ADVANCES IN WELDING SCIENCE AND TECHNOLOGY

TWR '86

Proceedings of an International Conference on Trends in Welding Research
Gatlinburg, Tennessee, USA
18–22 May 1986

Edited by
S.A. David
Oak Ridge National Laboratory
Oak Ridge, Tennessee, USA

Sponsored by
Joining Division Council

Co-sponsored by
American Welding Society and
Welding Research Council

Published by
ASM International

Library of Congress Catalog Card Number: 86-073083
ISBN: 0-87170-245-2
SAN: 204-7586

Printed in the United States of America

FOREWORD

In 1979 the ASM Joining Division Council organized a conference addressing the trends in welding research in the United States. The success of that conference prompted us to think in terms of global participation for the next welding science conference. Hence, the 1986 International Conference on Trends in Welding Research, was held on May 19-22, in Gatlinburg, Tennessee.

The conference was very significant for two reasons. First, it was truly an international event that focused mainly on the basic welding research trends in the world that will provide the necessary thrust in the future for the advancement of welding technology. It was also the first time a distinguished gathering of international welding researchers had ever taken place. Second, the proceedings of the conference demonstrated that welding is no longer an art but is truly a science—to be precise, an engineered science.

Welding science is unique for the wide variety of disciplines it brings together, such as arc and plasma physics, materials science, chemistry, chemical engineering, mechanical and electrical engineering, and economics. Worldwide scientists from all of these disciplines are currently contributing to welding research. In the past, lack of this multidisciplinary approach had left welding to be developed more as an art than as a science. Welding is such a complex field that this multidisciplinary approach is essential to significantly advance welding science. Now that it is being done, the future for welding science looks very promising.

Today because of this unique multidisciplinary approach to welding research, we have a better understanding of arc characteristics and arc-metal interactions. We are attempting to define better the heat and fluid flow problems in welds. With all the advanced microstructural characterization techniques and analytical tools now available, we can understand better the evolution of microstructures in weldments and develop better microstructure-property correlations. In addition, our capabilities to improve weld quality and productivity are increasing through the development of sensing and control devices.

The conference, co-sponsored by ASM International, The American Welding Society, and the Welding Research Council, featured eight parallel sessions on the most recent advances in welding science. In addition to the 100 technical papers presented by authors from 18 countries, 30 poster presentations were made. Seven keynote lectures were given by world renowned experts. Sessions were organized to cover several important aspects of welding, including heat and fluid flow modeling, solidification behavior, phase transformations, weldment properties, weld cracking phenomena, new process advancements, and sensing and control.

Finally, I would like to express my appreciation to the Organizing Committee for their help and support. A special note of thanks to Kathy Gardner and G.M. Slaughter of the Metals and Ceramics Division, Oak Ridge National Laboratory, Connie Marsch, Peg Ternovacz, and Joan Bellian, ASM staff members, who worked with me very closely on several aspects of this conference. Also my sincere thanks to the authors and attendees for making this conference a resounding success.

S.A. David
Conference
General Chairman

TABLE OF CONTENTS

Heat and Fluid Flow in Welds

Solidification Behavior of Welds

Phase Transformations in Weldments

Advances in Welding Processes

Keynote Address

Sensing and Control During Welding

Properties of Weldments

Welding and Weldability

Weldmetal Cracking

Poster Presentations

HEAT AND FLUID FLOW IN WELDS

KEYNOTE ADDRESS

THE MATHEMATICAL MODELING OF ARC WELDING OPERATIONS

Julian Szekely
Department of Materials Science and Engineering, MIT
Cambridge, Massachusetts 02139

INTRODUCTION

DURING THE PAST FEW YEARS there has been a tremendous growth in research efforts aimed at the mathematical modelling of arc welding operations. As sketched in Fig. 1, this effort is "pushed" by the more ready availability of computing facilities, software packages and computational fluid mechanics; while the "pull" is being provided by the industrial need for higher quality and more consistent welds, achievable by automation and robotization of these operations.

At this point it may be worthwhile to comment briefly on what a mathematical model is, and what mathematical models do in general and in the context of welding.

A mechanistic mathematical model represents a process or some aspect of a process by a set of differential or algebraic equations, which are derived from fundamental chemical and physical laws. A properly conceived mathematical model will provide quantitative relationships between the key process parameters and should be an invaluable component of any process study. As sketched in Fig. 2, a mathematical model is crucial in the planning and interpretation of experiments, and plays a pivotal role in process control and process optimization. Furthermore, mechanistic mathematical models are also needed as a part of artificial intelligence studies.

Within the welding context, the principal objective of process research as seen in the many papers given at this Conference is to relate the processing conditions to the structure and properties of the welds produced, hence to define the optimal welding conditions and ultimately devise the means to accomplish these through automation, process control, and robotization. Major advances have been made in this field in recent years in the U.S., largely through the support of the Office of Naval Research and the U.S. Department of Energy.

The purpose of this article is to trace briefly the evolution of mathematical modelling of arc welding systems, with emphasis on both the considerable accomplishments and the major tasks that still lie ahead.

Fig.3 shows a schematic sketch of an arc welding system which may be conveniently divided into three regions, namely:

- -- the arc;
- -- the weldpool and the surrounding solid;
- -- the arc-weldpool interface.

ARC BEHAVIOR

As seen in Fig.3 the potential between the cathode and the anode (the weldpool) causes current flow, which in turn generates thermal energy in the partially ionized gas. The passage of the current through the conducting medium generates a magnetic field, the interaction of which with the current produces an electromagnetic force field.

The net effect of these complex, mutually coupled phenomena is intense heat generation, high gas velocities and temperatures, and a rapid rate of heat transfer from the arc to the weldpool.

Welding arcs have been studied over the years and the maximum arc velocities may be estimated with the aid of Maecker's equation (1,2):

$$U_{max} = \sqrt{\frac{\mu_o}{2\bar{\rho}}} \quad R_c J_c \qquad (1)$$

where μ_o is the permittivity of free space;
$\bar{\rho}$ is the mean density;

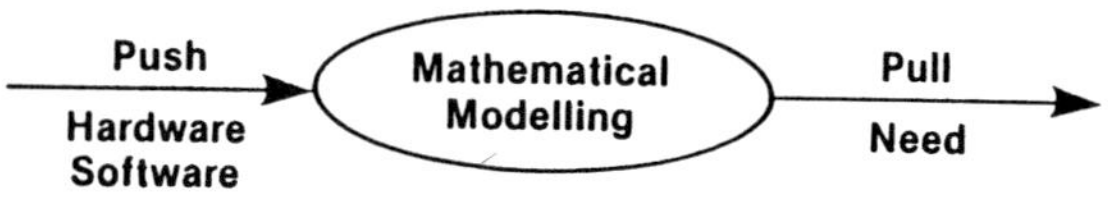

FIG. 1. The forces propelling mathematical modelling.

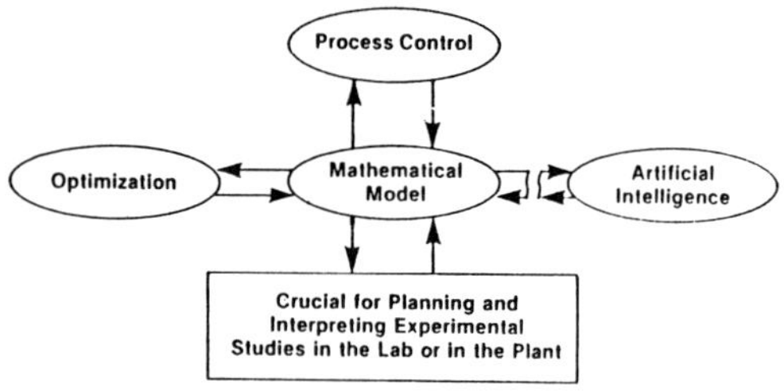

FIG. 2. The role of mathematical models.

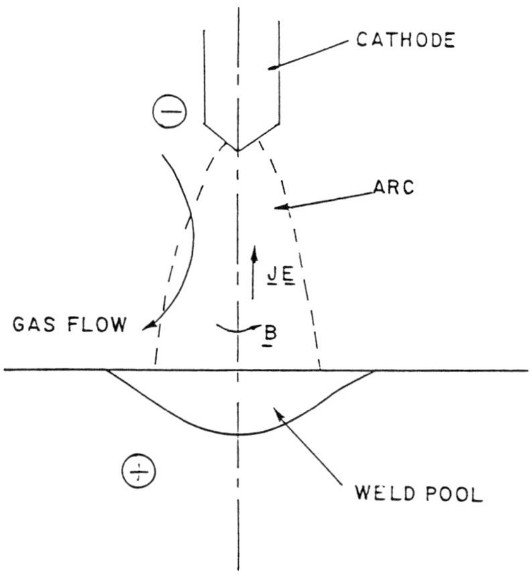

FIG. 3. Schematic representation of the arc and of the weldpool region.

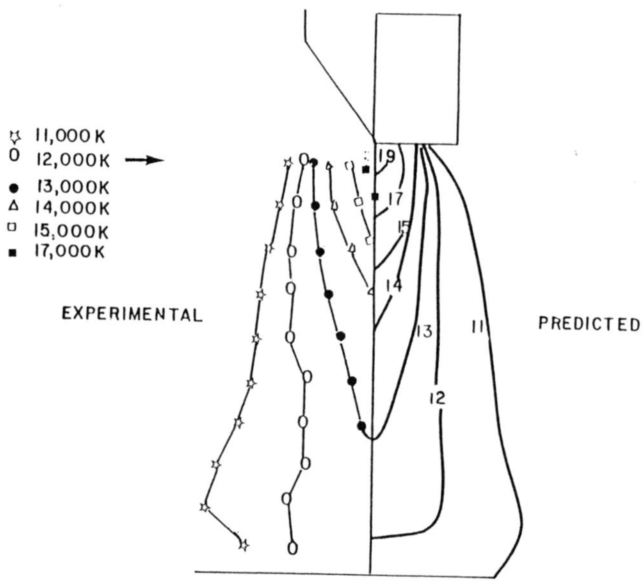

FIG. 4. Comparison of the theoretically predicted temperature profiles with measurements reported by Hsu and Pfender (8) for a 100 A, 10 mm arc (9).

R_C and J_C are the cathode spot radius and the corresponding current density, respectively

which would give velocities of the order of a few hundred meters per second for typical welding conditions.

A more rigorous representation of this system is available through the statement of the equation of continuity:

$$\nabla \cdot \rho \underset{\sim}{u} \quad (2)$$

equation of motion:

$$\underbrace{\nabla \cdot \rho \underset{\sim}{u}\underset{\sim}{u}}_{\text{convective transport}} = \underbrace{-\nabla p}_{\text{pressure gradient}} + \underbrace{\nabla \mu \nabla \underset{\sim}{u}}_{\text{viscous transport}} + \underbrace{J \times B}_{\text{Lorentz force}} \quad (3)$$

and the differential thermal energy balance equation.

$$\nabla \underset{\sim}{u} \rho C_P T = \nabla k \nabla T + S_R \quad (4)$$

The left hand sides of Eqns (3), (4), and (5) represent the convective transport of momentum and enthalpy due to fluid motion, while the source terms in the thermal energy balance equation are due to Joule heating, electron transport and thermal radiation.

Eqns (2)-(4) are readily integrated for the appropriate boundary conditions; some typical computed results are shown in Figs. 4-9.

Figs. 4 and 5 show a comparison between experimentally measured temperature profiles for 10 mm arcs operating at 100 and 200 A, respectively. The agreement between measurements and predictions is quite reasonable. However, from the standpoint of welding, knowledge of the heat and current fluxes to the anode are of rather greater interest. Fig.6 shows a comparison between the theoretically predicted maximum anode current density and the measurements of several investigators --using a water-cooled anode. The very good agreement between measurements and predictions is, again, readily apparent. As seen in Fig.7, the radial current density is also quite well predicted, while there is some scatter regarding the theoretical predictions and the experimental measured heat fluxes, as seen in Fig.8. Finally, Fig.9 shows both a comparison of the experimentally measured and the theoretically predicted radial distribution of the heat flux and the principal contributing components that make up the heat flux to the anode.

It should be noted that for welding, arcs, electron flow is the dominant contributor to the overall heat flux; thus, the precise knowledge of the current distribution is essential in model development.

One may summarize the results in this section by stating that a good quantitative understanding exists of the general behavior of plasma jets encountered in welding, although experimental verification of the computed velocity fields would be desirable. Most of the unresolved problems pertain to plasma jet stability, to the behavior of the near cathode region, and to the plasma jet-weldpool interface. This latter topic will be discussed subsequently.

WELDPOOL BEHAVIOR

The weldpool shown in Fig. 3 receives a spatially distributed current and heat flux and may undergo convective motion, driven by

-- electromagnetic
-- buoyancy and
-- surface tension forces.

From a practical standpoint we may distinguish between two types of operations, namely spot welds, where the heat flux is applied for a certain time period and during this time the weldpool expands. When the heat supply is disconnected, the process is reversed and the weldpool solidifies, giving the weldment. The other, more common mode of operation, involves a moving heat source, thus a moving weldpool.

Up to the present, the bulk of the work has involved spot welds, although recently, useful work has been initiated regarding moving heat sources (11).

The main motivation for studying weldpool phenomena is provided by the fact that weld penetration, weldpool shape and weldpool circulation will in general play an important role in determining the properties and hence the quality of the welds produced.

The early description of weldpools has ignored convection completely, but still provided a reasonable, often quantitative description of the weldpool shape. The first time that convection in weldpools was addressed in a realistic manner was through the adaptation of concepts from electro-slag refining to electroslag welding (12-14). The next major thrust came from adaptation of these ideas, first to static, idealized weldpools in arc welding (15-16) and then to the more realistic weldpool expansion (17-18) and collapse (19-20) in spot welding operations. This effort is also benefiting considerably from useful work in the related area of laser welding (21-24).

FORMULATION

Heat and fluid flow phenomena in weldpools are readily represented by writing down the conservation equations, which closely parallel those given previously for the arc. Thus we have

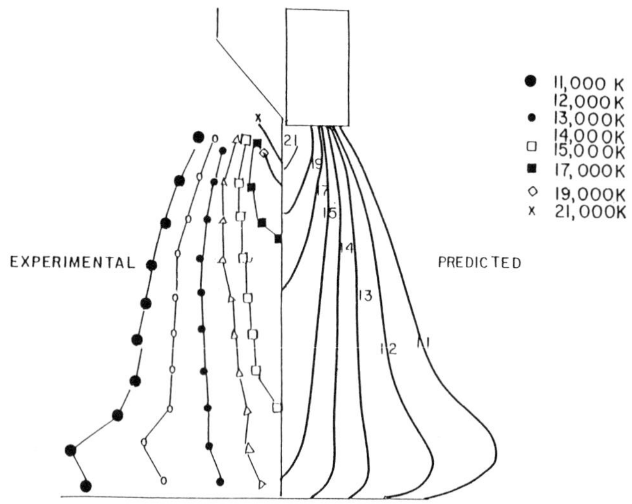

FIG. 5. Comparison of the theoretical predictions with measurements reported by Hsu and Pfender (8) for a 200 A, 10 mm arc (9).

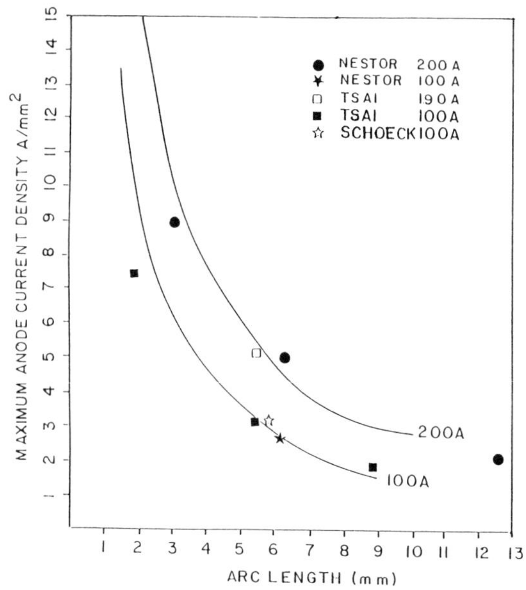

FIG. 6. Comparison of the experimentally measured and the theoretically predicted anode current densities, as a function of the arc length.

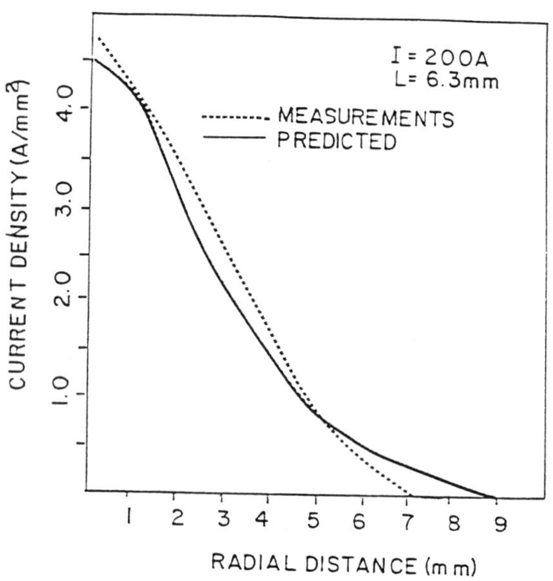

FIG. 7. Comparison of the experimentally measured and the theoretically predicted current density distributions.

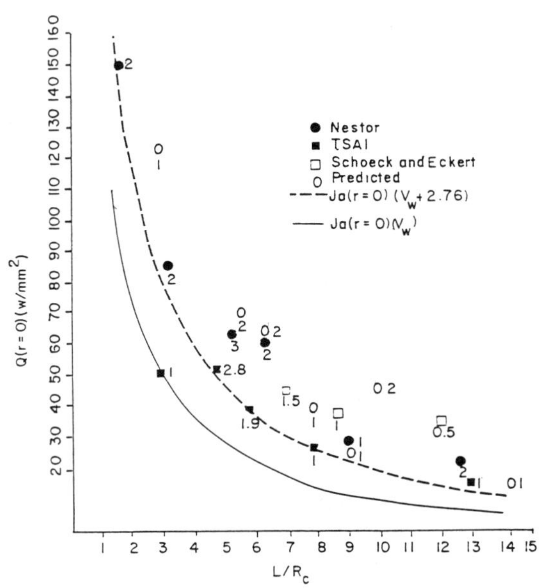

FIG. 8. Comparison of the experimentally measured and the theoretically predicted heat fluxes at the center of the target, using the experimental data of several investigators.

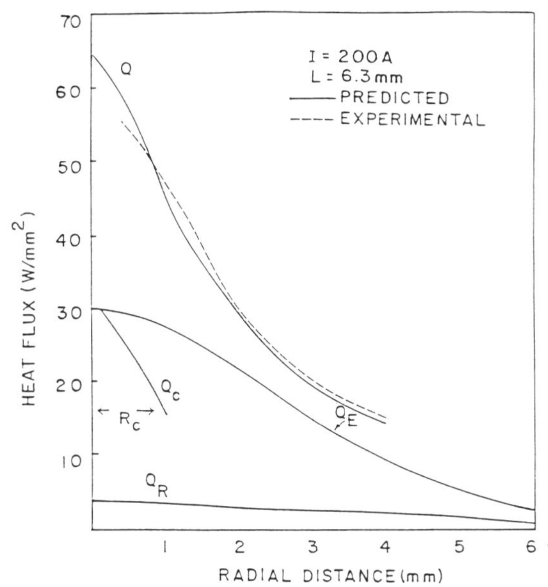

FIG. 9. The principal factors contributing to the radially distributed heat flux. Q_C = convective heat flux; Q_E = electron flow; Q_R = radiative heat flux; Q = total heat flux.

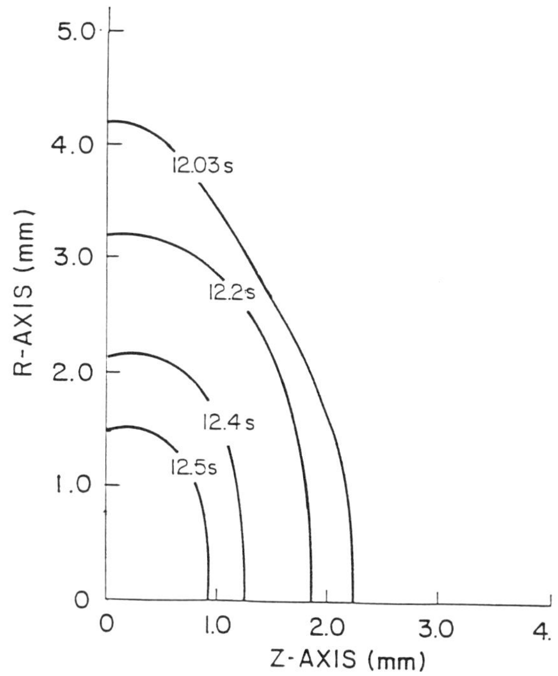

FIG. 10. The collapse of a weldpool in steel, showing the time dependence of the position of the solidification front.

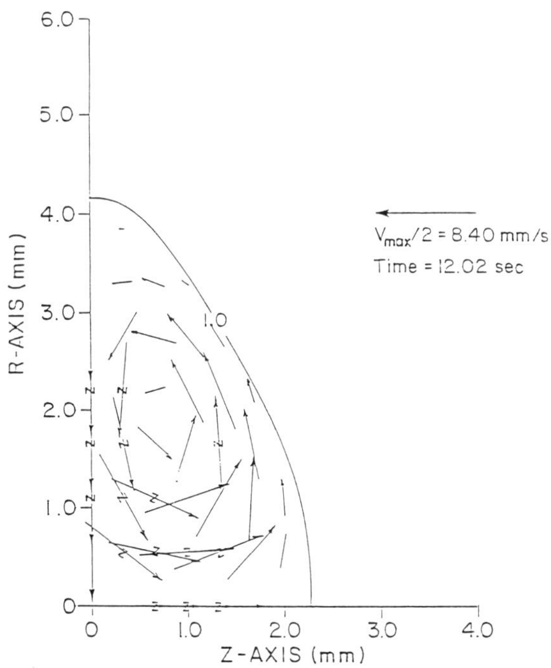

FIG. 11. The velocity map for a collapsing weldpool shown in Fig.10, just at the beginning of the solidification period.

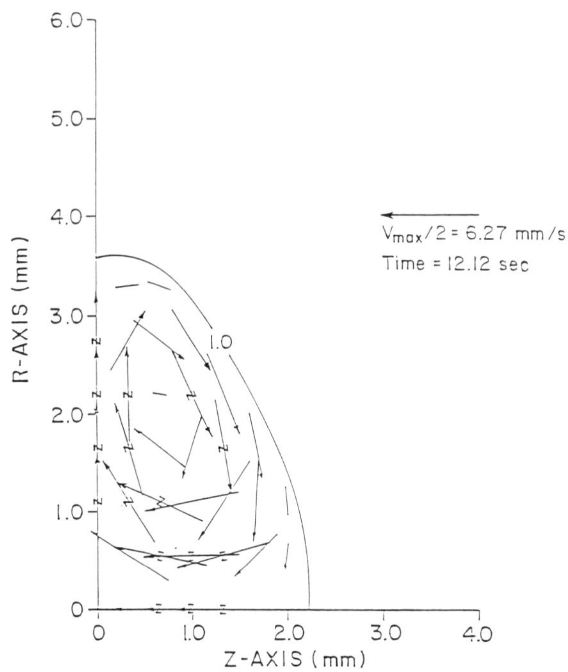

FIG. 12. The velocity map for a collapsing weldpool about 0.12 seconds into the solidification period.

$$\nabla \cdot \underset{\sim}{u} \quad (5)$$

(continuity)

$$\rho \underset{\sim}{u} \cdot \nabla \underset{\sim}{u} = -\nabla p - \nabla \mu \nabla \underset{\sim}{u} + \rho \underset{\sim}{g} + \underset{\sim}{J} \times \underset{\sim}{B} \quad (6)$$

(motion)

$$\underset{\sim}{u} \cdot \nabla \rho C_p T = \nabla k \nabla T \quad (7)$$

(thermal energy balance)

The boundary conditions for Eqn (5)-(7) have to express the fact that:

- -- the heat and current fluxes are specified at the free surface;
- -- the latent heat has to be supplied or released at the melt-solid interface;
- -- the velocities are zero at the solid surfaces; and
- -- surface tension gradients at the free surface (due to temperature gradients) may provide additional shear.

Numerous solutions to these equations have been presented in the literature. However, before proceeding with a discussion of these, some comment should be made on the order-of-magnitude estimates that may be made regarding the interrelationships between key process parameters. A full discussion of these issues has been presented in an earlier paper and only some key points will be made here.

THE EFFECT OF CONVECTION, THE PECLET NUMBER -- Let us define the Peclet number as

$$N_{Pe} = \frac{U_o \rho C_p L}{k} \quad (8)$$

where U is the characteristic velocity. When the Peclet number is small, convective heat transfer is unimportant; thus, the weldpool development can be described in terms of heat conduction alone. Such is the case for steel, when surface tension effects are unimportant and when the heat and current flux falling on the weld is broadly distributed. In contrast, when the Peclet number is large, convection (that is, weldpool circulation) plays an important role in determining the weldpool shape. It should be remarked here, however, that the principal resistance to heat flow in welding operations is in the solid phase, and for this reason good approximate estimates, that are good within a factor of two, of the weldpool dimensions may be made on the basis of heat conduction alone. However, in the majority of cases, knowledge of convection is important, because this will define the precise weldpool shape and may affect the structure and properties of the weldments produced.

ESTIMATION OF THE CHARACTERISTIC VELOCITIES -- The characteristic velocities in the weldpool may be estimated as follows. For surface tension driven flow and moderate velocities, we have:

$$U_{R,O} \sim \frac{\partial \gamma}{\partial T} \frac{\Delta T}{\mu} \left(\frac{L_z}{L_R}\right) \quad (9)$$

For high velocities, i.e. when a boundary is formed:

$$U_{R,O} \sim \left[\left(\frac{\partial \gamma}{\partial T}\right)^2 \Delta T^2 / \rho \mu L_R\right]^{1/3} \quad (10)$$

The corresponding expressions for buoyancy driven flow are:

$$U_{R,O} \sim \frac{g \beta \Delta T L_z^3 \rho}{\mu L_R} \quad (11)$$

and

$$U_{R,O} \sim (g L_z \beta \Delta T)^{1/2} \quad (12)$$

for low and high velocities, respectively. Analogous relationships may be developed for the electromagnetically driven flow component, which indicates that the velocity is proportional to the square root of the current density and is markedly dependent in the spatial distribution of the current flux.

Upon substituting the appropriate numerical parameters in Eqns (9)-(12) for typical welding conditions for steel, we would obtain velocities in the 100-1000 mm/s range in the presence of surface tension effects, while buoyancy or electromagnetic forces would give velocities of the order of 10 mm/s or less. The marked dependence of the velocity on the linear scale should also be apparent. For larger, deeper weldpools we should expect large Peclet numbers, hence a significant effect of convection on the weldpool shape.

THE TIME SCALE -- Spot welding is an inherently transient process, so it is important to estimate the time scale for solidification, once the supply of thermal energy has been discontinued.

This may be written as:

$$t_s \simeq \frac{L_z^2 C_p \rho}{k_s St} \quad (13)$$

while the time scale for melting is given as:

$$t_m \simeq \frac{L_z^2 \left[\Delta H + C_p (T_i - T_{mp})\right]}{k (T_{max} - T_{mp}) \rho} \quad (14)$$

Finally the time scale for establishing convection or for the inertial decay of convection is

$$t_i \simeq \frac{L_z^2 \rho}{\mu} \qquad (15)$$

Inspection of these relationships suggests that for typical spot welding conditions, the time scale for solidification is much shorter than the time scale for melting. Furthermore, the time for the inertial decay of convection would be much longer than the solidification time. The consequence of this behavior is that for spot welding, solidification will occur from an essentially circulating weldpool.

The other, perhaps more important, consequence of Eqns (9)-(15) is that one can estimate many of the weldpool characteristics from simple algebraic relationships, which is very helpful for the general appreciation of the system, and as a first necessary step before machine computation.

SOME COMPUTED WELDPOOL RESULTS -- In the following Figures, we shall show some computed weldpool results, taken largely from work in the author's laboratory, with emphasis on some of the newer developments.

Fig. 10 shows the position of the solidification fronts upon the collapse of a weldpool in a spot welding operation. It is seen that the time scale for the collapse (measured from t = 12 s), is quite short, of the order of 1/2 second, in contrast to the melting period, which took 12 seconds. Figs. 11 and 12 show the velocity maps computed for two times, just at the beginning of the solidification period, while the time dependence of the maximum velocity throughout the welding cycle is shown in Fig. 13. The velocity is seen to grow gradually with the development of the weldpool, as was to be expected, since the velocity is markedly dependent on the linear scale of the system; then the decay of the velocity is quite rapid as the weldpool collapses, but this is again due to the effect of the linear weldpool dimension.

Figs. 14, 15, 16, and 17 show the behavior of a steel system with a negative surface tension temperature coefficient, which results in an outward radial flow at the free surface. It is seen that at the time the weldpool collapse is initiated, the maximum velocity is of the order of 60 mm/s, which agrees with the order-of-magnitude estimates.

As seen in Figs. 14 and 15, growth of the weldpool took some 8 seconds, while the weldpool collapse is accomplished in less than 1/2 second--in agreement with the order-of-magnitude estimates given earlier. The trajectory of the maximum velocity shown in Fig. 17 is qualitatively similar to the behavior that was seen earlier in Fig. 13.

CONTINUOUS WELDING OPERATIONS

The behavior seen in Figs. 10-17 described spot welding operations, which are somewhat easier to analyze because these systems exhibit axial symmetry and are mathematically two-dimensional (with time dependence). Continuous welding operations are of rather greater practical importance and some interesting work has been initiated both by Kou (11) and by Chen and Mazumder (21) in this area. Fig.18, taken from a recent paper by Kuo, shows the computed isotherms for a continuous welding operation, which has been modelled allowing for full three-dimensional behavior. It should be remarked, however, that certain numerical problems still appear to remain in the tackling of this obviously difficult problem, and for this reason only slow welding speeds, which result in a relatively small distortion of the weldpool, have been modelled to date. Nonetheless, this is a very important endeavor in which major advances are likely in the future.

An alternate approach taken in our laboratory was to consider a two-dimensional slice in a travelling weldpool, as a first step in modelling problems of this type. Figs. 19, 20, 21, and 22 show the results of preliminary calculations depicting a travelling weld for progressively increasing travel speeds. The gradual distortion of the weldpool and the shallower penetration are readily apparent as the welding speed is increased. It is of interest to note that for higher travelling rates there is relatively little convection in the "melting zone" while the solidifying region is very strongly agitated. These preliminary calculations, which will be followed up by a three-dimensional study, are pointing to some very interesting effects that we are likely to encounter in the study of continuous welding operations.

One may summarize our current understanding of weldpool behavior by stating that a good quantitative understanding is emerging regarding the factors that affect weldpool development in the formation of spot welds, and important advances are being made regarding continuous welding operations. The important shortcoming of the work done to date is the lack of systematic correlation with experimental work and the as-yet largely unfulfilled need to relate weldpool behavior to the structure and properties of the welds produced.

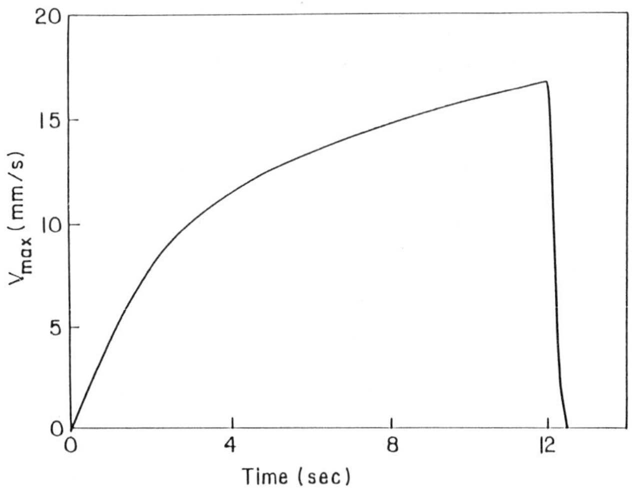

FIG. 13. The time dependence of the maximum velocity for the weldpool shown in Figs.10-12.

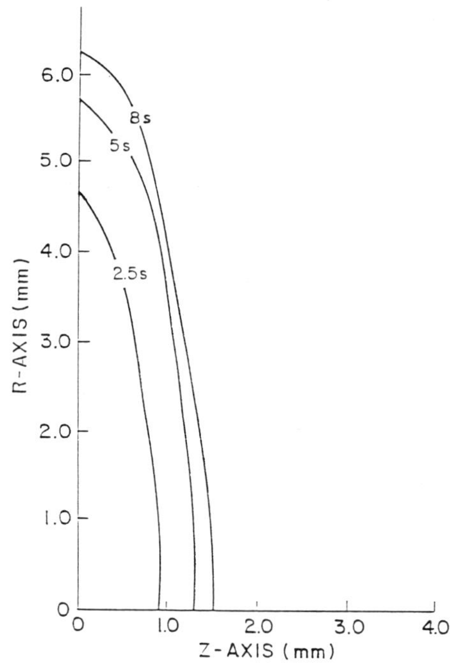

FIG. 14. The transient growth of a weldpool for steel, operated in the "normal mode" with a negative surface tension parameter.

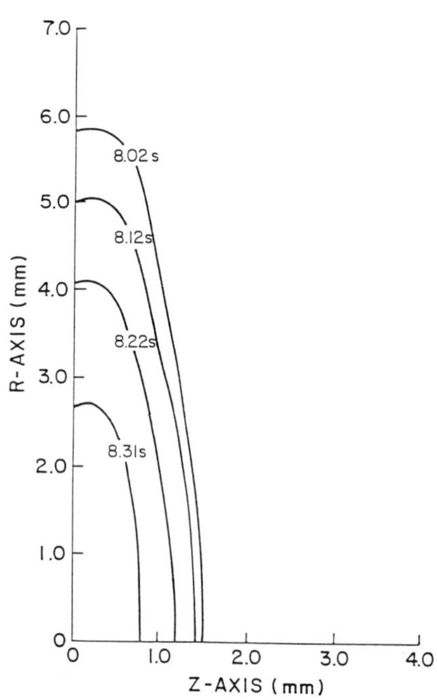

FIG. 15. The transient collapse of a weldpool for steel, corresponding to the growth shown in Fig.14.

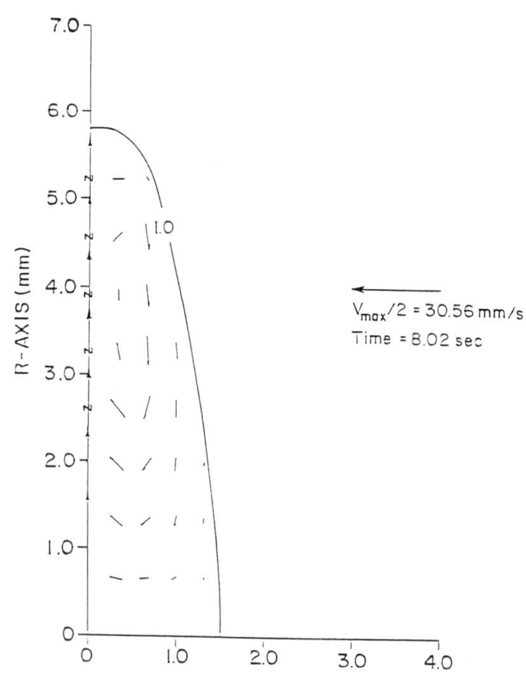

FIG. 16. The velocity map for the system shown in Fig.14, just at the commencement of the weldpool collapse period.

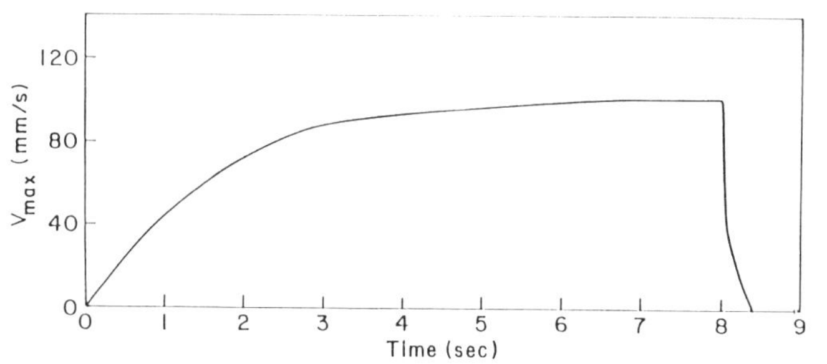

FIG. 17. The time dependence of the maximum velocity for the weldpool described in Figs.14-16.

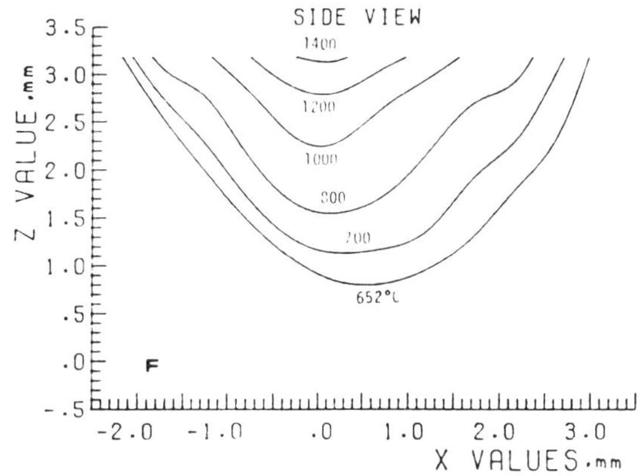

FIG. 18. Isotherms computed for a three-dimensional model of a continuous welding operation, by Kuo and Wang (11).

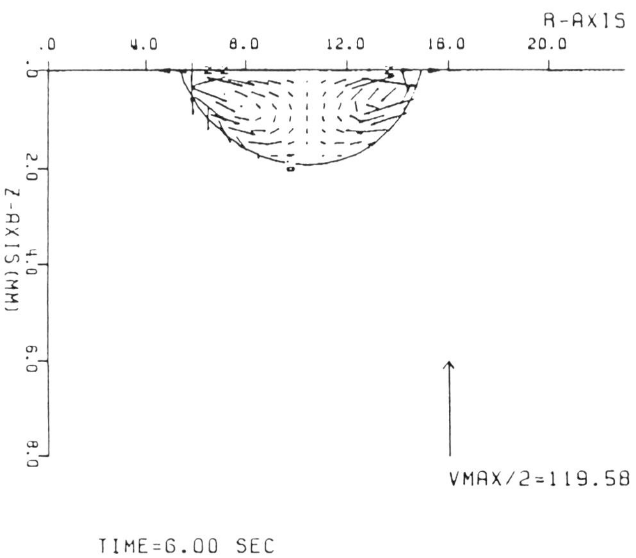

FIG. 19. Isotherms computed for a two-dimensional continuous welding model by Oreper and Szekely, for zero travelling speed.

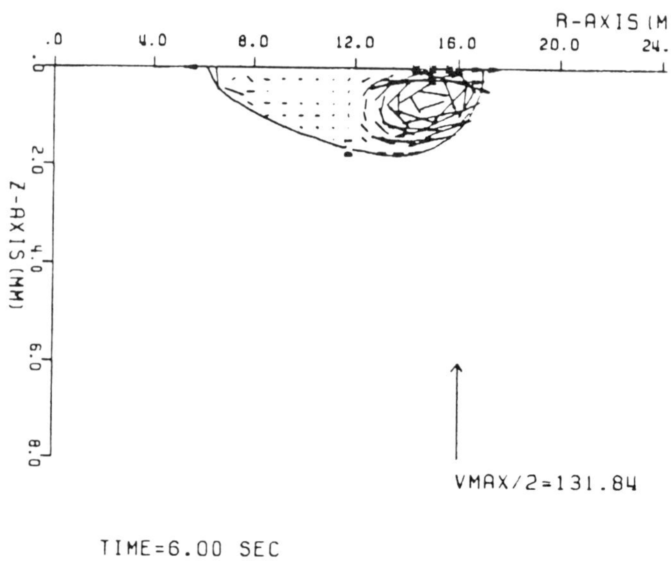

FIG. 20. Isotherms computed for a two-dimensional continuous welding model for a travel speed of 1 mm/s.

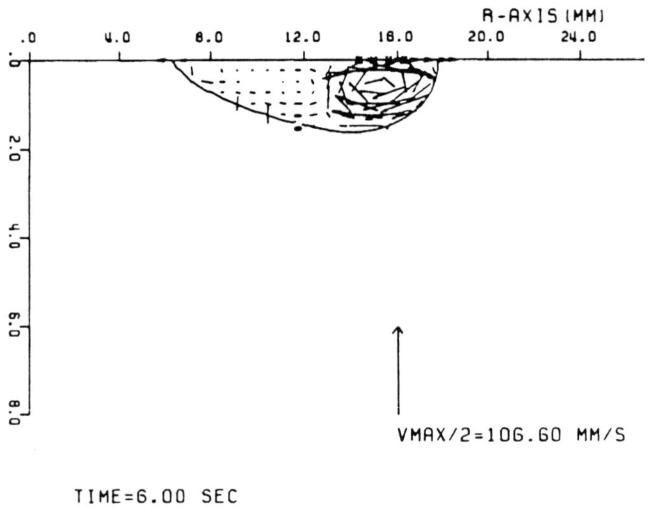

FIG. 21. Isotherms computed for a two-dimensional continuous welding model for a travel speed of 1.5 mm/s.

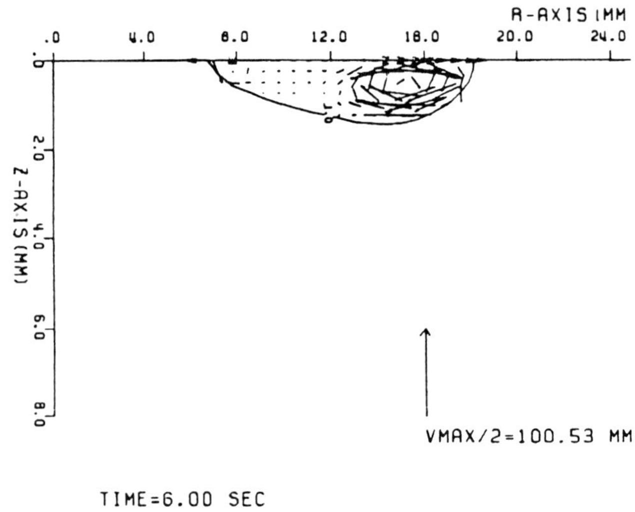

FIG. 22. Isotherms computed for a two-dimensional continuous welding model for a travel speed of 2 mm/s.

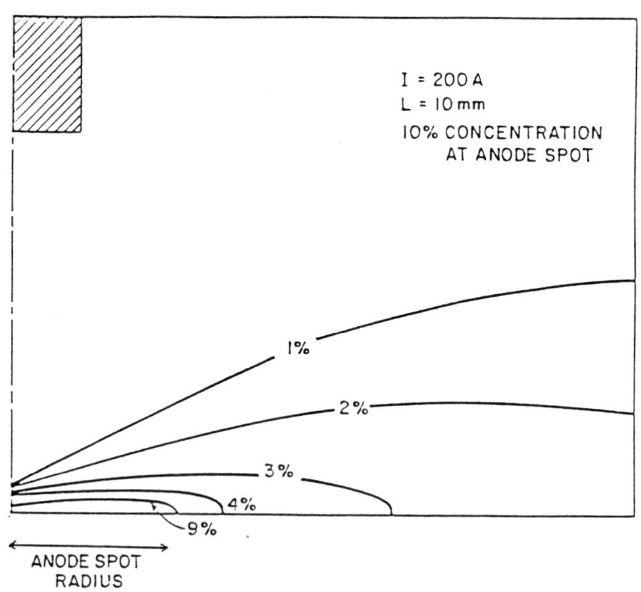

FIG. 23. The computed copper vapor distribution in the gas phase when a 10 volume % concentration of copper vapor was specified at the anode spot surface.

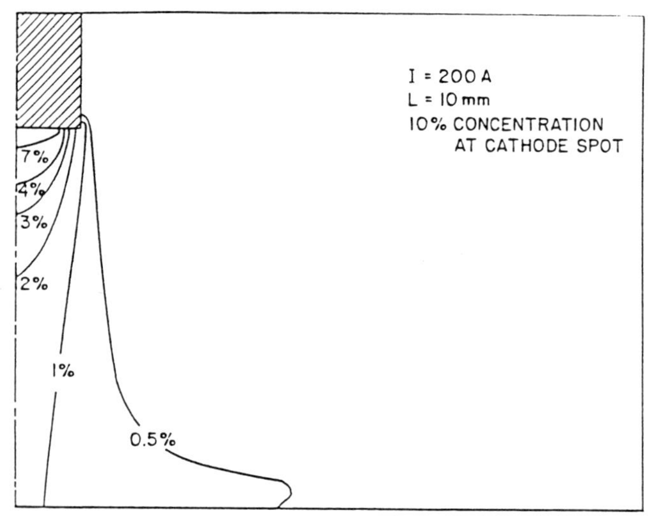

FIG. 24. The computed copper vapor distribution in the gas phase when a 10 volume % concentration of copper vapor was specified at the cathode spot surface.

Nonetheless, the developments made in the last five years should make this next step feasible.

WELDPOOL - ARC INTERACTIONS

The discussion up to the present was concerned with describing plasma arc and weldpool behavior. The interaction between these phenomena has entered these considerations implicitly, because the thermal and the current boundary conditions for the weldpool problem had to be derived from the plasma characteristics. Up to the present, all investigators examined the uncoupled problem; that is, the thermal boundary conditions for the weldpool were considered to be given.

Recent work has shown that the work done on plasma-surface interactions and in weldpool circulation is internally consistent, because the fluxes calculated for the former are quite close to the assumptions that were generally made in the weldpool circulations.

However, it is clear that beyond this zero order interaction, a host of more complex, yet equally important problems remain to be tackled.

METAL VAPORIZATION FROM THE WELDPOOL

The question has been raised whether the vaporization of metal from the pool is likely to affect the nature of the welding arc. Figs. 23 and 24 show the computed concentration isopleths for copper vapor, assumed to be at 10 vol% at both the anode spot and the cathode spot. It is seen that for an anode current of 100 A, the material is swept out of the anode region quite rapidly, so that its effect in modifying the plasma properties is not likely to be great. If the material were to be generated in the cathode region, it would be retained near the axis for larger distances, but even here the net effect would be quite minor.

Perhaps the most challenging problem that remains regarding weldpool interactions are:

- -- the weldpool deformation due to electromagnetic and surface tension forces and
- -- the formation of surfaces waves and instability in general.

All these present quite difficult problems, because we have to incorporate the tools of other disciplines, which are rather unfamiliar to workers in the welding field.

The existence of these much more complex problems should in no way invalidate the somewhat simpler analyses that have been presented in the previous Sections 2 and 3 of the paper. Indeed, all these more complex problems will have to be tackled in a stepwise manner, and the previously described work should represent a useful intermediate stage.

CONCLUDING REMARKS

Major advances have been made in the modelling of arc (and other) welding systems in recent years.

A good understanding exists of the factors that govern the behavior of welding arcs and arc-to-anode heat transfer, although more work needs to be done in order to define the anode region more precisely.

Excellent progress has been made in the past few years in modelling weldpool circulation, and more recently the solidification of weldpools for spot welding. Major work needs to be done, however, in the modelling of continuous welding processes and in relating the welding process parameters to the structure and properties of the welds produced. This latter area, together with a more precise definition of plasma-weldpool interactions, with special reference to deformation and surface stabilities, represents the major challenges ahead.

ACKNOWLEDGEMENTS

The author wishes to thank the Office of Naval Research for support of this work under Grant #N00014-84-K-0427. Thanks are also due to the U.S. Department of Energy (Basic Energy Sciences) which supported our earlier efforts in this field. Thanks are also due to my colleagues in the welding field, notably Professor T.W. Eagar for helpful discussions and to our co-workers, notably Drs. G. Oreper and J. McKelliget, for their contributions.

REFERENCES

1) H. Maecker, Z. Phys., 1955, vol. 141, pp. 198-216.

2) P. Schoeck and E.R.G. Eckert, 5th Int. Conf. Phen. Ionized Gases, Munich, 1961, vol. 2, pp. 1818-1829.

3) M. Ushio, J. Szekely, and C.W. Chang, Ironmaking and Steelmaking, 1981, vol. 8, pp. 279-286.

4) J.W. McKelliget and J. Szekely, Jnl. Phys. D: Appl. Phys., 1983, vol. 16, pp. 1007-1022.

5) H.A. Dinulescu and E. Pfender, Jnl. Appl. Phys., 1980, vol. 51, pp. 3149-3157.

6) N.A. Sanders and E. Pfender, Jnl. Applied Phys., 1984, vol. 55, pp. 714-722.

7) K. Hsu, K. Etemadi, E. Pfender, Jnl. Appl. Phys., 1983, vol. 54, pp. 1293-1301.

8) K. Hsu and E. Pfender, Jnl. Appl. Phys., 1983, vol. 54, pp. 4359-4366.

9) J. McKelliget and J. Szekely, Met. Trans., 1986, in press.

10) N. Tsai, Ph.D. Thesis, Department of Materials Science and Engineering, Massachusetts Institute of Technology, 1983.

11) S. Kuo and Y.H. Wang, Welding Journal, Research Supplement 63, 1986.

12) A.H. Dilawari and J. Szekely, Met. Trans., 1977, 8B, 123.

13) A.H. Dilawari, J. Szekely and T.W. Eagar, Met. Trans., 1978, 9B, p. 39.

14) A.H. Dilawari, J. Szekely and T.W. Eagar, The Welding Journal, 1978, 57, p. 1.

15) G.M. Oreper, T.W. Eagar and J. Szekely, Welding Journal, 1983, 62(11), pp. 307s-312s.

16) S. Kuo, and D.K. Sun, Met. Trans. A, 1985, vol. 16A(2), p. 203.

17) C. Chan, J. Mazumder, and M.M. Chen, Met. Trans. A, 1984, vol. 15A(12), p. 2175.

18) G.M. Oreper and J. Szekely, Jnl. of Fluid Mechanics, vol. 147(10), p. 53.

19) J. Szekely, G.M. Oreper, and J. McKelliget, Proc. Engineering Foundation Conference on the Modelling and Control of Welding and Casting Processes, (S. Kuo, ed.), 1986.

20) G.M. Oreper, J. Szekely, and T.W. Eagar, Met. Trans., 1986, to be published.

21) C.L. Chun, R. Zehr, J. Mazumder, and M.M. Chen, Proc. Engineering Foundation Conference on the Modelling and Control of Welding and Casting Processes, Santa Barbara, 1986.

22) C. Chan, J. Mazumder and M.M. Chen, Met. Trans. A, 1984, vol. 15A, 2175.

23) T. Chande and J. Mazumder, J. Appl. Phys., 1984, vol. 56(7).

24) C.L. Chan, M.M. Chen and J. Mazumder, AIME, Paper No. 85 WHA/HT-35.

COMPUTATIONAL HEAT TRANSFER FOR WELD MECHANICS

J. Goldak, M. McDill, A. Oddy, R. House, X. Chi, M. Bibby
Carleton University
Ottawa, Canada K1S 5B6

ABSTRACT

Developments in finite element analysis (FEA) techniques for computing transient temperature fields during welding are described in this paper. As an example, 3D and 2D transient thermal analyses of a root pass in a girth weld in a thick-walled pipe are presented. The joint geometry is modelled directly. It is shown that significant accuracy and negiligible computing cost are possible in many cases with a two-dimensional (2D) cross-sectional model but the important problems of run-on (weld start) and run-off (weld stop) are ignored entirely. However, weld start and stop can be included in an in-plane 2D analysis but this analysis is applicable only to relatively thin plates or high heat input welds in which through-thickness heat flow can be neglected. In reality, important heat transfer effects are usually three-dimensional and therefore a 3D FEA is required for most welding situations. With current commercial software and analysis techniques, the computing cost for 3D FEA can be punitive. Strategies that reduce computing time and cost to affordable levels and yet achieve the desired accuracy are presented.

IT IS NOW WIDELY RECOGNIZED that complex processes such as welding can be analysed by FEA. Today it is possible to analyse the temperature field in welds using temperature-dependent thermal properties, the latent heats of fusion and transformation, a realistic heat source model; i.e., the double ellipsoid model [**1**], surface heat loss by convection and radiation, and joint and workpiece geometry. The analysis applies not only to quasi-steady state conditions but also to the transients on starting and stopping the weld. This is important because thermal strain and thermal stress are almost always more severe at the beginning and the end of the welding process.

The goal of our research is to create and demonstrate accurate and efficient algorithms to compute transient temperature, stress, strain and microstructure changes during welding. Such algorithms will ultimately enable an engineer to quickly determine if a welding process and/or workpiece design will achieve the specified performance. The FEA technique is based on rigorous physics and mathematics; and is, therefore, more reliable than empirical approaches. Reference [**1**] is a detailed review of previous work in the area. Progress in computing thermal stress changes during the welding process is presented in a second paper at this conference.

Since 1970, it has been obvious that FEA should be an appropriate method for analysing welds. However, the problem has proved to be surprisingly difficult. Marcal and Hibbitt's monumental effort [**2**] achieved only limited success. A critical step was the development of the double ellipsoid model for the heat source [**1**], [**3**].

The need to achieve accuracy with a resolution of the order of 1 mm near the weld in a welded structure that is of the order of meters, creates the second critical issue. If it is not resolved effectively, the computing costs will be excessive. This probably is why, to the authors' knowledge, no serious attempts to perform 3D heat transfer analysis have been published. Andersson [**4**] stated that 3D analysis would be prohibitively expensive.

To reduce the computing cost, the problem has usually been reduced to a two-dimensional (2D) cross-sectional analysis which allows heat to flow only in the cross-sectional plane perpendicular to the weld [**1**]. In addition to this idealization error for steady state conditions, temperatures cannot be determined while starting and stopping the weld.

In other cases, a two-dimensional (2D) in-plane analysis is applied where the heat is allowed to flow in the plane of the plate but not in the through-thickness direction [**1**]. This can be more costly than the cross-sectional model because the cost is proportional to the length of weld analysed. Weld start and stop conditions can be analysed, but the in-plane analysis is only appropriate for high heat or thin plate welding situations in which heat flow in the through-thickness direction can be neglected.

The reality is that the heat flows in all three directions in most welds and 3D models are needed to compute accurate transient temperature fields. New FEA techniques and strategies for reducing the current cost of 3D FEA thermal

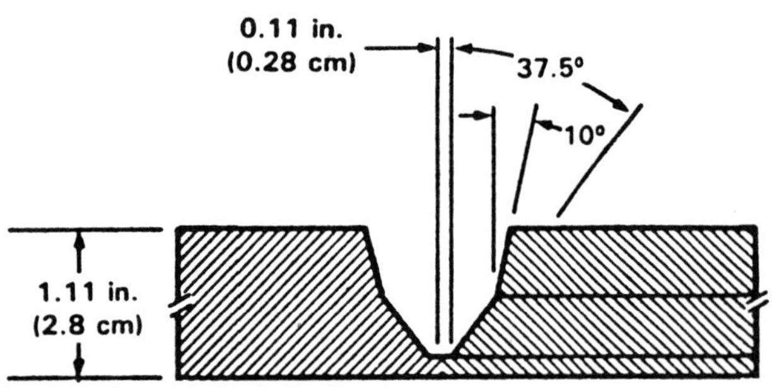

Fig. 1. Joint design for girth weld in a 60 cm (24 in.) diameter pipe.

analysis of the welding process are presented in this paper.

To demonstrate the effectiveness of the new algorithms, 3D transient temperature fields are computed for a root pass in a girth weld in a thick–walled austenitic stainless steel pipe. The joint preparation is shown in Fig. 1. Experimental data for this weld has been measured by the Battelle Pacific Northwest group [5]. An extensive comparison of experimental and computed results will be presented in a future publication. However, previous FEA analyses reported in [1] agreed with experimental results within 10 percent.

THEORY

The conservation of energy (the Fourier heat equation), together with geometry, the thermal properties, initial conditions and boundary conditions, define a unique solution of temperature, $T(x, y, z, t)$, as a function of position and time. A detailed description of the finite element analysis techniques and algorithm for solving this problem for the welding process is contained in [1].

The nonlinear transient finite element code for solving this problem was developed in our computing laboratory. This has enabled new algorithms to be implemented and tested rapidly.

DATA

Data submitted to the FEA program are categorized as follows:

(i) Geometry - Internal and external pipe diameter, length of pipe and joint design. In this presentation only a short length of the girth weld is analysed (sufficient to demonstrate the method).

(ii) Thermal Properties - Thermal conductivity, specific heat capacity and latent heat [7].

(iii) Initial Temperature Distribution - the temperature distribution when the welding process starts (i.e., $T_0 = 20\ C$ in this case).

(iv) Data Describing the Arc - current (100 amps), voltage (10.5 volts), travel speed (1.27 mm/s), position and estimates of the width, depth and length of the weld pool (width = 6 mm, depth = 3 mm, length = 12 mm in this case). It is preferable that the data be determined from experimental measurement. If experimental data on weld pool dimensions are not available, they can be estimated by the method of Christensen et al [6]. It is important to note that the size and shape of the weld pool are defined as data and must be input in order to run the analysis. If the computed weld pool differs from the weld pool data the problem is inconsistent.

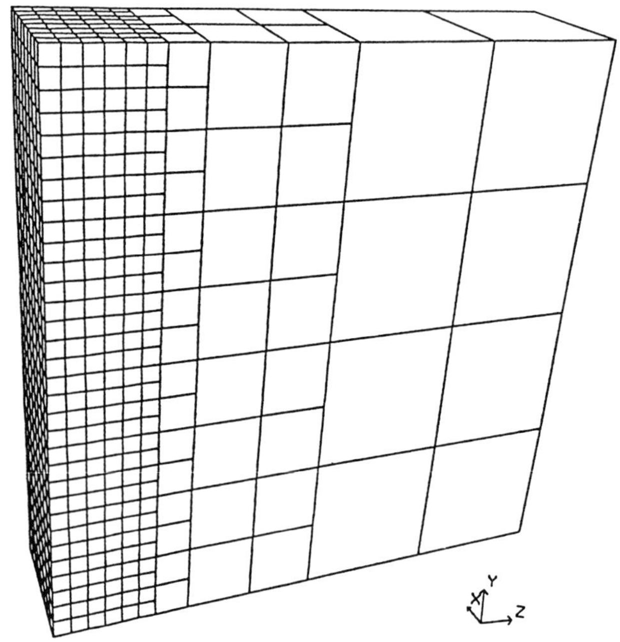

Fig. 2. An example of an edge–graded mesh that could be used for FEA of a weld.

COMPUTATIONAL STRATEGY

3D ANALYSIS - In 2D FEA, an effective mesh design is important. In the 3D case, it is a critical issue. Grading is necessary because small elements (e.g. of the order of 0.5 to 3 mm) are required near the heat source while larger (e.g. 30 cm × 28 cm × 30 mm in Fig. 7) elements are effective in areas of the workpiece remote from the arc. Far away from the heat source they can be as large as possible while still capturing the geometry and variations in the temperature field. Grading is not easy to achieve with current FEA programs. Global constraints, distorted elements, or tetrahedrons, the 3D equivalent of triangles, are often used. None of these alternatives is attractive for a variety of reasons.

Gupta developed the 2D grading elements used in the cross–sectional and in–plane analysis shown in Fig. 6, 7 and 8 [8]. The 3D grading elements shown in Fig. 3, 4 and 5, were recently developed by McDill et al. [9] for 3D analysis of welds. In these elements, the correct basis functions are computed directly. They are equivalent to eight noded bricks with constraints. However, computationally they are more efficient because the constraints need not be specified in input data; nor the extra basis functions constrained out. These elements are an important advance in computational weld mechanics.

While grading the mesh in one direction from the weld into the workpiece (Fig. 2) presents one level of cost reduction, grading the mesh in all directions (Fig. 3) from the heat source presents far greater cost savings. Algorithms are now being developed to move this fine mesh with the heat source so that long sections and eventually the entire

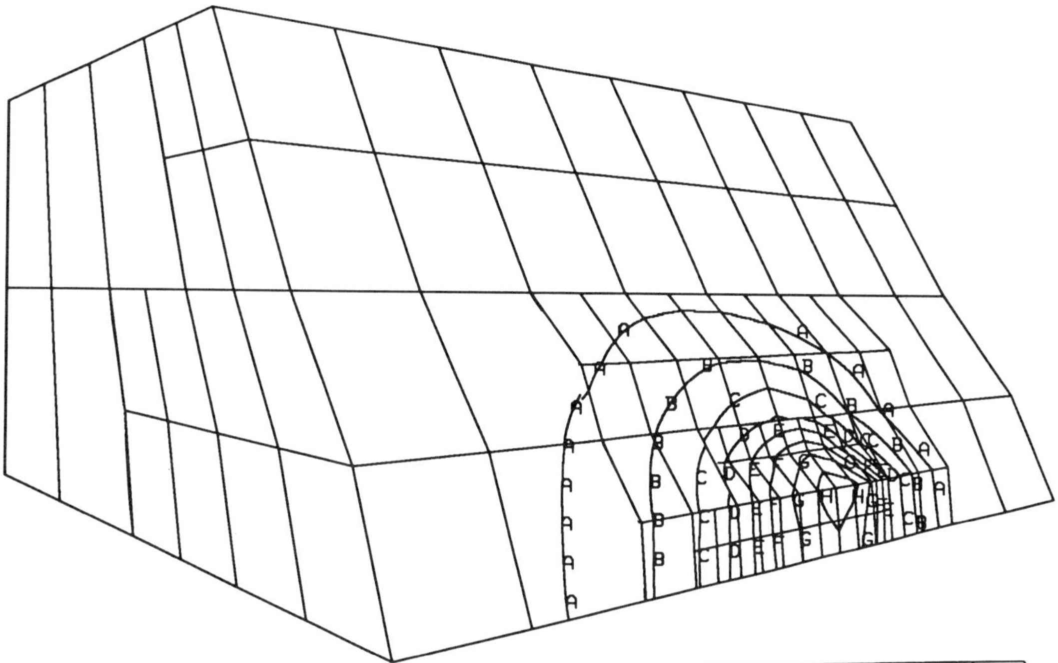

Fig. 3. Analysed section (50.8 mm (2 in.) × 50.8 mm (2 in.) × 28 mm (1.11 in.)) showing temperature contours for the root pass in a girth weld of a pipe 7.5 seconds after the arc started. See Fig. 1 for details of pipe geometry. Contour A = 200 C, B = 400 C, etc.

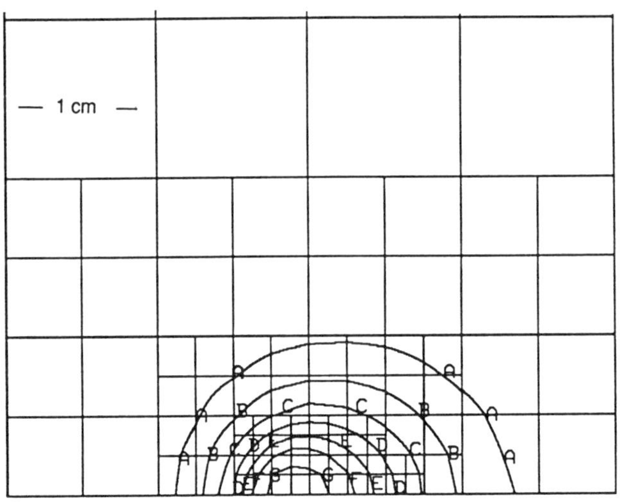

Fig. 4. Temperature contours for the root pass in a girth weld of a pipe viewed from the underside of the weld. This figure should be compared with the in-plane results in Fig. 8 but note that this is the bottom surface, not the mid-plane shown in Fig. 8. See Fig. 1 for details of pipe geometry and weld procedure. Contour A = 200 C, B = 400 C, etc.

girth weld can be analysed.

The temperature contours ranging from 200 to 1800 C are shown on the prepared surface (joint face) in Fig. 3. The isotherms on the bottom view are shown in Fig. 4 and in a cross-sectional view in Fig. 5. The fusion zone size

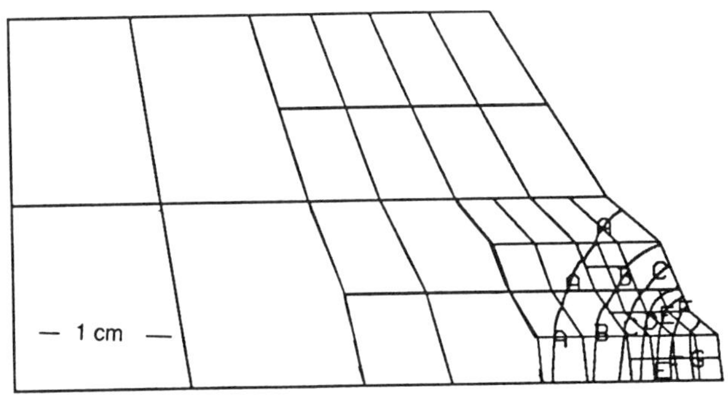

Fig. 5. Temperature contours for the root pass in a girth weld of a pipe. A cross-section is viewed through the center of the arc 7.5 seconds after the weld started. This figure should be compared with the cross-sectional results in Fig. 6. See Fig. 1 for details of pipe geometry and weld procedure. Contour A = 200 C, B = 400 C, etc.

is defined by the melting point isotherm of approximately 1450 C. This lies between the G and H isotherms.

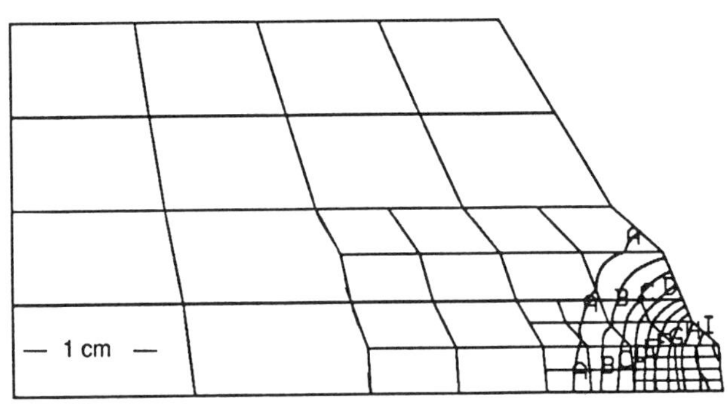

Fig. 6. Temperature field for a cross-sectional analysis of a root pass in the girth weld described in Fig. 1. A = 200 C, B = 400 C, etc.

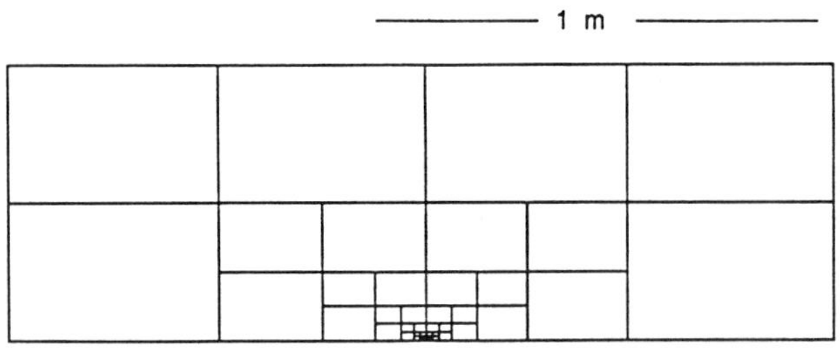

Fig. 7. In–plane analysis of girth weld in a 60 cm (24 in.) diameter pipe. The pipe has been unrolled to 194 cm (72 in.) × 60 cm (24 in). There are only 56 elements and 87 nodes in this mesh of one symmetric half of the pipe. Approximately 26 of the elements in this mesh are too small to be seen; they are shown in Fig. 8. The smallest elements have sides of only 2 mm but a thickness of 30 mm. The largest elements are 30 cm × 28 cm × 30 mm.

2D CROSS–SECTIONAL (XS) ANALYSIS - The fundamental principle, KEEP IT SIMPLE, remains good advice. The cross–sectional model is simplest and cheapest, and provides an excellent first approximation. Of particular importance, it indicates areas where the through–thickness temperature is not constant. In these areas a 3D analysis would provide the greatest gains in accuracy. The results of the analysis for this particular girth weld are shown as isotherms in Fig. 6. It is apparent that there are significant differences in the higher and lower temperature regions — a result of the fact that longitudinal heat flow is not allowed in the cross–sectional model. However, this analysis requires 90 times less computing time than the 3D analysis shown in Fig. 3.

IN–PLANE (IP) ANALYSIS - The results of an in-plane analysis for this welding situation are shown in Fig. 7. In the 3D analysis only a 5.08 cm × 5.08 cm section is analysed. In the cross–sectional analysis only a 5.08 cm deep slice is analysed. In Fig. 7, the entire pipe is analysed. These isotherms differ considerably from those on the bottom of the 3D model shown in Fig. 4. In fact, they are quite unrealistic in this case because they average the heat input and temperature over the entire weld thickness. However, this analysis can be quite accurate for thin section welds or farther from the arc where through–thickness heat flow is insignifiant. At the same time, transient temperatures that occur when the weld is started or stopped can be computed. This analysis is 200 times faster than the 3D analysis.

COMPUTATIONAL TIME AND COST

The computing cost will be measured in CPU seconds. The 3D analysis ran on an Apollo DN 660 workstation with 2 Mbytes of memory. The cross–sectional and in–plane analyses were done with an Apollo DN320 workstation with 1.5 Mbytes of memory. The current equivalent model, a DN3000, costs approximately US$10,000. Its speed is similar to a VAX-780.

If the complete girth weld was discretized with edge–grading, each step would require roughly 72 times as much cpu time. Stepping the weld around the entire circumference would require 72 times as many steps. Thus the cpu time to analyse the first pass of the complete weld using edge–grading would be roughly 5000 times greater. Assuming a Cray–2 supercomputer is 800 times faster, and the code were properly vectorized, a crude estimate of Cray–2 cpu time is 6 hours for a 3D analysis of the complete root pass. An algorithm which enabled the mesh to be graded from the arc might reduce this cpu time by a factor of 100.

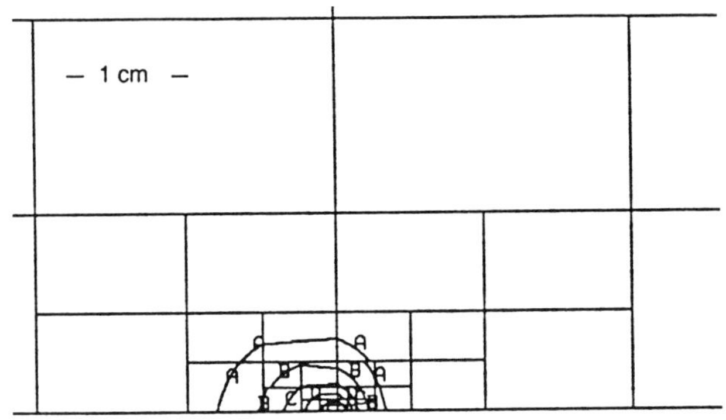

Fig. 8. In–plane analysis of girth weld in a 60 cm (24 in.) diameter pipe at a magnification 32X greater than Fig. 7. Only a 6 cm × 4 cm section is shown. The smallest element size is 2 mm. $A = 100$ C, $B = 200$ C, etc.

Table 1. Computing Times for Thermal Analysis

Analysis Type	*Num. of Elements*	*Num. of Nodes*	*Front Width*	*Num. of Increments*
XS	52	73	11	14
IP	56	87	12	4
3D	389	659	127	6

Table 1. Continued

Analysis Type	*Num. of Iterations Per Increment*	*CPU Time (Total) Seconds*	*CPU Time Per Iteration Seconds*
XS	3	536	12.8
IP	4	236	14.8
3D	6	23400	650

ERRORS IN COMPUTED TEMPERATURES

How accurate are the computed temperatures? The answer requires a discussion of the errors that arise from various sources. First, idealization errors are present in the cross–sectional and in–plane analyses. They can be evaluated by comparing these solutions with the 3D analysis which has no idealization error.

The temperature, $T(x,y)$, in the in-plane analysis averages the temperature in the 3D analysis, $T(x,y,z)$, through the thickness, h; i.e., $T(x,y) = \frac{1}{h}\int_0^h T(x,y,z)\,dz$. Because it ignores through-thickness heat flow, the temperatures are grossly in error near the arc. In fact, the midplane of the 3D analysis never reaches 200 C while the in-plane reaches 525 C. Therefore the in-plane is actually quite curvilinear, if the in-plane is defined to be the plane on which $T(x,y) = T(x,y,z)$.

Temperatures in the cross-sectional analysis are too high near the arc (2300 vs. 1800) because longitudinal heat flow has been neglected. Temperatures farther from the arc, say the 200 C isotherm, are too low. This is because the cross-sectional model does not permit the preheating effect due to longitudinal heat flow.

The second error is the discretization error; i.e., the error caused by using elements and time steps of finite size. This can be evaluated by repeating an analysis with a sequence of finer meshes and time steps. When the change in the temperature field (e.g. L_2 norm) becomes sufficiently small, the discretization error becomes small. This is the situation in the in-plane analysis, where the temperatures did not change substantially when each element was subdivided into four.

Far more difficult to control or estimate are errors in the data. Errors in geometry should usually be small. Errors in the temperature-dependent thermal conductivity, heat capacity and latent heats of transformation are more difficult to estimate. Perhaps the most important source of error is the data describing the welding arc, particularly the arc efficiency. However, voltage, current, and position are rarely accurately measured or controlled in the welding industry. This group of errors associated with the arc is best evaluated by comparing computed temperatures with experimentally measured temperatures.

If the efficiency is too high, or heat capacity is too low, the temperatures will be too high everywhere. If the thermal conductivity is too low, peak temperatures will be too high and temperatures far from the arc will be too low. If the weld pool size defined in the data is too small, peak temperatures will be too high. In each of the above cases, it was assumed only one error was present. In reality, all errors are present and vary with time. This requires a more sophisticated analysis. Space does not permit a detailed discussion of this important subject.

EXPERIMENT VERSUS COMPUTER SIMULATION

It is unfortunate that a perception has arisen that there is a conflict between experiments and computer analyses. In the long term, experiments will be needed and designed to test the accuracy of the computational analysis. The computed field allows the temperature to be determined at any point at any time. There is a pressing need for more accurate experimental data. Furthermore, the design of experiments can and should be optimized by computer simulation. Thus computer analyses and experiment should be complementary and synergistic.

In comparing experimentally measured temperatures with computed temperatures, it is useful to consider three welds, the design or target weld, the weld that was actually made and the weld data used in the computer model. Both the computer analysis and experiment are subject to error. The challenge is to estimate the errors that occur. The conclusion that a difference between experimental and computed results is due to error in the computed result is far from justified.

SUMMARY

1. Techniques and strategies for the accurate and effective computation of 3D transient temperatures in girth welds have been described.

2. The temperatures in 2D FEA models can be computed accurately in less than 3 minutes on a relatively small and inexpensive computer. This raises the possiblity of computer control based on the actual physics of the weld instead of some empirical function.

3. With the techniques and strategies presented, the accuracy of the computed temperatures is almost entirely a function of the accuracy of the data; i.e., geometric, thermal properties and heat input, and cooling effects.

ACKNOWLEDGEMENT

The financial support of NSERC A2992 (J. Goldak), NSERC Strategic Grant G1529 (J. Goldak), NSERC A4601 (M. Bibby) and AMCA International are gratefully acknowledged. Professors Goldak and Bibby also wish to acknowledge the contribution of several graduate students: M. McDill developed the mesh-grading system; A. Oddy investigated error measures; R. House did the 3D and cross-sectional analysis; X. Chi did the in-plane analysis. L. Meekisho implemented the front width minimization program. Our thanks to Professor N. Holtz for his assistance with computer methods. D. Downey implemented TAURUS, the post-processor, on our Apollo network. Special thanks to John Hallquist, Lawrence Livermore National Laboratory, for providing the post-processor, TAURUS, used to plot results.

REFERENCES

[1] J. A. Goldak, B. Patel, M.J. Bibby and J.E. Moore, *Computational Weld Mechanics*, Invited Opening Paper for AGARD Workshop — Structures and Materials 61st Panel Meeting, Oberammergau, Germany, September 8-13, 1985.

[2] H. D. Hibbitt and P. V. Marcal, *A Numerical, Thermo-Mechanical Model for the Welding and Subsequent Loading of a Fabricated Structure*, Computers and Structures, 1973, vol. 3, p. 1145–1174.

[3] J.A. Goldak, A. Chakravarti and M.J. Bibby, *A New Finite Element Model for Welding Heat Sources,* Trans. AIME., 1984, vol. 15B, p. 299–305.

[4] B.A.B. Andersson, *Thermal Stresses in a Submerged-Arc Welded Joint Considering Phase Transformations*, Journal of Engineering Materials and Technology, Trans. ASME, 1978, vol. 100, p. 356–362.

[5] D.G. Atteridge, S.M. Bruemmer, R.E. Page, *Evaluation of Welded and Repair Welded Stainless Steel for LWR Service*, NUREG/CR-3613, PNL-4941, Vol. 2, 1984.

[6] N. Christensen, L. de.V. Davies and K. Gjermundsen, *Distribution of Temperatures in Arc Welding*, British Welding Journal, 1965, vol. 12, p. 54–75.

[7] D. Peckner and J. Bernstein, *Handbook of Stainless Steel*, McGraw–Hill, 1977, pp21–47.

[8] A.K. Gupta, *A Finite Element For Transition Grading From A Fine to Coarse Mesh*, IJNME, 12, p. 35—45, 1978.

[9] M. McDill, J. A. Goldak, A. Oddy, M. J. Bibby, *Isoparametric Quadrilaterals and Hexahedrons for Mesh-Grading Algorithms*, Communications in Applied Numerical Methods.

WELD POOL GEOMETRY PREDICTIONS USING A TWO-DIMENSIONAL HEAT FLOW MODEL

H.A. Nied
General Electric Company
Schenectady, New York 12345

ABSTRACT

A mathematical model of the Gas Tungsten Arc Welding (GTAW) process for thin plates based on the two dimensional solution of the heat conduction equation for a distributed heat source moving at a constant velocity was formulated. A computer solution for the weld pool envelope was obtained from the temperature field by using the Newton-Raphson iteration technique. The process model was validated by conducting experiments on 1/16 inch thick type 304L stainless steel using a coaxial optical GTAW torch vision system to obtain video data of the weld pool. The video data of the weld pool geometry and shape were compared to predictions obtained from the model. It was shown that the predictions of the weld pool geometry and shape obtained by the mathematical process model compared favorably with the measured weld pool test data over a broad range of welding parameters.

INTRODUCTION

A simplified 2-D mathematical model of the weld pool was developed to simulate the GTA welding of thin sheet metal. This process model was to provide an inexpensive tool to conduct parametric studies over a large range of welding parameters. The first series of many experiments was conducted to determine the weld pool geometry for typical GTA welding conditions. This provided the data to compare the weld pool geometry obtained by experiment with the predictions from the model.

The objective was to determine if the process model could provide reasonable agreement with experimental results. If good agreement between the process model and tests could be established, then this process model would provide both a better understanding of the process and the necessary functional relations for developing the welding process control algorithms.

THEORETICAL MODEL

Many papers have been published in the open literature on the topic of heat flow analysis for welding processes. Essentially, two approaches for welding analysis have been used. The first method consisted of obtaining closed form solutions of the governing partial differential equation, while the second approach employed numerical methods such as finite difference, or finite element techniques. To obtain a solution in this study, only the first method was considered since it was desired to obtain a simple functional expression that could be used for an adaptive welding feedback control system algorithm.

The governing partial differential equation for either moving point or line heat sources seems to have been first derived by Wilson.[1] Three decades later Rosenthal[2-4] independently applied this formulation and solution to arc weld problems. These references are often quoted and have served as the starting formulation for papers that followed. Wells,[5] using this line source solution, showed fair agreement between theoretical predictions and experiments. In that paper, Wells seems to have been the first to have derived the implicit relationship for the maximum weld puddle width. Grosh, et al., [6] investigated the effect that temperature-dependent thermal diffusivity would have on the temperature distribution for moving heat sources. Their results for the 2-D solution showed approximately a 200°C difference in a region within 2 mm of the weld bead when compared to the experimental results obtained from the welding of thin stainless steel plates. Grosh and Trabant[7] found excellent agreement between experimental and analytical studies of quasi-stationary temperature distribution in both thick and thin arc-welded stainless steel plates using Grosh's earlier solution. This indicates, that for certain applications not involving complicated weldment geometries, analytical solutions do indeed provide good predictions of the arc welding process. Christensen et, al.,[8] conducted tests by welding single beads deposited on heavy sections of mild steel and aluminum. Their comparison of experimental results with the predictions obtained using the Rosenthal model showed a reasonable correspondence between weld puddle shape, width and length. However, their attempts to predict penetration were not successful. Their findings also indicated that the surface temperatures could be predicted much better than plate interior temperatures when using the Rosenthal model. Pavelic et, al.,[9] showed that their model provided good predictions of the peak temperatures in the heat-affected zone within 10% of experimentally measured values. They also showed that weld pool width and length predicted by line source theory, even though it did not exactly agree with experimental data, did plot parallel to the experimental results. Malmuth [10] determined that the latent heat of fusion exerted an almost neglible effect on the melting isotherm at low arc velocities, but had an appreciable effect on the trailing edge at high velocities. More recently Pavelic [11] used the line source theory as a building block for the prediction of weld pool width and length. The formulation derived was shown to predict the pool boundary within 3% of values obtained experimentally from GTA welding of carbon steel.

Pavelic[9] seems to be the first to include a finite area-distributed heat source into his welding analysis, formulated as a finite difference solution of the governing equations. He used a radially symmetric Gaussian heat flux distribution as the heat input for the electric arc. Tsai [12,13,14] extended this approach by including finite heat sources into his welding analysis models. The formulation of the process model to follow also uses a finite heat source which is considered to be most important if the model is to realistically simulate the GTAW process.

FORMULATION OF GOVERNING EQUATIONS

The GTAW welding process is depicted in Figure 1. For this formulation we are interested in the temperature field observed from a moving system attached to the moving arc where it impinges on the plate. The fixed Cartesian coordinate system *X, Y, Z* is oriented such that the *X*-axis is along the weld seam and points in the direction of the torch velocity. The *Z*-axis direction is through the plate thickness; while, the *Y*-axis is in a direction transverse to the welding direction. The origin is on the surface of the plate. The moving coordinate system, ξ, Y', Z' is oriented such that the origin is on the surface directly below the electric arc. The ξ axis points in the velocity direction. Figure 2 shows the weld pool shape and geometry viewed directly from above the plate sub-

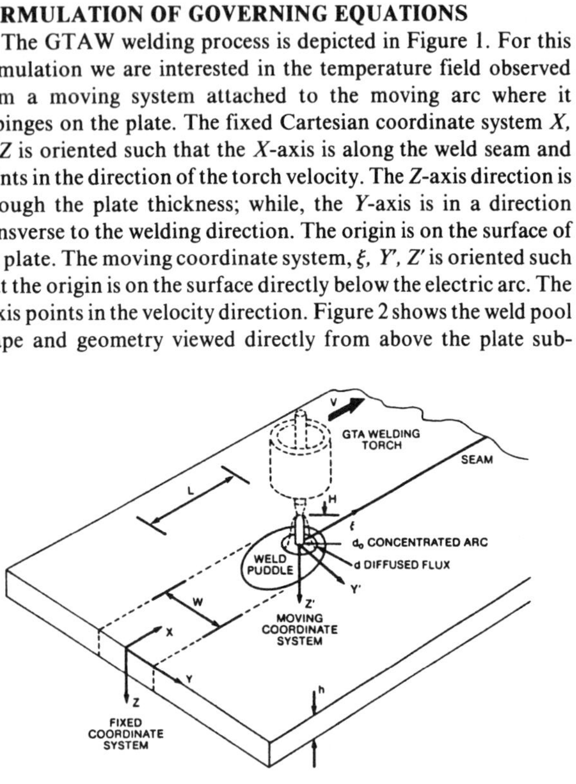

Figure 1. Gas tungsten arc welding (GTAW) process showing the reference and moving coordinate system.

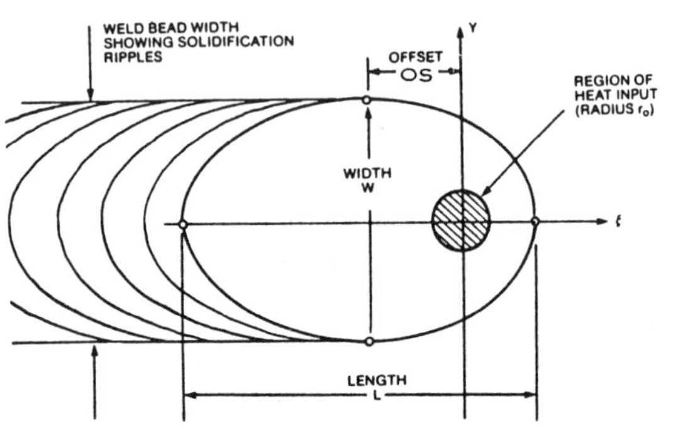

Figure 2. Weld pool shape showing geometric attributes.

jected to a finite heat source of radius (r_0). It is shown that the weld pool has a nearly elliptic shape of length (*L*) and width (*W*). The maximum width of the weld puddle is offset aft of heat input region, assumed to correspond with the electrode location, by the distance (0S). The three-dimensional heat conduction equation expressed in terms of the fixed coordinate system is given by:

$$\frac{\partial^2 T}{\partial X^2} + \frac{\partial^2 T}{\partial Y^2} + \frac{\partial^2 T}{\partial Z^2} = \frac{PC}{K}\frac{\partial T}{\partial t} \tag{1}$$

where (ρ), (*c*), and (*K*) are the material density, specific heat and thermal conductivity; respectively. The relationship between the fixed and moving coordinate system can be expressed in terms of the velocity (v).

$$X = \xi + vt \tag{2}$$

If it is assumed that the moving frame of reference is traveling at a uniform velocity, then the transient temperature effects can be removed from the formulation. Accordingly, when partial derivatives of the spatial coordinate and the time variable t obtained from Equation (2) are substituted into Equation (1), the following governing partial differential equation valid in the moving coordinate system is obtained for a quasi-steady state process.

$$\frac{\partial^2 T}{\partial \xi^2} + \frac{\partial^2 T}{\partial Y^2} + \frac{\partial^2 T}{\partial Z^2} = -\left(\frac{v}{\alpha}\right)\frac{\partial T}{\partial \xi} \tag{3}$$

With the thermal diffusivity (α) defined by:

$$\alpha = \frac{K}{\rho\, c} \tag{4}$$

Rosenthal [4] has shown that the form of the solution to Equation (3) can be expressed as:

$$T(\xi, Y, Z) = A_0 e^{-\lambda\xi}\, \Phi(\xi, Y, Z) \tag{5}$$

with the parameter (λ) defined as:

$$\lambda = \frac{v}{2\alpha} \tag{6}$$

When the solution given by Equation (5) is substituted into Equation (3), the Helmholtz equation in terms of (Φ) is obtained.

$$\nabla^2\Phi = \lambda^2\Phi \tag{7}$$

SOLUTION

The previous section describes the governing equations for a constant velocity frame of reference that are applicable for determining the temperature distributions in a three-dimensional heat conducting medium when the proper boundary conditions are specified. These governing equations can be specialized to simulate the welding of thin metal plates by reducing Equation (7) to a two-dimensional planar heat conducting material. Accordingly, the Helmholtz equation expressed in terms of Cartesian coordinates will reduce to:

$$\frac{\partial^2\Phi}{\partial\xi^2} + \frac{\partial^2\Phi}{\partial Y^2} = \lambda^2\Phi \tag{8}$$

It is more convenient to express this equation is terms of polar coordinates to obtain a model for the arc welding of thin sheet metal.

$$r^2\frac{\partial^2\Phi}{\partial r^2} + r\frac{\partial\Phi}{\partial r} - r^2\lambda^2\Phi = 0 \tag{9}$$

where the radius vector (r) is given as:

$$r = \sqrt{\xi^2 + Y^2}$$

The general solution for this equation is expressible in terms of modified Bessel functions.

$$\Phi = AI_0(\lambda r) + B\,K_0(\lambda r) \tag{10}$$

When this solution is substituted into the expression for the temperature (Equation 5), the two-dimensional form of the temperature field is obtained with the constants of integration (A) and (B).

$$T(\xi, r) = e^{-\lambda\xi}\,[AI_0(\lambda r) + B\,K_0(\lambda r)] \tag{11}$$

The constants of integration for this solution are determined by imposing the boundary conditions. It is assumed that the thin plate is infinite in size and that the temperature at distances far from the weld puddle remain at ambient. Consequently, if the temperature rise at infinity is zero, then the constant of integration (A) must be made equal to zero.

The constant of integration (B) is determined from the boundary condition used to simulate the heat flux input. Most solutions in the literature assume either a point or line heat source; i.e. [2,3,4,5,6,7,8] Pavelic[9] appears to be the first to incorporate a finite distributed heat source into the solution which provided a more realistic simulation of the welding process. In his formulation he assumed a radially symmetric Gaussian heat input distribution. A solution for that welding model was obtained by using the finite difference method. The introduction of a finite heat source makes the problem more difficult. Tsai[14] introduced the concept of a skewed finite heat source based on his experimental research. This form for the heat input seems to more realistically describe the heat flux distribution on the surface of the plate being welded. The welding model described in this report will follow his approach by assuming that the heat flux on the surface of the thin plate can be expressed as a Gaussian function modified by an exponential function in terms of the Helmholtz parameter. This separable form for the heat source distribution is appealing since it is functionally compatible with the solution and its derivatives to satisfy the partial differential equation and the governing boundary conditions. The heat input distribution used in this derivation is mathematically expressed as:

$$q = q_0\,e^{-Cr^2}\,e^{-\lambda\xi} \tag{12}$$

Figure 3 shows a plot of this heat input distribution in a plane coincident with the weld seam in the velocity direction. The implication of this distribution is that the maximum heat input is aft of the geometric center. At low velocities this offset effect is insignificant and a radially symmetric Gaussian flux distribution can be applied. However, at higher velocities, it is considered to have an appreciable effect on the mechanism for heat flow into the weld pool.

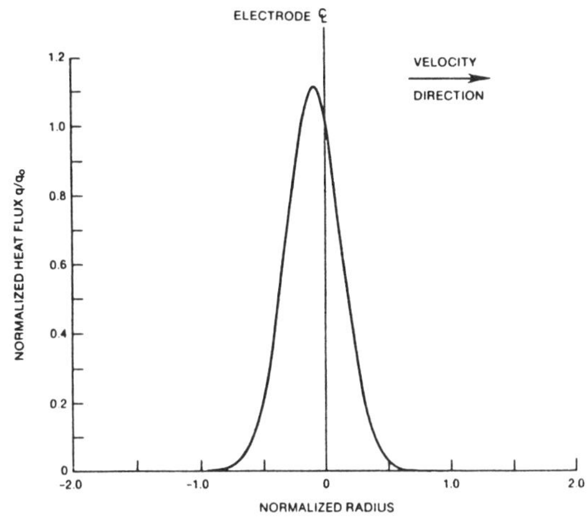

Figure 3. Skewed finite heat source distribution after C. Tsai.[14]

When the expression for the constant of integration is inserted into the solution given by Equation (11), the final solution for the temperature distribution was obtained as:

$$T(\xi,r) = \left(\frac{r_0 q_0}{2Ch\lambda K}\right) \times \left[\frac{1 - e^{-C}I_0(\lambda r_0)}{I_0(\lambda r_0)\,K_1(\lambda r_o) + I_1(\lambda r_0)\,K_0(\lambda r_o)}\right] \times e^{-\lambda\xi}\,K_0(\lambda r) \tag{13}$$

The chief purpose for modeling the GTAW welding process was to obtain the weld pool envelope and the associated shape and geometry. Once the temperature distribution has been determined, the weld pool solid-liquid interface can be determined. Since the temperature solution is expressed in a closed form, it is appealing to use the Newton-Raphson technique for obtaining the weld puddle envelope. This method was found to be computationally efficient for determining the points which describe the locus of this envelope. In this formulation, it is assumed that the latent heat effect is negligible based on Malmuth's[10] research. Ghent[15] showed that the width of the weld pool is independent of the latent heat, since there is neither melting nor solidification at the points of maximum puddle width shown in Figure 2, which also corresponds to the solidified weld bead width.

TEST DESCRIPTION

Test Facility

Process tests were conducted using a robot facility (CART) having a Cartesian motion with rotational capability about the vertical (*Z*-axis) of the torch holder.

The TIG Welding Vision System torch used for obtaining the process test data is shown in Figure 4. This torch design includes the optics for weld puddle geometry acquisition. The vision system torch had a 1-inch-diameter gas cup and a tungsten-thorium electrode 3/32 inch in diameter with a 30 degree included angle and a 0.020 inch flat on the tip. The 1-inch-gas cup diameter provided a large optical field of view that allowed one to obtain pool images over a broad range of welding parameters. This torch design proved to be an outstanding research tool to clearly observe the dynamic features of the weld pool during the welding run. The electrode height was set at 0.100 inch above the workpiece, for all of the process tests conducted.

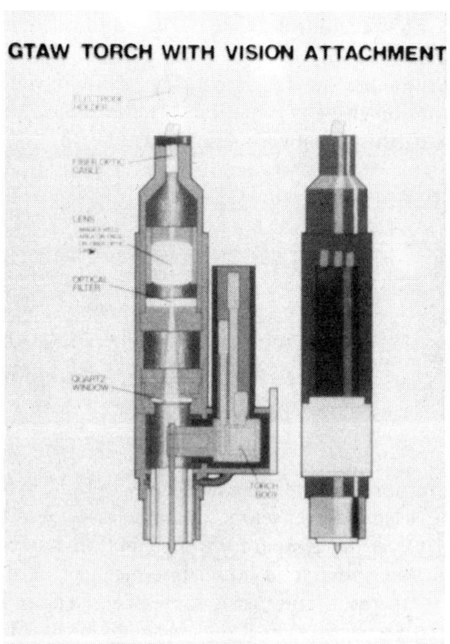

Figure 4. Vision system torch showing optics for viewing weld pool.

Test Specimen Configuration

The test specimen size was selected to minimize edge effects and welding distortion. Also the specimen was to be long enough to provide good, stable video data over a sufficient length without transient effects resulting from startup, shutdown, or welding parameter changes. The specimens for the test consisted of 2 pieces 1/16 inch thick, 3 inches width by 18 inches length.

Procedure for Conducting Process Test

A test matrix consisting of 12 welding runs was used to determine the relationship between the weld pool geometry and the welding parameters used. The arc velocity was varied from 8.9 to 13.7 cm./minute (3.5 - 5.4 inch/min) while the current was adjusted in steps of 10 amps, from 100 to 130 amperes. While the test was in progress the weld puddle image could be viewed on a TV monitor while the image was being recorded. In addition, during each test, the elapsed time (min:sec), arc power (kw), torch speed (IPM), voltage (VOLTS), current (AMPS) and heat input (KJ/IN) were continuously recorded.

VIDEO DATA ANALYSIS

Selected segments of the weld pool image recorded on video tape were loaded onto a frame accessible video disc. Each frame is digitized by a Grinnel module which is an electronics video digitizing hardware package that is interfaced to a VAX11/750 minicomputer. These images can be viewed on a video monitor to select a single image for analysis. Using the Talos digitizer pad and an intergraphics pen, the location of the electrode and the data points that define the envelope of the puddle can be established. Figure 5 shows a typical selected weld puddle image that was analyzed to determine geometric parameters.

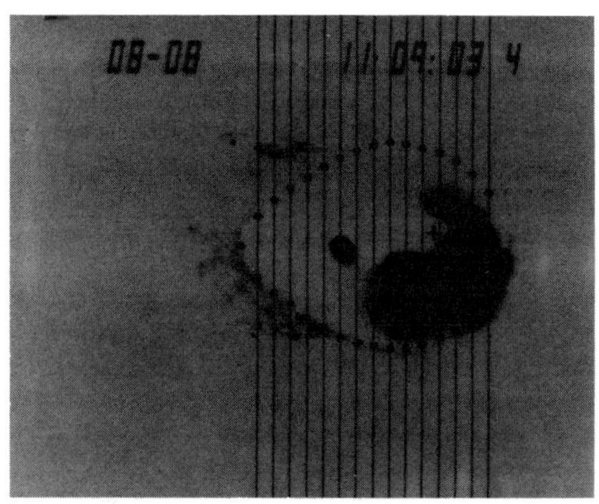

Figure 5. Weld pool video image.

For the image shown in this figure, the following weld puddle geometry was determined:

1. Puddle width = 304 mils
2. Puddle length = 420 mils
3. Puddle area = 95460.08 square mils
4. Puddle perimeter = 1121.44 mils
5. Aspect ratio L/W = 1.380

After this test data was extracted from the video images, a comparison was made with the analytical model predictions.

DISCUSSION OF RESULTS

A Fortran computer program of GTAW process model previously described was written which could be used to obtain a graphics display of the input data and the weld pool shape on a Tektronix 4014 graphics terminal. Figure 6 shows a typical plot of the weld puddle shape and geometric values superimposed on the graph.

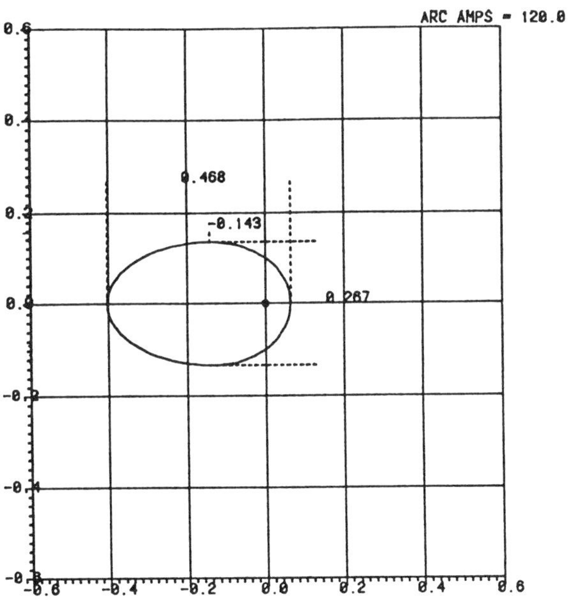

Figure 6. Typical computer plot of weld pool using solution.

This computer program had the feature of calculating the weld pool perimeter, area, maximum width, length and aspect ratio, which is the ratio of length to width. Figure 7 shows a typical computer printout listing the welding parameters, workpiece data, and the weld puddle geometry.

TIG WELDING ANALYSIS

Welding Parameters

Enter Arc Efficiency (%)
50
Enter Arc Voltage (VOLTS)
11.2
Enter Arc Current (AMPS)
55
Enter Arc Velocity (INCH/MIN)
3.5

Workpiece Geometry & Material Data

(Input Thermal Properties for Room Temperature Values)

Enter Metal Thickness (INCH)
.062
Enter Thermal Conductivity (BTU/HR. FT. DEG F)
14.4
Enter Density (LBS./CUB. FT.)
501
Enter Specific Heat (BTU/LB. DEG F)
.11
Enter Melting Temperature (DEG F)
2700

Weld Puddle Geometry

Perimeter = 0.719 inches
Area = 0.041 SQ. inches
Width = 0.215
Offset = 0.039 inches
Length = 0.242 inches
Aspect Ratio = 1.125
Fortran stop

Figure 7. TIG welding analysis program output.

Figures 8 and 9 are comparisons of predictions made with the GTAW process model and the video weld pool data. The results show that the weld pool width is nearly linear to the welding power; while the puddle area is a non-linear function of the power.

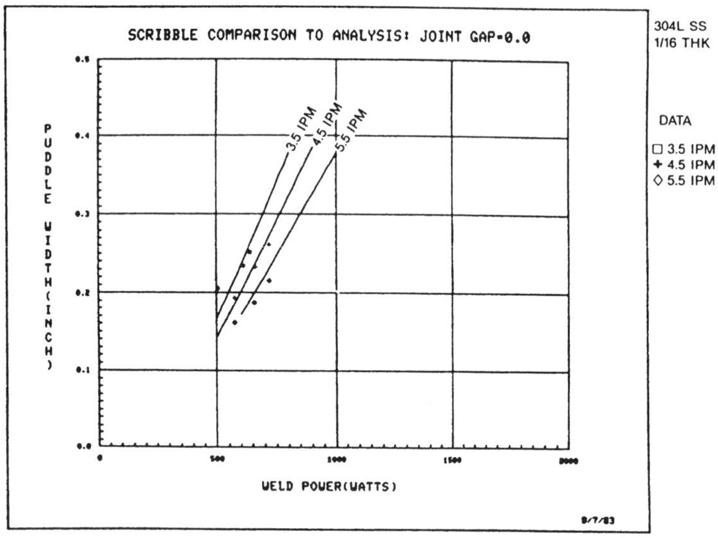

Figure 8. Puddle width vs. weld power, 1/16 inch thick 304L stainless steel, joint gap = 0.0 inch.

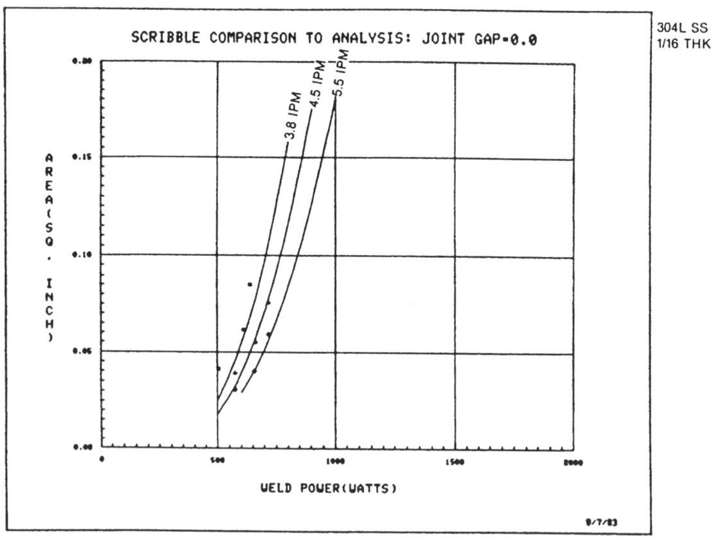

Figure 9. puddle area vs. weld power, 1/16 inch thick 304L stainless steel, joint gap = 0.0 inch.

The predictions obtained using the simple 2-D process model compared favorably with the tests conducted on thin sheet metal.

In addition to providing weld pool geometric data to validate the analytic process model, the process tests also established set points for conducting experiments for different materials and welding parameters. Once it was demonstrated that the process model could provide reasonable predictions of the weld pool geometry, it was used to determine those discriminants that could most effectively be used in a welding adaptive control system based on the optical sensing of the weld puddle. The studies revealed that the weld puddle maximum width and area were the most sensitive discriminants that could be extracted from the weld pool image for controlling the GTAW process.

REFERENCES

1. Wilson, H.A., "On Heat Convection," Proc. Cambridge Phil. Soc. *12* (1904), pp. 406-423.
2. Rosenthal, D., and Schmerber, R., "Thermal Study of Arc Welding — Experimental Verification of Theoretical Formulas," Welding Journal, *17*, April 1938, pp. 2-8.
3. Rosenthal, D., "Mathematical Theory of Heat Distribution During Welding and Cutting," Welding Journal, *20*, May 1941, pp. 220s-234s.
4. Rosenthal, D., "The Theory of Moving Sources of Heat and Its Application to Metal Treatments," Trans. ASME, *68*, Nov. 1946, pp.849-866.
5. Wells, A.A., 'Heat Flow in Welding," Welding Journal, *31* (5); May 1952, pp. 263s-267s.
6. Grosh, R.J., Trabant, E.A., and Hawkins, G.A., "Temperature Disbribution in Solids of Variable Thermal Properties Heated by Moving Heat Sources," Quar. Applied Math., Vol. *13*, No. 2, 1955, pp. 161-167.
7. Grosh, R.J., and Trabant, E.A., "Arc Welding Temperatures," Welding Journal, *35* (8); 1956, pp 396s-400s.
8. Christensen, N., de L. Davies, V., and Gjermundsen, K., "Distribution of Temperatures in Arc Welding," British Welding Journal, Vol. *12*, No. 2, Feb. 1965, pp. 54-75.
9. Pavelic, V., Tanbakuchi, R., Uyehara, O.A., and Myers, P.S., "Experimental and Computed Temperature Histories in Gas Tungsten-Arc Welding of Thin Plates," Welding Journal, *48* (7); 1969, pp. 295s-305s.
10. Malmuth, N.D., Hall, W.F., Davis, B.I., and Rosen, C.D., "Transient Thermal Phenomena and Weld Geometry in GTAW," Welding Journal, *53* (9); Sept., 1974, pp. 388s-400s.
11. Pavelic, V., and Tsao, K.C., "Weld Puddle Shape and Size Correlation in a Metal Plate Welded by the Gas-Tungsten-Arc (GTA) Process," ARC PHYSICS AND WELD POOL BEHAVIOR, Intl. Conf. Welding Institute, Cambridge, 1980, pp. 251-258.

12. Tsai, C.L., "Parametric Study on Cooling Phenomena in Underwater Welding," PhD Thesis, MIT (1977).

13. Tsai, C.L., "Modeling of the Thermal Behavior of Metals During Welding," TRENDS IN WELDING RESEARCH, Edited by S.A. David, American Society of Metals, 1981, pp. 91-108.

14. Tsai, C.L., "Finite Source Theory," Preceedings 1983 Engineering Conference; Modelling of Casting and Welding Process, New England College, Henniker, NH, July 31-Aug. 5, 1983.

15. Ghent, H.N., et al., "Heat Conduction Modelling of the Two-Dimensional TIG Weld Pools," ARC PHYSICS AND WELD POOL BEHAVIOR, Intl. Conf. Welding Institute, Cambridge, 1980.

HEAT TRANSFER AND FLUID FLOW IN LASER MELTED WELD POOLS

A. Paul, T. DebRoy
Department of Materials Science and Engineering
Pennsylvania State University
University Park, Pennsylvania 16802

ABSTRACT

A mathematical model was developed for the calculation of heat transfer and fluid flow in laser melted pools. Appropriate equations of change were solved numerically to obtain insight into the significance of various welding parameters. Furthermore, the model was used to examine the role of sulfur in determining the pool shape during laser welding.
The weld pool temperature obtained from vapor composition was in reasonable agreement with the predicted temperature values. From the analysis of the numerically predicted flow fields it was demonstrated that the heat transfer from the laser beam to the base metal via the molten weld pool was significantly improved by the presence of an appropriate surface active agent. Critical experiments were designed and conducted to determine the weld pool shapes with samples of pure iron and iron doped with sulfur. The pool shapes predicted by the model were found to be in good agreement with those obtained experimentally.

DURING LASER WELDING, the dynamics of heat transfer, fluid flow and mass transfer affect the characteristics and the quality of the fabricated product. The eventual properties of the weld zone are strongly influenced by the fluid flow in the weld pool, heat transfer and the transport of the alloying elements from the molten pool into the environment. Owing to the small size of the weld pool and the presence of plasma in its vicinity the physical measurements of the temperature and the velocity fields in the pool are not easy tasks. A recourse is to develop an understanding of the transport processes by mathematically modeling the essential physical features of the process.

These features must include appropriate considerations of the driving force for flow resulting from (a) the spatial variation of surface tension at the weld pool surface due to both composition and temperature dependance of surface tension and (b) the "body force" which originates from the relatively minor local variations of liquid density inside the weld pool. The possibility of enhanced heat and mass transport rates due to turbulence must also be considered. The extent of absorption of the laser beam by the workpiece in presence of plasma must be taken into account. The value of the laser beam absorptivity is crucial since the rates of all the important transport processes are profoundly affected by the extent of beam absorption.

The quality of the model predictions depends on our ability to understand and model these important physical features adequately. Although we can account for the attenuation of the beam energy due to the presence of the plasma by providing an experimentally determined value of absorptivity, this must be viewed as a sacrifice of the predictive capability of the model in our quest for accuracy. A more fundamental limitation is imposed by the lack of necessary data pertinent to the plasma-weld pool system. The primary driving force for fluid flow in the laser melted pools is the surface tension gradient. The interfacial tension values between molten metals and plasma are not available in the literature. The influence of plasma-weld pool heat exchange on the transport processes in the weld pool is also not clearly understood. Although the temperature of the plasma is known to be much higher than the temperature of the weld pool surface, the heat exchange between the weld pool surface and the plasma cannot be estimated with a reasonable degree of trustworthiness. At high laser powers additional complexities are introduced due to the presence of the keyhole. In view of these uncertainities in the modelling of transport processes, the model must be tested by designing suitable experiments before predictions can be accepted as reliable.

The velocity and temperature fields are predicted by numerical solution of the equations of change with appropriate consideration of the existence of surface tension gradient on the surface of the weld pool. An experimentally determined value[1] of the absorptivity of the laser beam was used for the calculations. To understand the nature of the heat transfer between the weld pool surface and its surroundings that contain the hot plasma, the predicted temperatures in the weld pool were compared with the temperatures determined experimentally. The experimental determination of the weld pool temperature was based on the evaluation of the ratios of the vaporization rates of various elements from the weld pool[2,3]. Critical experiments were designed and conducted to determine the weld pool shapes in samples of pure iron and iron doped with sulfur. The experimentally determined weld pool shapes were compared with the corresponding theoretically calculated values.

MATHEMATICAL MODELING

A schematic diagram of the computational domain is shown in figure 1. The equations of change for two dimensional, transient, incompressible flow were solved numerically for the shaded region of the diagram. Since these equations are well documented in standard textbooks[4,5], these are not reproduced here. Instead, only the salient features of the computation scheme are presented here. The weld pool was assumed to be symmetrical about the center of the computational domain. The curved weld pool boundary was approximated by a series of steps for simplicity. The turbulence in the molten metal region was represented by an effective viscosity model[6]. In the representation of the Marangoni effect, the temperature coefficient of surface tension was treated as a constant[7]. The laser power and the absorptivity was introduced in the computational scheme as a source term in the energy equation.

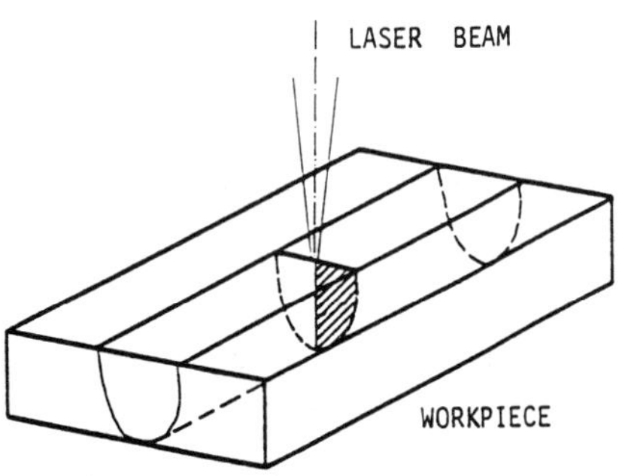

Fig. 1 - Schematic diagram of the computational domain

The important boundary conditions utilized for the calculations are illustrated in figure 2. At the surface of the weld pool, the shear stress was equated to the gradient of surface tension. Since the heat exchange between the surface of the weld pool and the surroundings that contain hot plasma is not clearly understood, the effect of allowing a negligible amount of heat exchange between the pool surface and its environment was examined. Heat exchange was only considered at the pool surface where the laser beam impinged on the material. The computatational domain was divided into 25 x 15 equally spaced grids. The equations of change were represented in explicit finite difference form. The pool shape was updated after each cycle. The details of the computations are given elsewhere[8]. The data used for the calculations are given in Table I.

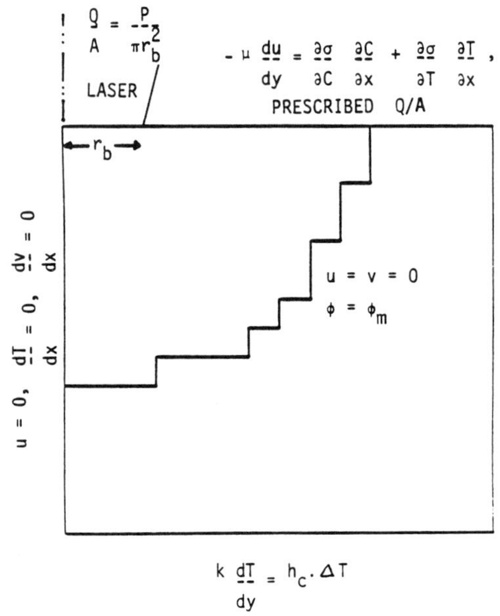

Fig. 2 - Important Boundary conditions used for calculations

Table I

Data used for the heat transfer and fluid flow calculations

Laser Power	400 watts
Absorptivity	12 %
Temperature coefficient of surface tension	-0.54 & 0.47 gm/sec^2 K
Effective viscosity of molten metal	5 poise
Effective thermal conductivity of molten metal	0.0745 x 5 cal/cm-sec-K
Heat transfer coefficient at the solid surface	$3.15\times10^{-4}[\Delta T]^{0.25}$ cal/sec cm^2 K

EXPERIMENTAL PROCEDURE

A schematic diagram of the experimental setup is shown in figure 3. A carbon dioxide laser, Coherent Model Everlase 525-1, capable of producing a maximum output power of 575 watts in the continous wave mode was used. Samples were placed on a remotely controlled, electrically operated table capable of providing linear motion. All welding was carried out inside a plexiglass box using a 2.54 cm diameter, 12.7 cm focal length Zn-Se lens with an anti reflection coating.

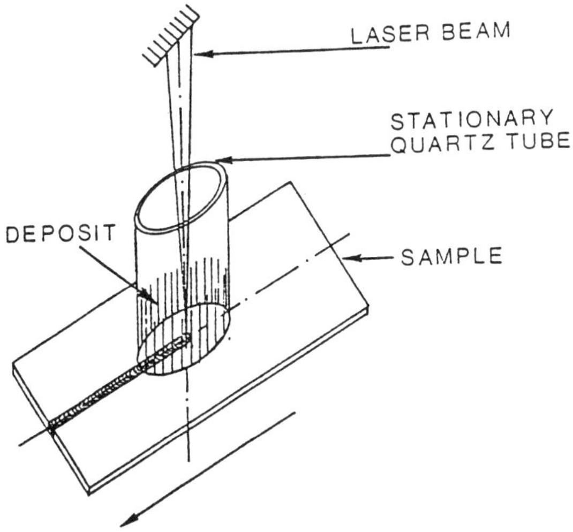

Fig. 3 - Schematic diagram of the laser welding setup

A portion of the vaporized material was collected as condensation on the inner surface of a hollow, cylinderical, both end open quartz tube which was held stationary and coaxial with the laser beam. The composition of the condensate was determined by electron probe microanalysis (EPMA) and atomic absorption (AA) techniques.

Experiments were conducted with three types of samples - high manganese stainless steel (USS Tenelon, 15% Mn, 18% Cr, 0.75% Ni, 0.12% C), puratonic iron obtained from Johnson Mathey Chemicals Limited containing less than 10 ppm total impurities and pure iron samples chemically doped with sulfur. The doping was achieved by exposing pure iron specimens to a predetertimed composition of H_2/H_2S gases isothermally at 1243 K for sufficient period of time. These samples were irradiated by a laser beam under identical welding conditions. After welding, transverse sections of the welds fabricated with pure iron and the iron samples doped with sulfur were examined under a microscope to study the effect of sulfur on the pool shape.

The technique utilizing the ratio of the vaporization rates of any two alloying elements was used to calculte pool temperatures in USS Tenelon. A brief description of this technique is given below.

Since the vaporization rates of individual elements are related to their equilibrium vapor pressures over the molten pool, the ratio of the vaporization rates of any two elements is a strong function of temperature. Alternatively, for a given composition of the molten pool if the ratio of the vaporization rates of any two elements is known, the average pool temperature can be determined. The part of the vaporized material collected in the quartz tube was used to determine the relative rates of vaporization of the individual elements. Details of this procedure are documented elsewhere[2,3].

RESULTS AND DISCUSSION

For the welding of pure iron, the computed velocity vector plot is shown in figure 4.

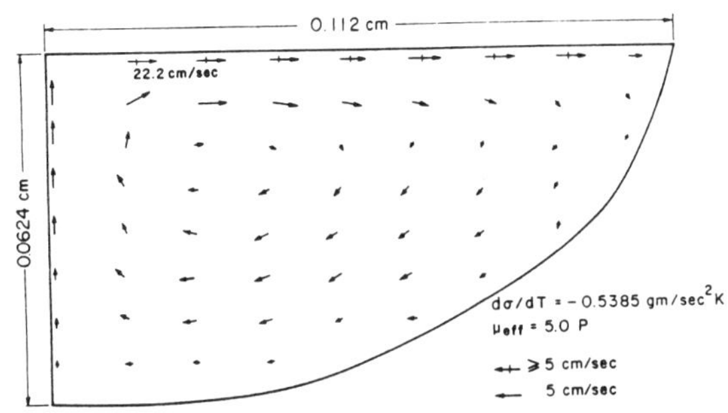

Fig. 4 - Velocity vector plot for the case of pure iron

The nature of the flow can be clearly seen to be radially outwards. The maximum velocity is about 22 cm/sec. Due to this radially outward flow the packets of fluid that absorb heat from the laser beam at the center of the pool are quickly transported to the edge of the pool. This results in widening of the pool and leads to a relatively lower depth to width ratio of the weld zone. Figure 5 is the corresponding temperature profile for the case of pure iron. Constant temperature

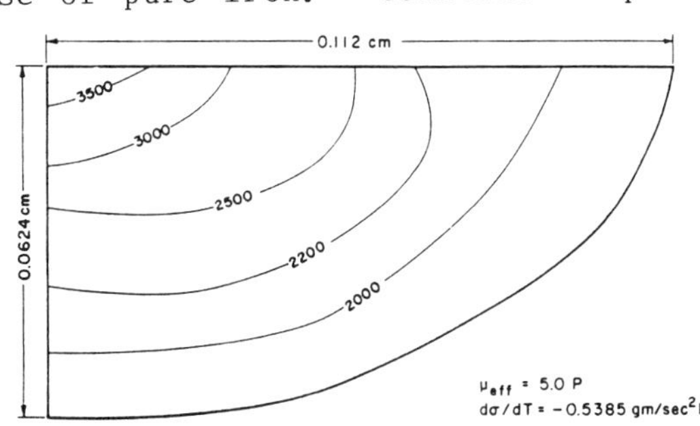

Fig. 5 - Temperature profile for the case of pure iron

lines are shown in this plot. It may be seen that the maximum temperature predicted by the model is slightly higher than the boiling point of iron. The existence of a strong temperature gradient in the weld pool is consistent with the absorption of a significant amount of energy in a small localized area at the centre of the weld pool.

It can be shown from theoretical considerations that the ratio of the vaporization rates is a strong function of temperature. Figure 6 is a plot of the ratio of vaporization rates of Fe and Mn from theoretical considerations versus temperature.

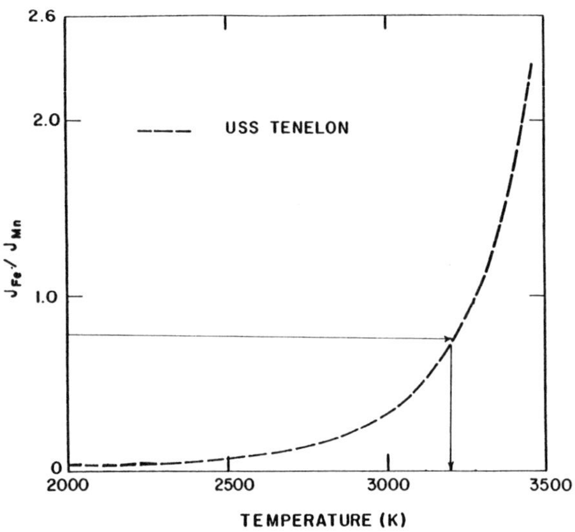

Fig. 6 - The ratio of calculated vaporization rates of Fe and Mn (J_{Fe}/J_{Mn}) as a function of temperature

Superimposed on this plot is the experimentally obtained value of the ratio of vaporization rates of Fe and Mn (shown by arrows). It is seen that the temperature corresponding to the observed ratio is 3200 K.

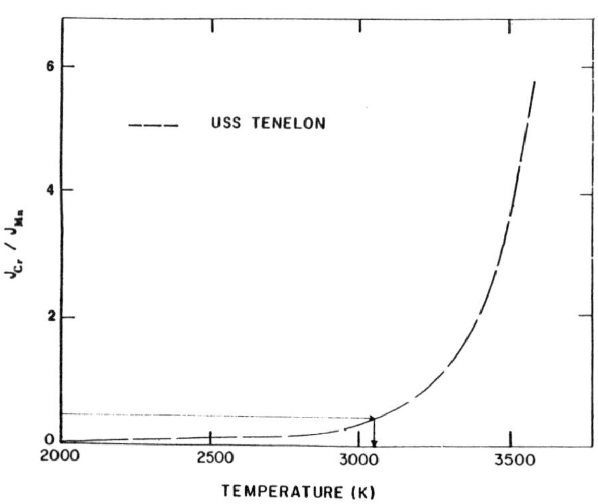

Fig. 7 - The ratio of calculated vaporization rates of Cr and Mn (J_{Cr}/J_{Mn}) as a function of temperature

Figure 7 is another plot of the ratio of vaporization rate of Cr and Mn from theoretical considerations versus temperature. Again, superimposed on this plot is the experimentally observed ratio. The temperature obtained is 3050 K. It may be noted that irrespective of the element pair chosen the temperature of the weld pool is the same. Since the vaporization rate is a strong function of temperature, the effective pool temperature estimated on the basis of the vaporization of various elements is expected to be close to the maximum temperature of the weld pool. Since the temperature coefficient of surface tension for most iron base alloys, that do not contain a strong surface active component, is roughly equal to that of pure iron, the computed temperature and velocity fields are also valid for stainless steels.

A comparison of the predicted and observed temperatures shows that the temperatures predicted by the model are roughly equal to those obtained experimentally. Since the heat exchange between the weld pool surface and the hot plasma is not well understood, the net heat transfer from the weld pool surface to its surroundings was assumed to be negligible. The good agreement between the temperature values obtained from the relative vaporization rates of various elements and the numerically computed results indicate that the assumption is not unduly restrictive.

Figure 8 is the temperature profile for the sulfur doped samples. It is observed from the constant temperature contours that the maximum temperature obtained in this case is lower than the value calculated for the fabrication of pure iron (figure 5). The reason for this may be explained from the data in figure 9 in which the velocity field predicted for the sulfur doped sample is shown. We can clearly see that the flow is radially inward and the maximum velocity in this case is higher than that for pure iron. The effect of an inward velocity is that the fluid packets that absorb heat at the center of the pool from the laser beam are quickly transported to the bottom of the pool. This

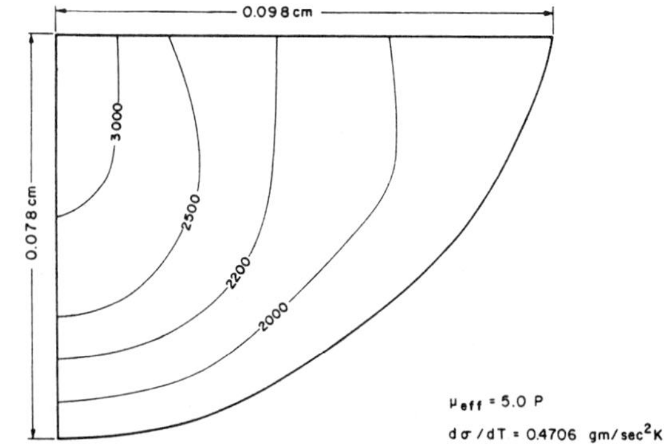

Fig. 8 - Temperature profile for the sulfur doped sample

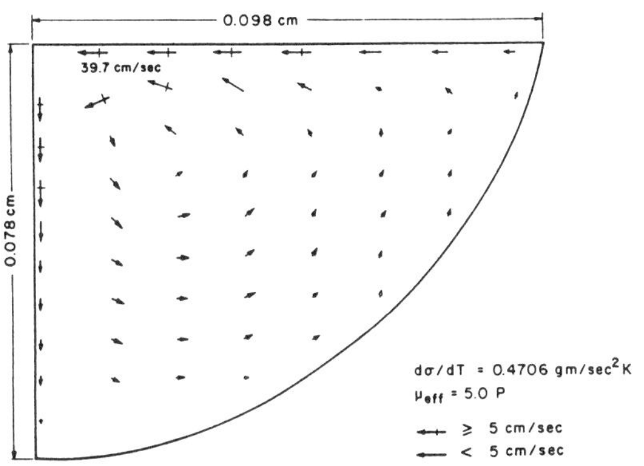

Fig. 9 - Velocity vector plot for the sulfur doped sample

results in a thermal drilling action and we get a higher depth to width ratio. Since packets of relatively cool liquid metal are transported from the edge of the pool to its center, the maximum temperature in the weld pool is lower than in the previous case where computations were presented for pure iron.

SEM micrographs of the weld zone for pure iron and sulfur doped iron are presented in figure 10. Both welds were prepared under identical conditions (laser power = 500 watts, welding speed = 3.5 mm/sec and helium flow rate = 6 liters/minute). The difference in the shape and size of the weld zone is clearly visible. The sulfur content of the weld zone in figure 10 was fairly uniform and was 0.39%.

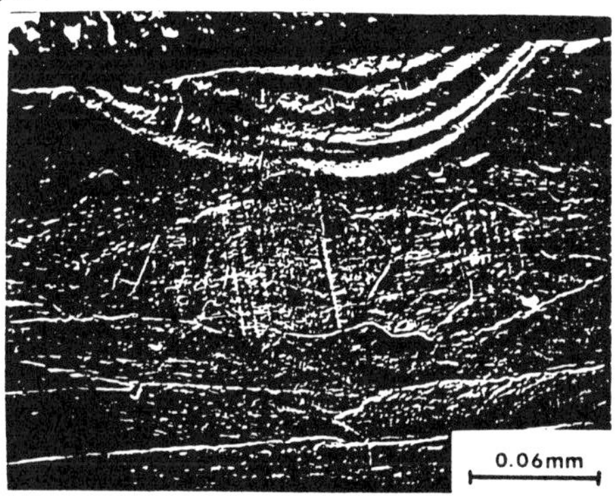

PURE IRON SAMPLE

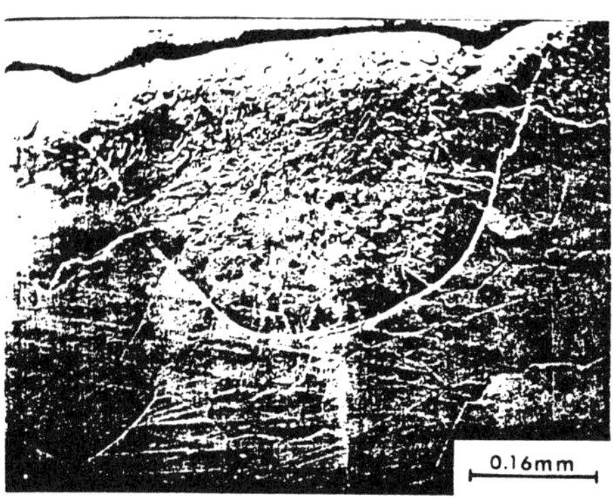

IRON DOPED WITH SULPHUR

Fig. 10 - SEM micrographs of pure iron and sulfur doped iron samples

It is apparent that during laser welding, sulfur increases the depth to width ratio of the weld pool which is consistent with the predictions presented in figures 4 and 8. A similar effect was observed by Heiple and Roper in their studies on the effects of minor elements during gas tungsten arc welding of stainless steels[9].

CONCLUSIONS

Although metal vapors in excited states are produced at relatively low laser powers, the values of the temperature coefficient of surface tension available in the literature seem to be adequate for the prediction of heat transfer and fluid flow in laser melted pools.

The computed values of the weld pool temperature ignoring heat exchange between the weld pool surface and its surroundings (other than direct absorption from the laser beam) was roughly equal to the measured weld pool temperature. Heat loss from the weld pool surface seems to be compensated, at least to a fair extent by heat gain from the hot plasma.

From the analysis of the numerically predicted fluid flow fields, it is demonstrated that the heat transfer from the laser beam to the base metal via the molten pool is significantly improved by the presence of an appropriate surface active agent. Critical experiments were designed and conducted to substantiate the predicted improvements.

ACKNOWLEDGEMENT

This work was sponsored by the U. S. Department of Energy, Office of Basic Energy Sciences, Division of Materials Science under grant number DE-FG02-84ER45158.

REFERENCES

1. Khan, P. A. A. and T. DebRoy, Metallurgical Transactions, 16B, 853-856, (1985)
2. Khan, P. A. A. and T. DebRoy, Metallurgical Transactions, 15B, 641-644, (1984)
3. P. A. A. Khan and T. DebRoy, "Laser Processing of Materials", p. 71, The Metallurgical Society of AIME, Pennsylvania (1985)
4. J. Szekely, "Fluid Flow Phenomena in Metals Processing", Academic Press, London, (1979)
5. R. B. Bird, W. E. Stewart and E. N. Lightfoot, John Wiley and Sons, Inc., (1960)
6. DebRoy, T., A. K. Majumdar and D. B. Spalding, Appl., Math. Modelling, 2, 146-150, (1978)
7. Tszin-Tan, V., R. A. Karasev and A. M. Samarin, Russian Metallurgy and Fuels, 2, 33-34, (1960)
8. Hirt, C. W., B. D. Nichols, and N. C. Romero, UC-34 and UC-79d, Los Alamos, (1975)
9. Heiple, C. R. and J. R. Roper, Trends in Welding Research in the United States, ed. S. A. David, p. 489-520, Metals Park, Ohio, Am. Soc. for Metals, (1982).

A THREE-DIMENSIONAL HEAT TRANSFER FINITE ELEMENT MODEL OF SUBMERGED ARC WELDING OF HSLA STEELS

Pat L. Mangonon, Mahesh A. Mahimkar
Mechanical Engineering Department
Florida Institute of Technology
Melbourne, Florida 32901

Abstract

A three-dimensional heat transfer finite-element model (FEM) of submerged arc welding was developed and applied to micro-alloyed HSLA steels. The model incorporates the temperature variable physical properties and the latent heat of fusion of steel. The model is validated by predicting the correct weld pool depths and shapes in 25 mm and 50 mm thick plates for three heat inputs. In order to simulate weld pool convection, a higher effective liquid thermal conductivity was used and the results compared with those using the normal liquid thermal conductivity. The results suggest that the weld pool depths are essentially independent of thickness and that weld pool convection is significant only at very high heat inputs. For an X-70-X-80 steel, the HAZ microstructures are predicted to be essentially acicular and/or bainitic.

SUBMERGED ARC WELDING (SAW) is primarily used in the construction of ships, offshore structures and pressure vessels. Recently, because of its inherent cost effectiveness, the trend is to use very high heat inputs to fabricate micro-alloyed high strength low alloy (HSLA) steel structures (1-4). The high heat inputs usher in toughness control problems for these steels at the heat affected zone (HAZ).

Micro-alloyed steel base plates attain their excellent structures and properties through controlled processings of steel making, soaking, hot-rolling and cooling (2). Extremely fine polygonal ferrite grains (5 μm), acicular ferrite or ultra-low carbon bainite structures can be formed. Additional strengthening is provided by very fine particles (2 - 10 nm) of vanadium, niobium and/or titanium carbides, nitrides or carbo-nitrides. These particle sizes precipitate in ferrite. Larger particle sizes (greater than 10 nm) are precipitated in austenite during hot-rolling and contribute little directly to the strength (5). Indirectly, they keep the austenite unrecrystallized (5,6) during hot-rolling to provide ferrite nucleation sites during transformation. Much larger particle sizes (greater than 0.1 μm) in the microstructure are left undissolved during the soaking operation.

The HAZ toughness control problems arise because of the number of processes that may occur during the short heating and cooling cycle in welding (1,8). Prominent among these processes are the dissolution and coarsening of the carbonitride particles as they influence solute redistribution in austenite, austenite grain coarsening, the subsequent transformation products, and reprecipitation of the particles. Furthermore, these particles may undergo modification in both their morphology and composition (9).

It is generally agreed that welding parameters of significance in the above processes are the peak temperatures, T_p, the time above the grain coarsening and particle dissoltution temperatures and the cooling time between 800-500°C (8). All of these parameters can be secured from temperature profiles of locations in the HAZ. To obtain the temperature profiles experimentally is tedious and can be time consuming. For this reason, a three-dimensional finite element model was developed for bead-on-plate submerged arc welding to predict the temperature profiles in the HAZ. Heat inputs of 1.58 KJ/mm (40 KJ/in), 3.15 KJ/mm (80 KJ/in) and 4.72 KJ/mm (120 KJ/in) were studied in welding 25 mm (1 inch) and 50 mm (2 inch) thick plates.

THE SAW MODEL

Figure 1 shows the stationary Cartesian coordinate system used with respect to the moving electrode. Admittedly, this is a more difficult problem in comparison to the quasi-stationary condition (where the coordinate system moves with the electrode). However, it gives a better overall physical picture of the process. For reasons associated with the finite element method (FEM), the start of the heat input was not done at Z = 0 but at the third internodal plane of the FEM model.

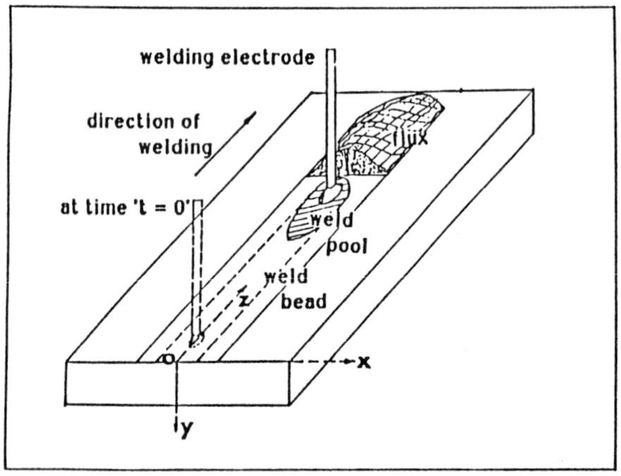

Figure 1 - Bead-on-plate welding model

The net heat input to the plate, Q_p, in Watts, is given by the expression

$$Q_p = \eta EI \tag{1}$$

where

E	=	arc voltage between the electrode and the plate
I	=	applied current, amps.
η	=	efficiency

The efficiency for submerged arc welding cited in the literature is somewhat variable. From peak temperature measurements, Barry, et al (10) suggest that 80 ± 3 per cent of the heat input enters the plate. Christensen, et al (11) claim the efficiency in the 90-99 per cent range. In this study, we used values 80-90 per cent as found by Barry, et al because of another study (12) which found that about 15 per cent of the arc energy is used to melt the electrode in consumable processes. Since energy is needed to partially melt the flux, 80 per cent was chosen for the two lower heat inputs. For the highest heat input studied here, 90 per cent efficiency was used because the energy to melt the electrode and the flux is a fixed amount and will not increase proportionately with heat input.

Following Pavelic, et al (13), the heat flux from the arc is assumed to be radially distributed on the plate surface as a normal Gaussian function.

$$q(r) = \left(\frac{3Q_p}{\pi r_e^2}\right) \exp\left[-3\left(\frac{r}{r_e}\right)^2\right] \tag{2}$$

where

q(r)	=	heat flux, W/m^2
r	=	heat arc radius
r_e	=	the effective radius at which 95 per cent of Q_p is enveloped

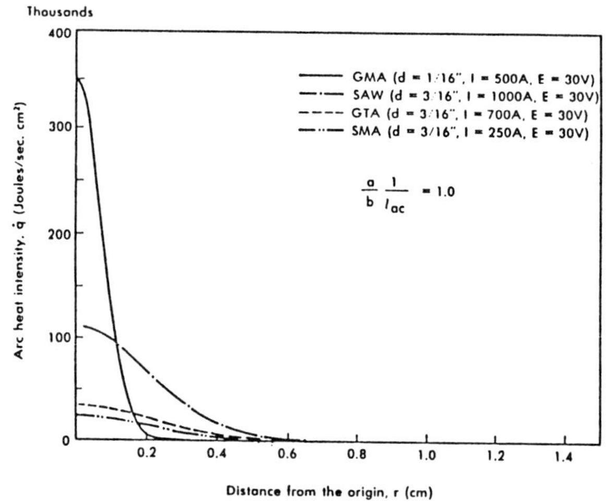

Figure 2 - Effective r_e value for arc welding

The value of r_e is derived from Figure 2 14 . Using the SAW case, the 1000 ampere - 30 volt power used with the 4.76 mm (3/16") electrode translates to an arc input of 3.54 KJ/mm (90 KJ/in) at an electrode speed of 8.47 mm/sec (20 in/min). This is very close to one of the heat inputs studied here, i.e., 3.15 KJ/mm, and the effective radius of 5.5 mm was therefore used. The r_e values for the 1.58 KJ/mm and 4.72 KJ/mm cases were taken as 4.50 mm and 6.50 mm, respectively. These estimates were made from a similar apparent linear variation of the arc size as the heat inputs were varied in gas tungsten arc welding (GTAW) (15).

The net Q_p to the plate is dissipated primarily to heat and melt the material through conduction. The familiar three-dimensional transient conduction equation for an isotropic material is given by

$$\kappa \nabla^2 T + Q = \rho c_p \frac{\partial T}{\partial t} \tag{3}$$

where

κ	=	thermal conductivity of the material ($W/m/^oK$)
T	=	temperature in ^oK
ρ	=	density of the material
t	=	time, in sec.
Q	=	heat source per unit volume (W/m^3)
c_p	=	specific heat ($J/kg/^oK$)
∇^2	=	$\frac{\partial^2}{\partial x^2} + \frac{\partial^2}{\partial y^2} + \frac{\partial^2}{\partial z^2}$

Solution of this transient conduction equation by the finite element method (FEM) (to be described later) uses the temperature variable properties of conductivity and heat capacity of the solid material. The latent heat of fusion was included. All the physical properties were those compiled by P.J. Wray (16). The influence of weld pool convection was examined by comparing the results of the static (case 1) and the dynamic (case 2) cases. Case 1 uses the normal liquid thermal conductivity of 45 W/m/°C while case 2 increases the liquid thermal conductivity from 45 W/m/°C to 120 W/m/°C over a range of temperatures.

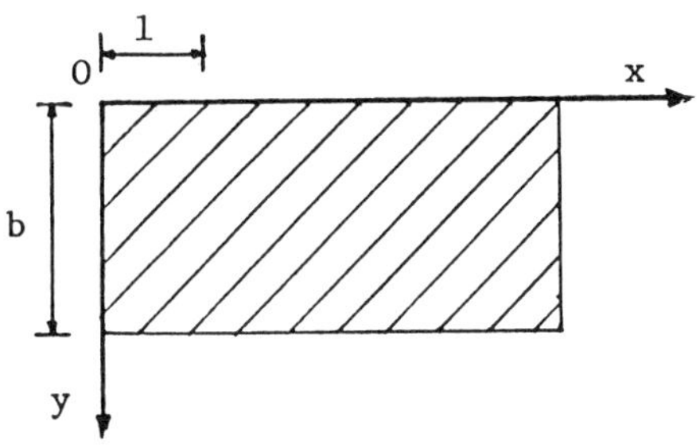

Figure 3 - Boundary conditions

The boundary conditions for the model as shown in Figure 3 are

a) No heat loss through the flux cover and to the other side of the plate at the weld center plane, i.e.

$$\frac{\partial T}{\partial x} = 0 \text{ at } x = 0 \text{ for } 0 \leq y \leq b \tag{4a}$$

$$\frac{\partial T}{\partial y} = 0 \text{ at } y = 0 \text{ for } 0 \leq x \leq \ell \tag{4b}$$

where b is the plate thickness and ℓ is the width of the flux cover.

b) Boundary convection and radiation, according to

$$Q_c = h_c A(T_s - T_o), \text{ and} \tag{5a}$$

$$Q_r = A\varepsilon\sigma(T_s^4 - T_o^4) \tag{5b}$$

where Q_c and Q_r are the convection and radiation heat losses at the top surface not covered with flux, at the bottom surface, and at the side of the plate.

h_c = coefficient of heat convection
ε = emissivity of surface
σ = Stefan-Boltzmann constant
T_s = surface temperature
T_o = ambient temperature
A = surface area

The energy losses to melt some of the flux and the electrode wire were already lumped as part of the original loss of the heat input.

THE FINITE ELEMENT METHOD

The complete details of the development of the finite element method (FEM) are described elsewhere (17,18). It is only sufficient to indicate here the nodal mesh arrangement of the linear brick element model used for the calculations. Figure 4 shows the y-z and the x-z internodal plane spacings to be closer together near the weld centerline, but the x-y plane spacing is equidistant along the z-direction. The positive axial directions are as indicated. A total of 2500-4000 nodes were used in the calculations.

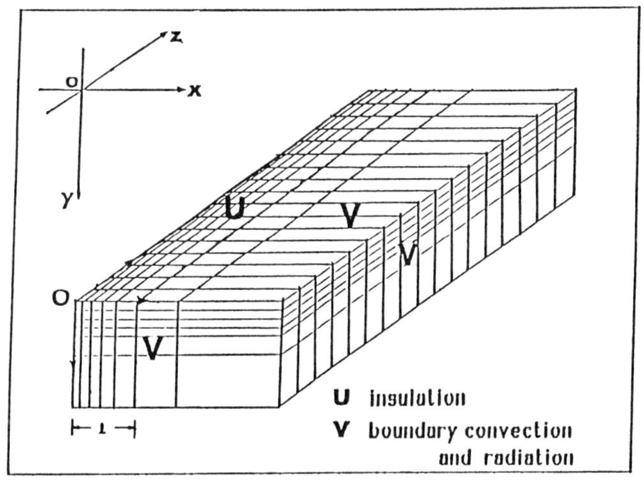

Figure 4 - Linear brick element FEM model showing the nodal mesh arrangements.

Figure 4 demonstrates also the insulation (U) due to the flux cover, and the convection and radiation heat losses from indicated areas of the plates in addition to the primary conduction mode within the plate. As mentioned earlier, some flux will melt but the flux cover over it keeps the heat loss through radiation and convection to a minimum. To keep the model simple, the assumption of insulation over the flux was taken to avoid using the properties of the flux.

The essential welding parameters used in the computational model are listed in Table 1. The effective arc radii have been discussed earlier. Very good agreement with empirical data were obtained using the arc efficiencies indicated. In order to save computational time on the Digital Equipment VAX 11/780 minicomputer, the total width of the plate was limited to 80 mm for the 1.58 KJ/mm heat input. The criterion for the width selections is simply that the temperatures at these distances away from the center of the welds do not deviate very much from the initial room temperature, T_c. (Because of symmetry, the widths of the plates analyzed are of course only half of the total widths.)

The computational FEM was stabilized, i.e., made to converge easily, by discretizing the energy inputs per second and applying them at fractions of a second. The latter are shown as time step increments in Table 1. Comparing these times with the internodal plane spacings, one sees that a distributed heat source is applied on one internodal plane along the z-direction for the 1.58 KJ/mm. For the 3.15 KJ/mm and 4.72 KJ/mm, two and three distributed sources, respectively, were applied according to the time increments in Table 1 in the internodal spacing. Considering the effective radii of the sources, the heat is seen also to be effectively distributed over 3 to 5 internodal plane spacings. To further contribute to the stability of the solution, the effective phase change range, $T_L - T_S$, was chosen to be 80^oK. This is about twice the equilibrium temperature range for a 0.1%C steel but can be rationalized by the non-eqiulibrium conditions during the welding cycle. These conditions can induce the well-known superheating and supercooling conditions during melting and solidification.

Table 1 - Welding Parameters used in Model*

Heat Input (KJ/mm)	1.575	3.15	4.724
Heat arc effective radius, r (mm)	4.50	5.50	6.50
Heat arc efficiency, η (%)	80.00	80.00	85.00
Width of plate modelled (mm)	80.00	100.00	100.00
Time step increment (sec)	0.25	0.20	0.15
Internodal plane spacing			
in terms of welding time (sec)	0.25	0.40	0.45
in terms of distance (mm)	2.12	3.39	3.81

* Electrode diameter = 4.76 mm (3/16")
Welding speed = 8.467 mm/sec (20.0 ipm)

The FEM model uses the lumped capacitance matrix instead of the consistent capacitance matrix because stable results were obtained. The Gauss-Siedel iterative temperature calculation procedure was employed and was continued until the difference between the old and the new calculated temperatures is within 2^oK. To start the computation, an initial guess of the solution is made. The closer the initial guess is to the real solution, the faster is the convergence. Therefore, the computer program was designed to estimate the temperature field based on the previous time step. As the quasi-steady condition is approached, the temperature field becomes repetitive.

RESULTS AND DISCUSSION

TRANSIENT CHARACTERISTICS OF SAW - The choice of the fixed coordinate system in Figure 1, instead of the moving coordinate system (19,20), enables one to determine the end effects in welding. In the moving coordinate system, the origin coincides with the point of application of the heat source and an observer moving with it notices no change of the temperature field around it. This condition is called quasi-stationary and eliminates the time factor in Eq. 3, i.e.

$$\frac{\partial T}{\partial t} = 0.$$

By far, the majority of investigations on welding have adopted the quasi-stationary approach and failed naturally to indicate the transient characteristics.

Figure 5 demonstrates the development of the steady-state weld pool depth for the heat inputs studied and for the case when the liquid thermal heat conductivity equals $45\ W/m/^oK$. As expected, one sees that it takes longer to attain the stable weld pool depths from the start of welding as the heat inputs increase. It takes

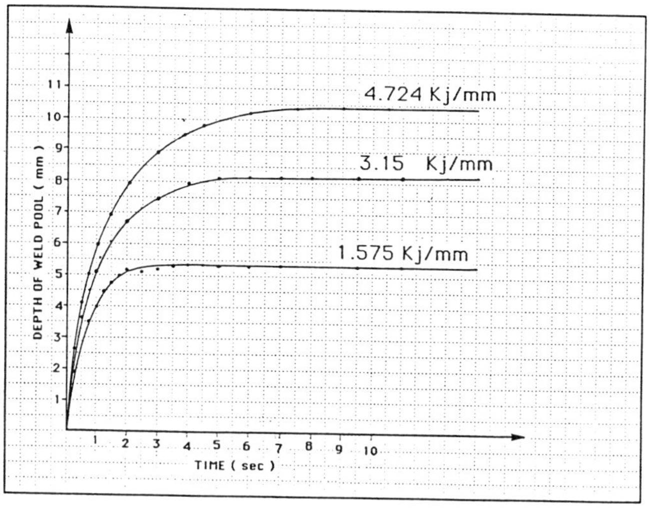

Figure 5 - Transient development of weld pool depth as function of time for different heat inputs ($\kappa = 45\ W/m/^oK$)

time to dissipate more thermal energy and to melt more material to larger depths. The results for both the 25 mm and the 50 mm thick plates indicate that the weld pool size is the same for both thicknesses. Therefore, within this thickness range at least, the weld pool depth is independent of thickness. This may be explained by the fact that the heat input rate to the plate is much faster than the heat dissipation rate within the plate. Thus, for a given heat input, a certain volume (mass) of material would melt independent of thickness (15).

The steady-state weld pool depths, the maximum HAZ depths, the times to form these depths and the maximum bottom plate surface temperatures are given in Table 2a. The weld pool depth of 5.25 mm in Table 2a for the 1.58 KJ/mm compare favorably with the values of 5.44 mm obtained from the average of two empirically regressed data (21,22) at the same heat input and 4.83 mm obtained for a heat input of 1.51 KJ/mm (23). However, in the same Table 2a, the weld pool depths of 8.06 mm and 10.27 mm at the higher heat inputs of 3.15 and 4.72, respectively, are lower than the corresponding average empirically regressed values of 9.09 mm and 12.50 mm. This discrepancy may be explained by the neglect of the effects of weld pool convection. Because of the well-known electromagnetic, buoyancy and surface tension effects, the weld pool is in fact stirred into motion rather than being static as assumed in Table 2a. Nonetheless, the good agreement of the calculated results for the 1.575 KJ/mm with three independent sources suggests that the convection effect at this heat input is negligible. In order to confirm this, a higher liquid thermal conductivity was used to approximate the convective effects.

There is no obvious way to estimate the "effective thermal conductivity" of the weld puddle to account for convection. It was found that 1.5 times the solid thermal conductivity gave a reasonable agreement with experimental results in gas tungsten arc welding (24) (GTAW). Since the thermal conductivity of iron varies from 80 at room temperature to 34.5 W/m/^{o}K at 1800°K, this translates to about 120 to 52 W/m/^{o}K as the effective thermal conductivity for the liquid. The higher value was chosen because other sources indicate that the effective conductivity can still go as high as 3 to 10 times the liquid conductivity (25). Table 2b tabulates the values for the same parameters (as in Table 2a) using the 120 W/m/^{o}K effective liquid thermal conductivity. It is obvious that the weld pool depth for the 1.58 KJ/mm heat input changed insignificantly from the static case. The calculated weld pool depth of 5.43 mm for the dynamic case is in better agreement with the 5.44 mm average empirically regressed value cited earlier. In contrast, for the 3.15 and 4.72 heat inputs, the weld pool depth are still lower than the empirical regressed values. Nevertheless, they demonstrate the influence of convection in increasing the weld pool depths. These results confirm the earlier suggestion that the influence of convection at the lowest heat input can be neglected. The increasing influence of weld pool convection with increasing heat input might be related to the Peclet number as suggested by Szekely (26). This number contains a characteristic velocity term which obviously increases as the heat inputs increase (27).

Table 2a: Calculated Results with 45 W/m/^{o}K Liquid Thermal Conductivity

Heat Input (KJ/mm)	1.58		3.15		4.72	
Plate thickness (mm)	25	50	25	50	25	50
Weld Pool Depth (mm)	5.25	5.25	8.06	8.06	10.27	10.27
Max HAZ Depth (mm)	7.64	7.64	11.47	11.31	14.92	13.99
Max Bottom Surface Temperature (^{o}K)	393	301	566	315	693	322
Initial Time to Form Weld Pool Depth (sec)	5.0	5.0	6.5	6.5	8.25	8.25
Initial Time to Form HAZ (sec)	6.0	6.0	10	10	15.0	11.1
Computer CPU Time (hr)	2.77	3.07	6.89	9.48	6.85	7.35

Table 2b: Calculated Results with 120 W/m/^{o}K Maximum Effective Liquid Thermal Conductivity

Heat Input (KJ/mm)	1.58		3.15		4.72	
Plate Thickness (mm)	25	50	25	50	25	50
Weld Pool Depth (mm)	5.43	5.43	8.81	8.81	11.46	11.37
Max HAZ Depth (mm)	7.60	7.6	12.02	11.82	17.09	15.45
Max Bottom Surface Temperature (^{o}K)	426.6	306.3	589.8	316.6	724.9	326.5
Initial Time to Form Weld Pool Depth (sec)	5.0	5.0	7.0	7.0	8.0	8.4
Initial Time to Form HAZ (sec)	6.5	6.5	10.0	10.4	15.0	10.5
Computer CPU Time (hr)	4.25	5.47	12.35	14.47	20.07	23.45

The maximum HAZ depth corresponds to the point at the centerline where the peak temperature is 1000°K. The results in Table 2 indicate that the HAZ depths in the 25 mm and 50 mm thick plates are only noticeably different at 4.72 KJ/mm, with the thicker plate having the lower value. This is of course expected because of the larger heat sink and steeper temperature gradient in thicker plates. This is depicted by the lower maximum bottom surface temperatures in the 50 mm thick plates. However, the HAZ depths are only noticeably different when the bottom surface temperatures differ by more than a factor of 2. The effect of increasing the liquid thermal conductivity on the HAZ depth is similar to its effect on weld pool depth, i.e., negligible effect at 1.58 KJ/mm and increasing HAZ depths as the heat input increases.

Initial transient times to form the weld pool and HAZ depths are shown also in Table 2 .When multiplied with the welding speed, these will indicate the minimum starting tab to be used for each heat input. Because of the oscillations during iteration and convergence, there could be a possible $\pm$ 0.5 second error in the selection of the values.

After the initial transient time, the welding process has supposedly attained the quasi-steady state condition. At this stage, Figure 6 illustrates the locations, relative to the heat arc, of the internodal planes containing the maximum weld pool and HAZ depths (1800°K and 1000°K peak temperatures, respectively) as well as an intermediate 1400°K peak temperature location. For the 25 mm thick plate, 45 W/m/^{o}K thermal conductivity and 1.58 KJ/mm heat input, it is seen that when the arc is at plane 27, the weld pool and the HAZ depths are observed at planes 22 and

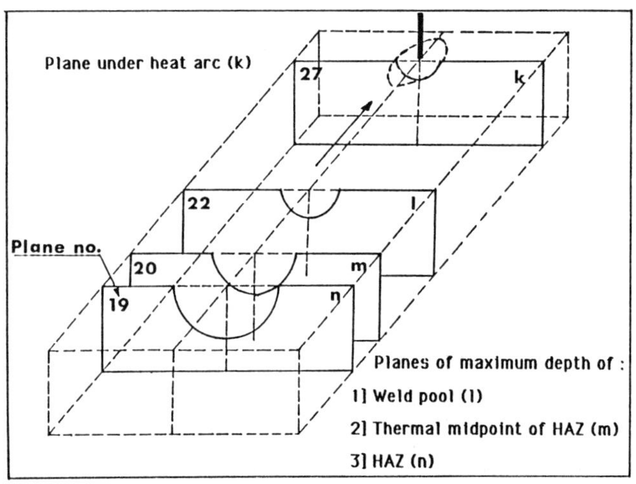

Figure 6 - Locations of internodal planes where maxima of weld pool depth (T_p = 1800°K), the HAZ depth (T_p = 1000°K) and an intermediate location (T_p = 1400°K) relative to instantaneous heat arc location.

19, respectively. As seen in Table 1, the internodal planes are 0.25 seconds apart and one might say that it takes 1.25 seconds and 2.00 seconds, respectively, to attain the weld pool and HAZ depths after the heat arc is applied. When the effective liquid thermal conductivity was increased to 120 W/m/$^{\circ}$K, the time elapsed, after the heat is applied to form the weld pool, is reduced to 1.0 second, while that for the HAZ remained at 2.0 seconds. At higher heat inputs, the effect of increased effective liquid thermal conductivity is to decrease both the time elapses to form the weld pool and HAZ after the heat application when compared to using the normal liquid thermal conductivity (18).

TEMPERATURE PROFILES AT THE HAZ - In order to obtain the cooling rate and the time spent above a certain temperature for a location in the HAZ, the temperature profiles were determined. Since locations in the HAZ are characterized by peak temperatures, the peak temperatures of 1800°K, 1000°K and 1400°K were chosen to correspond to the boundaries and the approximate midpoint of the HAZ, respectively. The temperature profiles of these three HAZ locations in the central y-z plane are shown in Figure 7 , curves A, B and C, respectively, for the three heat inputs using 45 W/m/$^{\circ}$K liquid thermal conductivity. At any one time, the locations are, however, not at the same x-y plane as the inset might suggest. Rather, they are in different planes as in Figure 6.

The profiles were determined by locating the 1800°K peak temperature at the maximum weld pool depth. At that instant, the 1000°K and 1400°K peak temperatures were located at other x-y planes. The temperatures at these locations before and after this time were then determined to depict the heating and cooling rate, respectively. One notes, however, that at the instant the location of the 1800°K was determined, this temperature has really peaked. The temperature

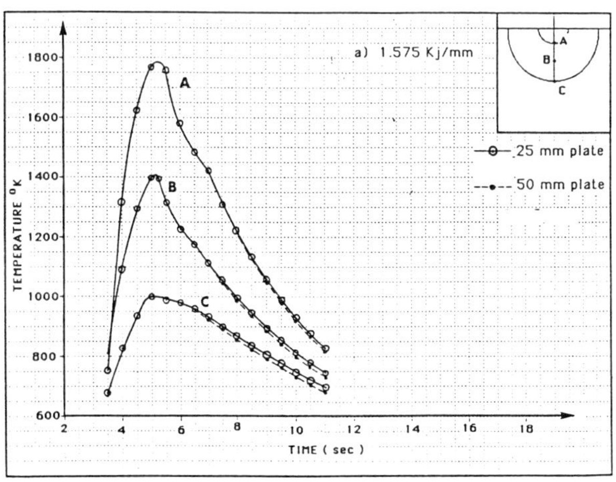

(a)

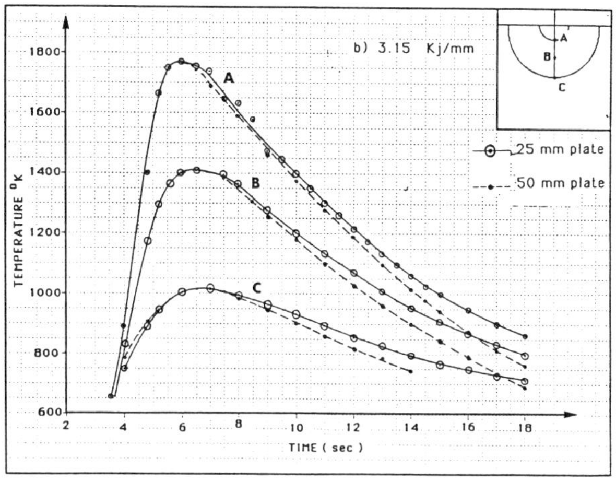

(b)

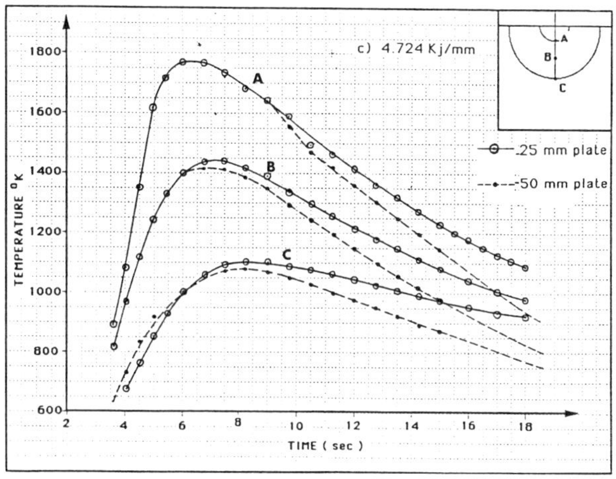

(c)

Figure 7 - Temperature profiles of three HAZ locations shown in inset corresponding to the boundaries and the midpoint of the HAZ (a) 1.58 KJ/mm, (b) 3.15 KJ/mm, (c) 4.724 KJ/mm.

at the other two locations may not have peaked yet, however. This is very noticeable in Figure 7c at 4.72 KJ/mm heat input where the B and C curves peaked at later times.

The results obtained from these temperature profiles are listed in Table 3. The peak temperatures of 1800°K and 1400°K compare with 1623°K (1350°C) and 1273°K (1000°C), which are commonly referred to as representing the grain-coarsened and the grain-refined areas, respectively, in the HAZ (1). The cooling times from 1123°K (850°C) to 773°K (500°C) was chosen instead of the commonly cited range of 1073°K (800°C) to 773°K (500°C) because of the availability of a continuous cooling transformation (CCT) diagram requiring cooling times from 850°C. It is obvious, as expected, that the cooling rates of the 50 mm thick plates are higher than

Table 3: Cooling Times and Average Cooling Rates From 1125°-773°K (850°-500°C) For Locations at Two Peak Temperatures

Plate Thickness (mm)	Heat Input (KJ/mm)	1.58		3.15		4.72	
25 mm	Peak Temperatures (°K)	1800	1400	1800	1400	1800	1400
	Δt (sec)	3.15	3.70	7.4	8.0	14.5	21.2
	Average Cooling Rate (°K/sec)	111.1	94.6	47.3	43.7	25	16.5
50 mm	Δt (sec)	3.00	3.40	5.0	5.55	5.5	6.9
	Average Cooling Rate (°K/sec)	116.7	102.9	70.0	63.1	64	50.7

those of the 25 mm plates. In addition, the higher peak temperature (1800°K) displays a higher cooling rate than the lower peak temperature (1400°K). While the difference in cooling rates between these two locations is about 10 per cent at the lower heat inputs, it is rather very significant at the highest heat input studied here, especially for the 25 mm thick plate. The latter needs to be qualified, however, because the cooling times were obtained by extrapolation of the cooling curves. The results at the lower heat inputs agree with those of Jhaveri, et al (28) who found that the difference in cooling rates at the weld centerline is only between 5 to 10 per cent. However, this may not be true at high heat inputs -- the uncertainty being in the extrapolation of the cooling curves in Figure 7c. The extrapolation was the alternative to the prohibitive computer time (see Table 2) on the VAX machine to run the program to 773°K (500°C). The cooling curves were extended with a french curve.

METALLURGICAL EFFECTS - Another important parameter that is derived from the temperature profiles is the length of time above a certain critical temperature. It relates to the kinetics of diffusional processes during the thermal cycle. For plain carbon steels, the diffusional processes of importance are the transformations of ferrite to austenite and pearlite to austenite and the dissolution and/or coarsening of individual cementite (may be spheroidized) particles. The rates of the transformations appear to be fast [8] but the dissolution of the cementite particles is sluggish [8,9] and naturally will depend on their size. These particles influence the austenite grain growth, another diffusion process, according to either Zener's [8] or Gladman's relation [29]. For all these diffusional processes, it is suggested that the area under the peak is related to the "kinetic strength" [8,30-33] of the diffusive processes.

In addition, the dissolution temperature of the particles will determine the pinning force [30] inhibiting grain growth. The cementite particles dissolve at much lower temperatures than the micro-alloyed carbides, nitrides or carbonitrides. For this reason, grain coarsening starts at a much lower temperature in plain carbon steels than in micro-alloyed steels. In micro - alloyed steels, titanium nitride has the highest dissolution temperature. Indeed it forms in liquid steel. For this reason, small titanium nitride particles are very effective in inhibiting grain growth in the HAZ [32] to improve toughness. The carbides and nitrides of vanadium dissolve at the lowest temperature among the micro-alloyed precipitate particles. Niobium carbo-nitrides start dissolving at about 1200°-1300°C depending on the concentration, as dictated by the solubility product. Not all the niobium carbo-nitrides will dissolve during the weld cycle and in fact some may coarsen [8].

The extent of particle dissolution, austenite grain coarsening and homogenization of composition at the high temperatures will determine the hardenability of the material in the HAZ and will influence the HAZ microstructures that will form during the cooling time between 850°C and 500°C. Composition of the austenite phase and its grain size are the primary factors determining the hardenability of steel. The higher the carbon content, the alloying elements, and the coarser the austenite grain size, the easier it is to form martensite on cooling. Along this line and as a matter of illustration, a continuous cooling transformation (CCT) diagram for an X-70-X-80 line pipe steel is shown in Figure 8 [33]. This figure is based on austenitizing the steel, with the composition shown, at 1100°C for three minutes. During the welding cycle, the austenite grain size at the grain-coarsened areas will be larger than that at 1100°C but the composition of the austenite will most likely be less

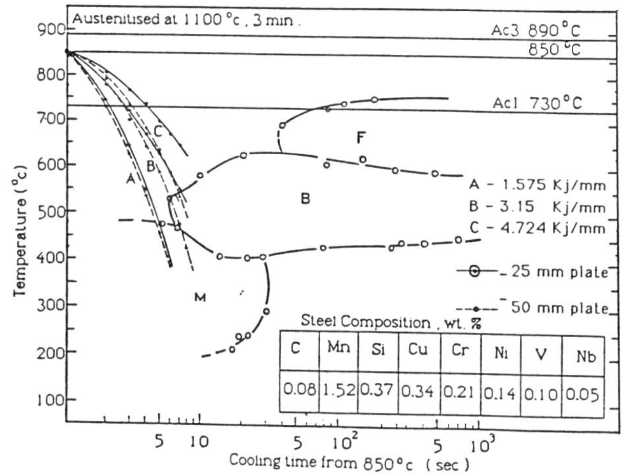

Figure 8 - Continuous cooling transformation (CCT) diagram of an X-70-X-80 line pipe steel with cooling curves from Figure 7 superimposed.

because of the much lesser time at temperature. It is suspected that the most probable CCT in the HAZ will be pushed slightly to the left (shorter times) and towards higher temperature relative to that in Figure 8. Thus, one might expect that the HAZ microstructures at all the heat inputs investigated here will result in predominantly bainitic. At much higher heat inputs, the cooling times from 850°C to 500°C will take longer and will most likely pass through the ferrite region. If this is the case, pro-eutectoid ferrite will form at prior austenite grain boundaries and such a microstructure is bad for toughness and ductility. These microstructures were observed in Gleeble simulated SAW specimens representing 4 KJ/mm and 8 KJ/mm heat inputs, respectively 1 . In addition to these basic microstructures, the cooling times will allow for the reprecipitation of the microalloy precipitates. Mangonon (9) found through thin foil transmission electron microscopy of GTA welded specimens that the morphology of the reprecipitated particles differ from those in the base plate. Direct observation of these reprecipitated carbo-nitrides have yet to be reported in SAW micro-alloyed steels.

SHAPE OF THE HEAT AFFECTED AREAS - Figure 9 shows the isotherms of 1000°K and 1800°K indicating the shpaes, respectively, of the HAZ 1 and the weld pool 2 for both the 25 mm and 50 mm thick plates and for the case of 40 W/m/°K liquid thermal conductivity. The weld pool shapes for both thicknesses are the same and are seen to be slightly off-semicircular. The weld pool shapes can vary depending on may factors but the shapes obtained here are very close to those predicted by Rosenthal's 19 three-dimensional case. According to Rosenthal, the weld pool shape should be semicircular for the 3-D case.

The effect of thickness is depicted only in the HAZ at the 3.15 and 4.72 KJ/mm heat inputs where the thicker plate shows a slightly lesser HAZ depth but tends to have a wider width. It is expected that the shapes will vary with the effective radius chosen.

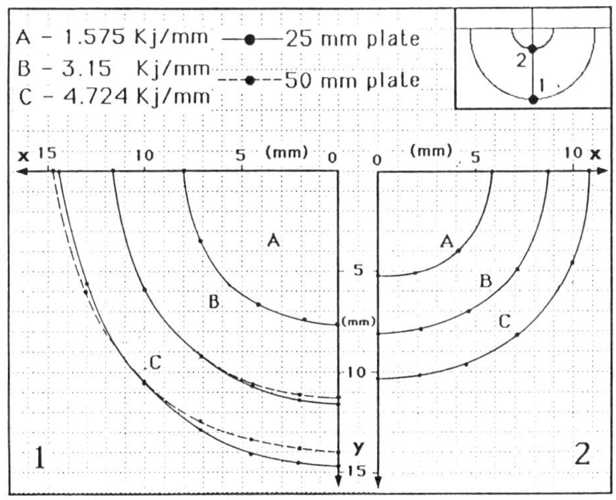

Figure 9 - Shapes of HAZ and weld pool in x-y planes.

SUMMARY AND CONCLUSIONS

A three-dimensional heat transfer finite element model of submerged arc welding was developed and applied to micro-alloyed HSLA steels. The model incorporates the temperature variable physical properties and the latent heat of fusion of steel. In order to simulate weld pool convection, a much higher effective liquid thermal conductivity was used to compare it to the simple conduction is a static weld pool. The essential conclusions from this investigation are the following:

1) The validity of the model is substantiated by predicting the correct weld pool depths and shapes.
2) The weld pool depths were essentially independent of thickness indicating the welding speed is much higher than the heat transfer.
3) The results suggest that weld pool convection effects are significant at high heat inputs but negligible at low heat inputs.
4) The choice of the fixed coordinate system, in lieu of the simpler moving coordinate system, brings out the transient time effects one needs to incorporate in practical welding situations.
5) The temperature profiles and the predicted cooling rates suggest that the primary microstructures

will be acicular and/or bainitic in the HAZ in an X-70-X-80 steel for all the heat inputs studied here.

ACKNOWLEDGEMENT

The authors wish to thank Florida Institute of Technology for the most generous allocation of computer time and to Drs. F.L. Stasa and R.P. Scaringe for their valuable suggestions and critical review of the model.

REFERENCES

1. J.T. Bowker, A.G. Glover, and R.B. Lazor, "The Structure and Toughness of the Weld Heat-Affected Zone of a C-Mn-V Steel," p. 679 in "HSLA Steels - Technology and Applications," American Society for Metals, 1984, Proceedings of International Conference, 3-6 October 1983, Philadelphia, PA.
2. I. Kozasu, "Recent Development of Micro-alloyed Steel Plate," p. 593, ibid.
3. "Colloquium on Micro-alloyed Steels," November 1983, Held at Naval Research Laboratory (NRL), Washington, D.C., Sponsored by NRL and David Taylor Ship R&D.
4. C.P. Royer, "Trends in Materials for Offshore Structures," p. 31 in "Materials for Future Energy Systems," Proceedings of Conference held in Washington, D.C., 1-3 May 1984, Published by American Society for Metals, 1985.
5. P.L. Mangonon, Jr. and W.E. Heitman, "Subgrain and Precipitation - Strengthening Effects in Hot-rolled, Columbium-bearing Steels," p. 59, Micro-alloying '75. Proceedings of Conference held in Washington, D.C., October 1-3, 1975. Published by Union Carbide Corporation, 1977.
6. R.K. Amin and F.B. Pickering, "Austenite Grain Coarsening and the Effect of Thermomechanical Processing on Austenite Recrystallization, p. 1, "Thermo-mechanical Processing of Micro-alloyed Austenite," Proceedings of Conference published by TMS-AIME, 1982. A.J. DeARdo, G.S. Ratz, and P.J. Wray, editors, TMS-AIME.
7. P.L. Mangonon, Jr., Unpublished results of work done at Inland Steel Research Laboratories.
8. K.E. Easterling, "Predicting Heat-Affected Zone Microstructures and Properties in Fusion Welds," Special lecture, Proceedings of International Trends of Welding Research Conference, Gatlinburg, TN, 18-22 May, 1986, ASM.
9. P.L. Mangonon, Transmission Electron Microscopy of Welded Cb-Microalloyed Steel, Welding J., 64, No. 1, 1985, p. 135-195.
10. J.M. Barry, Z. Paley and C.M. Adams, Jr., Heat Conduction from Moving Arcs in Welding, Welding J., 42, No. 3, 1963, pp. 97s-104s.
11. N. Christensen, V. de L. Davies, and K. Gjermundsen, "Distribution of Temperatures in Arc Welding," Brit. Weld. J., 12, 1965, p. 54.
12. C.E. Jackson and A.E. Shrubsall, "Energy Distribution in Electric Welding," Welding J., 29, 1950, p. 23s.
13. V. Pavelic, R. Tanbakuchi, O.A. Uyehara and P.S. Myers, "Experimental and Computed Temperature Histories in Gas Tungsten Arc Welding of Thin Plates," Welding J., 48, No. 7, 1969, pp. 295s-305s.
14. C.L. Tsai, "Modelling of Thermal Behavior of Metals During Welding," p. 91 in "Trends in Welding Research," ASM, 1982.
15. S.S. Glickstein and E. Friedman, Temperature Transients in Gas Tungsten ARc Weldments, Welding Review, May 1983, p. 72.
16. P.J. Wray, "Mechanical, Physical and Thermal Data for Modelling the Solidification Processing of Steels," p. 245 in "Modelling of Casting and Welding Processing," Conference Proceedings, TMS-AIME, 1981.
17. M. Mahimkar and P.L. Mangonon, to be published.
18. M. Mahimkar, M.S. Thesis, Department of Mechanical Engineering, Florida Institute of Technology, Melbourne, FL, 1986.
19. D. Rosenthal, The Theory of Moving Sources of Heat and Its Application to Metal Treatments, Trans. ASME, 68, 1946, p. 849.
20. E. Friedman, "Finite Element Analysis of Arc Welding," Report WAPD-TM-1430, January 1980, Bettis Atomic Power Laboratory, West Mifflin, PA 15122.
21. B.J. Bradstreet, "Effect of Welding Conditions on Cooling Rate and Hardness in the Heat Affected Zone," Welding J., 48, No. 11, 1969, pp. 499s-504s.
22. Welding and Brazing, Vol. 16, ASM Metals Handbook, 8th edition.
23. J.C. Ion, "Modelling the Microstructural Changes in Steels Due to Fusion Welding," Doctoral Dissertation, Lulea University of Technology, 1984: 36D, Sweden.
24. S. Kou and Y. Le, "Three-Dimensional Heat Flow and Solidification During the Autogenous GRA Welding of Aluminum Plates," Met. Trans. A., 14, No. 11, p. 2245.
25. G.M. Ecer, M. Downs, H.D. Brody, and A. Gokhale, "Heat Flow Simulation of Pulsed Current Gas Tungsten Arc Welding," p. 139 in "Modelling of Casting and Welding Processes," H.D. Brody and D.Apilian, editors. Conference Proceedings TMS-AIME, 1981.
26. J. Szekely, "Heat and Fluid Flow in Welds," Keynote lecture, International Trends in Welding Research Conference, Gatlinburg, TN, May 19-22, 1986.
27. G. Oreper and J. Szekely, "On Electromagnetically and Buoyancy Driven Flows in Weld Pools," Appendix to "A Basic Study of

Heat Flow in Fusion Welding," by J. Szekely and T.W. Eager, Progress report to U.S. Dept. of Energy. DOE/ER/04799-4, DE 83-002803.

28. P. Jhaveri, W.G. Mofatt, and C.M. Adams, Jr., "The Effect of Plate Thickness and Radiation of Heat Flow in Welding and Cutting," Welding J., 41, No. 1, 1962, 12s.
29. T. Gladman, "On the Theory of the Effect of Precipitation Particles on Grain Growth in Metals," Proc. Roy Soc., 266A, 1966, p. 298.
30. M.F. Ashby and K.E. Easterling, "A First Report on Diagrams for Grain Growth in Welds," Acta. Met., 30, 1982, pp. 1969-1982.
31. J.C. Ion, K.E. Easterling and M.F. Ashby, "A Second Report on Diagrams of Micro-Structure and Hardness for Heat Affected Zones in Welds," Acta. Met., 32, 1984, pp. 1949-1962.
32. L.J. Cuddy, J.S. Lally and L.F. Porter, "Improvement of Toughness in the HAZ of High Heat Input Welds in Ship Steels," p. 697 in Proceedings (see Reference 1).
33. T. Hashimoto, H. Ontani, T. Kobayashi, N. Hatano and S. Suzuki, "Controlled Rolling Practice of HSLA Steel At Extremely Low Temperature Finishing," p. 501 in "Thermomechanical Processing of Micro-alloyed Austenite," A.J. DeArdo, G.A. Ratz, P.J. Wray, editors, TMS-AIME, Conference Proceedings.

EFFECTS OF SURFACE DEPRESSION AND CONVECTION IN GTA WELDING

M.L. Lin, T.W. Eagar
Materials Processing Center
Massachusetts Institute of Technology
Cambridge, Massachusetts 02139

Abstract

Surface depression and convection have been studied with the goal of determining some of the physical characteristics of the weld pool which are changed by these factors. It is shown that, at currents in excess of 300 amperes, a deep surface depression may form. Experimental measurements of the angular velocity of circumferential flow in the pool indicate that the deep surface depression is due to formation of a vortex. When such a vortex depression forms, current distribution in the plasma may be radically altered, resulting in a different heat distribution on the weld pool surface.

ALTHOUGH THE GAS tungsten arc (GTA) welding process is capable of producing extremely high quality welds in a wide range of materials; its application is often limited to low current ranges due to formation of defects at welding currents in excess of 300 amperes.

Many researchers have studied heat flow (1-7), convection (8-17), and the effect of minor elements (18-23) in the base metal in order to predict more precisely the geometry of the GTA weld beads. However, most of this work has involved relatively low welding currents. In order to better understand the reason for the difficulties in the high current regime, a number of physical characteristics of the GTA process have been studied.

At low currents, the surface of a liquid pool in GTA welding is essentially flat. Many theories of the heat intensity distribution of the arc on the base metal have been based upon this assumption of a flat surface (24-26); however, at currents above 300A, a deep crater on the weld pool surface may form which makes these previous theories invalid. In order to account for the depressed surface of a weld pool, some modified heat distributions have been proposed (27,28). However, there is no physical basis for these proposed modifications. In addition, at high currents, the deep surface depression of the weld pool can cause formation of weld defects (29-31); hence, it was thought that an investigation of the influence of surface depression on weld formation at high currents would be a useful endeavor.

It has been reported that fluid motion in the weld pool can influence the heat distribution (8,10,32-34), homogeneity of minor elements (10-12), and defect formation (9,11,31,35). Since the strength of the fluid motion is expected to increase with increasing current, study of convection in the weld pool is also essential if one hopes to understand formation of the weld bead at high currents.

In this paper, a new study of surface depression in GTA welding is presented. This work is divided into two parts: (1) vortex formation and (2) current distribution in the arc.

VORTEX FORMATION

Although it is commonly believed that the deep surface depression which forms at currents above 300 amperes is caused by the arc force, this can easily be shown to be incorrect (29). The depression caused by the arc force can be estimated from a balance of the arc pressure with the metallostatic head and the surface tension curvature.

$$P_{arc} = \rho g H + \gamma\left(\frac{1}{r_1} + \frac{1}{r_2}\right) \qquad \text{Eq. (1)}$$

For a 300 ampere arc with a 30^0 electrode tip angle, $P_{arc} = 2.5$ kN/m^2 (36). For steel, $\rho = 7.8$ g/cm^3 and γ is approximately 1.2 J/m^2. If one assumes that $r_1 = r_2 = 0.1$ cm, the maximum depth of surface depression, H, is found to be 1.3 mm. A more elaborate analysis has been given in reference 37, but even from the simple analysis presented above, it is clear that a 5 mm surface depression on steel in a 300 ampere arc (29) cannot be explained by the impingement of the plasma jet.

Lin and Eagar (29) suggested that a weld pool could be approximated as a compound vortex; however, they did not explain the origin of this vortex. They proposed, but did not prove that the circumferential flow in the weld pool about the axis of the arc may be a result of toroidal flow in the liquid metal (38). It also may be due to a circular rotation of the anode spot on the weld pool. Other possible mechanisms are an asymmetric electrical ground position or tungsten electrode tip geometry, or inhomogeneous chemical composition of the base plates. Any such asymmetrical driving force in the weld pool, driven by either surface tension gradients or electromagnetic forces, may induce a circumferential flow in the liquid pool about the axis of the arc. In addition, the arc pressure can depress the surface of the weld pool slightly, causing the fluid particles rotating around the axis of arc to be drawn toward the center. If the angular momentum is conserved, this will cause an increase in the rotational velocity as the radial distance is decreased.

Regardless of the force driving the angular flow, such a flow should be observable on the surface of the weld pool. In order to test whether a circumferential flow does exist, high speed photography was used to measure the motion of aluminum oxide particles on the weld pool surface.

Although these markers revealed circumferential flow in the weld pool, no alumina particles on the surface of the weld pool were visible within 4 mm of the weld centerline because of the deep surface depression and the strong arc light intensity in this region. Occasionally, one or two of the alumina particles could be seen to swiftly accelerate as they moved radially inward toward the bottom of the deep crater. Outside the deep surface depression (i.e. beyond the 8 mm diameter central core), the circumferential flow velocity was measured to be on the order of 2 rps. Since it is difficult to measure the angular velocity inside the crater of the weld pool, conservation of angular momentum may be used to estimate the angular velocity in this inner region. Since the measured 2 rps angular velocity is about 4 mm away from the center of the weld pool, the angular velocity is estimated to be 32 rps at one millimeter radius. Using the previous described model of a free and forced vortex with a transition radius of 1 mm, the depth of surface depression caused by this vortex is about 3.6 mm (29) which is within the range of the experimental data for the actual weld pool surface depression. Hence, it is seen that the circumferential flow which is experimentally observed in the weld pool can explain the deep surface depression which is formed.

CURRENT DISTRIBUTION IN THE ARC

Since more than 80% of the heat input to the workpiece is carried by electrons (26,39,40), studies of current distribution in the arc can lead to a better understanding of heat distribution on the surface of the weld pool. Current distribution in GTAW has been measured by several authors (24-26) on flat water cooled copper anodes, and was found to be approximated reasonably well by a Gaussian (24). When a deep surface crater forms, this Gaussian approximation may need to be modified. An estimate of the effect of surface depression on modification of the heat source distribution is presented below.

The current density J in the arc column can be expressed as

$$J = \sigma_e E \qquad \text{Eq. (2)}$$

since the Hall effect (i.e. v x B) is very small in the arc plasma. Assuming that the electrical field strength, E, is irrotational, the field can be calculated from

$$E = -\nabla U \qquad \text{Eq. (3)}$$

This is valid when current of low frequency is employed.

The electrical potential U can be calculated from Gauss's Law

$$\text{div } D = \nabla \cdot D = \rho_e \qquad \text{Eq. (4)}$$

where ρ_e is the net charge density and D is the electrical displacement as expressed by $D = \varepsilon_0 K_d E$ or $D = -\varepsilon_0 K_d \nabla U$. Substituting this into equation 4, we obtain

$$\nabla \cdot (\varepsilon_0 K_d \nabla U) = \varepsilon_0 (K_d) \nabla^2 U + \nabla U \cdot \nabla K_d = -\rho_e \qquad \text{Eq. (5)}$$

Except for the region very close to the cathode and anode surface, the arc is essentially a neutral plasma and the net charge density ρ_e in the arc column can be assumed to be zero. Thus, equation 5 can be simplified as

$$K_d \nabla^2 U + \nabla U \cdot \nabla K_d = 0 \qquad \text{Eq. (6)}$$

The real plasma is an inhomogeneous medium with the temperature in the central part higher than in the surrounding region; however, in order to simplify this model and to obtain the relative distribution of equipotential lines in the arc, a homogeneous arc is assumed in this model.

With assumption of a homogeneous arc, equation 6 can be further simplified to the Laplace equation which is expressed as $\nabla^2 U = 0$.

Thus in order to calculate the current density in a homogeneous arc, Laplace's equation has to be solved to obtain the distribution of equipotential lines in the arc. Then both equation 2 and 3 can be applied to obtain the current distribution in the arc. Since the electrical conductivity of the metallic cathode

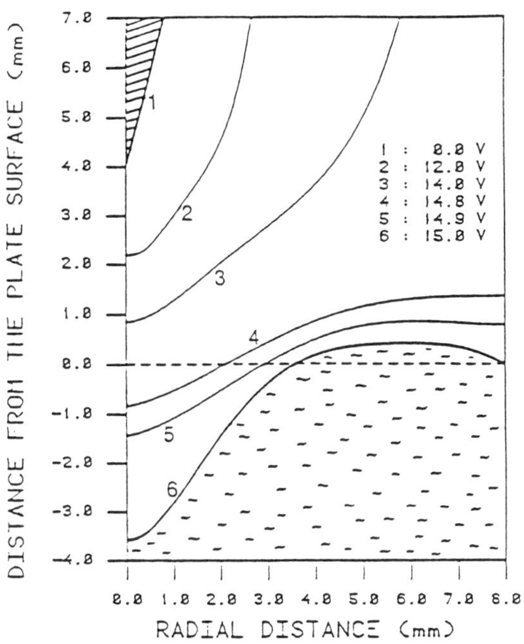

Figure 1: Equipotential lines between the 30° cathode tip and the depressed weld pool as calculated by the finite element model.

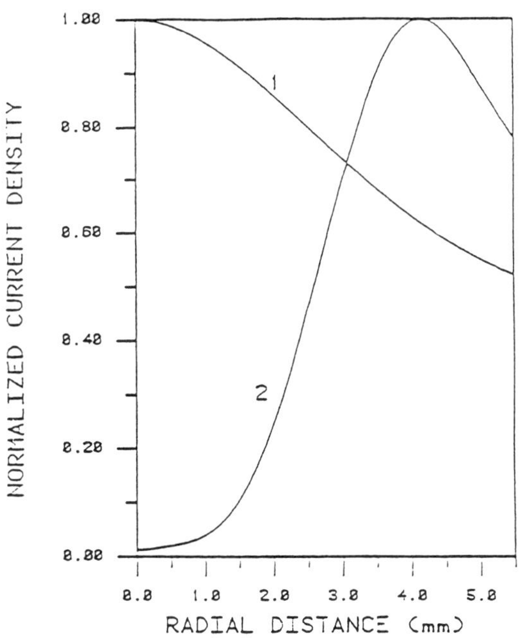

Figure 2: Normalized current density distribution on the surface of the weld pool. Curve 1: With the 30° cathode and the flat weld pool. Curve 2: With the 30° cathode and the depressed weld pool as shown in Figure 1.

and anode are much higher than that of the plasma, it can be assumed that both the cathode and the anode are equipotential surfaces.

Calculations were performed with three different cathode and five different anode geometries. Two of the cathodes were conical and had 30° and 105° included angles while the third was a paraboloid of revolution. The anodes include a flat surface and four distorted surfaces with maximum surface depression depths of 0.3 mm, 0.7 mm, 1.0 mm and 3.5 mm. A constant 4 mm arc length and an arc voltage of 15 volts were assumed. Laplace's equation was solved in cylindrical coordinates using a finite element method.

Since only the relative distribution of equipotential lines between the cathode and the anode is of concern, both the anode drop and cathode drop were not considered in this model. The current distributions at the anode were then calculated from these equipotential lines by using equations 2 and 3. Because a homogeneous arc is assumed in this model, and because the relative current density distribution is of concern, a normalized current density instead of an absolute current density on the anode surface was calculated.

Figure 1 shows the calculated equipotential lines between a 30° conical cathode and an anode of 3.5 mm surface depression. It was found that as the sharpness of the cathode increases, the equipotential lines tend to move toward the cathode, which gives a stronger electric field near the sharp electrode than near the dull electrode.

For a 30° cathode tip angle, the normalized current density distributions on the flat and the depressed surface of the 3.5 mm deep weld pool are shown in Figure 2. The normalized current density distributions on the other cathode geometries were found to be similar to those shown in Figure 2. Figure 3 shows the normalized current density distributions on the surface of the weld pool with shallower surface depressions. It can be seen that if the anode (base metal) surface is flat, the current density is the highest at the center of the weld pool, and the assumption of a Gaussian distribution of the current density on a flat weld pool surface (24) is reasonable. However, when the surface of the weld pool is distorted, even with surface depression depths as small as 0.3 mm, this Gaussian distribution assumption is no longer valid. When the maximum depth of the surface depression is about 1 mm, the highest current density is no longer at the center of the weld pool; rather, it shifts to the side walls of the weld pool crater. This results in a very significant change in the heat distribution. This redistribution of the heat source should be considered in high current arc welds.

Figure 4 shows the penetration, width, and area of a GTAW spot weld as a function of welding time at 285 amperes. It can be seen that as the welding time increases from 3 to 24

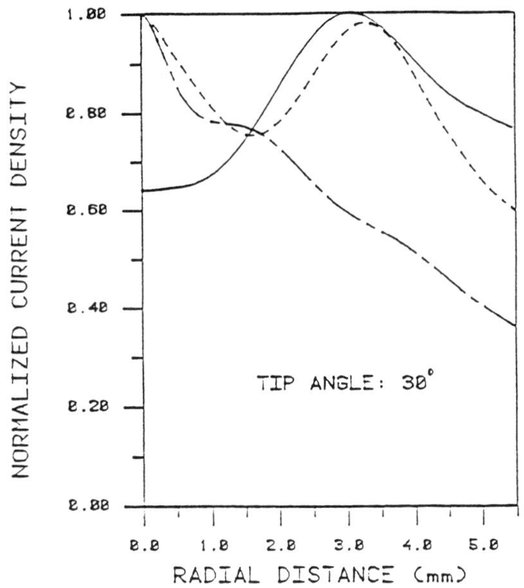

Figure 3: Normalized current density distribution on the surface of the weld pool with shallow depression. Electrode tip angle: 30°.
Maximum surface depression depth: 0.3 mm
Maximum surface depression depth: 0.7 mm
Maximum surface depression depth: 1.0 mm

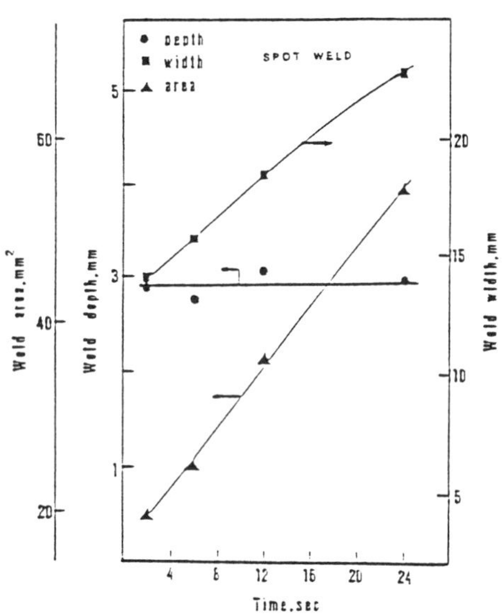

Figure 4: Variation of weld depth, width and area with welding time for a stationary 285 ampere argon arc with a 30° electrode tip angle.

seconds, there is no increase of the penetration depth while the width and area of the spot weld continue to increase with welding time. Since a deep surface depression is formed at 285 amperes, this surprising result is believed to be due to the redistribution of the current to the sides of the depressed weld pool. In effect, the deep crater separates the weld into two separate regions on either side of the depression.

The maximum current density at the side walls of the depressed weld pool shows that most of the heat generated in the arc is transferred to the side walls of the weld puddle and very little is transferred to the bottom of the crater. This is consistent with the experimental result that the 285A spot weld does not increase in depth as weld time increases. It is also consistent with the observation that the bottom of the deep surface depression is solid rather than liquid metal.

CONCLUSIONS

Using the principle of conservation of angular momentum, the maximum angular velocity of a compound vortex formed at 300 amperes is estimated to be about 30 rps at a radial distance of one millimeter from the weld pool centerline. This results in a calculated depth of surface depression which is of the same order as the experimental result.

Calculation of equipotential lines between the cathode and the anode permits one to estimate the current distribution, and therefore the heat distribution, in the arc. It is shown that the deep surface depression produced in the weld pool at high currents can cause significant redistribution of the heat source, with the maximum heat intensity shifting to the side walls of the weld puddle. Since very little heat is transmitted to the bottom of the weld pool, the penetration depth of a spot weld remains almost constant while the width and area of the spot weld increase with time.

Acknowledgement

The authors are grateful for the support of this work by the Office of Naval Research under Contract N00014-C-80-0384.

LIST OF SYMBOLS

- B : magnetic flux density (weber/m^2)
- D : electrical displacement (coulomb/m^2)
- E : electric field strength (V/m)
- g : acceleration of gravity (m/s^2)
- H : depth of the weld pool (m)
- I : current (A)
- J : current density (A/m^2)
- K_d : dielectric constant
- P_{arc} : arc pressure (N/m^2)
- r_1, r_2 : principal radii of the surface depression curve (m)
- U : electrical potential (V)

LIST OF SYMBOLS (continued)

v	: velocity of charged particles in the plasma (m/s)
γ	: surface energy of the weld pool (J/m^2)
ε_0	: electrical permittivity of free space (farad/m)
ε	: electrical permittivity ($=K_d\varepsilon_0$) (farad/m)
ρ	: density of the liquid metal (kg/m^3)
ρ_e	: net charge density ($coulomb/m^3$)
σ_e	: electric conductivity (mho/m)

REFERENCES

1. Rosenthal, D., Transactions of ASME, 43(11), 849-866 (1946)
2. Christensen, N., Davies, V. and Gjermundsen, K., British Welding Journal, 12(2), 54-75 (1965)
3. Adams Jr., C. M., Welding Journal, 737(5), 210s-215s (1958)
4. Pavelic, V. and Tsao, K. C., Arc Physics and Weld Pool Behavior, The Welding Institute International Conference, London, 251-258 (May 1979)
5. Friedman, E., Proceedings of the Conference on Numerical Modelling of Manufacturing Processes, ASME, Atlanta, Georgia, 35-47 (November 1977)
6. Kou, S., Kanevsky, T., and Fyfitch, S., Welding Journal, 61(6), 175s-181s (1982)
7. Rykalin, N. N., and Nikolaev, A. V., Welding in the World, 9, 112-132 (1971)
8. Willgoss, R. A., Arc Physics and Weld Pool Behaviour, The Welding Institute International Conference, London, 361-373 (May 1979)
9. Woods, R. A., and Milner, D. R., Welding Journal, 50(4), 163s-173s (1971)
10. Lawson, W. H. S., and Kerr, H. W., Welding Research International, 6(5), 63-77 (1976)
11. Bradstreet, B. J., Welding Journal, 47(7), 314s-322s (1968)
12. Houldcroft, P. T., British Welding Journal, 1, 468-472 (1954)
13. Mills, G. S., Welding Journal, 58(1), 21s-24s (1979)
14. Heiple, C. R., Roper, J. R., Stagner, R. T., and Aden, R. J., Welding Journal, 62(3), 72s-77s (1983)
15. Shercliff, J. A., Journal of Fluid Mechanics, 40, 241-250 (1970)
16. Andersson, D., I.I.W. Doc. 212-217-71
17. Oreper, G. M., Eagar, T. W., and Szekely, J., Welding Journal, 62(11), 307s-312s (1983)
18. Savage, W. F., Nippes, E. F., and Goodwin, G. M., Welding Journal, 56(4), 126s-132s (1977)
19. Bad'yanov, B. N., Davydov, V. A., and Ivanov, V. A., Auto. Welding, 27(11), 63-64 (1974)
20. Gurevich, S. M., Zamkov, V. N., and Kushnirenko, N. A., Auto. Welding, 18(9), 1-5 (1965)
21. Bennett, W. S., and Mills, G. S., Welding Journal, 53(12), 548s-553s (1974)
22. Glickstein, S. S., and Yeniscavich, W., WRC Bulletin, No. 226 (May 1977)
23. Block-Bolten, A., Eagar, T. W., Proceedings of the Conference on Trends in Welding Research in the United States, ASM, New Orleans, LA, 53-73 (November 1981)
24. Tsai, N. S., and Eagar, T. W., Welding Journal, 62(12), 346s-355s (1983)
25. Nestor, O. H., Journal of Applied Physics, 33(5), 1638-1648 (1967)
26. Schoeck, P. A., Modern Development in Heat Transfer, New York and London, Academic Press, 353-400 (1963)
27. Goldak, J., Chakravarti, A., and Bibby, M., Metallurgical Transactions B, 15B(2), 299-305 (1984)
28. Giedt, W. H., Proceedings of the Conference on Trends in Welding Research in the United States, ASM, New Orleans, LA, 109-131 (November 1981)
29. Lin, M. L., and Eagar, T. W., Transport Phenomena in Materials Processing, ASME PED 10, 63-69 (November 1983)
30. Ishizaki, K., Weld Pool Chemistry and Metallurgy, The Welding Institute International Conference, London, 65-76 (April 1980)
31. Paton, B. E., Mandel'berg, S. L., and Sidorenko, B. G., Auto. Welding, 24(8), 1-6 (1971)
32. Bukarov, V. A., Ishchenko, Yu. S., and Loshakova, V. G., Welding Productions, 25(11), 5-9 (1978)
33. Metcalfe, J. C., and Quigley, B. C., Welding Journal, 56(5), 133s-139s (1977)
34. Apps, R. L., and Milner, D. R., British Welding Journal, 10, 348-350 (1963)
35. Morgan-Warren, E. J., Arc Physics and Weld Pool Behavior, The Welding Institute International Conference, London, 325-336 (May 1979)
36. Lin, M. L., and Eagar, T. W., Metallurgical Transaction B, in press.
37. Lin, M. L., and Eagar, T. W., Welding Journal, 64(6), 163s-169s (1985)
38. Millere, R. P., Sharamkin, V. I., and Shcherbinin, E. V., Magnetohydrodynamics, 16(1), 66-69 (1980)
39. Wilkinson, J. B., and Milner, D. R., British Welding Journal, 7, 115-128 (1960)
40. Quigley, M. B. C., Richards, P. H., Swift-Hook, D. T., and Gick, A. E. F., Journal of Physics D: Applied Physics, 6, 2250-2259 (1973)

MINOR ELEMENT EFFECTS ON GAS TUNGSTEN ARC (GTA) WELD PENETRATION - WELD POOL PHYSICS

R.E. Sundell, H.D. Solomon, S.M. Correa
Corporate Research & Development
General Electric Company
Schenectady, New York 12345

ABSTRACT

This work is part of a cooperative industry/university project to develop a better understanding of the cause of variable GTA weld penetration and to develop techniques to overcome this variation. The work of Walsh and Savage[2] in these proceedings concentrates on determining the effects of minor elements on autogenous GTA weld bead shapes, whereas this work is directed towards developing a quantitative description of the weld pool flow. Surface tension measurements using the sessile drop technique are reported for heats of AISI 8630 steel and 316L stainless steel as a function of temperature and are correlated with weld bead D/W ratio. Results from a mathematical model of the weld pool flow are presented which indicate the importance of weld pool convection in determining weld bead shape. Also, experimental measurements of weld pool surface velocity and peak weld pool surface temperature are presented which agree well with the flow model. The results generally confirm the Heiple and Roper[1] model for variable GTA weld penetration.

A SIGNIFICANT PROBLEM IN THE GTA WELDING of many high-purity alloys is variable penetration caused by small differences in base metal chemistry (well within commercial specification). Heiple and Roper[1] have proposed that variations in weld penetration are the result of convective flow driven by surface tension gradients on the weld pool surface. These gradients exist due to temperature gradients on the weld pool surface. Small changes in minor element concentrations affect the magnitude and sign of the surface tension gradient and, thereby, the convective flow. Surface-tension-driven convection is sometimes termed Marangoni convection.

Pure metals exhibit a negative surface tension coefficient, i.e., the surface tension decreases as the temperature increases. Since the temperature is highest under the arc, the surface tension is highest at the edge of the pool. This gradient produces flow from the center of the pool outward, producing efficient heat transfer to the sides of the weld and a wide, shallow weld pool. Alternately, the addition of trace amounts ($\sim$50 to 100 ppm) of a surface-active element, such as sulfur, produces a positive surface tension coefficient. This produces a strong inward flow on the weld pool surface, efficient heat transfer to the weld root, and a narrow, deep weld pool.

This paper reports some of the results of a cooperative industry/university project to develop a better understanding of the cause of variable GTA weld penetration and to develop techniques to overcome this variation. Other results are reported in the paper by Walsh and Savage[2] in these proceedings. The work of Walsh and Savage concentrates on determining the effects of minor elements on autogenous GTA weld bead shapes, whereas this work is directed towards developing a quantitative description of the weld pool flow, including (1) measuring the surface tension of the alloys studied, (2) modeling the weld pool flow, and (3) measuring temperatures on the weld pool surface. More complete information detailing the results of this project may be found in the final report[3].

EXPERIMENTAL PROCEDURES

SURFACE TENSION MEASUREMENTS – The sessile drop technique[3–6] was used to measure the surface tension at temperatures as high as 1900 °C and as low as 1300 °C (where supercooling took place). All the data reported here was obtained in purified research-grade argon gas on ZrO_2 substrates. An optical pyrometer was used to measure the temperature of the sessile drops. See Reference 3 for more data concerning the experimental equipment and techniques.

The surface tension is determined from the shape of the drop. The shape is governed by the balance between gravity, which tends to spread out the molten drop, and surface tension, which tends to keep it a sphere. The result is a flattened sphere with a greater flattening the lower the surface tension. Photographs were taken of the drop and the drop shape was digitized by an image analyzer[3]. A computer program, based on one written by Butler and Bloom,[7] was developed[3] which determined the density and surface tension from the drop shape. It utilized the digitized drop shape data as input.

WELD POOL TEMPERATURE MEASUREMENTS – Weld-pool surface temperatures were measured using tungsten-rhenium thermocouples inserted from the underside of the test plate as shown in Fig. 1a. Intrinsic thermocouples (no welded junction bead) were used which consisted of 0.010-inch-diameter tungsten-rhenium thermocouple wire extending approximately 0.020 inch beyond the insulating 1/16-inch-diameter alumina rods. These were inserted to within 0.015 inch of the plate surface. The thermocouple junction was formed by the molten metal. Elimination of the thermocouple

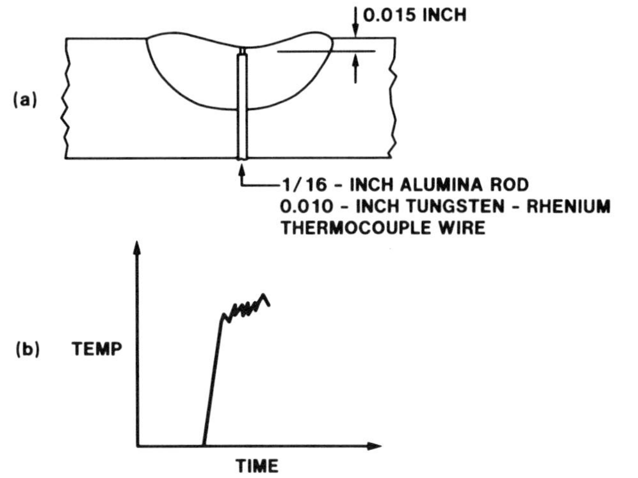

Fig. 1. (a) Intrinsic thermocouple setup; (b) typical thermocouple trace.

bead moved the effective junction closer to the workpiece surface by approximately 0.025 inch. Since the weld pool is depressed (<1 mm) directly beneath the welding arc[8], thermocouple traces usually registered (by an open reading) a point at which the thermocouple protruded through the surface, thereby indicating both the location of the thermocouple junction at the weld-pool surface and the temperature just prior to the open reading (see Fig. 1b).

The possibility of error due to nonthermal emf signals caused by the welding arc was investigated by passing the welding arc over thermocouples made with identical wire on the two sides of the junction. Measured values were always less than 5% of the temperature signal.

EXPERIMENTAL RESULTS AND FLOW MODEL

SURFACE TENSION – Figures 2 and 3 show the surface tension, γ, as a function of temperature, T, for the 316L stainless steel and the 8630 steel, respectively. Also compared on these figures are the important solute variations and values of the weld bead depth-to-width ratio (D/W) for the heats studied.* The general shape of these curves correlates with D/W and confirms the Heiple and Roper[1] model. Low D/W ratios are associated with a steep negative surface-tension gradient which is the behavior exhibited by relatively pure alloys. This is especially true of the highest purity 316L S.S. (heat J), which contained only 10 ppm sulfur and had the lowest D/W (0.17). As the impurity level (primarily S) is increased, γ is reduced and $d\gamma/dT$ at 1500 °C becomes positive. This gives rise to inward fluid flow and a high D/W. In the relatively pure 316L alloys where $d\gamma/dT$ is strongly negative, the flow is outward, producing a low D/W. In the 8630 heats, all of which contain a relatively large solute content, the γ-T behavior is actually more complex than the development of positive $d\gamma/dT$. The surface tension exhibits a positive shape until about 1700 °C, whereupon entropic mixing reduces the surface solute content and $d\gamma/dT$ becomes negative. This may result in a complex fluid flow.

The variation of $d\gamma/dT$ at 1500 °C and 1800 °C is shown more quantitatively in Fig. 4, which plots $d\gamma/dT$ for all of the experiments. There is, as expected, scatter but clear trends are shown. Low D/W ratios are associated with strongly negative

* The detailed composition of the heats used in this study along with welding parameters are reported in Walsh and Savage[2] in these proceedings.

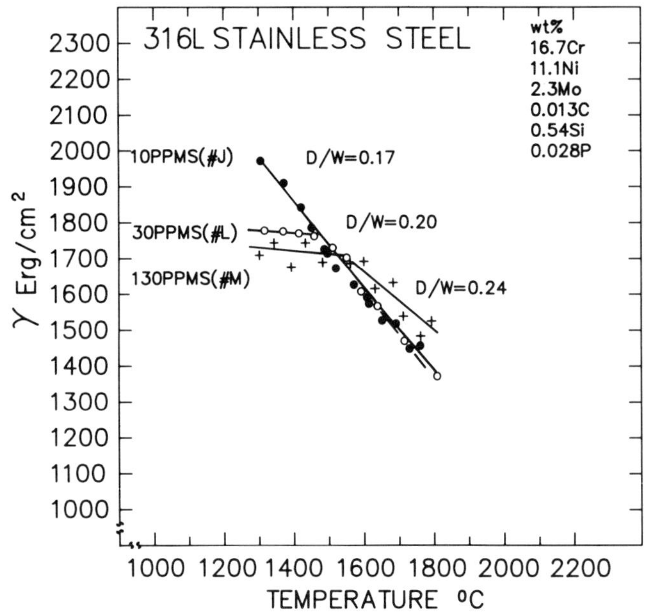

Fig. 2. Surface tension for heats of 316L stainless steel.

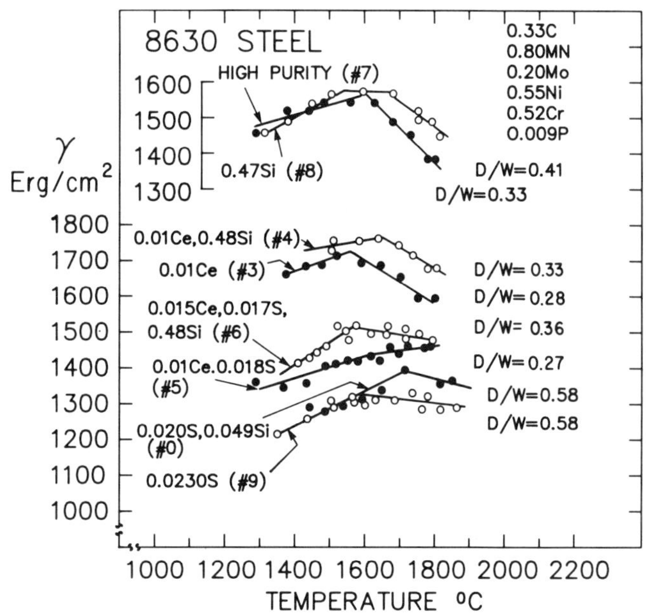

Fig. 3. Surface tension for heats of AISI 8630 steel.

$d\gamma/dT$ at both 1500 °C and 1800 °C. This negative $d\gamma/dT$ produces an outward Marangoni force throughout the weld pool, which produces the low D/W ratio. The 8630 heats have larger $d\gamma/dT$ than the 316L heats at both 1500 °C and 1800 °C. At 1500 °C $d\gamma/dT$ ranges from 0 to 0.9 $Erg/cm^2/°C$ compared to a $d\gamma/dT$ as low as -2.0 $Erg/cm^2/°C$ for the 316L heat J. This means that for the 8630 heats the Marangoni force at the edge of the weld pool ranges from 0 (where $d\gamma/dT$ is zero) to an inward force increasing in magnitude as $d\gamma/dT$ increases. At 1800 °C, however, $d\gamma/dT$ is reduced, becoming negative in most experiments. This negative $d\gamma/dT$ will produce an outward Marangoni force at the center of the weld pool, which will act to reduce D/W ratios. Those heats with highest $d\gamma/dT)_{1500}$ also have the highest $d\gamma/dT)_{1800}$, and the action at both temperatures serves to produce the highest D/W ratios. Likewise, those heats with the lowest $d\gamma/dT)_{1500}$ also have the lowest $d\gamma/dT)_{1800}$ and the action at both temperatures serves to produce the lowest D/W ratios.

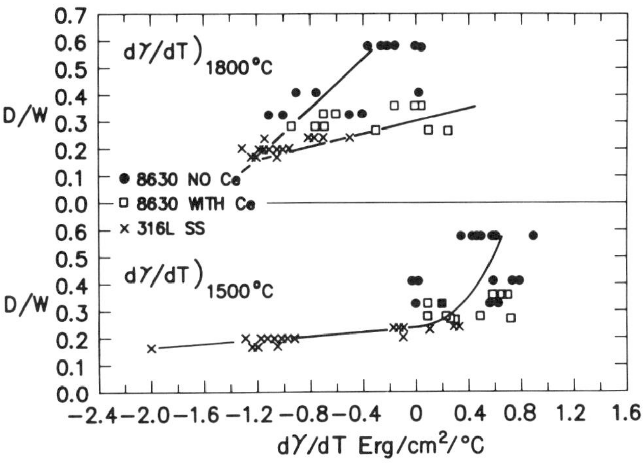

Fig. 4. Weld bead depth-to-width ratio (D/W) versus surface-tension gradient for heats of 316L stainless steel and AISI 8630 steel.

The $d\gamma/dT)_{1800}$ data shows more variation than $d\gamma/dT)_{1500}$. In particular, the 8630 heats which contain both Ce and S (heats 5 and 6) have lower D/W ratios than would be expected from the data taken on the heats not containing Ce. For this reason separate correlation lines have been drawn through the $d\gamma/dT)_{1800}$ data.

WELD POOL FLOW MODEL— A mathematical model has been developed for the fluid flow in gas tungsten arc weld pools. The fluid is taken to be incompressible, viscous, electrically conducting, and confined within a pool of specific shape. The flow is driven by surface-tension gradient (thermocapillary), electromagnetic, and buoyancy forces generally in this order of importance.

A nonuniform 35 × 35 curvilinear grid was used for the fluid flow computation (Fig. 5). Regions of steep gradient, such as the surface layer, are resolved by using a dense grid. The grid was extended to 70 × 70 for the electromagnetic field computation as shown in Fig. 5. The details of the computational technique are outlined more fully in Reference 9, where numerical inaccuracies caused by truncation in the finite difference schemes and by inadequate grid distribution are addressed.

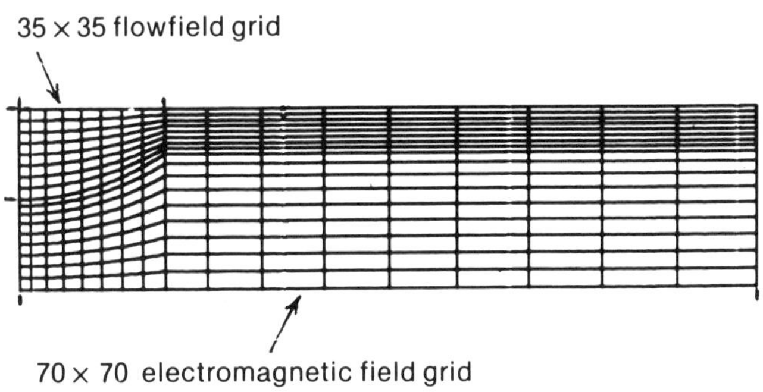

Fig. 5. Grid used for fluid flow and electromagnetic field computations; for clarity, not all grid lines are shown.

Weld pools with surface tension and MHD forces acting simultaneously (and including buoyancy effects) are considered using the same data (Table 1) as in the model of Oreper and Szekely[10]. Values chosen for the surface tension coefficient are typical of those measured in the Surface Tension section above.

Table 1. Parameters and Physical Properties

density, ρ (kg/m^3)	7200
viscosity, μ (kg/m-s)	6×10^{-3}
thermal conductivity, λ (W/m-K)	15.5
specific heat, Cp(J/kg-K)	753
volumetric expansion coefficient, β(K^{-1})	10^{-4}
surface-tension coefficient, $d\gamma/dT$ (N/m-K)*	$\pm 0.5 \times 10^{-3}$
heat input, Q (W)	2000
radius for heat input distribution, r_q (m)	3.5×10^{-3}
current input, I (A)	200
radius for current input distribution, r_i (m)	3.2×10^{-3}
radius of pool, r_1 (m)	5×10^{-3}
depth of pool, d_1 (m)	3×10^{-3}
radius of plate, R (m)	25×10^{-3}
thickness of plate, h (m)	6×10^{-3}
melting point, T_m	1750 K

*Note: 0.5×10^{-3} N/m−K = 0.5 $\frac{Erg}{cm^2-K}$

Contours of axial and radial velocity and of temperature for the case $d\gamma/dT > 0$ (Figs. 6a-c) clearly show the boundary layer on the free surface caused by the surface-tension forces. Because of the widely disparate velocities present in the flow field, velocity vector plots are not as revealing as contour plots taken in conjunction with one another. The flow is driven inwards along the surface and down along the symmetry axis to form a large counterclockwise cell. The largest velocities exist on the free surface and on the axis. Free surface velocities are close to 1 m/s.

A secondary clockwise cell on the bottom of the pool is also predicted. This cell is caused by the large pressure gradients set up where the downward flow along the axis must turn to follow the contour of the pool bottom. Such phenomena are well known from computational and experimental studies of laminar flow (Rhee et al.[11]). In fact, the prediction of such secondary flows is a critical test of the accuracy of finite-difference methods and the adequacy of the grid used[12]. When a coarse (15 × 15) grid is used, the secondary cell is not evident.

The temperature field (Fig. 6c) shows the deep penetration of heat caused by the flow down along the axis; however, the secondary flow reduces this heat flux somewhat at the center of the weld-pool bottom.

When the surface-tension coefficient is negative ($d\gamma/dT < 0$), the boundary-layer nature of the free-surface flow persists, but the flow in the interior is drastically altered. The velocity contours (Figs. 7a-b) show that the radially outward free-surface flow sets up a clockwise cell at the outer edge of the pool and then separates from the pool bottom to form an inner counterclockwise cell. The secondary flow at the pool bottom is still evident. The temperature contours (Fig. 7c) show very little heat penetration into the pool. Most of the heat is carried away laterally along the free surface, which would result in a broad, shallow weld pool. Some heat penetration along the axis is caused by the counterclockwise cell.

For flow with MHD forces alone, contours of the axial and radial velocity show a primary counterclockwise cell similar to that for the surface-tension driven flow with $d\gamma/dT > 0$; however, the velocities are smaller and the flow does not have a boundary-layer nature because the driving force in this case ($\vec{j} \times \vec{B}$) is a distributed body force rather than a surface force. Free surface velocities are on the order of 0.05 m/s. Reference 9 may be consulted for more details.

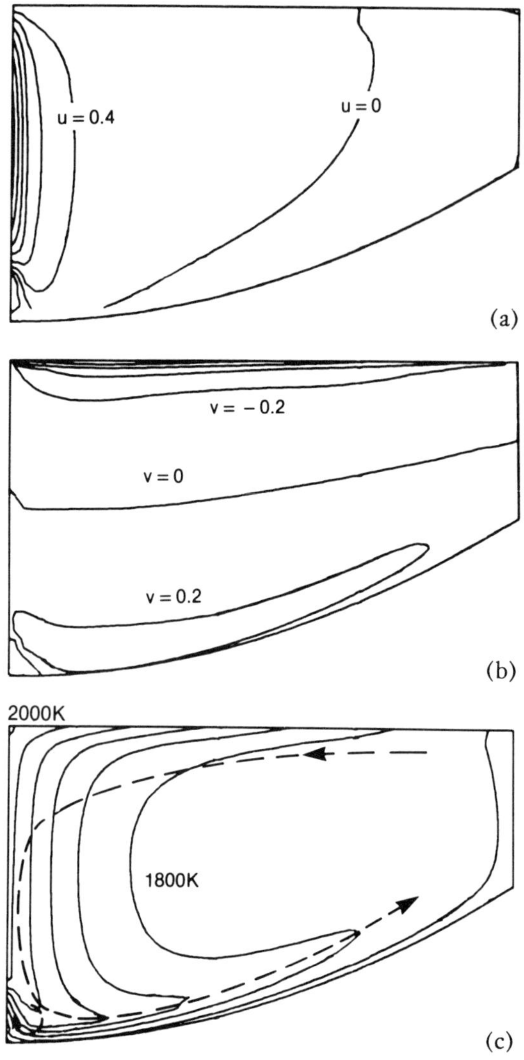

Fig. 6. Combined surface-tension driven and magnetohydrodynamic flow ($d\gamma/dT > 0$): (a) axial velocity contours spaced 0.4 m/s; (b) radial velocity contours spaced 0.2 m/s; (c) temperature contours spaced 40 K; dashed lines indicate flow direction.

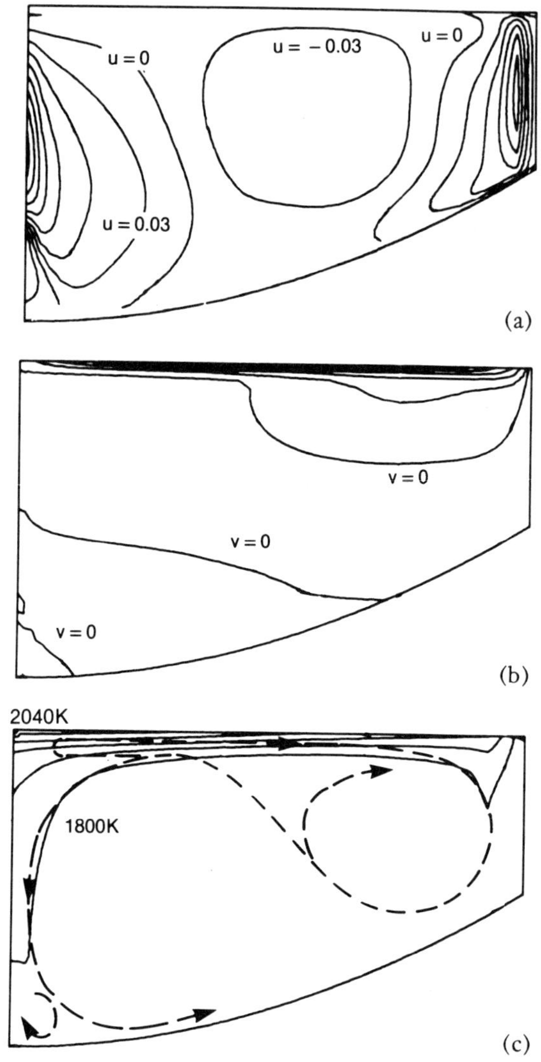

Fig. 7. Combined surface-tension driven and magnetohydrodynamic flow ($d\gamma/dT < 0$): (a) axial velocity contours spaced 0.03 m/s; (b) radial velocity contours spaced 0.1 m/s; (c) temperature contours spaced 60 K; dashed lines indicate flow direction.

FLOW AND TEMPERATURE EXPERIMENTS – Weld-pool flow visualization studies were conducted using a high-speed Kodak Spin Physics video-recording system at 1000 frames per second. A slurry of acetone and aluminum oxide tracer particles was employed to track weld-pool surface motion. Results agreed with the calculated flow streamlines for the $d\gamma/dT > 0$ and $d\gamma/dT < 0$ cases shown above. Aluminum oxide particles were pushed to the edge of the pool for low sulfur heats (~10 ppm) of stainless steel ($d\gamma/dT < 0$), whereas particles were swept to the center of the weld pool for heats with $d\gamma/dT > 0$. Weld-pool surface velocities for the inward radial surface flow were measured to be approximately 1 m/s, in agreement with the above calculations.

Weld-pool temperature measurements for three different heats of 8630 steel are shown in Fig. 8. The weld-pool width was measured to obtain additional data points for temperature gradient determination. Measured peak weld-pool surface temperatures agree well with those predicted by the computational model. Also, measured values are consistent with experiments conducted by Giedt et al.[13].

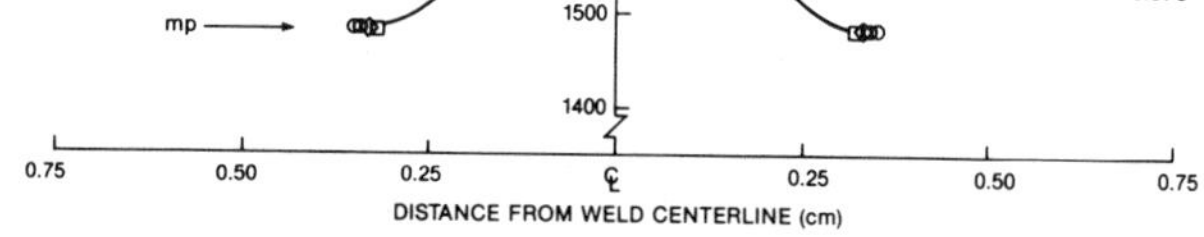

Fig. 8. Temperature across weld centerline; AISI 8630 low-alloy steel, 3/8-inch thick.

CONCLUSIONS

Surface tension measurements as a function of temperature for different heats of AISI 8630 steel and 316L stainless steel correlate with the weld bead D/W ratio and confirm the Heiple and Roper[1] model. Also, a computational study of weld pool flow indicates that convection driven by surface tension gradients plays a crucial role in determining the heat transfer to the solid-liquid interface. Experimental measurements of weld pool flow velocities and peak weld pool surface temperatures agree well with the computational flow model.

ACKNOWLEDGEMENT

This work was supported by National Science Foundation Grant No. MEA-8208950.

REFERENCES

[1] Heiple, C.R., and Roper, J.R., "Mechanism for Minor Element Effect on GTA Fusion Zone Geometry," *Welding J. 61*, p. 97s (1982).

[2] Walsh, D.W., and Savage, W.F., "The Effects of Minor Elements on Autogenous GTA Weld Bead Shape," (In these proceedings).

[3] Sundell, R.E., Correa, S.M., Harris, L.P., Solomon, H.D., Wojcik, L.A., Savage, W.F., and Walsh, D.W., *Minor Element Effects on Gas Tungsten Arc Weld Penetration*, Industry/University Cooperative Research, Final Report to the National Science Foundation, Production Research Program, Div. of Mechanical Engr. and Applied Mechanics, Contract No. MEA-8208950 (to be published).

[4] Brashforth, F., and Adams, S.C., *An Attempt to Test the Theories of Capillarity*, Cambridge U. Press, England, 1883, authorized facsimile, 1971 University Microfilms, Ann Arbor, MI, USA.

[5] White, D.W.G., *Trans ASM, 55*, pp. 757-777 (1962).

[6] Sangiorgi, R., Caracciolí, G., and Passerve, A., *J. Materials Sci., 17*, pp. 2895-2901 (1982).

[7] Butler, J.N., and Bloom, B.H., *Surface Science, 4*, pp. 1-17 (1966).

[8] Lin, M.L., and Eagar, T.W., "Influence of Surface Depression and Convection on Arc Weld Pool Geometry," *Transport Phenomena in Materials Processing (PED) 10*, ASME WAM, Nov. 13-18, 1983.

[9] Correa, S.M., and Sundell, R.E., "A Computational and Experimental Study of the Fluid Flow in Weld Pools," Proceedings of the 1986 Engineering Foundation Conference, *Modeling and Control of Casting and Welding Processes*, Santa Barbara, Calif., Jan. 12-17, 1986, pub. by Met. Soc. of AIME.

[10] Oreper, G.M., and Szekely, J., "Heat and Fluid-Flow Phenomena in Weld Pools," *J. Fluid Mech. 147*, pp. 53-79 (1984).

[11] Rhee, H.S., Koseff, J.R., and Street, R.L., "Flow Visualization of a Recirculating Flow by Rheoscopic Liquid and Liquid Crystal Techniques," *Exp. in Fluids 2*, p. 57 (1984).

[12] Schrieber, R. and Keller, H.B., "Driven Cavity Flows by Efficient Numerical Techniques," *J. Comp. Phys. 49*, p. 310 (1983).

[13] Giedt, W.H., Wei, X-C, and Wei, S-R, "Effect of Surface Convection on Stationary GTA Weld Zone Temperatures," *Welding J. 63*, pp. 376s-383s (1984).

THE MECHANISM OF MINOR ELEMENT INTERACTION IN AUTOGENOUS WELD POOLS

Daniel W. Walsh
California Polytechnic State University
San Luis Obispo, California

Warren F. Savage
Renssalaer Polytechnic Institute
Troy, New York

ABSTRACT

Substantial changes in the three-dimensional shape of autogenous GTAW weld pools were caused by both controlled changes in base plate composition and alterations to weld process parameters. Many treatments had significant effect on weld pool shape at the 95% confidence level. Typically, the main effects (single variable) were the greatest, but many multiple-factor interactions were large. Oxygen and nitrogen altered bead shape by a more complex mechanism than did sulfur. Many of the effects noted were common to both stainless and low-alloy steels. Several shielding gas additions were shown to minimize heat-to-heat variability in depth-to-width ratio. Varestraint testing has shown that weld pool sulfur concentrations at the 10 ppm level can produce hot-cracking. Pool flow velocities were measured by direct observation and recorded on a high-speed video device. Patterns and velocities observed are consistent with the Marangoni surface tension driven-flow mechanism. Pool depth-to-width ratio was unaffected by electrode diameter, but was directly related to tip apex angle. In general, sulfur and nitrogen additions to the pool increased depth-to-width ratios, oxygen additions increased depth, and both oxygen and nitrogen additions increased volume.

FABRICATORS OF MATERIALS HAVE NOTED that a wide variance in bead shape can exist between different heats purchased to the same material specification, despite conformance to a standard set of welding parameters during fabrication. The problem of variable bead geometry, although not a new one, has become critical because of tremendous pressure within the fabrication community to automate welding processes. The incentives for automation include the expansion of process applicability into hostile environments as well as cost savings and quality enhancement. The problem is most evident in autogenous (no filler added) welds made by the Gas Tungsten Arc Welding process (GTAW) at currents up to 250 amperes. However, the problem persists with higher current levels and with hot wire feed. The welds most sensitive to this effect are butt welds in pipe, seal welds, root passes in pipe of any diameter and wall thickness, and spot welds.

The critical variable that affects bead geometry is base metal composition. Existing standards allow ranges of composition that are a comromise between what is attainable with reasonable effort on the part of the producer and what is acceptable for final use by the customer. Often it is not the major components in the alloy , but rather the minor elements. These elements are typically those most difficult to control. Indeed, in many cases the causal element is sulfur, and variations in weld bead depth-to-width (D/W) ratios as great as 200% can result from changes in sulfur concentrations of less than 100 parts per million (ppm). The currently available refining processes which permit reducing the sulfur content to less than 100 ppm have actually aggravated the bead geometry problems associated with automatic welding . As a result some fabricators request that sulfur be added to material in order to augment weld penetration. It is ironic that some producers are being asked to add sulfur to production heats, thus abrogating achievements of 100 years of production research.

Many forces interact to shape an autogenous GTAW weld bead. The force of the plasma jet on the weld pool, the buoyancy forces within the weld pool, the surface tension differences over the pool surface and other forces have been considered by various investigators. Correlation between the plasma jet momentum and bead shape has not been found [1]. Buoyancy forces have been examined [2], but when compared to Lorentz effects [3,4], these were found to contribute only a small fraction of the total stirring within the pool (6% of that provided by Lorentz effects). Moreover, when surface tension effects are compared on a theoretical basis [5,6] or an experimental basis [7] with Lorentz effects in the range of currents normally used for thin-section autogenous GTAW welding , the flow velocities caused by surface tension are found to be an order of magnitude greater (10-20 cm/sec for Lorentz and 100-150 cm/sec for Marangoni).

Over the range of currents normally employed for autogenous GTAW, surface tension driven-flows (the Maragoni effect) seem to be the major agent of weld bead shape variations. Large differences in surface tension can be caused by slight composition differences. All pure metals and most alloys exhibit a decreased surface tension as temperature is increased; a manifestation of greater interatomic spacing and thus lower attractive forces at higher temperatures. In some alloys, however, most notably iron-base alloys containing group VI additions, surface tension actually increases with increased temperature. The surface tension gradient reversal is caused by the pronounced surface activity of group VI elements in iron-based alloys. Sulfur, as an example, segregates to the surface of the weld pool, drastically lowering the surface energy. The configurational entropy of the system is decreased by the concentration of sulfur atoms at this location, however, the net free energy change for the system is negative:

$$dG_A = dH - TdS + Ad\gamma$$

, H = enthalpy
S = entropy
G_A = Gibb's free energy @ constant area
γ = surface energy

and the process is thermodynamically feasible subject to the limits of kinetics. Note that as the temperature is increases, the decrease in configurational entropy becomes more critical, and at higher temperatures less sulfur will segregate to the surface. The surface tension is inversely related to the sulfur content in the range of interest. The high temperature regions of a weld pool, beneath the arc, will support lower surface adsorption of sulfur than will low temperature regions on the fringe of the weld pool, where the temperature is closer to the effective liquidus of the base metal. The surface tension in the pool center will be greater than that of the liquid remote from the center. If it is assumed that the temperature gradient in the weld pool is universally negative moving radially outward from the center, then the surface tension gradient in the weld pool is universally negative moving radially outward from the center, and a radially inward flow of fluid will be generated. This flow generates a deep narrow pool. In pools lacking surfactants the flow is reversed; broad, shallow pools are generated.

MATERIALS AND PROCEDURE

A study was made of the three-dimensional shape of autogenous GTAW pools. High-pressure inert gas jets were used to decant these pools in a controlled series of base materials, the composition of which provided the basis for a 2x3 designed experiment around the factors Si, Ce and S (Table 1). These factors were augmented by O_2 and N_2 additions to the shielding gas, generating a 2x5 matrix. The effects of the tip diameter, tip geometry and weld current were examined as well.

The main body of the experiment involved the placement of an autogenous GTAW bead on plate welds on 0.40" thick AISI 8630 material. All welds were made under identical conditions except when a process parameter was used as an experimental variable. Controlled amounts of O_2 and N_2 were added to the torch gas and made available to the pool. Active gas concentrations ranging from 10 to 5000 ppm were supplied by a mixing apparatus composed of a series of mass flow controllers. The weld was expunged by violently contacting the pool with a directed, high-pressure jet of inert gas. This occurred immediately after the arc was extinguished. using a method developed by Zanner [8], silicone rubber replicas of the decanted weld pool crater were made, and pool shape variables were determined from them.

The Varestraint apparatus [9,10] was used to determine the hot-cracking sensitivity of several heats of Type 316 austenitic stainless steel. Table 2 lists the composition of these materials. The cracking parameters recorded were total crack length (TCL), number of cracks (NC), and maximum crack length (MCL). Transverse and longitudinal sections of the weld bead were examined metallographically, and correlations between hot-cracking behavior and microstructure were made. Welding conditions for these tests are listed in Table 3.

High-speed vidoetape recordings of the motion of alumina (Al_2O_3) particles (placed on the plate prior to welding) on the pool surface were made with a Spin Physics SP2000 video device.

Table 1
Composition (wt. %) of AISI 8639 Alloy Heats

Heat	C	Ce	Cr	Mn	Mo	Ni	P	S	Si
37	.33	NA	.50	.79	.20	.56	.012	.007	.13
38	.33	NA	.52	.82	.20	.58	.010	.007	.47
39	.32	NA	.50	.80	.20	.58	.012	.023	.14
40	.33	NA	.51	.81	.20	.58	.010	.020	.49
43	.32	.010	.59	.81	.20	.52	.007	.004	.14
44	.32	.010	.50	.81	.20	.52	.008	.003	.15
45	.32	.010	.50	.81	.20	.52	.005	.018	.15
46	.32	.015	.51	.82	.20	.53	.005	.017	.48

Table 2
Chemical Composition of Type 316 Material, (Wt.%)

ID Code	1	2	3	4
C	0.011	0.015	0.012	0.013
Co	-	0.09	0.16	0.04
Cr	17.15	16.45	16.50	16.90
Cu	-	0.32	0.18	-
Mn	1.59	1.84	1.67	1.60
Mo	2.73	2.10	2.06	2.56
N	0.105	0.073	0.080	0.105
Ni	12.65	10.34	10.18	12.70
P	0.022	0.030	0.030	0.024
S	0.001	0.003	0.013	0.001
Si	0.60	0.40	0.62	0.59

Table 3
Welding Parameters

	AISI 8630	Type 316
Current	150 A	250 A
Voltage	11.0 V	12.6 V
Travel speed	3 ipm	4 ipm
Electrode diameter	3/32", EWTh-2, 90 tip	1/8", EWTh-2, 90 tip
Tip to work	0.090"	0.090"
Shielding gas	Argon, 35 cfh	Argon, 35 cfh

RESULTS AND DISCUSSION

VIDEOTAPE OBSERVATIONS- Weld pools in AISI 8630, heats 39 and 44, and Type 316, heats 1 and 3 were studied using the Spin Physics SP2000 high-speed videocamera. Alumina particles placed on the plate prior to welding were used to trace flow patterns. Flow in heats 39, 44 and 3 were all radially inward, while heat 1 exhibited outward flow. The flow velocities measured for heats 39 (200 ppm S) and 3 (130 ppm S) were similar, approximately 85 cm/sec. This number is consistent with values theoretically developed by Levich [5]. Heat 44 (10 ppm S) exhibited outward flow. The different flow velocities and directions are reconcilable with the work of Mills, et al [11], who postulated that a critical minimum concentration of surfactant is needed to support inward flow.

THE EFFECT OF MINOR ELEMENTS- A summary of the main effects of compositional variables, significant at the 95% confidence level, is given in Table 4. In addition to the main effects many interaction effects were present, the most prominent of which are listed in Table 5. Oxygen has a potent influence on the weld pool, increasing pool length, width, volume and section. However, it has no net effect on D/W. Further, oxygen additions increase the configuration factor and thus promote a stable, hemispherical pool. Sulfur, in contrast, decreases width and increases depth, greatly increasing D/W. Sulfur does not alter bead volume or cross-sectional area but does enhance pool symmetry. N_2 increases depth, volume, pool area and configuration factor in a manner similar to but more subdued than O_2. N_2 has no effect on pool length or width. The effect of gas additions was not only to increase pool depth, but to expand the molten volume as well. Interestingly, the arc voltage for the constant tip to work distance was increased by the addition of O_2 (1.5 to 2.5 V) and N_2 (0.75 to 1.25 V) to the shielding gas at the 400 ppm level. This implies a change in the drop regions of the arc and an improved transfer of energy to the work. No such increase in energy transfer was noted with sulfur additions.

Table 4
Major Effects of Composition on
Listed Geometric Variables as a Percent
of Grand Mean

	S	Si	Ce	O	N	G.Mean
Length	-	-	-	+11.5		.655 cm
Width	-6	-	-	+14.0	-	.598 cm
Depth	+12.5	+10.2	-24.3	+15.7	+8.3	.243 cm
Volume	-	+15	-25	+36	+11	cm^3
Area	-	+13.9	-9.8	+32.7	+10.3	.1252 cm^2
D/W	+19.3	+8.0	-27.8	-	+6	.407
Configuration factor	+7.7	-	-23.7	+15.5	+15	.124

Table 5
Significant Higher Order Interactions*

Depth	S-Ce	-8%
	S-O	+7%
	Ce-O	-9%
	S-Ce-Si-O	-7%
Volume	Si-O	+7%
	Si-N	-7%
	O-N	-11%
	S-Si-Ce-O	-15%
D/W	S-Ce	-13%
	S-Si	+8%
	S-O	-10%
	Ce-O	+11%
	Si-O	-5%
Fused Area	S-Si	-8%
	S-Ce-Si	-7%
	Ce-Si-O	-7%

*as a percentage of grand mean of particular geometric variable

Lower arc voltage was associated with cerium rich pools, indicating a reduction in the anode drop and a decrease in energy transfer for the weld pool. Cerium also forms stable oxides and sulfides (Ce_2O, CeS, Ce_2O_2S) with melting points in excess of the base material. The compounds would effectively remove surface active materials from the weld pool, and reduce the surface tension driven-flow fostered by these elements. Furthermore, large, refractory slag deposits were associated with cerium rich pools. These deposits appeared to interfere with pool flow and caused the anode spot to be unstable. Because of these factors, Ce diminishes pool depth, D/W, volume and configuration factor. Pool depth and D/W ratios are increased by the action of Si, as was pool cross-section and volume. The latter effect may be explained by the interaction of secondary electrons with the anode drop region, while the former can be explained by the reduction in melt viscosity caused by the presence of Si. The level of sulfur present in the low sulfur segment of the designed experiment is sufficient to cause mild flow reversal. Thus flow enhancing elements like Si reduce viscosity and seem to behave in a manner similar to S. The strong and positive effect the S-Si interaction has on D/W can be explained by the enhancement of the existing radially inward flow pattern as well. The multiple element interaction S-Si-Ce-O diminished pool depth and volume. This reduction is associated with the formation of a complex slag on the pool surface.

The effects of O_2 and N_2 additions on bead depth, averaged over a replicated set of welds on the entire test matrix is shown in Table 6. Note that O_2 and N_2, added alone or in combina-

Table 6
Effects of Shielding Gas Additions*

	Avg. bead depth	Standard deviation
No added gas	82.16 mils	20.8 mils
O_2	104.5 mils	14.5 mils
O_2/N_2	105.3 mils	13.5 mils
N_2	93.7 mils	22.5 mils

*Three beads on each of the eight heats of AISI 8639 with minor element differences

tion at the 400 ppm level increase the average bead depth, but only those additions involving O_2 decrease the standard deviation of the D/W distribution. The reactive gas content employed in the test matrix was the level that produced the maximum pool depth in the range of concentration between 0 and 1500 ppm. Both depth and D/W exhibit a maximum and then a decline in this range of composition. Interestingly, additions at levels in excess of 0.2% engender large increases in depth and D/W, attributable to blunting of the electrode tip by reaction with the active gas.

PROCESS PARAMETER EFFECTS- Electrode diameter, at fixed current, had no significant effect on pool geometry for the range of diameters examined. Electrode tip angle, however, had a strong influence on pool shape. The tip angle was found to be directly related to the depth of penetration. This finding is in accordance with the results of Savage et al [12], as well as the results of Burghardt [13]. The active level of sulfur present even in the low sulfur material promotes inward flow, and no reversal of this behavior is caused by the pool temperatures encountered at the current levels chosen. Interestingly, when current was increased pool length, width, depth and volume were enhanced, but D/W was not significantly affected, and no positive interactions between sulfur and current existed. Strong positive correlations between O_2 and current did exist for pool depth and D/W however.

VARESTRAINT RESULTS- Lippold and Savage [14], Cieslak and Savage [15], and others have shown that the mode of solidification in austenitic stainless steels is the critical factor in determining the susceptibility of these materials to hot-cracking. Typically, compositions that solidify with austenite as the primary phase are more sensitive to hot-cracking than welds solidifying as primary ferrite. Producers of austenitic stainless steel favor compositions that render austenite the primary phase during solidification. Thus, although common practice can reduce sulfur and phosphorus contents to less than 30 ppm, hot-cracking remains a problem when these steels are welded. Table 7 summarizes the Varestraint test results. At 1% augmented strain steel 2 shows a MCL of only 2 mils and a TCL of 6 mils. Steel 4 had the greatest MCL and the second largest TCL, steel 1 had the next to largest MCL and steel 3 had the lowest TCL and MCL. At the 2% and 4% levels of restraint, steel 3 is most resistant to cracking, 1 is least resistant, and materials 2 and 4 are intermediate in response.

The primary austenite solidification substructure typical of steels 1 and 4 is shown in Figure 1. Steels 2 and 3, however, solidify as primary ferrite, the ferrite being located in the cores of the cellular dendrites, Figure 2. Figure 3 shows a typical crack at the fusion boundary of a specimen of steel 1 at 1% strain. Note that the cracks extend across the fusion boundary and propagate along cell boundaries in the fusion zone. Although steel 3 has a sulfur content 13 times as great as steel 1, steel 1 is more crack sensitive.

Resulfurization is effective in homogenizing the weld bead geometry only when additions raise the sulfur level to 30 ppm, a level three times that found in crack sensitive heat 1. Resulfurization is a viable alternative only when the addition of sulfur is accompanied by the addition of some element that can mollify its behavior during the latter stages of solidification, without impeding the desirable effects sulfur has on pool flow dynamics.

Table 7
Summary of Varestraint Test Results

Identification Code	Augmented Strain (%)	Number of Cracks (NC)	Maximum Crack Length (MCL), mils	Total Crack Length, mils
1	1	13	8.75	45.0
2	1	4	2.0	6.25
3	1	8	5.0	22.5
4	1	9	12.5	40.0
1	2	18	14.0	102.0
2	2	21	8.75	59.0
3	2	16	6.25	37.0
4	2	16	10.5	68.0
1	4	34	22.5	246.0
2	4	45	14.0	195.0
3	4	27	10.0	114.0
4	4	49	11.0	153.0

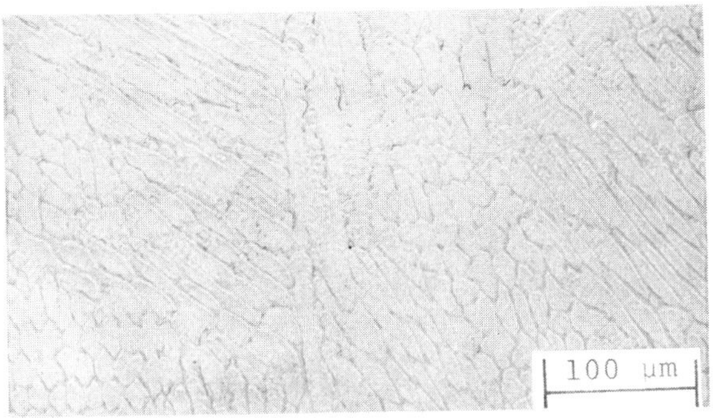

Figure 1. Primary austenite solidification

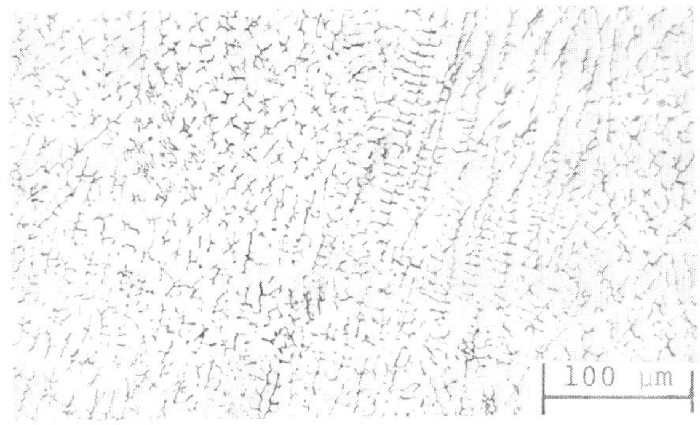

Figure 2. Primary ferrite solidification

Figure 3. Hot-crack in heat 1

CONCLUSIONS

1. Oxygen, nitrogen and sulfur increase pool depth and symmetry. Oxygen and nitrogen increase pool volume, while sulfur does not.
2. Arc voltage increases when oxygen or nitrogen is present at the 400 ppm level. These gases increase the energy transfered to the work.
3. Silicon increases the depth and D/W ratio of weld beads in the materials studied. The S-Si interaction increases penetration more than S or Si alone.
4. Cerium reduces the depth of penetration, D/W, symmetry, fused area and pool volume. The effect is large at the cerium level studied. Cerium reduces the anode drop, forms stable oxides and sulfides, and produces a refractory coating on the pool surface.
5. O_2 and N_2, when added to the shielding gas, individually or in combination, increased the average D and D/W for the series of alloys as a group. Furthermore, these additions decreased the standard deviation of the D and D/W distributions.
6. O_2 and N_2 exhibit a variable effect on D and D/W as their concentration in the shielding gas is changed. The D and D/W ratio initially increase, reaching a maximum at 400 ppm. At 400 ppm to 1200 ppm, D then increases slowly while D/W decreases. At levels 0.2 wt.% of gas (2000 ppm) D and D/W both increase again, because the tungsten tip erodes, becoming more blunt.
7. A direct relation exists between electrode tip angle and bead D/W for the alloys studied.
8. At the levels of current examined, electrode diameter has no effect on bead shape.
9. Sulfur concentrations produced in the weld pool by resulfurization techniques were shown to be quite sufficient to cause extensive hot-cracking.
10. The hot-cracking sensitivity of austenitic stainless steel was shown to be a stronger function of primary solidification mode than of sulfur content.

REFERNCES

[1] Converti, J., Ph.D. Thesis- Massachusetts Institute of Technology, June 1981.

[2] Mills, G.S., Arc Physics and Weld Pool Behavior, Welding Institute Proceeding, May 1979, pp. 279.

[3] Woods, R.A. and Milner, D.R., Welding Journal, 50(4), 1971, pp. 163.

[4] Atthey, D.R., Journal of Fluid Mechanics, 98, pp. 787.

[5] Levich, V.G., Physiochemical Hydrology, Prentice-Hall Publishers, 1962, pp. 388.

[6] LanVow, L. and Lifshitz, E., Fluid Mechnaics, Addison-Wesley Publishers, 1959, pp. 242.

[7] Heiple, G.R., Roper, J.R., Stanger, R.I., and Aden, R.J., Welding Journal, 62(3), March 1983.

[8] Zanner, F.J., Ph.D. Thesis- Rensselaer Polytechnic Institute, August 1967.

[9] Savage, W.F. and Lundin, C.D., Welding Journal, October 1965, pp. 433-442.

[10] Savage, W.F. and Lundin, C.D., Welding Journal, November 1966, pp. 497-502.

[11] Mills, K.C., Keene, R.F., Brools, R.F., and Olusanya, A., WI Conference Proceedings, The Effect of Minor Elements on Weld Pool Behavior, November 1983.

[12] Savage, W.F., Strunk, S.S., and Ishikawa, Y., Welding Journal, November 1965, pp. 457-464.

[13] Heiple, C.R. and Burghardt, P.R., verbal presentation at 1985 AWS Weld Show, Las Vegas, Nevada.

[14] Lippold, J.C. and Savage, W.F., Welding Journal, July 1978, pp. 257-269.

[15] Cieslak, M.J. and Savage, W.F., Welding Journal, May 1970, pp. 203-211.

DRIVING FORCES FOR CONVECTION IN WELD POOLS

Y.H. Wang, Sindo Kou
Department of Metallurgical & Mineral Engineering
University of Wisconsin-Madison
Madison, Wisconsin 53706

ABSTRACT

Weld pool convection can affect significantly the penetration and the structure and properties of the resultant welds. Three distinct driving forces for convection are considered in this study, i.e., the buoyancy force, the electromagnetic force and the surface tension force at the weld pool surface. The distributions of these three individual driving forces in the weld pool are presented, along with the weld pool convection patterns and penetrations they produce.

WELD POOL CONVECTION has been studied using computer modeling by Oreper et al. (1,2), Chan et al. (3) and Kou et al. (4). The computer models developed describe two dimensional convection in stationary weld pools in a semi-infinite workpiece. Recently, Kou et al. (5,6) have developed a weld pool convection model for a moving heat source and workpieces of finite thickness. Since the vast majority of the arc welding practices involve a moving heat source and weld pool, and the workpieces are finite in thickness, rather than semi-infinite, this convection model is much more useful.

Weld pool convection can strongly affect the structure and properties of the resultant welds. Variations in weld characteristics which are likely to occur from changes in weld pool convection are: weld penetration, macrosegregation, gas porosity, solidification structure, undercutting and surface smoothness. Apparently, the quantitative understanding of convection and heat flow in weld pools is of considerable practical interest. The present paper emphasizes on the driving forces for convection in weld pool.

Driving Forces

Fig. 1 shows a schematic sketch of an arc welding system, the heat source is moving at a constant speed U. The coordinate system (x-y-z) moves with the electrode at the same speed, and its origin coincides with the center of the electrode.

There are three distinct driving forces for weld pool convection, i.e., the electromagnetic force, the buoyancy force and the surface tension force at the weld pool surface. The first driving force is caused by the interaction between the divergent current path in the weld pool and the magnetic field it generates. The second driving force is caused by the temperature gradient within the weld pool, and the third by the temperature gradients at the weld pool surface and possibly by the surface active agents in the weld pool. These three driving forces are expressed as follows:

Electromagnetic Force

$$\vec{F}_{em} = \vec{J} \times \vec{B}$$

Buoyancy Force

$$\vec{F}_b = -\rho\beta\,\vec{g}\,(T-T_0)$$

Surface Tension Force

$$\vec{F}_\gamma = \nabla T \frac{\partial \gamma}{\partial T}\,\delta(z)$$

where $\delta(z)$ is the unit impulse function, indicating that the surface tension force exists only at the weld pool surface (z=o).

The current density vector $\vec{J}$ and the magnetic flux vector $\vec{B}$ can be obtained by solving Maxwell's equations with the MHD approximation (7-9) and the following two boundary conditions:

$$J_z = \frac{3I}{\pi b^2} \exp\left(\frac{-3r^2}{b^2}\right) \text{ at } z = 0$$

and

$$J_z = o \text{ at } z = g$$

Governing Equations

The three driving forces described above are substituted into the following equation of motion

$$\rho(\vec{V}.\ \nabla)\ \vec{V} = -\ \nabla P - [\nabla\ .\ \tau]+(\text{driving forces})$$

This equation is then solved simultaneously with the other two governing equations, i.e., the continuity and the energy equations, to get the weld pool convection patterns and penetrations.

The convection is considered laminar, in view of the small size of the weld pool. The weld pool surface is considered flat, which is reasonable since the GTA (gas tungsten arc) welding process is designed for use with relatively low currents, say around or below 200 A, to avoid melting of the tungsten electrode. It has been reported (10,11) that for a welding current not significantly higher than 200 A, the weld pool surface is flat.

Results and Discussion

The results to be shown have been calculated for a GTA weld in a 3.2 mm thick 6061 aluminum plate of 20.3 cm x 12.7 cm. The current, voltage and welding speed are 110 A, 10 V and 5.5 mm/sec, respectively. The arc efficiency, current flux radius and heat flux radius are 76 pct, 2.0 mm and 3.0 mm, respectively. These values were measured by using a calorimeter and a split-anode.

(i) Electromagnetic Force

Fig. 2a shows the distribution of the electromagnetic force in the weld pool. Because of the divergency of the electric current from the central part of the weld pool surface into the bulk weld pool and the surrounding workpiece material, the electromagnetic force points downward and inward. The resultant convection pattern in the weld pool is shown in Fig. 2b. The counterclockwise convection pattern is consistent with the direction of the electromagnetic force shown in Fig. 2a. As can be seen from Fig. 2b, the liquid metal near the weld pool surface flows inward, picks up the heat from the arc, and delivers it to the weld root. Consequently, full penetration is produced.

(ii) Buoyancy Force

Fig. 3a shows the distribution of the buoyancy force in the weld pool. Due to the heat input from the arc, the liquid metal near the center of the weld pool experiences higher temperatures than that near the pool boundary. This suggests that under the influency of the gravity force, the former is lighter and tends to float while the latter is heavier and tends to sink. The resultant convection pattern is shown in Fig. 3b. As shown, the convection pattern is clockwise and the penetration is shallow.

(iii) Surface Tension Force

Fig. 4a shows the distribution of the surface tension force. Due to the fact that the surface tension of liquid metals tends to decrease with increasing temperature, the surface tension at the weld pool surface is lower near the center and higher near the boundary. As a result, the surface tension force points outward. The resultant convection pattern is shown in Fig. 4b, indicating an outward convection pattern at the weld pool surface. The weld penetration is rather shallow. In the presence of surface active agents, the surface tension - temperature relationship can be reversed. This suggests that the surface tension force and the resultant convection pattern can also be reversed, as shown in Fig. 5. Full penetration is produced.

During welding all these three driving forces act simultaneously, the resultant convection patterns have been described elsewhere (5,6).

Conclusions

1. The electromagnetic force points downward and inward, producing deep penetration and a convection pattern in which the liquid metal flows downward near the center of the pool and upward near the boundary.

2. The buoyancy force points upward near the weld pool center, producing shallow penetration and a convection pattern opposite to that due to the electromagnetic force.

3. The surface tension force points outward, producing an outward surface flow pattern and shallow penetration. Both the driving force and the resultant convection pattern can be reversed by surface active agents, and the penetration can become significantly deeper.

Acknowledgement

The authors gratefully acknowledge support for this study from the National Science Foundation, under NSF Grant No. DMR 8419274.

References

1. Oreper, G. M., T. W. Eagar and J. Szekely, Weld. J. 62, 307s (1983).

2. Oreper, G. M. and J. Szekely, J. Fluid Mech. 147, 53 (1984).

3. Chan, C., J. Mazumder and M. M. Chen, Metall. Trans. A, 15A, 2175 (1984).

4. Kou, S. and D. K. Sun, Metall. Trans. A, 16A, 203 (1985).

5. Kou, S. and Y. H. Wang: Weld. J. 65, 63s (1986).

6. Kou, S. and Y. H. Wang: Metall. Trans. A, 17A (1986), in press.

7. S. Szekely, "Fluid Flow Phenomena in Metals Processing", p. 178, Academic Press, New York (1979).

8. W. F. Hughes and F. J. Young, "The Electromagnetic Dynamics of Fluids", chap. 7, John Wiley, New York (1966).

9. J. D. Jackson, "Classical Electrodynamics", 2nd ed., Chap. 3, John Wiley, New York (1976).

10. Lawson, W. H. S. and H. W. Kerr, Weld. Resh. Interl. 6, 1 (1976).

11. Lin, M. L. and T. W. Eagar, Weld. J. 64, 163s (1985).

Table I. Symbols and Definitions

Symbol	Definition
b	current flux radius
$\vec{B}$	magnetic flux vector
$\vec{F}_b$	buoyancy force
$\vec{F}_{em}$	electromagnetic force
$\vec{F}_\gamma$	surface tension force
g	workpiece thickness
$\vec{g}$	gravitational acceleration
I	welding current
J_z	axial component of current density vector
$\vec{J}$	current density vector
P	pressure
T	temperature
U	welding speed
$\vec{V}$	velocity vector
β	thermal expansion coefficient
γ	surface tension
$\partial\gamma/\partial T$	surface tension temperature coefficient
ρ	density
τ	shear stress

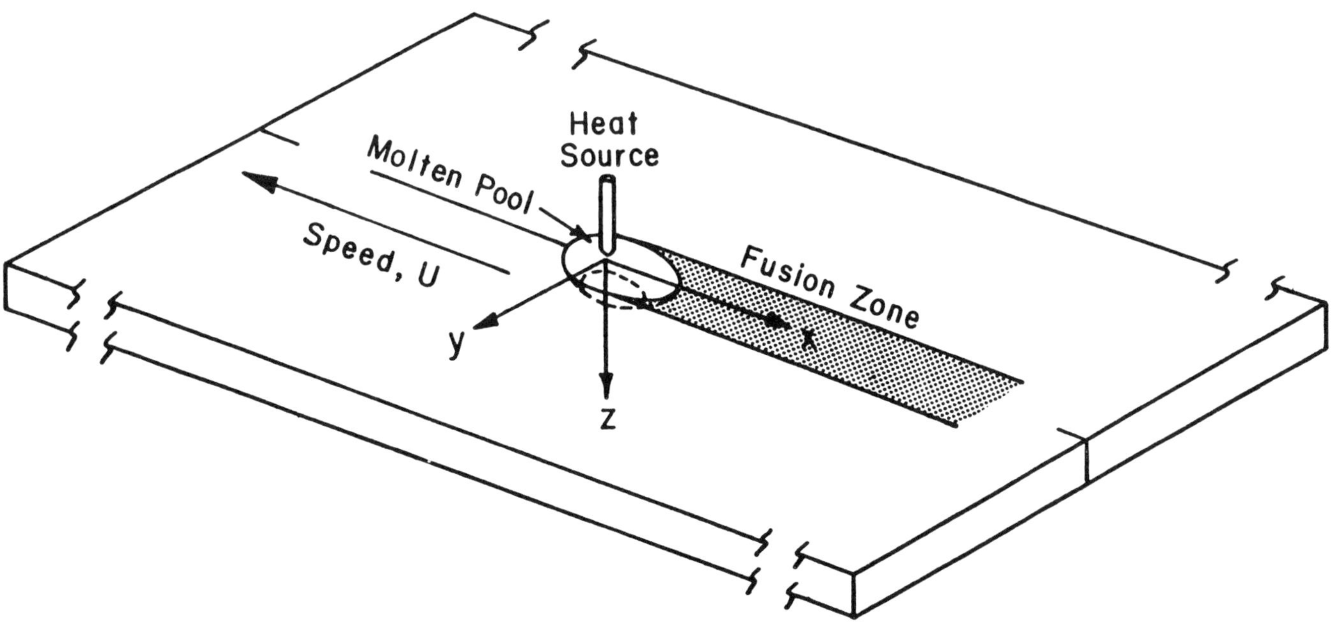

Fig. 1 Schematic sketch of the arc welding system.

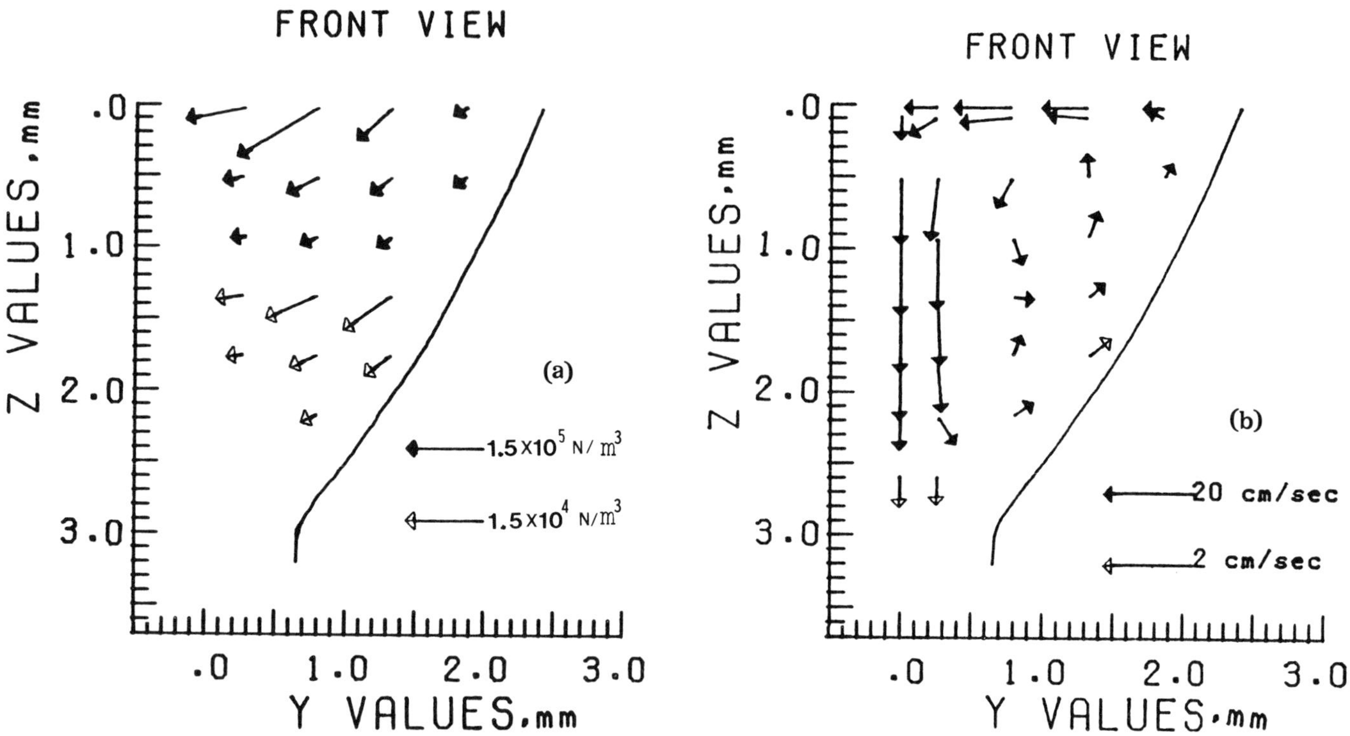

Fig. 2 Electromagnetic force: (a) distribution in the weld pool; (b) resultant convection pattern.

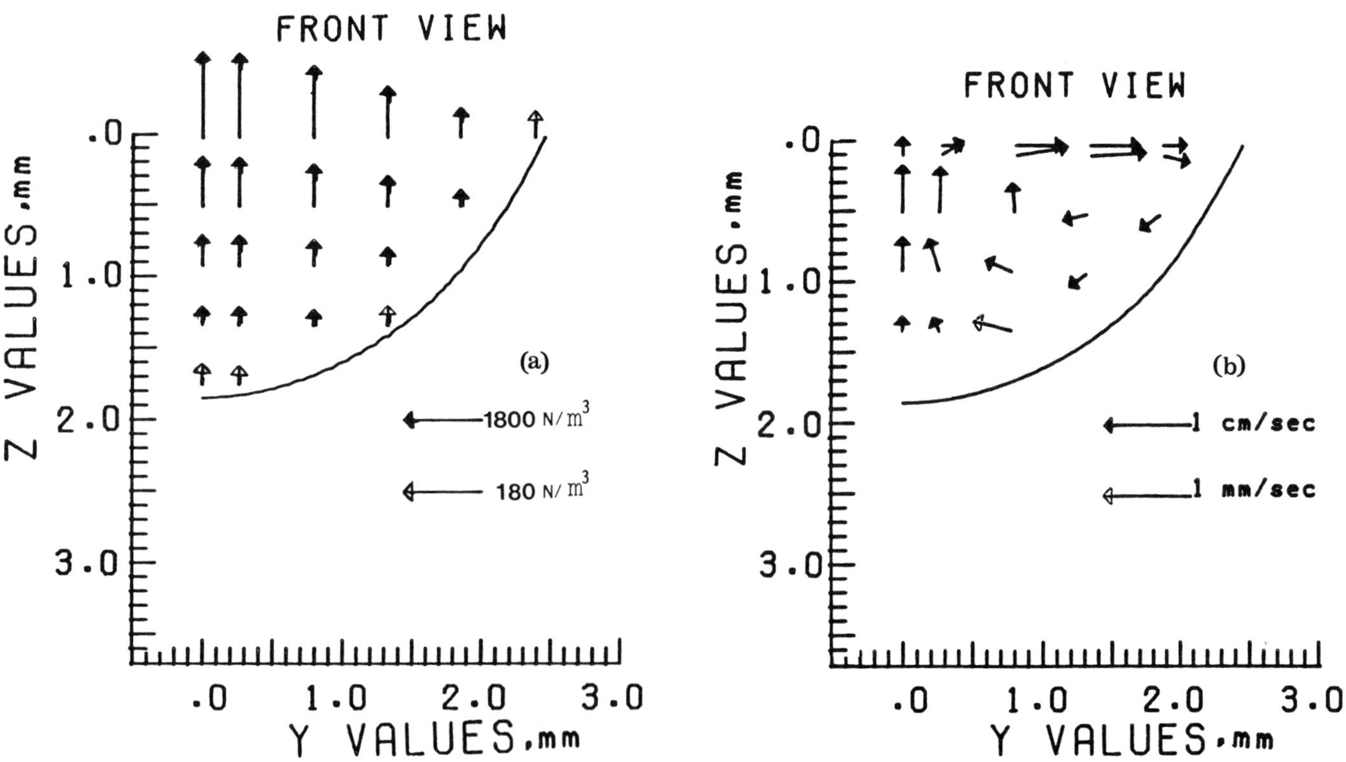

Fig. 3 Buoyancy force: (a) distribution in the weld pool; (b) resultant convection pattern.

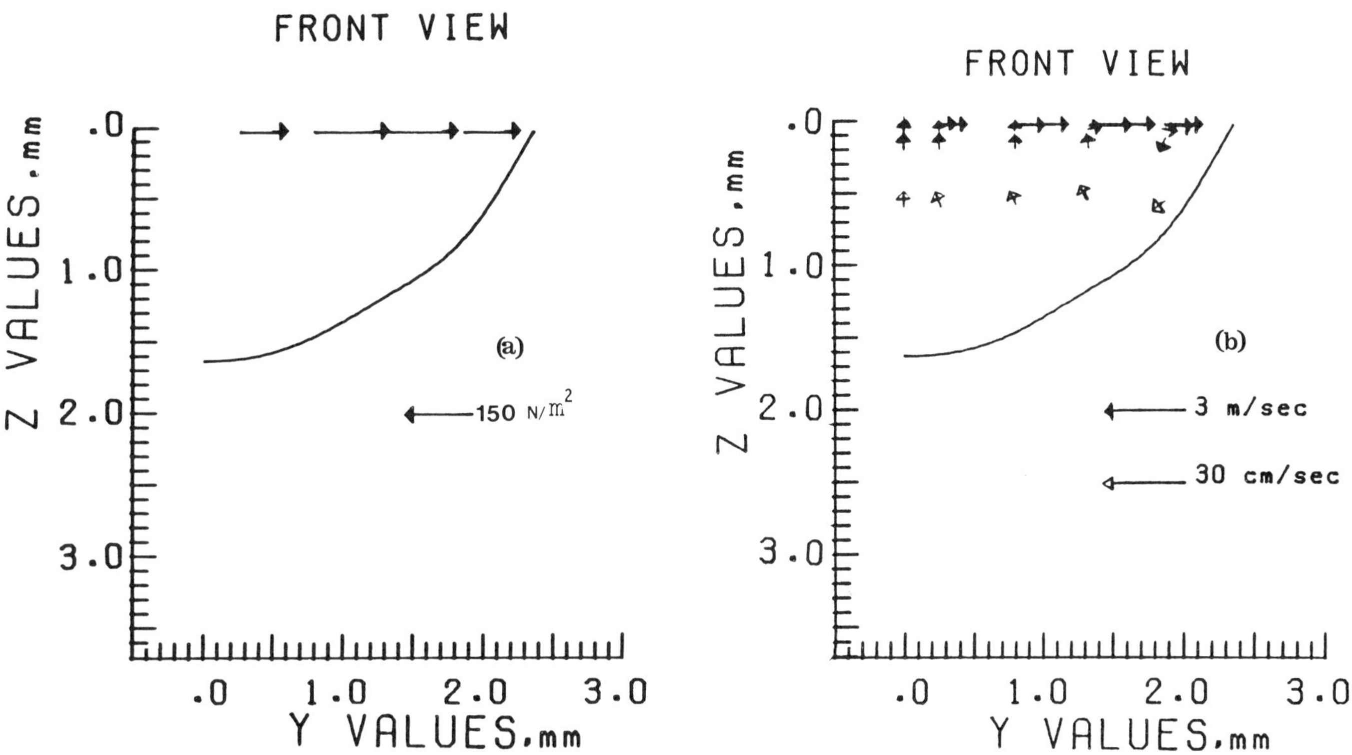

Fig. 4 Surface tension force: (a) distribution in the weld pool; (b) resultant convection pattern. $\partial\gamma/\partial T = -\ 0.35 \times 10^{-3}$ Kg/sec^2°C.

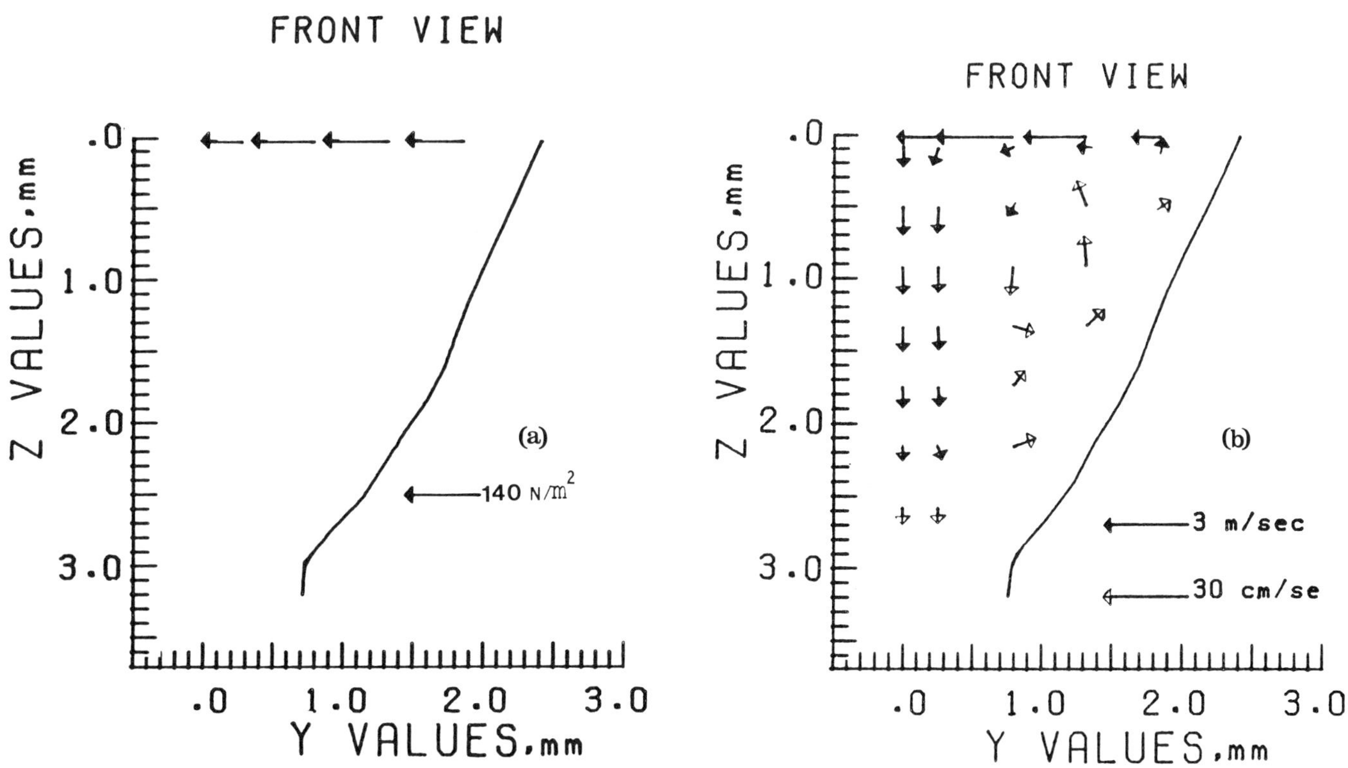

Fig. 5 Surface tension force: (a) distribution in the weld pool; (b) resultant convection pattern. $\partial\gamma/\partial T = 0.35 \times 10^{-3}$ Kg/sec^2°C.

FINITE ELEMENT MODELING OF ARC WELDING PROCESSES

P. Tekriwal, J. Mazumder
Department of Mechanical and Industrial Engineering
University of Illinois at Urbana-Champaign
1206 W. Green Street
Urbana, Illinois 61801

ABSTRACT

The finite element method has been applied to predict the thermal history of a weld joint produced by MIG welding process. The problem consists of one in which the size of finite element mesh is growing continuously in time. The procedure of how to incorporate the growth of the mesh in the analysis has been described. The model is a three-dimensional transient heat transfer analysis.

The finite element program ABAQUS along with a few user subroutines was employed to obtain the numerical results. Temperature-dependent thermal properties, effect of latent heat, and the convective and radiative boundary conditions are included in the model. Numerically predicted sizes of the heat-affected-zone and the melt-pool-zone are compared with the experimentally observed values.

LIST OF SYMBOLS

C_j	Incremental temperature of node j in an iteration, °K
Cp	Specific heat of metal, J/kg°K
Cp_{gas}	Specific heat of shielding gas, J/kg°K
e	Exponential (= 2.7)
D	Nozzle diameter, m
h	Heat transfer coefficient, W/m^2°K
H	Latent heat of fusion for metal, J/kg
k	Thermal conductivity of metal, W/m°K
k_{gas}	Thermal conductivity of shielding gas, W/m°K
L(T)	Differential operator
n	Iteration counter
$\underline{n}$	Unit vector normal to surface
N_i	Shape functions (i = 1,2,...,8)
NPD	Nozzle-to-plate distance, m
P	Power input, W
Pr	Prandtl number of shielding gas
q"	Surface heat flux, W/m^2
q''_s	Specified heat flux on the surface, W/m^2
Re	Reynolds number of shielding gas flow
dS	Elemental surface area, m^2
t	Welding time, s
Δt	Time step, s
T	Temperature of metal, °K
T_a	Ambient temperature, °K
T_i	Temperature at node i, °K
T_L	Liquidus temperature of metal, °K
T_S	Solidus temperature of metal, °K
u	Internal energy, J/kg
u_t	Internal energy at time t, J/kg
v	Welding speed, m/s
v_{gas}	Shielding gas flow velocity, m/s
dV	Elemental volume, m^3
x,y,z	Physical coordinates, m

Greek Symbols

α	Thermal diffusivity of metal, m^2/s
α_{gas}	Thermal diffusivity of shielding gas, m^2/s
δ_{jk}	Kronecker delta
$\partial/\partial t$	Partial derivative w.r.t. t, 1/s
$\underline{\nabla}$	$\underline{i}\ \partial/\partial x + \underline{j}\ \partial/\partial y + \underline{k}\ \partial/\partial z$, 1/m
ε	Thermal emmissivity
ε,η,ζ	Natural coordinates, m
$\varepsilon_i,\eta_i,\zeta_i$	Natural coordinates of node i,m
η	Arc efficiency
ν_{gas}	Kinematic viscosity of shielding gas, m^2/s
ρ	Density of metal, kg/m^3
ρ_{gas}	Density of shielding gas, kg/m^3
μ_{gas}	Dynamic viscosity, kg/m-s
σ	Stefan-Boltzman constant = 5.6697 x 10^{-8} W/m^2°K^4

WELDING TECHNOLOGY has been by and large empirical to date. Series of experiments and the experience have contributed to the present state-of-the-art of welding [1]. But with the advent of computer age, efforts are being made to make available the science behind welding to practicing engineers. The effort to quantify heat flow and residual stresses in welding goes back to as early as the 1930's when Rosenthal [2,3] and Bouton and Martin [4] developed analytical models for the theory of welding. These analytical models are of little practical use to design engineers as they are too idealized to be used in practice. Nevertheless, these methods have provided a foundation for the numerical methods which can take into account the non-linearities involved in the process and provide a practical solution. The finite element method has developed into a powerful tool to solve problems governed by differential equations [5], including transient heat conduction analysis of complex shapes [6,7]. While the technique has been used fairly well to simulate and model the gas tungsten arc welding (GTAW) process by Friedman [8,9,10], Krutz, et al. [11], and Stitt [13], not many attempts have been made to model a gas metal arc (GMA) welding process popularly known as the MIG (Metal Inert Gas) welding process. At the time of this writing, the authors saw the paper by Hibbit and Marcal [12] in which MIG welding was studied and a case study made for an axi-symmetric example. In this article the theory of finite element formulation will be presented and the solution method will be described for a three-dimensional MIG welding problem. The results of the analysis are evaluated in terms of the comparison of the size of the heat-affected-zone (HAZ) and the melt-pool-zone (MPZ) with the experimental observation. The numerical data used for the welding parameters were supplied by the U.S. Army Construction Engineering Research Laboratory, Champaign, Illinois and experiment was also conducted on samples at their facility.

MIG WELDING PROCESS

Figure 1 depicts the process in which a constant supply of the consumable electrode is maintained through the center of the welding nozzle. When the electrode comes close to the workpiece an arc is produced which causes the electrode tip to melt and join the V-groove between the two plates. A constant supply of the shielding gas (98 percent Argon and 2 percent Oxygen) is also maintained through the annulus around the feedwire inside the nozzle in order to protect the weld from atmospheric contamination.

With regard to modeling the process, the following may be noted:

(i) A part of the heat supplied by the arc is consumed to melt the wire electrode continuously,

(ii) A part of the heat is lost to the surroundings before it is received by the plates,

(iii) The heat of the arc and the molten metal joining the workpiece induces heat flow in all three dimensions in the workpiece. Consequently, complex metallurgical changes are caused in the fusion zone or the melt-pool-zone (MPZ) due to the severe thermal cycle the metal undergoes. Adjacent to the MPZ, the metal is heated to below the melting point, yet above the eutectoid temperature that causes significant change in the microstructure and the mechanical properties. This region is known as the heat-affected-zone (HAZ) and requires significant attention by the design engineers while trying to evaluate the joint strength. The high heat concentration near the center of the arc and rather sparse heat distribution away from the arc cause non-uniform heating and cooling of the weldment, thus generating plastic deformation and residual stresses in the weldment. This often results in permanent distortions of the weldment,

(iv) In the heat transfer process, heat is lost from the surfaces in form of convection and radiation, and

(v) The drift of the shielding gas causes enhanced convective heat loss from a part of the surface.

In this article the problem is solved for temperature distribution in the plates and no calculation has been made to predict the residual stresses and distortions. Vaporization is also neglected. However, results of this work (the temperature) can be

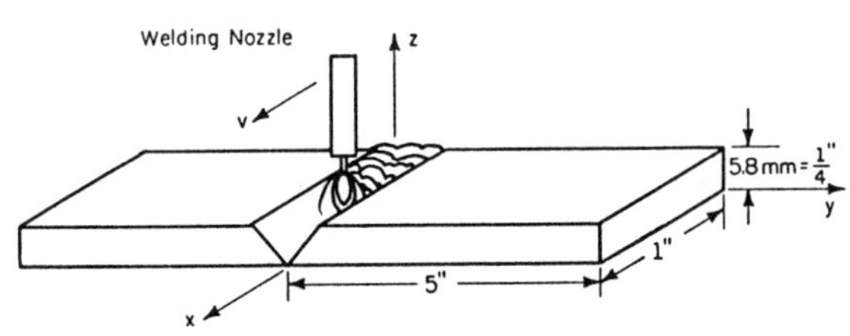

Fig. 1 - Illustration of MIG Arc Welding Process

used as input to the residual stress analysis and a complete thermo-mechanical model can be developed. The authors plan to do so in the future.

MODEL OF INPUT HEAT ENERGY TO THE WORKPIECE

The losses of energy that occur from the arc to the plate are extremely complex in nature and to avoid this complexity a rather obscure term called arc efficiency (η) is used to quantify the energy made available by the arc. $Q = \eta . V . I$ is the energy supplied by the arc. A value of 66 to 69 percent for η has been reported for MIG process of mild steel by Christensen, et al. [14], and a value of 71 percent (±3%) was calorimetrically determined by Essers, et al. [15], with M.S. wire diameter of 1.2 mm (0.05 in.) and argon plus 7 percent CO_2 as the shielding gas. It may be noted that any change in the welding parameters such as the shielding gas and its flow rate, electrode shape, nozzle-to-plate distance, welding volt and current, etc., can significantly affect the arc efficiency value. The data that were used in the present analysis are given in the APPENDIX and η = 67 percent was chosen. The temperature-dependent thermal properties such as conductivity (k) and specific heat (C_p) which were used as shown in Fig. 2(a) and 2(b) were taken from Ref. [11]. Waszink, et al. [23] quotes Villeminot [24] as determining the temperature of the filler wire droplet from melting point to 1000°K above melting point. For the analysis in this article a value of 2300°K (melting point plus 550°K) was chosen as the droplet temperature in consultation with the army research laboratory staff. Knowing the filler wire feed rate and its temperature (2300°K) at which it separates, we can determine the heat extracted from the arc by the filler wire electrode with the help of the (C_p - T) diagram of Fig. 2(a). It is based on the area under the (C_p - T) diagram between the room temperature (298°K) and the droplet temperature (2300°K). Now, $Q = \eta VI$ minus the heat required to melt the feed wire (up to 2300°K) was assumed to be received by the workpiece surface in a radially symmetric Gaussian profile. The Gaussian profile for the arc heat input has almost invariably been used in literature [10,11,13,16]. Smartt et al. [17] also found their experimental data of heat flux to fit the Gaussian profile.

$$q''(r) = q''_o \exp[-3 r^2/r_b^2]$$

$$\text{or, } q''(x,y,t) = q''_o \exp\{-3/r_b^2[(x-vt)^2+y^2]\}$$

where, $q''(r)$ = heat flux per unit area per unit time deposited at location r on the plate, q''_o = maximum q" (at r = 0), r = distance from the center of the arc, r_b = arc beam radius describing a region in which 95 percent of the total heat is deposited on the plate.

A user subroutine DFLUX was written to calculate the input heat distribution in the manner mentioned above and it was compiled along with the finite element package ABAQUS to obtain the temperature distribution in the weldment.

FINITE ELEMENT FORMULATION

The appropriate energy equation without any internal heat generation is:

$$\rho C_p \frac{\partial T}{\partial t} = \underline{\nabla} (k \, \underline{\nabla} T) \quad (1)$$

Boundary conditions are given by the Fourier equation

$$\underline{q}'' = - k \, \underline{\nabla} T \quad (2)$$

Heat flux q" consists of one or more of the following modes of heat transfer depending upon the boundary:

(i) $\underline{h} (T - T_a)$, the convective heat loss;
(ii) $\varepsilon\sigma (T^4 - T_a^4)$, the radiative heat loss;
(iii) $-q''_s$, if the surface is receiving a specified heat flux, q''_s. q" = 0 on y = 0 - plane due to symmetry.

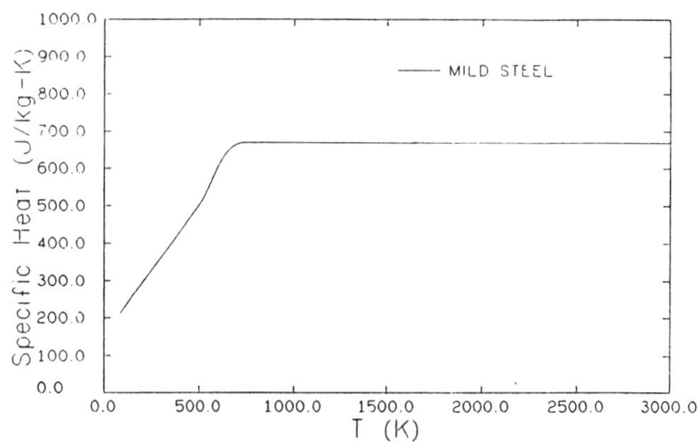

Fig. 2(a) - Specific Heat versus Temperature for Mild Steel

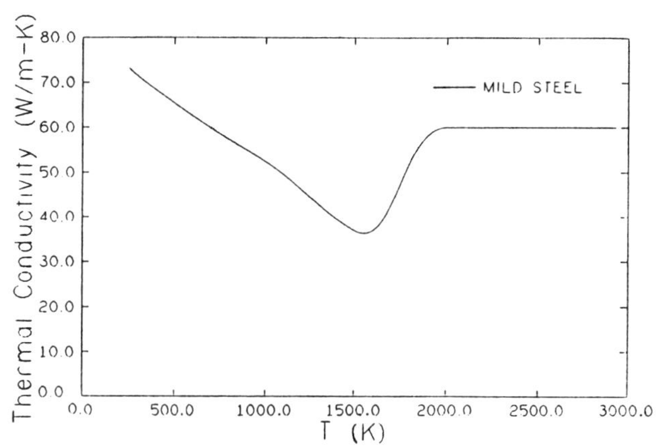

Fig. 2(b) - Thermal Conductivity versus Temperature for Mild Steel

All other surfaces incur the heat losses due to convection and radiation and only the top surface (z = thickness) facing the torch receives heat flux from the arc.

The Galerkin method dictates

$$\int_V N_i \, L(T) \, dV = 0$$

$$\text{or} \quad \int_V N_i \rho \, \dot{u} \, dV - \int_V N_i \, \underline{\nabla} \cdot (k \, \underline{\nabla} \, T) \, dV = 0 \qquad (3)$$

$\dot{u} = C_p \, \partial T/\partial t$, $T = N_i T_i$, and N_i are shape functions.

Note that summation convention is assumed on repeated indices unless mentioned otherwise.

Eight-noded rectangular linear finite elements (see Fig. 3) were used to discretize the plate. The shape functions are given by the equation $N_i = 1/8\,(1+\varepsilon\varepsilon_i)(1+\eta\eta_i)(1+\zeta\zeta_i)$ (no summation).

The highest order derivative that is allowable in Eq. (3) is one greater than the order of continuity in the interpolation equations. Since all the interpolation equations are of the order zero (continuity in T but not in the first derivative), derivatives greater than the first order cannot appear in Eq. (3). This restriction can be avoided by reducing the order of L(T) using integration by parts [5].

Integrating the second term of Eq. (3) by parts

$$\int_V N_i \rho \dot{u} dV + \int_V \underline{\nabla} N_i \cdot k \, \underline{\nabla} \, T dV = \int_S N_i k \, \underline{\nabla} \, T \cdot \underline{n} \, dS \qquad (4)$$

Substituting Eq. (2) and replacing T by $N_k T_k$,

$$\int_V N_i \rho \dot{u} dV + \int_V \underline{\nabla} N_i \cdot k \underline{\nabla} \, N_k dV \; T_k + \int_S N_i \, \underline{q}'' \cdot \underline{n} dS = 0 \qquad (5)$$

The time derivative of the internal energy term is taken by using the backward Euler's formula.

$$u_{t+\Delta t} = u_t + \left.\frac{\partial u}{\partial t}\right|_{t+\Delta t} \Delta t$$

$$\text{i.e.,} \qquad \dot{u} = \left.\frac{\partial u}{\partial t}\right|_{t+\Delta t} = \frac{u_{t+\Delta t} - u_t}{\Delta t} \qquad (6)$$

This method is unconditionally stable and is very sensitive to the grid size and the time step used [5]. Oscillations may occur in the numerical values even though the method itself is stable. The amount of oscillation is dependent on the material properties, the size of elements, and the time step. Reductions of the element size in high temperature gradient regions and reduced time steps decrease the seriousness of oscillations. But elements that are too small will result in too many degrees of freedom raising the cost of numerical analysis. This fact motivated us to use an expanding grid in the y- and z- directions and small elements near the weld centerline (Fig. 3).

Substitution of Eq. (6) in Eq. (5) yields

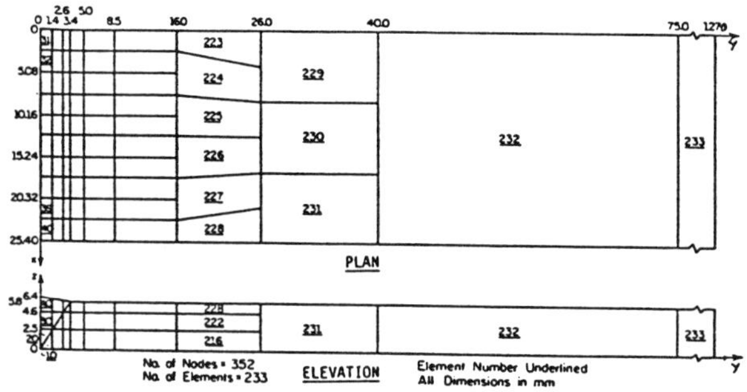

Fig. 3 - Orthographic Projection of the Mesh

$$\frac{1}{\Delta t}\int_V N_i \rho (u_{t+\Delta t} - u_t) dV + \int_V \underline{\nabla} N_i \cdot k \underline{\nabla} \, N_k dV \; T_k + \int_S N_i \, \underline{q}'' \cdot \underline{n} \, dS = 0 \qquad (7)$$

The non-linear system, Eq. (7), is linearized in the form

$$F_i \equiv K_{ij} T_j - b_i = 0 \qquad (8)$$

and is solved by ABAQUS using a modified Newton method [18]. In the Newton method, the tangent matrix (the Jacobian matrix) is formed by differentiating F_i with respect to $T_j\big|_{t+\Delta t}$. That is,

$$K_{ij} = \frac{\partial F_i}{\partial T_j\big|_{t+\Delta t}}, \qquad (9)$$

F_i = left-hand side of Eq.(7).

The method is a modified Newton method because the Jacobian matrix K_{ij} is not formed exactly. We now describe how the Jacobian is formed.

Jacobian Contributions:

The internal energy term contributes

$$\frac{\partial}{\partial T_j\big|_{t+\Delta t}} \left[\frac{1}{\Delta t} \int_V N_i \, \rho \, (u_{t+\Delta t} - u_t) \, dV\right]$$

$$= \frac{1}{\Delta t} \int_V \rho \, C_p\big|_{t+\Delta t} \frac{\partial T}{\partial T_j} dV$$

$$= \frac{1}{\Delta t} \int_V N_i \, \rho \, C_p\big|_{t+\Delta t} N_j \, dV \qquad (10)$$

$$C_p\big|_{t+\Delta t} = C_p(T_{t+\Delta t}), \text{ if } T_{t+\Delta t} < T_S \text{ or } T_{t+\Delta t} > T_L$$

$$= C_p + \frac{H}{T_L - T_S}, \text{ if } T_S < T_{t+\Delta t} < T_L$$

This term is small outside the solidus-liquidus temperature range and is stiff inside that rather narrow range. This may result in a numerical instability. To avoid the problem, this term is modified to a secant term during early iterations of the solution to a time step [19].

Jacobian contribution of the conductivity term in Eq. (7) is:

$$\frac{\partial}{\partial T_j}\bigg|_{t+\Delta t} [T_k \int_V \underline{\nabla} N_i \cdot k \, \underline{\nabla} N_k \, dV]$$

$$= \delta_{jk} \int_V \underline{\nabla} N_i \cdot k \big|_{t+\Delta t} \underline{\nabla} N_k \, dV$$

$$+ T_k\bigg|_{t+\Delta t} \int_V \underline{\nabla} N_i \cdot \frac{\partial k}{\partial T}\bigg|_{t+\Delta t} N_j \underline{\nabla} N_k \, dV$$

$$= \int_V \underline{\nabla} N_i \cdot k \big|_{t+\Delta t} \underline{\nabla} N_j \, dV \quad (11)$$

$\partial k/\partial T$ is usually very small, that is, thermal conductivity varies very slowly with temperature and hence is neglected. Note that if the second term is not negligible, it will make the Jacobian matrix nonsymmetric.

The boundary term contributions can be calculated as follows:

Note, $\underline{q}'' \cdot \underline{n} \doteq h(T-T_a) + \bar{\varepsilon}\sigma(T^4-T_a^4) - q_s''$

($\doteq$ sign has been used to designate that the equality stands with one or more terms on the right-hand side, depending upon the boundary). The Jacobian contribution is

$$\frac{\partial}{\partial T_j}\bigg|_{t+\Delta t} \int_S N_i \, \underline{q}'' \cdot \underline{n} \, dS$$

$$= \int_S N_i \frac{\partial \underline{q}'' \cdot \underline{n}}{\partial T}\bigg|_{t+\Delta t} N_j \, dS$$

$$\doteq \int_S N_i [\frac{\partial h}{\partial T}(T-T_a)+h+4\bar{\varepsilon}\sigma T^3]_{t+\Delta t} N_j \, dS \quad (12)$$

The modified Newton method then yields the following system:

$$C_j \{\frac{1}{\Delta t} \int_V N_i \rho \, C_p\big|_{t+\Delta t} N_j \, dV +$$

$$\int_V \underline{\nabla} N_i \cdot k\big|_{t+\Delta t} \underline{\nabla} N_j \, dV +$$

$$\int_S N_i \, [\frac{\partial h}{\partial T} (T - T_a) + h + 4\bar{\varepsilon}\sigma T^3]_{t+\Delta t} N_j \, dS\}$$

$$= - \frac{1}{\Delta t} \int_V N_i \rho (u_{t+\Delta t} - u_t) \, dV -$$

$$\int_V \underline{\nabla} N_i \cdot k \, \underline{\nabla} N_j dV \, T_j - \int_S N_i \, q_n'' \, dS \quad (13)$$

where, $$T_j^{n+1}\bigg|_{t+\Delta t} = T_j^n\bigg|_{t+\Delta t} + C_j \quad (14)$$

$(q_n'' = \underline{q}'' \cdot \underline{n})$.

SURFACE HEAT LOSSES

Natural convection heat transfer occurs on all surfaces of the plates except the plane of symmetry (y = 0). But the area directly beneath the nozzle of the torch experiences forced convection due to flow of the shielding gas. Based on previous models [9], h = 10 W/m^2°K was used for all the surfaces not influenced by the shielding gas and the following empirical relation initially suggested by Gardon and Cobonpue for air [20] and later modified by Steen for any gas [21] was used for a part of the top surface under the nozzle of the welding torch:

$$h = 13 \, Re^{1/2} \, Pr^{1/3} \, k_{gas}/NPD,$$

$$\text{Reynolds number } Re = \frac{v_{gas} \times D \times \rho_{gas}}{\mu_{gas}},$$

$$\text{Prandtl number } Pr = \frac{\nu_{gas}}{\alpha_{gas}} = \frac{\mu_{gas} \, C_{p_{gas}}}{k_{gas}}.$$

Program HCALC calculates Re and Pr based on the properties of the shielding gas. It then generates subroutine FILM which contains the expression for h. Subroutine FILM is compiled with the main program and calculates h for every integration point at a time step. DFLUX and FILM both transform the coordinate system to the moving arc and then calculate the heat flux and the heat transfer coefficient at any point appropriately. It may be noted that constant values of h are used for every point under the nozzle. There is no radial variation of h. The heat loss due to convection accounts for a very minor part (less than 5 percent) of the total heat involved in the process and hence a more accurate representation of h will not produce any difference in overall results [22].

Radiation heat losses are accounted from all the surfaces except the plane of symmetry (y = 0) by using the equation $q'' = \bar{\varepsilon}\sigma(T^4-T_a^4)$. The emmisivity value 0.08 has been used.

MESH GENERATION

The whole problem is symmetric about the x-axis and hence it is sufficient to discretize and analyze the plate and the filler metal geometry which is in the positive octant of the coordinate system shown in Fig. 1.

Choice of a suitable mesh is vital to the accuracy and economy of the finite element results. It may be noted here that the problem of MIG welding is one in which the mesh (or the solution domain, speaking mathematically) grows continuously with time due to the addition of filler metal until welding is completed. Ideally, in order to obtain accurate results the analysis should be carried out for an infinitesimal length of time and then the new element or elements (which will be infinitesimally small in size) corresponding to filler metal should be added to the mesh before carrying out the analysis for another infinitesimally small period of time. This alternate process of analysis and redefining the mesh along with new boundary conditions will continue until the weld length is covered. However, from a practical point, this is impossible and we have to opt for an approximate solution by adopting the same scheme of alternating between the analysis and the redefinition of the mesh on the level of finite time and finite size of the filler metal. For the present analysis the weld length (1 in.) was divided into ten parts and a 10-step analysis was carried out to complete one pass, each step adding a 0.1 inch weld length. Figure 4(a) shows the mesh for the first step of analysis in which 0.5 second welding time is completed. Then another set of elements is added and the new boundary conditions defined before the second step of analysis from 0.5 to 1.0 seconds is carried out. Figure 4(b) shows the mesh for the fifth step of the analysis. It may be pointed out that the volume of the additional elements at each step and their size in the welding direction (x-direction) are in agreement with the filler wire feed rate and the welding speed, respectively.

The second important requirement for the mesh is the fine size elements near the weld centerline. In order to pick the accurate peak temperatures in the heat input area, we must have certain minimum number of nodes under the arc beam radius. This number depends upon the magnitude of the heat input. The higher the heat input, the more severe is the temperature gradient in the region and hence more and more nodes (or degrees of freedom) are required in the region. The authors' experience recommends that at least 5 to 6 nodes should be taken under the arc beam radius for steel materials in order to track the peak temperatures accurately. The temperature distribution is not very sensitive to element size far away from the weld centerline and hence in order to reduce the cost of analysis, the size of the elements is increased as they get farther away from the region of high heat input.

Since the irregular shape of the boundary poses no threat to the finite element analysis, the boundary of the weld bead can more realistically be modeled. Based on experimental measurements, a 10° bulge (see Fig. 3) was allowed in the geometry of the filler metal. Since, first-order, 8-noded rectangular elements are more suited for the problems involving latent heat effects [18], the same were chosen to discretize the mesh. ABAQUS uses a 2 x 2 x 2 rule for numerical integration with the integration points located at the corners of the element. This renders only diagonal contribution of the internal energy term to the Jacobian matrix.

COMPUTED RESULTS: COMPARISON WITH OBSERVED HAZ AND MPZ

Figures 5(a) through 5(d) show the isotherms on the surface and on the lateral (xz-) plane. These curves give the pattern of heat flow in the weldment. These curves have been drawn for V = 28 Volts, I = 380 A, η = 0.67 and welding speed = 12 in/min (5.08 mm/s) as supplied by the research laboratory. From Fig. 5(c) we notice that the fusion boundary (T $\simeq$ 1750°K) is close to the bottom of the plate. Figure 5(d) shows that the region up to x = 10.6 mm (0.4 in.) is completely melted at t = 2 seconds after welding started. The weld length (Fig. 1) is 1 in. and it takes 5 seconds to weld the 1 in. length. Thus a melt-through is predicted from the numerical results unless there is some support underneath the plates. This was observed to be true experimentally at the army research facility. A lower energy input is sufficient to weld the pieces together.

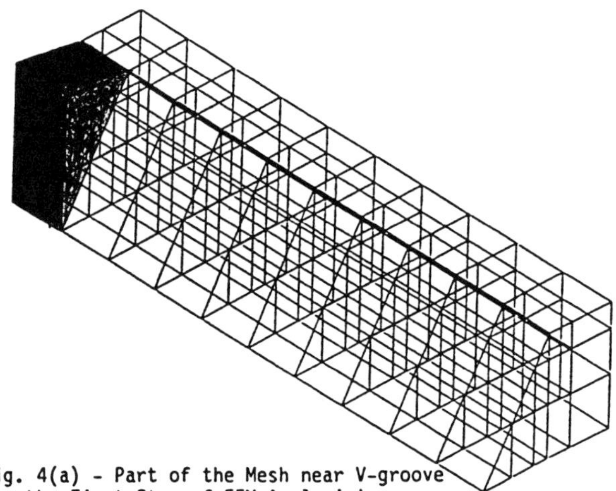

Fig. 4(a) - Part of the Mesh near V-groove for the First Step of FEM Analysis*

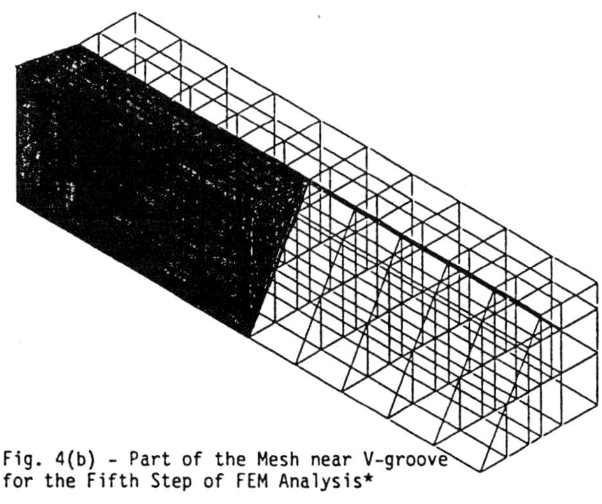

Fig. 4(b) - Part of the Mesh near V-groove for the Fifth Step of FEM Analysis*

* Filler metal elements are shaded.

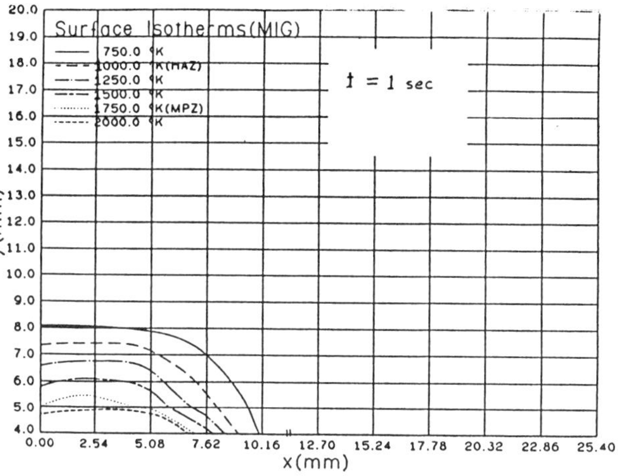

Fig. 5(a) - Isotherms on the Top Surface

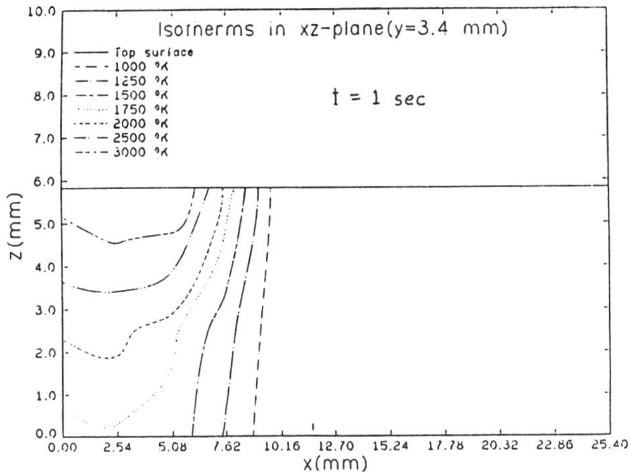

Fig. 5(c) - Isotherms on Transverse Plane

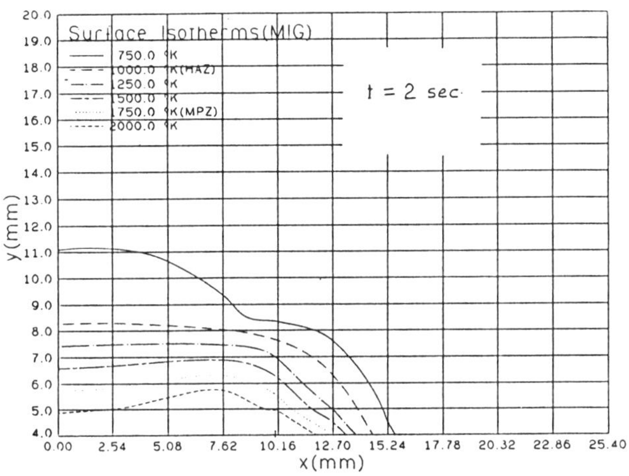

Fig. 5(b) - Isotherms on the Top Surface

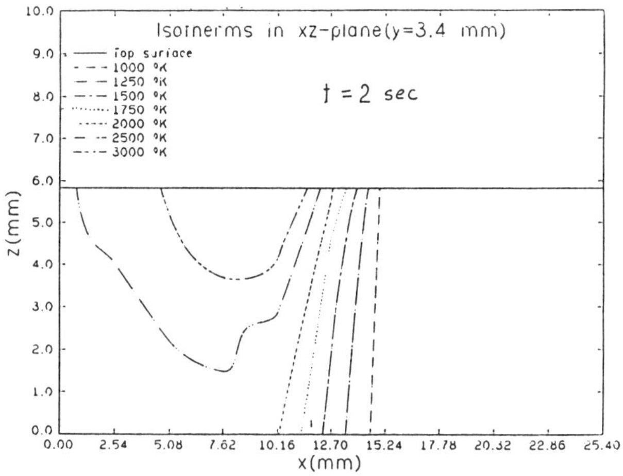

Fig. 5(d) - Isotherms on Transverse Plane

Thermal history of the material is very important in determining the microstructural changes the material undergoes and the strength of the weld joint. Any suspected critical region can be analyzed by plotting the thermal history in that region. Krutz and Segerlind's result [11] indicate that the cooling of metal adjacent to the weld pool is the critical metallurgical location. While the authors have not looked at the results very closely in all the regions in order to determine the critical metallurgical region, the thermal history of the two sample points was drawn in Fig. 6. One point belongs to the HAZ while the other to the melt-pool-zone (MPZ). Very high temperatures cannot be measured by simple techniques such as with thermocouples. Yet in the region away from the molten zone, the temperature can be measured and compared in order to get some idea of the accuracy of the numerical results. A thermocouple was attached at location of node 259 (12.7, 8.5) mm on the surface of the sample and a peak temperature of 1115°K above room temperature was recorded

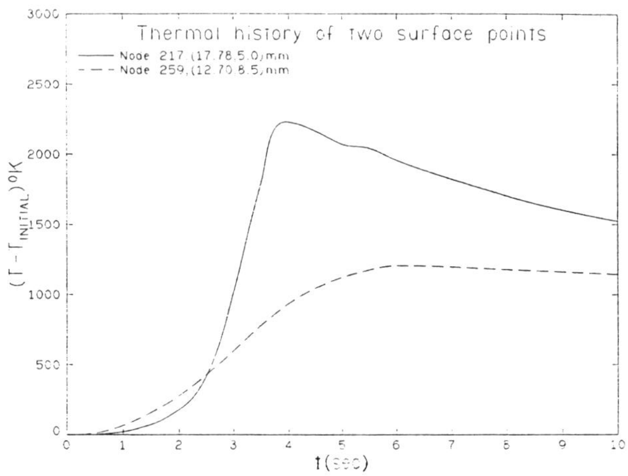

Fig. 6 - Heating/Cooling Curves of Two Sample Points

by the thermocouple. Figure 6 shows that the particular location should attain 1200°K above the room temperature. The difference is not very large considering the experimental error involved. The above experiment was not designed to record temperatures and compare with the numerical predictions, even though in the future this is planned along with parametric study of the heat input, welding speed, etc.

The aim of the experiment was to determine the size of the heat-affected-zone and the melt-pool-zone. Figures 7(a) and (b) plot the HAZ and MPZ boundary at different times and determine the extent of the two regions. The criterion for the HAZ boundary was 1000°K (mild steel) and for the MPZ boundary 1755°K. Figure 7(a) shows the time growth of the 1000°K isotherm on the surface. Growth rate is high up to 6 seconds of time (note that welding time is 5 seconds and after this time no arc heat is supplied to the plates) and then the growth is at a slower rate. The isotherms at t = 20 and 24 seconds are coincident with each other and then the HAZ decays due to the dominant atmospheric cooling. Thus the numerical results predict an extent of 16 mm from Fig. 7(a) in each plate for the HAZ. Experimental observation was found to be 1.25 in. (31.75 mm) when measured across both plates. The numerical prediction (16 x 2 = 32 mm) is in very close agreement with the experimental observation. Figure 7(b) predicts the size of MPZ 15.4 mm (7.7 mm in each plate) on the surface from finite element calculation. Experimentally the width of the MPZ was found to be 16.8 mm.

Figures 8(a) and (b) plot the temperature distribution along the top and bottom edge of the V-groove at different times after the start of the welding. We may notice in Fig. 8(a) that the welding speed is faster than the diffusion rate of heat in the material. The peak temperature distribution is not moving at the rate at which the arc is moving. For instance at t = 1 second, the arc is located at x = 5.08 mm but the highest temperature is located at about x = 4 mm on the top edge of the V-groove. Similarly, at t = 3 seconds, the highest temperature is attained by the point x ≈ 14 mm when the arc is passing through the x = 15.24 mm point. In Fig. 8(b) we notice that the peak temperature distribution moves at almost the same speed as the welding arc. These plots in Fig. 8 also give the estimate of the rate at which the points along the V-groove edge cool down.

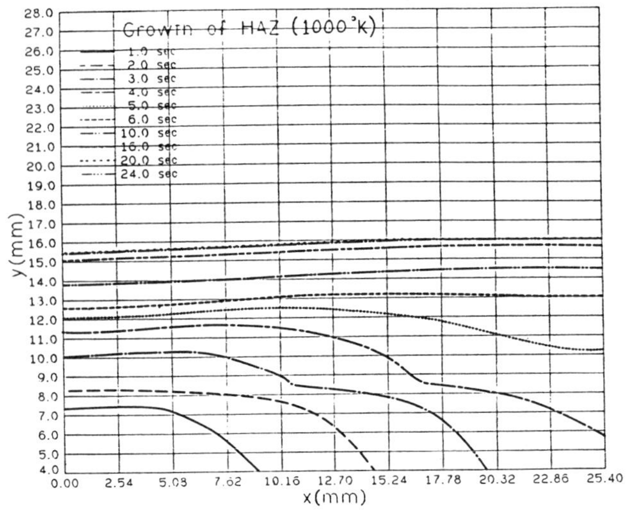

Fig. 7(a) - Numerical Determination of the Heat-Affected-Zone

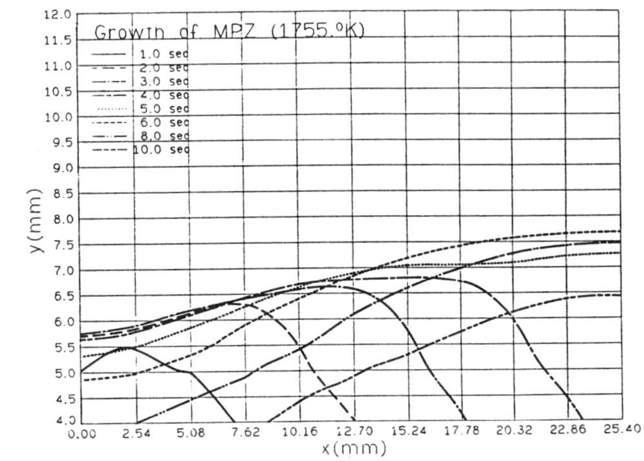

Fig. 7(b) - Numerical Determination of the Melt-Pool-Zone

CONCLUSIONS

1. This analysis proves that the fact that filler metal causes the mesh to grow continuously with time poses no major threat to the applicability of the finite element method to a MIG welding process. Of course, very good engineering judgment is required while discretizing the solution domain in order to get accurate and meaningful results at reasonable cost.

2. While the temperature distribution predicted by the numerical results were not completely verified experimentally, the size of the heat-affected-zone as calculated from the finite element method was found to be in close agreement with the experimentally measured value on one test piece. In future more experiments are

planned with varying welding parameters to check their effects on welding. This will also provide a parametric study of the numerical model, thus testing the accuracy of the model.

3. The analysis gives a very good spectrum of overall heat transfer phenomenon in a MIG welding process.

4. The temperature results can be used for thermomechanical analysis of the weldment.

5. This paper outlines the step-by-step procedure of alternating between the heat transfer analysis and changing the mesh geometry.

ACKNOWLEDGMENT

The authors gratefully acknowledge the sponsorship and the use of the facility of the U.S. Army Construction Engineering Research Laboratory, Champaign, Illinois.

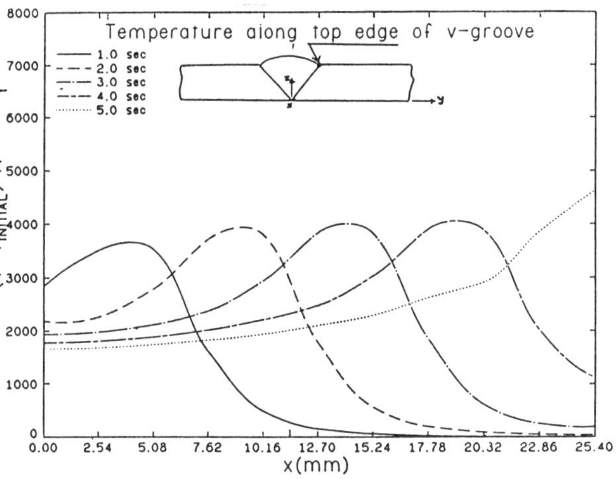

Fig. 8(a) - Transient Temperature Profile

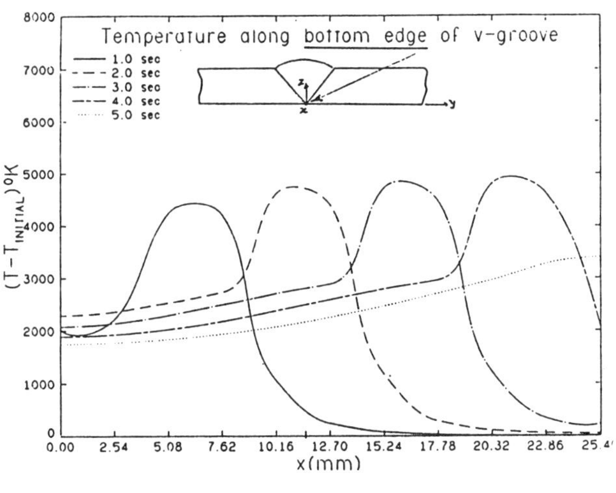

Fig. 8(b) - Transient Temperature Profile

REFERENCES

1. Masubuchi, K., Welding in The World 17, 11-24 (11/12, 1979)
2. Rosenthal, D., Weld. J. 20, 220s-234s (1941)
3. Rosenthal, D., Trans. ASME 68, 849-866 (1946)
4. Boulton, N. S. and H. E. Lance Martin, Proc. Inst. Mech. Engrs. (London) 123, 295-347 (1936)
5. L. J. Segerlind,"Applied Finite Element Analysis", John Wiley and Sons, Inc., (1976)
6. Zienkiewicz, O. C. and C. J. Parekh, Int. J. Num. Meth. Engr. 2, 61-71 (1970)
7. Wilson, E. L. and R. E. Nickell, Nuclear Engr. and Design 4, 276-286 (1966)
8. Friedman, E., J. Pressure Vessel Technology (Trans. ASME), 206-213 (Aug. 1975)
9. Friedman, E. and S. S. Glickstein, Weld. J. 55, 408s-420s (Dec. 1976)
10. Friedman, E., DOE Research and Development Report, Jan. 1980, Bettig Atomic Power Laboratory, West Mifflin, PA 15122
11. Krutz, G. W. and L. J. Segerlind, Weld. J. 57, 211s-216s, (1978)
12. Hibbitt, H. D. and P. V. Marcal, Computers and Structures 3, 1145-1174 (Sept.1973)
13. Stitt, M. C., M.S. Thesis (1983), Dept. of Mech. and Ind. Engr., University of Illinois at Urbana-Champaign, 1206 West Green St., Urbana, IL 61801
14. Christensen, N., V. de L. Davies and K. Gigermundsen, Br. Weld. J., 54-75 (Feb. 1975)
15. Essers, W. G. and R. Walter, Weld. J., 37s-42s (Feb. 1981)
16. Pavelic, V., R. Tanbakuchi, O. A. Uyehara and P. S. Myers, Weld. J.48, 295s-305s (July 1969)
17. Smartt, H. B., J. A. Stewart and C. J. Einerson, Amer. Weld. Soc. 66th Annual Convention, Las Vegas, NV (May 1985)
18. Hibbitt, Karlsson, and Sorensen, "ABAQUS Theory and User's Manuals", Hibbitt, Karlsson and Sorenson, Inc.,Providence, RI (1982)
19. J. F. Traub, "Iterative Methods for the Solution of Equations", Prentice-Hall, Inc., Englewood Cliffs, N.J., 1964.
20. Gardon, R. and J. Cobonpue, Int. Heat Transfer Conf., Part II, 454-460 (1961)
21. Steen, W. M., Ph.D. Thesis, Imperial College, London, 1976.
22. Mazumder, J. and P. Tekriwal, "Optimization of Welding Processes using a Mathematical Model", Progress report prepared for the U.S. Army Construction Engineering Research Laboratory, Dec. 1984.
23. Waszink, J. H. and L. H. J. Graat, Weld. J., 108s-116s (April 1983)
24. Villeminot, P., Soudage et Technique Connexes 21, 367-380 (9/10, 1967)

APPENDIX

The following data were used for numerical results:

Ambient temperature	= 298°K
Filler metal temperature when it joins the V-groove	= 2300°K
Solidus temperature	= 1700°K
Liquidus temperature	= 1755°K
Latent heat of fusion	= 273790 J/kg
Density of M.S.	= 7870 kg/m^3

Piecewise constant values of thermal conductivity of M.S. as follows:

K (W/m°K)	T(°K)
73.15	250.0
65.31	500.0
52.25	1000.0
36.92	1500.0
31.35	1950.0
60.00	2000.0
60.00	>2000.0

Specific heat as a piecewise constant function of temperature:

Cp (J/kg°K)	T(°K)
209.0	85.0
502.0	500.0
668.8	750.0
668.8	>750.0

Welding voltage	= 28 V
Welding current	= 380 A
Arc efficiency, η	= 0.67
Welding speed	= 5.08 mm/s (12 in/min)
Arc beam radius, r_b	= 5.56 mm
Filler wire diameter (1/16 in)	= 1.6 mm
Filler wire feed rate (131 in/min)	= 5.55 cm/s
Nozzle-to-plate distance (3/4 in.)	=19.05 mm
Nozzle diameter (3/4 in.)	= 19.05 mm
Nozzle angle with the vertical	= 0°

Shielding gas(98% Argon+2% Oxygen) data:

Flow rate (50 ft^3/hr)	= 0.3933 Litre/s
Thermal conductivity	= 0.0178 W/m°K
Specific heat	= 518.816 J/kg°K
Dynamic viscosity	= 2.217 x 10^{-5}kg/m-s
Density	= 1.7837 kg/m^3

SOLIDIFICATION BEHAVIOR OF WELDS

KEYNOTE ADDRESS

SEGREGATION AND STRUCTURE IN THE WELD ZONE

H.D. Brody
University of Pittsburgh
Pittsburgh, Pennsylvania

ABSTRACT

Control of the thermal conditions in the weld pool can be used to link welding parameters and weld zone structure and segregation. The undercooling at the weld pool solidification front is dependent on the alloy constitution and on the thermal parameters in the weld pool: temperature gradient, isotherm advance rate, and cooling rate. In turn, the undercooling at the solidification front (i.e. at the advancing dendrite or cell tips) is key in determining the potential for grain refinement, the solidification interface morphology, and the extent of microsegregation. The solidification structure and segregation influence the tendency for weld defects, such as cracking. Process control techniques that may be used to alter weld zone structure and segregation include molten pool oscillation and pulsed current welding.

A WELD ZONE may be considered as a small casting, often a continuous casting, formed under specialized conditions: an intensely stirred liquid pool, high temperature gradients, rapid growth rates, intimate contact between the molten alloy and the mold including partial melting of the mold. In this framework solidification principles may be used to predict and explain observations characteristic of welding processes. The understanding of solidification relations can be applied to the control of structure and segregation so as to improve weld quality. A few aspects of weld zone solidification have been selected for discussion within this review: grain structure, solid liquid interface morphology, solute redistribution, growth undercooling, grain refinement, and weld zone cracking.

GRAIN STRUCTURE

Grain refers to a region of a solidifying (or solidified) alloy in which the primary phase stems from a unique source and grows with a single crystallographic orientation.[1,2] Alloys seldom freeze to form only a single solid phase. The formation of secondary phases within a grain often are directly related to the primary phase in crystallographic orientation and distribution. Solidification substructure (dendritic, cellular) and the multiplicity of phases often make grains appear complex. As example, parts of two grains in a 310 stainless steel weld zone are shown in perspective view in Figure 1.[3] Within each grain numerous primary dendrite arms of the austenite phase are aligned in parallel arrays when viewed in longitudinal sections and in a regular arrangement when viewed in transverse sections.

Grains are characterized by their shape and their size.

Columnar grains are elongated in one direction which, locally, is close to the primary direction of heat flow. Columnar grains exhibit a preferred crystallographic orientation. A specific crystallographic direction tends toward alignment with the elongated axis of the grain and, thus, with the direction of heat flow.

Equiaxed grains have about the same width in all directions and have a random crystallographic orientation.

Any combination of columnar and equiaxed grains may form in a weld zone. Typically, weld zone grains are either totally columnar or columnar near the fusion line changing to equiaxed at the weld centerline. Depending on conditions, both equiaxed and columnar grains may be fine or coarse.

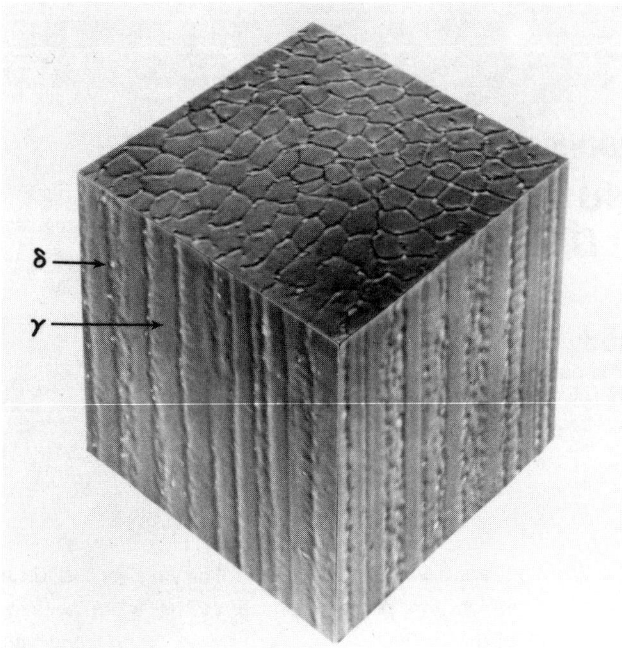

Figure 1: Three orthogonal views of weld zone structure in 310 stainless steel.[3]

Grain refinement is aimed at a reduction of grain size or an increase in the fraction of equiaxed grains. The objective of grain refinement may be an improvement in alloy weldability or an enhancement of weldment properties in service.

Grain formation in alloys usually is presented as a nucleation and growth process. In weld zone solidification the nucleation step is not required.[4] The molten alloy of the fusion zone freezes within a mold formed of the base metal. The alloy, as its heat of fusion is transferred to the base metal, may grow epitaxially on the base metal. Each base metal grain at the fusion line may serve as the growth site for a fusion zone grain. Thus, the initial size of the grains at the fusion line will be the same as the grain size of the base metal at the fusion line at the time freezing begins at that location. Also, the crystallographic orientation of each fusion zone grain will be the same as the orientation of the base metal grain on which it grows epitaxily.

The grains at the fusion line grow in a direction (roughly) perpendicular to the fusion line. As the grains grow, there is a competition between neighbors. Some continue to grow, others terminate. Generally, the surviving grains are the ones that have a preferred crystallographic axis closely aligned with the heat flow direction. In cubic alloys, the cube axis directions <100> are the preferred growth directions.[1-4] As a result of the competition, the grain size continually increases with increasing distance from the fusion line.

Epitaxial growth and columnar grain competition are illustrated in Figure 2 which is a photomicrograph of a weld zone in an iridium alloy.[3,5] The relation between base metal grains and weld zone grains at the fusion line is clearly shown. By drawing lines parallel to the fusion line at increasing distances and counting the number of grain boundaries intersected by each line, the reader can evaluate the grain coarsening that results from grain competition and spreading.

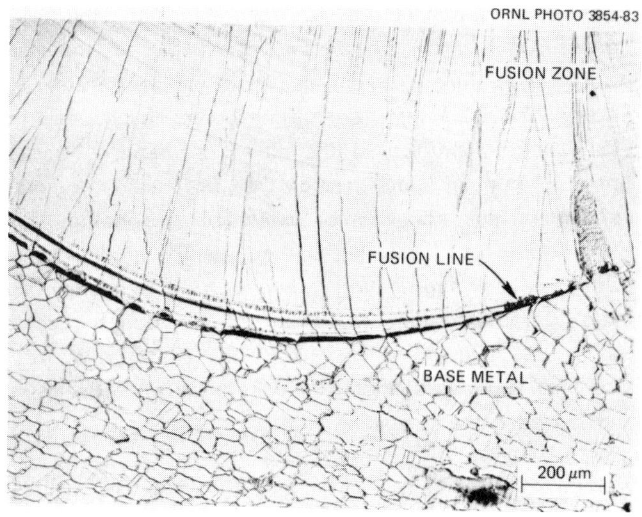

Figure 2: Epitaxial and columnar growth near the fusion line in an iridium alloy.[3,5]

A transition from columnar to equiaxed grains typically is observed in large weld zones and in alloys with a wide freezing range. Equiaxed grain formation may occur naturally as a result of solid fragments having a chance to grow sufficiently in the region ahead of the columnar growth. The solid fragments result from convection in the weld pool or settling due to gravity of small grains formed at upper surfaces. For the fragments to survive and grow there must be undercooling ahead of the columnar grains.[1,2,6]

SOLID LIQUID INTERFACE MORPHOLOGY

The mass rate of freezing of alloy welds is determined by the conditions of heat transport. The form of the freezing front is determined by the interaction of solute transport, heat transport, and crystallography. At the interface between solid and liquid, the alloy strives to achieve chemical equilibrium. Solid liquid interface morphology is a balance between the requirements to redistribute solute efficiently and to minimize interfacial energy. Three types of solidification interface morphology, dendritic, cellular, and planar, are illustrated in Figure 3 and discussed below.

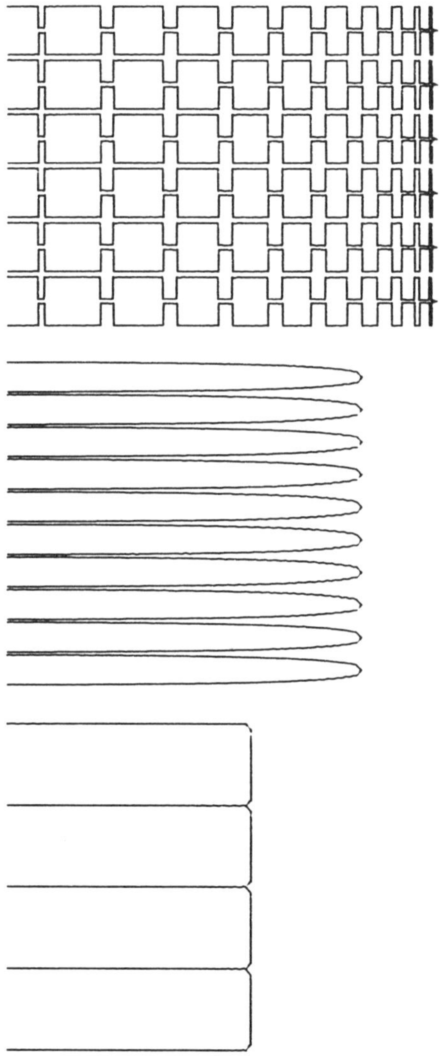

Figure 3: Illustration of dendritic, cellular, and planar growth morphologies.

Since most alloys freeze over a temperature range there is a thermodynamic requirement to redistribute solute during freezing. The distribution coefficient is a measure of the extent of solute redistribution required for a given alloy. The equilibrium distribution coefficient, k^*, is defined as the ratio of the interface composition of the forming solid, C_s^*, and the freezing liquid, C_l^*; i.e.,

$$k^* = C_s^*/C_l^*$$

For elements with k^* less than one, the alloying element or impurity will be rejected from the growing solid into the liquid phase. The liquid at the interface becomes enriched in the solute. For values of k^* greater than one, the liquid is depleted in solute.

A *dendritic* interface is highly branched and is an efficient means of redistributing solute. The high degree of branching creates substantial interfacial area and interfacial energy. Crystallography influences dendritic morphology in that the axes of the branches form along specific crystal directions. For example, in cubic alloys, the dendrite axes are aligned with the <100> directions, cube axes, which are the same as the preferred growth directions for columnar grains in cubic systems.[1-4]

The microradiograph of Figure 4 illustrates a Sn-13%Bi alloy in the process of freezing dendritically. The liquid pool is at the top and the completely solid region is at the bottom of the microradiograph. Due to partitioning of bismuth between phases, the liquid appears dark and the solid appears light. Figure 4 illustrates several features of dendritic solidification.

Going from the dendrite tips to the dendrite roots, the fraction of solid increases. However, the solid does not form linearly with distance. Typically, solid forms rapidly in the initial freezing region and then the incremental solidification rate decreases.

Numerous primary dendrite arms form within a single dendrite. The primary arms within a single columnar grain are parallel to the preferred crystallographic growth direction most closely aligned with the predominant heat flow direction; numerous secondary arms branch from each primary arm. Going from top to bottom, i.e. from regions of higher to lower temperatures within the freezing range, the spacing of the secondary arms *appears* to increase; the longer that

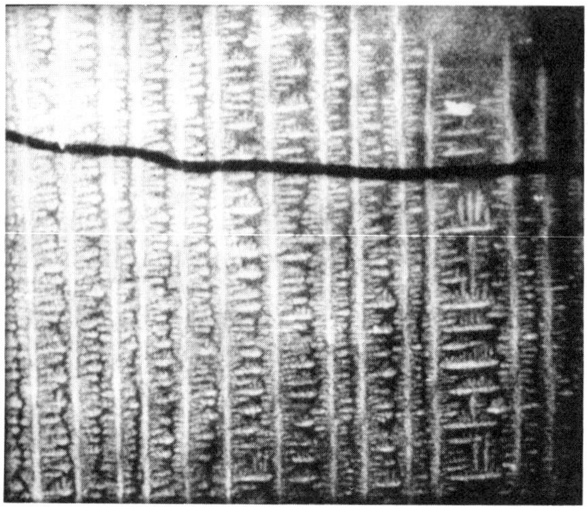

Figure 4: Microradiograph of dendrites solidifying in a Sn–13%Bi alloy.[7]

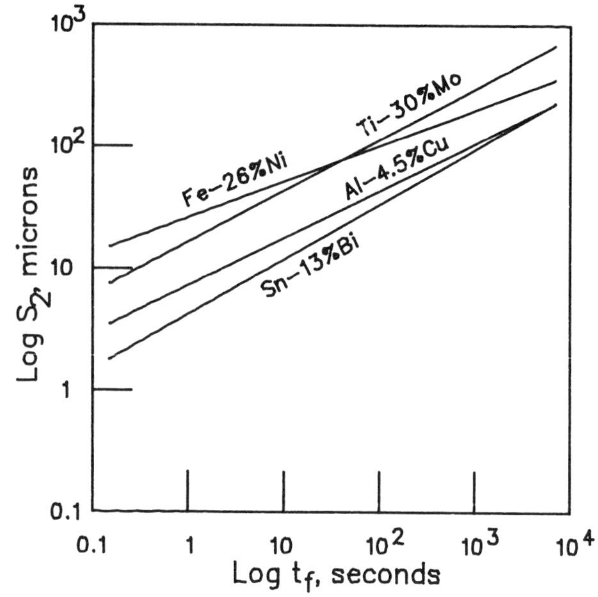

Figure 5: Variation of secondary dendrite arm spacing in four alloys as a function of local freezing time.[8]

the solid and liquid are in contact, the larger the apparent spacing of secondary arms.

Evidence of the effect of local solildification time on measured secondary arm spacings[8] is shown in Figure 5. for the four alloys shown fit a single straight line (on a log–log plot). Similar data have been obtained for numerous other alloys. Secondary dendrite arm spacing, S_2, increases with local freezing time, t_f, according to

$$S_2 = Ct_f^n \qquad 0.3 < n < 0.5$$

where the coefficient, C, and the exponent, n, are constants that depend on the alloy system.

The secondary dendrite arms effectively divide the liquid into isolated pools, and the redistribution of solute occurs within the locality of a pair of dendrite arms. Diffusion has little effect on transporting solute along the length of the freezing zone. Solute is redistributed from the thickening secondary arms into the neighboring liquid pools. Within the interdendritic pools the composition of the liquid is essentially uniform. For $k^*<1$ the composition of the liquid pools increases going from the region of the primary dendrite tips (higher temperature region) to the root of the mushy

Cellular morphology is differentiated from dendritic morphology in that there is no side branching.[1] As illustrated schematically in Figure 3, long liquid grooves are left between the growing cells. The cell cores and the liquid grooves within each grain are parallel. Going from the cell tips to the root of the mushy zone, the solid fraction increases by the thickening of the cells. For $k^*<1$ the thickening cells reject solute into the liquid of the cell grooves. Again the primary segregation direction is transverse to the growth direction.

The weld zone of a 321 stainless steel gas tungsten arc weld in Figure 6 illustrates characteristics of a cellular microstructure. The dark etching regions outline the intercellular grooves. Within a grain the cell axes are all aligned; at grain boundaries the orientation of the cell axes changes. The grain boundaries and cell boundaries etch deeply, indicating the pattern of the solute redistribution process. Although the cells are rod-like, they often appear as short ellipses in a polished cross-section when they are not cut on a plane parallel to the growth direction. Also, the cell spacing is not directly measured from each grain. The apparent spacing depends on the angle of the cut relative to the growth direction. Nonetheless, it is clear for the autogenous GTA weld of 321 stainless steel shown in Figure 6 that the cell spacing does increase going from the fusion line to the center of the weld bead.

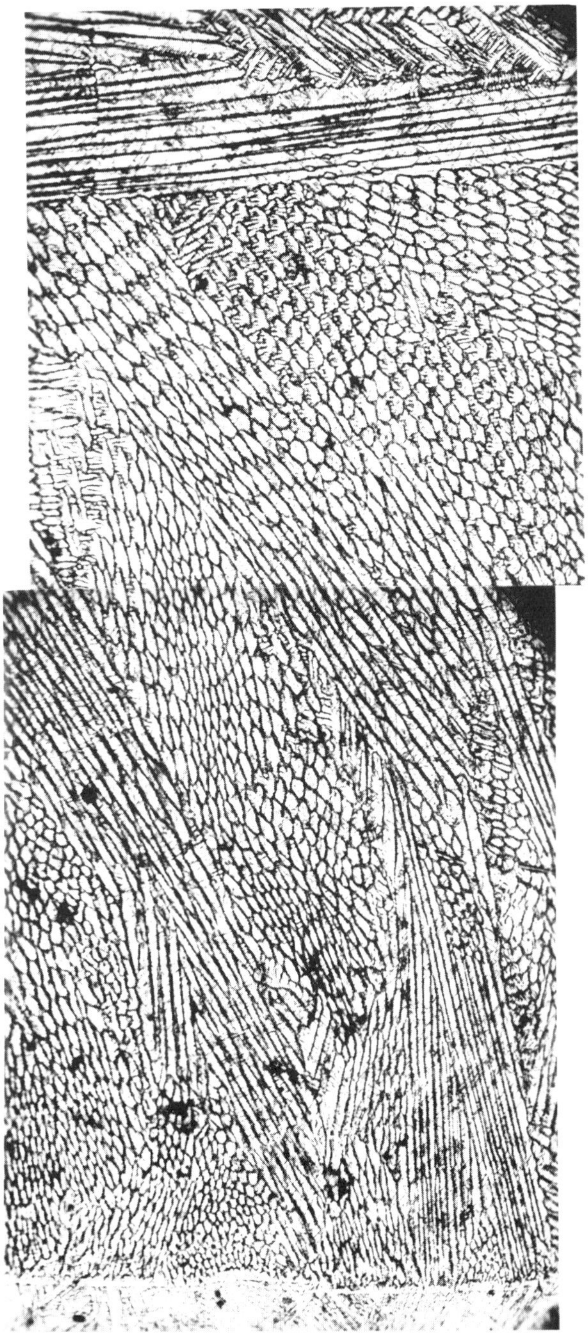

Figure 6: Cellular solidification structure of an autogenous 321 SS GTA weld zone.

In *planar* freezing the interface between solid and liquid is microscopically smooth. (On an atomic scale an alloy interface is rough.) Of course, planar interfaces have low curvature. The sketch in Figure 3 illustrates several grains growing with a planar interface. Grooves form at the grain boundaries. The solute is redistributed into the liquid ahead of the interface or into the grain boundaries. Predominantly, solute is transported in the same direction that the interface advances.

SOLUTE REDISTRIBUTION AND UNDERCOOLING

To look in more detail at the solute redistribution process, consider, as example, a binary alloy freezing with a cellular morphology. The phase relations are illustrated in Figure 7a. The composition of the weld pool, taking into account dilution, is C_0. Neglecting undercooling at the cell tips, if significant, the composition of the first solid to form at the liquidus temperature, T_l, would be K^*C_0. Accounting for undrcooling at the cell tips

$$\Delta T_t = T_l - T_t$$

the first solid to form at temperature T_t would be at composition k^*C_t. As the local temperature decreases, the composition of the liquid will increase as solute is rejected from the thickening cells to the intercellular liquid grooves. The composition of the forming solid, C_s^*, also increases. This continues until the eutectic temperature, which is the effective solidus temperature, is reached. Then, the remaining liquid transforms isothermally to the eutectic mixture.

A temperature distribution from base metal to weld pool is sketched in Figure 7b. The region that is above the liquidus temperature, T_l, (or T_t with undercooling) is still liquid. The region that is below the eutectic temperature, $T_s'=T_e$, is all solid, primary cells with eutectic mixture in the intercellular regions. The mushy zone, the region where cells of primary solid phase and liquid coexist, is bounded by the liquidus and eutectic isotherms at X_t and X_r, respectively.

Figure 7c combines the information in Figures 7a and 7b to depict the composition of the liquid in the mushy zone and the weld pool.[9] Ahead of X_t, the liquid is still at C_0. Between X_r and X_t the liquid is enriched in solute. The liquid near the roots of the cell grooves is more enriched, until at X_r the liquid reaches the eutectic composition. Diffusion of solute through the mushy zone as a result of the concentration gradient described by Figure 7c is one potential source of

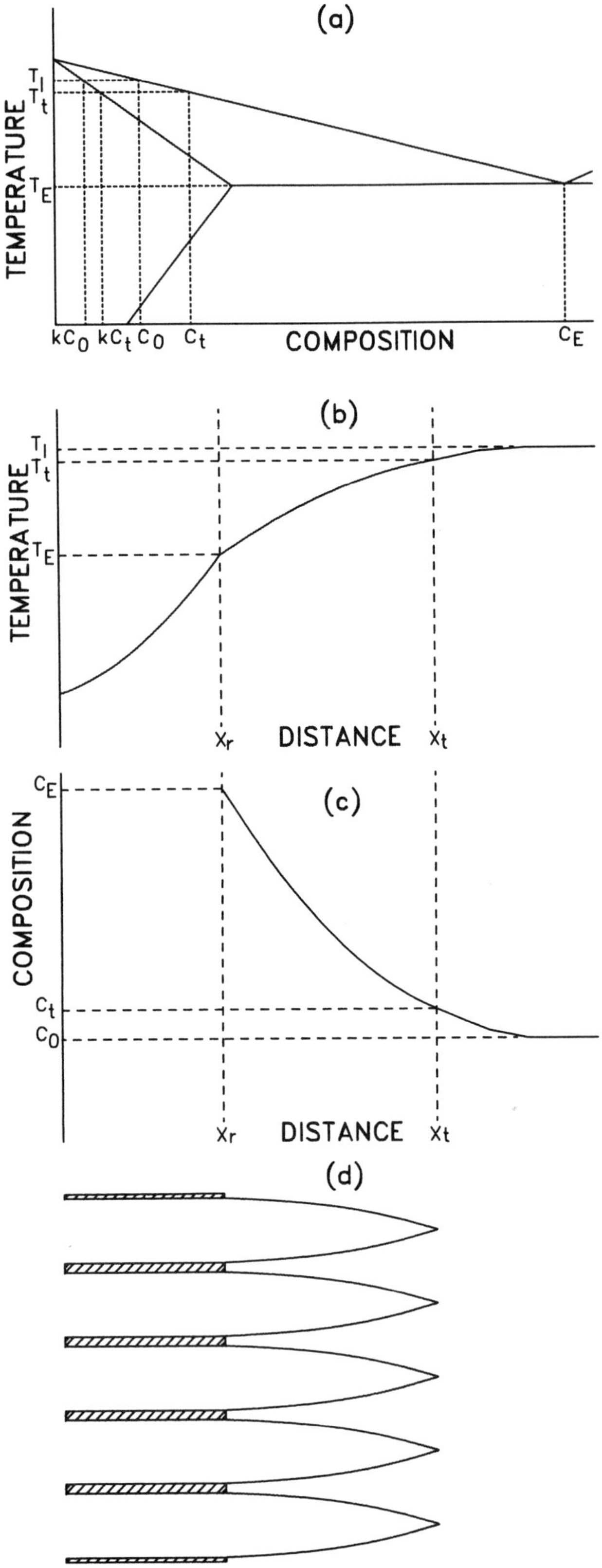

Figure 7: Elements of a model of alloy solidificatiion.[9]

undercooling at the cell tips that will be discussed below.

Figure 7d illustrates the formation of cells. The solid fraction increases continuously from X_t to X_r; and at X_r the remaining liquid transforms isothermally to the eutectic mixture.

The composition of the solid at the centers of the cells may deviate from k^*C_0 (which is the composition of the first solid to form under equilibrium conditions) because of two factors: undercooling at the cell tips and diffusion in the solid during freezing and on cooling to room temperature.

First, consider the undercooling. With reference to a model proposed by Burden and Hunt,[6] the undercooling at the cell tips may be estimated from

$$\Delta T_t = \Delta T_D + \Delta T_r + \Delta T_k$$

The diffusional undercooling is composed of two terms

$$\Delta T_D = D_l G_t / R_t \ - \ m_l R_t r (1-k^*) C_0 / D_l$$

where

m_l=slope of the liquidus, °/%

D_l=solute diffusion coefficient in the liquid, cm^2/s

G_t=thermal gradient at the tips, °/cm

R_t=rate of advance of the isotherm at cell tips, cm/s

r=radius of curvature of the cell tips, cm

The first term represents diffusion through the mushy zone due to the concentration gradient illustrated by Figure 7c.[8,9] The second term represents the point effect of diffusion at the cell tips.[6]

The undercooling due to curvature may be estimated from

$$\Delta T_r = 2\gamma / r$$

where γ is the solid liquid interfacial energy, $ergs/cm^2$.

The kinetic undercooling, ΔT_k, is the smallest contribution to undercooling for an alloy and may be neglected.[2]

Due to the undercooling at the cell tips, the composition of the liquid at the tips is enriched. The deviation of the composition of the liquid at the tips

$$\Delta C = C_t - C_0$$

will depend on the extent of undercooling of the cell tips; and from the above discussion, the important thermal parameters are G_t and R_t. (The local cooling rate, roughly the product of G and R, is also important but is related to the first two.) The composition of the first solid to form will be k^*C_t. As the undercooling approaches the magnitude of the freezing range, the dendrite tip temperature approaches the equilibrium solidus temperature, and the initial solid composition approaches C_0.

Second, consider the influence of solid state diffusion. As successive layers of solid form, they do so at lower temperature and at, for $k^* < 1$, higher composition. Thus, there is a gradient for diffusion within the solid cell. Solid state diffusion would lead to an increase of the composition at the center of the cells and a decrease in the extent of microsegregation. The contribution of solid state diffusion to the solute redistribution process can be estimated by evaluating the dimensionless parameter[9]

$$\alpha = D_s t_f / s_2^{\,2}$$

where t_f is the local freezing time, which is inversely related to the product of G and R, and s_2 is the secondary dendrite arm spacing. At values of α greater than one-tenth, solid diffusion during freezing begins to become an important factor in the solute redistribution process.

In terms of the thermal parameters, the solid composition at the center of the cells/dendrites will approach C_0 and microsegregation will be reduced for large values of R_t, large values of G_t/R_t, large values of $(G_tR_t)^{-1}$. Similarly, the amount of nonequilibrium second phases, such as eutectic mixtures, will be reduced.

GRAIN REFINEMENT

The formation of equiaxed grains, especially fine equiaxed grains, is often desired in welds and castings. For a given alloy three routes for grain refinement are available: control of heat input, application of convection, and addition of nucleating agents. The first approach is the most easily applied and the most often used.

The convection causes fragmentation of selected dendrite branches.[1,11] Some solid fragment are carried by the convective currents to regions of higher temperature. If carried to a region directly under an arc or in another region of high superheat, the fragments would remelt and have no direct effect on grain growth. In remelting, however, the solid fragments chill, to some extent, the local liquid. If the fragments were carried to an undercooled region ahead of the dendrite tips and were allowed to dwell there, they would survive. When the solid fragments accumulate sufficiently to interfere with the growth of the columnar grains, columnar growth would give way to equiaxed.

Weld pool convection is always present and often intense. Lorentz forces in arc welding stir the weld pool and the stirring can be intensified by current pulsing or by arc oscillation. The steep temperature gradients in the weld pool lead to convection driven by gravity forces and surface forces. The intense convection can cause fragmentation of dendrite branches and sweeping of the fragments into the liquid pool. However, unless the fragments are allowed some dwell time in the undercooled liquid at the dendrite tips, they will not be effective in providing growth centers for equiaxed grains. The steep temperature gradients and high freezing rates that are characteristic of welding have the potential to provide the required undercooling at the growth front. On the other hand, cellular microstructures are more predominant in weld zones, see Figure 6; and it appears cells are not as readily fragmented as are dendrites.

Grain refinement by chemical catalysis has been reviewed by numerous authors.[12] In welding epitaxial growth is predominant; undercooling prior to nucleation is not required. Growth undercooling is necessary before heterogeneous nucleation sites can be effective. A danger with innoculating the liquid, whether in casting or welding, is that the addition will be required in such large amounts that it harms the mechanical or corrosion properties of the weldment.

In perspective, it appears that application of electromagnetic forces together with tailoring of the thermal conditions, as with current pulsing, are the most promising approaches for cases difficult to control by standard techniques. As example, modified the weld zone grain structure from predominantly columnar to predominantly equiaxed in 321 stainless steel by selected application of pulsed gas tungsten arc welding.[13]

RESISTANCE TO CRACKING

One category of cracking in welds occurs during the final stages of freezing of the weld pool when the solid phases coexist with the final liquid film. This form of cracking, hot tearing, is more common in alloys with a wide freezing range. In a given alloy system, hot tearing susceptibility will be more pronounced in a critical band of composition and will be greatest for coarse columnar structures and least for fine equiaxed structures.[1,13,14]

Consider an alloy freezing with an equiaxed grain structure. As it begins to freeze, the solid and liquid mixture can flow to respond to the imposition of strains due to contraction of the workpiece or neighboring weld zone. However, at the coherency temperature, corresponding to around 30 - 40% fraction solid, the solid skeleton becomes interlocked; the mixture takes on the properties of a solid. Flow of liquid may continue through the intersticies between the solid grains. As the alloy cools, the stress on the solid skeleton can build up. The wider the range between the coherency temperature and the effective solidus temperature, the more likely will be hot tearing. Fine equiaxed grains are less susceptible to hot tearing than coarse equiaxed grains. Equiaxed grains are less susceptible than columnar grains.

For a given alloy, the load carrying capacity of the solid plus liquid mixture and the cracking tendency can be summarized in a chart such as Figure 8.[14] For a given strain rate, three regions can be defined. For low stresses and at low fractions of liquid, the skeleton is strained but the solid skeleton does not fracture. At higher stresses and higher liquid fractions, the solid skeleton tears along a path primarily defined by liquid films along dendrite arms. Above a critical fraction of liquid, the liquid can flow to fill incipient tears. At higher strain rates, as would be found in welding, higher liquid fractions would be required for refilling of tears.

The most common approach to avoiding weld zone cracking is to alter the shape of the weld pool.[15] With columnar grains and an elongated weld zone, the grains impinge in a nearly continuous surface at the weld centerline. The liquid film along that surface provides a continuous path for weld cracking. By changing the shape of the weld pool to elliptical, the grains are not aligned, the liquid film is spread over a larger area, and the liquid path is less continuous. In some cases a grain grows along the centerline.

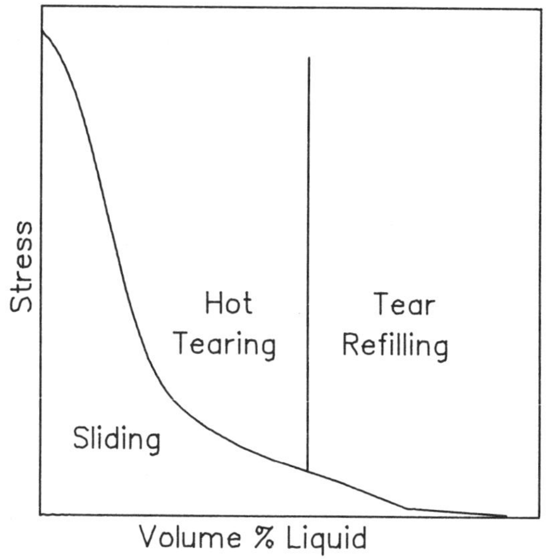

Figure 8: Relation of load and liquid fraction on deformation and fracture in Pb-Sn alloys.[13]

Another improvement would be to reduce the size of the columnar grains to spread the liquid film over more grain boundary area.[5]

Transforming the grain structure of the weld zone from columnar to equiaxed is effective in reducing cracking tendency. The application of pulsed welding to refine the grain structure of 321 stainless steel was noted above.[13] The influence of grain structure on cracking tendency in these was evaluated through varestraint testing.[13,16] Total crack length, number of cracks, and average crack length all decreased as the percent of columnar grain decreased, see as example Figure 9.

SUMMARY

The specialized conditions in the solidification of weld zones can be analyzed in terms of solidification principles. Understanding the solidification phenomena of weld zones can be used to alter the structure and segregation of the weld metal. In turn the quality of the welds can be enhanced.

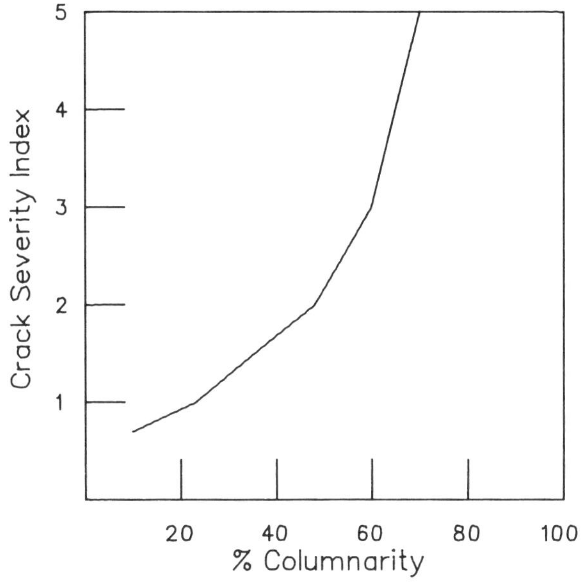

Figure 9: Influence of grain morphology on weld zone cracking of 321 SS.[12]

REFERENCES

1. M. C. Flemings, Solidification Processing McGraw Hill (New York, 1974).

2. B. A. Chalmers, Principles of Solidification, John Wiley (New York, 1964).

3. S. A. David, Oak Ridge National Lab, Personal Communication.

4. W. F. Savage, C. D. Lundin, and A. H. Aronson, Welding Journal, 44, 1965, pp175s.

5. S. A. David, and C. T. Liu, "Modification of Weld Fusion Zone Grain Structure in Thorium-Doped Iridium Alloys," Grain Refinement in Castings and Welds eds. G. J. Abbaschian and S. A. David, TMS-AIME (Warrendale, PA, 1983) pp. 249-258.

6. M. H Burden and J. D. Hunt, "Cellular and Dendritic Growth," J. Crystal Growth 22, 1974, pp. 109-116.

7. L. H. Cadoff, Westinghouse Electric Research And Development Center, Pittsburgh, PA. 15235, Personal Communication.

8. H. D. Brody, "Structure and Properties in Cast Non-ferrous Alloys," Solidification Technology, eds. J. J. Burke, M. C. Flemings, and A. E. Gorum, Brook Hill Publishing Co. (Chestnut Hill, MA., 1974) pp. 53-83.

9. H. D. Brody and M. C. Flemings, "Analysis of Solute Redistribution in Dendritic Solidification," Trans. TMS-AIME, 236, 1966, pp. 615-624.

10. W. Kurz, K. Giovanola, R. Trivedi, "Theories of Microstructural Development During Rapid Solidification,"" Acta Met. 34, 1986, pp. 823-830.

11. K. A. Jackson, J. D. Hunt, D. R. Uhlmann, and T. P. Seward III, Trans. TMS-AIME, 236, 1966, pp. 149.

12. G. J. Davies and J. G. Garland, International Metals Review, 20, 1975.

13. A. A. Gokhale, "Weld Pool Solidification Structure and Property Control in Pulsed Current Gas Tungsten Arc Welding," M. S. Dissertation, Metallurgical Engineering, University of Pittsburgh, Pittsburgh, Pa., 1980.

14. A. A. Gokhale, "Tensile Properties of Partially Solidified Pb-Sn Alloys in Relation to Hot Tearing," Ph. D. Dissertation, Metallurgical Engineering, University of Pittsburgh, Pittsburgh, PA. 15261, 1985.

15. W. F. Savage, Welding in the World 18, 1980.

16. A. A. Gokhale, A. A. Tzavaras, H. D. Brody, and G. M. Ecer, "Grain Structure and Hot Cracking in Pulsed Current Gas Tungsten Arc Welding of 321 Stainless Steel," Grain Refinement in Castings and Welds, eds. G. J. Abbaschian and S. A. David, TMS-AIME (Warrendale, PA, 1983) pp. 223-248.

WELD MICROSEGREGATION CHARACTERIZATION AND MODELING

J.A. Brooks, M.I. Baskes
Sandia National Laboratories
Livermore, California

IT IS WELL RECOGNIZED that microsegregation during weld solidification can have a strong influence on weld cracking susceptibility, phase equilibria, mechanical properties, and fracture behavior, all of which play a major role in weld performance. Still, little data are available in the literature in which weld microsegregation has been carefully studied. In our work we have chosen several binary alloy systems and weld processes in which to study the mechanics of weld solidification and microsegregation and to determine the important factors which lead to the final degree of alloy segregation.

EXPERIMENTAL DETAILS

GTA welds were made in 3 mm thick Al-Cu and Fe-Nb binary alloys. The alloy compositions and weld solidification conditions are shown in Table 1 for which cooling rates were obtained from thermocouple readings taken from near the weld center line.

Table 1 - Weld Solidification Conditions

Alloy	Weld Type	Weld Speed (cm/sec)	Cooling Rate (°C/sec)
Al 1.1% Cu	GTA	.43	375
Al 1.1% Cu	eb	3.2	
Al 2% Cu	GTA	.43	375
Fe-2% Nb	GTA	.21	100

Electron beam welds were made in 3mm thick Al-1.1% Cu alloy with a weld velocity of 3.2 cm/sec. In this case, weld cooling rates were not experimentally determined. Electron microprobe and analytical electron microscopy (AEM) were used to determine the degree of microsegregation. Analyses were taken from regions near the center of the welds where solidification occurred parallel to the weld direction. In this region the solidification velocity approximated the weld speed.

The Al-Cu weld samples were polished and etched with amonium persulfate for metallographic examination; while the Fe-Nb samples were etched with Vilella's reagent. TEM thin foils were prepared using a jet polishing technique in a solution of 8% nitric acid, 17% water, and 75% methanol with 20 volts at -40°C. Care was taken to ensure that redeposition of Cu on the foils did not occur because this condition would affect the STEM (scanning transmission electron microscopy) analysis.

CALCULATIONAL MODEL

We modeled microsegregation using a finite difference analysis in which solidification and microsegregation was represented in the volume element shown in Figure 1. We also modeled

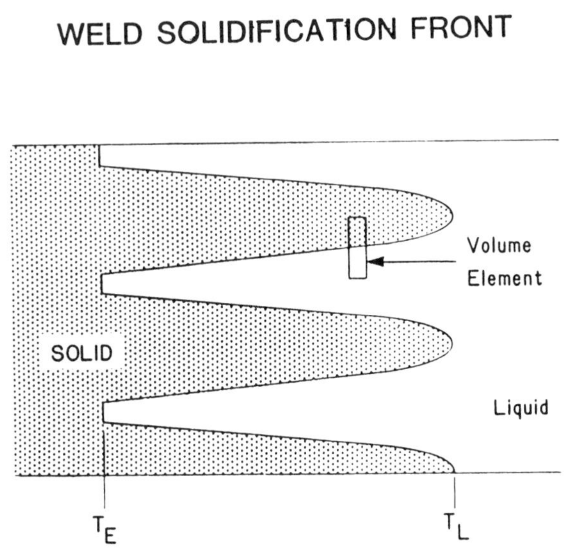

Figure 1 - Schematic of weld solidification front showing the volume element used to model microsegregation during cellular solidification. Note the first material solidifies at T_L with composition of the cell tip, and the last material solidifies at T_E, for the above schematic, at eutectic composition.

dendrite geometry using cylindrical coordinates with the axis along the dendrite length. The boundary conditions were dictated by the phase diagram as shown schematically in Figure 2. Under equilibrium conditions, the first material to solidify at the dendrite tip is of composition $C_o k$; where C_o is the nominal composition and k is the equilibrium partitioning coefficient defined

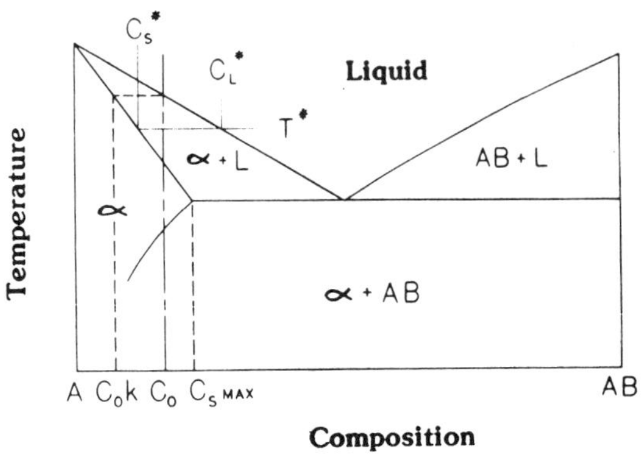

Figure 2 - Schematic of a binary eutectic phase diagram in which $k = C_s/C_L$ is constant and representative of Al-Cu and Fe-Nb.

as the ratio of the solid and liquid compositions at temperature T. At any temperature, i.e., position along the dendrite, the following conditions are assumed: (1) equilibrium exists between the liquid and solid at the solid-liquid interface; (2) k is constant; (3) complete mixing occurs radially in the liquid; (4) no diffusion occurs in either the solid or liquid along the dendrite; (5) the cooling rate is constant; and (6) the solute diffusivity in the solid exhibits an Arrhenius temperature dependence.

Thus, these conditions are the same as those for the Scheil equation (1), except that solid-state diffusion can occur during solidification and cooling. A similar analysis had been derived earlier by Brody and Flemings (2).

Al-Cu SYSTEM

The Al rich side of the Al-Cu phase diagram is a simple eutectic in which k is almost constant at 0.17 (3). The eutectic forms between Al (5.6 wt% Cu) and θ ($CuAl_2$) at 548°C. The optical micrograph in Figure 3 shows the cellular solidification structure typical of the Al-Cu welds. In this weld, which contains 2% Cu, the cell boundaries are decorated with θ resulting from some eutectic solidification, and θ` resulting from a one-hour, 350°C age. Microprobe results, taken across several solidification cells of an as-welded sample, are shown in Figure 3b. Note, the Cu content across the cell core is fairly uniform and exhibits a minimum of approximately 0.8 wt%. The maximum Cu concentration measured in the cell boundaries is very dependent upon the region analyzed. Since the eutectic product θ is present, the maximum Cu content can be as high as that of $CuAl_2$. However, the maximum in solid solution before aging would be expected to be near the maximum solid solubility, approximately

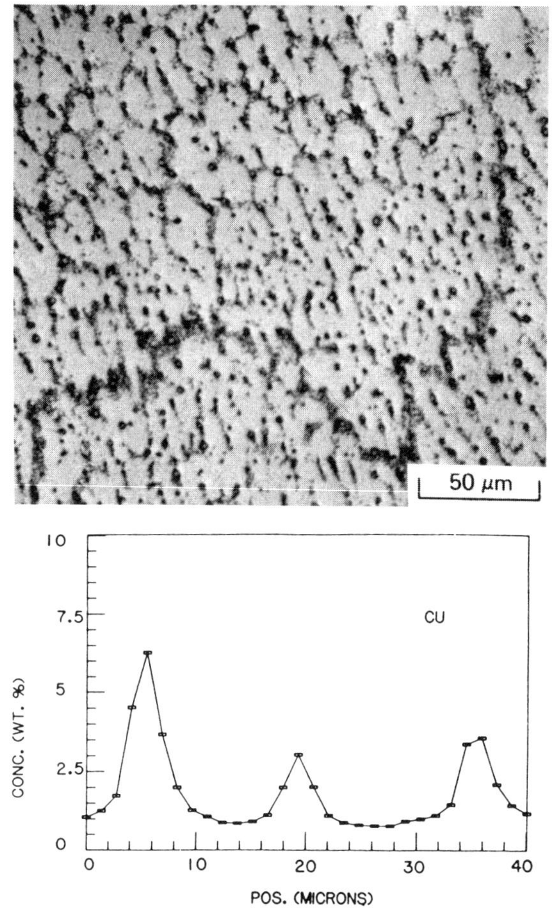

Figure 3(a) - Optical micrograph of an aged Al-2% Cu GTA weld showing the solidification cell boundaries decorated with θ and θ' precipitation. (b) electron microprobe analysis across several solidification cells of an unaged sample.

5.6% Thus, in the Al-Cu welds, we are mainly interested in the Cu concentration profile across the cell core and the quantity of eutectic θ within the cell boundaries.

In the TEM micrograph of Figure 4 the weld structure is seen more clearly. The eutectic product θ is present in the cell boundaries. As a result of aging at 350°C, two variants of θ` are also visible in the cell boundary region but not along the cell core. It is evident that in the core region the Cu content is below the θ` solvus at the 350°C aging temperature (4). However, the onset of θ` precipitation should correspond to the solvus composition.

A STEM image of a foil taken from an Al-1.1% Cu electron beam weld is shown in Figure 5. In this case the cell boundaries are decorated with a dislocation substructure in addition to a small amount of the eutectic product θ. Although a quantitative analysis was not conducted, the amount of θ resulting from eutectic solidification in the electron beam welds appeared to be considerably less than in the GTA weld in the alloy of the same composition. In Figure 5, contamination spots showing the region in which the AEM (analytical electron microscopy) analysis was conducted are also evident. Analysis at positions across the cellular structure were

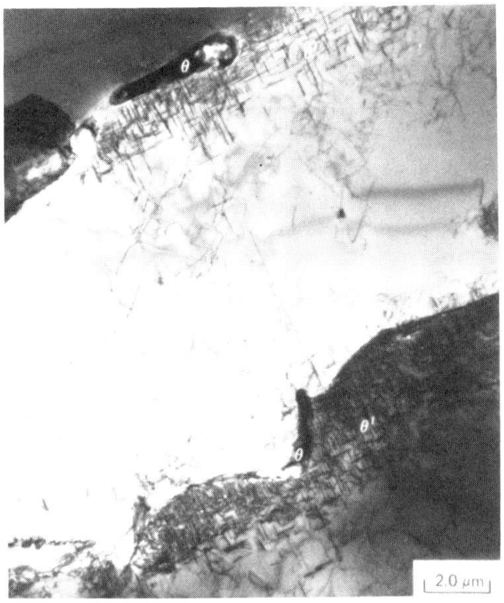

Figure 4 - TEM micrograph from the weld shown in 3(a). Note the eutectic constituent θ, and θ' precipitates in the cell boundary regions.

taken in three regions to increase the counting statistics while minimizing any errors due to contamination buildup. The results of the AEM analysis is shown in Figure 5b. In this case a spike in the Cu content is found at the center of each cell, but the maximum Cu content still occurs along the cell boundary. The maximum Cu concentration measured in the cell boundaries was approximately 2.3%; however, in regions surrounding the eutectic θ, the concentration is expected to be close to 5.6%. The maximum Cu concentration measured at the cell core was 1.1%, while the minimum was 0.6%. Although the spike in solute content at the cell center will not be discussed further in this paper, it must be noted that this phenomenon has been observed by Boettinger et. al. (5) in electron beam surface melting studies in the Ag-Cu system.

Of greatest interest in modeling the solidification behavior is the prediction of the solute concentration across the cell core. The minimum Cu concentrations measured for three welds, made under the conditions shown in Table 1, are given in Table 2.

Table 2 - Summary of Al-Cu Weld Segregation

Cell Size (Microns)	C_o (wt%Cu)	$C_o k$ (wt%Cu)	C core (wt%Cu)
15 (GTA)	2	0.34	0.8
10 (GTA)	1.1	0.19	0.35
3 (EB)	1.1	0.19	0.6

Our calculated Cu concentration profile for the Al-2% Cu GTA weld shown in Figure 6 is very similar to the probe analysis shown in Figure 3b, except that the calculated core concentration is only 0.38% Cu compared to the measured concentration of 0.8%. It should be noted that no evidence of an initial solidification transient exists in the experimental data. It

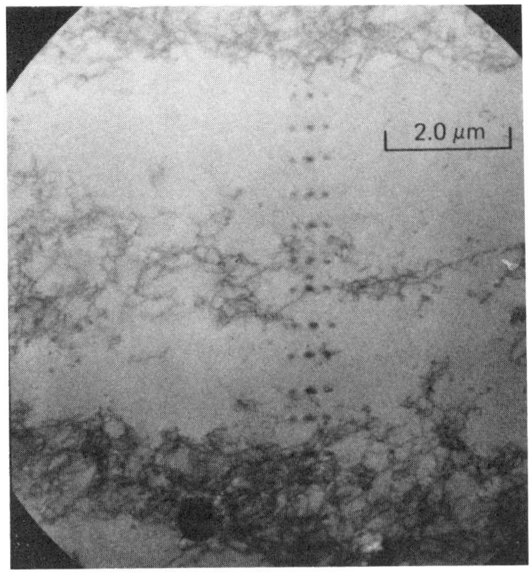

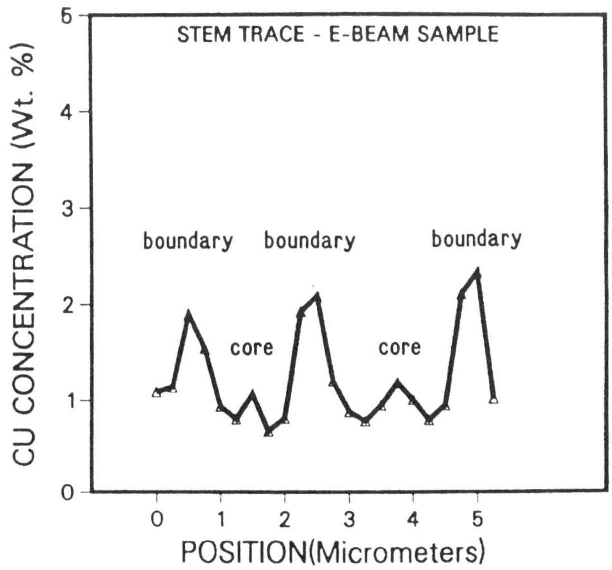

Figure 5(a) - STEM image of an electron beam weld of Al-1.1% Cu, cell boundaries are decorated with dislocations. (b) STEM analysis of sample shown in (a).

has been proposed by other invesitigators (6) who applied the solidification model developed by Smith, Tiller, and Rutter (7,8) that both initial and final transients occur during the cellular solidification of welds. The length of the initial transient is $X \sim \frac{D_l}{Vk}$, where D_l = diffusivity of solute in the liquid and V = the solidification velocity. However, when applying this model, it was assumed the solidification velocity, defined for the volume element in Figure 1, was about equal to the weld speed. The actual velocity depends upon other factors including the solidification temperature range and thermal gradient and is usually much slower than the velocity of the cell tip. The volume element of our Al-2% Cu GTA weld solidifies at an average velocity of ~ 3.9×10^{-5} cm/sec, the length of the volume element (7.5 microns) divided by the local solidification time (0.19 sec). In this case, the length of the initial

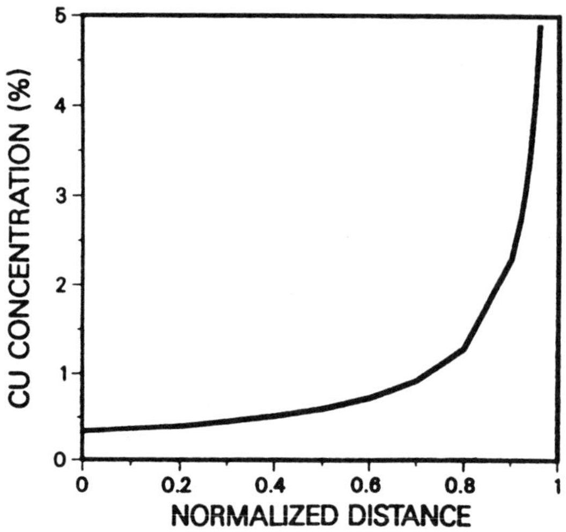

Figure 6 - Calculated Cu concentration of the Al-2% Cu GTA weld from the cell core to cell bound-boundary; the final 2% of liquid will solidify as a eutectic between Al (5.6%Cu) + $CuAl_2$.

transient would be ~ 380 microns on 25 cell diameters (using $D_1 = 2.5 \times 10^{-5}$ cm^2/sec) and not 3.4 microns that would be obtained by using the weld speed (9).

Initially, one may suppose that the high Cu concentration of the cell cores, being approximately 2 C_ok for the GTA welds and 3 C_ok for the electron beam weld, is due to solid-state diffusion. However, we calculate that solid-state diffusion only increases the core concentration a few hundreds of a percent, from 0.34 to 0.38% Cu in the 2% alloy, when using a value for the diffusivity of Cu in Al of $D_o = 0.084$ sq cm/sec, and Q = 32,600 cal/mole (10). The

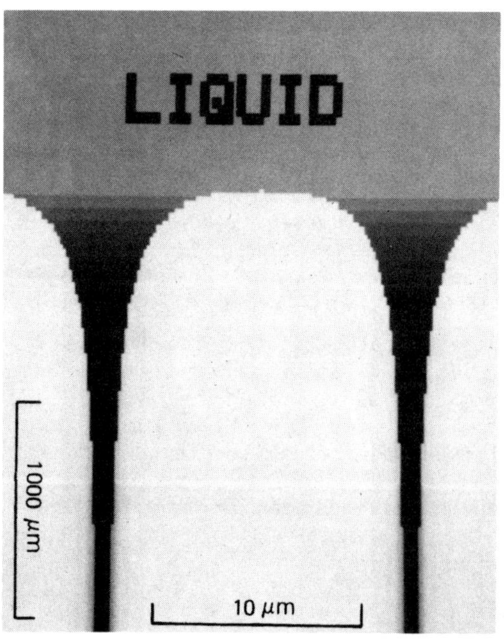

Figure 7 - Computer graphic results of Al-2% Cu GTA weld showing cell tips. Note length to width of cells have been reduced by a factor of 100.

computer graphics in Figure 7 shows the tips of solidifying cells extending into the liquid weld pool. It is easily seen that the Cu concentration within the liquid between the cells is uniform but increases from the cell tip to the final liquid to solidify at eutectic composition. By further examining the details of the concentration profile near the cell tip, we see that the Cu isoconcentration lines in the solid are nearly parallel to the cell axis illustrating the lack of a significant amount of solid-state diffusion. In this figure the relative length to diameter of the cell core has been reduced by a factor of 100 to demonstrate this point. In actuality the aspect ratio of the cells is approximately 250 to 1, i.e., the solidification cells appear as needles extending into the weld pool. To obtain, by diffusion, an increase in Cu concentration at the cell core from 0.38 to only 0.8% Cu, we would have to decrease the cooling rate by a factor of 300, increase the diffusivity by a factor of 300, or reduce the cell size by a factor of 17, none of which are reasonable.

Dendrite tip undercooling (11,12) has been used to explain why the concentration at the cell core is greater than C_ok. Burden and Hunt (11) have derived an expression for tip undercooling given by

$$\Delta T = \frac{GD_1}{V} + 2^{3/2}[-mV(1-k)C_o\theta]^{1/2}/D_1^{\;1/2} \qquad (1)$$

Where G = thermal gradient, V = tip velocity, D_1 = solute diffusivity in the liquid, m = liquidus slope, and θ = curvature undercooling coefficient.

The undercooling described by the first term results from an increase in solute content ahead of the solidification front due to diffusion down the concentration gradient between the solidification cells. In our case, this term is negligible. However, the dendrite tip curvature and a radial concentration gradient about the tip does result in undercooling which is given by the second term of Equation 1.

Table 3 lists the experimental data obtained from the three welds, along with calculated

Table 3 - Undercooling of Al-Cu Welds

C_o (wt%)	V cm/sec	C_ok (wt%)	Core Core (wt%)	ΔT Required (°C)	ΔT Burden Hunt (°C)
2	.4	.34	.8	8.5	2.8
1.1	.43	0.19	.35	3	2.1
1.1	3.2	0.19	.6	8	5.7

* T = Undercooling required to achieve measured core concentrations.

* ΔT = Undercooling calculated from Burden and Hunt--Eq 1 of text.

values of undercooling obtained from the experimental data, and predicted undercooling calculated using Equation 1. In the calculation, we used values for $D_l = 2.5 \times 10^{-5}$ cm^2/sec, and the Zener approximation for $\theta = \frac{\sigma}{\Delta S}$ with a surface energy of 121 erg/cm^2 and an entropy of melting of 1.63×10^7 erg cm^{-3} °C^{-1} (11). In Table 3, the amount of undercooling required to obtain the measured core concentrations in excess of $C_o k$ is obtained by $\Delta C\ m_s \Delta T$; where m_s is the solidus slope. The calculated amount of undercooling is less than that measured for all three welds. This result was also found to be the case by Boettinger et. al. (5) for electron beam surface melted Ag-Cu alloys. However, it is important to note that the amount of undercooling increases with increased alloy composition and with velocity, in agreement with Equation 1.

Fe-Nb SYSTEM

We selected the Fe-Nb system for study for several reasons: the Fe-Nb phase diagram, like Al-Cu, exhibits a simple eutectic on the Fe-rich side between Fe (4.5% Nb) and ε (Fe_2Nb) (13); k ~ constant, k being ~ 0.25 is similar to the value of 0.17 for Al-Cu; and since Fe-Nb is a b.c.c. system, we expected that solid-state diffusion may play a major role in reducing the final degree of weld microsegregation. Therefore, an interesting comparison in the degree of weld microsegregation can be made between two systems which exhibit very similar phase diagrams.

An optical micrograph showing the cellular solidification structure of a GTA weld in an Fe-2% Nb alloy is shown in Figure 8. The cell cores were found by microprobe analysis to contain

Figure 8 - Optical micrograph showing the cellular solidification structure of an Fe-2% Nb GTA weld.

approximately 1.8% Nb, close to the nominal composition of 2% and much greater than $C_o k$ = 0.5%Nb.

To our knowledge, the diffusion coefficient of Nb in γ-Fe has not been measured. Therefore, in modeling weld solidification, we considered the diffusivity of Nb to be the same as the self-diffusion coefficient of Fe, $D_0 = 6.8$ cm^2/sec Q = 55 kcal/mole (14). In dilute solutions this is a reasonable approximation (15). The results of a computer simulation using a cell spacing of 20 microns and a cooling rate of 100°C/sec is shown in Figure 9. The curved nature of isoconcentration lines in the solid is indicative of diffusion occurring during weld solidification and

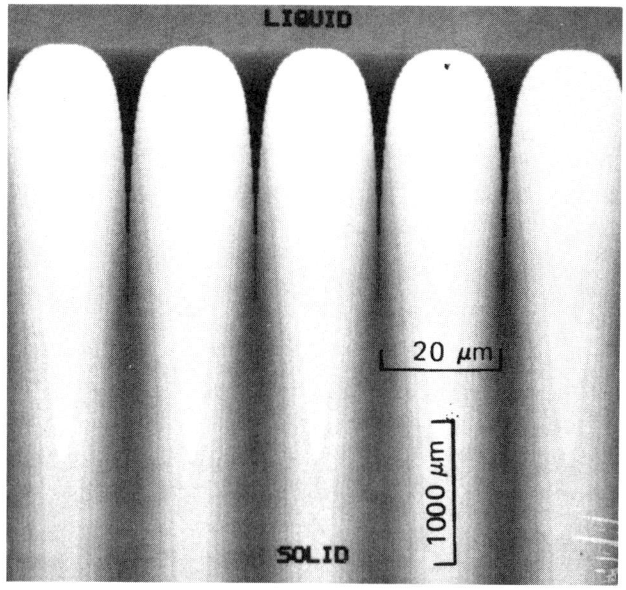

Figure 9 - Computer graphic results showing concentration profiles in a Fe-2% Nb GTA weld. The intensity of color represents the Nb content and shows the nature of solid-state diffusion.

cooling. Solid-state diffusion also results in solidification occurring at 1397°C, 37°C above the eutectic temperature of 1360°C. The final concentration of the cell core is calculated to be 1.4% Nb compared to the measured value of 1.8%. However, the higher core concentration of 1.8% can be achieved by doubling the pre-exponential of the diffusion coefficient, still a reasonable estimate for the diffusivity of Nb, or by invoking tip undercooling as discussed above. The diffusivity of Nb in γ-Fe has been shown to be greater than the Fe self-diffusion coefficient (15). If just the pre-exponential is doubled, the final solidification occurs at an even higher temperature, 1417°C.

Figure 10a shows the results of a microprobe analysis taken across a cellular structure of a GTA weld in an Fe 3.3% Nb alloy; while Figure 10b shows the calculated Nb content of a cell volume element during different stages of solidification and cooling. The solidification model shows that the first material solidifies at a composition of 0.82% Nb ($C_o k$). A large fraction (approximately 1/2) of the cell solidifies at a fairly uniform composition within the first 0.08 second. As solidification proceeds, the concentration of the cell core increases due to solid-state diffusion. Final solidification occurs after 1.4 seconds. At this point about 99% of the volume element (weld) solidifies as a eutectic, δ-Fe with 4.5% Nb and Fe_2Nb. The concentration of the cell core by this time has increased from 0.82% to 1.9% Nb. After 2.3 sec all the eutectic constituent has dissolved and the concentration of the cell boundary has been reduced to only slightly greater than 4%. Note in the next 3.3 sec, only a small amount of further diffusion occurs.

If we compare the measured results for the 3.3% Nb weld with the calculated values shown in Figure 10, we see the agreement is quite good. The measured core concentrations are 2.8% compared to the calculated value of approximately 2.3%. Both the measured and calculated cell boundary concentrations are 3.8%. In this analysis we have assumed the diffusion coefficient for Nb to be the same as that of δ Fe.

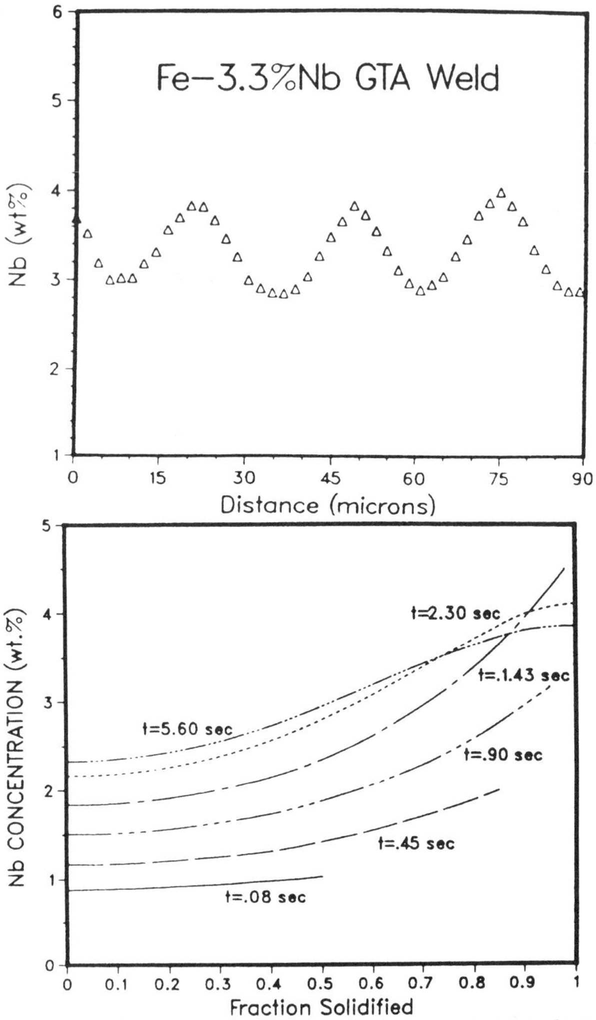

Figure 10(a) - Electron microprobe results from across a cellular weld structure in a Fe-3.3% Nb alloy. (b) calculated composition profiles for the volume element shown in Figure 1, plotted for different stage of solidification and cooling of the 3.3% Nb weld.

We must point out that we have shown, that even in GTA welds, dendrite tip undercooling will result in higher core concentrations. Thus, the calculated Nb concentrations of the core, being lower than those measured, can be accounted for by dendrite undercooling without having to adjust the diffusivity.

SUMMARY

We have characterized, and modeled using a finite difference analysis, the solidification behavior of welds in two alloy systems--Al-Cu and Fe-Nb. Although the constitutional diagrams are very similar for both alloy systems, the degree of microsegregation in the welds of similar solute content is very different. We have shown that in both GTA and electron beam welds for Al-Cu, dendrite tip undercooling occurs and increases with both increasing solidification velocity and composition. This serves to reduce the degree of segregation. However, solid-state diffusion is slow in the Al-Cu system so that little additional homogenization of solute occurs during solidification and cooling. Although the cell size of the electron beam welds were much smaller than those of the GTA welds, the increase in cooling rate still prevents any appreciable solid-state diffusion from occurring. Thus, in most welds in Al-Cu, some eutectic solidification will occur, but for a given composition, the amount of eutectic solidification will decrease with increasing velocity.

In Fe-Nb welds we found that solid-state diffusion not only greatly reduces the degree of microsegregation but greatly affects the final temperature and phases of solidification. For example, in an Fe-2%Nb weld, we find that as a result of solid-state diffusion, solidification occurs totally as single phase δ before the eutectic temperature is reached and the final structure is nearly homogeneous. In a 3.3%Nb alloy, some eutectic solidification ($L \rightarrow \delta Fe + Fe_2Nb(\varepsilon)$) occurs; however, the eutectic constituent ε in the cell boundaries is completely dissolved by the time the weld is cooled to room temperature.

These studies have provided much insight into the solidification mechanics and microstructural development of welds. It becomes very clear that the detailed mechanics of solidification are not necessarily apparent in the observed weld microstructure at room temperature. Analysis of this type will not only lead to a better understanding of the physical metallurgy of welding, including the evolution of weld structure, weld properties during solidification and cooling, and solidification cracking behavior, but they can be used as a valuable tool to control weld behavior, both through alloy design and welding practice.

REFERENCES

1. E. Scheil, Z. Metallk, 34, p.70, 1942.
2. H. D. Brody and M. C. Flemings, Met. Trans. Vol. 236 (5), 615-624, 1966.
3. M. Hansen and K. Anderko, "Constitution of Binary Alloys," McGraw Hill, p. 85, 1985.
4. E. Hornboken, "Aluminum," 43, PtII, 9-15, 1967.
5. W. J. Boettinger, Metallurgy Div. NBS, Washington, DC, unpublished research.
6. W. F. Savage, Welding in the World, Vol. 18, No. 5/6, 89-111, 1980.
7. V. G. Smith, W. A. Tiller, and J. W. Rutter, Canadian Journal of Physics, 33, 723-745, 1955.

8. W. A. Tiller, K. A. Jackson, J. W. Rutter, and B. Chalmers, Acta Met., 1, p. 428, 1953.
9. J. C. Lippold and W. F. Savage, TMS-AIME Proceedings Casting and Welding Process Modeling, Rindye, NH, Aug. 1980.
10. C. J. Smithells, "Metals Reference Book," 3rd Ed., Vol. II, p. 589, 1962.
11. M. H. Burden and J. D. Hunt, J. of Crystal Growth, 22, 109-116, 1974.
12. W. Kurz, B. Giovanola, and R. Trivedi, Acta Met., Vol. 34 (5), 823-830, 1986.
13. M. Hansen and K. Anderko, "Constitution of Binary Alloys," McGraw Hill, p. 676, 1985.
14. S. Kurokawa, J. E. Ruzzante, A. M. Hey, and F. Dyment, Metal Science, Vol. 17 (9), 433-438, 1983.

THE CONTRIBUTION OF WELDING TO THE UNDERSTANDING OF METALLURGICAL PHENOMENA

M. Dadian
Institut de Soudure
Paris, France

ABSTRACT

The work is concerned with a new approach which placed the solidification stage in the context of the thermal dynamics associated with the welding of iron and steel. The preliminary sequence of events taking place in the heat affected zone on heating, as the results of the successive $\alpha \rightarrow \gamma$ and $\gamma \rightarrow \delta$ transformations is described first. Then the genesis of the molten pool, the constitution of the $\delta \rightleftarrows$ liquid transition zone, the δ solidification, the $\delta \rightarrow \gamma$ and $\gamma \rightarrow \alpha$ transformations upon cooling are dealt with and analysed in detail. The movements of the fusion, solidification and allotropic transformation fronts, as well as the micrographic structures are detected thanks to a "memory effect" evidenced by appropriate procedures.

1. INTRODUCTION

Microstructures in welded joints are generally interpreted with the aid of equilibrium diagrams or continuous cooling transformation diagrams, taking into account factors such as chemical composition, maximum temperature attained, high temperature dwell time and cooling rate.

However, in order to fully understand the various microstructural features, together with their formation sequence, their metallographic appearance must be viewed in the light of the thermal dynamics of the weld cycle. Furthermore, the importance of the heating stage must be emphasized, for it is during this period that the physical-chemical changes are initiated in the heat affected zone while, close to the melting temperature, the structure of the transition region is determined adjacent to the fusion zone.

The purpose of the present study is to examine in detail the metallurgical processes which occur under the dynamic thermal conditions prevailing during the welding of iron and steel. Metallographic examination is carried out after completion of the weld cycle, using experimental techniques which enable reconstitution of the various stages: allotropic transformations, recrystallisation, formation of the melt pool and solidification.

It will be seen once again that fusion welding is an extremely fertile experimental technique for the study of metallurgical phenomena.

2. EXPERIMENTAL TECHNIQUES

2.1. Materials

The materials investigated ranged from pure iron to low carbon steel, a considerable amount of work being carried out on Armco iron (table I). In addition, compositions intermediate between iron and steel were obtained by controlled introduction of carbon into the melt pool of weld beads made on plates of Armco iron previously coated with a layer of carbon deposited with a carburising oxy-acetylene flame.

2.2. Thermal conditions

The fusion lines are made :

- either with an electric arc under an argon atmosphere (G.T.A. process):

  * 2 mm diameter thoriated tungsten electrode, connected to the negative terminal of a rectified generator,

  * electrical parameters: 120 amps, 11 volts,
  * welding speed : 20 cm/min.,
  * welding energy : 4400 J/cm ;

- or with an electron beam.

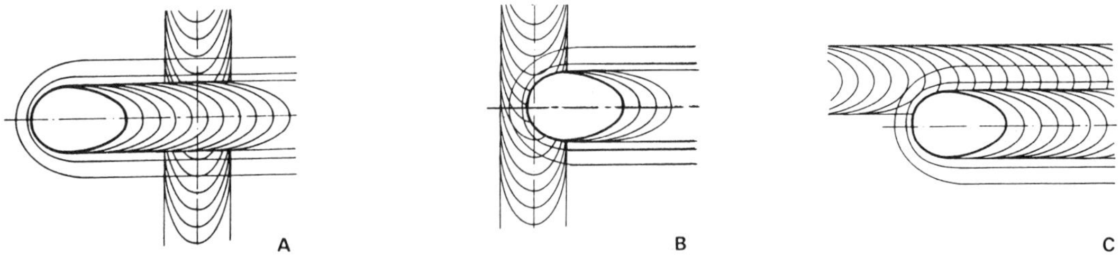

Fig. 1 - Different fusion line configurations for studying the effect of welding thermal cycles in a substrate with an annealed, cold-worked or as-solidified structure.

TABLE I

	C	Mn	Si	Cr	Ni	Mo	Cu	Ti	Nb	Co	V	Al	S	P	B	N	O
Pure iron (ppm)	3.0	7.0	-	1.1	0.5	0.04	2.5	-	-	2.4	-	20.4	1.2	-	-	11.0	13.0
Armco iron (annealed) (%)	0.006	0.050	0.020	0.016	<0.005	<0.005	0.005	<0.005	<0.005	<0.005	<0.005	0.009	0.007	0.010	<0.0005	0.005	0.028
Steel B * (%)	0.038	1.00	0.30	-	-	-	-	-	-	-	-	-	0.005	0.014	-	0.010	0.046

* Deposit obtained with a basic coated electrode
Welding energy : 10 000 J/cm.
Welding speed : 24 cm/min.

These techniques present the advantage of conserving a clean surface on the materials being studied. Furthermore, their automation ensures a regular melting rate.

The thickness of the base metal is limited to 2 mm in order to ensure a planar heat distribution.

Different configurations of the fusion lines (fig. 1) make it possible to study the effects of the thermal cycle on the substrate metal, either in the annealed or cold-worked condition or with an as-solidified structure.

Detailed information on the formation of the weld pool and epitaxial solidification can be obtained by means of micro melt spots and micro fusion lines made at high speed (100 cm/min.)

Thermal cycles were recorded by capillary chromel-alumel thermocouples and the results were processed by computer.

2.3. Metallographic techniques

Examinations were made either on as-welded or on polished and etched surfaces.

In the former case, the interest of the method is increased if the base metal is previously mechanically polished, or if it has a bright surface, either due to annealing in hydrogen (for pure iron) or to careful rolling. Benefit can be drawn in this way from high temperature thermal etching or from the variations in relief induced by the deformation (1,2). The choice of environment contributes to the success of this type of etching, which concerns not only the heat affected zone, but also the fusion zone.

In the second zone, the metallographic preparation of a cross section involves the following successive steps:

I) fine mechanical polishing, finishing with a 2 micron diamond paste;

II) chemical polishing; selective dissolution leads to an etching effects. Optimum conditions are obtained with an oxalic acid-hydrogen peroxide bath:

* Oxalic acid .... 3.0 g
* Hydrogen peroxide (30 %) .... 4.0 ml
* distilled water .... 100 ml

Duration: 5 to 25 minutes;

III) Etching in a saturated aqueous solution of picric acid, in order to reveal delta solidification structures and prior delta and gamma grain boundaries:

* Picric acid .... 2.0 g
* Distilled water .... 100 ml
* Wetting agent .... 0.5 ml

Bath temperature: 60 to 70°C
When this etch is prolonged it can faintly reveal alpha ferrite and structures resulting from the transformation of austenite;

IV) Alpha ferrite can sometimes be shown more clearly by an additionnal nital or sodium bisulfite etch:

* Sodium bisulfite .... 60 ml
* Distilled water .... 20 ml

The examinations were carried out by optical microscopy with normal or oblique lighting, or with interference contrast. Higher resolutions were obtained by scanning electron microscopy.

3. ANALYSIS OF THE THERMAL DYNAMICS

Consider the case of a thin homogeneous iron plate, of uniform thickness, subjected to the thermal action of a localised heat source of constant characteristics, endowed with a regular rectilinear movement. In these conditions, when the instantaneous heat distribution in the base metal is identical at all times, the isotherms around the melt pool take the form of oblong loops, close together in the frontal zone, but more widely spaced towards the rear.

The following isotherms will be considered in particular: $\alpha \rightleftarrows \gamma$ [911°C], $\gamma \rightleftarrows \delta$ [1392°C], $\delta \rightleftarrows L$ [1538°C], the latter defining the melt pool.

As the heat source is displaced, the translation of each isothermal loop generates in the metal a band of influence whose width is limited by that of the loop, the narrowest band being that of the fusion line (or weld seam) limited by the transition zone, locus of the singular points A and B (fig. 2).

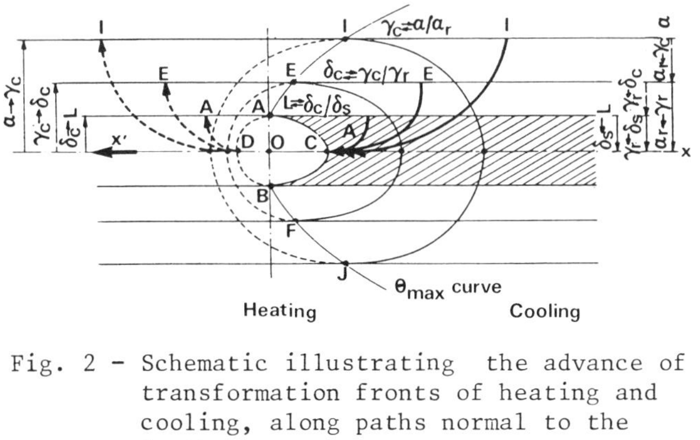

Fig. 2 - Schematic illustrating the advance of transformation fronts of heating and cooling, along paths normal to the isotherms.

In effect, these points represent the junction of the two branches of the $\delta \rightleftarrows L$ [1538°C] isotherm, the forward one corresponding to the fusion front ADB, where the $\delta \rightarrow L$ transformation takes place under dynamic heating conditions and the rear one corresponding to the solidification front ACB at which the reverse transformation $L \rightarrow \delta$ occurs under dynamic cooling conditions.

Finally, it is important to emphasize that the symmetrical points A and B occupy the most advanced positions on the solidification front. Since heat transport occurs along paths which represent the shortest distance between two successive positions of a given isotherm, the solidified grains begin to grow epitaxially at A and B, on the partially melted crystals of the transition zone, and at right angles to the latter. Their growth then continues along a path which curves progressively inwards in the welding direction, with a solidification rate V_s:

$$V_s = V. \cos \theta_s$$

where θ_s is the angle between the thermal gradient and the axis xx' during cooling. V_s varies along the curve ACB; it is very low at A and B and reaches a maximum, equal to the welding rate V, at C (fig. 3).

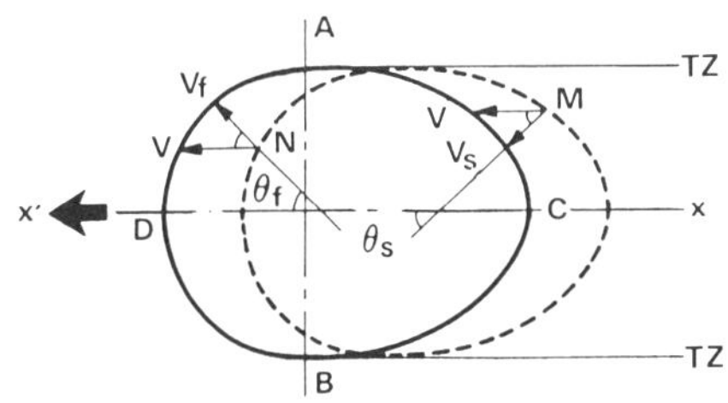

Fig. 3 - Schematic showing the fusion V_f and solidification V_s rates at points N and M on the fusion and solidification fronts, respectively ADB and ACB.

However, the epitaxial nucleation does not necessarily lead to equal chances of growth for all grains. Grains whose preferred growth direction (the [100] direction in the cubic system) is favourably oriented relative to the solification isotherm tend to develop more rapidly (3). This competetive and hence selective, characteristic of growth therefore leads to the appearance of a solidification texture during welding.

The thermal gradient also exerts an effect at the fusion front, where the fusion rate V_f at a point N on the isotherm ADB can be evaluated in a similar manner to V_s, by the relation :

$$V_f = V. \cos \theta_f$$

where θ_f is the angle between the thermal gradient and the axis xx' during heating.

The melting rate V_f is therefore very low at A and B. It tends towards the value V at D, the most advanced point on the melting front. It will be seen that at D, the heat flux, always perpendicular to the front ADB, follows the path xx'. However, because of the shape of the isotherm, for a neighbouring point this path gradually curves away relative to the axis xx' and reaches the transition zone at right angles, at point A or B (fig. 2).

TABLE II

DETAILS OF THE ALLOTROPIC TRANSFORMATIONS IN THE HAZ AND FZ	
ON HEATING	ON COOLING
Initial constituent α . From 911°C, γ_c invades α up to the $\gamma_c \rightleftarrows \alpha$ transition zone . From 1392°C, δ_c invades γ_c up to the $\delta_c \rightleftarrows \gamma_c$ transition zone	. From 1392°C, γ_r totally invades δ_c, then δ_s from $\delta_c \rightleftarrows \gamma_c/\gamma_r$. From 911°C, α_r invades γ_c, then γ_r from $\gamma_c \rightleftarrows \alpha/\alpha_r$ Final constituent α_r
BALANCE OF PHASES FORMED	
γ_c, δ_c (and liquid)	(δ_s), γ_r and α_r

It is possible to extend this analysis to the transformation isotherms $\alpha \rightleftarrows \gamma$ [911°C] and $\gamma \rightleftarrows \delta$ [1392°C], distinguishing:

- in the region of the base metal subjected to heating (subscript c), the $\alpha \rightarrow \gamma_c$ and $\gamma_c \rightarrow \delta_c$ transformation fronts, corresponding respectively to the progression of the γ_c and δ_c phases;

- in the region of the base metal subjected to cooling (subscript r), the $\gamma_c \rightarrow \alpha_r$ and $\delta_c \rightarrow \gamma_r$ transformations, corresponding to the progression of the α_r and γ_r phases.

The metallurgical effects associated with each family of loops occur within a band of influence whose maximum width is limited by the "transition zones" $\alpha \rightleftarrows \gamma$ and $\gamma \rightleftarrows \delta$.

Chronologically, the transformations propagate and succeed one another, during heating then during cooling, in the direction of thermal flow normal to each front. Thus, in the course of a complete transformation cycle $\alpha \rightarrow \gamma_c \rightarrow \alpha_r$ or $\gamma_c \rightarrow \delta_c \rightarrow \gamma_r$, the normal to the isotherm at the most advanced point on the $\alpha \rightarrow \gamma_c$ or $\gamma_c \rightarrow \delta_c$ transformation front is oriented in the welding direction. It then progressively curves inwards at a decreasing rate and finishes perpendicular to the xx' axis, in the $\alpha \rightleftarrows \gamma$ or $\gamma \rightleftarrows \delta$ transition zone. This is then immediately succeeded by the reverse transformation $\gamma_c \rightarrow \alpha$ or $\delta_c \rightarrow \gamma_r$ which progresses along a new inwardly curving perpendicular path, at an increasing speed, gradually approaching the xx' axis and finally merging with it at the end of the cycle. The allotropic transformations therefore begin in the heat affected zone, in the region where the base metal undergoes heating ; they then continue during cooling until they reach the fusion zone (see table II for details of the allotropic transformations in the HAZ or FZ).

REMARKS :

The above analysis of the thermal dynamics has deliberately been placed in an ideal context, where the thermal action of the heat source and the volume of the melt pool always remain constant, and where the L/δ, δ/γ and γ/α interfaces are considered to be isothermal surfaces.

In order to define the real, more complex, situation encountered during welding, it is necessary to make two preliminary remarks, one relative to the conditions under which the melt pool is formed and the other concerning the thermal variations along the L/δ, δ/γ and γ/α fronts.

Firstly, it should be remembered that, whatever the fusion welding technique employed and whatever the mode of operation used, a melt pool is an agitated medium, intense movements and oscillations being caused by both external and internal factors. It is therefore difficult, or even impossible, to keep the melting conditions strictly constant with time. For this reason, the melt pool continuously varies in volume, leading to corresponding fluctuations in the speed of the fusion and solidification fronts. If no sudden modifications of the external and internal factors disturb this situation, a quasi-periodic regime is finally established. One clear manifestation of this phenomenon is the appearance of solidification waves, representing successive positions of the solidification front. In this case, thermal fluctuations of the heat source are known to have a predominant influence (4 to 8) and their amplitude and periodicity can be traced in the waves.

In addition, except at very low welding

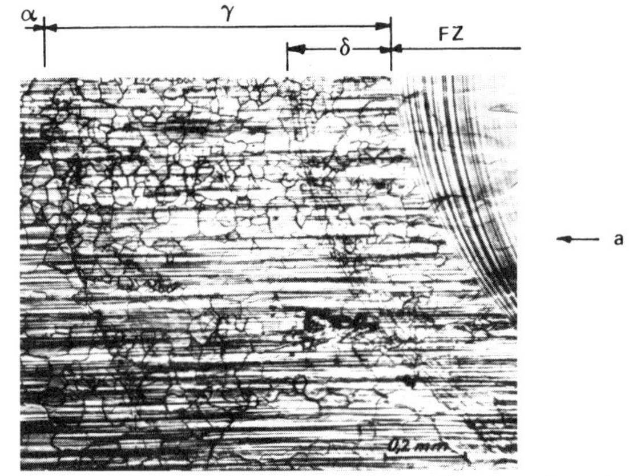

Fig. 4 - γ and δ transformation ripples in the heat affected zone of pure iron. The grain structure is revealed due to thermal etching, obtained by annealing the base metal in hydrogen. As electron beam welded surface.

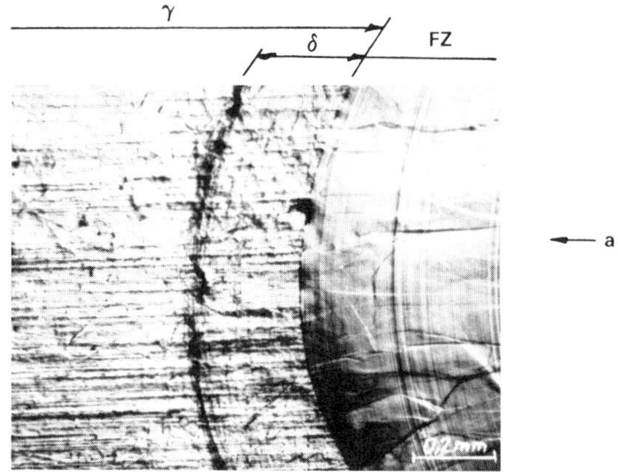

Fig. 5 - Transformation ripples in the δ zone of cold rolled pure iron. The arrow a indicates solidification waves.

speeds, the fronts can not be considered as isotherms, since the degree of superheating varies with solidification rate, while the transformation temperatures differ during heating and cooling, where the rate of heat evolution and the presence of chemical heterogeneities exert a marked influence. Thermal variations, the extent of which depend on the circumstances, can render the loops, which up to now have been considered as isotherms, either partially or even totally non-isothermal. In the real situation, it would be expected that the liquid-solid and solid-solid interfaces would tend rather to be spread over a variable temperature interval, depending on the welding speed and on the location. For a given loop, this interval would be practically zero at the points where the two fronts join and would be greatest at the xx' axis.

4. TRANSFORMATION CONFIGURATIONS IN THE HEAT AFFECTED ZONE

4.1. Preliminary metallographic indications

4.1.1. Pure iron

Metallographic examinations are usually carried out on polished and etched sections. However, a simple surface examination of as-welded samples can provide an additional source of information, the efficiency being increased if the base metal is previously polished or has a smooth surface, welding being carried out by a non-polluting process.

When the surface of a specimen of pure iron with an electron beam fusion line is examined by optical microscopy under oblique lighting, among a variety of relief effects, two sets of of fine ripples (or striations) can be discerned, similar to the solidification waves. The phenomenon is particularly clear at the beginning and at the end of the fusion zone (fig. 4 and 5). The ripples belonging to the first set run over slightly into the transition zone and are heavily marked. Those of the second family cover a wide area and are fainter, except in the peripheral region. These ripples are the traces left on the surfaces by the δ/γ and γ/α transformation fronts respectively, the successive positions of which are revealed by minute variations in relief brought about by thermal etching or by strain, and correspond to the thermal fluctuations during welding (fig. 6, 7, 8). In the case of the γ/α transformation front, the apparent absence of ripples (except at the extremities of the fusion line where the thermal conditions render them more readily detectable) can be explained by a lower temperature and wider transformation bands which reduce the effects of the thermal fluctuations. In the heat affected zone, the two sets of ripples leave in their wake two bands of influence, beginning at the $L \rightleftarrows \delta_c/\delta_s$ transition zone and limited respectively by the transition zones $\delta_c \rightleftarrows \gamma_c/\gamma_r$ and $\gamma_c \rightleftarrows \alpha/\alpha_r$.

At a sufficiently high resolution, each ripple appears to constitute the boundary of a perpendicularly developing oriented grain structure, the front being punctuated by discrete indentations wherever it meets non-metallic inclusions or transformations front-grain boundary junctions (fig. 8, 9) . The changes in curvature of the front at these points give it a characteristic convoluted shape. These observations also provide precious information on time dependent grain boundary migration.

4.1.2. Armco iron

In this case also, two sets of ripples are observed on the surface of specimens containing a fusion line made with the protection of an inert gas atmosphere. Examinations carried out on etched and polished sections reveal even more clearly the transformation bands bordering the fusion line, which can be distinguished due to remarkable final α_r grain structure (fig. 10).

The narrowest band corresponds to the region

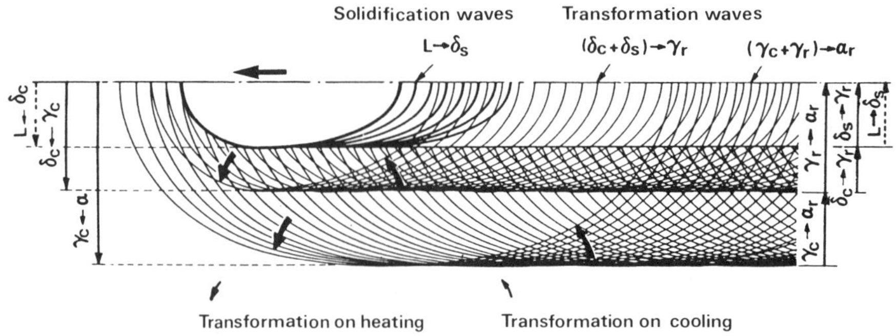

Fig. 6 - Extent and overlapping of transformation zones induced progressively during heating and cooling, in synchrony with the thermal fluctuations of the welding cycle.

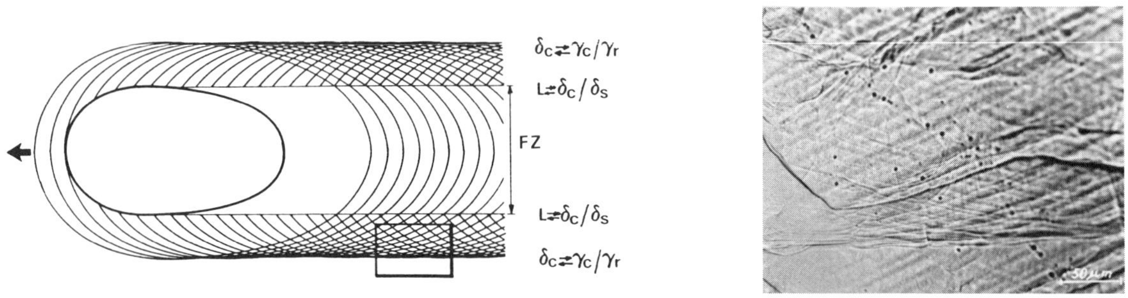

Fig. 7 - $\gamma_c \rightarrow \delta_c$ and $\delta_c \rightarrow \gamma_r$ transformation ripples in the vicinity of the $\delta_c \rightleftarrows \gamma_c / \gamma_r$ transition zone. Pure iron; as electron beam welded surface.

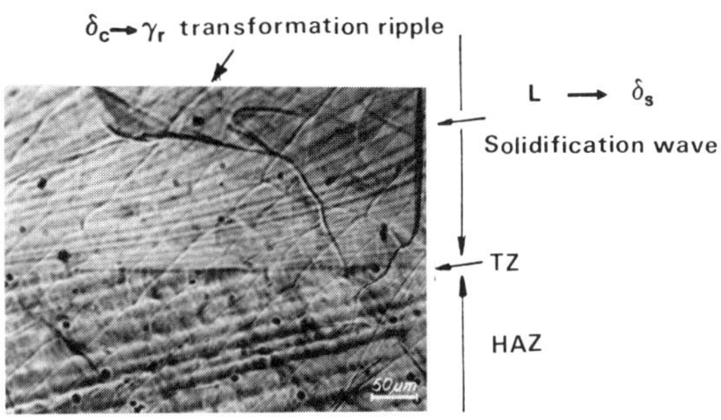

Fig. 8 - $\delta_c/\delta_s \rightarrow \gamma_r$ transformation ripples crossing the transition zone. Pure iron; as electron beam welded surface.

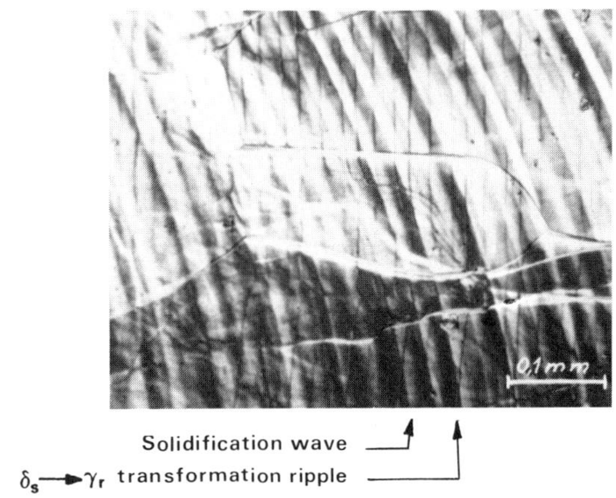

Fig. 9 - Progression of the $\delta_s \rightarrow \gamma_r$ transformation in the fusion zone and discontinuous growth of a γ_r grain in synchrony with the fluctuations of the thermal regime. Pure iron; as electron beam welded surface.

having undergone the δ_c transformation, within a wider band completely occupied by the γ_c transformation . The new α_r grain structure appears on cooling as a result of the respective reverse transformations $\delta_c \rightarrow \gamma_r \rightarrow \alpha_r$ and $\gamma_c \rightarrow \alpha_r$ with, for each phase, a crystal structure oriented in the direction of the thermal gradient growth being normal to the transformation fronts, and a grain structure influenced by the grain dimensions of the disappearing phase. The orientation of grains during the $\gamma \rightarrow \alpha$ transformation has been observed by LEHR (9) in high purity iron placed in a furnace with a longitudinal thermal gradient. It has also been found in the case of fusion lines in Armco iron (7, 8). It will be shown below that a similar orientation of the grain structure is also observed in the heat affected zone during the $\alpha \rightarrow \gamma_c$, $\gamma_c \rightarrow \delta_c$, $\delta_c \rightarrow \gamma_r$ and $\gamma_r \rightarrow \alpha_r$ transformations, due to nucleation and rapid growth in a thermal gradient (fig. 11). As in the case of solidification, this process promotes selective growth with preferred development of the most favourably oriented grains of the new phase. The transition zones $\gamma_c \rightleftarrows \alpha/\alpha_r$ and $\delta_c \rightleftarrows \gamma_c / \gamma_r$, which represent the limiting positions reached by the $\alpha \rightleftarrows \gamma$ and $\gamma \rightleftarrows \delta$ equilibria, are an example of this, distinct γ_r then α_r grain structures being formed by the $\gamma_c \rightarrow \delta_r$ and $[\gamma_c + \gamma_r] \rightarrow \alpha_r$ transformations during cooling.

In the case of pure iron and Armco iron, the nucleation and selective growth process also

Fig. 10 - Montage showing two contiguous areas, given different complementary etches after an initial mechanical polish. On the right, the preparation (I + II + III) reveals the γ and δ transformation zones. On the left, an additional sodium bisulfite etch reveals the oriented α_r structure. The latter appears coarser in the δ_c and δ_s zones.

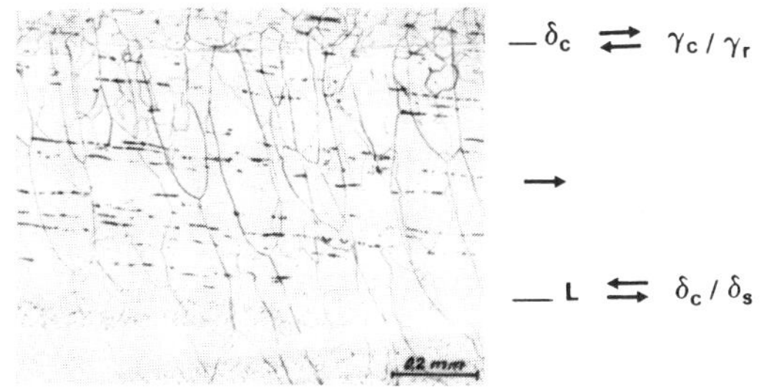

Fig. 11 - On cooling, a coarse oriented γ_r grain structure develops in the former δ_c zone, starting at the $\delta_c \rightleftarrows \gamma_r$ transition zone, in crystallographic prolongation of the γ_c grains partially transformed to δ_c. It then crosses the $L \rightleftarrows \delta_c / \delta_s$ transition zone and penetrates the fusion zone. Armco iron. Metallographic preparation (I + II + III).

occurs during heating. This can be shown by careful microscopical examination of the $\alpha \rightleftarrows \gamma$ transition zone, where the $\alpha \rightarrow \gamma_c$ transformation front appears, at high magnification, to be comprised of a succession of convex lines growing into the α grains, each curved unit belonging to an oblong γ_c cell oriented in the direction of the thermal gradient. The morphologies of the two phases do not seem to be related (fig. 12). The transformations appear to occur without any cristallographic relationship between the α and γ phases, as was observed by LEHR (9), and by FAIVRE and DONZE (10) in the case of pure iron. In thermal gradient conditions each phase grows with its own preferred orientation at the expense of the regressing constituent. It will be shown that a similar process also occurs during the $\gamma_c \rightarrow \delta_c \rightarrow \gamma_r$ transformations.

In contrast to high purity iron, the fine surface ripples in the prior δ region are only faintly visible in Armco iron. Only the $\delta_c \rightleftarrows \gamma_c / \gamma_r$ transition zone and a few points corresponding to halts or to decelerations of the transformation front during cooling are marked by a striation which fades out when it crosses the $L \rightleftarrows \delta_c / \delta_s$ transition zone . The trace of the transformation fronts (or striations) is particularly visible in polished and chemically etched metallographic sections. One especially noteworthy detail is that the $\gamma_c \rightarrow \delta_c$ heating transformation front sometimes has an irregular outline ahead of the melt pool, at islands of chemical segregation where the γ_c phase is stabilised sufficiently for the δ_c transformation to go round them as it advances. These islands are eliminated when the temperature increases locally as the fusion front approaches. Fig. 13 illustrates the spread of transformations within an interval of temperature in an area containing inclusions or chemical heterogeneities. These examinations shows, furthermore, that the outline of the prior γ grains, particularly in the zone affected by the δ_c transformation (where the structure is coarser) and in the fusion zone, is fairly clearly revealed, probably due to a sufficiently high concentration of impurities and physical defects in the grain boundaries. The situation is different in the case of the prior δ_c grain structure which is difficult to reveal, the limited stability at high temperature (1538-1392°C) and the short dwell time not being conducive to clear marking of the prior δ_c grains under the operating conditions employed.

If, on the contrary, the base metal is in a less stable physical chemical state than that of the equilibrium solid structure, it is possible to promote a reaction in the heat affected zone which enables the prior δ_c grain structure to be revealed more readily. As solidified structures (for example a cellular type structure) generally fulfill this requirement quite well, particularly since they contain fine segregation and a high density of physical defects.

4.2. Transformations in the thermally reaffected fusion zone

The technique consists in thermally affecting a first bead by the passage of a second one, as in multipass welding , but with a pass configuration better suited to the observation of the desired phenomena (fig. 1).

When examined after mechanical polishing and chemical etching, the thermally reaffected solidification structure appears to be strongly marked by the allotropic transformations. In particular, a dense network of striations entirely covers the $\gamma_c \rightarrow \delta_c$ transformation zone. Contrary to observations on annealed Armco iron, these striations, judging by their curvature, mark the successive positions of the $\gamma_c \rightarrow \delta_c$ transformation front on heating. After the passage of the melt pool, all that remains in a series of oblique rings within a band situated between the $L \rightleftarrows \delta_c / \delta_s$ and $\delta_c \rightleftarrows \gamma_c / \gamma_r$ transition zones on either sides of the fusion lines (Fig. 14 to 18).

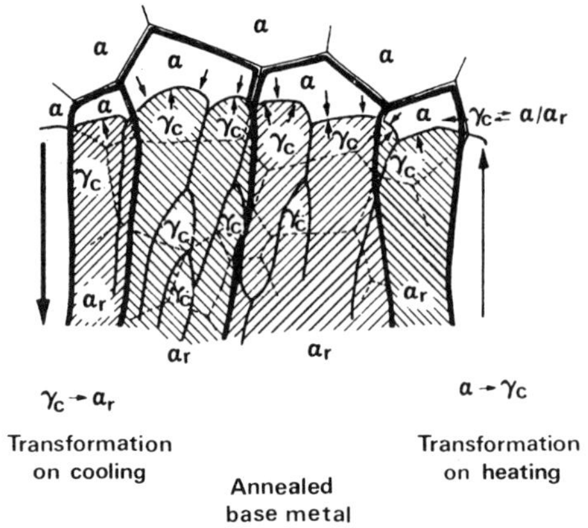

Fig. 12 - Constitution of the $\gamma_c \rightleftarrows \alpha/\alpha_r$ transition zone. On cooling, the α grains partially transformed to γ_c serve as nucleation sites for the new oriented α_r structures, which develops in crystallographic prolongation of the α grains, without coinciding with the prior γ_c grains.

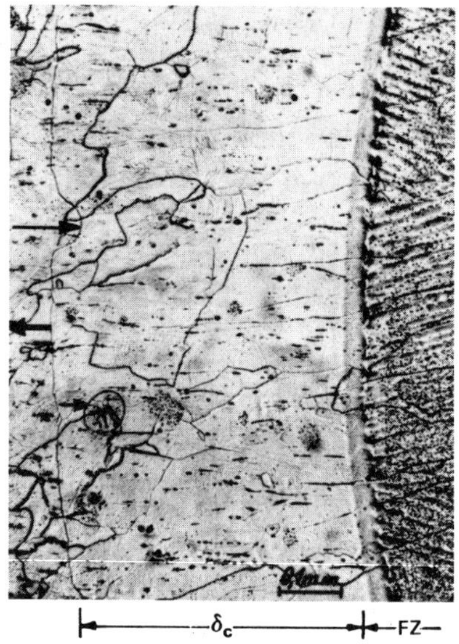

Fig. 13 - $\delta_c \to \gamma_c$ transformation front looping round islands of chemical heterogeneities during heating. Armco iron. Metallographic preparation (I + II + III).

Fig. 14 - Transformation in the thermally reaffected fusion zone at the edge of the terminal depression of a fusion line. Armco iron. Configuration B of fig. 1. Metallographic preparation (I + II + III).

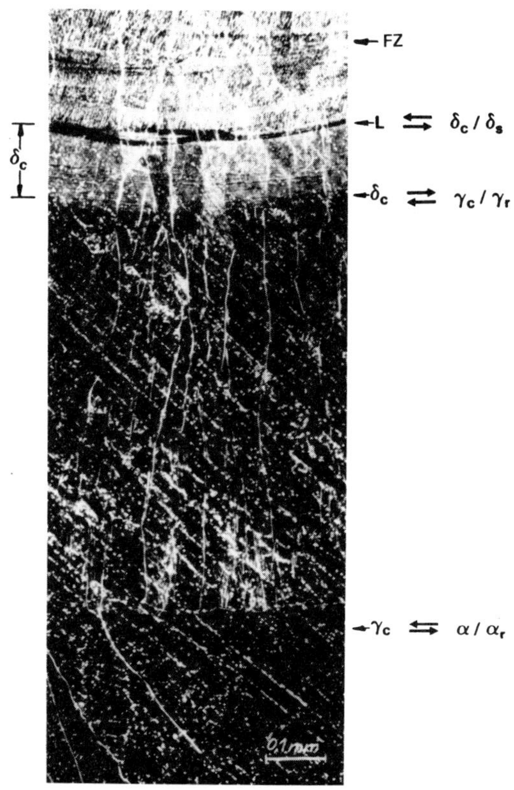

Fig. 15 - Detail of the micrograph in fig. 18 a. The presence of a network of striations in δ_c is clearly visible, and the $\gamma_c \rightleftarrows \alpha/\alpha_r$ transition zone can be seen in a background of microsegregation at solidification cell boundaries. Dark field micrograph.

At high magnification, each striation is seen to be comprised of a groove with a row of micropits, revealing the presence of persistent physical and chemical heterogeneities which have been left behind and which have been preferentially attacked by the metallographic etchant. This aspect is similar to the

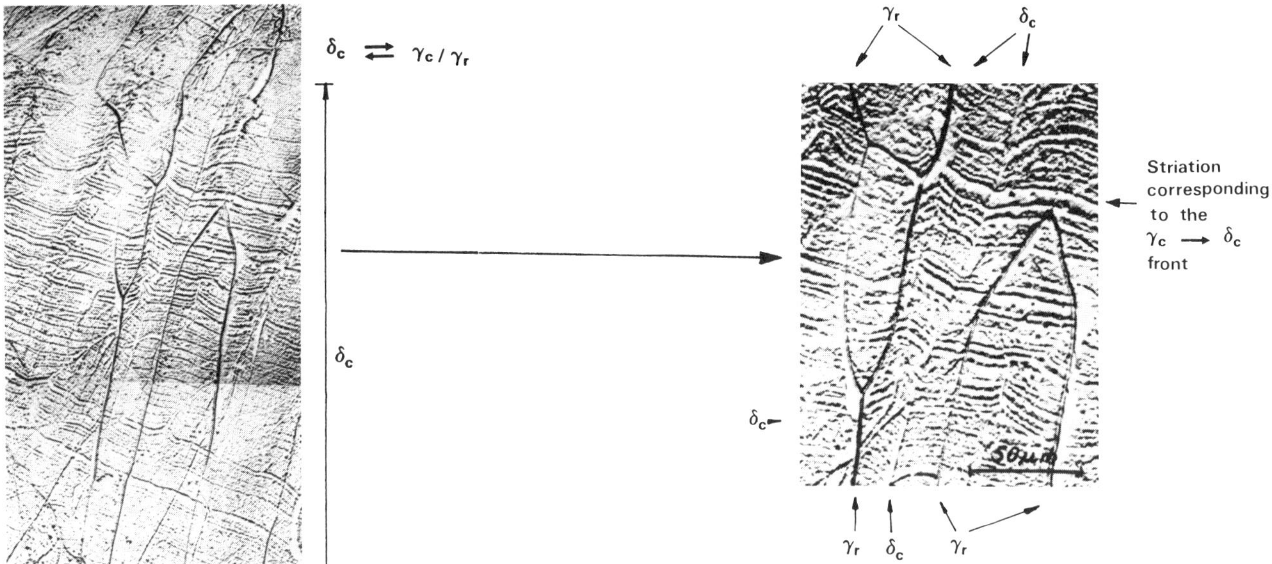

Fig. 16 - Striations formed in synchrony with the thermal fluctuations, revealing successive positions of the $\gamma_c \rightarrow \delta_c$ transformation front and enabling reconstitution of the oriented δ_c grain structure at the time of its formation from the junctions between the δ_c / γ_c interface and the δ_c grain boundaries. The γ_r grain structure forms on cooling, with its own kinetics, starting from the $\delta_c \rightleftarrows \gamma_c/\gamma_r$ transition zone. The deflection of a γ_r grain on crossing the γ_c / γ_r boundary, under the influence of the thermal gradient, can be seen, together with evidence of migration of some γ_r grain boundaries. Intercellular or interdentritic microsegregation is considerably reduced in the δ_c zone. Thermally reaffected fusion zone. Armco iron. Fusion lines in configuration A of Fig. 1. Metallographic preparation (I + II + III).

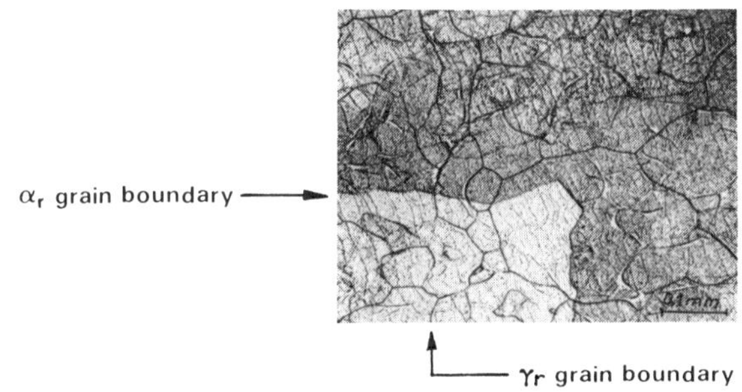

Fig. 17 - Longitudinal section through the δ_c zone. The $\gamma_c \rightarrow \delta_c$ striations, the γ_r grain structure and an α_r grain boundary are particularly clear. Armco iron. Thermally reaffected fusion zone. Metallographic preparation (I + II + III), completed by a nital etch.

observations reported by COULOMB (11) for pure iron, where the stages of the growth of α and γ grains were revealed by networks of impurities and imperfections which persisted in the metal after various heat treatments and which mark the sites of former grain boundaries. These networks exert an impeding action and retard grain boundary movement, making it discontinuous. Furthermore, EICHEN and DE JONG (12, 13) directly observed the discontinuous displacement of the $\gamma \rightarrow \alpha$ interface, using electron emission microscopy, also in pure iron. The phenomenon occurs whenever the movement of the interface is empeded by an obstacle, be it physical (e.g. a vacancy or a dislocation) or chemical. In the present case, the obstacles are clearly governed by the thermal fluctuations during welding whose consequences trigger two processes :

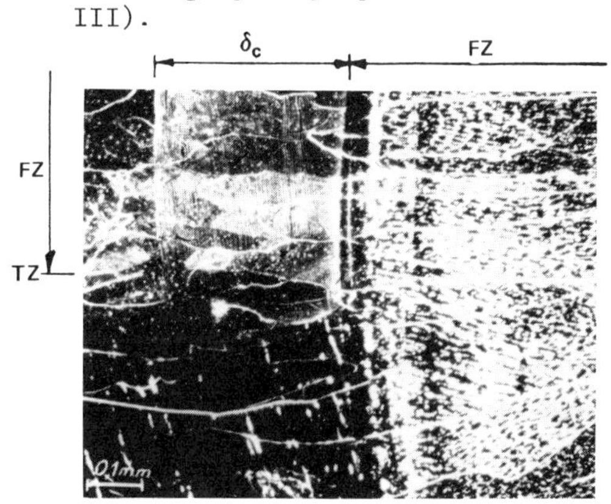

Fig. 18 - Illustration of the greater activation of physical chemical processes by the $\gamma_c \rightarrow \delta_c$ transformation in the thermally reaffected fusion zone. Fusion lines in configuration C of fig. 1. Armco iron. Metallographic preparation (I + II + III).

- collection and drainage of heterogeities along the moving transformation front;

- concentration and deposition of heterogeneities at halts and decelerations of the front.

In addition to variations in volume, the phase changes in iron are accompanied by a re-distribution of the impurities as the front passes, and this phenomenon is the basis of a process proposed by AILLOUD and LANGERON for purifying iron by displacing a zone of γ phase along a bar of α iron (14,15). However, because of unfavourable diffusion rates in the γ phase, the efficiency of the process is poor in the $\alpha \rightleftarrows \gamma$ transformation range, theoretically, the partition of foreign atoms should be more efficient during the $\gamma \rightleftarrows \delta$ transformation, for which diffusion is easier in the b.c.c. δ phase at high temperature. This analysis seems to be confirmed in the heat affected zone in the narrow δ_c transformation band. The high temperature of the $\gamma_c \rightarrow \delta_c$ transformation and the small temperature difference between the $\delta_c \rightarrow L$ fusion front and that of the $\gamma_c \rightarrow \delta_c$ transformation are factors which favour a stronger activation of the diffusion processes - as is confirmed by the tendency for the inter-cellular microsegregations to be taken back into solution (fig. 16). They also enhance physical chemical changes, since a network of "trans-formation waves" is formed, similar to the solidification waves, corresponding to the thermal fluctuations of the welding cycle. In spite of the abrupt changes in the properties of iron on crossing a transformation point, the distance from the $\alpha \rightarrow \gamma_c$ front explains the absence of visible striations, due to the reduced activity of the processes described above.

In the search for indications of former grain structures, every detail has its importance. In this respect, the trace of the $\gamma_c \rightarrow \delta_c$ transformation front contains in-formation which enables the δ_c grain structure to be reconstituted. In effect, on a metallographic section, each striation is found to be comprised of a more or less continuous line with multiple breaks or indentations at junctions between the γ_c / δ_c interface and δ_c grain boundaries (fig. 16). Between two indentations the curvature of the front is slightly convex in the direction of advance, as in pure iron. From one striation to another, the sequence of indentations reveals the position occupied by the boundaries and enables the initial form of the δ_c grains to be determined. In this way, it can be seen that the δ_c grain structure has developed at the expense of the γ_c phase, at right angles to the transformation front, by a selective growth process. The $\gamma_c \rightarrow \delta_c$ transformation zone is therefore composed of a δ_c grain structure elongated in the direction of the thermal gradient, up to the $\delta_c \rightleftarrows \gamma_c / \gamma_r$ transition zone. As is the case for the $\alpha \rightarrow \gamma_c$ transformation which precedes it, the $\gamma_c \rightarrow \delta_c$ transformation occurs in the region of base metal subjected to heating.

While the oriented growth of the δ_c grains is continuing to completion the gradual increase in temperature in the neighbourhood of the future $L \rightleftarrows \delta_c / \delta_s$ transition zone as the fusion zone approaches induces local re-crystallisation of the δ_c grains, from which, at the junctions A and B between the fusion and solidification fronts, a coarse δ_s solidification structure will grow epitaxially in the fusion zone during cooling.

The growth of the new γ_r grains in the δ_c zone also occurs epitaxially, beginning slowly along the $\delta_c \rightleftarrows \gamma_c / \gamma_r$ line from the limiting γ_c grains, partially transformed to δ_c, with a new preferred orientation, normal to the $\delta_c \rightarrow \gamma_r$ transformation front. This explains the presence of "breaks" in the grains where they cross the $\delta_c \rightleftarrows \gamma_c / \gamma_r$ line (fig. 16). A comparable situation occurs on the $\gamma_c \rightarrow \alpha_r$ transformation front during cooling whether the base metal is in the as-welded, annealed or cold-worked condition. In the latter case, the α recrystallisation, which takes place at a lower temperature (<911°C), precedes the $\gamma_c \rightarrow \alpha_r$ transformation. It is important to emphasize here the morphology of the recrystallised α grains which, in spite of the thermal gradient, conserve a typically equiaxed form up to the $\gamma_c \rightleftarrows \alpha/\alpha_r$ transformation front, where the oriented α_r growth initiates from the limiting α grains, partially transformed to γ_c (fig. 12 and 19). This relay plays an important part in determining the relationship between the sizes of the old and new grain structures.

In the light of this experience, a detailed analysis of the structures observed in as-welded or mechanically polished and chemically etched specimens shows that the transformation processes in both pure and Armco irons are

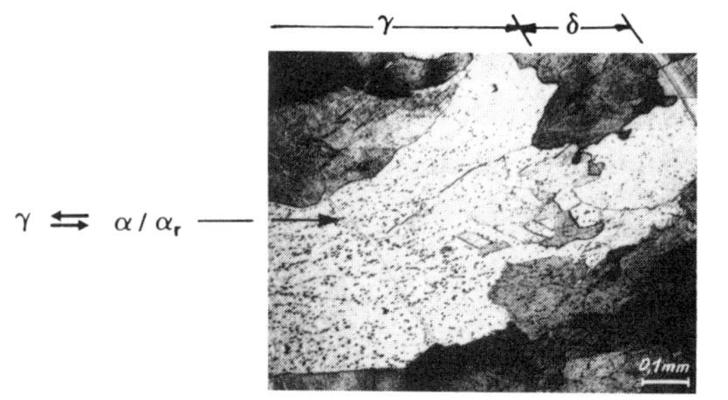

Fig. 19 - Oriented growth of the new α_r grain structure on cooling, in prolongation of the α grains partially transformed to δ_c on heating. Armco iron. Thermally reaffected fusion zone. Section transverse to the fusion line. Metallographic preparation (I + II + III).

comparable. During both heating and cooling, the growing phase develops its grain structure at the expense of the regressing component, in a direction at right angles to the transformation fronts .

Nevertheless, a significant difference exists in the physycal chemical activities at the γ/δ front. In pure iron, this activity appears to be equally great at both the $\gamma_c \rightarrow \delta_c$ and $\delta_c \rightarrow \gamma_r$ fronts. In the case of annealed Armco iron, the activity is greater at the $\delta_c \rightarrow \gamma_r$ transformation front. In contrast, when the base metal is in a metastable condition, the process of draining and concentration of heterogeneities is more active at the $\gamma_c \rightarrow \delta_c$ interface, on heating.

It is obvious that the successive transformations are unable to remove all traces of preceding transformations. In effect, the heat affected zone "memorises" the different stages, particularly the $\gamma_c \rightarrow \delta_c \rightarrow \gamma_r$ transformations in the former δ zone. The persistence of intergranular heterogeneities makes the contours of the γ grains easier to reveal.

4.3. Remarks concerning the heat affected zone of steels

Whereas oriented grains are detected in the heat affected zone of pure and Armco irons, in the case of steel the ferritic and austenitic grain structures do not show this feature. This behaviour could be attributed to the chronological disorder of the transformations due to the multiphase constitution of steels, impeding the oriented growth process, numerous nucleation sites being randomly activated in a variable interval of temperature and time. According to RASANEN and TENKULA (16), the growth of austenite grains in the neighbourhood of the transition zone of steels results from strain induced recrystallisation due to a reverse martensitic transformation on heating through the α ferrite $\rightarrow$ austenite transition. The transformation to δ ferrite would tend rather to inhibit growth due to the formation of intergranular films of liquid. In contrast, FEDOSEEV (17), simulating a thermal cycle during slow welding of a low alloy Cr-Mo steel, found the coarse austenite grains to be in the primary zone.

Grain growth in the heat affected zone is always observed to a greater or lesser extent, whether there is an allotropic transformation or not. It is related to the welding energy via the thermal kinetics and the high temperature dwell time. The phenomenon is amplified by any local strain, either previously existing or produced by phase changes during welding.

Austenite grain growth can thus take place with or without a δ transformation in the sequences :

- α ferrite $\rightarrow$ austenite $\rightarrow$ transformation of the austenite on cooling;
- α ferrite $\rightarrow$ austenite $\rightarrow \delta$ ferrite $\rightarrow$ austenite $\rightarrow$ transformation of the austenite on cooling.

It will be noticed, however, that in the first case, the austenite forms on heating. Grain growth therefore occurs during heating and can continue during cooling until the local temperature attains the austenite transformation point. In the second case, an additionnal austenite grain structure is formed during cooling, due to the δ ferrite $\rightarrow$ austenite transformation. Recrystallisation would be expected to be more strongly activated, due to the stimulating effect of the δ ferrite transformation, as in Armco iron, with a redistribution or concentration of chemical heterogeneities in the neighbourhood of the transition zone at very high temperatures, particularly at the boundaries of the austenite grain structure.

5. δ_S SOLIDIFICATION

Solidification takes place during cooling in the wake of the melt pool, initiating at the solidification front. It occurs in a band the width of which is determined by the breadth of the fusion front, whose extremities A and B trace out the $L \rightleftarrows \delta_c/\delta_s$ transition zones. The points A and B also represent the most advanced positions of the solidification front, where atoms in the liquid transfer to the solid support and initiate the very slow epitaxial growth of the new phase perpendicular to the transition zone. Furthermore, heat flow dynamics lead to a competitive situation, grains having their [100] crystal directions oriented parallel to the thermal gradient growing more rapidly than the others.

In the operating conditions chosen the δ_S solidification structure remains predominantly cellular except for a narrow planar band bordering the transition zone, formed at a very low solidification rate (fig. 13). Its grain structure can be distinguished due to the different orientation of cell blocks in neighbouring grains, intergranular segregation being insufficient to reveal the grain boundaries clearly. The δ_s grains can be distinguished by their greater width compared to the narrow δ_c grains formed in the heat affected zone by the $\gamma_c \rightarrow \delta_c$ transformation before the onset of overheating. This can not be attributed to selective growth, since the latter occurs far away from the $L \rightleftarrows \delta_c / \delta_s$ transition zone, beyond the planar zone. It must therefore be admitted that rapid solidification has occurred in a steep thermal gradient in the narrow δ_c band, on the approach of the fusion zone, before the epitaxial growth.

Experience has shown that the most favourable conditions for revealing the primary δ_s grains are obtained by promoting rapid solidification: rapid microspot melting or high speed (100 cm/min.) micro line fusion traverses, affecting only an extremely thin surface layer

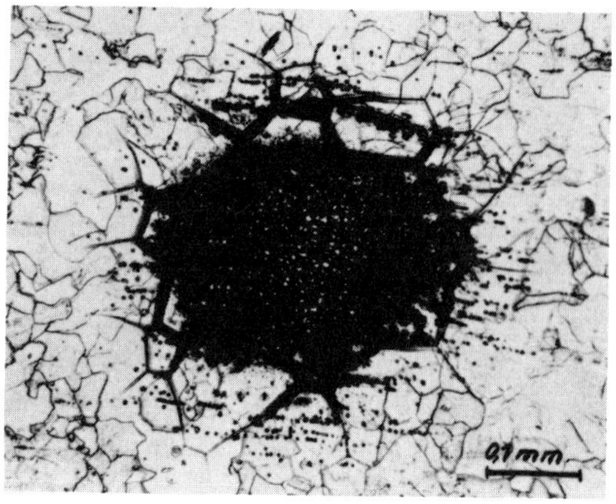

Fig. 20 - GTA micro-fusion spot in Armco iron. The polygonal δ grain structure is particularly clear in the precellular solidification zone. Cellular-dendritic transition in the centre. Metallographic preparation (I + II + III).

(fig. 20 to 22). In both operating conditions, the prior δ_s grain structure remains inconspicuous in the precellular zone, some distance away from the $L \rightleftarrows \delta_c / \delta_s$ transition zone, but becomes clearer in the cellular zone as the solidification rate increases.

In addition, the microline fusion technique provides a method for studying:

- the basic processes during the formation of a fusion line;

- the movements of the fusion and solidification fronts;

- the formation of solidification and transformation waves (fig. 21 and 22).

The technique is most successful when the pulsations of the thermal regime can be closely controlled. This is the case for the rectified current G.T.A. welding process used in the present study.

Under these conditions, the high speed employed makes it possible to decompose the fusion line into its constitutive components. In effects, the line appears to be made up of a series of crescents, a configuration which can be explained by a series of circular fusion spots, each overlapping the already solidified preceding one (fig. 21 and 22). Thus, each thermal pulse gives rise to a fusion spot which develops rapidly during heating, attaining a limiting dimension determined by thermal equilibria. When the fusion front ceases to advance, it then becomes a solification front which propagates in the reverse direction from the periphery to the centre of the spot. The partial overlap of successive spots leaves only a crescent-shaped part of the preceding spot

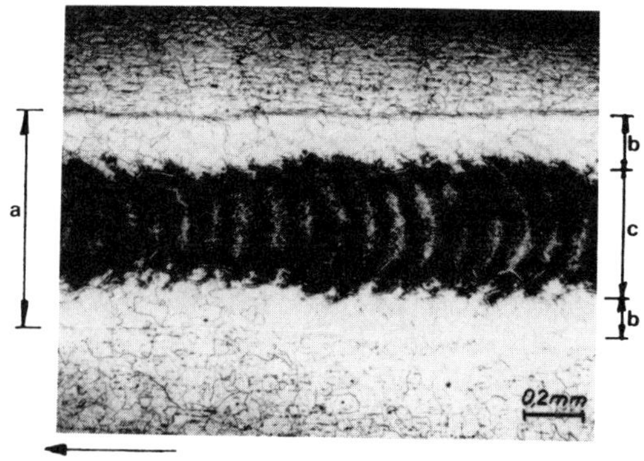

Fig. 21 - High speed (100 cm/min.) micro-fusion line on Armco iron. The thermal fluctuations of the heat source lead to periodic variations in the advance of the solidification front, the surface solidification waves being associated with structural variations in the underlying material. Time between successive waves: ∿ 6 ms. Metallographic preparation (I + II + III).
a : fusion zone; b : planar solidification; c : planar-cellular alternation.

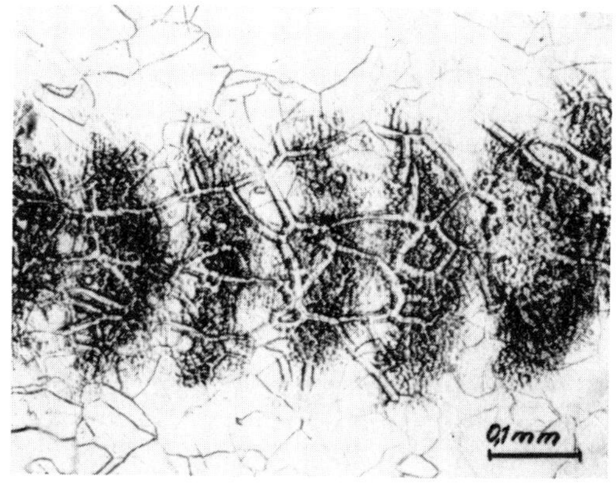

Fig. 22 - Micro-fusion line on Armco iron. Details of the planar-cellular alternations, showing a network of δ_s grains.

unmelted. The boundary separating two successive crescents constitutes a genuine transition zone where the epitaxial transition process occurs on contact with a thermally reaffected solidification structure. Since any irregularity in the advance of the liquid-solid interface leaves a trace in the solidification structure, the inversion of the front, which corresponds to a sharp disturbance in the thermal regime, marks the solidification front naturally, either in the form of a wave, on the surface, or as a line of heteregeneous structure within the metal. The waves become more accentuated and more closely spaced as the welding speed decreases. In the present case, their formation is due to a factor

TABLE III

POSSIBLE SUCCESSIVE MOVEMENTS OF THE SOLIDIFICATION FRONT	VARIATION OF THE EFFECTIVE SOLIDIFICATION RATE V_e
DECELERATION { . Increase of pool volume or . Flow of superhead liquid towards solidification front ↓ ACCELERATION { . Decrease of pool volume or . Reverse flow of liquid	$0 < V_e < V$ ↓ $V_e > V$
DECELERATION ↓ ARREST ↓ ACCELERATION	$0 < V_e < V$ ↓ $V_e = 0$ ↓ $V_e > V$
DECELERATION ↓ ARREST ↓ REVERSAL { . Abrupt increase in melt pool volume ↓ ARREST ↓ ACCELERATION	$0 < V_e < V$ ↓ $V_e = 0$ ↓ $(V_e < 0)$ { . Solidification front changes to fusion front ↓ $V_e = 0$ ↓ $V_e > V$

originating outside the melt pool, since the transformation rings remain clearly associated with the variations in the pulsating thermal regime. Random fluctuations probably have the same effect, provided that they occur for a time sufficient to enable the formation of a detectable imprint.

The high speed microline fusion traverse, by accentuating microscopic irregularities in the advance of the solidification front (whatever the liquid-solid interface profile), thus offers the opportunity of more clearly defining the notion of solidification rate.

In effect, the solidification rate V_s represents only an average rate, significant mainly on a macroscopic scale. On a finer scale, the reality is more complex. Table III indicates some examples of the variation of the effective rate V_e as a function of the successive movements of the solidification front at a point on the liquid-solid interface situated on the xx' axis, the welding speed V remaining constant.

Beyond the fusion and solidification fronts, the thermal fluctuations extend their influence to the movement of the $\gamma_c \rightarrow \delta_c \rightarrow \gamma_r$ and $\alpha \rightarrow \gamma_c \rightarrow \alpha_r$ transformation fronts, the intensity of the effects decreasing the distance.

6. $\delta_s \rightarrow \gamma_r$ AND $\gamma_r \rightarrow \alpha_r$ TRANSFORMATIONS IN THE FUSION ZONE

After the start of solidification at points $A_{\delta s}$ and $B_{\delta s}$, the most advanced positions of the $\delta_s \rightarrow \gamma_r$ and $\gamma_r \rightarrow \alpha_r$ cooling transformation fronts successively reach the $L \rightleftarrows \delta_c/\delta_r$ transition zone, at different oblique angles, in times (fig. 23):

$$t_{\gamma r} = A_{\delta s} A_{\delta r}/V = B_{\delta s} B_{\gamma r}/V \text{ and}$$

$$t_{\alpha r} = A_{\delta s} A_{\alpha r}/V = B_{\delta s} B_{\alpha r}/V$$

Since the metallurgical structure was established by epitaxial growth from the $L \rightleftarrows \delta_c/\delta_s$ transition zone, the latter does not constitute a barrier to the propagation of the allotropic transformations in the fusion zone. The transformation structures therefore continue across the transition zone into the fusion zone,

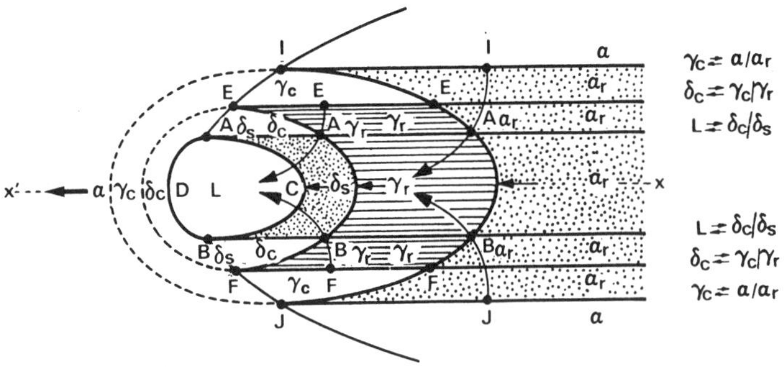

Fig. 23 - Sequence of allotropic transformations under dynamic cooling condition.

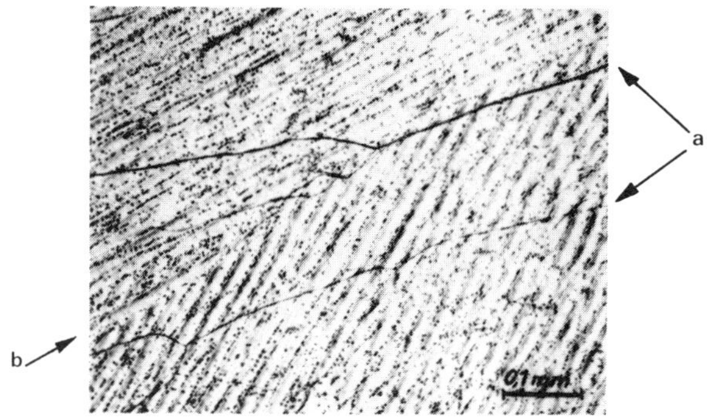

Fig. 24 - Intersection of δ_s and γ_r grain boundaries in the fusion zone of a fusion line on Armco iron. Metallographic preparation (I + II + III). a: γ_r grain boundaries; b: δ_s grain boundary.

advancing in a direction at right angles to each front, with selective oriented growth
Here also, the growth of each phase at the expense of the disappearing constituent develops a new grain structure (fig. 24) . The progression of the transformation fronts remains nevertheless discontinuous, being governed by the thermal fluctuations encountered during welding, whose effects decrease as the transformation takes place at lower temperatures.

In the case of steels, the greater the number of phase changes during cooling, the more numerous are the nucleation sites. As a result, not only is the size of the ferrite grains limited, but their growth in the direction of the thermal gradient is also strongly disturbed.

6.1. $\gamma_s \rightarrow \delta_r$ transformation

6.1.1. Pure iron

The transformation advances along a regular front in synchrony with the thermal fluctuations as is particularly well revealed on the as solidified surfaces (fig. 9 and 25), while behind the front, discontinuous recrystallisation occurs in a thermal gradient, governed by the same thermal pulsations (fig. 26a).

6.1.2. Armco iron

In spite of the rapid attenuation of the striations in the fusion zone, there possibly exists a less regular $\delta_s \rightarrow \gamma_r$ transformation front, as in the heat affected zone (fig. 26b). The transformation is therefore spread over a certain temperature interval $\Delta\Theta_1$.

6.1.3. Steel

The δ ferrite → austenite front becomes very irregular and the transformation by nucleation and growth takes place in a thermal gradient, under non-isothermal conditions (fig. 26c). In effect, the movement of this interface remains linked to the kinetics of the partition of the elements between the two passes, taking account of the local heterogeneities in chemical distribution, established principally during solidification; interdendritic or intergranular segregations or microinclusions, for example. Thus, areas which favour rapid austenite transformation facilitate movement of the front and austenite islands can even form ahead of it.

Other sites momentarily stabilize the front in certain positions and the δ ferrite → austenite interface takes up complicated configurations, with the appearance of numerous, rapidly growing protuberances, together with anfractuosities where the elements which retard the austenite transformation gradually concentrate, on the side nearest to the disappearing phase.

The phenomenon is comparable to the formation of persistent liquid stringers in the interdendritic spaces due to segregation during solidification, as is shown by the traces of certain arrest positions left by the front, revealed by appropriate metallographic etching, which indicate the presence in the fusion zone of a network of physical chemical heterogeneities, subsequent to the δ ferrite → austenite transformation (fig. 27 and 28). The indications concerning the configuration of the δ ferrite → austenite interface, obtained *a posteriori*, show an expected analogy with the α ferrite → austenite interface, observed directly at high temperature in a silicon rich steel by RATHENAU and BASS (18), using emission electron microscopy.

Furthermore, it is possible to estimate the extent of the austenite recrystallisation under the thermal gradient from the intersections of these traces with the final positions of the austenite grain boundaries, revealed by a network of intergranular α ferrite, formed at the beginning of the subsequent austenite→α ferrite transformation.

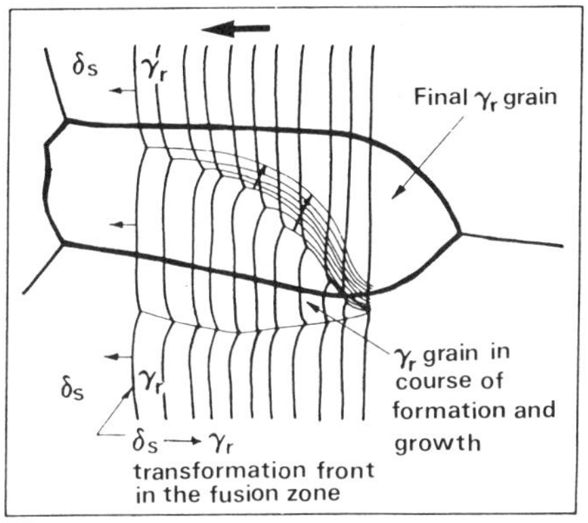

Fig. 25 - $\delta_s \rightarrow \gamma_r$ interface and discontinuous growth of a γ_r grain in the fusion zone of pure iron (see also Fig. 9).

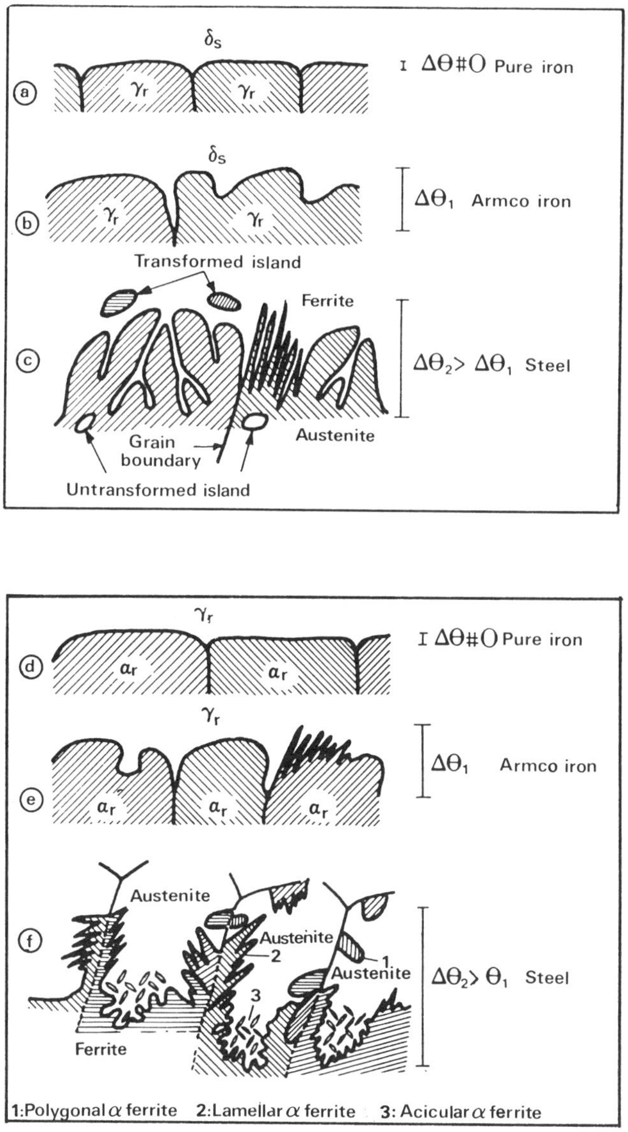

Fig. 26 - Transformation in the fusion zone.

In spite of the complex appearance of the transformation front on a microscopic scale, the heat flow conditions are such that the δ ferrite $\rightarrow$ austenite transformation is spread out within an incurving band, over a temperature interval $\Delta\Theta_2$. The intergranular or interdendritic chemical heterogeneities represent favourable sites for nucleation of the new phase. As for the growth of the austenite grains, it becomes more and more established that their development is determined by the dimensions and distribution of non-metallic microinclusions (19).

The austenite grains do not generally coincide with those of the delta solidification structure. They show a preferred orientation, contrary to the situation in the heat affected zone. This tendency is probably related to the oriented character of the parent solidification structure, developed under selective conditions.

6.2. $\gamma_r \rightarrow \alpha_r$ transformation

6.2.1. Pure iron

The phase change takes place along a less regular transformation front, nucleation ocurring preferentially at γ_r boundaries (fig. 26d). Growth continues over a long distance, producing a coarse, oriented α_r grain structure.

6.2.2. Armco iron

The $\gamma_r \rightarrow \alpha_r$ interface becomes more irregular and the front develops within a temperature interval $\Delta\Theta_1$. The α_r grains still conserve an oriented growth, but over a shorter distance. Nucleation remains preferentially at prior γ_r grain boundaries (fig. 26e).

6.2.3. Steel

The transformation front takes on an extremely irregular configuration, within a wide temperature interval $\Delta\Theta_2$ (fig. 26f). Abundant nucleation, initially intergranular, then intragranular, retards α_r grain growth by impeding their development in the direction of the thermal gradient.

In addition, compared to the austenite grains, the ferrite grain structure differs by its variety of morphologies - polygonal, lamellar or acicular - depending on the chemical composition of the steel, the austenite grain size, the solidification mode, the cooling rate and the inclusion content (fig. 26f)
The systematic definition of ferrite microstructures in the framework of the general identification of austenite decomposition products is presently of growing importance for the evaluation of the mechanical properties of weldments (19, 20, 21).

On completion of the $\gamma_r \rightarrow \alpha_r$ or austenite $\rightarrow \alpha$ ferrite transformations, continued cooling

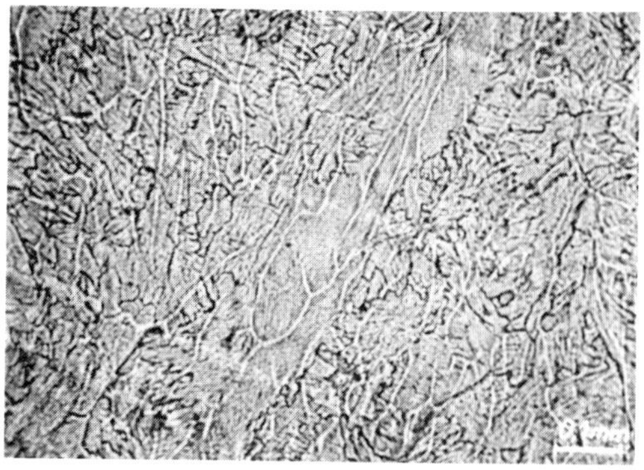

Fig. 27 - Network left by the physical chemical segregations at anfractuosities in the $\delta_s \rightarrow \gamma_r$ transformation front in the fusion zone of slightly carburized Armco iron. Metallogrphic preparation (I + II + III).

leads to some grain boundary migration and to intense polygonization, particularly in the coarse α_r structure.

7. CONCLUSIONS

Complex as they may be, the metallurgical phenomena which occur during welding can be analyzed in detail, the microstructure conserving traces of the previous states through which it has passed. This "memory" effect particularly concerns the sites of concentration of physical or chemical heterogeneities. In this way, the imprints of grain boundaries and former phase interfaces make it possible to follow structural changes and facilitate reconstitution of the various stages involved in the build-up of the solidification and allotropic transformation structures during the welding of iron and steel. Such reconstitution must take account of the dynamic conditions under which the metallurgical phenomena occur during the welding cycle. The search for metallographic indications can be aided by the use of simple but efficient experimental techniques such as the examination of previously polished as-welded surfaces, examination of as-solidified surfaces, appropriate metallographic etching, high speed micro-fusion lines, micro-fusion spots, chemical variations induced by the controlled addition of elements to a melt pool or promotion of a greater physical chemical activity by thermally re-affecting an as-solidified structure.

In particular, it has been possible in this way to show the following:

1) On both heating and cooling, the α/γ and γ/δ polymorphic transformations in pure and Armco iron propagate along fronts, developing new grain structures oriented in the direction of the thermal gradient, due to a selective growth process. Transition zones formed of partially transformed grains delimit the width of the bands swept out by successive transformations in the heat affected zone. The new phases grow

Ⓐ

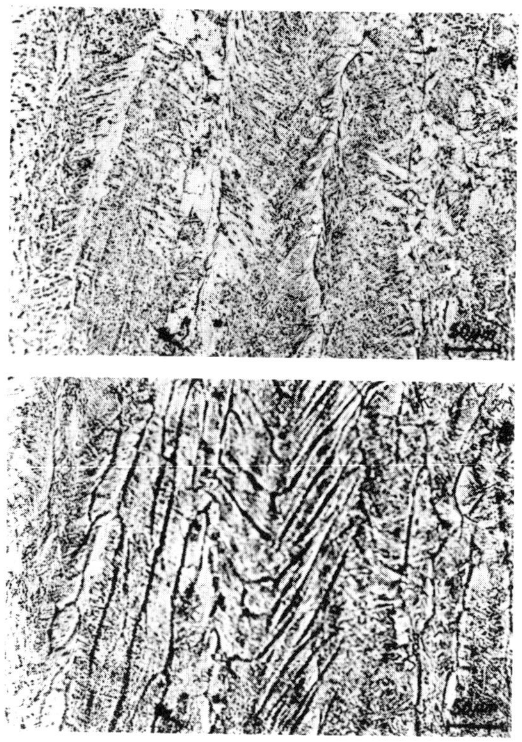

Ⓑ

Fig. 28 A - Microstructure revealed by a picric acid etch in the fusion zone of steel B. The oriented prior austenite grain structure can be seen in its final stage of development. A border of α ferrite outlines these grains.

Fig. 28 B - The same area examined with slight defocussing reveal both the δ ferrite solidification dendrites and the concentration of physical chemical heterogeneities at the anfractuosities of the δ ferrite $\rightarrow$ austenite transformation front.

epitaxially from these transition zones during cooling, in prolongation of the partially transformed supporting grains, with a varying growth rate, as for the formation and growth of the solidification grains from the partially melted crystals in the $\delta \rightleftarrows L$ transition zone.

2) In a base metal which has recrystallized due to a welding cycle, oriented growth is observed only in the transformation and fusion zones. This is explained by an abundant, but random, nucleation within a limited temperature interval which restricts growth of the recrystallized grains.

3) The consecutive development of the different transformation textures, all parallel to the thermal gradient, the selective growth and the progressive increase in the maximum temperature attained, lead to an increase in grain size from one structure to the next. Growth of the δ, γ, then α grains is particularly marked in the vicinity of the $\delta \rightleftarrows L$ transition zone, in the region affected by the δ transformation during heating.

4) In the case of steel, the absence of preferred orientation in the transformed grains of the heat affected zone can be explained by abundant nucleation spread over a wide interval of temperature and time, as for recrystallisation due to the numerous phases which occur.

5) The transformations advance discontinuously in synchrony with the thermal fluctuations which are inevitable in welding, along a front which is quite regular in pure iron, less so in Armco iron and highly irregular in steel. Any deceleration or arrest of the interfaces is marked by a persistent concentration of physical and chemical heterogeneities.

6) The high speed micro-fusion line technique is extremely fruitful for studying the conditions of formation of the melt pool and for analyzing the movements of the solidification front and the fluctuations of the solidification rate.

7) The physical chemical activity - which needs to be studied in detail - seems to be of a similar intensity on heating and cooling at each of the two transformation fronts in pure iron, whereas it appears greater on cooling in the case of annealed Armco iron. On the contrary, it is much higher at the heating transformation front in the thermally reaffected fusion zone of Armco iron. It appears to be especially intense in zones which have undergone two γ transformations or double austenizing, the first during heating and the second on cooling:

$\gamma \rightarrow \delta \rightarrow \gamma$, or austenite → δ ferrite → austenite.

8) Marked grain boundary migration is observed due to the thermomechanical effects of welding on the allotropic transformation structures, particularly in the γ and δ fields and in the austenitic fusion zone. In contrast, the α or coarse α ferrite structures undergo intense polygonisation.

Finally, the present study shows that a detail knowledge of the thermally activated metallurgical phenomena which occur in a temperature gradient in iron can lead to progress in understanding the mechanisms involved and can help to explain the behaviour in regions subjected to the thermal effects of welding in the more complex case of steel.

ACKNOWLEDGEMENTS

This work was carried out with the financial support of the French Ministry of Research and Technology (grant n° 81 S 0644). The author is particularly indebted to Miss H. GOUVEA, of the Portuguese Welding Institute and to Messrs. R. GAILLARD, J. DIJOLS and P. LEMOINE of the French Welding Institute in Paris, for their precious experimental assistance, and to Mr. J. CATTANI, also of the French Welding Institute, for drawing the diagrams.

REFERENCES

[1] P. COULOMB and P. LACOMBE. Revue de Métallurgie, 1958, n° 10, pp. 918-926.
[2] F.A. CALVO, K.P. BENTLEY et R.G. BAKER, B.W.R.A., march 1963.
[3] H. GRANJON and M. DADIAN. Soud. Tec. Conn. 1972, n° 5/6, pp. 181-195.
[4] J.G. GARLAND and G.J. DAVIES. Metal construction, 1970, vol. 2, p. 171.
[5] G.J. DAVIES and J.G. GARLAND. Intern. Met. Reviews, 1975, 20, pp. 83-106.
[6] D.J. KOTECKI, D.L. CHEEVER and D.G. HOWDEN. Welding Journal, july 1972, pp. 386s-391s.
[7] M. DADIAN, C. BOUCHER and H. GRANJON. Rapports Institut de Soudure n° 5464 and n° 5764, august 1973 and april 1975.
[8] M. DADIAN and H. GRANJON. Colloque Métallurgie de la Soudure, Marseille, 8 and 9 november 1973, SFM-SIS. Editions Métaux.
[9] P. LEHR. Thèse, Université de Paris, 22 june 1957.
[10] R. FAIVRE and G. DONZE. C.R. Ac. des Sciences 1957, n° 245, p. 2277.
[11] P. COULOMB. Métaux Corrosion Industries, january 1960, pp. 1-23 and february 1960, pp. 6-82.
[12] E. EICHEN and J.W. SPRETNAK. Trans. A.S.M., 1959, n° 51, pp. 454-475.
[13] M. DE JONG. Thèse, Amsterdam, october 1960.
[14] P. AILLOUD and J.P. LANGERON. Colloque C.N.R.S., Paris, 26/30 september 1966, Mem. Sc. Rev. de Métallurgie, 15 june 1968, pp. 73-84.
[15] P. AILLOUD and J.P. LANGERON. Journal of Nuclear Materials, 1968, n° 25, pp. 69-74.
[16] E. RASANEN and J. TENKULA. Scandinavian Journal of Metallurgy, 1972, n° 1, pp. 75-80.
[17] B.A. FEDOSEEV. Avt. Svarka, 1979, n° 12, pp. 19-21, Aut. Welding, 1979, n° 12, pp. 14-16.
[18] G.W. RATHENAU and G. BASS. Metaux Corrosion Industries, april 1954, pp. 139-150.
[19] R.C. COCHRANE. Doc. IIS/IIW : 713-82 - Soudage et Techniques Connexes, sept./oct. 1983, pp. 334-340.
[20] G.M. EVANS. Doc. IIS/IIW : II A 432-77 Oerlikon-Schweissmitteilungen, january 1978, n° 82, pp. 4-19.
[21] Doc. IIS/IIW : II 899-79.

TRANSFORMATIONS AND NEW PHACOMP ANALYSIS IN WELD METAL OF HASTELLOY C-TYPE ALLOYS

Michael J. Cieslak, Thomas J. Headley, Alton D. Romig
Sandia National Laboratories
Albuquerque, New Mexico 87185

ABSTRACT

The sequence of phase transformations occurring during solidification and cooling of weld metals in the Hastelloy C family (C-4, C-22, C-276) has been determined. An equivalent component chemistry model has been defined for these alloys which allows existing Ni-Cr-Mo ternary phase diagrams to be used to predict the transformation sequence (both liquid → solid and solid → solid) in these complex alloys. In the cases of Alloys C-22 and C-276, the solidification events terminate with the formation of a constituent containing a topologically-close-packed (TCP) phase (**σ** in C-22; **P** in C-276). The New PHACOMP formalism developed by Morinaga et al. for predicting TCP phase precipitation by solid state transformation in superalloys is applied for the first time to measured weld solidification segregation profiles. This formalism correctly predicts the occurrence of terminal solidification TCP phases in Alloys C-22 and C-276 and correctly predicts the absence of such phases as terminal solidification products in Alloy C-4. Implications to analysis of other nickel alloy weld metals are suggested.

HASTELLOYS* C-4, C-22, and C-276 are nickel-base alloys derived from the Ni-Cr-Mo ternary alloy system. They have excellent corrosion resistance in a variety of aqueous media and have good weldability when compared to other nickel alloys such as ALLCORR** or Inconel*** Alloys 625 and 718[1-3].

In the solution-annealed condition, the Hastelloys are single phase materials not hardenable by conventional aging heat treatments. Intermetallic TCP phases and carbides have been observed[4-9] in all of the alloys, the extent of precipitation being dependent upon both the particular alloy chemistry and the specific thermal cycle. Of the three, Alloy C-4 has the greatest thermal stability, experiencing partial transformation to **μ** phase only after greater than 1000 hrs. in the 800°C temperature range[4]. Precipitation of **μ**, **P**, and **M_6C** carbide may occur in Alloys C-22 and C-276 within only a few minutes exposure at elevated (>800°C) temperature[3,8,9].

*Trademark of Cabot Corp. **Trademark of Teledyne Allvac ***Trademark of Huntington Alloys

EXPERIMENTAL

The compositions of the alloys used in this study are given in Table 1. As can be seen, these alloys have low

Table 1. - Alloy Chemistries (Wt.%)

Element	Alloy C-4	Alloy C-22	Alloy C-276
C	0.004	0.006	0.003
Co	0.10	0.84	0.96
Cr	15.69	21.22	15.83
Fe	0.45	3.17	5.44
Mn	0.17	0.31	0.50
Mo	15.06	13.43	15.56
Ni	67.20	56.96	55.58
P	<0.005	0.010	0.014
S	<0.002	<0.002	0.002
Si	0.03	<0.02	0.03
Ti	0.23	0.03	<0.01
V	0.02	0.14	0.18
W	<0.10	3.29	3.93

residual element contents. Alloy C-4 is close to being a Ni-Cr-Mo ternary alloy, having no other elements present at the 0.5 wt% level or greater. Alloys C-22 and C-276 have Fe and W contents in excess of 3 wt.% and close to 1 wt.% Co. All alloys were in the sheet product form (~3mm thick) and in the mill annealed condition.

Welding was performed using the autogenous (no filler metal added) gas-tungsten-arc (GTA) process. All welding was done using direct current (100 ± 5 amps), electrode negative (straight polarity) with a machine voltage of ~13.5 volts at a travel speed of 3.4 mm/s (8 in/min). Welding grade argon was the shielding gas at a flow rate of 0.16 l/s (20 cfh). The welds examined in this study were from Varestraint Test[10,11] specimens (a fusion zone hot cracking test) welded according to the above parameters. The results of the hot-cracking analyses on these alloys are reported elsewhere[1].

In-situ water spray quenching of solidifying welds was used to suppress solid state diffusion and allow for a more accurate determination of the solidification segregation pattern with electron microprobe (EMP) analysis. The liquid → solid transformation sequence was examined with differential thermal analysis (DTA), using a Dupont Thermal Analyzer. Samples were heated under a flowing argon atmosphere to a temperature of 1450°C, where all of the alloys were completely liquid. The samples were then cooled at a rate of 0.17°C/s (10°C/min) through the solidification temperature range. Initial and terminal solidification events were recorded.

Analysis of weld metal was accomplished by using the scanning electron microscope (SEM), EMP (Cameca MBX, 15 kV, 20 nA, $\Phi(\rho,Z)$ data reduction[12]), transmission electron microscope (TEM) (JEOL 200CX), and analytical electron microscope (AEM) (JEOL 100C, 100 kV, equipped with a Tracor Northern TN2000 multichannel analyzer). Phase compositions were calculated from the AEM data using the standardless ratio technique[13]. Sensitivity factors were experimentally determined from a sample of Alloy C-276. Absorption corrections were not made to the data as the thin foil criterion[14] was not exceeded in the areas under analysis. The areas examined were typically 50-100 nm thick.

RESULTS

Figure 1 shows the weld metal microstructures found in the vicinity of solidification hot cracks in the three alloys. In the case of Alloy C-4 (Fig. 1a), which had the best hot cracking resistance, no secondary solidification constituent could be observed with the SEM. In the cases of Alloys C-22 and C-276 (Figs. 1b and 1c), fusion zone hot cracks were intimately associated with terminal solidification constituents which were readily discernible in the SEM. Figure 1d is a representative SEM photomicrograph of the lamellar eutectic-like morphology found along fusion zone hot cracks in Alloys C-22 and C-276.

Figure 2 shows the on-cooling DTA results for the three alloys. Each alloy has a relatively large exothermic peak corresponding to primary austenite (γ) solidification, but only Alloys C-22 and C-276 show evidence of a different terminal solidification constituent. (The labelling of these peaks as σ (Alloy C-22) and **P** (Alloy C-276) will be apparent from the discussion.) This is consistent with the SEM observations in that no secondary constituent was found associated with hot cracks in Alloy C-4. The existence of liquid at lower temperatures in Alloys C-22 and C-276 was supported by longer maximum crack length Varestraint Test data for these two alloys relative to Alloy C-4, as was reported earlier[1].

TEM analysis of weld metal thin foils revealed tthe identity of the minor phases in these alloys. Alloy C-4 had a very low volume fraction of widely dispersed **TiC**. The weld metal of Alloy C-22 contained three intermetallic TCP phases: σ (tetragonal, a = b= 0.908 nm, c = 0.475 nm) **P** (orthorhombic, a = 0.907 nm, b = 1.698 nm, c = 0.475 nm) and μ (hexagonal, a = 0.476 nm, c = 2.591 nm). Of the total minor phase content, ~80% was **P** phase, ~20% was σ phase and only a trace amount of μ phase was observed. Alloy C-276 weld metal contained **P** and μ phases in approximately equal amounts.

TCP phase chemistries obtained using the AEM procedures described previously are listed in Table 2. As can be seen, all of the possible TCP phases found in Alloys C-22 and C-276 are depleted in Ni and Fe relative to the γ matrix. In addition, these same phases are enriched in Mo and W relative to the the γ matrix.

The results of EMP analyses on water-quenched GTA welds from the three alloys are shown in Fig. 3. Care was

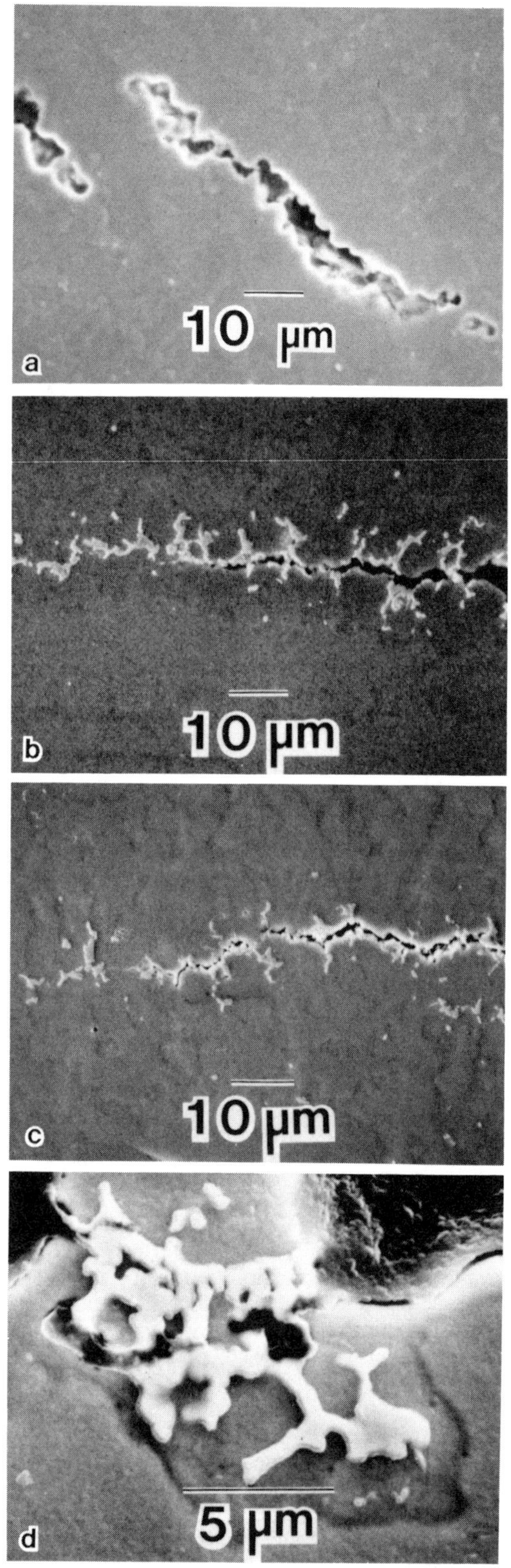

Figure 1. Weld metal microstructures associated with hot cracks in (a) Alloy C-4, (b) Alloy C-22, (c) Alloy C-276, (d) Higher magnification of constituent in Alloy C-22.

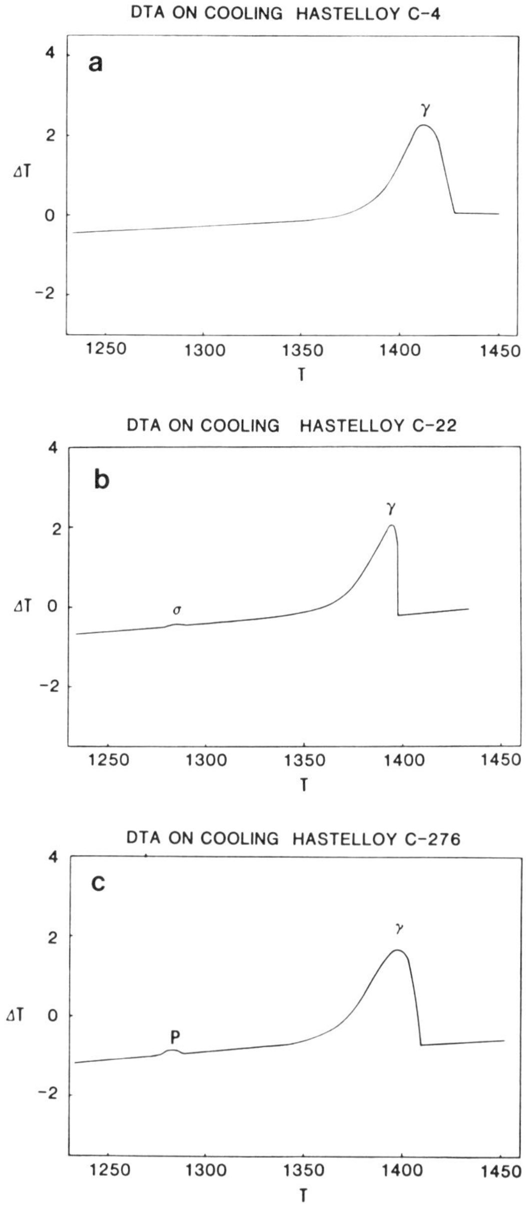

Figure 2. DTA on cooling profiles for the three Hastellc

taken to insure that TCP phases were avoided in the analyses. Each alloy is characterized by Ni depletion and Mo enrichment in interdendritic (ID) volumes. Conversely, the dendrite core (DC) regions are Mo depleted and Ni enriched. There is very little Cr segregation observed in these alloys. In the case of Alloy C-4, Fig. 3b, Ti segregation to interdendritic regions is observed. Amongst considerable scatter, Fe appears to segregate to dendrite core regions in these alloys. The pattern of W segregation was far more difficult to discern against the data scatter and no distinct pattern could be concluded.

DISCUSSION

ANALYSIS OF TRANSFORMATIONS - The sequence of phase transformations leading to the observed weld metal microstructures in these alloys will be explained using available isothermal sections of the Ni-Cr-Mo phase

Table 2. - Phase Chemistries (Wt.%)

C-22

	μ	P	σ	γ
Ni	33.1 (2.4)*	32.6 (0.8)	34.5 (0.6)	58.5 (1.2)
Mo	38.7 (0.2)	37.4 (1.0)	34.9 (0.4)	12.7 (0.8)
Cr	19.3 (0.7)	21.7 (0.7)	23.4 (1.3)	21.6 (0.5)
W	6.3 (1.1)	5.3 (1.1)	4.2 (1.2)	2.9 (0.7)
Fe	2.1 (0.2)	2.2 (0.2)	2.2 (0.1)	3.4 (0.1)
Co	0.6 (0.2)	0.9 (0.4)	0.9 (0.2)	0.9 (0.2)

C-276

	μ	P	γ
Ni	33.2 (0.8)	33.5 (0.6)	57.0 (2.3)
Mo	40.9 (0.6)	39.7 (1.5)	16.0 (1.9)
Cr	15.2 (0.9)	15.7 (0.6)	16.3 (0.4)
W	6.2 (1.6)	6.5 (1.6)	4.2 (0.8)
Fe	3.5 (0.3)	3.7 (0.2)	5.6 (0.3)
Co	1.0 (0.5)	0.8 (0.4)	1.0 (0.3)

* Value in parentheses equals 1 standard deviation, excluding counting statistics.

diagram. Bloom and Grant[15] and Raghavan et al.[16] have developed isothermal sections in this system. A solidus surface has not been experimentally determined, nor has a true liquidus surface. Liquidus isotherms have been determined[15], and a series of high temperature invariant reactions above 1250°C were proposed.

Bloom and Grant report[15] that **γ**, **P**, and **σ** are possible solidification phases in the region of this system relevant to the alloys under study. **γ** forms as the primary solidification phase at high Ni concentrations. As the Cr and Mo content of the ternary alloy increases (especially the Mo content), secondary solidification constituents (**σ**, **P**) are predicted to occur. **σ** is predicted at higher Cr and lower Mo concentrations. **P** is predicted at higher Mo and lower Cr concentrations. In the composition range where these TCP phases would be expected, the demarcation between **P** and **σ** formation is predicted at approximately 20 wt.% Cr. That is, **σ** would form in higher Cr regions, **P** in lower Cr regions.

Figure 4 shows the available ternary sections of the Ni-Cr-Mo system. The points of interest in these diagrams are: (1) all of the TCP phases are stabilized by increasing the Mo and Cr (especially Mo) concentrations; (2) **μ** exists only as a lower temperature (850°C) phase in this system; (3) **σ** is stabilized relative to the other TCP phases at all temperatures by increasing the Cr content at the expense of the Mo content; (4) where **μ** exists, it is stabilized by increasing the Mo content and decreasing the Cr content relative to the other two phases; (5) **P** is intermediate in Cr and Mo content relative to the other TCP phases at a given temperature where these phases are stable.

The average TCP phase chemistry data in Table 2 is consistent with these observations. That is, Mo is enriched and Ni is depleted in the TCP phases relative to the matrix. In addition, the partitioning behavior of Fe is similar to Ni and the partitioning behavior of W is similar to Mo. The consistency with the phase diagrams can be seen in better detail by examining the data from Alloy C-22, in which all of the TCP phases are observed. The average Cr content increases in t he following manner: **μ** < **P** < **σ** as the

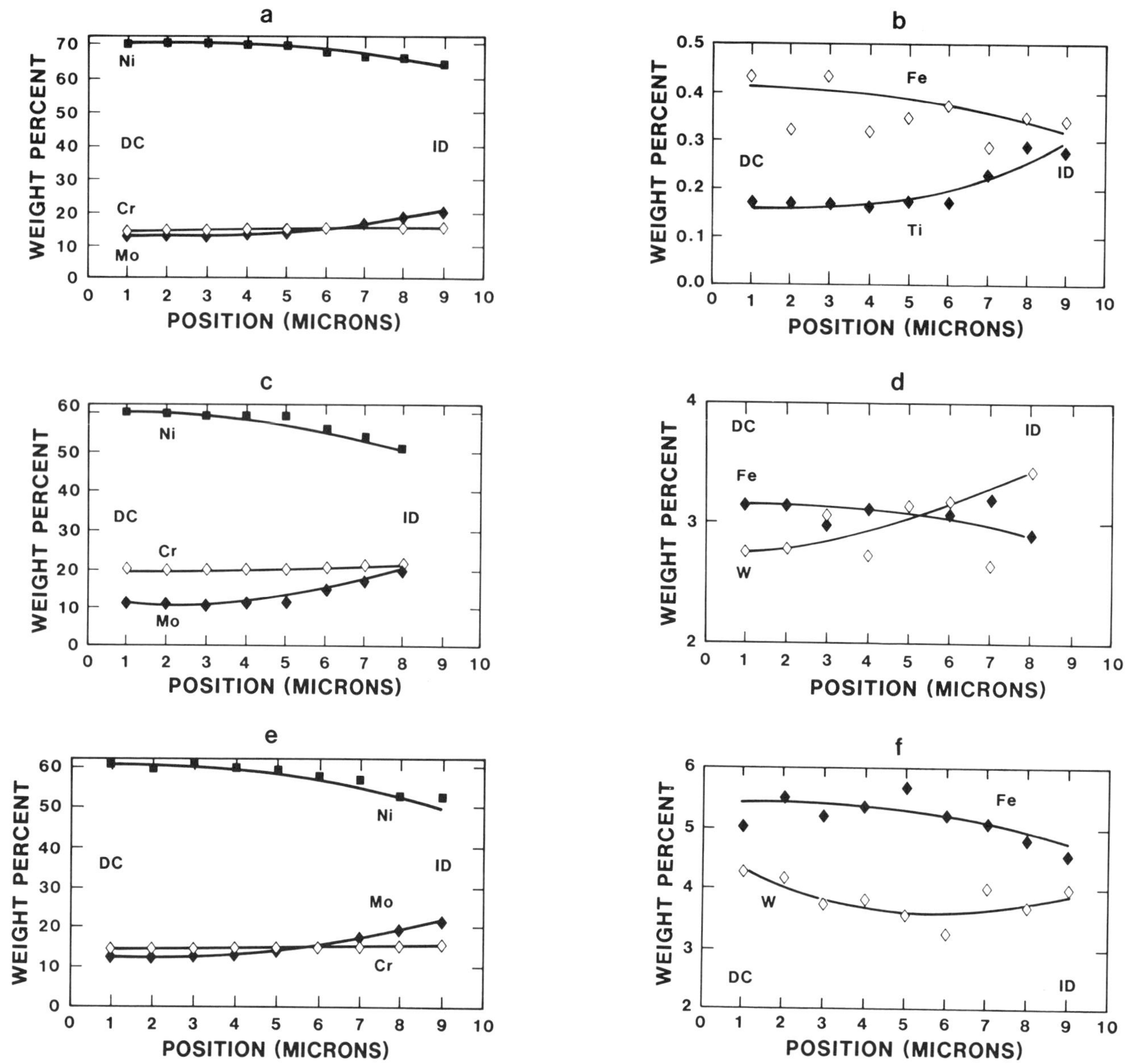

Figure 3. Electron microprobe analyses of water-quenched GTA weld metals on (a,b) Alloy C-4, (c,d) Alloy c-22, (e,f) Alloy C-276. Note Ni depletion and Mo enrichment in interdendritic (ID) regions.

ternary diagrams predict. The average Mo content increases in the following manner: **σ < P < μ**, also in accordance with the ternary phase diagram prediction. In addition, the partitioning behavior of W follows that of Mo in that the W content of the TCP phases increases in the same order as that found for Mo: **σ < P < μ**.

The similarity in partitioning behavior of Ni and Fe and Mo and W suggests an equivalence in chemical behavior among these two sets of elements. These similarities are not altogether surprising in that W and Mo are refractory elements and both have a body-centered-cubic (BCC) crystal structure. Hence, they would both tend to destabilize the face-centered-cubic (FCC) **γ** matrix. In the high temperature range under consideration, both Fe and Ni have FCC crystal structures and would tend to stabilize the **γ** matrix.

There was no apparent partitioning behavior of Co in this study. The low absolute weight fraction of Co present makes a determination of the partitioning behavior of this element difficult. Raghavan[17] investigated the isothermal transformation behavior in a heat of Alloy C-276 containing approximately 2 wt.% Co. Using similar AEM techniques, he measured a lower Co content in the TCP phases he observed (**μ,P**) than in the **γ** matrix, suggesting that Co acts in a manner similar to Ni.

Using the above data correlations, an equivalent composition model is proposed to account for the behavior of the minor alloying elements, relative to phase stability, in these alloys.

The proposed Mo equivalent, Mo_{eq}, is defined as:

$$Mo_{eq} = wt.\%\ Mo + wt.\%\ W\ . \qquad (1)$$

The proposed Cr equivalent, Cr_{eq}, is defined as:

$$Cr_{eq} = wt.\%\ Cr\ . \qquad (2)$$

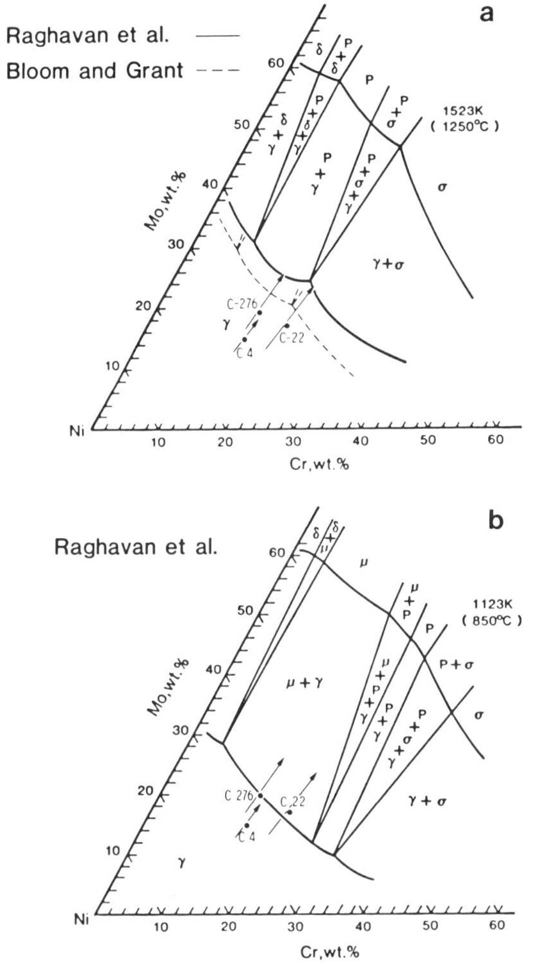

Figure 4. Isothermal sections of the Ni-Cr-Mo phase diagram at (a) 1250°C and (b) 850°C.

The proposed Ni equivalent, Ni_{eq}, is defined as:

$$Ni_{eq} = wt.\%\,Ni + wt.\%\,Fe + wt.\%\,Co + \Sigma\, wt.\%\,X_i \quad (3)$$

where Σ wt.% X_i is the remaining minor alloy content not included in the other equivalent definitions. At the present time, there is no data base for assigning these elements to other than the solvent element equivalent. Doing this also conveniently allows for the sum of the three equivalents to total to 100 wt.%.

If these definitions are applied to the segregation profiles shown in Fig. 3, the solidification paths for each alloy, in equivalent terms, can be projected on the ternary phase diagrams of Fig. 4. These are shown as arrows corresponding to each alloy. The tail of each arrow is the dendrite core equivalent composition and the head of each arrow is the equivalent interdendritic composition for the specific alloy. The bulk equivalent composition of each alloy is shown as the black dot next to its respective alloy designation.

As can be seen, each alloy follows a solidification path toward the regions of TCP phase stability with Alloys C-22 and C-276 having sufficient segregation to enter into these regions. The DTA data of Fig. 2 and the SEM micrographs of Fig. 1 indicate that solidification terminates in Alloys C-22 and C-276 with the formation of a secondary solidification constituent. Bloom and Grant[15] determined a minimum liquidus temperature of ~1275°C in the Ni-Cr-Mo ternary system. The DTA data shown in Fig. 2 indicates a terminal solidification event at ~1285°C for both Alloys C-22 and C-276.

Assuming that the presence of the other alloying elements does not affect the phase diagram much, the 1250°C section will be used as a close approximation of the solidus surface. (The phases present in this section are the same proposed by Bloom and Grant[15] as solidification phases in this region of the ternary system.) This suggests (from the position of the solidification path arrows) that the terminal solidification event in Alloy C-22 is the formation of a constituent containing **σ** phase. In the case of Alloy C-276, solidification terminates with the formation of a constituent containing **P** phase.

Alloy C-4 forms no terminal solidification TCP phase as the extent of alloying element segregation is not sufficient to exceed the maximum solubility in **γ** relative to TCP phase formation during solidification. Instead, it appears likely that solidification terminates in this alloy with the formation of a TiC-containing constituent. Figure 5 is a TEM micrograph of the TiC found in Alloy C-4. This dendritic-appearing morphology is identical to that identified by Lippold[18] and Lambert[19] as euetctic TiC in Alloy 800.

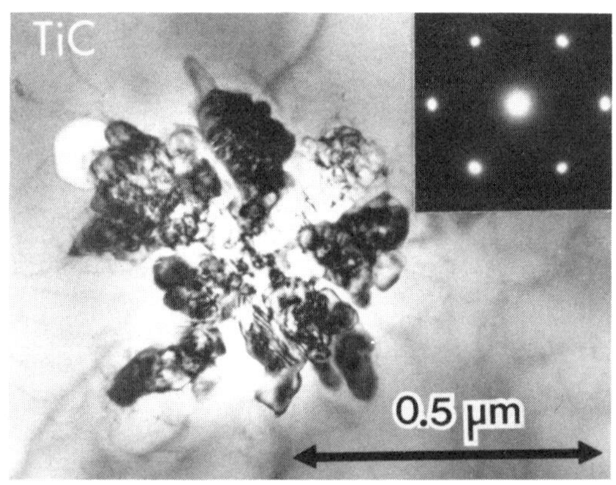

Figure 5. TEM photomicrograph of TiC constituent, [110] zone pattern shown, found in Alloy C-4 weld metal. Note the dendritic morphology of this constituent.

Figure 4b shows that the solidification paths of both Alloys C-22 and C-276 enter into the region of **μ** + **γ** stability as the weld metal cools to 850°C and below. This implies that the solidification TCP phases will transform to other TCP phases as the weld metal cools. In the case of Alloy C-276, **P** transforms to **μ**. In the case of Alloy C-22, **σ** transforms first to **P** phase as it must pass through a region of **P** + **γ** stability before it enters the region of **μ** + **γ** stability. This **P** subsequently transforms to **μ**. Because of insufficient time at temperature, none of these transformations go to completion.

These sequences of transformation are consistent with the phases observed during TEM analysis of each alloy's weld metal. Alloy C-22 contained all of the TCP phases. Alloy C-276 contained only **μ** and **P**. The relatively low Cr_{eq} in this alloy precluded the formation of **σ** as a terminal solidification constituent. Because of the shifting of the phase fields (with the introduction of **μ**) to higher Cr content as the temperature falls, there is no chance for **σ** to form as a solid state transformation product in this alloy.

NEW PHACOMP ANALYSIS - The intermetallic phases found in these alloys belong to the group of TCP phases which have historically[20] caused problems when they appear in superalloys during long term exposure in the high temperature operating range. A method has been developed for predicting their occurrence in these complex alloys. This formalism is known by the acronym PHACOMP[21] (for PHAse COMPutation). It is based upon the correlation between the average electron hole number of an alloy, as first defined by Pauling[22], and the precipitation of TCP phases. That is, when an alloy's average electron hole number exceeds a critical value, TCP phase precipitation can be expected. An excellent review of the development of this formalism has been written by Sims[23].

More recently, the PHACOMP formalism has been reformulated by Morinaga and coworkers[24,25] in terms of average Fermi energies. Quantum mechanical cluster variation calculations have been performed, with Ni_3Al as the standard zero-point Fermi energy structure (FCC). Alloying elements normally found in nickel-base superalloys have been substituted into this structure and the quantum mechanical calculations have been performed to determine the relative Fermi energies for these different alloyed clusters. In a manner analogous to the "old" PHACOMP, Morinaga and coworkers[25] correlated critical average Fermi energies, M_d^{crit}, with the formation of these same TCP phases in superalloys.

The methodology for determining the average Fermi energy, M_d, for an alloy is as follows. A net γ composition is determined after accounting for the formation of phases and compounds normally encountered during primary processing of the alloy. The atomic fraction of each element in this net γ composition is multiplied by its Fermi energy factor derived from the cluster variation calculations. Table 3 lists the values of these factors, m_d^i. These terms are summed to arrive at an average Fermi energy number for the residual γ matrix as follows:

$$M_d = \Sigma\, a_i\, (m_d^i) \qquad (4)$$

where a_i is the atomic fraction of element i and M_d and m_d^i are as defined above. If $M_d \geq M_d^{crit}$, σ or one of the other TCP phases would be expected to form.

Table 3. - Elemental m_d^i Factors[25]

Element	m_d^i Factor
Ni	0.717
Co	0.777
Fe	0.858
Mn	0.957
Cr	1.142
V	1.543
Mo	1.550
W	1.655
Al	1.900
Si	1.900
Nb	2.117
Ti	2.271
Ta	2.224
Zr	2.944

In the TCP phase precipitation temperature range, M_d^{crit} for σ formation was found[25] to have a linear temperature dependence, as such:

$$M_d^{crit} = 6.25 \times 10^{-5}\, T + 0.834. \qquad (5)$$

where T is the absolute temperature in Kelvin. **P** phase would be expected to form at similar M_d^{crit} values in the Ni-Cr-Mo system[25] (but at lower Cr concentrations).

The locus of M_d^{crit} values as a function of temperature is essentially a measure of the γ / γ + TCP solvus. That is, it determines the multi-component solubility limit of the γ matrix relative to TCP phase formation. When that limit is exceeded, TCP phases will form.

An analogous sequence of events happens during austenitic solidification. Segregation of alloying elements creates a continuous variation of chemistry from the dendrite core to the final solidification volumes. In the case of these nickel alloys, the segregation pattern of Ni depletion in interdendritic regions is such that M_d increases from the dendrite core to the associated interdendritic volumes. This can be seen in Fig. 6 which shows the results of pointwise calculating M_d, using Eq. 4, from the EMP data given in Fig. 3. When local critical values are exceeded, TCP phase-containing constituents should form as solidification products.

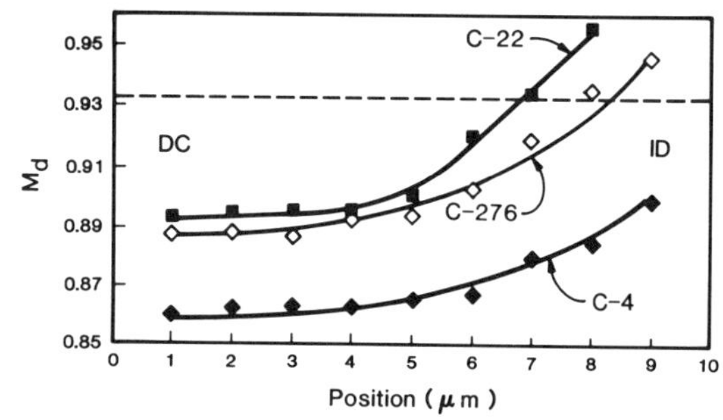

Figure 6. Calculated M_d values for the Hastelloy alloys as a function of position within an individual dendrite.

The horizontal dashed line is the M_d^{crit} value from Eq. 5 evaluated at 1285°C. This is the approximate temperature of σ solidification in Alloy C-22 and **P** solidification in Alloy C-276, as measured by the DTA experiments (Fig. 2). As can be seen in Fig. 6, the critical M_d is exceeded in the terminal solidification volumes of both Alloys C-22 and C-276 and TCP phases are observed. In the case of Alloy C-4, solidification segregation is not extensive enough to exceed the γ dendrite solubility limit and no TCP phases are observed.

CLOSURE- In the case of the alloys in this investigation, application of the New PHACOMP formalism correctly predicted the formation (or nonformation) of TCP phases as terminal solidification constituents. Other nickel alloy welds have shown the same consistent behavior[3,26] relative to correlation between M_d^{crit} and the formation of terminal TCP phase-containing solidification constituents.

The implication of this understanding is that weld metal microstructures in this class of alloys may be predicted *a priori* if the segregation patterns are known or can be deduced. In many nickel alloys, fusion zone hot cracking is associated with the formation of TCP phases[1,3,26] and the prediction of hot crack sensitive microstructures may be possible by applying the New PHACOMP formalism to these alloys.

CONCLUSIONS

• Ni, Cr, and Mo equivalents have been defined which, with the use of the Ni-Cr-Mo phase diagram, correctly predict the sequence of phase transformations (except for TiC formation in Alloy C-4) leading to the observed microstructures in Hastelloy Alloys C-4, C-22 and C-276.

• New PHACOMP calculations applied to measured segregation profiles in Hastelloy Alloys C-4, C-22, and C-276 correctly predict the occurrence (or non-occurrence) of terminal TCP solidification phases in these alloys.

• An understanding of austenite phase stability during solidification via a formalism such as New PHACOMP allows for a deeper comprehension of the development of weld metal microstructures and related properties (e.g. hot cracking susceptibility) in nickel base alloys.

ACKNOWLEDGEMENTS

The authors wish to acknowledge the experimental assistance given by R. E. Semarge and W. Hammetter. This work was performed at Sandia National Laboratories, Albuquerque, New Mexico, supported by the U. S. Dept. of Energy under contract number DE-AC04-76DP00789.

REFERENCES

1. M. J. Cieslak, T. J. Headley, A. D. Romig, Jr., *"The Welding Metallurgy of Hastelloy Alloys C-4, C-22, and C-276"*, submitted to *Met. Trans. A*, Dec. 1985.
2. M. J. Cieslak, G. A. Knorovsky, T. J. Headley, "Weldability/Microstructure Relationships of High Nickel Alloys for Header Applications", paper presented at the 1985 AWS Annual Meeting, April 29, 1985.
3. M. J. Cieslak, Unpublished Research at Sandia National Laboratories, Albuquerque, NM, 1986.
4. S. J. Matthews, *Superalloys* 1976, p. 215, The Metallurgical Society of AIME publication, Warrendale, PA.
5. H. M. Tawancy, R. B. Herchenroeder, A. I. Asphahani, *Jour. of Metals*, vol. 35, No.6, June 1983, p. 37.
6. N. Sridhar, J. B. C. Wu, P. E. Manning, *Jour. of Metals*, vol. 37, No. 11, Nov. 1985, p. 51.
7. F. G. Hodge and R. W. Kirchner, *Corrosion*, Aug. 1976, p. 332.
8. F. G. Hodge, *Corrosion*, Oct. 1973, p. 375.
9. R. B. Leonard, *Corrosion*, May 1969, p. 222.
10. W. F. Savage and C. D. Lundin, *Welding Journal*, Oct. 1965, p. 433-s.
11. W. F. Savage and C. D. Lundin, *Welding Journal*, Nov. 1966, p. 497-s.
12. G. F. Bastin, H. J. M. Heijligers, F. J. J. van Loo, *X-ray Spec.* 13: 91, 1984.
13. G. Cliff and G. Lorimer, *Proc. 5th European Congress on Electron Microscopy*, Institute of Physics, (1972), p. 140.
14. J. I. Goldstein, J. L. Costley, G. W. Lorimer, S. J. B. Reed, *SEM/1977*, p. 315.
15. D. S. Bloom and N. J. Grant, *Trans. AIME*, Feb. 1954, p. 261.
16. M. Raghavan, R. R. Mueller, G. A. Vaughn, S. Floreen, *Met. Trans. A*, vol. 15A, May 1984, p. 783.
17. M. Raghavan, B. J. Berkowitz, J. C. Scanlon, *Met. Trans. A*, vol. 13A, June 1982, p. 979.
18. J. C. Lippold, *Welding Journal*, March 1984, p. 91-s.
19. J. A. Lambert, Central Electricity Generating Board Report No. TPRD/M/1511/R85, Oct. 1985.
20. C. T. Sims, *Jour. of Metals*, 1966, vol. 18, No. 10, p. 1119.
21. L. R. Woodyatt, C. T. Sims, H. J. Beattie, Jr., *Trans. AIME*, 1966, vol. 236, p. 519.
22. L. Pauling, *Physical Review*, Dec. 1, 1938, vol. 54, p. 899.
23. C. T. Sims, *The Superalloys*, C. T. Sims and W. C. Hagel, eds., 1972. p. 276, John Wiley & Sons publication, New York, NY.
24. M. Morinaga, N. Yukawa, H. Adachi, *J. Phys. Soc. Japan*, 1984, vol. 53, No. 2, p. 653.
25. M. Morinaga, N. Yukawa, H. Adachi, H. Ezaki, *Superalloys 1984*, p. 523, The Metallurgical Society of AIME publication, Warrendale, PA.
26. M. J. Cieslak, G. A. Knorovsky, T. J. Headley, A. D. Romig, Jr., *"The Use of New PHACOMP in Understanding the Solidification Microstructure of Nickel Base Alloy Weld Metal"*, submitted to *Met. Trans. A*, March, 1986.

WELDING SIMULATION UP TO THE MOLTEN STATE - TESTING PROCEDURE AND FIRST RESULTS ON 10 MnMoNi 55 WELD METAL

K. Kussmaul, R. Schwab, J.N. Aoh
Staatliche Materialprüfungsanstalt (MPA)
der Universität Stuttgart
Stuttgart, Federal Republic of Germany

ABSTRACT

A method of weld thermal simulation up to the molten state has been developed. Using this method, a rectangular specimen made from weld deposit is heated by an electric current passing through it until it melts in the central region and is then cooled down. A crucible made of MgO-ceramics is used to confine the molten zone. By applying a TIG arc at low intensity to the specimen, the shrinkage and gas entrappment which would otherwise develop during solidification is avoided. The temperature is measured with a Mo-Re thermocouple. The temperature control is carried out by means of a closed-loop control system.
The solidification mechanisms under controlled conditions and the effect of temperature cycles on the phase transformations and the microstructure of the weld metal can thus be studied. First results of the simulation on 10 MnMoNi 55 weld metal are described. They indicate a good reproducibility of the simulated weld deposits. Impact testing is carried out on Charpy-V specimens machined from simulated weld metal. Furthermore, metallographic investigations of the microstructures are accomplished to interpret the toughness properties obtained from the impact testing.

THE CONVENTIONAL WELD THERMAL SIMULATION technique has been applied extensively to study the microstructures and mechanical properties of the heat-affected zone (HAZ) of welds, e.g. /1-4/. This technique offers a convenient way to produce a wide range of homogeneous synthetic HAZ microstructures under specific thermal cycles. Thus the specimen has a sufficient volume for mechanical testing. Weld thermal simulation has further been applied to investigate the relaxation behaviour of steels during stress-relief annealing /5,6/, the fracture toughness of the HAZ /7-9/, and the phase transformation during welding /10,11/.

Most of the work on weld thermal simulation has so far been carried out up to a peak temperature range below the melting point of the materials. Some attempts have been made to extend the peak temperature to the melting point /12-14/. Among these, a welding fusion simulation was carried out with a "Gleeble 510" machine on sheet specimens to investigate the surface reaction near melting and to assess the welding characteristics of special alloys /12/. Other simulations up to the molten state were developed to establish CCT-diagrams /13,14/. In all these cases the specimens were too small to represent the solidification structure in a real weld metal. Besides, a further assessment of the mechanical properties of the simulated fusion zone could not be realized because the dimensions of the specimens were too small to satisfy those specified for impact specimens.

In order to study the solidification mechanisms during the welding process and the microstructures of the weld metal developed under controlled conditions and in order to assess their corresponding toughness properties, the scope of this work is therefore to develop a suitable weld thermal simulation technique up to the molten state. It was of special importance to ensure that the size of the molten region should be large enough so as to represent the solidification structure of real weld metal as well as to satisfy the dimensions specified for impact specimens.

MATERIAL AND SPECIMEN PREPARATION

10 MnMoNi 55 submerged arc weld deposit (electrode material S3 NiMo 1, welding flux UV 14210) is selected as the first testing material. The chemical composition is given in Table 1. The dimensions of the testing specimen as well as the position of the later machined Charpy-V-specimen is shown in Fig. 1. The specimens are provided with holes drilled in both ends for injecting water as a coolant

Table 1 - Chemical composition of the testing material - 10 MnMoNi55 weld metal

Element	C	Si	Mn	P	S	Cr	Ni	Mo	V	Al	Cu	Co	W, As, Pb, Sb, Nb, Sb, Nb, Sn, Ti
Wt-%	0,10	0,16	1,30	0,009	0,004	0,06	1,00	0,59	0,01	0,02	0,03	0,03	≤ 0,01

during the simulation. The small inclined hole near the specimen center serves as an inlet for the jacket-thermocouple.

EXPERIMENTAL PROCEDURE

WELD THERMAL SIMULATOR - A thermal cycle simulator with a rated power of 26 kVA (type Smitweld) is used in conducting the simulation. The specimen is heated by an alternating current up to 5 kA passing through it until it melts in the central region. It is then cooled down by reducing or shutting off the electric current. During the whole procedure the specimen is provided with constant water cooling at both ends, Fig. 1. The specimen is held by slideable copper grips to compensate the thermal strains developed in the specimen during welding simulation, Fig. 2. The water-cooled grips are connected to the heating unit by flexible cables.

TEMPERATURE CONTROL - The heating transformer together with the test specimen, the thermocouple, the thyristor controller, the mA-mV converter, the PID-regulator and the programmer form a closed control loop. In the programmer, a potentiometer is turned in an optical-mechanical way according to the profile of a thermal cycle template attached to a turning drum. The instantaneous temperature of the test specimen is compared with the temperature dictated by the potentiometer. A signal proportional to the temperature difference is then supplied to the thyristor controller so that the difference is rapidly reduced. A Mo-Re thermocouple (Mo-41%Re, Mo-5%Re; wire diameter 0.5 mm) with a service temperature range up to 1900 °C is selected to carry out the temperature measurement of the test specimen during welding simulation. The thermocouple is protected by a thin double-hole pure alumina ceramic tube. The tube end is sealed up by using a TIG torch to prevent the thermocouple wires from direct contacting with the melt pool, which may cause dissolution of the thermocouple wires.

CRUCIBLE - In order to prevent the liquid steel from dropping, a crucible consisting of three pieces of porous MgO-ceramics together with a crucible holder is used, Fig. 2. The jacket-thermocouple is fixed on the holder to prevent it from sliding or floating while the specimen center is liquid.

PREVENTION OF SHRINKAGE FORMATION AND GAS ENTRAPPMENT - A TIG arc discharged at low intensity (current 25 - 30 A) towards the liquid surface before and while the melt begins to solidify can greatly improve the quality of the resolidified zone because shrinkage formation and gas entrappment are avoided. Due to the additional heat source from the arc the liquid surface would not first begin to solidify so that possible outgassing can take place. Furthermore, owing to the temperature gradient, the solidification front will be drawn towards the upper surface, so that the formation of a shrinkage in the center of the resolidified zone can be avoided.

During the welding simulation, an argon gas flow is continuously spread over the specimen through gas nozzles mounted on the crucible holder. This contributes to reduce the oxidation of the melt pool.

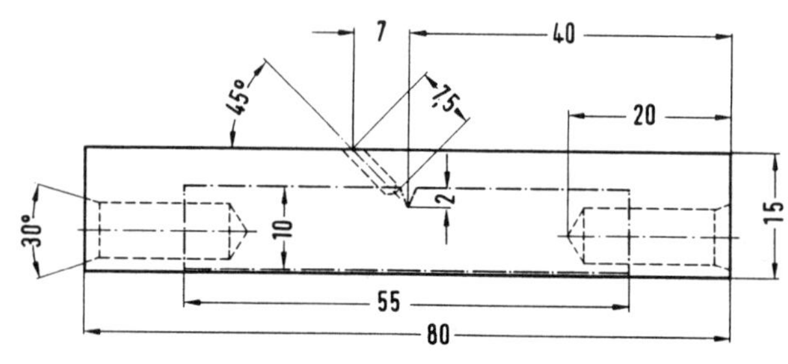

Fig. 1 - Specimen for weld thermal simulation

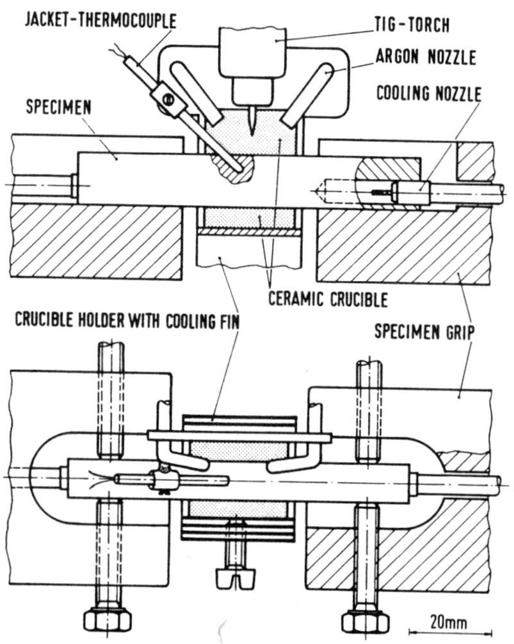

Fig. 2 - Crucible and grips system (a) front view (b) top view

IMPACT TESTING - Charpy-V impact specimens were prepared from the simulated specimens according to Fig. 1. A macroetching process should be carried out to reveal the mid-position of the solidification structure. Thus the correct position for the notch can be selected. In this study the notch was in the center line of the simulated weld.

RESULTS AND DISCUSSION

CHEMICAL COMPOSITION - The chemical composition of the specimen in the resolidified zone after the weld thermal simulation was analysed by means of an optical emission quantometer. It shows that the chemical composition is unchanged during the simulation process (gases like nitrogen, oxygen or hydrogen were not analysed). This implies that no vaporing of non-gaseous elements occurred during the melting procedure. The simulated weld metal was further analysed by means of an X-ray microanalyser along a longitudinal section and was proved to be free from macrosegregation in the resolidified zone.

METALLOGRAPHIC INVESTIGATION - In order to investigate the influence of the multi-pass effect on the solidification structure, the following three different thermal cycles were selected and applied in the simulation:

1) single cycle with a peak temperature (T_p) of 1480 °C
2) reheat thermal cycle with T_p of 1480 °C and 1150 °C
3) reheat thermal cycle with T_p of 1480 °C and 1350 °C

The cooling time between 800 °C and 500 °C ($t_{8/5}$) for all three thermal cycles was 20 sec. The temperature-time curves are shown in Fig. 3. The macrostructures of the resolidified zone along the longitudinal central section of the specimens simulated under these three thermal cycles are shown in Fig. 4 (a) to (c),

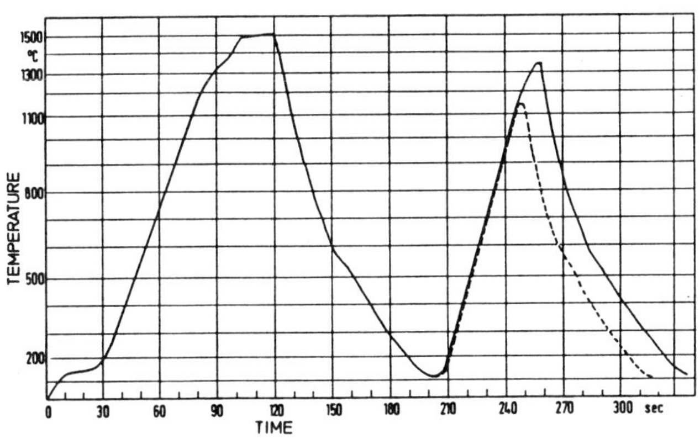

Fig. 3 - Temperature-time-curves

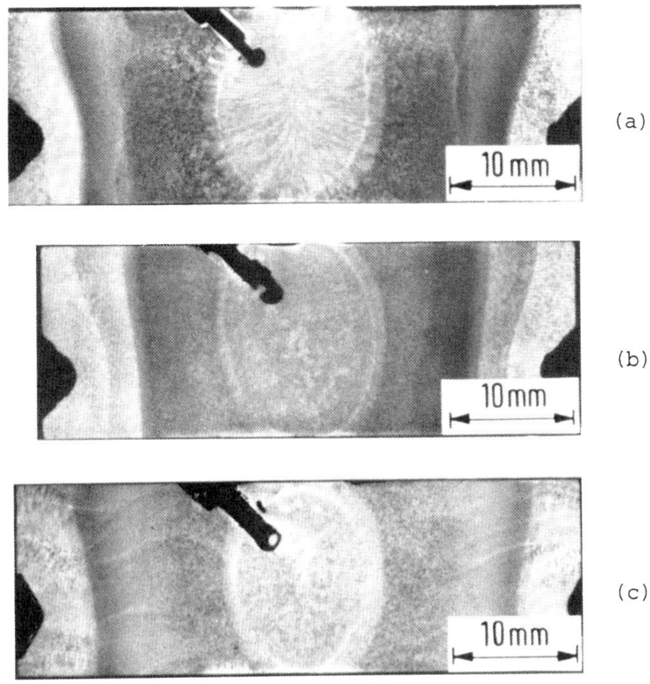

Fig. 4 - Macrostructure of the resolidified zone
(a) Peak temperature 1480 °C
(b) Peak temperatures 1480 °C and 1150 °C
(c) Peak temperatures 1480 °C and 1350 °C

respectively. All of the three photographs illustrate that the resolidified zone is free from shrinkage and gas entrappment. The length of the molten zone is about 12 mm.

Figure 4 (a) shows the macrostructure of a resolidified zone simulated with a peak temperature of 1480 °C. It reveals the columnar structure of the prior austenite grains. It should be noted that the prior grain size coincides with the grain size of the solid metal at the fusion boundary. This confirms the statement in /15/. The microstructure of the same region is shown in Fig. 5 (a). It consists of acicular ferrite, grain boundary ferrite with some aligned and dispersed martensite crystals and carbide particles. The grain size in this region corresponds to ASTM No. 6. Figs. 4 (b) and (c) show the macrostructures of the specimens simulated with reheat thermal cycles. The columnar solidification structure has disappeared due to the phase transformation. Both of the microstructures, Figs. 5 (b) and (c) consist of acicular ferrite and grain boundary ferrite with a small amount of martensite, which is very similar to the result from the single thermal cycle. The different grain sizes of the microstructures, namely ASTM No. 8 in Fig. 5 (b) and ASTM No. 6 in Fig. 5 (c) must be attributed to the different reheat peak temperatures of 1150 and 1350 °C respectively.

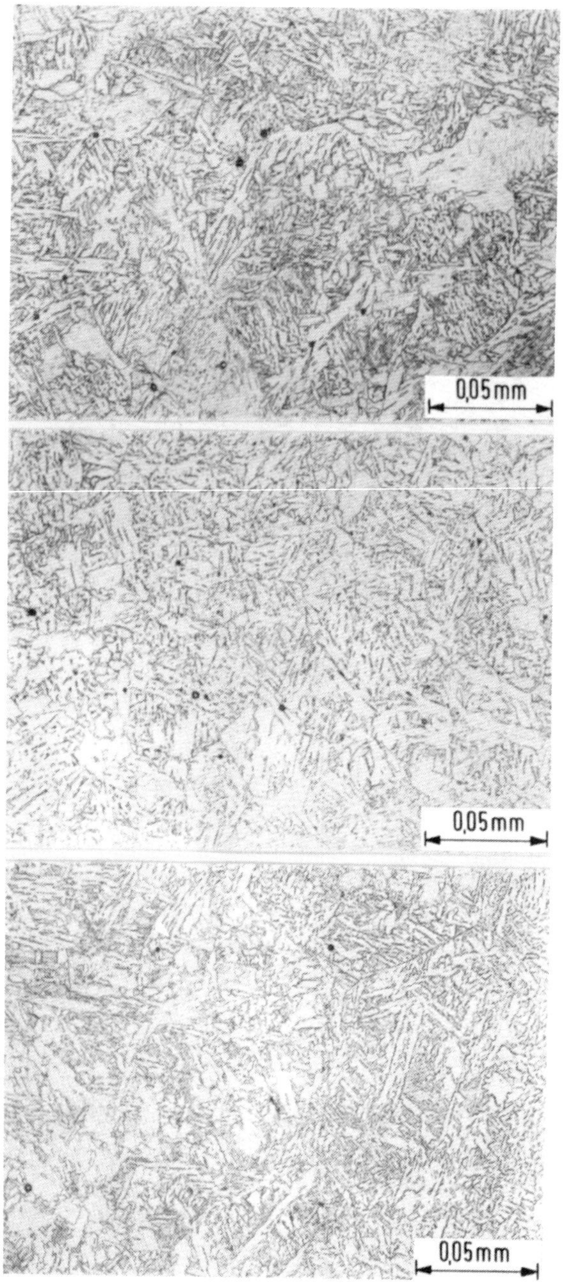

(a)

(b)

(c)

Fig. 5 - Microstructure of the resolidified zone
(a) Peak temperature 1480 °C
(b) Peak temperatures 1480 °C and 1150 °C
(c) Peak temperatures 1480 °C and 1350 °C

TOUGHNESS RESULTS - The Charpy-V impact curves obtained from the specimens which were simulated with three different thermal cycles are given in Fig. 6. It illustrates the influence of the peak temperature of the reheat cycle on the absorbed energy of the simulated weld metal. The upper shelf energy of the resolidified zone is improved by means of a subsequent reheat cycle with a peak temperature of 1150 °C. This result is in good agreement with the metallographic investigation with respect to

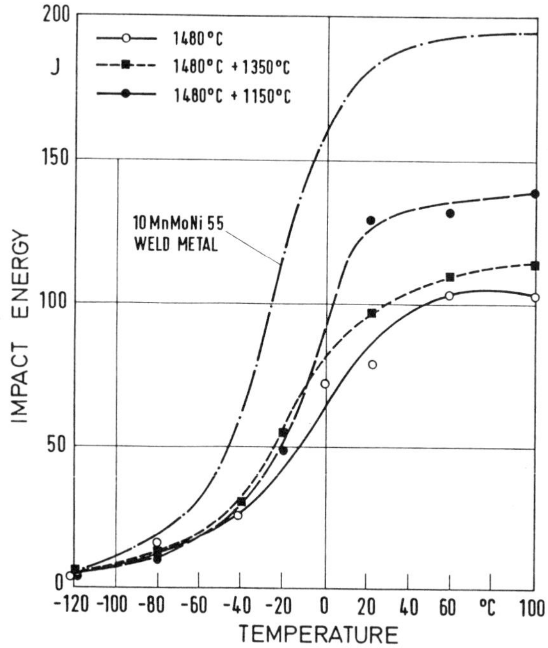

Fig. 6 - Charpy-V curves for 10 MnMoNi 55 simulated under various thermal cycles

the grain size. Less improvement in upper shelf energy was observed for reheating up to 1350 °C, since at this peak temperature grain coarsening would take place. In comparison with the simulated specimens, the original weld deposit shows a much higher C_V-toughness, Fig. 6 and finer microstructure, Fig. 7. The reason for this is not yet fully understood, but one of the reasons is certainly the large austenitic grain size for the single cycle. This is influenced by the large grain size along the fusion boundary, which is caused by the low heating rate in the simulation tests. Experiments with a more powerful current supply to obtain a higher heating rate will be performed in future and thus a smaller grain size would be expected.

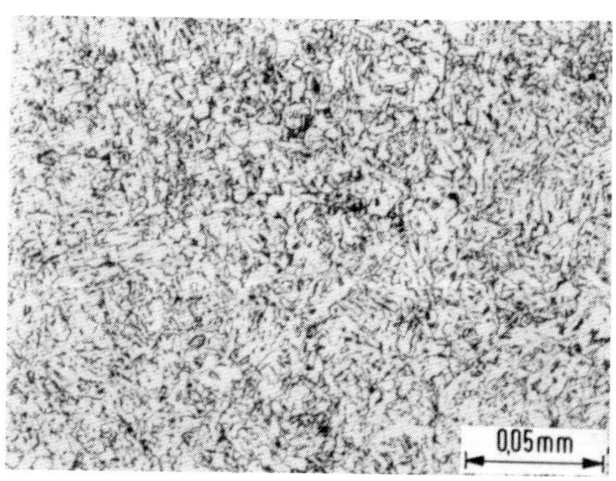

Fig. 7 - Microstructure of 10 MnMoNi 55 weld metal

The influence of the interpass temperature on the toughness properties was investigated as well. The simulation specimens were first heated up to 1480 °C, cooled at a rate of $t_{8/5}$ = 20 sec and then kept at different interpass temperatures between 200 °C and 450 °C for 7 minutes. Finally they were cooled down to RT. To reduce the number of specimens, the impact testing was only conducted at +60 °C. The results are illustrated in Fig. 8. The impact energy at +60 °C first increases with increasing interpass temperature up to 350 °C and then decreases. The interpass temperature at or below 300 °C has hardly any influence on the impact energy. An additional metallographic investigation was carried out to explain this phenomenon. It was found that, as the interpass temperature was under 300 °C, the microstructure was almost unchanged with respect to RT. At an interpass temperature of 350 °C, the acicular ferrite in the microstructure transformed into finer polygonal ferrite, which may have a beneficial effect on the toughness properties. A further increase in the interpass temperature results in coarsening of the polygonal ferrite, and thus a lowering of the impact energy would occur. A similar result was observed in /16/.

CONCLUSION

1. A weld thermal simulator (26 kVA) was extended to conduct a weld thermal simulation up to the molten state. By applying a TIG arc as an additional heat source during solidification and by using a suitable crucible and thermocouple the resolidified zone was proved to be free from shrinkage and gas entrappment. The resolidified zone was sufficiently large for further impact testing. The solidification mechanisms and the transformation of the microstructure during welding under controlled conditions can thus be studied.
2. A reheat cycle with a peak temperature of 1150 °C resulted in grain refinement and raised the upper shelf energy of the impact curve. No grain refinement occurred and the upper shelf energy hardly changed if the specimen was reheated at a peak temperature of 1350 °C.
3. Increasing the interpass temperature up to 350 °C had a beneficial effect on the impact value at +60 °C (upper shelf range), which was accounted for the formation of fine polygonal ferrite in the microstructure.

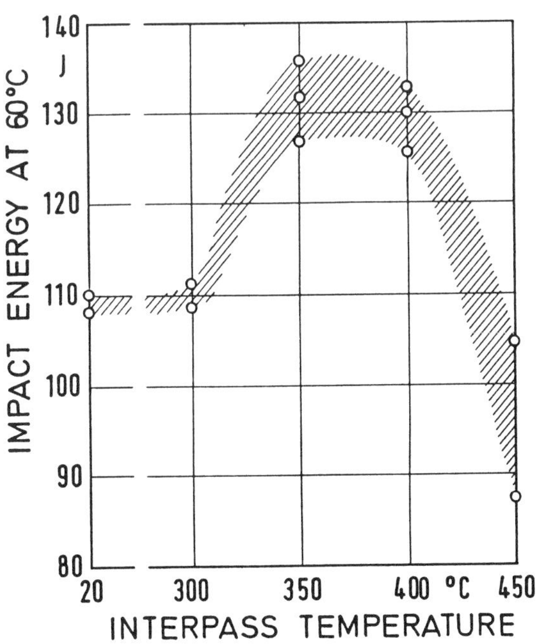

Fig. 8 - Influence of the interpass temperature on the impact energy at +60 °C

REFERENCES

1. Nippes, E.F. and W.F. Savage, Weld. J. 28, 534s-41s (1949)
2. Wu, K.C. and R.E. Herfert, Weld. J. 46, 32s-38s (1967)
3. Dolby, R.E. and D.J. Widgery, Weld. Inst. Report M/52/70; Weld. Int'l. Res. and Dev't. 1, 15pp. (1971)
4. Degenkolbe, J., H. Höhne and D. Uwer, Schweißen u. Schneiden 34, 137-41 (1982)
5. Kussmaul K., J. Ewald, and G. Maier, Schweißen u. Schneiden 28, 250-255 (1976)
6. Ewald, J., F. Beisswänger, J. Jansky and G. Maier, Schweißen u. Schneiden 29, 402-406 (1977)
7. Dolby, R.E., Metal Constr. 4, 59-63 (1972)
8. Schmidtmann, E. and W.D. Thinnes, Stahl u. Eisen 98 18-23 (1978)
9. Dhooge, A., K. Ostyn, W. Provost and A. Vinckier, WRC Bulletin 266 (1981)
10. Schmidtmann, E. and W.D. Thinnes, Stahl u. Eisen 97 1197-1201 (1977)
11. Berkhout, C.F. and P.H. van Lent, Schweißen u. Schneiden 20, 256-60 (1968)
12. Cremisio, R.S. and W.T. Koffler, Weld. J., 46, 259s-65s (1967)
13. Ruge, J., R. Müller and H.-J. Peetz, Schweißen u. Schneiden 33, 363-68 (1981)
14. Velkov, K. and A. Schelev, Schweißtechnik, Berlin, 32, 542-43 (1982)
15. Lancaster, J.F., "Metallurgy of Welding", 2nd Ed., P. 59, George Allen & Unwin, London (1980)
16. Evans, G.M., Oerlikon-Schweißmitteilungen 37, 17-31 (1979); (IIW-Doc. II-A-460-78).

FORMATION OF CENTRELINE WELD DEFECTS IN AUSTENITIC STAINLESS STEEL SHEETS AND STRIPS

Veli P. Kujanpää
Lappeenranta University of Technology
Lappeenranta, Finland

ABSTRACT

The formation of two kinds of centreline defect, solidification cracks and centre cavities, was studied by GTA welding of different austenitic stainless steels using various welding parameter combinations. The formation of solidification cracks is basically dependent on grain boundary phenomena, i.e. formation of solidification grain boundary and interfacial energy considerations. Centre cavities were found to be shrinkage cavities which form in the equiaxed zone in high-speed welds. They are analogous to cast pipes in these equiaxed zones and their formation is dependent on the probability of equiaxed zone formation and the ability of the melt to feed the zone.

AUSTENITIC stainless steel sheets and strips are generally found to have a good weldability, the only problem sometimes being penetration. They are generally easy to weld, provided that no severe restrictions are made as to the quality of the weld or productivity, but cracking is encountered when the sheets are severely restrained. Also, the demand for increased productivity, with automation and mechanization, has resulted in the use of the higher welding speeds, which is sometimes found to cause pore-like centreline defects. These can impair corrosion resistance in particular, and mechanical properties such as yield strength, ductility and creep and fatigue strength.

This paper is based on four studies dealing with the effect of welding parameters /1,2/ and composition /3,4/ on defect formation in GTA welds on austenitic stainless steel sheets and strips. The former studies showed that different kinds of defect can form in welds; cracks, centre cavities, ripple cavities and undercuts, depending on the welding parameters and the composition of the weld metal. The purpose here is to consider centreline defects, cracks and centre cavities and differences in their formation mechanisms.

EXPERIMENTAL

About 40 austenitic stainless steel sheets and strips of different types (thickness 2 or 1,5 mm) were welded by the single-electrode GTA method using a wide range of welding parameters (current 40...330 A, speed 25...1600 mm/min, arc shielding gas Ar+0...17 % H_2, He+20 % and Ar+0.3 % CO_2 and root shielding gas Ar and N_2+15 % H_2) and compositions (Cr = 16...22 %, Ni = 8... 25 %, Mo = 0...6 %, Cu = 0...1.5 %, N = 0...0.2 %, P = 0.006...0.04 % and S = 0.001...0.12 %). Experiments were also performed in a tube factory using the three-electrode GTA method with two strip compositions at several welding speeds and with various shielding gases.

RESULTS AND DISCUSSION

WELD DEFECTS - Two kinds of centreline defect were formed in the GTA welding of austenitic stainless steel sheets and strips: cracks and centre cavities. Their macroscopic locations are indicated in Fig. 1 and their shapes in Fig 2. The cracks were generally longer than the centre cavities, which were more numerous. The edges of the cracks were sharp, while those of the centre cavities

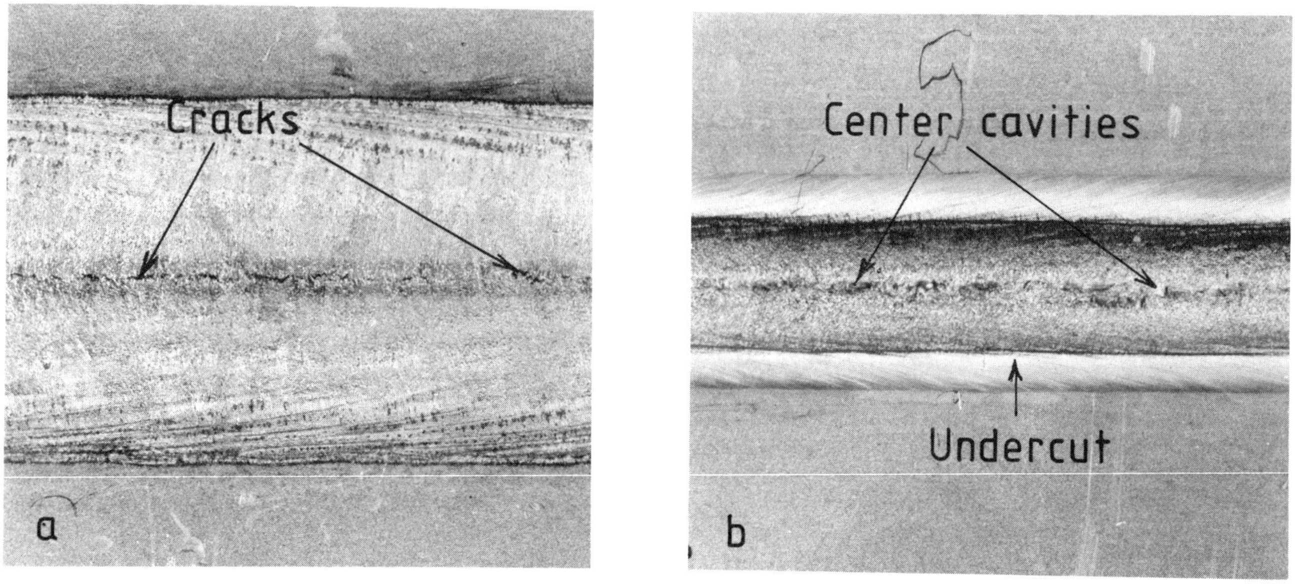

Fig. 1 - Centreline defects in seams welded with the following parameters: a) 190 A, 14 V, 400 mm/min b) 330 A, 18 V, 1200 mm/min /1/.

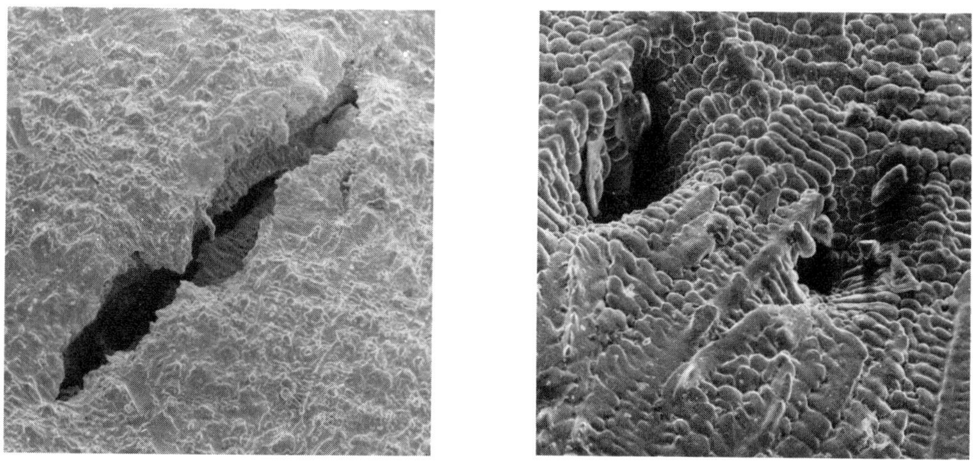

Fig. 2 - Types of centreline defect: a) crack, b) centre cavities /1/.

were rounded. In terms of microstructure, the cracks were situated in the last regions to solidify, i.e. at the grain boundaries, Fig. 3a, while the centre cavities were always situated in the equiaxed zone which formed typically on the weld centreline at high welding speeds, Fig.3b.

EFFECTS OF WELDING PARAMETERS - The effects of the welding parameters, including current and speed, on defect formation can best be shown on welding current - welding speed diagrams, Fig. 4. Here it can be seen that the weld was most susceptible to crack or centre cavity formation if the welding parameters were selected to be close to the hole formation field when the volume of the weld was high. The lower welding current values caused the depth of the defect to diminish.

The effect of welding speed was also considerable. Cracks were typically found at low speeds, Fig 4, and cracking became more pronounced as the speed was increased. At medium welding speeds ($\sim$ 400 mm/min) they were replaced by centre cavities, which increased at still higher speeds. The role of the shielding gases in centre cavity formation is seen in Fig. 5, which shows the dependence of the number of centre

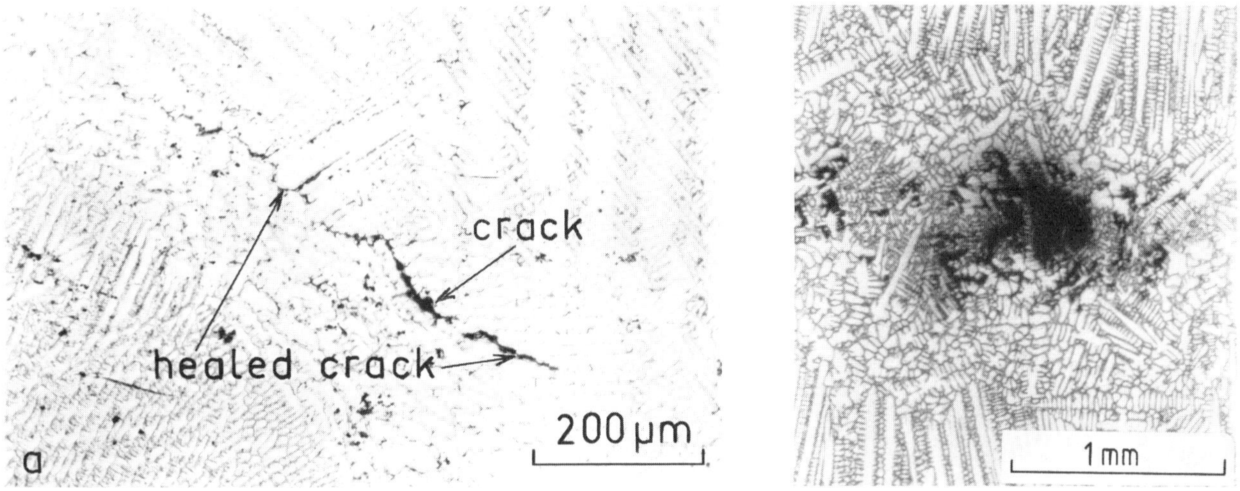

Fig. 3 - Location of centreline defects in weld microstructures: a) crack, b) centre cavities /1/.

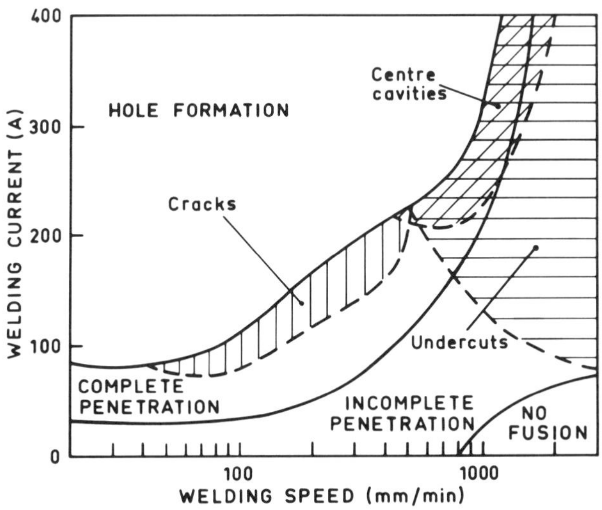

Fig. 4 - Fields of cracks and centre cavities on a welding current - welding speed diagram /1/.

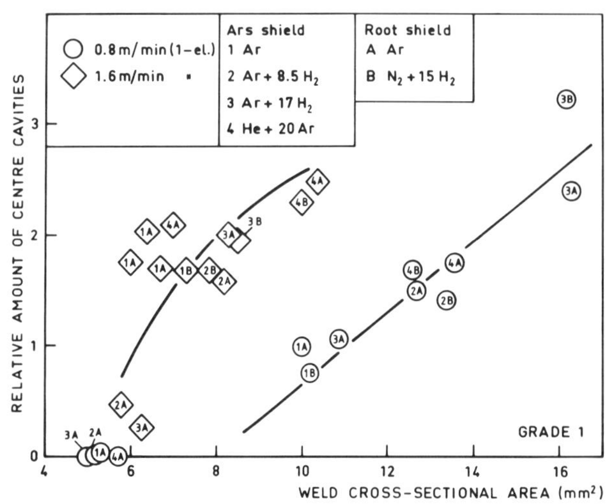

Fig. 5 - Dependence of centre cavities on the cross-sectional area of the weld /3/.

cavities on the cross-sectional area of the weld. This figure does not depend directly on the shielding gases themselves, but rather on the cross-sectional area caused by the gases, which is in turn affected by the type of gas. The effect of welding speed can also be recognized in Fig. 5, centre cavities forming more frequently at higher speeds.

Centre cavity formation was markedly reduced when three electrodes were used instead of the single-electrode GTA method. The general statements presented above concerning the effects of the cross-sectional area and welding speed are also valid for the multiple-electrode methods.

The shielding gases did not appear to affect crack formation in any way.

EFFECT OF COMPOSITION - The effects of sulphur content on the dimensions of the cracks and centre cavities is shown in Fig. 6. The tendency for crack and centre cavity formation increased with sulphur content, these defects being infrequent when the sulphur content was under 0.003 %. A reduction in the phosphorus content resulted in fewer cracks but did not greatly influence centre

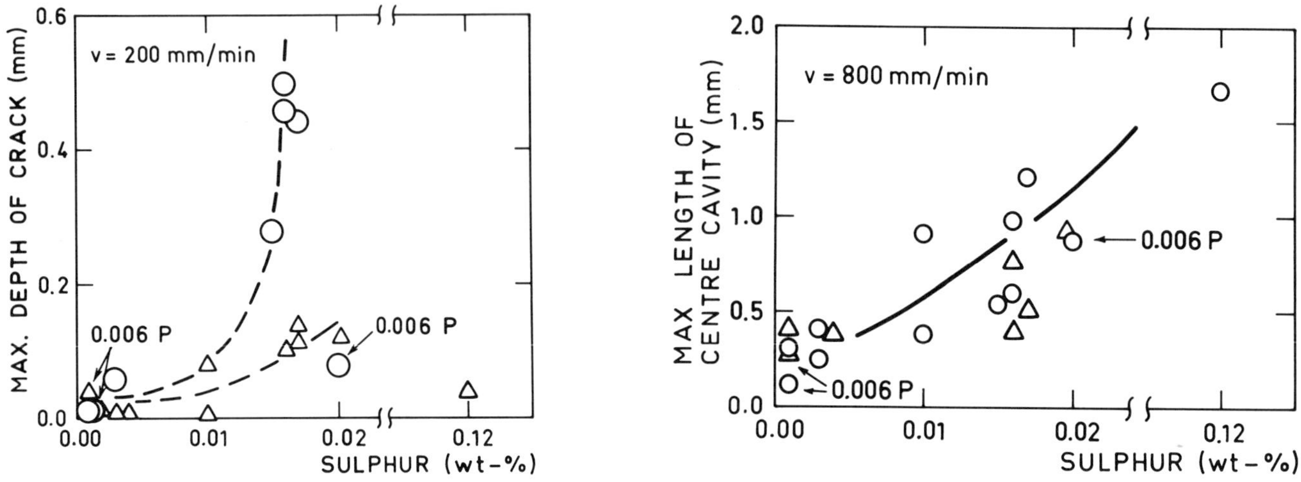

Fig. 6 - Relation between numbers of cracks and centre cavities and sulphur content. v = welding speed, o = primary austenitic solidification mode and Δ = primary ferritic solidification mode.

cavity formation.

The solidification mode of the weld is also a significant factor in Fig. 6, where welds which solidified in a primary austenitic mode are seen to have been more susceptible to crack formation than those which solidified in a primary ferritic mode. The number and size of the centre cavities did not seem to depend on the solidification mode at a welding speed of 800 mm/min.

FORMATION OF CENTERLINE WELD DEFECTS - Cracks formed on the weld centreline in the sheets when using moderate speeds (100 - 400 mm/min), being located microstructurally in the last part of melt to solidify, which confirmed them as solidification cracks.

The mechanism of solidification crack formation in welds on austenitic stainless steels and its dependence on solidification mode is discussed in many publications (e.g. Refs. 5-8), and it is found that the high susceptibility of welds solidifying in a primary austenitic mode is connected with the formation of solidification grain boundaries and with interfacial energy considerations.

Primary austenitic solidification gives rise to long, straight solidification grain boundaries, in which solidification cracks easily initiate and propagate, on the other hand primary ferritic solidification yields no solidification grain boundaries, since they are prevented by peritectic transformation and delta ferrite is formed between the austenitic dendrites /9/. Solidification cracking in welds of primary austenitic solidification is also enhanced by interfacial energy considerations.

In addition to the microstructure, the form of the weld puddle also affects solidification crack formation. At low welding speeds, the weld puddle is elliptical and cracking diminished, but when the welding speed is increased the puddle becomes drop-shaped /10/. With austenitic stainless steel sheets of 2 mm thickness, this transformation occurs at a speed of about 200 mm/min. The cracking level is increased in such cases, because the concentration of segregated impurities in the last droplets of the melt is probably greater than in elliptical puddles.

As the welding speed increases, the temperature gradient decreases and the

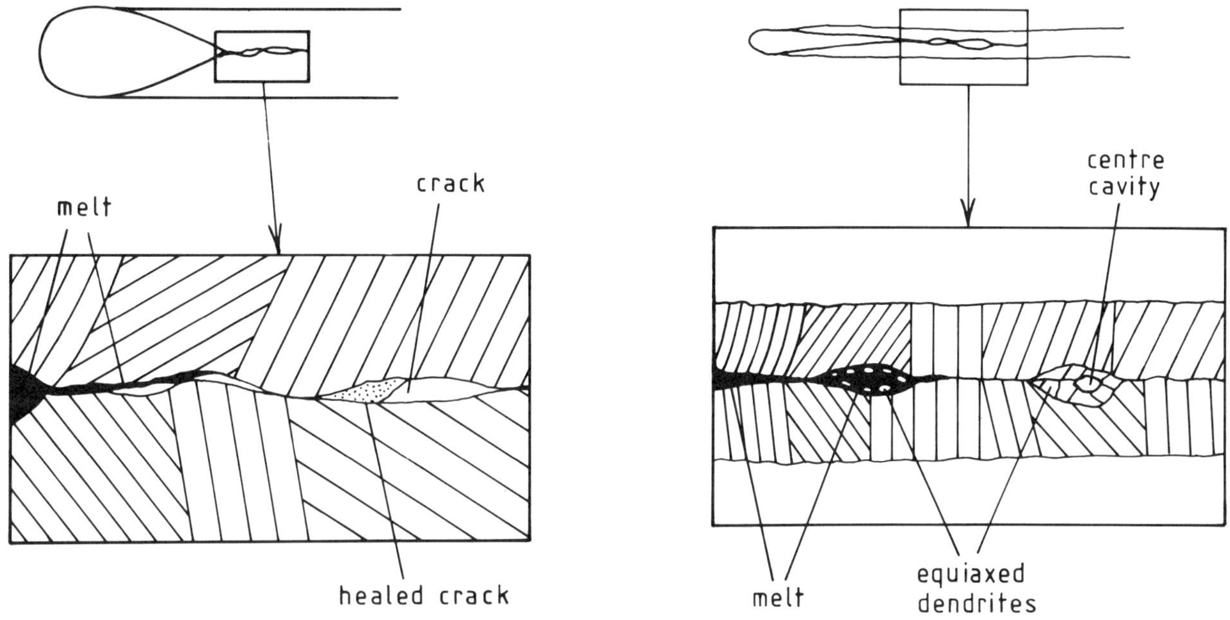

Fig. 7 - Schematic model for a) solidification crack formation and b) centre cavity formation in a weld.

solid-liquid range widens, so that back-filling or healing becomes more difficult, Fig. 7a.

Centre cavities are formed at the centreline of the weld at even higher welding speeds than solidification cracks. These can be differentiated from cracks by their morphology, in addition to which they are not affected by the solidification mode as solidification cracks are.

The centre cavities found here are attributed to shrinkage, for the following reasons:

1. Their number was found to be closely dependent on the cross-sectional area of the weld (Fig. 5), while an increase in the cross-sectional area implies greater shrinkage during the cooling of the weld pool.

2. No internal pores could be revealed radiographically or in the weld cross-sections studied microscopically even at the highest hydrogen content of the arc shielding gas.

3. The nitrogen, oxygen and hydrogen content of the weld did not increase during welding.

The formation of centre cavities is found to be connected with the formation of equiaxed zone in the weld at high welding speeds and with volume changes in the equiaxed zone. If the volume changes due to shrinkage cannot be offset by sucking liquid from the weld pool, the melt remaining between the equiaxed dendrites will not suffice and a centre cavity will be formed in the middle of the equiaxed zone, Fig. 7b. Centre cavites can thus be regarded as cast pipes in the equiaxed zones.

CONCLUSIONS

The formation of two kinds of centreline defect, solidification cracks and centre cavities, is studied here using GTA welding of different austenitic stainless steels with various welding parameter combinations. The results allow the following statements to be made:

1. The formation of solidification cracks is basically dependent on the formation of solidification grain boundary and interfacial energy considerations.

2. Centre cavities are shrinkage cavities which form in the equiaxed zone at high welding speeds. Their formation is dependent on the probability of equiaxed zone formation and on the ability of the melt to feed to the zone.

ACKNOWLEDGEMENTS

Financial support from the Foundation of Outokumpu Co. is acknowledged. The author also wishes to thank Mr. Malcolm Hicks, M.A. for revising the English of the manuscript.

REFERENCES

1. Kujanpaa, V.P., Weld.J. 62, 45-s to 52-s (1983).

2. Kujanpaa, V.P., L.P. Karjalainen and H.A.V. Sikanen, Weld.J. 63, 150-s to 155-s (1984).

3. Kujanpaa, V.P., Weld.J. 63, 369-s to 375-s (1984).

4. Kujanpaa, V.P., L.P. Karjalainen and H.A.V. Sikanen. Weld defects in austenitic stainless steel sheets - Role of steel type. Submitted for publ. in Weld.J.

5. Arata, Y. et al. Trans JWRI 5, 35-51 (1976).

6. Kujanpaa, V.P., et al., Weld.J. 63, Internat. 9, 55-76, (1979).

7. Brooks, J.A., et al. Weld.J. 63, 71-s to 83-s (1984).

8. Lippold, J.C. and Svage, W.F., Weld.J. 61, 388-s to 396-s (1982).

9. Kujanpaa, V.P., Davis, S.A. and White, C.L., Formation of hot cracks in austenitic stainless steel welds - Solidification cracking, to be published in Welding Journal.

10. Suutala, N. Metall. Transact. 13A, 2121-2130 (1982).

FORMATION OF THREE TYPES OF NEW GRAINS IN THE WELD METAL

Y. Le, Sindo Kou
Department of Metallurgical & Mineral Engineering
University of Wisconsin-Madison
Madison, Wisconsin 53706

ABSTRACT

It is well known that columnar grains in the weld metal grow epitaxially from the boundary of the fusion zone. New grains, however, can also form in the bulk weld metal, i.e., away from the boundary of the fusion zone. Three different types of new grains are discussed in this paper. The first type is columnar grains which grow epitaxially from grains in the bulk weld metal. This type of grains is often observed in oscillated arc welds and pulsed arc welds made with low frequencies of arc oscillation and pulsation. The second type is equiaxed grains which nucleate in the bulk weld metal, due to various possible nucleation mechanisms. The third type is in between the previous two types in geometry and is often called comet or stray grains, depending on their length. The mechanisms for the formation of the second and third types of new grains are discussed briefly.

IN a series of recent studies, Kou et al. (1-3) have been demonstrated that the grain structure of the weld metal can strongly affect: 1. its solidification-cracking susceptibility, and 2. its mechanical properties. In order to better understand and control the grain structure of the weld metal, it is essential to understand the formation of grains during weld metal solidification.

It is well known that columnar grains in the weld metal tend to grow epitaxially from the boundary of the fusion zone (4-7). This is the so-called epitaxial growth process. In the process, grain growth is initiated by arranging atoms from the liquid phase upon the existing crystalline substrate, thereby extending it without changing the crystallographic orientation. The continuity of crystallographic orientations across the boundary of the fusion zone was first confirmed by Savage et al. (5) using the Laue X-ray back-reflection technique.

Although the grain structure near the boundary of the fusion zone is characterized by epitaxial growth, in the bulk of the fusion zone the grain structure can be characterized by the formation of new grains in the weld metal. The phrase "new grains" refers to grains which have not originated from the boundary of the fusion zone by epitaxial growth, but have formed afterward. In the prsent study, the formation of three types of new grains in the weld metal will be addressed.

Experimental Procedure

The materials used for welding was 6061 and 1100 aluminum sheets of 1.6 mm (1/16 inch) thickness, and the welding process the GTAW (gas tunsten arc welding) process.

The power source was a programmable GTA welding machine, with regular and pulsed arc options. When arc oscillation was required, a magnetic arc oscillator was used.

After welding the specimen were sectioned, polished and etched with a solution containing 75 ml HCl, 25 ml HNO_3 and 5 ml HF to reveal the grain structure. For scanning electron microscopy, the specimens were electroplished using a solution of 200 ml phosphoric acid, 150 ml sulfuric acid, 275 ml water and 25 grams chromic oxide. An energy dispersive spectrometer (EDS) was also used.

Results and Discussion

Fig. 1 illustrate schematicaly the three types of new grains observed. They are discussed in the following sections.

I. Internal Epitaxial Columnar Grains

Fig. 2a shows the grain structure of a 1100

aluminum weld made with circular arc oscillation. The welding speed and the frequency of oscillation are 4.23 mm/sec (10 ipm) and 1 Hz, respectively. The welding direction was from right to left. As shown, many columnar grains grew epitaxially from those formed in a previous cycle of arc oscillation, the substrate being inside rather than at the boundary of the fusion zone and hence the adjective "internal". Of course, many columnar grains grown epitaxially from the boundary of the fusion zone are also visible.

Fig. 2b shows similar internal epitaxial columnar grains in an oscillated arc weld of 1100 aluminum. The welding speed was 2.54 mm/sec (6 ipm) and the frequency of pulsation 1 Hz.

II. Equiaxed Grains

The nucleation mechanisms for equiaxed grains in the weld metal have recently been discussed by Kou et al. (8). Fig. 3a shows the grain structure of a 6061 weld made with a current of 100 A and speed of 12.7 mm/sec (30 ipm). As shown, equiaxed grains nucleated near the centerline of the weld metal and blocked off the epitaxial columnar grains grown from the boundary of the fusion zone.

Fig. 3b and 3c show the SEM micrographs of one of the grains in the weld. As shown, a heterogeneous nucleus is clearly visible at the root of the equiaxed dendrite. The result of the EDS analysis is shown in Fig. 3d. As shown, the nucleus is rich in titanium, a typical element for heterogeneous nucleation of aluminum. Since the EDS analysis is not a quantitative analysis, the composition of the nucleus is not known. It is, however, believed to be some type of alumnum-titanium intermetallic compound, possibly with small amounts of other alloying elements.

This result is interesting since, unlike 2219 aluminum, commercial 6061 aluminum is not known to contain titanium. In fact, the evidence of heterogeneous nucleation of equiaxed grains due to a titanium-rich compound in 6061 aluminum welds is surprising. However, the chemical analysis of the alloy, shown in Table I, did indicate a low but significant level of titanium. Analysis of 6061 aluminum from a different producer showed similar results.

Generally speaking, equiaxed grains in the weld metal can also nucleate by mechanisms other than heterogeneous nucleation. Some of these mechanisms have been discussed recently by Kou et al. (8) and hence will not be repeated here. However, it should be emphasized that too often equiaxed grains in welds have been claimed to have nucleated by dendrite fragmentation without any proof at all. The equiaxed grains in the 6061 aluminum weld shown here could have been a perfect example, judging from the fact that this alloy is not known to contain titanium.

III. Comet and Stray Grains

Fig. 4 shows the grain structure of a 6061 aluminum weld made with a current of 70A and a speed of 5.08 m/sec (12 ipm). As shown, quite a few small grains scattered in the fusion zone. These small grains were not quite equiaxed but had a tail pointing toward the heat source during welding, hence the name "comet grains".

The maximum growth direction of a grain is in line with the direction of the maximum temperature gradient and the latter points toward the heat source. Consequently, when a small grain nucleates during welding, it can grow preferentially in the driection pointing toward the heat source. Of course, if the temperature gradient in this direction is not significantly higher than those in other directions, the grain will be more equiaxed in shape rather than comet.

Fig. 5 shows the comet grains previously observed in an Al-0.3 Fe-0.1 Si alloy (9). The sharp tails also pointed toward the heat source during welding.

Fig. 6 shows a comet grain in a 6061 aluminum weld. As expected, it pointed toward the heat source. The interesting thing about this micrograph, however, is that this comet grain nucleated heterogeneously from the dark particle located at the origin of the dendrite of the grain, i.e., the nucleus. This nucleus, like that shown previously in Fig. 3, is rich in titanium. Comet grains, like equiaxed grains, can also nucleate by mechanisms other than heterogeneous nucleation. But in this case, heterogeneous nucleation, rather than the often erroneously quoted dendrite fragmentation, is clearly the nucleation mechanism.

If the condition of competitive growth during welding is such that some grains nucleated inside the boundary of the fusion zone are able to grow a considerable distance, stray grains can form. These grains, which are columnar in nature, can be considered as the further extension of comet grains. A few stray grains are in fact visible in the micrograph shown in Fig. 4.

Acknowledgements

The authors gratefully acknowledge support for this study from the National Science Foundation, under NSF Grant No. 8419274.

References

1. Kou, S. and Y. Le, Weld. J. 64, 51 (1985).
2. Kou, S. and Y. Le, Metall. Trans. A, 16A, 1887 (1985).
3. Kou, S. and Y. Le, Metall. Trans. A, 16A, 1345 (1985).
4. Savage, W. F., C. D. Lundin and A. H. Aronson, Weld. J. 44, 175s (1965).
5. Savage, W. F. and A. H. Aronson, Weld. J. 45, 85s (1966).
6. Savae, W. F., C. D. Lundin and T. F. Chase, Weld. J. 47, 522s (1968).
7. Savage, W. F. and R.J. Hrubec, Weld. J. 51, 260s (1972).
8. Kou, S. and Y. Le: Weld. J., in press.
9. Nakqagawa, H., K. Mitsuaki, F. Matsuda and T. Senda, Trans. Japan Weld. Society, 1, No. 2 (1970).

Table I
Chemical Analysis of 6061 Aluminum

	Mg	Si	Cu	Fe	Mn	Cr	Ni	Zn	Ti	Zr	Al
wt. pct.	1.2	0.6	0.2	0.4	0.07	0.2	0.01	0.09	0.04	0	Balance

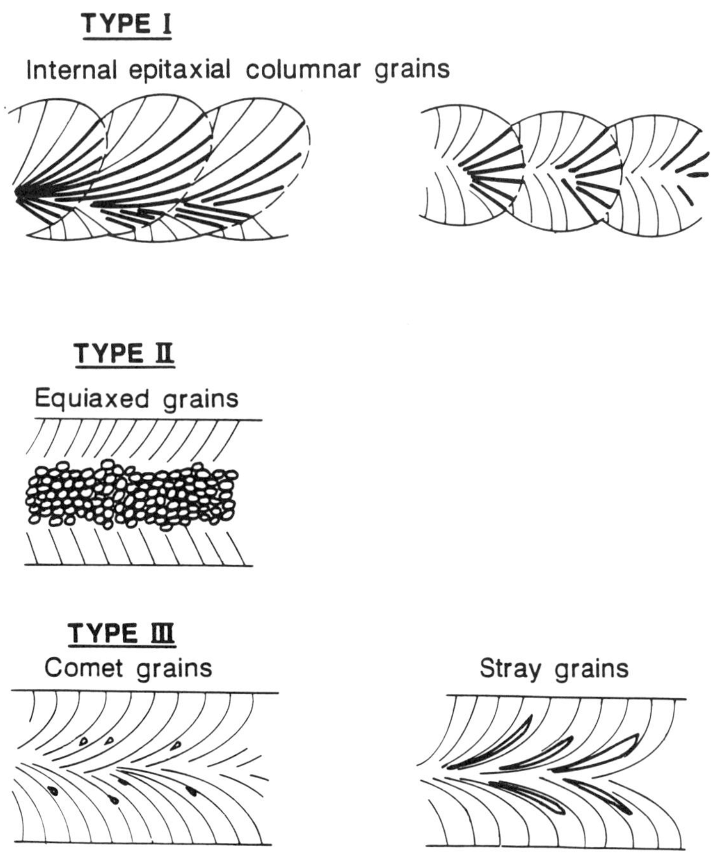

Fig. 1 Schematic sketch of three types of new grains in the weld metal: (a) Type I, internal epitaxial columnar grains; (b) Type II, equiaxed grains; (c) Type III, comet and stray grains.

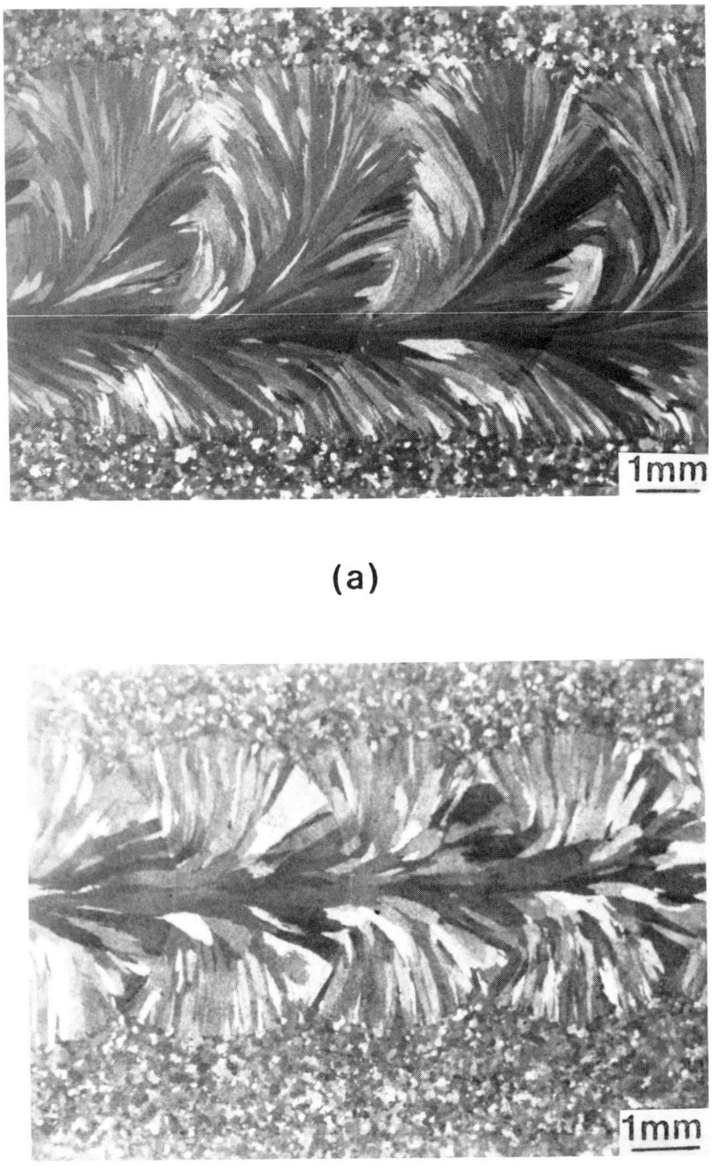

(a)

(b)

Fig. 2 Internal epitaxial grains in 1100 aluminum welds: (a) oscillated arc weld; (b) pulsed arc weld.

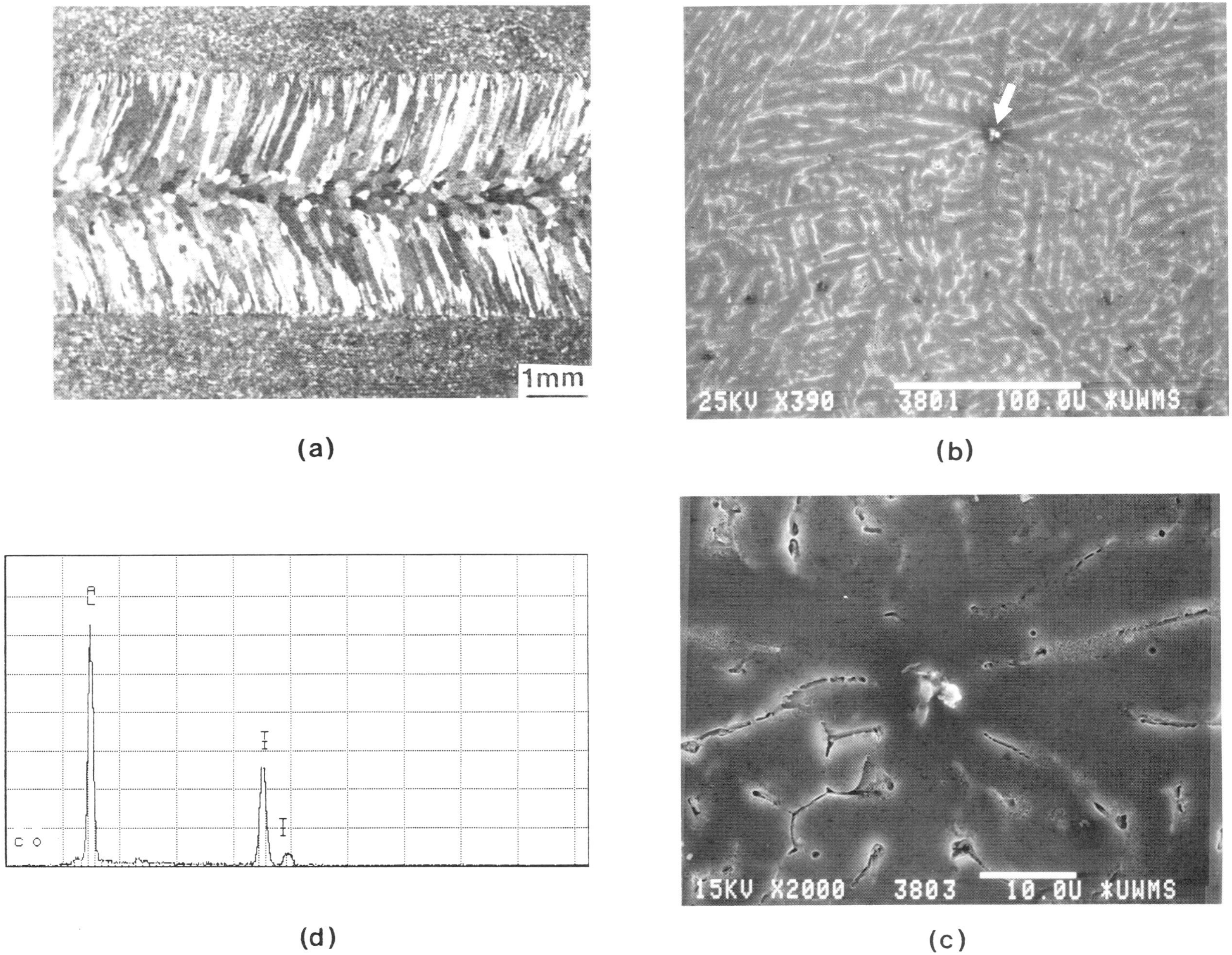

Fig. 3 Equiaxed grains in a 6061 aluminum weld: (a) grain structure; (b) structure near the origin of dendrite; (c) heterogeneous nucleus; (d) EDS analysis of the nucleus.

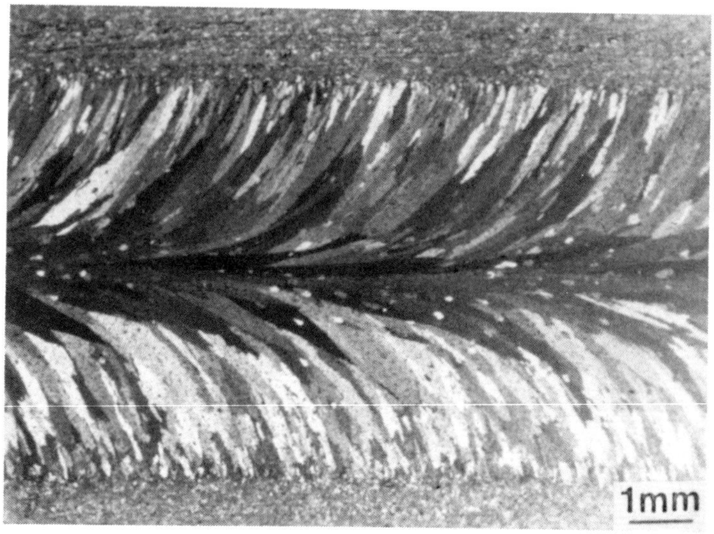

Fig. 4 6061 aluminum weld containing small comet grains.

Fig. 5 Comet grains in a weld of Al-0.3Fe-0.1Si alloy (From Nakagawa et al.[9]).

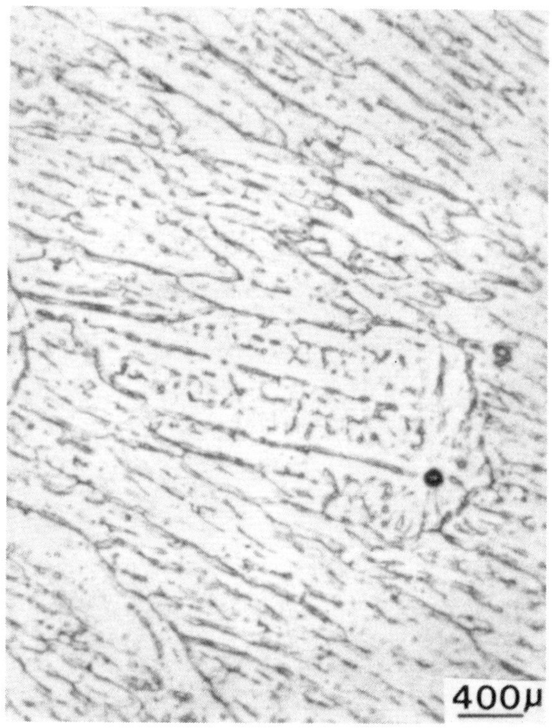

Fig. 6 A heterogeneously nucleated comet grain in a 6061 aluminum weld.

HOT CRACKING IN IRON BASE ALLOYS - EFFECT OF SOLIDIFICATION MODE

L. Pentti Karjalainen
University of Oulu
Oulu, Finland

Veli P. Kujanpää
Lappeenranta University of Technology
Lappeenranta, Finland

Niilo Suutala
Outokumpu Oy Stainless Steel Division
Tornio, Finland

Abstract

Austenitic stainless steels are resistant to hot cracking if their solidification mode is ferritic-austenitic. The weld hot cracking susceptibilities of Fe-C, Fe-C-Mn, low-alloy and Fe-Ni type steels are assessed on the basis of a literature review and transvarestraint tests carried out. It is observed that these steels are resistant to cracking if Nieq = Ni+22C+0.31Mn+14.2N+Cu (in wt%) is between 2.3 and 3.7, but prone to cracking if Nieq<2.3 unless S+P<0.04% or Nieq>3.7 unless S+P<0.02%. The high resistance is connected with primary ferritic solidification, which is followed by a peritectic reaction resulting in the formation of austenite, while ferrite still exists at the last stages of solidification, hence the presence of numerous ferrite-austenite phase boundaries.

CRACKING AT HIGH TEMPERATURES can cause troubles during continuous casting and welding. Where austenitic stainless steels are concerned it is now well established that the solidification mode plays an essential role in cracking susceptibility /1-5/. Steels solidifying primarily as austenite possess a great risk of hot (solidification) cracking unless their impurity content is very low, whereas those having ferritic-austenitic solidification are not prone to cracking even at a high impurity content. The cracking tendency also seems to increase also if the solidification takes place as single phase ferrite /4,6/.

Hot cracking does not often play a very serious role in the case of low carbon steels, but it does so more frequently in low-alloy steels, or higher-alloyed high strength steels, nickel steels over about 2.5% Ni and free-machining steels, especially at high welding energies. The detrimental role of the impurities sulphur and phosphorus and certain elements such as carbon, copper,and nickel and the beneficial effects of manganese, molybdenum, vanadium, calcium, REM, and oxygen have been noted and numerous regression formulae has been developed to assess the risk of cracking, as reviewed by Dixon /7/ and Jain et al./8/. These formulae are nevertheless often complex and of unknown significance and limited applicability.

The question of the role of the solidification mode may be arisen, as for austenitic stainless steels. The solidification mode of the Fe-C alloy, as deduced from the phase diagram (Fig.1), is single-phase ferritic if C<0.09% and single-phase austenitic if C>0.54% /9/. Between these values it begins as delta ferrite, which is followed by a peritectic reaction in which austenite is formed from liquid and ferrite.

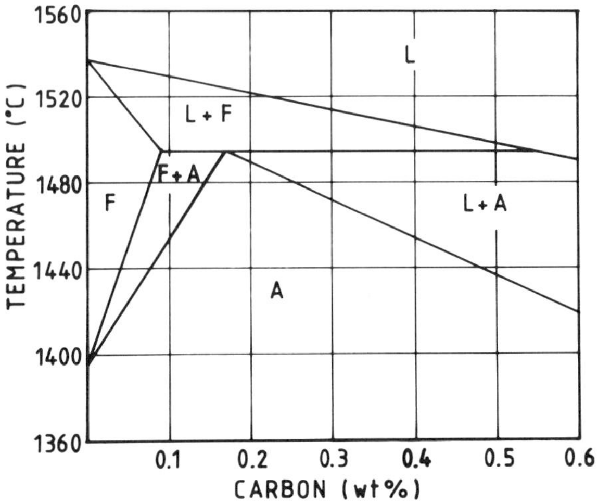

Fig. 1 - Part of the Fe-C phase diagram /9/

Masumoto et al./10/ and Masumoto and Imai /11/ found that Fe-C alloys containing more than 0.1%C are sensitive to hot cracking, as are Fe-Ni steels with more than about 2.5%Ni. This was explained as being due to the peritectic reaction with austenite formation and strong segregation of sulphur. Borland /12/ stated that the solidification should be ferritic to prevent cracking, and the beneficial effects of ferrite-promoting elements and the detrimental effects of austenitizers are emphasised by Bonomo /13/. Kihara and Matsuda /14/ deduces that the cracking susceptibility of high-strength steels must depend on their potential for forming austenite on solidification. Recently Kim et al. /15 / have pointed to the combined effect of C and Ni, finding an increasing risk of cracking with the promotion of austenitic solidification.

On some occasions cracking also seems to be able to take place also in steels with a carbon content below 0.1%, as shown by the results of Nakagawa et al /16 / and Jones /17/, for instance. Ohshita et al. /18/ also realized an increased cracking tendency in very low carbon steels (<0.07%) as the solidification is fully ferritic.This means that the hot cracking phenomenon has still not been explained. The present work extends the consideration of solidification mode developed earlier for austenitic stainless steels to other iron base alloys, in particular low carbon steels, low-alloy steels and nickel steels. Solidification cracking susceptibility is assessed in this respect on the basis of experimental data for welds obtained from the literature and from transvarestraint tests carried out.

EXPERIMENTAL

A series of transvarestraint tests was carried out on steels of the compositions given in Table 1, using the testing parameters in Table 2. The lengths of cracks were measured optically with a stereo microscope on a section parallel to the plate surface.

RESULTS

SOLIDIFICATION MODES-An extensive literature search was made on the effect of composition on the solidification mode of a steel, including iron base alloys from mild steels to high-alloy stainless steels. The results are given on a Creq-Nieq field in FIG.2, in which the solidification sequences are also presented

Table 2 - Conditions in Transvarestraint Tests

Welding current	170 A
Pole voltage	16-18 V
Arc length	2 mm
Welding speed	70 mm/min
Electrode	EWTh-2, diam 2.4 mm
Shielding gas	99.99% Ar
Bending time	0.5 s
Bending radius	150 mm
Augmented surface strain	3.2%

schematically. The results are derived from various laboratory experiments, production scale castings and welds. The criteria for inclusion are that the author has argued his proposition concerning the solidification mode and gives an exact composition.

The data in Fig.2 first confirm that simple equivalents Creq= Cr+1.37Mo+1.50Si and Nieq= Ni+22C+0.31Mn+14.2N+Cu (elements in wt%) can be used to predict the solidification mode of austenitic stainless steels, a fact established before /28,29/, and also show that these same equivalents can, more generally, be successfully extended to other steels, too. Three solidification regions are distinctly distinguishable in carbon steels: austenitic, ferritic-austenitic and ferritic. The demarcating Nieq's when the Creq is zero, are about 10 and 2.5, corresponding approximately to carbon contents of 0.54% and 0.09% with a coefficient of 22. The vertical projection of the composition, corresponding to 0.17%C in a binary Fe-C alloy, is shown schematically by the dashed line in Fig.2.

CRACKING SUSCEPTIBILITY-The steels were ranked into three cracking sensitivity classes, susceptible, somewhat-susceptible and non-susceptible, based on the same criteria as used previously for austenitic stainless steels /4/. The criteria are given in Table 3, including several hot cracking tests, the results of which are taken from literature.

Table 1 - Compositions of Test Steels (in wt%) and Maximum Lengths of Cracks in Transvarestraint Tests. Creq and Nieq are also given.

Code	C	Si	Mn	P	S	Cu	Cr	Ni	Mo	N(ppm)	l_{max}(mm)	Creq	Nieq
C01	.007	.00	.03	.011	.004	.01	.01	.03	.01	56	0.4	0.0	0.2
C11	.11	.16	.36	.016	.023	.08	.03	.07	.01	32	1.1	0.3	2.7
C15	.15	.22	.58	.015	.041	.03	.01	.01	.00		2.7	0.3	3.5
C30	.30	.29	.68	.026	.037	.30	.17	.16	.03		3.2	0.7	7.6
C50	.50	.24	1.07	.011	.022	.32	.16	.19	.02		2.8	0.6	11.8
C56	.56	.31	1.17	.019	.029	.22	.11	.11	.02	52	6.1	0.6	13.5
C42	.42	1.62	.76	.012	.032	.26	.15	.13	.04	76	2.7	3.0	9.8
C12	.12	.26	1.12	.012	.027	.31	.51	.21	.01	25	0.7	1.1	3.5
CNi	.03	.24	.67	.005	.005	.01	.02	8.97	.00	42	2.3	0.4	9.8
CS1	.15	.22	.82	.019	.013	.19	.16	.20	.02		1.4	0.5	3.9
CS2	.14	.25	.98	.024	.047	.29	.26	.18	.02		2.7	0.7	3.9
CS3	.15	.26	1.17	.026	.099	.30	.27	.15	.04		3.4	0.9	4.1
CS4	.12	.0	1.30	.051	.392	.01	.05	.01			3.6	0.1	3.0
CR	.08	.98	.82	.020	.019	.06	17.9	.11	.01	124	0.6	19.1	2.2

Table 3 - Assessment of Results of Hot Cracking Tests

Test	Quantity	Non-Susceptible	Somewhat susceptible	Susceptible
Transvarestraint Test	Brittleness Temperature Range	0-110 °C	110-145 °C	145 °C -
	Maximum Crack Length	0-0.5mm	0.5-0.8mm	0.8mm -
Circular Groove Test	Crack Angle	0-5	5-10	10 -
	Cracking	0-0.5 %	0.5-10 %	10 % -
Fissure Bend Test	Fissure Count	0	1-30	30 -
Test of Brooks	Rating	1	1.2-2.2	2.3-3.4
Test of Schmidtmann	Deformation Rate	0.2mm/s -	.15 -.20	0-0.5mm/s
Test of Sadowski	Number of GTA Weld Cracks	0	1-20	21-
Finger test	Cracking of Weld	1-10 %	10-20 %	21 % -
Patch Test	Cracking Angle	0-15	15-50	50 -
Murex Test	Cracking of Weld	0-10 %	10-40 %	41 % -
Huxley Test	Cracking of Weld	0-8	9-20	21 -
Zhdanov Method	Critical Rate of Deformation	11mm/min-	7-10.9	0-6.9mm/min

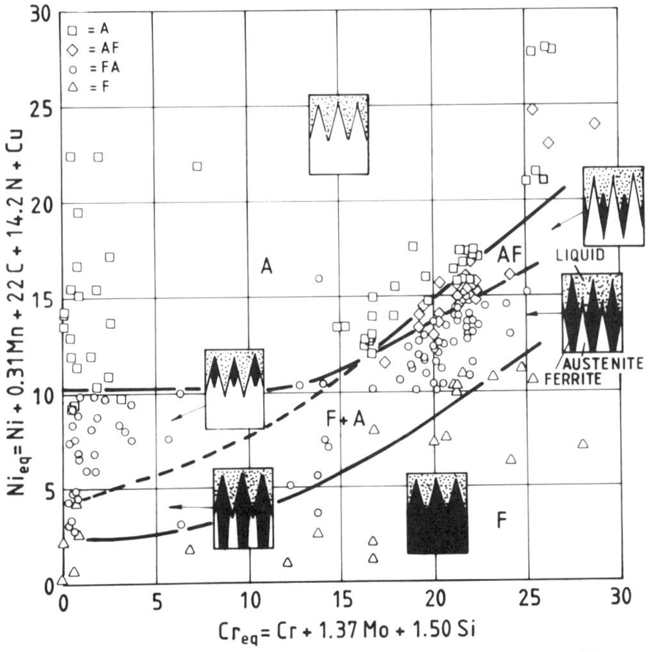

Fig. 2 - Solidification modes of iron base alloys (A = fully austenitic, AF = austenitic-ferritic, FA = ferritic austenitic, F = fully ferritic). Literature data from Refs 19-29, for stainless steels see Refs 27,28

The results from the literature and the transvarestraint tests here (for results, see Table 1) are plotted without any distinction on the Creq-Nieq field in Fig.3. For clarity, the low Creq region is enlarged in Fig.4. It must be noted that not every single data point is shown,e.g. regarding the very extensive data of Morgan-Warren and Jordan/30/, which cover more than one hundred low-alloy steels, only certain typical examples of the different steel groups are taken into account, though the whole body of data is quite consistent with the ideas to be expressed below.

Lines of demarcation can be drawn from the high equivalents region (stainless steels) down to the low equivalents region (low-carbon steels), separating the susceptible alloys quite distinctly from the non-susceptible ones. The interceptions with the Nieq axis are at about 4.0 and 2.5, the former value corresponding approximately to 0.17%C in a Fe-C alloy (Nieq=3.7) and falling slightly below 4.5%Ni for a Fe-Ni alloy (Nieq=4.5)/10,15/.

In order to take account of the impurity content of the steels, the data for steels with Creq <2, i.e. Fe-C, Fe-C-Mn, some low-alloy and Fe-Ni steels, are plotted in Fig.5. This includes steels with a low (<0.002%), moderate or high (>0.1%) sulphur content. The data contains also some alloys, for which the phosphorus content is not known (generally below 0.03%). These data points are marked by arrows.

Fig.5 shows that if Nieq is between about 2.3 and 3.7 the alloys are generally not prone to cracking unless they have a very high impurity content (S+P>0.12%). Exceptions are rare (mild steels C11 and C15 in Table 1) other than in the data of Bailey and Jones /34/, not included in Fig. 5, which concern several susceptible steels with 0.12...0.16%C and 1%Mn (Nieq=2.5...3.8). Conversely, if the Nieq isbelow 2.3 the steels are generally susceptible to hot cracking unless S+P<0.04...0.06% (S<0.03%). There are some exceptions in this region, involving

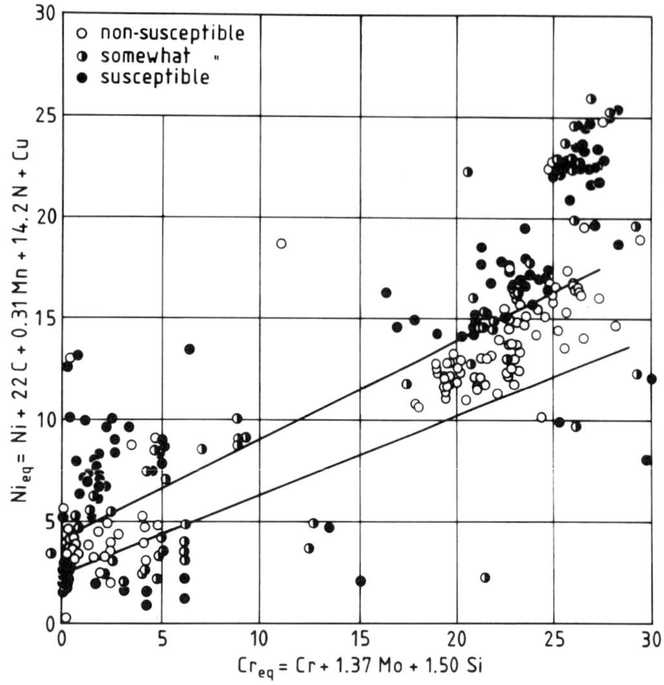

Fig. 3 - Solidification cracking susceptibility of iron base alloys. Data from Refs 4, 10-18, 30-35, for stainless steels see Ref. 4

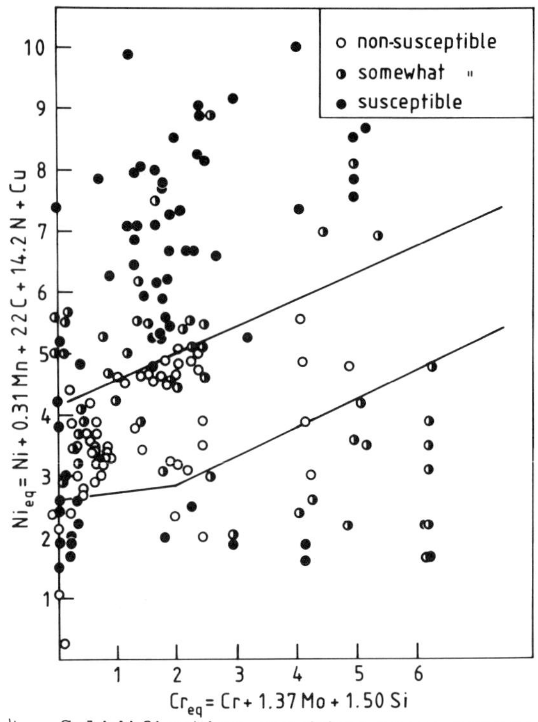

Fig.4 - Solidification cracking susceptibility of mild, low-alloy and nickel steels /10-18, 30-35/

non-susceptible alloys, but these are ones with a high manganese content (high Mn/S ratio). Similarly, if the Nieq is higher than 3.7 the steels are again susceptible to cracking unless S+P<0.01...0.02%.

DISCUSSION

As firmly established earlier in case of austenitic stainless steels, the susceptibility to solidification cracking is low as the solidification mode is ferritic-austenitic, i.e. the ratio Creq/Nieq is between 1.5 and 2.0 /4 /. In this case ferrite is the primary phase during solidification and the austenite is formed between ferrite dendrites through a three-phase reaction, which can be either peritectic or eutectic (see schemes in Fig.2). This superior resistance is attributed to the reduced segregation of impurities such as sulphur and phosphorus, their higher solubility in ferrite and the fact that the austenite, as it nucleates, increases markedly the inter-phase interfaces between ferrite and austenite. It is generally accepted that the interfacial energies of ferrite-austenite boundaries will be lower than that of austenite-austenite or ferrite-ferrite boundary, which means that they will be more slightly wetted by the last traces of liquid /6,35 /. Thus the initiation and propagation of solidification cracks is suppressed provided that a ferrite phase exists at last critical stages of solidification as is the case in austenitic stainless steels, in which the transformation from ferrite to austenite takes place relatively slowly.

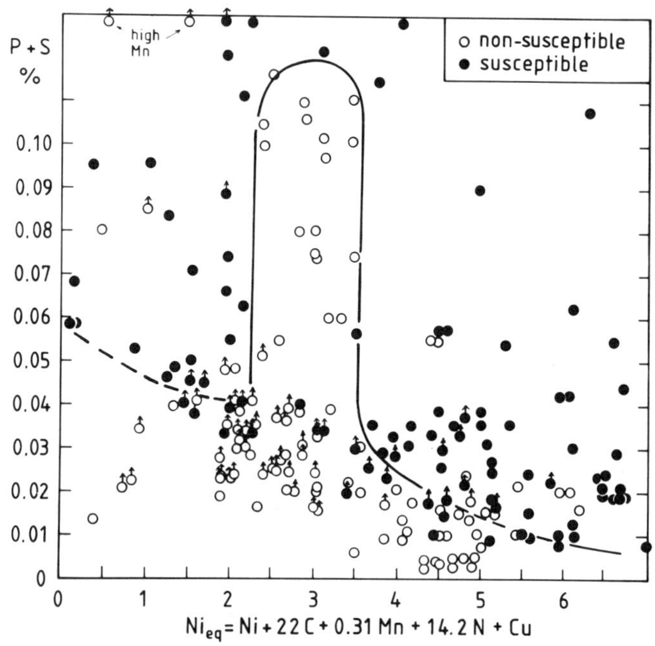

Fig. 5 - Effect of impurity content and composition on the cracking susceptibility of iron base alloys with Creq< 2.0 /10-18, 30-35/

The results for low-carbon, low-alloy and nickel steels shown in Figs. 3-5 also point out high resistance to cracking in the case of primary ferritic solidification if Nieq is between 2.3 and 3.7. Then both ferrite and austenite phases exist at the final stages of solidification (carbon content between 0.09 and 0.17% in the Fe-C alloy). Resistance is lower in cases of fully ferritic (Nieq<2.3) and austenitic

(Nieq>10) solidification, just as in austenitic stainless steels, but it is also poor during primary ferritic solidification if Nieq is between 3.7 and 10. Referring to Figs 1 and 2 for the Fe-C alloys, one realizes that in this case (carbon content between 0.17 and 0.54%) the peritectic reaction results in liquid and austenite, i.e. the ferrite disappears before complete solidification is achieved. Due to the high diffusion rate of carbon, the actual phases may well be those shown by the equilibrium diagram. Finally, only austenite is present at the critical stages for crack formation, as in the primary austenitic solidification of austenitic stainless steels. Hence, austenite-austenite grain boundaries will be formed, wetted by impurity films and with poor cohesion.

In low-carbon steels the relative content of Mn and S has been observed to control the structure and the shape of sulphides, which when film-like form weak paths along primary grain boundaries and contribute to cracking /12,16/. Phosphorus is also known to segregate, being harmful especially at low carbon levels /7,34/. As seen in Fig.5, the impurity content has an effect on the cracking risk, but the critical values obviously depend on the solidification mode, as for stainless steels /4 /. Manganese is a special element, however, as shown by the fact that high manganese steels have a higher resistance than could be predicted on the basis of the solidification mode (see Fig.5).

The Creq and Nieq adopted here are those observed to forecast the solidification mode for austenitic stainless steels most accurately /28,29/. Some slight modifications could perhaps give better results at low equivalents, e.g. for nickel coefficient relative to carbon of between 23 and 28 is suggested instead of the figure 22 used here /13,14,15,18/. This is hardly of any high significance for forecasting hot cracking risks, however, as several other factors such as restraint, joint design and welding parameters affect the cracking susceptibility in practical welding operations.

CONCLUSIONS

The solidification mode plays an essential role with regard to the weld hot cracking susceptibility of iron base alloys. For austenitic stainless steels the cracking resistance is high if the solidification mode is ferritic-austenitic, Creq/Nieq ratio (Creq=Cr+1.37Mo+1.5Si, Nieq=Ni+22C+.31Mn+14.2N+Cu) between 1.5 and 2.0. For low-carbon, low-alloy and nickel steels the resistance to cracking is highest if Nieq is between 2.3 and 3.7, while both ferrite and austenite phases exist at the final stages of solidification.

ACKNOWLEDGEMENT

The financial support provided by The Finnish Academy is gratefully aknowledged.

REFERENCES

1. Masumoto,I.,K.Tawaki and M.Katsuna, J.Jpn Weld. Soc. 41,1306-1314(1972)
2. Thier,H.,DVS-Berichte 33,69-75(1975)
3. Arata,Y.,F.Matsuda and S.Katayma, Trans. JWRI 5,31-51(1976)
4. Kujanpää,V.,N.Suutala,T.Takalo and T.Moisio, Weld.Res.Int. 9,55-76(1979)
5. Katayama,S.,T.Fujimoto and A.Matsunawa, Trans. JWRI 14,123-138(1985)
6. Hull,F.C., Weld.J. 47,399s-409s(1967)
7. Dixon,B., Austr.Weld.J.,Summer,23-30(1981)
8. Jain,S.C.,D.B.Goel and P.C.Gupta, Tool Alloy Steels 14,191-197(1980)
9. Metals Handbook, vol.8, 8th ed.,ASM, Metals Park USA,1975,p.276
10. Masumoto,I.,K.Tanaki,M.Kutsuna and H.Ozaki, Schweissen und Schneiden 27,450-454(1975)
11. Masumoto,I. and K.Imai, Trans. JWRI 39,104-111(1970)
12. Borland,J.C., Br.Weld.J. 8,526-530(1961)
13. Bonomo,F., Weld.Res.Int. 2,1-28(1972)
14. Kihara,H. and F.Matsuda, Trans. JWRI 2,83-95(1973)
15. Kim,J-G.,K-T.Rie and J.Ruge, Arch.Eisenhuttenwes. 56,639-644(1985)
16. Nakagawa,H.,F.Matsuda and T.Senda, Trans. Jpn Weld.Soc. 5,18-25,26-33,89-94(1974)
17. Jones,P.W., Br.Weld.J. 6.282-290(1959)
18. Ohshita,S.,N.Yurioka,N.Mori and T.Kimura, Weld.J. 64,129s-136s(1983)
19. A Guide to the Solidification of Steels, Jernkontoret,Stockholm Sweden, 1977, 162p.
20. Hammar,Ö. and G.Grunbaum, Scand.J.Metall. 3,11-20(1974)
21. Jacobi,H. and W.Pitsch, Arch.Eisenhuttenwes. 46,417-422(1975)
22. Jacobi,H. and K.Schwerdtfeger, Met.Trans. 7A,811-820(1976)
23. Bibby,P.A. and J.Bech, JISI 211,290-292(1973)
24. Edvarsson,T.,H.Fredriksson and I.Svensson, Met.Sci. 10,298-306(1976)
25. Fredriksson,H.,Met.Sci. 10,77-86(1976)
26. Loh,J., Im hochtemperaturbereich entstehende Risse beim Brammenstranggiessen nichtrostenden Stähle, dissertation, TH Aachen, 1984
27. Suutala,N., Solidification Technology in the Foundry and Cast House (Int.Conf.) 15-17 Sept. 1980, Coventry UK,p.310-314
28. Suutala,N., Met.Trans. 14A,191-197(1983)
29. Hammar,Ö. and U.Svensson, Solidification and Casting of Metals (Int.Conf.) 18-21July 1977, Sheffield UK,p.401-410
30. Morgan-Warren,E.J. and M.F.Jordan, Met.Technol. 6,271-278(1974)
31. Garland,J.G., Weld.Res.Int. 9,85-107(1979)
32. Matsuda,F.,H.Nakamma,H.Fujimaki and S.Saruwatari, Trans. JWRI 6,53-58(1977)
33. Sterenbogen,Yu.A., Weld Pool Chemistry and Metallurgy (Int. Conf.) 15-17April 1980, London UK,p.31-41
34. Bailey,N. and S.B.Jones, Weld.J. 58,217s-231s(1978)
35. Kujanpää,V.P.,S.A.David and C.L.White, will be published in Weld.J.

EFFECT OF Si ON THE WELDABILITY OF Fe-Mn-Al STEELS

D.K. Aidun, J. Suh, T. Zacharia
Mechanical Engineering Department
Clarkson University
Potsdam, New York 13676

Effect of Si on the Weldability of Fe Mn Al Steels

D.K. Aidun, J. Suh and T. Zacharia
Mechanical Engineering Department
Clarkson University
Potsdam, New York 13676

ABSTRACT

PRELIMINARY RESULTS of a study done on the impact of Si levels on the weldability of Fe Mn Al austenitic steels indicate that alternative welding procedures can provide good weldability and mechanical properties without sacrificing corrosion properties. The authors are optimistic that further research will lead to the development of Fe Mn Al austenitic alloys that will be feasible substitutes for some Cr-containing austenitic stainless steels at least in non-critical applications.

INTRODUCTION

THE PURPOSE of this paper is to outline an approach used to investigate the weldability of Fe Mn Al austenitic steels, specifically two compositions: 29.7 Mn 7.9 Al 1.04C, and 30.3 Mn 7.9 Al 1.05C with 1.5 Si added.

Research on mechanical properties and environmental effects on Fe Mn Al alloys has determined that replacing Cr with Al, Ni with Mn and C need not inhibit sulfidation or corrosion resistance (1,2). Furthermore, Casteletti and Spinelli (3) have shown that adding Si can, in fact, increase oxidation resistance. Attractive qualities albeit increasing Si decreases the weldability of the Fe Mn Al alloy, thereby limiting its current applicability to non-critical uses.

With welding, an important and required fabrication process, good weldability properties must be guaranteed before any alloy can be used in critical applications such as in the petroleum, aerospace and nuclear power plant industries. Weldability is then a primary concern in expanding the scope of applications of Fe Mn Al alloys to include those uses previously limited to the Cr containing alloys.

The research outline herein reports on preliminary results of weldability studies on two compositions chosen in part for their inherent oxidation resistance.

EXPERIMENTAL PROCEDURES

TWO HEATS of Fe Mn Al steel were selected with compositions shown in Table 1. The only difference between the two heats is the Si content. Specimens 15x1.88x0.32 cm were machined and butt-welded using autogenous GTAW process. The welding parameters are shown in Table 2. Flat tensile specimens 6.88x0.31x0.2 cm were machined out from the weldments such that they contained only the weld metals (or fusion zones). Mechanical properties of the welds versus the base metals were then compared. Optical and scanning microscopies were performed on the welds to identify microstructures in the weld zones.

RESULTS AND DISCUSSION

WELDABILITY STUDIES should include an investigation of the solidification phenomenon as it relates to hot-cracking and effects on mechanical properties. In a recent investigation (4), C was found to be the primary factor affecting solidification modes of four heats of a Fe Mn Al alloy similar in composition to that studied herein. In that investigation, and in an investigation on a different alloy conducted by Szumachowski and Kotecki (5), Mn was found to have a minimal effect on the deposit ferrite number in the welds. Thus, in this investigation emphasis is on the impact of increased Si on resistance to hot-cracking.

Because of the high Si content of heat Y68 (1.5 w/o Si) segregation was expected. Indeed, center-line cracking in Y68 was so extensive that propagation was visible to the naked eye. To avoid further excessive center-line cracking and weaving weld procedure, similar to that used by Kou and Le (6), was used on subsequent butt welds of Y68.

The "zig-zag" weaving passes eliminated center-line cracking and caused almost random grain growth. The distribution of Si within the Fusion Zone (FZ) was uniform, transversing the entire FZ. The base metal (BM) of Y68 consisted of a predominantly austenitic phase whereas the FZ showed regions of ferrite along austenitic cell boundaries. Figure 1 shows the photomicrograph of the Y68 weldment. The microstructures of the BM of Y66 and Y68 are similar due to identical C content. Solidification morphologies of the FZ in Y66 (Fig. 2) and Y68 are also similar, both showing ferrite along the austenitic cell boundaries.

Figure 3 shows the solidification structure of the Y68 weldment. Subject to torch oscillation, the dendrites in the FZ do not follow any particular direction toward the center of the weld. Without torch oscillation, the dendrites in the FZ do grow toward the center; see Figure 4a. Figure 4b clearly shows dendrites from both sides of the FZ merging at the center of the weld. The weld-center, being the last to solidify, will contain the highest concentration of solutes.

As shown in Figure 5a (top view of solidification structure of Y68 subjected to torch oscillation), the morphology is of a basket-weaved type, with fine austenitic cells bounded by ferrite. The area is enlarged in fig. 5b and the uniform distribution of the austenite cells is even more pronounced.

Figure 6 shows the fracture surfaces of Y68, with cracked during autogeneous GTA welding. Table 3 shows a "regional" EDAX analysis of the fracture surface (FS) and a "spot" EDAX analysis

Table 1 - Chemical Composition (Wt.-%) of the Heats

Heat	C	Si	Mn	Al	P	S	B	Fe
Y66	1.04	0.16	29.7	7.90	0.003	0.01	0.005	Bal
Y68	1.05	1.50	30.5	7.90	0.003	0.01	0.005	Bal

Table 2 - Welding Conditions and Parameters

Heat	Remarks	Voltage Volts	Current amps	Speed cm/min.
		15	215	22.5
Y66	Single Pass			
		14	185	22.5
Y68		13	175	22.5

Table 3 - Chemical Analysis (in Wt. %) of Heat Y68 Using EDAX

Location	Fe	Mn	Al	Si	P	S
FS	57.06	29.00	7.97	5.97	0.00	0.00
A	60.11	24.23	8.35	7.31	0.00	0.00
B	55.11	31.29	8.62	4.98	0.00	0.00
C	56.32	22.32	6.07	15.28	0.00	0.00

of locations A, B and C. The Analysis Speaks to the Large weight percentage of Si in the fracture surface. Specifically, locations A and C show ferrite formed along austenitic cell boundaries as well as excess level of Si, indicating formation of low-melting silicides, detrimental to weldability. In sum, EDAX analysis showed as was expected, large weight percentages of Si in both the ferrite and austenitic regions.

By oscillating the torch, i.e., "zig-zagging," Si was uniformly distributed from the FZ to the weld-center such that silicides did not concentrate in the interdendritic regions. Instead, silicides transversed the FZ.

After successfully welding Y68, mechanical testing was performed on both heats (Y66 and Y68) using the Instron with a cross head velocity of 0.051 cm/min.. Table 4 shows the tensile properties of both heats (Y66 and Y68) at room temperature. Y68 has a much higher yield and ultimate strength than does Y66, attributable to its higher Si content. The ductility of Y68 is, however, lower than that of Y66. While the weld zone strengths of both heats are lower than their respective base metal strengths, the ductility of both weld zones is better than that of the base metals.

Figures 7, 8 and 9 show the effect of weld heat input on the yield strength, ultimate strength, and ductility, respectively. The effect of heat input on the strength of both heats is not clear at this time and further investigation is necessary, However, for both heats Fig. 9 shows a decreasing trend in ductility with decreasing heat input.

Table 4 - Mechanical Properties of Heats Y66 and Y68 Base Metal and Weld Metal

Heat	Yield Strength	Ultimate Strength (MPa)	Elongation (%)
Y66 (BM)	759	1069	59
Y66 (WM)	503 - 579	883 - 952	50 - 64
Y68 (BM)	1104	1242	34
Y68 (WM)	683 - 724	1000 - 1035	47 - 51

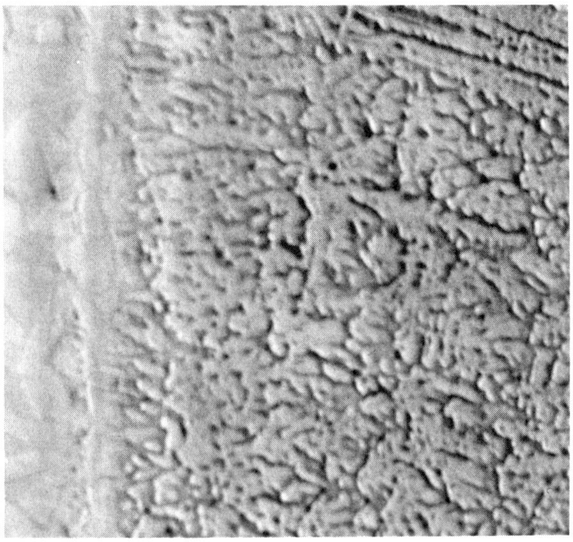

Figure 1- Microstructure of Y68 weldment
@ 375X.

Figure 3- Solidification morphology of
weaving Y68 weld zone @ 75X.

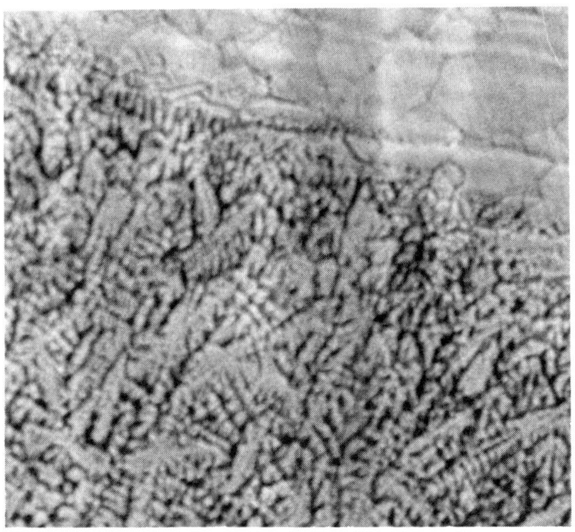

Figure 2- Microstructure of Y66 weldment
@ 375X.

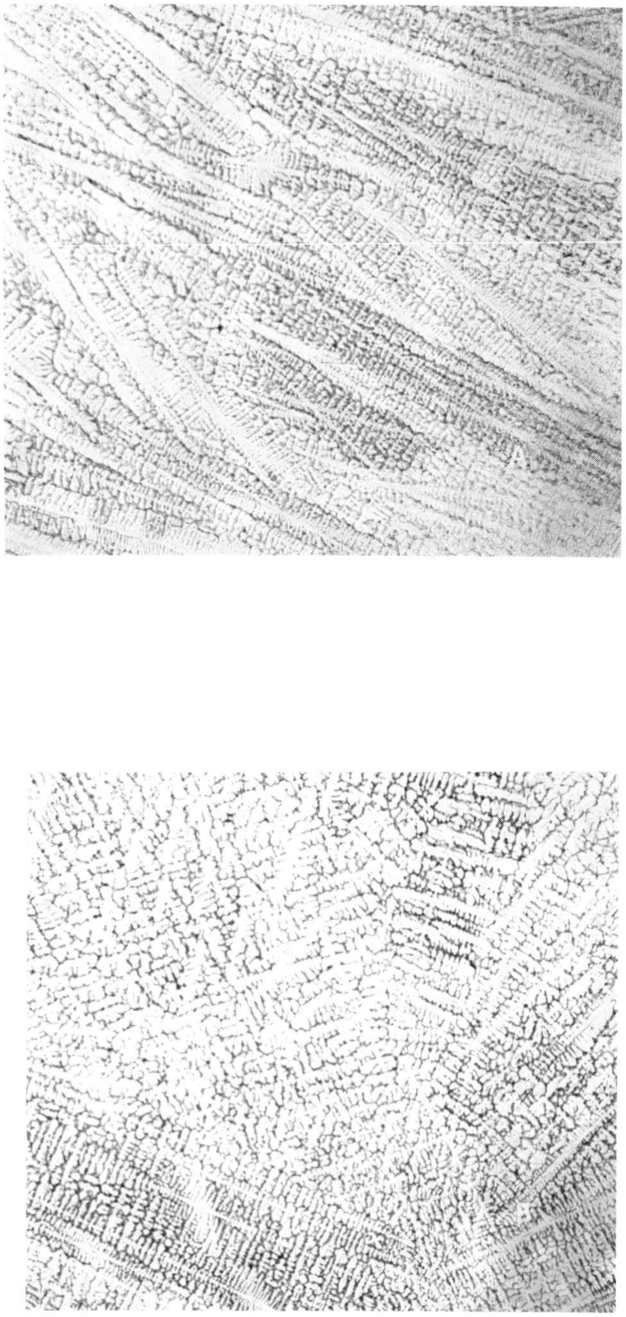

Figure 4- Solidification morphology of Y68 weld @ 150X.

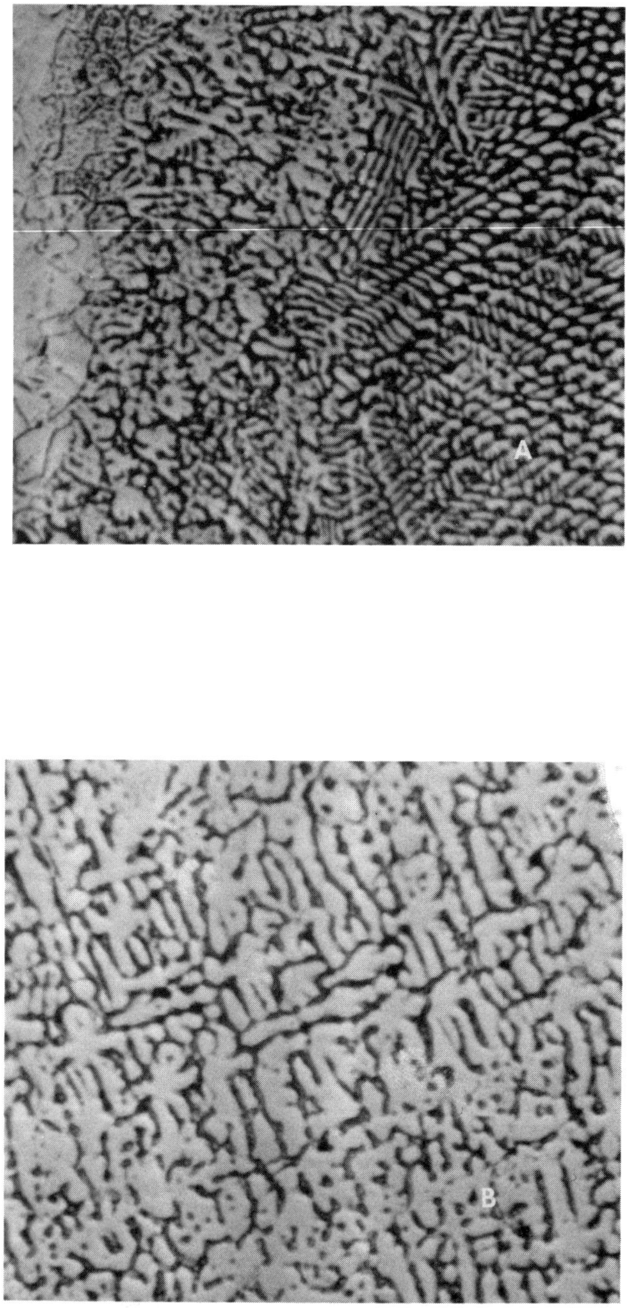

Figure 5- Solidification morphology of weaving Y68 weld zone @ 150X , top view.

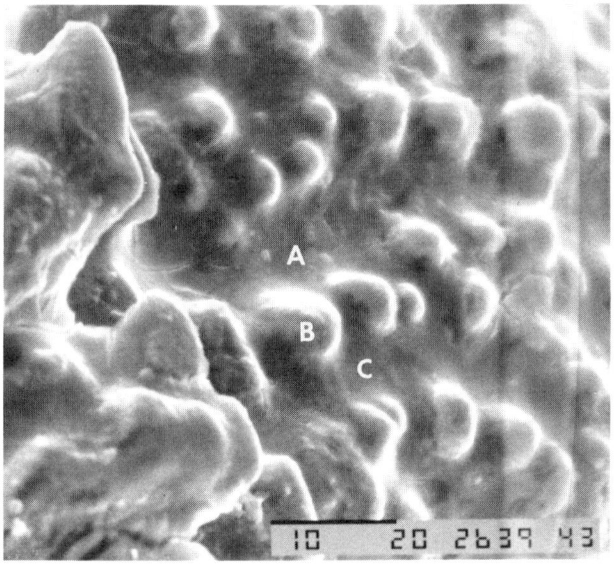

Figure 6- SEM picture of center-line cracking in Y68 weld @ 2000X.

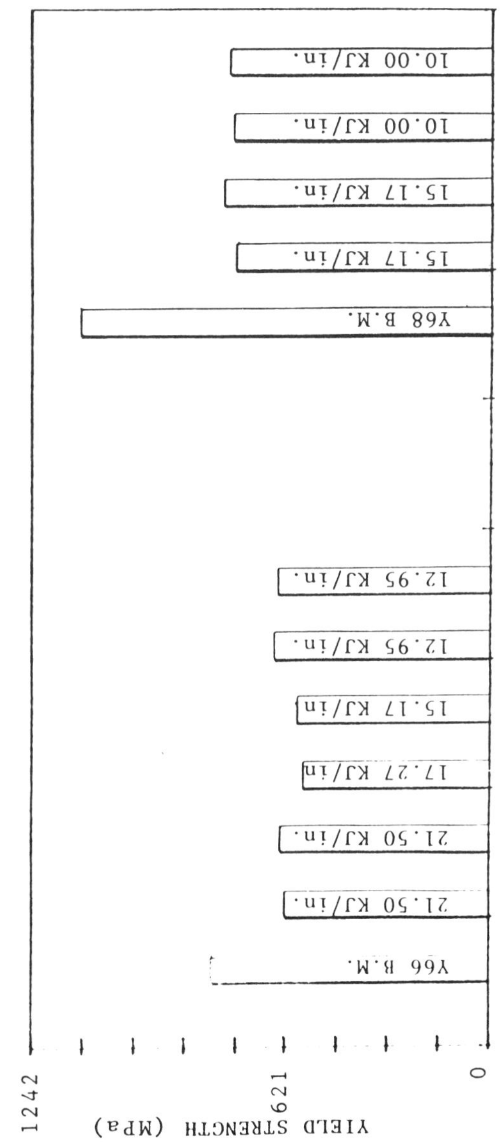

Figure 7- Effect of weld heat input on the yield strength of heats Y66 and Y68.

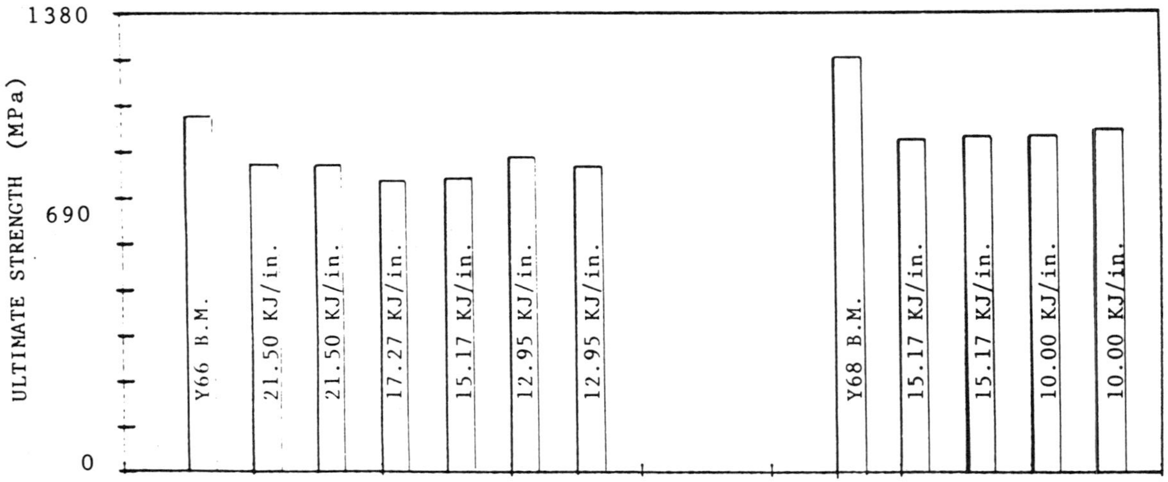

Figure 8- Effect of weld heat input on the ultimate strength of heats Y66 and Y68.

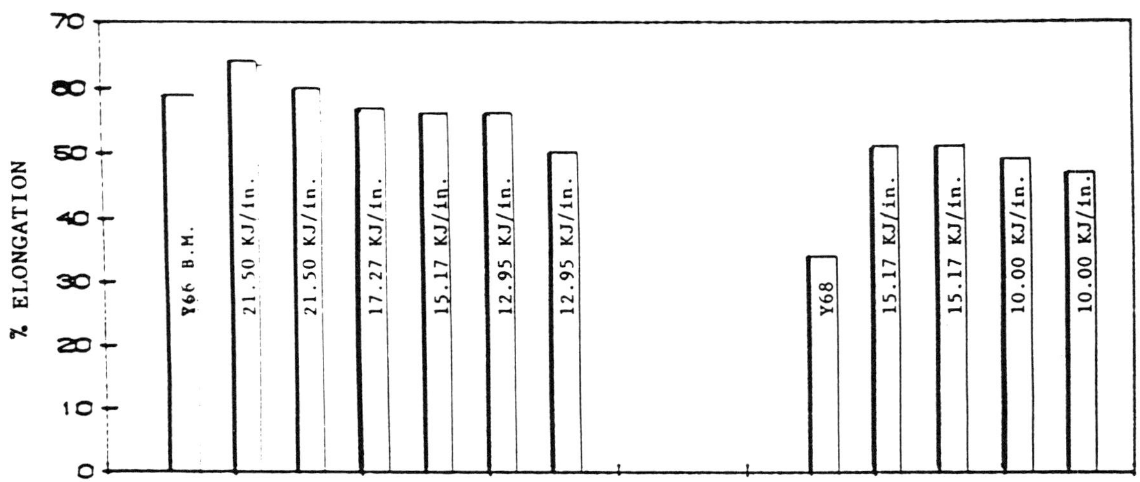

Figure 9- Effect of weld heat input on the ductility of heats Y66 and Y68.

SUMMARY

FROM THE ABOVE STATED PRELIMINARY RESULTS, it is clear that Fe Mn Al austenitic steels are weldable if properly cast and processed. High Si levels need not prohibit weldability. The cause of weld center-line cracking in heats used in this investigation is attributed to the formation of low-melting silicides along the austenitic phase boundaries, due in turn to the high degree of Si segregation. Although strengths of welded heats are lower than strengths of respective base metals (joint efficiency; 80-85%), ductility is better. Further tests are necessary to correlate mechanical properties to microstructures of welds for various cooling rates so as to better understand the mechanical and metallurgical properties of Fe Mn Al Weldments.

ACKNOWLEDGMENTS

The authors would like to thank Dr. G.R. Smolik of EG&G Idaho, Inc. for providing the materials.

REFERENCES

1. Banerji, S.K., "An Update on Fe Mn Al Steels," FASRAD 81-5, Foote Mineral Company, June 11, 81.

2. Banerji, S.K., "The 1982 Status on Fe Mn Al Steels," FASRAD #82-12, Foote Mineral Company, October 1, 82.

3. Castelleti, L.C. and Spinelli, "The Resistance to Oxidation and the Electrical Resistivity of an Fe Mn Al System Alloy," Dept. of Matls. Sci., EESC, Univ. of Sao Paulo, Sao Carlos, Brazil, Proc. XXXVI, Brazilian Soc. of Metals (ABM) Annual Congress, p. 249, held in Recife, Pernambuco, Brazil, July 5-10, 81.

4. Aidun, D.K. and Suh, J., "Effect of Carbon on the Solidification Morphology of Fe Mn Al Austenitic Steels," TMS/AIME Proceedings, March, 1986.

5. Szumachowski, E.R. and Kotecki, D.J., "Manganese Effect on Stainless Steel Weld Metal Ferrite," Welding Journal, Res. Suppl., p. 156-161, May 1984.

6. Kou, S. and Lee, Y., "Alternating Grain Orientation and Weld Solidification Cracking," Met. Trans. A, p. 1887-1896, October 1985.

PHASE TRANSFORMATIONS IN WELDMENTS

KEYNOTE ADDRESS

SOLID STATE PHASE TRANSFORMATIONS IN STEEL DURING WELDING

T.G. Gooch, P.H.M. Hart
The Welding Institute
Cambridge, England

ABSTRACT

Because the properties of welds in steel are critically dependent upon the microstructure developed in the weld metal and heat affected zone, it is essential that factors controlling transformation during the welding cycle are understood and quantified. Accordingly study has been carried out virtually since the inception of fusion welding. In recent years, a number of advances have been made in understanding metallurgical changes at welds both for ferritic materials and for high alloy grades in which the austenite phase is stable at room temperature.

In ferritic steels, attention has focused on the decomposition of austenite, especially the role of second phase particles in controlling the final transformation product, whereas in austenite stainless materials, changes during and immediately subsequent to solidification have been examined. The high temperature ferrite-austenite reaction has further been studied in regard to recently developed duplex stainless alloys. Superimposed on such microstructural studies, increasing consideration has been given to modelling the thermal cycle experienced during a welding operation.

These areas will be reviewed from both fundamental and applied viewpoints, with the aim of indicating the extent to which microstructural prediction is feasible at the present time.

INTRODUCTION

The use of fusion welding for fabrication results in the joint area being subject to localised heating and cooling with the thermal cycle experienced varying over a wide range depending on the material thickness and welding process. High energy density processes, such as electron beam welding, heat only a restricted volume of metal, but engender high heating and cooling rates, ca. 10^4°C/sec adjacent to the fusion boundary, in contrast to the much greater heating and protracted thermal cycle of electroslag welding with heating and cooling at perhaps 10^2°C/sec or less. Virtually all commercial metals and alloys display some metallurgical change between normal ambient and the melting point and inevitably imposition of a weld thermal cycle will mean that local microstructure, and hence joint properties, may differ significantly from those of the base metal. Moreover, because the time/temperature cycle experienced in welding can vary markedly, considerable differences can exist in final microstructure even for a given material (1).

Metallurgical transformation will take place during both heating and cooling, although for weld metal it is only the cooling cycle that is of concern. The extent of change will depend upon the position in the weldment, since this will determine the peak temperature and associated retention time, while, in multipass welding, a range of cycles of progressively lower peak temperature will be superimposed at any single point. In broad terms, the metallurgical changes during welding can be divided into two, namely those involving a major phase change and those involving second phase particles, whether dissolution during the heating cycle or precipitation on cooling. Because of the significance of microstructure in determining weldment properties, numerous studies have been undertaken to examine and quantify weld area transformation. In the past, these have largely followed conventional metallurgical investigations on parent materials; this is still the case, but, with increasing recognition of the industrial importance of welding, greater attention has been paid to specific study of transformations occuring under the transient thermal condition of a welding cycle, and the last decade has seen major advances in understanding.

The present paper considers the microstructural changes taking place during fusion welding of steels, with an emphasis on behaviour observed in practice. Attention is paid to conventional transformable ferritic types, to higher alloy austenitic materials, and to recently developed duplex austenitic-ferritic stainless grades, with the aim of indicating the current state of knowledge and of identifying areas where further work is required. The objective is to illustrate the extent to which microstructural prediction is possible at the present time. Hence, reference is made both to the various changes taking place during welding ferrous materials and to the extent to which thermal cycles experienced in welding can be predicted from knowledge of joint geometry and welding conditions.

FERRITIC STEELS

AUSTENITE DECOMPOSITION - The area of greatest concern in recent years regarding ferritic steels corresponds to a combination of material composition and thermal cycle which results in weld area microstructures covering almost the widest possible range of austenite transformation products, i.e. from proeutectoid ferrite to martensite. This interest arises directly from the widespread useage of microalloyed and lean alloyed C-Mn steels which produce such diversity of final microstructure under cooling conditions associated with the commonly employed welding processes. Particular reference is therefore made to these steel types in the present paper.

In transformable ferritic steels, transformation from $\alpha \rightarrow \gamma$ will occur on heating, with, depending on composition, the subsequent $\gamma \rightarrow \delta$ change, followed by the reverse reactions on cooling. Very largely, the final microstructure is dependent on the $\gamma \rightarrow \alpha$ transformation, and metallurgical welding research has been biased towards definition of factors controlling austenite decomposition. Two considerations in particular have contributed to this trend. First, there has been a demand for high weld metal toughness via achievement of a fine intragranular transformation product without the addition of nickel to impart cleavage resistance, and, second, concern has arisen that parent steels of apparently similar composition can show marked differences in hardenability and hence in risk of heat affected zone (HAZ) hydrogen cracking. In contrast, the $\delta \rightarrow \gamma$ change has been little studied, its practical effects usually being small although it may influence the austenite grain size achieved in the high temperature HAZ and weld metal, and also the relationship between the austenite grain boundaries and inclusion distribution.

For many years the transformation products during austenite decomposition have been described as being influenced just by steel composition in terms of major specified elements, austenite grain size and cooling conditions. In most metallurgical technology fields, this view has worked extremely well and has formed a satisfactory basis for classical assessment of the hardening behaviour of steels through Jominy-curves, H-curves, etc. However, in welding metallurgy, it has become increasingly clear that another factor needs to be considered. This is illustrated by the transformed microstructures in Figs 1a and 1b. Figure 1a is typical of the high temperature HAZ of a C-Mn-Nb steel composition typically 0.1C, 0.3Si, 1.4Mn, 0.010S, 0.015P, 0.015Nb, 0.025Al, 0.008N. Figure 1b is from a submerged arc (SA) weld deposit made with a basic flux and with a composition as described for the parent steel. Since the cooling rates in the HAZ and weld metal are effectively the same and the γ-grain sizes will be similar, "traditional" transformation concepts would predict similar microstructures, which is evidently not the case: the weld metal contains a high proportion of fine grained intragranular transformation product, which is virtually absent from the HAZ. More detailed consideration of the composition reveals an important difference in that the oxygen content of the steel will be around 20-30ppm and is an order of magnitude less than in weld metal made with a basic flux. It is now widely recognised in principle that this is a major cause of the observed microstructural difference.

Because of its higher oxygen content, the weld metal contains a very much higher volume fraction of inclusions that the steel. After intensive study in recent years, there is now general agreement that the difference in decomposition of austenite between the HAZ and weld metal arises from the influence of these second-phase particles, and in particular that they can promote the intragranular nucleation of non-martensitic transformation products. This ability of second-phase particles to influence austenite decomposition is not exclusive to weld metals. Indeed, the first reports in the welding field of second-phase particles influencing austenite decomposition were almost simultaneously related to both HAZ and weld metal (2-5). In the former case, it was observed that recently developed clean steels displayed increased HAZ hardenability, apparently because the formation of high temperature transformation products was suppressed, while in the latter case a correlation between oxygen level and transformation

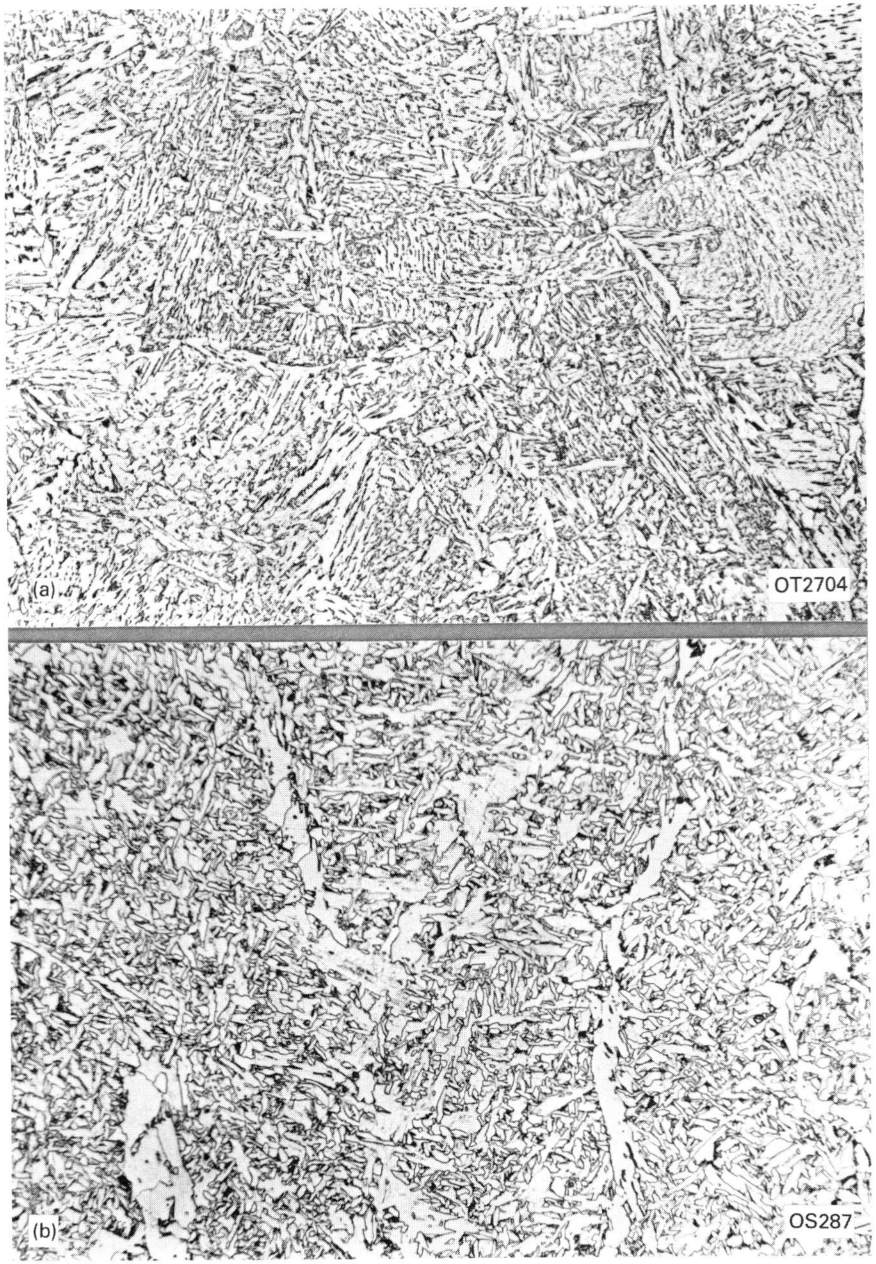

Fig.1. Transformed microstructures in a submerged-arc weld in a C-Mn-Nb steel, ×270:
a) HAZ   b) weld metal

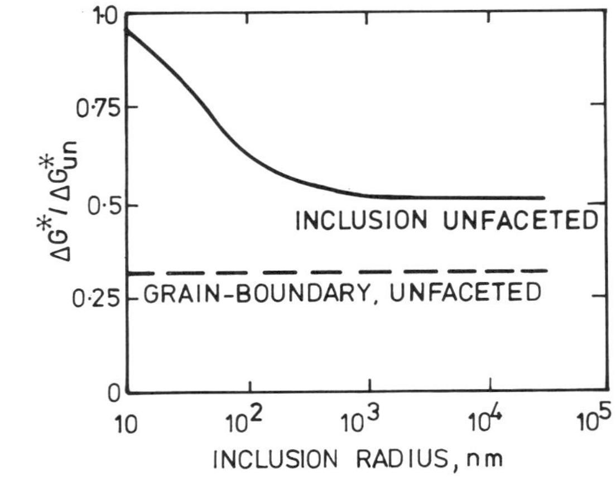

Fig.2. Predicted energy barrier to unfaceted ferrite nucleation in austenite as a function of inclusion radius: the energy barrier ΔG^* is normalised to that for homogeneous nucleation, ΔG^*_{un} (10).

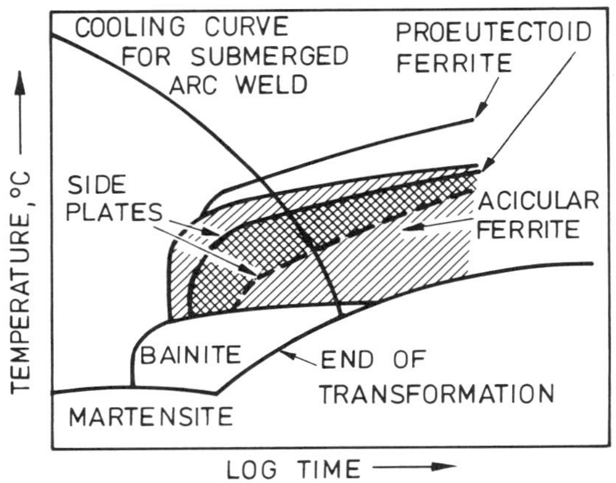

Fig.3. Schematic CCT diagram showing weld metal containing effective acicular ferrite nucleants (4).

product was noted, with direct evidence for nucleation on inclusions. Nonetheless, the major published effort since that time has been in respect of weld metals, especially with reference to the formation of acicular ferrite because of the marked beneficial effect of this constituent on toughness. Hence, the view has developed of an optimum weld deposit oxygen level for a given bulk composition and weld cooling rate. Perhaps in contrast, the trend in steel compositions, at least in respect of oxides and sulphides, has been to make steels cleaner with the aim of improving mechanical properties generally. Specialised attempts, it is true, have been made to use second-phase particles, most frequently TiN but also Ca-oxysulphides (6), TiN-MnS complexes (7, 8) and more recently, Ti oxides (9), for control of austenite grain growth in the HAZ.

ROLE OF SECOND-PHASE PARTICLES - At present, the mechanisms by which second-phase particles in weld metal produce a dramatic change in microstructure are not fully established, at least to the level necessary to give consistent prediction of transformation behaviour. Early efforts concentrated on the obvious potential aid to nucleation of the $\gamma \rightarrow \alpha$ reaction provided by second-phase particles, and today this is commonly held to be the prime cause of structure modification. This is an attractively simple concept, readily explaining for example the higher HAZ hardenability of clean steels in that a paucity of suitable nucleants will effectively suppress intermediate non-martensitic transformation products on cooling. Certainly, weld metal inclusions have been shown by calculation and observation to promote intragranular nucleation (10, 11), and it is also believed that they will enhance grain boundary nucleation when they are so located (12).

At the same time, the precise mechanisms of enhancing nucleation are not widely agreed. On energy considerations, and considering unfaceted inclusions, Ricks et al. (10), showed that, depending on the size of the inclusions, they can be intermediate between grain boundaries and the matrix (homogenous nucleation) in their effectiveness in promoting $\gamma \rightarrow \alpha$ transformation (Fig 2), and there seems little doubt about this in a general sense. However, in reality, the inclusions are highly complex, being mixtures of different sulphides (generally MnS and CuS (13)) and oxides (14), and can show complex faceted and unfaceted surfaces (15).

The calculations of Ricks et al. (10) omitted two important (but difficult to treat theoretically) aspects of inclusions which can have significant influence on their potency for aiding ferrite nucleation, namely differences in thermal contraction between inclusion and matrix (16), and the fact that lattice matching may occur (11). Both effects will aid nucleation, and this suggests at least the possibility that some inclusions may be as effective as grain boundaries in aiding ferrite nucleation. Experience from HAZ studies is that a number of inclusion types appear able to promote intragranular ferrite nucleation, namely, sulphides, silicates, TiN, BN on REM oxysulphides and Ti-oxides (2, 3, 9, 17). For weld metals, from the work of Terashima and Hart (18), it again seems that a range of types can be effective; in submerged arc weld metal of 350ppm O_2 and low Si, Al and Ti contents (i.e. oxide forming elements), intragranular acicular ferrite could in effect be "switched on" by individually increasing Si or Al or Ti level. Collectively, these findings indicate that the inclusion type may not be the most important factor in deciding the ability of second-phase particles to influence the $\gamma \rightarrow \alpha$ transformation.

Consideration of the transformation stages and other data strongly indicates that inclusions have a second, equally important, role to play in determining microstructure development, apart from aiding nucleation (and in addition to influencing austenite grain size). For weld metal and steel compositions of the general type considered here, transformation will begin at grain boundaries, probably at inclusions, and, depending on the precise cooling rate and composition, a grain boundary allotriomorph of ferrite will form, i.e. blocky ferrite or ferrite veining. In this example, Widmanstatten ferrite sideplates will subsequently develop from the boundary allotriomorph, and at this same stage, or shortly after, intragranular nucleation of acicular ferrite commences and then largely completes the transformation from austenite. The successful production of a predominantly acicular ferrite microstructure may thus require not only adequate intragranular nucleation (and of course adequate hardenability in the deposit for the cooling curve to intersect the acicular ferrite phase field (Fig 3)), but also conditions which prevent or reduce nucleation and growth of Widmanstatten ferrite despite a growth rate at least an order of magnitude greater than for grain boundary ferrite.

The possibility that second phase particles can indirectly influence acicular ferrite formation by retarding the nucleation or growth of Widmanstatten ferrite has received only little attention. However, Townsend and Kirkaldy (19) indicated that some instability of the advancing α/γ interface of grain boundary ferrite is necessary for nucleation of Widmanstatten ferrite, while Cochrane and Kirkwood (11) considered that inclusions in grain boundaries in weld metal, perhaps as a result of pinning, can promote "wavy" grain boundaries and give rise to the necessary instability. Thus, quite apart from any change in ferrite nucleation rate per se due to grain boundary inclusions, variation in the number, size and distribution of second phase particles may alter the nucleation and early stages of growth of Widmanstatten ferrite in particular (20).

Moreover, if the growth rate of Widmanstatten ferrite can be reduced, the final transformation product will be more dependent on the intragranular nucleants. Cuddy et al (21) reported a change from Widmanstatten ferrite to blocky ferrite in HAZ studies of Ti-treated steels compared to a Ti-free base material. They proposed that the fine TiN particles interact with the ledges involved in the rapid edgewise growth of ferrite side-plates and impede their motion, reducing the aspect ratio towards that of blocky ferrite. The larger TiN particles additionally changed the microstructure by promoting intragranular ferrite nucleation. Further, Purdy (22) suggested that small matrix precipitates may inhibit ferrite formation by producing a stasis of the α/γ boundary, and in this regard Terashima and Hart (18) considered that it was the production of fine oxides just prior to solidification which was the controlling factor on acicular ferrite production: the mechanism by which such fine particles was operating was not identified, but some retardation of Widmanstatten ferrite growth may well have been contributory.

Vanadium additions to weld metal are known to promote grain boundary ferrite and acicular ferrite and this seems to be at the expense of Widmanstatten ferrite (2, 3, 24), supporting the hypothesis that reduced nucleation and/or growth rate of Widmanstatten ferrite is important in the production of acicular ferrite. Indeed, vanadium seems to be effective in promoting acicular ferrite formation even at low weld metal oxygen levels (~ 100ppm) (25, 26) where general experience would indicate an inability to form acicular ferrite because of insufficient number of particles to provide nucleation. Even in HAZs, vanadium additions are capable of promoting intragranular ferrite development, cf. Figs (4a) and (4b). The mechanism by which vanadium acts has not been elucidated but could well arise from V(CN) precipitation either stabilising the α/γ interface of grain boundary ferrite and/or reducing the growth rate of Widmanstatten ferrite.

GENERAL COMMENTS - Overall, practical understanding of the austenite decomposition sequence in HAZs and weld metals has advanced significantly in recent years. The various allotriomorphic forms of ferrite are well defined, together with the appropriate temperature ranges of formation, at least qualitatively, and the major importance of second phase particles in determining the progress of transformation is widely recognised. Increasingly, direct experimental evidence is being gained of ferrite nucleation on inclusions of various types. Nonetheless, it would appear that study should be made of factors influencing the formation of Widmanstatten ferrite, at least as much as for those that might promote acicular ferrite.

The majority of properties required of welded joints in the materials under consideration can be enhanced by the production of finer grain structures, especially if this can be achieved without significantly increasing the level of alloying. Second phase particles appear to offer major advantages in this respect both in the HAZ and the weld metal. Although much of their use in HAZs is aimed at reducing γ-grain growth by pinning, it is already established that they are additionally beneficial through intragranular nucleation. The fine-grain microstructures which can be developed in weld metal, with good mechanical properties, indicate that the next stage of development for HAZs might well not be better grain growth control (after all, fine ferrite grain weld metal structures can be achieved in austenite grains of 200µm in size) but better control of Widmanstatten ferrite nucleation and growth, and the nucleation of intragranular ferrite.

AUSTENITIC STAINLESS STEELS

FERRITE-AUSTENITE TRANSFORMATION - The solidification behaviour of austenitic stainless steel weld metal has been studied in detail (27, 30), since solidification cracking sensitivity can be highly dependent upon the primary solidification phase (31). With weld metals of low Cr:Ni ratio, primary solidification will be to austenite with, if the Ni content is not too high, interdendritic ferrite forming as a result of a peritectic or eutectic reaction. At high Cr:Ni ratios, ferrite is the primary solidification phase, with austenite developing interdendritically, to give a range of final ferrite morphology, from vermicular to lath-like. In principle, a similar ferrite content at room temperature may result whatever the initial solidification phase, but higher final ferrite levels at ambient temperature normally reflect primary ferrite solidification. Over a wide range of compositions, austenite is the stable phase below, say, 1200°C, and thus, during cooling, solid state transformation of ferrite into austenite occurs in common 300 series weld metals (Fig 5). This has been examined by various workers (29, 32, 33), although attention has concentrated on transformation following ferritic solidification with very much less emphasis on the rate of decomposition of residual ferrite following

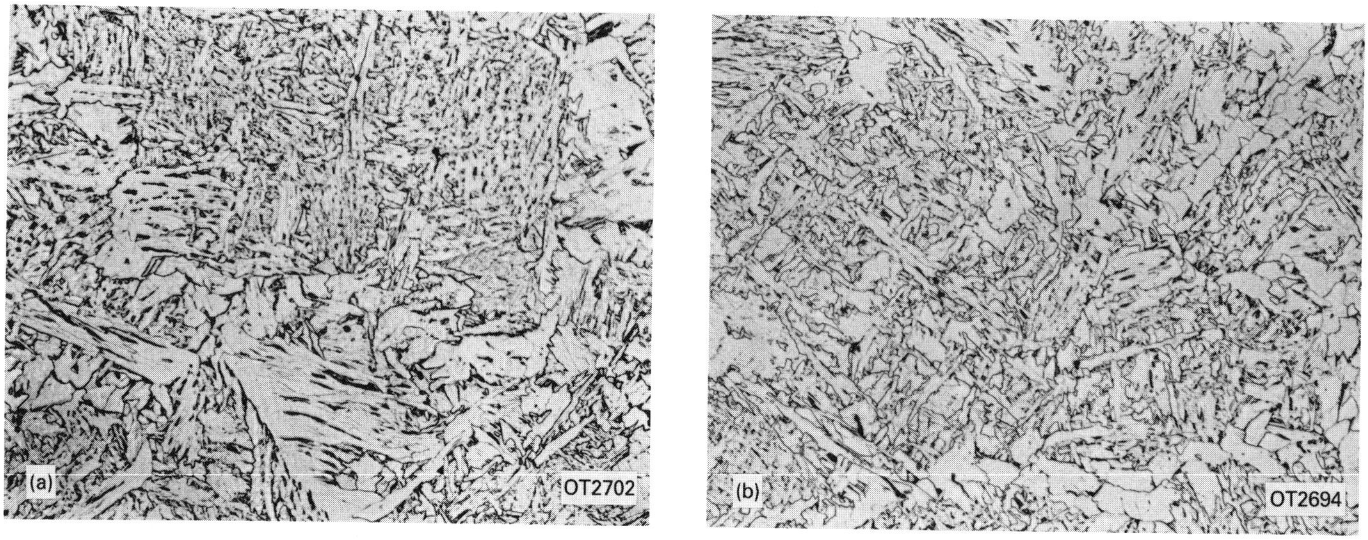

Fig.4. Effect of vanadium addition to a 0.07C-1.4Mn steel in promoting intragranular nucleation and reducing the extent of Widmanstätten ferrite, ×270:

a) no vanadium   b) 0.1% vanadium

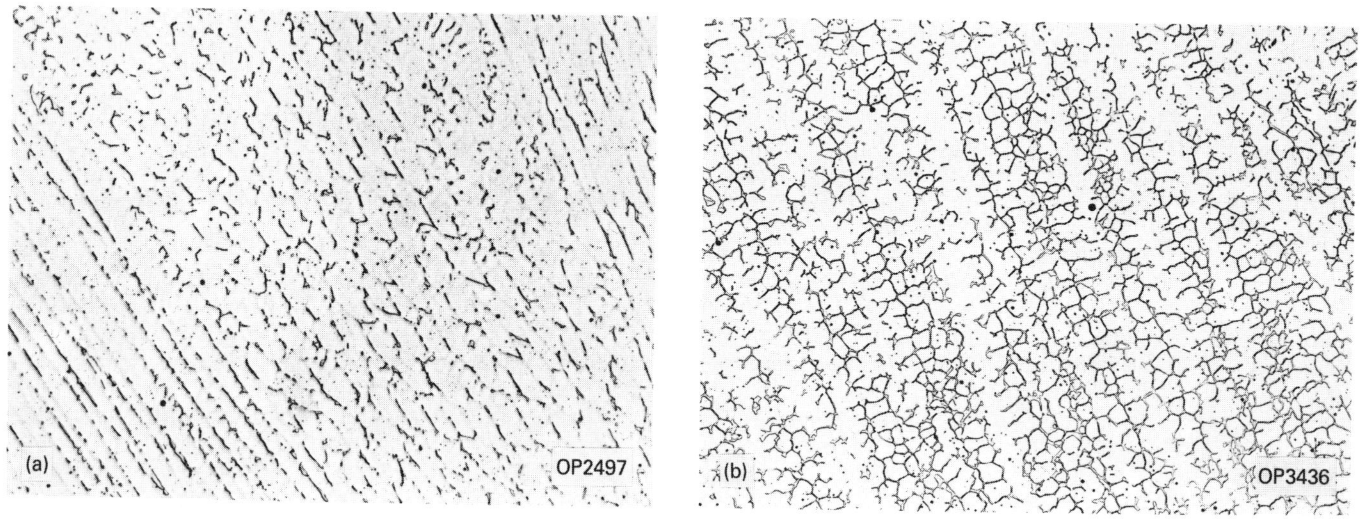

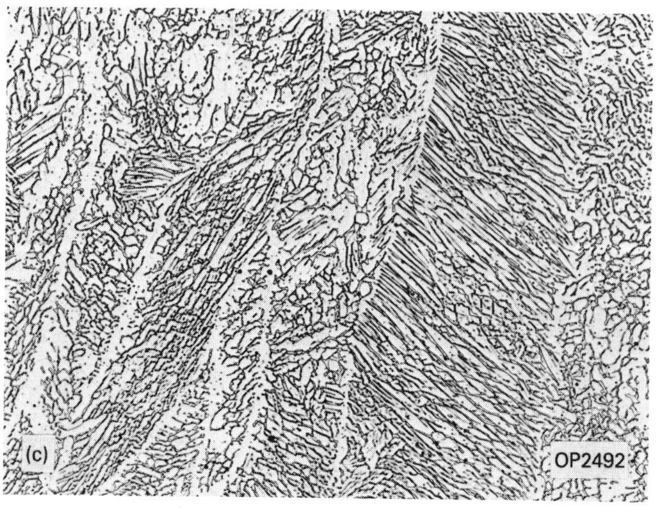

Fig.5. Variation in ferrite/austenite morphology in austenitic stainless steels:

a) primary austenitic solidification: residual interdendritic ferrite, ×340

b) ferritic-austenitic solidification: austenite from solidification and solid state transformation, ×210

c) primary ferritic solidification: austenite from solid state transformation, ×270 austenitic solidification.

Employing quenching studies on solidifying samples, Frederiksson (27) clearly showed that austenite present at room temperature could originate from the solidification sequence or by solid state transformation. This work was followed by a number of other investigations (28, 33) and has led to proposed modification to the Schaeffler diagram (31) to distinguish between deposit compositions solidifying primarily to austenite or to ferrite, with the aim of identifying compositional ranges preferred to obtain high solidification cracking resistance (Fig 6). In principle, therefore, whether or not major solid state transformation occurs can be assessed from the weld metal composition, and it is perhaps remarkable that for practical purposes the Schaeffler coefficients of individual elements appear similar for both the solidification and solid state reactions.

Austenite stainless steels of compositions covering those found commercially may, in fact, solidify virtually entirely as ferrite (28, 29). However, the final ferrite content can be as low as ca. 10% indicating a high reaction rate below the solidus temperature for the ferrite-austenite transformation. This consideration led some workers (29) to propose that the reaction was diffusionless, but microanalysis examination by Leone and Kerr (33) showed that the compositional gradients and elemental partition coefficients in austenite and ferrite differ between normally cooled weld metal and quenched samples, implying that significant diffusion can occur in the course of a weld cooling cycle. Leone and Kerr, moreover, calculated diffusion distances for Cr and Ni in ferrite from the relationship:

$$x = \sqrt{Dt} \qquad [1]$$

where x is the mean diffusion distance of an element, D is the diffusion coefficient, and t is the time. A fairly low heat input, rapid cooling rate situation was considered, and, by integrating $\sqrt{Dt}$ during the cooling cycle, it was concluded that total diffusion distances were commensurate with the scale of the observed microstructure, and hence that ferrite decomposition was by a conventional diffusion mechanism rather than by shear.

One possible reservation to the Leone and Kerr analysis is that transformation was assumed to commence virtually immediately on solidification, at over 1400°C. In fact, depending on the weld metal composition and segregation during solidification and on the cooling rate, the actual transformation start temperature could be well below 1400°C, when the total diffusion distance will be appreciably reduced. Nonetheless, the available evidence most strongly supports diffusional transformation (32, 33), and it can be remarked that diffusion rates at temperatures close to the solidus are not well defined, and may be greatly enhanced by the simultaneous strain experienced as a result of contraction on cooling.

There has been little effort specifically on nucleation of the $\delta \rightarrow \gamma$ reaction or to define the diffusional growth mechanism. Inclusions may play a role as in the $\gamma \rightarrow \alpha$ change in C-Mn steels, but in most cases, nucleation presumably takes place on austenite resulting from solidification. Brown et al, (34) concluded from the transformation in cast CF8M that a lattice correspondence existed between δ and γ, regardless of interfacial curvature, and was of the Kurdjumov-Sachs type, and from this work austenite growth was considered to occur by an interfacial ledge mechanism.

Accepting diffusional $\delta \rightarrow \gamma$ transformation, a dependence on weld cooling rate would be expected, rapid cooling leading to higher final ferrite content at room temperature. However, both French and Swedish workers (35, 36) have indicated final ferrite content to be fairly insensitive to changes in welding conditions over a wide range. Assuming Johnson-Mehl behaviour, i.e.

$$Vf = 1 - \exp(-\kappa t^{n}) \qquad [2]$$

where Vf is the proportion transformed to austenite, κ is a rate constant, t is the time and n depends on the growth mechanism, and provided that diffusion is sufficiently rapid for the reaction to be substantially complete even at low heat input conditions, the final ferrite content would be expected to be only slightly affected by cooling rate (Fig 7). Further, little additional transformation would be anticipated on reheating during multipass welding, unless, perhaps, the arc energy were greatly increased to give a very much more extended cooling time; this is consistent with general experience, in that the ferrite content in multipass welds in common 300 series alloys does not normally vary greatly within individual beads, or with distance from interrun fusion boundaries.

The ferrite-austenite reaction can take place also in the high temperature HAZ immediately adjacent to the fusion boundary. With sufficiently high Cr:Ni ratios, transformation to ferrite on heating will occur, followed by partial or complete reformation of austenite as in weld metal. More commonly, the extent of ferrite formation depends on segregation and banding in wrought products, and ferrite columns may form parallel to the plate surface in high Cr:Ni ratio areas (Fig 8), possibly in association with ferrite remaining after solution treatment in the as-

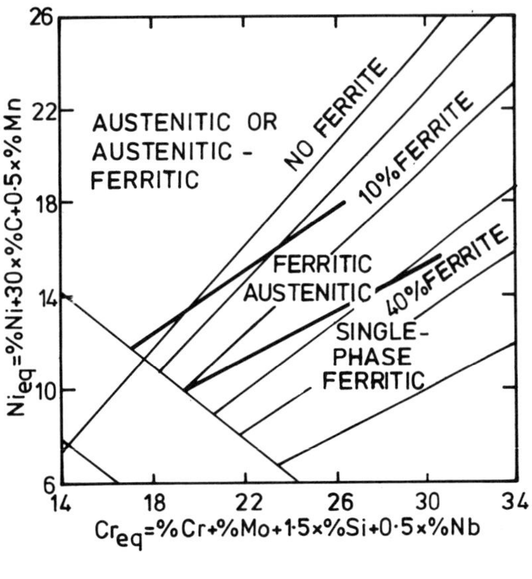

Fig.6. Schaeffler diagram with boundaries between austenitic, ferritic-austenitic or ferritic solidification modes (31).

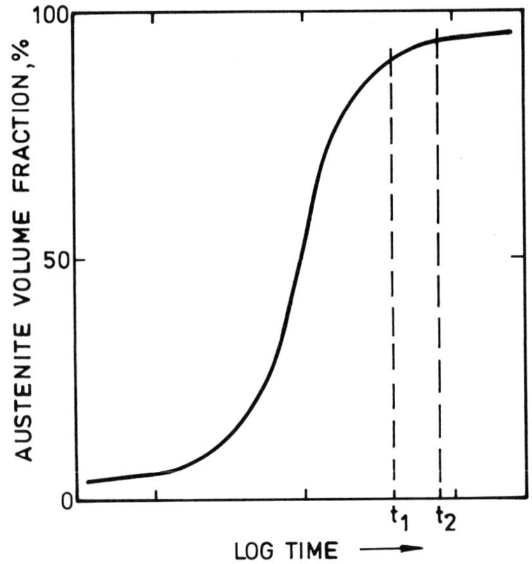

Fig.7. Schematic illustration of austenite formation from ferrite in a typical 18Cr-10Ni steel, assuming isothermal transformation. Time t_1 corresponds to a given weld cooling cycle; $t_2 = 2t_1$ and corresponds either to higher heat input (slower cooling rate) or to reheating. The final ferrite content is little affected by increasing heat input or by multipass welding.

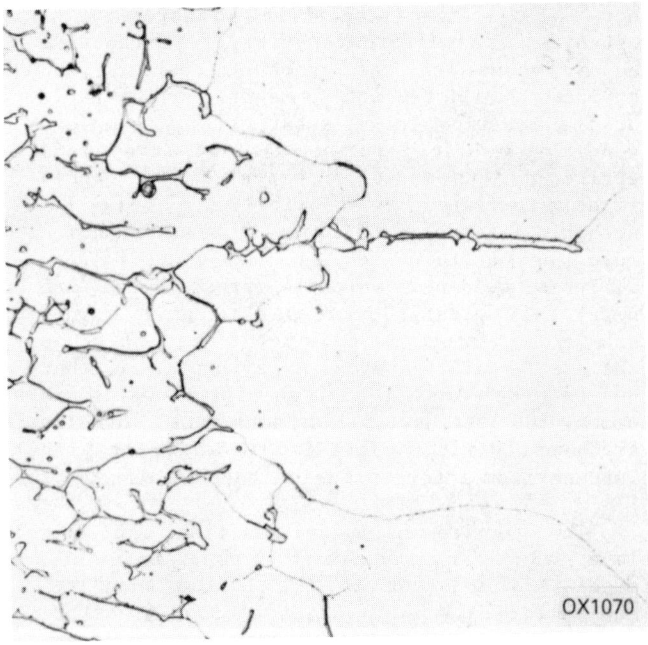

Fig.8. HAZ ferrite formation at a grain boundary in a segregation band in austenitic stainless steel plate, ×700.

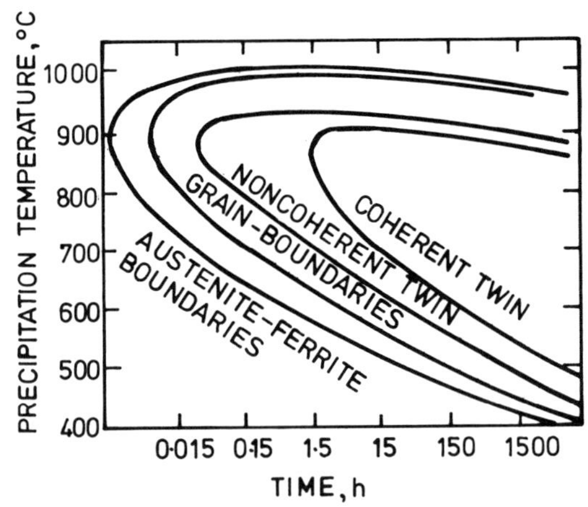

Fig.9. Precipitation kinetics of $M_{23}C_6$ carbide in 18Cr-9Ni stainless steel containing 0.038%C, originally quenched from 1260°C (grain size ASTM 1) (38).

received material. Ferrite may also form at grain boundaries close to the fusion boundary where local liquation arises, presumably in part due to the Rhebinder effect. Few data are available on HAZ transformation, but it can result in segregation of alloying elements between the phases, and hence influence corrosion resistance.

PRECIPITATION PROCESSES - During welding, austenitic stainless steels are subject to precipitation of a range of phases, including both carbides and intermetallic compounds. From the practical standpoint, concern attaches primarily to precipitation of chromium-rich carbides of the $M_{23}C_6$ type over the temperature range of ca. 550-850°C.

The main features of $M_{23}C_6$ precipitation have been well studied (37, 38) for many years, in the main because grain boundary precipitation (Fig 9) renders the material sensitive to intragranular corrosion via local depletion of the matrix chromium content. The effective activation energy for sensitization is ca. 210-240 kJ/mol, precipitate growth being controlled by chromium diffusion. Nucleation and growth rate at grain boundaries can be sufficiently high for a welding cycle to reduce corrosion resistance, but, following the introduction of AOD processing, the application of low carbon grades has increased, these greatly reducing the volume of carbide formed, and hence the degree of matrix chromium depletion relative to previous production practice.

The occurrence of intergranular stress corrosion in nuclear power generation pipework has stimulated further and more detailed investigation with the particular aim of defining material and welding conditions likely to cause unacceptable sensitisation to intergranular attack. Various workers have sought to quantify either the extent of precipitation or its influence on corrosion resistance, and Solomon (39) has presented data on critical HAZ cooling rates which must be exceeded to avoid sensitisation for a range of steel carbon contents (Fig 10), and has examined (40) the extent to which thermal cycles in multipass welding have an additive effect. Following such work, it is clear that sensitisation during continuous cooling can be more severe than indicated by isothermal studies, and that the simultaneous application of strain as experienced at welds can further enhance the effects of carbide formation.

Investigations in the USA (4) have shown that, even with weld cycles which do not cause sufficient carbide formation for corrosion resistance to be adversely affected, nucleation still takes place which can act to reduce corrision resistance via precipitate growth on long term ageing in service at fairly low temperatures. The full impact of this observation has yet to be assessed, but presumably the use of low carbon steel will be of benefit as in "traditional" sensitisation.

Comparable carbide precipitation can take place also in weld metal along ferrite-austenite interfaces. At least in the as-welded condition, this represents less of a practical problem because the high surface area means that the effect of precipitation in causing chromium depletion at any one point is reduced, and, indeed, given the presence of ferrite, nominally austenitic weld metals are much more tolerant to high carbon contents than wrought material.

In contrast to the substantial effort on carbide formation during welding, precipitation of intermetallic phases has received little attention, although, because of the marked effect on mechanical properties, work has been carried out relevant to heating after welding, either for stress relief or during service. For the common 300 series materials, intermetallic precipitation solely during welding is not of widespread importance, but small amounts of sigma and/or chi phases may form rapidly, most notably at ferrite in weld metals containing molybdenum (42). This can reduce corrosion resistance in highly oxidising media, while the tensile ductility of multipass 317 type weld metals (17Cr:12Ni:3.5Mo) is commonly below that of 308 deposits (18Cr:10Ni). Intermetallic precipitation may be more significant in high molybdenum steels recently developed for improved corrosion resistance. Liljas et al. (43) identified chi-phase in as-welded HAZs in a 20Cr:18Ni:6Mo:0.2N steel, and, while no adverse effect on material properties was observed, further study would seem required.

GENERAL COMMENTS - Prediction of residual ferrite levels is of major importance in welding austenite steels, especially in weld metal. For solidification and solid state transformation, composition can be seen from available information as more important than welding conditions in determining the final ferrite level, at least for the 300 series alloys, and a reasonable estimate of the ferrite content can be made from the various empirical composition/microstructure diagrams or formulae that exist. Nonetheless, for critical applications greater precision remains desirable; as yet the body of literature on the ferrite-austenite reaction during welding high alloy steels is less extensive than for austenite decomposition in C-Mn and low alloy steels, and further study of reaction rates and associated segregation behaviour is needed.

For many situations, the low carbon contents afforded by AOD processing have reduced concern regarding carbide formation and sensitisation to intercrystalline attack. However, this is not invariably the case, and the improved quantification of controlling factors in recent years must be noted, including the specific recognition of the effects of continuous weldment cooling on precipitation rate. Presumably an analogous approach would be of benefit also in terms of possible intermetallic formation in new higher alloy grades of steel.

DUPLEX FERRITIC AUSTENITIC STEELS

FERRITE-AUSTENITE TRANSFORMATION - The composition of these materials is so balanced that, as commercial products supplied typically in the 1050°C annealed condition, the microstructure is two-phase, ferrite and austenite being present in roughly equal proportions. On heating, transformation to ferrite takes place at high temperatures with concomitant grain growth, followed by reversion to austenite on cooling, and thus the structure in the HAZ of a weld differs markedly from that of the base steel (44). Austenite forms on cooling initially at ferrite grain boundaries, with subsequent growth within the grains, as Widmanstatten sideplates or allotriomorphs (Fig 11). A similar structure develops also in weld metal, although following epitaxial ferritic solidification, the grain size tends to be coarser than in the base metal.

The transformation sequence in duplex steels is thus essentially similar to that in nominally austenitic 300-type alloys, but with the significant differences that (a) ferrite is a stable phase from the melting point to room temperature, and (b) the $\gamma \to \delta \to \gamma$ transformation temperatures are appreciably lower. As a result, final ferrite contents (and grain sizes) in the weld area are much higher. For many years, available commercial materials gave predominantly ferritic HAZs and weld metals with a resultant deterioration in corrosion resistance and toughness. It is now appreciated that the properties of weldments in this type of steel depend critically on the ferrite-austenite balance achieved in HAZ or weld metal, and considerable effort has been paid in the last decade to ensuring adequate austenite reformation during cooling.

The phase changes in duplex steels have been examined by various workers (45 - 49), albeit mainly by conventional isothermal techniques, and the effect of composition on ferrite and austenite formation during heating and cooling is fairly well understood, although the coefficients ascribed to various elements may well differ from those in the Schaeffler diagram. Southwick and Honeycombe (45) pointed out the close parallels that can be drawn between the austenite morphology following the $\delta \to \gamma$ reaction and that of proeutectoid ferrite in ordinary steels, implying that similar growth mechanisms would be operative. As for CF8M, the ferrite-austenite orientation was close to the Kurdjumov-Sachs relationship, with growth kinetics intermediate between interface and volume diffusion control. Similar conclusions were reached by Mundt and Hoffmeister (49), who also found the growth exponent in the Johnson-Mehl equation to increase at lower temperatures, suggesting a ledge growth mechanism to become dominant. Given matching composition, weld metals normally contain somewhat lower austenite proportions than the HAZ, the effect possibly being associated with the coarser grain size common in weld metal, in that reduction in grain boundary area will effectively reduce the number of nucleation sites for initial transformation. If this is the case, transformation in weld metal may show a greater dependence on heat input and cooling rate than is the case for the HAZ.

Southwick and Honeycombe (45) also observed martensitic decomposition of ferrite to occur, but only in samples quenched to below 650°C. However, at least from reported optical metallographic examination, it does not seem that this reaction takes place during welding to any great extent.

In order to achieve high as-welded austenite contents, there has been increased use of alloying with nitrogen as an austenitising element, and current materials give substantial amounts of austenite over a wide range of welding conditions. Nonetheless, even with such alloys it is imperative that adequate time be allowed during the cooling cycle for austenite to occur. Nucleation per se seems to be an easy event (45) and, in general, it is growth rate that becomes important: the materials thus differ significantly from conventional austenitic alloys in that it is highly undesirable to use low heat input "cold" welding conditions. The situation is exemplified in Fig 12, showing an electron beam weld in which the rapid cooling after welding has led to retention of a virtually fully ferritic structure. Fortunately, current production materials are reasonably tolerant to changes in heat input and, to some degree, the material composition is dominant in determining the final austenite level. However, even within a single commercial grade, variation in HAZ transformation behaviour will exist between different heats (50) (Fig 13). This probably stems at least in part from differences in transformation start temperature as a function of composition, but, from the trends noted in C-Mn steels, it may be also that austenite nucleation and/or growth is influenced by second phase particles. Both effects would cause the differences in transformation

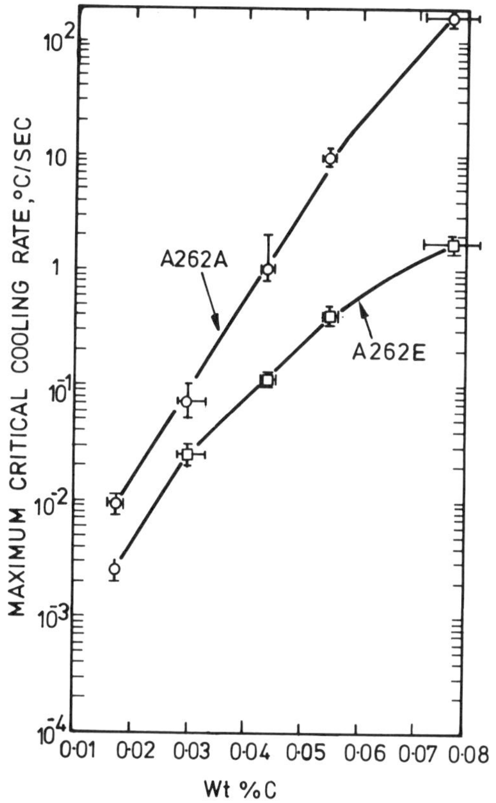

Fig.10. Maximum HAZ cooling rate required to avoid sensitisation in austenitic stainless steels of varying C content (39).

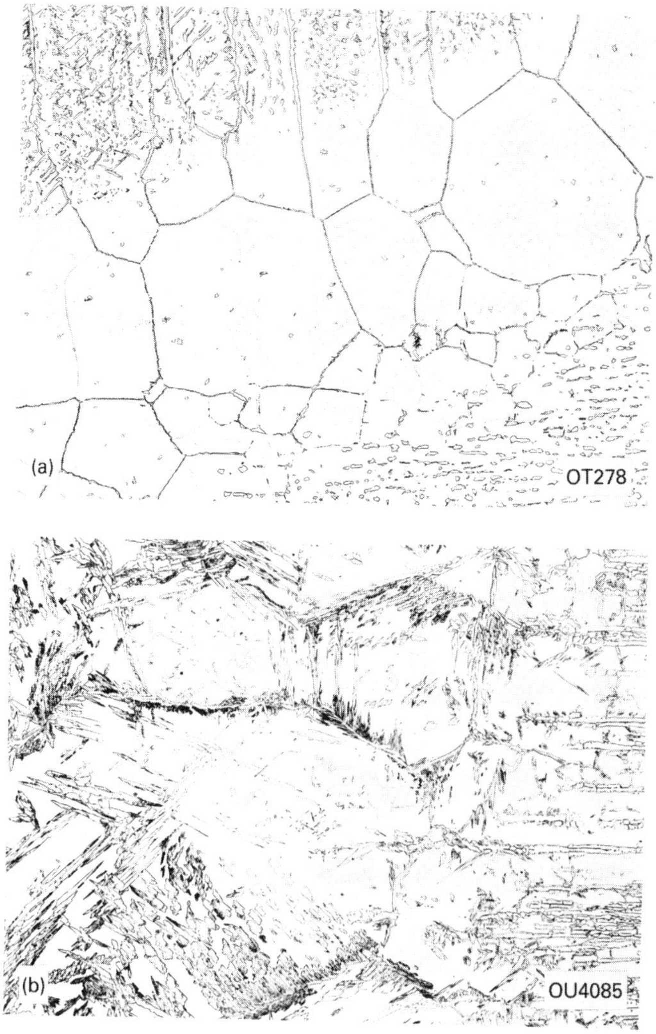

Fig.11. HAZ transformation in duplex stainless steels:
a) 26Cr-5Ni-1.5Mo steel, ×80 b) 22Cr-5Ni-2.5Mo-0.12N steel, ×70

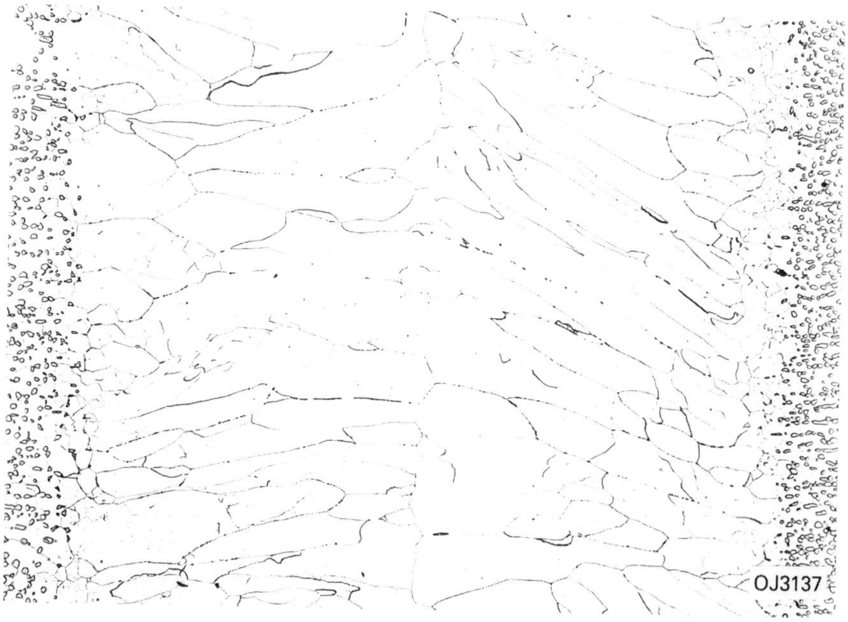

Fig.12. Retention of a ferritic structure in electron beam weld in 26Cr-5Ni-1.5Mo duplex stainless steel, ×90.

characteristics shown schematically in Fig 14.

At present, it is seldom possible to define the minimum arc energy required for a specific material/joint geometry to obtain substantial austenite formation. Van Nassau (51) has recommended that welding conditions should be such as to give a cooling time from 800°C to 500°C of 8 to 30 seconds; this would seem to be reasonably applicable, although such a criterion does not fully describe the high temperature part of the thermal cycle over which transformation actually takes place. This is an area where further work is required: on one hand transformation to austenite in commercial alloys during welding is clearly much faster than indicated by published isothermal TTT diagrams (Fig 15), while on the other a maximum arc energy must exist to avoid degradation of properties from precipitation processes (52).

In contrast to the case with nominally austenitic grades of stainless steel, the duplex alloys can show a marked response to reheating in multipass welding, with significant increase in austenite content. This is, however, dependent on a sufficiently high heat input being used for the fill passes, and also on the steel composition giving fairly high transformation temperatures. Again, further work is required from the practical standpoint to quantify the effort of reheating on HAZ and weld metal austenite content.

PRECIPITATION PROCESSES - As for austenitic grades, precipitation of carbides and intermetallics can occur in duplex materials in both HAZ and weld metal. The pattern of precipitation is dependent on the extent of the ferrite-austenite transformation, since the latter takes place at higher temperatures and thus occurs first on cooling (44). With predominantly ferritic structures, precipitation of carbides and carbonitrides is observed both in grain boundaries and intragranularly, a precipitate-free zone frequently being observed adjacent to grain or phase boundaries. Such precipitation can be especially pronounced in the HAZ if non-matching fillers of fairly high carbon level are employed as a result of carbon migration from the weld metal at high temperature. At higher austenite levels, intragranular carbide and nitride precipitation tends to be suppressed, but carbides may still form at the phase boundaries. Carbide formation occurs over a temperature range where ferrite is transforming to austenite, and in consequence takes place on a moving δ/γ interface, somewhat analogous (53) to the interphase precipitation on γ/α interfaces in microalloyed C-Mn steels, although the rate of movement will be fairly low at the temperatures concerned.

Few data have been reported on intermetallic formation during welding (44, 52). Precipitation of intermetallic phases clearly takes place in higher alloy duplex grades in HAZ and weld metal, and can adversely affect the service properties, but it is not presently possible to define the extent to which a practical problem is posed.

GENERAL COMMENTS - The fact that duplex steels have been successfully employed in welded fabrication for many years indicates not only recognition of the importance of ensuring local austenite reformation but also that the roles of composition and welding variables are fairly well defined. However, as well illustrated by the material variability in Fig 13, detailed understanding is very much less advanced, and further study of the $\delta\rightarrow\gamma$ reaction in commercial materials is essential from both material and welding procedure viewpoints. This is the case also for precipitation effects, where, apart from a general need for investigation of high arc energy conditions, the paucity of data on intermetallic formation will be noted.

WELD THERMAL CYCLES

HEAT FLOW - Study of the thermal cycle experienced during a fusion welding operation has been based predominantly on the classical Rosenthal approach:

$$\frac{\partial^2 T}{\partial x^2} + \frac{\partial^2 T}{\partial y^2} + \frac{\partial^2 T}{\partial z^2} = -\frac{v}{a}\frac{\partial T}{\partial t} \qquad [3]$$

where v is the welding speed and a is the thermal diffusivity. This has been employed with varying degrees of success (1, 54), although in recent years a number of alternative finite element analysis computer programs (55) have been developed for prediction of the thermal cycle experienced at a given point around a weld. By and large the approaches give predictions of the right order of magnitude, but for practical purposes they commonly give unacceptable errors,particularly in identifying the peak temperature experienced (56). The analyses have perhaps been most successful in predicting cooling rates at some 800-500°C. Thus while thermal cycle predictions have been usefully employed for examination of phase changes in transformable C-Mn and low alloy steels for example, greater difficulty may arise in terms of the high temperature part of the thermal cycle where dissolution of second phase particles can occur and where the $\gamma\rightarrow\delta\rightarrow\gamma$ transformation of significance in high alloy grades takes place. The inaccuracy no doubt stems in large part from the assumptions made, such as a point heat source, etc. (54) Equally, it is likely that the physical constants necessary for thermal cycle predictions are not sufficiently well defined over the high temperature regime. Moreover, process

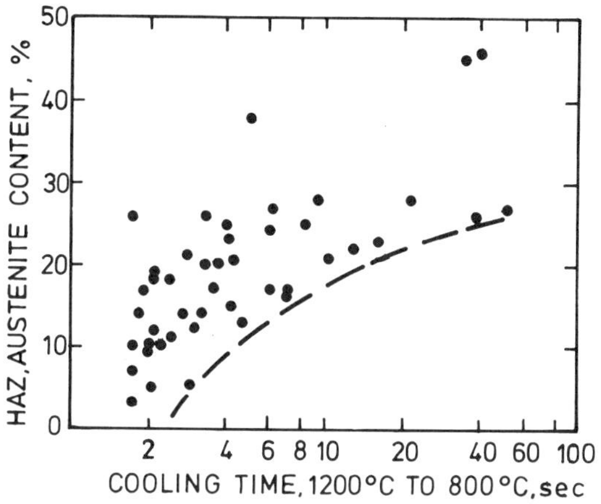

Fig.13. Variation in HAZ austenite content with cooling time for a range of commercial 22Cr-5Ni-2.5Mo-0.12N steels (50).

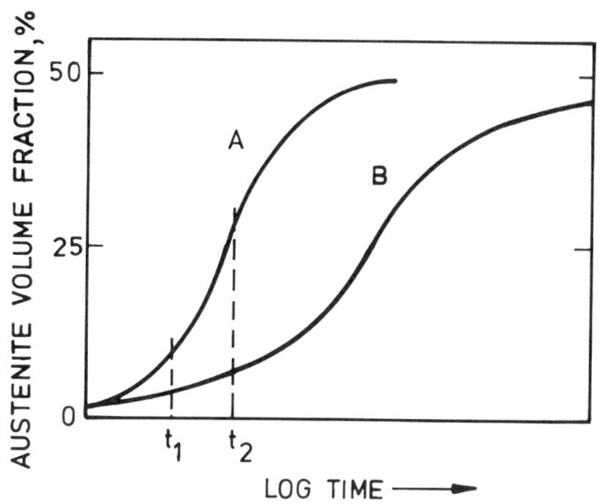

Fig.14. Schematic illustration of austenite formation from ferrite in two duplex stainless steels, assuming isothermal transformation. Time t_1 corresponds to a given weld cooling cycle; $t_2 = 2t_1$ and corresponds either to higher heat input (slower cooling rate) or to reheating. Steel A displays substantially more austenite formation than steel B when heat input is increased or multipass welding is employed.

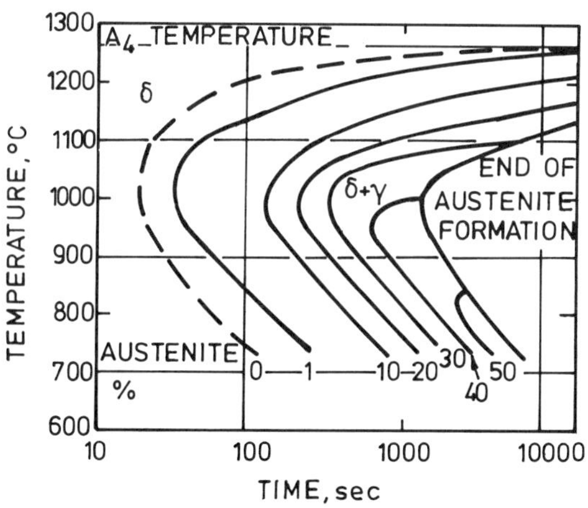

Fig.15. Isothermal TTT diagram for 26Cr-7Ni duplex stainless steel, heated to 1350°C for 1min (49).

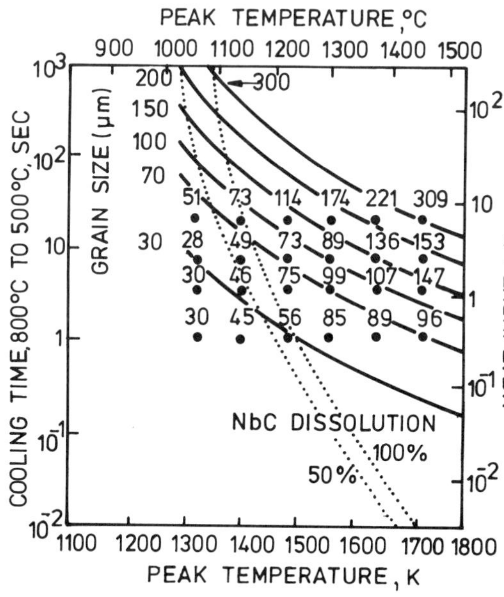

Fig.16. Grain growth diagram for a Nb-microalloyed steel. The points refer to the mean grain sizes around actual MIG welds (1).

efficiency can vary over a wide range so that the true heat input can seldom be defined with precision.

Although these analyses have differentiated between two and three dimensional heat flow (thin and thick plate), effort has concentrated on butt welds with little consideration to more complex joint geometries. The ability to define the weld thermal cycle is inherent in any work aimed at microstructural predicition. Accordingly, a wide range of nomograms has been produced to characterise specific situations (1, 58), both from direct measurements on a purely empirical basis, and based on heat flow analysis with suitable correction factors. Over the last decade, the database available has increased considerably, and it is possible to derive a reasonable estimate of thermal cycle in many instances, although some scatter in material behaviour must still be accepted. It can be expected that this trend will continue, with increased use of computerised acquisition, storage and manipulation of data.

TRANSFORMATION BEHAVIOUR - It is a characteristic of virtually all solid state transformations in steels that the extent of transformation is highly dependent on small changes in material composition and thermal cycle, and hence, at the present time, it is not possible to predict the final microstructure associated with a welded joint with exactitude. Nonetheless, in recent years, a number of studies have been made relating the progress of metallurgical changes to the total temperature-time cycle experienced on welding (1). The derivation of continuous cooling transformation data for ferritic steels was established many years ago, and can be paralleled by the more recent American work on sensitation of austenitic stainless steel. Ikawa et al. (59) and Allberry et al. (60) considered HAZ grain growth, by dividing the thermal cycle into small isothermal steps of hold time, Δt, and summing behaviour, namely:

$$D_f^{\,n} - D_o^{\,n} = A\, \Sigma_o^{\,f}[\exp(-\frac{Q}{RT})]\Delta t \qquad [4]$$

where D_f and D_o are the final and initial grain sizes after holding at temperature T, n is a grain growth exponent, Q is an appropriate activation energy and A is a material constant. The significance of this approach is in the concept of integrating the effects of the total time/temperature cycle. The approach has been extended by other workers to include allowance for grain boundary pinning, as exemplified by the "weldability diagrams" of Ashby and Easterling (1, 61), and offers considerable scope in providing a framework for quantitative microstructure prediction (Fig 16). Indeed, Alberry and his co-workers have successfully derived a computer model for prediction of HAZ microstructures in alloy steel weldments, with allowances for the effects of multiple thermal cycling in causing either re-austenitisation or sub-critical tempering (62, 63).

At the same time, direct experimental measurement remains necessary to provide quantitative input data to such computer models. In principle, and providing that the controlling reactions are adequately understood, it should be possible to describe most transformations during welding in terms of the total thermal cycle. For example, following Thomas and Keown (64), Eq. [2] can be expressed as:

$$\ell n\left(-\ell n(1-V_f)\right) = \ell n\ \kappa o - \frac{Q}{RT} + n\ell nt \qquad [5]$$

where κo is a rate constant. Hence, it should be feasible to derive the final austenite content in stainless steel welds by appropriate summation of the thermal cycle. However, this requires knowledge of κo, Q and n, and of their variation as functions of temperature and material composition: at present, such data are seldom available. Thus, precise characterisation of the phase changes occurring in a weld thermal cycle must presently be based on observations of actual welds made under conditions relevant to the practical situation.

GENERAL COMMENTS - From studies on heat flow and transformation behaviour under a weld thermal cycle, the possibility is emerging not only of more exact prediction of microstructure but also of microstructure control via optimisation of welding procedure. As demonstrated by Alberry et al (62), considerable success has already been achieved, and even though empirical input data generation remains necessary, the trend must be expected to continue. One point must be made regarding weld area microstructure. While considerable effort has been devoted to quantitative predictions of HAZ and weld metal microstructures, (e.g. Bhadeshia (65), Easterling (1, 61), Alberry (62)), to a large extent these currently ignore any effect of second-phase particles on transformation behaviour. At least in C-Mn steels, where second-phase particles demonstrably have a marked effect on $\gamma \rightarrow \alpha$ transformation, it is doubtful if a significant advance in microstructure prediction, for cooling rates relevant to welding, can take place without modelling their influence. (This is, of course, over and above the known effect upon austenite grain growth behaviour, where some allowance at least in respect of nitrides and carbides (1) has currently been made.) The need to include such modelling is likely to increase; the requirement is well-established for C-Mn weld metals, and there should be more controlled use of stable second-phase particle in the parent steels for manipulation of HAZ microstructures. Incorporation of second-phase particles effects on the $\delta \rightarrow \gamma$ change may well be essential also for austenitic and duplex steels, although in this case further work to clarify the situation is required.

CONCLUDING REMARKS - Solid state transformations occuring in ferritic, austenitic and duplex steels during a weld thermal cycle have been described, with particular reference to recent work. With regard to ferritic C-Mn and allied steels, the importance is emphasised of second-phase particles in determining the progress of the austenite decomposition reaction. For austenitic and duplex stainless alloys, attention has concentrated on the solid state ferrite/austenite transformation occurring at high temperatures, and on subsequent precipitation processes.

In all cases, it is clear that considerable advance has been achieved in understanding of the variables influencing behaviour, enabling more precise prediction and control of the microstructural changes that take place on welding. This has been paralleled by theoretical and empirical work on quantifying the thermal cycle experienced. At present, transformation behaviour cannot be predicted exactly solely on a fundamental basis, but to a large extent, controlling factors have been established and continued development of computer models with generation of appropriate quantitative input date can be foreseen.

ACKNOWLEDGEMENTS

The authors thank their colleagues at The Welding Institute for advice and assistance in producing this paper.

REFERENCES

1. K. Easterling, "Introduction to the Physical Metallurgy of Welding", Butterworths & Co. (Publishers) Ltd., London, 1983.
2. P. H. M. Hart, Proc. Conf. "Trends in Steel & Consumables for Welding", The Welding Institute, London, pp21-53, Nov 1978.
3. T. Fukanoshi, T. Tanaka, S. Ueda, M. Ishikawa, N. Koshizuka & K. Kobayashi, Trans. I.S.I.J., 17(7), pp419-427, Aug 1977.
4. D. J. Abson, R. E. Dolby & P. H. M. Hart, Proc. Conf. "Trend in Steel & Consumables for Welding", The Welding Institute, London, pp75-101, Nov 1978.
5. Y. Ito & M. Nakanishi, "The Sumitomo Search", (15)5, pp42-52, May 1974.
6. M. Nakanishi, Y. Komizo, I. Set, M. Nakamura & Y. Saitoh, 11W Doc 1X-1281-83.
7. N. Yurioka, Proc. Conf. "International Congress on Welding Research", IIW, Boston, pp184-185, July 1984.
8. K. Hatton, A. Tanaka, K. Okamoto, Y. Tomita, Y. Tsuda, R. Yamaba, T. Haze & K. Ito, Nippon Steel Tech. Report, No 24, pp35-52, Dec 1985.
9. H. Homma, S. Ohkita & S. Matsuda, "67th American Welding Society Annual Meeting", AWS, Atlanta, Paper 24C, April 1986.
10. R. A. Ricks, G. S. Barritte & P. R. Howell, Proc. Conf. "Solid-Solid phase transformations", AIME, Pittsburg, pp463-468, Aug 1981.
11. N. Mori, H. Homma, M. Wakabayashi & S. Ohkita, 11W Doc 1X-1196-81.
12. R. C. Cochrane & P. R. Kirkwood, Proc. Conf. "Trends in Steel & Consumables for Welding", The Welding Institute, London, pp103-121, Nov 1978.
13. S. G. Court & G. Pollard, J. Mat. Sci. Letters, 4(4), pp427-430, April 1985.
14. E. S. Kayali, J. M. Corbett & H. W. Kerr, J. Mat. Sci. Letters (2)1, pp123-128, Jan 1983.
15. B. R. Keville, Weld. J., 62(9), pp253S-260S, Sept 1983.
16. L. Devillers, D. Kaplan, B. Marandet, A. Ribes & P. V. Ribond, Proc. Conf. "The Effect of Residual Impurity of Micro-alloying Element on Weldability & Weld Properties", The Welding Institute, London, Paper 1, Nov 1983.
17. S. Kamazawa, A. Nakashima, K. Okamoto & K. Kamaya, Trans. I.S.I.J., 16(9), pp486-495, 1976.
18. H. Terashima & P. H. M. Hart, Proc. Conf. "The Effects of Residual Impurity of Microalloying Element on Weldability & Weld Properties", The Welding Institute, London, Paper 27, Nov 1983.
19. R. D. Townsend & J. S. Kirkcaldy, Trans. ASM Qtly, 61(9), pp605-619, Sept 1968.
20. P. R. Krahe, K. R. Kinsman & H. I. Aaronson, Acta. Met., 20(9), pp1109-1121, Sept 1972.
21. L. J. Cuddy & J. S. Lally, Proc. Conf. "Technology & Applications of High Strength Low Alloy (HSLA) Steels" ASM Metal Congress, Philadelphia, pp141-143, Oct 1983.
22. G. R. Purdy, Acta. Met. 26(3), Pt II, pp487-498, 1978.
23. D. J. Abson & P. H. M. Hart, The Welding Institute Members Report 162/1981.
24. R. E. Dolby, IIW Doc 1X-1213-81.
25. B. G. Kenny, H. W. Kerr, R. B. Lazor & B. Graville, Met. Con. 17(6), pp374R-381R, June 1985.
26. A. H. Koukabi, T. H. North & H. B. Bell, Met. Con. 11(12), pp639-642, Dec. 1979.
27. H. Fredriksson, Met. Trans., 3, 2989-2997, 1972.
28. Y. Arata, F. Matsuda & S. Katayama, Trans. J.W.R.I., 5(2), pp135-151, 1976.
29. J. C. Lippold & W. F. Savage, Weld. J., 58(12), pp362S-374S. Dec. 1979; 59(2), pp48S-58S, Feb 1980; 61(12), pp388S-369S,

Dec 1982.
30. S. A. David, G. M. Goodwin & D. N. Braski, Weld. J., 58(11), pp330S-336S, Nov 1979.
31. V. P. Kujanpaa, N. J. Suutala, T. K. Takalo & T. J. I. Moisio, Met. Con., 12(6), pp282-285, June 1980.
32. J. A. Brookes, J. C. Williams & A. W. Thompson, Met. Trans A, 14A, pp1271-1281, July 1983.
33. G. L. Leone & H. W. Kerr, Weld. J., 61(1), pp13S-21S, Jan 1982.
34. E. L. Brown, T. A. Whipple & G. Krauss, Proc. Conf. "Duplex Stainless Steels", pub. by ASM, Ohio, pp665-691, 1983.
35. T. Tsuk, M. Hubert & C. Messager, Rev. de Metallurgie, 68, pp829-837, Dec 1971.
36. S. Polgary, Schweiss Technik/Soudure, 56(9), pp364-375, Sept 1966.
37. P. Marshall, "Austenitic Stainless Steel: Microstructure and Mechanical Properties", pub. by Elsevirs Applied Science Publishers Ltd., Barking, 1984.
38. R. Stickler & A. Vinckier, Trans. Amer. Soc. Met., 54, pp362-380, 1961.
39. H. Solomon, Corrosion, 40(2), pp51-60, Feb 1984.
40. H. Solomon, Corrosion, 34(6), pp183-193, June 1978.
41. M. J. Povich, Corrosion, 34(2), pp60-65, Feb 1978.
42. R. A. Daeman & F. Dept, Weld. J., 35(1), pp33S-40S, Jan 1970.
43. M. Liljas, B. Holmberg & A. Ulander, Proc. Conf. "Stainless Steels 1984", pub. by Institute of Metals, London, pp323-329, 1985.
44. T. G. Gooch, vide ref. 34, pp573-602.
45. P. D. Southwick & R. W. K. Honeycombe, Met. Sci., 14(7), pp253-261, July 1980.
46. Y. Nakao, K. Nishimoto & S. Inoue, J. Jap. Weld. Soc., 50(5), pp86-92, May 1981.
47. H. Hoffmeister & R. Mundt, Schweissen und Schreiden, 38(11), pp573-578, Nov 1981.
48. Ibid, Archiv. fur das Eisenhutt, 52(4), pp159-164, April 1981.
49. R. Mundt & H. Hoffmeister, idem, 54(7), pp291-294, July 1983.
50. D. N. Noble, J. Honeycombe &. T. G. Gooch, unpublished work, The Welding Institute, U.K., 1985.
51. L. Van Nassau, Welding in the World, 20(1/2), pp23-30, 1982.
52. K. Ume, N. Seki, Y. Naganawa, T. Hyodo, K. Satoh & Y. Kuriki, "Corrosion 1986", NACE, Houston, paper 155, 1986.
53. P. D. Southwick & R. W. K. Honeycombe, Met. Sci., 61(10), pp475-481, Oct 1982.
54. P. S. Myers, O. A. Uyehara & G. L. Borman, WRC Research Bulletin No 123, July 1967.
55. J. A. Goldak, A. Chakravarti & M. J. Bibby, Met. Trans. B, 15B(2), pp299-305, June 1984.
56. C. L. Tsai, Proc. Conf. "Trends in Welding Research in the United States", ASM, Ohio, pp91-108, 1982.
57. J. E. Moore, M. J. Bibby & J. A. Goldak, 11W Doc. No 212-604-85.
58. A. Kohno & S. B. Jones, The Welding Institute Members Report 81/1978/PE, 1978.
59. H. Ikawa, Trans. Jap. Weld. Soc., 4(2), pp35-41, Sept 1973.
60. P. J. Alberry, B. Chew & W. K. C. Jones, Met. Technol., 4(6), pp317-325, June 1977.
61. M. F. Ashby & K. E. Easterling, Acta Met., 30(11), pp1969-1978, Nov 1982.
62. P. J. Alberry, R. R. L. Brunnstrom & K. E. Jones, Met. Technol., 10(1), pp28-38, Jan 1983.
63. P. J. Alberry & J. G. Feldstein, Proc. Conf. "Maintenance Welding in Nuclear Power Plants", AWS, Tennessee, 1985.
64. R. G. Thomas & S. R. Keown, Proc. Conf. "Mechanical Behaviour and Nuclear Applications of Stainless Steel at Elevated Temperatures", Metals Soc., London, pp30-38, 1982.
65. H. K. D. H. Bhadeshia, L.-E. Svensson & B. Gretoft, Acta. Met., 33(7), pp1271-1283, 1985.

KEYNOTE ADDRESS

PREDICTING HEAT-AFFECTED ZONE MICROSTRUCTURES AND PROPERTIES IN FUSION WELDS

K.E. Easterling
School of Materials Science & Engineering
University of New South Wales
Sydney, Australia*

ABSTRACT

Simple models of the temperature distribution due to arc welding and the resulting microstructural changes in microalloyed steel are developed. The phenomena included in the model are dissolution and coarsening of carbonitrides, phase transformations and grain growth. Estimations of the volume fractions of the various eutectoidal constituents and the resulting hardness distribution in the HAZ, are also made. Various ways of presenting the microstructure and hardness data, as plotted on a microcomputer, are presented. Finally, areas of modelling and experimentation needing further development are discussed.

IN ORDER TO PREDICT final HAZ microstructure and hardness, there are several processes occurring during the heating and cooling of the weld thermal cycle that need to be taken into account. These include,

(a) during the heating cycle:
- recrystallisation
- $\alpha \rightarrow \gamma$ transformation
- pearlite/cementite$\rightarrow \gamma$ transformation
- carbon redistribution in austenite
- carbonitride coarsening
- carbonitride dissolution
- austenite grain growth

(b) during the cooling cycle:
- continued austenite grain growth
- $\gamma \rightarrow \alpha + Fe_3C$ phase transformation
- reprecipitation
- continued coarsening.

Obviously, the extent to which any of these processes occurs depends on the welding process used, the geometry of the joint and plate thickness, and the composition of the steel. The starting microstructure of the steel also affects those phenomena in that cementite dissolution, carbon redistribution, austenite nucleation and grain growth, etc., all depend upon the distribution of cementite or pearlite in the ferritic structure.

Dissolution and/or coarsening of the carbonitrides in the case of microalloyed steels depends on their composition. Recent work has shown that the compositions of carbonitrides in microalloyed steels are fairly complex and highly dependent upon the prior thermal history of the steel. Since the temperature of dissolution of the particles, as well as their rate of coarsening, is associated with diffusion of the metallic components of the particles in the austenite, as well as the particles' overall solubility in austenite, prediction of particle stability during a rapid heating cycle as in welding is complicated. Possible ways of tackling this particular problem have recently been suggested, however(1,2).

In the present work, equations describing the various phenomena occurring during a weld thermal cycle are given, and ways of computing microstructural changes and hardnesses in the heat affected zone are summarised. The approach adopted is essentially one that allows calculations to be carried out on a microcomputer. The program is written in such a way as to allow a certain interaction between operator and computer concerning input (type of steel, welding process, joint geometry) and output required (optimum welding energy or process, microstructure or hardness, etc.).

WELD THERMAL CYCLE - The microstructural changes occurring to the base metal on welding depend on the amount of input energy of the weld as given by q/v (where q is the arc power and v is the speed of movement of the arc), and the heat absorbing characteristics of the metal being welded in terms of joint geometry and thermal properties, etc.

Consider a welding arc tracking across a plate in the direction as shown in Fig. 1. A point at (y,z) below the surface is subjected

* On leave from the University of Lulea, Sweden

to a thermal cycle of T(y,z,t), where t is time. At a point below the surface on the weld centre line (y = o), the temperature field is given by (3,4).

$$T(z,t) = T_o + \frac{2\pi\lambda q}{v[t(t+t_o)]^{\frac{1}{2}}} \exp - \left[\frac{(z+z_o)^2}{4at}\right] \quad (1)$$

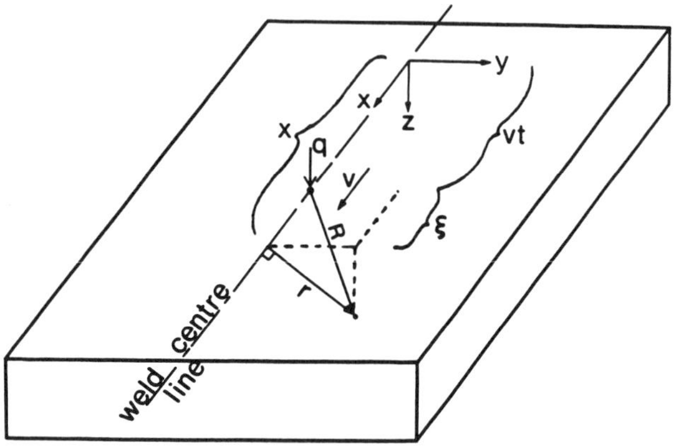

Fig. 1 - Welding geometry and coordinate system used.

where T_o is the temperature of the plate prior to welding, λ is the thermal conductivity, and a is the thermal diffusivity. The constant t is a measure of the time required for heat to diffuse over a distance corresponding to the source radius, such that

$$t = \frac{r_B^2}{4a} \quad (2)$$

Z_o is the distance that heat can diffuse during the interaction time r_B/v. Equation (1) is effectively a development of the familiar Rosenthal equation for a point source, but modified to take account of the size of the source used in practice. It thus gives a better measure of the shape of the temperature field about the moving electrode than the Rosenthal equation, particularly for the welding of thick plates as e.g. in bead - on - plate welding. However, it should be noted that if a point source is assumed, t and Z both become zero and eqn.(1) reduces to the Rosenthal equation.

Now eqn.(1) can be further refined by using an experimental measurement to adjust for unkown effects of e.g. alloying additions on the thermal properties of the steel, arc efficiency, etc. This is done in steel by measuring the position of the A_1 temperature in a bead- on plate weld and comparing its depth below the surface with that predicted from eqn.(1). This is illustrated in Fig. 2. It is observed that in this case no adjustment is apparently necessary since the position of the 723°C isotherm given by eqn.(1) corresponds well with the approximate position of the A_1 temperature, as indicated by the change in microstructure from the dark pearlitic regions to the yellow martensitic phase. Had it been necessary to adjust eqn.(1), in the modified form, it could still have been used for any other weld input energy or plate geometry for the same material.

THE $\alpha \rightarrow \gamma$ TRANSFORMATION For the 0.15 wt%C steel shown in Fig. 3, it is seen that under equilibrium conditions the $\alpha + Fe_3C \rightarrow \alpha + \gamma$ transformation would occur at the A_1 temperature.

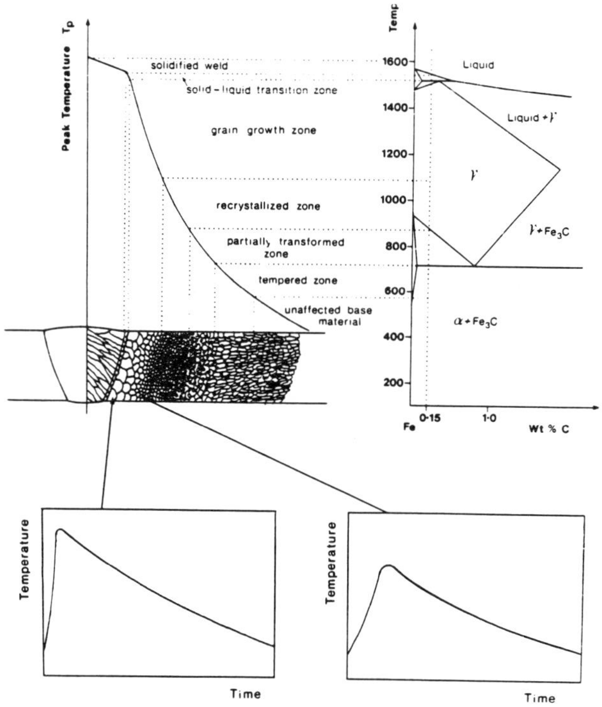

Fig. 3 - Schematic illustration of the various microstructures developed in the heat affected zone in terms of severity of the weld thermal cycle and the Fe-C phase diagram.

However, the heating and cooling rates in fusion welding are such that equilibrium conditions are extremely unlikely to hold. It is therefore an interesting fact that according to the evidence shown e.g. in Fig. 2, the transformation:

pearlite $\rightarrow$ austenite $\rightarrow$ martensite

occurs very close to the eutectoidal temperature even in the case of welding. This suggests that the nucleation barrier to austenite formation from pearlite must be very small. Other work done on laser treated steels containing spheroidised cementite (5,6) suggests that considerable superheating of the carbides can occur, but that pearlite transforms practically at the eutectoidal temperature. The reason for this is not obvious, though it could be due to the large interfacial area in pearlite between

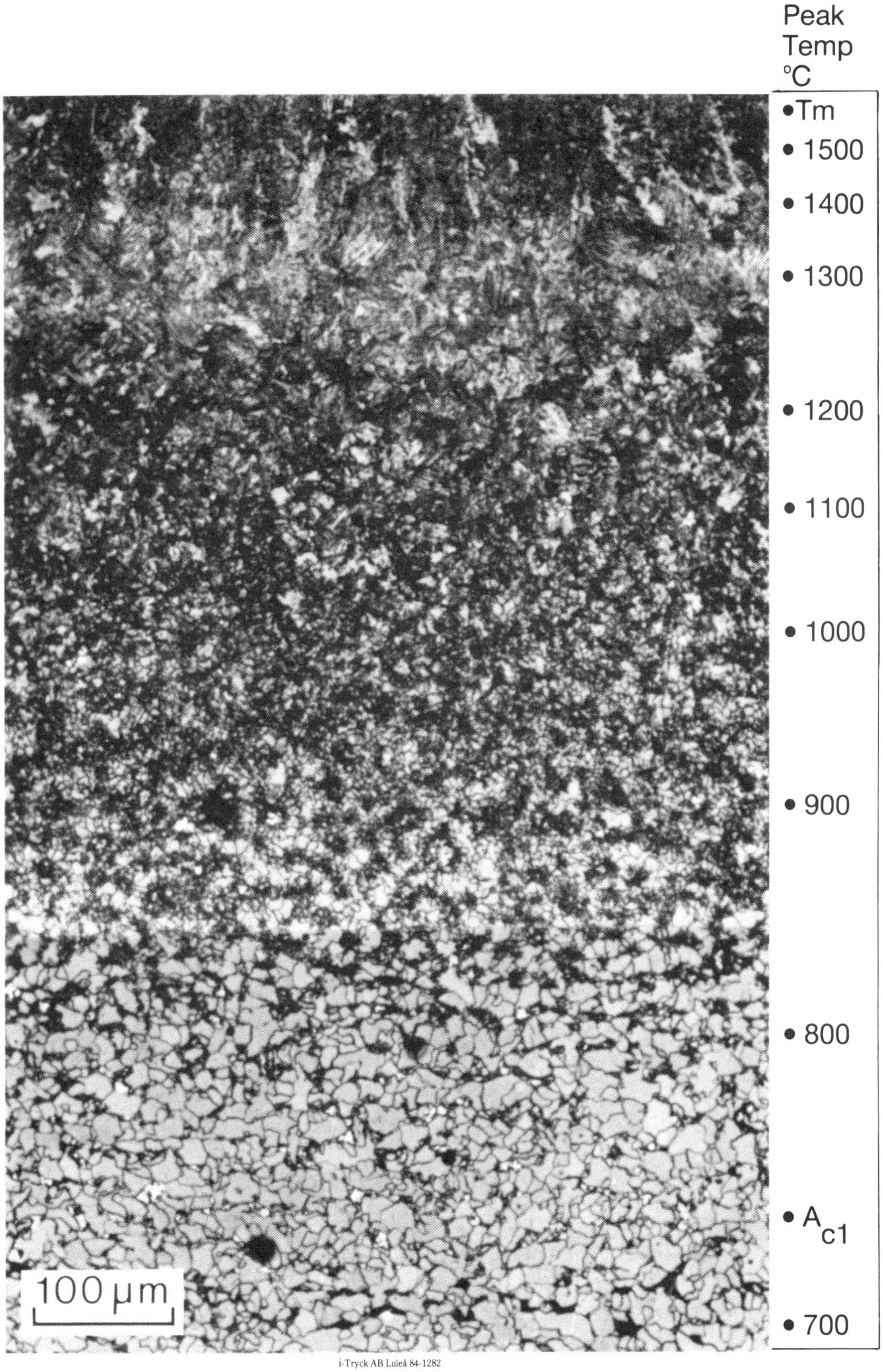

i-Tryck AB Luleå 84-1282

Fig. 2 - The microstructure of the heat affected zone of a metal-inert-gas weld of a Nb-micro-alloyed steel. Interferance microscopy is used to colour the various phases as: green ferrite, yellow martensite/bainite, brown pearlite, etc.

the ferrite and cementite larnellae, which provides good sites for the nucleation of austenite.

The ferrite → austenite transformation in pearlitic low carbon steels is thus a process in which the expansion of islands of austenite occurs. Experiments suggest that growth of the austenite, at least in the lower temperature range, is controlled by the rate of carbon diffusion into the ferrite. This implies in turn that as the ferrite surrounding the austenitic islands becomes sufficiently enriched in carbon, it is able to transform allowing the austenite to grow.

CARBONITRIDE COARSENING AND DISSOLUTION - In microalloyed steels, austenite grain growth is partially controlled by a dispersion of fine carbonitrides which effectively "pin" the grain boundaries. This is illustrated by the Zener equation, giving:

$$g = \frac{cp}{F} \quad (3)$$

where g is the austenite grain size, p is the mean precipitate size, F is the volume fraction of precipitates and c is a constant which depends on various factors such as precipitate shape. Actually, the Zener equation gives a final grain size as reached by an isothermal treatment. The Zener equation is not easy to apply in practice because of the difficulty of estimating F. A better experimental approach perhaps is to express g in terms of mean free spacing of precipitates, although the particles in question are only of the order of 10nm diameter and are hard to identify clearly in electron micrographs.

During the weld cycle, the particles coarsen or dissolve, depending on the peak temperature, the input energy of the process, and the composition of the particles. Assuming that the particles are of simple binary composition, e.g. NbC, TiN, VN, AlN, etc, their coarsening rate during an isothermal treatment is given by the Lifshitz-Wagnar equation:

$$p^3 - p_o^3 = \frac{Ct}{T} \exp - \left[\frac{Q}{RT} \right] \quad (4)$$

where P_o refers to the original mean precipitate size, t is time, Q is the activation energy for diffusion of the metallic component of the particle through the matrix volume, and C is constant that includes particle/matrix surface energy. In order to apply eqn.(4) to a welding (non - isothermal) thermal cycle, it has effectively to be integrated over this cycle(see later).

Above a certain temperature the particles disolve rather than coarsen. In the case of carbonitrides in steel, the isothermal solubility equation is:

$$\log\{ C_m^a C_c^b \} = A - \frac{B}{T_s} \quad (5).$$

C_m and C_c refer to the composition of the metallic and interstitial components of the compound, respectively, in weight percent, and a and b refer to the compound's stoichiometry. During the rapid heating of a weld thermal cycle, there is insufficient time for equilibrium dissolution to occur and the particles will be superheated. It can be shown(4,8) that the degree of superheating is well estimated by modifying eqn.(5) to:

$$\log[(\frac{C_m}{f})^a (C_c)^b] = A - \frac{B}{Ts} \quad (6)$$

where the factor f takes into account the time taken for the metallic component of the particle to diffuse into the surrounding austenite during a heating cycle as illustrated in Fig. 4. f is given by(7):

$$f = \frac{(6Dt)^{3/2}}{l^3} \quad (7)$$

where D is the diffusion coefficient of the

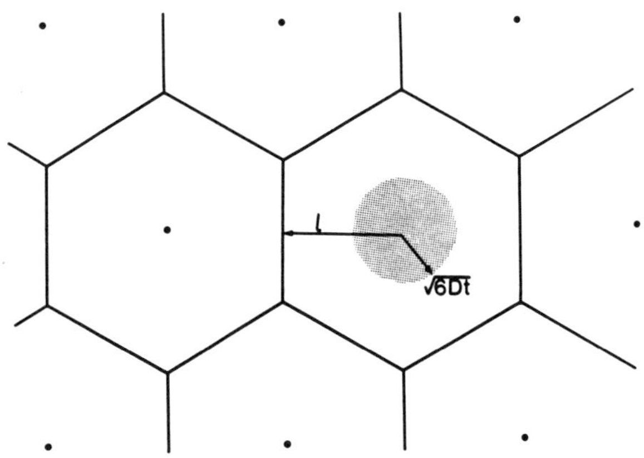

Fig. 4 - Cell model used for precipitate dissolution and the associated diffusion field.

metallic component in austenite and l is the radius of the volume of matrix needed to accommodate dissolution and is approximately equal to the interparticle spacing. If an experiment is made to measure the actual dissolution time, t*, then l can be better estimated from

$$l = (6Dt^*)^{\frac{1}{2}} \quad (8).$$

The extent of coarsening, dissolution or even grain growth, is given by the number of diffusive jumps that occur during the weld cycle. The kinetic strength of a complete heat cycle for a given diffusive controlled structural change can be defined as(7):

$$I = \int_0^\infty \exp - \left[\frac{Q}{RT(t)} \right] dt \quad (9)$$

where Q is the activation energy for the change

in question and R is the gas constant. This can be intergrated for the weld cycle to give:

$$I = 3\left(\frac{RT_P}{Q}\right)^{\frac{1}{2}}\tau \exp - \left(\frac{Q}{RT_p}\right) \qquad (10)$$

where τ is a thermal time constant. For the thick plate bead-on-plate solution to eqn(1), for example(3):

$$\tau = \frac{r^2}{4a} = \frac{Q}{2\pi\lambda ve}\left(\frac{1}{Tp-To}\right) \qquad (11)$$

From this it is thus possible to express the various diffusive processes in terms of the weld thermal cycle as calculated from eqn.(1).

GRAIN GROWTH - The change in grain size, g, assuming no barriers in the form of particle pinning, is given by:

$$g^2-g_o^2 = Ct \exp - \left(\frac{Q}{RT}\right) \qquad (12)$$

where C is a kinetic constant. Using eqns.(10) and (11),

$$g^2-g_o^2 = C\tau \exp - \left(\frac{Q}{RT_p}\right) \qquad (13)$$

where C can again be accounted for by an experimental measurement of grain growth in a real weld, as discussed earlier. For examples of how these modifications are carried out in detail, see ref.(8). It should be noted that eqn.(13) is used on the assumption that dissolution of particles is essentially complete. If coarsening occurs instead, as e.g. in the case of TiN particles in steel, eqn.(3) is used to estimate grain size.

HARDNESS OF THE HAZ - Hardness can be calculated if the volume fraction of the various constituents making up the zone can be estimated. By assigning hardnesses to each constituent, the net hardness at any point in the HAZ as a function of distance from the fusion line, can then be obtained from the rule of mixtures approach. Inagaki and Sekiguchi(9) have established the volume fraction of constituents (determined experimentally) as a function of the cooling rates, Δt, of various welds, Fig. 5. As examples, the curves defining the martensite and bainite regions in their diagram are well described by(4):

$$Fm = \exp - \left\{0,69\left(\frac{\Delta t_m}{\Delta t_{\frac{1}{2}}}\right)^2\right\} \qquad (14)$$

for the volume fraction of martensite, and

$$F_b = \exp - \quad 0,69\left(\frac{\Delta t_b}{\Delta t_{\frac{1}{2}}}\right)^2 - F_m \qquad (15)$$

giving the volume fraction of bainite. $\Delta t_{\frac{1}{2}}$ is the time in the diagrams for 50% of the transformation to occur. Similar volume fractions can be obtained for ferrite and pearlite. Changes in grain size are accounted for in each case by using a corrected time constant for transformation so that:

$$\Delta t_{\frac{1}{2}} = \frac{g}{g_o}(\Delta t_{\frac{1}{2}})_o \qquad (16)$$

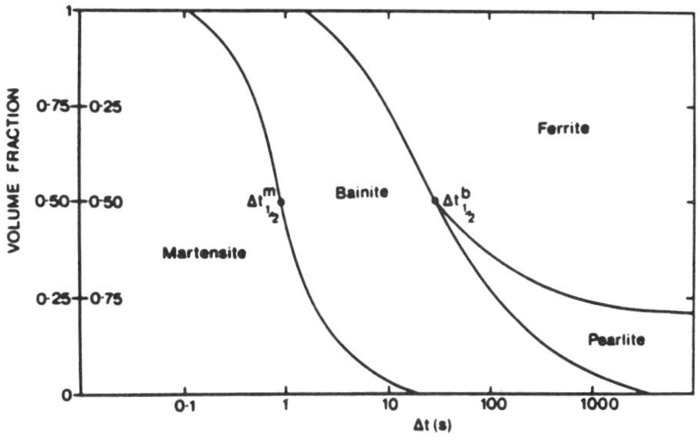

Fig. 5 - Volume fraction of eutectoidal constituent in steels as a function of $\Delta t_{8\to5}$, after Inagaki and Sekiguchi(9).

The amount of carbon in the martensite or bainite is interpolated directly from the Fe-C phase diagram to be:

$$C = 0,83 - \frac{(T_p-A_1)}{(A_3-A)}(0,8-C) \qquad (17)$$

$$\simeq \frac{C}{F} \quad \text{if } A_1 < T_p < A_3$$

Obviously if $T_p < A_1$, no bainite or martensite forms. The hardness of each point in the HAZ from the rule of mixtures is now estimated from:

$$H = H_m F_m + H_b F_b + \text{-----} \qquad (18)$$

Hardnesses of each of the phases, as a function of carbon content, can be obtained from the literature (see e.g. refs.4,8).

COMPARISONS BETWEEN PREDICTION AND EXPERIMENT

COMPARISON OF HEAT FLOW THEORY WITH EXPERIMENT - Comparisons of the thermal cycle prediction of eqn.(1) with experimentally measured thermal cycles in submerged arc and gas metal arc welds are shown in Fig. 6. The main discrepancy appears to be in the heating rate; this could be due to experimental problems of poor response from the thermocouples used for monitoring the change, on to the fact that in developing eqn.(1) no account is taken of changes in thermal properties as a function of temperature, or of heat losses from the surface. On the other hand the overall shape of the predicted thermal cycles are in good agreement with the experimental ones.

GRAIN SIZE COMPARISONS - These are illustrated in Fig. 7 for the thick plate welding of a Nb microalloyed steel. The shaded contours show the range of dissolution of NbC, from 50% to 100%. The slopes of the theoretical grain size contours have been adjusted slightly in this case to suit the experimental results, but the distance between contours is as predicted from theory. Agreement between the experimental measurements of grain size for a wide range of different welds with those predicted, is satisfactory considering the wide range of input energies involved in this case.

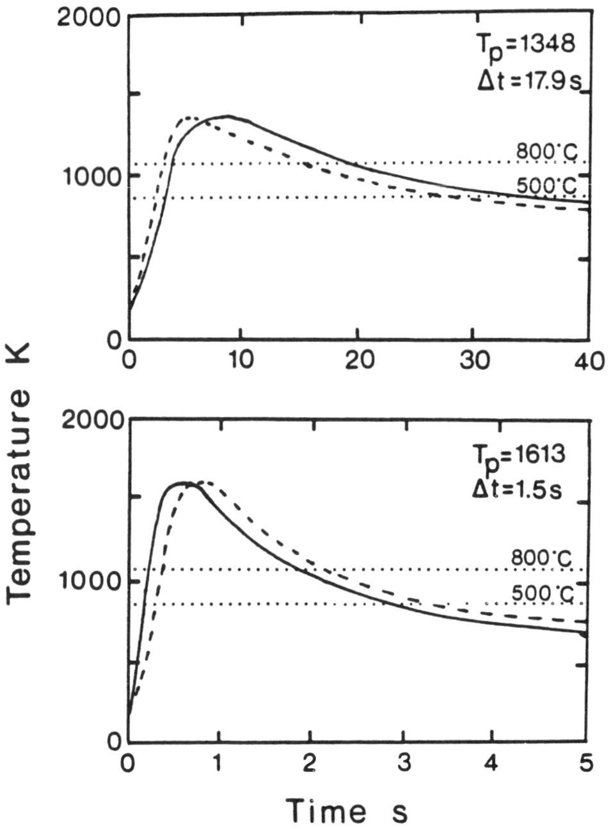

Fig. 6 - Comparison of experimental (full lines) and theoretical (dotted lines) weld thermal cycles for (a) a submerged-arc weld, and (b) gas-metal-arc weld.

An alternative way of presenting grain size contours, this time as a function of depth below the surface of the steel from the centre line of the weld, is illustrated in Fig. 8. In this more physical representation of the weld microstructure, the actual size of the various parts of the HAZ can be read off directly from the diagram.

An example of grain size changes in a Ti-microalloyed steel, in which coarsening of the TiN particles has occurred instead of dissolution, is given in Fig. 9. In this figure, experimental measurements of both particle size and grain size are included for comparison.

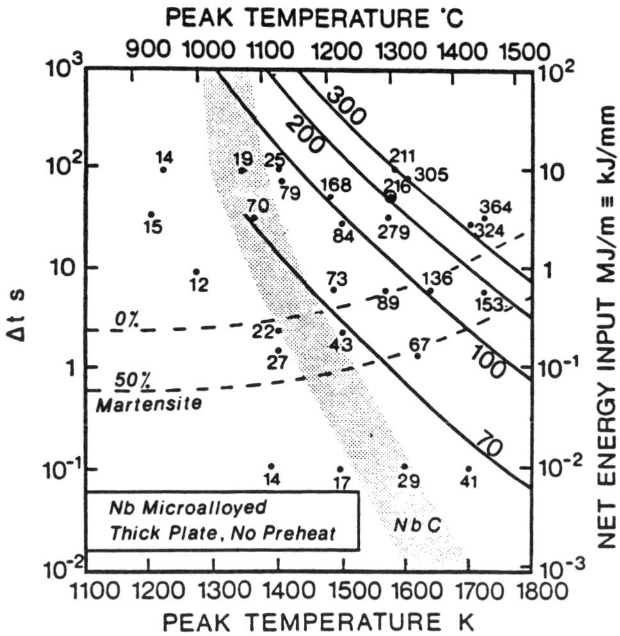

Fig. 7 - Diagramatic representation of HAZ microstructure. The shaded contour refers to NbC dissolution, the full lines are grain size contours and the dotted lines are martensite contours. The number refer to experimental measurements of austenitic grain size in various welds.

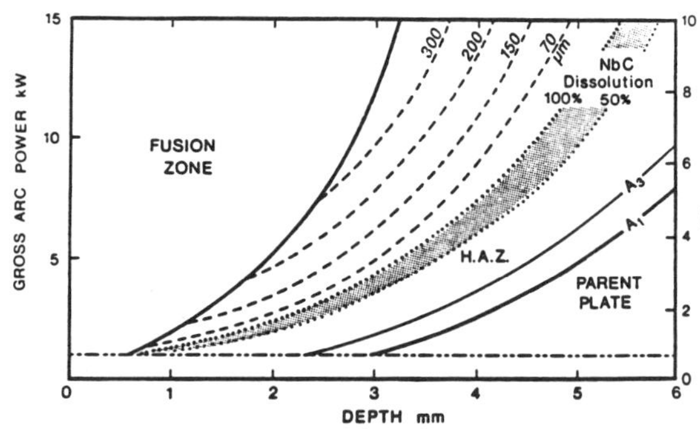

Fig. 8 - An alternative diagram of HAZ microstructure, showing a more physical picture of the various zones.

HARDNESS COMPARISONS - Comparisons between computed and experimentally measured hardnesses are shown in Fig. 10. No adjustments were made to the contours in this case, and as before agreement is surprisingly good.

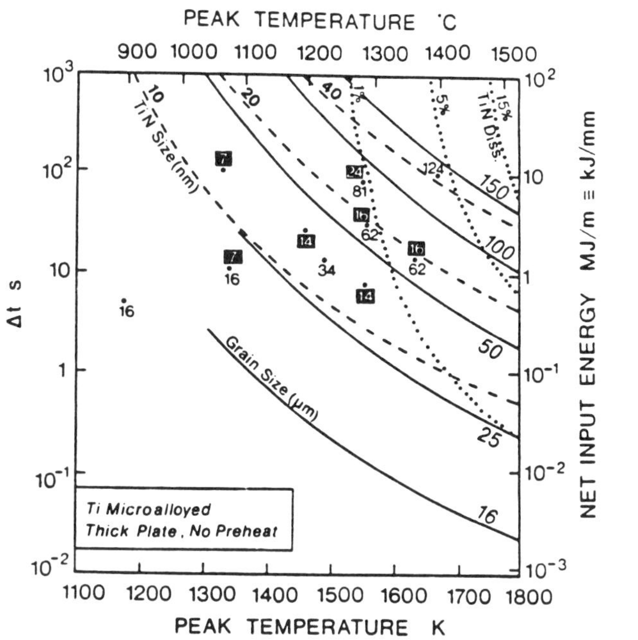

Fig. 9 - Microstructure diagram for a Ti-microalloyed steel in which TiN particle coarsening but not particle dissolution occurs. The boxed numbers are experimental measurements of particle size in welds and the other numbers refer to measured grain sizes.

An alternative way of presenting hardness and grain size data is shown in Fig. 11. In this case two different gas metal arc welds were made and the grain sizes and hardnesses measured across the HAZ. The dotted and full line curves are the theoretical predictions at these particular input energies of weld. As before, agreement is good.

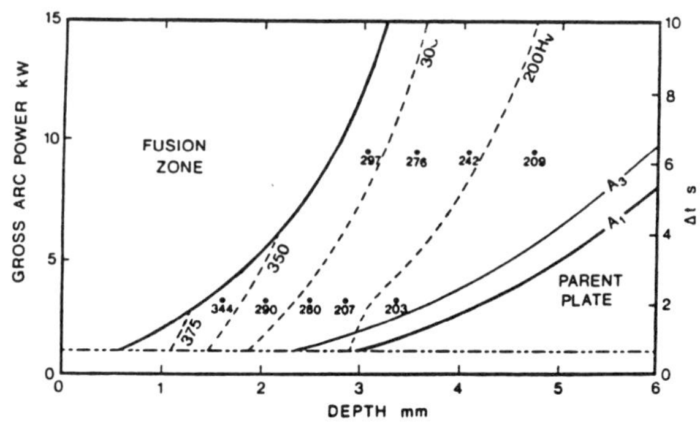

Fig. 10 - A digram of hardness contours (dotted lines), with associated experimental measurements made on various welds.

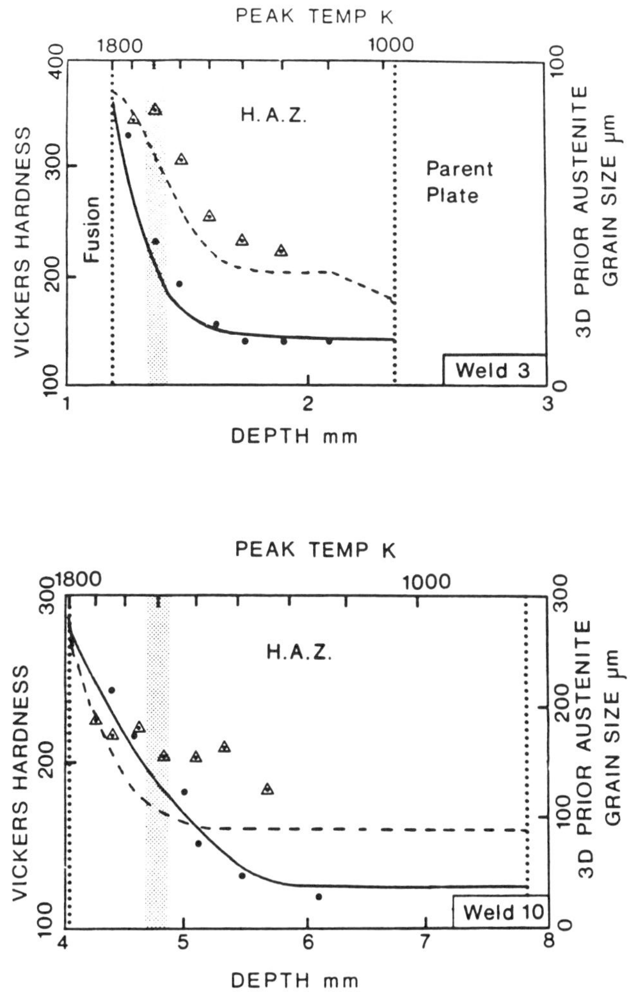

Fig. 11 - Grain size (full lines) and hardness (dotted) preditions of two different gas-metal-arc welds, with associated experimental plots. The shaded regions refer to particle (NbC) dissolution.

FURTHER DEVELOPMENTS

There are several aspects of the modelling of welding that could be worth working on. These include:

PRECIPITATE DISSOLUTION - In the present work, it has been assumed that carbonitrides are based on binary compounds. This is not correct in the case of practical steels as discussed earlier, particularly for the case of the microalloyed grades where it has been clearly demonstrated that the composition of particles is a complex function of steel composition and heat treatment(1,2). Since the solubility of particles is highly dependent upon their composition, this factor should obviously have been included when computing dissolution contours. A way to achieve this in the case of two metal ternary compound particles has been given(1,2). However, it would be useful to extend this analysis at least to quarternary particle compositions, to include two metal components plus carbon and nitrogen.

Apart from the problem of thermodynamic stability, the kinetic modelling of dissolution involving diffusion of two metallic components through austenite is fairly complex. For example, the extent of diffusion of one species compared to the other, depends on differences in their activation energies. However, as the particles dissolve their compositions alter due to these differences, thereby changing their thermodynamic stability. More theoretical work is needed to understand these effects better.

COARSENING - On the basis of the arguments given above, the effect of particle composition should also be included in precipitate coarsening models. Besides the problem of different rates of diffusion of the metallic components and the resulting changes in particle composition, an additional problem arises in this case in that the particle may be either growing or shrinking. This not only affects the particles composition but is also likely to affect the surface energy "constant" in the coarsening equation. As with dissolution of complex particles, the coarsening behaviour deserves more theoretical and experimental work.

BEAD AND JOINT GEOMETRIES - The hemispherical weld bead geometry cross-section, as predicted by a heat flow model based on a gaussian energy distribution, is found in only a limited number of welds in practice. Modelling the bead in terms of a finger and crater (Fig. 12) by employing two separate input energies, is one approach tried(8). However, a more general treatment to take account of the transfer mode of metal drops, surface energy, and heat loss effects on the shape of the melt is still awaited.

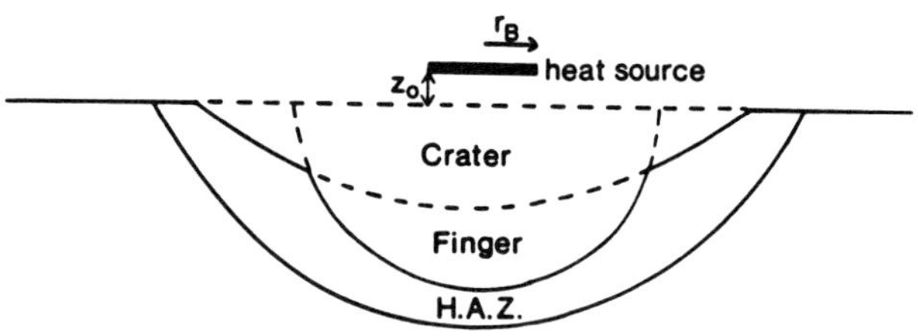

Fig. 12 - Modelling a submerged arc weld bead geometry using two overlapping gaussian energy distributions to simulate the "crater and finger" of the weld.

Another aspect of weld geometry concerns the type of weld joint being made, e.g. butt, fillet, edge, etc. In the present work, it is generally assumed that butt welds represent a two dimensional heat flow situation, while the others can be described by a three dimensional case. This approach may be too simple and it could be more appropriate to assign the various weld joints, plate thicknesses, etc., with a thermal severity grading based on the various heat flow paths possible.

MULTI PASS WELDS - Since each weld pass in multi-run welds effectively anneals the weld metal and HAZ of the previous pass, it might be desirable to incorporate these effects in the modelling procedure. Indeed, there have been attempts at doing this(10). In principle, the present approach could easily be adopted to account for multi-run thermal cycles, although a single pass weld of high heat input obviously provides the most severe changes in microstructure and properties possible.

HYDROGEN - Hydrogen induced cracking in welds is such an important phenomenon that more attention concerning the way hydrogen redistributes itself in the HAZ during the cooling cycle could be useful. Since the present work makes predictions of grain size, martensite volume fraction and hardness, incorporation of hydrogen levels into the model could provide a warning or recommendation in the computer program of the necessity for preheat, for example. More specifically, if it is assumed that cold cracking depends on a critical average hydrogen content in the HAZ, tied up with a certain minimum hardness in the grain growth zone, then for a given type of welding process and initial hydrogen level it should be possible to use diffusion theory to calculate the final hydrogen distribution in the HAZ. An attempt to do something like this has been made by Stenbecka(11). In principle it should be possible to adopt the same approach as used to handle diffusion phenomena in the present modelling, and to treat hydrogen diffusion and redistribution within the HAZ during any given weld thermal cycle.

ACKNOWLEDGEMENTS

I acknowledge with pleasure the work of my students, Doctors John Ion and Jan Strid, as well as invaluable cooperation with Professor Michael Ashby of Cambridge University.

REFERENCES

(1) J. Strid and K.E. Easterling, Acta Metall. 33, 2057 (1985).

(2) J. Strid, Ph.D. thesis, University of Lulea (1985).

(3) M.F. Ashby and K.E. Easterling, Acta Metall. 32, 1935 (1984).

(4) J.C. Ion, K.E. Easterling and M.F. Ashby, IBID, 32, 1949 (1984).

(5) W-B. Li, Ph.D. thesis, University of Lulea, 1984.

(6) W-B Li, K.E. Easterling and M.F. Ashby, Acta Metall., in print, 1986.

(7) M.F. Ashby and K.E. Easterling, IBID., 30, 1969 (1982).

(8) J.C. Ion, Ph.D. Thesis, University of Lulea, 1985.

(9) M. Inagaki and H. Sekiguchi, Trans. Nat. Res. Inst. Metals, Japan, 2, 102.

(10) L. Laroche, R. Blondeau, M. Palmier, J. Dollet, Soud.et Tech. Connexes, 33, 95 (1979).

THERMODYNAMICS OF THE ACICULAR FERRITE TRANSFORMATION IN ALLOY-STEEL WELD DEPOSITS

J.R. Yang, H.K.D.H. Bhadeshia
Department of Metallurgy and Materials Science
University of Cambridge
Pembroke Street
Cambridge CB2 3QZ, England

ABSTRACT

The mechanism by which acicular ferrite grows is not understood and this hinders the development of phase transformation models for predicting the microstructure of welds. In this work, we report experiments on the thermodynamics of the formation of acicular ferrite from austenite. It is found that on cooling below the B_S temperature, the degree of transformation to acicular ferrite increases, from zero at B_S, to an extent depending on the undercooling below the B_S temperature. However, at any given isothermal transformation temperature, the transformation ceases prematurely when the carbon concentration of the austenite reaches the level where diffusionless decomposition becomes thermodynamically impossible. The transformation thus exhibits the "incomplete reaction phenomenon" typical of the bainite reaction. Acicular ferrite is found to be similar to bainite, its growth being diffusionless, with carbon redistribution occurring after the initial formation of the ferrite plate. Acicular ferrite differs morphologically from bainite because it nucleates intragranularly on inclusions and because of hard impingement between plates nucleated on adjacent inclusions.

ACICULAR FERRITE (α_a) is a phase formed by the transformation of austenite (γ) during cooling of low-alloy steel weld deposits (1-3). It forms in a temperature range where diffusional transformations become relatively sluggish and give way to displacive transformations such as Widmanstätten ferrite, bainite and martensite. Its morphology consists of non-parallel plates of ferrite, which during the early stages of transformation nucleate on inclusions present in the columnar austenite grains typical of weld deposits. Subsequent plates may nucleate sympathetically on these inclusion-nucleated plates (4), so that a one-to-one correspondence between the number of active-inclusions and the number of α_a plates is not expected.

The term "acicular" means "shaped and pointed like a needle" but it is generally recognised that α_a has in three-dimensions the morphology of thin-plates. The true aspect ratio of such plates has never been measured but in two-dimensional sections, the plates typically are about 10μm long and perhaps 1μm wide (this implies that the true aspect ratio should be much smaller than 0.1). In wrought steels, the formation of plate-shaped ferrite (whether it be bainite or Widmanstätten ferrite) is always accompanied by a change in the shape of the transformed region (5). This shape change can be described as an invariant-plane strain with a significant shear component and is generally taken to imply a displacive transformation mechanism and the existence of an atomic correspondence between the parent and product phases, at least as far as atoms in the substitutional sites are concerned. Interstitial atoms, such as carbon may diffuse during transformation (especially when the ferrite forms at low-undercoolings where the partitioning of carbon is a thermodynamic necessity), without affecting the shape change or the displacive character of the transformation (5,6). The shape change accompanying the formation of α_a has never been determined experimentally.

The classification of microstructures on the basis of morphology is of considerable use in the study of structure-property relationships, but the prediction of microstructure requires a deeper understanding of the transformation mechanism. It is now well established that the presence of acicular ferrite in the primary microstructure of low-alloy steel weld deposits leads to improved toughness, but the mechanism of its formation is not understood. Attempts at modelling the primary microstructure thus rely on deducing the α_a content by difference rather than by direct prediction (3). In this work, our aim was to deduce fundamental information about the growth of acicular ferrite, focussing on whether the growth is diffusionless.

EXPERIMENTAL METHOD

To enable the study of acicular ferrite without substantial interference from other reactions, a high-hardenability manual metal arc weld was deposited at a speed of 2mm/s, using a welding current and voltage of 180A and 23V(DC+) respectively. The joint geometry was compatible with ISO2560, the weld consisting of 27 runs with 3 runs per layer, the interpass temperature being 250°C. The composition of the weld was found to be Fe-0.06C-0.27Si-1.84Mn-2.48Ni-0.02Mo wt.% and the deposit contained an oxygen level of 320ppm by wt. together with 0.01Al, 0.02Ti and 0.01Nb wt.%.

Specimens for dilatometric experiments were machined from homogenised (1200°C for 3 days while sealed in a quartz tube containing a partial pressure of pure Ar) weld metal. The specimens were 3mm diameter cylindrical rods with a 1.2mm diameter bore and a 20mm length, electroplated with a 7µm layer of nickel, to provide constraint and avoid surface nucleation or surface degradation effects. The experiments were carried out in a 'Theta' high-speed dilatometer, the specimens being austenitised at 950°C for 10 mins (or 1200°C for 30 mins) before helium gas quenching to the isothermal transformation temperature. The dilatometer was calibrated using pure Ni and Pt rods, so that absolute length changes could be measured.

The relative length change ($\Delta l/l$) as measured on a dilatometer can be converted into volume fraction of transformation; the method is described in detail elsewhere (7). The calculations, however, require the linear expansion coefficients and lattice parameters of austenite and of ferrite. The ferrite linear expansion coefficient e_α was determined by first tempering specimens at 600°C for 10 mins to decompose any retained austenite and then recording the length change during cooling at a rate of 0.063°C/s. The measurements do not therefore account for the presence of a very small amount of carbide. The γ expansion coefficient e_γ was measured while the specimen was in the γ single-phase field. The ferrite lattice parameter (a_α) at ambient temperature was measured using a Debye-Scherrer technique (CuK_α radiation) and Newton-Ralphson extrapolation using the 011, 002, 112, 022 and 013 lines. Having determined the volume fraction v of ferrite obtained by transformation, the carbon concentration of γ, x_γ is obtained from the relation:

$$x_\gamma = \bar{x} + [v(\bar{x} - s)/(1 - v)] \qquad (1)$$

where s is the amount of carbon in solution in the ferrite or tied up as carbides. $\bar{x}$ is the average carbon content of the alloy. Since carbide precipitation does not occur rapidly over the time scale of the experiments, s is taken to be 0.02 wt.%.

RESULTS AND DISCUSSION

PRIMARY WELD MICROSTRUCTURE AND TTT DIAGRAM - The primary weld microstructure, consisting mainly of acicular ferrite and microphases is illustrated in Fig. 1. Fig. 2 shows the calculated TTT curve for the weld (8); in such diagrams, the upper C curve represents the time taken for the initiation of diffusional transformations such as allotriomorphic fertite and pearlite, whereas the lower C curve represents the time taken for the initiation of displacive transformations such as bainite and Widmanstätten ferrite (8). It is notable, however, that in this alloy, the Widmanstatten ferrite start temperature (W_S) equals the bainite start temperature (B_S). This means (9) that Widmanstätten ferrite should not be observed at all and immediately implies that the acicular ferrite observed in the weld deposit is likely to be bainite if it is not a completely new transformation. Fig. 2 also shows that for any reasonable weld cooling rate, the cooling curve cannot intersect the diffusional C curve, so that the alloy only transform by a displacive mechanism. The alloy is therefore ideally suited to avoid interference from other transformation products during the study of acicular ferrite. In addition, transformation during the quench to the isothermal transformation temperature can easily be avoided. Of course, a weld deposit is necessarily chemically inhomogeneous (10), and some formation of allotriomorphic ferrite and Widmanstätten ferrite cannot thus be avoided in the solute depleted regions, as seen in Fig. 1.

GRAIN SIZE EFFECT - The problem of resolving the nature of α_a is made worse by the fact that it does not readily occur in wrought steels, due to the absence of suitable intragranular heterogeneous nucleation sites and because the γ grain size of such steels is deliberately kept small; transformations initiated at γ grain boundaries thus swamp the interior regions as well. However, if after deposition, a weld is reheated into the

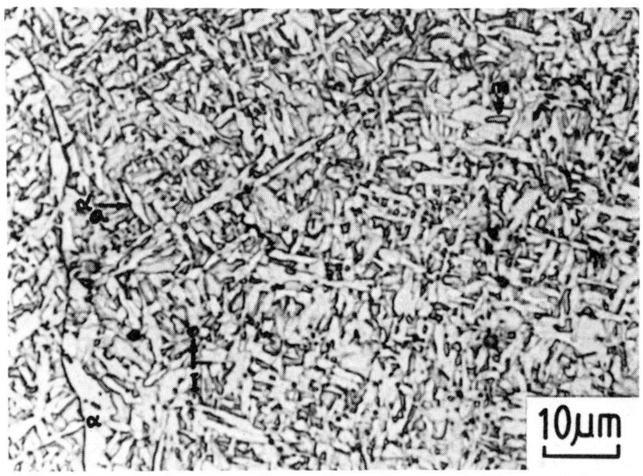

Fig. 1 - Primary microstructure of the weld deposit

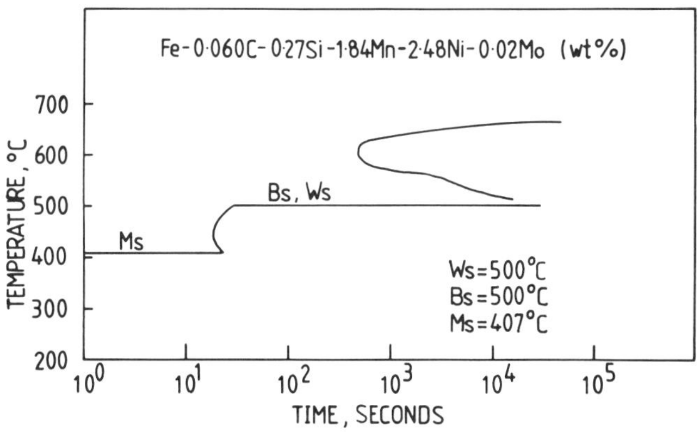

Fig. 2 - Calculated TTT curve for the Fe-Mn-Si-Ni-Mo-C weld

austenite phase field to generate large γ grains then these new grains still contain the inclusions which were present in the original deposit. Isothermal transformation of this structure in a temperature range where α_a usually forms in welds should then lead to the intragranular nucleation and growth of plates whose morphology is identical to that of the α_a observed in weld deposits. Fig. 3 illustrates the results of isothermal transformation experiments on reheated weld metal; in one case the weld was reaustenitised at a temperature of 950°C for 10 mins. (Fig. 3a) and another sample was austenitised at 1200°C for 30 mins prior to isothermal transformation at 460°C for 30 mins before quenching to ambient temperature. The higher austenitisation temperature has clearly led to a larger austenite grain size, and subsequent isothermal transformation has led to the intragranular nucleation of plates of ferrite which can be identified morphologically with acicular ferrite found in welds. On the other hand, in the specimen reaustenitised at 950°C, the small grain size has prevented intragranular nucleation, the ferrite plate nucleating at the austenite grain boundaries and growing by a sub-unit mechanism to give the morphology of sheaves of upper bainite. These results imply that acicular ferrite is not a new transformation product, but is simply intragranularly nucleated bainite. Its morphology differs from conventional bainite because it nucleates intragranularly, whereas in wrought steels, bainite nucleates initially at γ/γ or γ/α boundaries. α_a does not normally grow in sheaves because the development of sheaves is stifled by hard impingement between plates nucleated independently at adjacent sites.

BAINITE OR WIDMANSTÄTTEN FERRITE? - A further experiment is needed to clearly establish that acicular ferrite is similar to bainite and not to intragranularly nucleated Widmanstätten ferrite. Both these latter products grow by a displacive transformation mechanism, the details of which are discussed elsewhere (6-9,11). The formation of Widmanstätten ferrite involves the redistribution of carbon between the parent and product phases, but bainite initially forms by a diffusionless mechanism and the carbon later partitions into the remaining austenite (6-9,11). This difference can be exploited to deduce whether acicular ferrite grows like conventional bainite or like Widmanstätten ferrite.

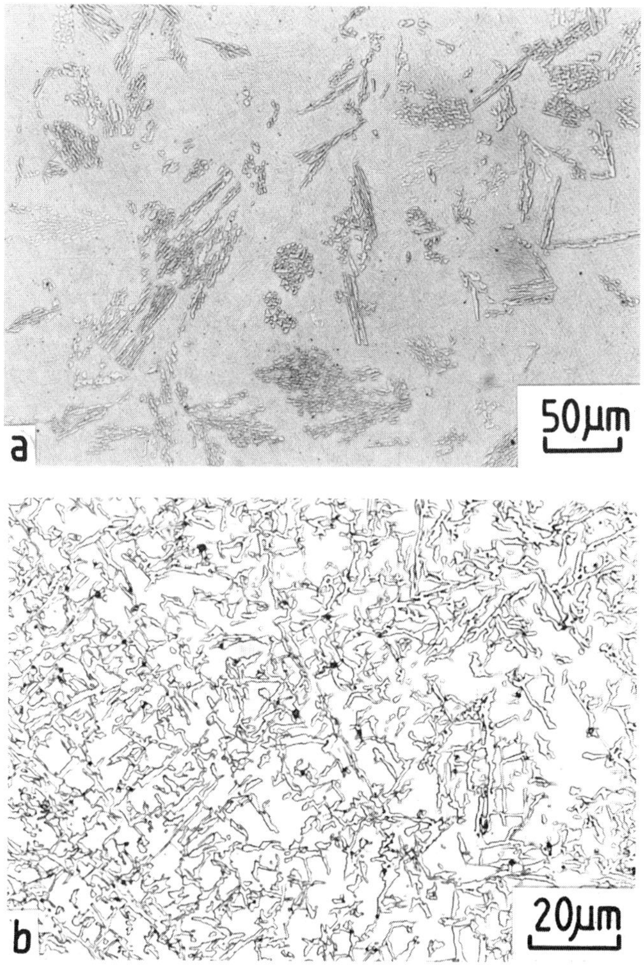

Fig. 3 - (a) Bainite obtained after austenitisation of weld metal at 950°C followed by isothermal transformation at 460°C. (b) Acicular ferrite obtained after austenitisation of weld metal at 1200°C followed by isothermal transformation at 460°C. Details of heat-treatments given in text.

Considering first the diffusionless formation of bainite, the reaction should go to completion since there is no diffusion necessary. In practice, the whole of the austenite grain does not transform instantaneously because of kinetic restrictions (e.g. heterogeneous nucleation); even if the first plate forms without diffusion, it has an opportunity to reject its excess carbon into the residual austenite. Any further increment of transformation is therefore associated with a lower free-energy change, due to the higher carbon content of the austenite from which it has to form. Eventually, a stage is reached where the transformation becomes thermodynamically impossible since the free energies of the residual (Fig. 4) austenite and bainitic ferrite of the same composition become identical (6-9, 11). The locus of such positions, as a function of isothermal transformation temperature defines the T_o' curve, where γ and α_b (with a certain amount of stored energy associated with transformation strains) of the same composition have equal free energies. The corresponding curve for stress-free γ and α of identical composition is conventionally called the T_o curve. The bainite reaction should therefore stop when the carbon concentration of the austenite (x_γ) reaches the level given by the T_o' for the isothermal transformation temperature concerned (6-9,11).

The Ae3′ curve may similarly be defined for the growth of α_w, which involves the para-equilibrium transformation with the partitioning of carbon. If the small amount of strain energy associated with the back-to-back formation of mutually accommodating plates of Widmanstatten ferrite is ignored (9), the α_w reaction should stop when x_γ reaches the level given by the Ae3′ curve for the isothermal transformation temperature concerned. For any given T, the carbon concentration given by the Ae3′ curve is much larger than that given by the T_o' curve, so that a measurement of x_γ at reaction termination can clearly distinguish between α_w and α_b. Hence, our aim was to isothermally transform a weld metal in a temperature range where it is known to transform to just acicular ferrite, and by studying the maximum extent of transformation in terms of the above criteria, to decide whether α_a resembles Widmanstatten ferrite or bainite with respect to its growth mechanism.

DILATOMETRY AND THERMODYNAMIC ANALYSIS - The dilatometric specimens were austenitised at 950 for 10 mins or at 1200°C for 30 mins before quenching to the isothermal transformation temperature. The isothermal transformation curves are presented in Fig. 5 and are seen to have a classical sigmoidal shape. It is immediately apparent that the maximum degree of transformation at any T increases as T decreases below the B_s temperature, showing that the weld metal exhibits the classical 'incomplete-reaction phenomenon' which is an identifying feature of bainitic transformation (6-9,11). From the data of Fig. 5, it is possible to calculate x_γ corresponding to the maximum degree of transformation, and these calculations are presented in Fig. 6, along with the relevant phase boundaries (calculated as in 6,9,11). The calculations use the following measurements: $e_\gamma = 1{,}15 \times 10^{-5}\ K^{-1}$, $e_\gamma = 1.769 \times 10^{-5}\ K^{-1}$, and the lattice parameter of ferrite in the weld, at ambient temperature = 2.8723 angstroms. It is clear that acicular ferrite formation ceases at the T_o' boundary, providing strong evidence that the formation of acicular ferrite is diffusionless, the redistribution of carbon into the residual austenite occurring at a stage after growth. The results also imply that the strain energy accompanying acicular ferrite growth is of the order of 400J/mol (9).

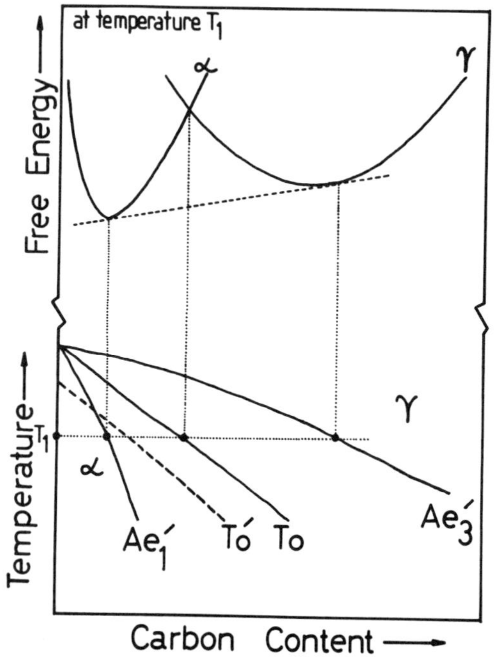

Fig. 4 - Schematic diagram showing the construction of phase boundaries from free energy data. The T_o' line includes the effect of 400/mol of strain energy due to transformation.

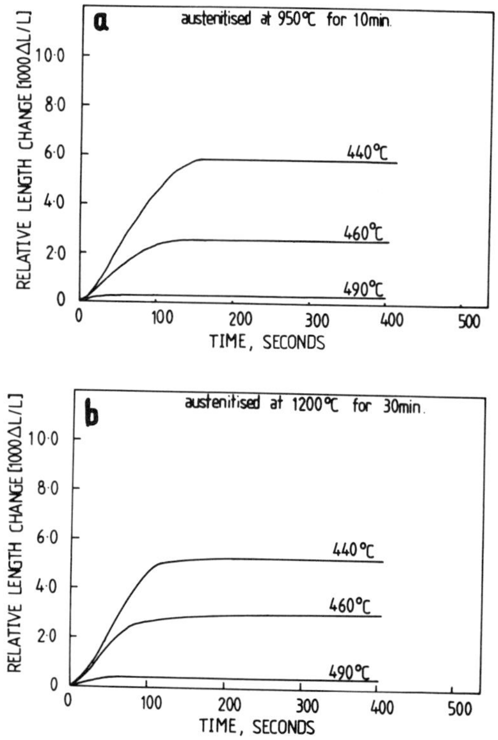

Fig. 5 - (a) Isothermal transformation curves for weld reaustenitised at 950°C. (b) Isothermal transformation curves for weld reaustenitised at 1200°C.

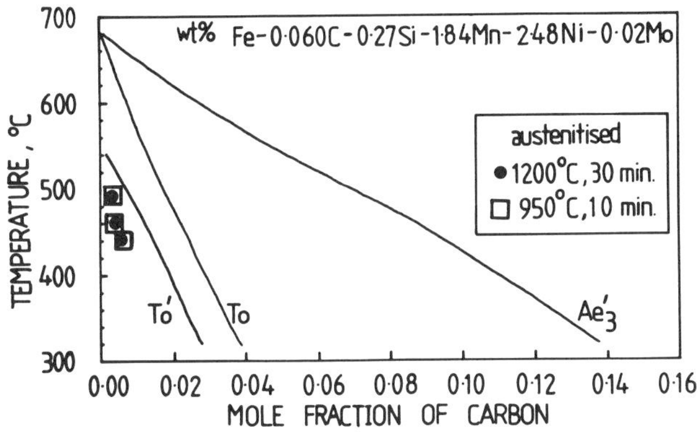

Fig. 6 - Phase diagram, with experimental data on the carbon concentration of austenite at the termination of isothermal transformation. The reactions stop well before the Ae3´ curves are reached.

CONCLUSIONS

It is found that the formation of acicular ferrite at any particular transformation temperature ceases as the carbon concentration of the residual austenite reaches the T_0' boundary, and that the transformation exhibits an "incomplete reaction phenomenon". These results provide strong evidence suggesting that the growth of acicular ferrite is diffusionless, with carbon partitioning into austenite after the transformation event. In this respect, acicular ferrite is identical to bainite. Evidence has also been provided to show that the morphology of acicular ferrite differs from that of bainite simply because the former nucleates intragranularly at inclusions within large austenite grains. It is concluded that acicular ferrite is in fact bainite which nucleates intragranularly in the large austenite grains typical of weld deposits.

ACKNOWLEDGEMENTS

The authors are grateful to Professor D. Hull for the provision of laboratory facilities at the University of Cambridge, to ESAB for the provision of weld samples and to the Chinese Ministry of Education for financial support.

REFERENCES

1) Abson, D. J. and R. E. Dolby, Weld. Inst. Res. Rep. 19/1978, Welding Institute, Abingdon, Cambridge, 1978.
2) Easterling, K., "Introduction to the Physical Metallurgy of Welding", Butterworths, London (1983).
3) Bhadeshia, H. K. D. H., L.-E. Svensson and B. Gretoft, Acta Metall. 33 1271-1283 (1985).
4) Barrite, G. S., R. A. Ricks and P. R. Howell, "Quantitative Microanalysis with High Spatial Resolution", p.112, Metals Soc., London (1981).
5) Bhadeshia, H. K. D. H., Mater. Sci. and Tech. 1 497-504 (1985).
6) Bhadeshia, H.K.D.H., Prog. in Mat. Sci. 29 321-386 (1985).
7) Bhadeshia, H. K. D. H., J. de Physique C4-43 443-449 (1982).
8) Bhadeshia, H. K. D. H., Met. Sci. 16 159-165 (1982).
9) Bhadeshia, H. K. D. H., Acta Metall. 29 1117-1130 (1981).
10) Gretoft, B., H. K. D. H. Bhadeshia and L.-E. Svensson, Acta Stereologica, in press.
11) Bhadeshia, H. K. D. H. and D. V. Edmonds, Acta Metall. 28 1265-1273 (1980).

MODELING OF THE HEAT AFFECTED ZONE OF ALUMINUM ARC WELDS

R.P. Martukanitz
Alcoa Laboratories
Alcoa Center, Pennsylvania

ABSTRACT

An arc welding heat transfer model, based on finite differencing techniques, was used in conjunction with kinetic relationships developed to represent precipitation during welding of 6009. These kinetic relationships were based on the additive characteristic of the precipitation reaction. The extent of the reaction, through loss in hardness, was then determined for the various thermal profiles in the heat affected zone. These produced hardness profiles that varied depending on the welding heat input. Welding experiments were conducted to compare measured heat affected zone hardness profiles to those predicted under identical conditions.

A Computer Aided Engineering (CAE) System is envisioned for designing welded aluminum assemblies. Such a system would allow joint strengths to be ascertained based on process parameters. It would also have the ability to perform structural analysis in a global sense by considering base metal properties, joint strengths, and loading conditions. Such a system is shown in Figure 1 and would utilize existing technology for heat transfer and structural analysis. Numerical models have been developed that accurately depict the various welding processes.[1-7] Structural analysis packages are commercially available and have been successfully applied to a wide variety of engineering problems.[8,9] The expertise which must be developed is the ability to predict welded joint strength based on choice of process and parameters.

In the case of welded aluminum alloys joint strength is usually dictated by the extent of base metal degradation in the heat affected zone (HAZ). This degradation is cause by microstructural modifications associated with elevated temperatures experienced in this zone. For precipitation-hardenable aluminum alloys, the HAZ represents dissolution or growth of precipitates. The response in the HAZ of 2XXX aluminum alloys follows dissolution; whereas, the primary modification for the 6XXX alloys is growth of precipitates.[10,11] The work-hardenable aluminum alloys, such as the 5XXX series, display a HAZ that is dominated by recrystallization and grain growth. Although the nature of these heat affected zones may differ they all are diffusion controlled and thermally dependent. Ideally, the HAZ may be described by heat transfer analysis and kinetics of the governing metallurgical reactions.

The purpose of this work is two-fold: to describe quantitatively the reactions governing the softening in the HAZ for aluminum alloys, and to use this information with the aid of an arc welding heat transfer model to predict HAZ property profiles. The kinetic relationships for these reactions may be obtained empirically through isothermal experiments. If these relationships describe the reaction in terms of both time and temperature, it is feasible to couple these relationships with thermal cycles of the HAZ to determine loss of mechanical properties.

MODEL DESCRIPTION

A model of the heat affected zone may be developed through the use of relationships describing the rate of metallurgical reactions associated with the elevated temperatures of this zone.

Isothermal kinetics for aluminum alloys may be defined by the equation:[12]

$$\zeta = 1\text{-exp}\ (-k_1 t) \qquad [1]$$

where: ζ = fraction transformed due to precipitation or recrystallization,

k_1 = constant that is related to the temperature and time required to form a specific amount of the reaction product, and

t = time.

Since strength is related to the amount untransformed Eq. [1] may be represented as:

$$\frac{\sigma}{\sigma_{max}} = \exp(-k_2 t) \qquad [2]$$

where: σ = strength,

σ_{max} = maximum strength, and

$k_2 = \ln(\sigma/\sigma_{max})$.

When the reaction rate is a function of the amount transformed and temperature, the reaction may be expressed as an additive quantity.[13] Hence, Eq. [2] may be transformed to represent the reaction during a continuous thermal cycle:[14]

$$\sigma - \sigma_{max} \exp(-k_2 \theta) \qquad [3]$$

and

$$\theta = \int_{t_o}^{t_f} \frac{dt}{C_t} \qquad [4]$$

where: θ = factor which represents the amount transformed during the thermal cycle,

C_t = time as a function of temperature from kinetic relationship,

t_o = time at start of thermal profile, and

t_f = time at end of thermal profile.

In practice θ may be approximated as a summation:

$$\theta = \sum_{i=1}^{i=F-1} \frac{\Delta t_i}{C_t} \qquad [5]$$

Through the use of Eqs. [3] and [5], provided the thermal cycle and kinetic relationship are known, the strength at any point in the HAZ may be determined. This is achieved by first calculating θ associated with the thermal cycle. If the kinetic relationship is expressed as time, the thermal cycle may be integrated numerically with θ being determined by the sum of the individual temperatures and C_t's. This procedure is shown graphically in Figure 2. Once θ is determined the resultant strength governed by the reaction for that thermal profile is calculated. A similar analysis was attempted by Smith, et al, to determine properties in the heat affected zone of aluminum alloy 6061-T6; however, the use of a simplistic thermal and kinetic model led to poor agreement between measured and predicted HAZ strengths.[15]

EXPERIMENTS

To determine the validity of the HAZ kinetic model, two sets of experiments were conducted. The first set involved isothermal aging experiments for developing kinetic data governing HAZ precipitation of 6009-T6 for the as-weld condition and 6009-T4 that would be post-weld aged. Isothermal aging experiments were utilized to obtain hardness data representing the precipitation reaction. This information was coupled with a heat transfer model to provide predictive abilities for loss in hardness in the HAZ. The second set of experiments entailed the production of welded panels under controlled conditions to obtain actual data regarding thermal histories and loss of hardness in the HAZ.

Isothermal aging, representing 6009-T6 for the as-welded condition, involved heating 2.0 mm thick 6009-T6 samples to various temperatures between 285°C and 430°C and holding for short periods of time (1, 3, 6, 10, and 90 seconds) followed by water quenching. Similar experiments were conducted on 6009-T4 samples followed by the normal aging practice of 205°C for one hour to represent 6009-T4 that would be post-weld aged. A Gleeble thermal/ mechanical simulator was utilized during the isothermal aging experiments. This simulator employs resistance heating and a computer controlled, closed-loop system to accurately simulate predetermined thermal cycles. Three hardness measurements, using a Rockwell tester and the E-scale, were taken on the surface of the samples after isothermal aging. The 6009-T6 samples were allowed to naturally age 30 days after isothermal aging prior to hardness testing. Hardness measurements for the 6009-T4 samples were made after the final aging of 205°C for one hour.

Welding experiments were conducted to obtain information concerning thermal histories and hardness profiles of the HAZ. Material used during welding was 2.0 mm thick 6009 sheet from the same lot of material used during the isothermal aging experiments. Autogenous, gas tungsten arc (GTA) welds using alternating current (AC) were made along the center of 160 mm wide by 180 mm long panels while varying heat input through changes in welding velocity. These produced linear heat inputs of 756 and 1128 J/cm. A third heat input of 504 J/cm was used to validate the heat transfer model during partial penetration but was not used in the HAZ analysis. Welds were made on 6009-T6 and 6009-T4 that were subsequently aged. Panels were fixtured using insulating material to prevent the occurrence of heat sinks. During welding temperatures at 1.0 cm and 2.0 cm from the weld centerline were obtained using 0.36 mm diameter thermocouples that were capacitor discharge welded to the bottom side of the panel. Temperatures were recorded on a laboratory data acquisition system. Welding parameters for each experiment are shown in Table 1. Hardness profiles of the HAZ were obtained after welding by hardness measurements. These hardnesses represented an average of three measurements in 0.25 cm increments from the weld centerline. Using identical parameters obtained during the welding experiments, a numerical heat transfer model was utilized to obtain temperature profiles in 1.0 mm increments from the weld centerline through the heat affected zone.

TABLE 1 - Welding Parameters Used During Experiments

Welding Amperage:	120 amperes
Welding Voltage:	14 volts
Welding Velocity:	0.67, 1.0, and 1.5 cm/sec
Shielding Gas:	9.4 l/min argon
Electrode:	2.4 mm diameter pure tungsten
Cup Size:	13.0 mm inside diameter
Electrode Stick-out:	6.3 mm
Tungsten-to-Work Distance:	1.5 mm

DISCUSSION OF RESULTS

Results of hardness measurements made after the isothermal aging experiments were statistically analyzed using multiple regression analysis. This analysis provided relationships describing loss in hardness in terms of time and temperature representing the T6 as-welded and T4 post-weld aged conditions. These relationships are shown graphically in Figures 3 and 4 as isohardness curves. These curves depict the time required to obtain a specific loss in hardness, based on the precipitation reaction, at various temperatures. The 6009-T6 material showed a fairly linear relationship to approximately 325°C, and at temperatures above 325°C curvature of the isohardness curves was exhibited. This curvature represents a change in the nature of the modification rather than a change in the rate of the precipitation reaction. It signifies dissolution of the strengthening phase present prior to welding as the solvus temperature is approached. This is further verified by viewing the isohardness curves for 6009-T4 material that was aged after isothermal heating (Figure 4). These curves show similar inflections but with a continuation of curvature at higher temperature to produce a characteristic "C-curve". The increase in hardness at the higher temperatures is due to the precipitation of dissolved particles into strengthening phases. Similar to the T6 material, dissolution occurred at these temperatures during isothermal aging and was made available to reprecipitate during the final aging at 205°C for one hour.

The statistical model used to generate this family of curves also made available relationships for specific curves. This information would ultimately be required for the kinetic model. The kinetic relationship for 6009-T6 representing the as-welded condition for 80% of the original strength was:

$$\ln C_t = \frac{-101.19 + 0.154\ (T)}{10.58 - 0.021\ (T)} \qquad [6]$$

and for 6009-T4 representing welding in the T4 temper and post-weld aging at 205°C for one hour:

$$\ln C_t = 106.96 - 42.28\ [\ln(530-T)] + 4.25\ [\ln(530-T)]^2 \qquad [7]$$

Figure 5 shows macrographs of the experimental welds made using three heat inputs. Superimposed on the macrographs are the liquidus and solidus isotherms obtained from the heat transfer model utilizing identical welding parameters. A process efficiency of 45% was employed in the heat transfer model to represent gas tungsten arc welding using alternating current with the power supply set in the balanced wave mode. Shown in Figures 6 and 7 are the measured and predicted thermal cycles for the full penetration welds made at 756 and 1128 J/cm heat input, respectively. These represent temperatures recorded during welding and those predicted by the heat transfer model (shown as dashed lines) at distances of 1.0 and 2.0 cm from the weld centerline. For both weld conditions the model predicted somewhat higher temperatures than those obtained experimentally.

The predicted hardness profiles for the two weld heat inputs for 6009-T6 and 6009-T4 that had been post-weld aged are shown in Figures 8 and 9. The predicted hardness points were obtained by utilizing the various thermal cycles from the heat transfer model in conjunction with the kinetic model for both cases. Good agreement between measured and predicted hardness values was observed with the greatest deviation occurring near the weld fusion zone. These anomolies may be explained by recalling that the kinetic model was constructed to represent the growth of precipitate. At positions close to the fusion zone temperatures experienced are such that dissolution, rather than growth, of precipitate will govern the modification.

SUMMARY AND CONCLUSIONS

An arc welding heat transfer model, based on finite differencing techniques, was used in conjunction with kinetic relationships developed to represent precipitation during welding of 6009. These kinetic relationships were based on the additive characteristic of the precipitation reaction. The extent of the reaction, through loss in hardness, was then determined for the various thermal profiles in the heat affected zone. These produced hardness profiles that varied depending on the welding heat input. Welding experiments were conducted to compare measured heat affected zone hardness profiles to those predicted under identical conditions.

The results of this evaluation indicate that the heat affected zone of aluminum-magnesium-silicon alloys may be accurately modeled through heat transfer and kinetic analysis. The model developed by the author provides accurate representation for the majority of the heat affected zone. The present kinetic model does not account for dissolution of precipitate which is found at distances close to the weld fusion zone.

REFERENCES

1. Friedman, E., and S. S. Glickstein, Welding Journal, 55, 408s-420s (1976)
2. Brody, H. D. and R. A. Stoehr, Journal of Metals, 24, 20-27 (1980)
3. Kou, S., and Y. Lee, Metallurigical Transactions A, 14A, 2245-2253 (1983)
4. Martukanitz, R. P., "A Three Dimensional Heat Transfer Model for Arc Welding Aluminum," Presented at the Annual Meeting of American Welding Society, Dallas, TX (1984)
5. Goldak, J., A. Chakrabarti, and M. Bibby, Metallurgical Transactions B, 15B, 299-307 (1984)
6. Blottner, F. G., AIME Conference Proceedings - Modeling of Castings and Welding Process II (1983)
7. Chan, C., J. Mazumder, and M. M. Chen, Metallurgical Transactions A, 15A, 2175-2184 (1984)
8. "Ansys-Engineering Analysis Systems," Swanson Analysis Systems, Inc., Elizabeth, PA (1977)
9. "ABAQUS," Hibbit, Karlsson, and Sorensen, Inc., Providence, RI (1984)
10. Dumolt, S. D., Ph.D. dissertation, Carnegie-Mellon University, Pittsburgh, PA (1983)
11. Engo, T., and T. Kuroda, Transactions of the Japanese Welding Research Institute, 11, 61-66 (1982)
12. Evancho, J. W., and J. T. Staley, Metallurgical Transactions, 5, 43-47 (1974)
13. Cahn, J. W., Acta Metallurgica, 4, 572-575 (1956)
14. Staley, J. T., Industrial Heating, 44, 6-9 (1977)
15. Smith, C. O., E. R. Funk, and H. Udin, Welding Research Council Bulletin Number 12, New York, NY (1952)

(4RPM7C)

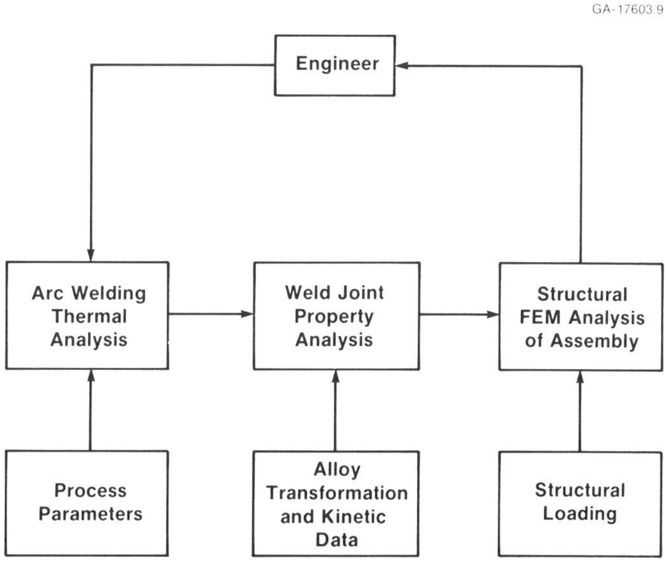

Computer Aided Engineering System
Figure 1

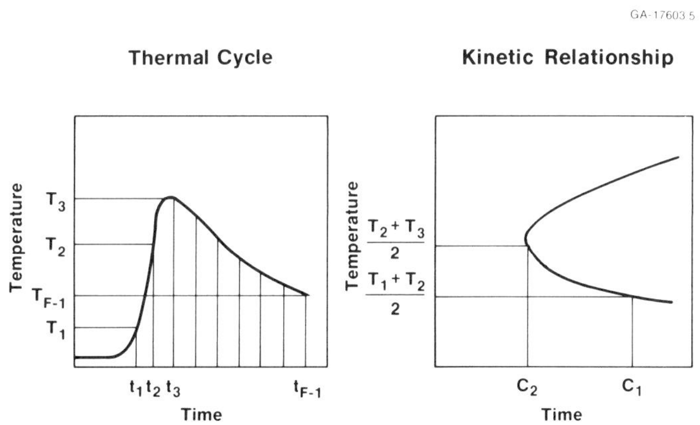

$$\tau = \frac{\Delta t_1}{C_1} + \frac{\Delta t_2}{C_2} + \ldots + \frac{\Delta t_{F-1}}{C_{F-1}}$$

Graphical Representation of Heat Affected Zone Model
Figure 2

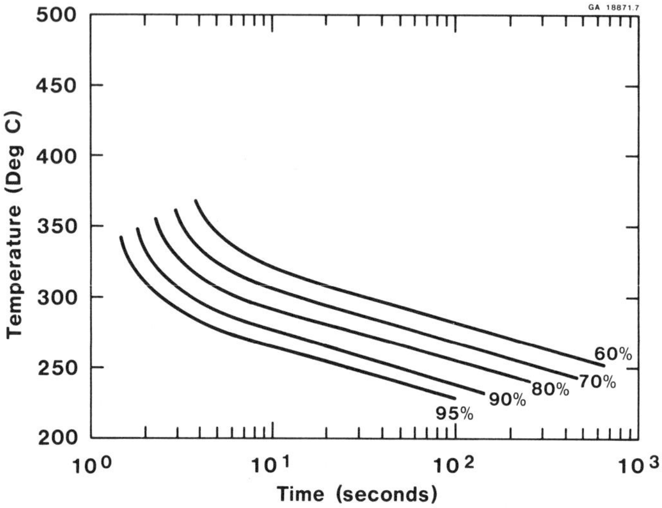

Kinetic Relationships for 6009-T6 in the As-Welded Condition
Figure 3

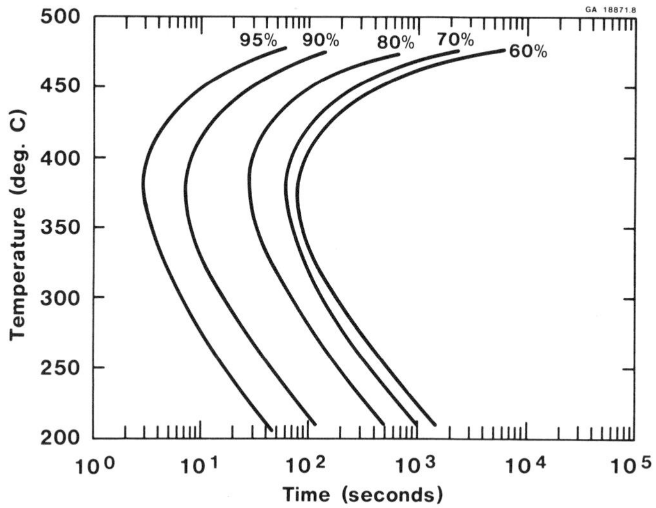

Kinetic Relationships for 6009-T4 that has been Post-Weld Aged
Figure 4

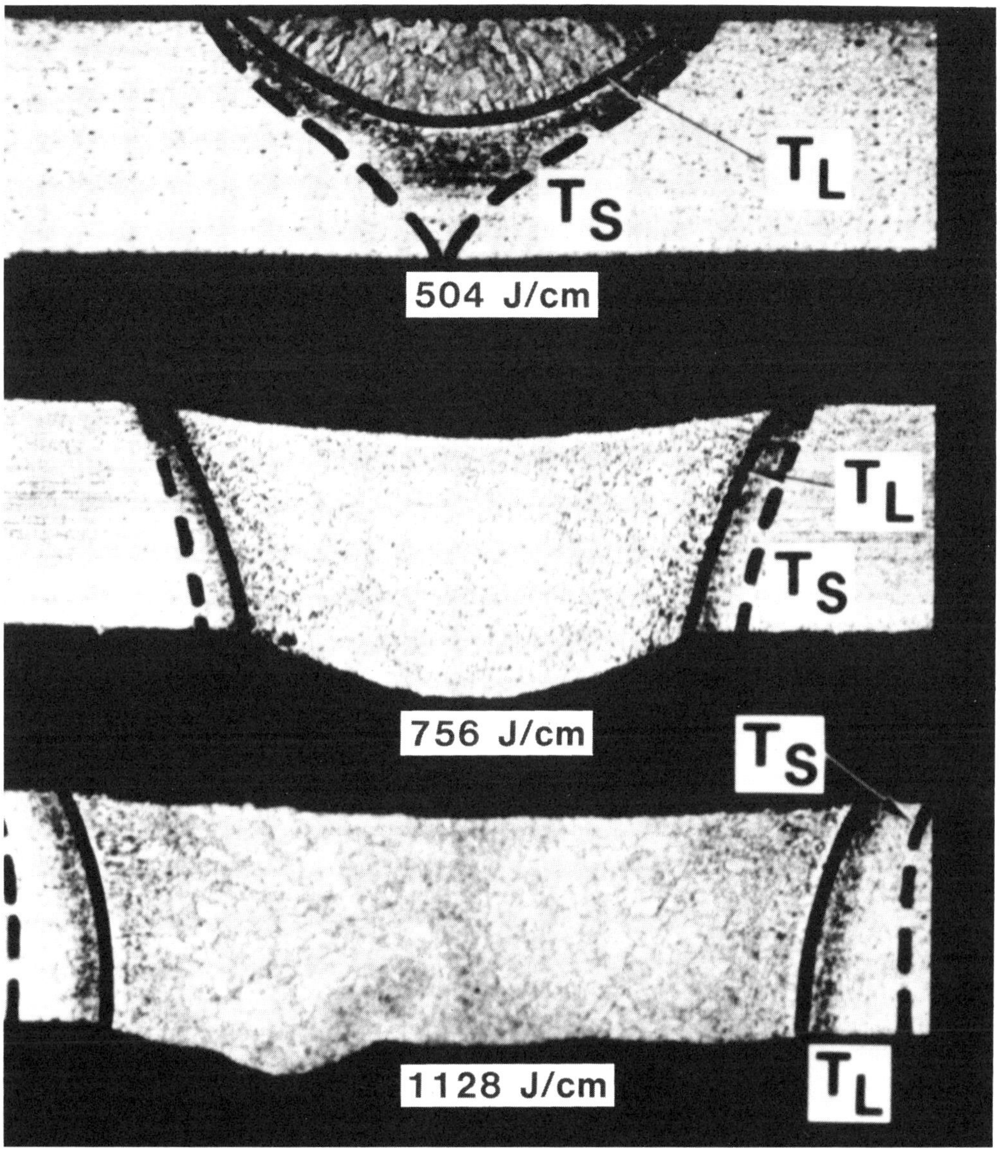

Macrographs of 2.0 mm Thick 6009 Welds Showing Comparisons of Fusion Zone and Model Predictions. (Graff-Sargent Etch)

Figure 5

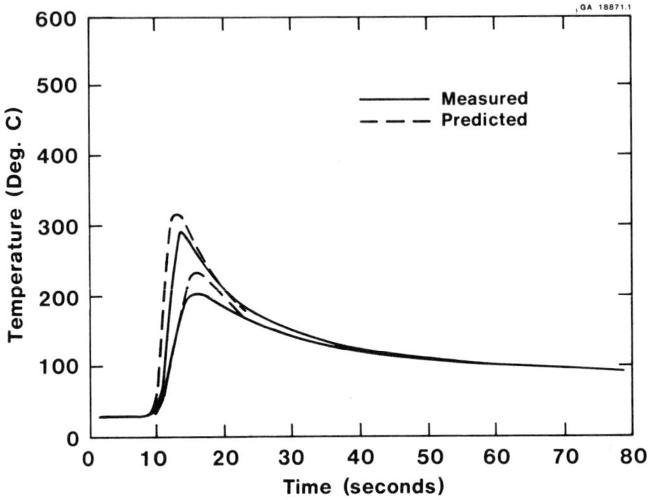

Measured and Predicted Thermal Cycles for 756 J/cm Heat Input
Figure 6

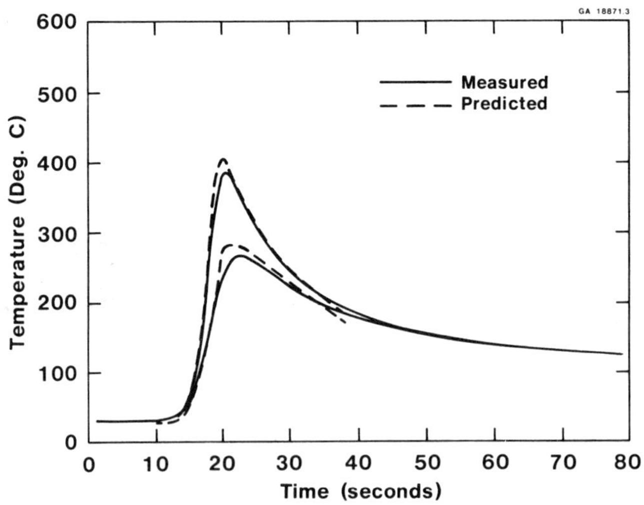

Measured and Predicted Thermal Cycles for 1128 J/cm Heat Input
Figure 7

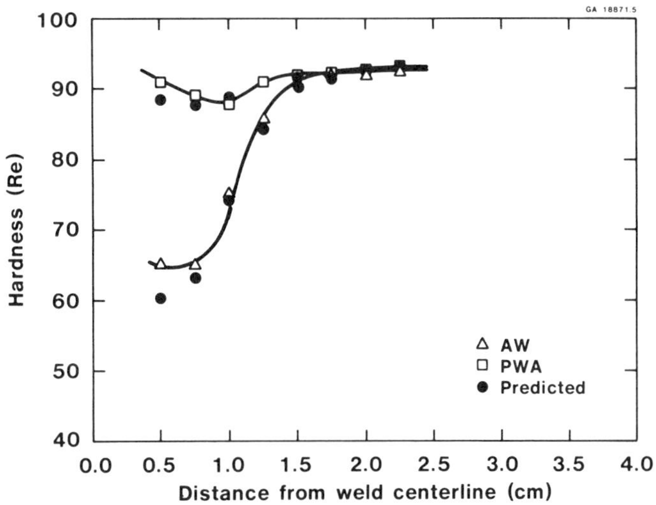

Measured and Predicted Hardness Profiles for Welds made with 756 J/cm Heat Input
Figure 8

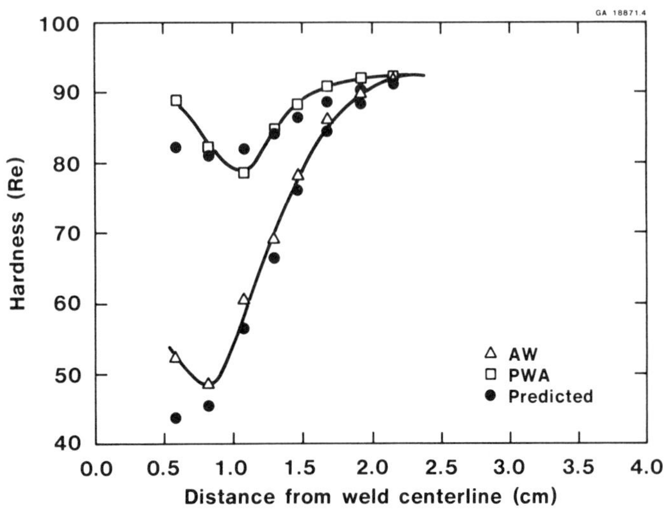

Measured and Predicted Hardness Profiles for Welds made with 1128 J/cm Heat Input
Figure 9

MICROSTRUCTURAL EVALUATION OF Fe-3Cr-1.5Mo-0.1V THICK-SECTION ELECTRON BEAM WELDS

J.M. Vitek, N.H. Packan, S.A. David
Oak Ridge National Laboratory
Oak Ridge, Tennessee 37831

ABSTRACT

The iron-base alloy 3Cr-1.5Mo-0.1V is intended to be used for high-temperature applications. The feasibility of welding heavy-section plates (102-mm thick) of this alloy by high-power, electron-beam welding was investigated. An evaluation of the mechanical properties of these welds, reported in a companion study, indicated that electron-beam welding is an attractive process for joining this type of alloy. The present study is concerned with the evaluation of the weld and heat-affected zone (HAZ) microstructures by means of optical and electron microscopy. The weld and immediately surrounding HAZ structures are bainitic with thin areas of retained austenite between the bainite laths. Further away from the weld, the HAZ structures consist of bainite and proeutectoid ferrite and correspond to a drop in the microhardness, which goes through a minimum. Beyond the minimum, the HAZ structure is tempered bainite. The various structures observed can be related to the maximum temperature experienced during the welding cycle and its relationship to the A_{c_1} and A_{c_3} temperatures of the alloy. Except for the small region containing bainite plus ferrite, the structures throughout the weld and HAZ are basically bainite tempered to various stages; these variations can be eliminated by a proper postweld heat treatment.

THICK-SECTION PRESSURE VESSEL STEELS are commonly used as components of energy systems; welding is an integral part of the fabrication of such components. Conventional welding processes such as shielded metal arc and submerged arc welding are commonly used techniques but they are time consuming and expensive. As an alternative to these processes, the feasibility of electron beam welding of thick-section steel was examined.[1] Such a process offers several advantages with regard to cost, speed, and the elimination of filler metals, although some inherent disadvantages are unavoidable.

In a companion study[1] to the present investigation, electron-beam welds of three ferritic steels were evaluated. Mechanical property measurements, including tensile testing, Charpy V-notch impact tests, and guided bend tests, were carried out. The test results indicated that electron-beam weldments had excellent properties and demonstrated that electron-beam welding is an attractive alternative process for heavy section steel welding. The intent of this paper is to examine in detail the microstructure of the thick-section electron beam welds and their associated heat-affected zones. Only one of the three ferritic steels previously evaluated[1] will be studied — namely, 3Cr-1.5Mo-0.1V.

The 3Cr-1.5Mo-0.1V ferritic steel examined in this study was developed as part of an effort to modify 2.25Cr-1Mo steels for better hydrogen embrittlement resistance and improved elevated-temperature properties.[2-5] The particular alloy variation used is based on the work of Wada and Cox.[2,6] This alloy has recently been investigated in the wrought condition by Klueh and Nasreldin.[7] The steel is typically utilized in the normalized-and-tempered condition, with a tempered bainitic microstructure.

EXPERIMENTAL PROCEDURE

The 3Cr-1.5Mo-0.1V steel used in this study was melted by means of the argon-oxygen decarburizing process by Carpenter Technology Company, Reading, Pennsylvania,[8] and was processed into 102-mm-thick plate by Lukens Steel Company of Coatesville, Pennsylvania.[8,9] The material was received in the normalized (955°C/2 h/air cool) and stress relieved (565°C/2 h) condition exhibiting the partially tempered microstructure shown in Fig. 1. The alloy composition is given in Table 1.

Full-penetration autogeneous welds were made by CBI Industries with the 102-mm-thick plate in the horizontal position using a Sciaky 100 kW electron beam welder. The welding parameters

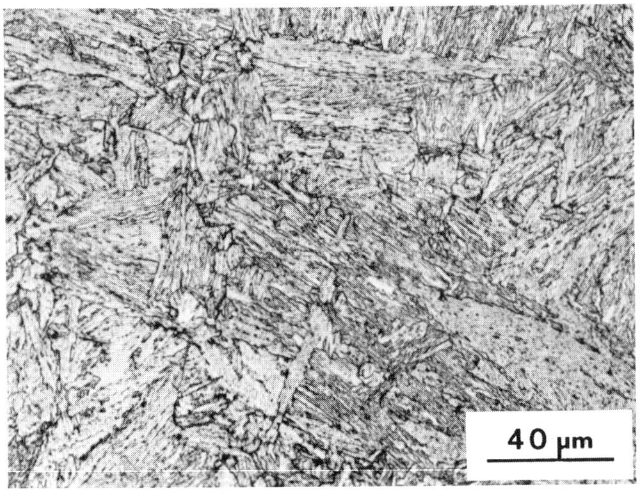

Fig. 1 — Optical micrograph of the base material showing a bainitic microstructure.

Table 1 - Alloy composition (in weight percent)

C	0.12	S	0.002
Mn	0.84	P	0.011
Si	0.27	Al	0.004
Ni	0.14	Sn	0.002
Cr	2.86	Sb	<0.001
Mo	1.48	Cu	0.06
V	0.09	N	0.0141

were: 95 kV accelerating voltage, 337 mA beam current, and 1.9 mm/s travel speed. Microhardness measurements were made across the fusion zone (FZ) as well as across the heat-affected zone (HAZ). Other mechanical property measurements were made but these results are reported elsewhere.[1]

Detailed structural analysis was performed using optical metallography and transmission electron microscopy (TEM). Disks for TEM were prepared from various regions of the weld and HAZ. Thin foils were prepared in a twin-jet polisher with a solution of 90% ethyl alcohol and 10% perchloric acid at −20°C and 22 V. Electron microscopy was performed on a JEOL 100C microscope with a special AMG low-field objective lens polepiece at 120 keV or a Philips EM430 microscope operated at 300 keV.

RESULTS

A cross-sectional view of the weld is shown in Fig. 2. The full-penetration welds were uniform and of high integrity. As noted earlier, the weldability of the 3Cr-1.5Mo-0.1V alloy has been found to be good. Detailed evaluation indicated the weld properties were favorable.[1] A microhardness trace across the weld and heat-affected zone is shown in Fig. 3. For convenience, the hardness profile is divided into regions I through V. Region I corresponds to the FZ itself. Region II is the area of the HAZ nearest the FZ where the hardness is comparable.

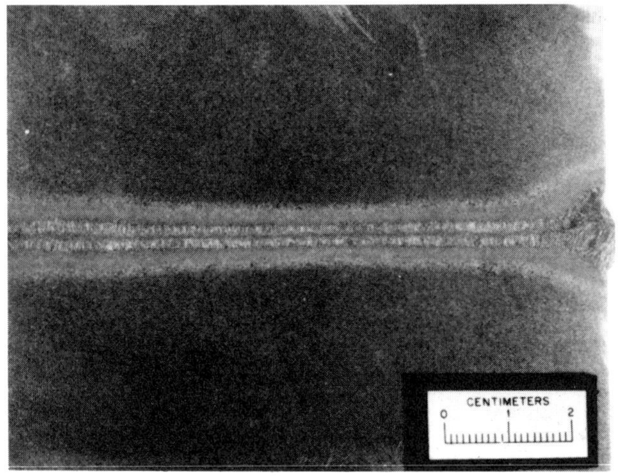

Fig. 2 — Transverse macroscopic view of the entire thick-section electron beam weld in the 3Cr-1.5Mo-0.1V alloy.

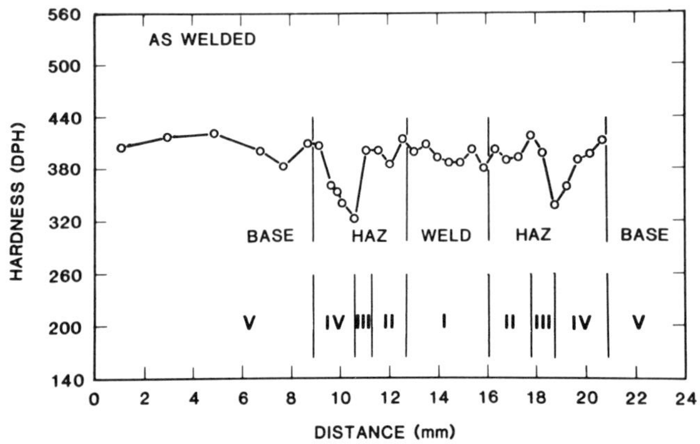

Fig. 3 — Hardness profile across the fusion and heat-affected zones with the five distinct regions superimposed.

In region III the hardness drops to a minimum value and in region IV the hardness increases to the base metal level. Region V corresponds to the base metal. The structure in each of these regions will be treated separately.

The FZ structure corresponding to region I in Fig. 3 is shown in Fig. 4. Figure 4(a) indicates the structure in region I is relatively uniform on a coarse scale. However, as shown in the TEM micrographs of Fig. 4(b) and (c), some areas are bainitic whereas other sections contained fine narrow fingers of retained austenite between the bainite laths.

The HAZ structure corresponding to region II is shown in Fig. 5. The area shown was located in the HAZ immediately next to the FZ. The micrograph clearly shows thin areas of retained austenite between the bainitic matrix, similar to the structure observed in the FZ.

An optical micrograph of the structure in region III is shown in Fig. 6. The matrix is bainite, as in regions I and II, but in addition there are clear, lightly etched areas which

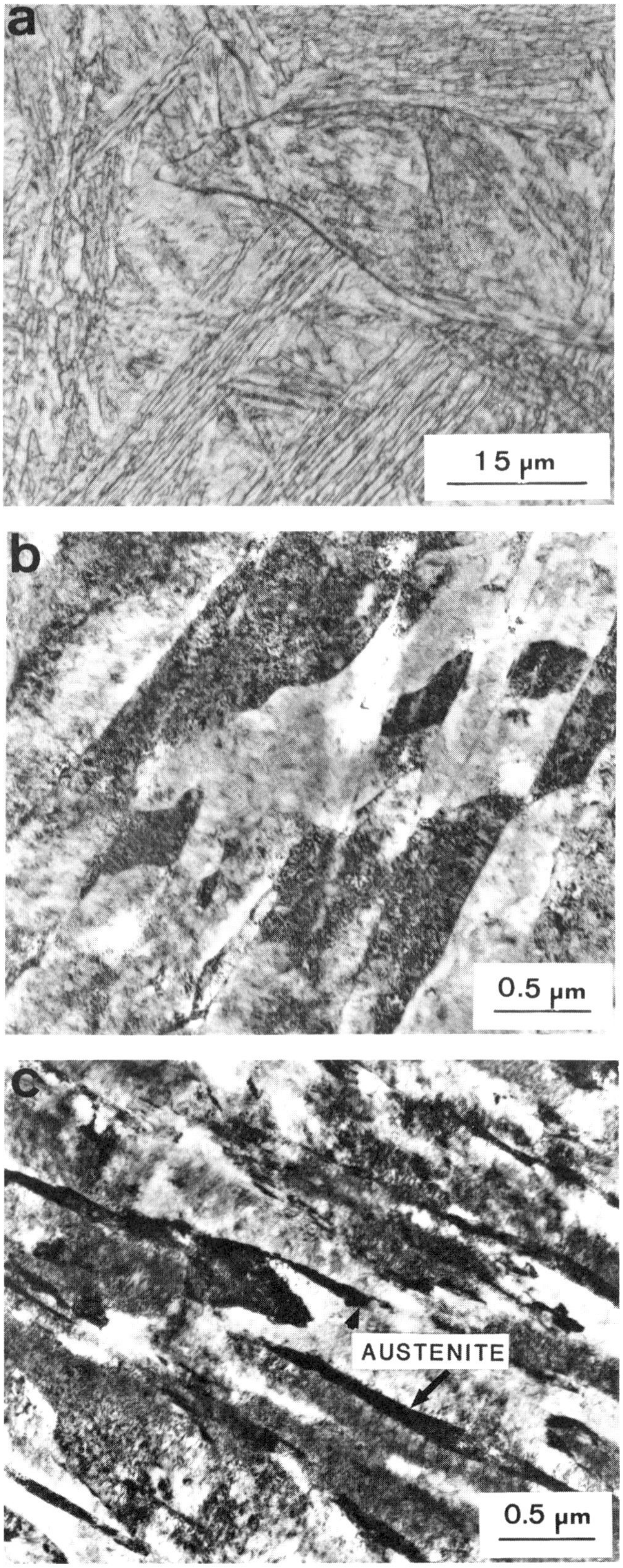

Fig. 4(a) — Optical micrograph of the bainitic fusion zone structure. (b) Electron micrograph showing the bainitic microstructure in the fusion zone. (c) Electron micrograph revealing thin areas of retained austenite between the bainite laths in the fusion zone.

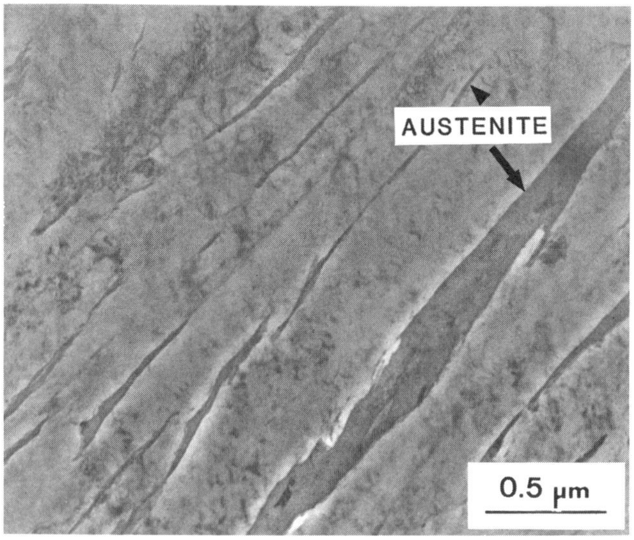

Fig. 5 — Electron micrograph showing the retained austenite between bainitic laths in the HAZ heated to above A_{c_3}.

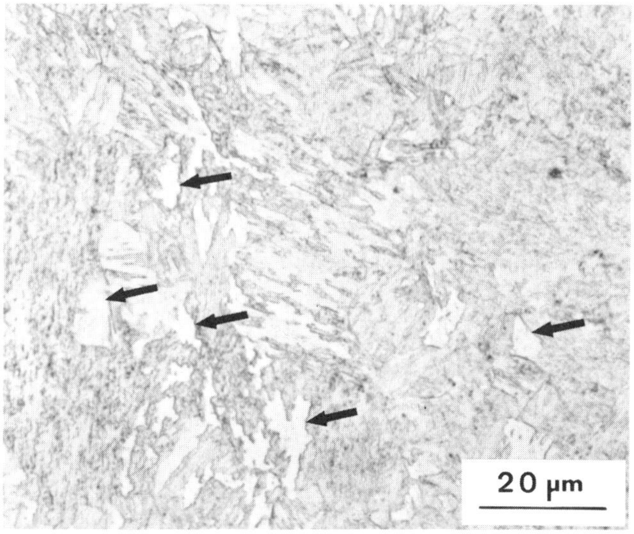

Fig. 6 — Optical micrograph of ferrite plus bainite microstructure in region III, corresponding to the HAZ heated above A_{c_1} but below A_{c_3}. Arrows indicate some of the ferrite areas.

correspond to proeutectoid ferrite. The extent of the region III microstructure is limited and somewhat irregular, thereby making the preparation of TEM specimens from this section difficult. The structure in region III was examined in detail by electron microscopy elsewhere,[10] using Gleeble thermomechanical simulation to produce larger areas containing the same structure.

The structure corresponding to region IV is shown in Fig. 7. The microstructure is that of tempered bainite. As shown in Fig. 7(b), extensive precipitation along bainitic lath boundaries as well as within the laths themselves has occurred.

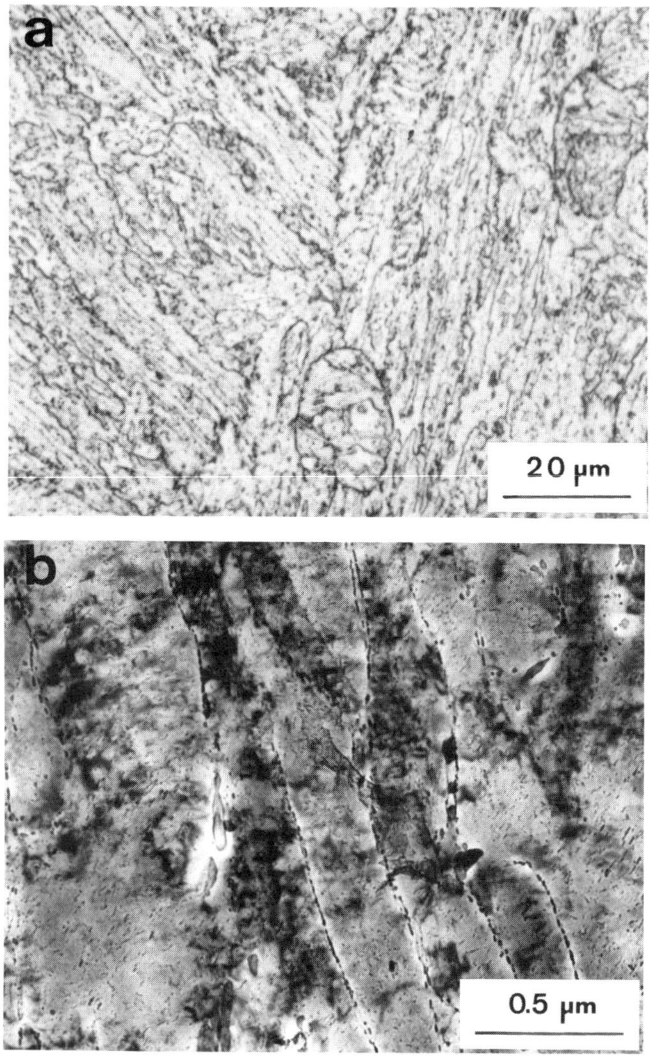

Fig. 7(a) — Optical micrograph of partially tempered HAZ heated to a maximum temperature less than A_{c_1}. (b) Electron micrograph of area similar to that in (a).

DISCUSSION

The presence of a range of microstructures in the FZ and HAZ may be understood by considering the phase stability of Fe-C alloys. Three distinct regions are of interest in the present work: (a) the single-phase austenite area at temperatures above the A_{c_3}, (b) the two-phase ferrite plus austenite phase field between the A_{c_3} and A_{c_1} temperatures, and (c) the two-phase field of ferrite plus carbide below the A_{c_1} temperature. The different structures observed in the weld and HAZ can be interpreted in terms of the maximum temperature attained during heating in the welding cycle. For materials exposed to temperatures above A_{c_3}, the structures developed are fully austenitic at elevated temperatures. Upon cooling, the austenite transforms to bainite. Before the transformation is complete, some limited diffusion of carbon to the bainitic lath boundaries occurs, thereby stabilizing some of the remaining untransformed austenite. This sequence yields a bainitic microstructure with retained austenite as observed. For the relatively high cooling rates prevalent during EB welding, formation of proeutectoid ferrite is avoided. The transformation mechanism is similar to that described in detail by Strangwood and Bhadeshia for the formation of acicular ferrite in welds.[11] This description of transformations and resultant structures for materials heated above A_{c_3} applies to both the FZ (region I) and the immediately neighboring HAZ (region II), where similar microstructures are observed. Apparently the details of the solidification process in the weld metal do not have a major impact on the final structures because the alloy is fully austenitic during cooling.

In the HAZ heated above A_{c_1}, but below A_{c_3}, a two-phase microstructure of proeutectoid ferrite and austenite is produced. When cooled, the ferrite remains stable, but the austenite is transformed to bainite. Consequently a mixture of ferrite plus bainite is expected in the region of the HAZ corresponding to this range of maximum temperature attained, as found for region III. Since the ferrite is a softer phase and since the temperature range between A_{c_1} and A_{c_3} is relatively small, one would expect a sharp drop in hardness over a limited range, as found in region III of the hardness profile (Fig. 3).

For the region of the HAZ exposed to a maximum temperature below A_{c_1} (region IV), the tempering reaction of the initial partially tempered bainitic structure simply continues toward a fully tempered state. The degree of additional tempering that occurs varies over the HAZ, thereby explaining the gradual increase in hardness in the HAZ from the minimum to the base metal value, as shown in Fig. 3. The minimum hardness found in the HAZ was approximately 330 DPH. This is considerably higher than the hardness of 200 to 225 DPH found for the same alloy in the fully tempered condition,[12] indicating that the heat cycle to which the HAZ was subjected was insufficient to fully temper the material.

Overall, the structures in the FZ and the HAZ were relatively uniform, except for the small area designated as region III. Zones I, II, IV, and V consisted primarily of bainite, with the major difference between zones being the extent to which the bainite is tempered. It should be possible to minimize or eliminate these differences with a postweld heat treatment (PWHT). Hardness measurements after a PWHT have indeed shown that a relatively constant hardness extends across the FZ and the entire HAZ.[1] The effect of a PWHT on the microstructure is being examined separately.[10] The results of the present study are promising in that they indicate that the details of the solidification reaction are not particularly significant in determining the final microstructures. Whether delta ferrite forms at the highest temperatures in the FZ or parts of the HAZ seems to play, at most, a secondary role. This is because the alloy passes through a single-phase austenite phase field upon cooling from very high temperatures and the alloy is effectively austenitized. These microstructural results show that uniform material properties can be expected across nearly the entire FZ and HAZ, particularly after an appropriate PWHT.

SUMMARY AND CONCLUSIONS

The microstructures of electron-beam welded thick-section 3Cr-1.5Mo-0.1V steel were examined in detail. It was found that the FZ structure was uniform on a macroscopic scale, consisting of bainite. In some areas, retained austenite was found at the interlath boundaries. The HAZ neighboring the FZ, which was heated to above the A_{c_3} temperature during welding, had a similar bainitic structure with areas of retained austenite. For the HAZ which was heated above the A_{c_1} temperature but below the A_{c_3} temperature, the microstructure consisted of proeutectoid ferrite plus bainite. In this region the hardness dropped to a minimum value. Further away from the FZ, in the HAZ heated to below the A_{c_1} temperature, the structure consisted of partially tempered bainite. The degree of tempering decreased with increasing distance from the FZ, resulting in a gradual rise in the hardness to that of the minimally tempered base material. Overall, the structure across the entire FZ and HAZ was fairly uniform, consisting primarily of bainite tempered to various levels. Thus, it is expected that the materials properties may be similar over the entire weld and HAZ, particularly after an appropriate PWHT.

ACKNOWLEDGMENTS

The authors would like to acknowledge the technical assistance of K. F. Russell, J. W. Jones, and R. W. Reed in the preparation of the specimens. The authors also wish to acknowledge the work of Frances Scarboro in the preparation of the manuscript. This research was sponsored by the Division of Materials Sciences, U.S. Department of Energy, under contract No. DE-AC05-84OR21400 with Martin Marietta Energy Systems, Inc.

REFERENCES

1. King, J. F. and S. A. David, "Electron Beam Welding of Heavy Section Ferritic Steels," this proceedings.
2. Wada, T. and T. B. Cox, "Research on Chrome-Moly Steels," pp. 77–93, MPC-21, ed., R. A. Swift, American Society of Mechanical Engineers, New York (1984).
3. Ishiguro, T., K. Ohnishi, Y. Murakami, S. Mima, and J. Watanabe, ibid., pp. 43–52.
4. Kozasu, I., H. Suzuki, M. Yamada, and H. Tagawa, ibid., pp. 53–76.
5. Parker, E. R., R. O. Ritchie, J. A. Todd, and P. N. Spencer, ibid., pp. 109–116.
6. Wada, T. and T. B. Cox, "Advanced Materials for Pressure Vessel Service with Hydrogen at High Temperatures and Pressures," pp. 111–121, MPC-18, ed., M. Semchyshen, American Society of Mechanical Engineers, New York (1982).
7. Klueh, R. L. and A. M. Nasreldin, "Microstructure and Mechanical Properties of a 3Cr-1.5Mo Steel," ORNL-6206, Oak Ridge National Laboratory, Oak Ridge, Tennessee, (September 1985).
8. Swift, R. A., "Research on Chrome-Moly Steels," ibid., ref. 2, pp. 95–107.
9. Edmonds, D. P., T. B. Cox, R. K. Nanstad, R. A. Swift, and R. W. Swindeman, "Evaluation of 3Cr-1.5Mo-0.1V Steel as a Gasifier Pressure Vessel Alloy," ORNL-6106, Oak Ridge National Laboratory, Oak Ridge, Tennessee (January 1985).
10. Vitek, J. M. and S. A. David, to be submitted to Welding Journal.
11. Strangwood, M. and H.K.D.H. Bhadeshia, "The Mechanism of Acicular Ferrite Formation in Steel Weld Deposits," this proceedings.
12. Swindeman, R. W., A. M. Nasreldin, R. L. Klueh, and J. F. King, "Metallographic Examination of Response of 3Cr-1.5Mo Steel to Heat Treatment," ORNL/TM-9691, Oak Ridge National Laboratory, Oak Ridge, Tennessee (November 1985).
13. "Metals Handbook," Vol. 8: Metallography, Structures and Phase Diagrams, 8th ed., American Society for Metals, Metals Park, Ohio (1973).

THE MECHANISM OF ACICULAR FERRITE FORMATION IN STEEL WELD DEPOSITS

M. Strangwood, H.K.D.H. Bhadeshia
Department of Metallurgy and Materials Science
University of Cambridge
Cambridge, England
CB2 3QZ

ABSTRACT

Detailed crystallographic measurements, surface relief experiments, microstructural observations and thermodynamic analysis reveal that acicular ferrite grows by a diffusionless mechanism, in which the parent and product lattices are related by an atomic correspondence. Acicular ferrite is found to be very similar to bainite and differs morphologically because it nucleates intragranularly on inclusions within the weld; the morphology is also modified by hard impingement between plates nucleated at adjacent sites.

ACICULAR FERRITE (α_a) in the primary microstructure of low-alloy steel weld deposits consists of non-parallel plates of ferrite within the austenite grains and its presence seems to correlate directly with improved toughness. The mechanism of acicular ferrite growth is not clearly established, but it apparently nucleates intragranularly at inclusions (1-4) within the large columnar austenite grains so typical of weld deposits. It is not known whether the formation of acicular ferrite leads to a change in the shape of the transformed region. Such a shape change would indicate the transformation mechanism. It is notable that the growth of all other plate-shaped ferrites in steels always leads to an invariant-plane strain-shape change which has a large shear component (5,6). The morphological development of acicular ferrite has also not been investigated and the carbon concentration of the ferrite during growth is unknown. The purpose of this work was to resolve some of these difficulties, focussing on the growth aspects of acicular ferrite.

EXPERIMENTAL

The study of acicular ferrite in welds is made difficult by the high degree of transformation which occurs during cooling. In such welds, most of the parent austenite has transformed by the time ambient temperature is reached so that crystallographic experiments are very difficult to conduct. The high degree of transformation influences the morphology of the ferrite by causing impingement between crystals growing from different sites; it is the morphology that exists during unhindered growth which usually provides clues to the mechanism of transformation. To avoid these difficulties, an unusual experimental weld with a high hardenability and carbon concentration was deposited in order to ensure a low-degree of transformation during cooling to ambient temperature, and to retain considerable quantities of austenite at the same time. The weld was deposited using a manual metal arc technique at a speed of 2mm/s, using a welding current and voltage of 180A and 23V(DC+) respectively. The joint geometry was compatible with ISO2560 and the weld consisted of some 24 runs (interpass temperature 250°C), with three runs per layer. The present work is concerned only with the primary microstructure of the top layer, which is not influenced by dilution or by the heat effect of other runs. The chemical composition of the weld was found to be Fe-0.201C-1.80Mn-0.44Si-0.03Ni-0.04Cr-0.01Mo-<0.01V wt.%, with 212 ppm (by wt.) of O and 110 ppm of N.

Samples from the top run were prepared for transmission electron microscopy (on a Philips 400T, operated at 120kV) by electropolishing in a 5% perchloric acid and 95% acetic acid solution at 50-75V and ambient temperature. Quantitative metallography was carried out on a Quantimet 720 image analysis system. A "Theta" high-speed dilatometer was used to measure the extent of reaction as a function of isothermal transformation temperature; the specimen preparation for this has been described elsewhere.

Some weld metal samples were polished and then sealed in quartz tubes, under a partial pressure of pure argon, and austenitised at

1200°C for 10 mins before continuously cooling to ambient temperature (by dropping the intact quartz tube into iced brine) with the specimen still sealed in the quartz tube. The samples were then examined for surface relief effects accompanying the formation of acicular ferrite, using Nomarski interference contrast microscopy.

RESULTS AND DISCUSSION

MICROSTRUCTURE - The as-deposited microstructure is illustrated in Fig. 1, and shows that the weld consists primarily of acicular ferrite plates with very little allotriomorphic ferrite and Widmanstaten ferrite. When sectioned on a random plane (Fig. 1,2), the acicular ferrite always presented a lenticular section, confirming the view that it has a thin-plate morphology in three dimensions. On such random sections, the thickness/length ratio of the α_a plates was found to be about 0.3, so that the true aspect ratio (after considering stereological effects) should be much smaller. Electron microscopy revealed considerable quantities of retained austenite between the acicular ferrite platelets (Figs. 3a,b). This austenite was used in crystallographic measurements discussed below. Some of the plates contained inclusion particles (Fig. 2) which were presumably responsible for the heterogeneous, intragranular nucleation of α_a (of course, the probability of observing such nucleation sites is rather small, so that inclusions need not be observed in all plates even though they may be present).

The plates have an apparent length of a few microns and the α_a/γ boundary gently curves to generate the lenticular outline of the plates (Fig. 2). The tips of the plates were also found to be smoothly curved. These morphological observations are consistent with a displacive transformation mechanism in which the lenticular plate shape of α_a arises through the need to

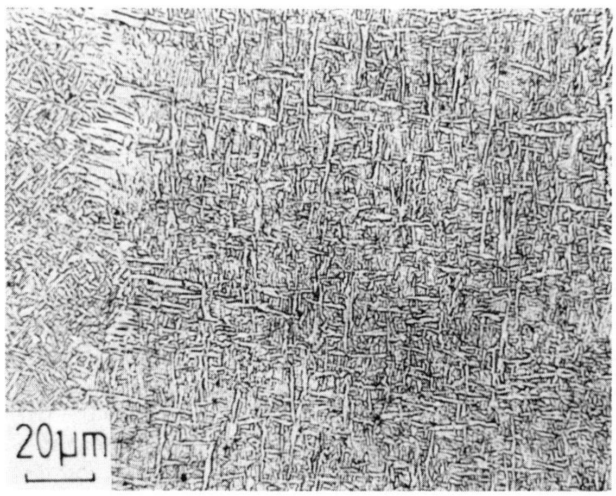

Fig. 1 - Optical micrograph showing the primary microstructure of the weld deposit.

Fig. 2 - (a) Electron micrograph illustrating the lenticular morphology of acicular ferrite. (b) Montage illustrating the detailed tip morphology of acicular ferrite.

minimise the strain energy associated with the displacive transformation. The observations are inconsistent with growth involving a ledge mechanism since the α_a/γ interface is certainly not found to be facetted in any way: more often than not, it exhibits clear curvature. The acicular ferrite plates were on a few occasions found to consist of "sub-units" (Fig. 4); such sub-units are found during the formation of bainite, which grows by the martensitic propagation of discrete platelets (sub-units) which together constitute a bainite "sheaf".

SURFACE RELIEF - Pre-polished specimens were transformed to examine (using Nomarski interference microscopy) the surface relief accompanying the transformation of austenite to acicular ferrite. The results are presented

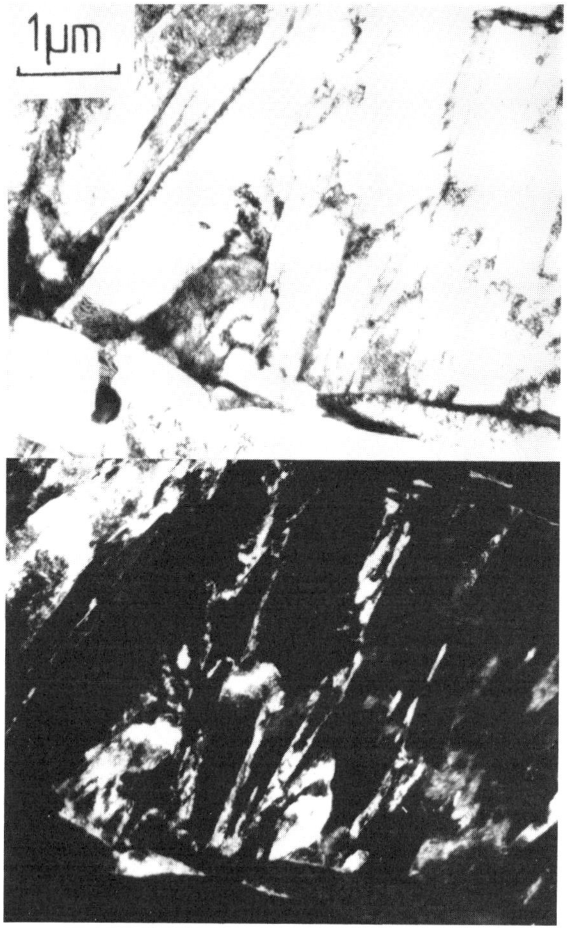

Fig. 3 - Bright and Dark field images of retained austenite in the primary weld microstructure.

Fig. 4 - Electron micrograph illustrating the sub-units which make up what macroscopically appears to be a single plate of acicular ferrite.

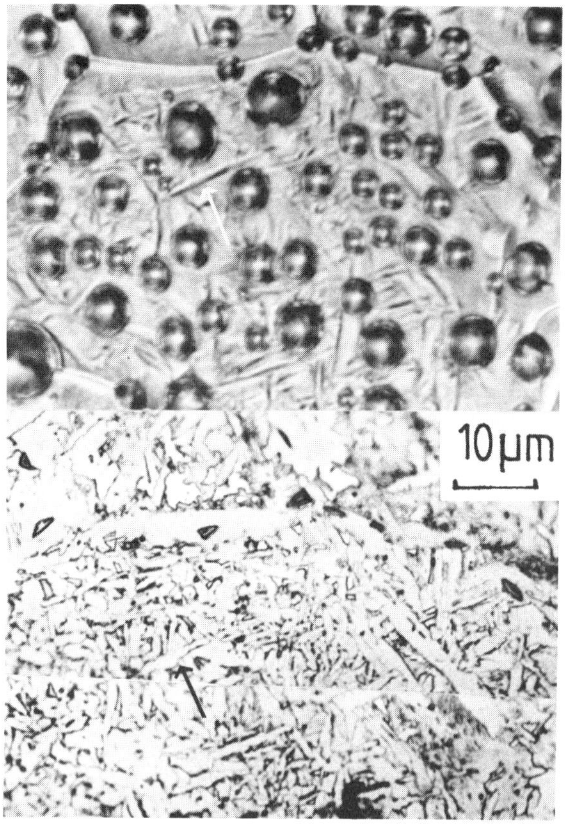

Fig. 5 - (a) Nomarski interference micrograph illustrating the surface relief accompanying the formation of acicular ferrite. (b) Corresponding optical micrograph.

in Figs. 5a,b and indicate that the formation of plate-shaped acicular ferrite is accompanied by a change in the shape of the transformed region. The uniformity of the contrast indicates that the shape change can be described as an invariant-plane strain with a significant shear component. This shape change can be taken to imply a displacive transformation mechanism and the existence of an atomic correspondence between the parent and product phases, at least as far as the atoms in the substitutional sites are concerned.

CRYSTALLOGRAPHY - For displacive transformations where the parent lattice changes into that of the product by some kind of a deformation of the former, the two lattices must always be intimately related (6). For steels, this means that the γ/α orientation relation must fall within the "Bain Region" (7-9). The homogeneous deformation which accomplishes the γ-α transformation is the Bain strain, which does not rotate any plane or direction by more than about 11°, so that any set of corresponding planes and directions can be made parallel after this strain by a rotation of not more than 11°. This 11° region about the Bain orientation is called the Bain region; for example, the Kurdjumov-Sachs and Nishiyama-Wasserman orientation relations fall within this region. For displacive transformations the γ/α orientation should always fall within this region. Diffusional growth on the other hand, occurs by the uncoordinated transfer of atoms across the interface so that the transformation is not limited by the presence of parent phase grain boundaries. It follows that the orientation relationship between α and the γ grain in which it grows, will not necessarily fall within the Bain region if the nucleation event occurs in one grain but growth in another adjacent grain.

The large amount of retained austenite found was used to advantage to determine the crystallography of acicular ferrite. All of 32 measurements of the α_a/γ orientation measurements were found to fall within the Bain region, the majority of which were within 6° of the Kurdjumov and Sachs relationship (Table 1).

Table 1 - Orientation relationship between γ and α_a. θ is the angle (degrees) between the closest packed planes of the two lattices, $\{111\}_\gamma$ and $\{011\}_{\alpha_a}$. ϕ is the angle between corresponding close-packed directions $<01\bar{1}>_\gamma$ and $<11\bar{1}>_{\alpha_a}$.

θ	ϕ	θ	ϕ	θ	ϕ
6	6	6	6	3	3
10	10	8	8	7	7
0	8	0	0	6	6
6	6	7	7	5	5
5	5	0	2	0	0
3	7	6	6	6	6
2	2	7	7	0	0
4	4	0	0	4	4
0	0	10	10	0	0
10	10	9	9	7	7
9	9				

Single-surface trace analysis to determine approximately the habit plane indices of acicular ferrite, with respect to the parent austenite was also carried out using electron diffraction. The results of some sixty experiments, some of which are illustrated in Fig. 6, indicate that the habit plane is near $\{0.117\ 0.675\ 0.729\}_\gamma$, which is very close to the $\{3\ 10\ 15\}_\gamma$ habit plane of many conventional martensites (10).

THERMODYNAMIC ANALYSIS - After reaustenitisation at 1200^oC for 1 hour, the reheated weld metal was isothermally transformed for various time periods and at several temperatures prior to quenching to ambient temperature. The amount of acicular ferrite formed was measured on the Quantimet 720 image analysis system (the measurements refer to the α_a plates alone and do not include microphases as is conventional practice). The results are presented in Fig. 7a and show that the degree of transformation increases with undercooling below the B_s temperature, and thermodynamic analysis (the detailed interpretation of which is presented in refs. 11-13) of these data (Fig. 7b) showed that the formation of acicular ferrite ceased when the carbon concentration of the residual austenite reached the T_0' phase boundary. These results strongly suggest that acicular ferrite is really bainite which nucleates

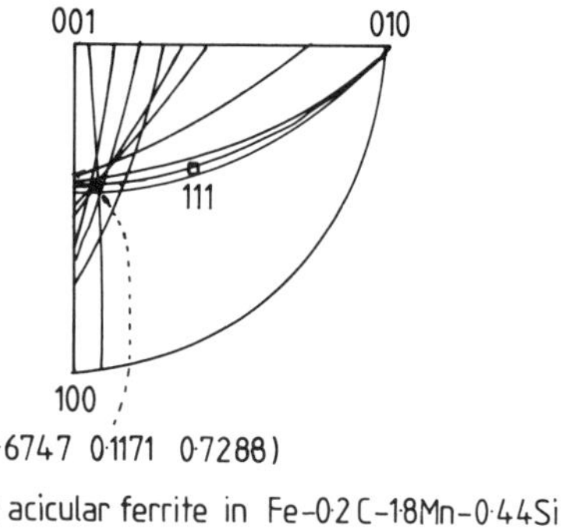

Fig. 6 - Single-surface habit plane trace analysis.

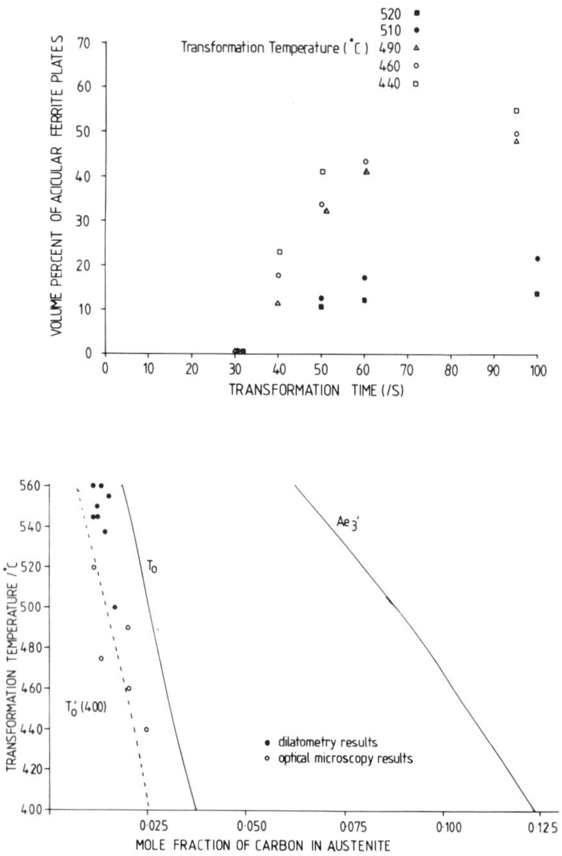

Fig. 7 - (a) Results of quantitative metallography on specimens partially isothermally transformed to acicular ferrite. (b) Thermodynamic analysis of results presented in (a), and further results obtained by dilatometry.

intragranularly; the growth of acicular ferrite is thus diffusionless and the carbon is redistributed between α and γ after the transformation event.

CONCLUSIONS

Our results suggest that the growth of acicular ferrite is diffusionless and occurs by a displacive transformation mechanism. The formation of α_a is accompanied by an invariant-plane strain shape change which implies the existence of an atomic correspondence between the parent and product phases. The crystallographic results are also consistent with this in that the orientation relationship between α_a and γ is always found to lie within the expected Bain region. The detailed shape of acicular ferrite also suggests a displacive transformation mechanism which does not involve interface advance by a step mechanism.

ACKNOWLEDGEMENTS

The authors are grateful to Professor D. Hull for the provision of laboratory facilities at the University of Cambridge and to the SERC and ESAB (UK) for financial support.

REFERENCES

1) Abson, D. J. and R. E. Dolby, Weld. Inst. Res. Rep. 19/1978, Welding Institute, Abingdon, Cambridge, 1978.
2) Easterling, K., "Introduction to the Physical Metallurgy of Welding", Butterworths, London (1983).
3) Bhadeshia, H. K. D. H., L.-E. Svennson and B. Gretoft, Acta Metall. 33 1271-1283 (1985).
4) Barrite, G. S., R. A. Ricks and P. R. Howell, "Quantitative Microanalysis with High Spatial Resolution", p. 112, Metals Soc., London (1981).
5) Bhadeshia, H. K. D. H., Mater. Sci. and Tech. 1 497-504 (1985).
6) Bhadeshia, H. K. D. H., Prog. in Mat. Sci. 29 321-386 (1985).
7) Ryder, P. L. and W. Pitsch, Acta Metall. 14 1437 (1966).
8) Ryder, P. L., W. Pitsch and R. F. Mehl, Acta Metall. 15 1431 (1967).
9) Crosky, A., P. G. McDougall and J. S. Bowles, Acta Metall. 28 1495 (1980).
10) Dunne, D.P. and C. Wayman, Metall. Trans. 2 2327 (1971).
11) Bhadeshia, H. K. D. H. and D. V. Edmonds, Acta Metall. 28 1265-1273 (1980).
12) Bhadeshia, H. K. D. H., Acta Metall. 29 1117-1130 (1981).
13) Yang, J. R. and H. K. D. H. Bhadeshia, this conference.

ASSESSMENT OF BOUNDARY CONDITION LIMITATIONS ON THE DEVELOPMENT OF A GENERAL COMPUTER MODEL FOR FUSION WELDING

K.W. Mahin
Sandia National Laboratories
Livermore, California 94550

A.B. Shapiro, J. Hallquist
Lawrence Livermore National Laboratory
Livermore, California 94550

ABSTRACT

This paper reviews the progress we have made in developing a general computer model for fusion welding. Both transient and steady state phenomena were studied for gas tungsten arc welds in 304 stainless steel. A vaporization model was incorporated into the computer code to improve temperature predictions under the arc. In addition, a method was developed to emulate the effect of fluid flow on weld pool geometry, without actually computing fluid flow. With these modifications, the accuracy of the temperature predictions were significantly improved. However, accurate prediction of residual stresses was limited by the lack of an adequate materials model, coupled with the lack of sufficient high-temperature material property data.

A GENERAL MODEL FOR FUSION WELDING requires the integration of sophisticated heat transfer, stress, and microstructural models to adequately represent the complexities of the process. Although research has been underway for a number of years to model the fusion welding process [1-7], most of this work has been limited by the lack of accurate experimental or mathematical definition of the complex physics of the process. For ease of computation, computer simulations typically assume 2- or 3-dimensional quasi-steady state conditions [5-30]. Agreement between code prediction and experiment is often an iterative process, involving adjustable parameters tailored for a specific material or geometry. As a result, errors can arise in the weld areas not as well specified by the code. How accurately we can predict changes in microstructure and properties in a weld will depend on how accurately we can define the boundary conditions of the heat transfer and stress analyses.

A schematic of a stationary gas tungsten arc (GTA) experiment, defined in terms of the input required for the computer model, is shown in Figure 1. The critical boundary conditions for the heat transfer include the following: (i) the heat flux distribution due to the arc; (ii) the interaction of the arc with the surface of the material; (iii) convective heat transfer due to fluid flow in the molten pool; and (iv) thermal conduction into the solid. The steep thermal gradients generated in the welding process produce nonuniform thermal strains [8-17]. To accurately incorporate the transient, three-dimensional nature of actual weldments requires sophisticated computations.

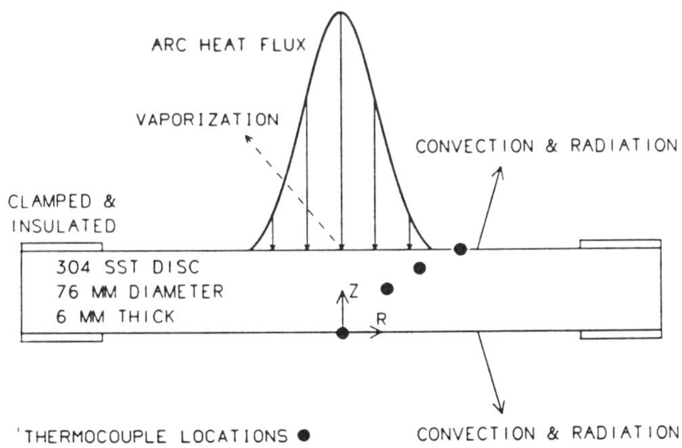

Fig. 1. Schematic of Stationary GTA Weld Experiment showing the heat transfer boundary conditions and thermocouple locations.

Although definitive data on the shape and the arc efficiency of the heat flux can be obtained experimentally [31-36], the physics of the interaction of the arc with the molten pool is not as well-defined [31]. The temperatures in the anode region of the arc (just above the weld pool surface) have been determined spectroscopically to be in excess of 6000 K [31,37-42], while experimental measurements of the weld pool surface temperature indicate temperatures below the boiling point of the material (e.g. 3080 K for 304 SS) [43]. The factors influencing this temperature drop include vaporization, fluid flow, radiation, and natural convection, none of which are easy to model. Experimental data on fluid flow forces and vaporization effects in molten alloys is only now becoming available [28-29,31,44-48], and the theoretical models are still being developed [49-53].

In the area of stress analysis, predictions of residual stress and distortion are often obtained by iterating between computer predictions and the experimental results [11-17]. The lack of accurate high temperature mechanical property data, as well as the corresponding lack of constitutive models describing the thermomechanical response of the material limits more general approaches [3,7-9,15-16,54]. In addition, actual verification or "calibration" of the computer predictions experimentally is difficult. This is particularly true for single pass or small diameter welds, due to probe size limitations (1 - 2 mm for X-Ray Diffraction) or texture effects in the material [55-58].

The ongoing research at Sandia National Laboratories, Livermore and Lawrence Livermore National Laboratory is directed toward translating our phenomenological understanding of the welding process into numerical models for a general computer-based weld model. Our approach has been to couple experiment with simulation to examine the effects of various modeling assumptions on the accuracy of the computer predictions. A stationary gas tungsten arc weld on 304 stainless steel was selected to provide input data to the computer codes, as well as experimental verification of the computer predictions. Traveling arc experiments tested the code's ability to handle the 3-D, transient problem.

This paper will cover the computer modifications required to incorporate the "physics" of the welding problem, the accuracy of the computer predictions, and the applicability of the current computer codes to actual metallurgical problems.

EXPERIMENTAL APPROACH

Stationary gas tungsten arc welds were made on disks of 304 SS, 76.2 mm in diameter by 6.35 mm thick. The speciman was restrained in a stainless steel fixture insulated with Zircar to prevent heat transfer to the fixture. Four small screws made contact with the plate through the insulation, providing a symmetric ground. Table I lists the welding parameters for the experiments.

The heat flux distribution for a 30 degree thoriated tungsten electrode was determined experimentally by H. B. Smartt and colleagues at Idaho National Engineering Laboratory (INEL) [33]. Using a water-cooled split copper anode, Smartt found the distribution to be Gaussian with a 2-sigma radius of 2 mm. Arc efficiency, defined as the total fraction of arc energy deposited into the workpiece, was measured with

Table 1 - Welding Parameters for Stationary Gas Tungsten Arc Welds

Welding Current	135A
Arc Voltage	16-17 V
Arc Length	3.175 mm
Dwell Time	5-7 sec
Shielding Gas	100% Argon
Flow Rate	25 cfh
Electrode	Thoriated-W
Diam.	3.175 mm
Tip Angle	30 deg.

dry calorimetry [33]. In this technique, you determine the heat content in the plate by measuring the total change in mass due to boiling of a given volume of liquid nitrogen. Transfer times between arc-off and immersion of the welded sample in the liquid nitrogen were less than 1 sec. Arc efficiencies of 65-69% were obtained for an initial arc gap of 3.175 mm at 135A.

S-type (Pt-Rh) thermocouples, 0.127 mm in diameter, monitored the thermal history of the specimen. The temperature data was recorded on high speed Nicolet storage oscilloscopes at a rate of 1 reading every 10 or 200 msec. Six thermocouples were used for each test. Three thermocouples, embedded at various depths within the sample, measured internal temperatures. To insure good contact with the specimen, the thermocouples were percussion welded to the bottom of the drilled holes. The locations of the embedded thermocouples after welding were determined by radiographing the specimen and generating a "3-D" picture. A top surface thermocouple, placed 5.4 mm from the weld center, was shielded from the arc in an aluminum oxide sheath.

A high temperature infrared viewing system, calibrated to 1800 K, monitored the temperature distributions for the traveling arc experiment. The infrared camera was shielded from the arc by surrounding the torch and the arc with a quartz tube lined with graphite foil.

Residual stress distributions were determined using X-Ray diffraction techniques [56,57]. The specimens were prepared by grinding the top surface to a 600 grit finish, followed by a vacuum anneal, prior to welding.

FINITE ELEMENT COMPUTER MODELS

The gas tungsten arc (GTA) welding problem was modelled using two LLNL-developed finite element computer codes: (i) TOPAZ, for the heat transfer analysis [59,60]; and (ii) NIKE, for the prediction of thermal distortion and residual stresses [61,62]. Both of these codes are capable of handling transient problems in 2- and 3-dimensions. For the 2-D axisymmetric stationary GTA weld, the finite element mesh consisted of 1490 elements. Small elements were used in the region directly under the arc to accurately represent the Gaussian input flux distribution and capture the high temperature gradients that exist there. A minimum of ten elements were required to capture the shape of the Gaussian heat flux distribution, with respect to a 2-sigma radius of 2 mm. To conserve computation time, a staggered grid was used outside of the melt region.

Table 2 - Boundary Condition Values Used for the Heat Transfer Analysis

Heat Flux Distribution	Gaussian 2σ radius = 2 mm
Efficiency Factor*	75%
Surface Emissivity	0.7 [63]
Laminar Free Convection Heat Transfer Coeff.	$8.72\ W/m^2 \cdot K$
Latent Heat of Fusion	2.65×10^5 J/kg
Latent Heat of Evaporation	7.35×10^6 J/kg
Thermal Conductivity (W/m • K)	
Solid ($T \leq 1670$ K):	$k = 8.116 + 1.618\times10^{-2}\ T$
Liquid ($T \geq 1730$ K):	$k = 12.29 + 3.248\times10^{-3}\ T$
Specific Heat (J/kg • K)	
Solid ($T \leq 1670$ K):	$c = 465.4 + 0.1336T$
Liquid ($T \geq 1730$ K):	$c = 788.$

*Effic. Factor = Arc Effic. + Loss due to Vapor.

The boundary conditions and material property data for the heat transfer analysis are listed in Table 2. The temperature-dependent thermal properties for 304 stainless steel were obtained from a report published by Argonne National Laboratory [43]. The properties were represented in the computer code as explicit equations [59,60]. We modeled the heat flux distribution from the arc in one of two ways: (i) using a distributed Gaussian heat flux for the case of the stationary arc; or (ii) as a temperature boundary condition, fixing the upper temperature of the surface at 3080 K. The second option significantly reduced the computation times required for the more complex 3-D calculation for the traveling arc. Fluid flow effects were approximated by enhancing the thermal conductivity of the melt in a linear fashion, such that the thermal conductivity at 3000K was 10 times greater than that calculated in Table 2. With an isotropic thermal conductivity, the weld pool geometry has a depth-to-width ratio of one-to-two (Figure 2a). Using anisotropic properties, we were able to emulate both shallow (Figure 2b) and deep penetration (Figure 2c) welds, which can arise due to bulk changes in fluid flow direction [23-29, 44-48].

The latent heat of fusion was represented as sensible heat, Q, in the energy balance equation given by:

$$\rho c \frac{\partial T}{\partial t} = \nabla \underset{\sim}{K} \cdot \nabla T + Q \qquad \text{Eq. (1).}$$

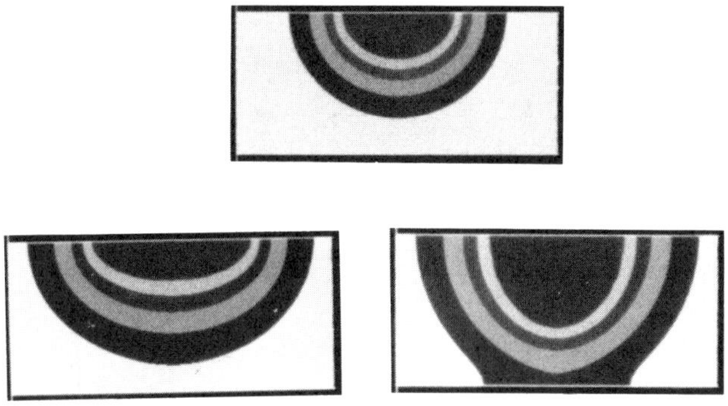

5X Enhancement in x-direction

10X Enhancement in z-direction

Fig. 2. Three different weld pool geometries generated by enhancing the thermal conductivity of the melt equally in each direction (a) and with a bias in a given direction (b) and (c).

where T = temperature [K]
K = temperature dependent thermal conductivity tensor
Q = internal heat generation term or source term
ρ = material density (kg/m**3)
c = material specific heat at constant strain (J/kg K)
t = time (seconds)

Heat flux, radiation and convection boundary conditions were specified at the surface. Since temperatures under the arc were high, a kinetic vaporization algorithm was added to the code to handle the boundary condition in this region. The model, based on the work by Block-Bolten and Eagar [49-51], assumes regular solution behavior, non-coupling of the vaporizing elements, and very low vaporization pressures. A complete description of the numerical methods required to incorporate this boundary condition is provided in another publication [64]. Addition of the vaporization boundary condition increased computation times on the CRAY from 44 seconds to 4 hours for the 2-D stationary gas tungsten arc problem.

In the stress analysis, we used a zero-displacement boundary condition to represent the restraint from the fixture. Material behavior was approximated with a temperature-dependent elastoplastic model, which incorporates elements of both kinematic and isotropic hardening [61]. The material properties, such as yield strength, the elastic and plastic moduli, the thermal expansion coefficient, and Poisson's ratio were input as a function of temperature. For temperatures up to 1100 K, we extracted the data for moduli and Poisson's ratio from reference material [65] . Property data at higher temperatures had to be obtained by linear extrapolation. To avoid the accumulation of unrealistic plastic strains [3,9,12,66], the plastic modulus was set to zero. Thermal distortion was computed using a temperature

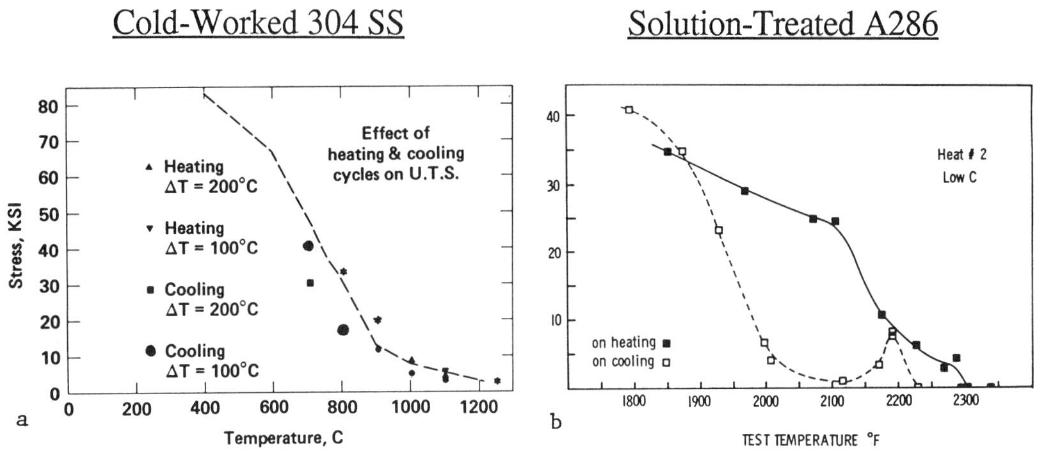

Fig. 3 The Effect of Rapid Heating and Cooling Cycles on the Ultimate Tensile Strength of (a) cold-worked 304 SS and (b) solution-treated A286.

dependent coefficient of thermal expansion coefficient.

Yield strength data was obtained by subjecting the material to thermal cycles similar to those obtained in the actual weld. Testing with a Gleeble 1500 revealed that the yield strength and the ultimate strength varied between the on-heating and the on-cooling cycle of the weld. The variation in ultimate tensile strength is illustrated in Figure 3a for cold-worked 304 stainless steel and in Figure 3b for solution treated A286. During the on-cooling cycle, the material strongly reflects its previous thermal history and is unable to regain the strength at temperature predicted by the on-heating peak temperature tests. Since there is no constitutive materials model in the present version of NIKE to describe this type of cyclic high temperature behavior, only the on-heating properties were used in the computer input.

RESULTS

HEAT TRANSFER ANALYSIS

Stationary Gas Tungsten Arc Weld (2-D) - For the stationary gas tungsten arc welding experiments on 304 stainless steel (SS), excellent agreement was obtained between the computer predicted weld penetration at 5 seconds and the experimental cross-section (Figures 4a and b). An upper bound on weld pool temperature was determined for the Gaussian heat flux distribution by equating the peak heat flux with the evaporative cooling loss due to vaporization. For an arc power of 1620W, the temperature under the arc varied from 3216 K to 2879 K for arc radii ranging from 1.5 mm to 3.0 mm. Enhancement of the actual isotropic 304 SS liquid thermal conductivity by a factor of 10 decreased the temperature directly under the arc from 3049 K to 2920 K and brought the temperature profile for the top surface into closer agreement with experiment (Figure 5). At a radial distance of 2 mm, which would represent the edge of the arc itself, the predicted temperature was 2250 K. This calculated temperature agrees exceptionally well with experimentally reported measurements of weld pool temperatures near the arc [31]. At temperatures lower than the melting temperature, a temperature-dependent isotropic thermal conductivity was used to calculate conduction within the solid region.

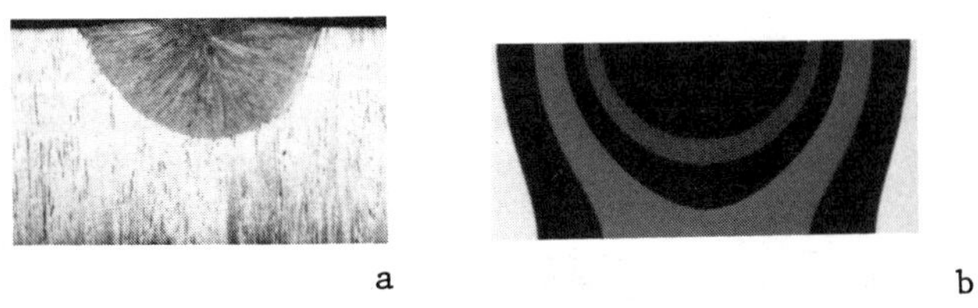

Fig. 4 Comparison of the Weld Pool Geometries in the actual weld cross-section (a) and the computer predicted cross-section (b). The center region in (b) denotes the melt zone.

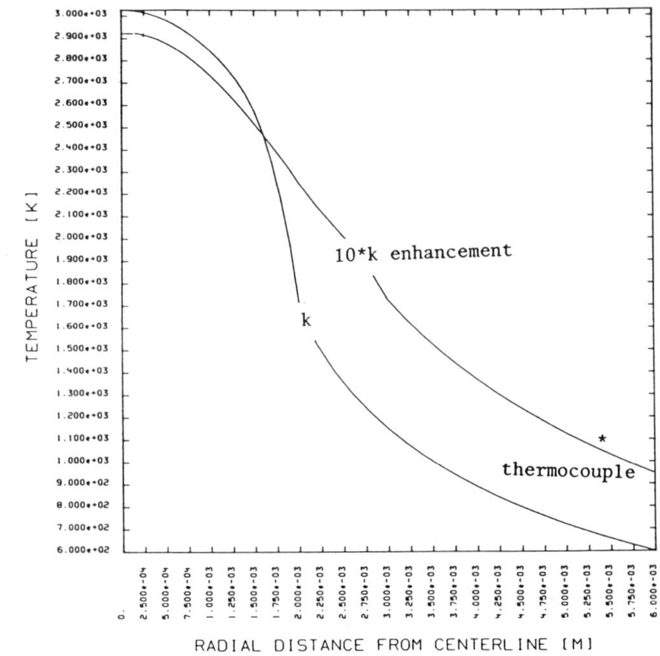

Fig. 5 Enhancement of the weld pool liquid thermal conductivity, k, brought the top surface temperature profile into closer areement with the experimental thermocouple measurement.

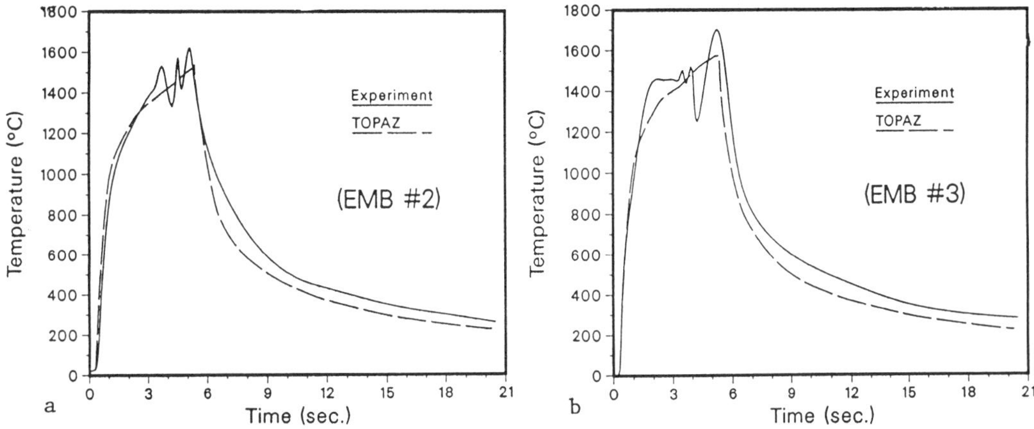

Fig. 6 Comparison between the temperature history predicted by TOPAZ (dashed) line and the history measured with the embedded thermocouples (a) 2 and (b) 3.

Using radiography to determine the experimental location of the embedded thermocouples, we found that the thermal history predicted by the corresponding location within the TOPAZ2D mesh correlated well with the actual thermal history (Figures 6a and 6b). The temperature gradients at the fusion boundary were on the order of 120 C (393 K)/0.2 mm, where 0.2 mm is the distance between nodes for the finite element grid in this region. The experimentally recorded peak temperature fluctuated considerably after the fusion zone boundary passed the thermocouple location. While the heating rate prediction agrees well with the experimental results, the cooling rate predicted by TOPAZ2D is slightly steeper than measured. However, the difference does not appear to be significant. The code predictions are less accurate, however, for the back surface thermocouple, as shown in Figure 7. In this case, the prediction overestimated the peak temperature by 120 C (373 K). As the gradients in this region are not steep, the error is probably caused by inadequate representation of the boundary conditions on the back surface, coupled with any accumulative error generated by assuming, rather than computing, fluid flow heat transfer effects.

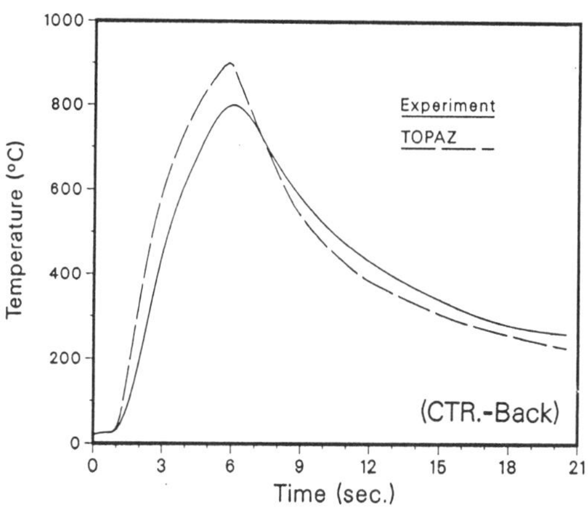

Fig. 7 Comparison between the TOPAZ-computed and the experimentally-measured thermal histories for the back surface.

Traveling Gas Tungsten Arc (3-D) - Temperature profiles for the case of the traveling arc on 304 SS were calculated using both the 2-D and the 3-D versions of TOPAZ. The weld pool was established in the code by allowing the heat source to remain stationary for 2 seconds before moving the heat source along the workpiece, similar to actual practice. Figure 8a shows the 3-D projection of thermal profiles generated for an arc traveling at a theoretical speed 20 in/min. In reality, it would be very difficult to maintain a weld pool at such a high travel speed. However, the 3-D simulation showed us that the elliptical weld puddle, which is found experimentally at the higher travel speeds, could be duplicated in the conduction code, without enhancing the thermal conductivity to account for fluid flow. In comparison, at slower travel speeds (1 in/min), the 3-D calculation produced more spherical weld pools.

In the 2-D calculation, large variations in travel speed produced significant instabilities in the calculations. We found that a travel speed of 1 in/min in the computer simulation produced a 50% penetration weld (Figure 8c), similar to that measured experimentally with an

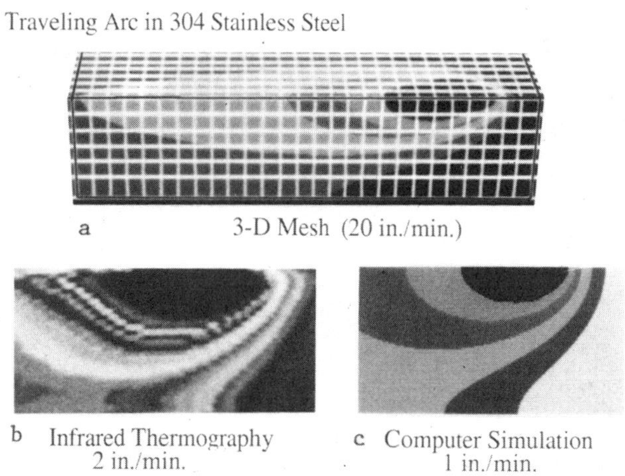

Fig. 8 Computer-predicted temperature contours for the traveling arc at travel speeds of (a) 20 in/min and (c) 1 in/min. Infrared thermography measurement of an actual weld at 2 in/min (b).

infrared camera for a traveling arc weld on 304 SS at 2 in/min (Figure 8b).

To minimize computational time, a temperature boundary condition, rather than a flux boundary condition, was used to simulate the arc source in the case of the traveling arc. While the shape of the thermal profiles looks reasonably accurate for the traveling arc, more work is required to determine the accuracy of the temperature predictions in terms of rate and location.

STRESS ANALYSIS (2-D) - NIKE2D was used to compute the thermal distortion and the resultant residual stress patterns in the stationary GTA weld. Evaluation of the stress state proved inconclusive from both a computer modeling and an experimental standpoint. Figure 9 shows the type of output generated by NIKE2D using temperature dependent material properties and a thermoelastoplastic materials model. The thermal distortion (Figure 9a) corresponds well to quantitative measurements on the actual sample. Comparison of stress and strain calculations with experiment, however, showed that a satisfactory quantitative agreement was not yet possible. Figure 9b is a plot of the effective plastic strain distribution in the 304 SS plate after cooling to room temperature.

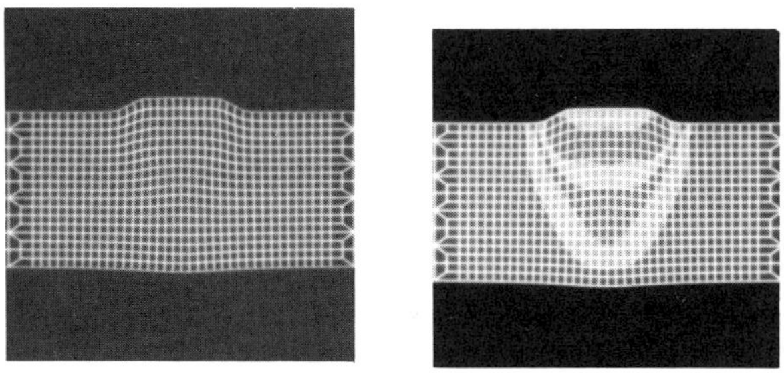

a THERMAL DISTORTION b EFFECTIVE PLASTIC STRAIN

Fig. 9 Computer generated plots of thermal distortion (a) and effective plastic strain (b), calculated using NIKE2D. The plot shows the finite element mesh in the region of the weld.

Calibration of the code results with experimental measurements proved to be difficult. X-Ray diffraction measurements were scattered due to a number of factors: (i) the small size of the heat affected zone, relative to the 2 mm probe size; (ii) the limited penetration depth of the probe; and (iii) possible grain size and texture effects [56]. The experimental measurements of the radial stress component indicated a 20 ksi residual stress level in the fully annealed sample, while in the hoop direction the stress level in the base metal measured zero ksi [67]. The transition from tension to compression for the hoop stress component occurred at 8 mm from the center of the weld. In the NIKE calculation, however, the transition point for the hoop stress component occurred 12 mm from the weld center.

DISCUSSION

HEAT TRANSFER ANALYSIS (2- AND 3-D) - Three of the boundary conditions shown in Figure 1, specifically:

(i) the experimentally determined heat flux distribution;

(ii) the vaporization kinetics under the arc; and

(iii) the convection effects in the weld pool;

were not solidly defined in the heat transfer simulation. In order to evolve towards a more general computer model for fusion welding, it is necessary to understand the limitations of these boundary conditions and their effect on the accuracy of the simulation.

Heat flux distribution - It is very time-consuming and difficult to obtain experimental measurements for the heat distribution in a gas tungsten arc weld [8,33]. The distribution varies with electrode geometry, arc gap, and material [8,32-34,38,40-41], as well as metal vaporization [31,37-38]. If fluid flow or vaporization significantly alters the heat flux distribution, experimental determinations on unmelted substrates may not be representative. Inaccuracies in experimental measurements will limit the accuracy of the computer predictions. For example in the TOPAZ simulation, a change in arc radius from 1.5 mm to 2.0 mm for a constant arc power reduced the top surface temperature by 200 degrees. However, the sensitivity of surface temperature to arc radius decreased with increasing radius. How accurately do we need to define the heat source distribution in the computer models? This will depend to a certain extent on how accurately other boundary conditions are accounted for within the simulation. For the case of the stationary GTA weld, small changes in the adjustable "fluid flow" parameter produced much larger changes in the D/W ratio of the weld pool than did changes in arc radius from 1.5 mm to 2.5 mm. Other considerations include the amount of computer time available and the degree of accuracy required to solve the problem. If the temperatures of interest are some distance from the fusion boundary, the accuracy of the actual heat source distribution becomes less significant [5-7].

For the case of the traveling arc, a distributed temperature boundary condition was used to model the heat source. This provided two main advantages: (i) it reduced the degree of mesh refinement required, i.e. the number of elements; and (ii) avoided the need to calculate vaporization. These simplifications reduced the computation time of the 3-D simulation by a factor of 5. An alternative approach would be to specify the heat flux distribution on the solid-liquid interface directly. This representation is the basis for the power density model proposed by J. Goldak [3-4,8]. In this model, an empirical 3-D representation of the heat source is mathematically approximated from experimental data on weld pool geometry. The approach avoids the the complexity of the fluid flow region and significantly reduces computation times for the 3-D traveling arc [3]. In both cases, accuracy of the heat source distribution is exchanged for more reasonable computation times.

Further work is required to determine the effect of such simplifying assumptions on the accuracy of the temperature distributions, especially in the region of the fusion zone boundary.

Vaporization and Fluid Flow - It is not possible to decouple vaporization from fluid flow. The high velocities predicted by recent fluid flow models [23-29] have been correlated with a significant cooling of the top surface of the weld pool. However, the predicted temperatures of 2200 C (2473 K) [29] have been calculated without considering vaporization. The current vaporization model assumes that all of the surface cooling in the region of the arc occurs by vaporization. It further assumes that none of the vaporized material recondenses on the surface. However in reality, up to 20% of the material may be returned to the surface through molecular collisions alone [49]. With these limitations, the model still predicted top surface temperatures at the periphery of the arc of around 2200 C (2473 K). At the center of the weld pool, use of the adjustable "liquid thermal conductivity" parameter to simulate fluid flow effects only reduced the temperature by 100 degrees.

At this point we need to integrate the actual fluid flow calculations with the vaporization and conduction calculations. Although the approach we took for the stationary GTA weld on 304 SS proved adequate for this particular case, this might not be the case for other materials or other configurations.

STRESS ANALYSIS - Evaluation of the residual stress distribution in the 304 SS plates using the NIKE code was limited by:

(i) the lack of accurate high temperature property (>1073 K) data;

(ii) the inability of the code to account for microstructure or history-dependent material properties; and

(iii) the uncertainty of how to handle work hardening and work softening.

Calibration of the code results with experimental data would have allowed us to perform a series of parametric studies to assess the effect of each limitation. However, X-Ray diffraction techniques were unable to provide conclusive data on the residual stress distribution in the 304 SS plate [67].

Accurate measurements of Young's modulus and Poisson's ratio at elevated temperatures requires the ability of the test setup to quickly ramp the material to peak temperature and take the measurement before any significant creep effects occur [66]. This can be both time-consuming and expensive to do for actual production materials. The alternative is to use existing data and make reasonable extrapolations for the higher temperatures.

The importance of using transient versus isothermal yield strength data will depend largely on the sensitivity of the material to thermomechanical cycling. For precipitation hardening materials, such as 7050-T6 Al, small variations in time at temperature can lead to large variations in both microstructure and mechanical properties [68]. For other materials, such as 5083Al, the effect may be less significant. In the case of cold-worked 304 SS, the major differences between the transient and the isothermal properties occurred at temperatures below 1073 K. Since residual stresses develop in this temperature range [54], the code accuracy will be limited by the accuracy of the input data.

A more significant limitation than material property data, however, is the inability of the code to account for history-dependent material property changes. As was shown in Figure 3, a considerable change in properties can occur between the on-heating and the on-cooling portion of the weld cycle. Since the residual stresses form during the on-cooling cycles, accurate data for this portion of the cycle is critical. However, the current code structure allows for only one set of temperature dependent material properties. Although we chose to use the on-heating data for the simulation, a better approach in hindsight would have been to use the on-cooling data.

The inability of the NIKE code to quantitatively predict thermal strain distributions during welding limits our ability to evaluate metallurgical phenomena such as Bauschinger effects, creep, cracking phenomena and complex phase transformations. In order to improve the code predictions, we need to develop sophisticated materials models, which will account for path dependent thermomechanical properties. In lieu of the this, there may be ways to directly incorporate on-heating and on-cooling mechanical properties into the code, until more general models become available.

CONCLUSIONS

Significant progress has been made toward obtaining reliable thermal histories for the case of a stationary gas tungsten arc weld. Using a sophisticated finite element heat transfer code, it was possible to model the thermal history of the weld with a minimum number of assumptions. Determination of the shape of the heat flux distribution and the arc efficiency experimentally improved the accuracy of the computer results. However, the computer calculations showed that small variations in arc radius produced changes up to 200 degrees in top surface temperature. Thus experimental accuracy is a necessity. Incorporation of a kinetic model for vaporization reduced top surface temperatures to within experimentally predicted ranges. Although fluid flow effects were accounted for using an adjustable "liquid thermal conductivity", the effect of high fluid velocities on surface temperatures was not modeled.

The thermal distortion values calculated in the stress analysis code agreed quantitatively with the experimental measurements. However, the calculated stress and strain distributions were only qualitatively correct. The accuracy of the stress analysis was limited primarily by the lack of a history dependent materials model, as well as by available experimental data. Experimental work is currenting underway to develop a first order materials model, which would incorporate the experimental on-heating and on-cooling properties of the material within the framework of the thermoelastoplastic materials model.

REFERENCES

1. Glickstein,S.S. and E.Friedman, Weld. Jnl., 55, (12), 408s-420s (1976)
2. Paley,Z. and P.D.Hibbert, Weld. Jnl., 54, (11), 385s-392s (1975)
3. Goldak,J., B.Patel, M.Bibby and J.Moore, "Computational Weld Mechanics", submitted to 1985 NATO AGARD meeting on "Advanced Joining of Metallic Materials", Bavaria, West Germany, 1-30 (Sept. 1985)
4. Goldak,J., A.Chakravarti, and M.Bibby, Met. Trans. B, 15B, (6), 299-305 (1984)
5. Glickstein,S.S. and E.Friedman, DOE Research and Develop. Report, WAPD-TM-1428, Oct. 1979.
6. Friedman,E., "Simulation for Materials Applications", Nuclear Metallurgy, 20, Part 2, 1160-1170 (1976)
7. Cacciatore,P., Modeling of Casting and Welding Processes, ed. H.Brody and D.Apelian, publ. AIME, 113-128 (1981)
8. N.S.Tsai, Ph.D. Thesis, Dept. of Mat. Sci. and Eng., MIT, Cambridge, Mass. (1983)
9. B.Patel, Ph.D. Thesis, Dept. of Mech. and Aeronautical Eng., Carleton Univ., Ottawa, Ontario, Canada (1985)
10. D.B.Duncan, M.S. Thesis, Univ. of Calif., Davis, Lawrence Livermore National Laboratory Report, UCRL-52891, (Jan. 1980)
11. Krutz,G.W. and L.J.Segerlind, Weld. Jnl., 57, (7), 211s-216s (1978)
12. Jonsson,M., L.Karlsson, L-E.Lindgren, Num. Methods in Heat Transfer, Vol.III, ed. R.W.Lewis and K.Morgan, publ. Wiley, N.Y., (1984)
13. Nagaraja Rao,N.R. and L.Tall, Weld. Jnl, 40, (10), 468s-480s (1961)
14. Ueda,Y. and T.Yamakawa, Trans. Jpn. Weld. Soc., 2, (2), 90-100 (1971)
15. K.Masabuchi, Modeling of Casting and Welding Processes, ed. H.Brody and D.Apelian, publ. AIME, 223-237 (1981)
16. R.A.Chihoski, Weld.Jnl., 58, (1), 9s-18s (1979)
17. Rybicki,E.F., R.B.Stonesifer, E.A.Merrick, and J.R.Haulter, Weld. Jnl., 65, (4), 99s-104s (1986)
18. Malmuth, N.D., W.F.Hall, B.I. Davis, C.D. Rosen, Weld. Jnl., 53, (9), 388S-400s (1974)
19. Kou,S., Modeling of Casting and Welding Processes, ed. H.Brody and D.Apelian, publ. AIME, 129-138 (1981)
20. Kou,S., Met. Trans. A., 13A, (3), (1982)
21. Kou,S., T.Kanevsky, S. Fyfitch, Weld. Jnl., 61, (6), 175s-188s (1982)
22. Myers,P.S., O.A.Uyehara, G.L.Borman, WRC Bull. #123, 1-46 (July 1967)
23. Chan,C.L., R.Zehr, J.Mazumder, M.M.Chen, 3rd Eng. Found. Conf. on Modeling and Control of Casting and Welding Processes, ed. by Mehrabian, publ. by AIME (1986)
24. Kou,S. and D.K.Sun, Met. Trans. A., 16A, (2), (1985)
25. Oreper,G.M. and J.Szekely, J. Fluid. Mech., 147, 52-79 (1984)
26. Chan,T., J.Mazumder, and M.M.Chen , Met. Trans. A., 15A, (12), 2175-2184 (1984)
27. Chande,T. and J.Mazumder, Jnl. of Applied Physics, 57, (6), 2226-2232 (1985)
28. Sundell,R.E., S.M.Correa, L.P.Harris, H.D.Solomon, L.A.Wojcik , NSF Interim Report, G.E.-Schnectady, Rep. #85SRD008, Contract No. MEA-8208950, (Oct. 1983-Sept. 1984)
29. Sundell,R.E., S.M.Correa, and H.D.Solomon, 3rd Eng. Found. Conf. on Modeling and Control of Casting and Welding Processes, ed. R. Mehrabian, publ. by AIME (1986)
30. Giedt,W.H., X.-C.Wei, S.-R.Wei, Weld. Jnl., 63, (12), 376s-383s (1984)
31. Int'l Inst. of Welding (IIW), Physics of Welding, ed. J.F.Lancaster, publ. Pergammon Press, New York (1984)
32. Smartt,H.B., J.A.Stewart, and C.J.Einerson, "GTA Arc Efficiency", presented at 66th Annual AWS Convention, Las Vegas, (May 1985)
33. Smartt,H.B., J.A.Stewart, and C.J.Einerson, "GTA Heat Input Distribution Measurements", presented at 66th Annual AWS Convention, Las Vegas (May 1985)
34. Key,J.F., J.W.Chan, H.B.Smartt, M.McIlwain, Weld Jnl, 62, (7), 185s-199s (1983)
35. Pavelic,V. R.Tanbakuchi, O.A.Uyehara, P.S.Meyers, Weld. Jnl., 48, (7), 295s-305s (1969).
36. Niles,R.W. and C.E.Jackson, Weld. Jnl., 54, (1), 25s-32s, (1975)
37. Lancaster,J.F., Phys. Technol., 15, 73-79 (1984)
38. Johnson,D.C. and E.Pfender, Plasma Chem. and Plasma Processing, 3, (2), 259-273 (1983)
39. Shaw,C.B. Jr. , Weld. Jnl., 54, (2) and (3), 33s-44s and 81s-86s (1975)
40. Key,J.F., J.W.Chan, M.E.McIlwain, IIW Doc. 212-549-83 (1983)
41. Key,J.F., J.W.Chan, M.E.McIlwain, Weld. Jnl., 62, (7), 179s-184s (1983)
42. Glickstein,S.S., Arc Physics and Weld Pool Behavior, Int'l. Conf., May 1979, London, Welding Institute Reprint, 1-16 (1979)

43. Leibowitz,L. "Properties for LMFBR Safety Analysis", Argonne National Laboratory Report #ANL-CEN-RSD-76-1 (1976)
44. Matsunawa,A., IIW Doc. 212-618-85, Trans. of JWRI, 11, (2), 145-156 (1982)
45. Burgardt,P. and C.R.Heiple, presented at 67th Annual AWS Convention, Atlanta, GA (April 1986)
46. Yokoya,S., Y.Asako, A.Matsunawa, Trans. of JWRI, 14, (2), 51-58 (1983)
47. Glickstein,S.S. and E.Friedman, Weld. Jnl., 60, (6), 110s-112s (1981)
48. Heiple,C.R.,J.R.Roper,R.T.Stagner,R.J.Aden, Weld. Jnl., 62, (3), 72s-78s (1983)
49. S.Dushman, Scientific Foundations of Vacuum Technique, publ. John Wiley and Sons, 2nd ed., New York (1962)
50. Block-Bolten,A. and T.W.Eagar, Met. Trans. B, 15B, (9), 461-469 (1984)
51. Block-Bolten,A. and T.W.Eagar, Trends in Weld. Research in U.S., ed. S.A.David, publ. Amer. Soc. for Metals, 53-58 (1982)
52. Khan,P.A.A. and T.Debroy, Met. Trans. B, 15B, (12), 641-644 (1984)
53. Lin,M.L. and T.W.Eagar, submitted to Met. Trans. A, (1985)
54. K.Masubuchi , Analysis of Welded Structures, Int'l Series on Mat.Sci. and Tech., publ. by Pergamon Press, Vol. 33, (1980)
55. Masubuchi,K., IIW Doc. X-675-72, (May 1972)
56. Ruud,C.O. and D.J.Snoha, Jnl. of Metals, 36, (2), 32-38, (1984)
57. Boyd,D.M. and J.F.Shackleford, Lawrence Livermore National Lab. report, UCID 19417, (Aug. 1982)
58. MacEwen,S.R., J.Faber,Jr., and A.P.Turner, Scripta Metall., 18, (1984)
59. A.B.Shapiro, "TOPAZ - A Finite Element Heat Conduction Code for Analyzing 2-D Solids", Lawrence Livermore National Lab. report, UCID 20045, (1984)
60. A.B.Shapiro, "TOPAZ3D - A Three-Dimensional Finite Element Heat Transfer Code", Lawrence Livermore National Lab. report, UCID 20484, (1985)
61. J.O.Hallquist, "NIKE2D - A Vectorized, Implicit, Finite Deformation, Finite Element Code for Analyzing the Static and Dynamic Response of Solids", Lawrence Livermore National Lab. report, UCID 19677, (1983)
62. J.O. Hallquist, "NIKE3D - An Implicit, Finite Deformation, Finite Element Code for Analyzing the Static and Dynamic Response of Three-Dimensional Solids", Lawrence Livermore National Lab. report, UCID 18822, (1981)
63. Y.S.Touloukian and C.Y.Ho, Thermophysical Properties of Selected Aerospace Materials, Part 1: Thermal Radiative Properties, Cindas, Purdue University, W. Lafayette, IN.
64. Shapiro,A.B. and K.W.Mahin, submitted to 1986 ASME Winter Annual Meeting, Anaheim, Dec. 1986; Lawrence Livermore National Lab. report, UCRL-94627, (May 1986)
65. Aerospace Structural Metals Handbook, Mechanical Properties Data Center, Battelle-Columbus Labs, Columbus, Ohio (1982)
66. B.Thomas, Ph.D. Thesis, Dept. of Metallurgical Eng., Univ. of British Columbia, (March 1985)
67. Ruud,C.O. and Ivkovich,D.P., Unpublished report on "Residual Stresses on a 304 Stainless Steel GTA Spot Weld Sample" (March 1986)
68. Mahin,K.W., Lawrence Livermore National Lab. Report, UCID 20329 (1985)

PREDICTION OF MICROSTRUCTURE OF THE FUSION ZONE OF MULTICOMPONENT STEEL WELD DEPOSITS

H.K.D.H. Bhadeshia
University of Cambridge
Cambridge, U.K.

Lars-Erik Svensson, B. Gretoft
ESAB AB
Gothenburg, Sweden

ABSTRACT

A model based on phase transformation theory, has been developed to allow the prediction of steel weld metal microstructures. It requires the input of just chemical composition and welding conditions; this facilitates the calculation of the appropriate part of the phase diagram (for steels containing C, Mn, Si, Ni, Cr, Mo and V in any combination), needed to define the relevant paraequilibrium tie-lines for detailed kinetic analysis. TTT and CCT curves are also computed, together with various transformation start and finish temperatures. These data in turn lead to the estimation of the volume fractions of allotriomorphic ferrite, Widdmanstätten ferrite and acicular ferrite. The theory is found to be in good agreement with experimental results.

WELDING is a complex metallurgical process involving many variables; in this work we present a recent model (1-5) for the calculation of the primary microstructure of low-alloy steel welds. The model has been shown to be capable of predicting the primary microstructure of Fe-Mn-Si-C and Fe-Mn-Si-Ni-C manual-metal-arc (MMA) welds as a function of heat input and chemical composition (1-5). Here we demonstrate its application by analysing quantitatively, published data (6) on the effect of relatively large variations in Si and Mn concentrations, on the primary microstructure of MMA welds.

THEORY FOR THE PREDICTION OF THE MICROSTRUCTURE

In low-alloy steel welds, solidification begins with the epitaxial nucleation of δ-ferrite at the fusion boundary. The melt then solidifies by the cellular growth of δ (7) which in turn transforms into a columnar-austenite (γ) grain structure. The morphology of the γ grains can be represented as a honeycomb of hexagonal prisms, the side length of each prism being 'a' and the length of the prism being 'c' (3,8). The c-axes of these prisms are approximately parallel to the direction of maximum heat flow during solidification; if $\bar{L}_{tn}$ is the mean lineal intercept measured on a transverse section of the weld, in a direction normal to the major axes of the γ grains, then (3)

$$\bar{L}_{tn} = \pi a \cos(30^{\circ})/2, \qquad (1)$$

so that 'a' can be determined experimentally. The parameter 'c' is not needed for the microstructure calculations because c>>a so that any effect due to the ends of the hexagonal prisms can be neglected (1,3). On further cooling, at $T=T_h$, the γ begins to transform, to layers of allotriomorphic ferrite (α) which grow by a diffusional transformation mechanism, at the γ/γ boundaries. As the temperature falls, diffusional growth becomes more difficult and Widmanstätten ferrite (α_w) plates nucleate at the α/γ boundaries and grow into the γ by a displacive transformation mechanism (9-12), at a rate controlled approximately by the diffusion of C in the γ ahead of the plate tips (12). At the same time, acicular ferrite (α_a) nucleates on inclusions (and also by autocatalytic effects) within the γ grains and grows in the form of thin plates. As the M_s temperature is approached, the small amount of remaining austenite decomposes either into degenerate pearlite and/or mixtures of martensite and retained austenite; because the volume fraction of these phases is relatively small, they are called "microphases".

T_h is determined by using an additive reaction rule (1), the cooling curve and a computed Time-Temperature-Transformation (TTT) diagram for the alloy concerned (10). The TTT diagram consists of two 'C' curves, the upper one giving the time for the isothermal initiation of diffusional transformations such as α and pearlite, while the lower C curve represents the initiation times for displacive transformations such as α_w, α_a and bainite (10). The cross-over point of the two C curves is the temperature T_1 below which displacive transformations are assumed to be kinetically favoured, so that the growth of α ceases and gives way to α_w and α_a formation. In the temperature range T_h-T_l, the α layers at the γ/γ boundaries are assumed to thicken by a paraequilibrium mechanism, at a rate controlled by the diffusion of carbon in the γ ahead of the moving α/γ interface. During isothermal transformation at a temperature T, the half thickness of each such layer is given by:

$$q = \alpha_1 t^{\frac{1}{2}} \qquad (2)$$

where α_1 is called the one-dimensional parabolic thickening rate constant, obtained by solving (see for example, ref. 11)

$$\{2(x^{\gamma\alpha}-\bar{x})/(x^{\gamma\alpha}-x^{\alpha\gamma})\}\{(\underline{D}/\pi)^{\frac{1}{2}}\}$$
$$= \alpha_1\{\exp[\alpha_1^2/(4\underline{D}/\pi)]\}\{1-\mathrm{erf}[\alpha_1/(2\underline{D}^{\frac{1}{2}})]\} \qquad (3)$$

and t is the time, defined to be zero when q=0. $x^{\alpha\gamma}$ and $x^{\gamma\alpha}$ are paraequilibrium carbon concentrations of α and of γ respectively, calculated as in Ref. 1. The diffusion coefficient D of carbon in γ is a sensitive function of the carbon concentration in austentite (x_γ) and since x_γ varies with distance ahead of the α/γ interface, a weighted average diffusion coefficient $\underline{D}$ is used in equation 2, and is defined as

$$\underline{D} = \int_{x^{\gamma\alpha}}^{\bar{x}} D\, dx/(\bar{x} - x^{\gamma\alpha}). \qquad (4)$$

Since the α grows anisothermally during cooling of the weld deposit, q is actually obtained by numerically integrating the function

$$q = \int_{t=0}^{t=t_1} 0.5\alpha_1\ t^{-0.5}\ dt \qquad (5)$$

where t_1 is the time taken for the weld deposit to cool from T_h to T_1. It should be noted that the parabolic rate constant is a function of time and temperature. The volume fraction v_α of α is then derived as a function of a and q:

$$v_\alpha = [4qC_3(a-2qC_3)/a^2] \qquad (6)$$

where $C_3 = \tan(30^\circ)$. A better fit is obtained by empirically correcting this to give a corrected volume fraction V^α (and this gives a corresponding corrected value of allotriomorph thickness q') where

$$V_\alpha = 2.04v_\alpha + 0.035, \qquad (7)$$

the calculated value thus being smaller than the actual value by a factor of ~ 2, although the correlation between v_α and the actual volume fraction of α is typically found to be 0.95 (1).

At $T=T_1$, α_w formation begins, but the growth rate of α_w in welds is generally so high that growth ceases within a fraction of a second, and α_w formation can therefore be considered to occur isothermally at T_1. However, the volume fraction v_w of Widmanstätten ferrite in general correlates badly with its lengthening rate (G); impingement with acicular ferrite has to be taken into account. If the time available for unhindered α growth is t_2, then v_w is given by:

$$v_w = 3.34G(a-2q'C_3)t_2^2/a^2 \qquad (8)$$

G is calculated for $T=T_1$, by the method described in (12). In the absence of impingement with α_a, the time available for α_w to grow right across the γ grains is t_3, given by:

$$t_3 = 2[a \sin(60^\circ) - q']/G \qquad (9)$$

If t_3 is less than a critical time t_c, then α_w can grow without impingement with acicular ferrite and t_2 in Eq. (9) is set to equal t_3. On the other hand, if $t_c<t_3$ than α_w growth is terminated by impingement with α_a, t_2 is set equal to t_c. t_c is experimentally found to be 0.211s for welds containing inclusions which nucleate acicular ferrite; although t_c is not sensitive to oxygen concentration, it is recognised that it may vary, depending on the ability of inclusions to nucleate acicular ferrite. The microphase volume fraction v_m can also be estimated thermodynamically, so that:

$$v_a = 1 - V_\alpha - v_w - v_m \qquad (10)$$

where v_a is the volume fraction of α_a. It is usual to include the microphases in the v_w and v_a measurements, since v_m is generally small; hence, microphases are not measured or calculated separately in this work.

The high cooling rates associated with arc welding inevitably lead to solute segregation during solidification. The biggest effect of such segregation is to alter T_h (3); if solidification occurs with δ as the primary phase, then the partition coefficient k_i where i is the subscript identifying an alloy element) which gives the ratio of the mole fraction of i in δ to that in liquid is given at a temperature T by:

$$k_i = \exp\{\Delta^o G_i/(RT)\} \qquad (11)$$

where $\Delta^o G_i$ is the Gibbs free energy change per mole in transforming the pure i from δ to liquid, and R is the gas constant. It is assumed that the segregation arising during solidification is not influenced by subsequent diffusion during rapid cooling from the liquidus. The composition of regions which contain the lowest substitutional content is then given by $k_i\bar{x}_i$ where $\bar{x}_i$ is the average concentration of i and T is set to the melting point in Eq. (11), assumed to be 1520°C. Carbon is assumed uniform everywhere since its activity gradients should rapidly be eliminated by diffusion during cooling. It follows that the compositions of the solute-depleted regions can be calculated, and these regions enhance ferrite nucleation. New TTT curves calculated for these solute-depleted regions can then be used to obtain better estimates of T_h. This procedure works well in calculating the microstructure of welds (3).

THE EFFECT OF Si ON WELD METAL MICROSTRUCTURE

From our knowledge of isothermal transformations in steels, it should at first sight be concluded that Si should enhance the formation of α in welds. Silicon is known to accelerate the nucleation and growth of ferrite (13). It should therefore increase V_α at the expense of, say, α_a. However, the development of microstructure during welding does not involve isothermal growth, and there are numerous other complicating and contradictory factors to be taken into account. It will become apparent later that information from isothermal experiments can be misleading in the context of weld deposits.

Evans (6) has reported quantitative microstructural measurements on a systematic series of Fe-Mn-Si-C alloys, in which the carbon concentration was kept approximately constant (0.065 wt.%). The aim of his work was to study the effect of silicon variations (∿0.2-0.95 wt.%) at two levels of Mn (0.6, 1.4 wt.%). The compositions of the alloys concerned are given in Table 1. It is unfortunate that C varies significantly in these welds, because such variations are known to have a powerful effect on microstructure (1). This makes it difficult to isolate the effect of Si, since only welds 2 and 4, and welds 5 and 8 are comparable in terms of C and Mn.

Table 1 - Chemical compositions (wt.%) of welds studied by Evans (6). The austenite grain size 'a' is calculated as discussed in the text.

Weld	C	Mn	Si	2a,μm
1	0.060	0.60	0.20	138
2	0.070	0.66	0.38	132
3	0.065	0.65	0.61	135
4	0.070	0.64	0.95	133
5	0.064	1.40	0.20	125
6	0.066	1.41	0.36	124
7	0.073	1.44	0.62	120
8	0.063	1.38	0.93	126

The welds (6) were deposited using manual metal arc welding, the joint geometry being compatible with ISO-2560. Welding was carried out in the flat position using 4mm diameter electrodes and 20mm thickness base plate. The current and voltage used were 170A and 21V (DC+) respectively, the weld consisting of 27 runs with 3 beads per layer. The electrical energy input (Q) was stated as being approximately 1kJ/mm, the interpass temperature (T_i) being fixed at 200°C. Evans (14) has separately reported the time taken for his welds to cool over the temperature range 800-500°C, and this, together an analysis of the cooling curves for MMA welds (2) shows the cooling rate (dT/dt, °C/s) for this temperature range is well approximated by the equation:

$$dT/dt = [2155(T-T_i)^{1.6}]/[Q\eta] \qquad (12)$$

where η is the arc transfer efficiency, taken to be 0.775. Evans (6) did not report any grain size measurements, but analysis (2) of some of his other work (14-16) on Fe-Mn-Si-C welds suggests the following relationships, with a correlation coefficient of 0.95 (standard error 6.01):

$$\bar{L}_{tn}\ (\mu m) = 108.2 - 348C(wt.\%) - 9.6Mn(wt.\%) + 13.24Q(kJ/mm) \qquad (13)$$

The earlier experiments of Evans (14-16) do not include significant variations in Si concentration (0.27-0.41 wt.%), so that the above regression equation ignores Si as an independent variable; its inclusion would make the equation unreliable for Si values outside of the range of Evans' data. The excellent correlation coefficient obtained indicates this procedure to be satisfactory. Calculated grain size values using Eq. (13) are included in Table 1. Having established the necessary data, primary weld microstructures for welds 1-8 were calculated using the theory discussed earlier; the results are presented in Tables 2-3 and in Fig. 1. Bearing in mind that the cooling curves and grain sizes are estimated, agreement between theory and experiment is rather good. The correlation coefficient for a plot of the calculated versus experimental data (Fig. 1) is found to be 0.96. The mean experimental values of V_α, v_w and v_a are 0.23, 0.22 and 0.60 respectively, and the mean differences between calculated and experimental data are 0.05, 0.06, and 0.09 respectively. The good agreement confirms that the model (1) can be used satisfactorily to account for the relatively high Si concentrations used in Evans' experiments.

Table 2 - Calculated parameters for welds 1-8

Weld	q´,μm	t_1,s	t_2,s	G,μm/s	T_h-T_1,°C
1	10.7	3.61	0.211	218	804-600
2	9.0	3.31	0.211	147	800-610
3	10.0	3.41	0.211	177	808-610
4	9.5	3.51	0.211	151	816-610
5	7.1	2.64	0.211	79	735-600
6	7.0	2.68	0.211	72	738-600
7	6.5	2.71	0.211	59	740-600
8	7.7	2.77	0.211	105	760-610

Table 2 shows that for all the welds, t_2=t_c; the growth of Widmanstätten ferrite is limited by hard impingement with intragranularly nucleated acicular ferrite. This is consistent with the micrographs published by Evans (6). When α_w is limited by impingement with α_a, v_w should depend to a large extent on G; this is indeed the case, the correlation coefficient relating them being 0.90.

Defining $\Delta=T_h-T_l$ (the temperature range over which α forms), Table 2 shows that Δ is slightly larger for weld 4 when compared with weld 1. In spite of this, t_l, q' and V_α are higher for weld 1. This is because ferrite formation begins at a higher temperature for weld 4 (T_h higher than weld 1) and since the cooling rate increases with T, there is less time available for α to grow. Experience from isotheraml experiments would indicate that weld 4 should have a higher V_α (since its Si level is higher), and hence is misleading for the continuous cooling conditions that arise during welding. Keeping Δ constant but increasing T_h has the effect of reducing the amount of allotriomorphic ferrite. In such circumstances, Si should not influence toughness through microstructure, assuming that all else is constant.

On the other hand, if Si increases Δ, V_α should also increase. Comparison of welds 5 and 8 (which contain similar C, Mn and have similar grain size) illustrates this; q', t_l and V_α are all higher for the higher Si weld 8. In these circumstances Si should be detrimental to toughness.

Finally, it has already been pointed out that the experimental data (6) are not entirely suitable for the study of Si effects, since the other alloying additions have not been kept sufficiently constant. Indeed, precise control of weld composition is a major experimental difficulty. However, this is easy to do theoretically, and a major theoretical study of alloy effects on primary weld microstructure will be published later.

Table 3 - Comparison of calculated and experimental volume fractions of α, α_w and α_a. The superscripts 'c' and 'e' on the volume fractions indicate calculated and experimental data, respectively.

Weld	V_α^e	V_α^c	v_w^e	v_w^c	v_a^e	v_a^c
1	0.32	0.32	0.40	0.39	0.28	0.29
2	0.22	0.29	0.33	0.29	0.45	0.42
3	0.24	0.31	0.37	0.33	0.39	0.36
4	0.30	0.30	0.18	0.29	0.52	0.41
5	0.23	0.24	0.08	0.17	0.69	0.59
6	0.13	0.24	0.15	0.16	0.72	0.60
7	0.20	0.23	0.10	0.13	0.70	0.64
8	0.18	0.26	0.10	0.22	0.72	0.52

CONCLUSIONS

It has been demonstrated that a recent model for the prediction of the primary microstructure of low-alloy steel weld deposits (1-5) gives good agreement with published experimental data (6) on the effect of large variations in Si and Mn on weld microstructure. The model predicts, and it is experimentally observed, that information obtained from simple isothermal experiments can be misleading in predicting alloying element effects.

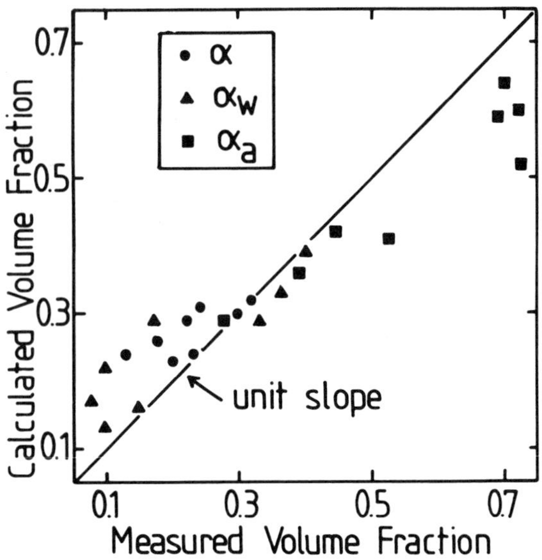

Fig. 1 - A plot of calculated α, α_w and α_a volume fractions versus the experimental data of Evans, for the weld deposits listed in Table 1.

ACKNOWLEDGEMENTS

The authors are grateful to Professor Hull for the provision of laboratory facilities at the University of Cambridge, and to ESAB AB (Sweden) for the provision of laboratory facilities and for financial support for this project.

REFERENCES

1) Bhadeshia, H. K. D. H., L. -E. Svensson and B. Gretoft, Acta Metall. 33 1271-1283 (1985).
2) Svensson, L.-E., B. Gretoft and H. K. D. H. Bhadeshia, Scand. J. of Metall. (1986) in press.
3) Gretoft, H. K. D. H. Bhadeshia and L.-E L.-E. Svensson, Acta Stereologica (1986) in press.
4) Bhadeshia, H. K. D. H., L.-E. Svensson and B. Gretoft, J. Mat. Sci. Let. 4 305-308 (1985).
5) Bhadeshia, H. K. D. H., L.-E. Svensson and B. Gretoft, Proc. 1st Int. Conf. on Computer Tech. in Welding, Welding Institute, Cambridge, England (1986), in press.
6) Evans, G. M., IIW document II-A-630-84 (1984).
7) Davies, G. J. and J. G. Garland, Int. Metall. Rev. 20 83-120 (1975).
8) Bhadeshia, H. K. D. H., L.-E. Svensson and B. Gretoft, J. Mat. Sci. (1986), in press.
9) Bhadeshia, H. K. D. H., Acta Metall. 29 1117-1130 (1981).

10) Bhadeshia, H. K. D. H., Met. Sci. 16 159-165 (1982).
11) Bhadeshia, H. K. D. H., Prog. in Mat. Sci. 29 321-386 (1985).
12) Bhadeshia, H. K. D. H., Mater. Sci. and Tech. 1 497-504 (1985).
13) Kinsman, K. R. and H. I. Aaronson, Metall. Trans. 4 959 (1973).
14) Evans, G. M., IIW document II-A-546-81.
15) Evans, G. M. IIW document II-A-469-79.
16) Evans, G. M., IIW document II-A-490-79.

THE KINETICS OF PHASE TRANSFORMATIONS IN HSLA-80 WELDMENTS

R.A. Vandermeer, C.L. Vold
Naval Research Laboratory
Washington, DC 20375-5000

ABSTRACT

The decomposition of austenite to polygonal ferrite was studied in HSLA-80 - an ASTM A710 grade A commercial steel. Isothermal transformation (IT) behavior was measured dilatometrically in the temperature range 716 to 793°C. The reaction kinetics were described analytically and converted to continuous-cooling-transformation (CCT) kinetics by numerical calculation using the additivity principle. A partial analytical CCT diagram was established and compared to data determined from independent experimental measurements.

A GOAL OF WELDING science is to develop quantitative, analytical models of the welding process capable of predicting weld zone microstructures and properties. Even for a specified welding process and an available, predictive heat flow model this is an enormously difficult task because of the wide range of peak temperatures and thermal histories experienced by the various volume elements in the fusion (FZ) and heat-affected (HAZ) zones. These difficulties are especially apparent in HSLA steels where solid-state phase transformations (reversion and transformation of austenite to ferrite, the solution and reprecipitation of second phases and austenite grain growth) can occur. These reactions along with the liquid to solid transformation ultimately determine the microstructure of the weldment. A comprehensive, analytical model of these transformations <u>individually</u> is a prerequisite to the overall goal of modeling the welding process.

This paper describes the progress of our effort to model the austenite to ferrite transformation in the HSLA-80 steel (ASTM A710), a high strength ferritic steel recently introduced in the construction of U.S. Navy ships. To be applicable to welding, phase transformation models and kinetics must be directed toward a continuous-cooling-transformation (CCT) perspective. Currently the CCT diagram is the most commonly available guide to understanding the resulting microstructures. Unfortunately the basic mechanistic understanding of phase transformations is, with the exception of martensitic transformations, usually garnered from isothermal transformation (IT) investigations rather than CCT studies. Since the resulting HAZ microstructures are directly dependent upon the operative transformation mode, IT characteristics must be known and subjectively adapted to aid in the interpretation of CCT behavior. In the past this methodology has not always been successful. Furthermore detailed open literature IT and CCT information on HSLA steels is scarce. This paper addresses some of these important issues.

The interrelationship of IT and CCT diagrams has been recognized for a long time. The additivity principle which began with the work of Scheil [1] has achieved modest but not always decisive success calculating CCT diagrams from IT diagrams. The theoretical studies of Avrami [2] and Cahn [3] suggested that the additivity principle was justified only for certain restricted nucleation and growth rate behavior but recently Sharma and Kirkaldy [4] argued for its application under far more general circumstances. A resurgence in the use of additivity has found application in steel transformations primarily the kinetics of the pearlite reaction [5,6,7]. In this study the validity of the additivity principle is tested for the austenite to ferrite transformation in HSLA-80.

EXPERIMENTAL METHODS

The starting material was a commercial, mill-heat-treated plate 16 mm thick of ASTM A710 grade A class 3 HSLA steel. The pertinent penultimate heat treatment consisted of austenizing the plate at 900°C for 2.7 ks, water quenching, aging at 593°C for 1.8 ks and air cooling. The chemical composition of the plate in wt% was; 0.036 C; 0.51 Mn, 0.0045 P, 0.0085 S, 0.34 Si, 0.66 Cr, 0.95 Ni, 0.21 Mo, 1.25 Cu, 0.04 Nb, 0.037 Al, 0.0092 N, <0.002 Ti, <0.007 V, 0.016 Sn, <0.0001 Ca, <0.002 Ce, balance Fe. Dilatometry specimens in the form of hollow cylinders 10 mm long by 5 mm outside diameter with a wall thickness of 1 mm were machined from the plate. The axis of each specimen was parallel to the rolling direction in the plate.

The IT experiments were conducted at the Oak Ridge Y-12 Plant using a Theta Industries metallurgical research dilatometer. A run typically consisted of heating the specimen inductively in a vacuum <0.1 mPa to the austenite state at 900°C at a rate of 7.3° C/s. After an austenitizing anneal of 1.8 ks, the specimen was rapidly cooled to a preset isothermal transformation temperature, held until transformation was complete or 57.6 ks duration whichever came first and helium gas quenched to room temperature. An S-type thermocouple spot welded to the outer cylindrical surface of the specimen monitored the temperature. Seven different transformation temperatures between 716 and 793°C were studied. During the run specimen length was measured contiuously using a calibrated linear-variable-differential transformer (LVDT). The LVDT is especially sensitive to the volume expansion associated with the austenite to ferrite phase transformation and thus its output can be used to detect and track the course of the transformation as a function of time and/or temperature.

The CCT experiments were performed under contract by AMAX Materials Research Center under the direction of Thomas B. Cameron. Specimens from the same commercial plate were austenitized at 900°C for 1.8 ks and cooled at ten different cooling rates ranging from 127°C to 0.02° C/s* in a programmable Materials Measuring Corporation quenching dilatometer. Rapid quenches were accomplished using a regulated flow of helium gas. Phase transformation information was obtained by analyzing the length change versus temperature data in a manner similar to that described by Eldis [8].

After the transformed specimens were removed from the dilatometer, they were sectioned and prepared for microstructural examination in both the optical and scanning electron microscopes using conventional metallographic practice.

*The cooling rates quoted in this paper have been estimated based on the measured times required to cool the specimens from 800°C to 500°C.

RESULTS

An example of data obtained during a typical isothermal transformation run is shown in Figs. 1 and 2. The consequences of the quench from austenite to T_t, the isothermal transformation temperature, and the subsequent hold at T_t are illustrated in Fig. 1. The initial contraction of the specimen with decrease in temperature corresponds to the expected thermal contraction of austenite based upon its coefficient of thermal expansion. The subsequent expansion after reaching T_t is a manifestation of the transformation of austenite to ferrite. Figure 2 depicts the dilation characteristics for the combined cool/hold/cool sequence. In this example complete

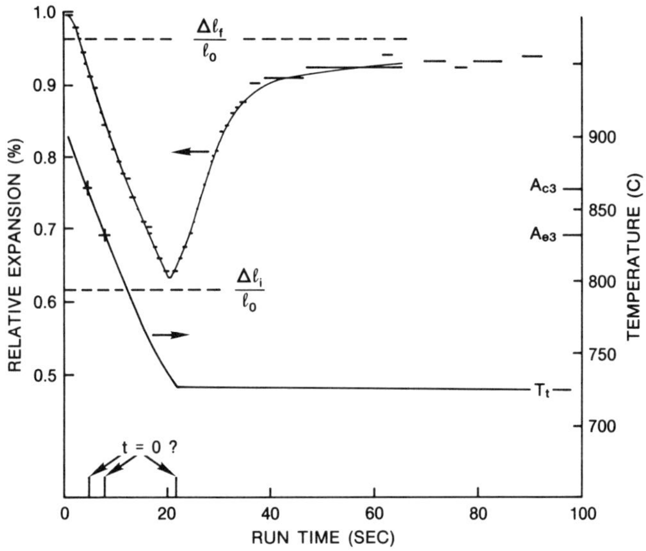

Figure 1 - Relative expansion and temperature plotted versus run time during quench from 900°C and isothermal transformation at 729°C.

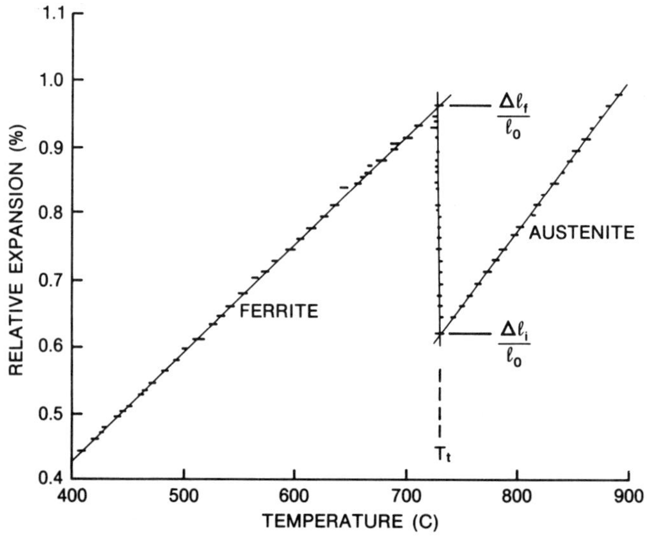

Figure 2 - Relative expansion versus temperature for the entire quench/hold/quench sequence.

transformation was accomplished at T_t = 729°C. The slopes of the high and low temperature segments of the curve in Fig. 2 are approximately the coefficients of thermal expansion of austenite and the transformed product respectively. Microscopy confirmed that the microstructure of the product consisted largely of polygonal ferrite.

To analyze data such as that shown in Fig. 1, the assumption is made that the length change fraction and the austenite fraction transformed, ζ, are equal so that

$$\zeta = \frac{\Delta l_t/l_o - \Delta l_i/lo}{\Delta l_f/f_o - \Delta l_i/l_o} \tag{1}$$

where $\Delta l_t/l_o$ is the relative isothermal expansion at transformation time, t, and $\Delta l_i/l_o$ and $\Delta l_f/l_o$ are the relative expansions corresponding to the start and finish of transformation as obtained from Fig. 2.

For transformation times to be meaningful theoretically and in calculations, a correct zero point for time must be established. Three possibilities corresonding to the instants when the specimen reached the temperatures Ac_3, Ae_3 or T_t were considered and these are shown in Fig. 1. Steel industry practice generally uses Ac_3 as the t = o point because it is readily measureable in a dilatometer [8]. Theoretically Ae_3 is more appropriate as the starting time coordinate because it represents the equilibrium temperature above which no transformation of austenite is possible. The determination of Ae_3 is difficult but a method of calculating it from thermodynamic principles with reasonable accuracy has been suggested by Bagannis and Kirkaldy [9]. The method includes contributions from seven of the most widely used alloying elements in HSLA steels. Using the Bagannis and Kirkaldy equation, Ae_3 was estimated to be 828°C (1101 K) for the composition of the HSLA-80 steel in this study. Throughout the analysis crossing the Ae_3 temperature was adopted as the instant where t = o. This choice gave the most consistent results.

Many investigators seek to describe isothermal phase transformation kinetics analytically using the equation

$$\zeta = 1 - \exp - Bt^n \tag{2}$$

where for a given T_t, B and n are constants. Use of this equation is restricted to transformations involving a single mechanism which can be modeled with simple nucleation and growth characteristics. In some cases t in Eq. 2 is replaced by $t-t_o$ where t_o may be viewed as a nucleation incubation period. Hawbolt et al. [5,6] invoked the latter form to represent the decomposition of austenite to pearlite and ferrite in plain carbon 1025 and 1080 steels. For 1025 the best data fit occurred when n was independent of temperature with values 1 and 4/3 for pearlite and ferrite respectively. Efforts to analyze the present results in terms of Eq. 2 proved unsuccessful. In Fig. 3 the transformed fraction is plotted according to the double logarithmic form of Eq. 2. At no temperature was a linear relationship over the measured range of ζ evident. Nor at any given value of ζ could the slope which is equal to n be construed as being independent of temperature. This implies n = f(ζ, T), a complexity which becomes a severe problem in applying Eq. 2 mathematically. Incorporation of an induction period tended to improve the situation at some temperatures but not at others and the same difficulties persisted. As LeBlond and DeVaux [10] pointed out Eq. 2 suffers from yet another deficiency in that it does not take into account austenite transformation to a two phase $\alpha + \gamma$ equilibrium microstructure. Apparently the decomposition kinetics are sufficiently complex that the simple nucleation and growth rate models leading to Eq. 2 are not valid for HSLA-80.

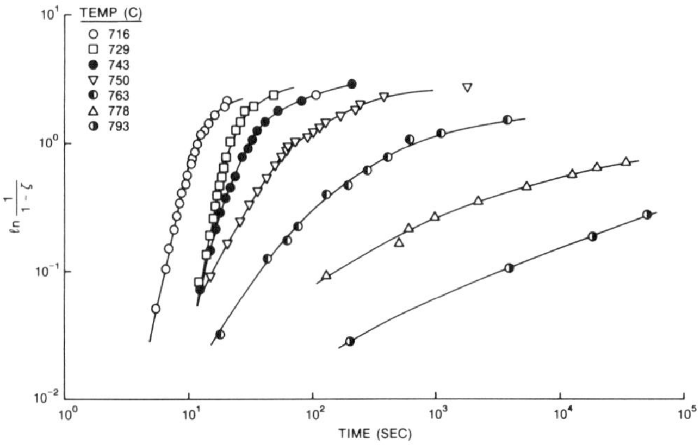

Figure 3 -Log-log plot of ln 1/(1-ζ) versus time for isothermal transformation.

In view of these shortcomings the Eq. 2 approach was discarded and an alternate, empirical representation of IT kinetics was sought. Upon testing of the data with several trial functions, Eq. 3 evolved.

$$\zeta = \frac{C_3}{2} + \frac{C_4}{\pi} \cdot \tan^{-1} \left(\frac{\ln t - C_1}{C_2}\right) \tag{3}$$

this equation was chosen because of its capacity to describe the complete data set at each of the temperatures studied using the C's as adjustable parameters. It also accounts for the partial transformation cases in which some equilibrium austenite remains. Figure 4 which plots ζ versus lnt shows the degree of fit that was achieved with Eq. 3 when a non-linear least squares computation was carried out on the data at each temperature. The temperature dependence in Eq. 3 is through the C's. To describe the temperature dependence of the C's analytically a cubic polynominal was selected.

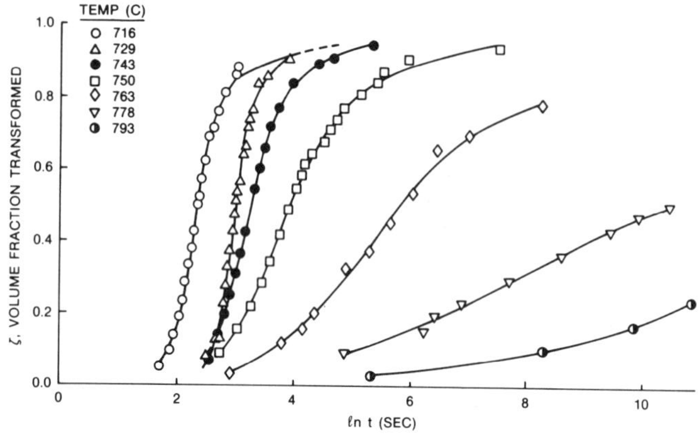

Figure 4 - Fraction transformed versus ln (time) for isothermal transformation.

$$C_i \;=\; a_i + b_iT + c_iT^2 + d_iT^3 \qquad i = 1 \text{ to } 4 \qquad (4)$$

The constants a_i, b_i, c_i and d_i were computed using Eq. 4 and regression analysis. When these 16 constants are combined with Eqs. 4 and 3 an analytical description of the isothermal austenite to ferrite transformation kinetics between 716 and 793°C is established.

A conjecture as to why the kinetics are complex can be offered based on an interpretation of the microstructures presented in Fig. 5. In Fig. 5a two microconstituents are apparent-light etching and dark etching in contrast. Quantitative point counting procedures indicate that 0.682 volume fraction was light etching polygonal ferrite. Examination of the dark etching regions at higher magnification, as in Fig. 5b, showed them to consist of ferrite (elongated) and an aligned second phase. This specimen after austenitization had been isothermally transformed at 774°C for 86.4 ksec and then quenched to room temperature. Dilatometry revealed that while much of the austenite decomposition occurred at 774°C, it was not complete and further transformation took place during the final quench to room temperature. A calculation of the transformation fraction expected at 774°C using Eqs. 3 and 4 gave a value of 0.665. Thus based on the optical metallographic results it is concluded that the light etching polygonal ferrite microconstituent formed at 774°C whereas the dark etching, apparently two phase regions, formed during the quench. If austenite decomposes to ferrite by a diffusional mechanism and partitioning of carbon and the other alloying elements is required, the last austenite to transform would be highly alloyed and the formation of other phases besides ferrite would be likely. The nucleation and growth kinetics in such diffusional, partioning systems might be expected to be very complex and Eq. 2 would find only limited applicability.

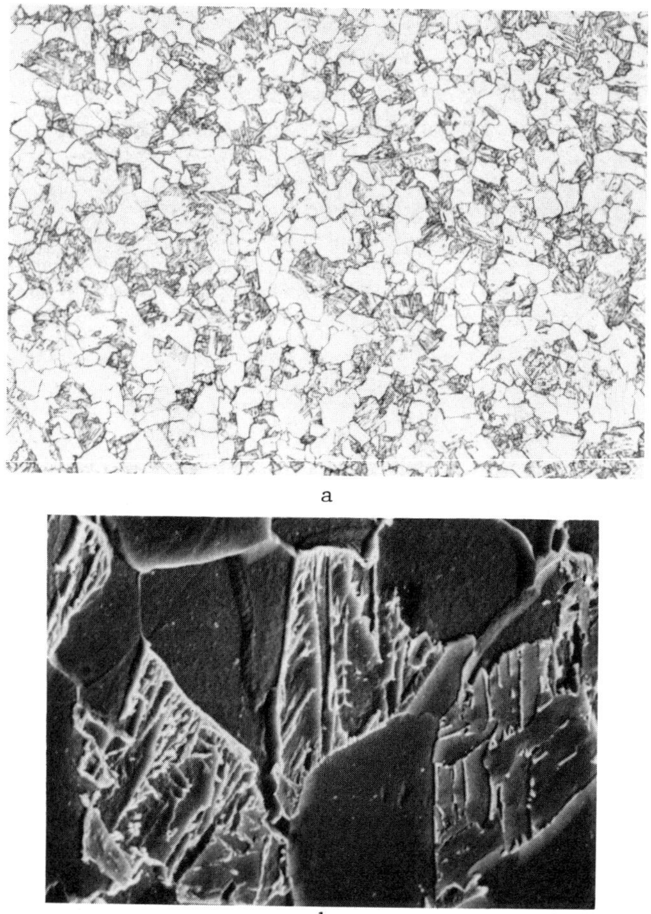

a

b

Figure 5 - Microstructure of specimen isothermally transformed at 774°C for 86400 sec and quenched - a (500X); b (5000X).

The conversion of IT kinetics to CCT kinetics can be accomplished using the additivity principle which is stated mathematically in Eq. 5.

$$\int_{A_{e3}}^{T(\zeta)} \frac{dT}{t(\zeta,T)\dot{T}} = 1 \qquad (5)$$

Where $T(\zeta)$ is the temperature and fraction transformed of interest, $\dot{T}$ is the cooling rate and $t(\zeta,T)$ is the isothermal t from Eq. 3. We can simplify the calculations by assuming $\dot{T}$ to be approximately constant and equal to $[T(\zeta)-A_{e3}]/\tau(\zeta)$ where $\tau(\zeta)$ is now the CCT time required to attain ζ fraction transformed at $T(\zeta)$. Inserting this aproximation into Eq. 5 and solving for $\tau(\zeta)$ yields

$$\tau(\zeta) = \frac{A_{e3} - T(\zeta)}{\displaystyle\int_{T(\zeta)}^{A_{e3}} \frac{dT}{t(\zeta,T)}} \qquad (6)$$

The integration of the denominator in Eq. 6 does not have an analytical solution when Eq. 3, rewritten in terms of t, is inserted.

Evaluation of $\tau(\zeta)$ therefore required numerical integration. To map out τ as a function of ζ and T numerous calculations using Eq. 6 were performed. The computed τ's were then represented mathematically by the equation

$$\ln \tau = \ln(A_{e3}-T) - \{B_1+B_2(A_{e3}-T) + B_3(A_{e3}-T)^2 + B_4(A_{e3}-T)^3\} \quad (7)$$

where the B's are functions of ζ only. The B_i dependence on ζ was evaluated using cubic polynomials analogous to the C_i dependence on T (Eq. 4) and regression analysis. Once the various cubic polynomial coefficients of the B_i were known, Eq. 7 contained all the information necessary to calculate $\tau = f(T,\zeta)$ which when plotted is the CCT diagram. The results for ζ = 0, .25, .5 and .95 are depicted in Fig. 6. The CCT diagram calculated by this methodology is limited to the temperature range over which the isothermal transformation data were measured (716 to 793°C). Any extension of the lines in Fig. 6 to lower or higher temperatures would be highly speculative and unwarranted unless a transformation model based on mechanism can be developed.

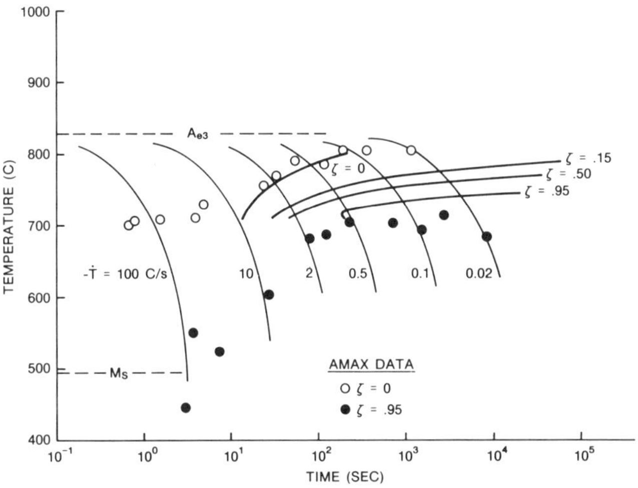

Figure 6 - Calculated and experimental CCT diagram for α to γ transformation in HSLA-80.

In order to test the validity of the additivity principle and assess its usefulness in calculations, the experimental CCT data independently developed by AMAX for this steel was evaluated. Transformation temperatures at ζ = 0 and .95 were estimated for each cooling rate investigated using a methodology similar to that described by Eldis [8]. The results are plotted in Fig. 6 also. In general, reasonable agreement was achieved at low ζ. An apparent systematic deviation at the higher ζ may be evident but as Eldis [8] has pointed out there can be an experimental uncertainty of ± 35°C in locating the transformation completion temperature from dilatometric data. Thus it is not certain whether the deviation at high ζ is an experimental difficulty or one due to a breakdown of the additivity principle.

Our subjective opinion is that broadly speaking, the use of the additivity principle to calculate CCT diagrams from IT data has much merit. A note of caution needs to be raised, however. When analyzing the IT data for this purpose, it is important to choose the proper t = 0 origin. The A_{e3} temperature provided the most consistent results and is suggested to be the best choice.

The present IT information does not allow calculation of the CCT diagram at temperatures below 716°C with reliability. This is due to the inability of the current dilatometer technique to obtain accurate isothermal transformation data at lower temperatures. Rapid transformation rates, recalescence due to the release of the heat of transformation and inability to cool at fast enough rates while preventing transients in the measurement signals all contribute to the problem. Currently another approach is being pursued to overcome these difficulties. The new approach utilizes a derivation by Kirkaldy and Sharma [4] whereby the additivity principle can be inverted to extract IT kinetics from CCT information instead of visa versa. Thus low temperature IT kinetics could be determined by analyzing the CCT data at the higher cooling rates. By combining such calculated low temperature IT information with that measured previously at the higher temperatures, a more complete transformation picture would emerge from which a plausible model could be developed. Once this model is established the additivity principle would be invoked to calculate the CCT diagram. Research currently in progress involves computations of this nature along with model development.

ACKNOWLEDGEMENTS

The authors wish to thank Dr. T. Kollie and Martin-Marietta Energy Systems for making time available on their dilatometer. Our sincerest appreciation goes to William Elliott for performing the isothermal experiments.

REFERENCES

1. Scheil, E., Archiv. fur Eisenhuttenw. 12, 565-67 (1935).
2. Avrami, M., J. Chem. Phys. 8, 212-24 (1940).
3. Cahn, J. W., Acta Met. 4, 572-75 (1956).
4. Kirkaldy, J. S. and R. C. Sharma, Scripta Met. 16, 1193-98 (1982).
5. Umemoto, M., N. Komatsubara and I. Tamura, "Solid-Solid Phase Transformations", p. 1111-15, The Metallurgical Society-AIME, Warrendale, PA (1982).
6. Hawbolt, E. B., B. Chau and J. K. Brimacombe, Met. Trans A, 14A, 1803-15 (1983).
7. Hawbolt, E. B., B. Chau and J. K. Brimacombe, Met. Trans A, 16A, 565-78 (1985).
8. Eldis, G. T., "Hardenability Concepts with Applications to Steel", p. 126-48, The Metallurgical Society-AIME, Warrendale, PA (1978).

MICROSTRUCTURAL CHARACTERISTICS OF AUSTENITIC STAINLESS STEEL INERTIA FRICTION WELDS

John C. Lippold
Edison Welding Institute
Columbus, Ohio

Ben C. Odegard
Sandia National Laboratories
Livermore, California

ABSTRACT

The microstructural and fractographic features of inertia friction welds in two austenitic stainless steels have been examined. In general, the bond region in these materials consisted of a refined austenitic grain structure resulting from dynamic recrystallization in the vicinity of the bondline during welding. The presence of ferrite along the bondline in these welds was a consequence of both a solid-state phase transformation at elevated temperatures and fragmentation of parent material ferrite. A significant level of ferrite along the bondline in high ferrite potential materials was found to influence both the mechanical properties and fracture behavior of the inertia friction welds.

INTRODUCTION

Inertia friction welding has proven to be a reliable and economical method for joining a variety of similar and dissimilar material combinations (Ref. 1-3). Since bonding occurs in the solid-state, the process is often selected as an alternative to fusion welding in applications where solidification defects such as porosity and weld hot cracking are encountered. Despite the utility of the inertia friction welding process, it is likely that this process will not become more widely used, particularly in critical applications, until the metallurgical and mechanical properties of the bond region are better understood.

Austenitic stainless steels are frequently used in critical applications where the level of ferrite in the weld region must be minimized to avoid property degradation during service, such as in corrosive environments (Ref. 4) or at cryogenic temperatures (Ref. 5). In fusion welds, the level of ferrite can be predicted based on chemical composition using diagrams such as those published by Schaeffler (Ref. 6) and DeLong (Ref. 7). This investigation evaluates the effect of composition on the microstructure and subsequent tensile properties of the bond region of inertia friction welds in two austenitic stainless steels of different ferrite potential.

EXPERIMENTAL PROCEDURES

Material. The materials evaluated in this investigation included a Type 304L stainless steel of composition 19.6Cr, 10.6Ni, 0.64Si, 1.8Mn, 0.013C, 0.038N, 0.004S, and 0.012P and a compositional variant of this alloy containing 20.9Cr, 9.0Ni, 0.62Si, 1.7Mn, 0.024C, 0.007N, 0.008S, and 0.025P (all values in weight percent). The high-Cr variant was designated alloy 21-9. The Type 304L was supplied in the form of 25 mm (1 in.) diameter bar, while alloy 21-9 was in the form of cross-rolled plate. Both materials were in the annealed condition and exhibited an ASTM grain size in the range from 5 to 7. All weld blanks machined from these materials were nominally 12.7 mm (0.5 in.) in diameter and 50.8 mm (2.0 in.) long.

Inertia Friction Welding. Welds were made using a MTI model 100 inertia welding machine. The welding conditions used during this investigation are listed in Table 1 and were selected on the basis of acceptable axial upset and lateral "flash" previously demonstrated to be indicative of a good weld in austenitic stainless steels (Ref. 8). Cleaning of the weld blanks consisted of an acetone rinse within 48 hours prior to welding.

TABLE 1

Welding Conditions

Flywheel Moment of Inertia, lb_m-ft^2 (kg-m^2)	1.5 (0.063)
Axial Force, lb_f (kN)	6000 (26.4)
Rotational Speed (rpm)	6600

Mechanical Testing. Smooth and notched tensile samples were machined from the welded blanks. The diameter of the smooth bar gage section was 4.95 mm (0.195 in.). Notch depths of 1.23 mm (0.048 in.) and 0.825 mm (0.032 in.) were used in order to measure the weld strength as a function of radial distance from the center of the welded blanks. Tensile samples were tested at a crosshead velocity of 0.02 mm/s (0.05 in/min). Tensile tests were performed at ambient temperature in both the as-welded condition and after thermally aging at 300 °C for 2 weeks in high pressure hydrogen.

RESULTS

Metallographic Evaluation. Welded specimens were sectioned axially to allow examination of both the bondline and the adjacent heat-affected zone (HAZ). The bond region in a Type 304L inertia friction weld sectioned in this manner exhibited a refined austenitic grain structure relative to the parent material. A radial transition in microstructure within this refined region was evident from the center to the outer diameter of the weld.

The bond region from the center to the mid-radius of the weld contained an extremely fine austenitic grain structure, as evidenced by the micrograph in Figure 1A. The width of this region was relatively uniform. Radially outward from the mid-radius point, the grain size progressively increased, reaching a maximum near the outer diameter of the weld (Figure 1B). The axial width of the grain-refined bond region also increased from approximately the mid-radius to the outer diameter.

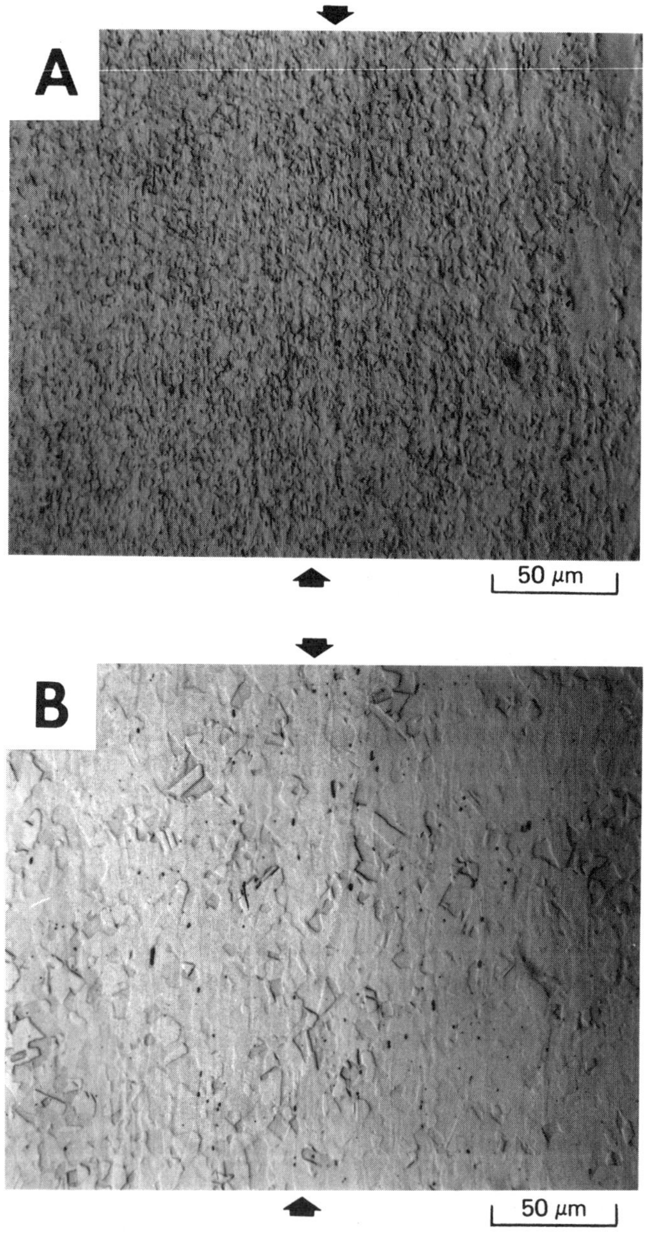

Figure 1. Microstructure along the bondline in a Type 304L inertia weld, A) center, B) near the outer diameter. Arrows indicate approximate bondline location.

Ferrite was not observed in any region along the bond line. Inspection of the extruded metal, or "flash", revealed a fine dispersion of ferrite in a refined austenitic matrix, as shown in Figure 2. Note the defect (arrow) associated with this two phase region. The localized presence of ferrite in the weld flash suggests that this ferrite formed via a solid-state reaction at near-solidus temperatures.

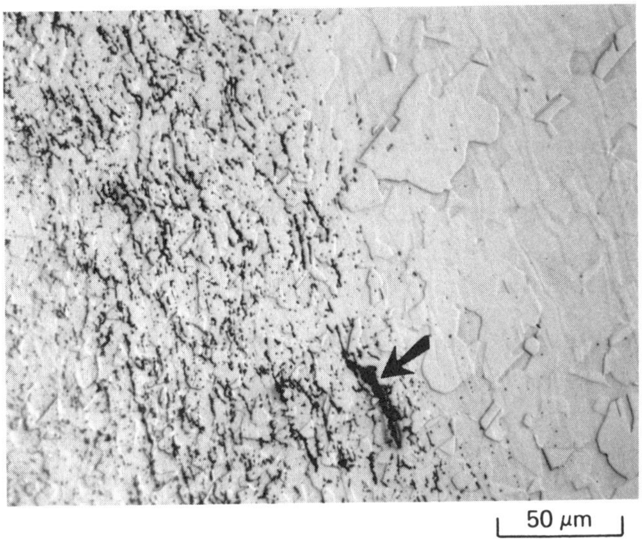

Figure 2. Microstructure in the "flash" region of a Type 304L inertia weld. Note the presence of a defect in the two-phase, austenite/ferrite region.

Examination of the bond region in Alloy 21-9 revealed a wide range of ferrite distribution and morphology. A narrow region extending radially along the bondline from the center to the outside diameter of the weld exhibited an extremely fine dispersion of ferrite, as shown in Figure 3A. The fine dispersion and large volume fraction of ferrite disguised the extent of austenite grain refinement. Outside this region a bimodal distribution of ferrite was evident, as illustrated by the micrograph in Figure 3B. Since the 21-9 parent material contained approximately 8 volume percent ferrite, it is likely that the larger ferrite particles in Figure 3B were remnants of the parent metal ferrite. The finer ferrite observed along the bond region may result from both fragmentation of the original ferrite and a solid-state transformation from the austenite at elevated temperatures along the bond line.

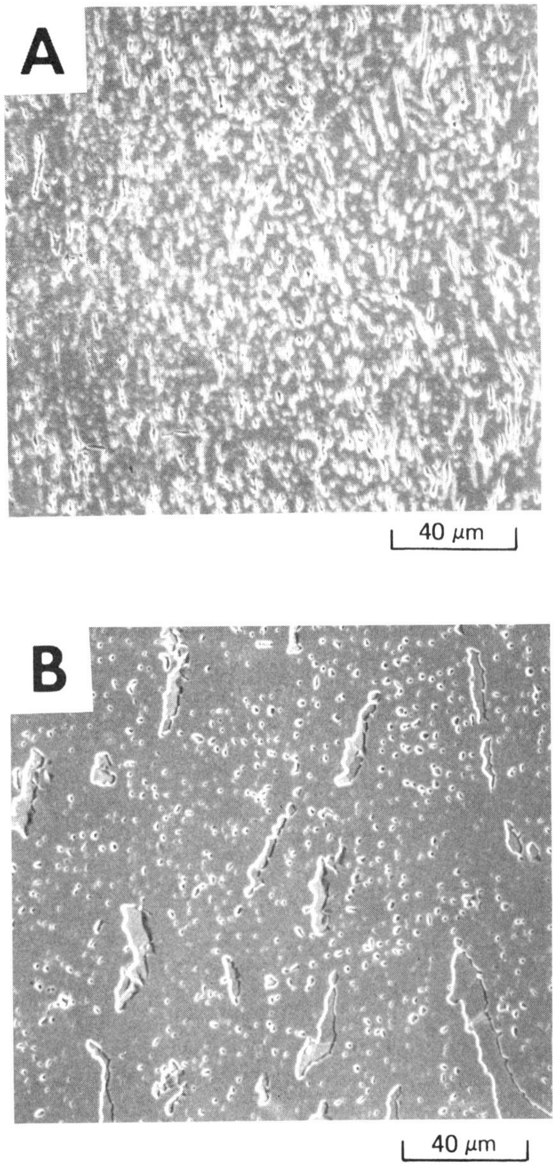

Figure 3. Microstructure of an Alloy 21-9 inertia weld, A) along the bondline, B) adjacent to the bondline. SEM micrographs.

Examination of the flash region in the 21-9 welds revealed an increase in the volume fraction of the ferrite and corresponding change in ferrite morphology relative to the bond region. A large portion of the weld flash exhibited a nearly continuous network of ferrite situated at austenite grain boundaries, as shown in Figure 4A. In the region of the flash which was extruded directly from the bond line (i.e. the region at the highest temperature), the microstructure consists of a ferrite matrix with continuous grain boundary austenite and Widmanstatten sideplates (Figure 4B). The presence of this microstructure suggests that this region was fully ferritic at some time during the welding process.

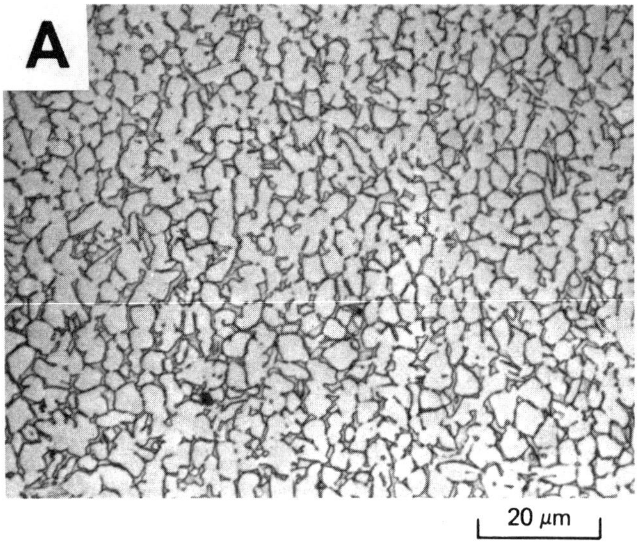

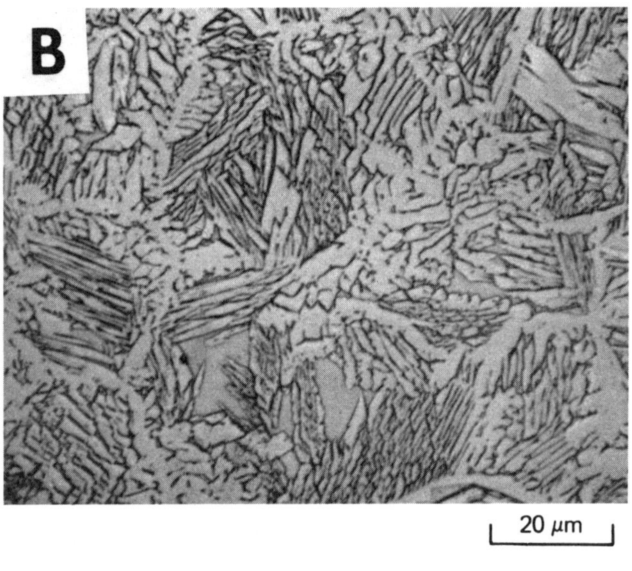

Figure 4. Microstructure in the "flash" region of a Alloy 21-9 inertia weld, A) ferrite along austenite grain boundaries, B) Widmanstatten austenite in a ferrite matrix. Ferrite is the dark-etching constituent.

Microhardness Traverse. A microhardness traverse along the bond line in a Type 304L weld is shown in Figure 5. Note that peak hardness occurs precisely at the center of the weld and decreases rapidly upon moving radially outward along the bondline. At approximately the mid-radius point, the hardness drops below the level of the parent material. This corresponds to the region of the bond line where grain coarsening was observed metallographically (Figure 1B).

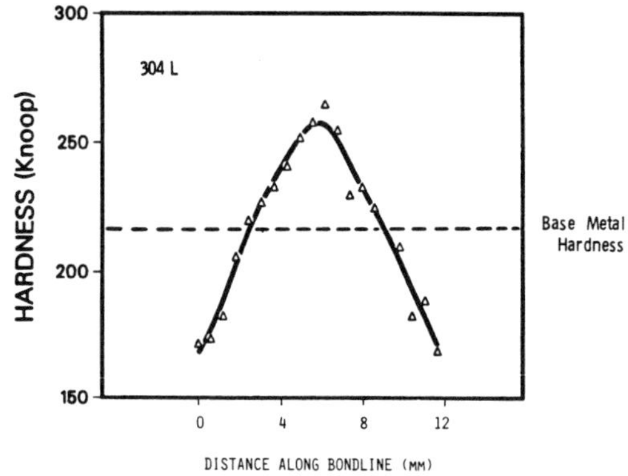

Figure 5. Microhardness traverse along the bondline in a Type 304L inertia weld. Note that bondline hardness exceeds base metal hardness in the center of the weld between the mid-radius points.

Microhardness surveys transverse to the bondline in the Type 304L inertia friction welds revealed a narrow (2 mm) band of increased hardness centered on the bondline at the weld center. At the mid-radius location, a softened region 1 mm in width was observed at the bondline. This region was bounded by a narrow band exhibiting increased hardness.

Hardness traverses both along and across the bondline in Alloy 21-9 were similar to those for the Type 304L inertia friction welds. However, the hardness gradient along the bondline was not as steep as that observed for Type 304L.

Mechanical Testing. The results of both the smooth and notched bar tensile tests are listed in Table 2. The use of notched tensile specimens served a dual purpose in that notching forced failure in the bond region and provided some indication of the bond properties as a function of location in the bond. Reference to Table 2 reveals that of the smooth bar tests, only the 21-9 sample which was hydrogen-charged failed in the bond region. All other smooth bar samples failed in the base material outside the bond region.

The large volume fraction of ferrite along the bond line in the 21-9 was

TABLE 2

TENSILE PROPERTIES OF INERTIA FRICTION WELDS

Material	Smooth Bar			Notched Bar(1)	
	YS(MPa)	UTS(MPa)	Elong.(%)	Ratio A	Ratio B
304L (air)	287	607	>50*	1.61	2.14
304L (H_2)	362	695	44*	1.53	1.98
21-9 (air)	326	689	>50*	1.50	1.97
21-9 (H_2)	413	732	19	1.38	1.77

* failure outside bond region
(1) bars notched at bondline, ratio = UTS(notched)/UTS(smooth)
notch depth A = 0.825 mm, notch depth B = 1.23 mm

most likely the cause for the bondline failure in the hydrogen charged smooth bar tensile sample. Similar ferrite-induced ductility losses have been previously observed in fusion welds in austenitic stainless steels containing up to 10 volume percent ferrite (Ref. 9). The slight increase in the yield strength of the charged samples relative to the uncharged samples has been attributed to solid solution strengthening (Ref. 9,10).

Notched bar tests were used to discriminate bond strength variations radially along the bond line. The increase in the notch/smooth bar ratio as notch depth increases is typical behavior in ductile materials and indicates that the bond strength is uniform or increases as a function of radial distance from the outer diameter. The increase in hardness near the center of the bond (Figure 5) suggests that the increase in ratio is due to bondline strengthening near the center of the bar.

Fractography. The scanning electron microscope (SEM) was employed to characterize the bondline failure mode. A macrofractograph of a notched Type 304L tensile bar is shown in Figure 6A. Note the spiral pattern on the fracture surface indicative of the metal flow pattern during inertia friction welding (Ref. 8). At higher magnification (Figure 6B), the fracture surface exhibits elongated dimples oriented parallel to the direction of metal flow. The relatively few particles observed on the fracture surface were identified as manganese sulfides. There was no evidence of ferrite on the fracture surface.

A macrofractograph of an Alloy 21-9 notched tensile bar is shown in Figure 7A. This fracture surface appears macroscopically flatter than the Type 304L fracture surface and does not manifest the metal flow pattern evident in Figure 6A. At higher magnification (Figure 7B), the dimples are larger and more uniform than those shown previously in Figure 6B. Small particles frequently associated with the dimples in Alloy 21-9 were enriched in chromium and undoubtedly correspond to the fine ferrite observed metallographically (Figure 2A).

Transmission Electron Microscopy(TEM). TEM analysis of thin foil samples removed from the bond region in the Type 304L inertia friction welds revealed a highly dislocated structure with a tendency for dislocation cell formation. No evidence of either ferrite or intermetallic particles (such as carbides) was found in the Type 304L welds (Ref. 12). TEM analysis of the Alloy 21-9 welds revealed a similar dislocation structure interspersed with small ferrite particles. It was difficult to determine the nature of this ferrite, i.e. forming via a solid-state transformation or by fragmentation. Carbides were not detected in 21-9 welds.

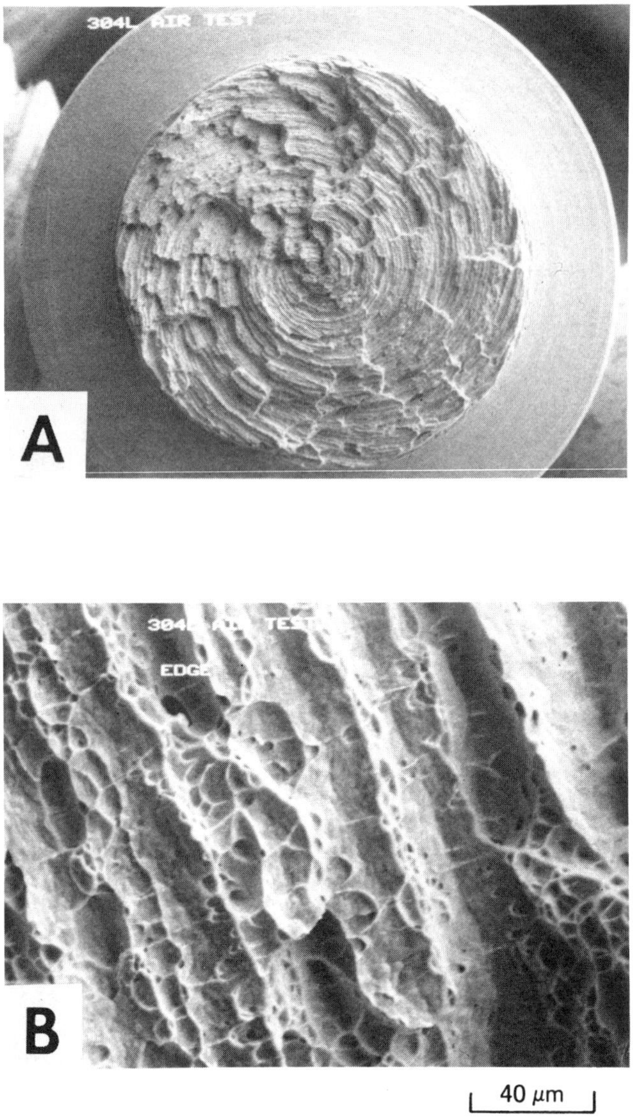

Figure 6. Fracture morphology of a Type 304L inertia weld tensile bar notched along the bondline, A) low magnification, B) high magnification showing elongated dimples.

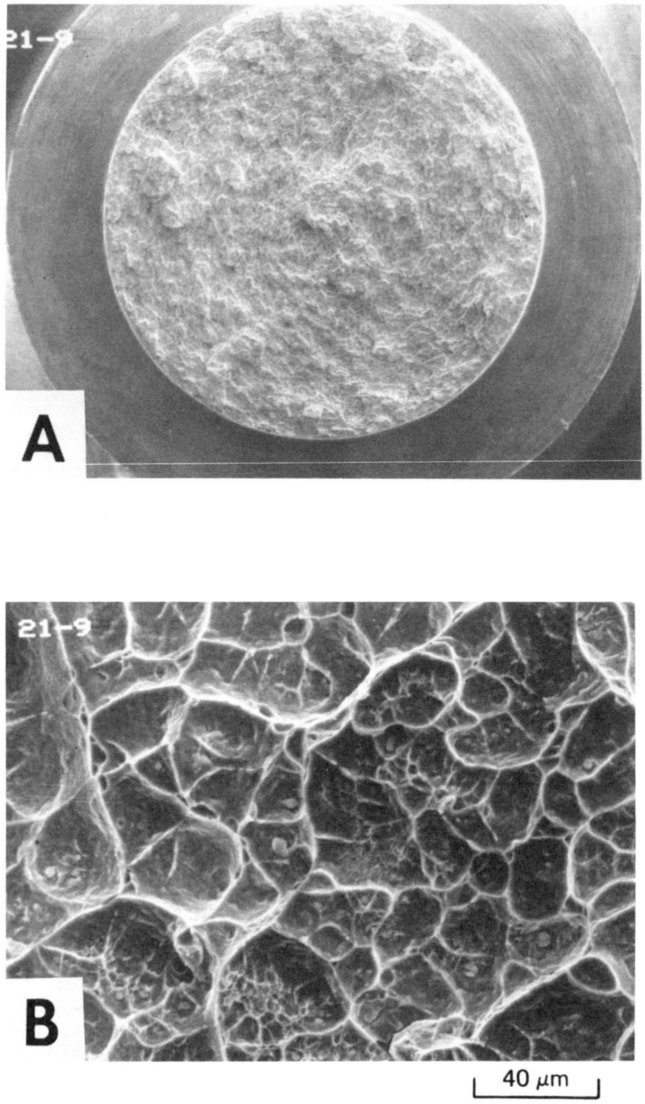

Figure 7. Fracture morphology of Alloy 21-9 inertia weld tensile bar notched along the bondline, A) low magnification, B) high magnification showing ferrite particles associated with the dimples.

DISCUSSION

High temperature deformation along the bondline in austenitic stainless steel inertia friction welds results in a number of metallurgical phenomena including dynamic recrystallization, grain growth, particle and phase dissolution, and phase transformations. All of these reactions may in some way influence the properties of the weld region.

Dynamic recrystallization results in refinement of the austenite grains (Figure 1A) and a subsequent increase in the hardness (Figure 5) and strength of the bond region. In the outer portion of the bond, the time duration at elevated temperature is greater (Ref. 11) and growth of the recrystallized grains (Figure 1B) softens this microstructure relative to the parent material. Despite this localized softening, the overall bond region remains stronger than the annealed parent material, as evidenced by the smooth bar tensile results reported in

Table 2.

The presence of ferrite along the bondline in austenitic stainless steel inertia friction welds influences both the mechanical properties and fracture behavior of these welds. The ferrite results from either a solid-state transformation or by fragmentation of parent metal ferrite. Ferrite formation via a solid-state reaction can be rationalized with the aid of an Fe-Cr-Ni pseudo-binary phase diagram, as shown in Figure 8.

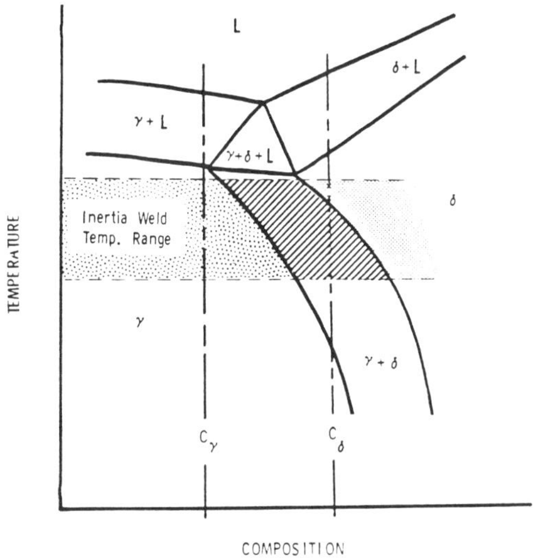

Figure 8. Fe-Cr-Ni pseudo-binary phase diagram with inertia welding temperature range superimposed.

Within the inertia weld temperature range, material at or adjacent to the bondline flows plastically under the applied axial load. Alloys with a low ferrite potential, such as C_γ in Figure 8, remain fully austenitic upon heating into this temperature range. As a result, no ferrite would be expected to form along the bondline. At higher ferrite potentials, such as C_δ , the bondline temperature lies within the two phase field and a transformation from austenite to ferrite would be expected. This composition would closely approximate the behavior of Alloy 21-9 during inertia friction welding and explains both the ferrite formation along the bondline (Figure 3A) and the various ferrite morphologies in the weld flash (Figure 4).

The Type 304L inertia friction weld behavior would be approximated by a composition midway between C_γ and C_δ in Figure 8. Note that in this case ferrite would only be expected to form in regions of the weld heated to sub-solidus temperatures, such as in the flash extruded along the bondline. The fine dispersion of ferrite observed in the flash of Type 304L welds (Figure 2) is indicative of this transformation behavior in low ferrite potential welds.

Severe deformation concomitant with temperature excursions into the two phase region accelerate the rate of transformation relative to strictly isothermal behavior. Plastic flow in the narrow band along the bondline enhances diffusional transport, thereby promoting the normally sluggish transformation of austenite to ferrite.

CONCLUSIONS

1. Inertia friction welds in austenitic stainless steels exhibit a weld region consisting of extremely fine austenite grains resulting from dynamic recrystallization along the bondline.

2. The microstructure and properties of the weld vary locally along the bondline. Grain refinement, hardness, and strength are maximum in the center of welds between solid bars.

3. Ferrite in austenitic stainless steel welds results from both a solid-state phase transformation and fragmentation of parent material ferrite.

4. Ferrite along the bondline influences both the mechanical properties and fracture behavior of the weld region.

ACKNOWLEDGEMENTS

Financial support for this investigation was provided by the U.S. Department of Energy, DOE, under Contract No. DE-AC04-76DP00789.

REFERENCES

1. T.L. Oberle, 1970, Inertia Welding, Metal Construction, 2(5), 193-195.
2. Anon., 1979, Inertia welding: simple in principle and application, Welding and Metal Fabrication, 47(8), 585-590.
3. R.A. Bell, J.C. Lippold, and D.R. Adolphson, 1984, An Evaluation of Copper-Stainless Steel Inertia Friction Welds, Welding Journal,

63(11), 325s-332s.

4. W.A. Baeslack, D.J. Duquette, and W.F. Savage, 1979, Effect of Ferrite on Stress Corrosion Cracking in Duplex Stainless Steel Weld Metals at Room Temperature, Corrosion, 34, p. 46.
5. E.R. Szumachowski and H.F. Reid, 1978, Cryogenic Toughness of SMA Austenitic Stainless Steel Weld Metals: Part 1 - Role of Ferrite, Welding Journal, 57(11), p. 325s.
6. A.L. Schaeffler, 1949, Metals Progress, 56, p. 680.
7. W.T. Delong, 1974, Ferrite in Austenitic Stainless Steel Weld Metal, Welding Journal, 53(7), 273s-286s.
8. J.C. Lippold and B.C. Odegard, 1984, Technical Note: Microstructural Evolution During Inertia Friction Welding of Austenitic Stainless Steels, Welding Journal, 63(1), 35s-38s.
9. J.A. Brooks and A.J. West, 1981, Hydrogen Induced Ductility Losses in Austenitic Stainless Steel Welds, Met. Trans., 12A, 213-223.
10. A.J. West and J.H. Holbrook, 1980, Hydrogen in Austenitic Stainless Steels: Effect of Phase Transformations and Stress State, AIME Conference Proc. Hydrogen Effects in Metals, Jackson Lake Lodge, Wyoming.
11. K.K. Wang and P. Nagappan, 1970, Transient Temperature Distribution in Inertia Welding of Steels, Welding Journal, 49(9), 419s-426s.
12. Unpublished research performed by F.A. Greulich, Sandia National Laboratories, Livermore, CA. 1986.

EFFECTS OF ALUMINUM AND TITANIUM ON THE MICROSTRUCTURE AND PROPERTIES OF MICROALLOYED STEEL WELD METAL

J.K. Brownlee, D.K. Matlock, G.R. Edwards
Center for Welding Research
Department of Metallurgical Engineering
Colorado School of Mines
Golden, Colorado

ABSTRACT

A series of submerged arc welds with systematic variations in aluminum and titanium contents were prepared at a constant heat input on ASTM A656 base plate. Standard quantitative metallographic techniques were used to evaluate the weld metal microconstituents, and transmission electron microscopy of extraction replicas was used to evaluate the weld metal inclusions. The results showed that with aluminum as the primary deoxidant a critical aluminum concentration of approximately 0.025 wt. pct. was required to produce a maximum in the acicular ferrite content and a minimum fracture transition temperature. The maximum in acicular ferrite content correlated with a maximum in the density of inclusions with diameters greater than 0.5μm. The primary role of titanium was to participate in the processes which minimize grain boundary ferrite.

INTRODUCTION

The increased use of high strength low alloy (HSLA) steels has required improved weld metals with mechanical properties equivalent to those of the base material. The welding consumables which have been developed generally depend on balanced additions of hardenability agents, complex deoxidants, and microalloying elements for their final properties. Microalloyed weld metals derive their best final properties from a refined ferrite microstructural component known as acicular ferrite.

Acicular ferrite formation depends on microalloy/deoxidant additions, most notably aluminum and titanium, and is influenced by other factors such as flux basicity and hardenability (1). However, the mechanisms by which aluminum and titanium affect weld metal microstructure remain unclear. This investigation examines the effects of systematic additions of aluminum and titanium to submerged-arc weld metal with emphasis on the relationships between weld metal deoxidation and microalloying practice, microstructural development, and mechanical properties.

EXPERIMENTAL PROCEDURES

Nineteen submerged arc bead-on-plate welds were made using a series of experimental welding wires with nominal compositions, in weight percent, of 0.1C, 1.3Mn, 0.04Si, 0.32Mo, 0.013B and with aluminum and titanium concentrations between 0.02 to 0.08 and 0.00 to 0.04 weight percent respectively (2). All welds were prepared on 19mm thick ASTM A656 base plate using one commercial basic flux, designated OP in this study, at a constant heat input of 3.5 MJ/m. The base plate composition in weight percent, was 0.14C, 1.52Mn, 0.023P, 0.004S, 0.23Si, 0.081V, 0.029Nb, and 0.028Al. The nominal flux composition in weight percent was 10.7SiO_2, 17.3Al_2O_3, 31.7MgO, 6.6CaO, 24.1CaF_2, 1.1MnO, 0.9TiO_2, 0.8Na_2O, 1.7Fe_2O_3, and 0.35C.

The changes in weld metal microstructure and properties associated with variations in the weld metal aluminum and titanium concentrations were quantified with light microscopy, transmission electron microscopy, and charpy V-notch impact testing with sub-size (5mm thick) samples.

RESULTS

The chemical compositions of the experimental welds are summarized in Table I. Table II summarizes the results of quantitative metallography and shows that wide variations in weld metal microstructures were obtained. The welds consisted mainly of acicular ferrite (AF) and grain boundary ferrite (GBF) with welds 4/05, 8/0, and 8/4 containing significant amounts (10-28 percent) of ferrite with aligned carbide (FAC). Fractions of Widmanstatten ferrite (WF) and polygonal ferrite (PF) are also included in Table II. All samples of weld metal exhibited hardnesses between 94 and 98 Rockwell B.

TABLE I: Summary of Al/Ti Matrix Weld Metal Chemical Compositions, in Weight Percent

Weld Name	C	Mn	Si	P	S	Cr	Ni	Mo	B	V	Nb	Cu	O	N	Ti	Al
Al/Ti 2/0	0.082	1.41	0.190	0.017	0.008	0.037	0.040	0.120	0.002	0.055	0.019	0.018	0.032	0.0089	0.003	0.019
Al/Ti 2/05	0.090	1.45	0.200	0.017	0.008	0.035	0.031	0.110	0.002	0.058	0.021	0.014	0.032	0.0091	0.007	0.023
Al/Ti 2/1	0.088	1.49	0.190	0.017	0.008	0.036	0.042	0.130	0.002	0.058	0.021	0.014	0.032	0.0110	0.006	0.022
Al/Ti 2/2	0.085	1.44	0.210	0.017	0.008	0.038	0.036	0.110	0.002	0.058	0.021	0.014	0.031	0.0100	0.012	0.022
Al/Ti 2/4	0.083	1.43	0.210	0.017	0.008	0.036	0.034	0.110	0.002	0.056	0.020	0.018	0.032	0.0089	0.019	0.022
Al/Ti 3/0	0.085	1.69	0.180	0.019	0.008	0.038	0.056	0.200	0.004	0.058	0.021	0.018	0.044	0.0240	0.002	0.023
Al/Ti 3/2	0.082	1.75	0.180	0.018	0.009	0.038	0.063	0.220	0.003	0.053	0.019	0.021	0.045	0.0360	0.005	0.021
Al/Ti 3/4	0.080	1.91	0.180	0.018	0.008	0.039	0.084	0.280	0.004	0.054	0.020	0.020	0.040	0.0350	0.008	0.023
Al/Ti 4/05	0.090	1.46	0.200	0.017	0.008	0.035	0.043	0.120	0.003	0.056	0.020	0.014	0.031	0.0087	0.003	0.026
Al/Ti 4/1	0.088	1.43	0.200	0.017	0.008	0.036	0.033	0.110	0.002	0.058	0.021	0.014	0.033	0.0089	0.009	0.029
Al/Ti 4/2	0.091	0.46	0.210	0.017	0.008	0.035	0.040	0.120	0.002	0.058	0.022	0.014	0.030	0.0092	0.018	0.032
Al/Ti 4/4	0.091	1.46	0.210	0.017	0.008	0.035	0.033	0.110	0.002	0.058	0.021	0.014	0.032	0.0098	0.025	0.027
Al/Ti 6/0	0.088	1.45	0.190	0.018	0.008	0.038	0.031	0.100	0.002	0.058	0.020	0.018	0.040	0.0086	0.006	0.029
Al/Ti 6/1	0.086	1.39	0.180	0.018	0.008	0.038	0.030	0.093	0.001	0.058	0.021	0.018	0.042	0.0085	0.007	0.022
Al/Ti 6/2	0.083	1.38	0.180	0.017	0.008	0.038	0.029	0.100	0.001	0.057	0.019	0.018	0.042	0.0086	0.010	0.020
Al/Ti 8/0	0.090	1.42	0.210	0.017	0.008	0.034	0.034	0.100	0.002	0.057	0.020	0.018	0.031	0.0091	0.003	0.041
Al/Ti 8/05	0.092	1.42	0.180	0.017	0.008	0.038	0.037	0.110	0.002	0.057	0.020	0.018	0.043	0.0087	0.004	0.026
Al/Ti 8/1	0.083	1.39	0.160	0.017	0.008	0.038	0.047	0.130	0.001	0.054	0.018	0.019	0.037	0.0086	0.002	0.018
Al/Ti 8/4	0.087	1.43	0.220	0.017	0.008	0.036	0.033	0.110	0.002	0.059	0.022	0.018	0.027	0.0088	0.026	0.045

TABLE II: Quantitative Metallographic Measurements in Volume Percent

Weld	GBF	WF	AF	PF	FAC
Al/Ti 2/0	21.2	5.6	68.6	4.2	0.4
Al/Ti 2/05	19.0	4.0	73.6	3.0	0.2
Al/Ti 2/1	15.8	5.4	77.4	1.4	0.0
Al/Ti 2/2	16.7	2.9	78.3	2.1	0.0
Al/Ti 2/4	10.0	4.6	83.8	1.6	0.0
Al/Ti 3/0	8.0	17.4	70.6	1.2	2.8
Al/Ti 3/2	6.8	9.8	82.0	1.0	0.4
Al/Ti 3/4	7.2	5.4	86.8	0.6	0.00
Al/Ti 4/05	16.4	12.0	58.8	3.2	9.6
Al/Ti 4/1	20.8	6.0	71.5	1.3	0.4
Al/Ti 4/2	15.6	5.6	77.8	1.0	0.0
Al/Ti 4/4	11.6	7.4	80.4	0.6	0.00
Al/Ti 6/0	25.5	5.6	64.3	1.6	3.0
Al/Ti 6/1	18.6	6.6	73.0	0.8	1.0
Al/Ti 6/2	19.8	8.6	69.0	1.6	1.0
Al/Ti 8/0	16.7	16.1	51.1	1.2	14.9
Al/Ti 8/05	14.0	6.2	76.2	3.2	0.4
Al/Ti 8/1	15.0	15.6	64.8	2.4	2.2
Al/Ti 8/4	4.2	18.8	42.8	6.4	27.8

Figure 1 shows the volume percent AF plotted as a function of the weld metal aluminum concentration, and indicates that the AF content was maximized near an aluminum concentration of 0.025 weight percent. From Table II, the decrease in AF above 0.025 weight percent aluminum is shown to be accompanied by a simultaneous increase in the volume fraction of FAC. The volume fraction of GBF was strongly influenced by the weld metal titanium and manganese concentrations. Figure 2 shows that additions of titanium or manganese caused the amount of ferrite veining to decrease. The data separate according to manganese content as summarized in Table I. The 1.45% Mn band shown on Fig. 2 presents microstructures for samples between 1.39 and 1.49 wt. pct. Mn while the 1.75% Mn band presents results for samples between 1.69 and 1.91 wt. pct. Mn. As discussed in an earlier paper (1), the effect of manganese shown in Fig. 2 and Table II is consistent with previous observations which have demonstrated that Mn lowers the transformation temperature of austenite to ferrite and enhances acicular ferrite formation at the expense of grain boundary ferrite.

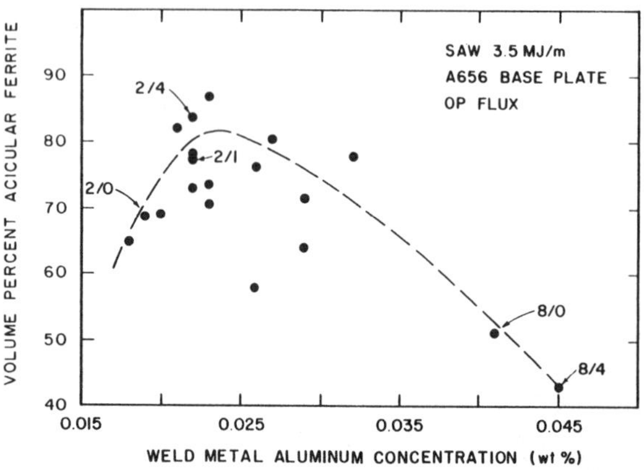

Fig. 1. - Summary plot of the variation in acicular ferrite as a function of weld metal aluminum concentration.

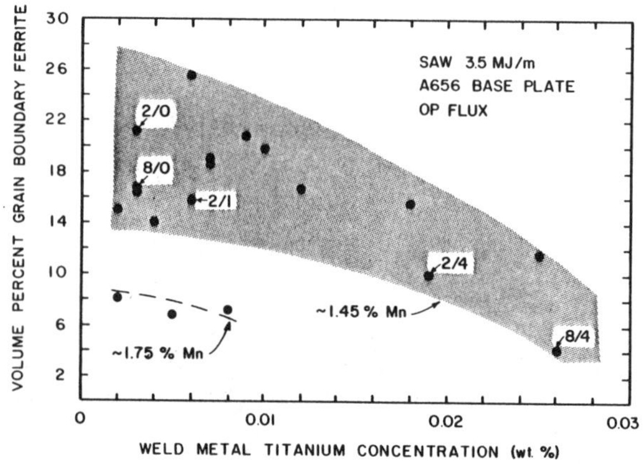

Fig. 2 - Summary plot of the variation in grain boundary ferrite as a function of the weld metal titanium concentration.

Weld metal microstructures, representative of the data in Figs. 1 and 2, are presented in Figs. 3 and 4. In Fig. 3 the effects of aluminum are shown for weld metal centerline micrographs of welds 2/0, 2/4, 8/0 and 8/4. Note that the data for these welds are identified in Figs. 1 and 2 and represent the complete range in microstructures. Comparison of the low and high aluminum welds in Fig. 3 show that increasing the aluminum concentration caused the primary intragranular transformation product to change from AF to FAC. Figure 4 illustrates the effect of titanium in welds 2/0 and 2/4 along with an intermediate titanium concentration weld, 2/1. The low magnification micrographs in Fig. 4 show that the main effect of increasing the titanium concentration at both low and high aluminum concentration was to reduce the fraction of ferrite veining while having little effect on the morphology of the intragranular transformation products.

The Charpy V-notch impact data, presented as percent ductile fracture versus test temperature for the four welds considered in Fig. 3 are shown in Fig. 5. Consistent with the variations in microstructure discussed above, the curves in Fig. 5 show that increasing the AF volume fraction and/or decreasing the GBF decreased the 50% fracture appearance transition temperature (FATT, i.e. temperature corresponding to 50% ductile fracture) and thereby increased the weld metal toughness. It is significant to note that while weld 8/4 had less AF than weld 8/0, the reduced GBF content of weld 8/4 caused it to exhibit a 50% FATT lower than that for weld 8/0. Thus, the results shown in Fig. 5 confirm the importance of refining all microstructural components when optimal toughness is required.

DISCUSSION

Aluminum Effects on Weld Metal Microstructures

The effects of aluminum in this study were quantified by correlating the observed changes in the weld metal inclusion distributions with the aluminum content in welds with an essentially constant oxygen content of 320 ppm. Six welds were selected for analysis. These include the four welds discussed in Fig. 3 along with welds 4/4 and 6/1 (a weld with 420 ppm oxygen). The results are summarized in Table III. The number of inclusions per unit volume with diameters greater than 0.5 microns exhibited a maximum near an aluminum content of 0.022 wt. pct., as shown in Fig. 6. The corresponding effect of inclusion distribution on the weld metal microstructure is shown in

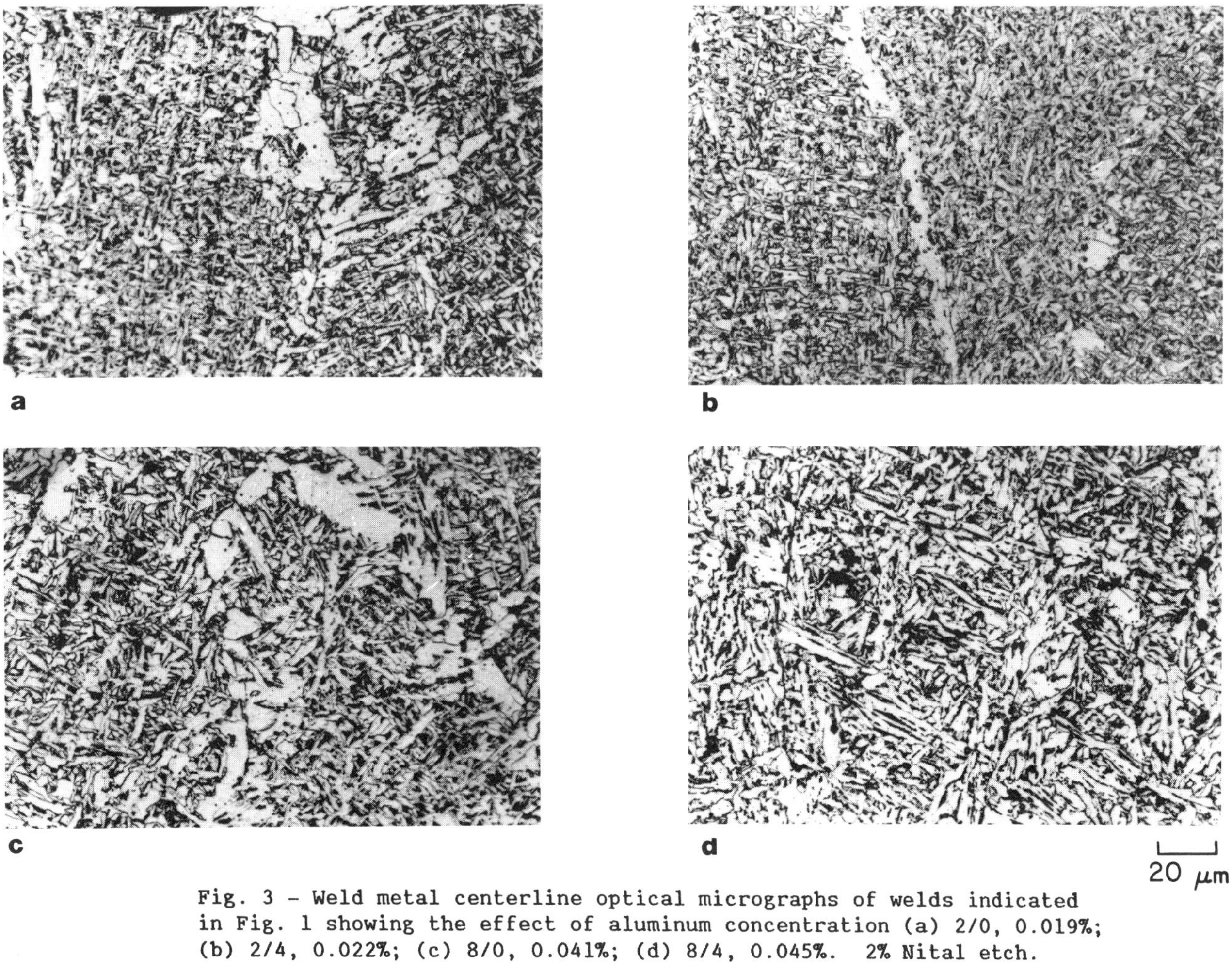

Fig. 3 - Weld metal centerline optical micrographs of welds indicated in Fig. 1 showing the effect of aluminum concentration (a) 2/0, 0.019%; (b) 2/4, 0.022%; (c) 8/0, 0.041%; (d) 8/4, 0.045%. 2% Nital etch.

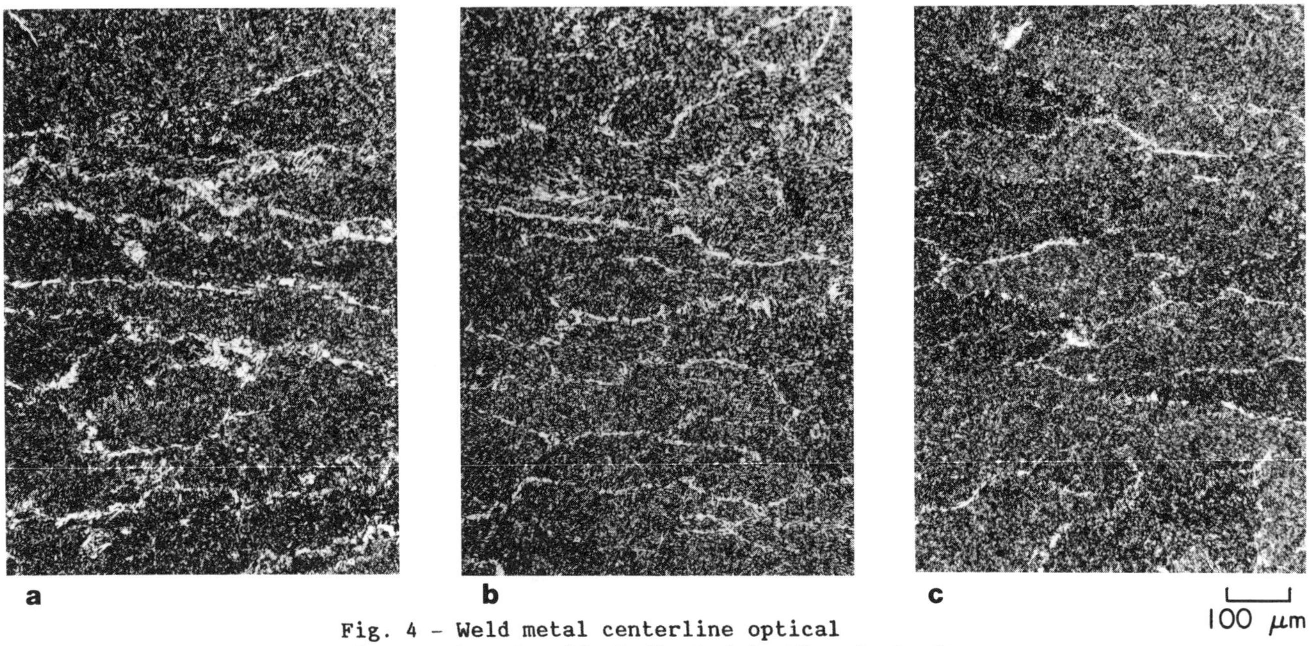

Fig. 4 - Weld metal centerline optical micrographs of welds indicated in Fig. 2 showing the effects of titanium concentration. (a) 2/0, (b) 2/1, (c) 2/4. 2% Nital etch.

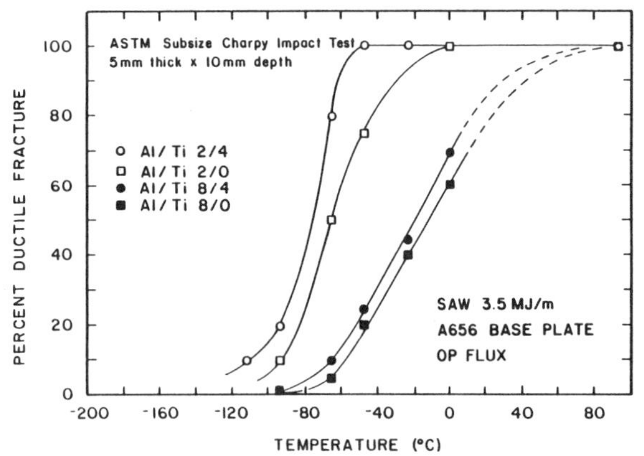

Fig. 5 - Fracture appearance transition behavior for welds shown in Fig. 3.

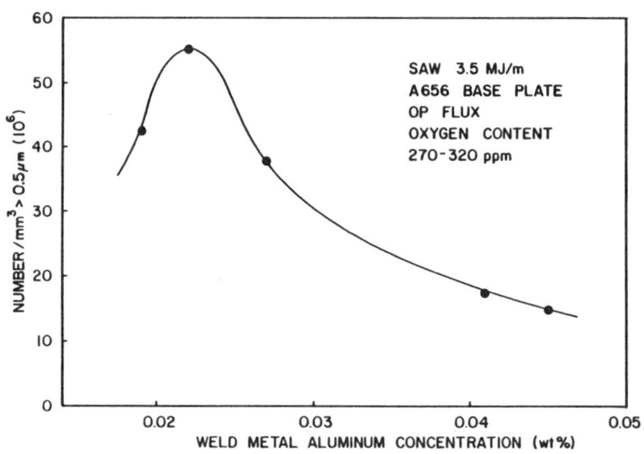

Fig. 6 - The effect of aluminum content on the number of inclusion with diameters greater than 0.5 μm.

TABLE III: Summary of Inclusion Distribution Data Obtained From Transmission Electron Micrographs of Extraction Replicas

Weld	Av. Dia. (μm)	Mode (μm)	Nv (#/mm^3)	Nv>.0.5μm (#/mm^3)	Al (w/o)	Ti (w/o)	O w/o)
Al/Ti 2/0	0.307	0.350	3.130E+08	4.23E+07	0.019	0.003	0.032
Al/Ti 2/4	0.352	0.350	2.870E+08	5.52E+07	0.022	0.019	0.032
Al/Ti 4/4	0.373	0.350	1.690E+08	3.78E+07	0.027	0.025	0.032
Al/Ti6/1	0.227	0.50	4.600E+08	4.51E+07	0.022	0.07	0.042
Al/Ti 8/0	0.352	0.150	1.250E+08	1.72E+07	0.041	0.003	0.031
Al/Ti 8/4	0.358	0.350	9.450E+08	1.50E+07	0.045	0.026	0.027

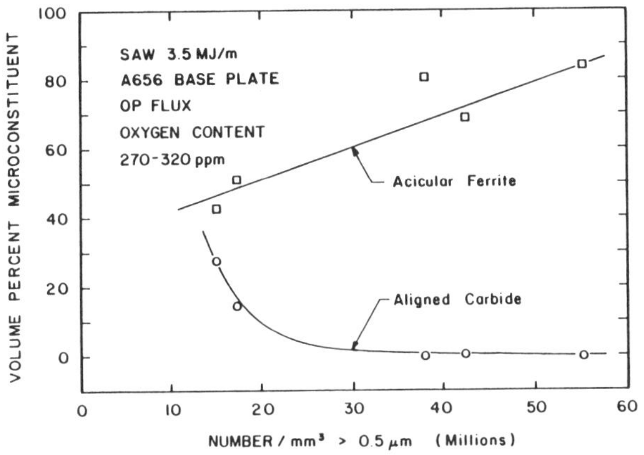

Fig. 7 - Summary plot of the effect of the weld metal inclusion distribution on the volume fractions of acicular ferrite and aligned carbide in the final microstructure.

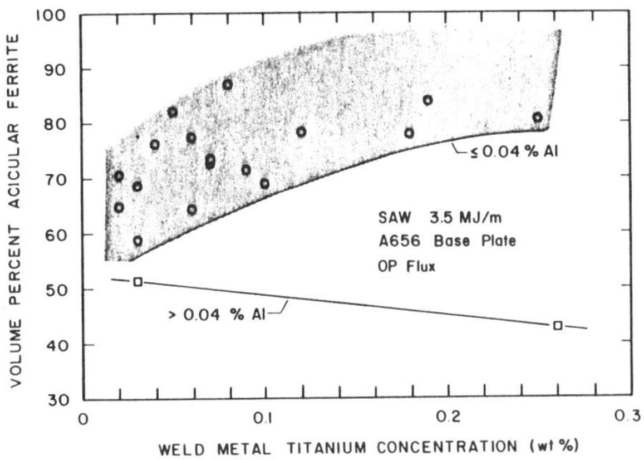

Fig. 8 - The volume percentage of grain boundary ferrite plotted as a function of the weld metal titanium concentration. Note that the effect of Ti changes depending on the Al concentration.

Fig. 7, a plot of the volume percent of intragranular microconstituents versus the number of large inclusions per unit volume. Data presented in Fig. 7 are for the five welds in Table III with 270-320ppm oxygen. Increasing the number of large inclusions caused the AF volume fraction to increase, and the FAC volume fraction to decrease. This result is in general agreement with the theory of Ricks, Barrite and Howell (3) and clearly shows that high AF concentrations are favored by large numbers of critically sized inclusions. Large inclusions have been shown to control acicular ferrite volume fraction primarily by limiting the volume fraction of grain boundary ferrite through control of the prior austenite grain size (4). Specifically, large prior austenite grain sizes correlate with small volume fractions of grain boundary ferrite. The relationship between the Al:O ratio and the size distribution of weld metal inclusions is not fully understood but it is thought to be related to the role of aluminum in the deoxidation process and to the kinetics of precipitation of oxides from the weld pool (1).

Titanium Effects on Weld Metal MIcrostructure

Numerous researchers (5,6) have proposed that boron is effective in reducing the volume fraction of GBF in weld metals by segregating at grain boundaries and retarding the formation of ferrite allotriomorphs. Mori, et. al. (5) have proposed that titanium additions are important in protecting the boron solute from nitrogen. Thus, according to this model, a decrease in ferrite veining depends primarily on the nitrogen gettering capability of titanium.

Terashima and Hart (6) showed that increasing the titanium content of SA welding consumables caused the titanium content of weld metal inclusions to rise, and others (7-11) have suggested that the presence of titanium in weld metal inclusions may enhance the nucleation of AF on those sites. Terashima and Hart (6) suggested further that such a mechanism could only be operative when the overall level of deoxidation in the weld metal was such that titanium, which has less affinity for oxygen than aluminum, could enter the deoxidation process. Figure 8, a plot of the AF volume percent as a function of the weld metal titanium concentration, supports the latter contention by showing that the effect of increasing titanium concentration at high aluminum concentrations was to slightly decrease the volume fraction of acicular ferrite in the weld metal. If the effect of the titanium on AF nucleation was independent of the weld metal deoxidation process, the results shown in Fig. 8 could not be rationalized. Therefore, it was concluded that while titanium in the absence of aluminum can have a direct beneficial effect on the nucleation of AF, its main effect in this investigation was to reduce the amount of ferrite veining through the Ti-B model proposed by Mori, et. al. (5).

CONCLUSIONS

The results of this investigation led to the following conclusions:

1. The density and size of nonmetallic inclusions in microalloyed steel weld metal was controlled by the concentration of aluminum when that element was the primary deoxidant.

2. The maximum acicular ferrite volume fraction was associated with a high density of inclusions in a critical size range, approximately 0.5 μm.

3. When both aluminum and boron were present in the microalloyed steel weld metals, the beneficial effects of titanium resulted from a reduction in the volume fraction of grain boundary ferrite.

ACKNOWLEDGEMENTS

The authors acknowledge the support of the Office of Naval Research, Contract N00014-83-K-0079, and the Bethlehem Steel Corporation.

REFERENCES

1. O. Grong and D.K. Matlock, Int. Metals Reviews, 31, No. 1, 27-48 (1986).

2. J.K. Brownlee, "Effects of Aluminum and Titanium on the Microstructure and Properties of Microalloyed Steel Weld Metal", M.S. Thesis No. T-3064, Colorado School of Mines (1985).

3. R.A. Ricks, G.S. Barrite and P.R. Howell, "Proc. Int. Conf. on Solid-State Transformations", TMS-AIME, Warrendale, PA, 463-468 (1983).

4. N.A. Fleck, O. Grong, G.R. Edwards, and D.K. Matlock, Welding Journal, 65, No. 5, 113s-121s (1986).

5. N. Mori, H. Homma, S. Ohkita, and M. Wakabayashi, "Mechanism of Notch Toughness Improvement in Ti-B Bearing Weld Metal", Int. Inst. of Welding, IIW Doc.-1196-81 (1981).

6. H. Terashima and P.H.M. Hart, "Proc. of Int. Conf. on the Effects of Residual Impurity, and Microalloying Elements on Weldability and Weld Properties", The Welding Institute, Cambridge, England, Paper #27 (1983).

7. N. Bailey, Welding Institute Research Report, No. 221/1983, 1-5 (1983).

8. R.J. Pargeter, Welding Institute Research Report No. 154/1981, 1-9 (1981).

9. A.H. Koukabi, T.H. North, and H.B. Bell, Metal Construction, 11, No. 12, 639-644 (1979).

10. T.H. North, H.B. Bell, A.H. Koukabi, and I. Craig, Welding Journal 58, No. 12, 343s-354s (1979).

11. H. Terashima and P.H.M. Hart, Welding Journal, 63, No. 6, 173s-183s (1984).

HYDROGEN INDUCED WORM-HOLE POROSITY IN STAINLESS STEEL WELD METAL

David G. Howden, Lucy S. Yen
Department of Welding Engineering
The Ohio State University
Columbus, Ohio

ABSTRACT

A new hypothesis is introduced for the mechanism of hydrogen induced worm hole porosity formation in stainless steel weld metals. The hypothesis is that hydrogen gas, required for pore propagation, enters the pore from solidified weld metal by diffusion rather than from the molten weld metal, as is popularly believed. The mechanism was tested experimentally and a simultaneous heat and mass flow analysis was used to show that mass flow rates of hydrogen into the pore were adequate to support pore growth at the welding speed. Experimental and theoretical determinations of hydrogen levels in the arc atmosphere for pore formation and the effect of welding variables on pore diameter were in fair agreement.

INTRODUCTION

HYDROGEN IS OFTEN ADDED to the argon shielding gas for the Gas Tungsten Arc Welding (GTAW) of austenitic stainless steels to increase heat input, weld penetration and productivity. In some cases, the weld develops gas porosity in the form of elongated pores (worm holes) which range in diameter up to about 2 mm and which extend along the weld to lengths of several centimeters. In the case of a full penetration weld in relatively thin sheet, the pores originate at the fusion line and tend to propagate perpendicular to the solidification front finally running along the weld centerline. They are characterized in Figure 1 which shows a longitudinal vertical section at the weld centerline and the corresponding radiograph of the weld prior to sectioning. This particular weld was made in 304 stainless steel using the GTAW process and a partial pressure of hydrogen in the arc of 0.15 atmosphere. It is interesting to note that there are multiple overlapping pores which, on the radiograph, appear to be just one.

A similar type of porosity has been observed in type 316 stainless steel (1) and in nickel and nickel alloys (2), although in the latter case a somewhat higher level of hydrogen partial pressure in the arc was needed (40 percent).

For a gas bubble to exist in the liquid metal, the gas pressure in the bubble must equal the combined pressures applied by the general atmosphere, that of the liquid surface energy and the "hydrostatic" head due to the depth of the bubble in the pool, Eq. (1).

$$P_b = P_{at} + 2\ T/r + \rho g h \qquad \text{Eq. 1.}$$

where;

P_b is the pressure in the bubble
P_{at} is the atmospheric pressure
T is the liquid surface tension
r is the radius of the bubble
ρ is the density of the liquid metal
g is the gravitational acceleration
h is the depth of the bubble beneath the surface of the pool

For a 1 mm diameter bubble in the depth of a weld pool, the surface tension and gravitational pressures are so small that they may be ignored.

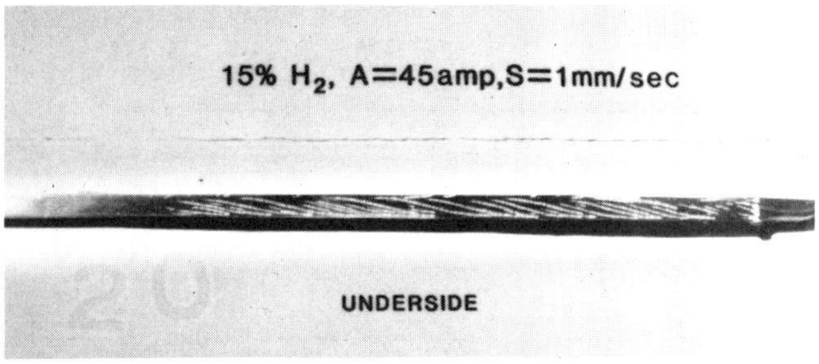

Fig. 1 - Longitudinal - vertical cross section of a stainless steel weld with gross worm hole porosity and the corresponding radiograph.

This means that the pressure of hydrogen in the bubble must be approximately equal to one atmosphere and that the surrounding metal must be supersaturated with hydrogen (with respect to the hydrogen pressure of 0.15 atmosphere in the arc) for the bubble to remain stable. Further, for the bubble to form a worm hole, each pore must grow at the same rate as the solidification front proceeds, being supplied with hydrogen to satisfy the rate of increase of volume. Traditionally, it has been assumed that the hydrogen is supplied by the diffusion of gas from the supersaturated liquid in the weld pool across the liquid/gas interface into the gas pore. To investigate this mechanism, Willgoss and Atthey (3) carried out a mathematical analysis of this kind of model (Figure 2), and, making some assumptions, including a supersaturation of hydrogen in the liquid metal, found good agreement for the occurrence of pores in AISI types 316 and 304 stainless steels. Chew and Willgoss (1) in their study of hydrogen absorption by weld pools, explained that the necessary supersaturation of hydrogen in the liquid metal could be produced by hydrogen rejection during the solidification process.

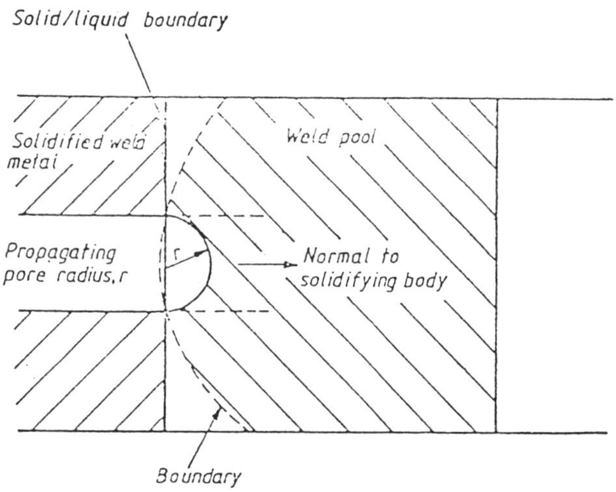

Fig. 2 - Diagram to illustrate the Willgoss and Atthey (1) model for the formation of worm hole porosity.

This hypothesis is not supported by other work. Howden and Milner (4) reported that the mean concentration of hydrogen in the molten weld pool of type 304 stainless steel was not grossly in excess of the amount predicted by Sievert's Law for a temperature slightly above the melting point. Further, Howden (5) found that hydrogen was not rejected on solidification according to hydrogen solubility relationships between liquid and solid. It was hypothesized that, on solidification, the hydrogen concentration in the liquid was retained at essentially the same level in the solid. Consequently, due to the change of solubility from liquid to solid, the solid metal becomes supersaturated with respect to the prevailing hydrogen pressure in the arc atmosphere. This leads to an effective increase in the internal solution pressure in the solidified metal which, theoretically, can exceed that necessary for sustaining a gas bubble. This hypothesis also implies that, for hydrogen bubbles to form and propagate, the hydrogen gas has sufficient mobility in the solid metal to provide the mass flow rate required for pore formation and growth at the welding speed. Also implied is that gas would in fact diffuse from the gas phase into the liquid of the weld pool; opposite to that of the earlier hypothesis.

Some evidence is available to support the idea that hydrogen from the solidified weld metal is influential in the formation of weld metal porosity. Both Ekerot et alia (6) and Pokhodnya and Demchenko (7) have reported that a nucleated pore can grow by diffusion of hydrogen from the solidified material near the pore. In neither case was a detailed analysis carried out, however.

The present work was designed to investigate the second hypothesis, namely, whether it is possible to provide sufficient hydrogen mass flow rate by diffusion of the gas from the solidified weld metal into the propagating pore to sustain its growth and to account for any additional loss of gas through the gas/liquid interface into the pool. The model is shown in Figure 3. The approach was to determine experimental welding conditions necessary for pore formation and growth and then to determine mass flow rates using a simple, simultaneous two dimensional heat flow and mass flow computer analysis. Only the steady state pore growth condition was considered at this time; the pore initiation and termination events were not investigated.

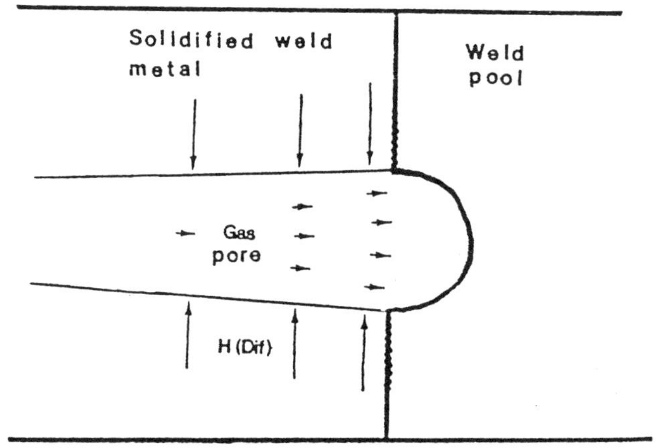

Fig. 3 - Diagram to illustrate the model used in the present investigation.

EXPERIMENTAL DETAILS

The experimental effort was directed to provide a check and calibration of the mathematical analysis to be explained later.

Full penetration welds were made with a square, butt joint configuration, in both AISI 304 stainless steel and nickel sheet of thickness 3 to 4 mm with no filler metal addition.

The GTAW process was used with argon shielding gas, to which additions of hydrogen up to 15 percent (by volume) were made. The shielding gas flow rate was 12 liters per minute. The same gas mixture as that in the arc was used for back up shielding. Travel speeds ranged up to 2.5 mm/sec and currents up to 100 amps with an arc length of 2 mm.

The base metal chemical compositions are given in Table 1.

Table 1 - Chemical Composition of the Base Metals used in the Investigation (wt. percent)

	AISI 304	NICKEL
Carbon	0.05	0.05
Manganese	1.96	0.24
Silicon	0.47	0.02
Phosphorus	0.025	-
Sulfur	0.008	0.001
Nickel	8.20	Rem.
Chromium	19.10	-
Molybdenum	0.26	-
Copper	0.34	0.01
Aluminum	0.003	

Each weld was radiographed to document the extent of worm hole porosity prior to sectioning of the weld longitudinally at the centerline for further examination and measurement of the pore dimensions. A sample was shown in Figure 1. Of particular interest to the investigation was the presence of pores and their diameters.

RESULTS

The experimental results may be broken down into two areas:

1) Characteristics of the worm hole pores
2) Effect of welding variables on pore dimensions

CHARACTERISTICS OF WORM HOLE PORES - Certain features of the pores can be seen in Figure 1. The pores are able to propagate for relatively large distances along the weld compared to the thickness of the sheet. They originate, at least in the full penetration welds produced in this study, at the fusion line adjacent to the base metal, follow perpendicularly to the solidification front to the centerline of the weld and progress along the weld. The pores behave in a manner similar to that of metal crystals which are suitably oriented with respect to the maximum temperature gradient that growth is unrestricted. The pores, of course, have no preferred growth direction and therefore propagate with ease. It was noticed that the pores oriented themselves at a an angle with respect to the horizontal centerline of the weld, sometimes diving to the root of the weld, sometimes rising towards the top, and, at othertimes, remaining parallel to the welding direction. The direction was dictated by the heat flow and solidification direction, as indicated by the columnar grain and sub structure. In a transverse section, the pores were observed to be round.

Of significance is the fact that, occassionally, two adjacent pores merged into one. At this point, there was a rapid decrease in pore diameter to a size similar to that of the original pores which merged (Figure 4). Apart from this merging event, termination of the pores normally took place near a free surface at the face or root of the weld.

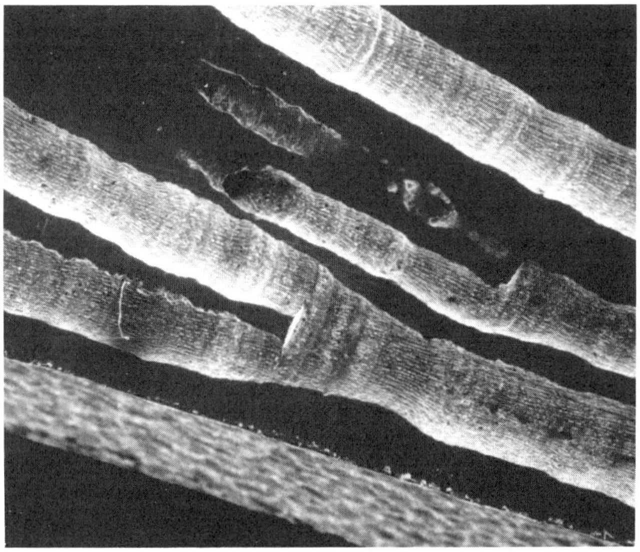

Fig. 4 - Enlarged view of the merging of two worm hole pores.

Close examination of the pore internal surface, showed that the surface was rippled in much the same way as the surface of a weld reflects the current oscillations of the arc in forming surface ripples. This indicates that the pore diameter is able to reflect rapid changes in weld solidification rate, as does the pool surface.

EFFECT OF WELDING VARIABLES - The major welding variables studied and their effect on porosity were:

Hydrogen partial pressure in the arc
Arc current
Travel speed

Most of the experimental work was carried out on the stainless steel because it was found not possible to create pores in the nickel with up to 15 percent hydrogen additions to the shielding gas.

Hydrogen Partial Pressure - Above a certain level, increase in hydrogen pressure increased both the quantity of pores and the diameter of the individual pores. The situation may be seen in several of the figures referenced in the following sections. The threshold pressure for pore formation was between 5 and 10 percent hydrogen in the arc.

Arc Current - For given travel speeds and hydrogen additions to the arc, the diameter of the worm hole pores increased with increasing current. The magnitude is shown in Figure 5.

Travel Speed - The pore diameter increased with travel speed for conditions of constant arc current and hydrogen level. The trend is shown in Figure 6.

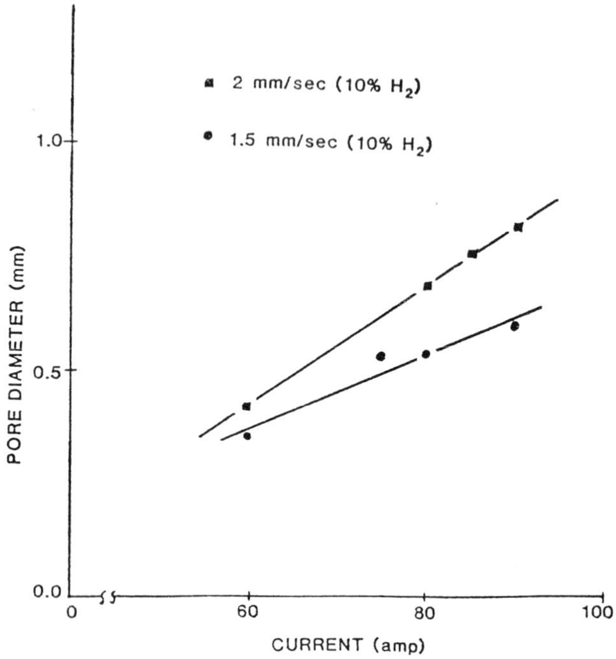

Fig. 5 - Relationship of pore diameter to arc current for different travel speeds and hydrogen levels.

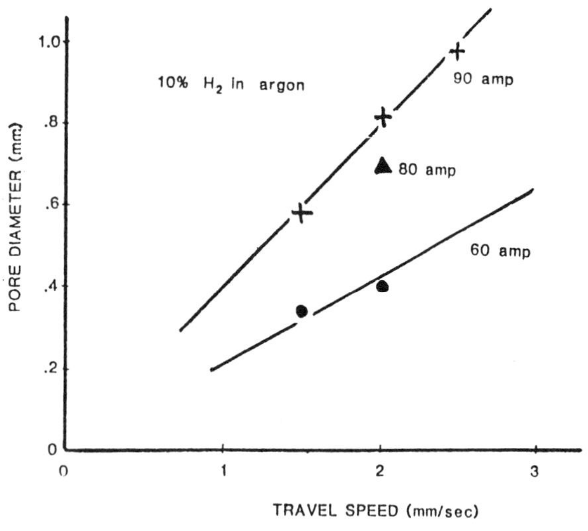

Fig. 6 - Relationship of pore diameter to travel speed for different arc currents and hydrogen levels.

MATHEMATICAL MODEL

In order to examine the mechanism of pore propagation, it was assumed that hydrogen gas, absorbed into the pool according to Sievert's relationship at a temperature of approximately 2000°C (4), was retained in the solidified weld metal on solidification (5). Due to the decreased solubility of hydrogen in the solid metal, the internal solution pressure of the gas increases for this constant concentration. This increased activity was assumed to be the driving force for pore formation and propagation with the hydrogen diffusing from the surrounding solid metal into the pore. For steady state propagation to occur, the rate of gas flow into the pore must equal the sum of that required to satisfy the volumetric increase rate and that amount of gas being lost through the gas/liquid interface into the pool (Figure 7). This may be expressed mathematically as follows,

Mass flow to pore from solid metal. ml/sec	Flow needed for pore growth. ml/sec	Mass flow to liquid. ml/sec

$$F_1 \cdot \pi \cdot d \cdot l = \pi d^2/4 \cdot s + \pi d^2/2 \cdot F_2 \qquad \text{Eq. 2}$$

and

$$d = \frac{4 \cdot F_1 \cdot l}{s + 2F_2} \qquad \text{Eq. 3}$$

where

F_1 is the flow rate from solid to gas phase per unit area
F_2 is the flow rate from gas to liquid phase per unit area
d is the diameter of the pore
s is the weld travel speed
l is the length of the pore providing the flow from solid to gas phases

C_L and C_S (Figure 7), the concentrations of hydrogen in the liquid and solid phases respectively, are assumed to be equal except that C_S decreases with distance (time) behind the solidification front due to diffusion of hydrogen into the surrounding solid metal and into the gas phase. Gas will diffuse into the pore only when the concentration of hydrogen in the solid metal is greater than the hydrogen solubility at the particular location (temperature). This condition, C_S (or C_L) > S_T, where S_T is the solubility of hydrogen in the solidified metal, defines the the minimum condition for the stability of the pore. As shown in Figure 8, only when C_S is greater than S_T will a positive gradient occur for hydrogen diffusion to the pore. It should be noted that both C_S and S_T will change with distance along the pore; C_S because of hydrogen diffusion and S_T because of temperature changes. This situation lends it self well to a combined heat and mass flow analysis.

HEAT AND MASS FLOW ANALYSIS - The analysis consisted of a simple, point heat source, two dimensional analysis to plot temperature and hydrogen concentrations for the steady state conditions of heat flow and hydrogen diffusion. (Eqs. 4. 5. 6.). Details are given by Yen (8).

$$\frac{d^2T}{dX^2} + \frac{d^2T}{dY^2} = \frac{1}{\kappa}\frac{dT}{dt} \qquad \text{Eq. 4}$$

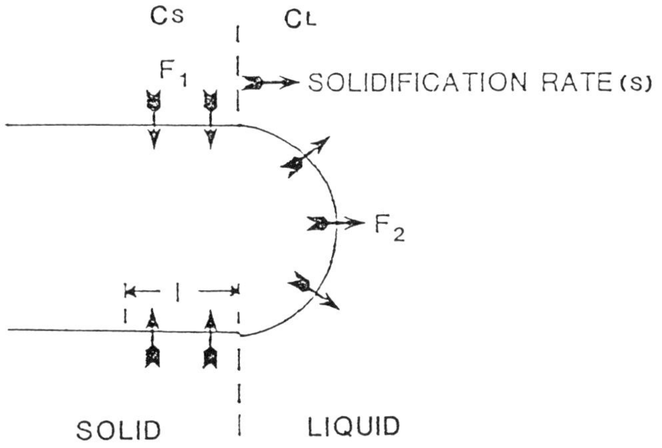

Fig. 7 - Flow of hydrogen assumed for the mathematical model.

$$\frac{dc}{dt} = \frac{d\,D}{dX} \cdot \frac{dc}{dX} + \frac{d\,D}{dY} \cdot \frac{dc}{dY} \qquad \text{Eq. 5}$$

For a moving coordinate (w,y), which has a defined temperature distribution, equation (5) is reduced to be a time independent equation.

$$\frac{d\,D}{dw} \cdot \frac{dc}{dw} + \frac{d\,D}{dy} \cdot \frac{dc}{dy} = 0 \qquad \text{Eq. 6.}$$

From the above analysis, a series of conditions were generated to study the flow of hydrogen to the pore as a function of distance (time) behind the solidification front. An example is shown in Figure 9, where the conditions of 60 amp. arc current, 2 mm/sec travel speed and a hydrogen pressure in the arc of 15 percent hydrogen, were used. The hydrogen concentration perpendicular to the welding direction is shown. Attention should be paid to the plot of triangles (second curve down from the top) which represents a location in the solid metal, close to the solidification front. Here the positive concentration gradient of hydrogen will lead to a flow of gas into the pore.

By knowing this concentration gradient and assuming that hydrogen would flow into the pore from all directions around the pore, it was possible to integrate to find the total mass flow rate of hydrogen (ml/sec/mm^2) to the pore over the length of the pore which showed a positive concentration gradient.

This plot is shown in Figure 10 for conditions of 10 percent hydrogen in the arc, 90 amp. arc current and a travel speed of 2 mm/sec for 304 stainless steel, with diffusion and thermal conductivity constants (4,9,10,11) assumed for austenitic stainless steel. The flow rate is much less than that required for pore propagation according to the calculated flow rate needed and shown in Table 2, under the heading of austenitic stainless steel. A similar calculation for the experimental results of Ohno and

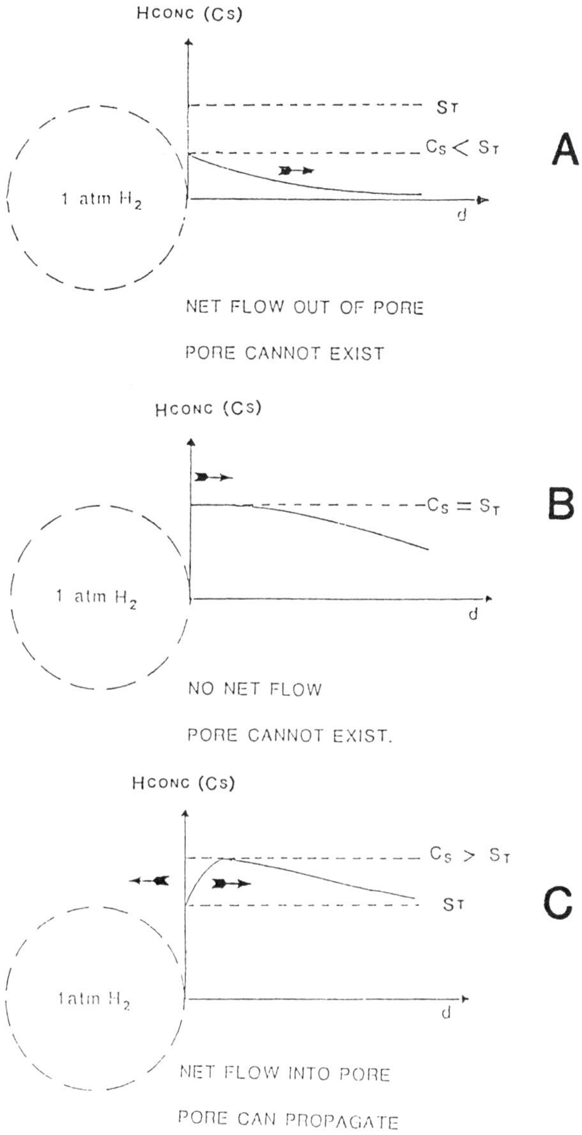

Fig. 8 - Conditions prevailing for pore growth.

a) $C_S < S_T$ No flow to the pore. Pore cannot exist.

b) $C_S = S_T$ No flow to the pore. Pore cannot exist.

c) $C_S > S_T$ Flow can proceed to the pore because of the positive temperature gradient. Pore can exist and propagate

Uda (2), using 50 percent hydrogen in the arc and inserting the solubility diffusion and thermal conductivity coefficients appropriate for this metal showed a relatively good correlation between theoretical and experimental values of mass flow rate and an adequate supply of hydrogen to propagate a pore. (Table 2. under the heading of Nickel) Typical solubility and diffusion constants, used for the generation of data in Table 2, are given in Table 3.

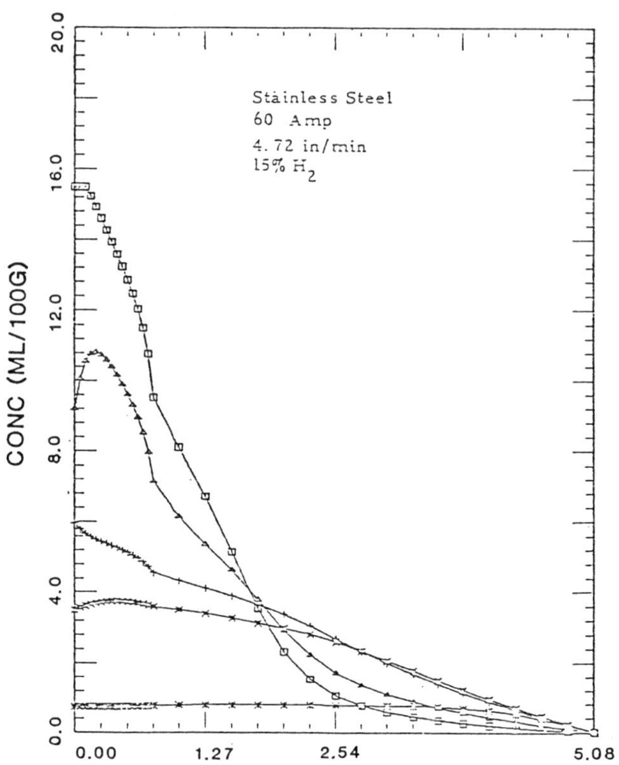

Fig.9 - Plot of hydrogen concentration perpendicular to the welding direction for various distances behind the weld pool center.

From the top down, at the weld pool center, 3 mm, 6.4 mm, 12.7 mm and 57 mm.

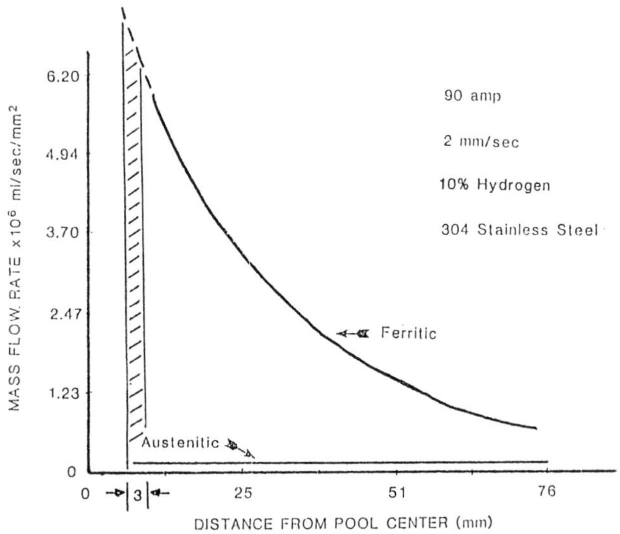

Fig. 10 - Hydrogen mass flow rate (ml/sec/mm^2) as a function of distance behind the weld pool center.

DISCUSSION

The discrepancy of flow rate predicted for austenitic stainless steel, and that needed, is disturbing, whereas the flow rate for nickel is in reasonably good agreement with experimentation.

In order to explore the situation further, the solidification characteristics of 304 stainless steel were examined. Depending upon the composition, it is possible that the 304 steel may solidify as delta ferrite. This is shown in Figure 11, which shows the pseudo-binary phase diagram of the Fe-Cr-Ni system with a constant iron content of 70 percent. Acknowledging the fact that the fixed points on the diagram may vary appreciably with small changes in composition, it appears that an alloy of chromium:nickel ratio of 2:1 solidifies primarily as ferrite and that this phase may remain in the structure until the temperature has been reduced to about 1150°C. Because of the lower solubility of hydrogen in the ferrite phase and a higher diffusivity than in the austenite phase, Table 3, there would be major differences in the hydrogen diffusion rate and hence the mass flow rate to the pore if ferrite were present.

For this reason, the mass flow rate to the pore was recalculated using the solubility and diffusion coefficient of delta ferrite. The heat flow program was not modified since, it was

Table 2 - Calculated Versus Needed Flow Rate of Hydrogen to Propagate a Pore.

CURRENT amp.	TRAVEL SPEED mm/sec	HYDROGEN IN ARC percent	CALCULATED FLOW RATE ml/sec x 10^6	FLOW RATE NEEDED
		AUSTENITIC STAINLESS STEEL		
90	1.5	10	1.08	72
90	2.0	10	1.46	185
90	2.5	10	1.83	326
		FERRITIC STAINLESS STEEL		
90	1.5	10	205	72
90	2.0	10	150	185
		NICKEL		
150	2.0	50	205	280
150	3.0	50	150	420

Table 3 - Selected Hydrogen Solubility and Diffusion Coefficients used for the Development of Table 2 Data.

	Solubility ml/100g		Diff. Coeff. cm^2/sec x10^3	
	1200°C	1400°C	1200°C	1400°C
Austenitic Stainless	14.1	15.1	0.08	0.14
Ferritic Stainless	7.3	9.9	3.31	3.48
Nickel		13.6		3.19

believed, that the presence of a small amount of ferrite would not change the temperature profile to any great extent. The recalculated values have been inserted in Table 2 for total mass flow rate and in Figure 10 for flow rate per

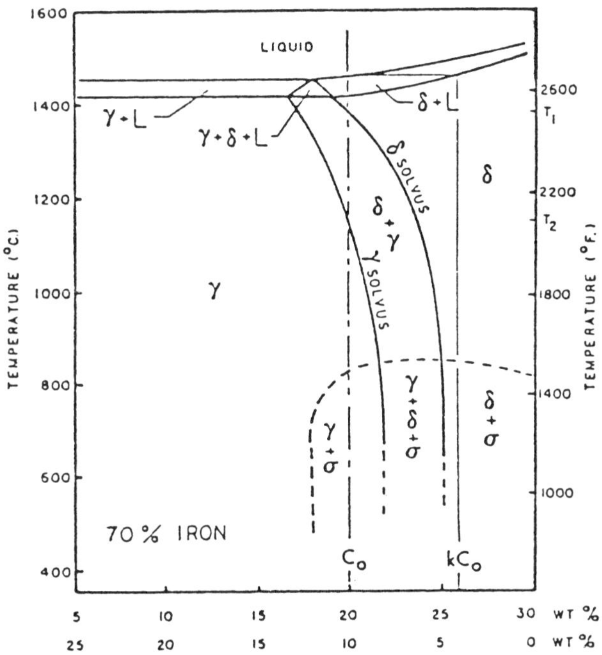

Fig. 11 - Pseudo-binary phase diagram of the Fe-Cr-Ni alloy system with a constant iron content of 70 percent.

unit area as a function of distance along the pore.

Since the flow rate obtained from a ferritic structure is so much greater than from an austenitic structure, any flow from the austenite might be conveniently ignored and the "length of delta ferrite" along the pore behind the pool be used for the distance "l" in Eq. 3. A simple calculation for the conditions given in Figure 10, shows that a 3 mm length of delta ferrite behind the solidification front, for a pore of 0.81 mm diameter, would provide a total mass flow rate of $72 \cdot 10^{-6}$ ml/sec, a value predicted in Table 2 for propagation of the pore.

The heat flow analysis was therefore used to predict the longitudinal temperature gradient along the centerline of the weld behind the pool so that the distance from the solidification front to the 1150°C isotherm could be found. An example of the plot is shown in Figure 12. The distance was calculated for various arc currents at given travel speeds (Figure 13) and for various travel speeds at a given current (Figure 14).

Clearly, the distance "l" increases with current. When the pore diameter is plotted with respect to the length of delta ferrite, Figure 15, there is a direct relationship, as predicted by Eq. 3. When the same maneuver is applied to the effect of travel speed, however, "l" decreases with increasing speed suggesting that the pore diameter should decrease with increasing travel speed, according to Eq. 3. In the experimental work quite the opposite occurred.

The reason for this discrepancy is not readily apparent. One possible explanation

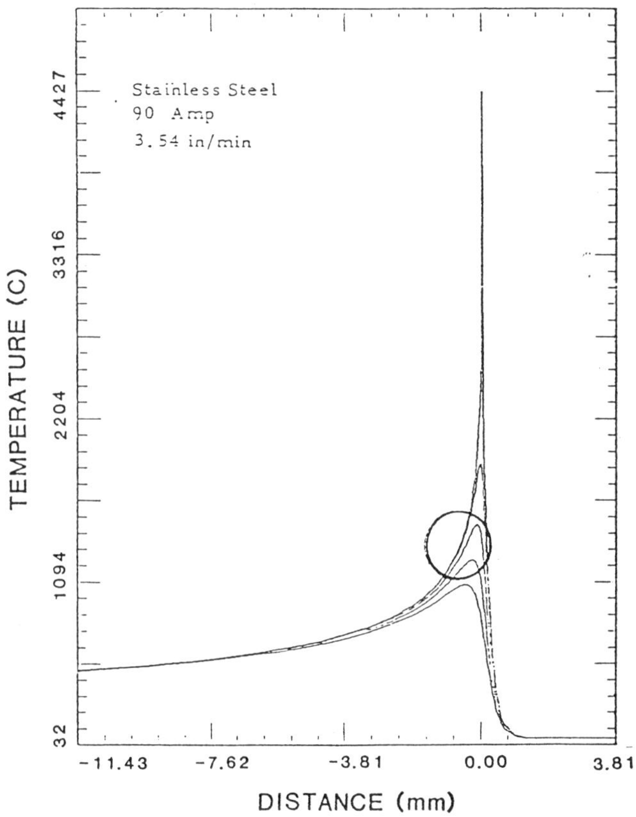

Fig. 12 - Example of the temperature cycles experienced by a weld in stainless steel. Temperature is plotted as a function of distance behind the center of the pool for various distances from the centerline of the weld. The highest curve is at the centerline.

could involve the over-simplified heat flow analysis which did not include a consideration of the latent heat of fusion. A more sophisticated analysis, performed by Tsai (12), shows that the effect of increasing travel speed at constant arc current does not change the longitudinal temperature distribution along the weld centerline but does, of course, reduce the width of the isotherms perpendicular to the welding direction. Another possible effect, on which little information exists, is that the kinetics of the ferrite to austenite transformation may allow some supercooling of the ferrite to occur, thereby extending "l". It would be anticipated that this effect would be greater for the higher cooling rates associated with higher welding speeds.

Outside of this perturbation, the model proposed for the pore growth mechanism explains most of the experimental observations. According to Eq. 3, the diameter of the pore should depend on the flow rate per unit area from solid to gas phase, F_1, the effective length along the pore over which gas diffuses into the pore, 1, the travel speed, s, and the flow rate of hydrogen per unit area, from the pore to the liquid metal, F_2. F_1 increases with C_S or C_L which are both related to the hydrogen pressure in the arc. Further F_1 increases as S_T, the solubility of the solidified metal for hydrogen, decreases, and for an increasing pore diameter. "1" is affected by the heat flow conditions (both arc current and travel speed). F_2 should increase for increasing pore diameter and for decreasing pool hydrogen concentration, C_L, which is related to the hydrogen pressure in the arc.

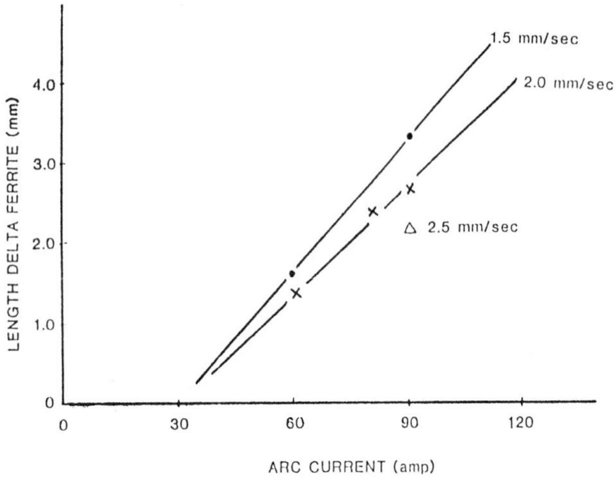

Fig. 13 - Calculated length of delta ferrite as a function of arc current for given travel speeds.

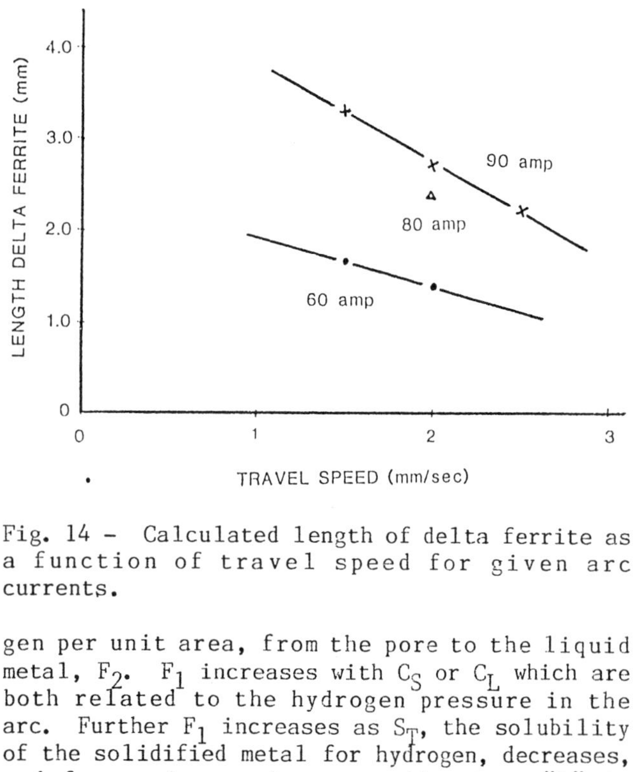

Fig. 14 - Calculated length of delta ferrite as a function of travel speed for given arc currents.

The pore diameter must then adjust to these complex controlling factors. How this is accomplished physically, is indicated in Figure 16 and relates to the geometry of the end of the gas pore protruding into the molten pool. In

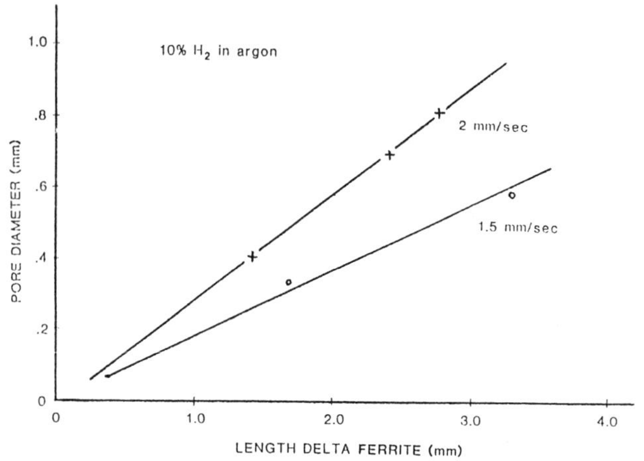

Fig. 15 - Relationship of pore diameter with the calculated length of delta ferrite.

Figure 16a, the rates of gas flow are balanced at this diameter and the protrusion is hemispherical so that the pore propagates with the same diameter. If the gas flow into the pore is increased, for some reason, Figure 16b, the diameter of the bubble is increased and the pore diameter increases. By increasing the bubble diameter, two things happen to compensate for the increase of flow rate into the pore. First, the volumetric rate needed for pore growth increases and, second, the area of the bubble in contact with the molten metal increases and flow rate to the pool is increased. The diameter of the pore then stabilizes at the new size. Conversely, Figure 16c, a reduced influx of gas to the pore would reduce the diameter. Apparently, these adjustments take place very rapidly such that the high frequency arc heat input changes caused by the power supply ripple are reflected by pore surface ripples.

What happens to the pore diameter after two pores merge, may help to substantiate the mechanism that gas enters the pore from the solidified metal rather from the molten pool. Unfortunately, the experimental evidence is not absolutely conclusive. It appeared in many cases, that the diameter of the newly formed pore was similar in size to that of each of the original pores, at least after a short distance needed for the re-establishment of steady state flow. Using the geometrical features of the pores, i.e. cylindrical in length and with a hemispherical bubble protruding into the molten pool, when the pores meet, the cross sectional area is initially maintained constant. This, in turn, means that the diameter of the new pore must increase by 40 percent, that the hemispherical area remain constant and that the unit circumferential area of the pore decrease by 30 percent. For the Willgoss and Atthey mechanism (1), this would mean no change in the hydrogen flow rate to the pore and that the diameter should remain stable at the enlarged diameter. For the presently proposed mechanism, the flow rate of hydrogen from the solid metal at the cicumferential surface of the pore should see a 30 percent decrease and the diameter of the pore should adjust to approximately the same as that of the original pores. The experimental studies tended to bear this out.

Further evidence that gas must be supplied to the pore from the solidified weld metal,

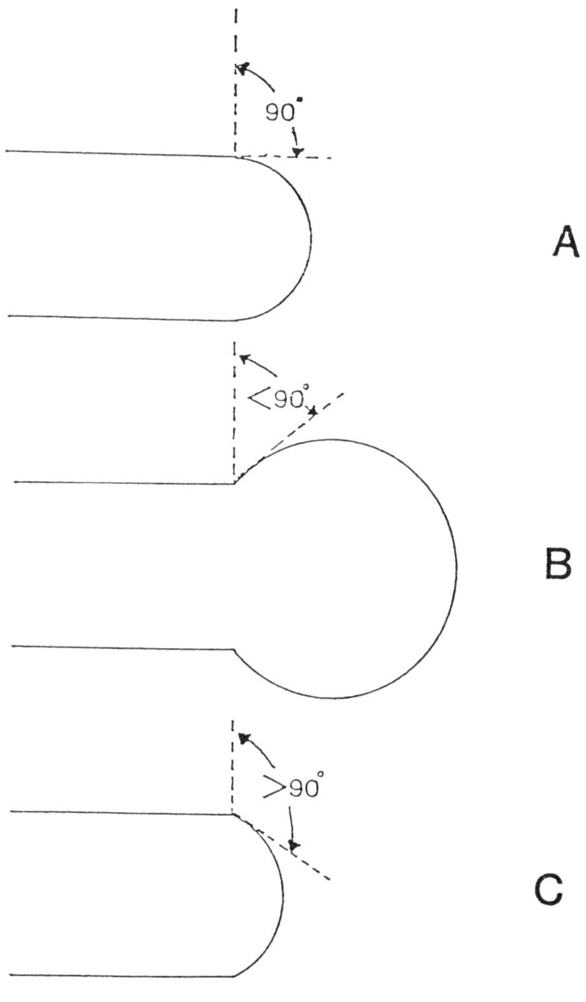

Fig. 16 - Self adjustment of the pore diameter.
a) Constant hydrogen flow to the pore
b) Transient increase of hydrogen flow
c) Transient decrease of hydrogen flow

rather than from the liquid pool, to induce growth was provided experimentally. When small holes were drilled in the faying surfaces of the joint to act as nuclei for pores as the weld pool passed, growth was not observed. Since the gas in the holes was at atmospheric pressure at the time the pool passed, any gas from the molten pool would be expected to increase the volume of the pore and initiate growth. Hydrogen was not available in the solid base metal surrounding the pore in this case, however, and growth did not take place.

CONCLUSIONS

Worm hole porosity may be induced in AISI 304 stainless steel weld metal with less than 10 percent hydrogen in the arc shielding gas during GTA welding. The pores grow perpendicularly to the solidification front reaching diameters of 2 mm. and may proceed for a length along the weld of several times the base metal thickness.

A model is proposed to show that the worm hole porosity can be propagated by hydrogen flow from the solidified weld metal by a solid state diffusion process, rather than from the molten pool, as is popularly believed.

A simplifed heat and mass flow analysis predicts, to a large extent, pore occurrence and behavior. The diameter of the pore adjusts to the welding variables, increasing with hydrogen potential in the arc, arc current and travel speed. Some discrepancy between the model prediction and the experimental data exists with respect to travel speed.

The crystal structure of the solidified weld metal, austenite or ferrite in AISI 304, apparently affects pore growth rate greatly, based on the differences in hydrogen solubility and diffusivity. The role of delta ferrite on the behavior of gases during solidification is not well understood and should receive more attention in future research efforts.

ACKNOWLEDGEMENTS

The authors wish to thank Dr. Chon L. Tsai, Associate Professor, Ohio State University, for his valuable discussions and assistance with the mathematical analysis. Mr Sun Ming Park, Graduate Research Assistant, The Ohio State University, was very helpful in his discussions of the work and his assistance in organizing the contents of the paper. Gratitude is also expressed to L. T. V. Steel Corporation, Cleveland, OH. and Huntington Alloys, Inc., Huntington, WV. for providing the materials used in the investigation.

REFERENCES

1) Chew, B. and R. A. Willgoss, Proceedings of International Conference - Weld Pool chemistry and Metallurgy, p. 155. London, Apr 15-17, 1980. The Welding Institute, Abington, England.

2) Ohno, S.and M.Uda, Trans. of National Res. Inst. for Metals. 23, 243, 1981.

3) Willgoss, R. A. and D. R. Atthey, Proceedings of International Conference - Weld Pool chemistry and Metallurgy, p. 121 - 131, London, Apr 15-17, 1980. The Welding Institute, Abington, England.

4) Howden, D. G. and D. R. Milner, Brit. Welding Journ., 10, 304 - 310, (1963)

5) Howden, D. G.,
Proceedings of International Conference - Weld Pool chemistry and Metallurgy, p. 205 - 215. London, Apr 15-17, 1980. The Welding Institute, Abington, England.

6) Ekerot, S. and H. Fredriksson,
Svetsaren, No. 1, 1974.

7) Pokhodnya, I. K. and V. F. Demchenko,
Proceedings of International Conference - Weld Pool chemistry and Metallurgy, p. 107. London, Apr 15-17, 1980. The Welding Institute, Abington, England.

8) Yen, L. S., Master of Science Thesis, 1984
The Ohio State University, Columbus, Ohio.

9) Robertson, W. M., Z. f. Metall. K.,
64, 436, 1973.

10) Hobson, J. D., J. of Iron and Steel Inst.
164, 342, 1950.

11) Smialowski, M., J. of Iron and Steel Inst.
164, 100, 1950.

12) Tsai, C. L., Private Communication.
Ohio State University, Dept. of Welding Eng. Columbus, Ohio.

MICROSTRUCTURAL EVOLUTION IN THE HAZ OF A SUBMERGED ARC WELDED MICROALLOYED STEEL

R.A. Nichting, E.L. Brown
Colorado School of Mines
Golden, Colorado

Abstract

Microstructural evolution in the HAZ of a submerged arc welded Cb-V steel was studied. Thermal and microstructural models were experimentally established. A statistical-empirical model was required to establish HAZ thermal cycle as a function of heat-input and position. Austenite microstructural evolution was found to be characterizd by abrupt grain coarsening, controlled by dissolution of a V-based microalloy precipitate. Grain growth in the HAZ during welding was predicted with reasonable accuracy by integrating the grain growth equations over the experimentally established thermal cycle. Continuous cooling transformation behavior in the HAZ was primarily controlled by prior austenite microstructure in this range of heat-input.

Introduction

Microalloyed HSLA steel base-plate derives its balanced package of high strength and toughness via optimized thermomechanical processing to produce a fine-grained, microalloy precipitation strengthened ferrite. The delicate balance of microalloy precipitation and microstructure in the base-plate is significantly perturbed by the heat of welding processes, especially those associated with high heat-input. The microstructure evolved in the heat-affected-zone (HAZ) is primarily a function of the thermal cycle imposed by welding on the prior structure and chemistry of the base-plate. The microstructure (s) of the HAZ play an extremely important role in determining the toughness of the HAZ and the overall toughness of the weldment. The formation of a coarse grained austenite region may be especially detrimental with regard to toughness.

It would be desirable to formulate and confirm a model of HAZ microstructural evolution with which predictions could be made based on base-plate properties and welding parameters. A complete model actually consists of three parts: a) a thermal model capable of predicting the thermal cycle as a function of position in the HAZ and welding parameters such as heat-input, b) an austenite microstructural model to predict austenite grain growth and state of microalloy precipitation as a function of thermal history and, therefore, as a function of welding parameters and position in the HAZ, c) a transformation product microstructural model to predict the continuous cooling transformation behavior in the HAZ as a function of prior austenite microstruture and thermal history, namely cooling rate through the austenite-to-ferrite ($\gamma \rightarrow \alpha$) transformation. It is the fundamental interrelationship of these models that allows for the establishment of an overall processing - structure - property model and a rational choice of welding parameters.

The objective of the current study was to investigate the physical metallurgical basis underlying evolution of HAZ microstructure during submerged arc welding of a Cb-V microalloyed steel. The effect of variation in heat-input on austenite and transformation product microstructural evolution in the HAZ was studied. The effects of welding on the state of microalloy precipitation were also studied in an effort to provide a fundamental basis for microstructural evolution.

Material and Experimental Procedure

A controlled rolled 16mm thick Cb-V microalloyed base-plate, with chemical composition described in Table 1, was employed in this investigation. A single 45° bevel weld

Table 1 - Steel Composition

C	Mn	Si	Al	Cb	V	N	w/o
0.14	1.48	0.24	0.045	0.038	0.080	0.016	

preparation was employed to facilitate instrumentation for temperature measurement during welding. Small diameter blind holes, drilled perpendicular to the bevel face, provided wells for fine wire thermocouple installations. Thermocouples were welded to the hole bottoms. Rapid, high resolution digital thermal cycle data acquisition was accomplished with a digital storage oscilliscope. Single pass submerged arc weldments were produced at heat inputs of 19.7, 33.5 and 49.2 KJ/cm.

Transverse cross-sections of the weldments at thermocouple installations were metallographically prepared and observed via light microscopy in order to reveal the microstructure of the HAZ and the exact locations of thermocouple installations.

Simulation of microstructures at various HAZ locations was performed via a programmed thermal cycle test apparatus, the Gleeble 1500, utilizing previously obtained thermal cycle data. Dilatometry data obtained during microstructural simulation was utilized to assess the continuous cooling transformation behavior during welding. The Gleeble 1500 was also used to generate isothermally heat treated and quenched specimens in order to quantitatively specify austenite grain growth behavior.

Light microscopy for austenite microstructural characterization was performed on specimens etched in a saturated aqeous picric acid solution. Etching for transformation product microstructure, was accomplished with a 2% Nital solution. Grain size measurements were performed via a linear intercept technique. The state of microalloy precipitation and fine scale structure was monitored with transmission electron microscopy (TEM) of carbon extraction replicas and thin foils. Thin foils were produced by twin jet electropolishing in a 5% perchloric acid-acetic acid solution at 40V and room temperature. Precipitate size distributions were assessed from carbon extraction replicas per Ashby and Ebeling (1). The state of microalloy precipitation was also determined by X-ray diffraction of extracted precipitate residues.

Results and Discussion

The base-plate possessed a banded ferrite-pearlite microstructure with 17% pearlite and a 5-6 μm grain size, Fig. 1(a). TEM of thin foils and carbon extraction replicas and x-ray diffraction of extraction residues revealed a duplex dispersion of microalloy precipitates composed of V-based and Cb-based precipitates, Fig. 1(b). The V-based dispersion was characterized by a finer particle size (160A) and higher volume fraction than the Cb-based dispersion.

The nature of heat flow during welding has been theoretically considered by Rosenthal (2) and analytical solutions to the problem have been formulated with modifications due to finite-plate thickness, geometry of heat source

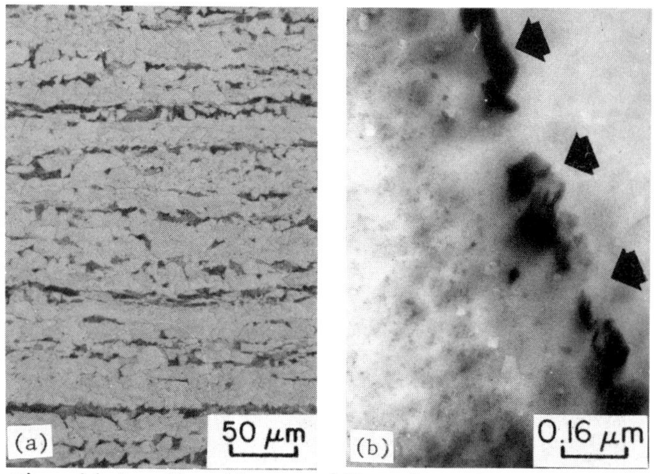

Fig. 1(a) - Ferrite-pearlite microstructure of base-plate, light micrograph.

Fig. 1(b) - Coarse Cb-based precipitates (arrow) and fine V-based precipitates of base-plate, thin foil. TEM micrograph.

and rate of movement and strength of the heat source (3). These solutions do not adequately account for variations of thermal parameters with temperature nor do they adequately account for complexities due to weld pool shape, extent and phase changes. Numerical methods have been employed to circumvent some of these problems.

Attempts were made to fit the thermal cycle data to the theoretically-based analytical solutions due to Rosenthal but were unsuccessful. In general, the cooling porition of the theoretical thermal cycle for a given heat-input and HAZ location significantly underestimated the corresponding region of the experimentally determined thermal cycle, Fig.

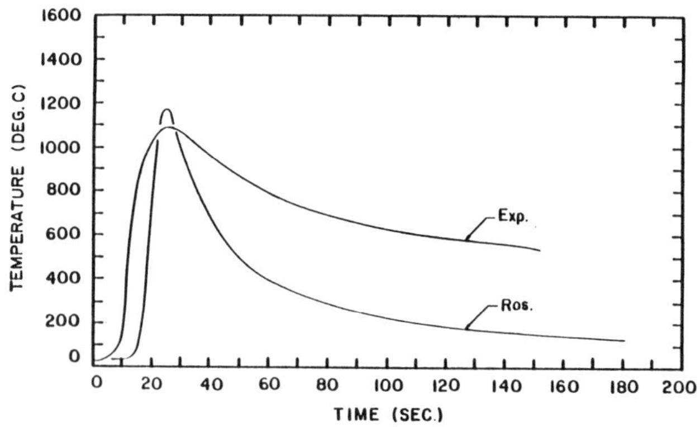

Fig. 2(a) - Comparison of experimental and Rosenthal-type theoretical HAZ thermal cycles; q = 49.2 KJ/cm, r = 0.152 cm.

2(a). The observed discrepancy could not be accounted for with realistic variations in thermal parameters or with finite thickness corrections. It was therefore necessary to empirically fit data separately in the heating, peak and cooling portions of the thermal cycle, Fig. 2(b), and as a function of heat-input and position in the HAZ, Table 2.

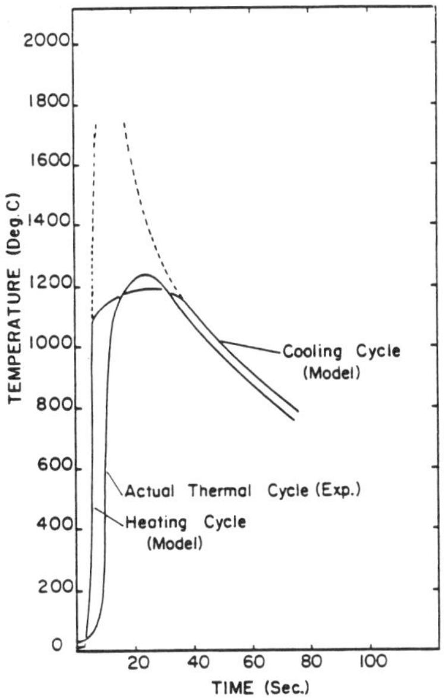

Fig. 2(b) - Typical comparison between results of empirical thermal model and experimental thermal cycle.

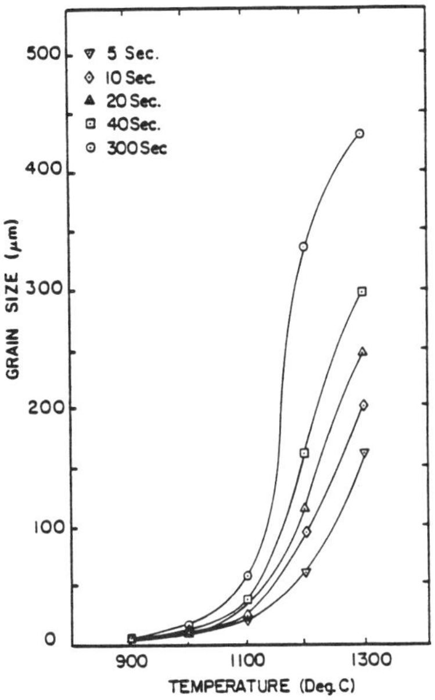

Fig. 3 - Results of isothermal austenite grain coarsening study.

Table 2 - Thermal Model Parameters

Heating:	$\log T = A - B/t$
	$\log A = 1.599 - (0.891)\ r$
	$B = 5.61 + 0.038\ (q/r)$
Peak Temperature:	$T_p = 452.90\ (q/r)^{0.16}$
Cooling:	$\log T = C - 0.5 \log t$
	$C = 1.069 \log (q) + 2.047$
	q = heat input
	r = distance from fusion zone boundary

Table 3 - Austenite Grain Growth Model Parameter

$$D = Kt^n$$

log K:	900-1000°C	$1.73 - (1402.99/T_A)$
	1000-1300°C	$8.21 - (9623.56/T_A)$

T_A = absolute temperature

n:	900-1100°C	$(6.7 \times 10^{-4})\ T - 0.506$
	1100-1200°C	$(2.37 \times 10^{-3})\ T - 2.381$
	1200-1300°C	$(1.7 \times 10^{-3})\ T + 2.503$

The empirical fit displayed greater scatter for conditions of high heat-input and positions close to the fusion zone boundary. That could be due to the aforementioned complexities associated with the weld pool. The empirical fit was employed in subsequent microstructural modeling efforts.

Isothermal heat treatments were utilized to elucidate the general austenite grain coarsening behavior of this steel and its relation to the state of microalloy precipitation. Austenite evolution was characterized by abrupt grain coarsening between 1000 and 1100°C and at least two temperature regimes of behavior, a low temperature slow growth regime and a high temperature rapid growth regime, Fig. 3. The grain growth data in each regime was fitted to an equation of the form, $D = K(T)\ t^n$ in which T = absolute temperature (°K), t = time(s) and K and n are temperature regime dependent constants. The constants are summarized in Table 3.

TEM and x-ray diffraction of microalloy precipitate extractions indicate significant dissolution of the V-based precipitate at temperatures above 1000°C while low volume fraction remnants of the Cb-based precipitate dispersion remained. The variation of mean precipitate size with time and temperature was complex, at a given temperature generally increasing through a maximum with time, Fig. 4.

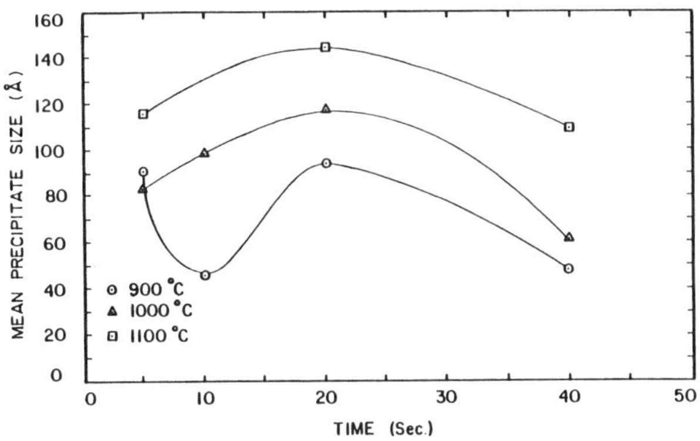

Fig. 4 - Mean precipitate size as a function of temperature and time for isothermal heat treatments.

All mean particle sizes were less than that of the base-plate. This behavior is due to perferential dissolution of fine V(C,N) and/or fine Cb(C,N) particles followed by eventual dissolution of coarser particles. The initial increase in mean particle size with dissolution could be confused with overall particle coarsening but size distribution analysis indicates that dissolution is actually occurring. This has been observed by others (4).

It would therefore seem that abrupt grain coarsening in this steel is associated with dissolution of the V-based precipitate dispersion. This was consistent with calculations of the respective equilibrium solution temperatures of V(C,N) and Cb(C,N); ≤1075°C and ≥1150°C, respectively.

Theoretical treatment of grain growth in the presence of a dispersed precipitate phase (5,6) has indicated that a critical density of dispersed particles is required to pin boundaries of grains of a given size, i.e. a critical precipitate size exists that is proportional to f D where D is the grain size and f is the volume fraction of particles. If the actual particle size exceeds this critical size grain growth will ensue. Therefore, two mechanisms for the onset of abrupt grain coarsening via grain boundary unpinning are possible. The volume fraction of particles may be reduced by dissolution or the actual particle size may be increased via coarsening at a constant, equilibrium volume fraction.

Austenite grain growth in microalloyed steels has been shown to be intimately associated with the scale and stability of the microalloy precipitate dispersion (7-9). Dependent upon steel composition and precipitate stability boundary unpinning may be controlled by either dissolution or coarsening.

In the current study the calculated critical particle size was always greater than the actual particle size so that precipitate coarsening could not account for the observed abrupt grain coarsening. Besides, particle coarsening via a capillarity-driven mechanism, such as Ostwald ripening, would not be expected prior to attainment of the equilibrium volume fraction of precipitate by dissolution. The modification of the base-plate precipitate dispersion must first be associated with precipitate dissolution due to the greater solubility of austenite for microalloy precipitates. Precipitate coarsening control of austenite grain growth is probably only viable for precipitates that are stable to higher temperatures where equilibrium is more easily attained and solute diffusivities are greater. It is interesting that the Cb-based precipitate dispersion appeared to have little effect on grain growth control under these conditions.

The formulation of a completely theoretical description of the grain coarsening process in this steel requires a comprehensive theory of dissolution of a size distribution of microalloy precipitates. The dissolution model of Ashby and Easterling (10) does does not appear to be adequately detailed.

Models for HAZ austenite microstructural evolution have been proposed (10-12). These models assume that the mechanisms that govern the processes of isothermal microstructural evolution may be employed to determine microstructural evolution during an athermal process, such as a HAZ thermal cycle. The integration technique is based on the so-called rule of additivity for thermally activated processes (13) and does not take into account the effects of the ferrite-to-austenite ($\alpha \rightarrow \gamma$) phase transformation during the heating portion of the thermal cycle on austenite microstructural evolution. That is, the evolution of austenite microstructure is considered solely as a growth process.

The thermal and austenite microstructure models, formulated in this study, were thus coupled in order to predict austenite grain size as the result of the athermal weld thermal cycle. The grain growth equations were integrated over thermal cycles as a function of heat-input and positions in the HAZ and compared with measurements of austenite grain size in the actual welds, Fig. 5. The predicted austenite

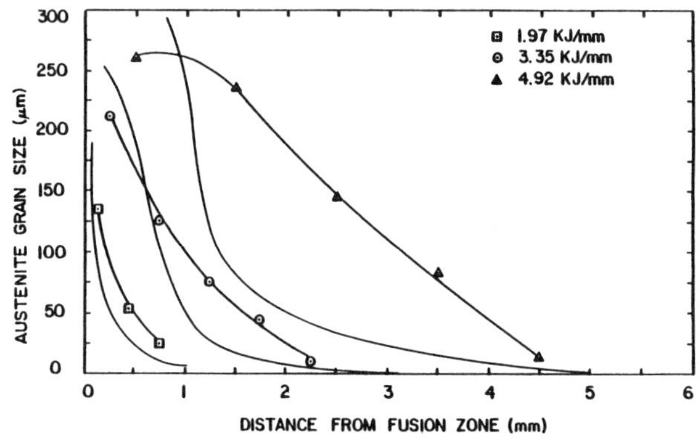

Fig. 5 - Comparison between austenite grain sizes in actual welds and those predicted with model.

grain size is close to that of the actual austenite grain size with the differences being greater for high heat-inputs and HAZ positions close to the fusion zone. This is probably due to the aforementioned problems in fitting thermal data for these conditions. It would appear, however, that overall, the integration of isothermal grain growth laws over the weld thermal cycle is an adequate means for predicting austenite microstructural evolution during submerged arc welding of this steel.

Austenite microstructural evolution in the HAZ of this Cb-V microalloyed steel is controlled by precipitate dissolution rather than coarsening, consistent with the results of isothermal grain growth studies and with the theory of Ashby and Easterling (10). The modeling approach of the present study does differ from other modeling approaches cited in its employment of an empirical thermal model rather than a Rosenthal-type thermal model. It would appear that use of the latter for microstructural modeling could result in error for high heat-input processes.

Evolution of transformation product microstructure in the HAZ is controlled by composition, prior austenite grain size and cooling rate. The latter are specified via thermal history. The isothermal and continuous cooling transformation behavior of steels of homogeneous composition with simple imposed thermal cycles has been extensively studied. However, as yet, no fundamental theory exists for predicting transformation behavior. Empirical models for transformation behavior have been proposed. Austenite structural and chemical inhomogeneties in the HAZ as well as the complicated thermal response as a function of position serve to exacerbate the problems of modelling transformation behavior in the HAZ.

The continuous cooling transformation behavior as a function of heat-input and HAZ location was adequately simulated by the Gleeble 1500 so that dilatometry data was applicable to actual submerged arc welding. The transformation from austenite was found to be a strong function of prior austenite grain size and less a function of cooling rate in this range of heat-input, Fig. 6.

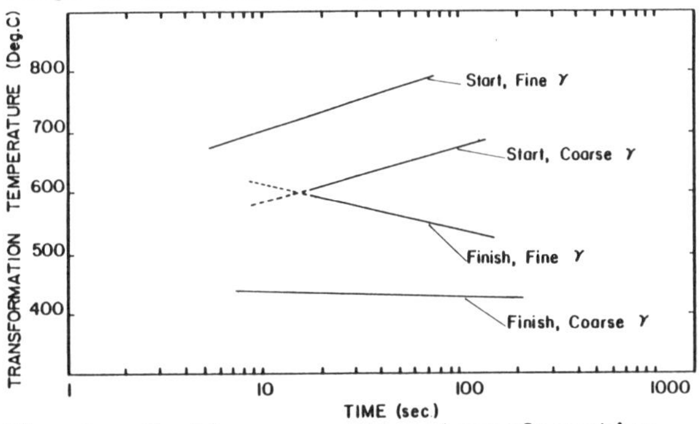

Fig. 6 - Continuous cooling transformation diagram for submerged-arc weld HAZ of Cb-V steel.

The transformation products of the coarse grained austenite region were generally mixtures of martensite and bainite associated with complete dissolution of microalloy precipitates. The cooling rates in this region are high enough to preclude reprecipitation. The measurement of ferrite grain size indicated close correlation with prior austenite grain size, Fig. 7. In fact, a minimum in ferrite grain size less than that of the base-plate occurs in association with peak temperatures just above the Ac_3.

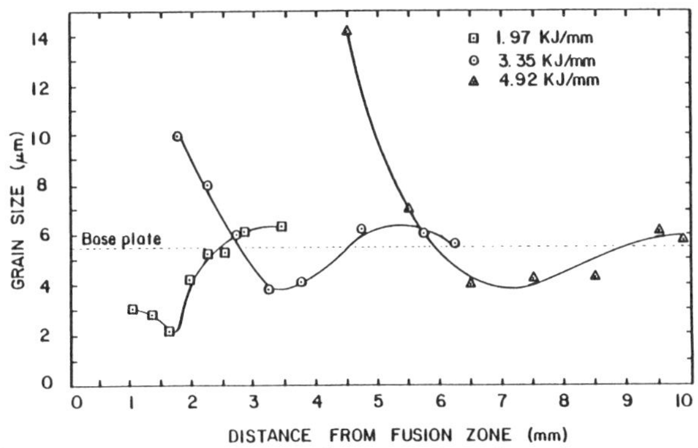

Fig. 7 - Measured ferrite grain size in actual welds.

Conclusions

1. A statistically derived empirical model was required to describe the thermal cycle as a function of heat-input and location in the HAZ of a single-pass submerged arc weld. Rosenthal-type closed form thermal models proved inadequate, probably due to inadequacies in modeling the geometry and phase transformations of the weld pool.

2. The evolution of austenite microstructure was controlled by the dissolution of V-based microalloy precipitates and grain growth during welding could be described by a simple integration of isothermal equations over the weld thermal cycle if caution was taken in applying the grain growth laws appropriate to specific regions of the thermal cycle.

3. The transformation product microstructure is mainly controlled by prior austenite grain size in this range of heat input. It is further characterized by essentially microalloy precipitate-free mixtures of martensite and bainite in the austenite grain coarsened regions of the HAZ and very fine ferrite grain sizes in the region of the HAZ associated with peak temperatures just above Ac_3

References

1. M.F. Ashby and R. Ebeling: Trans. AIME., 236, (1966), p. 1396.

2. D. Rosenthal: Welding J., 20, (1941), pp. 2205-2345.

3. N.N. Rykalin: "Calculation of Heat Flow in Welding", Moscow, (1951).

4. A.R. Jones and B. Ralph: Metallography, 10, (1977), pp. 469-480.

5. M. Hillert: Acta Met., 13, (1965), p. 227.

6. T. Gladman: Proc. Roy. Soc., 294A, (1966), p. 298.

7. T. Gladman and D. Dulieu: Metal Sci. J., 8, (1974), pp. 167-176.

8. T. Gladman and F.B. Pickering: JISI, (1967), pp. 653-664.

9. L.J. Cuddy and J.C. Raley: Met. Trans. A, 14A, (1983), p. 1989.

10. M.F. Ashby and K.E. Easterling: Acta Met., 30, (1982), p. 1969.

11. H. Ikawa, H. Oshige, S. Noi, H. Date, K. Uchikawa: Trans. Jap. Welding Soc., 9, (1), (1978), p. 47.

12. P.J. Alberry and W.K.C. Jones: Metal Technology, 9, (198), p. 419.

13. J.W. Cahn: Acta. Met., 4, (1956), p. 572.

THE INFLUENCE OF LOW ALLOY FILLER METAL ON SOLIDIFICATION AND FERRITE TRANSFORMATION IN ESW

D.W. Yu, H.S. Ann, J.H. Devletian, W.E. Wood
Oregon Graduate Center
19600 N.W. Von Neumann Drive
Beaverton, Oregon 97006-1999

ABSTRACT

Substantial changes in both the solidification and solid state microstructures resulted from small additions of Cr, Ni and Mo to the filler metal when joining A36 and A588 steels with the consumable guide electroslag welding (ESW) technique. The combined presence of Cr and Mo in the weld metals, being below 0.7% Cr-0.3% Mo, suppressed grain-boundary ferrite and side-plate ferrite and favored the formation of acicular ferrite. Above that content, the formation of acicular ferrite was also suppressed. Conversely, the combined presence of Ni and Mo, up to 1.5% Ni-0.3% Mo, continuously decreased the volume fractions of both grain-boundary ferrite and side-plate ferrite while giving rise to a matrix which was predominantly fine acicular ferrite. An attempt was made to explain this alloying mechanism by studying the influence of those alloying elements on the primary solidification mode and subsequent solid state transformations. The effects of those alloy additions on CVN toughness were also discussed.

FOR YEARS, a great deal of research has been devoted to overcoming the low toughness problem in electroslag weldments deposited on mild steels. Improvements in weld metal toughness have resulted from narrow-gap techniques [1], alloyed filler metals [2], powder metal additions [3,4], etc. In this investigation, low alloyed (Cr, Ni and Mo) filler metals were deposited on 50 mm and 75 mm thick A36 and A588 steels using the narrow-gap ESW technique, in order to improve both the microstructure and toughness of the weld metal.

Acicular ferrite, owing to its positive contribution to toughness, is the most sought constituent in the as-welded mild steel microstructure. Mn, Cr, Mo, and Ni are known to promote the formation of acicular ferrite. However, no quantitative correlation between alloy content and toughness has been systematically studied. And surprisingly, little was fully understood about their roles in weld metal transformation kinetics [5].

The main objective of this investigation was to study the influence of Cr-Mo and Ni-Mo additions on the solidification structures, solid state transformation products and the CVN impact toughness of as-welded electroslag welds on A36 and A588 steels. In addition, the application and effects of powder-cored filler metal were evaluated.

EXPERIMENTAL PROCEDURE

MATERIALS - The base metals welded in this investigation were ASTM A36 and A588 commercial steels having 50 mm (2") and 75 mm (3") thicknesses. The filler metals were all 2.4 mm (3/32") diameter and included (1) mild steel solid wires, (2) Cr-Mo steel solid wire, (3) Ni-Mo steel solid wire, and (4) Ni-Mo steel powder-cored tubular wire. The consumable guide plates were pre-welded in a webbed shape for dual-wire feeding and made of AISI-1010 mild steel. The chemical compositions of all plates and welding consumables are given in Table 1. The flux used throughout this study was Hobart-201 flux.

WELDING PROCESS - ESW was performed in the DC reverse polarity mode with current and voltage in the ranges of 1000-1300 A and 32-40 V, respectively. A 19 mm (3/4") gap was utilized. Slag depth was maintained constant at about 25 mm (1") by continuous flux addition during operation.

METALLOGRAPHY - Stead's reagent (mixed 10 g $CuCl_2$, 40 g $MgCl_2$, 20 ml HCl and 1000 ml methanol) was used for revealing the primary solidification structure, while 1% nital solution was used for the solid phase transformation products.

MECHANICAL PROPERTIES - Rockwell B hardness and standard CVN impact toughness tests (notches vertical to the welding direction) were performed on weldments according to ASTM E23 and D1.1 Structural Welding Codes.

RESULTS AND DISCUSSION

SOLIDIFICATION MODE - Weld metal deposited on A36 and A588 steels using the 25P and WS filler metals and having the weld metal compositions given in Table 2 solidified with a cellular dendritic mode near the weld edges and columnar dendritic near the weld center. It was shown in the author's previous paper [6] that cellular dendrites gave rise to the commonly observed coarse columnar grain region, while columnar dendrites gave rise to the thin columnar grain region of the weld. Increased welding speed deepened the weld pool, decreased the form factor and increased the volume fraction of weld metal cellular dendrites. It also enhanced the susceptibility of centerline hot cracking.

A small amount of Cr, Ni and Mo added into the weld metal (which already contained over 1% Mn) produced a major change of solidification mode to equiaxed dendritic at the weld center. With a form factor value of about 1.2-1.4, a mild steel filler typically produced columnar dendrites at the weld center (Fig. 1a). But, a low Cr-Mo or Ni-Mo steel filler (AS521, AX90 or TW8544) brought about a substantial equiaxed dendritic region in the weld center (Figs. 1b and 1c). Some typical weld metal compositions are given in Table 2. It is recognized [7] that the increase of solute content in the weld metal substantially influenced the transition of solidification mode from columnar dendritic to equiaxed dendritic growth. That was analogous to a decrease in the solidification parameter, G/R (where G is the local thermal gradient and R is the growth rate). The central equiaxed dendrite zone induced by alloy additions showed superior resistance to hot cracking when the welding speed was increased since the simultaneously nucleated equiaxed dendrites are believed to disperse remaining impurities to a comparatively large area other than a narrow centerline, thereby reducing detrimental segregation.

In addition, due to the rightward displacement of maximum δ-Fe composition in the Fe-Cr, Fe-Ni, and Fe-Mo phase diagrams compared with the Fe-C diagram, the primary dendrites in alloyed mild steel weld metal solidified as δ-ferrite, which had a higher impurity solubility and less interfacial segregation than γ-austenite during primary solidification. Thus, Cr, Ni and Mo additions provided a relatively beneficial (crack resistant) solidification structure.

FERRITIC MORPHOLOGY - In addition to their effects on the solidification structure, the additions of Cr, Ni and Mo significantly governed the ferrite morphology of the weld metal. Fig. 2 indicates that the volume fractions of microstructural constituents varied with increasing alloy content in both A36 and A588 weld metal. (Data in the extreme left of each diagram represented the case of mild steel filler).

With a constant Mn content of about 1.0% to 1.2%, Cr-Mo additions effectively reduced grain-boundary ferrite and side-plate ferrite, and increased acicular ferrite as long as the alloy content was below 0.7% Cr-0.3% Mo. When the Cr-Mo content exceeded those values, the volume fraction of acicular ferrite decreased while bainitic ferrite increased. Further increasing the alloy content above 1.1% Cr-0.4% Mo resulted in a fully bainitic microstructure (Fig. 3a). The Ni-Mo addition, within the limit of this investigation (1.5% Ni-0.3% Mo), not only greatly reduced both grain-boundary and side-plate ferrite, but also substantially promoted the formation of acicular ferrite. In both A36 and A588 steel welds containing above 0.8% Ni-0.2% Mo, acicular ferrite became the predominant weld metal constituent, as shown in Fig. 3b.

In determining the reason why Cr-Mo and Ni-Mo additions caused such sharply different morphologies of ferritic transformation, it is important that the influence of each alloying element on the ferritic transformation products be discussed.

FORMATION OF GRAIN-BOUNDARY AND SIDE-PLATE FERRITE - The volume fraction of ferrite during transformation, f, can be expressed as

$$f = 1 - \exp\left[-\pi/3\ \dot{N}\ V^3\ t^4\right]$$

where $\dot{N}$ is the nucleation rate, V is the growth rate, and t is time [8]. $\dot{N}$ is known to be directly proportional to the transformation driving force, ΔG. Considering the change in carbon concentration of austenite around ferrite nuclei (Fig. 4), the driving force for ferrite nucleation can be written as:

$$\Delta G = RT\ \{C_0\ \ln(C_\gamma/C_0) + (1-C_0)\ \ln[(1-C_\gamma)/(1-C_0)]\} \qquad [9]$$

where C_γ is the enriched carbon concentration in the austenite around the ferrite nuclei, and Co is the nominal carbon content of the base metal. Due to low nominal carbon content (i.e., $C_0 << 1$), the above equation reduces to:

$$\Delta G = RT\ \ln\{[(C_\gamma)^{C_0}\ (1-C_\gamma)]\ /\ (C_0)^{C_0}\}$$
$$= RT\ \ln\ (1-C_\gamma)$$

This meant that, in the case of mild steel welds, the driving force for ferrite nucleation was mainly governed by the enriched carbon concentration around the ferrite nuclei, C_γ. According to the well-known effects of Cr, Ni and Mo on the A3 line and the eutectoid temperature in steels [10], low Cr (with small undercooling) and Ni additions in the mild steel weld metal lowered C_γ, thus decreasing the ferritic nucleation rate, while Mo raised C_γ, increasing this rate. But, it should be noted that, in fact, the site saturation of energetically favorable austenite corners occurs before about 20% transformation [10], so, with a small undercooling, the actual nucleation rate was not as important as the growth rate of ferrite. The growth rate of grain-boundary ferrite (Vgb) forming at a small undercooling, according to the change of carbon concentration across the growing interface, can be expressed as:

$$Vgb = [D_C^{\gamma} (C_{\gamma}-C_0)] / [L (C_{\gamma}-C_{\alpha})] \quad [11]$$

where Dc^{γ} is the carbon diffusion coefficient in austenite, L is the carbon-enriched length ahead of the growth interface, and C_{α} is the carbon content of ferrite.

If L could be considered as the square root of Dc^{γ} t, where t is time, then

$$Vgb = [(D_C^{\gamma})^{1/2} t^{-1/2} (C_{\gamma}-C_0)] / (C_{\gamma}-C_{\alpha})$$

By similar reasoning, the growth rate of side-plate ferrite (Vsp) forming at an intermediate undercooling is:

$$Vsp = [D_C^{\gamma} (C_{\gamma}-C_0)] / [A \;\; r (C_{\gamma}-C_{\alpha})] \quad [11]$$

where A is a constant and r is the tip radius of the side plate. An approximation in the case of low carbon weld metal,

$$C_0 \rightarrow C_{\alpha}, \text{ thus } (C_{\gamma}-C_0) \rightarrow (C_{\gamma}-C_{\alpha}), \text{ so}$$

$$Vgb \rightarrow [(D_C^{\gamma})^{1/2} t^{-1/2}] \text{ and } Vsp \rightarrow [D_C^{\gamma}/A \; r]$$

It follows that the growth rates of both grain-boundary ferrite and side-plate ferrite in low carbon steel welds were principally governed by the diffusion coefficient of carbon in the austenite matrix. Cr, Ni and Mo, being dissolved in austenite-Fe, raised the diffusion activation energy of carbon and reduced the diffusion coefficient of carbon in austenite, thereby effectively depressing the growth of both grain-boundary ferrite and side-plate ferrite. Furthermore, Ni belonged to the group of elements which, during ferrite growth, caused a partition between austenite and ferrite under local equilibrium conditions [10]. Hence, the diffusion of Ni atoms also governed the growth rate of ferrite, resulting in a further restraint of grain-boundary ferrite and side-plate ferrite.

FORMATION OF ACICULAR FERRITE AND BAINITIC FERRITE - The formation of acicular ferrite has been related to the existence of weld inclusions [12]. Finely dispersed inclusions inside austenite grains provide a large number of energetically favorable sites for the nucleation of acicular ferrite, due to the strain energy caused by the difference in thermal contraction between inclusions and the austenite matrix.

Generally, in the absence of those additional alloying elements, grain boundaries are more favorable sites for ferrite nucleation than inclusions. In mild steel electroslag welds, the formation of grain-boundary ferrite raised the carbon content within the thin columnar austenite grains, thereby reducing the supersaturation and making intergranular nucleation even more difficult. As a result, less acicular ferrite formed in the thin columnar grain region. Only in the untransformed interior of coarse columnar austenite grains, where the enrichment of carbon played a less significant role, and with more undercooling, which made more inclusion sites become energetically favorable, could acicular ferrite form. Since Cr and Mo, both ferrite stabilizers, displaced the upper part of the austenite decomposition curve (C-curve) upward to the right and raised the ferritic transformation temperature, the following process is proposed. Above a certain Cr-Mo content, the formation of acicular ferrite is less energetically favorable and there is a decreasing or even no chance for intersection between the cooling curve associated with ESW and the upper C-curve. Thus, the austenite matrix can bypass the upper C-curve and transform to a fully bainitic microstructure. On the contrary, the presence of Ni, an austenite stabilizer, shifts the whole C-curve downward to the right and causes the ferritic transformation temperature to decrease. Therefore, for relatively wide ranges of Ni content and transformation temperature, the cooling curve associated with ESW continuously intersects the ferritic transformation curve, resulting in a fine acicular ferrite-predominating microstructure throughout the weld metal. No doubt, excessive Ni will further delay the ferritic transformation and eventually produce bainite or even martensite. In this investigation, the beneficial role of Ni in promoting acicular ferrite has extended to at least 1.5% Ni (plus 0.3% Mo and 1.1% Mn). Of course, the comprehensive function of multi-alloy elements is often not a simple linear summation. In this investigation, the concurrent addition of Mo plus Cr exerted too much of a negative effect on ferritic transformation, but a small amount of Mo could help Ni to promote the formation of acicular ferrite.

HARDNESS AND CVN TOUGHNESS - Excessively overmatched hardness in the weld metal due to alloying was not desired since maintaining high CVN toughness was the major concern. In comparison with the hardness of the base metals (Fig. 5), the Cr-Mo addition, due to its greater hardenability, raised the weld hardness for more than the Ni-Mo addition. Referring to a 20 Joule (15 ft-lb) minimum CVN impact energy at -18°C (0°F) and at the quarter-thickness/centerline location stipulated by AWS code D1.1, the experimental data very clearly demonstrated the main benefit to the impact toughness gained by Ni-Mo additions and the unsatisfactory behavior resulting from Cr-Mo additions (Fig. 5). On the basis of the foregoing, low toughness of the weld metal containing more than 1.1% Cr-0.4% Mo was attributed to the bainitic matrix which exhibited poor resistance to cleavage fracture, and that substantial improvement in the weld metal containing more than 0.9% Ni-0.2% Mo was brought about by the fine and interlaced acicular ferrite substructure which was advantageous in reducing the mean free path of cleavage fracture. For the mild steel electroslag weld metal, toughness at the half-thickness/centerline location was usually much worse than for the quarter-thickness/centerline. Ni-Mo additions significantly made both the hardness and toughness of the entire weld more uniform. Fig. 6 demonstrates that the transition curves for impact tests of both half-thickness and quarter-thickness locations in the Ni-Mo added weld metal overlapped.

Obviously, keeping a certain Ni-Mo content was the key point to obtaining a satisfactory toughness level in the weld. A strictly maintained narrow joint gap became quite critical to obtain reduced base metal dilution and rapid welding speeds. Consequently, TW8544 tubular filler metal (Table 1) allowed higher welding speed, required less melting power and resulted in reduced penetration. Tubular filler metal effectively decreased the heat input (and base metal dilution) of welds and, thus, maintained higher percentages of alloying elements in the weld metal. Figs. 7 and 8 show these effects of tubular wire on the heat input and base metal dilution, respectively.

A NEW CHARACTERISTIC PARAMETER FOR ACICULAR FERRITE STUDY - Conventionally, the width of acicular ferrite plates was initially used as its characteristic parameter. But, due to an uneven distribution of weld inclusion sizes and spacings, the acicular ferrite width varied greatly. In view of this fact, another characteristic parameter, the density of acicular ferrite plates (i.e., the number of acicular ferrite plates in a unit area), was used for the acicular ferrite study. Fig. 9 shows that the density of acicular ferrite plates was almost linearly proportional to the Charpy V-notch impact toughness. Fig. 10 indicates that this density increased with increasing weld metal Ni-Mo content.

This new characteristic parameter has already been proved more convenient and more logical. For example, the relatively low toughness in those acicular ferrite matrixes induced by Cr-Mo additions (about 0.7% Cr-0.3% Mo) (Fig. 9) could correlate to their low density of acicular ferrite plates and larger mean free path for cleavage fractures.

CONCLUSIONS

This ESW investigation led to the following conclusions:

(1) Small additions of Cr, Ni and Mo tended to promote equiaxed dendritic growth at the weld center of mild steel welds. The equiaxed structure possessed substantially improved resistance to hot cracking.

(2) The presence of Cr, Ni and Mo in the weld metal effectively inhibited the formation of grain-boundary ferrite and side-plate ferrite.

(3) At an alloy content below 0.7% Cr-0.3% Mo, Cr-Mo addition promoted the formation of acicular ferrite, and above that resulted in bainitic ferrite with low impact toughness.

(4) Ni additions produced weld metal microstructures rich in fine acicular ferrite with improved toughness. 0.9-1.5% Ni plus 0.1-0.3% Mo was the optimum alloy content in A36 and A588 weld metals for best toughness.

(5) Alloy powder-cored filler metal effectively reduced the heat input and the base metal dilution.

ACKNOWLEDGMENT

The authors are grateful for the support of the DoT/Federal Highway Administration (Contract DTFH61-83-R-00062).

REFERENCES

1. Ann, H. S., Yu, D., Devletian, J. H. and Wood, W. E., "Narrow-Gap High Deposition Rate ESW", presented at the 66th Annual AWS Convention, Las Vegas, NV, 1985.
2. "Technology Forecast '86", Metals Progress, Vol. 13, January, 1986, pp. 13.
3. Eichhorn, F., Hirsch, P. and Wubbels, B., IIW Document KII-J-76-80, IIW, 1980.
4. Andreev, V. P. and Novikova, D. P., Auto. Weld., Vol. 10, 1981, p. 46.
5. Grong, O. and Matlock, D. K., Int. Met. Rev., Vol. 31, No. 1, 1986, p. 27.
6. Yu, D., Ann, H. S., Devletian, J. H. and Wood, W. E., "Welding: The State of the Art", E. Nippes (ed.), ASM, 1986, pp. 21-32.
7. Savage, W. E., Lundin, C. D. and Aronson, R. H., Welding Journal, Vol. 44, No. 4, 1965, p. 175-s.
8. Porter, D. A. and Easterling, K. E., "Phase Transformation in Metal and Alloys", Van Nostrant Reinhold, London, 1981, p. 29.
9. Hsu, Z. Y., "Thermodynamics of Metallic Materials" [in Chinese], Science, Peking, 1981, p. 280.
10. Honeycombe, R. W. K., "Steel: Microstructure and Properties", Edward Arnold, London, 1981, pp. 32-75.
11. Verhoeven, J. D., "Fundamental of Physical Metallurgy" John Wiley and Sons, New York, NY, Chapter 12, 1975.
12. Ricks, R. A., Howell, P. R. and Barritte, G. S., Journal of Materials Science, Vol. 17, 1982, pp. 732-740.

Table 1 - The Compositions of Base Metal, Electrodes and Guide Plate (wt.%)

Material	C	Mn	Si	Cr	Ni	Mo	Cu	S	P	V
A36	0.17	0.92	0.20	0.02	0.10	--	0.20	0.02	0.01	0.06
A588	0.18	1.20	0.37	0.56	0.16	--	0.32	0.02	0.01	0.08
Hobart-25P	0.11	1.12	0.50	--	--	--	0.32	0.02	0.02	--
Linde-WS	0.09	0.50	0.30	0.55	0.50	--	0.30	0.03	0.02	--
Page-AS521	0.10	0.80	0.50	2.25	--	1.00	--	0.03	0.02	--
Airco-AX90	0.08	1.40	0.46	0.06	2.10	0.40	--	0.02	0.02	0.007
Stoody-TW8544*	0.03	1.20	0.45	--	2.30	0.45	--	0.02	0.02	0.05
1010	0.10	0.45	--	--	--	--	--	0.05 (max)	0.04 (max)	--

* Stoody Powder Cored Tubular Wire (Balance: Fe)

Table 2 - The Composition (wt%) of As-Deposited Weld Metal

Plate	Filler	C	Mn	Si	Cr	Ni	Mo	O
A36	25P	0.18	1.10	0.40	--	--	--	0.0240
(50 mm)	AS521	0.16	1.10	0.40	1.20	0.10	0.40	0.0246
	AS521+25P	0.14	1.01	0.30	0.50	0.10	0.20	0.0240
	AX90	0.17	1.10	0.36	0.03	0.91	0.15	0.0230
	AX90+25P	0.16	1.10	0.36	0.03	0.50	0.10	0.0227
	TW8544	0.13	1.31	0.35	0.03	1.13	0.21	0.0260
A588	WS	0.18	1.05	0.35	0.40	0.36	--	0.0230
(50 mm)	AS521	0.16	1.02	0.40	1.25	0.10	0.40	0.0221
	AS521+WS	0.16	1.05	0.36	0.79	0.07	0.30	0.0233
	AX90	0.13	1.20	0.36	0.32	0.94	0.16	0.0260
	AX90+WS	0.14	1.12	0.34	0.41	0.53	0.08	0.0236
	TW8544	0.11	1.21	0.35	0.15	1.30	0.20	0.0280

(Balance: Fe)

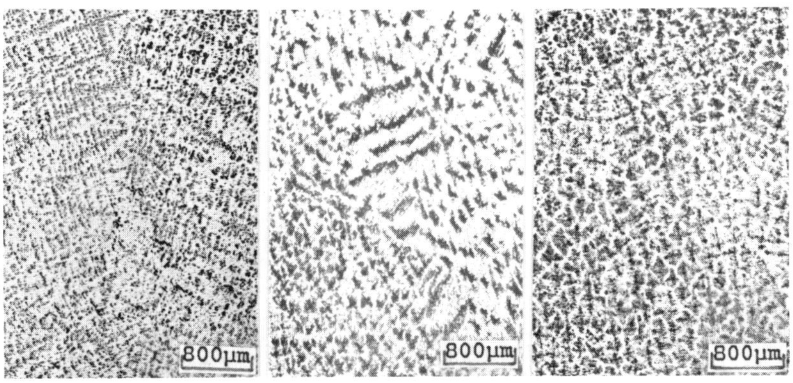

Fig. 1. Primary Dendrite Structures in the Weld Center of 50 mm A36 Electroslag Weldments using the Filler Metals of (a) 25P, (b) AS521, and (c) AX90.

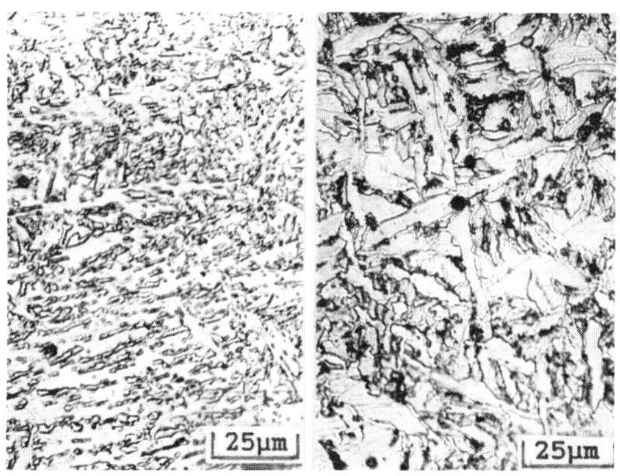

Fig. 3. Morphologies of Alloyed A36 [50 mm Thick] Electroslag Weld Metals: (a) Bainitic Ferrite Generated by Cr-Mo Addition, and (b) Acicular Ferrite Generated by Ni-Mo Addition.

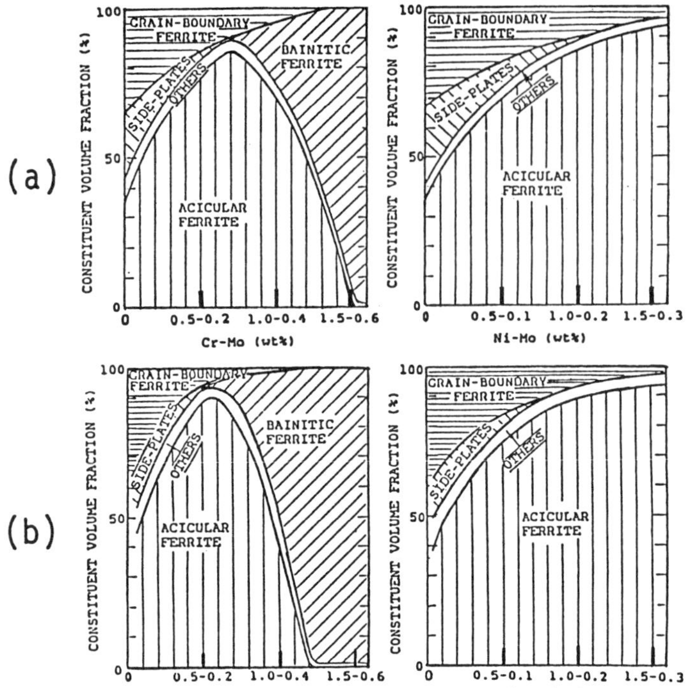

Fig. 2. Effect of Alloy Additions on the Constituent Volume Fractions of (a) A36, and (b) A588 [50 mm Thick] Electroslag Weld Metal.

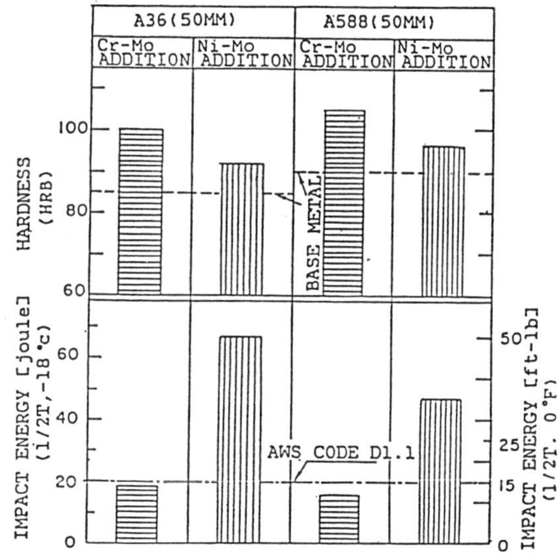

Fig. 5. Comparison of Mechanical Behavior Between Weld Metal Containing Cr-Mo and Ni-Mo Steel Filler Metal.

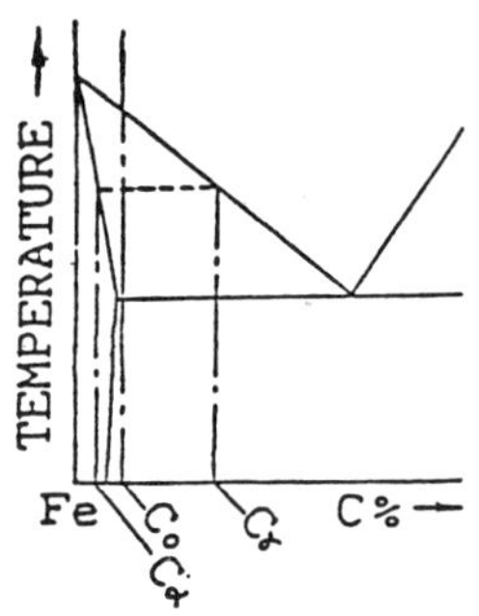

Fig. 4. Variation of Carbon Concentration During Austenite Decomposition.

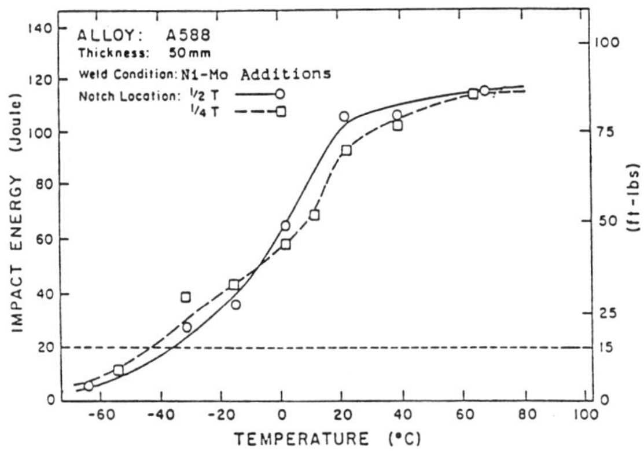

Fig. 6. CVN Impact Energy Transition Curves for Weld Metal Deposited on A588 Using TW8455 Filler.

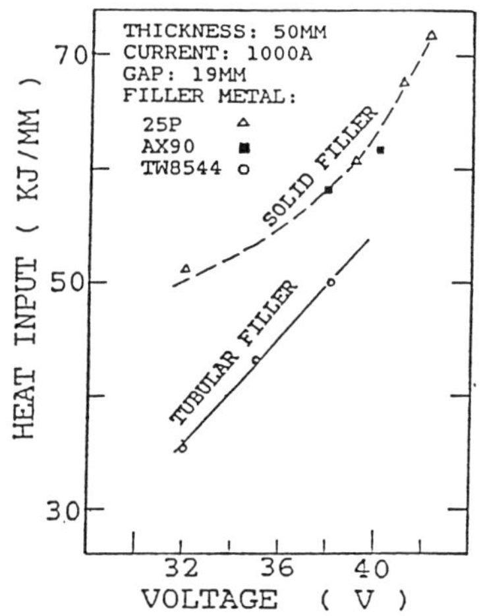

Fig. 7. Effect of Tubular Filler Metal on Heat Input of A36 Electroslag Welds.

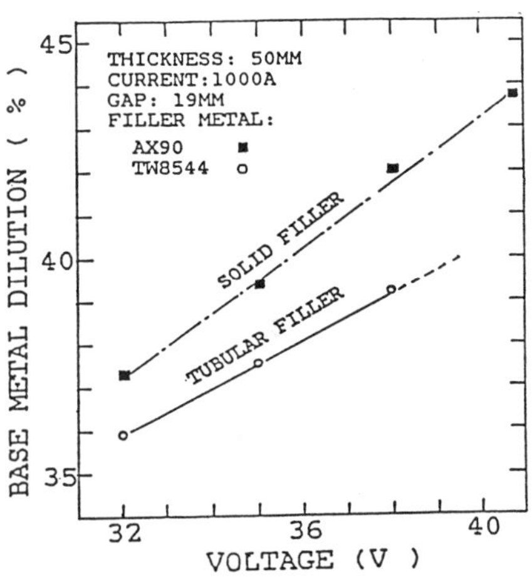

Fig. 8. Effect of Tubular Filler Metal on Base Metal Dilution of A588 ESW.

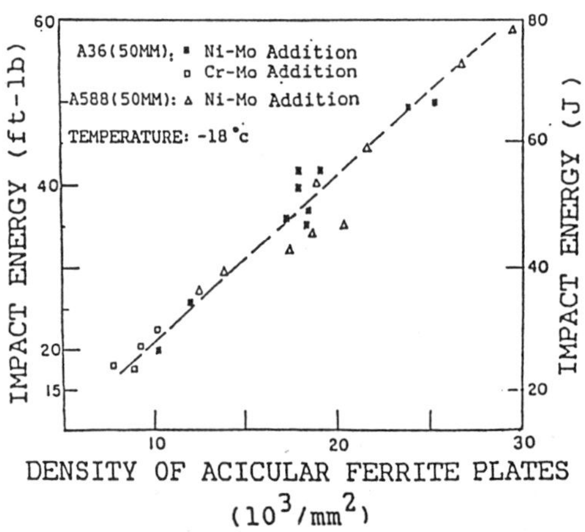

Fig. 9. Correlation Between the Density of Acicular Ferrite Plates and CVN Impact Toughness.

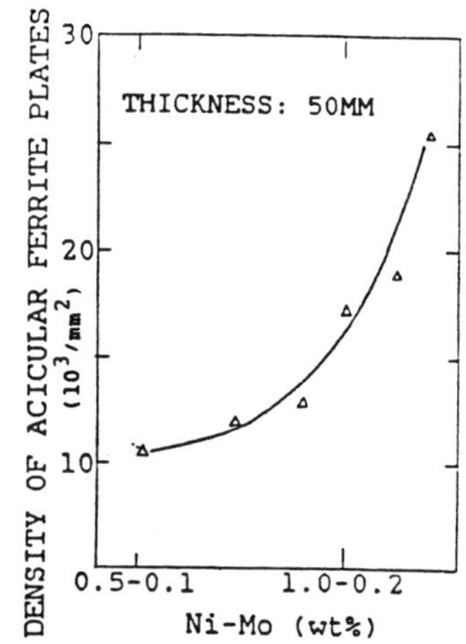

Fig. 10. Correlation Between Alloy Contents and the Density of Acicular Ferrite Plates in A36 Electroslag Weld Metal.

MICROSTRUCTURAL REFINEMENT THROUGH MULTI-PASS GTAW PROCESS

H.J. Kim
Hyundai Heavy Ind. Co., Ltd.
Ulsan, Korea

J.W. Morris, Jr.
University of California, Berkeley
Berkeley, California

ABSTRACT

The present work was undertaken to illustrate and clarify the microstructural features of martensitic Fe-Ni weld deposits made by a multipass welding process. To simulate this, rapid thermal treatments were applied to an Fe-12Ni-0.25Ti alloy using (1) an induction furnace and (2) bead-on-plate welding. The rapid thermal treatment refines the microstructure significantly by destroying the previous packet structure. Hence, the term 'packet refinement' is used to distinguish its microstructural features from the grain refinement associated with recrystallization of austenite. During multipass GTAW process with an 11% Ni ferritic filler wire, a coarse columnar grain structure is formed after solidification but this structure is refined repetitively by the rapid thermal cycles of subsequent passes. The result is a fully packet refined and extremely fine structure, in the order of several microns, throughout the weldment.

THE THERMAL HISTORY of a weld thermal cycle is highly dynamic. Hence, the solid-solid phase transformations experienced in a welded region are different from those during conventional furnace or thermomechanical heat treatment. The weld cycle accomplishes either a very rapid austenitizing cycle or a very short time tempering treatment depending on its peak temperatures. Previously, the present authors [1,2] applied rapid thermal cycles to 9 and 12% Ni steels to simulate the microstructural changes occurring in the HAZ. These studies demonstrated that the simulated HAZ microstructures are quite different from those obtained by corresponding furnace heat treatments. For instance, a rapid austenitizing cycle refines the microstructure by destroying the previous packet structure (packet refinement) while an austenitizing treatment performed in the furnace for a longer time refines the size of the prior austenite grains (grain refinement) through the recrystallization process. It was also demonstrated that the multi-cycle treatment was more effective in refining the effective grain size and, thus, suppressed the ductile-brittle transition temperature (DBTT).

The present research was performed to extend these microstructural studies to actual multipass weld deposits. The first part of the paper will described the effect of simulated weld thermal cycles in an induction furnace. The microstructural features obtained after this treatment have been published elsewhere [2] but some of them are quoted here to help in the understanding of the characterisitcs of "packet refinement". Later, the microstructural changes due to actual weld thermal cycles applied in the HAZ of bead-on-plate weldment and in the multipass weld deposit are presented. In the multipass weldment, each bead is expected to experience weld thermal cycles due to subsequent passes and is thus packet-refined repetitively to yield a very fine structure.

MATERIALS AND EXPERIMENTAL PROCEDURE

RAPID THERMAL CYCLES - The alloy used for cyclic treatment has the nominal composition of Fe-12Ni-0.25Ti with less than 0.01C. A plate of this alloy was annealed at 1200C for three hours furnace cooled to 900C and quenched in water. Its condition is designated as 'AN'. The 'AN' plate was machined and sliced to slightly oversized Charpy specimens. Single and double austenitizing cycles were applied to these specimens in an induction furnace. During this treatment, the specimen was heated to 790C within 15 seconds and then immediately dropped into an oil bath. A single such cycle constitutes the 'C' treatment: a double cycle is designated 'CC'.

BEAD-ON-PLATE WELDMENT - In order to investigate the effect of peak temperature on microstructure, a GTAW bead-on-plate weld was made on the annealed Fe-12Ni-0.25Ti alloy (AN). Its HAZ microstructure was analyzed by both optical and fractographic means. To perform the fractographic analysis, the welded plates were sliced into standard Charpy specimens and then temper embrittled at 450C for 500 hours. After inserting a 0.2 mm saw cut within the weld metal as shown in Fig. 1, the embrittled specimens were hydrogen charged in a solution of 1 N sulfuric acid at a current density of 10 mA/mm². The hydrogen-charged specimens were bolt loaded immediately after charging to initiate and to propagate a crack through the HAZ. After the crack ceased propagating into the base metal, the half broken specimens were fractured into two pieces. This method is quite effective in

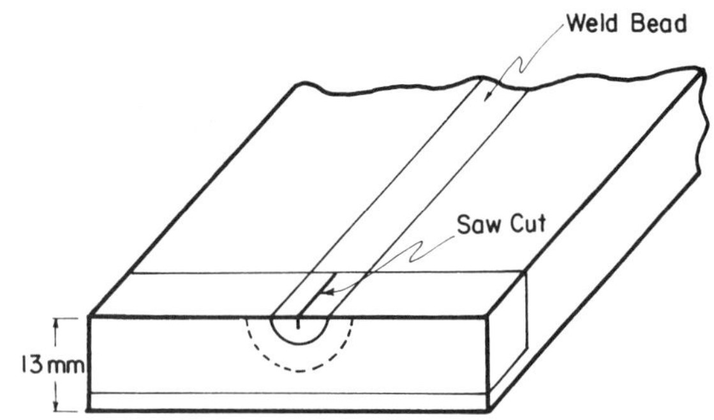

Fig. 1 - Schematical drawing of bead-on-plate weldment and location of specimens taken for the fractographic analysis. XBL 847-7186

promoting intergranular fracture in the annealed or the recrystallized specimens [2]. The fractographic information obtained was correlated with the corresponding microstructure revealed on the etched specimen.

MULTIPASS GTAW WELDMENT - A fully weld deposit material was fabricated by building up weld beads on the 24.5 mm thick 9% Ni steel plate. Using a fully automated GTAW process, weld beads were deposited side by side and layer by layer until the size of the weld deposit reached about 11 x 38 x 3.6 cm. The chemical compositon of the filler wire and weld deposit are shown in Table 1. The weld deposit contains 11% Ni and a very low content of interstitial elements.

Weld deposit samples for optical microscopy were polished and etched with a 5% Nital or an acidified ferric chloride solution. Samples for transmission electron microscopy were spark-cut from chemically-thinned slices of weld deposit. They were electropolished in a twin jet polishing apparatus using a chromic-acetic acid solution. The specimens were examined in a Philips 301 electron microscopy at 100 kV.

RESULTS

MICROSTRUCTURE OF THE THERMALLY CYCLED SPECIMENS - The microstructure of the 12Ni alloy in the annealed and quenched (AN) state is illustrated by the optical micrograph in Fig. 2a. This micrograph shows coarse grains of the prior austenite phase, each of which is subdivided into several packets of lath martensite. The boundaries of the packets are irregular, as evidenced by their jagged appearance in the plane of the micrograph and by the appearance of islands of one packet in the interior of its neighbor. Detailed TEM studies performed on this microstructure [2] showed that a single martensite variant is present within a packet. The packet size is the effective grain size governing transgranular cleavage since each packet shares a (100) cleavage plane. The microstructure of the alloy in single and double, cycled conditions ('C' and 'CC', respectively) are shown in Figs. 2b and 2c. In the 'C' condition the prior austenite grain boundaries are roughened and partly obscured by the martensite transformation products formed across the boundaries. In 'CC' condition prior austenite grain boundaries can hardly be found and the original packet structure was completely and homogeneously destroyed into very fine scale. There is no evidence of recrystallization, i.e., nucleation of new austenite grains and growth in the 'C' and 'CC' microstructures. Following the fractographic and TEM evidence, the present authors [2] have used the term 'packet refinement' to distinguish this kind of microstructural change from the 'grain refinement' associated with the conventional austenitizing treatment.

HAZ MICROSTRUCTURE OF THE BEAD-ON-PLATE WELDMENT - The bead-on-plate welding process generates high rates of heating and cooling in the parent metal adjacent to the weld and produces a band of HAZ surrounding the fused weld metal. The HAZ extends from the fusion line, which experiences the liquidus temperature, to an outer location heated to a lower critical temperature. By investigating the HAZ microstructure from fusion line to outer limit, the microstructural changes due to successively lower peak temperatures can be assessed. The

Table 1. Chemical Composition of Filler and Weld Material.

Material	C	Ni	Mn	Si	P	S	Fe
Filler Wire	0.02	11.1	0.39	0.02	0.003	0.002	bal.
Weld Metal	0.01	11.0	0.39	0.02	-	-	bal.

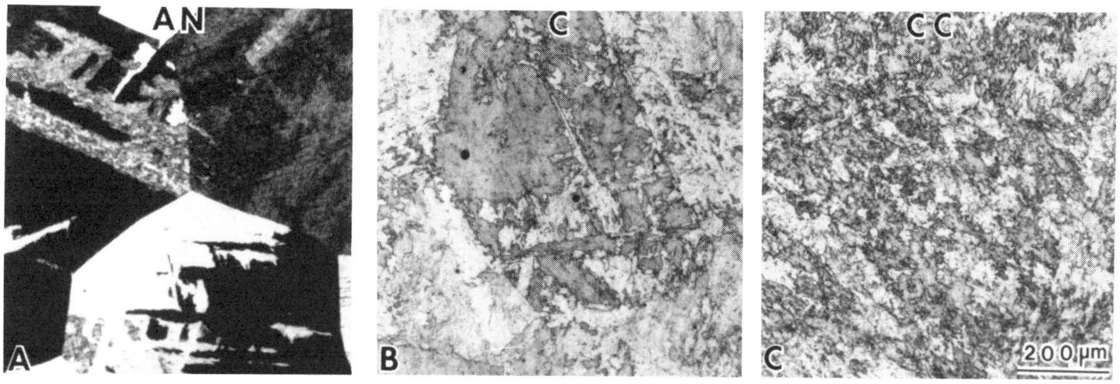

Fig. 2 - Optical micrographs of 12Ni alloy: (a) anneal and quenched (AN), (b) a single cycled, and (c) double-cycled conditions. XBB 855-4037

macroscopic view of a bead-on-plate weld made on an annealed 12Ni plate is shown in Fig. 3. As can be seen in this figure, the HAZ structure, which extends about 5mm from the fusion line, has a much finer grain structure than the base metal. This feature becomes more clear in the magnified picture (Fig. 4) taken across the HAZ. To visualize the small grains formed along the fusion boundary, fractographic analysis was performed using the method described

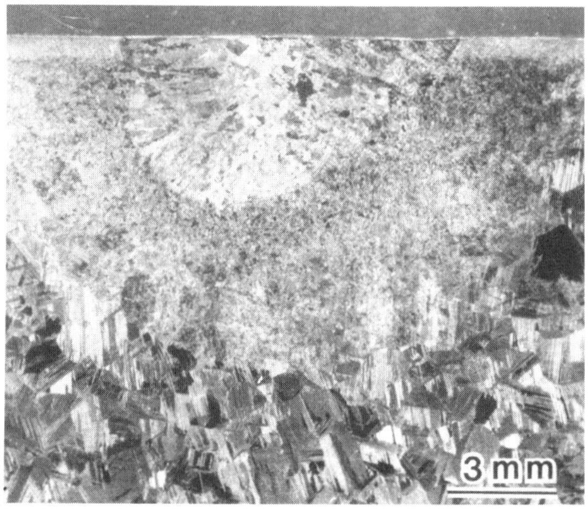

Fig. 3 - Microstructure of the bead-on-plate weldment. XBB 843-2117

earlier. The result is represented in the SEM fractographs shown in Fig. 5. As shown in Fig. 5a, the columnar grains in the weld metal and the equiaxed grains in the HAZ immediately adjacent to the fusion line fracture in an intergranular mode. However, the fracture mode suddenly changes to ductile dimple in the outer side of HAZ and then to intergranular again in the base metal. Figure 4b clearly illustrates the equiaxed HAZ grains next to the fusion boundary, as well as the columnar grains grown on the half-melted HAZ grains. These grains are much smaller than the base metal grains as observed in the optical means. It follows that grain refinement took place in the area surrounding the weld metal to a width of about 100 μm during the weld thermal cycle. It is also clear from Fig. 4 that the grains become smaller with increasing distance from the fusion line, i.e., with lowering of peak temperature.

In contrast to the grain refinment near the fusion boundary, the microstructural refinement in the outer region of HAZ seems to be accomplished by the destruction of the original packet structure, i.e., by 'packet refinement' [2]. The packet refinement is illustrated in the bottom of Fig. 4. In the boundary area between the HAZ and the base metal, a portion of the base metal grain is destroyed but the other half of the grain

Fig. 4 - Optical picture of bead-on-plate weldment showing the distinguishable change in microstructure across the HAZ. XBB 843-2118A

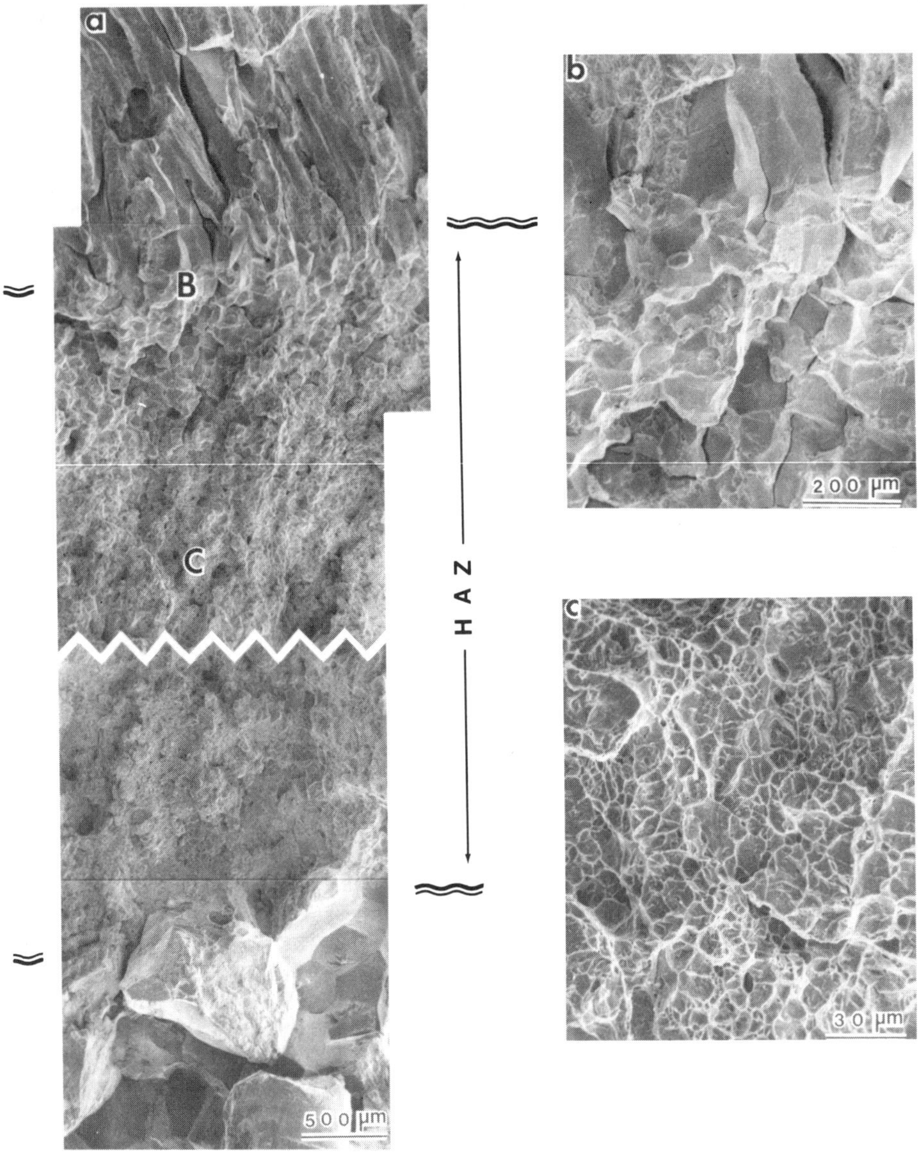

Fig. 5 - SEM fractographs showing (a) the fracture mode change occurring from weld metal to base metal, (b) and (c) are the magnification of the areas maked B and C, respectively in (a). XBB 845-3296

preserves its original shape. In the packet-refined region, the microstructure is quite similar to that in the 'C' specimen (Fig. 2b). Optical examination of the cross-section of the broken specimen shows that the ductile dimple region of the fracture surface shown in Fig. 5a matches well with the packet-refined part of the HAZ.

Figure 6 shows partially overlapped bead-on-plate weldments. Detailed examination of the weld metal structure shows the elongated columnar grains of the previous equiaxed grains next to the fusion line and the 'packet-refined' structure in the outer HAZ. The columnar grains formed in earlier passes did not coarsen but were mostly refined into fine grains or packets during the weld thermal cycle of subsequent passes.

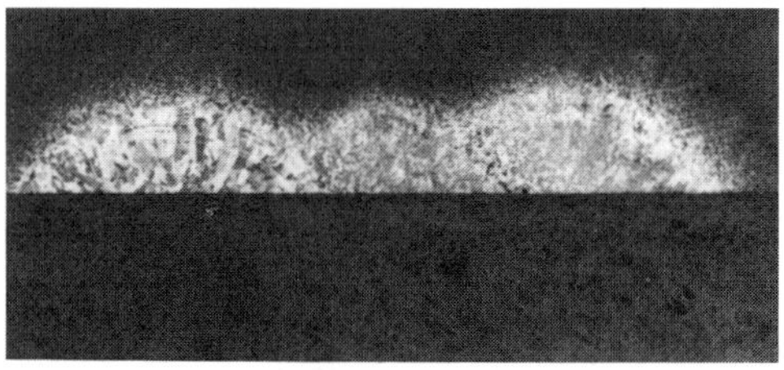

Fig. 6 - Microstructure of all weld deposit speciman etched with (a) 5% Nital and (b) acidified $FeCl_3$ solutions. XBB 805-6225

MICROSTRUCTURE OF MULTIPASS WELD DEPOSIT - As described in section II, an all-weld deposit plate was fabricated and its microstructure studied with optical and transmission

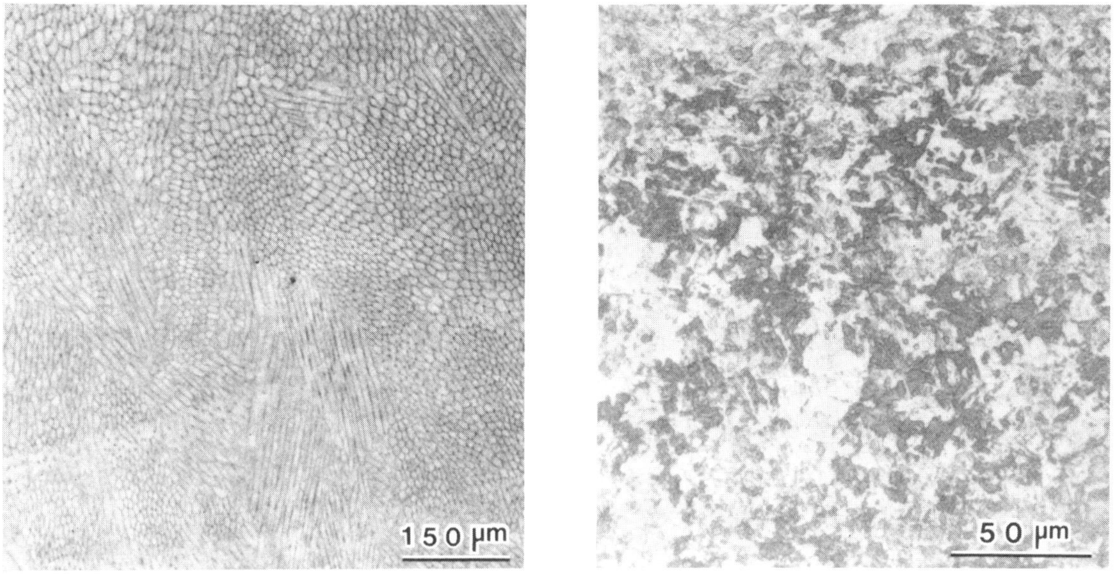

Fig. 7 - Microstructures of all weld deposit specimens etched with (a) 5% Nital and (b) acidified $FeCl_3$ solutions. XBB 846-4528

electron microscopy. The characterization specimens were sliced perpendicular to the welding direction. Figures 7a and 7b show optical microstructures after etching with 5% Nital and acidified $FeCl_3$ solutions, respectively. Figure 7a shows the cellular solidification structure that is typical in the Fe-(9-12)Ni system [3,4]. Even though the solidification structure is preserved, the transformed structure of the weld deposit is severely packet refined, as shown in Figure 7b. The microstructure is quite similar to that of double-cycled specimen ('CC') shown in Fig. 2c. The results of transmission electron microscope (TEM) are presented in Fig. 8. The area shown in the bright field micrograph contains at least three different orientations that yield distinct martensite diffraction patterns as shown in the figure. The distinct variants can be separated with dark field techniques as shown in Figures 8c, d and e. The boundaries between areas having different orientations are jagged and have sharp edges indicating that they are

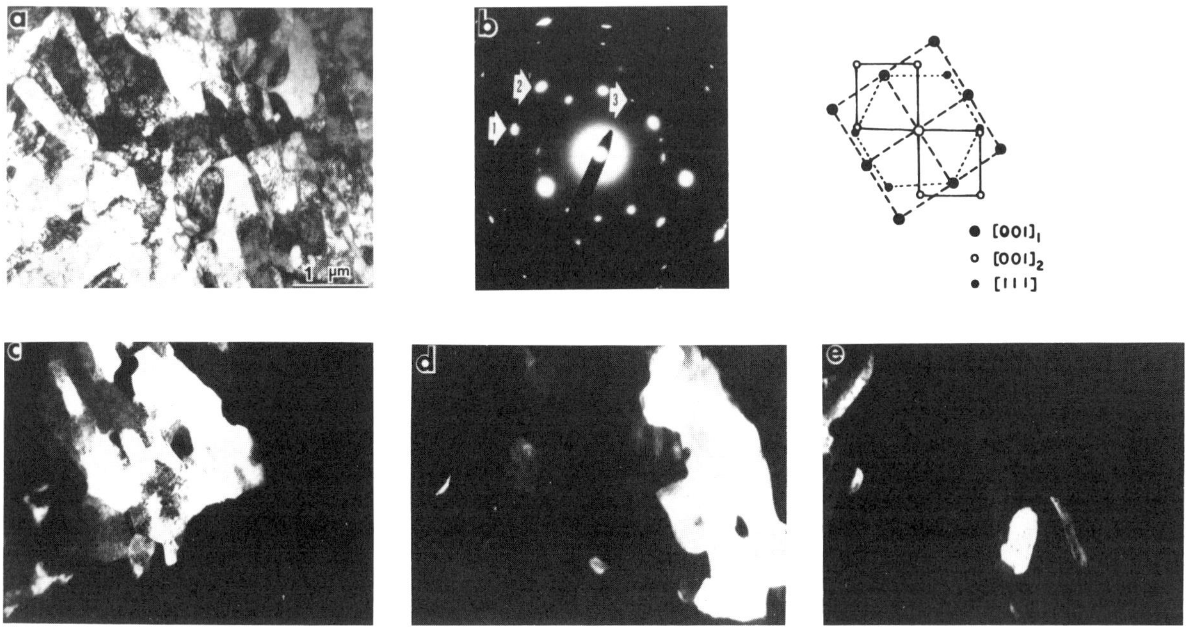

Fig. 8 - Transmission electron micrographs of the multipass weld deposit: (a) bright field image, (b) selected area diffraction pattern, (c), (d) and (e) are dark field images taken from the diffraction spots marked with arrows 1, 2 and 3, respectively in (b). XBB 847-5313

more like packet boundaries than grain boundaries. No grain boundary-like feature, i.e., smooth boundaries, was observed throughout the TEM examination. The TEM study demonstrates that different martensite variants are intermixed on a very fine scale resulting in an extremely fine effective grain size, in the order of several microns.

DISCUSSION

It is clear from the above that the HAZ in Fe-9-12Ni steels contains two regions with distinctly different microstructures: (a) a grain-refined region and (b) a 'packet-refined' region. The grain-refined region borders the weld metal while the packet-refined region constitutes most of HAZ. The results indicate that the type of refinement depends on the peak temperature of the weld thermal cycle. Immediately adjacent to the fusion boundary, the parent metal is raised to a temperature well above the austenite transformation temperature. Substantially complete transformation to austenite takes place, and the temperature is high enough for nucleation and growth of new austenite grains to occur. Maki [5] and others [6] have proposed that the driving force for recrystallization is the strain energy in the austenite phase after reverse shear transformation. If the recrystallized grains are smaller (or larger) than the original grains, the term 'grain refined (or coarsened) region' can be used to describe the resulting microstructure. Further away from the fusion boundary, the peak temperature is lower. Where the transformation temperature is exceeded but the recrystallization temperature (or time) is not, the austenite formed by the reverse shear transformation has a high residual defect density. Given the high density of defects and the propensity for the heterogeneous nucleation of martensite on cooling, it appears that the cooperative growth of adjacent laths is suppressed so that martensite laths from in small elements whose variants are more randomly chosen from among the 24 variants permitted by the Kurdjumov-Sachs (K-S) orientation relationship. This results in the destruction of the previous packet and in the 'packet-refined' structure.

In the case of a multipass weldment the as-cast columnar grains can be grain-refined and/or packet refined by subsequent passes. Moreover, if the HAZ of successive packets overlaps, the refined structures are further refined into fine packet structures by the rapid thermal cycles associated peak with still latter passes. This results in a fully refined packet and extremely fine structure throughout the weldment (effective grain site of several microns), as shown in Figs. 7 and 8. The fine effective grain size and the absence of clear prior austenite boundaries suggest that the multipass weld deposit may have exceptional resistance to low temperature cleavage and to embrittlement by tempering or hydrogen. The result of appropriate tests will be presented separately.

CONCLUSION

The microstructural refinement in the HAZ of weldments in 9-12Ni steel consists of grain refinment near the fusion boundary and packet refinement in the bulk of the HAZ. In the case of multipass weldment, the fine grains formed near the fusion boundary are further refined into fine packets by the weld thermal cycles of subsequent passes to lower peak temperatures. If the successive packets overlap, this results in a wholly packet refined structure.

ACKNOWLEDGEMENTS

This work was supported by the Director, Office of Energy Research, Office of Basic Energy Science, Material Science Division of the U.S. Department of Energy under contract No. DE-AC03-76SF00098.

REFERENCES

1. Kim, H.J., H. Shin and J.W. Morris, Jr., Proc. of Int. Cryo. Mat. Conf., Kobe, Japan, pp. 129-133 (1982)
2. Kim, H.J., Y.H. Kim and J.W. Morris, Jr., Lawrence Berkeley Laboratory Report #17617, Berkeley, CA 94720 (1982)
3. Devletion, J.H., J.R. Stephen and W.R. Witzke, Welding J., 56:4, 97-102S (1977)
4. Agusa, K., M. Koshio, N. Nishiyama, A. Kamada and Y. Nakano, Kawasaki Steel Technical Report No. 6, pp. 1-11 (1982)
5. Maki, T., H. Morimoto and I. Tamura, Trans. ISIJ, 20:10, pp. 700-706 (1980)
6. Nakazawa, K., Y. Kawabe and S. Muneki, Trans. ISIJ, 23:4, pp. 347-356 (1983)

WELD HAZ THERMAL CYCLES MEASUREMENT VS. COMPUTATION

Hugo S. Ferguson, David L. Hollinger
Duffers Scientific, Inc.
Troy, New York

ABSTRACT

Modern computer controlled simulators reproduce HAZ thermal cycles with good precision. The thermal cycles used are derived from two main sources - measured values and calculated values. In this investigation two methods of calculation are used; one initiated by Rosenthal and later modified by Rykalin, and the other due to Hannerz. The Hannerz computation is idealized and contains only peak temperature, preheat temperature and cooling time from 800 to 500 C.

Thermal cycles computed by these two methods are compared with thermal cycles directly interpolated from F(s,d) tables due to Nippes and Savage. The F(s,d) tables are obtained directly from measured values. The HAZ thermal cycles are now easily obtained from the F(s,d) tables by newly developed HAZ software.

Recent Continuous Heating Transformation (CHT) studies indicate more attention must be paid to analyzing HAZ rapid thermal cycles and resulting microstructures, particularly in the areas of the A1 and A3 in plain carbon and low alloy steels. Substantial depression of both the A1 and A3 are seen at high heating rates.

In steels the austenite formed by these high heating rates is very unstable. Processes, which have high heating rates, will yield transformation products at temperatures well below the equilibrium A1.

THE HEAT-AFFECTED-ZONE (HAZ) has been the subject of hundreds of major investigations. As welding designs are begun or welding problems occur, each situation usually requires further study of the particular HAZ. A knowledge of the thermal cycle seen at critical regions of the HAZ is required to predict the behavior of the material or to expose the material to thermal conditions which exist in the HAZ, for subsequent testing. Frequently, as in multipass welding, several different thermal cycles are required to evaluate the HAZ at a particular location. The temperature at which a critical region begins or ends is also important. Dynamic heating and cooling, present in welding, can change these temperatures substantially.

With the aim of developing thermal cycles, most studies of heat flow in the HAZ have been undertaken in two discrete areas: computation and direct measurement. There are distinct advantages in both methods. The beginning of heat flow studies, which resulted in a substantial following throughout the welding industry, is attributed to Rosenthal's work (Ref. 1) in 1935. Reviews of the early solutions to HAZ heat flow were done by Spraragen and Claussen (Ref. 2), but the first analytical solution is that of Rosenthal. The welding parameters - current, voltage, welding speed, material variables and geometry, were also first used in the Rosenthal solution. Extensive detail on the study of HAZ heat flow is presented by Rykalin (Ref. 3), which is the most complete book on the subject.

Every study carried out on the calculation of HAZ heat flow must use some source of information to check the correlation between the calculated and measured values. The degree of correlation varies widely in work presented to date by investigators. Some calculations work well for a certain set of conditions and not for others. While computer calculations are easily performed regardless of complexity, the lack of suitable boundary conditions and exact material constants always leads to questionable solutions. This requires that the investigator check with reality by making a few measurements.

The measurement method of HAZ studies has been done extensively by Hess, Nippes and Savage (Refs. 4 - 7). These studies used buried thermocouples in plates subjected to bead-on plate welds. The thermocouple measurements were curve fit using the equation

$$(1) \quad T-To = \frac{F(s,d)}{1000} \text{ X Energy Input X Input Factor}$$

For constants, units, and details, see Appendix I and for further explanation of this equation see Refs. 4 and 6.

The curves were then correlated by a parameter F(s,d), where s is time and d is distance from the weld centerline. The values T-To were then calculated by multiplying the heat input times the value of F(s,d) appropriate for the time s and distance d from the weld centerline under consideration. This approach uses experimental (measured) data to define the temperature profile and avoids the complications of thin or thick plate assumptions. It also avoids the variations in the temperature dependent material constants.

COMPARISONS BETWEEN MEASURED AND CALCULATED THERMAL CYCLES

The measured thermal cycles are derived from the F(s,d) data tables of Nippes and Savage. HAZ software, which interpolates data points from the tables for chosen energy inputs, initial plate temperature, and peak temperatures, was used to generate thermal cycles. The software gives heating and cooling rates at any point on the thermal cycle. The time above a chosen temperature is also available, as is the time between two chosen temperatures. An energy function is also applied to the area of the thermal cycle above a chosen temperature and may be used to compare qualitatively the effect of the time at temperature for different thermal cycles. Once the F(s,d) data tables are available the generation of thermal cycles is fast and accurate. If the data tables are representative of the same general range of energy input, then the final thermal cycle is always a good solution and no other checking is necessary. For additional information on the HAZ software, see Appendix II.

Rosenthal in 1946 showed that the thermal cycle can be analytically derived from Fourier's equations by considering the heat flow outward from a moving point source. This work may be further simplified by considering all energy dissipated in a plane at zero time and diffusing outward from that plane. For the two-dimensional case it is now possible to avoid the Bessel function solution and use the basic solution.

$$(2) \quad T-To = \frac{Q}{2d\sqrt{\pi \lambda c \rho t}} e^{-\frac{x^2}{4\alpha t}}$$

Variables, constants and units are listed in Appendix I.

This equation has been used by some investigators, the first of whom appears to be Rykalin. This equation has now been used by Hannerz in a very interesting simplification.

An idealized thermal cycle for HAZ work has been derived by Hannerz (Ref. 8) for steel. The following assumption was made:

> The cooling time between 800 and 500 C is independent of the distance from the weld centerline.

Measurements by many investigators show this assumption to be reasonable. The Hannerz derivation starts with the two-dimensional case used by Rosenthal and simplified by Rykalin, which is equation (2).

Assuming that the time between 800 and 500 C is independent of x, Hannerz eliminates all coefficients in equation (2) and replaces them by peak temperature, Tmax., preheat or interpass temperature, To, and delta T (800 - 500 C).

The final Hannerz equation is:

$$T - T_o = \sqrt{\frac{A}{300\,B \cdot t}}\; e^{-\left[\frac{A}{2e \cdot 300 \cdot B \cdot (T_{max} - T_o)^2 \cdot t}\right]} \qquad (3)$$

where :

T_{max} is the peak temperature
T_o is the preheat temperature
Δt is cooling time from 800 C to 500 C

$A = \Delta t(500 - T_o)^2 \cdot (800 - T_o)^2$
$B = 1300 - 2T_o$

Using the simplified equation (2) and the Hannerz equation (3) along with thermal cycles generated using the HAZ software and data from the Nippes/Savage work, thermal cycles generated are shown in Figs. 1 - 3 for a peak temperature of 1300 and Figs. 4 - 6 for a peak temperature of 900 C. The advantage of eliminating the "constants" is clearly shown in the plots of the Hannerz equation vs. the data from the F(s,d) tables. The peak temperatures always match, as do the cooling rates between 800 and 500 C. There is some discrepency in the time at temperature.

The first most noticable feature of the Hannerz plots is that they correspond very well with the F(s,d) measurement based curves even though no constants are required in the application of the equation. This is a substantial accomplishment. The

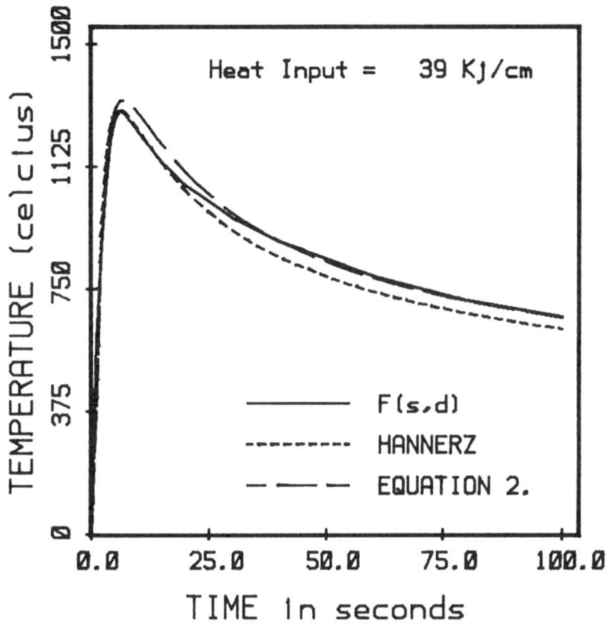

Fig. 2

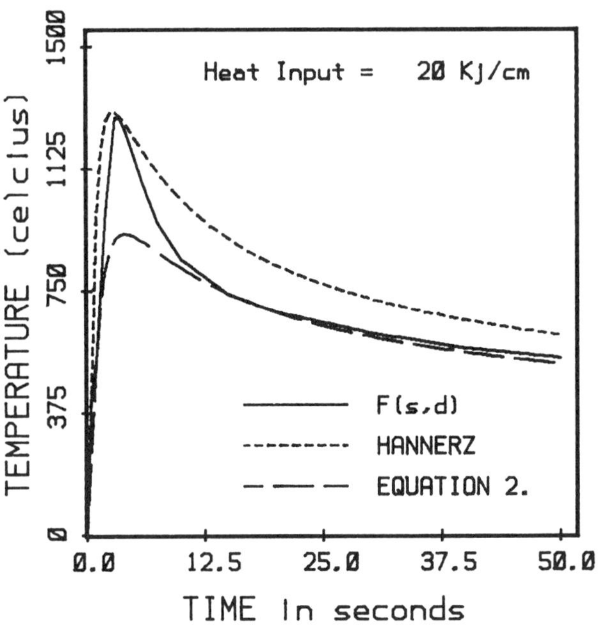

Fig. 1

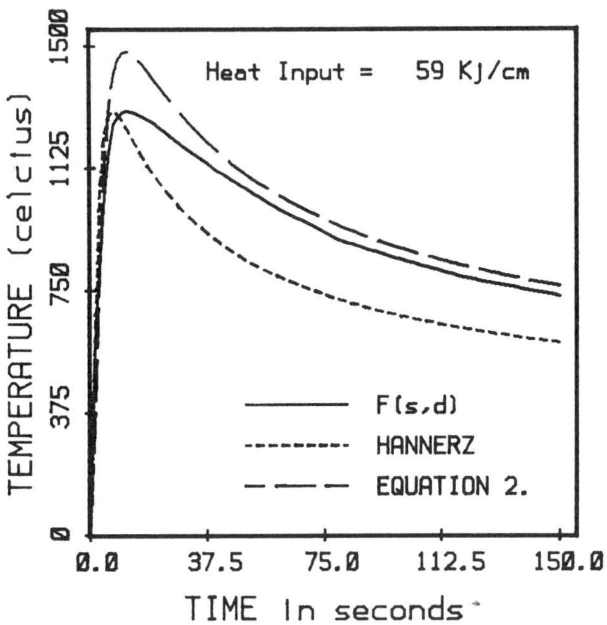

Fig. 3

constants used for the plot of equation (2) are:

thermal conductivity = 0.5 J/cm s C
thermal diffusivity = 0.07 cm/sec
density * specific heat = 6.7 J/cm C

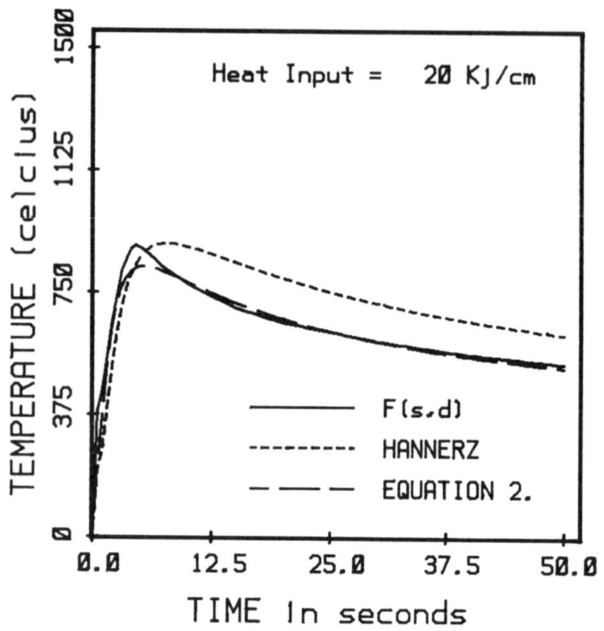

Fig. 4

The correlation of equation (2) is good only for an energy input of 39 Kj/cm. and a peak temperature of 1300 C, where its peak temperature is only slightly higher. For a peak temperature of 1300 C and 20 Kj/cm. the peak temperature is over 300 C too low and for a 59 Kj/cm. input the peak temperature is nearly 200 C too high. The time at temperature for high temperatures is critical for most materials. Errors of several hundred degrees is certainly not acceptable. Considering the plots of equation (2) for a peak temperature of 900 C similar problems are seen. The peak temperature is low for the 20 Kj/cm. plot and high for both the 39 and 59 Kj/cm. plots. Other values may be used for the constants or the equation expanded to accommodate the changes in the constants as a function of temperature for the material under study as a means of obtaining a better fit. Such changes complicate the thermal cycle generation.

The peak temperature for the Hannerz equation correlates very well for changes in energy input and changes in peak temperature. The peak temperatures plotted for equation (2) do not correlate close enough for practical use. Changing the constants will cause the plot to be close for one set of conditions, but all other

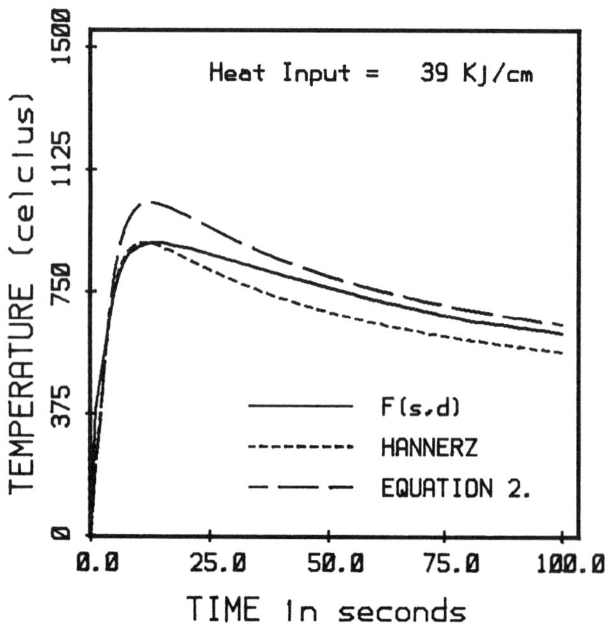

Fig. 5

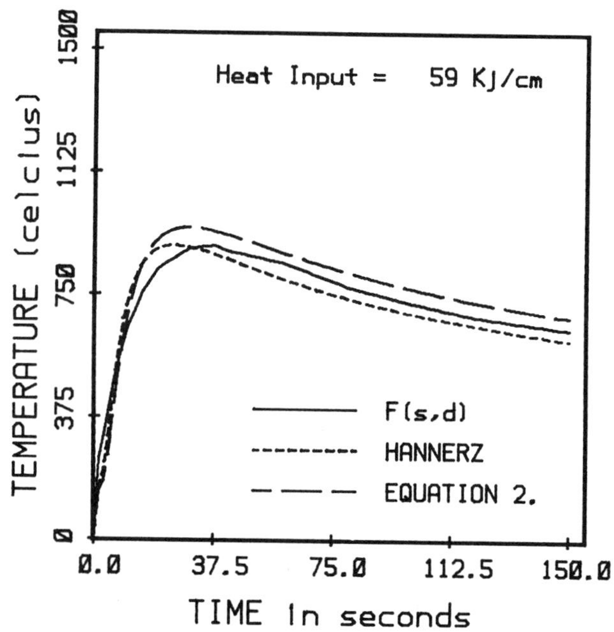

Fig. 6

conditions will then be unsatisfactory. There are too many variables left to choice in equation (2).

Considering the time at temperature or the area under the curve, discrepencies are substantial for all the plots except for a peak temperature of 1300 C and an energy input of 59 Kj/cm. For this one plot all of the curves are reasonably close. The Hannerz solution yields much more time at temperature for low energy inputs and less time at temperature for high energy inputs. Aside from the incorrect peak temperatures, equation (2) yields much less time at temperature for low energy inputs and much more time at temperature for high energy inputs. For materials which are sensitive to time at temperature, care must be exercised in using any thermal cycle which is different from the measured thermal cycle.

Of the two equations evaluated the Hannerz solution is far more practical, since no constants are required. The Hannerz solution also correlates very well at peak temperature and in cooling from 800 to 500 C.

TRANSFORMATION TEMPERATURES

In the study of steel HAZ work many investigators use beginning- and end-of-transformation data from the equilibrium diagram. In work by Paley and Hibbert (Ref. 9) contours of the HAZ are superimposed on a macro-section of the actual HAZ. The contour outlines are advantageous in the analysis of the exact HAZ location. However, the authors show the A1 to be 720 C and refer to the use of the iron-carbide equilibrium diagram. In this case and many others the use of equilibrium data in describing the limits of the HAZ frequently result in substantial errors. Both the Continuous Heating Transformation (CHT) and Continuous Cooling Transformation (CCT) data must be used to accurately describe the HAZ. Basically, welding is NOT an equilibrium process.

Figs. 7 - 9 present data for CHT of a low carbon steel and 4140. Contrary to popular belief, the CHT is substantially depressed at higher heating rates for these materials. Rates up to 15,000 C/sec. are shown. Heating rates of a few thousand centigrade degrees/sec. are possible in welding with low energy inputs and thin sheet. In laser and electron beam welding the heating rates may be even higher. The large depression of the A1 and A3 for high heating rates shows that a region heated at a 10,000 C/sec. rate to a temperature below the equilibrium A1, would be completely austenite for either material.

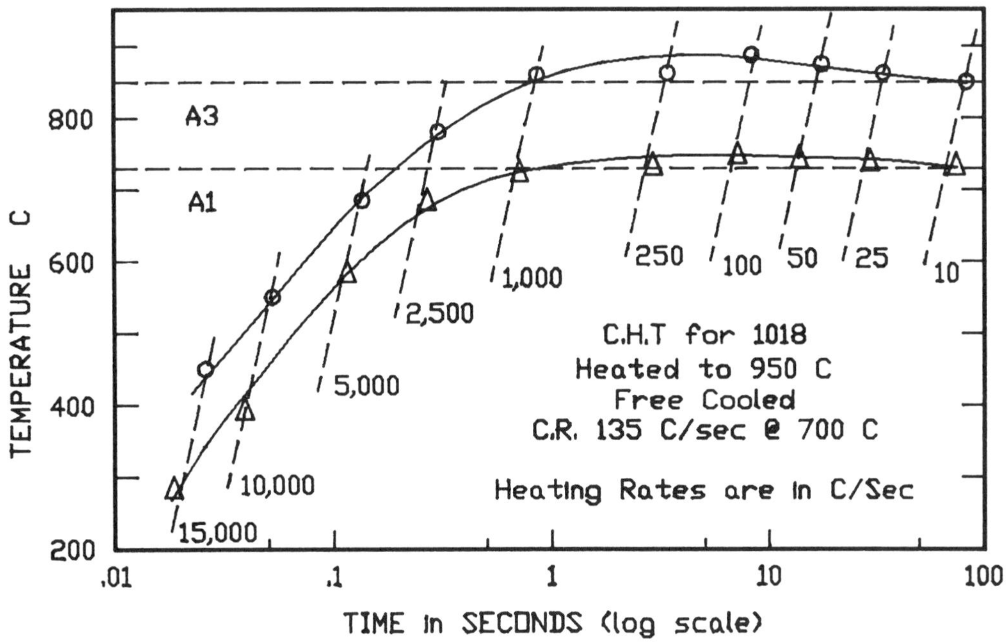

Fig. 7

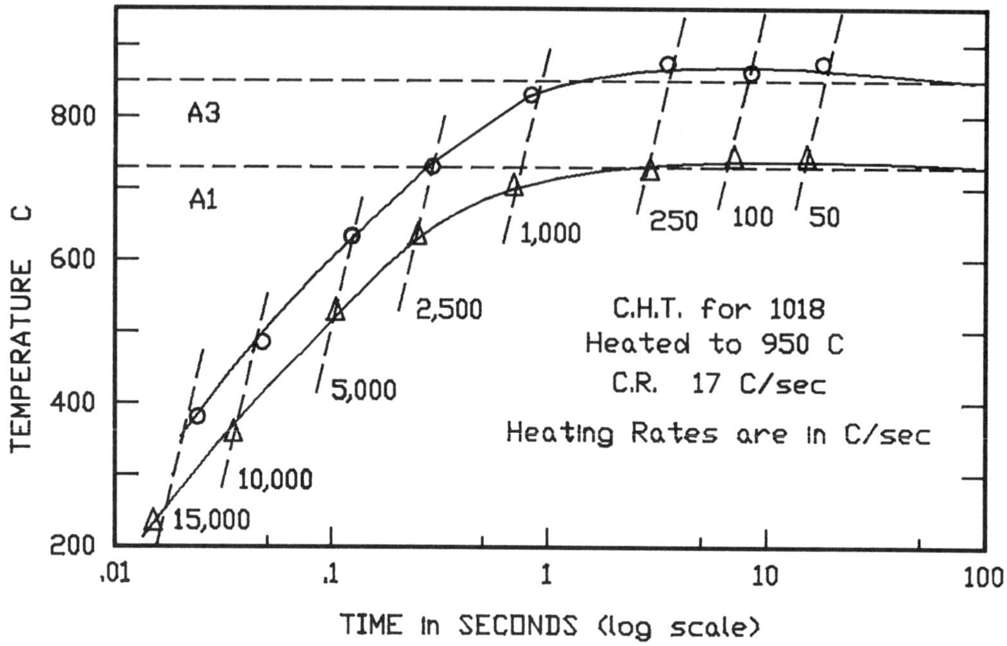

Fig. 8

The austenite formed below the A1 is very unstable. If the material is heated at 10,000 C/sec. to 650 C and then at 200 C/sec. to 950 C, no further transformation occurs during the 200 C/sec. heating period. If the material is heated rapidly to 650 C and held, the material is unstable and transforms within a few milliseconds to a few seconds.

High speed dilatometry was used to confirm the thermal coefficient of expansion for the b.c.c. and f.c.c. structures below and above the transition temperatures respectively. The volume change was also compared to

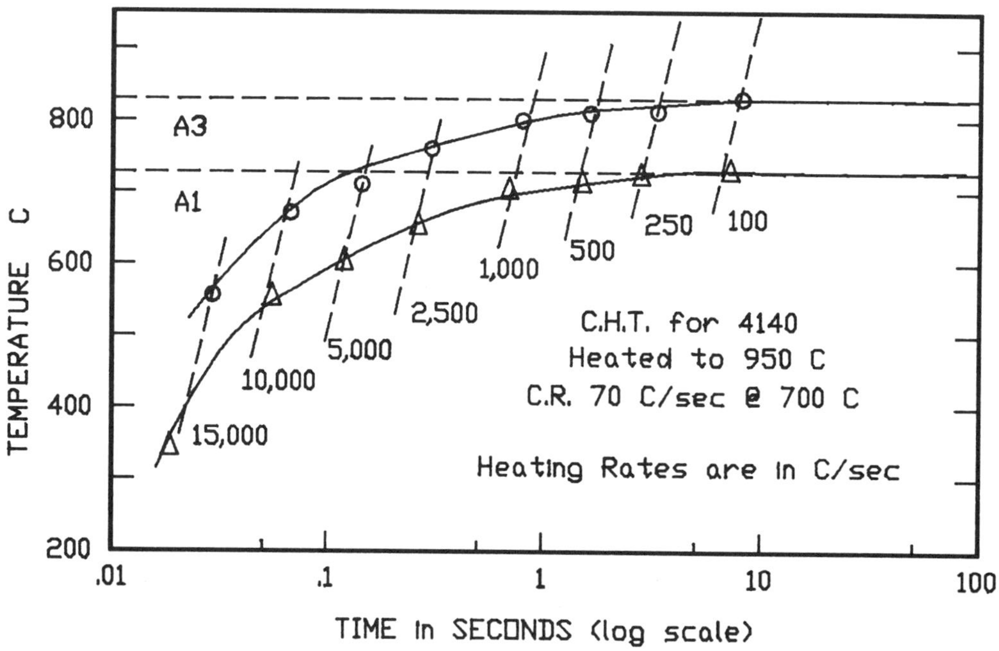

Fig. 9

the slow speed 10 C/sec. value. These comparisons of both thermal coefficients and volume expansion were within 10% when counted for temperature, except at the two highest heating rates, which were somewhat more. Due to the substantially depressed temperature of the austenite present for high heating rates the transition time on arresting the temperature below the equilibrium A1 is very short. This makes metallographic examination after quenching very difficult. Specimens used for this high speed CHT work were 12.7 mm in diameter reduced to 5 mm for a length of 10 mm at the mid-span. The reduced mid-span increases the thermal and electrical conductivity for the specimen over the use of plain 5 mm diameter bars. This increases both the heating and cooling rate, which is desirable for dynamic CHT and CCT work.

DYNAMIC THERMAL/MECHANICAL MEASUREMENTS

One of the primary purposes of obtaining thermal cycles representative of the HAZ is to be able to test materials exposed to these thermal cycles. Synthesizing HAZ microstructures and performing tests on these structures requires careful experimental procedures. Dynamic thermal work can only be done in the presence of a thermal gradient. If no gradient exists, no heat will flow. The specimen configuration shown in Fig. 10 allows the heating and cooling of a work zone, while maintaining isothermal planes perpendicular to the specimen axis at the middle of the work zone. By controlling the thermal gradient axially along the specimen, the thermal heating and cooling rates may be controlled without affecting the isothermal planes in the work zone. The measurement of CHT and CCT data at high rates requires that an isothermal section of specimen be

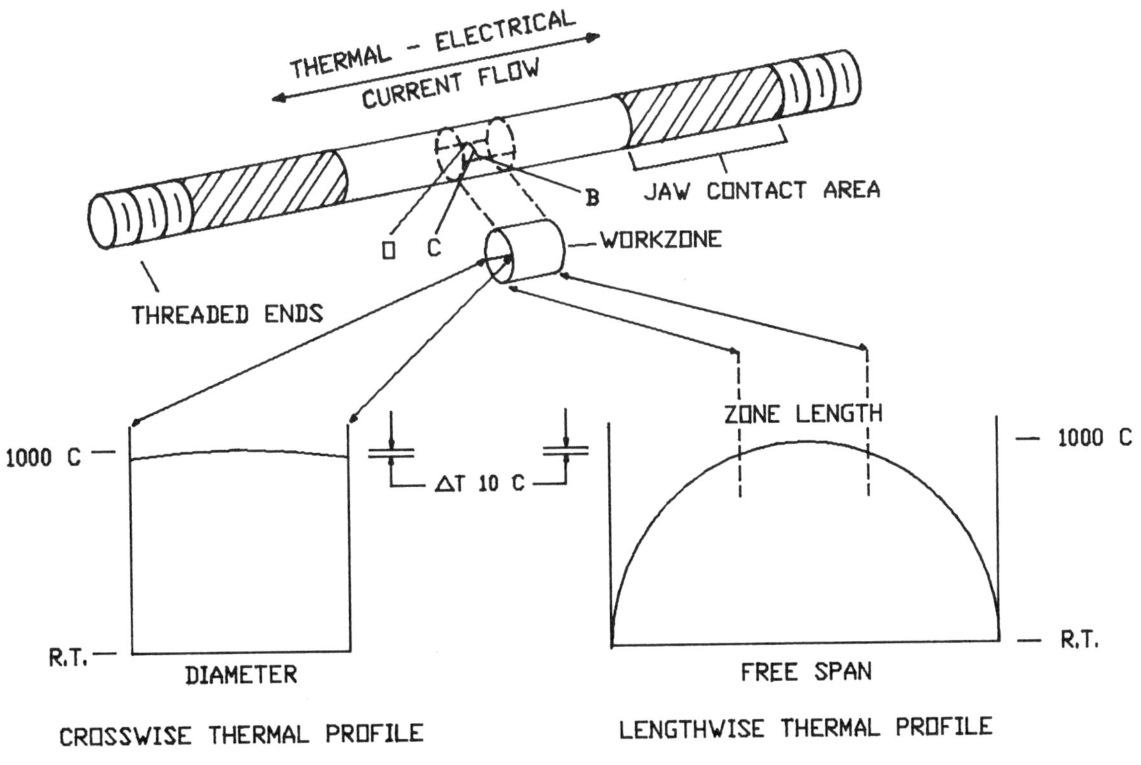

Fig. 10

present, which precisely follows the programmed temperature/time curve. The dilation measurements may then be made on the isothermal plane at the mid-section of the work zone. Reduced work zone cross sections should be used for high speed work.

The argument that any chosen point in the HAZ has a higher temperature on one side and a lower temperature on the other side except at the weld centerline may be accommodated in the specimen of Fig. 10 by choosing an isothermal plane to one side or the other side of the work zone mid-point. The width of the synthesized HAZ can be controlled by adjusting the free span of the specimen. The non-centered isothermal plane is then programmed for the thermal cycle of the position chosen in the HAZ. This procedure yields a point in the simulated HAZ, where the temperature on one side of the point is higher and the other side is lower, similar to the HAZ in the work piece. The only requirement is that the mid-point of the work zone must not exceed the melting temperature of the specimen. The use of a non-centered isothermal test plane is suitable for the generation of microstructures and dilation work. It is not suitable for hot ductility or other mechanical work.

CONCLUSIONS

1. The idealized thermal cycle equation of Hannerz yields thermal cycles which are accurate in peak temperature and 800 to 500 C cooling rate.
2. The most exact thermal cycles are obtained from measurements.
3. With HAZ software and F(s,d) measurement based tables precise thermal cycles are easily obtained. No further checking is necessary as to the accuracy of thermal cycles obtained by this method.
4. Welding is NOT an equilibrium process. CHT and CCT data must be carefully observed, when analyzing the HAZ.

REFERENCES

1. Rosenthal, D., 2eme Congres National des Sciences, Brussels, 1935, pp. 1277-1292.
2. Spraragen, W. and Claussen, G. E., Welding Journal, 16, Sept. 1937, pp. 4-10.
3. Rykalin, N. N., "The Calculation of Thermal Processes in Welding" Mashgiz 1951; translated into English by Z. Paley and C. M. Adams, Jr. 1963.
4. W. F. Hess, L. L. Merril, E. F. Nippes, and A. P. Bunk, Welding Journal, 22, 1943, pp. 377s-422s.
5. E. F. Nippes, L. L. Merril, and W. F. Savage, Welding Journal, 28, 1949, pp. 556s-564s.
6. E. F. Nippes, H. Wawrosek, and W. L. Fleischmann, Welding Journal, 34, 1955, pp. 169s-196s.
7. E. F. Nippes, and W. F. Savage, Welding Journal, 28, 1949, pp. 599s-616s.
8. Private communication, N. E. Hannerz, Royal Institute of Technology, Stockholm, Sweden.
9. Z. Paley and P. D. Hibbert, Welding Journal, 54, 1975, pp. 385s-392s.

APPENDIX I

The following variables, constants, letters, and units apply to all equations used in this paper:

T is temperature at time t in degrees C.

To is initial, preheat or interpass temperature in degrees C.

Δt is cooling time from 800 - 500 C in seconds.

F(s,d) is a function of distance, d, from the weld centerline, and the time, s, after the electrode passes the point.

EI is weld energy input expressed in Joules/cm (inch).

IF is a multiplying factor, which accounts for the effect of preheat temperatures. The value of this factor is 1.0 for an initial plate temperature of 20 C and other values are employed for other initial plate temperatures.

A is added energy/unit length of weld in Joules/cm.

d is plate thickness in cm.

Q/d is added energy/unit surface in Joules/cm squared.

λ is heat conductivity in Joules/cm * s * C.

C is specific heat in Joules/g * C.

ρ is density in g/cm cubed.

α is thermal diffusivity in cm squared/sec.

α is $\lambda/\rho C$ in cm squared/sec.

APPENDIX II

HAZ software generates data points which define temperature vs. time curves. Each curve represents the temperature of a point some fixed distance from the centerline of an arc weld. For each different material there is a table of values which are used to generate the points. These tables are values for a function of two variables, namely time and distance. The value of the function at some time and distance is used to generate the temperature at that point with the following relationship:

T = To + HI * F(s,d) * IF

Where:

s is the time

d is the distance from the weld centerline

T is the temperature of the point distance D from the weld centerline

To is the preheat temperature of the plate

HI is the heat input in Kjoules/inch (cm)

IF is an input factor based on the preheat

The values of the function F(s,d) have been collected experimentally using thermocouples and bead on-plate welds. The HAZ software interpolates the data in the F(S,D) tables to generate a temperature vs. time (thermal cycle) curve to match the user input.

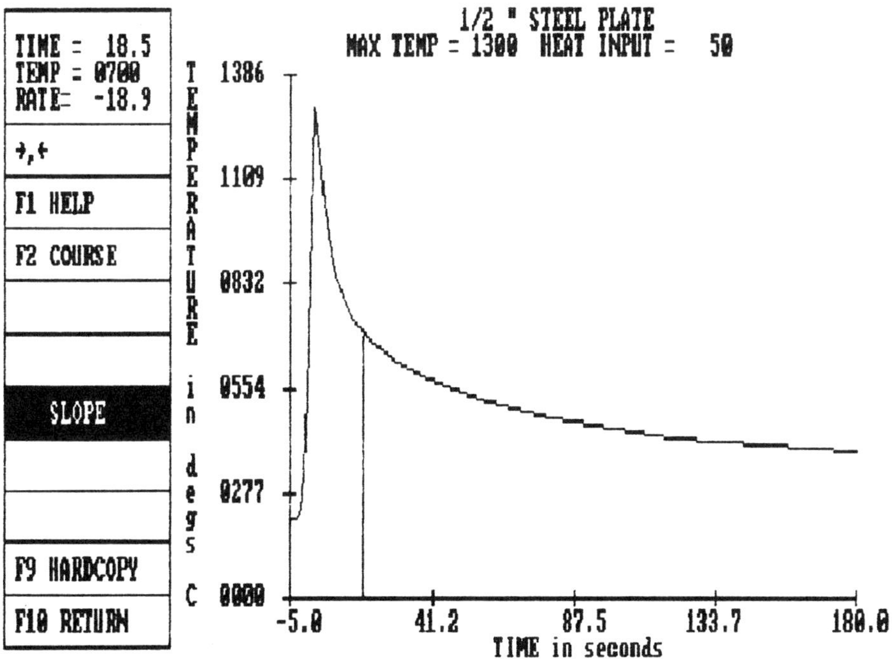

Fig. A-1

The user must input the peak temperature, preheat temperature, heat input, type of material and thickness.

Fig. A-1 is a thermal cycle for 1/2 inch thick steel using a preheat temperature of 200 C, a peak temperature of 1300 C, and an energy input of 50 Kjoules/inch. The plot is shown as it appears on the screen of the IBM PC. The resolution is that of the video screen, when output for use the curves are very smooth. The heating and cooling rates are easily obtained for any temperature on the curve, as shown by the example of Fig. A-1.

Fig. A-2 is the same thermal cycle, where the area above 1000 C is shown as a hatched region. The area between 700 and 1000 C is shown as a second hatched region. This feature allows the study of the time at temperature for the thermal cycle. An energy function is also available to estimate the effect of time at temperature on diffusion processes.

There are currently 6 sets of data available, corresponding to the following materials:

- 1/2" Steel Plate
- 1" Steel Plate
- 1-1/2" Stainless Steel
- 1/4" Titanium A55 Plate
- 1/2" Titanium
- 1/4" Zircaloy 2 Plate

Other sets of data will be incorporated into the program as the data becomes available.

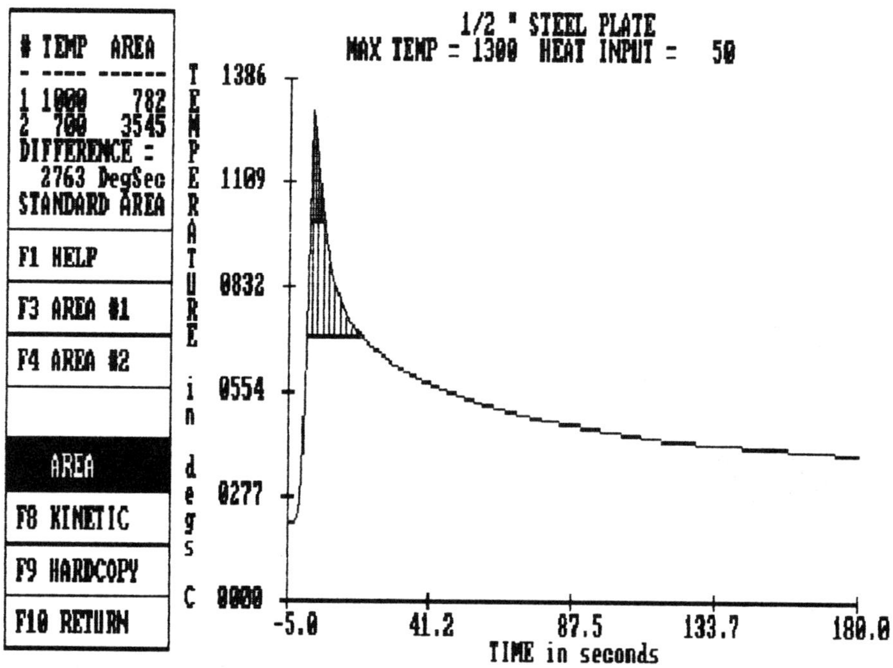

Fig. A-2

ADVANCES IN WELDING PROCESSES

KEYNOTE ADDRESS

THE PHYSICS AND CHEMISTRY OF WELDING PROCESSES

Thomas W. Eagar
Materials Processing Center
Massachusetts Institute of Technology
Cambridge, Massachusetts 02139

Abstract

Historically, welding processes have developed empirically and have been applied to technological problems with unusual quickness. As a result, a large number of processes have evolved, many of which are not well characterized from a scientific point of view. Many attempts have been made to categorize these processes, mostly with disappointing results; however, there have been a few categorizations which have proven useful. The first considers the thermodynamic stability of surface contaminants and describes how the temperature, pressure and chemical potential of the welding system can be changed to eliminate these contaminants. The second describes the heat intensity on the surface of the material and relates this to the maximum weld travel speed, the heat affected zone width, the equipment cost and the minimum sampling frequency necessary to control the process.

THE NEED FOR WELDING AND JOINING is ubiquitous; there are very few manufactured products that do not rely on welding or joining in some form. Indeed, only monolithic parts can be made without joining. Unfortunately, our methods are generally imperfect, either in properties or in affordability; and there is a constant search for improved processes.

The perfect joint is one which is indistinguishable from the material surrounding it. Some processes, such as diffusion bonding, come very close to this ideal; however, such processes seem to be either cost intensive or restricted to a narrow group of materials. It is clear that there is no universal process which will perform adequately on all materials and in all geometries. As a result, a welding engineer must be able to wisely select the best process for a particular material in a given application. In order to do this, it is useful to categorize the various welding processes in a systematic, scientifically based manner. If this can be achieved, it may be possible to determine the strengths and weaknesses of a given process a priori.

Given the great economic importance of welding to any manufactured product, one might ask why a fundamental science has not developed around this field? There are several answers to this question. Firstly, welding itself is not a discipline but is a process. As with any process, it involves scientific principles from many different disciplines, e.g. physics, chemistry, mechanics, electronics, materials and the like. In this sense, there is a science base for welding already in existence. It is up to the welding engineer to search out the knowledge available in other fields and to apply it, judiciously, to the problems of welding. The study of welding cannot be expected to create new science. Rather, it is an eminently practical field of study which can bring the promises of science to each of us in our daily lives. It is this practical application of science which makes the study of welding so exciting for many of us.

Ironically, this great need for improved welding processes is the second reason why a fundamental science has not been built around welding. Virtually any new process which might be applicable to welding is tested on welding immediately after it is discovered. Some of the earliest experiments on electric arcs during the nineteenth century involved welding. Electron beams and lasers were used to weld small wires together within a few years after these new heat sources were discovered. It is even rumored that particle beams have been used to make welds. Some of these new heat sources were adapted for practical use in welding long before we understood the fundamentals of how the heat source interacts with the material. In many cases, our use of a given process has far exceeded our science base. As a result we often experience unexpected problems. Progress is being made, but it is slow. The papers in this conference illustrate the areas of greatest concern in welding and joining today.

WELDING PROCESS FUNDAMENTALS

It would behoove us to develop a scientific framework which will allow us to evaluate existing and potential new welding processes. To do this, we must look back at what we are trying to achieve. As noted previously, we wish to form a joint which has indistinguishable properties from the material surrounding it. Ideally, one would like to merely place two parts in contact and achieve a perfect bond. There are two fundamental problems which prevent this. First, the surfaces of any material in the atmosphere is contaminated with either oxygen or water vapor or carbon dioxide. Second, the parts to be joined are solid and the surfaces to be joined do not mate perfectly. Diffusion bonding overcomes these chemical and geometric incompatibilities by application of heat and pressure. The heat allows the surface contamination to diffuse into the bulk. Further, the heat permits sufficient deformation or shape change to promote perfect mating of the surfaces. In materials such as iron, titanium or copper, which dissolve all or most of their usual surface contaminants, a nearly perfect bond can be made; but in aluminum or other metals with refractory oxides which are insoluble in the base metal, the process does not work so well.

Indeed, diffusion or other forms of solid state bonding achieve remarkable results. Even materials which are immiscible in both the solid and liquid states, such as iron and silver, can form interfacial bonds of exceptional strength. O'Brien, Rice and Olson achieved tensile strengths of nearly 100 ksi when bonding maraging steel with silver. (1) One of the reasons for this seemingly exceptional bond strength is that dislocations do not propagate across the bond interface; hence, near theoretical bond strengths are obtainable in principal. The theoretical iron-iron bond strength in the absence of dislocations is in excess of 1000 ksi; hence, even iron and silver which may not have a very high intrinsic bond strength can produce excellent interfacial tensile strength provided the interface is not contaminated or otherwise flawed.

Clearly then, we can achieve excellent results if we can eliminate the surface contamination. Since these are chemical bonds, we can look to chemical thermodynamics to determine how to eliminate them. The free energy of any bond is related to the temperature, pressure and chemical potential of the system, i.e.

$$G = G\ (T, P, \mu_i)$$

hence, to break a bond we must manipulate these fundamental variables. In fact this is clearly what we do in current practice, even if it is not a conscious decision. For some materials, notably silver, with a low chemical affinity for oxygen, a few hundred degrees centigrade will cause the surface oxide to decompose. Silver will also dissolve the oxygen, which is one reason why O'Brien et. al. found such good diffusion bonding at low temperatures. (1) Nonetheless, the temperature must be maintained (and other contaminants such as sulfur must be kept out of the system) if one is to avoid reformation of the surface film. The monolayer time-pressure is on the order of 10^{-8} atm-seconds, so it is not practical to clean the surface at high temperatures and later cool it unless one is operating at extremely high vacuums.

In materials other than silver with somewhat more stable oxides, such as copper or iron, higher temperatures are necessary to decompose the oxide. One can estimate the temperature necessary for decomposition of the oxide by reference to Ellingham diagrams (see Figure 1); however, one must be cautious and remember that such diagrams are based on all solids and liquids being in their standard states (which is usually the pure form) and all gases, such as oxygen being at one atmosphere pressure. If the material is alloyed, the relative activity of each metal component must be known in order to calculate a decomposition temperature. Since the atmospheric oxygen pressure is not very different from one atmosphere (at least in the logarithmic sense of thermodynamics), correction for the oxygen activity difference is not very important if one is operating in the atmosphere. However, if the system is operated in an inert gas or a vacuum, the oxygen potential may vary markedly. In such cases, we are using the second variable in the free energy function to assist in decomposition of the surface oxide. Reducing the partial pressure of oxygen and elevating the temperature can reduce oxides of intermediate stability. These effects of temperature and pressure on reduction of the oxide of any metal is readily calculated by chemical thermodynamics.

For more stable oxides, the partial pressure can be further reduced by addition of a reactive gas. The AWS Brazing Manual provides a useful graph, reproduced in Figure 2, of the relative stability of a number of metal oxides as a function of temperature and oxygen pressure, in the presence or absence of a reducing gas such as hydrogen. Clearly, any welding process engineer should understand the chemistry behind such a graph when considering diffusion bonding or brazing as a joining process for a given material.

The third variable in the free energy equation is the chemical potential of a species. The common method of altering this variable involves both temperature and the use of a flux. The flux provides a system of very low chemical potential for the surface contaminant; hence, the surface film will be absorbed into the flux. This is perhaps the most common method of removing surface contamination, yet, the specific chemistry of most flux systems is poorly understood.

In his classification of welding processes, Houldcroft describes this requirement of elimination of surface contamination as the "shielding method." He lists vacuum, inert gas, reactive gas, flux, no shielding and mechanical exclusion as possible methods. (3) It should be recognized that each of these shielding methods (with perhaps the exception of the last one) can be quantified in terms of the free energy of the system, as defined by the temperature, pressure and chemical potential of the surface contaminant. In this way, one can determine whether a shielding method is appropriate for a given material.

It will be noted that temperature is a major variable in nearly all methods of removing surface contaminants. This suggests another fundamental ranking for joining processes, i.e. the method of heating. In his classification Houldcroft called this the "source of heat", yet again, this can be quantified to yield more information about a given joining process. As before, one should choose an intensive, rather than an extensive variable to describe the system. A number of different authors have suggested that the relevant quantity is the power density on the surface of the material.

One of the primary welding processes is fusion welding since this solves both the chemical and the geometric problems of joining two surfaces. Melting the base material permits the surface contamination to float or dissolve away while the liquid metal conforms perfectly to the solid mating surface. Hence, heat used for welding usually involves melting the surface.

If one considers a planar heat source diffusing into a very thick slab, the surface temperature will be a function of both the surface power density and the time. Figure 3 shows how this temperature will vary on steel with power densities from 400 watts/cm^2 to 8000 watts/cm^2. (4) At 400 watts/cm^2, it takes two minutes to melt the surface. If the 400 watts/cm^2 heat source were a point on the flat surface, the heat flow would be divergent and it might not even be possible to melt the steel; the solid metal might be able to conduct away the heat as fast as it is being introduced. Generally, it is found that heat source power densities of approximately 10^3 watts/cm^2 are necessary to melt most metals.

At the other end of the power density spectrum, it is found that heat intensities of 10^6 or 10^7 watts/cm^2 will cause vaporization of most metals within a few microseconds. Above these power densities, all of the solid interacting with the heat source is vaporized and no fusion welding can occur. Thus it is seen that the heat sources for all fusion welding processes lie between approximately 10^3 and 10^6 watts/cm^2 on the power density spectrum. The spectrum with the locations of several common joining processes is shown in Figure 4.

Inspection of Figure 1 shows that the power density is inversely related to the interaction time of the heat source on the material. Since this is a transient heat conduction problem, one can expect the heat to diffuse into the steel to a depth which increases as the square root of time, i.e. from the Einstein equation

$$x \propto \sqrt{at}$$

where: x is the distance that the heat diffuses into the solid, in cm
a is the thermal diffusivity of the solid, in cm^2/sec, and
t is the time in seconds

For the planar heat source on a steel surface as represented by Figure 3, one finds the time in seconds to produce melting on the surface, t_m, is given by:

$$t_m = \left[\frac{5000}{H.I.}\right]^2$$

where H.I. is the heat intensity in watts/cm^2. If we consider the time to melting to be equal to the necessary interaction time, t_I, of the heat source with the material, one can generate the graph shown in Figure 5. One can see that heat sources of the order of 10^3 W/cm^2, such as oxyacetylene flames or electroslag welding require interaction times of 25 seconds in steel while laser and electron beams at 10^6 W/cm^2 need interaction times on the order of 25 microseconds. If we divide this interaction time into the heat source diameter, d_H, we obtain a maximum travel speed, V_{max} for the welding process, as shown in Figure 6. From this, it is clear why welders begin their training with the oxyacetylene process as it is inherently slow and does not require rapid response time in order to control the size of the weld puddle. Greater skill is needed to control the more rapid fluctuations in arc processes, while no human can control the pool of the high heat intensity processes such as laser and electron beam. It is an inherent fact that these processes must be automated in order to control them. This need to automate leads to increased capital costs for these high heat intensity processes. One can approximately replace the number of watts/cm^2 of a process with the dollar cost of the capital equipment.

For constant total power, a decrease in the spot size will produce a squared increase in the heat intensity. This is one of the reasons why the spot size decreases with increasing heat intensity as shown in Figure 6. It is easier to make the spot smaller than to increase the power rating of the equipment. In addition, one generally wishes only to melt a small volume of material. If the spot size were kept constant and the input power were squared in order to obtain higher power densities, the volume of fused metal would increase dramatically, with no beneficial effect. However, this decreasing spot size, coupled with a decreased interaction time at higher power densitites, compounds the problem of controlling the higher heat intensity process. A shorter interaction time means that the sensors and controllers for automation must operate at higher frequencies. The smaller spot size means that the positioning of the heat source must be even more precise. This positioning accuracy must be on the order of the heat source diameter, d_H, while the control frequency must be greater than the travel velocity divided by the diameter of the heat source. For processes operating near the maximum travel velocity, this is the inverse of the process interaction time, t_I (see Figure 5).

Thus we see that not only must the high heat intensity processes be automated due to an inherently high travel speed, but the fixturing requirements become greater and the control systems and sensors must have ever higher frequency response. Both of these factors lead to increased costs of high heat intensity processes, which is one reason that these processes, which are very productive, have not found wider utilization.

Another important welding process parameter that is related to the power density of the heat source is the width of the heat affected zone. Using the Einstein equation one can estimate a heat affected zone width from the process interaction time and the thermal diffusivity of the material. This is shown in Figure 7 with one slight modification. Above about 10^4 W/cm^2 the heat affected zone width becomes roughly constant. This is due to the fact that the HAZ grows during the heating stage at power densities below 10^4 W/cm^2 but it grows during the cooling cycle at higher power densities. Thus at low power densities, the HAZ width is controlled by the interaction time, while at high power densities it is independent of the heat source interaction time. In this latter case, the HAZ width is controleed by the cooling time necessary to remove the heat of fusion from the weld metal. In such a case the HAZ width is proportional to the fusion zone width.

The change of slope in Figure 7 also represents the heat intensity at which the heat utilization efficiency of the process changes. At high heat intensities, nearly all of the heat is used to melt the material and little is wasted in preheating the surroundings. As the heat intensity decreases, this efficiency is reduced. For arc welding, it may be only one-half of the heat entering the plate and for oxyacetylene it may be 10 percent or less.

Finally, the heat intensity also controls the depth to width ratio of the molten pool. This can vary from 0.1 in low heat intensity processes to more than 10 in high heat intensity processes.

CONCLUSION

It is seen that there are fundamental chemical and physical limits to our selection of welding processes. The chemical stability of surface contamination influences our choice of shielding or surface removal during the process. A knowledge of the power density on the surface of the material produces quantifiable parameters which provide criteria for maximum travel speed, minimum frequency of process control systems, heat affected zone width and equipment cost. Although these criteria are only approximations and will not define any process precisely, they do provide guidelines from which to predict how a new process or a new material will behave. Intelligent application of such guidelines can often help in the selection or application of a wide range of welding processes on a variety of materials.

Acknowledgements

The author wishes to express his sincere appreciation to Dr. Bruce MacDonald of the Office of Naval Research and Dr. Robert Reynik of the National Science Foundation for a decade of support on the physics and chemistry of welding processes. This generalized approach to welding process could only have been developed through such a long term committment to the study of a number of different welding processes.

REFERENCES

1. M. O'Brien, C. R. Rice and D. L. Olson, Welding Journal, 25 (January 1976)

2. D. R. Gaskell, Introduction to Metallurgical Thermodynamics, Second Edition, McGraw Hill, 1982

3. P. T. Houldcroft, Welding Process Technology, Cambridge University Press (1977)

4. N. S. Tsai and T. W. Eagar, Selection in Railroad Rail Welding, 421, Railway Systems and Management Association, Northfield, NJ (1985)

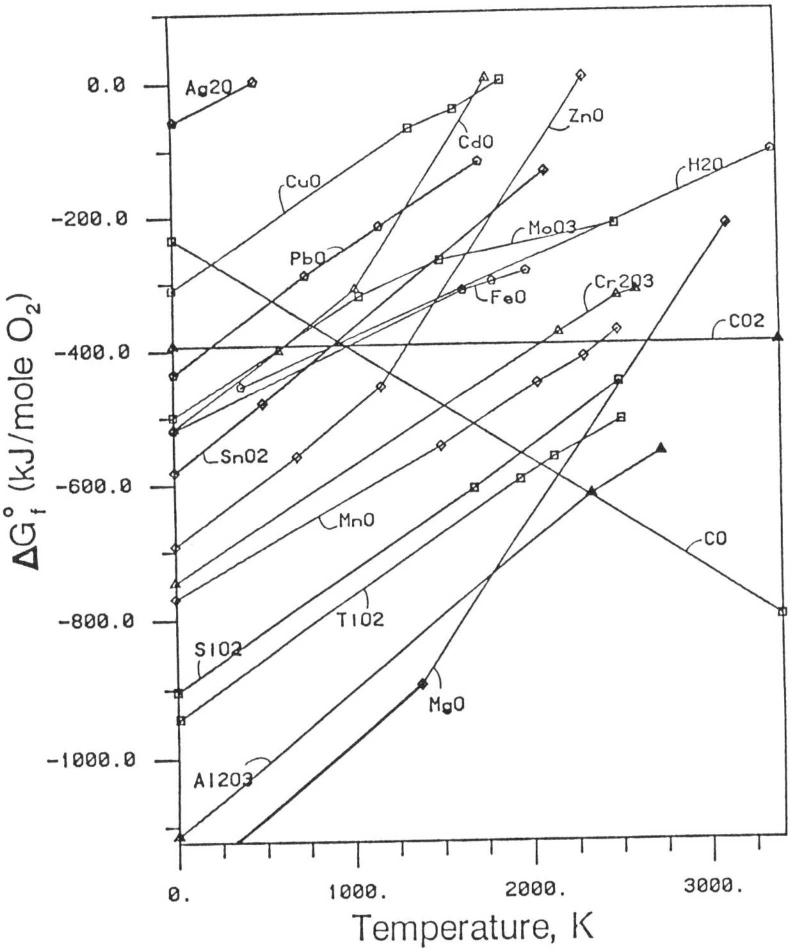

Figure 1

Ellingham diagram illustrating the relative thermodynamic stability of a number of metal oxides.

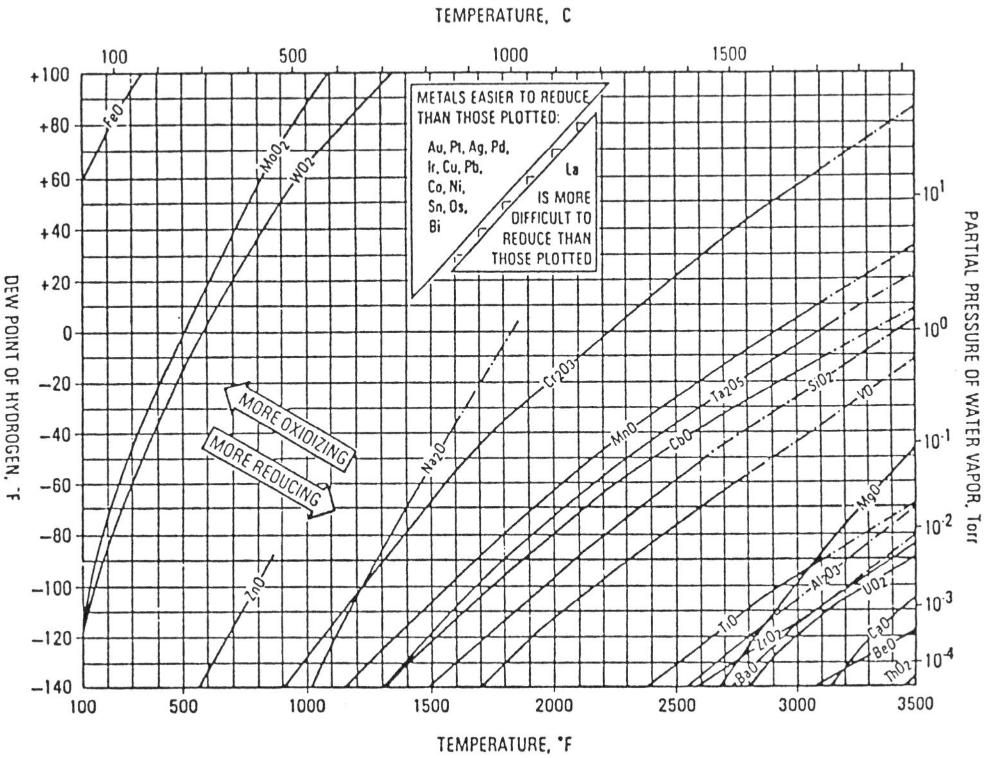

Figure 2

Metal-metal oxide equilibria in hydrogen atmospheres. From the AWS <u>Brazing Manual</u>, used by permission.

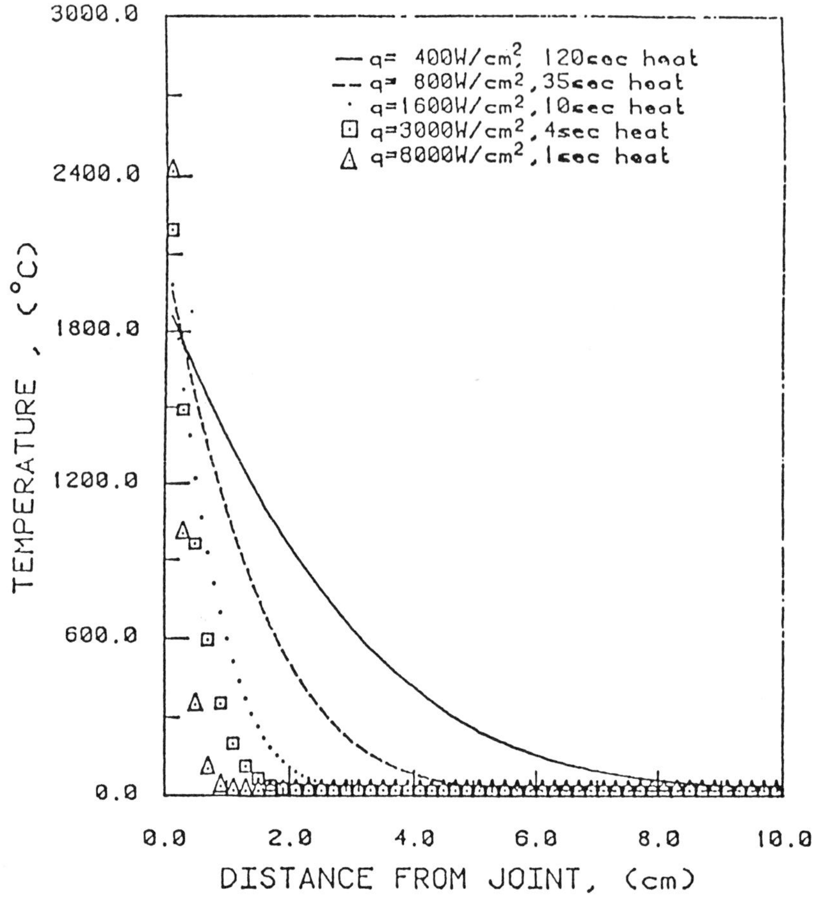

Figure 3

Temperature distribution after a specific heating time in a thick steel plate, heated uniformly on one surface as a function of applied heat intensity. The initial temperature of the plate is 25°C.

Figure 4

Spectrum of practical heat intensities used for fusion welding.

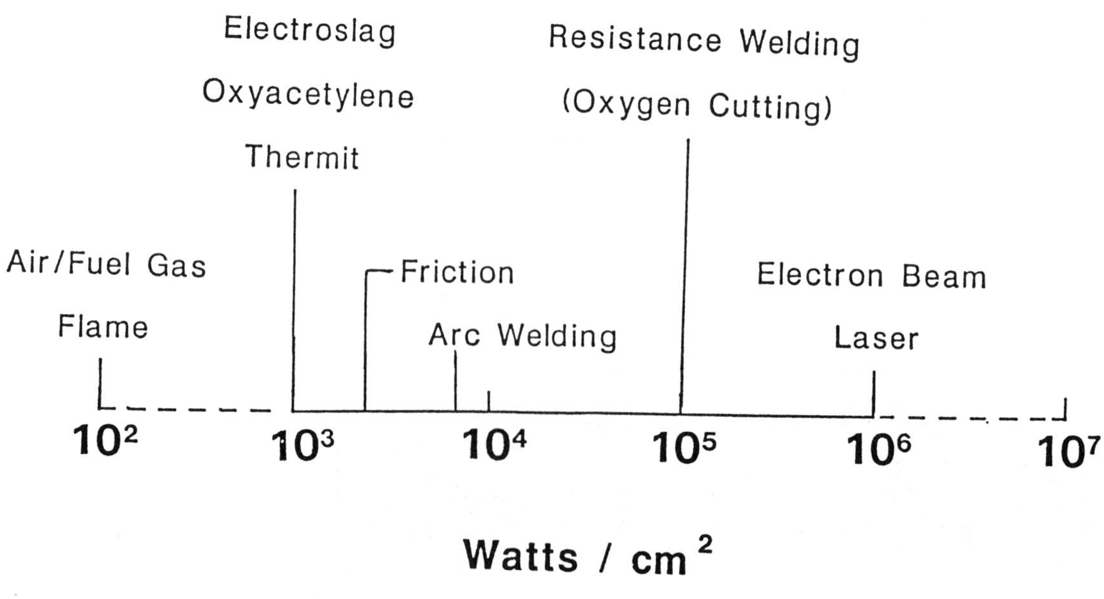

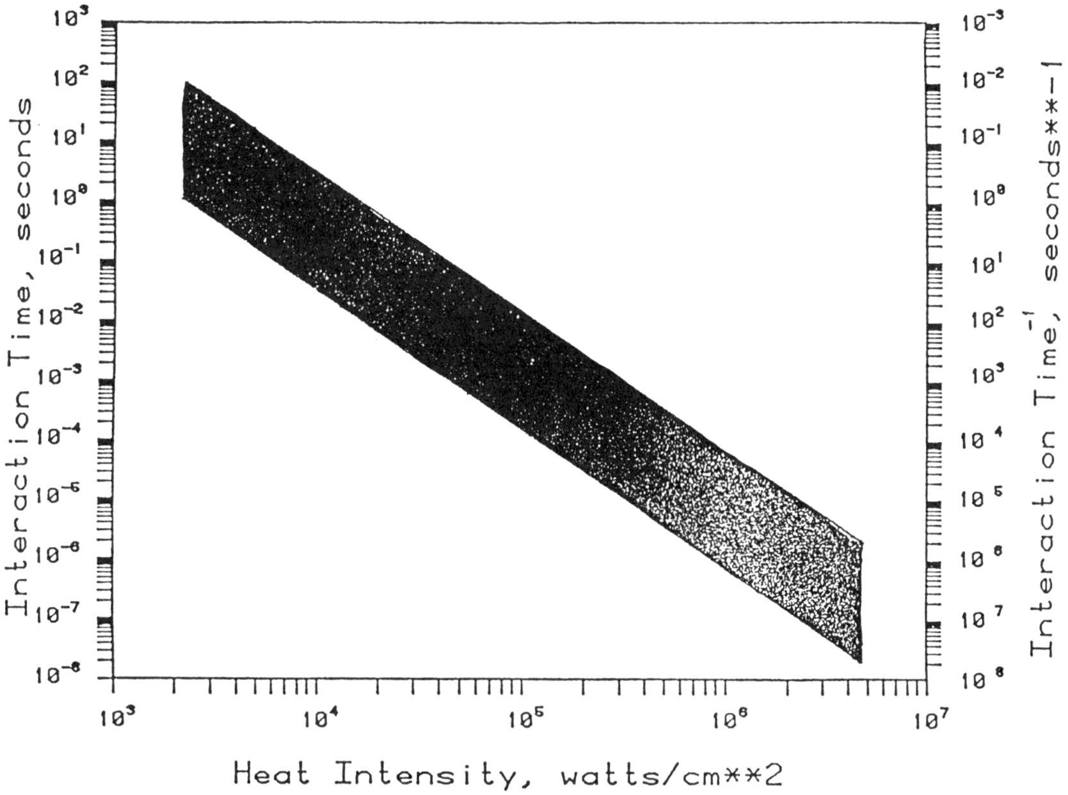

Figure 5

Typical weld pool-heat source interaction times as a function of source heat intensity. Materials with a high thermal diffusivity, such as copper or aluminum would lie near the top of this band, while steels, nickel alloys or titanium would lie in the middle, and uranium and ceramics, with very low thermal diffusivities, would lie near the bottom of the band.

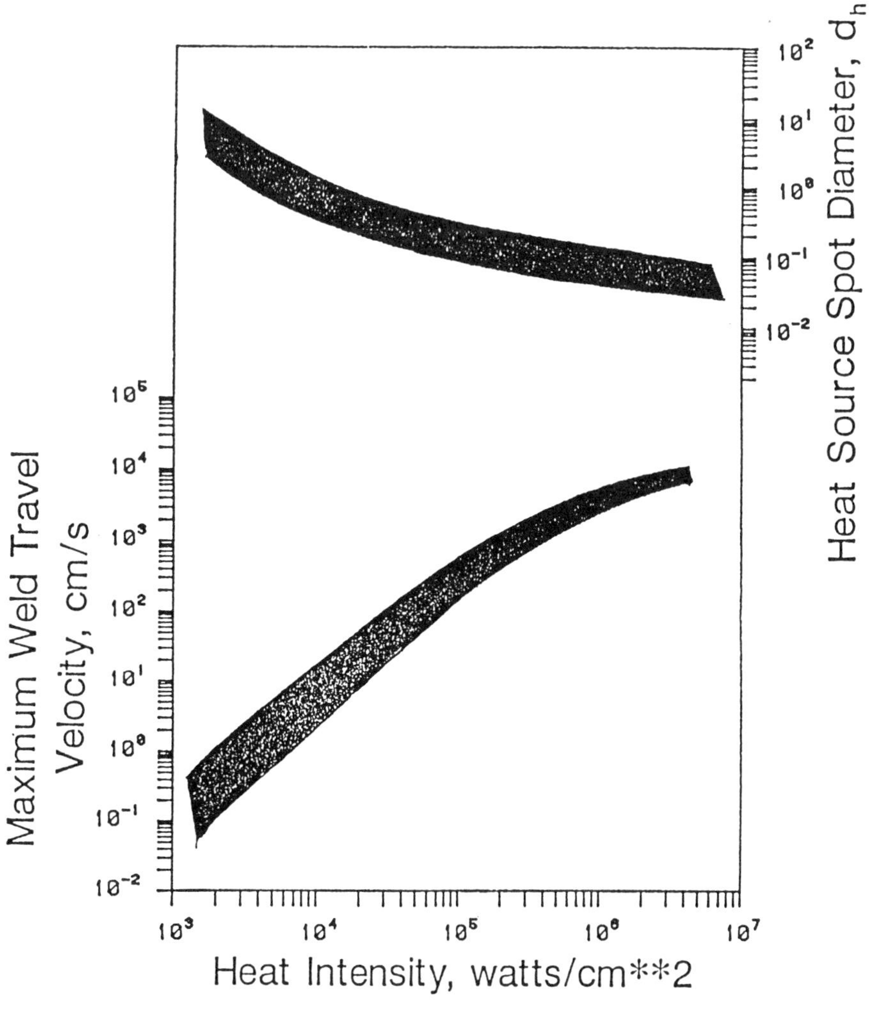

Figure 6

Maximum weld travel velocity as a function of source heat intensity based upon typical heat source spot diameters.

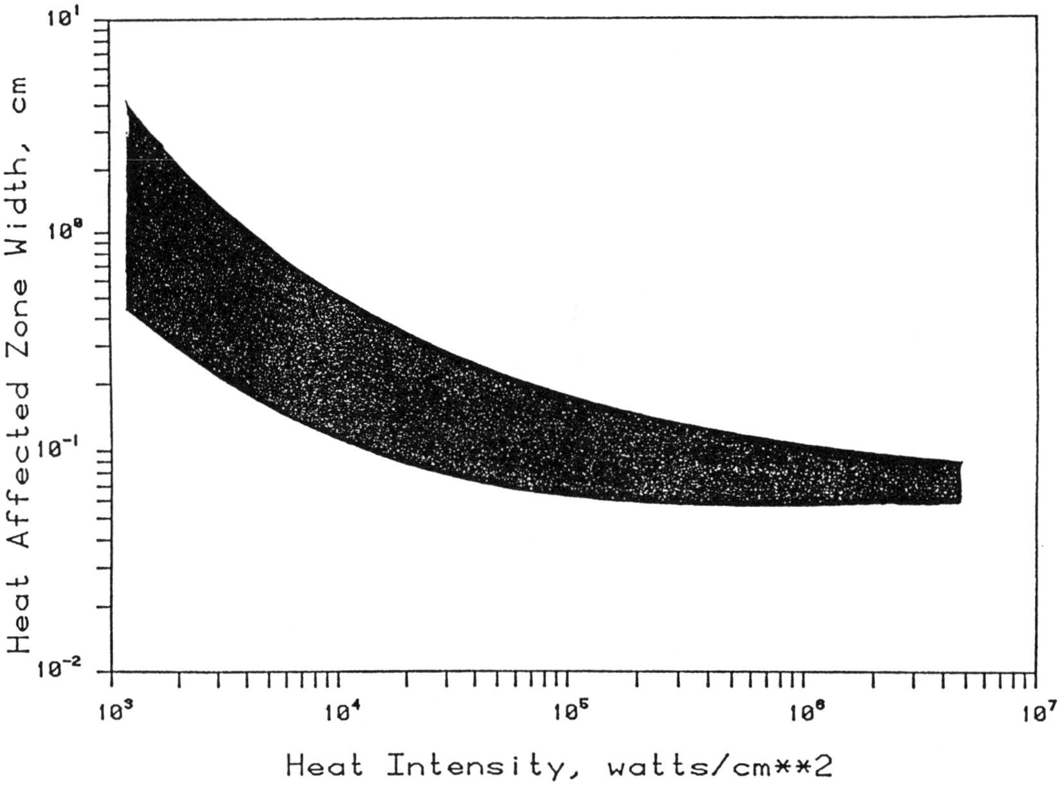

Figure 7

Range of weld heat affected zone widths as a function of source heat intensity.

THE EFFECT OF POWER SUPPLY RESPONSE CHARACTERISTICS ON DROPLET TRANSFER OF GMA WELDS

G. Kohn*, T.A. Siewert
Fracture and Deformation Division
National Bureau of Standards
Boulder, Colorado

ABSTRACT

The droplet transfer in gas metal arc welding was studied with a digital oscilloscope, laser backlighting, and a video camera. Fast Fourier transform of the digital current and voltage signals showed droplet transfer rates that were within 10% of the values determined with the video-tape record. A modified current controller with a variable upper current limit reduced the spatter in short circuiting welding by more than half.

MOST STUDIES of gas metal arc (GMA) welding have been performed with simple, constant-voltage power sources that have controls for maximum output (open-circuit) voltage and wire feed rate. More sophisticated power sources have controls for slope and inductance as well. The V/A response curves of these constant-voltage power supplies vary; a representative curve is shown in Figure 1. Its slope is usually defined at the intersection of the response curve and the arc characteristic line, which is given by

$$V = 20 + 0.04\ I, \qquad (1)$$

where I is the current in A and V the potential in V [1].

The slope determines the short-circuit current; small slopes result in large short-circuit currents and considerable spatter. Large slopes result in insufficient short-circuit current to melt the electrode and, therefore, the arc cannot reignite. The optimum slope for constant arc power is shown in Figure 1, but for stable arc length, the slope must be minimized [2].

* Guest Worker; on sabbatical leave from the Nuclear Research Centre-Negev, Beer-Sheva Israel.

The inductance determines the rate at which the current rises to its maximum value during the short-circuit period. Fast rise times (low inductance) are associated with excessive spatter; increasing the inductance increases the rise time. Another way to increase the rise time is to decrease the slope. The time constant T (in seconds), is the time required for the current to reach 63% of its peak value, given by

$$T = L/R, \qquad (2)$$

where L is the inductance of the circuit and R its resistance.

Modifications to a welding current controller gave us the ability to select the slope independently of the short-circuit current and open-circuit voltage and to control the time constant independently of the slope. Thus we could separate the effects mentioned above and investigate them independently.

METHODS

The GMA welding system consisted of a commercial power source connected to an automatic torch via a modified current controller. A motorized single-axis table carried the strip-shaped workpiece at a constant speed under the stationary torch. Welds were made with 0.9-mm-diameter AWS E70S-3 electrodes and Ar-5% CO_2 shielding gas on the surface of a mild-steel base plate.

By means of separate potentiometers on the control board of the current controller, we could independently regulate the maximum current of the power supply (up to 300 A), adjust the open- and short-circuit voltage, and control the slopes of two portions of V-A response curve. The first portion of the response curve extended from 0 to 100 A and was fixed at a slope that promoted easy arc initiation. The second portion extended from 100 to 300 A at a constant slope and was the region used for this study. With a special built-in filter, we could determine the response time (inductance) and vary the controller's response to any voltage changes during welding. After the parameters were set, a conventional x-y plotter was used to check the resulting V-A characteristic. With this system we could determine the effects of small variations in voltage, slope, and current on arc behavior.

Other advantages of this modified current controller are reduced electrical noise and elimination of the characteristic primary-voltage ripple from the output.

DATA AQUISITION SYSTEM - A digital oscilloscope recorded voltage vs. time and current vs. time during welding. Recordings were typically 800 ms long with 0.1 ms between data points. Voltage-vs.-time and current-vs.-time plots produced by the digital storage oscilloscope were analyzed for droplet transfer. A fast Fourier transform (FFT) method was used to characterize the frequency of arc instabilities caused by the droplet transfer.

A video camera and recorder were used to record droplet size and transfer rate at up to 300 frames per second. Backlighting of metal droplets with a 15-mW HeNe laser produced a shadowgraph. Recording time was limited only by the length of the workpiece and, therefore, was relatively long.

RESULTS AND DISCUSSION

The use of a video camera has the advantage of instantaneous viewing of the welding process and monitoring the effect of changing the welding parameters on the droplet size and transfer rate. It also permitted a single record up to 2 hours in length. This was not possible with the high-speed film cameras used in previous studies [3-5]. Another study used a 2-mW laser for backlighting the droplets [6]. The more powerful 15-mW laser used in this study increased the ratio of laser to arc light and permitted better resolution of the image.

A typical oscillogram showing current and voltage vs. time for a steady globular transfer mode is shown in Figure 2a. The droplets are represented by peaks of uniform size that are equally spaced in the current vs. time plot. Increasing wire speed from 125 to 200 cm/min (and current from 110 to 120 amperes) increases the droplet transfer rate from 18 drops per second (dps) to about 78 dps Figure 2 (b). The droplet transfer rate agreed within 10 percent when measured by fast Fourier transform of the oscillogram current and voltage records and from video tape (Figure 3). Stability of the arc and the shape of the instantaneous voltage vs. current curve were unaffected.

Increasing the operating voltage increased both the current and droplet size, as noted in the high-speed video pictures, but decreased their transfer rate. The change in droplet size as a function of current as the voltage is increased from 22 V (100A) to 36 V (150 A) is shown in Figure 4. In the lower voltage range, the arc was stable; above 26 V at 110 A the stability deteriorated until, at 30 V and 120 A, the arc lost its globular transfer characteristics.

Increasing the power supply time constant via the built-in filter decreased the short-circuit current because the droplet usually detached at lower levels of current (Figure 5). The smaller short circuit current also decreased the instantaneous power of the arc (evident in the power vs. time curves). The height of the power peak at the end of the short circuit portion of the droplet transfer process was related to the volume of spatter encountered in GMA welding.

At wire speeds greater than 150 cm/min and filter settings resulting in a slewing rate lower than 4000 A/s, the spatter could not be reduced by slope changes. Under these conditions, establishing a current limit was found to reduce the spatter. An average current of 110 A and an average voltage of 22 V were found to well represent short circuiting transfer conditions. During the actual short circuit, the current momentarily reached 210 A (similar to Figure 2b). In three subsequent tests, the maximum current was limited to 180, 150 and 135 A. Figure 6 is a plot of these maximum currents and the maximum power for each of these cases. Again, the height of the power peak at the end of the short circuit portion of the droplet transfer process was related to the volume of spatter. Since the wire feed speed was kept constant, there was no change in the volume of metal transfered across the arc. Yet the reduction in the maximum power in each droplet-transfer cycle due to the imposed current limit reduces the spatter significantly. A reduction in spatter of more than half was estimated by comparison of a video tape of the spatter in the four cases and subsequent examination of the specimen surface.

CONCLUSIONS

1. A technique was developed that used a video camera, higher energy laser, and digital oscilloscope to resolve the arc transfer characteristics.

2. Establishing a limit for the maximum short-circuit current reduced the spatter in GMA welding by more than half.

3. The height of the power peak at the end of the short circuit portion of the droplet transfer process was associated with the volume of spatter encountered.

ACKNOWLEDGMENTS

The authors appreciate the assistance of D. Mitrakovic in designing and constructing the power source controller, the assistance of

J. Moulder in providing the laser and its accessories, the assistance of R. D. Kriz in computer graphics, and the assistance of M. Stieg in organizing the manuscript.

REFERENCES

1. Manz, A. F., Welding Journal, vol. 59, September 1969, pp. 707-712.

2. Lancaster, J. F., "The Electric Arc in Welding," Lancaster, J. F., editor, Pergamon Press, London England (1984), pp. 134-203.

3. Quigley, B. C. and Webster, J. M., Welding Journal, vol. 61, November 1971, pp. 461s-466s.

4. Matsuda, F., Ushio, M., Kuwayama, N. and Koyama, K., Transactions of JWRI, vol. 12, No. 1 1983, pp. 35-41.

5. Essers, W. G., and Van Gompel, M. R. M., Welding Journal, vol. 63, June 1984, pp. 26-32.

6. Allemand, C. D., Schoeder, R., Ries, D. E., and Eagar, T. W., Welding Journal, vol. 64, January 1985, pp. 45-47.

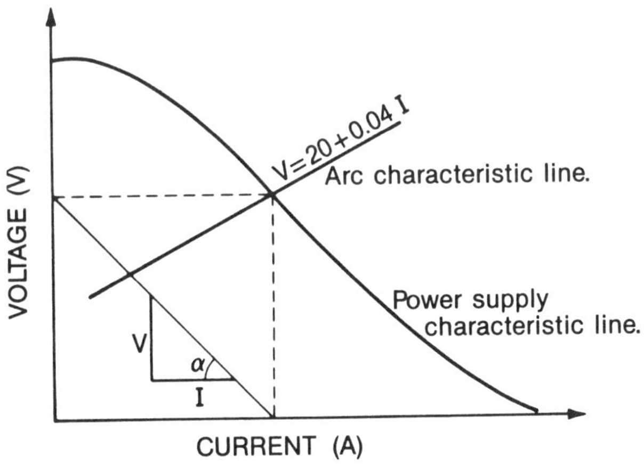

Figure 1. Volt-ampere response curve of a typical power supply. The slope of the curve at the intersection of the arc characteristic line is given by:

$$\text{Slope} = \tan \alpha = V/I = R = \text{Resistance}$$

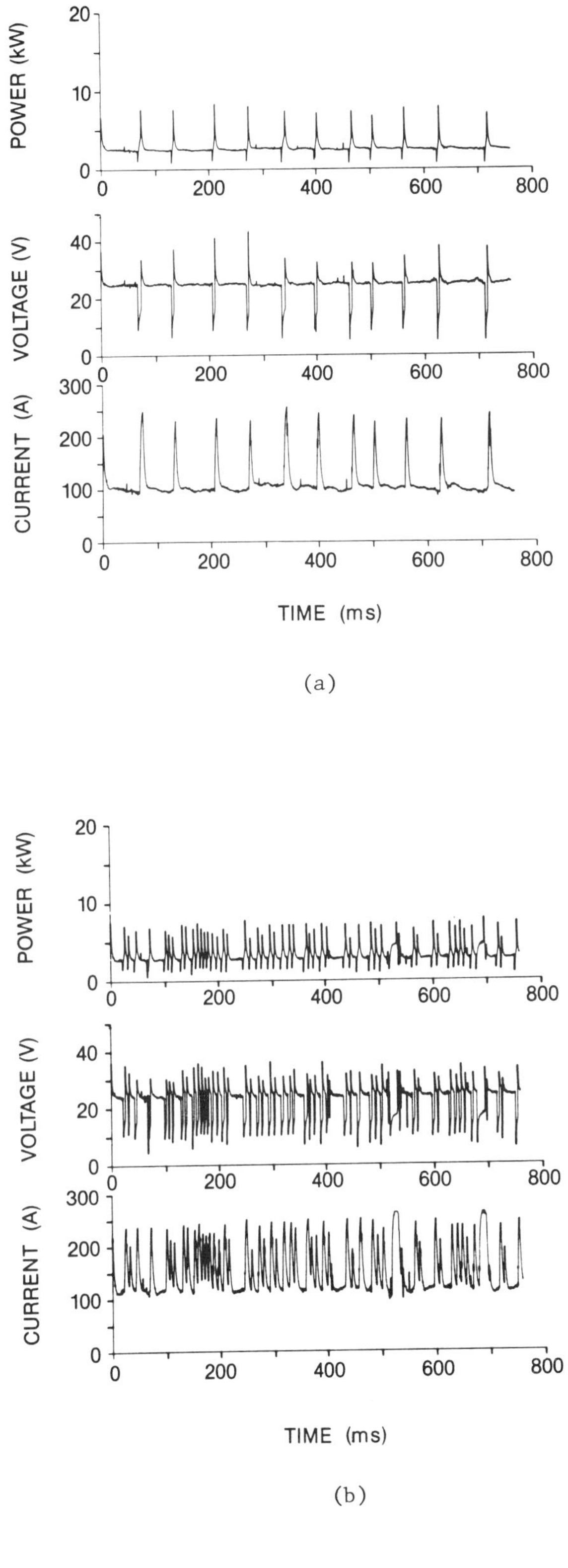

Figure 2. Oscillogram of current, voltage and power vs. time for two wire speeds; (a) 125 cm/min, (b) 200 cm/min.

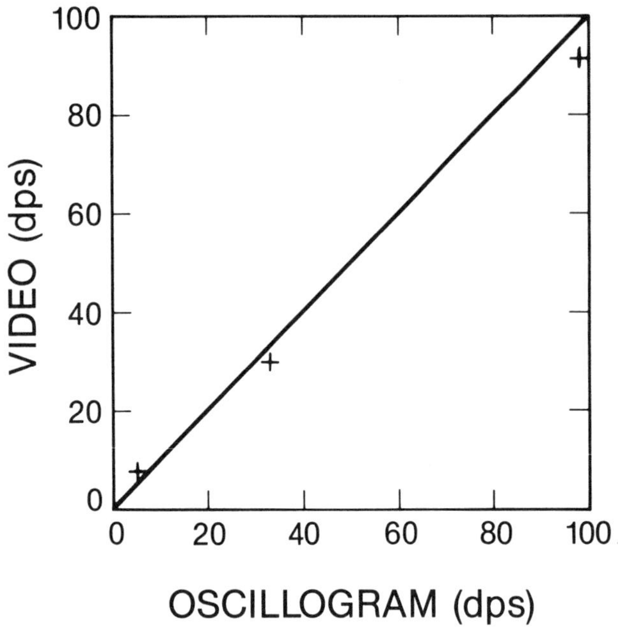

Figure 3. Droplet transfer rate in drops per second as measured by both video tape and oscillograms.

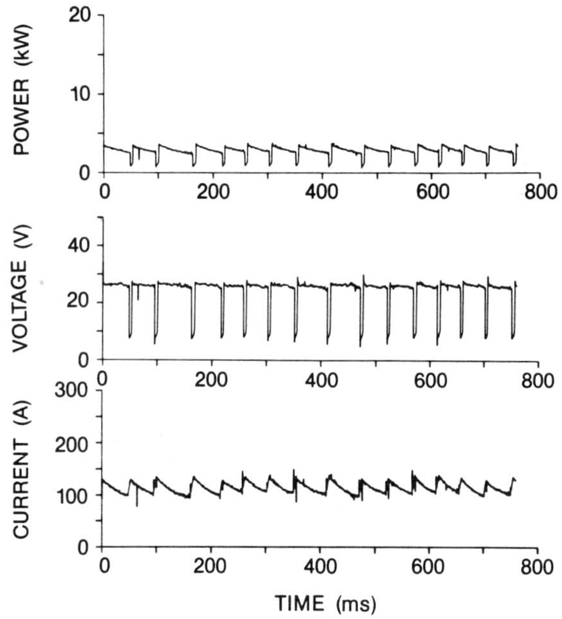

Figure 5. Oscillogram of current, voltage and power vs. time for similar welding parameters as in Fig. 2 (a) but with increased time constant (inductance).

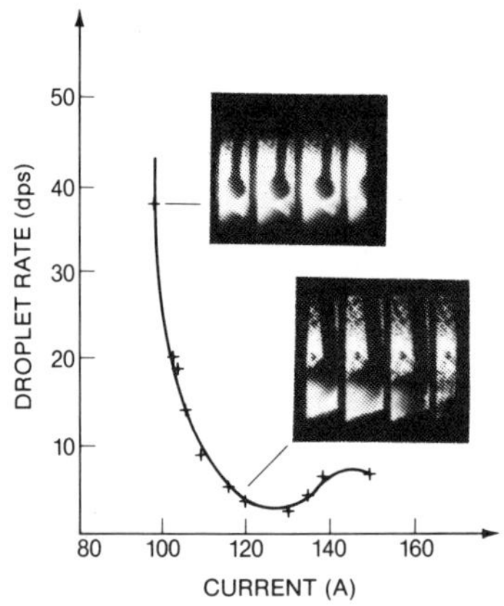

Figure 4. Droplet rate (drops per second) vs. current obtained for a wire speed of 100 cm/min and increasing voltage. The photographs indicate the change in droplet shape.

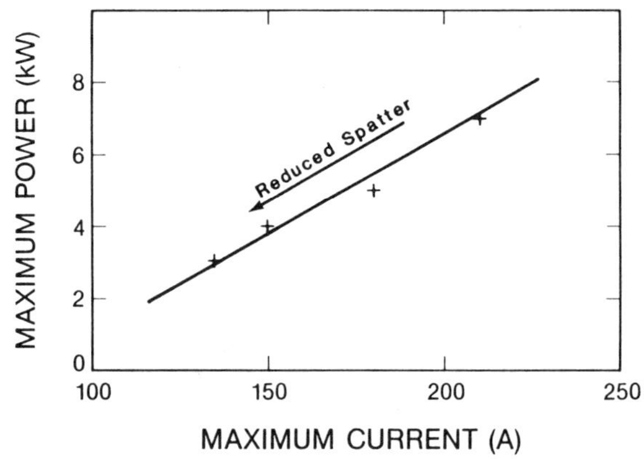

Figure 6. Maximum power vs. maximum current as the maximum current was limited.

FLUX CORED ELECTRODE DEVELOPMENTS FOR SHIPBUILDING APPLICATIONS

Lee G. Kvidahl
Ingalls Shipbuilding
Pascagoula, MS

ABSTRACT

Effective shipbuilding fabrication requires the use of efficient, economical welding methods while maintaining high levels of quality. Ship construction demands a high percentage of out of position welding making the flux cored arc (FCAW) welding process one of the most efficient processes for this type of construction.

Recent developments in flux cored electrodes have expanded the applications for the process to include high strength steels with low transition temperatures. These electrodes provide the necessary mechanical properties for the weldment design while possessing excellent weldability.

As a result of the electrode improvements, fabricators can use a more economical joining process than those available in the past.

FLUX CORED ARC WELDING is not a recent development. The first use of the process at Ingalls Shipbuilding was during 1968. Although used to some extent then, process constraints due to the electrode characteristics and weldability restricted the applications and the process was one of minor utilization. However, during the past five years, many improvements in the electrodes have resulted in FCAW now being the most widely used process in the shipyard. These improvements include weldability, availability of various alloy chemistries and as deposited mechanical properties.

The filler material manufacturing industry has performed much research and development work in flux cored electrodes. This work has resulted in electrodes with excellent strength and toughness that can be welded in all positions. Improvements in the manufacturing process controls and raw material selection ensures consistent high quality electrodes. Technological advances such as the advent of microalloyed flux cored weld deposits, provide fabricators with consistent mechanical properties and joint quality.

These developments have resulted in the expanded use of flux cored welding in shipbuilding applications.

SHIPYARD ROLE

As a fabricator Ingalls does not perform filler material research, rather, the identification of applications that require improvements in electrode properties or performance and the thorough investigation of electrode characteristics are the principle roles of a shipyard. The information collected through the testing and evaluation processes is discussed with the filler material manufacturers for their continued research and development efforts.

It is this cooperative effort which has accelerated the developments of the newer high strength electrodes. For example, a specific application was identified by the shipyards-flux cored arc welding of HY-80.* Representatives of the various flux cored electrode manufacturers were contacted and the applications were reviewed. The electrode manufacturers responded in a relatively short period of time with products which were evaluated in laboratory and production environments. The results of these test welds revealed that acceptable welds could be made with this new family

*MIL-S-16216, Steel Plate, Alloy Structural, High Yield Strength (HY-80 and (HY-100

TABLE I MIL-101TM/C ELECTRODE MECHANICAL PROPERTIES

	Tensile Strength Mpa (ksi)		Yield Strength Mpa (ksi)		Dynamic Tear			
					-29°C J	-20°F ft-lb	-1°C J	30°F ft-lb
MIL-101TM	751	(109)	689	(100)	502	369	782	575
MIL-101TC	477	(97)	599	(87)	477	351	721	530

of electrodes. After the necessary contractual approvals were obtained, the electrodes were placed in production.

The net result of this effort is a more economical fabrication method for constructing ships without any ensuing loss of quality.

EVALUATION PROCEDURES

Standard shipyard procedures have evolved for evaluating flux cored arc welding electrodes. The first step of the evaluation and the most important to the welder is the weldability of the electrode. With few exceptions, it is essential that an electrode operate in all positions. Shipbuilding demands that a welder adjust to out of position work since structural assemblies become too large to always place in the flat position. Similarly welders demand that their electrodes operate efficiently and easily in all positions.

A proper evaluation of weldability requires that the laboratory simulate the production environment to the maximum extent. This requires that the welding equipment, base materials and gases be the same as those used in production situations. During this evaluation, all positions of welding are utilized. Both fillet welds and butt welds are made with the sole purpose of watching the arc transfer and the weld deposit geometry.

This preliminary testing is mandatory to properly evaluate a welding electrode. There have been cases of weld materials which provide the necessary arc characteristics in the flat or horizontal fillet position but can not be considered "all position". There have also been cases where vertical or horizontal welds are possible but overhead becomes impossible for production environment applications. The surface tension characteristics of these products result in material falling from the puddle.

The reason for utilizing both fillet and butt weld configurations is that each is unique and both are used in construction of a ship. Even though a welding electrode can be used for "all position" welding of a fillet joint, this does not always indicate that butt welding is equally achievable. There have been instances where electrodes produce acceptable welds for those applications having a groove edge to assist in supporting the puddle. The final passes of a butt weld can be difficult for electrodes with low surface tension properties and slow freezing slag systems. Although the slower freezing slag systems can provide improved mechanical properties, welding in non-flat positions with these systems can result in arc transfer and puddle stability problems.

During this phase of the evaluation process, it is essential that the arc transfer be spray droplets. When the transfer is a fine spray, better puddle control and weld quality will be achieved by the welder in out of position applications. The spatter will be reduced and the welder discomfort and fatigue level will be minimized. The average welder prefers a stable spray transfer to a more globular transfer.

Upon satisfactory completion of the weldability evaluation, the mechanical properties are tested. Butt welds are made in the vertical and overhead positions. Each weld is tested for tensile, bend and notch toughness properties. Typical values for the MIL-101TM/C electrodes are shown in Table 1.

Tests are run in accordance with MIL-STD-248* which is the military standard for welding procedure qualification. The criteria for evaluation of mechanical properties are contained in MIL-E-24403/2.**

* MIL-STD-248, Welding and Brazing Procedure and Performance Qualification

**MIL-E-24403/2, Electrodes-Welding, Flux Cored, Low Alloy Steel

Both vertical and overhead positions are used because overhead is more difficult for electrodes with minimal weldability while vertical has a tendency to reduce mechanical properties due to the higher heat input required to deposit welds in the vertical position. Because of differences in cooling rates and the resulting microstructures, it has proven prudent to evaluate electrodes in both positions. The slower cooling rates typical in vertical welds can permit grain growth and self tempering which may reduce yield strength and toughness. The overhead position provides a finer grain microstructure which can result in good impact toughness and strength but the weldability of the electrode may be unacceptable.

CARBON DIOXIDE SHIELDING

Most of the development work to date has been with an argon/carbon dioxide mix shielding gas. Because of economic considerations, it is beneficial for a fabricator to use carbon dioxide for all flux cored arc welding rather than utilizing an argon mix. For this concept to be practical it is essential that the electrode be acceptable in weldability and mechanical properties. Several electrodes can provide the weldability requirement but may not be able to satisfy all of the mechanical properties.

Much of the present work at Ingalls has been related to the application of carbon dioxide shielding to FCAW of HY-80. Our work to date follows the trends of data reported in previously published work.* During the procedure qualification testing of the electrode, it was discovered that the microstructure is coarser with the carbon dioxide shielding than with the mixed gas shielding. This has resulted in a reduction in tensile properties and toughness.

The weldability of the electrode is different due to the change in shielding. There is a slight increase in spatter and improvement in penetration. However, with the necessary arc stabilizers the electrode provides a welding arc that is being used in an economical fashion.

*Lathabai, S. and Stout, R.D., Welding Journal, November 1985, pp. 303-S to 313S

A similar program has been successfully completed for MIL-71T1-HY* class electrodes. Although acceptable mechanical properties have been achievable with this type of electrode for several years, until recently, the weldability with carbon dioxide shielding has been unsatisfactory. The ability to obtain a spray transfer without a percentage of argon in the shielding gas has expanded the applications for this electrode with the resultant economic savings.

Table 2 contains a comparison of mechanical properties that resulted from weld tests when carbon dioxide shielding was substituted for the argon/carbon dioxide mixture. In all cases the properties are acceptable, however, the difference in values because of the change in shielding gas must be considered before altering a weld procedure.

APPLICATION FLEXIBILITY

Another factor which must be considered is the flexibility available with an electrode that will permit its use in more than one application. This becomes an important consideration as the welded products become more varied and an inordinate number of filler materials may be required. To assure proper quality control in filler material selection and to minimize the warehouse stockage, an electrode that can provide the necessary weldability and mechanical properties for more than one application may be selected rather than utilizing two or more electrodes that may be "textbook" optimums.

Much of the laboratory effort is spent evaluating the suitability of the flux cored electrodes for a range of applications. This research work requires electrodes to be tested in other than the seemingly typical applications. Test plates are welded using alternate gas mixtures, heat inputs, weld positions or other variables to determine the feasibility of the wire for use in multiple production situations.

*MIL-E-24403/1, Electrodes - Welding Flux Cored, Ordinary Strength and Low Alloy Steel

TABLE II MIL-71T1-HY ELECTRODE
MECHANICAL PROPERTIES

Shielding gas	Tensile Strength Mpa (ksi)	Yield Strength Mpa (ksi)	CVN -29°C J	CVN -20°F ft-lb	CVN -18°C J	CVN 0°F ft-lb
Argon/CO_2	661 (96)	606 (88)	44	32	56	41
CO_2	599 (87)	531 (77)	58	43	54	40

The decision to use one electrode for multiple applications requires a laboratory analysis of consumables to determine what properties can be consistently achieved using shipyard production welding techniques. It is not uncommon for the shielding gas used for specification certification to be different from that used in the shipyard. Therefore, test plates must be welded using the shielding gas supplied in the shipyard to determine the mechanical properties that will be obtained in actual construction.

Several of the chemical elements in a steel weld deposit are likely to be affected by a change in gas mixture. It is known that the manganese and silicon levels will vary with the percentage of carbon dioxide in the shielding gas. As the carbon dioxide increases, the manganese and silicon levels are reduced in the weld deposit.

For this reason, it is important for a fabricator to perform test welds using the shielding gas that will be used in production. Electrodes may be developed for specific gases or can be adaptable to suit alternate mixes. Although the weld appearance may not be altered significantly, the mechanical properties can differ with various shielding gas combinations.

The cooperative research and development effort between electrode manufacturers and fabricators is essential for continued progress in the field of welding. As specifications change, the interface between the two parties becomes very important. Recent specification changes in maximum diffusible hydrogen levels and minimum toughness values could have had a major impact on the shipbuilding industry but for a cooperative effort.

CONCLUSIONS

1. A cooperative effort by industry and filler material manufacturers can accelerate research and development progress.
2. Flux cored electrodes are available in both low and high strength grades that meet the requirements of MIL-E-24403/1 and MIL-E-24403/2 when using carbon dioxide shielding.
3. The microstructure of the weld deposit is more coarse and less uniform with carbon dioxide shielding than with an argon/carbon dioxide shielding.
4. The mechanical properties of the weld deposits will be altered when carbon dioxide shielding is used in lieu of mixed gas.
5. Only after thorough testing should an electrode be used with both gases rather than a single more optimum gas.

METALLURGICAL ASPECTS AND APPLICATION OF FRICTION WELDING

H. Kreye, G. Reiners
Universität der Bundeswehr
Hamburg, Federal Republic of Germany

ABSTRACT

Dissimilar material combinations including aluminum/copper, aluminum/steel, copper/steel, austenitic stainless steel/carbon steel and mechanically alloyed aluminum/steel were friction welded using continuous drive friction welding machines. Metallographic and mechanical tests were performed to study the effect of the process variables on the microstructure and the properties of the welds. In all instances welds could be produced having a strength similar to the strength of the weaker component.

FRICTION WELDING is a solid state welding process which utilizes the frictional heat generated between a rotating and a stationary workpiece to produce a metallurgical bond. The localized heating and the absence of any melting make friction welding better suited for joining dissimilar metal combinations than other welding processes. This applies especially to combinations with a large difference in melting temperature or thermal conductivity and for those with a tendency to form brittle phases at the interface. Such combinations, including aluminum/copper, aluminum/steel, copper/steel, stainless steel/low alloy steel, and mechanically alloyed aluminum as a rather new type of material as well, have been used for this study. The influence of the welding variables on the microstructure and mechanical properties has been examined in order to improve the basic understanding of the metallurgy of the process and to elucidate the great potential of friction welding.

EXPERIMENTAL PROCEDURE

MATERIALS - Aluminum/copper and aluminum/steel welds were chosen to study the formation of brittle intermetallic phases during the welding process and their control through the welding variables. Aluminum/copper welds were made from 1050A (Al99,5) and OFHC-Cu, bar to bar with a diameter of 20, 40, 60 and 100 mm. For the aluminum/steel welds 20 mm bars of 1050A (Al99,5), austenitic stainless steel 316Ti (X10CrNiMoTi 1810) and carbon steel C1060 (C60) have been used.

Copper/steel and austenitic stainless steel/low alloy steel welds were chosen to study the formation of martensite due to a high cooling rate and due to mechanical mixing, its control through the welding variables and its influence on the mechanical properties. Copper/steel welds were made from 20 mm bars of OFHC-Cu and carbon steel C1034 (Ck35), and stainless steel/low alloy steel welds from 30 mm bars of 316Ti (X10CrNiMoTi1810) and M1044 (St60-2).

The mechanically alloyed aluminum used in this study was a 1050A aluminum containing 14 vol.-% of a fine dispersion of aluminum oxides and carbides, tradenamed Dispal2 /1/. Due to the high thermal stability of these particles the material exhibits high strength at elevated temperatures. However, heating above the melting point causes degassing and formation of pores. This behavior strongly impairs the weldability of the material for any fusion welding technnique. Friction welds of Dispal/Dispal, Dispal/ C1060 and Dispal/316Ti were made with 20 mm bars.

FRICTION WELDING - Three different commercial type friction welding machines (Kuka, continuous drive) capable of 120, 450 and 2500 kN maximum load were used in the experiments. The machines were equipped with powerful load systems and fast acting hydraulically operated brakes. The start of the forging stroke and the brake application could be controlled independently. Therefore, the forging could be performed onto the rotating spindle and the slowed down spindle as well (fig. 1). This method, which combines some features of both continuous drive and inertia welding, is also termed as hybrid friction welding /2/.

The main welding variables of rotational

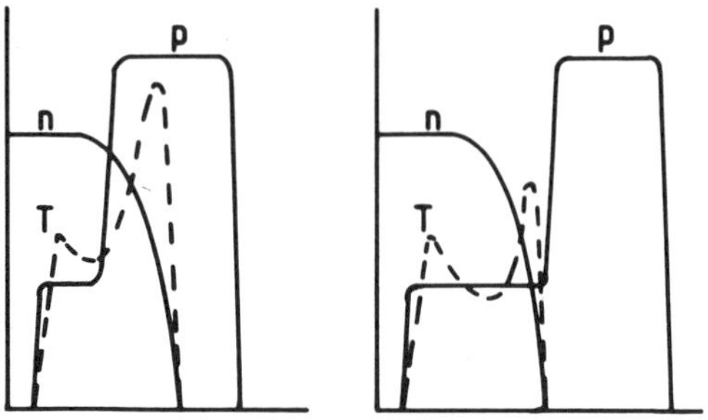

Fig. 1 - Variation of speed n, pressure p and torque T during the welding cycle. Forging onto the rotating spindle (left) and forging onto the slowed dcwn spindle (right).

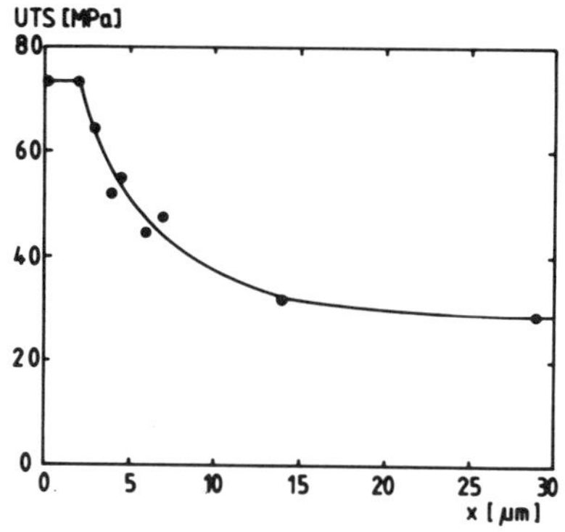

Fig. 2 - Width cf the diffusion zone x vs. strength (UTS). Aluminum/ccpper welds annealed at different temperatures.

speed n, friction time t_1, forging time t_2, friction pressure p_1, forging pressure p_2 and the start of the forging stroke were investigated for each of the combinations in order to determine the nominally optimum conditions which provided acceptable bend test results. Prior to welding the faying surfaces were dry-turned and degreased.

EVALUATION - The microstructure of the welds showing acceptable bend test results was analysed by metallographic techniques. Based on this analysis the welding conditions were further improved in order to minimize undesired diffusion and formation of brittle phases. Such welds were than subjected to further mechanical testing including tensile, shear and fatigue tests.The fatique tests were performed by starting with a load of 80 % of the yield stress of the weaker component and increasing the load in steps of 5 % of the yield stress after all $2 \cdot 10^5$ cycles. As a criterion for a sound bond strength properties of the welds similar to the strength values of the weaker component were required.

RESULTS

With all combinations of materials investigated, welds could be produced having a strength similar to the strength of the weaker component. The welding conditions were similar to those reported in the literature /3-6/. However, for combinations with a tendency to form intermetallic phases at the interface shorter friction times were used and the forging was performed onto the rotating spindle.

ALUMINUM/COPPER - The strength of the welds is limited by intermetallic phases which form in a narrow zone at the interface as a result of mechanical mixing and diffusion. The influence of this zone on the tensile strength of the weld is shown in figure 2. Welds with a diffusion zone, ≤ 2 μm wide, did not fail at the interface in tensile and fatigue testing. Such welds

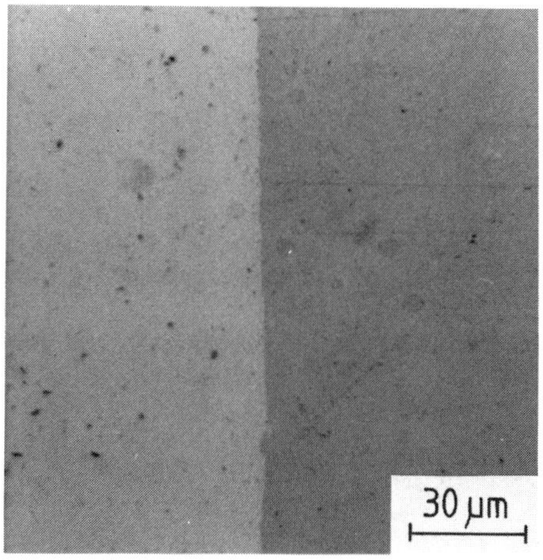

Fig. 3 - Microstructure of an aluminum/ copper weld.

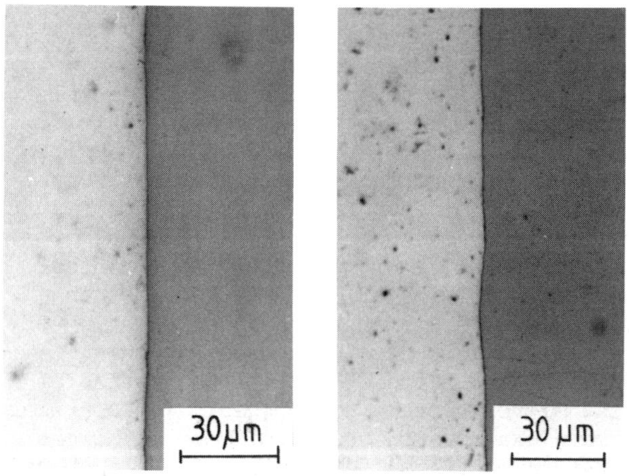

Fig. 4 - Microstructure of aluminum/steel welds. 1050A/316Ti (left) and 1050A/C1060 (right).

could be produced with all bars having a diameter of between 20 and 100 mm. Figure 3 shows a micrograph with no intermetallics visible at the interface. In this case the diffusion zone is only 0,3 µm wide as revealed by transmission electron microscopy /6/.

ALUMINUM/STEEL - The strength of the welds is limited by the formation of intermetallic phases, too. In both examples, aluminum to austenitic stainless steel and aluminum to carbon steel, the width of the diffusion zone could be reduced to a value ≤ 1 µm, not visible under an optical microscope (fig. 4). Such a narrow zone turned out to have no detrimental effect on the strength of the weld. In all testing procedures failure occured within the aluminum base material.

COPPER/STEEL - During the welding the ferritic-pearlitic steel is heated to a temperature at which it transforms to austenite and then to martensite upon cooling. The formation of martensite does not affect the ductility of the weld, when the width of the heated and transformed zone is kept narrow and when the formation of martensite is limited to the former pearlite grains. Such a favourable ferrite-martensite duplex structure could be achieved through the variation of the welding conditions as shown in figure 5.

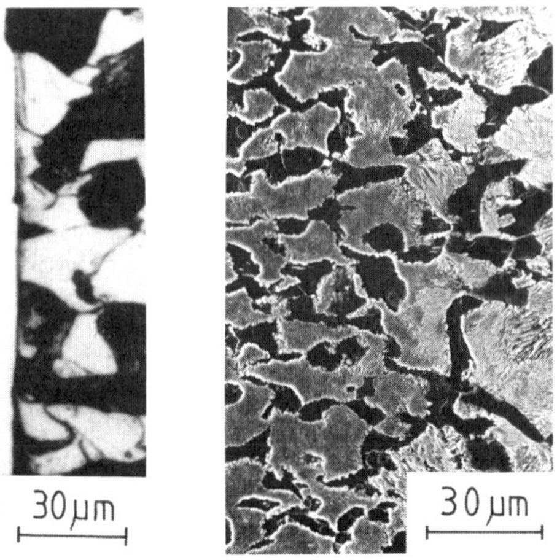

Fig. 5 - Microstructure of a copper/steel weld. Interface (left) and martensite-ferrite duplex structure (right).

AUSTENITIC/FERRITIC STEEL - Diffusion and mechanical mixing at the interface of these two materials always leads to an alloy composition, which transforms to martensite even at very slow cooling rates. Figure 6 shows two samples with different degrees of mixing of the two components. In this combination the increase in friction and forging pressure have proved most effective in reducing the width of the martensite zone. The bond strength of both samples shown in figure 6 was higher than the strength of the base material. However, in the fatigue test only the sample with less martensite did not fail at the interface.

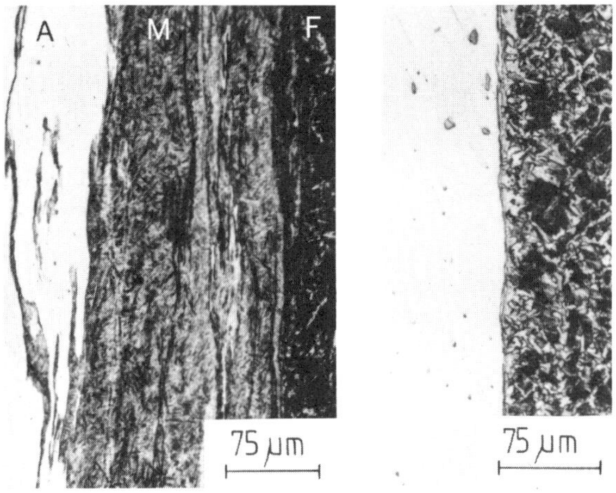

Fig. 6 - Microstructure of austenitic/ferritic steel welds with different amounts of martensite at the interface.

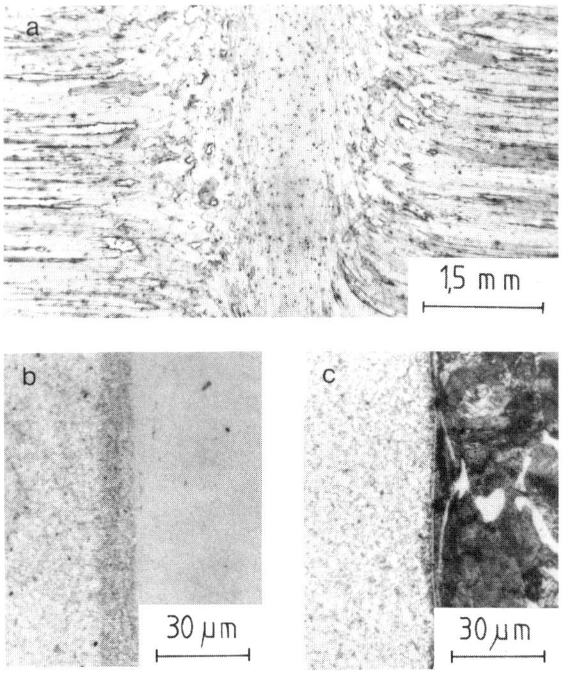

Fig. 7 - Microstructure of a Dispal/Dispal (a), Dispal/316Ti (b) and Dispal/C1060 (c) welds.

MECHANICALLY ALLOYED ALUMINUM - The forging of the material during friction welding turned out to be very effective in refining the distribution of the dispersed particles of Dispal (fig. 7a). A bond strength of 300 MPa could be achieved for Dispal/Dispal and for Dispal/316Ti and Dispal/C1060 welds. In the Dispal/steel welds the width of the diffusion zone at the interface containing intermetallic phases could be kept ≤ 1 µm (figs. 7b, c), so that the intermetallic phases did not affect the bond strength.

DISCUSSION

The results demonstrate that friction welding is a useful method in joining dissimilar materials which have a tendency to form brittle phases at the interface. Such phases have no detrimental effect on the weld properties, when the width of the diffusion zone containing these phases can be kept below a critical value, which is in the order of a few microns. In the process of welding aluminum to copper and aluminum to steel this could be achieved by applying short welding times (< 1s), high friction and forging pressures and by forging onto the rotating spindle. These parameters cause a fast heating of the material to a homologous temperature of $> 0.7\ T_m$ where the intermetallic phases become ductile and can be squeezed out of the weld zone into the collar. The narrow diffusion zone remaining at the interface experiences a grain refinement as a result of a dynamical recrystallization during the forging stroke.

In the welds of copper to steel and ferritic steel to austenitic steel which are also difficult to perform the formation of martensite must be limited to a narrow zone. In the case of copper to steel welds the width of this zone is determined by the depth to which the ferritic steel is austenized during the welding. A narrow martensite zone can mainly be achieved by applying short welding times. The short duration of the heating phase has another beneficial effect. Only the pearlite grains transform to martensite, while the ferrite grains preserve their structure. In the case of ferritic to austenitic steel welds the martensite zone is caused by a change in composition due to a mechanical mixing of the components. This mechanical mixing can bei minimized by applying high friction and/or forging pressures, leading to a more homogeneous flow of the material at the interface.

All examples given in this study show that in friction welding the microstructure of the bond zone can be controlled very effectively through the welding variables. The sound welds described were performed on commercial type welding machines and turned out to be highly reproducable.

ACKNOWLEDGEMENT

The welding experiments were carried cut at the Kuka, Schweißtechnik und Roboter GmbH, Augsburg. The authors wish to express their thanks to the staff of Kuka for technical assistance. Financial support granted by the Arbeitsgemeinschaft Industrieller Forschungsvereinigungen (AIF) is also greatly acknowledged.

REFERENCES

1. Arnhold, V. and J. Baumgarten, Powder Metall. Int. 17, 168-172 (1985)
2. Dunkerton, S. B., Int. Conf. on Developments and Innovations for Improved Welding Production, Birmingham 46.1-46.9 (1983)
3. Hartwig, H., Aluminium 57, 615-618 (1981)
4. Sada, N.B. and A.S. Bahrani, The Joining of Metals Vol. 2 p 38-46, The Institution of Metallurgists, London (1981)
5. Ishibashi, A., S. Ezoe and T. Tanaka, Bull. Jap. Soc. Prec. Engin. 17, 125-126 (1983)
6. Kreye, H. and G. Reiners, Z. Metallkde. (to be published)

THE EFFECT OF DEOXIDATION OF BASE METAL AND SHIELDING GASES ON THE PLASMA ARC WELDING OF CONSTRUCTIONAL STEELS

T. Moisio
Lappeenranta University of Technology
Lappeenranta, Finland

ABSTRACT

The chemistry of welding gases and the method of deoxidation of base metal affect the behaviour of the weld pool. In killed steels aluminium occurs mainly as a high melting oxides. It forms a solid slag film on top of the solidifying weld. The surface tension between the slag film and the melt makes bulges to form on the weld bed.

In calsium-silicon treated silicon-aluminium killed steels, aluminium exists as a low melting aluminates and slag on top of the melt forms droplets on the solidifying weld. The surface is smooth.

The choice of plasmagas has a detrimental effect on the weld quality.

INTRODUCTION

Industrial plasma arc welding is used today to join austenitic stainless steel and perhaps titanium. Keyhole plasma arc welding of constructional low yield or high tension steels is unusual and not much dealt with even in the literature. It is considered that the behaviour of the unalloyed steel weld pool is uncontrollable compared with that of the stainless steel weld pool.

The present interest in the possibility of the keyhole plasma arc welding of plain carbon constructional steels is due to the development of the control systems and the expansion of automatic welding as well to the lack of a proper oneside welding method.

Rather few research reports concerning the issue have been published. In the reports there are differences even in basic matters such as the choice of plasma and shielding gases. The purpose of this work is to investigate the effect of the deoxidation of the base metal and the shielding atmosphere on welding and weld pool behaviour in the keyhole plasma arc welding of constructional steels.

Materials

Materials used were sheets of steel of 3, 6 and 8 mm. Bead on plate tests were normally used. But weld trials were used just to check the differences between buttwelds and beads on plate tests. The analysis of the materials is shown in table 1. In the whole series deoxidation was allowed only on the 3 mm sheets. Silicon killed 6 or 8 mm sheets are not produced any more. Tests sheets were shot blasted before welding to prevent contaminated surfaces from affecting weld pool behaviour.

Shielding atmosphere

Argon and argon +5 % H_2 were used as plasma gases. Shielding gases were argon, argon +2 % O_2 and argon +20 % CO_2. The root side of the weld was shielded by argon or argon +5 % H_2 in the backing groove of the welding jig.

Table 1 Analysis of test materials

N % of the material	Deox. meth.	Analysis					
		C	Si	Mn		S	Al
1	Al	0.09	0.02	0.37	0.017	0.020	0.036
2	Si	0.11	0.20	0.50	0.017	0.021	0.001
3	Al-Si	0.09	0.22	0.31	0.011	0.015	0.028
4	Al-Si (TN)	0.09	0.20	0.47	0.017	0.008	0.035
5	Al-Si (TN)	0.11	0.20	1.58	0.022	0.007	0.032
6	Al-Si	0.14	0.33	1.31	0.030	0.016	0.044
7	Al-Si (TN)	0.17	0.36	1.36	0.018	0.009	0.026
8	Al-Si (TN)	0.12	0.18	1.42	0.022	0.009	0.038
9	Al-Si	0.13	0.52	1.36	0.014	0.015	0.048
							+0.002 B

Filler-material

As a filler-material Ø 0.8 mm wire purported for the MAG-welding of unalloyed steels was used. The list analysis is shown in table 2.

Equipment

Power source: Rectifier type constant current power source I max 400 A.

Torch: For mechanized plasma arc welding I max 300 A.

Experiments

It was noticed in preliminary experiments with 3 mm thick plates that undercutting was the main factor limiting welding speed. With thicker materials the turbulence of the pool could cause unstability in the pool but here too the undercutting was still the limiting factor. In each case the speed was increased just enought to prevent undercutting. Other parameters were kept constant. During welding the pool behaviour was observed visually and recorded by cine camera. Ready made welds were checked visually and by radiography.

WELDING PARAMETERS

Results

Welding parameters used and speeds achieved are shown in table 3.

Visual inspection showed that the welds made in the argon atmosphere had a smooth surface with perhaps a shallow dent along the center line when the base metal which were used is Si -killed. In Al- and Al-Si -killed steels there were local bulges rather lustreless compared to the bead surface. The rootsides of the welds were satisfactory and regular in the Al-Si -killed steels but some were irregular in the silicon killed base materials. The fusion line indicated that the wetting of the base metal by the melts could be better.

Table 2. Analysis of the filler-wire

C	Si	Mn	P	S
0.07...0.14	0.07...1.0	1.3...1.6	0.025	0.025

Table 3

Material	Plasmagas Argon			Argon +5 % H_2			Argon +10 % H_2		
3 mm				Q_p	I	v			
1				4.5	170	49			
2				4.5	170	52			
3				4.5	170	49			
4				4.5	170	49			
5				5.6	246	95			
6 mm	Argon			Argon +5 % H_2			Argon +10 % H_2		
5	5.2	270	30	5.6	268	38	4.8	270	43
6	5.6	246	27	5.2	238	29			
7	5.5	258	28	5.4	265	38	4.8	270	44
8	5.5	258	28	5.4	265	38	4.8	270	44
9	5.1	264	31	5.2	268	43			
8 mm	8 mm								
9				5.5	270	29			

Q_p = plasma gas flow l/min
I = welding current amp
v = welding speed cm/min

When Ar +20 % CO_2 was used as the shielding gas the wetterbility as well as the welding speed were improved and about the same effect was noticed with the addition of either 5...10 % H_2 or 2 % O_2 to the shielding gas.

It is also possible to increase the welding speed by adding H_2 to the plasmagas. However if the Ar + CO_2 mixture is used as shielding when the Ar + H_2 mixture is used as a plasmagas there are large blowholes both with Si- and Si-Al -killed steels and also the possibility of pores in other materials too. The surface of low sulfur (TH-treated) steel welds is extraordinarily smooth in comparison with material No. 6 for instance.

Some slag droplets were noticed on the top of the welds when argon atmospheres were used but none when Ar + H_2 plasmagas was used. The welds of materials 4.5 and 8 were especially clean. More slag was noticed on the root side of the welds with all steels. The amount of slag on the root side was clearly greater than that on the top of the weld. It was found to be most extended when Ar + CO_2 was used as a shielding gas.

DISCUSSION

It was found during this work that weld pool behaviour is strongly dependent on the interaction of the welding gases and the base metal. The formation of bulges is due to the aluminium-containing slag film on the top of the weld. The slag film is formed from Al or Al-containing residual slag in the base metal. The melting point of slag containing Al-oxide is considerably higher than that of the melting point of steel. A solid slag film forms on the molten metal and the surface tension between the slag film and the metal disturbs the surface of solidifying bed and forms the bulges.

In the Si-killed steel there is practically no Al and the slag consists of low melting silicates. No slag formations were noticed on the top of the weld.

In TN types of steel the aluminium forms low melting calcium aluminates. The slag forms small liquid droplets on the melt and does not disturb the surface tension of the solidifying metal.

The slag formed on the root side was a dark greenish colour and according to the analysis should be something of the (Fe, Mn)O-SiO type. The amount of slag was high when a CO_2 mixture was used as a shielding gas. It could be due to the dissociation of CO_2 to CO + O and the dissovling of O. It is supposed that due to the high energy and temperature of the plasma silicates in the residual slag dissociate and perhaps the degass on the top, where the temperature is highest in the melt, the O_2 reacts with Mn-Si-Fe and the surface tension fixes it on the surface of the root side.

It is proposed that H_2 -containing plasmagas and CO_2 containing shieldinggas gives the hotest weld and pores formed are due to the deoxidation process taking place in the melt.

In this work it is specified that when utilizing plasma arc welding more specific details must be considered than when conventional arc welding is used.

REFERENCES

F. Cuny: Soudage Plasma des aciers au carbone. Soudage et Techniques Connexes 34 (1980) 3-4, pp. 95-101.

C. Messager and F. Cuny: Reliability of Plasma Process in Welding Carbon Steels. Fundamentals and Practical Approaches to the Reliability of Welded Stuctures. Conferences Proceedings. Osaka, Japan, November 24-26, 1982. Publ. Japan Welding Society. Tokyo 1982. pp. 91-96.

J.P. Larue and M. Thomsen: Welding of Carbon Steels, Joining of Metals Conference Proceedings, Helsingor, Denmark, August 9-12, 1981. Publ. Ingenior hojskolen Helsingor Teknikum, Helsingor, Denmark 1981. pp. 28-35.

M. Giraud: Essais semi-industriales de soudage plasma a' plat des aciers au carbone realises en chantier naval. Soudage et Techniques Connexes 34 (1980) 3-4, pp. 102-110.

W. Lucas: TIG- and Plasma Welding of the 80's Part 4 Plasma and Pulsed Current Applications. Metal Construction 14 (1982) 12, pp. 659-667.

K.-E. Nielsen: Plasma Welding of Unalloyed Steel. Joining of Metals Conference Proceedings. Helsingor, Denmark, April 15-18, 1984. Publ. Ingenior hojskolen Helsingor Teknikum, Helsingor, Denmark, 1984, pp. 107-113.

K. Born: Plasmaschweissen. WT-Zeitschrift für industrielle Fertigung 71 (1981) 1, pp. 3-7.

W.F. Bethlehem: Einfluss des Wasserstoffanteils auf Schutzgas beim Plasmastichlochschweissen. Maschinenmarkt 87 (1981) 23, pp. 429-432.

NARROW GROOVE WELDING OF STEAM TURBINE ROTORS

Dennis R. Amos
Westinghouse-Power Generation
Charlotte, North Carolina USA

Robert E. Clark
Westinghouse-Power Generation
Orlando, Florida USA

Abstract

As a method to improve the availability of older low pressure rotors, a unique welding technology has been developed to maximize utility repair options. These methods allow each weld to closely match the original rotor properties with corresponding nondestructive testing requirements that significantly exceed modern forging acceptance standards. Subsequent post weld processing results in negligible dimensional deviations while attaining freedom from undesired residual stresses.

Because of the variable needs of older rotors, one weld technique is not sufficient. For example, where one blade attachment groove is damaged, the high degree of gas tungsten arc weld process control ensures reliability in fatigue critical areas. Conversely, where rotor bore ultrasonic examination may result in the retirement of a forging, removal and replacement of the defective area is feasible using a narrow gap welding process. Finally, larger repairs such as rubs and transverse shaft cracks utilize a high quality submerged arc method significantly advanced from today's standard methods.

MODERN LOW PRESSURE ROTOR alloy processing utilizes technology advances such as vacuum carbon deoxidization, ladle refining, computer controlled forging pressing, and ingot solidification design methods to obtain a forging with maximum toughness and minimum segregation (1 - 6). Unfortunately, these current state-of-the-art processes do not apply to older units currently in service; hence, a significant number of rotors are showing signs of aging such as journal wear, exterior cracking, rubs, and other degradation. These "wearout" phenomena are more common today because of the trend for unit life extension and upgrading rather than replacing these machines with new, more efficient turbogenerator sets.

To meet the need for greater reliability in older low pressure rotors, a series of rotor weld repairs have been developed to address specific requirements depending on the distressed location of the forging. The weld repair program takes into account unit design requirements, welding residual stresses, original forging metallurgy, and applicable non destructive testing methods to determine that the repair meets or exceeds the original material requirements.

The philosophy of weld repair requires that all welding processes use one filler metal chemistry compatible with most low pressure metal chemistries. This requirement avoids inventorying multiple welding consumables and reduces the risk of mixups resulting from human errors. Table 1 lists all substrate materials repairable by our techniques (1).

Table 1 - Forging Alloys Compatible with the Optimized Repair Methods

Component	Generic Alloy Name	ASTM Identification
LP Rotor	2.5 NiMoV	A470, Class 2
LP Rotor	2.5 NiMoV	A470, Classes 3 & 4
LP Rotor	3.5 NiCrMoV	A470, Classes 5 to 7
LP Disc	3.5 NiCrMoV	A471, Classes 1 to 3
LP Rotor	2.0 NiMoV	A293, Classes 2 & 3
LP Rotor	2.5 NiMoV	A293, Classes 4 & 5
LP Disc	2.8 NiMoV	A294, Grades B & C
Generator Rotor	2.8 NiMoV	A469, Classes 2 & 3
Generator Rotor	3.2 NiMoV	A469, Classes 4 & 5
Generator Rotor	3.5 NiCrMoV	A469, Classes 6 to 8

Additionally, every repair is "mocked-up" prior to actual welding. Items such as weld shrinkage control, stress relieving practice and ultrasonic test configuration are all addressed during this initial phase. All welding is done by automatic methods, thus, variables are well understood prior to component repair. New weld geometry/process combinations are metallurgically tested on the "mock-up" to obtain microstructure and mechanical properties. All critical process steps are measured, reviewed, and/or tested to document each weld repair.

WELD PROCESS EVALUATION

To adequately repair low pressure rotor steels, several requirements must be met. First, the weld must be applied soley by automatic methods. Second, the weld must show superior characteristics such as freedom of porosity and cracking, excellent fluidity, and bead shape. Third, it must be suitable for specific geometries and groove shapes as required for the area requiring repair. Finally, the metallurgical properties of the weld must be comparable with the properties of the low alloy rotor steel.

To meet these requirements a screening program was established in 1977 to evaluate all commerical welding processes. The initial review stressed that the welding process must be fully automatic. Prior work on critical component welding of similar low alloy steels had proven that the welding operator was the single most important variable. By automating a process, the risk of human skill variation is minimized.

Those weld processes considered during the screening program were:

- Gas Tungsten Arc - Hot Wire
- Flux Cored Arc Welding
- Submerged Arc Welding
- Narrow Gap Gas Metal Arc Welding
- Gas Tungsten Arc - Cold Wire

As a result of these initial investigations, both hot wire gas tungsten arc welding and flux cored arc welding were eliminated from consideration. In the former case, metallurgical properties of hot wire gas tungsten arc filler deposits were inferior to the rotor steel. Flux cored arc welding did not result in the required non-destructive examination quality, or mechanical properties. In this latter case, weld defects were often associated with the powder distribution in the filled wire.

The three remaining processes: Submerged Arc Welding (SAW), Narrow Gap, Gas Metal Arc Welding (GMAW), and Gas Tungsten Arc Welding (GTAW) showed great promise. Each had advantages and disadvantages which made them more feasible for a specific type repair weld.

Submerged arc welding is used for the repair of rotor areas involving a wide and/or deep groove, such as, the repair of a crack that is not completely in a radial-transverse orientation relative to the component axis. This method of welding is also applicable for large area or volume restoration such as replacement of worn metal resulting from hard part rubs. The main advantages of the submerged arc welding process is that is has a very high deposition rate; however, it requires a preheat and produces the largest grain size of the three processes. Because of the excellent heat input and bead size control, the resulting grain size is still small compared to most commercial submerged arc methods. For this reason, no problem occurs in meeting the original rotor tensile and impact specification requirements, even though these properties are the lowest of the three welding methods. During the screening of the submerged arc welding technique, each variable that could influence weld quality was investigated to assure overall process control (Table 2).

Table 2 - A Summary of Process Qualification Welding Parameters

Submerged Arc Weld (SAW)	Gas Metal Arc Weld (GMAW)	Gas Tungsten Arc Weld (GTAW)
wire type	pulse voltage	gas mixtures
flux type	background voltage	gas flow rates
wire diameters	amperage	tungsten sizes
nature of current, AC or DC	travel rate	cup sizes
power sources	gas mixtures	power sources
joint geometries	gas flow rates	wire diameters
amperage	root opening	wire type
voltage	root geometry	amperage
travel rate	wire tensile strength	voltage
nozzle design	wire extension	wire feed speed
preheat	torch design	travel rate
	preheat	joint geometries

Narrow gap, gas metal arc welding is generally utilized for joining of thick wall rotor forging components. This process was selected because the narrow groove geometry results in a relatively small deposited weld volume, thus distortion from shrinkage effects causes minimal dimensional changes. Gas metal arc welding requires preheating of the component and produces a grain size larger than gas tungsten arc welding. Its tensile and impact properties are intermediate between gas tungsten arc and submerged arc techniques. The major process qualification variables are also given in Table 2.

Cold wire gas tungsten arc welding is utilized for repairing individual rotor attachment grooves, cosmetic, or shallow groove repairs to correct minor surface defects or to allow multiple buildup of blade attachment groove locations (i.e., for a 360° application). Of the three welding process, gas tungsten arc welding offers the greatest number of advantages. This method has the highest ultrasonic quality, requires no preheat, has tensile and impact properties which exceed rotor material specification requirements by a large margin, and develops the finest microstructural grain size. A disadvantage of this process is the low deposition rate, about one and one quarter pounds (0.56 Kg) per hour. Again, weld process variables were analyzed and are given in Table 2.

WELD EVALUATION PLAN

During the initial weld process investigation phases of the program test plates were welded to provide material for rating nondestructive examination quality, evaluating the ease of welding, and generating mechanical test specimens. After the filler metal, flux (submerged arc weld) and shielding gas (gas tungsten arc weld), (gas metal arc weld) were selected, additional qualification tests were run on small mock-up components simulating the actual various configurations that would be repair welded. These latter tests were utilized to determine process variable differences associated with actual component geometries rather than conventional weld preparations. This work did show some filler metal alloy dilution associated with exposed surfaces as compared to deep grooves, and a series of corrective actions were undertaken to eliminate this problem. As a final qualification, all weld processes were verified on full-sized turbine components with mechanical tests actually removed from these welds. On production welds, final verification tests include nondestructive inspection, dimensional verification, hardness, and other metallurgical methods (if geometrically practical).

MATERIAL

Rotor and Disc Materials - The materials shown in Table 1 are all suitable for repair welding. These alloys are from either the NiMoV or NiCrMoV groups. Repair welds are easily deposited on both compositions. Yield strength (0.2%) levels through 115 ksi (793 MPa) are compatible with the Westinghouse filler metal composition. Repair welding of components having yield strengths greater than a 115 ksi yield strength must be reviewed on an individual basis.

FILLER METALS

During the development stage of this program, a single filler metal was desired that would meet or exceed all low pressure rotor material minimum specification requirements to 115 (793 MPa) ksi yield strength and at the same time meet or exceed the original forging nondestructive examination requirements. To achieve this, numerous gas tungsten arc weld filler metal compositions were evaluated with the result being a single filler metal that far exceeds the minimum Table 1 ASTM rotor and disc specification mechanical property requirements. Filler metal fluidity and other weldability issues are unaffected by the generic base metal alloy (i.e. NiMo or NiCrMoV).

A caution required when using this process applies to the original rotor forging residual alloy element distribution. If the base metal is high in sulfur or copper, then hot shortness of the weld may result in weld deposit cracking along interdentritic paths. Conversley, if temper embrittling elements exist, then underbead cracking becomes a possibility. Modern rotor forgings do not generally exhibit these phenomena, because of desulfurization and scrap selection during melting practice but these cracks are distinct possibilities in older steels. Some remedial measures may be available that still allow weld repair, but the entire repair must be carefully evaluated.

For gas metal arc welding several compositions were initially reviewed. The filler wire qualified for gas tungsten arc welding has proven to be excellent with regards to metallurgical properties and nondestructive quality.

For submerged arc welding thirty-five flux filler metal compositions were evaluated during test plate welding. The majority of these were quickly eliminated due to poor ultrasonic quality. The selected flux-filler metal combination meets the base metal nondestructive examination and mechanical property requirements. Again, the prior composition is suitable for this process.

POSTWELD HEAT TREATMENT

The postweld heat treatment temperature used is selected on an individual basis to meet the following requirements:

1. High as practical to minimize weld stresses.
2. High enough to temper the weld and heat affected zone hardness.
3. Low enough to prevent over-tempering the unaffected base metal.
4. Optimum temperature to obtain the required weld strength.

Lowering the weld stresses has been verified by residual stress measurements taken before and after welding and after the postweld heat treatment.

The compromise between high and low temperatures has been achieved by heating as close as practical to the rotor or disc final tempering temperature. Because we know the weld metal tempering kinetics response to heat treatment, we can select tempering parameters to obtain the desired weld strength and base metal properties specific to the forging alloy chemistry. In every case for Table 1 alloys, the heat treatment is lower for the weld metal than for the forging; hence the forging is unaffected by the postweld heat treatment.

Rotors are postweld heat treated locally adjacent the repair weld area. This local stress relief consists of a 360° heating at the repair area and axially along the rotor to meet prescribed requirements for axial and radial temperature gradients. These requirements have been established by shell calculations based on both theoretical and finite element analysis. All postweld heat treatment conducted to date have met these requirements.

DIMENSIONAL STABILITY

Dimensional instability arises from uneven residual stresses caused by welding shrinkage, by postweld heat treatment methods, and by machining techniques (2). Because machining methods for the weld repair are no different than the original rotor fabrication, these stresses are well known to use and are easily controlled to desired levels.

Weld shrinkage stresses are limited by the amount of filler metal added to the joint per pas (i.e. bead size) as well as the rapidity of solidification. The variables given in Table 1 specifically avoid significant dimensional variations arising during welding, but some residual stress will remain at a moderate level.

The postweld heat treatment is conducted for two specific reasons. First, the prior induced shrinkage stresses can be eliminated by relaxation of remnant shrinkage stresses at elevated temperature. Next, some weld metal and the adjacent heat affected zone has been quenched during welding and contains a hard untempered martensitic structure. Some residual stress occurs from the volumetric mismatch of the tempered/untempered phases. The postweld heat treatment eliminates these volumetric strains by uniformly tempering the weld/heat affected zone microstructure while relaxing the weld shrinkage stresses and allowing local yielding to occur. When done properly, the total residual stress pattern of the weld repair should be closely similar to the new rotor forging.

Dimensional stability during the postweld heat treatment can be compromised if care is not taken during the process. For example, the same relaxation phenonema that reduces weld shrinkage stresses can allow a shaft to bow if the component is not properly fixtured in the heat treatment furnace. Also, excessive heating and cooling rates will cause large thermal differential stresses in the weldment and locally yield material at the forging interior.

Dimensional studies are made on all full sized weld mock-ups plus on all production repairs. Disc and rotor blade groove serrations have been welded and stress relieved with dimensional checks made before welding, after welding, after stress relief and after machining. Measurements made on adjacent blade grooves have not evidenced distortion. When repairs are made on shafts/or rotors, truth indicator readings are made before welding, after welding, and after stress relief to verify that no significant radial movement ("run out") has occurred.

NON-DESTRUCTIVE EXAMINATION

In order that repair welding be considered, we must be able to verify by nondestructive examination that the quality of the weld equal or exceed the original unwelded forged material. Several nondestructive examination techniques are used to evaluate the weld and heat-affected zone. Inspections following welding and to prior postweld heat treatment are used to detect porosity, lack of fusion and cracking resulting from the welding process. The postheat treatment inspections should detect flaws caused by delayed cracking or stress relief effects. A summary of these techniques are as follows:

A. Ultrasonic - Used both before and after postweld heat treatment to the rotor and disc acceptance standards. These standards are rejection of 50% or larger of an equivalent flat bottomed 1/16th inch (1.6 mm) diameter hole. In addition, highly stressed welds such as blade groove serrations are ultrasonically inspected to a 1/64th inch (0.04 mm) diameter flat bottomed hole criteria.
B. Radiography - Used to inspect for GTAW tungsten inclusions before postweld heat treatment.
C. Direct Current Flourescent Magnetic Particle Inspection-Used before and after postweld heat treatment (after final machining) to the original forging acceptance standards.

PROPERTIES

Test Plate data has confirmed that all-weld postweld heat treated tensile properties of the three weld processes show excellent strength and ductility. Typical postweld heat treated properties are shown in Table 3.

Table 3 - Tensile and Charpy V-Notch Test Plate Properties

Property of Interest	Gas Tungsten Arc	Submerged Arc	Gas Metal Arc
0.2% Yield Strength ksi (MPa)	90 - 115 (621 - 793)	85 - 100 (586 - 689)	85 - 100 (586 - 689)
Ultimate Tensile Strength, ksi (MPa)	100- - 125 (689 - 862)	95 - 120 (655 - 827)	95 - 120 (655 - 827)
Elongation, %	26 - 30	24 - 28	25 - 29
Reduction of Area, %	72 - 76	70 - 74	70 - 75
Charpy "V" Notch 50% FATT, ^{o}F (^{o}C)	-50 to -80 (-46 to -62)	-10 to -30 (-23 to -34)	-20 to -40 (-29 to -40)
Room Temperature Charpy Energy, (j) FT-LB	160 - 210 (218 - 286)	100 - 120 (136 - 163)	110 - 130 (150 - 177)

Investigation results for postweld heat treated gas tungsten arc welds deposited on a conventional test plate to simulate build-ups are shown in Table 4.

Table 4 - Tensile and Charpy V-Notch Properties
Gas Tungsten Arc Serration Welds

Tensile

Location	0.2% YD, KSI (MPa)	UTS, KSI (MPa)	% EL	% RA
R1	105 (723)	108 (745)	30	77
R2	107 (738)	109 (752)	29	77
R3	104 (717)	109 (752)	27	75
F1	109 (752)	113 (779)	28	76
F2	104 (717)	110 (758)	26	75
F3	108 (745)	114 (786)	26	76
T-1	103 (710)	108 (745)	23	77
T-2	105 (723)	108 (745)	24	77

CVN

Location	75F FT LB (J)	$FATT_{50}F(C)$	Upper Shelf, FT LB (J)
4 specimens	205 (279)	-50 (-46)	205 (279)

Again, excellent yield strengths and ductility values plus excellent toughness are shown. An additional indication of quality is that all weld tensiles failed in the weld. No heat affected zone or fusion zone failures have been observed in our transverse tensile tests. Also transverse weld gas tungsten arc weld axial high-cycle fatigue tests run at zero and forty ksi mean loads in air and in low oxygen steam (less than 20 ppb) showed endurance limits equal to or higher than NiCrMoV base metal of equal strength levels. Charpy "V" notch toughness data show that 75F impact strength and the $FATT_{50}$ values are typical of modern NiCrMoV rotor material.

In summary the tensile and toughness properties of welds meet base metal properties at comparative base metal strength levels.

MOCK-UP AND PRODUCTION REPAIR WELDS

Various welds have been made on full size turbine components. These have been for mock-ups of production components and on components destined for service. Figure 1 shows the narrow-groove submerged arc welding of an 8 inch (203 mm) deep groove to repair a shaft crack. A complete view of this shaft repair is seen in Figure 2. Figure 3 shows the narrow-groove gas metal arc welding of a new shaft-end to an old rotor.

CONCLUSIONS

1. A wide scope of proven options in welding technology can be tailored to meet many types of service repair for low pressure rotors.
2. The repair welds can meet the highly restrictive nondestructive requirements of modern advanced melting technology forgings.
3. Tensile and impact properties of all three methods described in this paper meet or exceed rotor specification requirements irrespective of NiMoV or NiCrMoV alloy group, using one filler metal.
4. All types of repairs are qualified by mock-ups prior to application on an actual rotor or disc to minimize the risk of production component weld problems. Correlation with test plate properties is excellent.
5. Production welds made to date have been successful in that there is no significant distortion, excellent properties, and ultrasonic quality equal to rotor requirements.

ACKNOWLEDGEMENTS

The authors would like to thank the Westinghouse Electric Corporation for permission to publish this work. Specifically, the assistance and support of A. F. Axt, A. Hodgson, and D. Taylor from the (W) Turbine Plant in Charlotte, N.C., is greatly appreciated. Mr. R. Pigott, Dr. R. L. Novak, and Mr. J. D. Conrad of the (W) Steam Turbine Headquarters in Orlando, Florida are also acknowledged.

FIGURE 1 - 9" (23mm) Deep Groove Weld Repair

FIGURE 2 - Submerged Arc Welding of Rotor Shaft

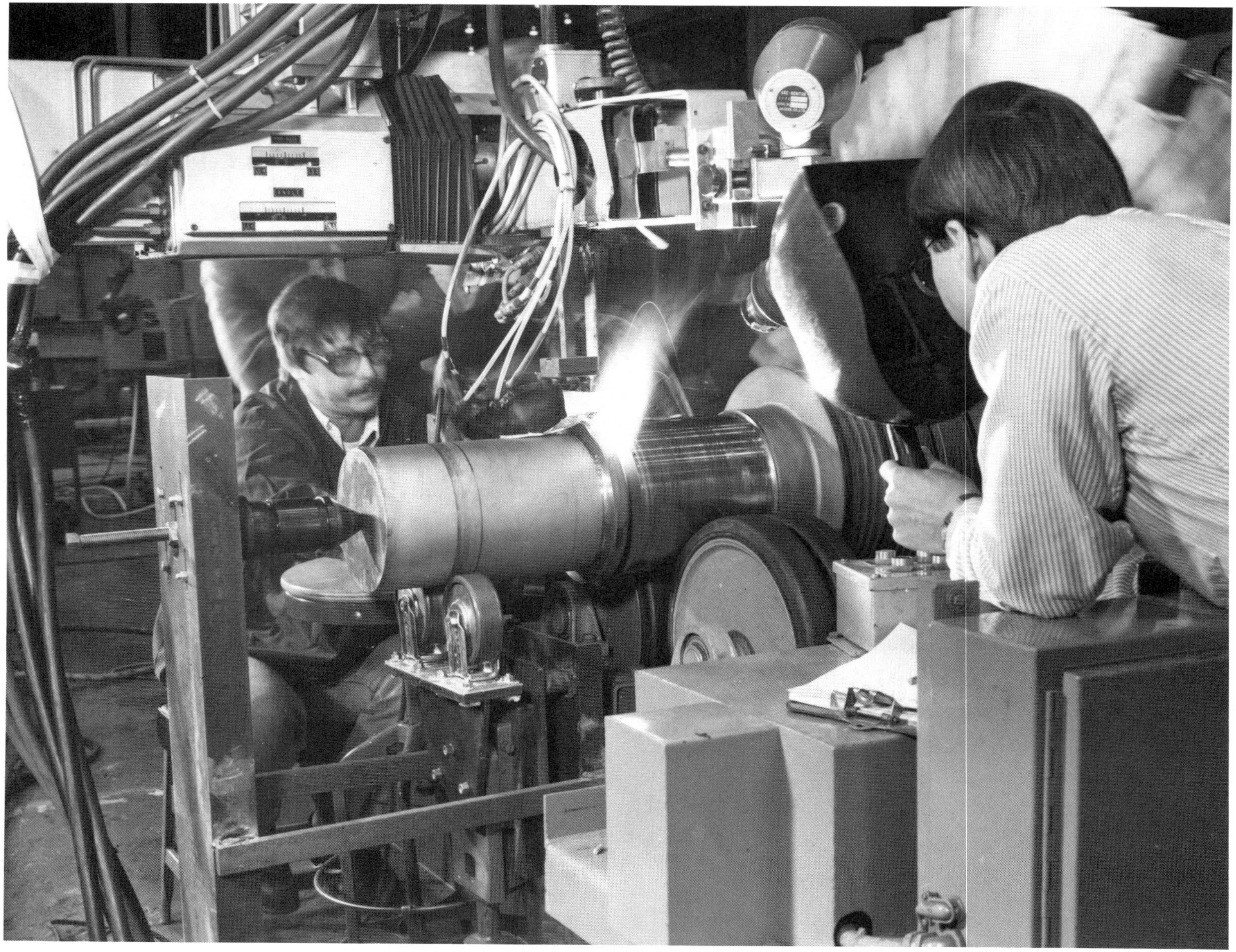

FIGURE 3 - Narrow Groove GMAW of a new forging end on a boiler feed pump turbine rotor.

REFERENCES

1. "Steel-Bars, Forgings, Bearings, Chain, Springs," ASTM 1984 Annual Book of Standards, Volume 01.05.

2. K. Masubuchi, "Control of Distortion and Shrinkage in Welding," Welding Research Council Bulletin 149, April 1970.

3. P. S. Prerey, M. Field, "Variation in Surface Stress Due to Metal Removal", Annuals of the CIRP, Volume 24, 1975.

4. J. Watanabe, T. Iwadate, T. Karaushi, "Recent Development of the Strength Evaluation of Large Forgings," Eighth International Forgemasters Meeting, Kyoto, Japan, October 23, 1977.

5. K. Tokeda, "Recent Developments of Steelmaking Technology and Situation of Special Steel in Japan," Special Lecture at the 95th ISIJ Meeting, April 1978 University of Tokyo.

6. J. E. Steiner, et al, "Advanced Steelmaking Processes for Rotor Forgings, "EPRI RD-3336, December 1983.

7. R. A. Wood, "Status of Electroslag Processing for Production of Large Rotor Forgings," Electric Power Reserach Institute. FP-799 July 1978.

8. Koreaki Suzuki, Tokehiro Meyamoto, "On the Formation Mechanism of the V Segragation in Steel Ingot," Transactions of the Iron and Steel Institute of Japan, April 1972.

9. K. Suzuki, K. Taneguchi, and S. Watanabe, "Control of Macro-Segregation in Large Steel Forging Ingots by Vacuum Carbon Deoxidation," Sixth International Vacuum Metallurgy Conference, April 23, 1979, San Diego, CA.

CRACKING IN PULSED LASER WELDS IN NICKEL-BASED ALLOYS, STAINLESS STEELS, AND THEIR COMBINATIONS

J.P. Reynolds, H.W. Kerr
Department of Mechanical Engineering
University of Waterloo
Waterloo, Ontario, Canada N2L 3G1

P.J. Fehrenbach, L. Bourque, R.D. Davidson
Fuel Materials Branch
Chalk River Nuclear Laboratories
Atomic Energy of Canada Ltd.,
Chalk River, Ontario, Canada KOJ 1J0

ABSTRACT

Cracking in pulsed laser welds in austenitic stainless steels, some Ni-based alloys and combinations of the two types has been investigated using both single and overlapping welds. Scanning Auger analysis showed both S and Ti on dissimilar metal fracture surfaces. Regression analysis also indicated that cracking increased with S content. Various other factors which may influence cracking were also examined, including weld joint gap, other alloy additions, minor phases and differential expansion.

THE PRESENT INVESTIGATION was undertaken because of intermittent cracking in pulsed laser welds of stainless steels and nickel-based alloys. In general, it is expected that the relative weldability of different alloys will be comparable in pulsed laser welds and arc welds, but differences have been observed (1). These differences may be due to one or more of several factors, including lack of mixing (2), the spacing of microsegregation, non-equilibrium phases (3,4), and differences in expansion coefficients. Intermediate compositions may crack more than either base material (5,6).

EXPERIMENTAL MATERIALS AND PROCEDURES

Chemical analyses of the materials investigated are given in Table 1. Both commercial and experimental stainless steels were studied, as well as 3 Ni-based alloys. A Nd-YAG pulsed laser with a beam diameter of about 0.5mm was used to make the welds, normally with an energy input of 5.2 J per pulse. Small bars approximately 5mm in diameter were butt welded. The initial welds studied were overlapping spots. Later pieces were tack welded to fix the restraint, and then a series of individual spot welds made along the joint.

Scanning Auger microanalysis (SAM) was carried out on certain fracture surfaces using a 5kV electron beam, with sputtering by argon ions to remove surface layers.

RESULTS

OVERLAPPING WELDS - A summary of the material combinations studied, and the observed intensity of cracking, is shown in Table 2, for overlapping welds.

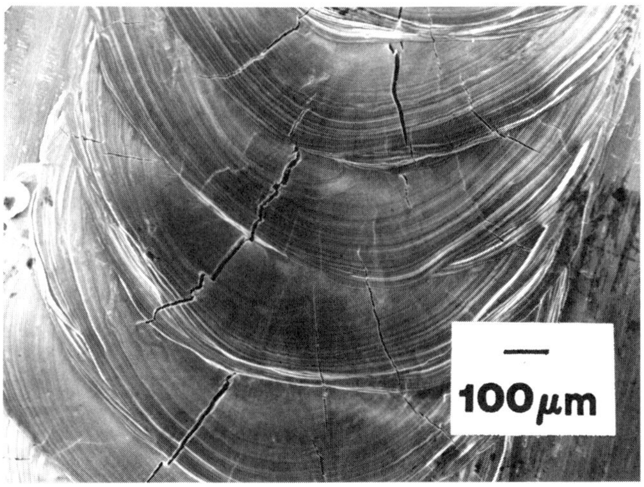

Fig. 1 Cracking in weld between Inconel 82 and 304A stainless steel.

Except for Inconels 600 and 718, materials joined to themselves did not crack. More cracking was found in dissimilar welds of stainless steels with either Inconel 82 or Inconel 600 (Fig. 1). The cracks were found to be mainly intergranular (Fig. 2), consistent with solidification cracking, (7,8).

Table 1: Composition of Alloys Studied (wt%)

Alloy	Fe	Ni	Cr	Mn	Cu	Mo	Si	S	P	C	Nb	Ti
304a	bal.	8.16	18.23	1.58	.53	.430	.42	.025	.035	.055	--	--
304b	bal.	10.64	19.48	1.69	.182	.072	.54	.013	.015	.025	--	--
308L	bal.	9.69	20.37	1.67	.095	.059	.50	.006	.016	.008	--	--
316	bal.	9.5	17.2	1.4	.27	2.11	.38	.009	.014	.06	--	--
316 Fill	bal.	12.5	19.6	1.15	.1	2.55	.32	.017	.010	.011	--	--
347	bal.	9.8	17.8	1.55	.102	.129	.59	.009	.014	.05	.61	<.02
347 Fill	bal.	9.2	20.4	1.13	<.03	.04	.34	.011	.010	.015	.54	<.02
Inconel 82	.75	bal.	20.6	2.99	.158	.014	.144	.010	.004	.008	2.4	.57
Inconel 600	8.4	bal.	15.4	.2	.141	.106	.30	.003	.010	.05	<.04	.39
Inconel 718	16.4	bal.	19.4	.16	.16	4.3	.16	.003	.010	.04	5.5	.73
Nickel 200	.2	bal.	<.01	.16	<.01	<.03	.16	.003	.010	.08	<.04	<.01
304c	bal.	10.02	18.43	1.68	--	.23	.42	.015	.022	.019	--	--
304d	bal.	8.25	18.24	1.78	.28	.18	.45	.010	.023	.070	--	--
304e	bal.	8.4	18.34	1.71	.25	.24	.56	.005	.027	.062	--	--
SSf	bal.	14.8	18.0	.70	--	--	.47	.002	.011	.015	--	--
SSg	bal.	11.3	20.0	.68	--	--	.52	.003	.007	.03	--	--
SSj	bal.	9.39	17.62	<.001	<.001	<.001	<.001	<.001	<.001	<.006	--	--
Alloy I	34.38	40.82	20.04	2.34	0.17	.043	.342	.009	.013	.016	1.2	.29

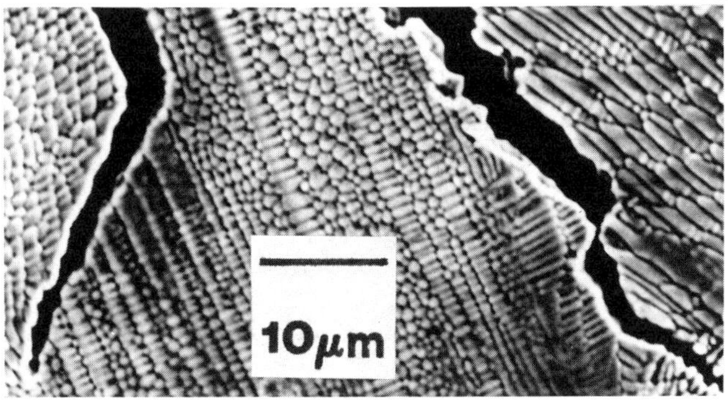

Fig. 2 Intergranular crack in dissimilar metal weld.

SAM composition-depth profiles from fracture surfaces are shown in Figs. 3 and 4. High levels of carbon and oxygen at the fracture are believed to be contamination effects (9). Segregation of S and Ti was observed just under the contamination on both fracture surfaces. Adjacent ductile fracture regions showed no evidence of S, and only a trace of Ti.

Cracks in one spot weld often propagated into a subsequent overlapping weld (Figs. 1 and 5). Subsequent specimens were made using single spots to examine crack initiation. Reduction of initiation would clearly reduce total cracking.

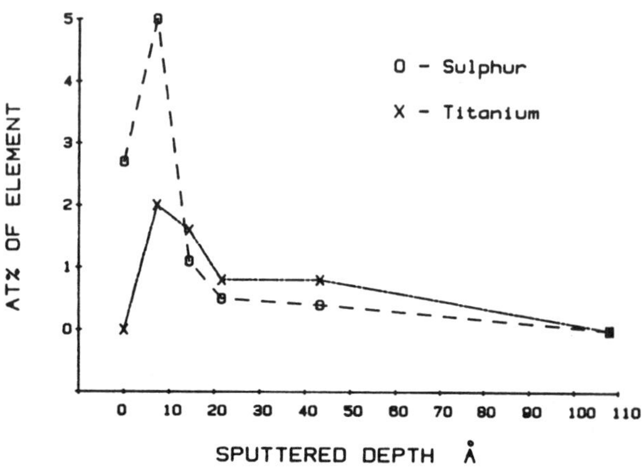

Fig. 3. SAM analysis from fracture surface of Inconel 600- stainless steel A weld.

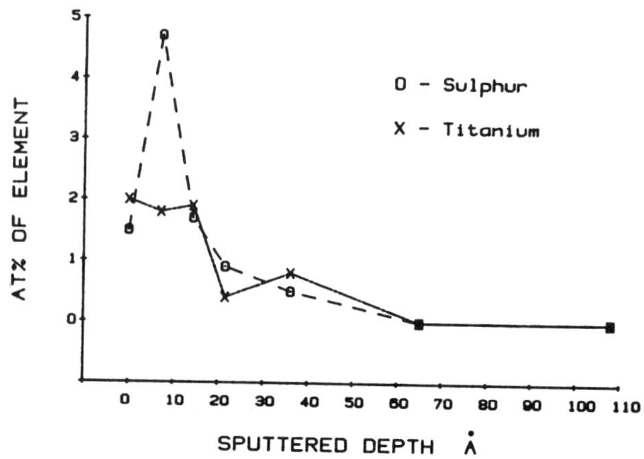

Fig. 4 SAM analysis from fracture surface of Inconel 82- stainless steel A weld.

Table 2: Summary of Cracking Observed in Overlapping Welds

	304a	304b	308L	316	316 Fill	347	347 Fill	In 82	In 600	In 718	Ni 200
304a	0	0	0	0		0	0	X	X		
304b		0	0	0		0	0	X	X		
308L			0	0		0		X			
316				0	0		0	X			
316 Fill					0						
347						S	0				0
347 Fill							0				
In 82								0	0	S	0
In 600									S	S	0
In 718										S	
Ni 200											0

0 = zero cracking; s = small cracks; x = extensive cracking

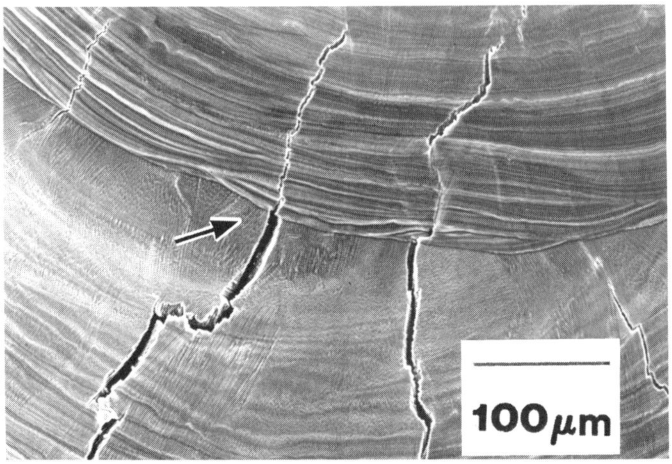

Fig. 5 Crack in original weld propagating into subsequent weld (at arrow).

SINGLE DISSIMILAR METAL LASER WELDS - Austenitic stainless steels with a range of C and S levels were welded to Inconels 82 and 600. The matrix of C and S levels in the materials is given in Table 1. The total crack length at the surface of each weld and the adjacent weld joint gap, were measured in a scanning electron microscope. For a given specimen with several spot welds, considerable variation of total crack length was often found.

Multiple regression analyses were carried out of crack length (L) in individual welds versus gap and various elements (S,P,C,Mn). The most significant element in all cases was sulphur. The maximum crack length (MCL) for a specimen of given alloy combination increased

Table 3. Regression Analysis Results of Cracking in Dissimilar Spot Welds

(i) Alloy 82

L = -92 + 42047 S + 5.8 x Gap (R^2 = 0.48)
MCL = 252 + 67791 S (R^2 = 0.63)
MCL = 77 + 59723 S + 7.8 x Gap (R^2 = 0.83)

(ii) Alloy 600

L = -209 + 59843 x S (R^2 = 0.72)
MCL = -174 + 78592 x S (R^2 = 0.81)

where L = Length of crack in individual welds.

MCL = Maximum crack length for a given alloy combination (all dimensions are μm), S = wt% S.

R^2 = the fraction of variation predicted by the equation.

as the S level increased, as shown in Fig. 6 for both weld types. Regression analyses also indicated that crack length increased with gap in welds containing Inconel 82, but not Inconel 600 (Table 3). No clear effect of C was found. For example, examination of Fig. 6 shows the increased C in 304e versus stainless steel f did not noticeably affect maximum crack length, for a given S level.

For welds containing both Inconel 82 and 600, very little cracking was seen for S levels below 0.003 wt% (Fig. 6). These crack-free points at low S represent 17 different welds, with gaps up to 60 µm and C levels up to 0.032 wt%. Hence it appears that at least 0.003 wt% S is necessary to cause cracking.

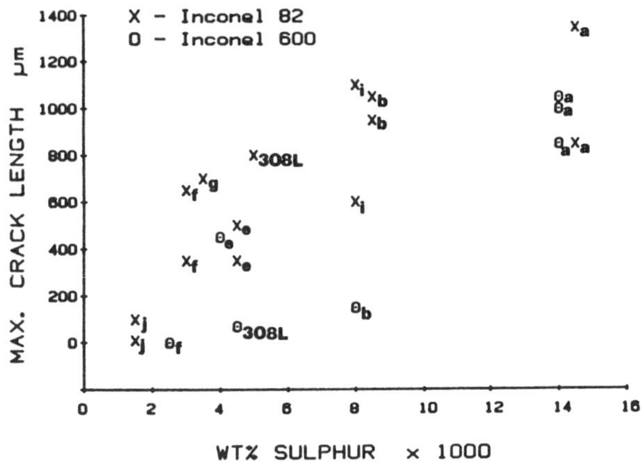

Fig. 6 Maximum crack length versus sulphur for dissimilar metal welds. Letters represent steels in Table 1.

Fig. 6 also shows greater maximum crack lengths in dissimilar welds with Inconel 82 than with Inconel 600. This difference was also evident in the overlapping welds.

OTHER WELDS - Differential thermal expansion between alloys could contribute to stresses and cracking. To examine this possibility, two types of experiments were carried out.

Specimens made from an ingot of about 50% each of Inconel 82 and stainless steel B were welded together. Cracking was again observed. The observed cracking is consistent with the effect of S (0.008 wt%).

Mini GTA (Gas Tungsten Arc) and EB (Electron Beam) welds, each 2-4 mm in width, were used to join Inconel 82 to stainless steels 304a and 304b. No large cracks were observed after welding with either of these processes. Single and overlapping laser welds were then superimposed on the original welds. Although some individual spot welds did not crack, cracking was not eliminated in any of the specimens.

DISCUSSION

More cracking was observed in dissimilar stainless steel to high nickel alloy welds than in welds of either stainless or high nickel alloys alone. Both statistical regression and Auger analyses indicate that sulphur is the most significant element with respect to cracking. In high Ni alloys, cracking is usually prevented by low S (<.004 wt%). Stainless steels, often contain more S (Table 1), but ferrite helps to avoid cracking (10). In the overlapping stainless steel welds, a thin band containing ferrite was observed at the weld edges (Fig. 7) presumably due to the lower transient solidification velocity there (4,11). No cracking was observed, despite the absence of ferrite at the centre of these welds.

Fig. 7 Overlapping stainless steel welds, with ferrite (dark) at outer edges.

Laser spot welds cool much faster than higher heat input welding processes. The cellular dendrites (Fig. 2, 8) had a spacing of 1µm for the laser welds, whereas the EB and miniGTA welds contained branched dendrites with a spacing of ≈ 10µm (Fig. 8) which permit localized microvoids. Fig. 8 shows a crack in a laser weld which has propagated from an existing microvoid in the EB weld.

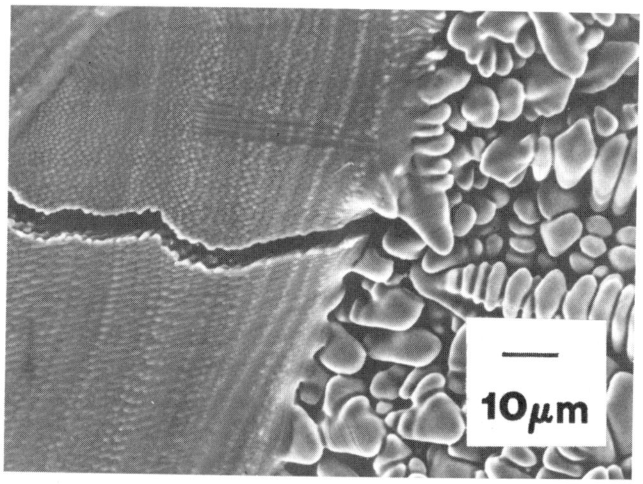

Fig. 8 Crack in laser weld propagating from microvoid in previous EB weld.

There was considerable scatter in the amount of cracking observed in individual laser spot welds. This may be due to several factors. First, the small weld size may lead to significant differences of S level if there is macrosegregation of S in the base material.

Second, the exact weld composition depends on the position of the spot weld relative to the butt joint. Others (5,6) have noted increased cracking of intermediate Ni compositions, which is not well understood. Third, the short welding time does not allow complete mixing of the weld pool in laser welds (Fig. 9). The cracking seems to be contained within a certain compositional band, and does not propagate into the darker (higher Ni) regions.

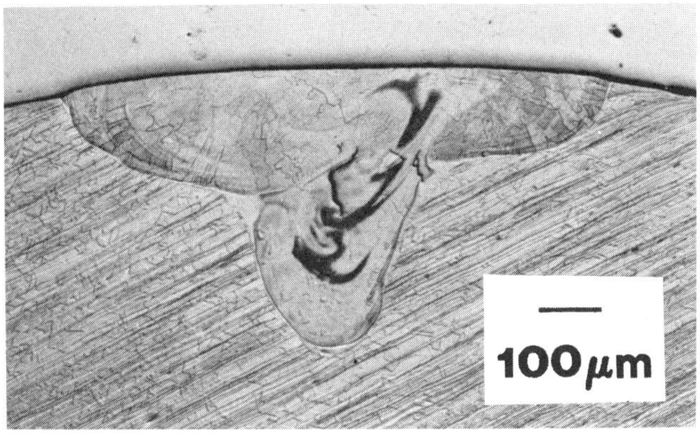

Fig. 9 Section in dissimilar metal weld showing lack of mixing.

Fig. 6 shows greater maximum crack length in dissimilar welds made with Inconel 82 than those made with Inconel 600. Interdendritic phases were found in the Inconel 82 laser welds. Other authors have noted cracking due to such phases (12).
We are continuing work to identify the phases, as well as clarify other aspects.

SUMMARY AND CONCLUSIONS

Extensive solidification cracking was observed in Ni-based to stainless steel combinations, but much less or none in stainless steel or Ni-based alloys in single and overlapping laser spot welds.

SAM analyses on crack surfaces indicated segregation of S and Ti. Multiple regression analyses of crack length indicated that S was the most significant factor, but that joint gap was also significant in welds containing Inconel 82. Dissimilar stainless - Ni alloy welds with less than 0.003 wt% S did not crack.

Welds involving only stainless steels contained a small amount of ferrite in the outer region of the weld.

Laser spot welds both in specimens of an intermediate composition and on prior mini GTA and EB dissimilar metal welds also cracked, indicating that differential thermal expansion is not a necessary condition for weld cracking.

Dissimilar welds with Inconel 82 exhibited more cracking than those with Inconel 600, for comparable S levels, possibly due to the presence of interdendritic phases.

ACKNOWLEDGEMENTS

This work was supported by the Natural Sciences and Engineering Research Council of Canada, and by Atomic Energy of Canada Limited. INCO and Atlas Steels Ltd. supplied materials.

REFERENCES

1. Weeter, L.A., Albright, C.E. and Jones, W.H., Welding J. 65, 51-s to 62-s (1986).
2. Seretsky, J. and Ryba, E.R., Welding J. 55, 208-s to 211-s (1976).
3. David, S.A. and Vitek, J.M. in "Lasers in Metallurgy", Met. Soc. AIME, Warrendale, Pa. (1981).
4. Vitek, J.M., Dasgupta, A. and David, S.A., Met. Trans., 14A, 1833-1841 (1983).
5. Payne, B.E., Met. Constr. and Br. Weld. J. 1 79-87, (1969).
6. Patterson, R.A. and Milewski, J.O., Welding J., 64, 227-s to 231-s (1985).
7. Matsuda, F., Nakagawa, H., Minehisa, S., Sakabata, N., Ejima, A. and Nohara, K., Trans. J. W.R.I., 13, 241-247 (1984).
8. Arata, Y., Matsuda, F. and Katayana, S., Trans. J.W.R.I., 6, 105-116 (1977).
9. Lippold, J.C., Welding J., 63, 91-s to 103-s, (1984).
10. Brooks, J.A., Thompson, A.W. and Williams, J.C., Welding J., 63, 71-s to 85-s (1984).
11. Lippold, J.C., Welding J., 64, 127-s to 136-s (1985).
12. Ritter, A.M. and Savage, W.F., Met. Trans., 17A, 727-737 (1986).

APPLICATION OF ELECTRON BEAM WELDING ON AUSTENITIC STAINLESS STEEL FOR PWR NUCLEAR POWER PLANT COMPONENTS

Shuhei Kuri, Yoshimi Uratani
Kobe Shipyard & Engine Works
Mitsubishi Heavy Industries, Ltd.
Kobe, Hyogo Pref.,
Japan

Genta Takano
Takasago Technical Institute
Mitsubishi Heavy Industries, Ltd.
Takasago, Hyogo Pref.,
Japan

ABSTRACT

In fabricating nuclear power plant components resultant safety is of most importance, and it is desirable to use highly reliable welding methods. The authors studied the electron beam weldability of austenitic stainless steel from the viewpoint of weld defects and mechanical properties of welded joints. Then, on the basis of the results obtained, they applied the electron beam welding to nuclear power plant components in practice.

NUCLEAR POWER GENERATION in the energy supply is becoming increasingly important. Therefore, the assurance of safety in the nuclear power plant is an absolutely essential condition. In the fabrication of related nuclear power plant equipment improvements have been sought in every aspect in order to improve reliability.

For the more common welding techniques used in general manufacturing, low welding heat input and narrow gap procedures are being developed and improved, especially for austenitic stainless steels used for PWR nuclear power plant components. Low heat input welding helps to prevent stress corrosion cracking and reduce welding distortion.

One of the most suitable joining techniques for stainless steel is electron beam welding and its application can do much toward improving the quality and reliability of nuclear power plant components.

Accordingly, the authors studied the electron beam weldability of austenitic stainless steel thoroughly from the viewpoint of weld defects and mechanical properties of welded joints. Then, on the basis of the results obtained they applied electron beam welding for the nuclear power plant components.

CHEMICAL COMPOSITION FOR PREVENTING WELD DEFECTS

CRACK SUSCEPTIBILITY - For welding austenitic stainless steel, in which prevention of hot cracking is important, weld metal with a ferrite level of above 5% is generally used.[1)2)]

However, the ferrite level of weld metal for electron beam welding, for which filler metal is not usually added, is determined by the ferrite level of the base metal. We examined a series of alloys to determine the influence of ferrite on cracking in electron beam weldments. Because (P+S) also contributes to hot cracking susceptibility, we included two (P+S) levels in the materials to be examined : 0.03% (ordinary material) and 0.07% (upper limit of standard). The ferrite level varied at 5 levels of 0, 1, 3, 5 and 8%. Further, a total of 11 types of materials with ferrite level of 0% and (P+S) level of 0.045% were applied to two kinds of material, AISI 304 and AISI 316, to provide a total of 22 types of materials for testing. Table 1 lists the chemical compositions and mechanical properties of materials for testing.

The shape of test plate is shown in Fig.1. For a thickness of 20mm, the FISCO type cracking test was performed. The slit-type cracking test was performed on 80mm plate. Welding conditions are shown in Table 2. For the plate thickness of 20mm and 80mm, crack examination was conducted on 6 sections and 5 sections, respectively.

Shown in Fig.2 are the test results on AISI 304. In materials with (P+S) levels of 0.045% and 0.03%, no cracks were observed for changes of 0 to 8% in ferrite level. In materials with (P+S) level of 0.07%, however, fine hot cracks were detected inside weld metal at a plate thickness of 20mm for ferrite level of below 1.1%. Fig.3 indicates the test result on AISI 316. As in the case of AISI 316, no cracks were observed in materials with (P+S) level of below 0.045%, while not cracks were detected in materials with (P+S) level of 0.07% and ferrite level of below 3.8%. Fig.4 shows microstructure of weld metal.

Table 1 Chemical compositions and mechanical properties of materials used

Test	Material		Thickness (mm)	C	Si	Mn	P	S	Ni	Cr	Mo	0.2%Y.S (kg/mm²)	T.S. (kg/mm²)	El. (%)	Remarks	
				Chemical compositions (wt. pct.)								Mechanical properties				Ferrite (%)
Cracking	Forge	AISI 304	20, 80	0.08	0.43	0.82	0.020	0.003	9.82	18.01	–	26	58	65	(P+S)≒ 0.03%	0
				0.08	0.52	0.89	0.026	0.005	10.35	18.28	–	24	55	69		0.7
				0.07	0.45	0.83	0.020	0.006	9.07	18.09	–	27	59	67		2.7
				0.08	0.47	0.85	0.027	0.012	8.64	18.23	–	25	59	70		4.4
				0.06	0.58	0.95	0.021	0.002	8.16	18.76	–	24	61	68		8.4
				0.08	0.40	0.96	0.036	0.035	10.26	18.31	–	25	27	62	(P+S)≒ 0.07%	0
				0.07	0.40	0.77	0.036	0.035	9.83	18.26	–	26	58	66		1.1
				0.08	0.43	0.75	0.036	0.037	9.01	18.24	–	28	60	64		2.7
				0.07	0.44	0.77	0.038	0.033	8.59	18.11	–	25	60	66		4.2
				0.08	0.66	0.98	0.039	0.040	8.17	18.65	–	28	64	64		7.4
		AISI 316		0.08	0.86	1.04	0.020	0.005	12.93	17.55	2.34	28	58	61	(P+S)≒ 0.03%	0
				0.08	0.86	1.06	0.020	0.005	12.32	17.50	2.34	27	57	59		1.9
				0.07	0.84	1.06	0.019	0.005	11.37	17.35	2.37	28	57	64		3.9
				0.07	0.80	1.06	0.018	0.005	10.49	17.19	2.35	26	57	68		5.4
				0.07	0.80	1.03	0.020	0.010	10.14	17.90	2.35	28	58	64		8.3
				0.07	0.87	1.04	0.036	0.037	12.76	17.33	2.36	28	58	55	(P+S)≒ 0.07%	0
				0.07	0.88	1.05	0.036	0.035	12.18	17.18	2.36	27	58	60		1.8
				0.07	0.85	1.05	0.036	0.035	11.33	17.28	2.36	27	58	64		3.8
				0.07	0.84	1.04	0.039	0.037	10.49	17.27	2.37	27	59	63		5.8
				0.07	0.83	1.03	0.038	0.041	10.12	17.86	2.38	29	59	62		8.4
		AISI 304	20	0.05	0.49	1.93	0.022	0.023	10.88	18.07	0.01	–	–	–	(P+S)≒ 0.045%	0
		AISI 316	20	0.04	0.50	1.85	0.023	0.022	12.48	16.65	2.45	–	–	–		
															O (ppm)	N (ppm)
Porosity		AISI 304	50, 80	0.05	0.62	1.00	0.030	0.007	9.06	18.22	–	23	58	65	71	320
				0.04	0.61	0.92	0.029	0.005	9.22	18.25	–	25	57	67	41	560
				0.04	0.64	0.99	0.030	0.002	9.23	18.19	–	29	60	63	60	1160
				0.03	0.59	0.87	0.026	0.003	9.22	18.47	–	22	57	74	80	320
				0.04	0.58	0.96	0.031	0.014	9.23	18.35	–	23	57	68	87	480
				0.04	0.63	1.02	0.028	0.004	9.27	18.36	–	29	61	63	85	1260
				0.03	0.59	0.87	0.026	0.003	9.22	18.47	–	22	55	69	93	320
				0.04	0.58	0.99	0.029	0.006	9.29	18.30	–	24	55	68	146	570
				0.04	0.63	1.02	0.028	0.004	9.27	18.36	–	29	60	60	98	1260
			20, 50	0.04	0.61	1.18	0.029	0.023	9.15	18.32	–	26	58	64	33	990
			20, 50	0.04	0.52	0.91	0.028	0.008	9.21	18.31	–	28	58	65	112	930
			20	0.04	0.64	0.99	0.030	0.002	9.23	18.19	–	28	61	64	60	1160
			20	0.04	0.63	1.02	0.028	0.004	9.27	18.36	–	27	59	60	98	1260
Joint Properties	Plate	AISI 304	20	0.04	0.56	1.07	0.031	0.013	9.18	18.33	–	26	62	58		
			50	0.04	0.61	1.01	0.029	0.009	9.22	18.29	–	24	58	65		
			80	0.05	0.60	0.94	0.029	0.007	9.16	18.43	–	22	56	71		

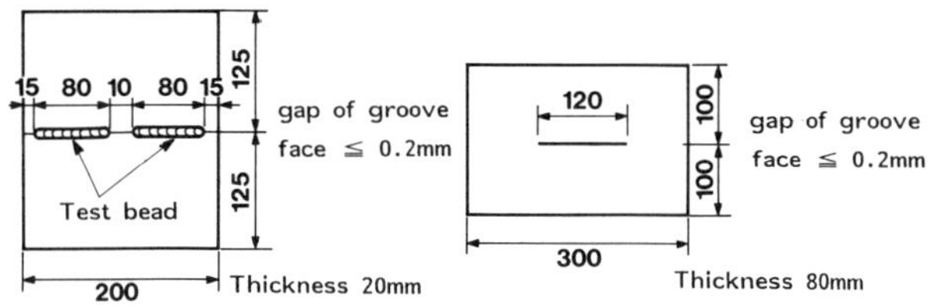

Fig.1 Shape of the weld cracking test plate

Table 2 Welding Conditions

Thickness (mm)	Welding position	Accelerating voltage (kV)	Vacuum pressure (torr)	Beam current (mA)	Welding speed (mm/min)	Focus current (A) (Focal length (mm))	Work distance (mm)
20	Horizontal	60	1×10^{-2}	300	700	0.903 (220)	320
80	Horizontal	60	1×10^{-2}	500	200	0.755 (670)	470
80	Flat	60	1×10^{-2}	350	550	0.750 (570)	450

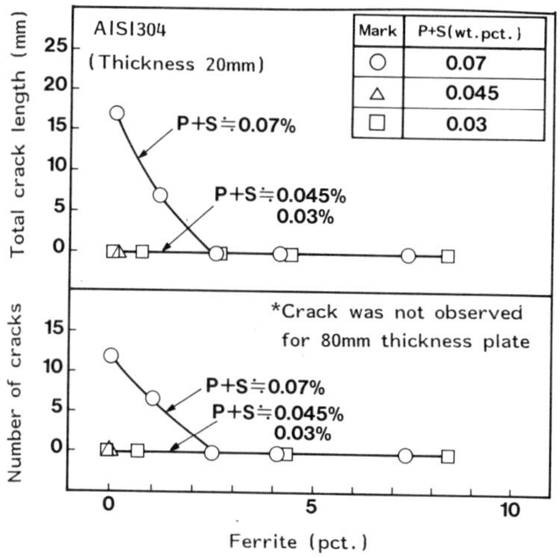

Fig.2 Weld cracking test results (AISI304)

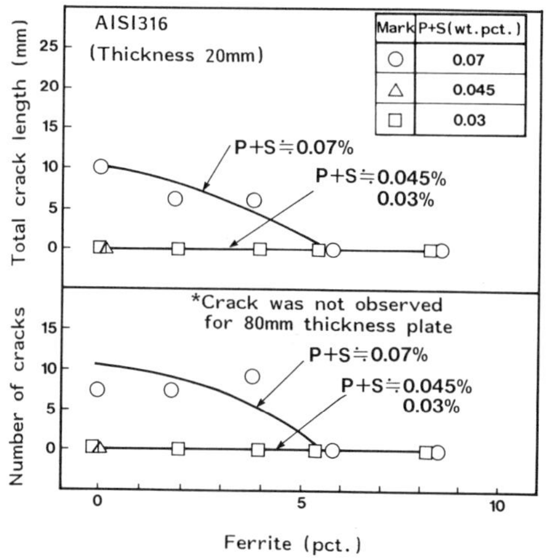

Fig.3 Weld cracking test results (AISI316)

This difference in crack susceptibility between AISI 304 and AISI 316 is thought attributable to a solidification structure. That is, all the AISI 304 materials, whose weld metal microstructure is considered to have the primary crystal of δ, had a structure of ferrite being present in austenite matrix, not being influenced by ferrite level. On the other hand, AISI 316 materials mostly had an austenite structure at a ferrite level of below 2%, and a structure of ferrite in an austenite matrix at ferrite levels of above about 4%.

The weld cracking test results are compared with Moisio's diagram[3] in Fig.5. Cr_{eq}/Ni_{eq} level to occur hot cracking is almost equal between electron beam welding and arc welding. But concerning (P+S) level to occur hot cracking, critical value of electron beam welding is higher than that of arc welding. The difference in crack susceptibility between electron beam welding and arc welding is thought the following reason. Electron beam welding is performed by 1 pass with 2 dimensional heat flow, consequently induced strain to weld metal is small.

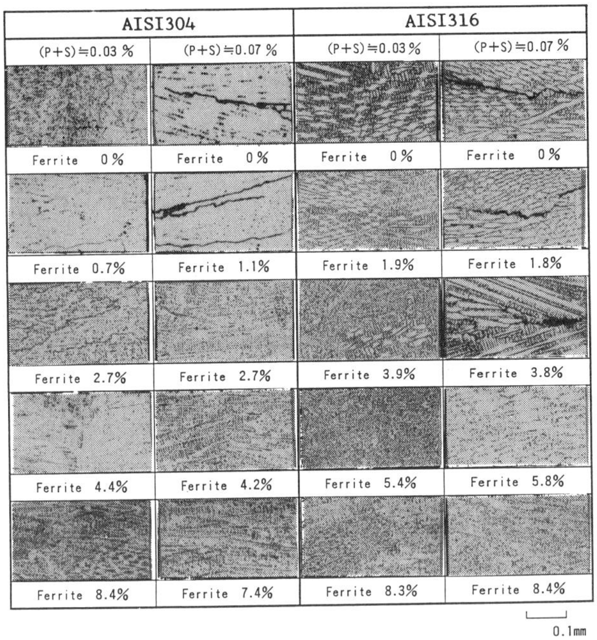

Fig.4 Microstructure of weld metal

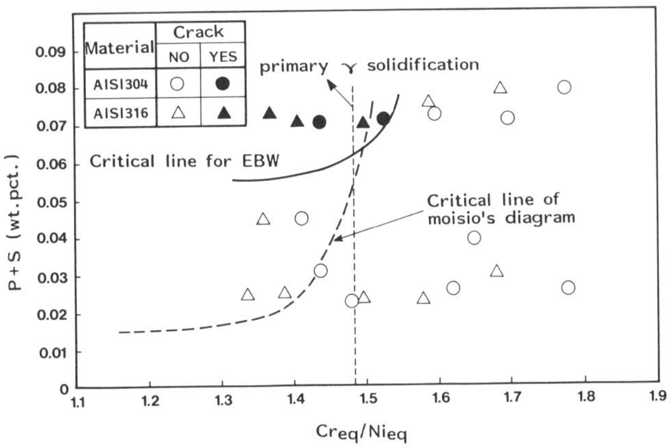

Fig.5 Effect of Cr_{eq}/Ni_{eq} and (P+S) on hot cracking susceptibility

The above results revealed that electron beam weld metal also requires its (P+S) or ferrite level to be limited so as to prevent hot cracking as in the case of arc welding. In actual welding, it is considered to be more effective to control (P+S) level.

OCCURRENCE OF POROSITY - With electron beam wedling, high gas contents of the base metal lead to porosity inside the weld metal occurring. To prevent this, for steel materials, the oxygen and nitrogen levels are required to be limited to approximately 100 ppm or below. Austenitic stainless steel usually contains about 400 ppm of nitrogen and about 75 ppm of oxygen in the austenite structure, which allows larger solubility of nitrogen than ferrite structure. AISI 304 materials whose gas contents have been varied at nitrogen levels of 300, 600, 900 and 1200 ppm and oxygen levels of 50, 75 and 100 ppm were used for testing. In addition, to investigate the effects of plate thickness and welding position, plate thickness of 20, 50, 80 mm were employed for horizontal position welding and 80 mm plate (depth of penetration about 40 mm) was used for flat position partial penetration welding.

Testing was conducted by meltrun welding using conditions representative for each plate thickness.

The results of examination are given in Fig.6 on possible porosity occurrence.

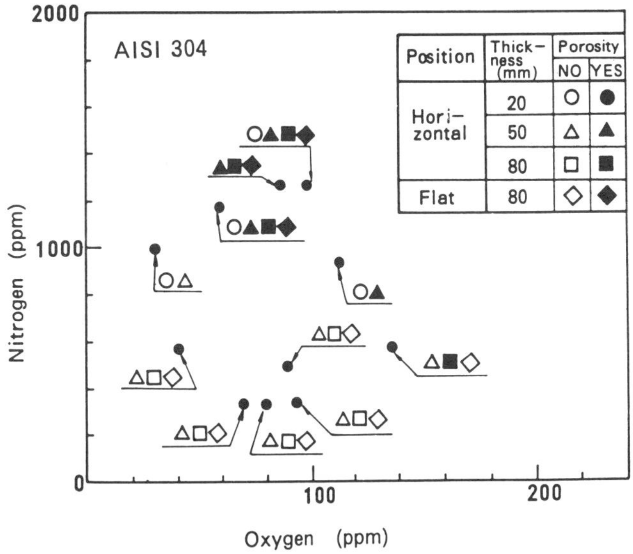

Fig.6 Relations between gas contents and porosity

The gas contents that caused no porosity to occur were nitrogen levels of below 1200 ppm and oxygen levels of below 100 ppm for the plate thickness of 20 mm, nitrogen levels of below 900 ppm and oxygen levels of below 50 ppm, and nitrogen level of below 600 ppm and oxygen levels of 50 to 100 ppm for 50 mm plate thickness, and nitrogen level of below 600 ppm and oxygen level of below 100 ppm for 80 mm plate thickness. No definite difference was observed in porosity occurrence between horizontal position welding (50 mm) and flat position welding in relation to the testing conditions used in the present tests, although the latter is considered to be more advantageous.

The difference in nitrogen level between base metal and weld metal was 20 to 126 ppm when the nitrogen level of base metal was about 300 to about 600 ppm, and 216 to 336 ppm when the base metal nitrogen level was about 1200 ppm, indicating that the larger the nitrogen content of the base metal, the larger the reduction. On the other hand, the oxygen level of weld metal decreased to 23 to 49 ppm, about below half of 43 to 150 ppm for base metal.

As mentioned above, both nitrogen and oxygen levels of weld metal were lower than those of the base metal. This suggests that nitorgen gas and CO gas result in porosity.

The above results revealed that austenitic stainless steel also needs gas contents to be controlled according to plate thickness to prevent porosity.

ELECTRON BEAM WELDED JOINT PROPERTIES

BASIC WELDED JOINT PROPERTIES - With selected combinations of welding position, accelerating voltage and plate thickness, testing was carried out using two types of materials, AISI 304 and AISI 316, to examine the basic welded joint properties.

Chemical Compositions of Weld Metal - Table 3 lists an example of analyses of chemical compositions of weld metal for 80 mm thick material. In relation to C, Si, P, S, Ni, Cr and Mo of elements analyzed, no marked difference was observed in their levels between base metal and weld metal. With regard to N and O, their levels were lower for weld metal than for base metal. Mn level showed a tendency to decrease in the front. Regarding elements other than Mn, no marked difference was given in their levels inside plate thickness. This tendency was also observed in the other plate thickness.

Table 3 Chemical analysis of weld metal

Welding position	Material	Accelerating voltage (kV)	Thickness (mm)	Position	C	Si	Mn	P	S	Ni	Cr	Mo	N	O
Horizontal (full penetration)	AISI 304	60	80	surface	0.041	0.56	0.77	0.032	0.007	9.26	18.2	0.15	0.0348	0.0024
				middle	0.047	0.57	0.88	0.032	0.007	9.29	18.2	0.15	0.0371	0.0025
				back	0.044	0.57	0.84	0.034	0.008	9.26	18.2	0.15	0.0369	0.0022
	AISI 316		80	surface	0.034	0.62	0.75	0.030	<0.005	10.5	16.9	2.12	0.0276	0.0021
				middle	0.038	0.62	0.87	0.034	<0.005	10.5	16.7	2.11	0.0274	0.0024
				back	0.037	0.62	0.87	0.033	<0.005	10.5	16.9	2.14	0.0293	0.0022

Tensile Properties - The results given in Fig.7 are of full thickness joint tension tests conducted at room temperature, and the results at room temperature and high temperature joint tensile tests on 5 mm sampled from the middle of plate thickness in relation to AISI 304. The room temperature joint tensile strength was slightly higher for full thickness joint than for a round bar joint. The tensile strength for all of these welded joints satisfied the base metal standard value of Japan. The tensile strength ratio of base metal to welded joint varied in the range of 0.93 to 1.0 with the strength of welded joint being silghtly lower than that of the base metal. The joint strength ratio, which showed no difference between that at room temperature and at high temperature up to 430°C, was almost constant (welded joint tensile strength/base metal tensile strength).

The strength of stainless steel is determined by chemical compositions (in particular C and N levels) and grain size. On the other hand, for electron beam weld metal, whose nitrogen level decreases from that of the base metal, the joint strength ratio is considered to become below 1 with weld metal subjected to fracture in a welded joint tensile test.

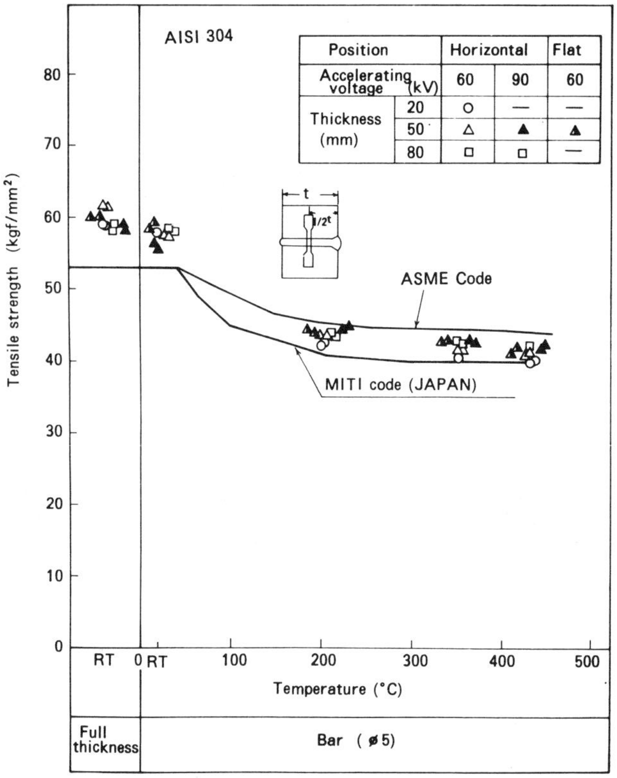

Fig.7 Tensile properties of butt-welded joint

Fig.8 shows the relation between (C+N) level and tensile strength of welded joint. Tensile strength of electron beam welded joint shows a good coincidance with the regression curve for small ingot, from which the strength of electron beam weld metal can be estimated by chemical composition and grain size.

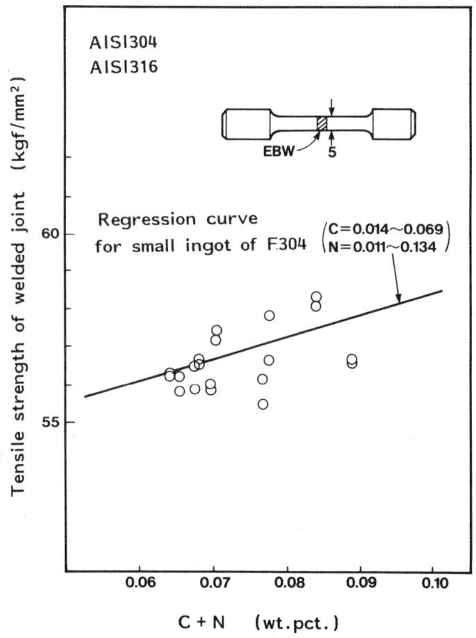

Fig.8 Relationships between (C+N) and tensile strength of welded joint

STRESS CORROSION CRACKING SUSCEPTIBILITY - Thermal cycles of the sensitizing zone is shown in Fig.10. The cooling rate of electron beam welding and GTA welding is almost same. But electron beam method is single pass welding against multi-pass welding of GTA method. Then electron beam welding is superior to arc welding from the view point of sensitizing of HAZ. The degree of sensitizing measured by EPR (Electrochemical Potentiokinetics Reactivation) method for AISI 304 (C = 0.05%), is about 20 $C/cm^2 \cdot GBA$ for electron beam welding (lap portion) and about 40 $C/cm^2 \cdot GBA$ for GTA welding, respectively.

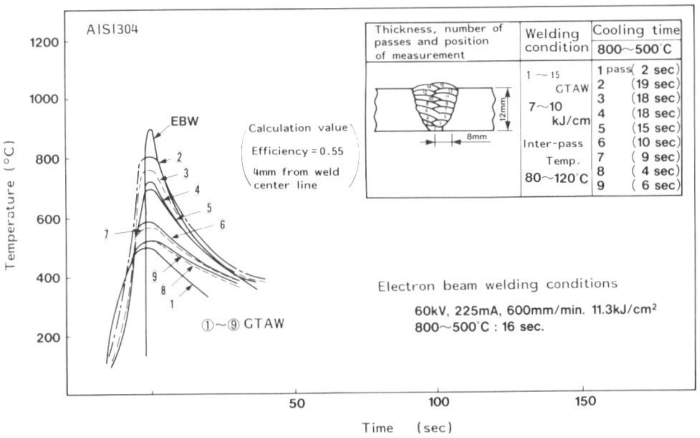

Fig.9 Thermal cycles of the sensitizing zone of pipe inner surface

Fig.10 Appearance of core internal structure after longitudinal and circular seam welding

APPLICATION OF ELECTRON BEAM WELDING FOR PWR NUCLEAR POWER COMPONENTS

Since the electron beam welding technology for austenitic stainless steel has been established by the various tests mentioned above. We started to apply the electron beam welding to major PWR nuclear power plant components. The completed components already made with this method are core internal structures (Fig.10) and auxiliary tanks (5 m or less in O.D, 6 m or less in height, 50 mm or less in plate thickness), which have longitudinal and circular seams welded with the electron beam welding.

Consequently not only is the anti-SCC characteristic improved in the quality field but also weld defects due to the human factors that have been a severe problem are eliminated.

The welding time is reduced sharply and the cost is reduced to about half, even when cost of the expensive electron beam welding machine is taken into account.

It is now being planned to put this method to practical use by expanding its application to circular seam welds of piping in the factory as well as to welds of pipe brackets fitted on various vessels using recent electronic control techniques.

SUMMARY AND CONCLUSIONS

The present research has conducted studies on the appropriateness of electron beam welding austenitic stainless steels for nuclear power plant components. A wide range of many-sided testing was conducted and the following results were obtained.

(1) In order to prevent hot cracking, the (P+S) level should be controlled to below 0.045%, and porosity may be prevented by limiting the nitrogen level to within 600 to 1200 ppm and the oxygen level to within 50 to 100 ppm depending on the plate thickness.

(2) N and O levels are lower for weld metal than for base metal. Therefore, the strength of weld metal is slightly lower than that of the base metal.

(3) Electron beam welding is superior to arc welding from the view point of stress corrosion cracking.

REFERENCES

1) F. C. Hull, Welding Journal, vol.46 (1976), 399S ∿ 408S

2) J. A. Brooks, A. W. Thompson and J. C. Williams, Welding Journal, vol.63 (1984), 71S ∿ 83S

3) T. Takalo, N. Suutala and T. Moisio, Metallurgical Transactions A. vol.10A Aug. (1979), 1173 ∿ 1181

THE ELECTRON BEAM WELDING OF THICK STRUCTURAL STEEL PLATES

L. Devillers, D. Kaplan
Institut de Recherches
de la Siderurgie Francaise
France

Productivity requirements have prompted the use of new joining techniques with high energy density. The volume of the weld metal (that means energy input, labour cost, filler material) is then significantly reduced as compared to conventional welding processes. For sheets (up to 5 mm) Laser Beam Welding has been developped, whereas for thick plates (over 40-50, up to 300 mm), Electron Beam Welding (EBW) has gained much audience [1, 2].

Numerous applications of this last technique have been put forward : pipes, pressure vessels, exchangers... Most of the publications relative to EBW have aimed at demonstrating applicability of this process to a given material and a set of welding conditions.

In this paper, we review briefly some general metallurgical characteristics of this process when applied to the welding of steel plates over 40 mm thick. We will insist on the toughness and the soundness of the EB welded joints. Thick plates users are indeed concerned with these two properties, in order to avoid the risk of sudden failure, eventually initiated from a weld defect which could propagate in an unstable mode, due to the inadequate toughness of the weld metal.

Experience has shown that most of the problems in EBW arise from the weld metal. Thus, we will not discuss here the properties of the Heat Affected Zone (HAZ) which can be considered as safe, due to narrowness and the very low level of hydrogen associated to the process. Yet we will mention the similarities between the HAZ and the weld metal microstructures observed in EBW.

I - The composition of the weld metal - The problems of porosities

EBW can be schematically described as a short vacuum treatment of a small volume of liquid metal. In spite of the fact that equilibrium is far to be reached, thermodynamics may be helpful to predict the occurence of some reactions. For example, elements with a high equilibrium vapour pressure at elevated temperatures are likely to escape from the molten metal. For structural steels, this holds mainly for chromium and manganese for which we have measured losses from 5 up to 20% in the welding of 50 mm thick plates, under a vacuum of 10^{-2} - 10^{-3} torr.

Moreover spontaneous degassing will occur for elements such as hydrogen or nitrogen, whose activities in the liquid metal are drastically reduced.

The loss of hydrogen in the weld metal is difficult to appreciate, in as much as the initial level of hydrogen in the base metal is itself very low.

Figure 1 indicates that a mean loss of 10% of nitrogen occurs for the welding conditions above.

Furthermore, the low partial pressure above the liquid metal yields the dissociation of oxides of manganese, silicon, or aluminium (stable at atmospheric pressure), to form gaseous oxides, mainly CO.

These degassing phenomena may induce porosities when gas bubbles are trapped during solidification. This can be enhanced by some elements which lower the surface tension of liquid steel and promote homogeneous and spontaneous degassing : figure 2 indicates that increasing the sulphur level from 2 to 14×10^{-3} w % increases significantly the risk of blowholes formation. It is therefore possible to reduce this phenomenon by selecting base materials with low nitrogen, oxygen and sulphur levels.

Nevertheless, modern steels can always be EB welded.

Too high levels of the above elements simply reduce the "weldability range", that is to say the range of effective welding parameters, such as tension, intensity, focusing, vibration.

II - Thermal cycles in EBW and their metallurgical effects

Mechanical properties of EB welds depend on cooling thermal cycles, and thus on operating conditions. For the HAZ, it has been shown elsewhere [3] that the maximum temperature distribution can be simply derived from data such as initial temperature, fusion temperature and width of the molten zone. This arises from undimensionnal heat flow analysis.

The cooling rate of the weld, as expressed by parameters such as Δt_{500}^{800} or Δt_{300}^{700}, may be inferred from welding heat input E, and plate thickness e, as shown in figure 3. This type of presentation indicates that the thermal efficiency of the process is of the order of 80%.

The weld metal and HAZ are submitted to similar thermal cycles but for their peak temperatures. Neglecting the minute losses of some elements in the weld metal, it may be assumed that the coarse grained zone of the HAZ and the weld metal will display similar microstructures after the γ to α transformation and therefore similar mechanical properties.

It is generally difficult (particularly in EBW) to assess the HAZ properties. This is the reason why we used a simulation technique on "Gleeble" machine to study HAZ whose widths were sufficient to be compared to EB weld metal. The simulation cycle included a 1300°C peak, followed by different cooling rates, identical to those observed on actual welds. Metallographic observation indicate that the microstructures of both zones are very much alike (figure 4) as well as hardness (figure 5) or toughness (figure 6).

This important conclusion indicates that the search for satisfactory properties in EB weld metal should be guided by experience obtained in HAZ metallurgy.

For example, the influence of cooling rate on the EB weld metal toughness may be foreseen : figure 7 indicates that good toughness properties are achieved through cooling rates yielding self-tempered martensite or lower bainite microstructures. It is to be noted, that, unlike in conventional processes, it is unnecessary to avoid martensitic structures, as cold cracking does not occur in EBW.

On the other hand, inferior toughness properties are associated with coarse ferrite-carbide microstructures.

Thus, for a given steel composition, the selection of the heat input (control of the cooling rate) should be a compromise between toughness requirements (met by low heat input) and practical efficiency (high heat input yielding a wide bead).

III - Influence of some residual or alloying elements on the toughness of weld metal

Figure 7 shows that for a given cooling rate, the proper selection of steel composition is very important. We will review briefly the influence of some alloying (Mo, B, Zr) or residual (S, P, N) elements on the toughness weld metal. For sake of simplicity, toughness is here assessed with the 28 Joules transition temperature of the Charpy V test (TK_{28J}).

Figure 8 indicates that nitrogen is very detrimental to toughness, especially in solid solution : TK_{28J} increases by 5°C for each extra 10^{-3} % W nitrogen in the case of rapid cooling rates. At usual levels, sulphur (figure 9) has very little influence on the TK_{28J}, although it does reduce the upper shelf energy.

On the opposite, phosphorus (figures 9 and 10) is detrimental, especially for low cooling rates.

This effect is further increased by a post weld heat treatment. As seen previously, the low heat input generally yields improved toughness. Nevertheless, the use of a high heat input may be necessary when magnetic deviation of the beam is suspected. In this case, some alloying elements can increase hardenability, thus improving the weld metal toughness.

Molybdenum or nickel additions can be considered, possibly as foils inserted between the plates before welding. Figure 11 shows that an addition of 200×10^{-3} % Mo, alone or along with minute amounts of boron may significantly lower the TK_{28J} for cooling rates Δt_{300}^{700} under 50 s.

For slower cooling rates, it might not be desirable (for economic reasons, or because the plate may be welded with another process) to increase the amount of elements promoting quenchability.

Good results can still be achieved, by using steels containing elements such as zirconium (figure 12), capable of removing free nitrogen and forming fine carbonitride precipitates.

IV - The Post Weld Heat Treatments (PWHT) of Electron-Beam Welds

According to regulations, PWHT are compulsory for the thicknesses considered in this paper. These regulations do not take into account the specificity of the various welding processes but only the steel grade and thickness of the plates. As a general rule, the opportunity of PWHT can be appreciated by two criteria :

- a mechanical aspect which is refering to the lowering of welding residual stresses. This is the primary purpose of the PWHT. MACHERAUCH and WOLFAHRT [4] works, as well as our own measurements, indicate that residual stresses in the as-welded condition are much lower in EBW than those observed in arc welding (less than 100-150 MPa). In this respect, it is obvious that PWHT on EB welded structures has a very limited interest ;

- a metallurgical aspect arising from structural modifications in the different zones of the weld. These possible modifications of toughness after PWHT will often be the result of complex phenomena and are, in general, similar to those observed in the HAZ.

For C-Mn steels with sufficiently low impurity levels, PWHT have generally a limited or a beneficial effect on toughness properties. It is likely that the PWHT beneficial intrinsic effect is opposed to adverse mechanical evolutions (see next item). For steels containing microalloying elements, PWHT may result in precipitation of nitrides or carbonitrides. No toughness reduction was found for niobium contents up to 45×10^{-3} W % ; on the opposite higher vanadium (50 to 100×10^{-3} W %) or aluminium (100×10^{-3} W %) contents seem to impair toughness after PWHT.

In the case of low alloy (Mo, N, Cr, Cu) steels the situation is less clear and depends on possible complex precipitation or grain boundary segregation phenomena.. High purity steels are thus required to meet stringent toughness requirements in the stress relieved state.

V - Some problems pertaining to toughness evaluation of EB welds

The assessment of EB weld toughness is most frequently conducted on bending specimens, whose notch is located in the HAZ or the weld metal. These zones are very narrow, and often harder than the surrounding zones.

We have shown elsewhere [5] that the toughness measured in these conditions is in general not intrinsic to the central zone, but depends on surrounding zones properties. Figure 13 shows for example, that changing the base metal properties around a central zone can cause drastic toughness variations. This is because stress level and constraint factor in the specimen closely depend on parameters such as the central zone width or the difference of hardness between base and weld metal.

Figure 14 summarizes the influence of some geometrical or mechanical factors on the toughness of EB weld metal. So, it must be kept in mind that a notched bending test in EBW gives almost never an exact measurement of the notched zone properties, but only an image of the "sandwich" : base and weld metal.

Another phenomenon frequently encountered in EB bending test, is the deviation of the fracture towards base metal (figure 15). This phenomenon is temperature dependent : at low temperatures, cleavage occurs, following a "normal" path of fracture in the weld metal. For higher temperature, fracture occurs after ductile initiation and follows slip (or maximum shear) lines. At this point there is a competition between a cleavage mode after ductile initiation ("normal" mode) and a mode where fracture still follows slip lines outside of the weld metal. In this last situation, it is obvious that the measured toughness is metallurgically meaningless.

When testing temperature is further raised the deviated fracture mode becomes the only stable mechanism.

This phenomenon is favorised by two parameters :

- the narrowness of weld metal. We have shown (everything else being constant) that the broadening of the central zone diminishes the risk of deviation ;

- the difference between stress-strain law (as a first approximation : hardness difference) of the weld and the base metal. The greater the difference, the more likely is the deviation to occur (figure 16).

Therefore, these results suggest some techniques which could be applied to avoid deviation : broadening of the central zone by testing side by side beads, increase of base metal hardness by heat treatment...

Conclusion

We have reviewed some aspects of the EB welding of structural steels, emphasizing the relations between steel composition and the toughness and soundness of the weld metal.

In order to avoid porosity, low O-N-S steels must be used. Too high levels of these elements reduce the range of useful welding parameters.

On a metallurgical point of view, EB weld metal and HAZ are very much alike. It is therefore important to keep impurity (P-N) level as low as possible. Elements acting as scavangers for nitrogen or promoting the formation of fine carbonitride precipitates can be used for high heat input welding.

The effect of Post Weld Heat Treatments on EB weld metal is the complex result of adverse mechanical or metallurgical phenomena. As a general rule, stress relieving is less beneficial for EB welding than for other welding processes.

Finally, particular features of the toughness evaluation of EB welds have been discussed.

References

[1] H. GRANJON - Travaux sur le soudage par Faisceau d'Electrons. Bilan et Perspectives. Soudage et Techniques Connexes. May-June 1981.

[2] R. BAKISH - Twenty five years Electron Beam Welding in the U.S. 3ème Colloque International sur le Soudage et la Fusion par Faisceau d'Electrons et LASER. Lyon, September 1983.

[3] L. DEVILLERS, D. KAPLAN - Aspects thermiques et métallurgiques du soudage FE des tôles de fortes épaisseurs en acier de construction. 3ème Colloque International sur le Soudage et la Fusion par Faisceau d'Electrons et LASER. Lyon, September 1983.

[4] E. MACHERAUCH, H. WOHLFAHRT - Different sources of residual stresses as a result of welding. Conference Residual Stresses in Welded Construction and their Effects. The Welding Institute, London, November 1977.

[5] L. DEVILLERS, D. KAPLAN - Evaluation de la ténacité d'éprouvettes composites. IIW Document IX 1376-85 presented at the International Institute of Welding, Strasbourg, September 1985.

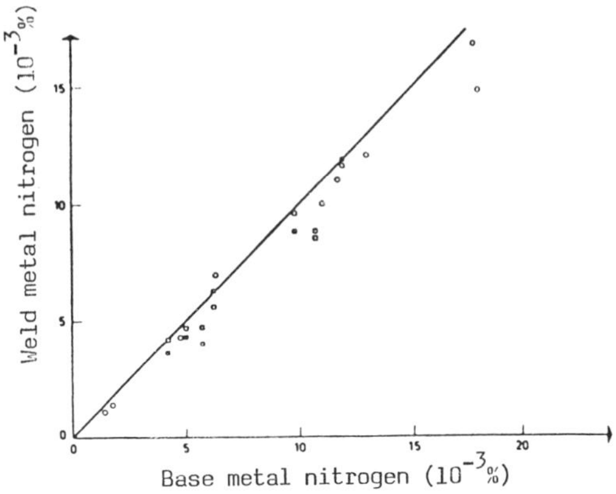

Figure 1 - Weld metal nitrogen content as a function of nitrogen base metal content.

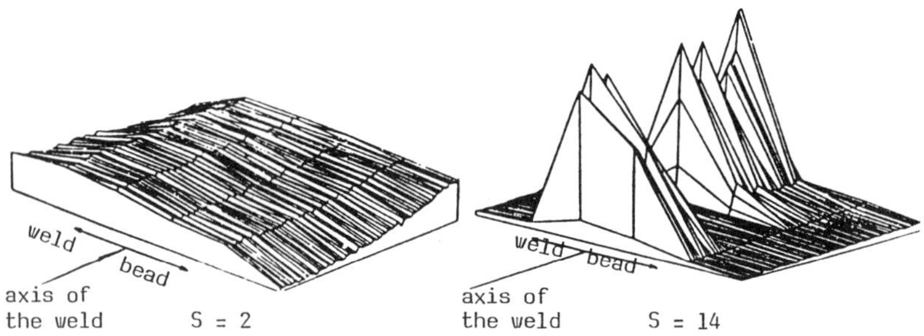

C	Mn	Si	Al	P	N	O
165	1350	300	40	9	8	2

(en 10^{-3} %)

Figure 2 - Three dimensional representation of density measurements in EB welds for two steels with different sulphur levels (2 and 14×10^{-3} W %).
Note : "peaks" indicate porosities in the weld metal.

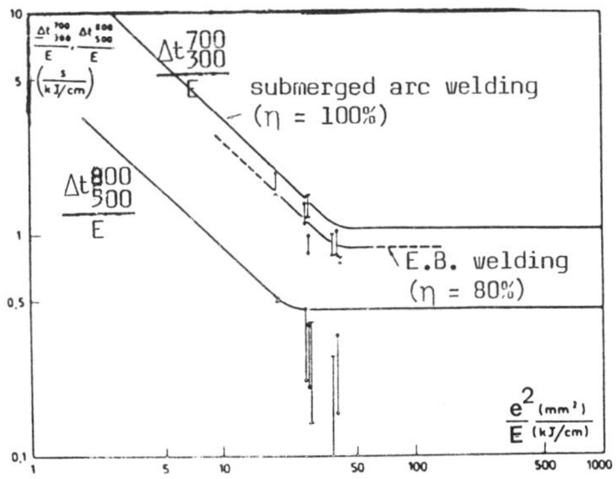

Figure 3 - Correlation between cooling parameters Δt_{300}^{700}, Δt_{500}^{800}, heat input E and plate thickness e.

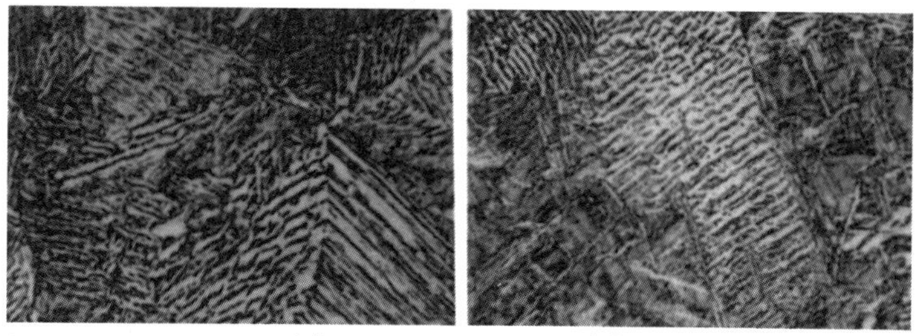

(a) (b)

Figure 4 - Comparison of microstructures observed in HAZ (a) and EB weld metal (b) for the same cooling rate. Δt_{300}^{700} = 80 s.

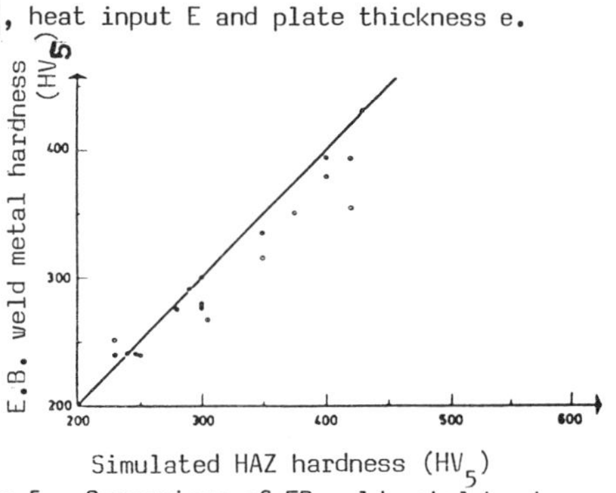

Figure 5 - Comparison of EB weld metal hardness with simulated HAZ hardness for the same cooling rate.

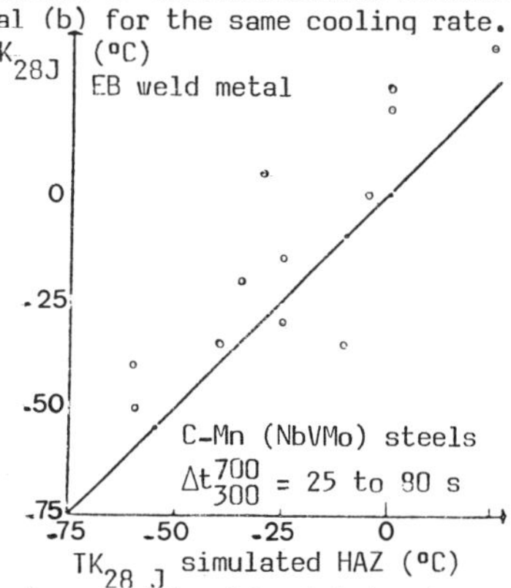

Figure 6 - Comparison of EB weld metal toughness with simulated HAZ toughness for the same cooling rate.

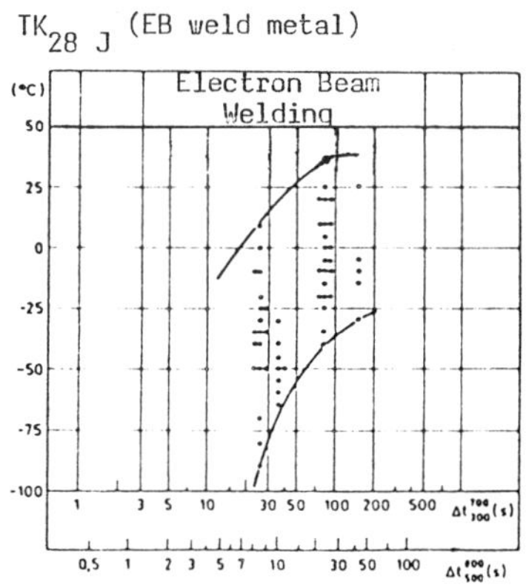

Figure 7 - Transition temperature at the 28 J level of EB weld metal, as a function of cooling rate after welding. (C-Mn, C-Mn-Nb, V, Mo steels).

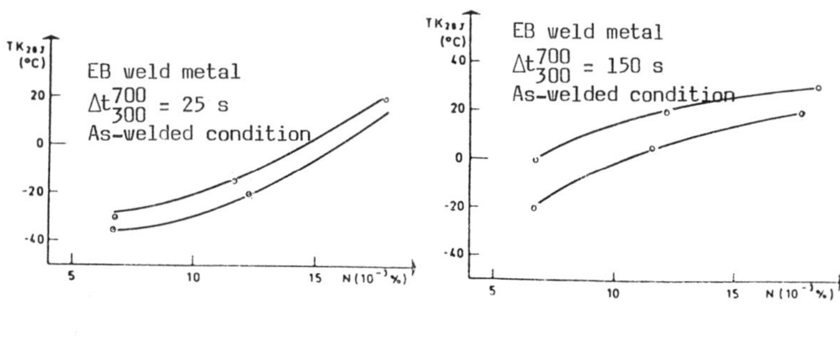

(a) (b)

Figure 8 - Influence of nitrogen on the toughness of EB weld metal for cooling rates of Δt_{300}^{700} = 25 (a) or 150 s (b).

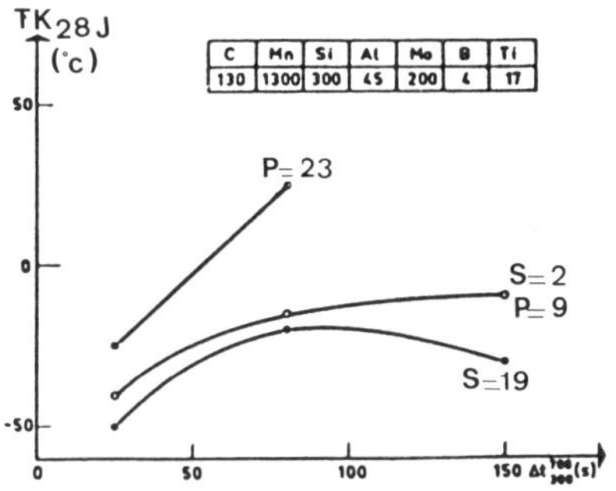

Figure 9 - Influence of sulphur and phosphorus on the toughness of EB weld metal for different cooling rates (as welded condition).

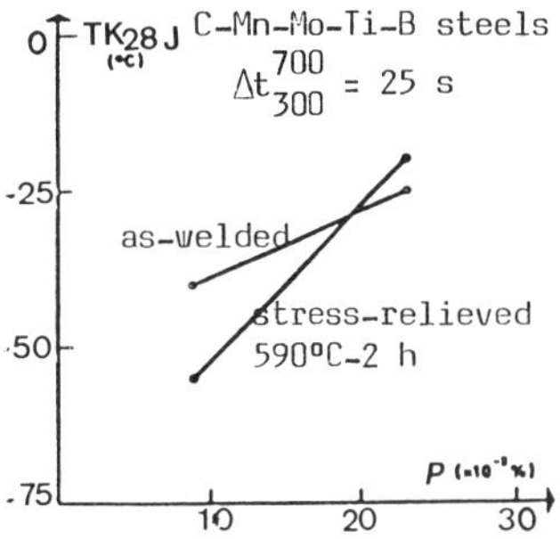

Figure 10 - Influence of phosphorus on the toughness of EB weld metal in the as-welded or stress relieved conditions.

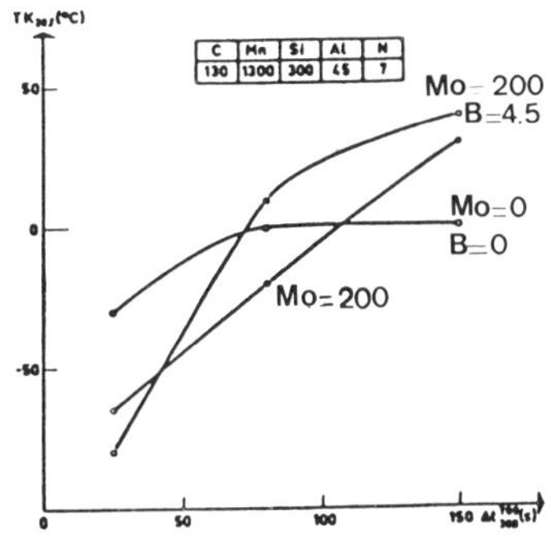

Figure 11 - Influence of molybdenum and boron on the EB weld metal tougnness (as-welded condition).

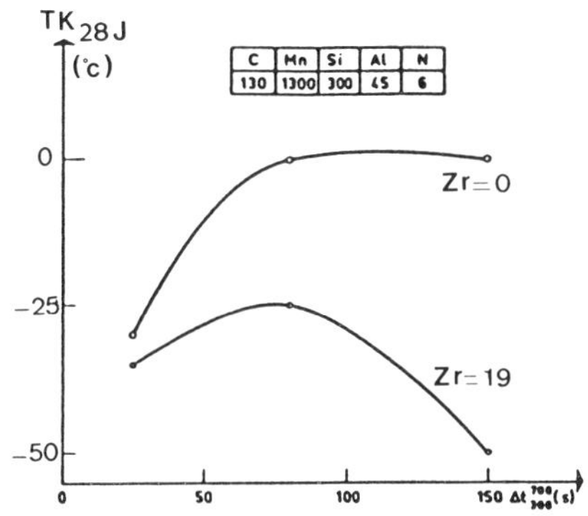

Figure 12 - Influence of zirconium on the toughness of EB weld metal (as-welded condition).

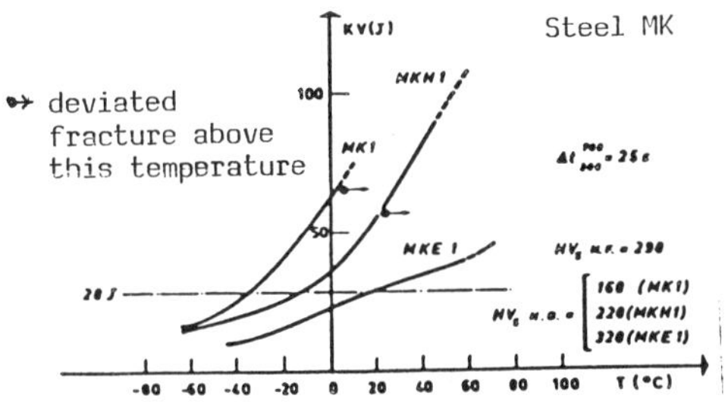

Figure 13 - Charpy V transition curves of the same EB weld metals with different heat treatments of the base metal (MKE : water quenched, MKH : oil quenched, MK : normalized).

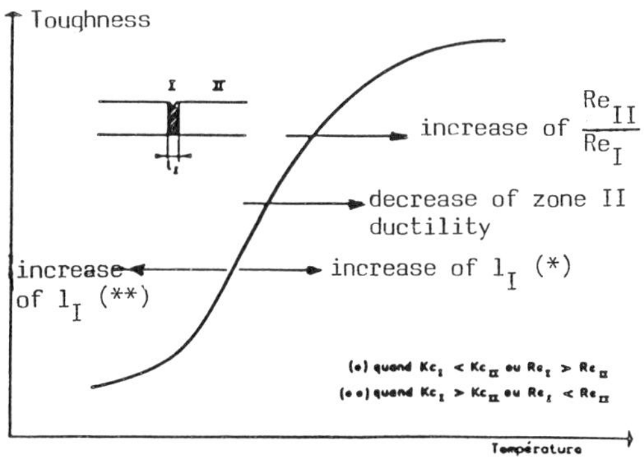

Figure 14 - Influence of some geometrical or mechanical factors on the apparent toughness of EB weld metal.

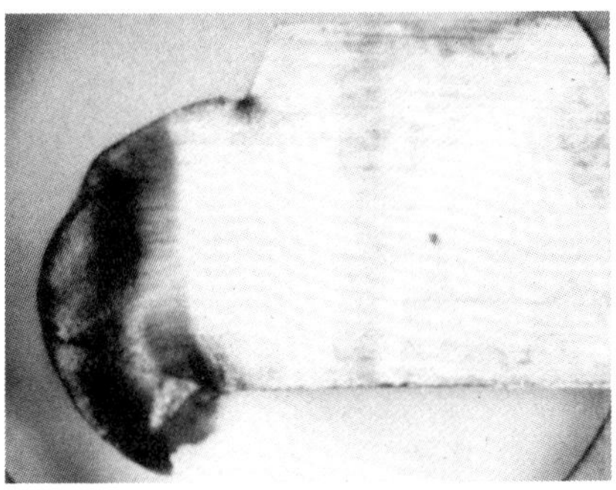

Figure 15 - Example of deviated fracture from notch located in the weld metal towards the base metal.

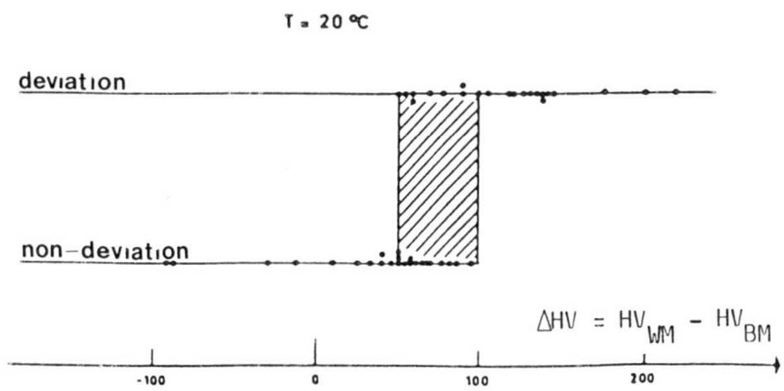

Figure 16 - Apparition of deviation in Charpy V at 20°C testing after ductile initiation in EB weld metal, as a function of weld and base metal hardness difference.

NEWLY DEVELOPED MOBILE EB GUN/COLUMN CONFIGURATION

Russell K. Colegrove, Donald E. Powers
Leybold-Heraeus Vacuum Systems Incorporated
Enfield, Connecticut

ABSTRACT

Development testing on a new configuration EB gun/column assembly has recently been completed. This assembly contains a self-accelerated, triode style electron gun that is capable of generating up to 45kW of beam power at 60kV, plus all of the auxiliary operational devices needed to make up a totally self-contained electron beam producing unit. As such, this system provides users of vacuum EB processing with a completely integrated and fully mobile EB producing unit that is ideally suited for accomplishing tasks which require a number of simultaneous axes of beam-travel motion.

THE PRIMARY INTENT of this paper is to review all the various unique features of a newly developed mobile EB gun/column configuration - and to describe the overall funtional capabilities of a gantry version, robotic style motion-producing device specifically designed for use with this new EB gun/column assembly.

For over 30 years now, the electron beam has been successfully employed as a highly versatile means for delivering energy to a workpiece - and, as such, has been universally used to perform a wide variety of different types of production material processing tasks (1/2/3/4). During this fairly extensive period of actual production application, a continued effort at enhancing the technology of EB processing has resulted in today's user having the capability to employ the process either in or out of vacuum, at high or low beam voltages, under manual or automatic operation, with intermittent or continuous beam output, etc.

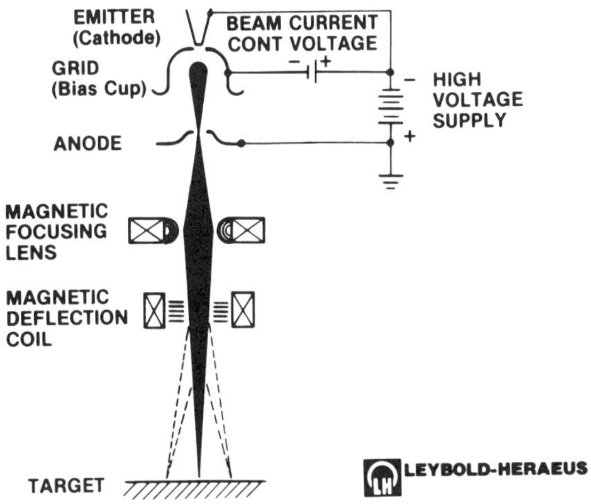

Fig. 1 - Schematic drawing of EB producing system

However, although present day users of EB have this much broader range of process application capability, the true flexibility of this method of energy delivery can still only be fully realized when the process is applied under reasonably good vacuum conditions - since it is only in this type of an environment that the greatest degree of beam focusing capability, widest range of beam focal plane variance, and the highest level of beam deflection versatility will be achieved (Fig. 1). Thus, as the size of workpieces being processed with EB have steadily increased, the need for a highly mobile EB producing unit (to provide a direct means for reducing the size of the vacuum enclosure required to EB process large workpieces under vacuum) has also grown.

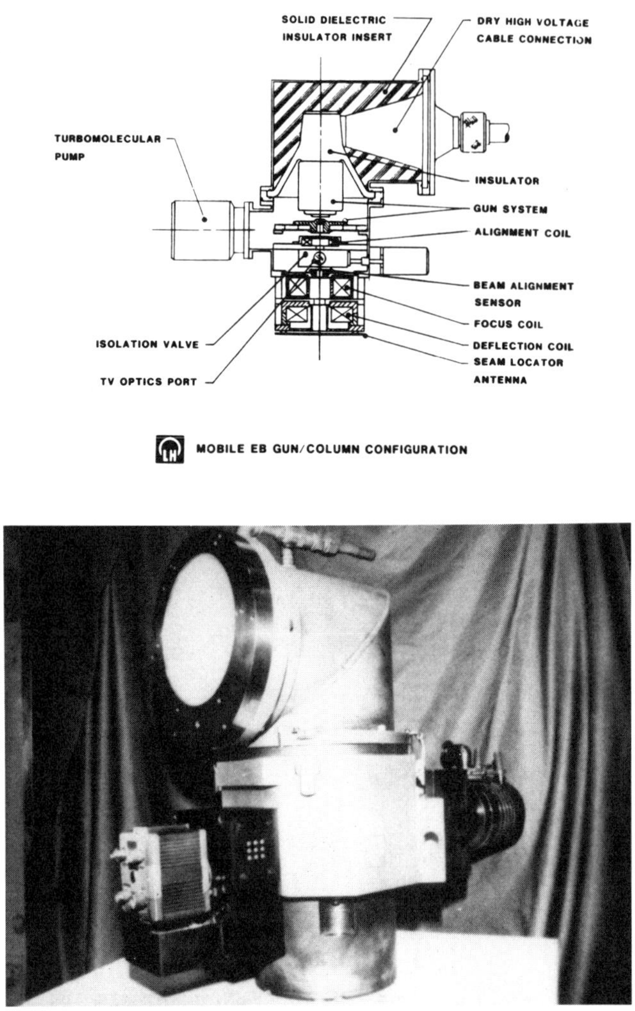

Fig. 2 - Photo and schematic of new mobile EB gun/column assembly

NEW MOBILE GUN/COLUMN ASSEMBLY DESCRIPTION

A relatively compact, highly mobile EB gun/column assembly that L-H has recently developed for satisfying this need is shown in Fig. 2. This unit contains a self-accelerated, triode style electron gun which uses a directly heated, ribbon variety filament as its source of electrons. This gun system is designed to reliably generate an electron current output of up to 0.75A at an operating voltage level of 60kV. Both gun voltage and current are independently adjustable, and both are controllable directly from the operator's work station.

The electron current output being generated emerges as a fairly collimated, but slightly divergent beam of high energy electrons that passes down through the lower segment of the gun/column assembly - exiting the assembly through the bottom of the gun/column structure. Modern computer plotting techniques (Fig. 3a) were employed during gun development, to allow the potential field configuration and beam profile shape that would result from various gun geometries intended for testing to be examined - thus ensuring that the most optimum beam profile attainable (at the 60kV/750mA operating level) was achieved.

Located in the lower segment of this overall gun/column structure are electromagnetic focusing and deflection devices. By varying the driving current of the electromagnetic focusing lens assembly employed, the beam generated can be focused into a fine spot over a range of distance out from the bottom of gun/column structure (up to a minimum distance of some 4 inches away from the bottom of the structure). The electromagnetic deflection coil assembly provided, just below the focusing lens, allows the focused beam spot produced in this fashion to be moved about with respect to its normal gun/column structure centerline position. This deflection coil assembly is a dual axes unit, and provides the capability for independent deflection along either the "X" or "Y" axes - or simultaneous deflection along both these axes; the driving current employed can be either AC or DC - so that both stationary and/or oscillatory beam deflection patterns can be produced.

The upper gun portion of this new gun/column assembly is independently pumped with a turbomolecular pump, in order to ensure that the best vacuum level possible is maintained in this area during beam operation. A high vacuum valve is also provided in this gun/column assembly, allowing the gun area to be isolated (and maintained under vacuum) when the gun/column assembly is exposed to atmospheric pressure - thereby minimizing the possibility that gun components might inadvertently become contaminated from being exposed to atmosphere. In addition, an electromagnetic alignment coil, located just below the gun region, allow precise alignment of the gun output beam (with the column electron optical devices in the lower segment) to be easily accomplished - directly from the operator's work station.

Other unique features that have been included, as an integral part of this newly developed mobile gun/column configuration, are:

Dry Cable Connection Design - The cable providing power to the gun is terminated in a resiliant potting material, and this potting is configured in a manner which allows it to plug directly into the gun/column assembly (without any need for having to employ some form of oil immersion interfacing scheme).

Direct Gun Cooling Design - A special, closed loop/self circulating/high dielectric strength oil cooling method is used as a means for continuously extracting heat from the gun area, thereby allowing gun temperatures to be maintained under control during lengthy periods of full power operation. The heat being extracted in this manner is continuously removed from the oil by a water cooled heat exchanger, thru which the oil circulates during its closed loop flow.

Quick Change Gun Design - The filament and grid cup portions of the gun system are held in place by an easily releasable mechanical mechanism, which allows the cartridge style assembly (Fig. 4) holding these component portions of the gun to be quickly removed (thereby facilitating gun maintenance). The employment of a spare, pre-assembled cartridge unit additionally reduces the time required for accomplishing a normal filament replacement task.

Beam Passage Protection Circuit - The current being picked up by a sensing diaphragm located in the lower portion of the gun/column structure is continuously monitored. The indication of an excessive current draw by this sensor, usually indicative of some form of undesirable beam position and/or beam profile condition, can be used to automatically shut the beam down should either of these undesirable conditions inadvertently occur during a Beam ON period - thereby protecting the gun/column assembly from accidental beam damage. In addition, this sensing diaphragm can be used as an assist device during performance of a beam alignment procedure.

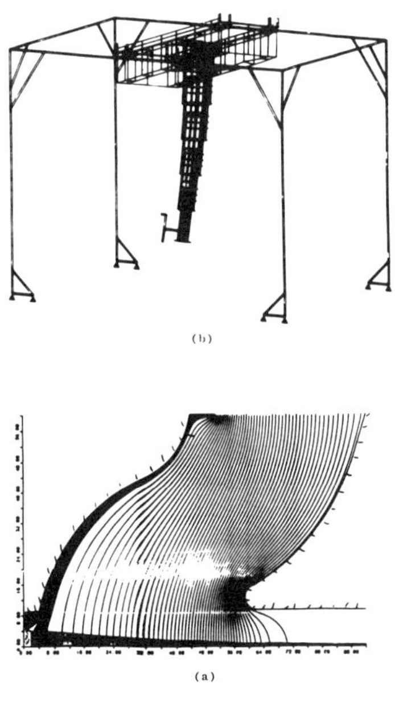

Fig. 3 - Computer plotted graphic representations

TV Viewing Capability - A miniaturized TV camera mounted to the gun/column structure displays the joint seam on a TV screen located at the control console, thereby allowing operator to have a clear "beam's eye view" of the joint under the gun - regardless of the gun's position (and/or orientation) with respect to the operator's location.

NEW ROBOTIC STYLE, GUN MOTION-PRODUCING SYSTEM

A robotic style, motion-producing device was required to provide this new mobile EB gun/column assembly with the highest degree of gun/column motion flexibility possible, and the

Fig. 4 - Cartridge style grid/filament assembly unit

system specifically designed for providing this type of multi-dimensional motion capability is shown in Fig. 5. The total (gun motion-producing) system concepted combines a telescoping Z-travel column, suspended from a gantry type X/Y - travel motion device, with a gun/column assembly that has a power operated tilting and pivoting capability. In this fashion, five axes of beam travel motion are available. Again, modern computer plotting techniques (Fig. 3b) were employed to enhance the final design provided. As such, a finite element analysis of the gun motion-producing system design concept was conducted, for both static and dynamic conditions, and required design corrections determined by this analysis were incorporated prior to start of manufacture of the system. One especially unique feature of this particular motion-producing system is that the telescoping Z-travel column design concept employed is one that allows this structure to collapse to a length that is only about one-half the value of its total Z-travel stroke length (thereby facilitating a complete minimization of the vacuum enclosure height needed for accommodating the length of Z-travel stroke provided).

The X, Y, and Z axes are simultaneously contouring axes, powered by variable speed DC servo motors - through precision ballscrew or high precision rack and pinion drives, which travel on hardened (precison ground) guide rails. To assist in performing three-dimensional (contouring or straight line) welding applications, the power operated gun tilting and pivoting feature allows the gun to be CNC positioned for the most effective weld processing capability. Contouring axes speed and position are established under direct control of the Computerized Numberical Control and can be programmed for fully automatic contouring operation. All axes can be integrated into a weld sequence program or operated manually by means of the pendant station jog pushbottons or operator station joystick control. This multiple axes gun motion-producing system is capable of following a programmed beam travel path with an accuracy of ± 0.025mm (0.001 in.), per foot of individual axes travel.

Fig. 6 is a photograph of a system presently being manufactured for one specific application of this newly developed mobile EB gun/column assembly. This system employs a roughly 50 cubic meter (approx. 1800 cubic foot) vacuum enclosure, constructed from nominally 30mm (1¼ in.) thick stainless steel (to simultaneously provide both a mechanically strong and a radiation tight structure), that is used to contain a gantry version, robotic style motion-producing device that provides 3.05 meters (120 in.) of gun X-motion/ 2.34 meters (92 in.) of gun Y-motion/1.79 meters (70 in.) of gun Z-motion/± 180 degrees of gun rotational motion/+ 90, - 45 degrees of gun tilting motion. This system pictured,

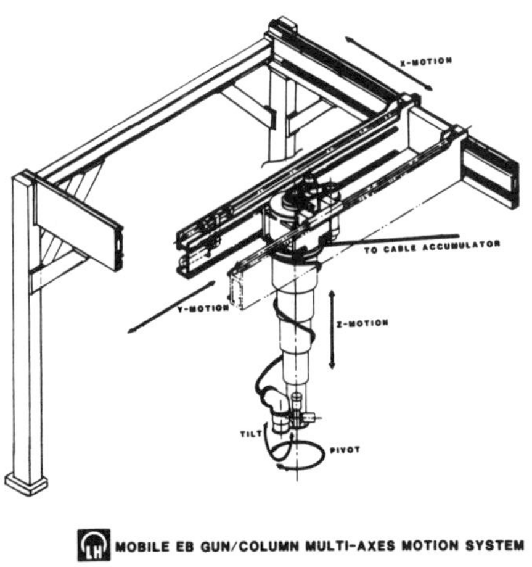

Fig. 5 - Schematic of motion system device

Fig. 6 - Mobile EB gun/column under construction

when programmed to function automatically under CNC control, will accurately weld three-dimensional parts requiring contouring beam-travel motion speeds of up to 42mm/s (about 100ipm), while also providing varying degrees of automatic gun operating parameter adjustment during executing of this beam travel motion.

SUMMARY

An EBW machine assembly utilizing a recently developed, highly mobile EB gun/column system and a newly fabricated, highly versatile EB gun/column motion-producing system has been described - and the various unique features these new systems provide, to present day users of vacuum EB processing who require an expanded capability for handling a greater variety of part sizes and shapes, have been outlined in detail.

ACKNOWLEDGMENT

The authors gratefully acknowledge the efforts of Dieter Fritz (L-H, Hanau, W. Germany) and Gary LaFlamme (L-H, Enfield, Conn.), for technical assistance provided during design concepting and final testing of the EB gun.

REFERENCES

1. Colegrove, R. and M. Carroll, "EBW System for Fully Automatic Welding of Large, Heavy Section Pieces", p 97, Proceedings of Bakish Laser vs Electron Beam - State of the Art Conference (1985).

2. Powers, D. and W. Dietrich, "Application of Electron Beam to Glazing and Transformation Hardening", p 44, Proceedings of Australasian Institute of Metals - 1st National Heat Treatment Conference (1984).

3. Dietrich, W. and G. Schubert, "Exploiting Electron Beam Welding in Production", Proceedings of The Welding Institute Conference on the Application of Energy Beams for Joining (1984).

4. "Electron Beam Welding", p 609, ASM Metals Handbook - 9th Edition, 6 (1983).

ELECTRON BEAM WELDING OF HEAVY SECTION FERRITIC STEELS

J.F. King, S.A. David
Oak Ridge National Laboratory
Metals and Ceramics Division
P.O. Box X
Oak Ridge, Tennessee 37831

ABSTRACT

A study was undertaken to investigate the feasibility of using a high power electron beam welding machine to weld heavy section steel of three different compositions. These were 3 Cr-1.5 Mo-0.1 V, 2.25 Cr-1 Mo, and modified 9 Cr-1 Mo steels. Emphasis in this work was placed on determining the mechanical properties of the weldments, characterizing the microstructure of the various weldment regions, and comparing these results with those from other welding processes. Characterization of the electron beam weldments was made by microhardness measurements, Charpy V-notch impact tests, tensile tests, and guided bend tests. The testing results showed the electron beam weldments had excellent properties which makes it an attractive process for heavy section steel welding.

WELDING OF THICK-SECTION STEELS is a common practice in the fabrication of energy systems. Components such as pressure vessels are large and must be economically fabricated with high integrity. The fabrication cost is strongly influenced by the speed at which these large components can be welded. Conventional welding processes such as shielded metal arc (SMA) and submerged arc (SA) are time consuming and expensive. Hence there is a great need to reduce welding time and the tonnage of weld metal deposited.

Electron beam welding (EBW) is a process that could be used to achieve dramatic reduction in the welding time and hence the cost. It has several attractive characteristics for the welding of thick sections. The penetrating ability of the beam produces welds with high depth-to-width ratios at relatively high travel speeds, making it possible to weld thick sections with one or two passes. The joints welded are usually autogenous square butt welds, eliminating the use of costly filler metals and other consumables. There is an additional saving in the labor and time required because the welding can be completed at a fast rate relative to the processes mentioned above. The limitations of the process include highly critical joint preparation, lack of flexibility in the joint fit-up, and the requirement to operate in a vacuum to achieve the high depth-to-width ratios in the weld.

The paper describes a study that was undertaken to investigate the feasibility of using a high-power EBW machine to weld thick-section steel. Also, the microstructural characteristics and the properties of the weldment are described and compared with weldments produced by conventional welding processes.

EXPERIMENTAL PROCEDURE

Materials selected for the heavy section electron beam welding study were 3 Cr-1.5 Mo-0.1 V, 2.25 Cr-1 Mo, and modified 9 Cr-1 Mo steel alloys with the compositions listed in Table 1. Emphasis was given to the 3 Cr-1.5 Mo-0.1 V steel alloy developed by Wada and co-workers for high temperature, high pressure hydrogen service as a possible replacement for 2.25 Cr-1 Mo pressure vessels.[1–4] The plates used in this study were normalized at 954°C for 2 h and air cooled, and then stress relieved at 565°C for 2 h and air cooled to produce a 100% bainitic microstructure.

The 2.25 Cr-1 Mo steel alloy conformed to ASME specification SA-387 grade 22 class 2 plate. Plates 100-mm-thick for the electron beam welding study were sawed from the top and bottom surfaces of the original 330-mm-thick material which had been heat treated by austenitizing at 954°C for 6.5 h, water quenching and then tempering at 663°C for 6.5 h. These 100-mm-thick plates from the surfaces of the larger plate were uniform in properties and microstructure through this thickness.

Modified 9 Cr-1 Mo steel plates for this study had received the standard heat treatment for this material which is to normalize at

1038°C, air cool and temper at 760°C. The starting material was 200-mm thick so the holding time at each heat treatment temperature was 4.5 h. After heat treatment the plate was cut to produce the nominal 100-mm thick weld plates.

Table 1 - Composition of Steels Welded by Electron Beam Process

Element	Content (wt %)		
	3 Cr-1.5 Mo V	2.25 Cr-1 Mo	9 Cr-1 Mo
C	0.12	0.12	0.089
Mn	0.84	0.40	0.47
Si	0.27	0.23	0.28
Ni	0.14	0.04	0.16
Cr	2.86	2.27	9.24
Mo	1.48	1.04	0.96
V	0.09	0.00	0.21
S	0.002	0.002	0.006
P	0.011	0.009	0.021
Al	0.004	0.00	0.002
Sn	0.002	0.00	0.004
Cu	0.06	0.08	0.08
Nb	0.00	0.00	0.054
N	0.014	0.00	0.035

WELDING

Full-penetration autogenous welds were made with a 100 kW electron beam welder. The welds were made with the 100-mm-thick plates of each material in the horizontal welding position. Welding parameters used to produce full penetration were: 95 kV accelerating voltage; 377 mA beam current; and 1.9 mm/s travel speed. A macro cross sectional view of a typical weldment is shown in Fig. 1. The weld microstructures were characterized by using standard optical and scanning electron microscopy techniques.

WELDMENT PROPERTIES

Electron beam weldments were evaluated by microhardness, Charpy V-notch impact, tensile, and guided-bend tests. The microhardness test was conducted with a Shimadzu diamond pyramid microhardness tester. Charpy V-notch impact tests were conducted in accordance with the procedure described by Nanstad.[5] The tests were conducted over a temperature range in accordance with ASTM procedure E23.[6] Charpy V-notch specimens were tested for the electron beam weld metal and the heat-affected zone (HAZ). Specimens were taken with the HAZ and weld metal perpendicular to the specimen axes and notched for the cracks to propagate in the longitudinal welding direction. Charpy V-notches were placed in the center of the weld for the weld metal tests and in the center of the HAZ for its tests.

RESULTS AND DISCUSSION

MICROSTRUCTURE - The transverse section of the weldment in Fig. 1 shows the fusion zone (FZ) and the HAZ were uniform and essentially parallel-sided through most of the thickness. A few voids were found in the weld metal of the 3 Cr-1.5 Mo and 2.25 Cr-1 Mo steels during sectioning, but these were rare and presented no significant problems for the weldment in terms of mechanical properties. The modified 9 Cr-1 Mo steel however contained large voids, 2 to 3 mm in diameter, near the center of the weld and over the entire length of the plate. The voids were attributed to the relatively high nitrogen content of the alloy. No welding parameter variations were evaluated to reduce or eliminate this condition. A sufficient amount of sound weld metal was present above and below this line of defects to permit weldment characterization.

The base metal microstructure for the 3 Cr-1.5 Mo alloy was predominantly bainitic. The FZ and HAZ were mixed bainite and martensite with some retained austenite present in the FZ. The as-welded microstructures for these three regions are shown in Fig. 2. Scanning electron microscopic examination of the microstructures revealed that the weldment contained no carbides or other precipitates in the as-welded condition. Although the base metal received a tempering treatment, the absence of clearly visible carbide precipitates in the base metal microstructure may be attributed to the low-temperature tempering treatment. Typical microstructures of the weldment after a postweld heat treatment of 690°C for 5 h showed the presence of precipitates which should be $M_{23}C_6$ as observed by Klueh[7] for a similar tempering treatment. Also, a microstructural comparison was made between the electron beam

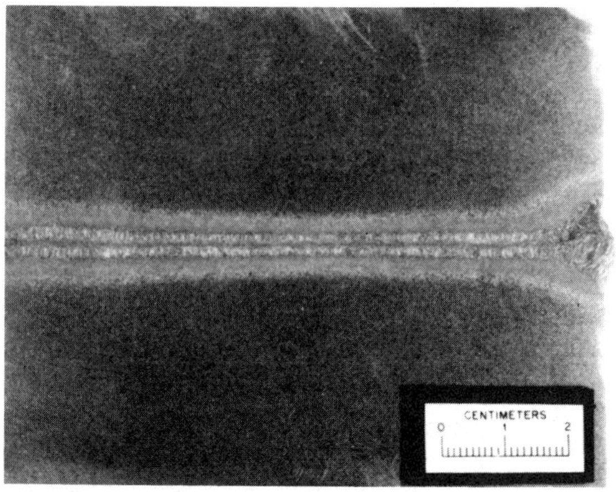

Fig. 1 - A transverse section of the electron beam weldment in 102-mm-thick 3 Cr-1.5 Mo-0.1 V steel. The fusion zone and HAZ were uniform and parallel-sided through most of the thickness.

weld metal and that of a submerged arc weld. The submerged arc weld metal (Fig. 3) in 3 Cr-1.5 Mo-0.1 V steel is predominantly granular bainitic.

Weld metal microstructures for the 2.25 Cr-1 Mo and modified 9 Cr-1 Mo steel electron beam weldments are shown in Figs. 4 and 5, respectively. The microstructure of the 2.25 Cr-1 Mo steel weld was martensitic and the modified 9 Cr-1 Mo weld was martensite with some ferrite.

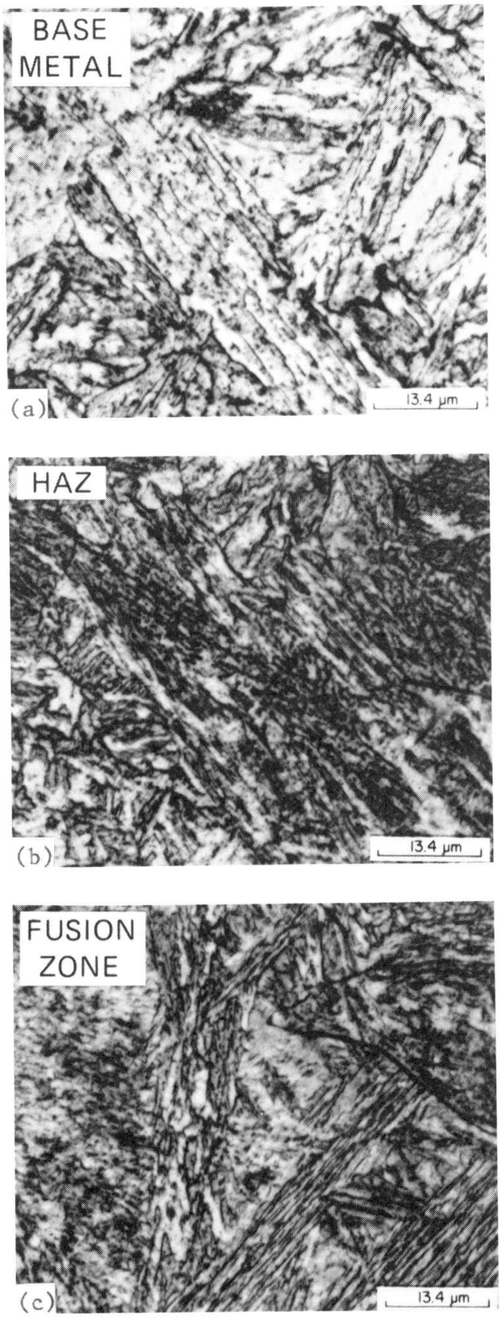

Fig. 2 - The microstructures of the electron beam weldment in 3 Cr-1.5 Mo-0.1 V steel were identified as (a) base metal-bainite, (b) HAZ mixed bainite and martensite, and (c) FZ mixed bainite and martensite with some retained austenite.

MICROHARDNESS - Microhardness traverses across the 3 Cr-1.5 Mo-0.1 V steel electron beam weldment both in the as-welded and in the post-weld heat treated (PWHT) conditions are shown in Fig. 6. Because the base metal had only received a stress relief heat treatment (565°C for 2 h), it shows a high average hardness of 400 DPH. The regions, (FZ and part of the HAZ) exposed to temperatures above austenitic temperatures during the weld thermal cycle are transformed to martensite during cooling to room temperatures. They have the same hardness as the base metal. The hardness profile in the as-welded condition exhibits a pronounced "sawtooth" effect in the weldment. This is typical of ferritic steels, and similar observation has been made earlier in conventionally welded ferritic steels.[8] The variation of hardness with location in the weldment reflects the varying amounts of tempering and stress relief produced in these regions by the weld thermal cycle. The lowest value in hardness of 320 DPH was observed in the HAZ on both sides of the FZ at approximately 2 mm from the fusion line. The hardness drop as in other ferritic steels may be related to the overtempering of the base metal microstructure at a temperature just below the Ac_1. Microstructural examination of the soft zone shown revealed tempered martensite and overaged carbides. The presence of untempered martensite in both the FZ and most of the HAZ of this particular steel after welding precludes the use of this alloy in the as-welded condition, so a PWHT of 690°C for 5 h was given to the weldment. After the 5 h of PWHT, the base metal has a hardness of about 210 DPH with the FZ showing an average hardness of 230 dph. The tempered zone hardness is 200 DPH, which is about the same as the base metal. The transformed zone has a hardness of 230 DPH, about 30 DPH higher than the base metal.

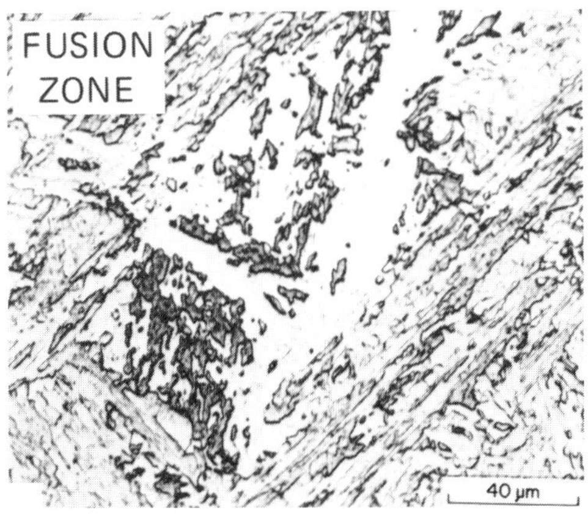

Fig. 3 - Submerged arc weld metal in 3 Cr-1.5 Mo-0.1 V steel has a granular bainitic microstructure.

Fig. 4 - Electron beam weld metal in 2.25 Cr-1 Mo steel.

Fig. 5 - Electron beam weld metal in modified 9 Cr-1 Mo steel.

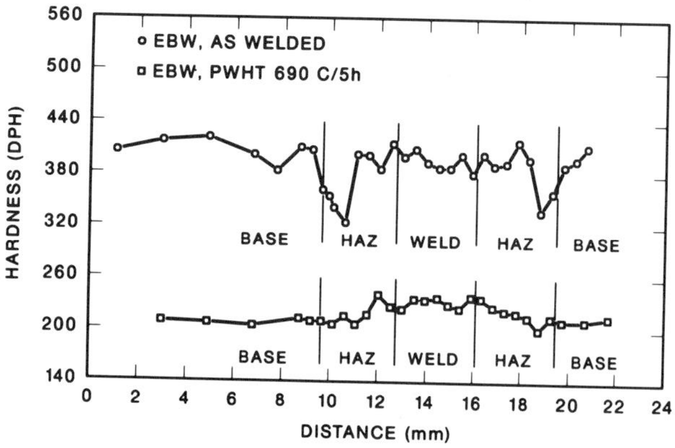

Fig. 6 - Hardness traverse in 3 Cr-1.5 Mo-0.1 V electron beam weldments as welded and after postweld heat treatment.

Hardness profiles for the 2.25 Cr-1 Mo and modified 9 Cr-1 Mo steels were similar to that of the 3 Cr-1.5 Mo weldment shown in Fig. 6 except for base metal hardness. These steels had been tempered to hardnesses ranging from 200 to 220 DPH. After PWHT of 677°C for 2.5 h the 2.25 Cr-1 Mo steel weld had a hardness of 260 DPH and the base metal remained the same. Modified 9 Cr-1 Mo steel received a PWHT of 760°C for 2.5 h to produce a weld metal hardness of 250 DPH. Both of these steels, as did the 3 Cr-1.5 Mo steel, had electron beam weld hardnesses higher than the base metal after their PWHT.

TENSILE PROPERTIES - Tensile testing was conducted on 3 Cr-1.5 Mo specimens taken transverse to the welding direction. The gage length of these specimens contained weld metal, base metal, and HAZ. The weldments used for these tests received a PWHT of 690°C for 5 h prior to machining and testing the specimens. This PWHT condition yields a temper parameter of 19.93 as calculated from

$$TP = T(20 + \log t) \times 10^{-3},$$

where T is the temperature in Kelvin and t is the time in hours. The tensile results are presented in Figs. 7 and 8 as a function of the test temperature. Base metal data for other tempering parameters are shown for comparison. The trend for both yield strength and ultimate tensile strength is the same in that both decrease with an increase in tempering parameter. Also it was found that the yield strength and ultimate tensile strength of the electron beam weld metal is higher than that of the base metal because all fractures in the tensile specimens occurred outside the weld. This correlates with the slightly higher hardness values measured for the weld.

Modified 9 Cr-1 Mo and 2.25 Cr-1 Mo steels exhibited similar tensile behavior to the 3 Cr-1.5 Mo steel. Testing was performed in both the as-welded and PWHT conditions at room temperature. All fractures in the tensile specimens occurred in the base metal outside the weld region which again indicates higher tensile strength for the electron beam weld metal.

CHARPY IMPACT TEST RESULTS - Figure 9 shows the results of the Charpy V-notch impact test on 3 Cr-1.5 Mo. The results for the base metal with a similar tempering parameter are included for comparison. The base metal and weld metal properties are nearly identical, with only a slight reduction in the upper-shelf energy of the HAZ. The original criteria for the 3 Cr-1.5 Mo alloy specified that it have good toughness, with Charpy V impact energy exceeding 54 J at or below -18°C. The data for the electron beam weld metal and HAZ both exceed this minimum requirement as does the base metal. The Charpy data for the weld metal in the as-welded condition are compared with the PWHT data in Fig. 10. The impact energy data for the

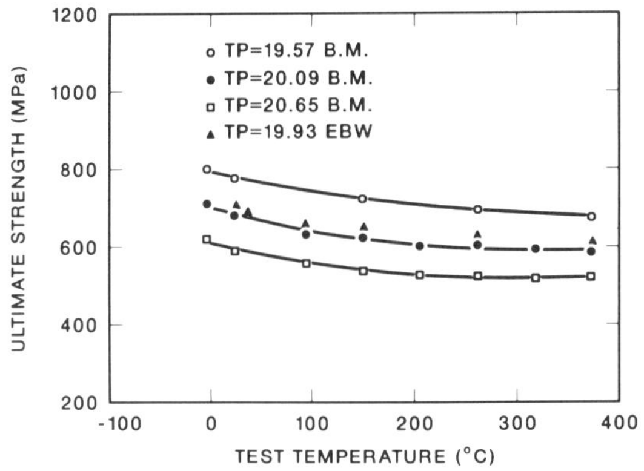

Fig. 7 - Electron beam weld metal and 3 Cr-1.5 Mo-0.1 V steel base metal temper parameter study of ultimate tensile strength.

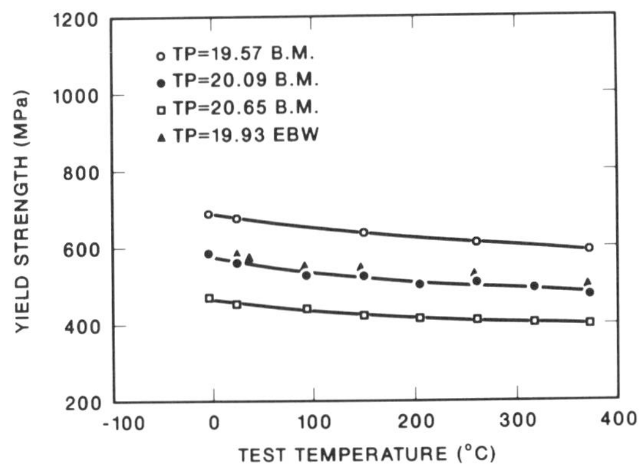

Fig. 8 - Electron beam weld metal and 3 Cr-1.5 Mo-0.1 V steel base metal temper parameter study of yield strength.

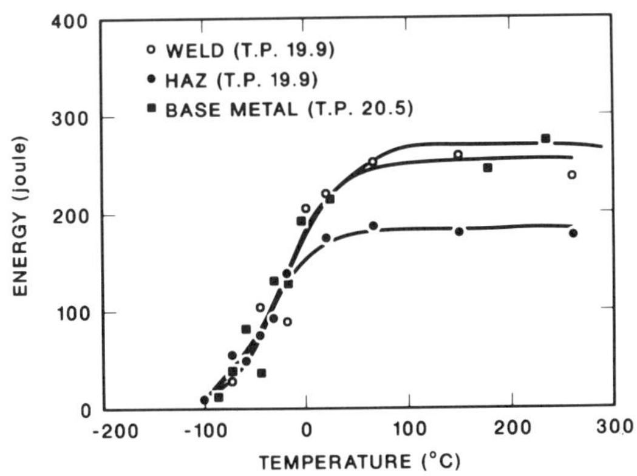

Fig. 9 - The Charpy V-notch impact test results for the electron beam weld metal are the same as those for the base metal after a PWHT of 690°C for 5 h.

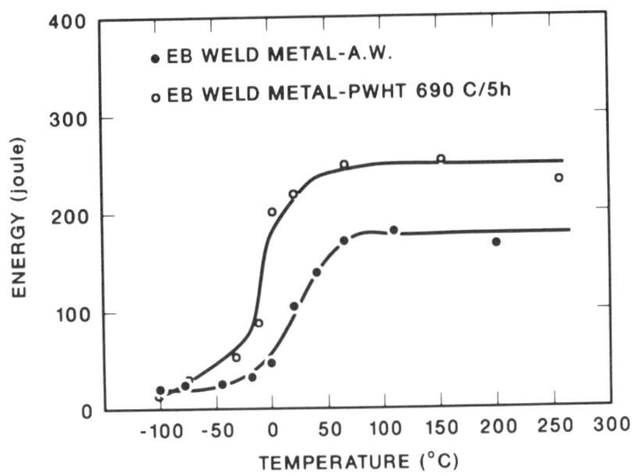

Fig. 10 - Charpy impact energy test results for 3 Cr-1.5 Mo-0.1 V electron beam weld metal.

as-welded weld metal appears to be exceptionally high (120 J) at room temperature. The room-temperature impact energy value is improved from 120 J to 220 J by a PWHT of 5 h at 690°C. Overall, the as-welded weld metal property falls only slightly below the criteria for this steel with a 54-J transition temperature of 3°C, which is very respectable. The HAZ in the as-welded condition had a 54-J transition temperature of -23°C. Hence, Charpy impact properties of electron beam weldments in this steel appear to be excellent even in the as-welded condition.

Charpy V-notch test results for the 2.25 Cr-1 Mo steel welds are presented in Fig. 11. In the as-welded condition, the HAZ and weld metal had similar transition temperatures with the weld having a higher upper shelf energy. After a PWHT of 677°C for 2.5 h the two regions again had similar transition temperatures but they were over 50°C lower than the as-welded data and the weld upper shelf energy increased to over 300 J. These are very acceptable impact properties.

Weld metal and the HAZ in electron beam welded modified 9 Cr-1 Mo steel were tested in the as-welded condition and after a PWHT of 760°C for 2.5 h. Results from these Charpy impact tests (Fig. 12) show the as-welded fusion zone in this steel has a high transition temperature compared to the other steels in this study but PWHT reduces this to a temperature below 0°C. This is similar to results obtained for gas tungsten arc welds in this steel.[8]

GUIDED-BEND TESTS - Guided-bend tests were conducted on the electron beam weldments in all three steels by preparing three specimens through the thickness. Approximately one-third of the weldment was present in each side-bend specimen. The bend tests were conducted according to ASME Boiler and Pressure Vessel Code, Section IX, Welding and Brazing Qualifications, using the guided-bend wrap-around jig specified in QW-466.3. Each specimen was 9.5 mm thick and 32 mm wide. Specimens were tested to a bending angle of 180° satisfactory results for each steel. Weldments were tested only in the PWHT condition.

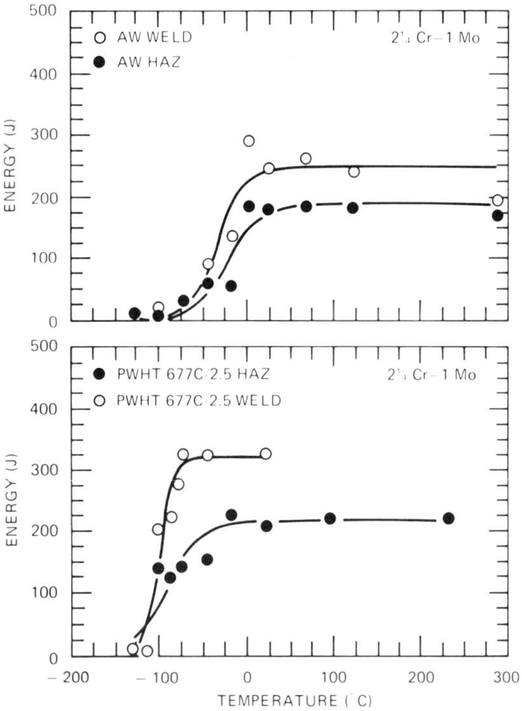

Fig. 11 - Charpy V-notch test results for electron beam welded 2.25 Cr-1 Mo steel as welded and after PWHT.

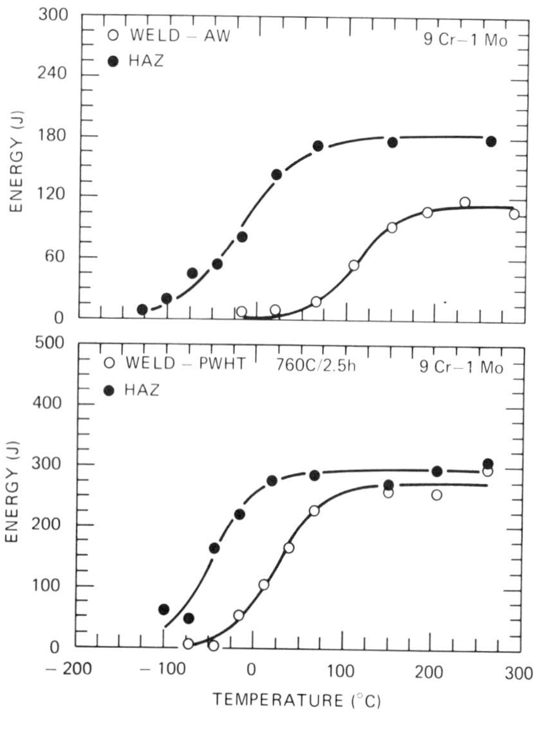

Fig. 12 - Charpy V-notch test results for electron beam welded modified 9 Cr-1 Mo steel as welded and after PWHT.

COMPARISON OF EB WELDMENTS WITH OTHER WELDMENTS - Properties of electron beam weldments in 3 Cr-1.5 Mo steel were compared with those of weldments made by two other welding processes being developed, namely, the submerged arc (SAW) and shielded metal arc (SMAW) welding processes.[9] It is recognized that weld metal properties can vary depending on the deposit composition from the consumables used in these processes. The properties used for this comparison represent those obtained from the currently ongoing welding development of this steel.

The Charpy V-notch 54-J transition temperature for the electron beam weld metal is slightly below the values for SAW metal in the 3 Cr-1.5 Mo-0.1 V steel. The test results from two weldments made at two different laboratories indicated that the EB weld has a lower transition temperature than either of the SA weldments. The Charpy V-notch upper-shelf energy of the EB weld was compared with that of a SMAW deposit in this steel. As shown in Fig. 13, the upper-shelf energy of the EB weld is the same as that of the base metal but is much higher than that of SMAW metal. These comparisons show that the EBW metal in 3 Cr-1.5 Mo-0.1 V steel has generally superior properties to the weld metal of the other commonly used processes and, as stated previously, the as-welded properties are very acceptable.

Modified 9 Cr-1 Mo and 2.25 Cr-1 Mo electron beam weld properties compare very favorably with those of other welding processes. The weld tensile strength exceeds that of the base metal for both alloys. The Charpy V-notch impact properties of modified 9 Cr-1 Mo electron beam weld metal fall in the range of data from gas tungsten arc weldments.[8] These have lower transition temperatures than welds produced by the submerged arc or shielded metal arc processes. The 2.25 Cr-1 Mo electron beam weld impact properties are similar to those of submerged arc deposits in this steel which were reported by Swindeman et al.[10]

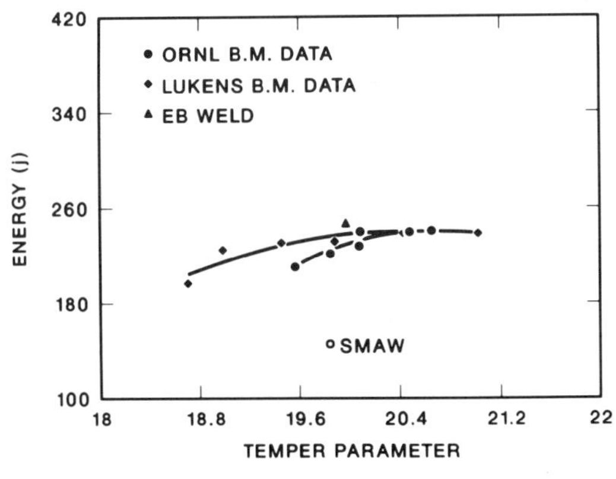

Fig. 13 - Charpy V-notch upper shelf energy of electron beam weld metal is superior to that of shielded metal arc weld metal and near that of 3 Cr-1.5 Mo-0.1 V base metal.

CONCLUSIONS

As a result of this study, several conclusions can be made. It has been shown that single-pass, full-penetration electron beam weldments of high integrity can be made in 102-mm thick plates of 3 Cr-1.5 Mo-0.1 V and 2.25 Cr-1 Mo steels. Weldments in modified 9 Cr-1 Mo steel contained large porosity presumably from its relatively high nitrogen content. Nitrogen content has been shown to affect the occurrence of porosity in other steels.[11] Welding parameter variations may reduce or eliminate this porosity but it was not evaluated in this study. The weld metal produced by the electron beam process has a slightly higher hardness and tensile strength than the base metal for these steels after their respective PWHT. Charpy impact properties of the weld metal in the 3 Cr-1.5 Mo-0.1 V steel were found to match those of the base metal and were better than the results obtained from weldments produced by the SAW and SMAW processes. Electron beam weldments in 3 Cr-1.5 Mo-0.1 V and 2.25 Cr-1 Mo steels readily pass the bend tests of Section IX of the ASME Code. These results indicate that electron beam welding offers an attractive alternative for heavy section steel joining.

ACKNOWLEDGMENTS

The authors are grateful to R. W. Reed, R. H. Baldwin, and T. D. Owings for their technical assistance in testing, C. W. Houck for metallography, to M. L. Santella and R. W. Swindeman for reviewing the manuscript, and to K. Gardner for preparing the manuscript. The research was sponsored by the Division of Materials Sciences, U.S. Department of Energy, under contract DE-AC05-84OR21400 with Martin Marietta Energy Systems, Inc.

REFERENCES

1. Wada, T. and T. B. Cox, "3 Cr-1.5 Mo Steel for Pressure Vessels in Hydrogen Service," pp. 111-21 in Advanced Materials for Pressure Vessel Service with Hydrogen at High Temperatures and Pressures, MPC-18, ed. M. Semchyshen, American Society of Mechanical Engineers, New York, 1982.
2. Wada, T. P. J. Grobner, D. L. Sponseller, and N. B. Heckler, "Hydrogen-Resistant Steel for Service in Thick-Walled Pressure Vessels at Elevated Temperatures," pp. 81-98 in Applications of Materials for Pressure Vessels and Piping, MPC-10, American Society of Mechanical Engineers, New York, 1979.
3. Wada, T. and T. B. Cox, "A New 3 Cr-1.5 Mo Steel for Pressure Vessel Applications," pp. 77-93 in Research on Chrome-Moly Steels, MPC-21, American Society of Mechanical Engineers, New York, 1984.
4. Swindeman, R. W., R. D. Thomas, R. K. Nanstad, and C. J. Long, Assessment of the Need for an Advanced, High-Strength Chromium-Molybdenum Steel for Construction of Third Generation Gasifier Pressure Vessels, ORNL/TM-8873, October 1984.
5. Nanstad, R. K., Characterization of Four Prestressed Concrete Reactor Vessel Liner Steels, ORNL/TM-7480, December 1980.
6. Standard Methods for Notched Bar Impact Testing of Metallic Materials, ASTM E23-72 Part 10, American Society for Testing and Materials, Philadelphia, 1978.
7. Klueh, R. L. and A. M. Nasreldin, Microstructure and Mechanical Properties of a 3 Cr-1.5 Mo Steel, ORNL-6206, September 1985.
8. Sikka, V. K. and P. Patriarca, Analysis of Weldment Mechanical Properties of Modified 9 Cr-1 Mo Steel, ORNL/TM-9045, May 1984.
9. Swindeman, R. W., J. F. King, and R. K. Nanstad, Final Report on 3 Cr-1.5 Mo Steels for Gasifier Pressure Vessel Construction, ORNL/TM-9731, September 1985.
10. Swindeman, R. W., R. K. Nanstad, J. F. King, and W. J. Stelzman, "Effect of Tempering on the Strength and Toughness of 2 1/4 Cr-1 Mo Steel Weldments, ORNL/TM-9307, October 1984.
11. Shimoyama, T., Y. Urantani, G. Takano, and S. Kuri, "Study on Weldability of Electron Beam Welding for Thick Plate of Austenitic Stainless Steel (Report 1)," Trans. of the Japan Welding Society, 14(2), October 1983.

ELECTRON BEAM WELDABILITY OF A RAPIDLY SOLIDIFIED ALUMINUM ALLOY

W.A. Baeslack III, S. Krishnaswamy
Department of Welding Engineering
The Ohio State University
190 W. 19th Avenue
Columbus, OH 43210

ABSTRACT

The electron beam weldability of a rapidly solidified/powder metallurgy Al-8Fe-2Mo alloy has been studied, with the influence of electron beam energy input on the weld zone structure, mechanical properties and fracture behavior specifically investigated. Results indicate that higher weld cooling rates associated with lower energy inputs promote the formation of finer fusion zone primary intermetallic and dendritic alpha structures. These finer structures promote increased weld fusion zone hardness, strength and ductility. The presence of coarse intermetallics near the fusion boundary in low energy input welds and throughout the higher energy input welds promotes lower strength levels and ductility. These reduced properties are attributed to fracture of the intermetallics or delamination of the intermetallic/alpha matrix interfaces.

RECENTLY, THE APPLICATION OF RAPID SOLIDIFICATION/POWDER METALLURGY (RS/PM) technology has led to the development of a new generation of aluminum alloys. These alloys are basically hypereutectic Al-Fe compositions with ternary additions of Mo, Ce or Ni. Through rapid solidification, the formation of equilibrium primary intermetallic phases can be suppressed, with transformation instead occurring to a microeutectic or solute supersaturated dendritic alpha. Subsequent consolidation and thermomechanical processing promote decomposition of these highly metastable structures by the formation of semi-coherent dispersoids which are structurally stable for long times at temperatures up to 230°C. The moderate to high room-temperature strengths provided by these fine microstructures, combined with excellent elevated temperature properties, make these alloys ideal candidates for a variety of structural applications.

If these alloys are to be utilized most effectively, the development of suitable joining techniques will be required. Considering the strong dependency of Al-Fe alloy properties on the metastable nature of the microstructure, it is apparent that the microstructure must be retained and/or recreated during joining if high joint efficiences are to be obtained. A previous weldability study of an Al-8Fe-1.7Ni alloy (1) showed that the rapid solidification and cooling rates associated with electron beam welding offer a strong potential for achieving these requirements. This earlier work, however, was limited by a high base alloy gas content and the resulting formation of excessive fusion zone porosity. The objective of the present study was to extend this initial work through characterization of the structure, properties and fracture behavior of electron beam welds in a low hydrogen Al-8Fe-2Mo RS/PM alloy.

EXPERIMENTAL

The Al-8Fe-2Mo alloy studied in this investigation originated as rapidly-solidified powder produced by Pratt and Whitney's "RSR" rotary atomization process (2,3). This process utilizes a rapidly spinning disk to produce powder by centrifugal atomization and helium forced convective cooling to provide droplet cooling rates between 10^4 and 10^6 °C/s. Subsequent inert compaction of the powder and thermomechanical processing provided sheet 1.27 mm in thickness with a chemistry of 8.0 wt%Fe, 2.3 wt%Mo, and less than 1 ppm hydrogen.

Autogenous, full-penetration electron beam welds were produced on the sheet using a range of energy inputs from 15 to 27.6 J/mm. (Specific welding parameters are provided in Table 1.) Following radiographic inspection, the welded coupons were sectioned, mounted, polished and etched (Keller's etch) for metallographic examination using light and scanning-electron microscopy and hardness testing (DPH, 500 g load) Additional coupons were machined into longitudinal and transverse weld tensile specimens

Table 1 - Electron Beam Welding Parameters[1] and Weldment Characteristics

Weld #	Travel speed mm/sec(in/min)	Energy input J/mm(J/in)	DPH at weld center	Threshold bending strain %	UTS MPA (ksi) Trans.	Long.	EL % Trans.	Long
1	12.7(30)	27.6(701)	110	3.3	--	--		
2	14.8(35)	23.6(600)	116	3.3	--	--		
3	16.9(40)	23.6(600)	116	4	330(46.5)	445(64.2)	1.6	5.8
4	16.9(40)	20.7(526)	119	4	--	--		
5	23.3(55)	19.3(490)	120	4	370(53.5)	470(67.7)	5.6	10.0
6	19.1(45)	18.4(467)	121	4	--	--		
7	21.2(50)	16.5(419)	126	4	--	--		
8	23.3(55)	15(381)	130	5	--	--		

[1]Voltage - 100 KV, current 3.5 ma except for #3 (4.0ma) and #5 (4.5ma)

(gage length = 12.7mm) and longitudinal weld bend specimens. Finally, scanning-electron microscopy (SEM) and energy dispersive X-ray analysis (EDX) were employed for fractography of bend and tensile specimens.

RESULTS AND DISCUSSION

WELD INTEGRITY - Previous studies on the fusion welding of RS/PM aluminum alloys have shown hydrogen-induced porosity to be a significant problem (4). The source of hydrogen is the hydration of aluminum oxide on powder surfaces during and subsequent to atomization. The oxide thickness and the degree of hydration are dependent on the atomization process (i.e. air or inert gas) and the manner of powder storage, handling and consolidation. Although hydrogen in the consolidated product can be reduced by hot vacuum degassing prior to final consolidation, such degassing temperatures for the Al-Fe alloys are limited by the coarsening and softening of the metastable dispersoid microstructures. In the present study, total inert atomization, handling and processing of the powder minimized hydration and provided a base alloy with below 1 ppm hydrogen. Correspondingly, radiographic examination of the weldments and metallographic study found minimal evidence of porosity.

WELD STRUCTURE - Characterization of the electron beam weld macrostructure using light microscopy (Fig. 1A) showed two distinct regions within the fusion zone: a dark-etching central region and a light-etching region adjacent to the fusion boundary. Examination of these structures at increased magnification using SEM (Fig. 1B) showed the central region to consist of extremely fine, nearly equiaxed primary intermetallics (Al_3Fe type) in a matrix of fine dendritic alpha (Figs. 2B and D) and the fusion boundary region to consist of much coarser, high aspect ratio intermetallics in a matrix of coarser dendritic alpha (Figs. 2D and D). Although not observed with light or SEM microscopy, previous scanning-transmission electron microscopy studies(4) suggest the presence of fine intermetallic at dendrite boundaries resulting

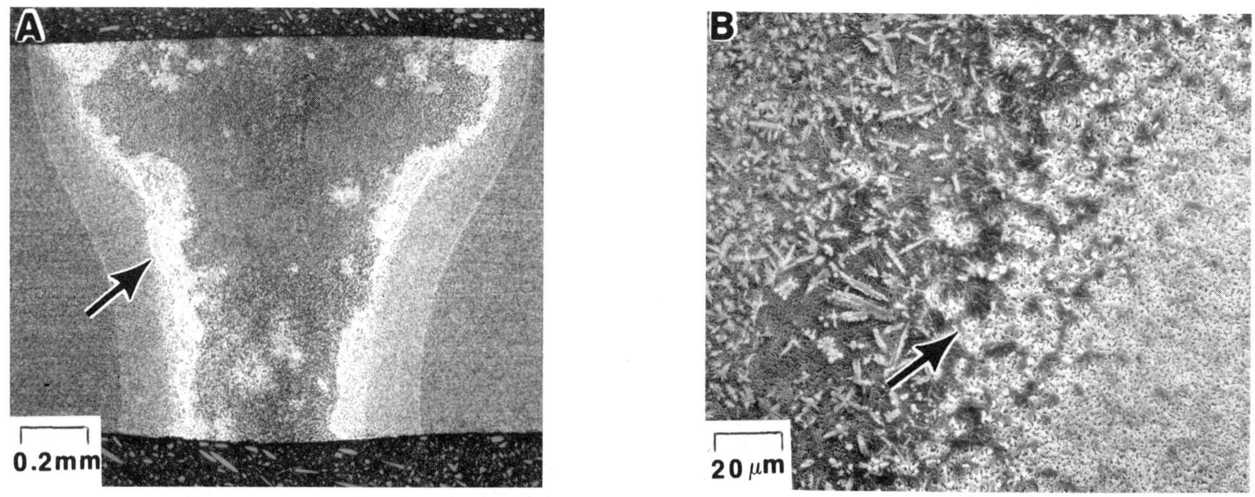

Fig. 1 - Light macrograph (A) and SEM micrograph (B) of electron beam weld produced with energy input of 15 J/mm. Arrows indicate fusion boundary.

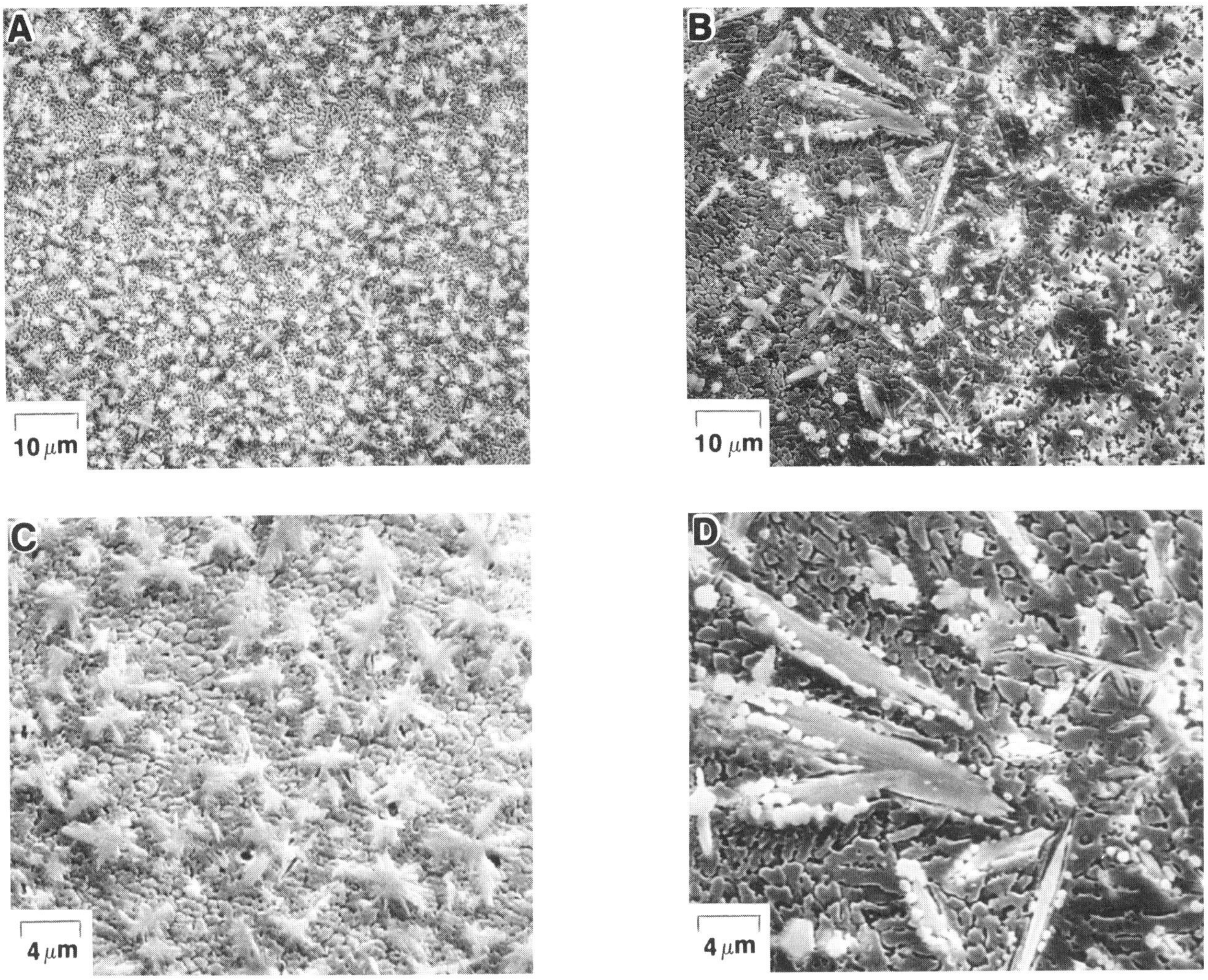

Fig. 2 - SEM micrographs of region in Fig. 1B at increased magnification showing electron beam weld fusion zone center (A,C) and fusion boundary region (B,D)

from eutectic decomposition of the final liquid to solidify. An increase in the weld energy input was shown to promote an increase in the proportion of light to dark-etching regions and also in the microstructure coarseness of both regions. Correlation of dendrite spacings in the central fusion zone regions with previous dendrite spacing/cooling rate relationship data for aluminum alloys (5) indicated cooling rates of between 10^3 and 10^4 °C/s, which were consistent with cooling rates calculated using an analytical 2D heat flow model. Despite these moderately rapid cooling rates, it was not possible to suppress intermetallic nucleation even for the lowest energy inputs in the present study.

Solidification of the electron beam welds occurred primarily by the equiaxed growth of fine alpha aluminum dendrites from intermetallic nuclei. Considering the melting point differences between the Al_3Fe intermetallic (≃ 1150°C) and the alpha aluminum matrix (≃ 650°C), the possibility that base metal dispersoids remained unmelted and provided these nucleation sites in the fusion zone becomes apparent. The potential for this phenomena was examined analytically using particle dissolution equations (6,7) and experimentally by producing identical electron beam welds on base metals exhibiting a wide range of initial dispersoid sizes. Although details of these analyses are not presented in this paper, results presently suggest that the base metal dispersoids were completely dissolved in the central fusion zone region during welding.

In contrast, dissolution near the fusion boundary was found to be incomplete (Fig. 1B). In this region, the dispersoids likely served as growth centers for the coarser intermetallic particles observed and possibly promoted the stagnation of fluid flow. Such stagnation may explain the slower cooling rates experienced in this region versus the weld center, as indicated by coarser intermetallic and alpha dendrite sizes.

It is of interest to note that the coarse fusion boundary structure contrasted previous results on Al-8Fe-1.7Ni (1), which found an absence of primary intermetallic formation at the fusion line (epitaxial solidification to microeutectic) and a gradual increase in the quantity and size of intermetallics toward the weld centerline. This structure was attributed to the steep temperature gradient and rapid cooling rates experienced near the fusion boundary of these high depth-to-width ratio electron beam welds which inhibited nucleation and growth of the intermetallics. A possible explanation for the differences observed in the present study is that the electron beam welds on the thin Al-8Fe-2Mo sheet exhibited much shallower temperature gradients, which was also suggested by the observation of nearly entirely non-epitaxial, equiaxed grain growth in the fusion zone. Indeed, high depth-to-width ratio electron beam welds produced on 12.5 mm thick Al-8Fe-2Mo plate have shown a structure more comparable to that of the Al-8Fe-1.7Ni welds (4).

Also evident in Fig. 1A is the presence of a heat-affected zone adjacent to the fusion boundary. SEM analysis of this region found a noticable coarsening of the dispersoid structure as compared to the unaffected base metal.

WELD PROPERTIES - Results of mechanical property tests are summarized in Table 1 and Figures 3 and 4.

Figure 3 shows a hardness traverse across an electron beam weld produced at the lowest energy input (15 J/mm). As indicated, the fine intermetallic/alpha dendrite structure in the center of the fusion zone exhibited a hardness comparable to the unaffected base metal (DPH 130 versus 126). However, in the weld HAZ and more significantly in the fusion zone adjacent to the fusion boundary, the hardness dropped appreciably which was consistent with the coarser intermetallic and dendrite structures observed in this region. As Table 1 indicates, the hardness of the fine central fusion zone regions increased with decreasing weld energy input and increasing cooling rate, which corresponded to the increase in microstructure fineness.

The tensile strengths of the longitudinal weld tensile specimens were found to increase with increased cooling rate (Fig. 4), with the more rapidly cooled specimen exhibiting a tensile strength approximately 90% of the typical base metal strength for this alloy (3). Consistent with the hardness traverses, transverse weld tensile specimens failed at lower strength levels through the coarse fusion boundary region in the weld fusion zone. However, weld efficiency values for the more rapidly cooled welds exceeded 73%, which was relatively high for aluminum alloy weldments.

Table 1 summarizes bend ductility test results, and shows an improvement in ductility with increased cooling rate. It is important to note that crack initiation in these specimens occurred exclusively in the coarse fusion zone microstructure directly adjacent to the fusion line. Tensile ductilities for both longitudinal and transverse specimens were also higher for the more rapidly cooled welds.

FRACTURE CHARACTERISTICS - Figure 5 illustrates the longitudinal (A and B) and transverse (C and D) tensile specimen fracture surfaces for a moderate energy input electron beam beam weldment (E in = 23.6 Kj/mm). Differences in the fracture behavior between the unaffected base metal and the fusion zone were visible even at low magnification (Fig. 5A), and ranged from a flat topography in the base metal to a much more irregular topography in the fusion zone. Figure 5B reveals that at higher magnification three different fracture surfaces could be observed on the longitudinal weld specimens. The unaffected base metal exhibited a ductile-appearing, dimpled fracture which was associated with the fine dispersoid structure in this region

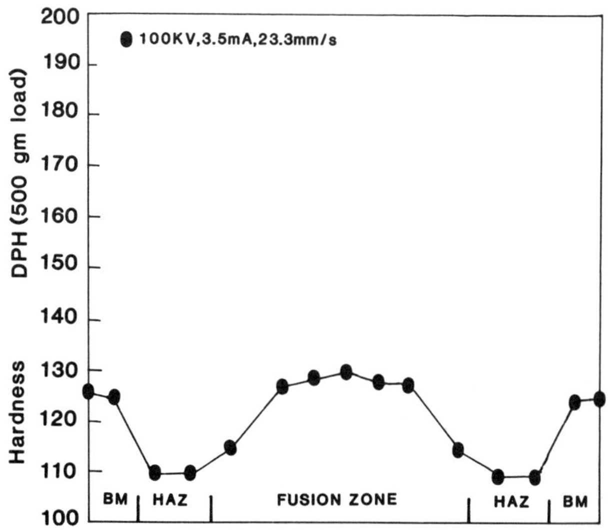

Fig. 3 - Diamond pyramid hardness (DPH) traverse across electron beam weld (E_{in} = 15 J/mm).

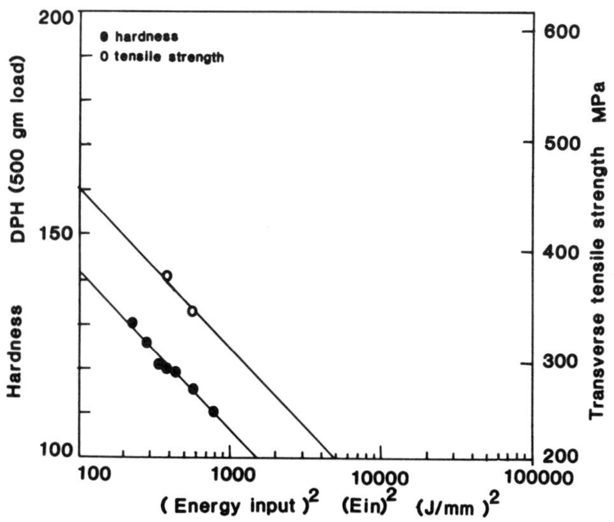

Fig. 4 - Diamond pyramid hardness at center of weld fusion zone and transverse weld tensile strength versus electron beam weld energy input.

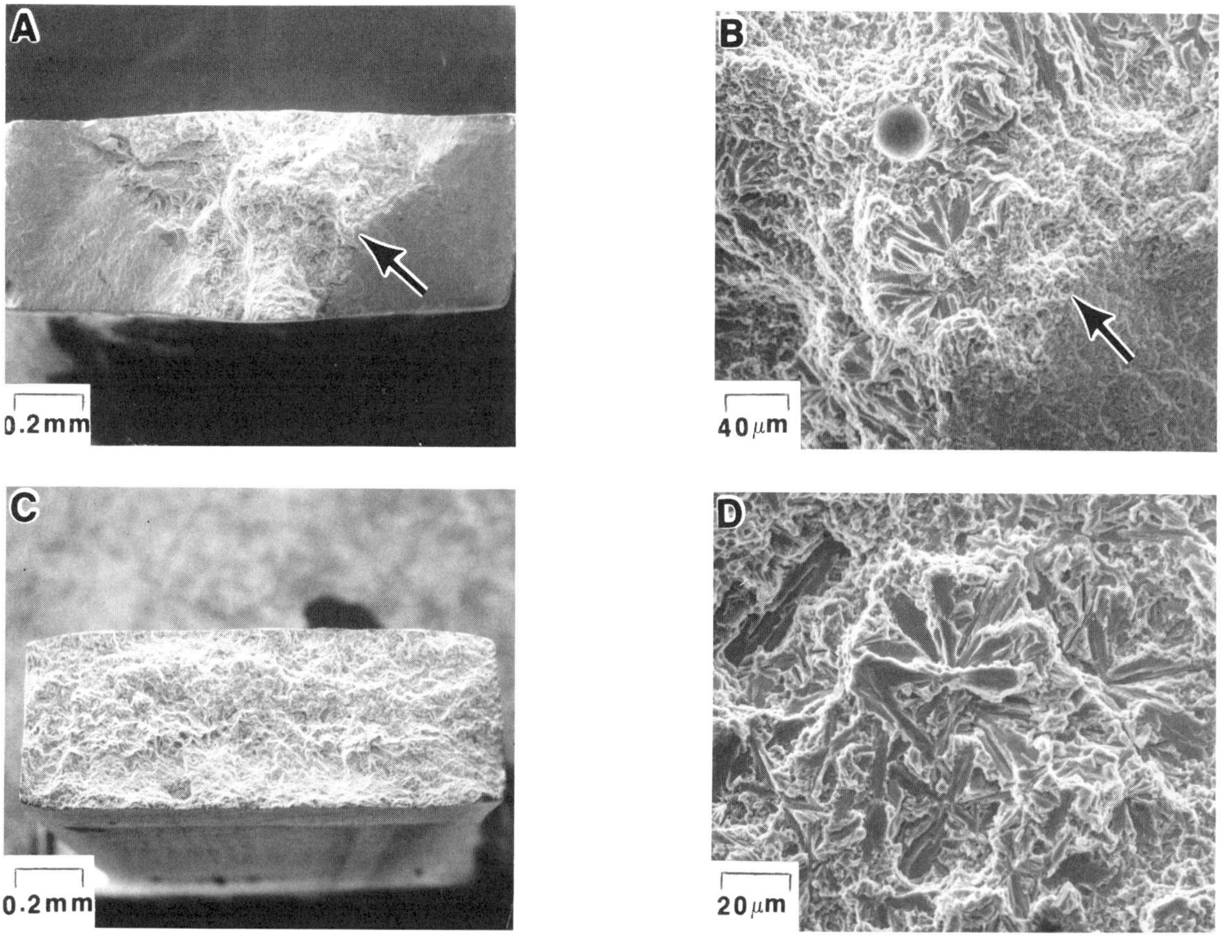

Fig. 5 - Fracture surfaces of longitudinal (A,B) and transverse (C,D) weld tensile specimens (E_{in} = 19.3 J/mm). Arrows in (A) and (B) indicate approximate fusion boundary.

(Fig. 6A and D). In the central fusion zone, where the intermetallic structure was coarser, considerable ductile-appearing fracture was again observed (Figs. 6B and E). However, SEM/EDX analysis also revealed the presence of intermetallics on the surface, suggesting possible local fracture by delamination of the alpha/intermetallic interfaces. Evidence of intermetallic fracture and delamination was most apparent in the fusion zone adjacent to the fusion boundary, where the coarse intermetallic structure promoted low ductility fracture (Figs. 6C and F). Relatively little evidence of ductile fracture was observed in this region. In the transverse weld tensile specimens, the entire fracture surface exhibited this third type of fracture, since failure occurred entirely in the fusion zone adjacent the fusion boundary.(Figs. 5C and D).

SUMMARY

The results described above clearly show the importance of maintaining a fine intermetallic/alpha aluminum dendrite structure in providing optimum weld strength and ductility. It is also evident that the weld fusion zone adjacent to the fusion boundary represents the "weak link" in the weldment, and that if properties are to be improved this region must be eliminated. Continuing work on the electron beam welding of the Al-8Fe-2Mo alloy sheet and plate is optimizing parameters such that depth-to-width ratios are increased and optimized. The steeper temperature gradients and more rapid cooling rates associated with these welds will promote finer fusion zone structures and potentially suppress primary intermetallic formation, thereby providing enhanced weld properties.

REFERENCES

(1) Baeslack, W.A. III, Metallography, 18, 73-82 (1985).

(2) Adam, C.M. and Bourdeau, R.G., "Rapid Solidification Processing, Principles and Technologies II," p. 246, Mehrabian, R., Kear, B.H. and Cohen, M., Eds., Claitors Publishing, Baton Rouge, Louisiana (1980).

(3) Adam, C.M., "Rapidly Solidified Amorphous and Crystalline Alloys," p. 411, Kear, B.H., Giessen, B.C. and Cohen, M., Eds., Elsevier Science Publishing Co., New York (1982).

(4) Baeslack, W.A. III, Unpublished Research, The Ohio State University, Columbus, OH 43210, 1982 - 1986.

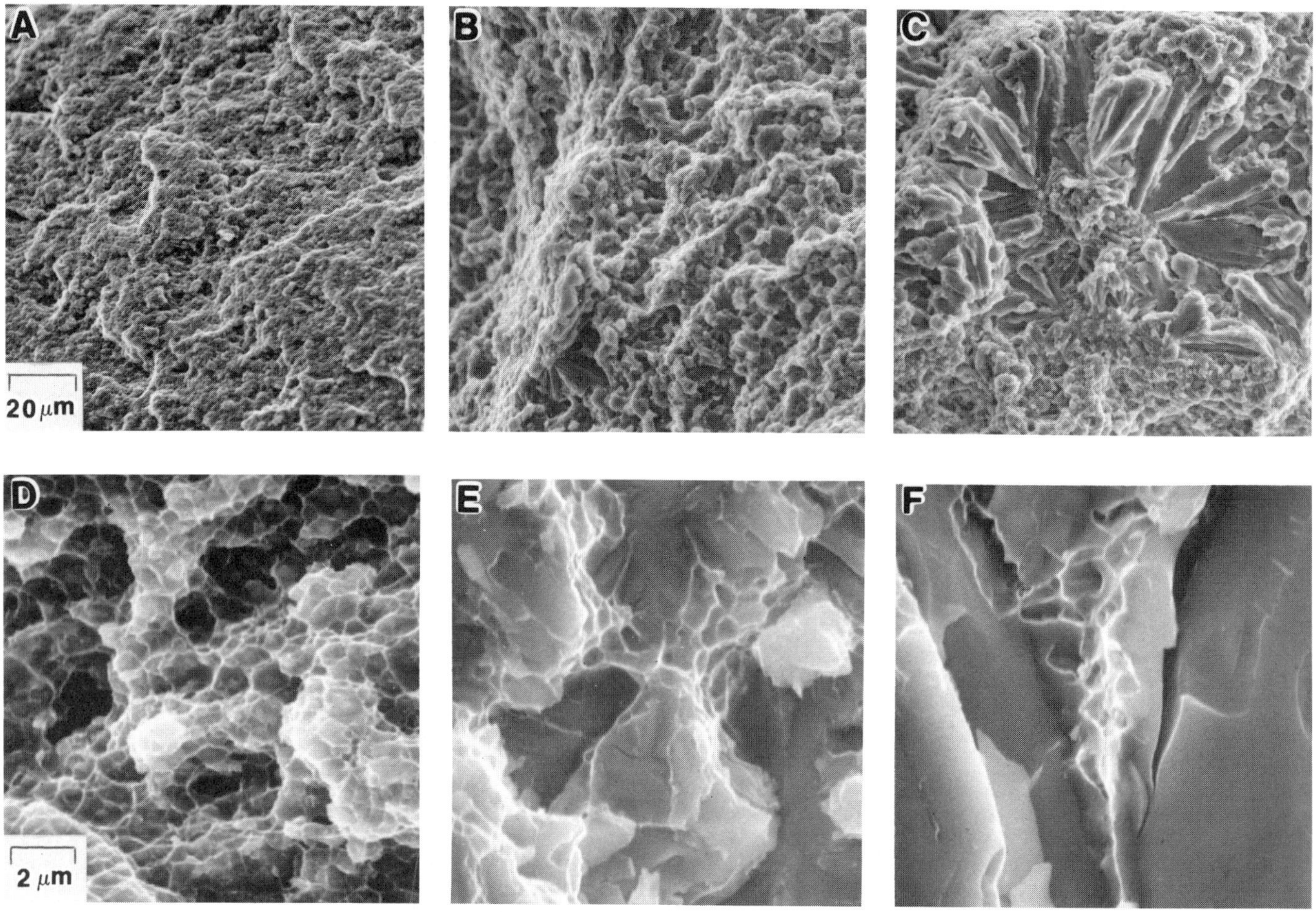

Fig. 6 - Fractographs of longitudinal weld tensile specimen in Fig. 5B at increased magnification showing unaffected base metal (A,D), fusion zone center (B,E) and fusion zone adjacent to the fusion boundary (C,F).

REFERENCES (continued)

(5) Jones, H., "Rapid Solidification of Metals and Alloys," p. 42, Institution of Metallurgists, London (1982).
(6) Doremus, R.H., "Rates of Phase Transformations," p. 18, Academic Press Inc., Orlando, Florida (1985).
(7) Geiger, G.H. and Poirier, D.R., "Transport Phenomena in Metallurgy," p. 541-542, Addison Wesley Publishing Co., New York (1973).

ACKNOWLEDGEMENT

The authors express their gratitute to Dr. Andrew Crowson and the Army Research Office for their financial support. In addition, we are indebted to the Materials Laboratory/Air Force Wright Aeronautical Laboratories for providing the alloy studied in this investigation.

ELECTROCHEMICAL EFFECTS ON WELD POOL CHEMISTRY IN SUBMERGED ARC AND D.C. ELECTROSLAG WELDING

M. Blander
Argonne National Laboratory
Argonne, Illinois

D.L. Olson
Colorado School of Mines
Golden, Colorado

ABSTRACT

Electrochemical reactions could be an important factor governing the chemistry of weld pools in D. C. welding. The anodic reaction at the weld wire-slag interface leads to a relatively high P_{O_2} which leads to the formation of an oxide

$$nO^{2-} + M(metal) \rightarrow MO_n + 2ne \quad (1)$$

where M is a metal at the weld wire-slag interface and n is related to the valence of M in the oxide. After the molten weld wire forms a droplet which separates from the wire, the electrochemical reaction ceases and the oxide dissolves in the flux. The cathodic reactions at the weld pool lead to the electrodeposition of metals

$$M^{2+}(slag) + 2e \rightarrow M(metal) \quad (2)$$

$$Si^{4+}(slag) + 4e \rightarrow Si(metal)$$

where M can be Fe^{2+}, Mn^{2+}, or other metals which are electrodeposited at the interface. The relative amounts of all these deposits will be readjusted by chemical reactions such as

$$Si(metal) + 2MO(slag) \rightarrow SiO_2(slag) + 2M(metal) \quad (3)$$

$$Mn(metal) + FeO(slag) \rightarrow MnO(slag) + Fe(metal)$$

The resultant changes in the compositions of the weld metal and the slag depend on the rates of the electrochemical reactions Eq. (2) relative to the rates of the back reactions Eq. (3). If the proposed mechanism is correct, experimental data would indicate that both the electrochemical and back reactions appear to be important. Analogous electrochemical reactions can occur at metal-plasma interfaces.

THE CONTROL OF WELD METAL COMPOSITION is important for producing high quality welds and requires an understanding of the fundamental physicochemical mechanisms for altering metal chemistry. Some of the possible mechanisms we have proposed[1,2] involve partial diffusion control of the chemistry at a slag-metal interface, nucleation constraints on reaction rates, enrichment of alloying elements upon melting and subsequent freezing and the use of precise thermodynamic analyses of exchange reactions and of the chemical driving forces. In this paper we discuss electrochemical effects on weld pool chemistry in D.C. welding. Electrochemical effects can occur at the interfaces of the slag or plasma with weld metal or weld wire. The relative transference numbers of ions and electrons in the slag or plasma partly govern the importance of these effects which could be major factors controlling weld chemistry. Experimental investigations are planned to determine the relative importance of electrochemical reactions in altering metal compositions in electroslag and submerged arc welding.

ELECTROCHEMICAL MECHANISM

When liquid slag has an interface with the weld metal and/or the weld wire, the electrochemical reactions are relatively simple to understand. This type of interface is present in electroslag welding and is present, at least in part, in submerged arc welding. Current densities in submerged arc welding (SAW) are higher at the weld wire interface than at the weld pool interface so that the likelihood of the presence of a slag-metal interface is higher at the weld pool than at the weld wire. The weld wire is generally anodic and the weld pool is cathodic. If we consider, as an example, welding with SiO_2-FeO-MnO fluxes containing 40 wt% SiO_2,[1,2] then the anodic reaction

$$nO^{2-} + M(metal) \rightarrow MO_n + 2ne \quad (1)$$

(where M is a metal at the weld wire-slag interface and n is related to the valence state of M in the oxide) will lead to a relatively high P_{O_2} at the weld wire. This reaction produces oxides at the interface. For steel M would be largely Fe. Because of the driving force for the reaction of product FeO with less noble metals such as Mn (or Cr or Si), reactions such as

$$FeO + Mn \rightleftarrows MnO + Fe \quad (2)$$

will lower the chemical potentials (concentrations) of metallic Mn (or Cr or Si) at the metal-oxide interface. The manganese (or Cr or Si) in the bulk metal will then diffuse to the interface and react with FeO to form the oxide. Thus, the molten weld wire will preferentially lose the more oxidizable metals to the interfacial oxide phase. After the molten weld wire forms a droplet which separates from the wire, the electrochemical reaction ceases but reactions such as (2) continue. The oxides produced will diffuse into both the metal and the flux. The rate of dissolution will depend on kinetic factors as well as on the driving force which is related to the differences between the chemical potential of the oxide at the interface and in the slag or metal phase. Thus, a slag which is relatively low in iron oxide will be more effective in dissolving the majority FeO product and in minimizing the increase in oxygen content of weld metal. On the other hand, only the product FeO will have any significant tendency to diffuse into the metal because of the very large activity coefficients and low solubilities for MnO (or Cr_2O_3, or SiO_2) in the metal.

The cathodic reactions lead to the electrodeposition of metals.

$$M^{2+}(slag) + 2e \rightarrow M(metal) \quad (3)$$

$$Si^{4+}(slag) + 4e \rightarrow Si(metal)$$

where M^{2+} can be Fe^{2+}, Mn^{2+} or other metal ions in the flux. The products are formed metastably because the voltages used in welding are much larger than the decomposition voltages of all the melt components. Consequently, the relative amounts of all these deposits will be readjusted by chemical reactions such as

$$Si(metal) + 2MO(slag) \rightarrow SiO_2(slag) + 2M(metal) \quad (4)$$

$$Mn(metal) + FeO(slag) \rightarrow MnO(slag) + Fe(metal)$$

Because of the large imposed overpotential used in welding, the deposition reaction will tend to deposit metals with similar diffusion coefficients in the approximate ratio of their equivalent fractions in the slag. The back reaction will shift these quantities by an amount commensurate with the equilibrium constants and the diffusion coefficients. Heat transfer considerations suggest that metal-slag interfacial temperatures are not much above the melting temperature of the metal (~1800 K) where, at equilibrium, there would be a high ratio of the concentrations of MnO to FeO in a slag in contact with a metal which contains *e.g.*, one percent of Mn. For the particular SiO_2-FeO-MnO slag we considered previously[1], the molar ratio of MnO to FeO in the slag at equilibrium is greater than five when the Mn content of the metal is 0.9 mole % . Thus the electrochemical reaction with MnO containing slags at the cathode will usually increase the Mn content of the metal. The degree of back reaction cannot be easily determined experimentally because the anodic oxidation reaction tends to preferentially remove Mn from the weld wire. A similar consideration of the back reaction of cathodically electrodeposited Si leads to the conclusion that the diffusion controlled back reaction (4) will be relatively independent of the MnO-FeO ratio for a fixed content of SiO_2. In addition to the back reaction of cathodically produced metals with slag, deposited metals which are less noble than Fe will react with anodically produced iron oxides on or near the surfaces of weld wire droplets which reach the weld pool. The oxides formed are relatively insoluble in the metal. Thus there is a mechanism for back reactions of electrochemically produced cathodic and anodic materials.

Detailed reaction mechanisms are considerably more difficult to deduce at plasma-metal interfaces. The relative contribution of electrons to total charge transport is likely to be larger in the plasma than in most molten slags so that there is likely to be less of an electrochemical reaction at the interfaces. Nevertheless, since the plasma operates at about atmospheric pressure and is relatively dense, the mean free paths and mobilities of electrons should not be extremely different from those of ions. Consequently, because of overall charge neutrality of the plasma, ions could transport a significant fraction of the current. Of the possible metallic cations, Mn^+ is likely to be more abundant than Fe^+. Oxygen can be present as O^+ or perhaps O^- if, as is likely, electron energies are not higher than the electron affinity of oxygen of about 1.5 eV. Oxygen atoms in the plasma near the weld pool can capture electrons since the kinetic energy of the electrons must be smaller than the electron affinity until they have travelled a fractional distance towards the weld wire commensurate with a voltage drop of at least 1.5 V. Oxycations and oxyanions derived from e.g. SiO, SiO_2 and FeO vapor molecules, may also be present. The ions thus produced would be transported in the electric field between the two electrodes and would participate in electrochemical reactions at the metal-plasma interface. Negatively charged ions will be oxidized at a weld wire anode and positively charged ions will be reduced at the weld pool cathode to deposit Mn, Fe, O and Si at the electrodes. Although the resultant chemical changes of the weld metal due to the plasma undoubtedly parallel those of the slag, the absolute and relative reaction rates for different elements are different in a manner which can only be guessed at present. Our expectations are that relatively more Mn than Fe will be deposited at the cathode and that oxygen can be deposited anodically or cathodically from a plasma. The deposition of silicon will be controlled by the concentrations of particular silicon containing species in the plasma. Consequently, electrochemical reactions at plasma-metal interfaces are likely to be qualitatively similar to those at the slag-metal interfaces. However, the transport of electrons in plasmas will lead to a smaller electrochemical effect with plasmas than with slags.

CONSEQUENCES OF THE MECHANISM

The mechanisms we propose at a SiO_2-FeO-MnO slag-metal interface lead to weld metal compositional changes in oxygen, manganese, and silicon. The oxide produced at the anode should be largely FeO which tends to dissolve away as the molten weld wire droplets fall through the slag. The rate of dissolution will depend on the driving force related to the difference between the activity of FeO at the metal surface (essentially unity) and that in the slag. In experimental results for SiO_2-FeO-MnO slags with 40 wt% SiO_2 and FeO contents ranging from 60 down to 0 wt%, the measured increase in the oxygen content of weld metal went from about 0.12 wt% at 60 wt% FeO to 0.09 wt% at 30 wt% FeO to 0.07 wt% at 0 wt% FeO[1]. The magnitude of these changes is commensurate with the variations of the activities of FeO with composition and with the assumption that some of the oxide which had formed on the weld wire droplets dissolves in the metal.

More direct evidence for the anodic oxidation can be deduced from the measurements of Lau, Weatherly and McLean[3] which indicate high oxygen contents of weld wire tips and weld wire droplets using CaO-Al_2O_3 based fluxes under an argon atmosphere. Reported oxygen contents ranged from 844 to 1943 ppm for the tips and from 1913 to 2987 ppm for the droplets. These quantities were always much larger than the oxygen contents of the weld metal which ranged from 321 to 1039 ppm. The ratios of the oxygen concentrations in weld metal to that in the droplets was always much less than the base plate dilution factor (0.5). This suggests the possibility of a reaction of oxides in or at the surface of the droplets with metals less noble than Fe (e.g. Ca, Al, Si in this case) which had formed electrochemically at the cathode interface. The oxides thus formed are insoluble in the metal but can dissolve in the flux. Similar reductions of droplet oxides may have also taken place in the experiments with the SiO_2-FeO-MnO flux discussed above.

The increments in manganese contents of weld metal with the SiO_2-FeO-MnO slags range from −0.26 at 60 wt% FeO to about 0.17 at 0 wt% FeO in the slag with a neutral point (which has a zero increment) at 23 wt% FeO[1]. With 0.9 wt% Mn in weld metal, the electrochemical effects will increase the Mn content of the weld metal for all concentrations of FeO less than about 59.5 wt% FeO by an amount which is approximately linear with the MnO content of the slag. The back reaction (4) will remove much of this Mn. In addition, the reaction with FeO at the weld wire and weld wire droplets will lead to a preferential loss of Mn from the wire and the weld pool by oxidation and subsequent dissolution in the slag. The result is a combination of these positive and negative increments in the manganese contents which is consistent with the observations.

A similar argument can be made for silicon. The electrochemical cathodic deposition will tend to increase the Si content of weld metal and the back reaction at the cathode as well as anodic oxidation of silicon in the weld wire will tend to decrease the silicon content of weld metal. Since, for a constant silica content of the slag, the kinetic factors do not vary greatly with the MnO/FeO ratio one expects an increment which is essentially constant. The experimental results indicate an essentially constant increment averaging about +0.08 wt% Si.

Thus we see that electrochemical effects at slag-metal interfaces can be correlated with measured compositional changes. Because electrons carry part of the current in plasmas, the total electrochemical effects at plasma-metal interfaces are likely to be smaller overall and to result largely from reactions at the cathode. This stems from the presence of more positively charged than negatively charged ions in plasmas. Manganese metal is much more volatile than iron and the volatility of manganese relative to iron from the slag is likely to be greater than unity albeit much less than from the metal. These vapor species partly ionize to form Mn^+ and Fe^+. Consequently, electrochemical effects in the plasma, if significant, should usually lead to positive increments in the manganese content of weld metal. In addition, the presence of oxygen and metallic oxyspecies such as SiO will produce some O^+ and SiO^+ species which will deposit some oxygen and silicon in weld metal. If there is a significant concentration of O^-, oxygen will also be deposited at the weld wire anode. The resultant increments in composition are difficult to quantify at present but are likely to be smaller than similar increments at a slag-metal interface.

CONCLUSIONS

Compositional changes in weld metal in SAW experiments with SiO_2-FeO-MnO slags as well as with CaO-Al_2O_3 based slags are consistent with electrochemical mechanisms involving anodic metal oxidation and cathodic metal ion reduction and deposition at a slag-metal interface. Back reactions of product metals with the slag and with anodically produced oxide and the subsequent dissolution of product oxides in the slag moderate the total effect. Electrochemical reactions at plasma-metal interfaces should lead to similar chemical effects. Further experiments are needed which define the relative importance of electrochemical effects involving slag and a plasma.

ACKNOWLEDGMENTS

This work was supported by the Office of Naval Research under Navy Order Number N00014-85-F-0097 with the Argonne National Laboratory and by the Army Research Office at the Colorado School of Mines.

REFERENCES

1. J. E. Indacochea, M. Blander, N. Christensen, and D. L. Olson, Metall. Trans., **16B** 237-245 (1985)

2. M. Blander and D. L. Olson, "Thermodynamic and Kinetic Factors in the Pyrochemistry of Submerged Arc Flux Welding of Iron Based Alloys", in Second International Symposium on Metallurgical Slags and Fluxes, eds., H. A. Fine and D. R. Gaskell, TMS-AIME, Warrendale, PA (1984) pps 271-277

3. T. Lau, G. C. Weatherly and A. McLean, Welding Journal, **64**(12) 343s-347s (1985)

A COMPARISON BETWEEN THE SPOT WELDING AND WELDBONDING OF ALUMINUM ALLOYS

Ahmed H. ElSawy, Adel S. Salama
Mechanical Design & Production Department
Faculty of Engineering
Cairo University
Giza, Egypt

Zakaria B. Nan
Helwan Plane Factory
Helwan, Cairo, Egypt

ABSTRACT

In the present work, the weldbonding of aluminum alloys has been studied and compared with the spot welding technique. Spot welding of aluminum alloys requires surface preparation by pickling and may be followed by mechanical brushing while in weldbonding, a corrosion resistant film of complex chromate is deposited on the faying brushed surfaces. Acceptable welds in weldbonding are obtained by using a relatively low current values with low slope rates and higher forces. Meanwhile, the tensile shear strength is increased by several folds compared with that for spot welded joints depending on surface preparation and welding conditions.

THE PURPOSE of this paper is to study the parameters and strength of the weldbonded joints of aluminum alloys compared with that of the ordinary spot welded ones. Spot welding and weldbonding of aluminum alloys are of prime importance in aircraft and aerospace industry (1,2). Weldbonding is characterized by a combination of a spot weld and adhesive bond. The adhesive is applied between joint surfaces either before spot welding (weld thru technique) or after spot welding (flow-in technique).

Joint performance is affected by : surface preparation, contact and dynamic resistance, welding current, welding force and welding time. In weldbonding, a stable film is deposited on the faying surfaces of the joint either by anodizing or by using Alodine 1200 process (3-6). Surface roughening prior coating reduces contact angle and consequently improves wettability (7).

Joint defects in spot welding are eliminated by using either a high welding force or a low welding current (8). In weldbonding, it is preferable to apply a high welding force with a relatively low value of welding current of low slope rate due to high electrical resistance of the joint (9-14).

The weldbonded joints has a more uniform stress distribution compared with the spot welded ones. Meanwhile, the maximum stress in spot welded joints shows to be higher than that occurs in the weldbonded joints (15).

EXPERIMENTAL WORK

Sheets of 1.6 mm thickness of aluminum alloy designated as 2017-T3 were used (16). Joint surfaces were pickled and mechanically brushed by a stainless steel wire brush. In weldbonding, this prosess was followed by the application of Alodine 1200 process where a corrosion resistant film of complex aluminum chromate was deposited on the faying surfaces of the joint. The flow-in method is adopted where an adhesive of a commercial name Vary Bond and another one of a commercial name of Super Bonder were applied. The contact resistance was measured by a pickling controller instrument.

A capacitor discharge welder with a water cooled electrode of 18 mm diameter and 75 mm radius of the dome shaped end was used (17). The peak value of the welding current was measured by a toroidal coil. The welding conditions were selected according to the aircraft specifications for resistance spot welding (18). The welding time was kept constant during allover the tests

Joint strength was evaluated by tensile shear test. Spot welded and weldbonded specimens were subjected to radiographical and metallographical examinations for defect detection.

RESULTS AND DISCUSSION

The increase of welding current causes the nugget diameter and penetration to increase in both spot welding and weldbonding, figs. 1,2 . The tensile shear strength is also increased up to a maximum value where heavy expulsion occurs, then it is decreased, fig. 3 .

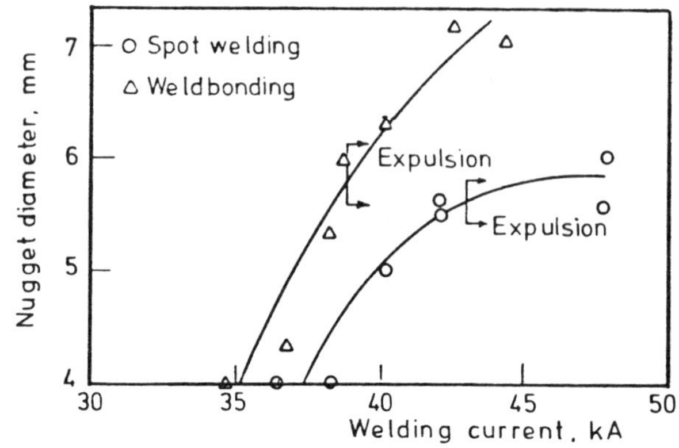

Fig.1- Effect of welding current on nugget diameter.

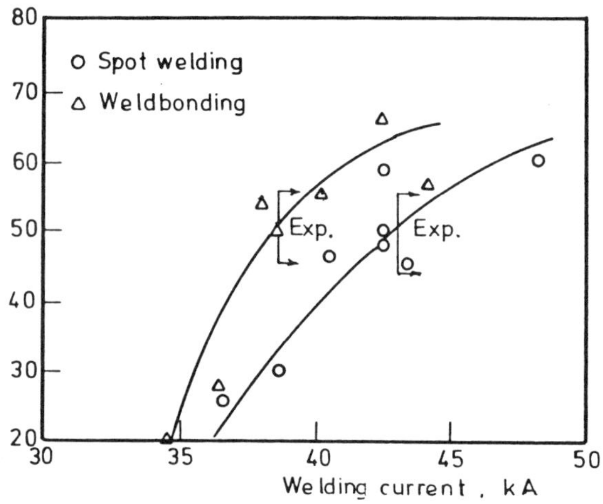

Fig. 2- Effect of welding current on penetration.

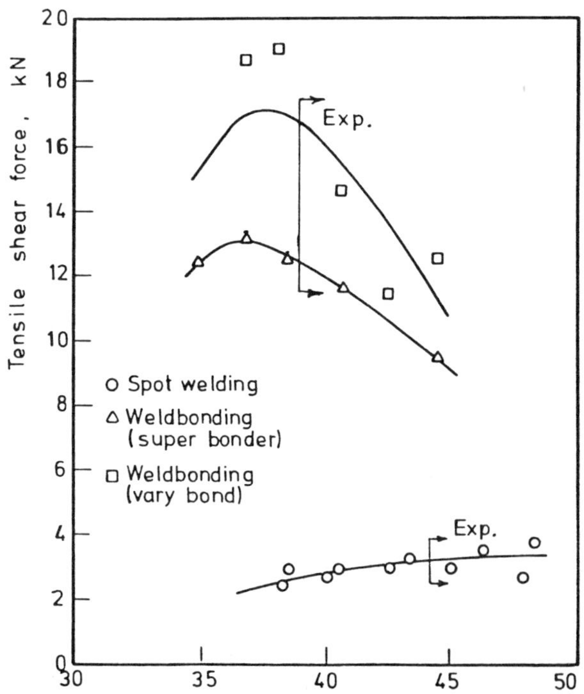

Fig. 3- Effect of welding current on tensile shear force.

This may be attributed to the presence of defects which are formed in case of heavy expulsion. The tensile shear strength of weld-bonded specimens appears to be higher when compared with that of the spot welded ones by several times. This increase may be attributed to the effect of adhesive bond.

The electrode force affects the total resistance between the electrodes and consequently the current flowing through the weld. The effect of welding force on nugget diameter and penetration is shown in fig. 4,5 .

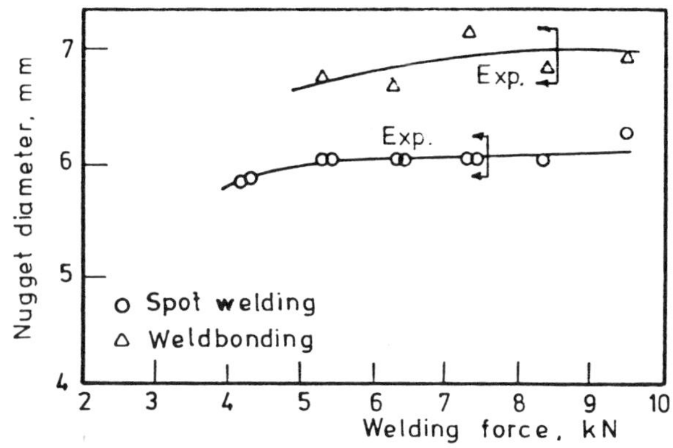

Fig. 4- Effect of welding force on nugget diameter.

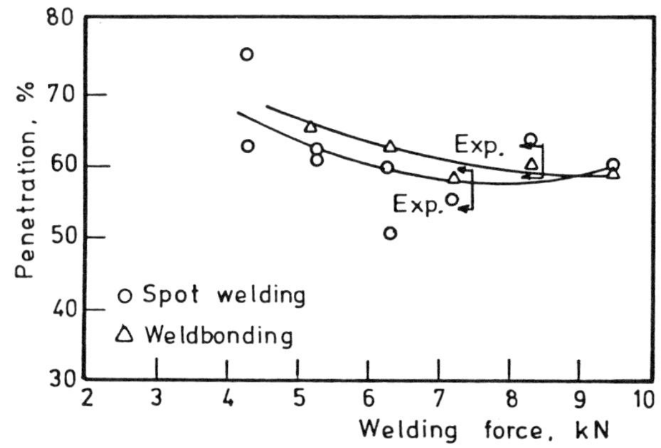

Fig. 5- Effect of welding force on penetration.

Fig. 6. shows the effect of welding force on the tensile shear strength of both spot welded and weldbonded joints. It could be shown that the tensile shear strength in weldbonding is higher than that of the spot welded specimens.

It is noted that the decrease in tensile shear strength as expulsion starts may be due to the decrease in the faying adhesive layer, cracks and porosity formation, decrease in the nugget diameter and decrease in penetration.

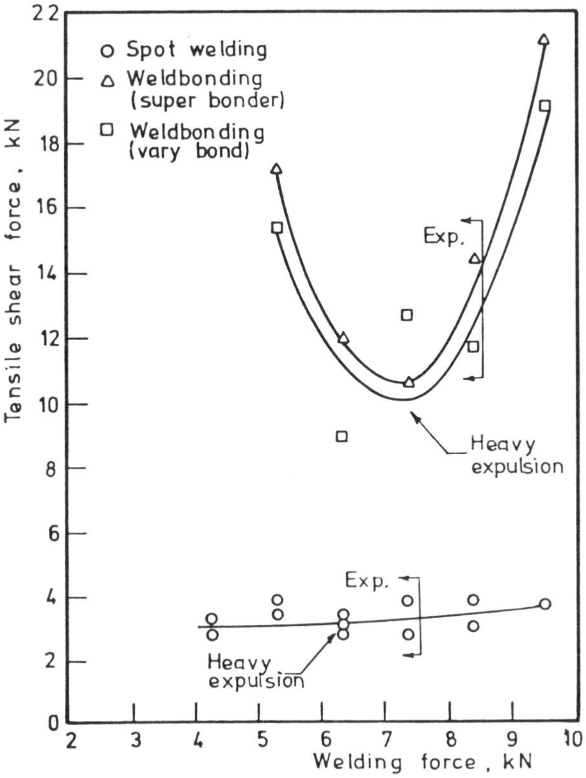

Fig.6-Effect of welding force on tensile shear force.

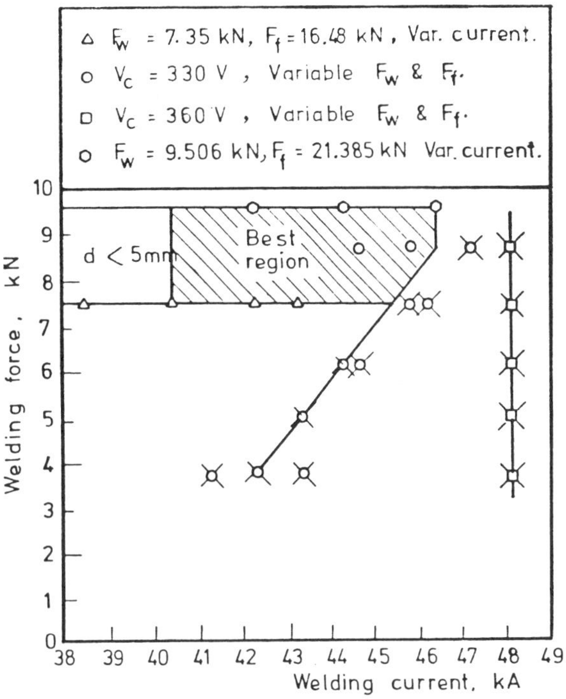

Fig.7- Best welding region for spot welding.
(d: Nugget diameter. – F_f : Forging force)
(F_w: Welding force. – ⨯: Expulsion)

Fig. 7. represents the best region for spot welding of 2017-T3 aluminum alloy. The presence of copper and magnesium alloying elements cause a secondary recrystallization of the phase Cu Mg Al_2 which gives rise of more serious defects due to coarse and hetrogeneous crystalline structure.

Fig. 8. represents the best region for weldbonding of 2017-T3 aluminum alloy using a relatively high welding force and low welding current with low slope rate.

CONCLUSIONS

The following conclusions could be withdrawn:

1. Heavy expulsion is generally accompanied by defects such as cracks and holes due to the secondary recrystallization of Cu Mg Al_2 phase. The defects can be eliminated by increasing the forging force.

2. The increase in welding current causes an increase in nugget diameter and so the tensile shear strength. After certain current value depending on contact resistance, dynamic resistance and applied force, expulsion occurs causing a decrease in shear strength.

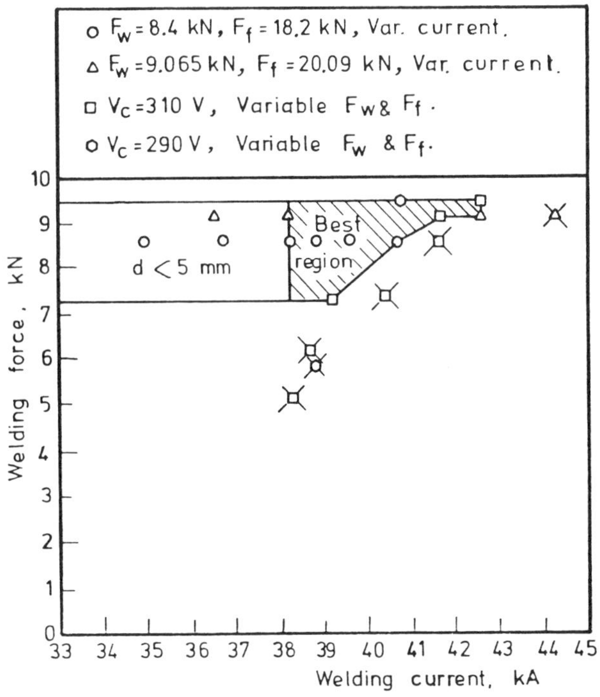

Fig.8 - Best welding region for weldbonding.
(d :Nugget diameter – F_f : Forging force)
(F_w :Welding force – ⨯ : Expulsion)

3. The weldbonded joint is stronger than the spot welded joint and the adhesive bonded one.

REFERENCES

1 . Griffing, L., "Welding Handbook, Metals and their Weldability", A W S (1972)

2 . Reinhart, T.J. Jr., Proc. Int. Conf. on Computer Simulation for Material Applications- Nuclear Metallurgy, 20, 201 - 207 (1976)

3 . Wu, K.C., Weld. J.- Res. Suppl. 55, 238s - 243s (1977)

4 . Thomson Company, "Specifications for Pickling"

5 . Mitchell, R.A., Wooly, R.M. and Baker, S.M., Proc. Int. Conf. on Computer Simulation for Material Applications- Nuclear Metallurgy, 20, 1039 - 1050 (1976)

6 . Marcel Dassault Brequet Aviation, "Documentation d'Avion", I.C.O.4.186

7 . Milner, D.R., Brit. Weld. J. 5, 89 - 96 (1958)

8 . Schwartz, M.M., "Metal Joining Manual", McGraw Hill, New York (1979)

9 . Bhattacharya, S. and Andrews, D.R., Sheet Metal Ind., 400 - 466 (1972)

10. Roberts, W.L., Weld. J. 30, 1004 - 1019 (1951)

11. Savage, W.F., Nippes, E.F. and Wassel, F.A., Weld. J. 57, 43s - 50s (1978)

12. Dikinson, D.W., Franklin, J.E. and Stanya,A., Weld. J. 59, 170s - 176s (1980)

13. Mallya, V.D., Weld. J. 63, 41 - 44 (1984)

14. Kimchi, M., Weld. J. 63, 58s - 63s (1984)

15. Dalley, J.M. and McClaren, S.W., Proc. Int. Conf. on Computer Simulation for Material Applications - Nuclear Metallurgy, 20, 615 - 623 (1976)

16. ElSawy, A. and Messler, R.W. Jr., Technical Report CMP 84-070-01, RPI, New York (1984)

17. Capacitor Discharge Spot Welder, Type MTK - 6301, Technical Description and Service Manual.

18. Nom Air 3383, "Conditions de Controle des Fabrications Soudees par Resistance par Points ou a'la Molette", Cedocar - 6767 - 24 - 473 (1973)

JOINING TECHNIQUES FOR A FERRITIC OXIDE DISPERSION STRENGTHENED ALLOY

Lewis E. Shoemaker
Inco Alloys International, Incorporated
Huntington, West Virginia

ABSTRACT

INCOLOY alloy MA 956 is a mechanically alloyed, oxide dispersion strengthened alloy exhibiting excellent strength and resistance to corrosion at temperatures up to 2300°F (1260°C). Because of the complex nature of the alloy's structure, joining procedures must be carefully designed to produce joints with optimum properties and avoid degradation of the base metal properties. With proper design, fusion welds can be used to join alloy MA 956. Solid state processes are being explored and seem to offer a good potential for success. Brazing also looks quite promising; research programs are underway to develop improved brazing alloys and procedures.

AS OUR TECHNICAL SOCIETY approaches the twenty-first century the material engineer seems to be entering a design revolution. Engines, turbines, and other such equipment run at higher pressures and speeds to increase their fuel efficiency. Materials must be stronger and more heat resistant so they may be used in thinner sections, and at higher stresses and temperatures. Materials, including metals, ceramics, and composites not dreamed of 10 years ago are in use today.

To the metallurgist the challenge of this period is to develop alloys for operation at temperatures well above 2000°F (1100°C). These materials must be capable of exhibiting strength as well as resistance to corrosion at these heretofore considered extreme temperatures. Alloying can only partially meet these requirements. Simple alloys, even those solid solution strengthened and precipitation hardened reach the limits of their use below the requirements of many applications. To increase the usage range of existing alloys, new technological advances are required.

Oxide dispersion strengthening (ODS) is one method that can be employed to extend the usage range of metals. By introducing a small, stable dispersoid into the metal matrix, the high-temperature strength of the materials is increased. The fine dispersoids act to pin dislocation movement much in the manner that second phases do in precipitation hardened alloys. However, where the results of the precipitation reaction are erased by solutioning of the phase, the oxide dispersoids are stable at much higher temperatures. Thus, these alloys exhibit strength at temperatures where conventional alloys cannot operate. Additionally, alloy systems which lack high-temperature strength but exhibit other desired properties (e.g., high-temperature oxidation resistance) can be strengthened for high-temperature service. These alloys combine a high grain aspect ratio, coarse grained microstructure, and a stable oxide dispersoid to produce properties previously unattainable by conventional alloys.

The initial efforts at dispersion strengthening of high-temperature alloys resulted in the invention of thoria dispersion strengthened nickel and nickel-chromium alloys by DuPont. These alloys were produced by a chemical process by which thoria and nickel carbonate were co-precipitated; reduction in hydrogen produced the nickel powder containing a thoria dispersoid. Sheritt Gordon uses a similar process to make a commercial thoria strengthened alloy, "DS Nickel"*. Since these alloys are produced by chemical processing, there is a limit to the number of species

*DS Nickel is a trademark of the Sherritt Gordon Company

allowable in the reaction and, thus, a limit to the number of elements which may be alloyed (1).

The mechanical alloying process developed by International Nickel in the 1960's removed the limitations of the chemical processes for producing ODS alloys (2). In this process, alloyed powders are produced in a high energy driven ball mill which facilitates continuous cold welding, work hardening, and fracturing elemental powders and the oxide dispersoid to form a uniform powder composition; the composition of each particle is equal to that of the bulk. Proper thermal/mechanical processing of these powders produce alloys with unique properties. In this manner alloys with both good strength and resistance to corrosion at high temperature can be produced (3,4).

Optimum utilization of these materials often require joints between sections of the alloy or dissimilar materials. The joints must exhibit properties suitable for the service but must be produced without degrading the properties of the ODS alloy. Excessive heating of the ODS alloy can cause oxide coalescence to the point that the size of the oxide dispersoid is no longer effective in pinning the dislocation (as happens when precipitation hardened alloys are overheated). Melting the ODS alloy results in "slagging off" of the oxide dispersoid so it is no longer present for strengthening. Melting also results in an equiaxed grain structure. Since most ODS alloys derive much of their strength from an elongated grain structure, this disruption in the structure results in reduced properties. Thus, it is seen that the development of procedures for joining these materials offers a unique challenge to the welding engineer.

INCOLOY* alloy MA 956 is a ferritic, ODS superalloy. The nominal composition of the alloy is given in Table 1. The elongated grain structure of the material is seen in Figure 1.

Table 1

Nominal Chemical Composition
INCOLOY alloy MA 956 (%)

Element	(%)
Chromium	20
Aluminum	4.5
Titanium	0.5
Yttria	0.5
Iron	Balance

*INCOLOY is a trademark of the Inco family of companies.

The design of the alloy is quite simple. The iron-chromium-aluminum alloy system is known for its resistance to oxidation and sulfidation at high temperature. However, at the temperatures where this corrosion resistance is needed, the alloy exhibits relatively low strength. By the addition of a yttrium oxide dispersoid by mechanical alloying, a new material, alloy MA 956, is produced. The alloy exhibits the corrosion resistance of the conventional alloy but also excellent strength up to 2300°F (1260°C) (5). The properties of the alloy have made it desirable for many high-temperature applications often requiring joining. Considerable research has and continues to be conducted to define suitable joining procedures.

For the purpose of this paper, the joining processes to be discussed are divided into three categories:

a) fusion

b) solid state

c) brazing

Specific processes under each category will be analyzed as to their applicability for joining the alloy. Experience with each process will be described. In keeping with the spirit of the conference, directions of joining development programs for this type alloy will be described and the potential for success projected.

It is noted that mechanical fastening techniques such as riveting are also used to join alloy MA 956 components. These techniques are relatively well established and, thus, do not require significant research to optimize procedures and properties. Consequently, this type of joining technique will not be included in this discussion.

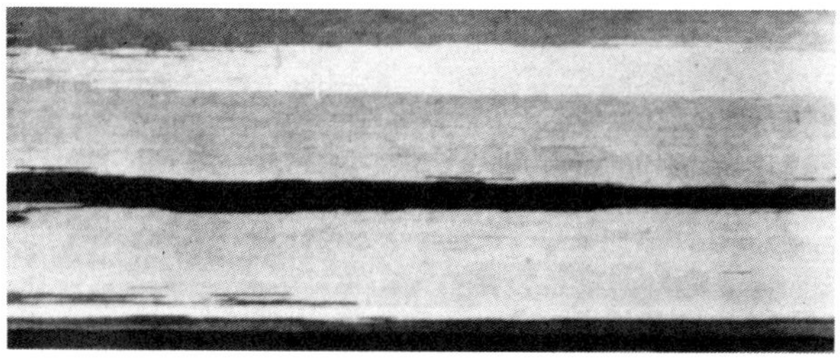

Figure 1. Highly elongated grain structure of INCOLOY alloy MA 956 extruded bar.

FUSION PROCESSES

Fusion processes, that is, processes which result in melting of the base metal, are probably the most widely used joining processes. Consequently, these processes have been widely explored for joining alloy MA 956.

Gas tungsten arc welding (GTAW) is commonly used for joining sheet metal and, thus, is used for joining ODS alloys. Both autogenous and filler metal weldments have been made in alloy MA 956 sheet. Alloy MA 956 fixtures, furnace muffles and tubes, and other equipment for the heat treating industry have been fabricated by GTAW either autogenously or by using an iron, chromium, aluminum, yttrium filler metal, BRIGHTRAY* alloy ICA. Such structures exhibit excellent life and economy as compared to conventional alloy structures. Fabrication and repair are accomplished by procedures similar to their conventional heat resistant alloys. It must be remembered that the alloy is ferritic and, thus, exhibits a ductile-brittle transition temperature. Consequently, a preheat of about 400°F (200°C) is recommended. Slow cooling may be required for complex welded structures to avoid post weld cracking (6). INCONEL* Filler Metal 617 and INCO* Filler Metal HX have also been used for joining the alloy for similar applications.

The cross section of an autogenous GTA weldment in alloy MA 956 is seen in Figure 2. The weldment contains porosity. Also the grain structure of the weldment is equiaxed as compared to the elongated grain structure of the base metal. A weld centerline perpendicular to the direction of the original grain structure is seen. Closer examination has shown that the dispersoid floats out of such weldments while molten ("slagging off"). Thus, such weldments exhibit strength well below that of the base metal (7). In many cases design considerations can account for the lesser strength by placing weldments in areas of lower stress and/or temperature. GTAW is also used for applications in which the alloy's resistance to corrosive attack at high temperatures is required rather than the limits of its strength.

High energy beam processes such as electron beam (EBW) and laser have been evaluated for use for joining ODS alloys. Since the concentrated heat of these processes melts a minimum of base metal, weldments are of better quality than those produced by GTAW. Porosity and dispersoid agglomeration are significantly reduced.

Figure 2. Autogenous GTA weldment in INCOLOY alloy MA 956 sheet.

Alignment of the energy beam with the grain orientation of the base metal has been shown to significantly increase the strength of weldments in alloy MA 956. As the beam angle is reduced, the grain misalignment is reduced and, thus, the properties are significantly improved. The macrostructure of such weldments are seen in Figure 3 (8,9).

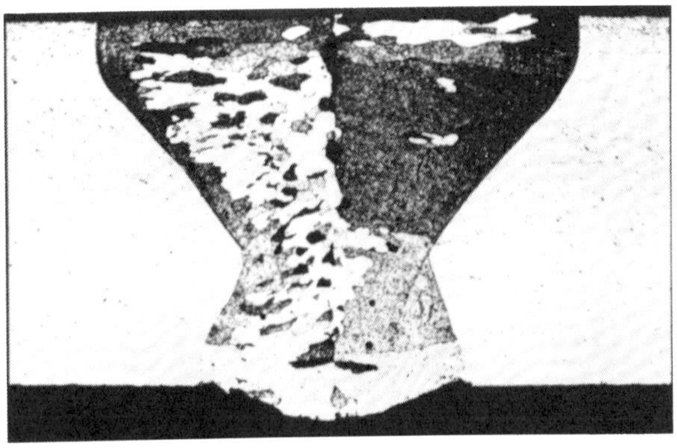

a) Beam at 90° to workpiece.

b) Beam at 20° to workpiece.

Figure 3. Electron beam weldments in INCOLOY alloy MA 956 sheet.

---

*BRIGHTRAY, INCONEL, and INCO are trademarks of the Inco family of companies.

Variations of these fusion processes, "TIG" brazing or beam brazing have shown promise for some joining applications for alloy MA 956. In these processes an insert is placed between the two sheets to be joined. The energy column (arc or beam) is focused directly on the insert so the amount of base metal melted is minimized. Such joints seem to be optimum for fusion processes in which filler metals are added.

Resistance spot welding is accomplished by imposing a current between two sections. The gap between the two sections causes resistance heating and melting at the interface. The process has been used with success for joining alloy MA 956. By adjusting the parameters so that the ratio of the width of the weld nugget to its thickness is increased, the strength of the weld is optimized. Pressure should be as low as possible to avoid weld cracking. A well designed spot weldment is shown in Figure 4. As with the other processes, the grain growth in the weldment is opposite to the direction of the grain structure of the base metal. Thus, the properties of the joint are less than the sheet (9).

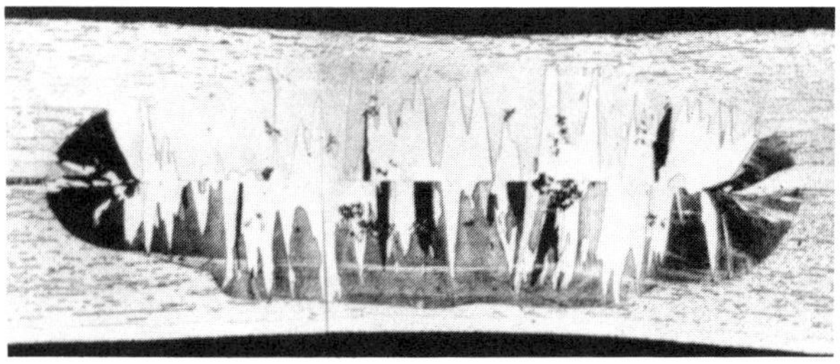

Figure 4. Resistance spot weld in INCOLOY alloy MA 956 sheet exhibiting desired high width to thickness ratio.

Resistance seam welding takes place by producing a series of overlapping spot welds. While this process has proven quite useful for making seal welds in conventional alloys, its use for ODS alloys (e.g., MA 956) results in multiple melting of the overlapping regions of the weld nugget. This increases dispersoid agglomeration and rejection and, thus, decreases weld properties.

Flash butt welds are made by imposing a current between two separated surfaces. Resistance heating melts the two surfaces which are then upset. Ideally, all the melted metal is expelled producing a near solid state joint.

A flash butt weldment in alloy MA 956 is seen in Figure 5. The structure is somewhat comparable to GTA or resistance weldments (9). Thus, this process should be applicable to similar fabrications where the other processes are used.

To summarize, fusion weldments have been successfully used to join alloy MA 956 for several applications. While the weldments exhibit strength properties significantly less than the base metal, designing so that the weldments are in low stress areas has allowed use of the alloy's excellent corrosion resistance to high temperatures.

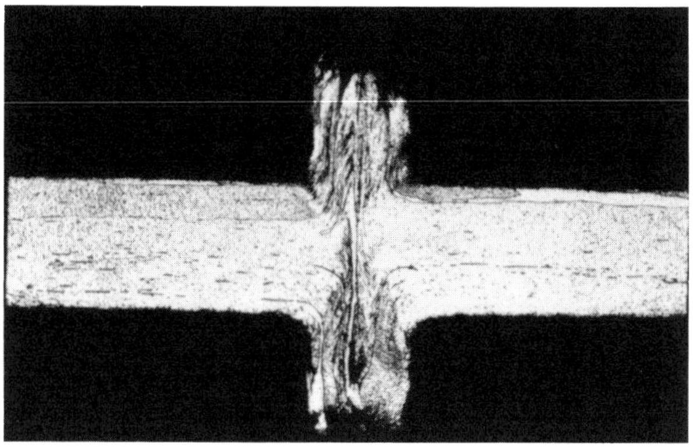

Figure 5. Flash butt weldment in INCOLOY alloy MA 956.

SOLID STATE PROCESSES

Solid state joining processes operate without melting of the base metal usually by the application of high pressure. Bonding is achieved by the introduction of mechanical, electrical, or thermal energy and/or diffusion.

Inertia (or friction) welding is a rather specialized process for joining a cylindrical part to a stationary section. The cylindrical unit is spun at high speed. Light contact between the parts creates heat at the interface and cleans the surfaces to be joined. Axial pressure is applied causing the parts to rotate against each other forming the joint. The cleaning and solid state nature of the bond make the process attractive for joining alloy MA 956.

A program is being designed to define procedures for inertia welding alloy MA 956 to itself and to dissimilar alloy. Initial results from this and past studies are positive. Preheat and post weld heat treatment are used to improve weld properties and avoid post weld cracking.

Explosive bonding is achieved between two metallic surfaces by a directed shock wave produced by detonation of an explosive charge. While the process is most often used for cladding of steel plate by conventional alloys, the process has more recently been applied to joining complex or dissimilar materials. Perhaps the latest application is the joining of tubes to tubesheets in heat exchangers.

A study is in progress at a European research laboratory to identify applications and procedures for explosively bonding alloy MA 956. As with many of the other processes preheat and a post weld heat treatment are utilized to produce a sound weld zone with an acceptable microstructure.

Microstructures of two explosive bonds are seen in Figure 6. The first micrograph shows a bond made at room temperature without post weld heat treatment. Newmann bands are evidence of adiabatic shearing. However, in the lower figure which was preheated to 400°F (200°C) prior to bonding and post weld heat treated at 2000°F (1100°C) the microstructure is clear (9).

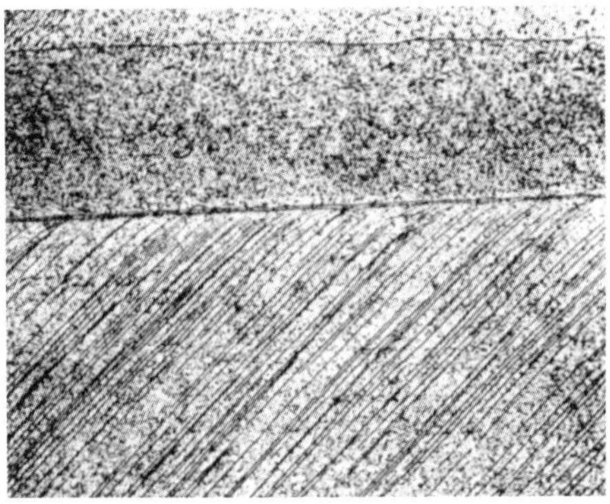

a) Bonded at room temperature; no post weld heat treatment.

b) Bonded at 400°F (200°C); heat treated 2000°F (1100°C) for 112 hours immediately after bonding.

Figure 6. Explosive bonds in INCOLOY alloy MA 956.

Diffusion bonding takes place when two faying surfaces are subjected to pressure at elevated temperature for an extended period of time. Diffusion takes place across the interface and a bond is formed. An interlayer of some rapidly diffusing element is sometimes used to decrease bonding time.

Work by Moore and Glasgow (10) has shown that diffusion bonding of a nickel-base ODS alloy, INCONEL alloy MA 6000, is feasible. While the process has not been applied to joining alloy MA 956, it would seem to offer great potential for suitable applications. Surface cleanliness requirements for MA 956 would be critical. A technique such as that required for brazing (cool wet grinding to a mirror finish and storage in alcohol until bonding) should be suitable.

Results thus far indicate that solid state welding processes are excellent candidates for joining alloy MA 956. Since no melting of the base metal takes place, the problems of grain structure alteration and slagging off of the dispersoid are avoided. Accordingly, solid state joints in alloy MA 956 exhibit properties superior to those of fusion weldments.

BRAZING

Brazing, like the solid state processes, offers a good potential for success for joining ODS alloys since melting of the base metal is avoided. Several commercial braze alloys have been evaluated for joining alloy MA 956. Additionally, development programs are in progress at Inco research facilities to develop suitable brazing alloys.

Preparation of the surface of alloy MA 956 prior to brazing is essential to assure flow of the molten braze metal. The high aluminum content of 4-5% in the alloy is required for resistance to high-temperature oxidation. The tenacious oxide formed, however, impairs brazeability by inhibiting the wetting and flow of brazing alloys. Consequently, the surface oxide layer must either be covered or removed to allow acceptable brazements to be made.

Depositing a layer of about 0.001 inch (0.025 mm) of nickel on the surfaces to be brazed was found to produce acceptable braze alloy flow and wetting. Electroplating is an acceptable means of depositing the layer of nickel. While no experience is known, vapor deposition should also be a suitable method of producing a barrier layer. There should also be other suitable barrier materials than nickel. It should be possible to choose a barrier material which will diffuse into the base metal at the brazing temperature without causing any alteration of the base metal properties (7).

Mechanical abrasion is used to remove the oxide from the surface to be brazed. This is accomplished by polishing on a slow, water cooled wheel or belt to a 3 micro-inch finish followed by rinsing in ethyl alcohol. If the sample is to be stored for more than one hour before brazing, it should be kept immersed in alcohol (9).

Because of complexities in shape, it is often desirable to chemically clean the surface of an alloy to be brazed. Research is currently being conducted to determine a suitable solution; caustic, acidic pickling, acidified chloride, and mixed acid solutions are being evaluated. Molten salt baths also may be useful in conjunction with the above solutions.

Furnace brazing in a vacuum is the brazing process most commonly used for alloy MA 956. Pressures of 10^{-4} - 10^{-6} torr are generally suitable. Brazing temperatures are usually quite high 2150-2300°F (1180-1260°C) because of the high melting point brazing alloys required.

The use of magnesium as a getter increases the cleanliness and brightness of the braze. The flow of a brazing alloy was shown to significantly increase when magnesium vapor was used; the contact angle was decreased from 50° without the getter to 18.5° with its use (7).

Properties matching the base metal have not been achieved with the commercial braze alloys presently available. Brazements generally suffer from insufficient high-temperature strength or corrosion of the matrix or secondary phases. Thus, research programs are in progress to develop suitable brazing alloys. Alloys showing promise utilize a matrix of iron, nickel, and/or cobalt. Chromium up to 30% and aluminum up to 5% are added for resistance to hot corrosion. Molybdenum, tungsten, and columbium are being evaluated for their effect on high-temperature strength. Manganese, germanium, and/or palladium are used as melting point depressants. Boron and silicon are limited to avoid base metal erosion and formation of brittle phases in the brazement, Figure 7a.

An iron-base, nickel, chromium, manganese, aluminum, braze alloy developed and patented by Inco Alloys International, Inc. is of particular interest, Figure 7b. This system uses manganese as a melting point depressant. Because of the high vapor pressure of manganese, the element evaporates during the brazing cycle or in a subsequent heat treatment; thus, the remaining braze metal has a greatly increased melting point (11).

The use of such filler metals with a volatile melting point depressant, transient liquid phase brazing, seems especially promising and is the focus of present developments. The brazing alloy in the patent exhibits excellent strength and with modification to increase oxidation resistance should provide optimum brazement properties.

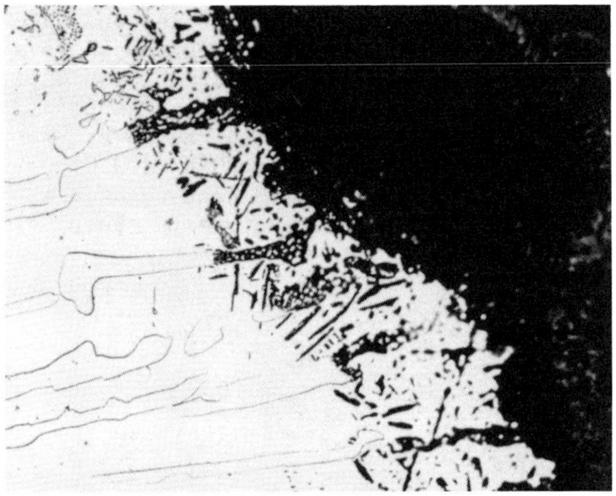

a) Preferential oxidation of chromium boride phase in conventional alloy brazement.

b) Microstructure - experimental alloy brazement.

Figure 7. Brazements in INCOLOY alloy MA 956.

CONCLUSION

Joining alloy MA 956 presents a series of unique requirements and problems to the welding engineer. For a joint to provide optimum properties, the grain morphology of the base metal must be maintained. It must also retain the dispersoid in an effective distribution. Porosity and cracking must be avoided.

Solid state joining processes seem most likely to meet these goals. Since melting of the base metal does not occur the above requirements are most nearly met.

Furnace brazing also offers excellent potential for joining alloy MA 956. The definition of a suitable chemical cleaning procedure will be a great stride toward making the use of process industrially feasible. The braze alloys commercially available produce brazements of excellent quality and properties. However, to date none have provided properties matching the base metal. Thus, work will continue on developing a suitable brazing alloy. To exhibit strength matching alloy MA 956 at high temperatures, the brazement will likely have to contain a dispersoid like the base metal. Certainly achieving this will open a new frontier for both brazing alloys and processes.

Fusion processes can be adapted to joining alloy MA 956. By designing so that weldments are in low stressed or low temperature areas, fusion welds can be readily and economically used to join the alloy and provide excellent service.

REFERENCES

1. L. P. Rice, "Metallurgy and Properties of Thoria Strengthened Nickel," DMIC Memorandum 210, October 1, 1965, Defense Metals Information Center 1 Battelle Memorial Institute.

2. J. S. Benjamin, "Dispersion Strengthened Superalloys by Mechanical Alloying," Metallurgical Transactions, Vol. 1, Oct., 1970, pp 2943-2951.

3. G. A. J. Hack, "Fundamentals of Mechanical Alloying", Frontiers of High Temperature Materials II, London, May 22-25, 1983, Inco Alloys International, Inc.

4. J. S. Benjamin, "Mechanical Alloying", Scientific American, May, 1976, Vol. 234, No. 5.

5. Technical Bulletin: "IncoMAP Mechanically Alloyed Products: INCOLOY alloy MA 956", Inco Alloys International, Inc., 1981.

6. D. T. Busby, Unpublished Research, Wiggin Alloys Limited, Hereford, England.

7. J. E. Franklin, Unpublished Research, Paul D. Merica Research Laboratory/ International Nickel Company, Suffern, NY.

8. T. J. Kelly, "Welding of Mechanically Alloyed ODS Materials", Trends in Welding Research, American Society for Metals, November 16-18, 1981.

9. T. J. Kelly, Unpublished Research, Paul D. Merica Research Laboratory/International Nickel Company, Suffern, NY.

10. T. J. Moore, T. K. Glasgow, "Diffusion Welding of MA 6000 and a Conventional Nickel-Base Superalloy", Welding Journal, August, 1985.

11. T. J. Kelly, United States Patent No. 4,444,587, April 24, 1984.

EFFECT OF WELDING PARAMETERS AND GROOVE ANGLE ON THE SOUNDNESS OF ROOT BEADS DEPOSITED BY THE SAW PROCESS

R.S. Chandel,
PMRL / CANMET
Energy, Mines and Resources
Canada

S.R. Bala
AMCA International
Research and Technology Center
Kanata

ABSTRACT

The effects of submerged-arc-welding parameters (current, voltage, wire diameter, electrical stickout, heat input and polarity) and groove angles (60°, 52.5°, 45° and 37.5°) on the soundness of root beads has been assessed. The results show that as the groove angle decreases, the frequency of defect occurrence increases. It has been observed that center-line cracking and lack of penetration are the main problems when depositing root beads in narrow grooves. Center-line cracking is mostly associated with the use of high current and high travel speeds while most of the lack of penetration occurs when low current and low travel speeds are used. Although some welds do suffer from undercuts, excessive reinforcement and lack of side-wall fusion, these problems are not considered critical. The results show that it is possible to deposit sound beads in grooves of 37.5 degrees.

IT IS WELL KNOWN that considerable savings can be made both in terms of consumables and labour cost by reducing groove angles when welding thick plates (1). The total heat input per unit length is also reduced, which is an advantage when mechanical properties and distortion are considered). However, when such an attempt is made, the probability of the occurrence of defects such as center-line cracking, lack of side-wall fusion and lack of penetration, etc., increases. These defects are related to weld-bead geometry and to a certain extent can be controlled by obtaining an appropriate bead profile. In the literature (2-6) numerous references can be found correlating the welding parameters and weld-bead geometry of submerged-arc welds, however, most of the work is concerned with bead-on-plate type of welds. Although some attempts have been made to assess the influence of welding parameters on weld bead geometry and soundness of groove welds, the results to-date are inconclusive and of little significance to shop welding fabrication (7).

The present industrial practice of butt welding plates thicker than 25 mm is to prepare 'single vee' or 'double vee' grooves of 60°. Although the narrow-gap process is becoming popular for welding thick sections with very little or no edge preparations, it is only economic for sections thicker than 100 mm (8). Therefore, in order to increase the productivity and decrease the weld metal volume requirements for thicknesses between 25 mm and 100 mm, the obvious choice is to reduce the groove angle. The aim of this work is to assess the possibility of depositing sound welds in narrow grooves (<60°) i.e., in 52.5, 45, and 37.5° grooves by using the submerged-arc-welding process.

EXPERIMENTAL TECHNIQUES

MATERIAL - The base material used for the experimental work was a 38 mm thick ASTM A36 steel plate. This plate was cut into 160 mm x 150 mm pieces and grooves of 30°, 26.25°, 22.5° and 18.75° were machined to obtain V-grooves of 60°, 52.5°, 45°, and 37.5° for butt welding. A root face of 12.5 mm and a root gap of 1.5 mm was used for each joint. Carbon-Manganese (AWS EL 12) electrode wires of 2.4 mm and 4.0 mm diameter were used along with an AWS F64 flux to prepare the experimental welds. This wire-flux combination was used since it is applicable to a wide range of current and voltage of both polarities.

WELDING - The machined plates were tack welded with a root gap of 1.5 mm and 100 mm-long run-on and run-off tabs were welded onto them. The run-on tabs allowed enough time to adjust the welding current and voltage while the runoff tabs absorbed the crater of the bead. Thus for each weld a 150 mm-long deposit with the required welding parameters was obtained.

The submerged-arc-welding system consisted of a DC transformer/rectifier with a constant current output along with a control unit and wire feeder.

Heat inputs of 3.1, 3.9 and 4.7 kJ/mm were used. For each heat input various combinations of current, voltage and travel speed were employed. All welds were made under the same conditions of external restraint and for each weld the current and voltage were recorded on a dual chart recorder. The travel speed, electrode stickout, and electrode polarity were preset for each deposit before the commencement of welding. In groove welding, the most critical beads are the root beads, therefore, only the root bead was deposited in each groove type.

BEAD MEASUREMENTS AND CHARACTERIZATION OF WELDS

To measure the bead characteristics the transverse sections of each weld were removed from the mid-length position, polished and etched by 2% nital. Using a Reichart microscope the welds were projected on a screen with a magnification of 5X and their features were traced on a transparent paper. From these traces weld bead features such as crack length, root penetration, bead width and bead height as shown in Fig. 1 were measured.

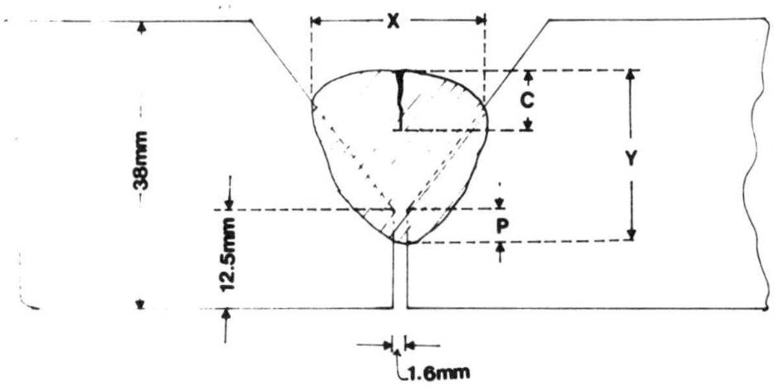

Fig. 1 - Weld features that were measured

Other features such as lack of side-wall fusion, excessive reinforcement and undercuts were judged qualitatively.

RESULTS AND OBSERVATIONS

A factorial experimental design was used so that the effect of current, travel speed, voltage, electrode diameter, electrode polarity, and electrode stickout could be evaluated

Table 1 - Occurrence of Weld Defects in Groove Welds

Process	Groove Angle	No. of Welds	Welds with Undercut	Welds with Lack of Sidewall Fusion	Welds with Centreline Cracks	Welds with Lack of Penetration	Welds with Excessive Reinforcement	No. of Acceptable Welds
2.4mm wire DCSP 32mm E.S.O.	60°	27	0	0	0	5	0	22
	52.5°	27	0	0	0	16	0	11
	45°	27	0	0	0	18	0	9
	37.5°	27	0	0	0	22	0	5
2.4mm wire DCSP 76.2mm E.S.O.	60°	27	0	0	0	5	0	22
	52.5°	27	0	0	2	5	0	20
	45°	27	0	0	2	6	0	19
	37.5°	27	0	0	3	14	0	10
2.4mm wire DCSP 32mm E.S.O.	37.5°	9	0	0	3	1	0	5
4mm wire DCSP 32mm E.S.O.	60°	36	1	0	2	1	0	32
	52.5°	36	0	0	1	0	2	33
	45°	36	0	2	10	1	1	24
	37.5°	36	0	0	15	3	0	18
4mm wire DCSP 32mm E.S.O.	37.5°	12	0	0	4	2	1	6

on the soundness of root beads deposited in 60°, 52.5°, 45° and 37.5° grooves.

The range of parameters used in the present investigation are those that are common for submerged-arc-welding.

To analyze the results a "tolerance box approach" proposed by Jones (9) was used. Welds free from defects such as: 1. Centerline cracking, 2. Lack of penetration, 3. Lack of side-wall fusion, 4. Undercut, 5. Excessive reinforcement, were judged to be sound.

The results are given in Table 1 and Fig. 2-6.

CENTER-LINE CRACKING - Table 1 and Fig. 2-6 show that the proportion of welds with centerline cracks increases as the groove angle decreases. For instance, Table 1 shows that when the groove angle decreases from 60° to 37.5°, the number of welds with center-line cracks increases from 2 to 15. Similar trends were observed for other welds produced by a range of welding process variables. Figures 2-6 also reveal that for a

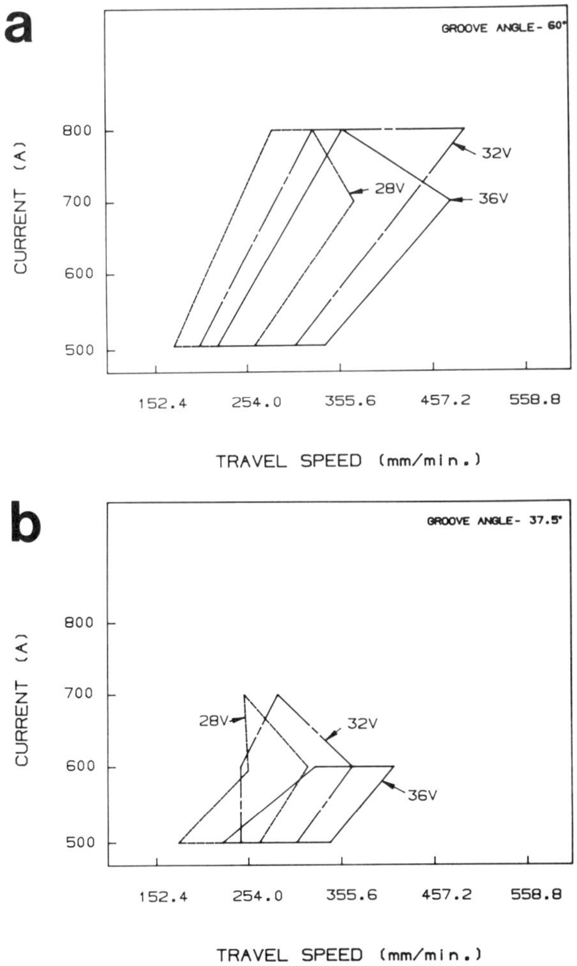

Fig. 2 - Represents combinations of welding currents, travel speeds and arc voltages for depositing Sound root beads [wire dia - 4 mm, E.S.O.-2 mm, polarity - DCSP] (a) 60° and (b) 37.5° grooves

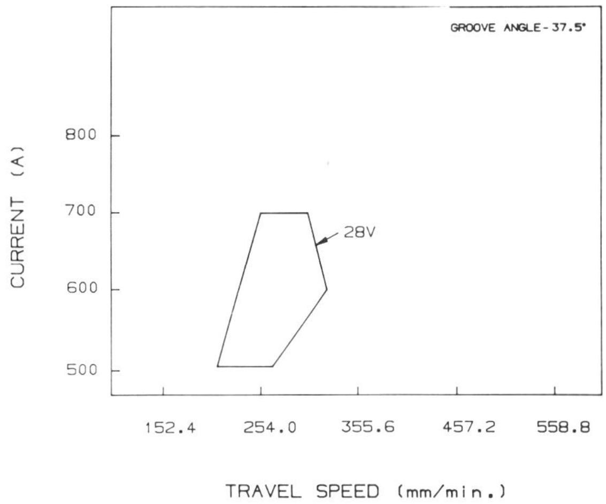

Fig. 3- Represents combinations of welding currents and travel speeds for depositing Sound root beads [wire dia - 4 mm, E.S.O.-32 mm, polarity DCRP] in groove of 37.5°

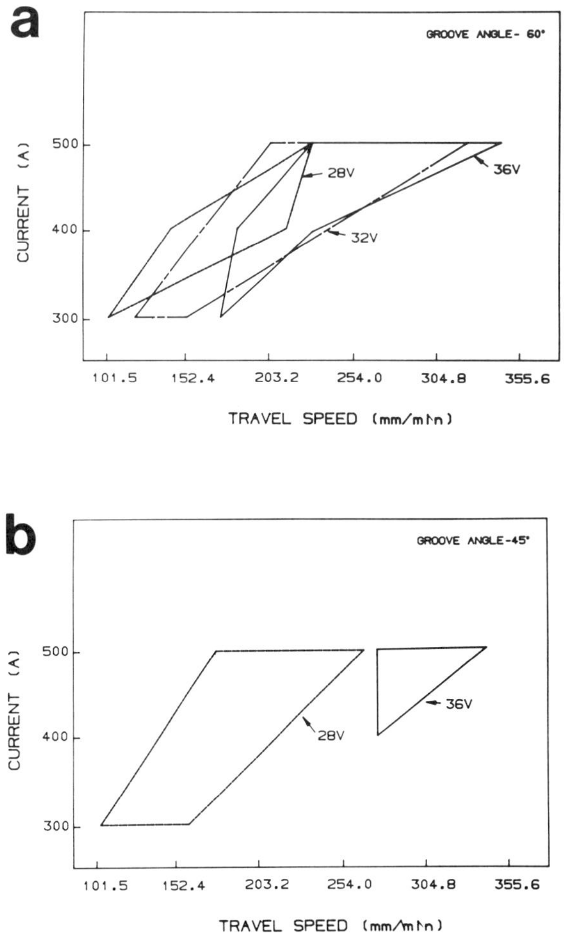

Fig. 4 - Represents combinations of welding currents, travel speeds and arc voltages for depositing Sound root beads [wire dia - 2.4 mm, E.S.O.-2 mm, polarity - DCSP]. (a) 60°, (b) 45° grooves

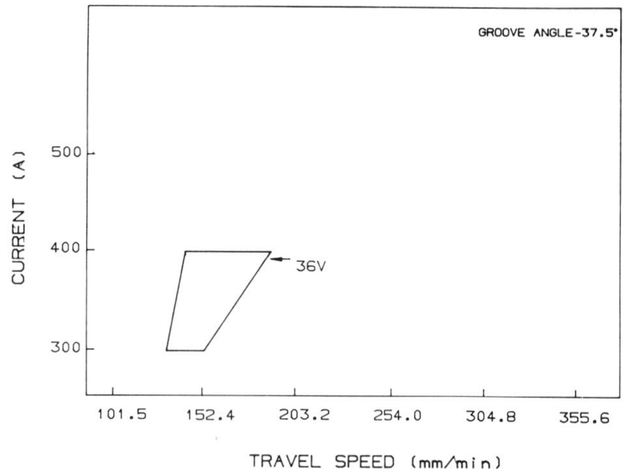

Fig. 5 - Represents combinations of welding currents and travel speeds for depositing Sound root beads [wire dia 2.4 mm, E.S.O.-32 mm, polarity - DCRP] groove of 37.5°.

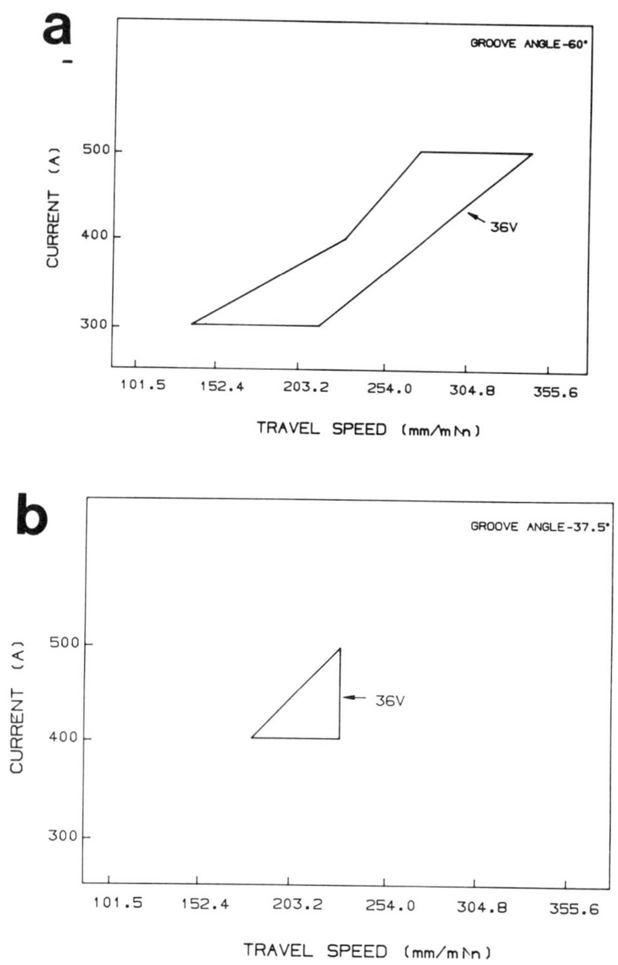

Fig. 6 - Represents combinations of welding currents, travel speeds and arc voltages for depositing Sound root beads [wire dia - 2.4 mm, E.S.O.-76.2 mm, polarity - DCRP] in (a) 60°, (b) 37.5° grooves

given groove angle, center-line cracking always occurs when high current and high travel speeds (for the same heat input and other welding variables) are used. Voltage seems to have no effect on the occurrence of center-line cracking. It is further observed that for all welding conditions, welds produced with the 4.0 mm diameter electrode are more susceptible to center-line cracking compared with those produced by the 2.4 mm diameter electrode (Fig. 2,4). For welds deposited by using a 4.0 mm diameter electrode, polarity does not seem to have any influence. The welds produced with a 76 mm stickout are more susceptible to center-line cracking compared with those produced by a 32 mm stickout and 2.4 mm electrode.

LACK OF PENETRATION - Table 1 shows that the proportion of welds that suffer from lack of penetration increases as the groove angle decreases. Furthermore, the welds with inadequate penetration are those that are produced by using low current and low travel speed conditions. Varying the voltage between 28V and 36V for any of the welding conditions seems to have no influence on the occurrence of lack of penetration. It can also be observed that irrespective of other welding parameters, the chances of producing a weld without adequate penetration increase when small diameter electrodes are used. Welds made with the electrode +ve suffer less from lack of penetration compared with those deposited using a -ve electrode for the same welding parameters. It is further observed that susceptibility to lack of penetration is higher when the electrode stick-out is increased.

LACK OF SIDE-WALL FUSION - From the results given in Table 1 it is apparent that lack of sidewall fusion was not a major problem. Only two welds made with a -ve electrode suffered from this defect.

UNDERCUTS - None of the welds had any significant undercut along the length.

EXCESSIVE REINFORCEMENT - Table 1 shows that 24 welds out of a total of 381 deposited had excessive reinforcement. Most of the welds that had excessive reinforcement were deposited with a 2.4 mm diameter electrode and 76 mm electrode stickout.

DISCUSSION

The results show that during the deposition of root beads in grooves of various angles the main problems are center-line cracking and lack of penetration. Other defects such as lack of side-wall fusion, undercut and excessive reinforcement did occur, but their occurrence is of minor importance. It is worth noting that center-line cracking and lack of penetration occurred in a systematic manner and can be explained on the basis of submerged-arc-welding parameters.

CENTER-LINE CRACKING - Three factors, namely; (a) restraint, (b) bead shape, and; (c) weldmetal composition are responsible for causing center-line or solidification cracking. These factors may act individually or in combination.

The experimental results have indicated that center-line cracking occurred when welds were deposited by using high current and high travel speed, large diameter electrodes and short stickouts. The probability of the occurrence of center-line cracking also increased when groove angle decreased.

It is well known that in submerged-arc-welding for a given heat input an increase in welding current (hence a corresponding increase in travel speed) results in a decrease in weld bead width (x) to weld bead height (y) ratio (5). Since weld beads with smaller x/y ratio are more prone to centerline cracking, this is probably the reason that most of the center-line cracking occurred when welds were deposited by using high current and high travel speed. Figure 7 shows a relationship for a limited number of representative welds between critical x/y ratios and groove angles to obtain welds without center-line cracks. It can be seen that as the groove angle decreases, the critical value of x/y ratio required to avoid center-line cracking increases. It varies from 0.75 for a 60° groove to 0.85 for a 37.5° groove. This is clearly because as the groove angle decreases, the restraint imposed upon the weld beads by the side-walls increases, hence cracking can take place even with higher x/y ratios since the two factors become effective. Therefore, it

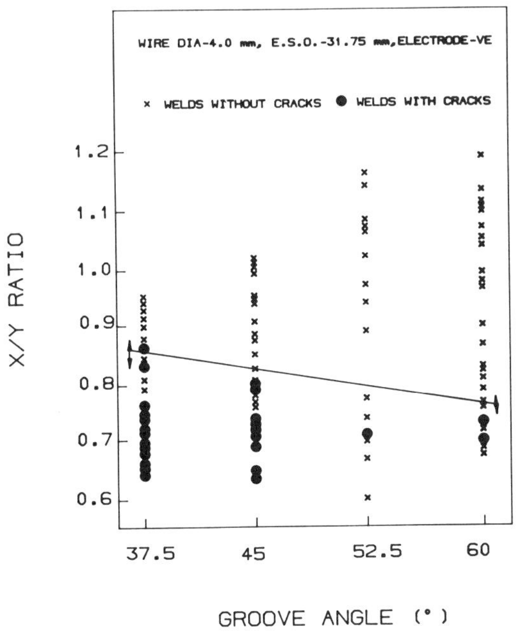

Fig. 7 - Effect of X/Y ratio and groove angle on the centre line cracking.

can be suggested that in order to avoid center-line cracking in narrower grooves, welds with larger x/y ratios should be deposited.

The voltage did not seem to have any influence on center-line cracking. Generally an increase in voltage increases the x/y ratio and hence should decrease the chance of cracking. It seems that compared with the effect of groove angle and current and speed, the effect of voltage was negligible for these tests.

The results also show that the welds produced with larger diameter wire (4 mm) were more susceptible to center-line cracking. In a separate work Chandel (1) has shown that for the same welding conditions, welds produced with smaller diameter electrodes had larger x/y ratios. Hence it can be assumed that because of their smaller x/y ratios, the welds produced by larger diameter electrodes were more susceptible to cracking.

Because of the limited number of welds deposited under conditions of maximum susceptibility to center-line cracking (37.5° groove) the effects of stickout and polarity on such cracking could not be derived.

LACK OF PENETRATION - The results show that the 'lack of penetration' in groove welds was influenced by current, travel speed, groove angle, wire diameter, electrode stickout and electrode polarity. It was further shown that current, travel speed and groove angle had the greatest influence on penetration.

It is well established that in arc welding, for the same heat input, current is the single most important parameter influencing penetration (2-4). To maintain the same heat input, lower travel speed means lower welding current. This is probably the reason for most of the lack of penetration occurring when lower current and lower travel speeds were used. The penetration of the root welds is caused by the digging power of the arc. Whenever the arc operates in restricted areas such as grooves, its digging power is lost because of the hydrostatic pressure imposed by the side-walls. The loss in digging power is directly proportional to the narrowness of the grooves. Thus, lack of penetration was more prevalent when welds were made in narrower grooves.

Normally an increase in voltage decreases the weld penetration. However, in the present work voltage seemed to have no influence on penetration. This is probably because compared with the combined effect of current and groove angle, the effect of voltage was unmeasurable. Because of the limitation on number of welds deposited with DCRP, the effect of polarity on the penetration was very difficult to assess. There was an indication that welds produced with longer stickouts suffered more from lack of penetration. This is probably because with longer stickouts the deposition rates are higher and larger weld puddles are formed which decrease the digging power of the arc and hence reduce penetration.

LACK OF SIDE-WALL FUSION - In arc welding lack of side-wall fusion usually occurs when cold arcs and fluxes that produce viscous slags are used. In the present investigation only two welds suffered from this defect, and no relationship with the welding parameters could be established.

EXCESSIVE REINFORCEMENT - Excessive reinforcement is defined as weld metal deposited in excess of the quantity required to fill the joint (10). In general this defect is associated with improper welding conditions such as high current and low travel speed. As can be seen in Fig. 8, excessive reinforcement occurred mostly when high currents and slower travel speeds were used. It was further observed that most of the welds with excessive reinforcement were those that were made using smaller wire diameters and longer stickouts (Table 1). This was expected because smaller wire diameters and longer stickouts and higher current enhance deposition rates. Since the travel speed was also low, the weld metal was deposited over a shorter distance hence excessive reinforcement occurred. Whenever higher travel speeds were used, this defect was eliminated because the same weld-metal volume was spread over a longer distance.

TOLERANCE BOXES

Tolerance boxes for various groove angles, wire diameters, stickouts, polarities and voltages are shown in Fig. 2-6. These boxes show the boundaries of current and travel speed combinations that can be used to deposit sound root beads. These boxes also show that

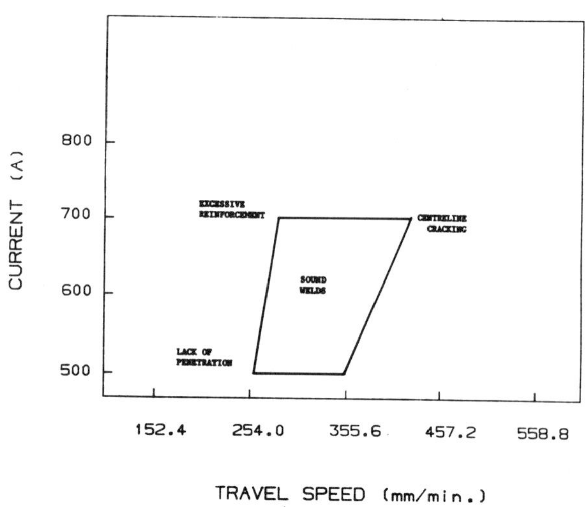

Fig. 8 - Schematic diagram showing the effect of current/travel speed on the occurrence of weld defects.

as the groove angle decreases, their size shrinks i.e., the number of possible current-travel speed combinations for depositing sound beads decreases. A noteworthy feature of these boxes as summarized in Fig. 8 is that the occurrence of a particular type of defect is always at a particular location, e.g., the lack of penetration is always towards bottom left.

CONCLUSIONS

1. The results show that as the groove angle decreased, the frequency of the occurrence of defects increased.
2. Center-line cracking and lack of penetration were the major problems when depositing root beads.
3. Center-line cracking was associated with high current and high travel speed while lack of penetration occurred when low current and low travel speeds were used and with narrower groove angles.
4. The results show that it is possible to deposit sound root beads in grooves of 37.5° by using the consumables employed in this study with 4.0 mm diameter electrodes, 32 mm long stickout and straight polarity.

ACKNOWLEDGEMENTS

Funding of this research program by the National Research Council of Canada is gratefully acknowledged. The work was undertaken in the Corporate Research and Technology Center of AMCA International, who gave permission for the publication of this paper.

REFERENCES

1. Chandel, R.S. "Optimization of Groove Angles", AMCA International Report No. 1-R78-016/9 (1982).

2. Apps, R.L., Gourd, L.M. and Nelson, K.A. "Effect of Welding Variables Upon Bead Shape and Size in Submerged-Arc-Welding", Welding and Metal Fabrication, 31(11) 453 (1963).

3. Drayton, P.A. "An Examination of the Influence of Process Parameters on Submerged-Arc-Welding", The Welding Institute Report No. PE/4/72 (1972).

4. Doherty, J. and McGlone, J.C. "Relationships Between Process Variables and Bead Geometry", The Welding Institute Report No. 52/1977/PE (1977).

5. McGlone, J.C. and Chadwick, D.B. "The Submerged Arc Butt Welding of Mild Steel: The Prediction of Weld Bead deometry from the Procedure Parameters", The Welding Institute Report No. 80/1978/PR (1978).

6. Doherty, J. and Salter, G.R. "Development in Procedure Selection Techniques", The Welding Institute Research Bulletin, 18*6), June (1977).

7. McGlone, J.C. "An Investigation into the Effect of Joint Angle on Bead Geometry for the Submerged-Arc-Welding of Mild Steel", The Welding Institute Report No. 53/1977/PE (1977).

8. Chandel, R.S., Orr, R.F. and Gianetto, J.A. "Submerged Arc Narrow-Gap Welding of Heavy Sections", Physical Metallurgy Research Laboratory, Energy, Mines and Resources, Ottawa, Report No. ERP/PMRL 83-42(TR) (1983).

9. Jones, S.B., Doherty, J. and Salter, G.R. "An Approach to Procedure Section in Arc Welding", Welding Journal, 56(7) 19 (1972).

10. ASM Handbook, Volume 6 "Welding, Brazing and Soldering"; 9th Edition, 14 (1983).

CONTROL OF GTA WELD POOL SHAPE

C.R. Heiple, P. Burgardt, J.R. Roper
Rockwell International
Golden, Colorado

ABSTRACT

The surface tension driven fluid flow model originally proposed to explain the effect of trace elements on GTA weld penetration has been used to understand the interaction between base metal chemistry and welding parameters in determining weld shape. The proposed model is supported by measurements of the response of weld pool shape to beam power density in an electron beam weld on two heats of 304 stainless steel with different sulfur levels. The response of weld shape to base metal sulfur level and input power density at moderate power densities was similar to that observed in GTA welds. There are a number of possible approaches to eliminating GTA weld shape variability, including doping the shielding gas or filler wire with surface active elements, changing welding parameters to make the process less sensitive to chemistry variations, and changing material specifications to require a minimum as well as a maximum in surface active residual element content. Promising preliminary results are being obtained with all these approaches.

PREDICTABLE AND REPRODUCIBLE WELD SHAPE is critical in high precision automated welding applications. Unfortunately there have been numerous instances of variations in GTA weld shape among different heats of material having the same nominal composition and using identical welding parameters. The weld shape as measured by depth/width ratio (d/w) can change by more than a factor of two. The variation in weld shape has been traced to differences in residual element content of the materials being welded (1). Heiple and Roper (2) proposed that GTA weld pool shape is determined largely by fluid flow patterns in the weld pool. The dominant force driving weld pool fluid flow, under normal GTA welding conditions, is a surface tension gradient. The surface tension gradient is caused by interaction of the large temperature gradient on the weld pool surface with the surface tension temperature dependence. Surface active trace elements which segregate to the liquid surface lower the surface tension and change the temperature dependence of the surface tension and thereby modify the surface tension gradient. This in turn changes the fluid flow in the weld pool and the weld shape.

Surface tension gradients can also be altered by changing temperature gradients on the weld pool surface. The magnitude of the surface temperature gradient is determined by the input power to the weld as well as the input power distribution so that the temperature gradients can be varied by changing welding parameters. Because the surface tension temperature dependence is a function of surface active residual element content, the response of weld pool shape to a given change in welding parameters can depend strongly on weld pool chemistry. Measurements of the interaction between base metal chemistry and welding parameters in determining weld shape are summarized here.

The strong dependence of GTA weld shape on residual element content suggests several strategies for improving weld shape and eliminating variable weld shape. These include intentional additions of appropriate elements into the weld, changing material specifications to include a minimum requirement on the appropriate surface active residuals (primarily sulfur in steels), changes in welding parameters to make the weld shape less sensitive to the base metal chemistry, and a certification program to determine the weld characteristics of the materials. Promising preliminary results for all these approaches will be described.

INTERACTION BETWEEN WELDING PARAMETERS AND BASE METAL CHEMISTRY IN DETERMINING WELD SHAPE

The surface tension driven fluid flow model

which has been used previously to explain the effects of surface active residual elements on weld shape can also be used to understand the effect of changing welding parameters on weld shape (3). The model is illustrated in Figure 1. [Note that the schematic flow pattern illustrated for high purity materials, Figure 1A, differs from that previously proposed (2).

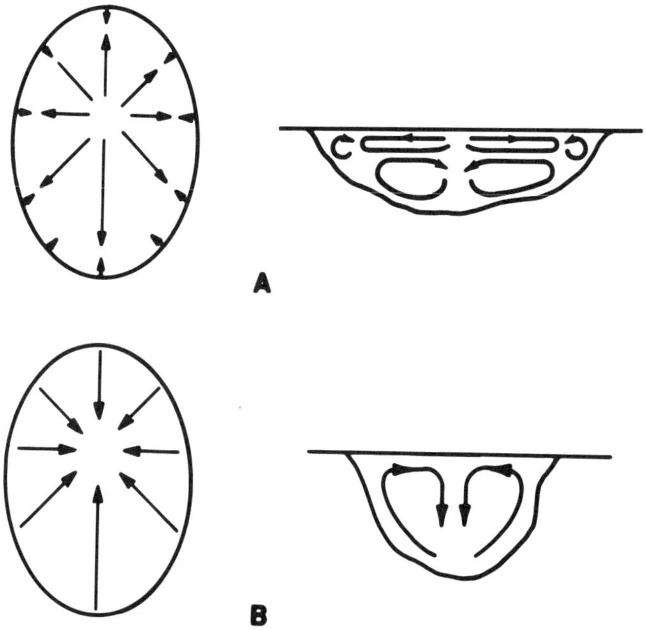

Fig. 1. Schematic surface and subsurface fluid flow in the weld pool. A - surface tension temperature coefficient negative; B - surface tension temperature coefficient positive.

Current three dimensional traveling weld numerical simulations (4) indicate the presence of a secondary circulation loop at the bottom of the weld pool similar to that shown. The presence of the secondary circulation loop does not qualitatively alter the predicted response of weld pool shape to changes in surface tension gradients because the upper circulation loop transfers heat to the edge of the weld pool in essentially the same way as the single loop originally proposed.] The effect of a change in welding parameters which increases the temperature gradient (and thus the surface tension gradient) from center to edge of the weld pool is to drive the existing fluid flow pattern more strongly. Thus, the same change in welding parameters can produce relatively wider and shallower welds in one material (Figure 1A), but relatively deeper, narrower welds in another material (Figure 1B).

There is an additional complication when surface active elements are present. The positive surface tension temperature dependence (which produces the fluid flow pattern shown in Figure 1B) only exists over a limited temperature range. Thus, as the central temperature of the weld pool is raised, eventually a critical temperature is reached above which a region near the center of the weld pool has a surface tension gradient opposite to that in the rest of the pool. The fluid flow pattern necessary for deep, narrow welds is disrupted and, with increasing weld pool central temperature, the weld pool becomes relatively wider and shallower. The temperature at which the positive surface tension temperature dependence ends can be estimated by extrapolation of surface tension measurements on two heats of 316 stainless steel (5), as shown in Figure 2.

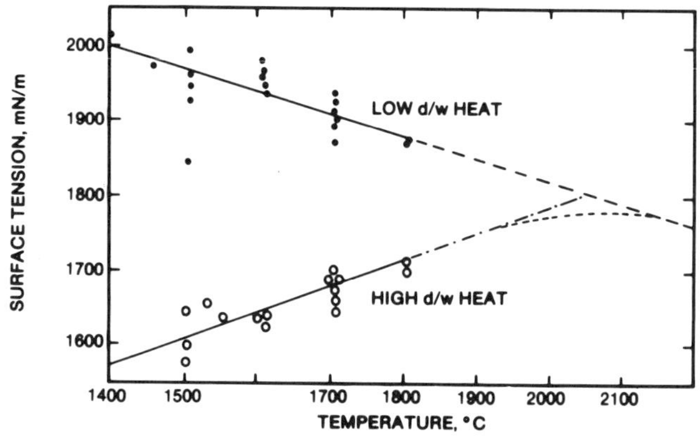

Fig. 2. Surface tension versus temperature for two heats of 316 stainless steel (Ref. 5). The high d/w heat had approximately 160 ppm more sulfur than the low d/w heat. The dashed lines indicate the expected behavior at higher temperatures.

This temperature may be quite different for other iron base alloys. Preliminary measurements have placed the maximum in surface tension at about 1700°C in AISI 8630 steel (6).

Tsai (7) has recently measured the effects of current, arc length, electrode tip angle and helium additions to the argon torch gas on the arc input power density. He found that increasing current, shorter arc length, sharper electrodes and increasing helium content all increased the central input power density in the arc and would therefore increase the weld pool peak temperature. Weld shape measured as a function of current is shown for high purity, sulfur doped, and selenium doped 304L stainless steel in Figure 3. Details of this measurement as well as measurements of the effects of the other welding parameters are given elsewhere (3). The response of weld shape to current and to the other welding parameters is consistent with the surface tension driven fluid flow model.

WELD POOL TEMPERATURE

From Figure 2, it would appear that (at least for 316 stainless steel) the weld pool

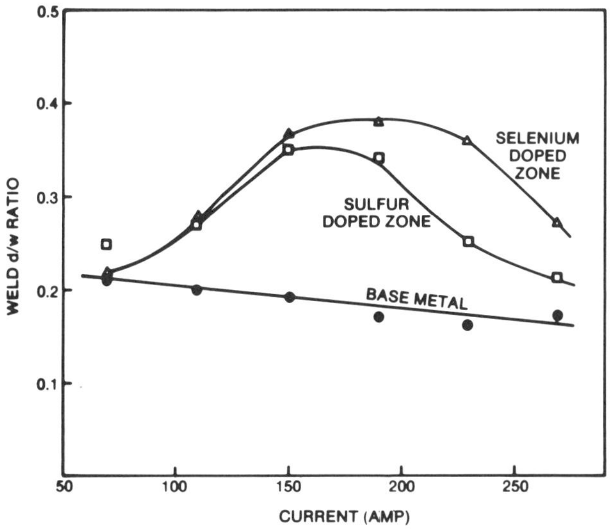

Fig. 3. GTA weld d/w ratio versus weld current for low sulfur 304L stainless steel (20 ppm) base metal as well as sulfur and selenium doped zones (Ref. 3).

maximum temperature would have to exceed about 2050°C for the maximum in d/w ratio to be observed. Recent efforts to measure weld pool temperatures by infrared (8) and thermocouple (9) techniques have predicted a maximum surface temperature of 1700-2100°C. Block-Bolten and Eagar (10) have estimated that the maximum weld pool temperature in stainless steel will not exceed about 2500°C because of heat loss due to evaporation of the more volatile constituents. In view of the uncertainties in instrumental measurements of weld pool temperature, we devised a simple experiment to estimate the weld pool central temperature. Holes 1 mm in diameter were drilled perpendicular to the surface of a 304 stainless steel plate. The holes were filled with wires of different melting point materials and GTA welds (180 amps; 15 cm/min. travel speed; pure argon cover gas) were passed across the filled holes. Tungsten (melting point = 3370°C), Ir (2450°C), Rh (1960°C), and Pt (1770°C) wires were tested. The welds were cross-sectioned at the wire locations and traces of the weld cross-sections and remaining wire were made and are shown in Figure 4.

As expected, the tungsten was melted only by the approximate pool depression caused by the cathode jet, while the other wires with lower melting point were melted progressively deeper under the surface. The difficulty with this analysis is that Ir, Rh and Pt are all completed soluble in liquid iron so that some dissolution as well as melting may have occurred. Therefore, the same experiment was repeated with a tube of 99.9% aluminum oxide (2000°C melting point). Aluminum oxide is not significantly soluble in liquid iron. Figure 4 shows the result of this experiment and shows that the

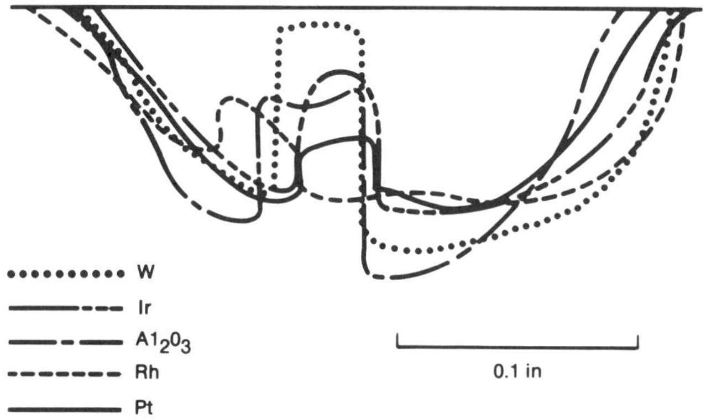

Fig. 4. Superimposed cross sections of GTA welds and unmelted portions of high melting point wires where the weld pool passed over the buried wires. Base metal 304L stainless steel.

alumina was melted to essentially the same depth as the Rh, which has a comparable melting point. Since the 2000°C isotherm is fairly deep in the pool, it appears that the surface temperature can be significantly above 2000°C, which is consistent with the surface tension driven fluid flow model.

ELECTRON BEAM WELD POOL MEASUREMENTS

It is possible that welding parameter data, such as that shown in Figure 3 can be interpreted as some complex arc - weld pool interaction. However, the surface tension driven fluid flow model is required to explain material chemistry effects on welds made with

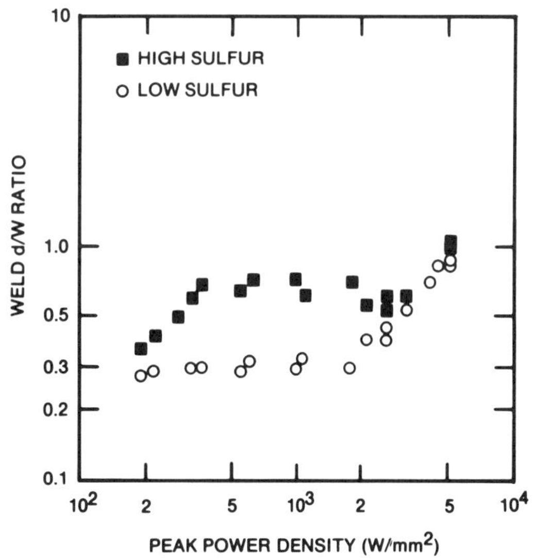

Fig. 5. Electron beam weld pool d/w ratio versus electron beam power density for low sulfur (20 ppm) and high sulfur (>120 ppm) 304L stainless steel. Keyhole formation begins at about $2x10^3$ w/mm^2. Welding parameters 100 KV, 3.5 mA, 1 cm/sec; power density varied by changing focus.

non-arc processes. As part of another study, Burgardt et al. (11) have investigated the effect of changes in beam focus on weld shape in electron beam welds. The results for low and high sulfur material are shown in Figure 5. Observe that the high sulfur material exhibits a maximum in d/w ratio at a moderate power density away from sharp focus. (The low and high sulfur exhibit deep and nearly identical penetration near sharp focus. Near sharp focus where the power density is high, penetration is by a keyhole mechanism, which is unrelated to surface tension driven fluid flow.) The maximum away from sharp focus is known to not be a result of any anomalous beam property, such as a beam width maximum, and is proposed to originate from exactly the same mechanism as for increasing current with GTA welds.

Measurements of electron beam power distribution were made as a function of beam focus and were used to calculate the beam power density at the d/w maximum of 700 W/mm^2. The power density at d/w maximum for electron beam is much higher than the equivalent value, 100 W/mm^2, obtained for GTA welds. However, the electron beam welds were made at much higher travel speed, 1 cm/sec. compared to 0.25 cm/sec. for the GTA welds. Calculation of weld pool surface temperature immediately under the heat source using a traveling distributed heat source model (12) showed the central surface temperatures, at the d/w maximum for the electron beam and GTA processes, to be essentially identical. The actual surface temperatures calculated using the traveling distributed heat source conduction approximation are much too high because most of the heat transport in the weld pool is from mass flow rather than conduction. However, the equality of calculated weld pool peak temperature at the d/w maximum provides a strong confirmation that the mechanism responsible for the presence of the maximum in weld d/w with increasing input power density is independent of heat source and therefore is not a result of an arc phenomenon.

STRATEGIES FOR CONTROLLING GTA WELD SHAPE

WELD POOL DOPING - A reduction in the surface active element content of the weld pool will cause a lowering of weld d/w ratio until the fluid flow pattern indicated in Figure 1A is produced. In the composition range where the transition from one fluid flow pattern to the other occurs, quite small differences in trace element content can produce substantial changes in weld shape. This leads to variable penetration, often within the same heat of material and even at different locations on the same part. Poor penetration and variable weld shape can be eliminated by increasing the surface active element content of the weld pool so that the fluid flow pattern shown in Figure 1B is obtained independent of base metal composition. We have reported (13) preliminary tests in which SO_2 or oxygen additions to the shielding gas improved penetration and suppressed variable penetration for GTA welds on 304 stainless steel. Oxygen doping of the torch gas is being used to produce 316 components for the Italian fast breeder test reactor (14) and SO_2 doping is being used on components for the French Super Phoenix breeder reactor (15). We have also demonstrated that sulfur additions to the weld pool on 21-6-9 and 304 stainless steels by doping the weld filler wire also increased weld d/w ratio and suppresses variability.

MATERIAL SPECIFICATION CHANGES - A second approach to insuring a high enough surface active element content in the weld pool is to change the material specification so that there is a range in the content of relevant trace elements rather than simply a maximum. We are presently evaluating 20000 Kg of 304L purchased to an 80-150 ppm sulfur specification. The evaluation involves circular GTA and EB welds on test blocks, as well as forgeability studies, using material with a sulfur content at the high end of this specification range. So far no cracks have been observed, even in deeply penetrating, high speed EB welds (although the calculated ferrite numbers have also been high, ranging from 6.0 - 7.5). The sensitivity to hot cracking (16) as measured by a Varestraint test is not appreciably different for 304L stainless steel containing 40 ppm S or 150 ppm S, provided the ferrite content is similar (Figure 6). More weldability tests using actual parts are planned.

A similar approach would be to use steels with very low levels of residual surface active element content, for example, a maximum of 20 ppm each sulfur and oxygen. While most of the 304L stainless steel we hvae recently received would meet such a criteria, it is likely that producers would find it difficult or impossible to guarantee such low sulfur and oxygen levels. GTA welds on such steels are wide and shallow, but they are consistent. Variability is generally a more difficult problem than low penetration since changes in joint design and welding parameters can often be used to compensate for poor penetration.

MATERIAL CERTIFICATION - Another approach is to categorize the weld penetration characteristics of incoming heats of material using a standardized weld test as part of the certification procedure. High and low penetration heats have been welded and we are currently in the process of statistically evaluating d/w data taken from multiple sections from multiple welds to determine how many welds and sections must be made to accurately characterize each incoming heat. Preliminary results with three heats of 304L stainless steel indicate that to characterize the d/w ratio to ± 0.04 with 95% confidence, four sections of one weld on each of four randomly selected bars from a given heat are required (17). Characterization of our stock of two other types of stainless steel is underway.

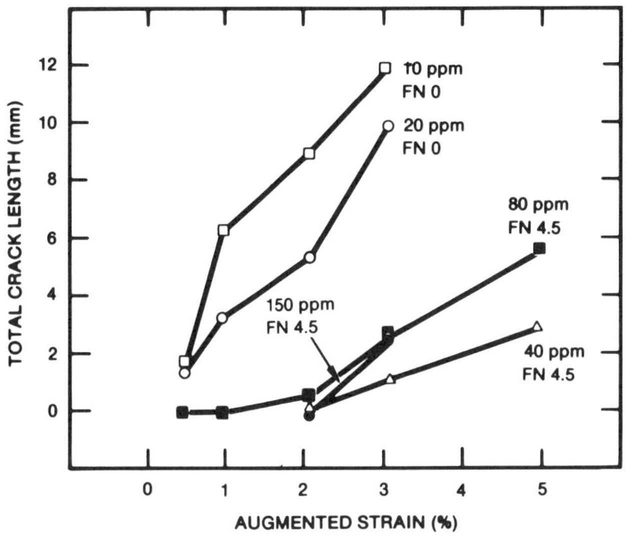

Fig. 6. Hot cracking susceptibility of several heats of 304L stainless steel, as measured by a Varestraint test. Heats are labeled by sulfur content and ferrite number (FN). (Ref. 16).

It is intended that following weld shape characterization, the material be subject to a comprehensive system of material scheduling for specific types of parts. The objective of these tasks is to ensure that drastic mismatches in composition do not occur and that the penetration of each component is similar to that which was typical of development parts. This can be difficult in situations where a large variety of parts are produced from the same material stocks and where lead times from material selection to final welding may be long (1-2 years in our case).

CONCLUSIONS

The response of weld pool shape to changes in welding parameters arises primarily from changes in weld pool fluid flow. Different welding parameters produce different temperature gradients on the weld pool surface, which lead to different surface tension gradients and thereby variations in fluid flow. The change in weld pool shape for a given welding parameter change can depend on material composition because the temperature dependence of the surface tension is strongly dependent on the surface active element content of the weld pool. A simple measurement of weld pool temperature indicates that high enough temperatures (significantly above 2000°C) are achieved in stainless steel weld pools to account for the dependence of weld pool shape on increasing input power density with the surface tension driven fluid flow model. The depth/width ratio of both GTA and conduction mode electron beam weld pools passes through a maximum with increasing power density. A simple conduction calculation indicates that the peak weld pool temperature at which this maximum occurs is essentially the same for both heat sources.

Variations in weld pool shape with changes in material chemistry can be controlled by intentional doping of the weld pool with appropriate surface active elements, such as sulfur or oxygen in iron-base alloys. Controlling the sulfur content of certain stainless steels to roughly an 80-150 ppm range also eliminates poor and variable penetration with a minimal increase in hot cracking susceptibility. Controlling the sulfur and oxygen contents to very low values produces consistent, but low, penetration.

ACKNOWLEDGMENT

This work was supported by the Department of Energy, Albuquerque Operations Office. Their support is gratefully acknowledged.

REFERENCES

1. Heiple, C. R. and J. R. Roper, "Trends in Welding Research in the United States", pp. 489-520, American Society for Metals, Metals Park, OH (1982).
2. Heiple, C. R. and J. R. Roper, Weld. J. 61, 97-s to 102-s (1982).
3. Burgardt, P. and C. R. Heiple, "Interaction Between Impurities and Welding Parameters in Determining GTA Weld Shape", accepted for publication in Weld. J. (1986).
4. Kou, S. and Y. H. Wang, Weld J. 65, 63-s to 70-s (1986).
5. Keene, B. J., K. C. Mills, and R. F. Brooks, Mater. Sci. Technol. 1, 568-571 (1985).
6. Solomon, H. D. Unpublished research. General Electric Co., Schenectady, NY (1985).
7. Tsai, N. S., "Heat Distribution and Weld Bead Geometry in Arc Welding", Ph.D. Thesis, Massachusetts Institute of Technology, Cambridge, MA (1983).
8. Giedt, W. H., X. C. Wei, and S. R. Wei, Weld. J. 63, 376-s to 383-s (1984).
9. Sundell, R. E., S. M. Correa, L. P. Harris, H. D. Solomon, L. A. Wojcik, W. F. Savage, and D. W. Walsh, "Minor Element Effects on Gas Tungsten Arc Weld Penetration", 85SRD008, General Electric Co., Schenectady, NY (1985).
10. Block-Bolten, A. and T. W. Eagar, "Trends in Welding Research in the United States", pp. 53-73, American Society for Metals, Metals Park, OH (1982).
11. Burgardt, P., "Minor Element Effects on Electron Beam Welding", to be published, Rockwell International, Rocky Flats Plant, Golden, CO (1986).
12. Eagar, T. W. and N. S. Tsai, Weld. J. 62, 336-s to 355-s (1983).
13. Heiple, C. R. and P. Burgardt, Weld. J. 64. 159-s to 162-s (1985).
14. Casarini, R., F. Franchini, and B. S. Tranchina, "Seminar on Factors Influencing

the Penetration in TIG Welding of Austenitic Materials", Paper II-2, Italian Commission for Nuclear and Alternative Energy Sources (ENEA), Brasimone, Italy (1985).

15. Schickel, R., G. Schickel, and H. Vaugoyeau, "Seminar on Factors Influencing the Penetration in TIG Welding of Austenitic Materials", Paper II-3, Italian Commission for Nuclear and Alternative Energy Sources (ENEA), Brasimone, Italy (1985).
16. Brooks, J. A., Unpublished research, Sandia National Laboratories Livermore, Livermore, CA (1985).
17. Greer, D. A., Unpublished research, Rockwell International, Rocky Flats Plant, Golden, CO (1986).

CALORIMETRY OF PULSED VERSUS CONTINUOUS GAS TUNGSTEN ARC WELDS

G.A. Knorovsky, P.W. Fuerschbach
Sandia National Laboratories
Albuquerque, New Mexico 87185

Abstract

A technique to measure heat input to welds is used in conjunction with metallography to compare the heat inputs, penetrations, volumes melted, and arc and melting efficiencies of pulsed versus conventional Gas Tungsten Arc autogenous standing edge welds in thin gauge 304L stainless steel. These results are a necessary link between computer models and real weld machine settings. Results show that machine output and heat input correlate better with RMS values of current than with average values. Arc efficiency increased slightly and melting efficiency increased dramatically for increased travel speed at nominally constant heat inputs. Also, for two conditions of average current and travel speed corresponding to similar nominal heat inputs, one an efficient combination (81 amps at 12.2 mm/s) the other an inefficient combination (42 amps at 6.3 mm/s), current pulsing increased penetration only for the inefficient case. Comparing the processes on the basis of penetration divided by heat input, the continuous weld made under efficient conditions was best. However, some nonreproducible pulsed welds gave an exceptionally high value for this parameter. It is believed that resonance between the liquid pool motion and the pulse frequency may explain this phenomenon.

*This work performed at Sandia National Laboratories supported by the U.S. Department of Energy under contract number DE-AC04-76-DP00789.

CALORIMETRY EXPERIMENTS on Gas Tungsten Arc (GTA) welds were performed for several reasons. First, computer modeling of welds can reduce experimental effort in developing weld schedules for specific components; however, it is necessary to relate heat fluxes used in the models to real welding machine settings. Measurements of arc efficiency (defined as the ratio of heat input to the part divided by the energy output of the arc) are necessary to accomplish this. Second, it was desired to evaluate some of the claims made for pulsed current GTA welding: reduced heat input, increased penetration, etc. (1-3). Third, we wished to determine optimum conditions for pulsed GTA welding, where the optimum was defined as maximum penetration for minimum heat input. Many components designed by our organization have thermally-sensitive features (such as glass-to-metal seals or pyrotechnic materials) in close proximity to closure welds. Quite often the most severe requirement imposed on the weld is hermeticity, with strength of little importance. No calorimetrically-determined heat input data are available for continuous GTA welds of the arc current/travel speed/material thickness combinations employed in these parts. No calorimetric data is known to the authors for pulsed GTA welds. From the data that are available (4-6), melting efficiency tends to increase with current (at constant speed), and increases with travel speed at low speeds but decreases with travel speed at high speeds (at constant current). Arc efficiency tends to decrease with increased currents and arc lengths.

EXPERIMENTAL PROCEDURE

Calorimetry experiments were conducted using a Seebeck Envelope Calorimeter. The specimen is mounted in the calorimeter, which is a box with internal dimensions 150 mm x 150 mm x 75 mm. The box lid is left open during welding and closed immediately after the weld is complete. The walls of the box contain a double layer of thermocouple junctions which effectively integrate the heat flux through the walls of the box. The walls of the calorimeter are maintained at a constant temperature by a controlled temperature water source. This provides isolation from the environment, and gives a reference temperature for the outer layer of thermocouple junctions. Heat is thus measured by integrating the output of the calorimeter with respect to time and multiplying by a calibration constant. This type of calorimeter has the advantage that quenching of the specimen is not necessary, and if metallurgical examination of the weld is required, the microstructure will not be altered. Also, the calorimeter is robust and insensitive to shock, only one electrical signal needs to be monitored, and there is no constraint other than size on the weld geometry which may be studied. Electrical output of the calorimeter was measured on a digital storage oscilloscope, using a differential input, to reduce common mode noise. The output vs. time trace was then integrated on the oscilloscope via an internal program.

Voltage measurements of the arc were made between contacts on the torch electrode holder, and on the ground return cable. A relay was used to begin voltage measurement after the arc started to prevent high voltages from reaching the oscilloscope. To enable closure of the calorimeter, the ground cables were made of four symmetrically-placed flat stranded copper leads. After exiting the calorimeter, these leads were connected to a conventional welding cable. Current was measured at the ground return cable using a wide frequency response Hall-effect device. This device also insulated the oscilloscope from high voltages during arc starts. Arc energy was measured by multiplying the voltage and current waveforms together and integrating the product with respect to time (using an internal oscilloscope data manipulation program). The weld geometry studied consisted of a standing edge weld made between two pieces of 1.3 mm thick x 25 mm wide x 127 mm long 304L stainless steel. This geometry was chosen because it is representative of actual welds in miniature electromechanical components. Welds were about 100 mm long. A transistorized 300 Amp power supply was used. W-2% Th electrodes of 1.6 mm diameter with a machine ground 90 degree conical tip were used. An initial arc gap of 1 mm was set. Argon gas flowing at 11 l/m shielded the welds. Metallography was used to determine fusion zone cross sectional area (using a planimeter) and centerline penetration at four locations in each weld. Preliminary welds showed that a nominally-calculated heat input of approximately 60 J/mm provided a reasonable amount of melting. All subsequent welds were made with a similar nominal heat input/unit length. Calorimetric measurements showed that the actual heat inputs for the continuous welds did remain nearly constant. Omar et.al. (1) have stated that average current is the pulsed GTA weld parameter controlling the degree of melting, therefore, pulsed weld parameters were chosen to maintain constant average current while varying both peak current and the percentage of time at peak current (hereafter called the duty cycle). The parameters used are listed in a later section. A true square wave pulse shape was obtained. The frequency of pulsing was chosen to obtain approximately 70% spot overlap; this corresponded to about a 0.8 mm spot spacing. This spot spacing has been found necessary to ensure hermeticity in pulsed welds. However, it is not the maximum possible as recommended by Becker et.al. (2).

EXPERIMENTAL RESULTS AND DISCUSSION

EFFECT OF WELD PARAMETERS ON WELD FUSION ZONE AND ARC EFFICIENCY - A series of welds were made at constant nominal heat input using speeds and currents as listed in Table 1. The centerline penetrations for these welds are shown in Fig. 1. Penetration increased with travel speed until a plateau was reached at ≈ 0.8 mm, at a travel speed of 8.3 mm/s.

TABLE 1 - CONTINUOUS WELD PARAMETERS

Current	Travel Speed
29 A	4.2 mm/s
55 A	8.3 mm/s
81 A	12.2 mm/s
107 A	16.1 mm/s
132 A	19.9 mm/s

The fusion zone cross-sectional area increased continuously with travel speed (Fig. 2); however, it became kidney-shaped as higher currents and speeds were used (Fig. 3). Arc efficiency increased very slightly over the same range (Fig. 3), while melting efficiency (which is defined here as the amount of heat necessary to melt the fusion zone (7) divided by the heat input to the part, rather than having the total arc energy output as the denominator) increased quite markedly (Fig. 4). Average current and travel speed conditions corresponding to penetration levels of ≈ 0.8 mm and ≈ 0.4 mm for continuous GTA (12.2 mm/s at 81 Amps, and 6.3 mm/s at 42 Amps, respectively) were chosen to investigate the effect of pulsed current. Duty cycles ranged between 20% and 100% at the 42 Amp level, and 40% to 100% at the 81 Amp level

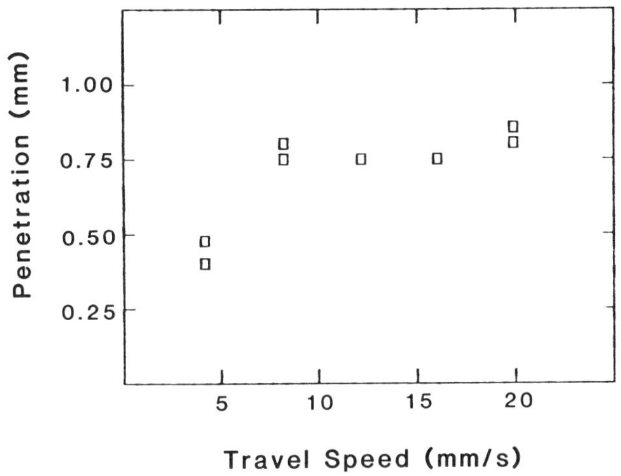

Fig. 1 Centerline penetration vs. weld travel speed at constant nominal heat input for continuous GTA welds.

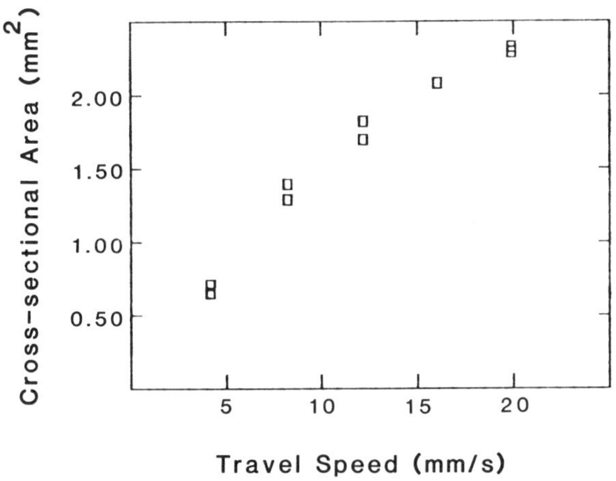

Fig. 2 Cross-sectional area vs. weld travel speed at constant nominal heat input for continuous GTA welds.

(lower duty cycles resulted in humped, non-uniform welds). Table 2 shows the current and time settings. Penetration (Fig. 5) increased with decreasing duty cycle for the slower travel speed condition, and remained essentially constant for the faster travel speed condition. Penetration produced by the 20% duty cycle at the 42 Amp level nearly equalled the best found for the 81 Amp welds.

HEAT INPUT VS CURRENT - For the continuous GTA welds, the heat inputs measured remained essentially constant for the currents and travel speeds used. However, this was not the case when duty cycle was varied at constant average current for the pulsed GTA welds. Plotting heat input versus RMS current (analogous to heat input in resistance heating) seems to indicate a better correlation (Fig. 6); however, the lines drawn through the data do not go through the origin as would be expected if RMS current alone were the controlling parameter.

MELTING EFFICIENCY VS DUTY CYCLE - The plot of Melting Efficiency versus Duty Cycle (Fig. 7) demonstrates the relative merits of pulsed versus continuous GTA welds for the conditions of this experiment. First of all, melting efficiency can be either very dependent or nearly independent of duty cycle for pulsed welds. When one is welding under efficient conditions (high travel speed), there does not seem to be any advantage to pulsing; when one is welding under inefficient (slow travel speed) conditions, the shorter duty cycle welds generate higher melting efficiencies. In the latter case, the melting efficiency nearly equals that obtained under efficient weld conditions. This is of particular interest in welding a specimen which tends to hot crack at travel speeds corresponding to efficient conditions. Since melting efficiency is proportional to the melt volume, it is clear that average current does not necessarily control the degree of melting.

PENETRATION/HEAT INPUT VS SPEED - When welding near heat sensitive parts, the most

TABLE 2 - PULSED CURRENT WELD PARAMETERS

Duty Cycle	Peak		Background	
	Current	Time	Current	Time
6.3 mm/s Travel Speed, 8.3 Hz:				
20%	181 Amps	0.024 s	7 Amps	0.096 s
40%	94 Amps	0.048 s	7 Amps	0.072 s
60%	65 Amps	0.072 s	7 Amps	0.048 s
80%	51 Amps	0.096 s	7 Amps	0.024 s
Cont.	42 Amps	-	-	-
12.2 mm/s Travel Speed, 16.7 Hz:				
40%	191 Amps	0.024 s	6 Amps	0.036 s
50%	153 Amps	0.030 s	6 Amps	0.030 s
60%	131 Amps	0.036 s	6 Amps	0.024 s
70%	112 Amps	0.042 s	6 Amps	0.018 s
80%	99 Amps	0.048 s	6 Amps	0.012 s
Cont	81 Amps	-	-	-

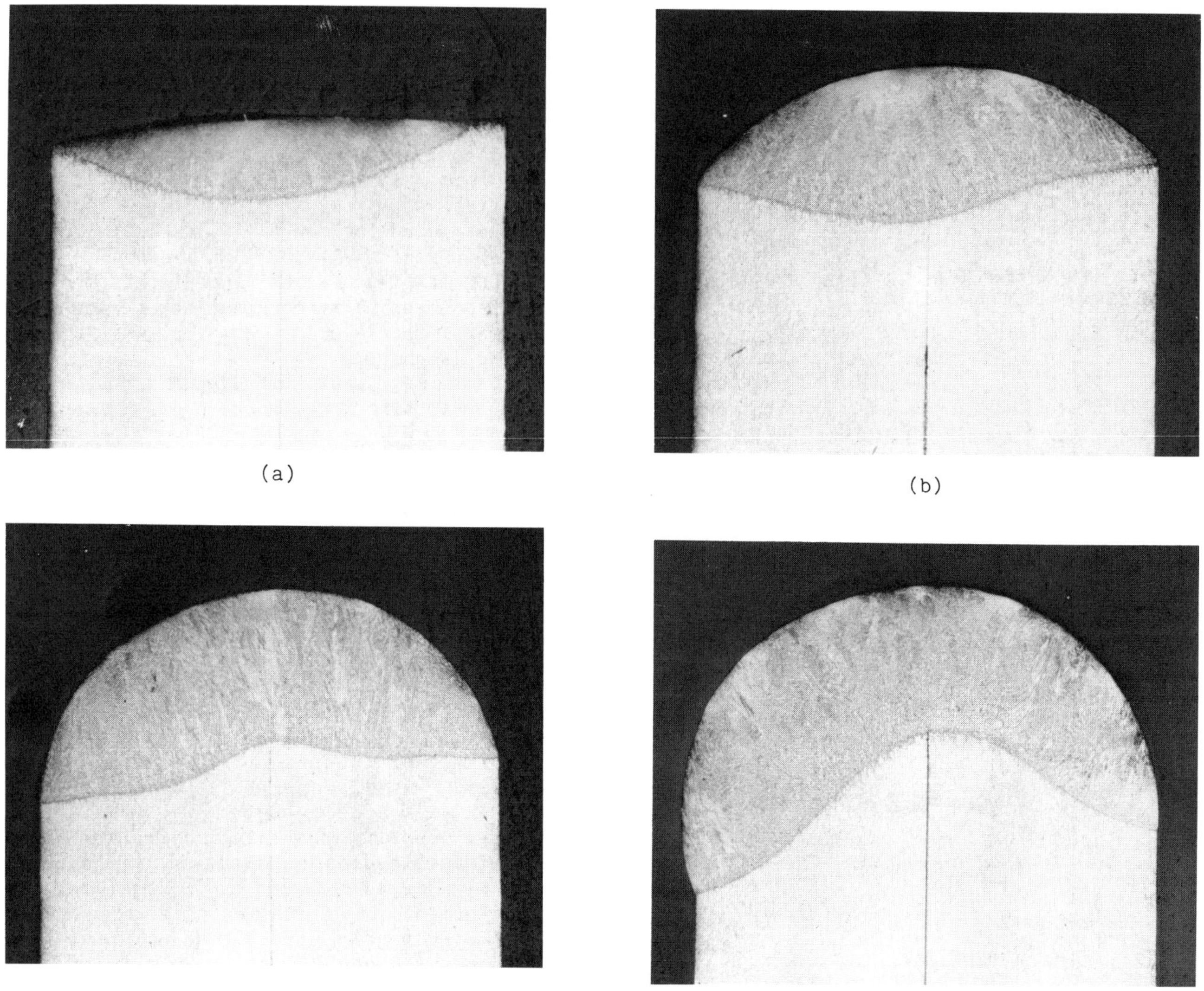

Fig. 3 Photos showing weld cross sectional area vs weld travel speed for continuous GTA welds at constant heat input. (a) 4.2 mm/s, (b) 8.3 mm/s, (c) 12.2 mm/s, (d) 19.9 mm/s. (31.6x)

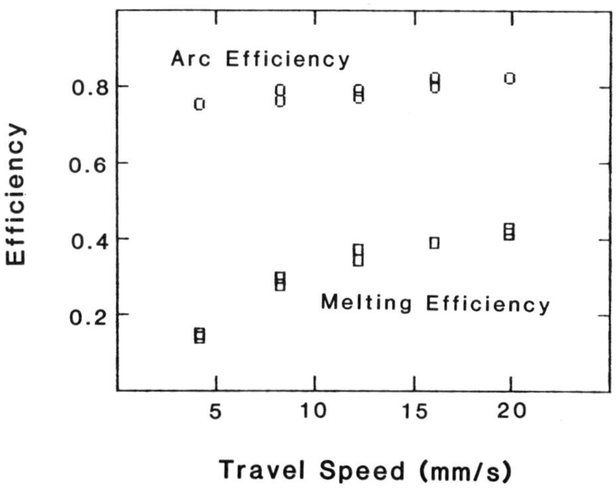

Fig. 4 Arc efficiency and melting efficiency vs. weld travel speed at constant nominal heat input for continuous GTA welds.

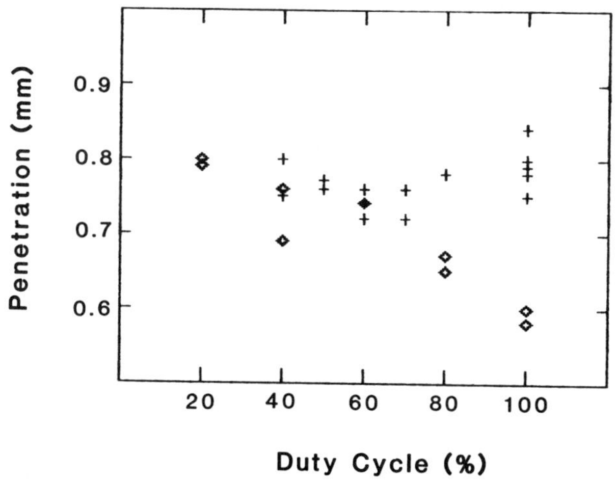

Fig. 5 Centerline penetration vs. duty cycle for pulsed GTA welds. Crosses correspond to 81 Amps average current at 12.2 mm/s; diamonds correspond to 42 Amps average current at 6.3 mm/s.

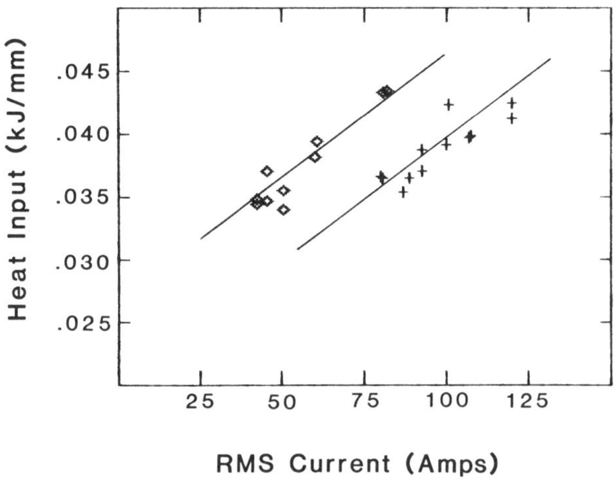

Fig. 6 Heat input vs RMS current for pulsed GTA welds. Crosses correspond to 81 Amps average current at 12.2 mm/s; diamonds correspond to 42 Amps average current at 6.3 mm/s.

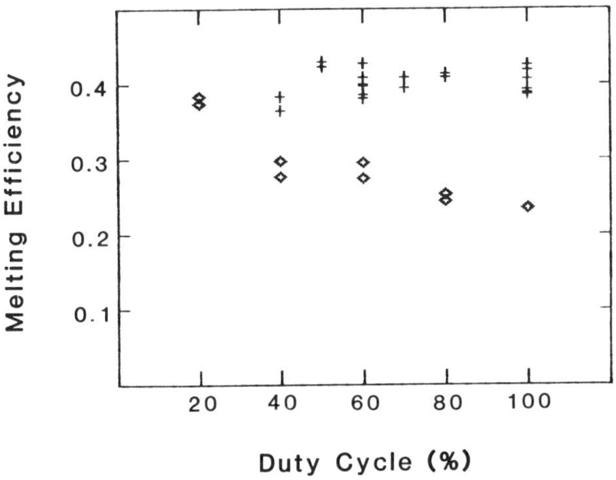

Fig. 7 Melting efficiency vs. duty cycle for pulsed GTA welds. Crosses correspond to 81 Amps average current at 12.2 mm/s; diamonds correspond to 42 Amps average current at 6.3 mm/s.

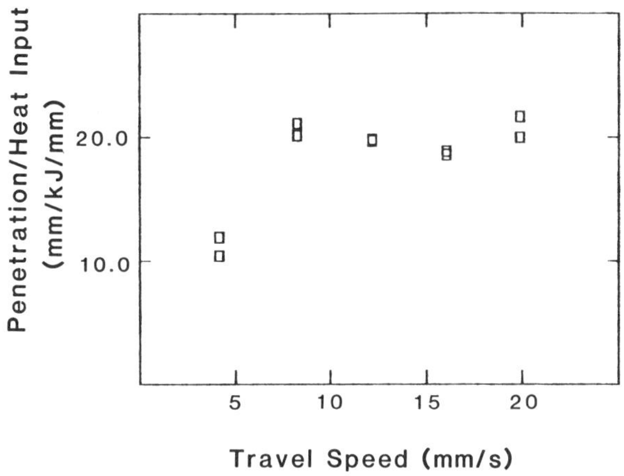

Fig. 8 Penetration/heat input vs. weld travel speed for continuous GTA welds at constant heat input.

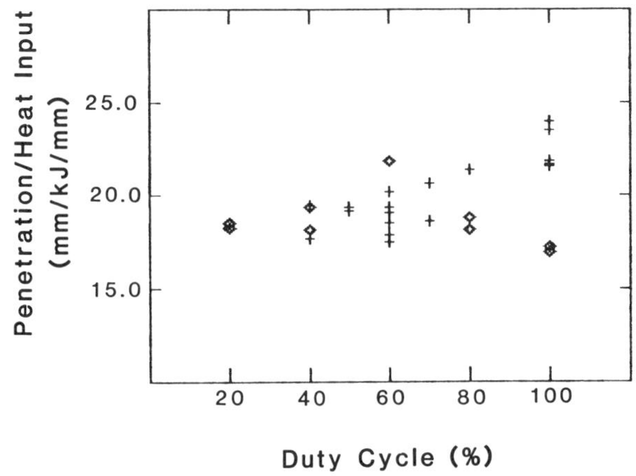

Fig. 9 Penetration/heat input vs. duty cycle for pulsed GTA welds. Crosses correspond to 81 Amps average current at 12.2 mm/s; diamonds correspond to 42 Amps average current at 6.3 mm/s.

prudent approach is to minimize heat input for a given required strength level. Thus, an optimal weld condition is one which maximizes weld penetration at minimal heat input. Plotting heat input normalized by penetration shows where this condition is reached for continuous GTA welds in this study. Fig. 8 demonstrates a rather insensitive behavior once a critical travel speed/current condition is reached. By adding the pulsed current feature (Fig. 9), one sees that there is no advantage to pulsing when operating under efficient conditions, but that an improvement can be made at intermediate duty cycles when operating at inefficient conditions. Comparing Fig. 9 with the earlier figure showing melting efficiency vs duty cycle (Fig. 7) shows that the condition of maximal melting efficiency does not correspond to the condition of minimum heat input for maximum penetration for inefficient conditions. Even though penetration and melting efficiency increase with decreasing duty cycle, the increase in heat input is faster.

GENERAL DISCUSSION OF OBSERVED BEHAVIOR - Quigley et.al (8) have discussed the various mechanisms contributing to heat transfer to the workpiece in gas tungsten arc welds. Among the factors discussed are 1) electron work function, 2) the electrons' thermal energy, 3) heat dissipation in the anode fall region, 4) conduction, 5) convection effects of the arc plasma, 6) radiation, 7) re-radiation from the arc and workpiece respectively, and 8) metal vaporization effects. By increasing the travel speed at constant arc energy per unit length, the time available for the weld is made shorter, which minimizes factors 7 and 8. Also, by increasing the arc current, factor 5 should be increased. This combination of factors may explain the slight increase in arc efficiency with travel speed. In Quigley's theoretical treatment, factors 5, 7 and 8 accounted for only about one fifth of the total heat transferred. Factors 1 and 3 accounted for the largest contributions (60% and 21% of the total, respectively).

The increased melting efficiency with increased weld speed is most probably explained by time considerations; if time is shortened for conduction, less heat is extracted from the weld pool, enabling it to remain larger for a given heat input. Berezovskii (9) calculated that melting efficiency due to a given size surface heat flux would increase as the welding speed increases, until the velocity increased to the point where the heat flux size became larger than the molten zone size, after which the efficiency would drop. We are not in this latter regime for the conditions of this study.

For the pulsed current experiments, the inefficient condition led to increased melting efficiency at short duty cycles, while the efficient condition was insensitive to duty cycle. Under already efficient conditions of heat input, it seems likely that reducing the arc current for part of the time may counteract any benefit derived by being at a higher current the rest of the time. Under inefficient conditions, the more efficient heat transfer while at high arc current apparently can overcompensate for the reduced time at high current.

ANOMALOUS PENETRATION RESULTS - Some early welds made in this study were found to exhibit much deeper penetration than others made at nominally identical conditions. Figure 10 illustrates the cross section of one such weld. Part of the irreproducibility was attributed to having hand-rather than machine-ground tungsten electrodes. However, a brief study was conducted with electrodes ground to sharper tip angles (which seemed to accentuate the phenomenon somewhat). A high speed video system was used to record the motion of the weld pool surface. He:Ne laser cross illumination was used in conjunction with narrow bandpass optical filters to create a shadow on the video camera's lens. The standing edge weld configuration lends itself admirably to this type of observation. It was found that the pool motion was quite complicated, involving a frequency corresponding to the pulse frequency as well as a number of others. When these other frequency motions interfered constructively with the fundamental, a much larger puddle motion occurred. We attribute the deeper penetration to this motion "uncovering" the pool root and allowing arc heat input at a position much nearer to the liquid/solid interface than would otherwise be the case. Since the frequencies and modes of vibration are functions of many parameters

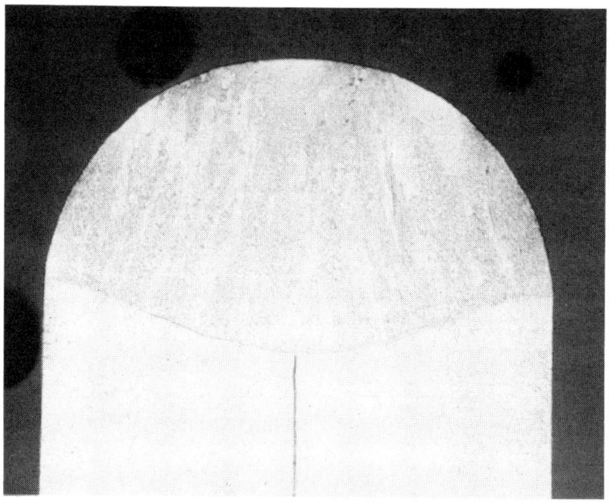

Fig. 10 Anomalous penetration cross-section photo (31.6x) made with 60% duty cycle, 81 Amps average current, 12.2 mm/s.

(including the puddle size and shape, the weld schedule parameters, and the vibration of the table, which was driven by a stepping motor), it is clear that a complicated resonance phenomenon may be involved. If one could consistently reproduce this resonance condition the penetration/heat input parameter could be almost doubled.

SUMMARY

A calorimetric technique has been developed to obtain heat inputs of welds in a geometry representative of miniature components. These results will be useful in attempts to use computer modeling to develop weld schedules. We have applied the calorimetric technique to GTA welds which range from 29 to 132 Amps at travel speeds from 4.2 to 19.9 mm/s. By varying current and travel speed proportionately, an approximately constant heat input of 37 J/mm was maintained. Arc efficiency increased slightly and melting efficiency increased dramatically with travel speed over the above range. Pulsed current GTAW welds at average current/travel speed combinations within the above range have also been investigated. Melting efficiencies have been determined, and the optimum penetration for minimal heat input condition identified. The effect of varying duty cycle at constant average current has also been investigated. It was noted that heat input correlates better with RMS current than with average current, and that pulsing is beneficial to penetration and melting efficiency only when the average current and travel speed correspond to an inefficient combination. At these inefficient conditions, decreasing duty cycle led to penetration and melting efficiency values which approached those of a highly efficient combination for continuous GTA. An anomalous high penetration condition occasionally obtained for certain pulsed current welds was tentatively attributed to resonant motion of the weld pool.

ACKNOWLEDGEMENTS

We would like to thank Mr. P. D. Thatcher for his help with calibration and instrumentation, and Mr. A. J. Romero for his metallographic assistance.

BIBLIOGRAPHY

1. A. A. Omar, C. D. Lundin, "Pulsed Plasma-Pulsed GTA Arc: A Study of the Process Variables," Welding Journal Research Supplement, (58), 1979, p. 97-s.

2. D. W. Becker, C. M. Adams Jr., "Investigation of Pulsed GTA Welding Parameters," Welding Journal Research Supplement, (57), 1978, p. 134-s.

3. R. E. Leitner, G. H. McElhinney, E. L. Pruitt, "An Investigation of Pulsed GTA Welding Variables," Welding Journal Research Supplement, (52), 1973, p. 405-s.

4. R. W. Niles, C. E. Jackson, "Weld Thermal Efficiency of the GTAW Process," Welding Journal Research Supplement, (54), 1975, p. 25-s.

5. N. Collings, K. Y. Wong, A. E. Guile, "Efficiency of Tungsten-Inert-Gas Arcs in Very-High-Speed Welding," Proc. IEEE, (126), 1979, p. 276.

6. H. W. Ghent, D. W. Roberts, C. E. Hermance, H. W. Kerr, A. B. Strong, "Arc Efficiencies in TIG Welds," in Arc Physics and Weld Pool Behavior, The Welding Institute, Cambridge. 1980, p. 17.

7. L. Leibowitz, "Properties for LMFBR Safety Analysis," Argonne National Laboratory, ANL-CEN-RSD-76-1, April 1976.

8. M. B. C. Quigley, P. H. Richards, D. T. Swift-Hook, A. E. F. Gick, "Heat Flow to the Workpiece from a TIG Welding Arc," J. Phys. D: Appl. Phys., (6), 1973, p. 2250.

9. B. M. Berezovskii, "The Thermal Efficiency of the Process of Penetrating Metals with a Welding Arc at the Surface," Automatic Welding (English translation), (10), 1979, p. 15.

PERFORMANCE EVALUATION OF METAL POWDER CORED STRIP ELECTRODES FOR HIGH SPEED ELECTROSLAG WELDING

Stephen Liu, Chwen Tzeng Su
Department of Industrial Engineering
The Pennsylvania State University
University Park, Pennsylvania 16802

ABSTRACT

The high heat input of electroslag welding is the factor responsible for the high deposition rate and welding speed. However, large heat affected zone with thick coarse grained region also results with low toughness, which sometimes prohibits the process from being used in critical component fabrication. Recent efforts in electroslag welding research are directed to minimize the heat input effect by using high speed processes which distribute heat more evenly along the weld.

In this work, a metal powder cored strip electrode was designed for electroslag welding. The powder to strip weight ratio was 3:1. Flat guide plates were used on both sides of the electrode. Due to the geometry and dimensions of the electrode and guide, much higher current and melt rate were reached. The average rate of weld metal deposition was approximately two to three times higher than the solid wire process. Combining experimental observation and theoretical calculations, it was possible to determine a range of welding parameters with satisfactory base metal penetration for the cored strip electrode.

The temperature distribution in the weld heat affected zone was also affected by the cored strip electrode. The joint edges of the base metal experienced higher temperatures than the conventional electroslag welds. Even though the welding current is higher than that used in the conventional electroslag welding processes, the specific heat input of the welds made with the metal powder cored strip electrodes actually decreased. Macrographic and micrographic evidences also showed that the final weld structure was refined by the application of higher welding current. Thus it is possible to utilize high speed electroslag welding process with the metal powder cored electrode to produce sound welds with good structural and mechanical properties.

ELECTROSLAG WELDING is an excellent method for the joining of thick components. Since its development in the early 1950s (Ref.1), it has been used in shipbuilding and structural welding of heavy machineries, pressure vessels, etc. It has the characteristics of large weld deposit, high heat efficiency, and is able to join vertical or near vertical plates in one single pass. As compared with other welding processes, only minimum joint preparation is required with little weld distortion.

The process is initiated by an electric arc between the electrode and the bottom of the joint Pieces of steel wool are generally introduced to facilitate the arc establishment. A starting flux is then added to form rapidly a molten slag layer covering the liquid metal. Gradually the slag pool will rise and extinguish the arc. During the process, the heat necessary for welding is generated by the resistance of the molten slag layer to electric current passing between the electrode and the work. The inherent high heat input makes the process unsuitable for some applications. This restriction can be related to the metallurgical transformations that occur in the heating and cooling cycle of the weldments. The extremely slow cooling rate will result in coarse weld solidification structure with large columnar grains and severe tramp element segregation (Ref.2-3). In the case of low alloy steel electroslag welding, it is not unusual to observe $\Delta t_{8/5}$ of 200-400 seconds (Ref. 4). The final microstructure is generally constituted of large volume fraction of grain boundary ferrite, ferrite sideplates, bainite, and microconstituints (Ref.5). Some acicular ferrite may also be seen. As result, electroslag welds present poor mechanical properties with low strength and toughness.

The same cooling rate effect can also be seen in the base metal next to the fusion line. Extended period of heating at high temperatures coarsens the inclusions of second phase particles allowing the austenite grains to grow. Subsequent slow cooling will not be able to refine the grain structure. Large volume fraction of allotriomorphic structures with carbide precipitation and ferrite sideplates will also reduce the toughness of the coarse grained heat affected zone of the weld (Ref.2,5). Post weld heat treatments are often specified to improve the notch toughness of the weld metal and the heat affected zone such that they satisfy the mechanical requirements. This is sometimes uneconomical, and other times, impracticable.

More recent efforts in electroslag welding are concentrated in the development of processes which improve productivity as well as enhanced toughness of the as welded joints. The most effective way is to decrease the overall heat input of the process by decreasing the volume of the weld deposit and the dimension of the root gap. This will lead to an increase in the heating and cooling rates, with shorter $\Delta t_{8/5}$, refining the weld metal and heat affected zone microstructure.

TECHNIQUES OF HIGH SPEED ELECTROSLAG WELDING

It was shown that in order to achieve the goal of large volume of deposited metal and lower heat input, the electrode melt rate and welding speed had to be increased. Responding to this technological challenge, many processes were developed attempting to maximize the welding speed and minimize the heat input. The individual characteristics of these processes are reviewed below.

Early in the 1970s, Avramenko, Lebedev and Bozhko (Ref.6) suggested that the most expedient way of increasing the productivity of the electroslag welding process is the reduction of the root gap with a simultaneous increase of electrode wire feed rate. The process decreases mainly the amount of weld metal deposited in the weld joint and the time needed to complete the weld. The extent to which the cross sectional area of the gap can be reduced is limited by the stability of the process and the size of the electrode and consumable guide tube inserted into the gap. The increase of electrode wire feed rate actually corresponds to an increase of the welding current. Contrary to the arc welding processes, an increase of welding current does not necessarily increase the heat input. This is because current affects both the electrode melt rate and weld travel speed. The wire feed rates of higher welding currents will reduce the specific heat input and decrease the weld penetration (Ref.7). Watanabe, Sejima, Kokura, Taki and Miyake (Ref.8) took a similar approach and developed the method of narrow gap electroslag welding. Figure 1 shows the electrode setup for the process. Strip electrodes of cross sectional dimensions of 20mm x 0.8mm (3/4in. x 1/32in.) were used. Blocks of solid flux were inserted between the electrode and base metal to avoid arcing. For welds made with 800 amperes, 34 volts and a root gap of 15mm (5/8in.), the welding speed was about 0.6 inch per minute. The specific

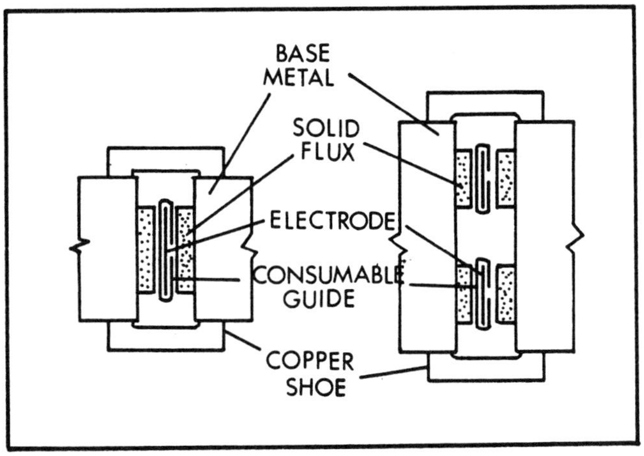

Figure 1. Schematic diagram showing the electrode configuration with the consumable guide and solid flux.

heat input was approximately 10^5 joules per square centimeter. Improved weld mechanical properties were also observed.

The concept of adding powder metal filler to increase the deposition rate was introduced in the early 1960s with the Bulkweld Process (Ref.9) More recently, Reynolds & Kachelmeier (Ref.9) reported some of their experience of metal powder filler in hardfacing and cladding. They indicated very successful results of incresed deposition rate, decreased heat input, and reduced consumption of fluxes. Fine powder can be easily melted by the excess process heat of the arc and deposited into the joint. Mild steels, low alloy steels, low and high temperature service steels are some of the materials that they welded using the method described. The powder to electrode weight ratio was typically 1:1. As compared to welds without powder addition, no microstructural change in the weld metal could be observed. However, the heat affected zone size was observed to decrease and could be related to the lower heat input of the process. Metal powder additions also provide known and exact chemical composition filler to the weld. Smirnov & Efimenko (Ref.10) reported that powdered filler metal in electroslag welding refined the weld metal and heat affected zone microstructure. Low temperature impact strengths of the weldments in the as welded condition were also observed to have doubled. They further concluded that with powder to electrode weight ratio of 1.7:1, heat input could be reduced by sixty percent. As result, the volume of molten metal pool and the time of solidification were also reduced by fifty percent, Figure 2. Eichhorn, Remmel, and Wubbels (Ref.11-13) mentioned in their work that the powder metal filler without electric current will increase the deposition rate while reducing the overheating of the weld pool. They used different electrode configuration and powder delivery location to successfully increase the welding speed to approximately nine inches per minute. A simultaneous reduction of energy input was also observed. The metal powder is

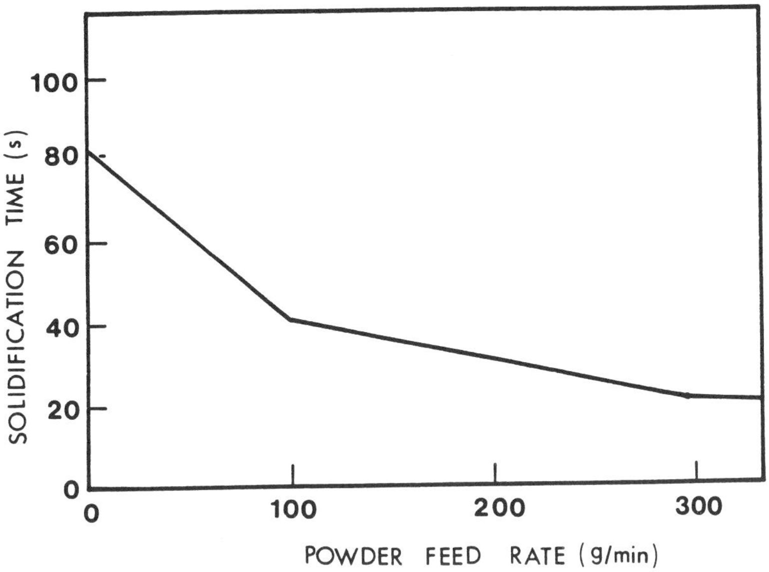

Figure 2. Plot of solidification time as a function of powder feed rae (Ref.10)

supplied to the weld by a current of argon gas, as shown in Figure 3. Once the powder reach the root gap, it will adhere to the electrode and move towards the weld pool. The quantity of metal powder transferred was 120%. Notch impact testing results measured with samples taken at a distance of 1 and 2 mm (0.04 and 0.08in.) from the fusion line showed significantly high Charpy impact toughness.

Atteridge, Ventkataraman, and Wood (Ref.7) initiated an extensive research program in electroslag welding and investigated the effects of root gap, welding current, electrode and consumable guid tube geometry, non-consumable mullite and quartz shroud, and electrode oscillation. They found that plate grounding and electrode position were critical in the control of weld symmetry. Narrow gap welds

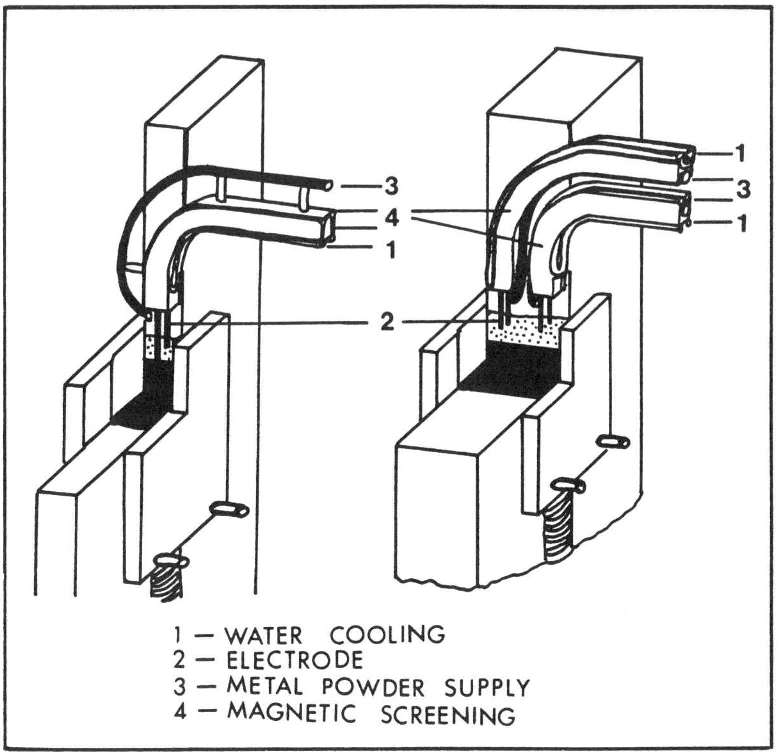

Figure 3. Powder metal filler addition without electric current to increase the welding speed and reduce the specific heat input (Ref.11-13).

(19mm root distance) resulted in higher base metal dilution than standard gap welds (32mm) and dilution increased linearly with decreasing joint spacing. Welding current was found to be influenced by the geometry of the electrode or guide tube, but it showed minimum influence on penetration. For a wing shaped guide tube, lower voltage was required to obtain complete penetration. However, increased current was necessary to achieve specific weld heat input reduction. The wing shaped guide tube eliminated the need for electrode oscillation. Combining high current (1000A), narrow gap, and wing shaped guide tube, the welding speed was increased significantly (from 0.67 to 2 inches per minute). The specific heat input was reduced by fifty percent. The introduction of non-consumable shrouds into the slag pool refined the as welded microstructure without the need of mechanical vibration. This was achieved because of the perturbation of the solidification front (Ref.14) and the intense weld pool stirring action generated due to the sharp thermal gradient established in the slag pool, Figure 4. Columnar

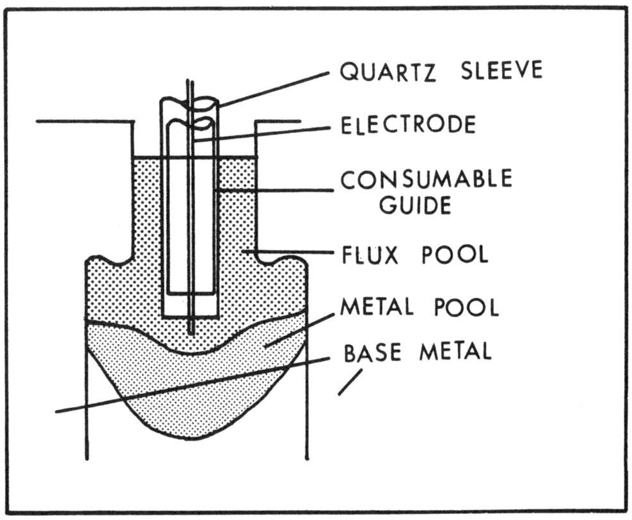

Figure 4. The introduction of a non-consumable quartz shroud modifies the weld pool form factor and refines the solidification structure.

grain width was reduced approximately by ten times.

In a more recent article, Eagar (Ref.15) examined the applications and trends of electroslag technology in Japan. He reported that the limitation of electroslag welding is the poor fracture toughness of the weld heat affected zone. Even though many methods had been tried to improve the as welded toughness of the weld metal and heat affected zone, there was not much success achieved.

It is the objective of this work to report some of the efforts in the development of a high speed electroslag welding process at the Penn State University. It involves the design of a special cored strip electrode which add metal powder to the weld pool in a more efficient way. It also incorporates the ideas of reduced root gap distance, electrode geometry, chemical composition modification, and high welding current and electrode melt rate.

EXPERIMENTAL PROCEDURE

ASTM 588 grade structural steel was chosen for the experiment. Plates of 305mm x 305mm x 38mm (12in. x 12in. x 1-1/2in.) were used. During the initial phase of the program, 2in., 4in., and 6in. high plates were also used to explore the ranges of welding parameters. The electrode has the following cross sectional dimensions: 35.4mm x 4.5mm (1-5/16in. x 11/64in. The width of the electrode covers almost the entire thickness of the base plate used. Figure 5 illustrates the electrode to joint geometry.

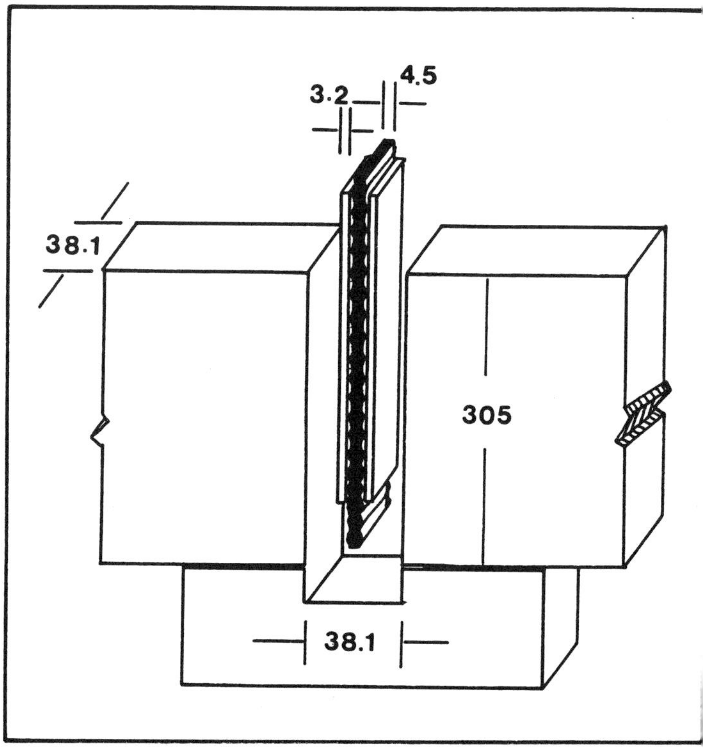

Figure 5. Schematic drawing showing the electrode to joint and guide plate geometry. (Unit: mm)

A high alloyed electrode was used to perform the first experiments and a low carbon manganese steel electrode completed the rest of the weld specimens. No flux was added. The chemical composition of the two electrodes and the base plate are given in Table I.

Table I. Chemical composition of the electrodes and base metal used in the welding experiments.

(%)	Electrode #1	Electrode #2	Base Metal
C	0.05	0.12	0.26
Mn	0.80	1.00	0.99
Si	0.40	0.40	0.20
Cr	10.00	-	-
Ni	5.00	-	-
Mo	0.70	-	-

As shown in Figure 6, the surface of the electrode is corrugated for the feeding mechanism and also the retention of powder in case the end of an electrode is cut open. The electrode feeder was custom designed and fabricated for the research project. The guide plates were also made of low carbon steel of approximately 1/8in. of thickness. Figure 5 also shows the relative position of the strip electrode with respect to the guide plates. The root gap size was held constant for all the welds (38.1mm) and is also shown in Figure 5. For the balance of the electric current path, split grounding was used. There was no electrode oscillation during welding.

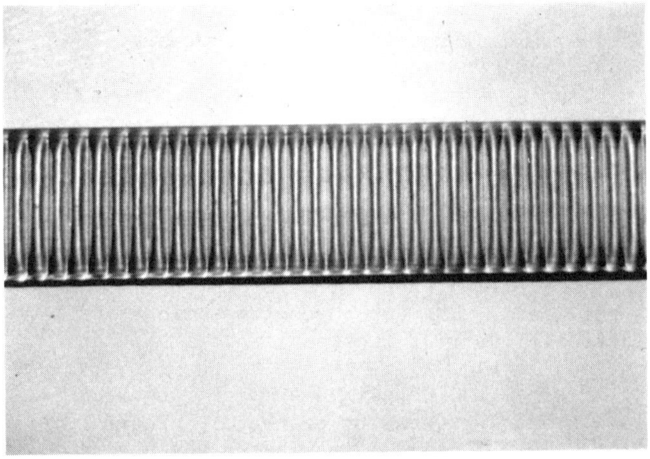

Figure 6. Photograph of a cored strip electrode showing the corrugated surface.

Comercial starting and running fluxes were used to carry out the welding. The chemical composition and the basicity of the running flux is shown in Table II.

Table II. Typical chemical composition of a running flux for electroslag welding.

CaO	12.2%
MgO	2.3%
MnO	22.5%
CaF_2	8.6%
SiO_2	33.0%
Al_2O_3	8.3%
TiO_2	8.0%
K_2O	0.9%
Na_2O	0.6%
FeO	1.8%
B.I.	0.9

Metallographic analyses and Rockwell hardness testing were performed on each weldment to determine the weld structure and hardness. The heat affected zone (particularly the coarse grained region) was also measured for comparative studies.

For the temperature distribution and heat extraction conditions in the heat affected zone, thermocouples were inserted into drilled holes in the plates and monitored throughout the heating and cooling cycle. Temperature profiles of some of the welds were obtained and compared to data available in the literature.

RESULTS AND DISCUSSION

PROCESS VARIABLES - The primary variable which influence the degree of base metal melting, weld metal dilution, and the heat affected zone microstructures and properties is the power input per unit length of a weld. Particularly in electroslag welding, it is imperative that the heat input be controlled to a minimum level. This is also the reason for the development of many research projects in the area of high speed electroslag welding.

To evaluate the performance of the metal powder cored strip electrodes, the operating parameters for fully penetrated welds need to be established. The model developed by Edwards, Frost, and Jones (Ref. 16) was followed, with some modifications, to derive the functional relations between the various process parameters. Conditions for adequate penetration of the base metal as a function of voltage, current, and the geometry of electrode, guide plate, and weld joint were identified. Experimental data of welds and theoretical predicted results were used for the determination of the range of optimum welding current and voltage.

First the relationship between the electrode feed rate (F) and the welding speed (S) was determined by the mass balance equation,

$$S = \left[\frac{w_e t_e f}{(Lg - 2w_g t_g)}\right] F \qquad (1)$$

where w_e is the width of the electrode, t_e is the thickness of the electrode, f is the compacting factor of the metal powder in the electrode, L is the plate thickness, g is the root gap, w_g and t_g are the thickness and width of the guide plate, respectively.

It is known that for a given constant voltage, current increases with electrode feed rate. Several investigators (Ref.1,5,17-19) reported a linear relation, while others (Ref.19) found a non-linear behavior. In the case of metal powder cored strip electrodes, the following relationship was observed.

$$i = AF^{0.35} \qquad (2)$$

where i is the welding current and A is a proportionality constant. Figure 7 is a plot of current versus electrode feed rate for welds made with the cored strip electrode at a constant potential of 40 volts.

Further analysis showed that electrode feed rate is also a function of welding potential. Using experimental data, a constitutive equation between current, electrode feed rate and welding potential was determined. Figure 8 is a plot of this equation. Good correlation was observed with $r^2 = 0.8813$.

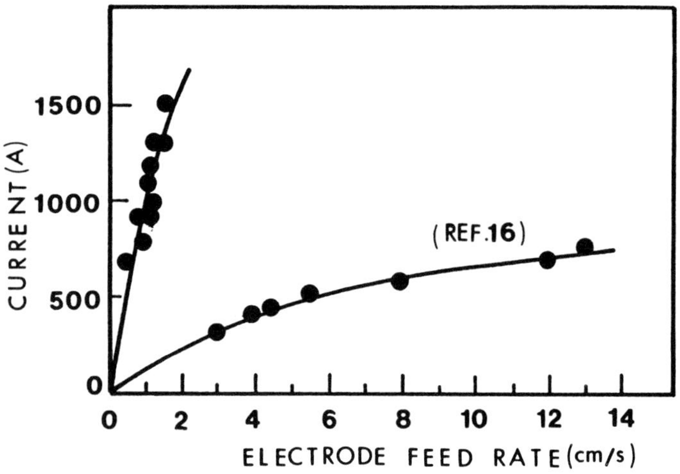

Figure 7. Welding current plotted as a function of electrode feed rate.

$$i = A' F^{0.35} V^{1.03} \qquad (3)$$

A' is a constant in this equation.

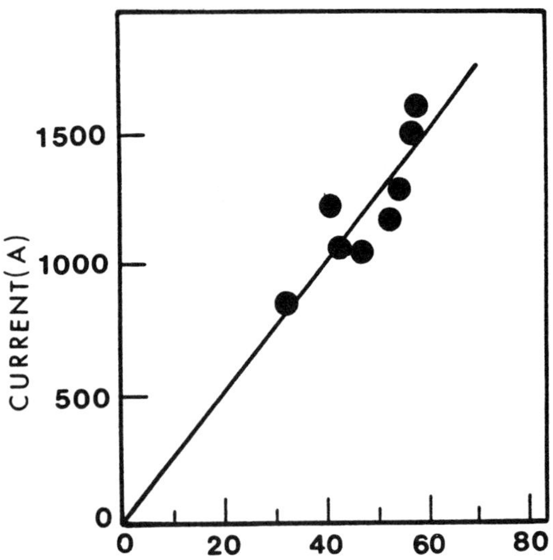

$F^{0.35}V^{1.03}$

Figure 8. Welding current plotted as a function of electrode feed rate and welding potential.

As mentioned earlier, the problem of electroslag welding is the great amount of heat absorbed by the base metal resulting in large coarse grained heat affected zone. Therefore, the exposure time of the plate surface to the flux pool becomes extremely important in the determination of weld metal and heat affected zone mechanical properties. The exposure time, t, can be determinedby the following equation,

$$t = \frac{d}{S} \qquad (4)$$

where d is equal to the depth of the flux pool. The total heat input per unit area of weld surface, E, can be defined as:

$$E = \frac{iVt}{2[L+(g-2t_g)]d} \qquad (5)$$

where $(g-2t_g)$ is the "effective" gap size. Substituting equation 3 and 4 in equation 5, E can be determined as a function of current, potential, and weld geometry.

$$E = B\frac{(Lg-2w_g t_g)}{[L+(g-2t_g)](w_e t_e)} \cdot \frac{V^4}{i^2} \qquad (6)$$

B is a proportionality constant. This equation indicates that for a cored strip electrode, the energy input is proportional to the fourth power of the welding potential, and inversely proportional to the square of the current. This is reasonable because an increase in the electrode feed rate will increase the pool velocity, decreasing the overall heat input. Similarly, a faster cooling rate will result, generating fine microstructure in the weldments. Comparing the constitutive equation derived for the strip electrode to that developed by Edwards et al. (Ref.16), the effect of current was more significant. The decrease of heat input with increasing current becomes larger at higher current levels. This is due to the large volume of molten metal deposited at higher current levels and the geometry of the electrode/guide plate assembly which distributes the heat generated and molten metal more evenly to a longer section of weld.

In order to determine the current-voltage operating parameter range, energy balance of the process was performed. The power contributions required to melt the electrode, to fuse the base plate surfaces, and to provide for the various heat losses were included in the analysis. A significant portion of the power input is lost to the water cooled copper retaining shoes, flux pool radiation, guide plate resistive heating, and weld puddle superheating. Figure 9 is a plot showing the operating range for the metal powder cored strip electrode in electroslag welding. Boundary A was determined using energy balance equations and represents the voltage threshold for base metal fusion at lower energy inputs. Boundary B represents the constitutive equation for adequate penetration at high power levels. Empirical results for weldments with unwetted joints were used in the computation. Boundary C represents the maximum power output of the welding equipment, which is a practical limit of the machine. Boundary D represents a limit on the electrode feed rate. Experimental values of current levels at which the elctrode feeding actually stopped were used to determine this boundary. The four boundaries enclose a process operating space in which adequate weld penetration can be obtained. Some experimental data are superimposed on the figure showing the welds with and without successful penetration. When compared with the conventional solid wire electroslag welding data (Ref.16). A significant

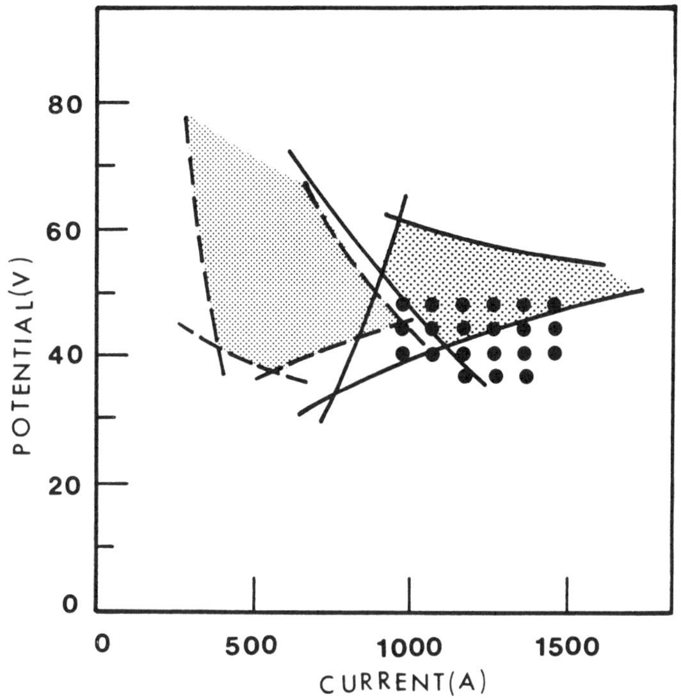

Figure 9. Operating space for electroslag welding with metal powder cored strip electrode.

increase in the current level is observed. In addition, the operating space is more limited, reminding of the importance and the need of process control in the strip electrode welding.

From the heat balance calculations, the role played by various factors in the consumption and transport of energy could also be determined. Table III shows the effect of each of the heat terms.

Table III. Energy balance of electroslag welding with the metal powder cored strip electrode.

Power contribution	Energy dissipated (%)
Heat input to the base metal	32.9
Resistive heat loss to guide plates	27.8
Heat input to melt metal powder cored strip electrode	22.1
Heat loss to copper shoes	14.0
Radiation heat loss by flux pool	1.9
Conduction heat loss by flux pool to weld pool	1.1
Total	100.0

WELD METAL MICROSTRUCTURE AND PROPERTIES - As mentioned before in this work, some of the major concerns in electroslag welding are the high heat input, coarse as welded structure and extended coarse grained heat affected zone. Therefore, the structure and properties of the welds made with the metal powder cored strip electrode need to be examined.

Standard metallographic techniques were used to prepare the macrographs shown in Figures 10 and 11. As predicted, the high current welds

Figure 10. Macrograph of a 1500A, 40V electroslag weld with metal powder cored strip electrode.

Figure 11. Macrograph of a 1000A, 40V electroslag weld with metal powder cored strip electrode.

exhibited fine grains. For the low current welds the solidification structure is extremely coarse with large columnar grains. By increasing the welding current from 1000 to 1500A, the columnar zone size was much reduced. The thickness of the heat affected zone also varied with the current level. At 1000 A, the coarse grained heat affected zone measured 4-5mm. Increasing the welding current to 1500A, the coarse grained heat affected zone decreased to approximately 2mm. This represents a fifty percent reduction in the coarse grained heat affected zone. When compared with data from other research work (Ref.18), welds produced in the present work showed much finer heat affected zone. This is because of the faster heating and cooling cycle of the high current weld which limits the austenite grain growth. The cross sectional area of the weld also decreased with increasing current. Using the same samples of 1000 and 1500A, the reduction of weld size was approximately thirty percent. Again, this is a result of higher welding speed and shallower penetration.

Hardness profiles were obtained to illustrate the effect of welding current, Figure 12. High and low current welds showed slightly different heat affected zone hardness readings. The average hardness of the weld metal with the refined solidification structure was approximately ten percent lower than that of the coarse structure weld. Ventkataraman et al. (Ref.7,14) reported similar findings that the hardness of the quartz shroud refined weld metal was lower than that of the standard electroslag welds. On the first sight, the refinement of the weld metal grain size in the high current welds would not explain the decrease in hardness as observed. However, after the micrographs shown in Figure 13 and 14 were examined , there was no contradiction found. The lower current cored strip electrode weld, as well as the conventional solid wire welds (Ref.7), showed extremely large austenite grains, delineated by the ferrite veins. In the intragranular region, large volume fraction of ferrite with aligned carbide (essentially bainite) could be seen. The high current welds exhibited small austenite grains, large volume fraction of grain boundary ferrite and smaller amount of lower temperature transformation products. The observed hardness decrease can be explained in the following way. The higher cooling rate refined initially the dendrite arm spacing which limited the austenite grain size. The higher surface area to volume ratio of a fine grained structure will provide large number of nucleation sites and promote the austenite decomposition. As result, large anount of grain boundary ferrite was observed in the high current welds, lowering the average hardness of the weld metal. The shorter heating and cooling cycle will also limit the coalescence and growth of inclusions and precipitates, which in turn, will control better the austenite grain growth, providing large surface area to volume ratio. This again makes grain boundary ferrite nucleation more favorable.

Figure 15 and 16 show the heat affected zone microstructure of the high and low current welds. Significant difference in the average grain size and structure of the coarse grained regions of the high and low current welds could be observed. This agrees with the shorter heating and cooling cycle observation made in the temperature profile determination. In addition, the heat affected zone hardness of the welds made with the cored strip electrodes were generally lower than that of the solid wire welds which could have beneficial effects concerning embrittlement.

HEAT INPUT, DILUTION AND DEPOSITION RATE - Jones, Olson, and Martins (Ref.20) suggested that for a particular welding flux system, the weld morphology is controlled by the current and voltage, since the heat for the welding process is obtained from resistive heating of the flux pool. The present results support the above statement and also point out that welding current is the more important factor. In order to further emphasize the advantage of the metal powder cored strip electrode process, the microstructure of the welds will be analyzed as a function of process parameters, current, and specific heat input. Figure 17 is a plot of welding current vs specific heat input. For the experimental welds produced with metal powder cored strip electrodes, the specific heat input was found to decrease with increasing current, which ranged from 1.6 x10^5 joules/cm^2 at 0.5mm per second of welding speed to 0.8 x 10^5 joules per square centimeter at the speed of 1.0 mm per second. The two welds were made at the current

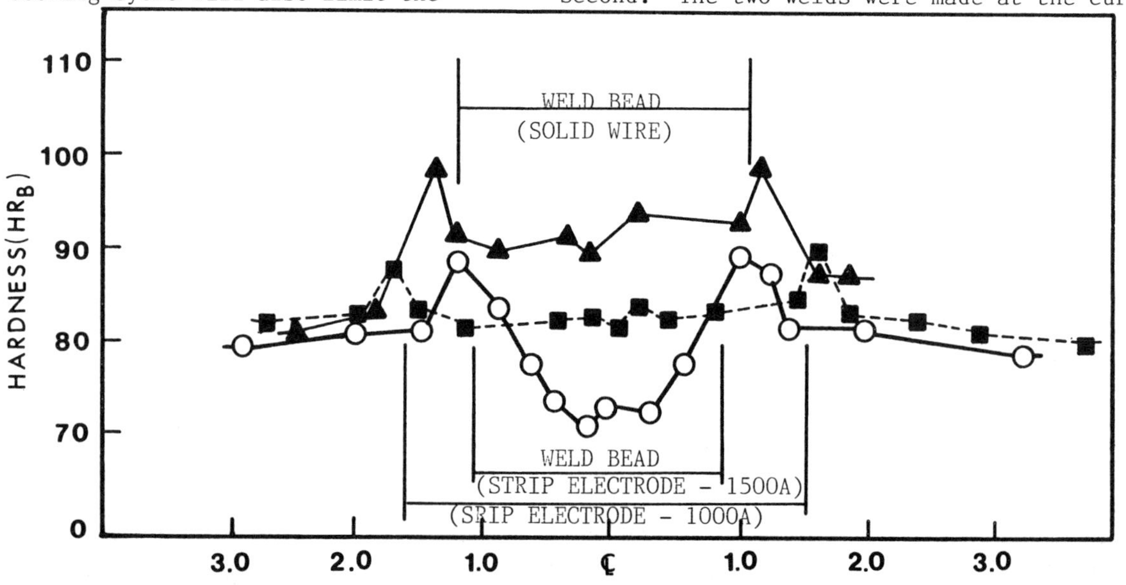

Figure 12. Hardness profiles of electroslag weldments, showing the level of hardness and the size of the weld beads.

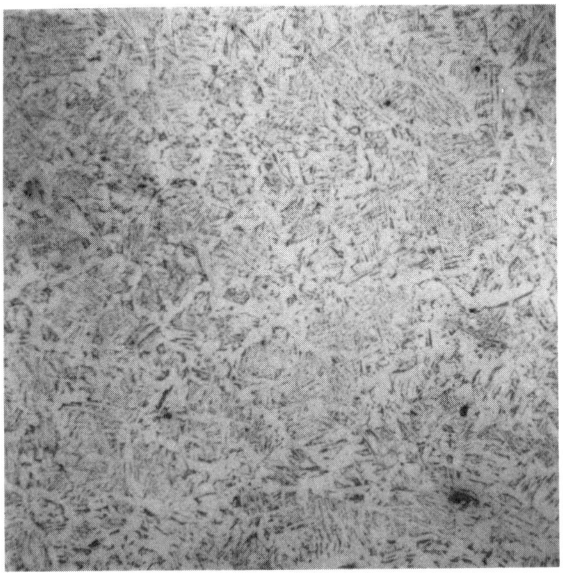

Figure 13. Light micrograph of the weld metal of a high current weld (1500A and 40V).

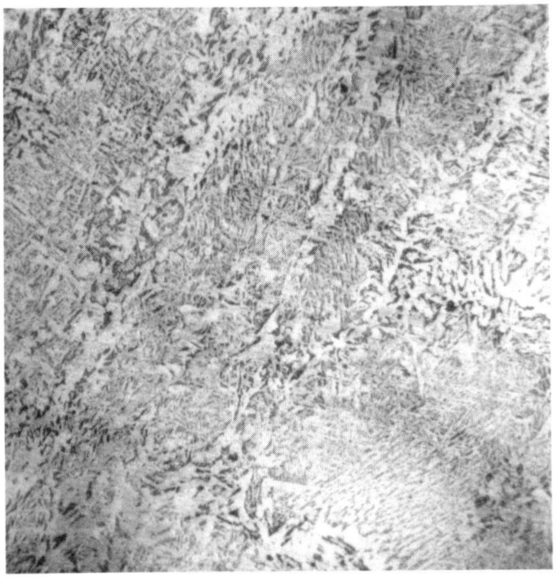

Figure 14. Light micrograph of the weld metal of a low current weld (1000A and 40V).

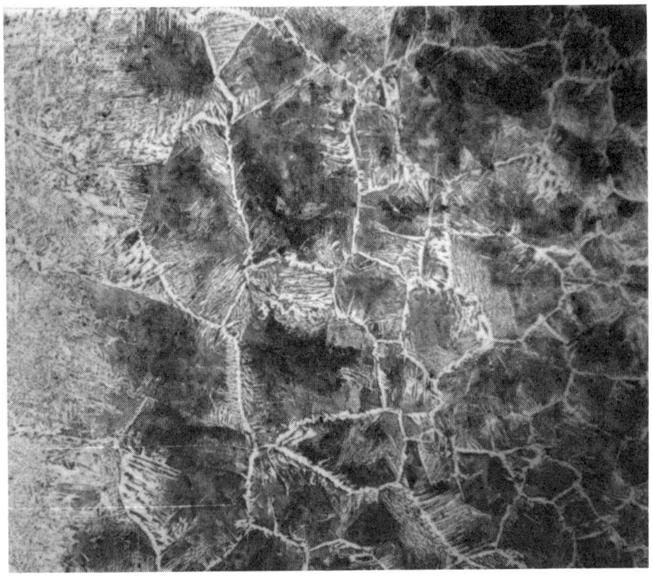

Figure 15. Light micrograph of the heat affected zone of a high current weld (1500A and 40V).

Figure 16. Light micrograph of the heat affected zone of a low current weld (1000A and 40V).

of 1000 and 1500A respectively.

The average welding speed reached in this work is faster than most of the reported results of 0.3-2.0inches per minute (Ref.7,8,18,23,24). It is however lower than that reported by Eichhorn et al. of 9.0 inches per minute (Ref.11-13).

Figure 18 shows the effect of welding current on deposition rate. By increasing the current from 1000 to 1500A, the deposition rate was almost doubled going from 30 to 85 pounds per hour. The amount of metal deposited is very large with respect to the conventional solid wire process demonstrating the effect of the metal powder cored strip electrode. An additional advantage is the reduction in dilution with current. The average dilution obtained in this study is approximately 30%, which is smaller than the welds made by the conventional solid electrode process (40-60%) (Ref.18,19,21). The major reason for this feature is the electrode configuration, welding speed, and the heat distribution.

In summary, the effects of high speed electroslag welding with metal powder cored strip electrode on weld microstructure and properties can be explained as follow. The geometry of the electrode and guide plates enable them to carry much higher currents than conventional solid wire process. The higher current increases the strip electrode melt rate and deposition rate. At the same time, the specific heat input and

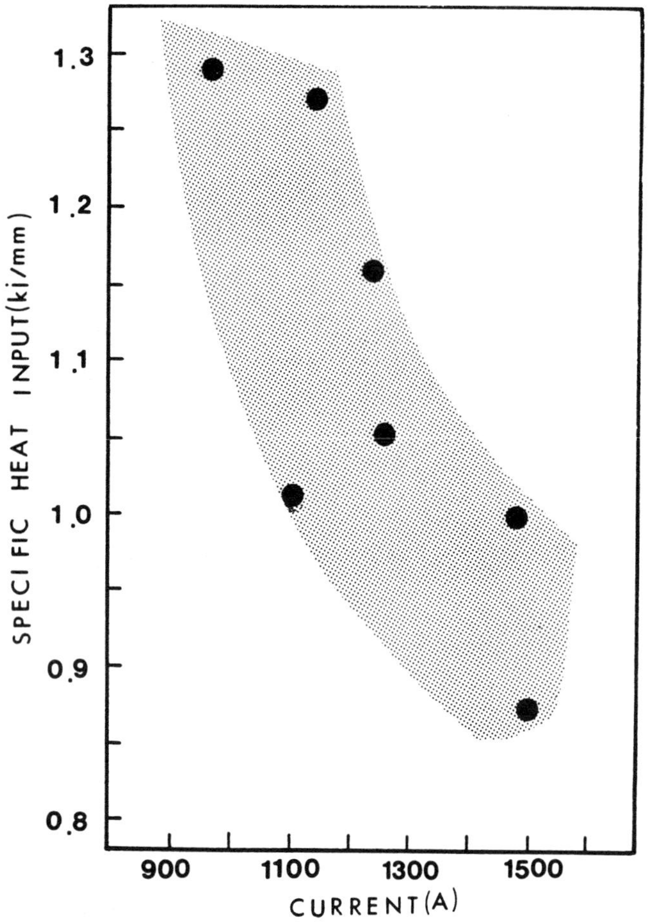

Figure 17. Specific heat input is plotted as a function of the welding current.

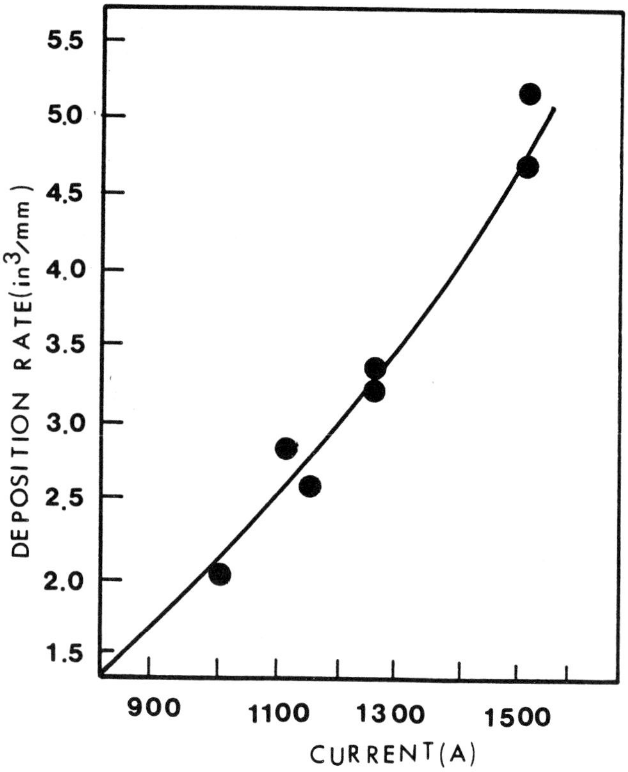

Figure 18. Deposition rate is plotted as a function of the welding current.

dilution are decreased. For a constant root gap and plate thickness, this results in an increase of the welding speed. The solidification rate is increased, refining the as welded structure and the austenite grain size. Solid state transformation reactions will produce the final microstructure of grain boundary ferrite, ferrite sideplates, ferrite with aligned carbide, and other microconstituints. As result of the shorter heating and cooling cycle, the heat affected zone size, particularly the coarse grained region, is also reduced. For the same reason, the average grain size of the heat affected zone size is also reduced.

CONCLUSIONS

The major results of this investigation led to the following conclusions:

1. The metal powder cored strip electrode with consumable guide plates demonstrated stable and high speed welding operation.
2. Functional relations for the electroslag welding process with metal powder strip electrode (i,V,F,S, weld and consumable geometry) could be developed.
3. It was possible to determine the electroslag welding process boundaries for successful penetration for the metal powder cored strip electrode.
4. Due to the geometry of the cored electrode and guide plates, higher current could be used which result in higher deposition rate, faster welding speed, lower dilution, lower specific heat input, shorter heating and cooling cycle, refined as welded structure, and reduced coarse grained heat affected zone.
5. The metal powder cored strip electrode ESW process showed superior performance compared to many other high speed electroslag processes.

ACKNOWLEDGEMENT

The authors acknowldege the research support of the National Science Foundation, the equipment and material support of the Hobart Brothers Co. and Alloy Rods, Inc. The assistance of the PSU IE laboratory technicians during the different phases of the welding experiments is much appreciated.

REFERENCES

1. B.E.Patton. 1962. Electroslag welding. AWS.
2. P.J.Konkol & W.F.Domis. 1979. Cause of grain boundary separations in electroslag weld metals. Welding Journal. V.58, p.161s.
3. M.Okumura et al. 1985. Grain boundary segregation of solute elements in embrittled 2-1/4Cr-1Mo electroslag weld metals. Trans. Japan welding society. V.16, No.1.

4. R.E.Dolby. 1976. The HAZ toughness of electroslag welds in C-Mn steels. The welding Institute.
5. A.L.Liby & D.L.Olson. 1974. Quarterly of the Colorado School of Mines. V.I, p.41.
6. V.I.Avramenko, B.F.Lebedev & V.I.Bozhko. 1973. Some ways of increasing the productivity of electroslag welding. Svar. Proiz. No.10, p.16.
7. D.Atteridge, S.Ventkataraman & W.E.Wood. 1982. Improving the reliability and integrity of consumable guide electroslag weldments in bridge structures. Research report - Oregon graduate center.
8. K.Watanabe, I.Sejima, S.Kokura G.Taki & H.Miyake. 1975. Problemas and improvement of large heat input electroslag welding. Japan welding society.
9. G.H.Reynolds & E.J.Kachelmeier. 1978. Adding powdered metal filler speeds deposition rate. Metals construction, p. 426.
10. S.A.Smirnov & L.A.Efimenko. 1973. Special structural features and mechanical properties of electroslag welded joints made using powdered filler metal. Svt. Svarka. No.9, p.46.
11. F.Eichhorn & J.Remmel. 1983. Situation of research in electroslag welding - A tendency of further development. Indian welding Journal. No. 4, p.37.
12. F.Eichhorn, J.Remmel & B.Wubbels. 1984. High speed electroslag welding. Welding journal. V.62. No.1, p.37.
13. F.Eichhorn & J.Remmel. 1985. Efficient fillet welding in the vertical welding position with electrogas and electroslag welding methods. IIW-DOC XII-908-85.
14. S.Ventkataraman, J.H.Devletian, W.E.Wood & D.G.Atteridge. 1982. Grain refinement dependence on solidification and solid state reactions in electroslag welds. Conf. Proc. in "Grain refinement in castings and welds". AIME, p.275.
15. T.W.Eagar. 1985. Applications and trends of electroslag technology in Japan. Welding research abroad. V.XXXII. NO.2, p.27.
16. G.R.Edwards, R.H.Frost & J.E.Jones. 1982. Advances in electroslag welding metallurgy and process development. Conf. Proc. "Welding technology 82". Australian welding Institute.
17. R.D.Thomas. 1960. Electroslag welding. Welding journal. V.39. No.2, 0.111.
18. M.Solari & H.Biloni. 1977. The effect of wire feed speed on the structure in electroslag welding of Low carbon steel. Welding journal. V.56. No.9, p.274s.
19. H.b.Cary. 1970. Porta-slag welding. Hobart Brothers Co.
20. J.E.Jones, D.L.Olson & G.P.Martins. 1980. Metallurgical and thermal characteristics of non-vertical electroslag welding. V.59. No.9, p.245s.
21. J.D.Culp. 1979. Electroslag weldments: Performance and needed research. Welding journal. V.58. No.7, p.27.
22. W.S.Ricci & T.W.Eagar. 1982. A parametric study of the electroslag welding process. V.61. No.12, p.397s.
23. Y.Asai & U.Nakamura. 1975. Electroslag welding of steels and slabs with plate electrodes. Advanced welding technology. Japan welding society. p.525.

DEVELOPMENT OF WELDING MATERIALS FOR LOW TEMPERATURE SERVICE

Yukihiko Horii, Shigeru Ohkito, Masakuni Wakabayashi
Nippon Steel Corporation
Sagamihara, Kanagawa
229 Japan

Masaru Namura
Nippon Steel Welding Products
& Engineering Company Ltd.
Japan

ABSTRACT

Preferably high toughness and CTOD property at low temperatures are demanded for steel welds of offshore structures and storage tanks. Based on the mechanism of notch toughness improvement of Ti-B containing weld metal, various factors which affect notch toughness and CTOD properties have been studied. The main results are as follows;
(1) There is an optimum carbon equivalent and tensile strength for toughness according to heat input in Ti-B containing welds.
(2) More attention should be paid to elements such as Al,Ti,O and N; they are not described in CEQ, but these elements largely affect weld metal toughness and other properties.
(3) Current type and polarity of power source are also important factors in SAW and SMAW metals.

IN ORDER TO SATISFY the requirement on welds, new welding consumables and welding procedures have been developed[1,2,3]. And it is known Ti-B type weld metal shows the best notch toughness of the various types of weld metals.

When designing the alloying elements to improve the properties of weld metals, it is very important to clarify the influence of factors that affect the weld metal properties. The benefits of Ti and B on weld metal toughness and the behavior of Ti and B have been clarified[4,5,6,7,8] and various welding consumables containing Ti and B have been developed. According to Mori et al, the mechanism of notch toughness improvement in the Ti-B type weld metal could be summarized by considering the metallurgical reaction between Ti, B, O and N during welding process.

The factors controlling the CTOD property have also been studied; Many equations on the correlation between the CTOD property and the other weld metal properties have been presented[9,10]. These equations have clearly showed high CTOD can be obtained by reducing TS and lowering vTrs.

Based on this information, the effects of the important factors controlling weld metal toughness and TS and some additional points which should be noted for the development of welding consumables in some welding processes are described in this study.

EXPERIMENTAL PROCEDURE

It was expected the weld metal microstructure and its toughness would be affected by such elements as Ti, B, N, O, and Al from the mechanism mentioned above. Impurities yielded from welding materials or steel plates are also an important factor that affects weld metal toughness.

In this study, the submerged arc (SA) welding process was mainly employed using wires listed on Table 1. To make Ti-containing welds, TiO_2-Al_2O_3-MgO type agglomerated flux was used. The amount of the alloying elements such as Mn, Mo, Cu and Ni was controlled by changing ferroalloys and oxides in fluxes or welding wires. Oxygen content was increased from 70 ppm to 450 ppm by changing the amount of calcium fluoride in flux. And heat input was increased from 4 to 11 kJ/mm. Standard Charpy V-notch specimens were sectioned and notched in so-called as welded region. Also, specimens of tensile test and chemical analysis were machined from same area. The results were applied to SMA, MIG, EG, and SA weldings. These testing conditions were described in each figure.

Table 1 Chemical composition of wires.

(wt %)

	Composition		C	Si	Mn	P	S
SAW	Wires	D	0.10	0.03	1.51	0.010	0.011
		E	0.11	0.05	1.92	0.018	0.010

RESULT and DISCUSSION

REACTION BETWEEN Al, Ti, B, N, AND O

Table 2 shows the chemical compositions of weld metal used. Microstructures of these weld metals are shown in Fig. 1. The intragranular microstructure changes by weld metal oxygen contents. Specificly, the microstructure of low oxygen deposit is ferrite with aligned second phase but the microstructure of adequate and higher oxygen deposits show acicular ferrite.

Figure 2 a)-d) shows the relative contents of Al, Ti, Si, and Mn in nonmetallic inclusions calculated semiquantitatively. Si and Mn contents in inclusions decrease drastically with a decrease of oxygen content. Ti shows large fluctuation, but exists in nonmetallic inclusions. This result clearly shows Al is the most active element to deoxidize in molten weld metal among Mn, Si, Ti, and Al. Most oxygen is combined with Al and, thus, there is no oxygen to combine with the other elements in welds of low oxygen content. Considering the composition of oxide inclusions and the amount of nitride, the optimum contents of Al, Ti, B, N, and O for toughness can be calculated. Figure 3 shows the effect of oxygen content on the composition of oxide inclusions and on the state of B and N. It is made by considering the order of deoxidation reaction and the result of the characterization of Ti, B, Al, N and O. Al_2O_3, TiO, SiO_2, B_2O_3 and Mn-Oxide are assumed as deoxidation products. It can also be assumed from the characterization that the residual of Ti and B react with N and form TiN and BN.

As shown in Fig. 3 a), the suitable oxygen range for toughness will be about 120-180 ppm. If oxygen content is lower than the suitable range, toughness decreases because of the existence of excess free B and the lack of Ti containing oxide inclusions. On the other hand, if oxygen content becomes higher than that level,

Table 2 Typical weld metal composition.

(wt %)

	(kJ/mm)	C	Si	Mn	P	S	Mo	Al	Ti	B	N	O
Ti-B	10.0	0.082	0.26	1.43	0.010	0.005	0.17	0.018	0.014	0.0025	0.0036	0.0232
Ti-B	4.5	0.080	0.21	1.45	0.011	0.007	–	0.008	0.018	0.0029	0.0026	0.0252

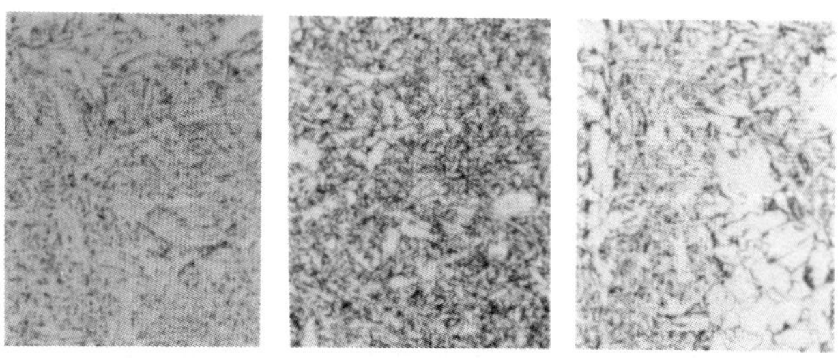

Fig. 1 Effect of oxygen content on the microstructure of Ti-B containing weld metals.

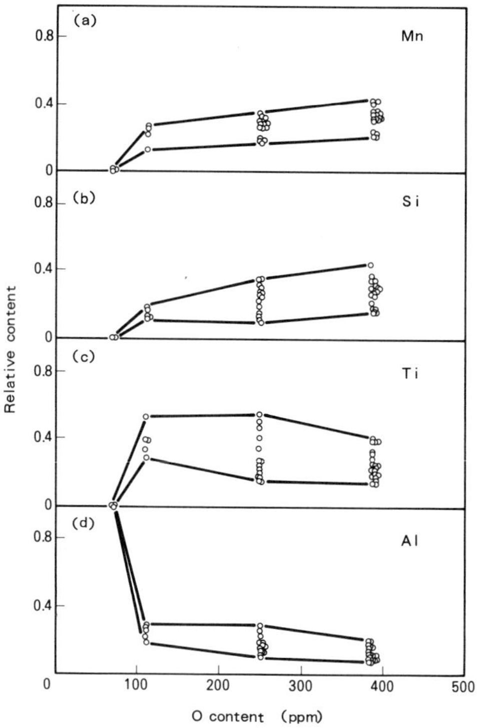

Fig. 2 Effect of O content on chemical composition of oxide inclusions.

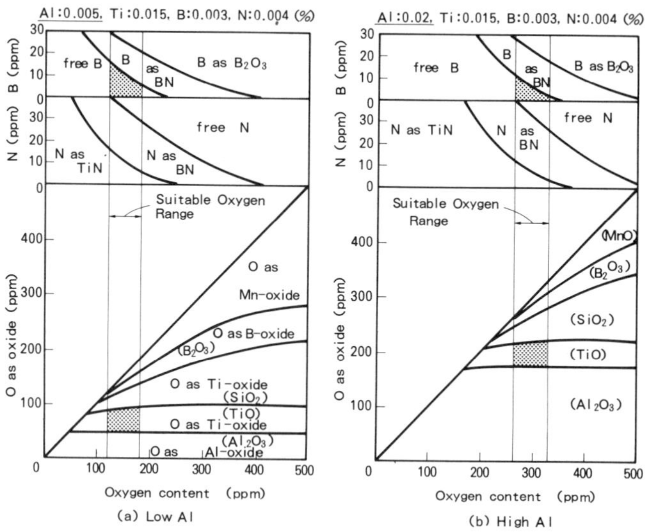

Fig. 3 Effect of O and Al content on the oxide composition and the state of B and N.

toughness also decreases because of the lack of free B and the increase of ineffective oxide inclusions. Figure 3 b) shows the suitable oxygen range increases to 250-300 ppm with the increase of Al content up to 0.015%.

Figure 4 also shows the effect of Al content in steel plates on the weld metal toughness. Lower vTrs is shown at higher Al content of 0.02%. This can be explained by added Al combines with oxygen, and protects Ti and B from oxygen.

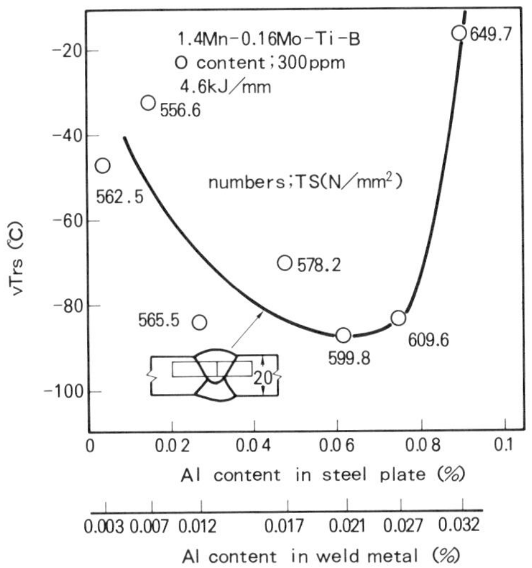

Fig. 4 Effect of Al content on weld metal toughness.

From the above results, it can be said free oxygen at deoxidation temperature (free oxygen = total oxygen - oxygen as Al_2O_3) and free B at austenite / ferrite transformation temperature (free B = total B - B as oxide - B as BN) are necessary for the control of weld metal microstructure and toughness. Also control of Al, Ti, B, N, and O are as important as the alloying elements of weld metals.

ALLOYING ELEMENT AND IMPURITIES

A tensile test was performed to clarify the relationship between chemical composition and tensile strength of welds made by one pass SA welding from both sides using heat input of 10 kJ/mm. Equation 1 shows the result of regresson analysis for this data.

$$TS = -28.13+1680*C+674*Si+175*Mn+62.7*Ni+191*Cr+411*Mo+192*Cu+375*V \quad (n=45,\ R=0.86) \qquad (1)$$

where : TS, 534-703 (N/mm^2)
C , 0.065-0.097 ; Mo, 0.0-0.50 (wt%)
Si, 0.20 -0.36 ; Cu, 0.0-0.30
Mn, 0.87 -1.61 ; V , 0.0-0.05
Ni, 0.0 -2.80

Evidently, the most effective element for strengthening weld metal is C followed by Si and then Mo, V, Cr, Mn, and Ni in order of decreasing efficiency. Depending on heat input, coefficients of these elements change, however, based on this data alloying design for weld metals TS can be done.

Figure 5 shows the effect of Mn content on toughness. There is an optimum range in Mn content: about 1.5 to 2.0% with the microstructure of homogeneous fine acicular ferrite. At 1.0% Mn content, the microstructure shows a large amount of GBF. On the other hand, as Bu increases, vTrs becomes high at Mn content higher than 2.0%.

Figure 6 shows the effect of Ni content on vTrs. vTrs become low with the increase of Ni content at 0.6% Mn, but vTrs increase with the increase of Ni content from 0.6 to 1.0% at 1.2% Mn weld. This suggests there is an optimum range of TS for toughness. And it is remarkable the vTrs is lower than -60°C at large heat input conditions, choosing good combination of Ni and Mn content. As shown in Fig. 6, Ni is an effective element to lower vTrs. However, it should be remembered shelf energy decreases with the increase of Ni content even at high testing temperature and it emphasizes microsegregation during solidification.

Mo shows the same effect on toughness as Mn and the optimum range is 0.1-0.2% at heat input of 8.5 kJ/mm. The reason why a relatively small amount of Mo is effective for toughness is the high hardenability of Mo. But an excessive addition of Mo seems to deteriorate the toughness because of the formation of upper bainite. Therefore, it should be used only for large heat input or high TS welds.

The suitable addition of Cu is effective to improve toughness because of the reduction of GBF. Cu is also effective for increasing the hardenability of weld metals. But, it should be restricted to less than 1.0%.

Nb and V have been widely used for strengthening steel plate. Nb is also effective for refining base metal microstructure. Figure 7 shows the effect of Nb and V on vTrs and TS of welds. Up to 0.02% of Nb and V show the same effect on TS, and Nb has a higher efficiency of reinforcing welds than V at higher than 0.02%. The addition of Nb or V is profitable to weld metal toughness because of the reduction of GBF: but a large amount of Nb or V deteriorates the toughness. Furthermore, it is well known that they increase the susceptibility of PWHT embrittlement.

Nitrogen is also an impurity element in weld metal. It comes from welding consumables, steel plate, and the atmosphere. Figure 8 shows the effects of B and N on toughness. It clearly shows there is an optimum ratio of B/N. The reasons for the embrittlement are explained by the increase of free-N and the amount of GBF in the lower range of B/N and by the existence of excessive free-B at austenite grain boundary in the higher range of B/N. And, as the increase of free N causes hot-strain-embrittlement, the selection of welding consumables and the welding procedure is important.

From this information, weld metal toughness cannot be explained by only chemical composition, but it shows there is an optimum TS for toughness. The best combination of alloying elements should be chosen for TS, Charpy, and CTOD properties.

HEAT INPUT

Figure 9 shows the effect of heat input on oxygen content, amount of GBF, size of acicular ferrite, and cooling time from 800 to 500°C. With the increase of heat input, oxygen content is reduced when using the same welding materials. This is because, with the increase of heat input,

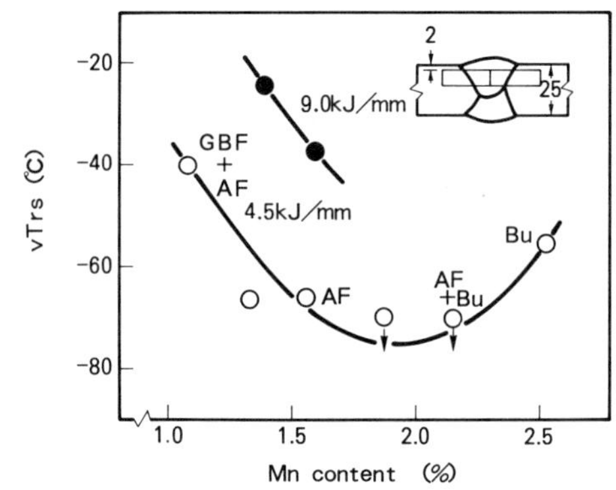

Fig. 5 Effect of Mn content on vTrs.

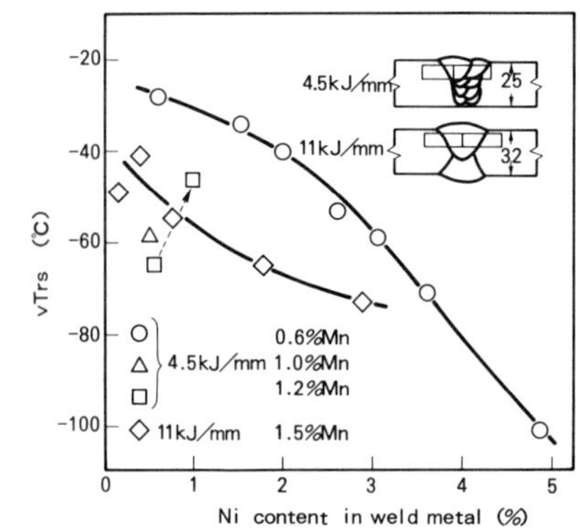

Fig. 6 Effect of Ni content on weld metal toughness.

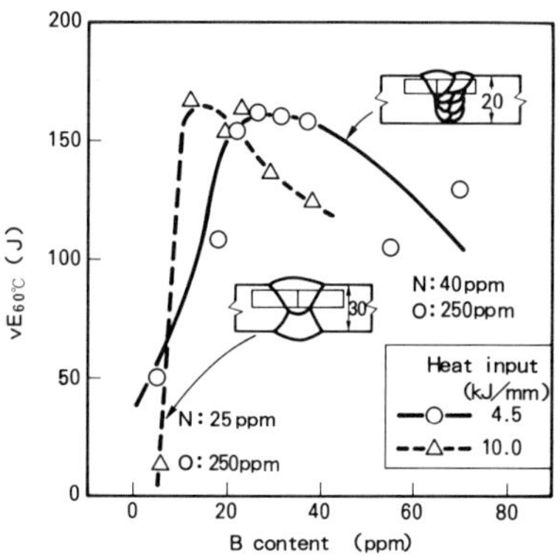

Fig. 8 Effect of N and B content on toughness.

the period from melting to solidification becomes longer and larger amounts of oxide inclusions are separated at slag/metal intersurface. With the increase of heat input, the amount of GBF and the size of acicular ferrite are increased; therefore, weld metal toughness and TS drastically decrease. This is because the cooling rate decreases and austenite/ferrite transformation rates decrease with the increase of heat input. Consequently, intra- and intergranular microstructure at large heat input detrimentally effects weld metal toughness. Based upon this relationship between the cooling time from 800 to 500°C, carbon equivalent (CEQ) to give absorbed energy of 50J at -60°C and TS of 637-657 N/mm^2 is summarized in Fig. 10. With the increase of cooling time between 800 and 500°C, from 50 to 220 seconds, the suitable CEQ range for toughness changes from 0.30 to 0.48.

In this study, Mn, Mo, and Ni are used to increase CEQ. However, Cu, V, and other elements also seem to have the same effect on weld metal toughness.

This information was obtained by using the SA welding process. However, it should be applied for the various welding processes, considering the peculiarities of each process.

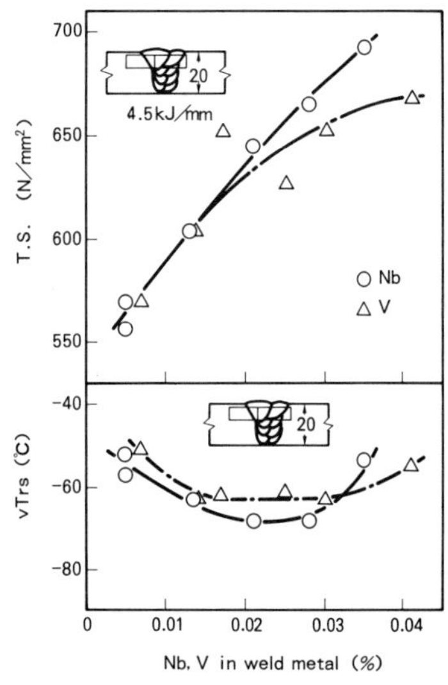

Fig. 7 Effect of Nb and V on vTrs and T.S.

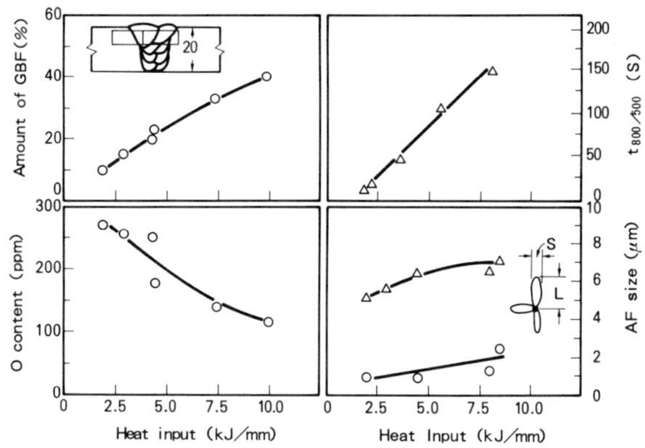

Fig. 9 Effect of heat input on the O content, amount of GBF and acicular ferrite size.

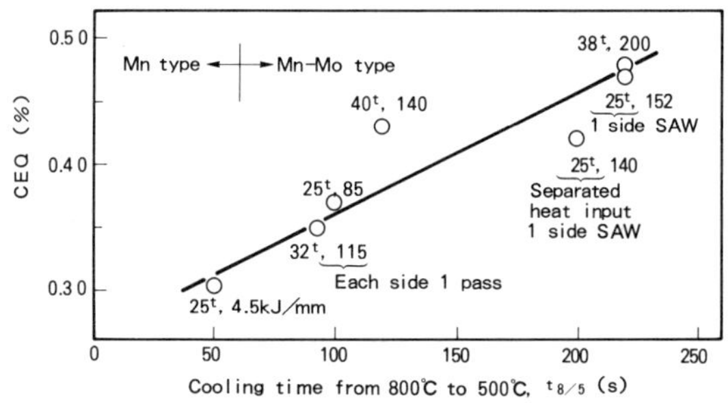

Fig. 10 Relation between $t_{8/5}$ and CEQ giving 50J at -60°C and 650-670 N/mm^2.

SOME EXAMPLES OF PRACTICAL APPLICATION

Based on the above philosophy, many welding materials have been developed in the various welding processes. However, there are indispensable points for the development of welding consumables in each welding process.

SMAW - The heat input range of SMAW is not so high; but nitrogen content in weld metal which depends on arc length and electrode polarity is as high as more than 70 ppm. Then the quantity of Ti and B added is a little larger than in the other processes. The CTOD properties at -10°C are sufficiently high if nitrogen content is less than 200 ppm and less than 150 ppm in the weld metal with intentionaly long arc. Test results for the one-side welding with SMAW are shown in Fig. 11. They show the CTOD properties of one-side welds are influenced by the electrode polarity and the electrode type for root pass. Generally, DC-EN polarity is more preferable than DC-EP polarity for one-side welding using a DC power source from the point of workability for URANAMI welding because the penetration by DC-EP at root is large and molten metal is too much. Otherwise, DC-EP polarity is more preferable from the point of weld metal toughness because the nitrogen absorption is large in the DC-EN polarity. However, Ti-B bearing electrode shows as high a CTOD as 1 mm at -10°C even by DC-EN is employed only for root pass and DC-EP for successive passes. URANAMI welding by AC using an electrode designed for AC can perform as good as DC-EN.

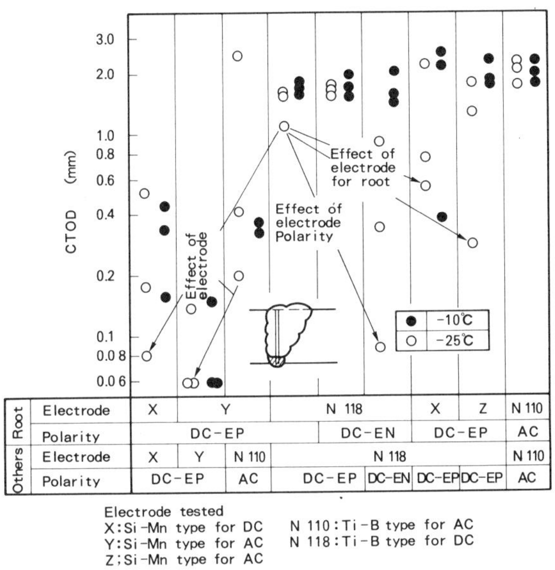

Fig. 11 CTOD test result for one side weld metal (W=2B).

MIG - MIG welding wire containing Ti and B is developed and can be applied for temperatures as low as -80°C. But it is necessary to watch for nitrogen absorption, especially on the root pass, the same as SMAW. An alternative current MIG welding technique with thick wire has been developed for high efficiency welding. It gives lower oxygen and hydrogen contents in weld metal whose toughness is excellent at less than -80°C. This welding method is effective for the welding of high tensile strength heavy plate which requires high toughness.

ELECTRO GAS WELDING WITH FINE WIRE - An automatic vertical-up electro gas welding process, more attention should be paid to steel composition of such elements as Al, N, V, and Nb than in the other welding processes.

SAW - Several kinds of Ti-B bearing fluxes have been developed for various purposes such as small, medium, and large heat input multi pass welding and heavy heat input one side welding. In accordance with heat input, Si-Mn, Si-Mn-Mo, and Si-Mn-Mo-Ni wire are combined with them. Figure 12 shows the influence of electrode polarity on the SA weld metal properties. Weld metal toughness with DC-EP is lower than with AC. This is because the oxygen contents with DC-EP is larger than that with AC, and the increase of oxygen in weld causes the decrease of soluble B, and the increase of GBF.

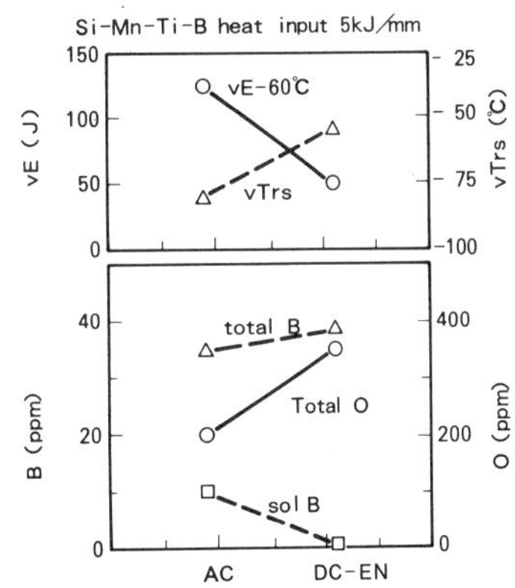

Fig. 12 Influence of electrode polarity on notch toughness and chemical composition of weld metal.

CONCLUSION

Many ideas for designing the composition and essential point for design welding consumables of Ti-B type weld metal were described. In conclusion, the main points to be noted are as follows:

i) The Ti-B type weld metal provides excellent CTOD properties.

ii) There is an optimum CEQ and TS to get maximum toughness in certain heat input.

iii) More attention should be paid for such elements as Al, Ti, B, N, and O. They are not described in CEQ, but they largely affect weld metal properties.

iv) Excess additions of Mn, Si, Nb, and V reduce weld metal toughness. But by controlling their amounts, high toughness can be maintained.

REFERENCES

1) K. Shinmyo et al., IIW Doc. XII-A-138-77.
2) K. Nagano et al., IIW Doc. XII-E-40-79.
3) Y. Horii et al., The forth JIW-11-31. Nov. Osaka.
4) D.J. Abson et al., Proceeding of International Conference. "Trends in steel and consumables for welding." TWI. 1978.
5) N. Mori et al., IIW Doc. IX-1196-81.
6) H. Terashima et al., Trend in welding consumables. 1983, Paper 27.
7) N. Mori et al., IIW Doc. II-82, IX-82.
8) R. Ricks et al., J. Material Science. 17 (1982) 732-.
9) R.J. Pargeter, TWI report 15/1981.
10) R.E. Dolby, IIW Doc. X-960-80.

RELIABLE WELDING OF HSLA STEELS BY SQUARE WAVE PULSING USING AN ADVANCED SENSING (EDAP) TECHNIQUE

C. Connelly, G.J. Fetzer, R.G. Gann, T.E. Aurand
Applied Fusion Technologies, Incorporated
Fort Collins, Colorado

Abstract

Fundamental movements in the molten weld puddle can be observed without the addition of sensors located in the hostile environment of the welding arc. These puddle movements directly correlate to weld puddle size. In addition full or partial penetration welds can be identified. The weld puddle is excited into motion by an arc current, square wave pulse. Small signal arc voltage fluctuations are sensed. Precise penetration and root widths can be maintained by continuous, automatic welding parameter corrections.

Pulsed GTA welding using square wave MILLI-PULSE, MICRO-PULSE and POLY-PULSE rates was studied as a part of this work. A wide range of pulse rates to control nugget shape, penetration, and improve quality were employed.

Background

Weld puddle oscillations and their relationship to puddle size, were discovered in the early work by Kotecki et al (Ref 1). They were interested in the mechanisms involved in the formation of ripples, during the solidification of the weld puddle. Continuing work was published by Zacksenhouse and Hardt (Ref 2) in which a theoretical model of the puddle is offered. Renwick and Richardson (Ref 3) recently completed experiments extending this to traveling partial penetration welds. These experiments indicated that through control of the weld puddle oscillation frequency via automatic control of the BASE current, the bead width could be maintained.

THE PRIMARY PURPOSE of this research was to develop an automatic control system to improve both the reliability and predictability of welding. Weld nugget size and particularly root bead width, has been successfully used by the welding engineer as a means of predicting weld joint strength. Therefore, a system that accurately controls these conditions substantially improves the reliability of the welding process.

There are a great many events that take place in the arc and molten weld puddle that directly relate to final nugget geometry and quality. Characteristically these events are physically small, very short in duration and sometimes temporal. Unfortunately, the welding operator finds this information out of his range of observable parameters. Therefore, if real time control of nugget size and penetration is to be accomplished, a precise method of sensing and control is required.

The proposed method of welding process control is the **EDAP** system (Electro Dynamic Arc and Puddle Control), based on sensing the fundamental motions in the weld puddle. This initial research was conducted with the objective of determining the feasability for producing the final hardware and software for such a system. The research method employed the use of advanced computer techniques and enhanced, high resolution, high speed video (**ERV**).

While conducting this research, some valuable information on pulsed GTA welding was developed. This included the discovery of a weld nugget shape termed a "TRANSITION" bead.

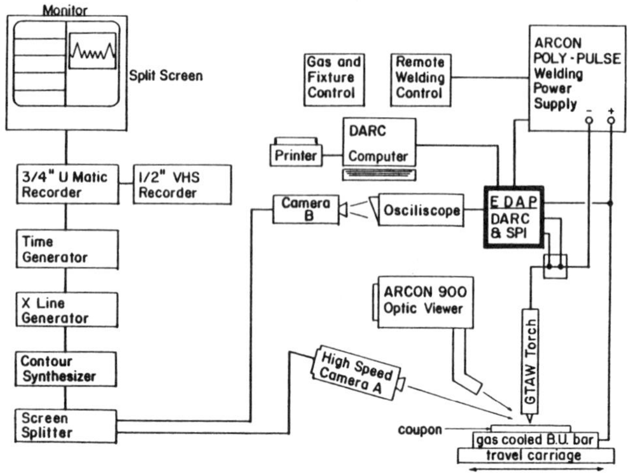

Figure 1 - Final Research System Block Diagram

Figure 1 is a schematic diagram of the research set up. The arc voltage was monitored, discretized and stored using the Data Acquisition and Control (DARC) through a custom designed Special Purpose Interface (SPI). Data from each test weld is stored on a separate diskette. In addition, test welds were monitored using an oscilloscope to observe the arc voltage and using Extended Range Video (**ERV**).

The welding power supply is set at a BASE current level and a narrow current pulse is added to the BASE level. Between the pulses the arc voltage is monitored by the EDAP system which is illustrated in Figure 2. Arc voltage measurements are low pass filtered to prevent aliasing, then digitized by the computer.

The samples taken between pulses are input to a frequency estimation algorithm which is implemented in software. The frequency estimate is compared to a reference frequency and any resulting error is transformed to a control signal. This can be fed back to the power supply and the BASE or pulse current adjusted accordingly.

The frequency estimator used in this research was a weighted least squares approach based on a single resonant frequency signal model (Ref 5).

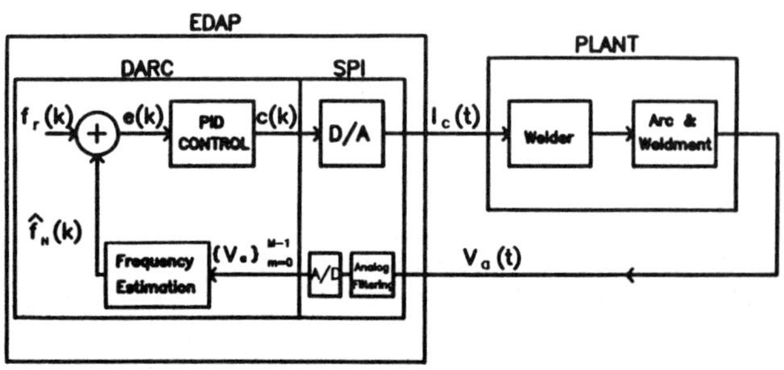

Figure 2 - EDAP System Diagram

PRELIMINARY RESEARCH

Previous researchers limited themselves to welding a maximum of 1/8" thick metals. This research was for steel thicknesses of 3/16", 1/4" and 3/8". The objective was to duplicate the full penetration root pass using a tight square butt on heavier construction steels. This required large weld puddles and high current ranges. Preliminary research was undertaken to prove the puddle dynamics theory for these larger puddles.

Based on the work of earlier researchers, it was expected that the technique might be sensitive to arc length and electrode shape. This was found not to be true with the EDAP system.

a. The electrode shape was important only as it related to penetration, travel speed and arc stability. (A 5/32" dia., 2% Thoriated Tungsten electrode, 30° apex angle with a .050" truncation was used for most of the final weld testing.)

b. An interesting discovery was the effect of arc length collecting data. Very short arc lengths (11-13V), desireable for maximum weld penetration, provided the strongest voltage signals. However, these short arc lengths signals contained so many variations that accurate frequency estimations were difficult. Using longer arc lengths (14-16V) solved this perturbation problem, with the arc acting as its own filter and averaging device.

Both Argon and Helium, and various mixes of these gases were tested and found acceptable. Due to the full penetration requirements a mix of 30 cfh He plus 15 cfh Ar was used. Flow of more than 80 cfh completely dampened puddle oscillation.

Travel speeds were initially limited to 2 IPM, due to the interference of the front wall of the weld puddle with the electrode. By designing and installing a magnetic arc positioner, travel speeds were increased to 4 IPM. This was accomplished by using the electromagnet to pre-position the arc ahead of the electrode, in the direction of travel.

There appears to be a current plateau for penetration (Ref 4) and EDAP implementation. Average currents above 350 Amps resulted in very wide puddles. These large puddles produced inconsistent signal responses.

PROCESS LIMITATIONS

Multiple weld puddle motions causing different arc voltage fluctuations were encountered. They could be controlled, attenuated or eliminated by several techniques. Puddle oscillations are also affected by pulse duty cycle. To assure consistent puddle oscillations, process limits were determined.

1. "Secondary" wave forms were observed at higher frequencies than the primary wave forms. It is expected that these disturbances exist primarily on the surface of the weld puddle. It was established that this "noise" could be reduced significantly, by superimposing a MICRO-PULSE (1 - 16KHz) on the BASE current.

2. **The most important discovery in this research was what we believe to be a "Traveling Wave"** (or what is sometimes called a bow wave). A large wave sometimes moves toward or away from the electrode, distorting frequency wave form plots and frequency estimations. This is a high energy wave which is interpreted by the computer as a very low frequency oscillation. When the amplitude of the current pulse was above 90 Amps the wave was almost always present. When the amplitude was below 70 Amps a "Traveling Wave" was not observed.

3. It was determined that excitation pulse width was causing problems in developing consistent puddle oscillations. Pulse widths of between 2 ms and 4 ms provided the most consistent frequency response. Longer excitation pulse widths had a tendency to dampen puddle oscillations.

4. The sensitivity of the EDAP system to various base current times was considered. Good, clear wave forms were produced when BASE current times were under 100 ms. Wave forms become slightly distorted beyond 40 ms with considerable variances 100 ms after the pulse.

Figure 3 is a waveform overlay of the arc voltage signals following the excitation pulse. The period of the waveform is approximately 22 ms which corresponds to a frequency of 45 Hz. The repetition is very consistent with each pulse.

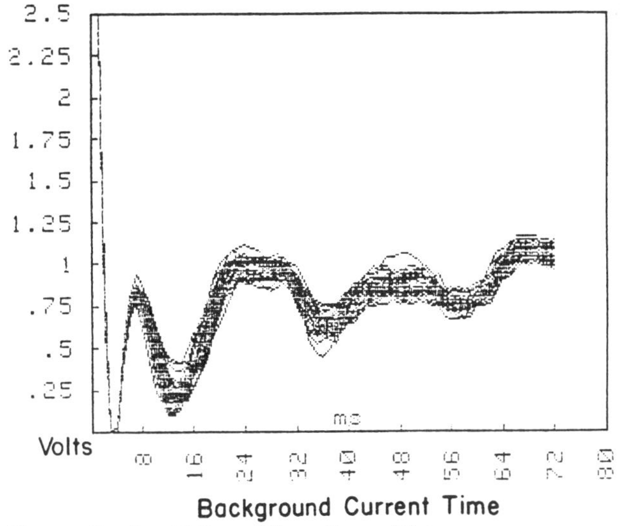

Figure 3 - Arc Voltage Wave Form Plot

The EDAP system was used in an "observation" mode during the following three welds. They verify the inverse relation between nugget size and oscillation frequency estimates.

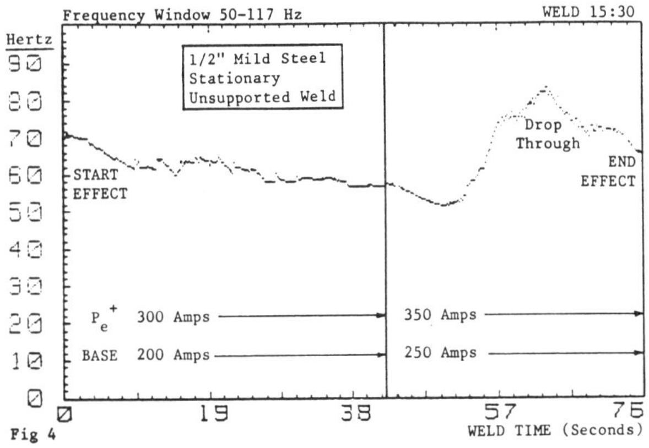

Fig 4

WELD 15:30 - Figure 4 - The BASE current was maintained at 200 Amps for 40 seconds, then increased to 350 Amps. The weld reached full penetration at 42-55 seconds. The arc dropped through the unsupported puddle at 56 seconds. **Amplitude of the puddle oscillations increased when the weld puddle reached full penetration,** then continued to get larger until drop through.

Figure 4 is a plot of the estimated weld puddle oscillation frequencies, as the weld progresses in time. Note the gradual decrease in frequency as puddle size increases. The vertical bar at 40 seconds into the weld indicates the area the current change was made. At this time frequency estimates dip indicating increasing puddle size. The point the weld burns through is obvious, shown by the sudden increase in frequency. This is probably because the puddle is not uniform, if indeed the concept of a puddle is appropriate at burn through.

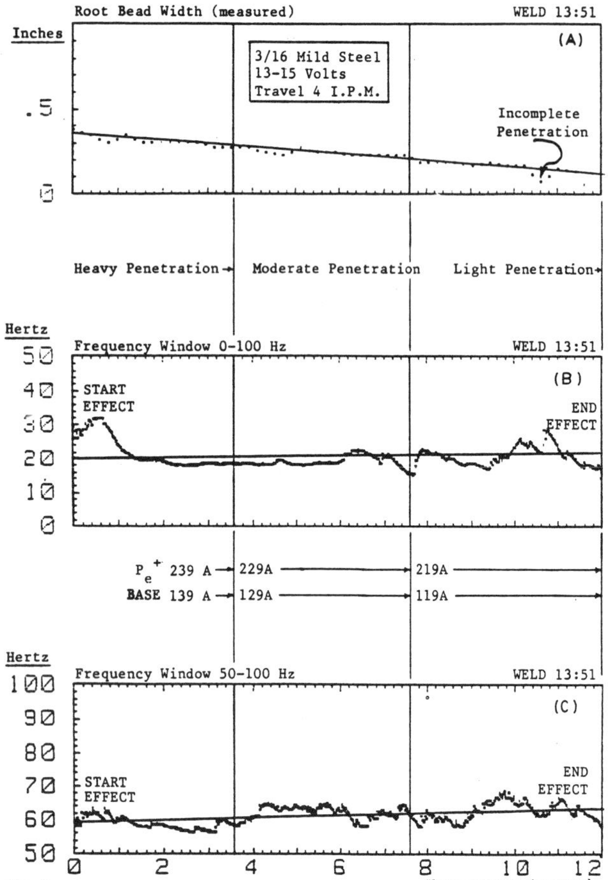

Fig 5

WELD 13:51 - Figure 5 - The BASE current was decreased with small step changes from 139 Amps to 119 Amps. Pulse levels were reduced from 239 to 219 Amps. There was full penetration for the length of the weld with one incomplete spot at 10.6".

Figure 5A is a plot of root bead width. The other figures show corresponding puddle frequencies for windows of 0-100 Hz (Fig 5B) and 50-100 Hz (Fig 5C).

Notice the linear variation of the root width with distance. This is reflected in the weld pool oscillation data in the direction expected. **A reduction in puddle size corresponds to an increase in frequency.** Another characteristic of the increasing trend is that it is nearly identical in both frequency windows. Average frequencies from the two estimators are much different however.

The data at two different frequencies indicates the weld puddle may have more than one resonant frequency. Numerous modes of oscillation, such as those observed in many resonant cavities as microwave wave guides, lasers, and musical instruments, may exist.

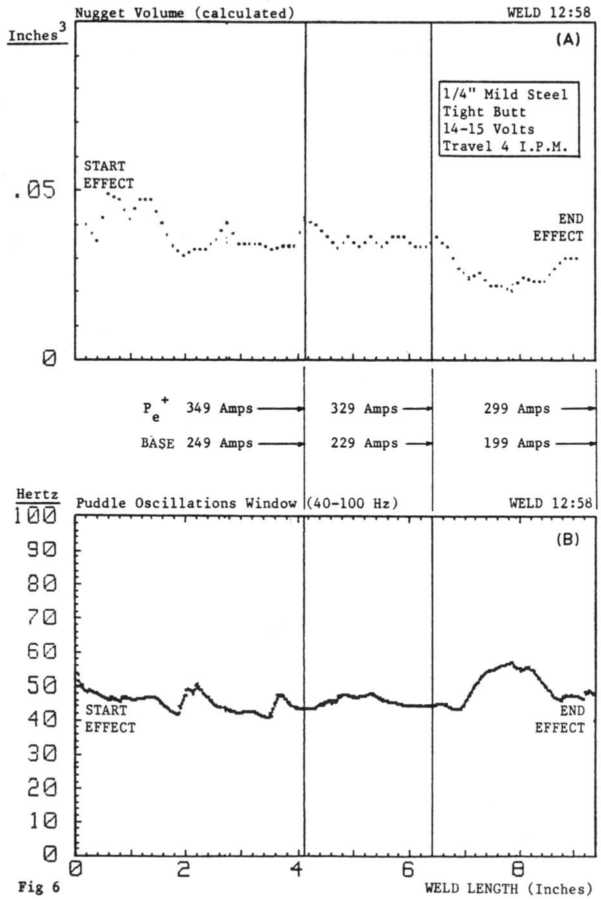

Fig 6

WELD 12:58 - Figure 6 - The initial current of 249 Amps was reduced in three steps to 199 Amps. The excitation pulse was started at 349 Amps and was simultaneously reduced to maintain a 100 Amp pulse amplitude.

The results of this test illustrate the perfect setting for a controlled weld scenario. This weld had good penetration decreasing for the entire length, with some deviations in top and root bead widths.

Figures 6A and 6B are plots of the nugget volume and puddle oscillation frequencies. It is apparent that not only can the general trend of the puddle size be detected but relatively short perturbations in the weld size can be sensed and compensated for.

This is the same weld that the wave form data in Figure 3 was collected. This data was taken at approximately 6 inches into the weld. Note the excellent agreement between the temporal estimate in Figure 3 and the estimates plotted in Figure 6B.

RESULTS AND DISCUSSION

Just prior to the completion of this paper, an inverted pulsing scheme was developed in which the increasing current pulse (P_e^+) was replaced by a decreasing current pulse (P_e^-). Following the diminishing current pulse, the puddle oscillations are monitored. Very stable and strong small signal oscillations were observed. Initial tests suggest that this inverted pulse technique may be less sensitive to process variables. Weld speeds of up to 6 I.P.M. were successfully employed using this pulse technique.

The broad range of pulse rates tested in this study resulted in an interesting observation for weld nugget shapes. Using MILLI-PULSE (0-999 Hz) or MICRO-PULSE (1 KHz-50 KHz) produced the expected metallurgical improvements and increases in depth of penetration.

However, using the POLY-PULSE mode, superimposing MICRO-PULSE on MILLI- PULSE, resulted in uneven relationships of pulse rates to bead shape, as compared to steady D.C. welds. Specific combinations of pulse rates will either decrease or increase penetration substantially.

A new nugget shape was identified. This nugget is irregular, appearing to be developed by a combination of convection flows in the weld puddle, that were both inward flowing and outward flowing. This is now identified as a "Transition" bead.

As MICRO-PULSE rates may be used for smoothing of the weld puddle in final **EDAP** implementation, these various nugget shapes need to be considered in final application.

CONCLUSIONS

A system (**EDAP**) has been developed which estimates the frequency of oscillation of a molten weld pool excited by a current pulse. The technique used is one of monitoring small signal arc voltage fluctuations through the arc plasma. The signals are collected directly at the electrode, eliminating the need for external sensors for back side access. The frequency estimation algorithm utilized, has been demonstrated to be both robust and stable.

Results of testing the system in an "observation" mode indicate that puddle oscillation frequencies are inversely related to the puddle size including root bead width. The correlation appears to be strong enough to effectively use the frequency estimates to control the size of the weld nugget. This includes identifying effects caused by short perturbations, such as varying joint gap size, base metal defects, and differential heat sinking. A discernable step (increase) in both voltage level and amplitude is seen when full penetration occurs in a partially or unsupported weld puddle. Amplitude and arc length rapidly increase just prior to full penetration and therefore may possibly be used to predict burn thru.

Also recognized was the multimode nature of the puddle oscillation. Final **EDAP** implementation may exploit this factor and monitor a set of frequency bands to add some redundancy to the calculations. There is a probability that different frequency ranges, along with amplitude changes, will be used to control different welding parameters.

Based on the work of previous researchers and this work, it can be concluded that there is a close correlation between induced puddle oscillations and nugget size for both very small and large weld beads. Steel, low alloy steel, stainless steel and aluminum all exhibit this same relationship.

ACKNOWLEDGEMENTS

The authors wish to express their appreciation to the Office of Naval Research, Department of the Navy, for funding this research under the SBA-SBIR program, Contract No. N00014-85-C-0759.

REFERENCES

1. Kotecki, D. J., Cheever, D. L., Howden, D. G., "Mechanism of Ripple Formation During Weld Solidification," Welding Journal, 51, (8), August, 1972.

2. Zacksenhouse, M., Hardt, D. E., "Weld Pool Impedance Identification for Size Measurement and Control," Journal of Dynamic Systems, Measurement and Control, Vol. 105, Sept. 1983.

3. Renwick, R. J. and Richardson, R. W., "Real Time Control of GTA Weld Size by Controlling Pool Oscillation Frequency," CWR Technical Report, June 1983.

4. Chihoski, R. A., "The Rationing of Power Between the TIG Arc and Electrode," Martin Marietta Corp., October, 1968.

5. Connelly, C., Fetzer, G. J., Gann, R. G., Aurand, T. E., "Reliable Welding of HSLA Steels by Square Wave Pulsing Using An Advanced Sensing (EDAP) Technique," Technical Report, Applied Fusion Technologies, Inc., April, 1986.

SENSING AND CONTROL DURING WELDING

KEYNOTE ADDRESS

SENSING AND CONTROL OF ARC WELDING

Hirokazu Nomura, Yuji Sugitani
Tsu Laboratories
Nippon Kokan K.K., Japan

- Abstract -

In Japan, arc sensors utilizing arc characteristics and optical sensors with image processing techniques have increased. Research is aimed primarily on sensor aided welding parameter control rather than just seam tracking.

A seam tracking system in which the direction of the GTAW torch with high response AVC is reversed when it reaches a pre-determined height has been developed. The deposition amount is controlled by utilizing data obtained from the tracing of the torch. A similar system has been developed for GMAW with high response ACC.

A new welding process has been developed with a high speed rotating arc, that has been used on horizontal fillet welding as well as NGW. By analysing the arc voltage changes in each rotating cycle precise seam tracking can be achieved.

A welding current control system has been developed for one side SAW. The welding current feedback control is based on detecting the intensity of arc light passing through the back side.

In FCB one side SAW, when the welding process is stable, the voltage between the base plate and the copper backing can be detected. This voltage is fed back to the welding current control.

1. Introduction

Developments of arc welding processes and automation systems have been promoted by the developments of sensors and sensor systems. The inventions of various sensors for arc welding in Japanese patents laid open are classified according to the principle and the purposes as shown in Fig.1. In the principle, contact sensors, optical sensors and sensor utilizing arc characteristics or phenomena are mostly proposed. Especially in the last five years, arc characteristics sensors and optical sensors have increased.

In purpose although the sensors for seam tracking make up a large portion, sensors for welding parameter control have been increased in the last five years as well.

Especially, the application of seam tracking systems utilizing arc characteristics has been increased remarkably in Japan. For example, its functions have been used mainly in commercial arc welding robots. The application of optical sensors with laser beam sectioning techniques has been also increased. Thus, it can be said that the seam tracking control has nearly finished the stage of development and entered into the stage of application. From now, development will be aimed primarily on sensor aided welding parameter controls rather than just seam tracking. In this report, sensors utilizing arc characteristics and those control systems for welding parameters as well as seam tracking systems developed by the authors are introduced.

2. Seam tracking control by arc sensor system

An arc sensor is the control system for the automatic seam tracking system utilizing arc characteristics as a sensor. The most important advantages which distinguish the system from other sensing system are ; no need of sensing device and real time control.

2.1 Weaving GMA welding

In consumable electrode welding with constant wire feeding rate, current-voltage characteristics of an arc changes with the change of an electrode-base plate distance. This is caused by the change of the melting rate of the wire in relation to the well known "self regulation characteristics of an arc". This characteristics can be utilized for seam tracking. Fig.2 shows the principle of seam tracking control of weaving GMA welding. Welding current or arc voltage changes with the weaving motion in relation to the inclined angle of the groove surface. Wave form is divided into left and right half cycle of weaving. Each integrated value is compared. The integration process prevents disturbance of noise such as instantaneous short circuiting of the wire. This process, therefore, can be applied to short circuiting GMA welding (2). Fig.3 shows comparison of "ΔS" (subtraction of integrated value) and "ΔE"(subtraction of peak value at weaving ends) in relation to torch deviation in short circuiting GMA welding.

2.2 Weaving GTA welding

Constant arc length control (AVC) is generally used in automatic GTA welding. It is achieved by torch height control in order to keep the arc voltage constant. Developed automatic seam tracking process uses a precise AVC control. As shown in Fig.4, weaving direction (X axis) is reversed when torch height reaches to a preset value "eo", which is constant distance from the surface of base plate. According to this system, optimum weaving width can be controlled in change of groove width. Another advantage obtained by this system is application to bead height control by using weaving width and weaving area. It is introduced in the next chapter.

In this seam tracking system, the high response of torch height movement is necessary. Fig.5 shows the dynamic movement of the electrode at inclined groove surface. Deviation of arc length (Δl) after t sec is given by

$$\Delta l = t \cdot v_X \tan\theta - \int_0^t v_Y \cdot dt$$

and the solution is

$$\Delta l = 1/\beta \cdot v_X \tan\theta \, (1 - e^{-\beta t})$$

where, θ is the inclination angle of the groove wall and β is the gain value of Y axis driver ($Vy=\beta \cdot \Delta l$).

An example of the control result is shown in Fig.6. A U-shaped groove with varied root width is used. Precision control of both seam tracking and weaving width has been achieved (3).

2.3 High speed rotating arc

High speed rotating arc process which was originally developed by NKK has the advantage of adoption of arc sensor control as well as improvement of the bead shape. Fig.7 shows the principle of the process applied for the NGW process. The welding wire is fed through a little eccentric hole at a contact tip which is rotated mechanically with a high speed (4). In accordance with the conventional weaving method, 4 to 5Hz is the practical limit for the weaving frequency. However, by this process, for example, 100Hz rotation can be easily obtained by a commercial DC motors. High speed rotation increases the sensitivity of the arc sensor and also improves the bead shape and the side wall fusion. Fig.8 shows the principle of seam tracking in horizontal fillet welding (5). Arc voltage wave form becomes asymmetrical at the center front (Cf) of the arc rotation when the torch deviates from the center axis of the groove. The deviation in transverse is detected by comparing the area of the wave forms (SL and SR) as shown in the figure. Torch height guidance is carried out by detecting average welding current. Seam tracking control by this system can be applied to each electrode individually in tandem electrode welding. Fig.9 shows a control result of horizontal fillet welding.

3. Bead height control by arc sensor

According to the arc sensor system which utilizes constant arc length control as shown in Fig.4, variation of the groove area and the width can be detected from the trace of the weaved torch. Two methods for bead height control have been developed.

3.1 Control with weaving width

Fig.10 shows the control system for GMA welding using only the weaving width. The turning point of the weaving is determined by the system to keep a constant height from the surface of the base plate. The variation of the groove width, that is (G-Go) in the figure, can be represented by that of weaving width (W-Wo). When a preset bead height "ho" is given, necessary deposition amount to keep a constant bead height can be easily obtained. Thus, an optimum welding speed can be controlled at each cycle of weaving according to the equation shown in the figure.

Fig.11 shows a control result in pulsed GMA welding (300A, 200Hz) with varied groove width. Welding speed, weaving width and bead height are precisely controlled as can be seen. This system has been adopted in multipass GMA welding robot which was developed by NKK, and already been applied with a satisfactory results (6).

3.2 Control with weaving area and width

Weaving area in GTA welding, which is obtained by the weaving trace and turning level (eo), closely corelated to the change of groove area and necessary deposition amount. This is caused by the phenomena of that the tip of the electrode moves keeping almost constant distance from the surface of the weld bead (7). Based on this phenomena, the deposition amount control to keep the bead height constant has been developed, where the weaving area and weaving width are used as the parameter to be detected. Fig.12 shows the principle. In the case of changing the groove shape, weaving trace changes as shown. In order to keep a constant bead height, deposition amount to be controlled corresponds to the area ΔS. Since the height displacement of the weaving is so small comparing to the weaving width in practice, ΔS can be given by an approximate equation shown in the figure. Control of the deposition amount can be made at each cycle of weaving by changing both the welding speed and the filler wire feeding rate.

4. Bead shape control system in one side SAW

Multi-electrode one side SAW is a widely used welding process which is a key technology in the most Japanese shipyards. In this process, a welding arc of a leading electrode forms a key hole as far as in case that a stable back bead is formed. Fig.13 shows a X-ray fluoroscopic observation result of a molten pool shape in two electrode one side SAW. The key hole formed under the leading electrode repeats open and shut with 2 to 3Hz vibration. This seems to be the natural vibration frequency of the molten pool. According to this key hole forming phenomena, important information which is closely related to the back bead width can be obtained. Both back bead and face bead shape control systems have been developed using this information (8)(9).

4.1 Control by arc light intensity

Arc light, which is passed through the key hole and backing glass fiber tape, can be detected by photo sensors mounted on a sliding copper shoe as shown in Fig.15. The copper shoe is driven synchronously with the travelling arc. Fig.14 shows the close correlation between the detected arc light intensity and the back bead width. The experiment has been carried out by welding current change with three kinds of groove conditions. Based on the result, back bead width control system (BBWC, shown in right side in Fig.15) has been developed, where, welding current of leading electrode is controlled to keep constant arc light intensity.

By using the BBWC system, the back bead width is controlled constantly in varied groove conditions. Therefore, for example, when the groove gap is widened, the welding current is decreased. Consequently, the deposition amount decreases and the bead height decreases. The block diagram shown on left side of Fig.15 is the face bead height control system (FBHC) which solves this problem. Fig.16 shows the relation between the nugget height and the welding current. Where, BBWC is carried out with different groove shapes. As is seen in the figure, when the back bead width is controlled constantly, the nugget height has close correlation with the welding current not being affected by the groove condition. This indicates ; in one side welding of key hole forming type, the welding arc reaches to the backing material and the arc pressure is just balanced to gravity head of molten pool. The developed FBHC system is carried out by controlling filler wire feeding rate in order to keep the constant welding current which is decided by plate thickness, namely, the nugget height (8).

4.2 Control in FCB process by detected voltage (Vd)

Small voltage (Vd) can be detected between the base plate and the backing copper plate in FCB one side SAW, in which, a layer of flux covered on fixed copper plate is used for the backing material. Fig.17 shows supposition of detecting Vd. Fig.18 shows comparison between Vd and arc voltage wave form. Polarity and phase of Vd agree with those of arc voltage, however its value is larger in electrode positive, being asymmetric in each polarity. In addition, Vd repeats appear and disappear with some frequency synchronized to the vibration of the molten pool (namely, synchronized to open and shut of key hole). Based on the experiment facts, it can be considered that ; Arc plasma passes through the key hole and comes in contact with the copper backing plate and then reflects to the base plate, therefore, it is a part of the arc voltage, as shown in Fig.17. It is supposed that Vd is a similar physical value with above mentioned arc light intensity passed through the key hole.

Fig.19 shows the relation between

the detected voltage Vd and the back bead width. In region "a" where key hole is formed, close correlation between Vd and the back bead width can be recognized. On the other hand, in region "b" where the key hole is not formed, Vd does not appear. The back bead width control system has been developed for the FCB one side SAW process. Where, the welding current of the leading electrode is controlled to keep the constant detected voltage "Vd" (9).

5. Conclusion

1) It is predicted by the authors that arc characteristics sensors and optical sensors will become major aids, and welding parameter control systems will become the most essential technology for arc welding automation.
2) A bead height control system combined with seam tracking control has been developed for GTAW and GMAW processes respectively.
3) A bead shape control system which consists of back bead width and face bead height controls have been developed for one side SAW.
4) The development of these sensing control systems are based on the observation and analysis of arc characteristics and molten pool phenomena.

Reference

1) H. Nomura, "Sensor and Sensor System of Arc Welding" (in Japanese) WELDING TECHNIQUE 1985 Jun. Vol.33 No.6
2) H. Nomura, Y. Sugitani, N. Tamaoki, "Investigation of Arc Sensor in All Positional Short-Circuiting Arc." (in Japanese) Welding Process Comm. No.88, JWS, SW-1371-82, 1982
3) H. Nomura et al, "The Development of Automatic Seam Tracking with Arc Sensor" Nippon Kokan Technical Report (Overseas) No.37 (1983)
4) H. Nomura, Y. Sugitani, "Narrow Gap Welding Process with High Speed Rotating Arc" Doc. IIW SG212-527-82, XII-C-033-82, 1982
5) H. Nomura, Y. Sugitani, M. Murayama, "Development of Horizontal Fillet Welding with High Speed Rotating Arc" (in Japanese) JWS, Arc Physics Comm, No. 85-594 (1985)
6) H. Nomura, Y. Sugitani, N. Tamaoki, "Development of Pulsed MAG Arc Welding Robot with Adaptive Control by Arc Sensor" Intl Conf. TWI, June, 1986 (to be published)
7) H. Nomura, Y. Sugitani, Y. Suzuki, "Automatic Real Time Bead Height Control with Arc Sensor in TIG Welding" Doc. IIW SG212-606-85, XII-884-85, 1985
8) H. Nomura et al, "An Automated Control System of Weld Bead Formation in One Side Submerged Arc Welding" Doc. IIW XII-C-032-82, 1982
9) H. Nomura et al, "Back Bead Width Control in One Side SAW Using Flux Copper Backing" Doc. IIW SG212-607-85, XII-883-85, 1985

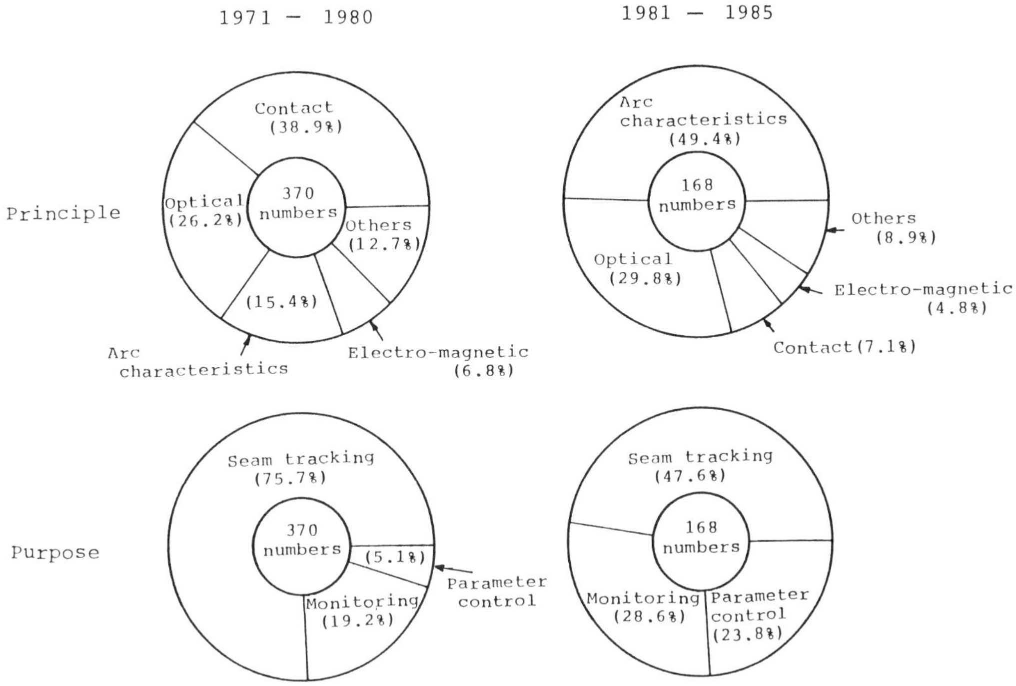

Fig.1 Inventions of sensors for arc welding (from Japanese patent laid open)

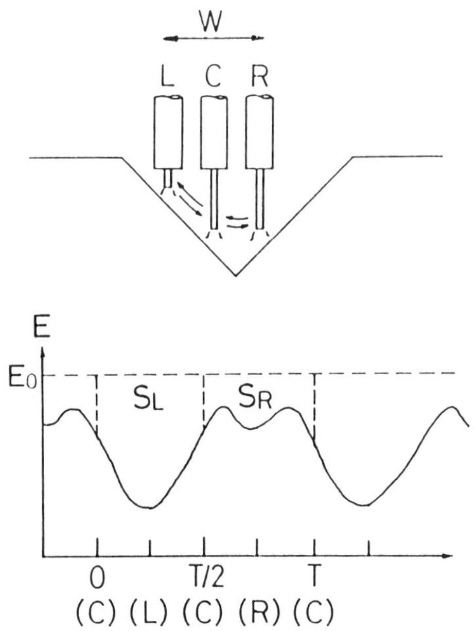

Fig.2 Principle of seam tracking with arc sensor in GMAW

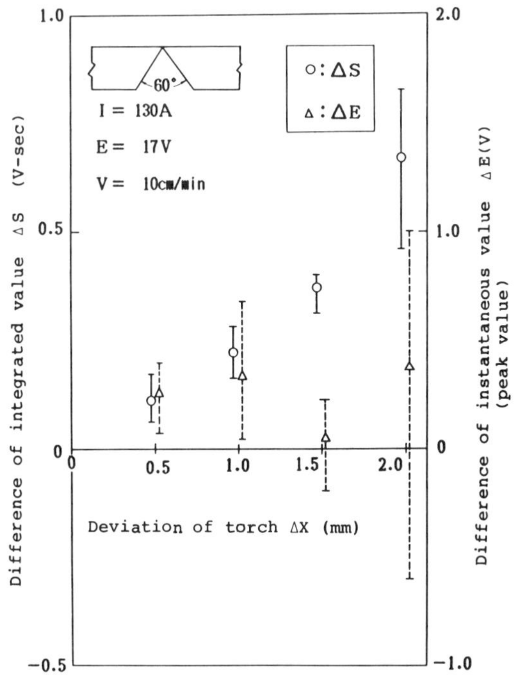

Fig.3 Comparison of ΔS and ΔE obtained at various deviation of torch (ΔX)

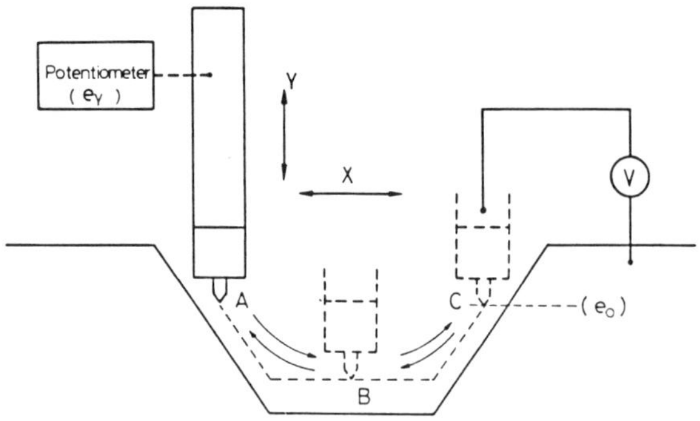

Fig.4 Principle of seam tracking with arc sensor in GTAW

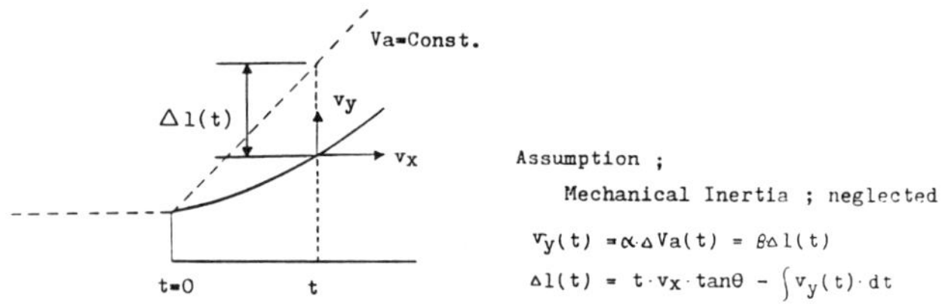

Fig.5 Dynamic characteristics of electrode movement

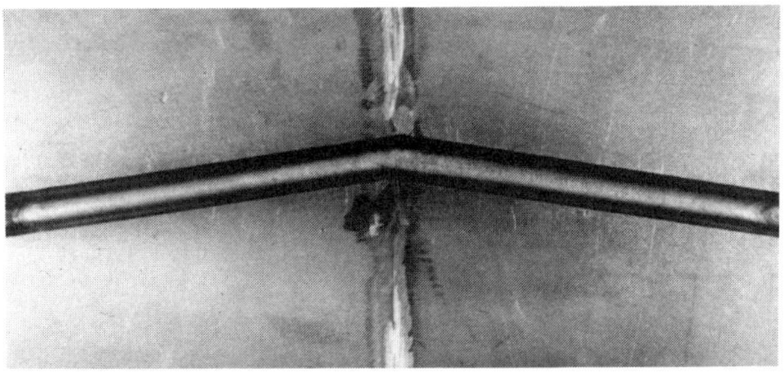

Fig.6 Test result of seam tracking

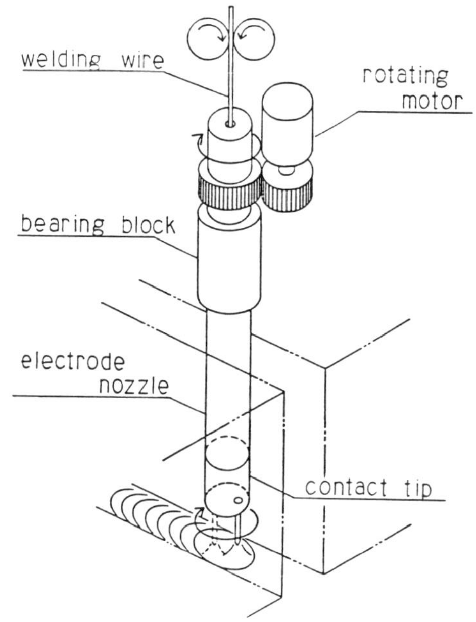

Fig.7 Principle of high speed rotating arc NGW process

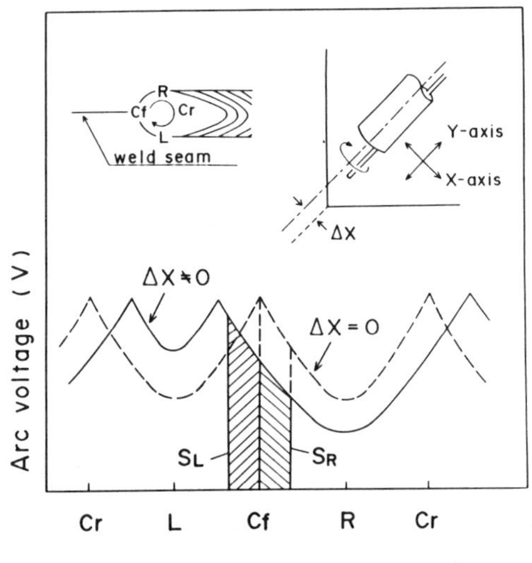

Fig.8 Principle of arc sensor in fillet welding

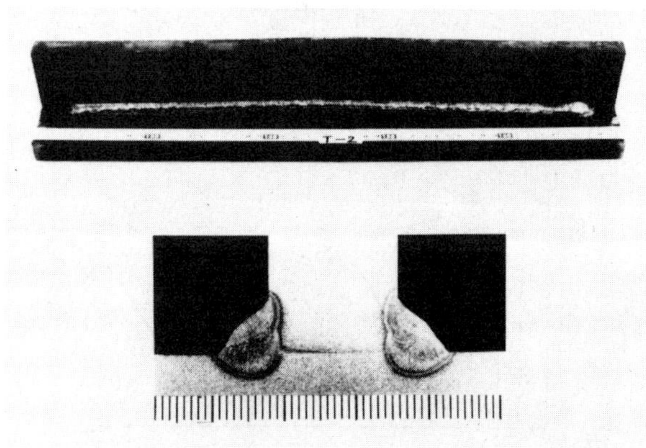

V=100cm/min,20%CO2-Ar

leading electrode	I=320A,1.2mm solid wire PLL-pluse,N=50Hz,D=3mm
trailing electrode	I=250A,1.2mm flux cored wire constsnt current,N=50Hz,D=2mm

Fig.9 Result of automatic seam tracking with arc sensor

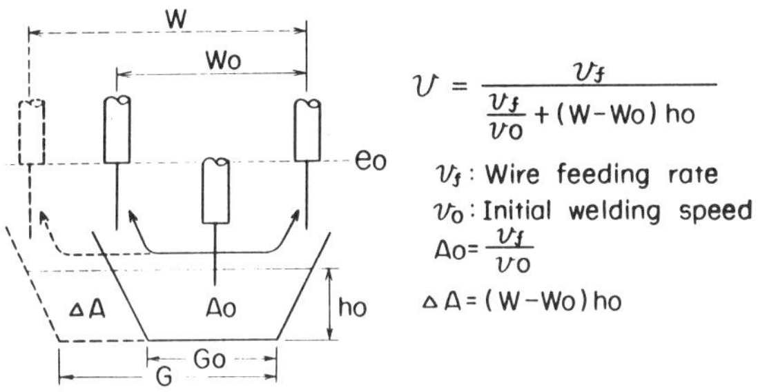

Fig.10 Principle of welding speed control in pulsed MAG welding

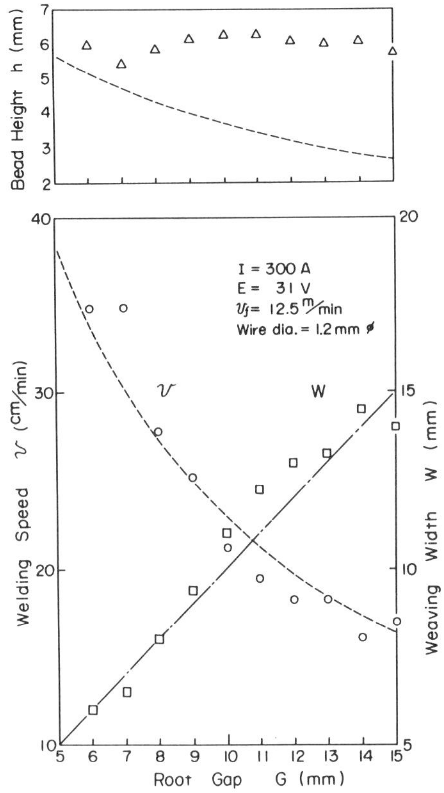

Fig.11 Result of welding speed control in various groove width

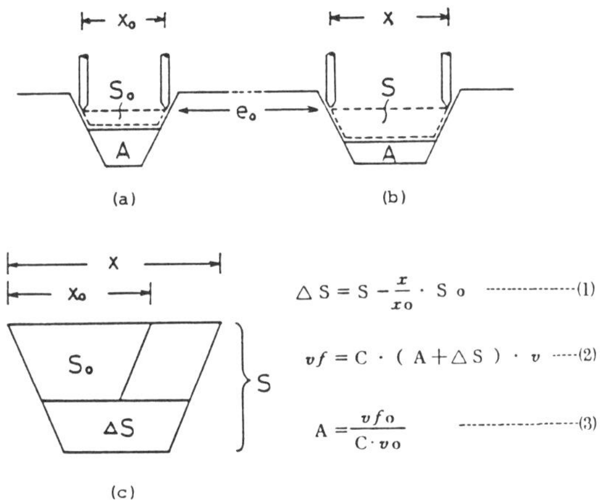

Fig.12 Principle of detecting deposition area (ΔS) to be controlled

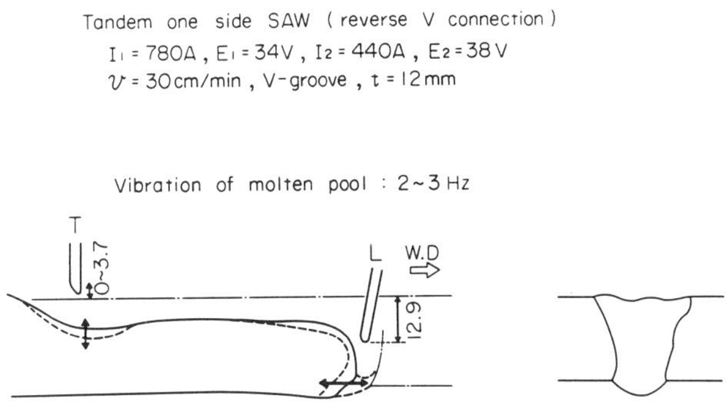

Fig.13 Observation of molten pool in one side SAW

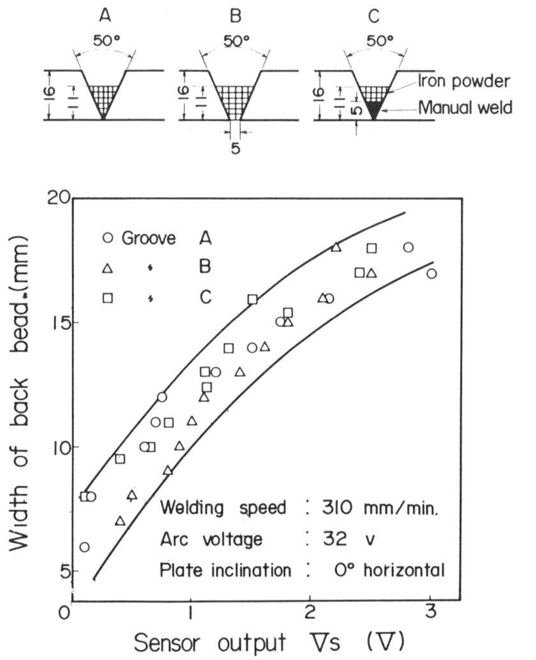

Fig.14 Correlation between arc light intensity and back bead width

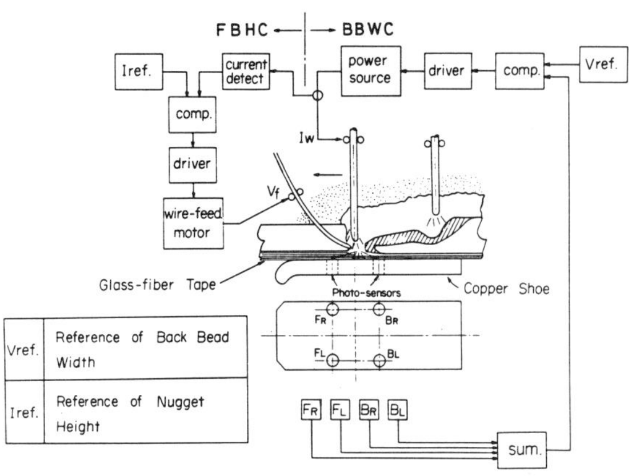

Fig.15 Block diagram of bead shape control system in one side SAW

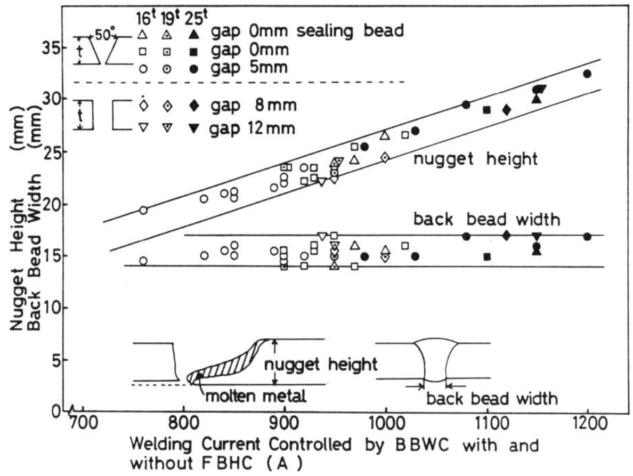

Fig.16 Correlation between nugget height/back bead width and controlled welding current under BBC

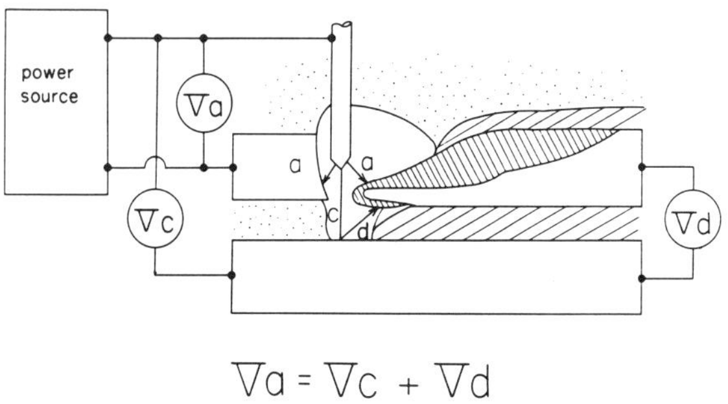

Fig.17 Voltage distribution of arc voltage

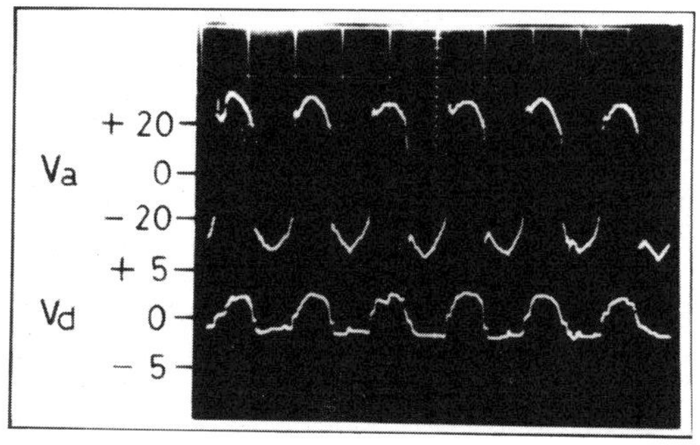

Fig.18 Wave form of detected voltage Vd

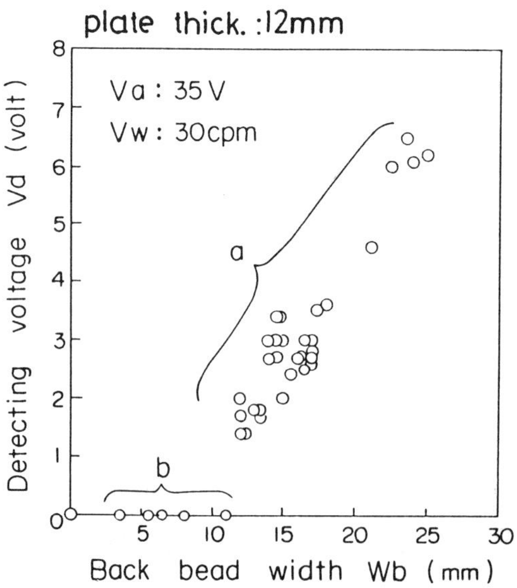

Fig.19 Correlation between back bead width Wb and the detecting voltage Vd.

ADAPTIVE SUBMERGED ARC WELDING CONTROL

George E. Cook
Vanderbilt University
Nashville, Tennessee

Milton D. Randall
CRC-Evans Automatic Welding
Houston, Texas

Mark E. Shepard, Liu Yizhang
Vanderbilt University
Nashville, Tennessee

Abstract

A digital data acquisition system has been used to study the effects of variations in the contact-tip-to-work-distance (CTWD) on the electrical arc signals for the submerged arc welding (SAW) process. Statistical analysis of the recorded digitized waveforms has been performed to obtain a more fundamental understanding of the electrical arc signal variations, that occur with cross-seam oscillation in through-the-arc positioning and process control. Small-signal dynamic models have been developed, and computer simulation has been used to obtain a better understanding of the mechanisms which change arc voltage and current in response to changes in CTWD or arc length. It has been observed that the current and voltage variations that result from the cross-seam oscillation are very robust and quite indicative of weld joint geometry and electrode placement relative to the joint. The dynamics of the SAW process are substantially different than that of the other consumable electrode processes. As a result, the voltage and current variations as a function of changes in the CTWD are substantially greater as compared to the other processes.

SENSING AND CONTROL of the electrode tip position relative to the joint is important for automation of the SAW process just as it is for the other consumable electrode processes. There are more limitations, however, on the sensing methods available for use with the SAW process due its inherent flux covering. The purpose of the research reported in this paper was to obtain a more fundamental understanding of the use of variations in the voltage and current waveforms as a functions of changes in the CTWD for positioning control. This general methodology has come to be called through-the-arc sensing [1-4].

Through-the-arc sensing is based on variations in the electrical signals as the electrode is oscillated back and forth across the welding joint. The variations in the electrical signals occur in response to the variations in the CTWD, which varies as a function of the joint geometry. For a sudden change in the CTWD, there will be virtually an instantaneous change in the arc length and voltage, and a corresponding change of opposite direction in the current. The new operating point that results from this sudden change will exist only momentarily. The wire feed rate (assumed constant) no longer equals the melting rate, and the system acts to re-establish an equilibrium. Assuming a constant potential, self-regulating system, the operating point will, in time, return to within a few percent of the voltage and current values that existed prior to the sudden change in the CTWD [1]. The difference between the new steady state operating point and the old operating point will be reflected primarily in a small change in the electrode extension with the steady-state arc length remaining essentially constant.

The rate at which the system re-establishes the new steady-state operating point is a function of the power source characteristic, the arc characteristic, the electrode extension, the wire size, and the properties of wire material. For most applications of gas-metal-arc welding (GMAW) and flux-cored-arc welding (FCAW), the desired welding conditions are such that the time constant of the self-regulating process is shorter than the oscillation rate. Hence, the steady-state operating point "follows" the change in CTWD as the electrode traverses across the joint. The operation of these systems thus depends on the relatively small change in steady-state operating conditions due to the change in the electrode extension.

For the SAW process, however, the larger size of the electrodes normally used, and the resulting lower current density results in a longer time constant for the self-regulating process. Indeed, the tests conducted in the research reported in this paper have shown that the self-regulating time constant for the SAW process may be expected to be comparable to or slower than the oscillation rates that would typically be employed. As a result, electrical signal variations that result from oscillation of the SAW electrode back and forth across the joint are transient in nature and hence, considerably more robust than those that occur in steady-state. This is advantageous from the standpoint of signal processing and maintenance of reliable control. However, to fully exploit the capabilities of this phenomenon, a more thorough understanding of the system and signal dynamics was required. This is particularly true of narrow gap SAW [5], where the rates of change in CTWD are high and the positioning tolerances are small.

MATERIALS

The steel base material purchased for the welding experiments was 19.0 mm-thick A516 Grade 70-S4.1 PVQ, made by U.S. Steel Corp. The back-up bar material was 9.5 mm by 38.1 mm hot-rolled A36 steel.

The submerged-arc flux and filler wire were purchased from Oerlikon. The combination flux-wire is designated F72EM12K. The filler wire is identified as 2.4 mm diameter SD3. The submerged-arc flux was Oerlikon OP121TT, which is a basic flux of the approximate composition: $CaO+MgO$ - 40%, CaF_2 - 20%, SiO_2 - 15%, and $TiO_2+ZrO_2+Al_2O_3$ - 5%.

WELDING EQUIPMENT

The submerged-arc welding power supply was a Lincoln Electric Co. Idealarc DC-600, 600-A constant-voltage machine. The air-cooled welding torch was mounted on a motor-driven, X-Y slide fixture to provide lateral movement and oscillation to the torch and vertical control over CTWD.

PREPARATION OF TEST PLATES

The 19.1 mm thick, A516 steel base plate was flame cut into 203.2 by 609.6 mm pieces for the weldments. Some of the pieces were machined to produce two joint designs; one joint preparation consisted of a 45-degree included angle with a 5.3 mm root opening, and the other preparation consisted of a narrow-gap 15-degree included angle with a 9.5-mm root opening. The A36 steel weld backup bars were 9.5 mm thick by 38.1 mm wide. The balance of the test pieces were welded in the flame-cut condition. The plates were fitup and tack welded to produce the 406.4 by 609.6 mm weldments.

DATA ACQUISITION AND SIGNAL PREPROCESSING CONSIDERATIONS

A major objective in this project was the development of a computer-based data acquisition system to sample and store large amounts of welding data for later analysis using digital signal-processing methods.

ACQUIRED SIGNALS - The four signals chosen for acquisition were voltage, current, cross-seam position, and wire feed rate. These provided both of the arc signals (voltage and current) as well as information about changes in the electrode position to correlate with the arc signals. Voltage was measured between the torch and the workpiece. Current was obtained from a 100mV/500A shunt in the workpiece ground lead. Cross-seam position was measured with a linear variable differential transformer (LVDT) coupled to the oscillating cross-slide carrying the torch. Wire feed rate was obtained with an optical encoder which was capstan-driven from the electrode wire.

ANALOG-TO-DIGITAL CONVERSION - The analog-to-digital converter (ADC) was a model DT2818 system manufactured by Data Translation Inc. The DT2818 provides for simultaneous sampling (at a maximum throughput rate of 27,500 samples per second) of 4 analog inputs. The DT2818 consists of a single board which fits into one of the expansion slots on an IBM PC/XT/AT computer.

CALIBRATION - The ADC produces an output between 0 and 4095 corresponding to input voltages of 0 and 10.00 volts. By knowing the gains and attenuation of the various elements in the signal path, a theoretical calibration factor for each channel was determined. Actual calibration tests were done with voltage references on the voltage and current inputs, and by measuring the LVDT displacement with micrometers. The calculated scale constants produced recorded data values accurate to 0.5% of the actual voltages and currents.

RAW DATA - Raw data was stored in files, each of which contained a contiguous set of samples from each of the four data signals. The size of a data file was chosen to fit onto one floppy disk, which holds 360 kbytes. Data is acquired in the computer in blocks of 64 kbytes; five blocks are collected, giving a total of 327,680 bytes of data in a file. A 512 byte header is prepended to the data, for a total of 328,192 bytes in each raw data file.

SAMPLE RATE - With a fixed-length data file, choice of sample rate becomes a tradeoff between bandwidth and total time of acquisition. On the one hand, it is desirable to have as many millimeters of weld traverse as possible in digitized data; on the other hand, undersampling can result in aliasing problems, time skew from analog filters, and loss of important information. In this project, it was desired to acquire some data at the widest practical bandwidth to permit evaluation of the high frequency characteristics of the arc signals, and some data at a much slower sample rate to provide long records for correlation with position information. A 6-kHz, high-frequency sample rate (with a 2500 Hz analog antialiasing filter) yielded data records slightly less than seven seconds long, but with excellent high-frequency resolution.

Because of the ripple components introduced by the SCR-based power source, a sampling rate of 1 kHz (with a 250 Hz analog prefilter) was chosen for the lower frequency sampling rate. This combination of sampling rate and antialiasing filter was necessary to prevent aliasing of the 360 Hz ripple component with lower frequency signal information.

DATA REDUCTION - Since the torch cross-seam oscillation rate is typically concentrated between 0.1 to 10 Hz, a sample-rate-reduction procedure was developed and used to prepare the data for the relatively low frequency positional information analysis.

The data reduction was done in two stages of filtering and decimation. The first stage reduced the effective sample rate from 1 kHz to 200 Hz. The second stage reduced the sample rate by a further factor of four by a similar filtering/decimation operation. The resulting overall decimation factor of 20 yielded a data file with two 1024 sample records, simplifying fast Fourier transform (FFT) operations on the data. The 25 Hz bandwidth proved adequate for analysis of the seam-tracking information. By digitally performing the sample-rate reduction operation on the 1 kHz sampled data, a data set was produced which retained the true time-relationship of all components of the signals.

ANALYSIS AND RESULTS

To determine the effect on the instantaneous voltage and current signals to variations in the joint width, CTWD, and parallelism between the axis of travel and the weld joint, a controlled series of welds were made with different spatial relationships between the welding electrode and the joint. Welds were made with the axis of weld torch travel aligned with the weld joint and other welds skewed to the weld joint. Welds were also made with varying widths of cross-seam torch oscillation, and with fixed lateral offsets. Some welds were made to study the effect of inclining the weld joint on CTWD variation.

Waveforms of voltage, current, instantaneous power, and instantaneous impedance were analyzed and correlated with the lateral cross-seam position of the electrode on a cycle-by-cycle basis. Power spectral analysis was used to determine the frequency content of these waveforms, permitting their frequency spectra to be compared with that of the cross-seam position waveform as well.

A number of algorithms were applied to the cycle-by-cycle variations of voltage, current, power, and impedance to study the ability to generate reliable control signals indicative of positional offset of the electrode relative to the joint.

WAVEFORM FEATURE CHARACTERIZATION ON SKEWED JOINT - A series of welds were run in which the travel path of the electrode was skewed with respect to the joint centerline. With this condition, it would be expected that the arc signals would initially indicate good centering of the electrode in the joint, and that the signals would indicate a gradual deterioration of proper centering as the weld progressed.

Figure 1 shows the current, voltage, cross-seam position, and wire speed of the third pass of one such test weld broken down on a cycle-by-cycle basis. The weld was made with the joint skewed with respect to the line traversed by the electrode by approximately 2.5 mm over the 600 mm length of the weld. The nominal current was 550 amperes, the nominal voltage was 31 volts, the CTWD was 13 mm, the wire speed was 66 mm/sec, the travel speed was 279 mm/min, the peak-to-peak oscillation width was 11.25 mm, and the oscillation frequency was 1.25 Hz.

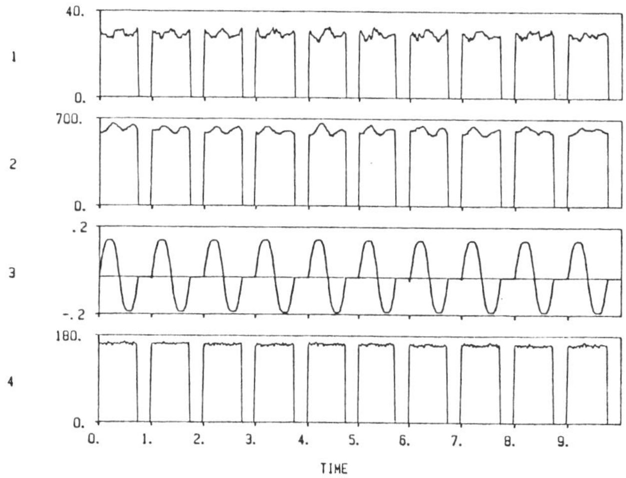

Fig. 1 - Cycle-by-cycle variations of voltage (1), current (2), position (3), and wire speed (4) for first ten cycles of skewed joint test

The data shown in Figure 1 was originally acquired at a sample rate of 1 kHz. Hence, the complete data file was approximately 40 seconds long and contained 50 cycles of cross-seam oscillation. The first ten cycles are shown in Figure 1. Since the complete weld time was approximately 131 seconds, the recorded data represents only 31% of the weld. The skew during this period of the weld was therefore approximately 0.75 mm.

Typical cyclic variations of the weld current are plotted in Figure 2 for the first ten cycles of oscillation. The current peaks at either sidewall are clearly evident; the peaks at the right sidewall are slightly greater than those at the left indicating a slight bias in the centering at the beginning of the weld, as was indeed the case.

An example of the cycle-by-cycle variations of instantaneous power is shown in Figure 3. The plot shown is for the last 10 cycles of oscillation, showing a distinct bias in the centering to one side of the weld. For this test the instantaneous power decreased for a decrease in the CTWD. Further testing, under a variety of conditions, will be required, however, before a general conclusion may be reached relative to the power variation as a function of change in CTWD.

The cycle-by-cycle variations of instantaneous impedance is shown in Figure 4 for the cycles numbered 11-20. In this region of the weld, the centering was proper, as indicated by the distinct minima at either sidewall. Like the instantaneous power, the impedance function decreases as the CTWD decreases at the sidewalls. This is to be expected because the voltage decreases and the current increases at the sidewalls. Both of these variations tend to make the impedance decrease. From the standpoint of signal robustness, this is a desirable feature.

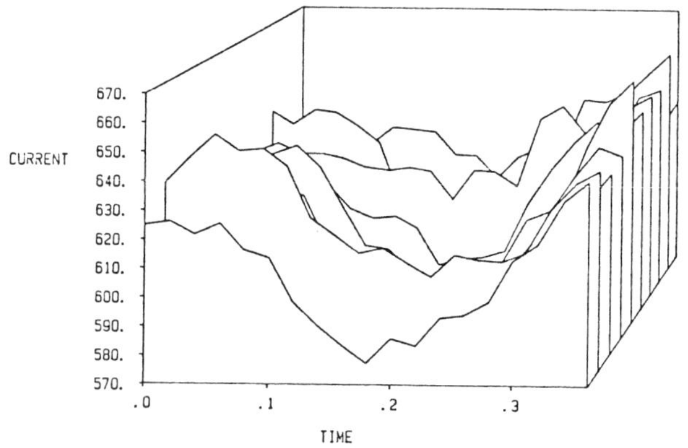

Fig. 2 - Cycle-by-cycle variations of current for first ten cycles of skewed joint test

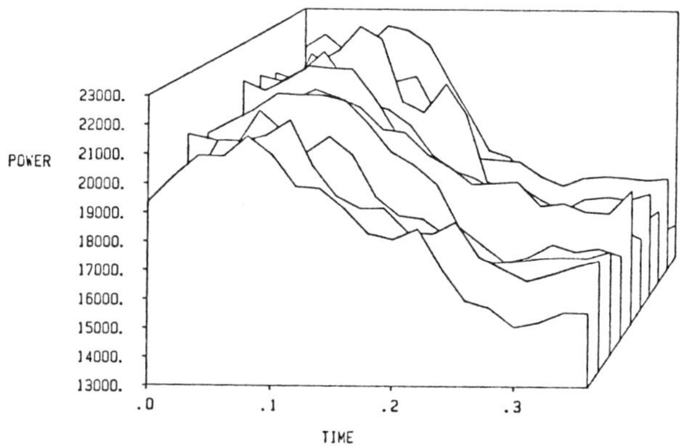

Fig. 3 - Cycle-by-cycle variations of instantaneous power for cycles number 41-50 of skewed joint test

To further substantiate the effect of the small skew between the joint centerline and the path of the electrode, power spectra were computed and plotted for current, power, and impedance waveforms over three different regions of the weld. The power spectra were computed using the FFT with 128 points corresponding to data time spans of 2.3 seconds. The specific regions chosen for computing the power spectra were t = 0 to 2.3 seconds, t = 19 to 21.3 seconds, and t = 36 to 38.3 seconds. Based on knowledge that the joint skew was such that the electrode was centered near the beginning of the weld and progressively deviated from the joint center as the weld was made, it was concluded that these conditions should be reflected in the power spectra functions.

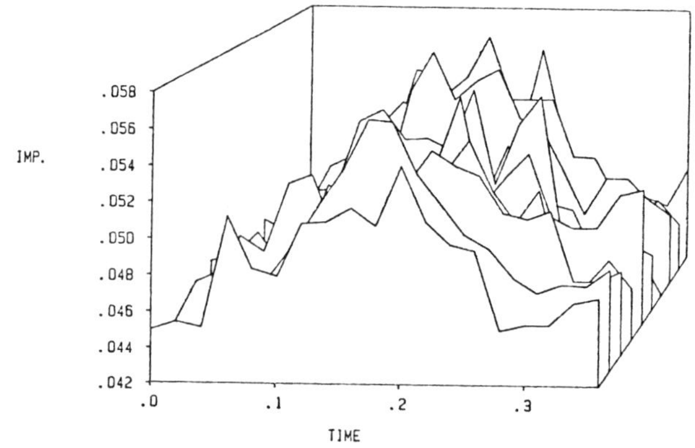

Fig. 4 - Cycle-by-cycle variations of instantaneous impedance for cycles number 11-20 of skewed joint test

The three power spectra are shown in Figure 5. At the beginning of the weld, the power spectrum for the time span [0-2.3s] showed a pronounced peak at twice the cross-seam oscillation frequency. This was due to symmetrical peaks in the current signal and symmetrical minima in the voltage signal at each extreme of the oscillation excursion, producing a current variation at twice the frequency of cross-seam oscillation. Near the center of the weld, the electrode had become progressively biased toward one side, hence the peaks in the current signal and the minima in the voltage signal at the extremes of oscillation were no longer symmetrical. As a result, the power spectrum in this region of the weld [18-20.3s] not only exhibited a strong peak at twice the oscillation frequency, but it exhibited a strong peak at the fundamental frequency of oscillation as well. Finally, at the end of the weld, the skew was sufficient to make the current and voltage variations essentially one-sided, and the power spectrum [36-38.3s] exhibited a strong peak at the fundamental frequency of oscillation and only a relatively small component at twice the oscillation frequency.

Power spectra of the instantaneous power waveform and the instantaneous impedance waveform, computed over the same three regions of the weld, showed similar results as conveyed by the power spectra computed for the current waveform.

The conclusion that can be drawn from this test and from the signal analysis presented, is that even for a total centering offset of less than half the wire diameter the signals are very robust and very accurately predict the precise positioning of the electrode. It is emphasized that the small condition of centering offset tested in this experiment would normally be considered within acceptable tolerances for producing sound welds. This was further substantiated by metallographic sections taken from the three sections of the weld.

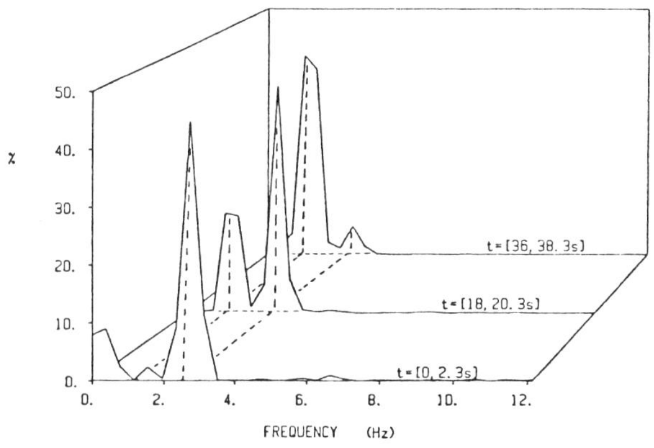

Fig. 5 - Power spectra of current for skewed joint test

PATTERN RECOGNITION AND CONTROL ALGORITHMS

The objective of this part of the research program was to investigate control algorithms that could be used to make reliable corrections in the vertical and lateral position and in other process parameters as a function of the processed signals obtained from the voltage and current waveforms. Generally, it was found that the signal changes experienced with the SAW tests were of equal or greater magnitude than those obtained from the GMAW or FCAW processes. Therefore, all of the algorithms that have been used with the GMAW or FCAW processes are potentially useful in controlling the SAW process as well.

Perhaps the simplest algorithms are those based on the peak values of the smoothed current waveform at each extreme of oscillation and on the value of the current at the center of the oscillation. These three values of current, measured per half-cycle of lateral oscillation, may be used to control the electrode position relative to the joint in several different ways. For simple centering purposes, the peak value at the left extreme of the oscillation excursion may be compared to the peak value at the right extreme of the oscillation excursion, and the difference of these two signals used to determine the direction and the amount of lateral correction called for on the next oscillation cycle.

The current difference between the left- and right-hand sides is plotted in Figure 6. The effect of the plate being skewed relative to the line of travel of the electrode is clearly evident. The horizontal line at 0 represents the centerline of the joint; the slanting line represents the path of the electrode. As previously explained, data used to produce the results shown in Figure 6 represents only 40 seconds of the approximate 131 seconds required to make the complete weld of 600 mm length. Hence, the skew between the joint and the electrode's line of travel, for the 40 seconds of data shown, is approximately 0.75 mm.

The data that is plotted in Figure 6 was obtained from the filtered and decimated data (corresponding to a 50 Hz sampling rate). Hence, as compared to the original data, considerable smoothing has taken place and Figure 6 effectively represents a smoothing of 1000/50 or 20 data points obtained at the original sampling rate. Further smoothing could be implemented to reduce the variance of the peak current estimates.

The algorithm that has been described is but one of several algorithms that could be implemented based on the same peak current measurements. The sample taken at the center of the oscillation excursion may be compared against a fixed reference to obtain an error signal for the CTWD positioning. This error signal may be used to drive a servo in the proper direction to maintain the CTWD constant.

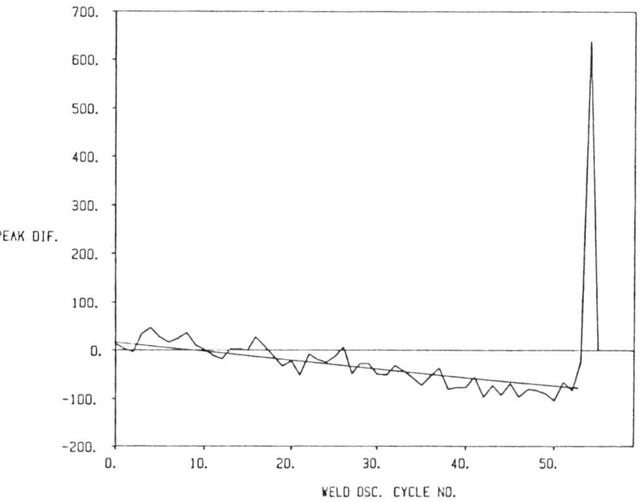

Fig. 6 - Peak current difference between left and right sidewalls for skewed joint test

For adaptive width control, the peak currents, measured at the extremes of oscillation excursion, may be compared against a constant value corresponding to desired sidewall fusion. If the measured peak current is less than the reference, then the oscillation width is increased; if the measured peak current is more than the reference, then the oscillation width is decreased.

A slight modification of the peak-current-measurement approach may be used to implement centering and width control on an instantaneous cycle-by-cycle basis. With this approach, the oscillation mechanism is directed to move the electrode toward a sidewall at a desired selectable rate until the current reaches a preset value corresponding to a desired condition of sidewall fusion. The direction of lateral oscillation is then reversed, and the process is repeated on the opposite side. By so doing, the automatic adjustment of electrode centering and width control is simultaneously achieved. To avoid driving the electrode into the sidewall, under conditions of noisy current measurements, position windows may be placed on the amount of increase or decrease in oscillation width permitted from one cycle to the next.

In addition to the algorithms discussed, a number of other algorithms were studied and compared in relation to one another [6]. Waveforms of instantaneous power and impedance were considered as well as waveforms of current. The experiments conducted, to date, however, were too limited in number to reach any definitive conclusions.

For the particular test described in this section, metallographic sections taken from the first, middle, and last part of the weld showed good fusion characteristics for all three sections. This serves to demonstrate the robustness and excellent sensitivity of the through-the-arc signal analysis methodology for SA welding. The signal analysis clearly showed that there was a skew between the electrode path and the joint centerline, even when the skew was well within tolerance levels that would normally be taken as acceptable for producing sound welds.

MODELING & SIMULATION

Preliminary results have been obtained from research efforts aimed at modeling and computer simulation of the consumable-electrode welding processes. The objective of the modeling has been to better understand the mechanisms which change arc voltage and current in response to changes in CTWD or arc length. These changes are used to infer joint profile information during cross-seam torch oscillation in seam-tracking systems using through-the-arc sensing [7].

The model used for these results is similar to one developed by Wu and Richardson at Ohio State [8], [9]. It consists of two coupled first-order nonlinear differential equations: the first describes the electrode burnoff rate in terms of current and stickout; the second describes changes in current in terms of stickout and CTWD. The simulation is implemented on a VAX-11/785 computer, using an adaptive-step-size 5th/6th-order Runge-Kutta numerical integrator.

The burnoff-rate equation uses results developed by Halmoy [10], and based on previous work by Wilson et al [11], Lesnewich [12], Amson [13], and Manz [14], which describe the burnoff rate as:

$$v_b = k_1 J + k_2 L J^2 , \tag{1}$$

where L is the stickout length, J is the current density in the electrode, and k_1, k_2 are constants of the electrode material. The first term describes the component of melting rate due to anode heating, and is linear in current; the second term describes the component due to ohmic heating in the electrode extension. This static model is strictly valid only in steady-state, as the ohmic heating term assumes that the stickout has been subjected to a constant current density J during its travel from the contact tube to the arc. The error introduced by using this model with time-varying current should remain small for small current perturbations or when the ohmic heating contribution is small.

The current equation results from applying Kirchoff's Voltage Law around the loop containing the power source, the electrode extension, and the arc. The power source is modeled as a resistance and inductance in series with a constant voltage source; the arc voltage is modeled as a constant voltage drop plus a term proportional to arc length; the voltage drop along the stickout uses the nonlinear model from Halmoy [10].

Examples of the computer simulation are shown in Figure 7. The curves represent simulations of the frequency response of welding current to CTWD changes for two consumable-electrode processes with very different current densities, but with the same deposition rate. Curve (a) shows a SAW process with 2.39 mm diameter wire at an average current of 471 A; curve (b) shows an argon-shielded spray transfer GMAW process with .889 mm wire at an average current of 222 A. The current density of the SAW process is 105 A/mm^2, compared with 358 A/mm^2 for the GMA. Both processes are steel-on-steel with 5.3 Kg/hr deposition rate, 25 mm average CTWD, and 1.0 mm peak CTWD perturbation. The SAW average arc length is 4.6 mm. and the GMA average arc length is 3.2 mm; both processes use the same power supply, with a current rolloff at 24 Hz due to series inductance. The curves are normalized to the average welding current, and represent the percentage change in current in response to a 1 mm change in CTWD.

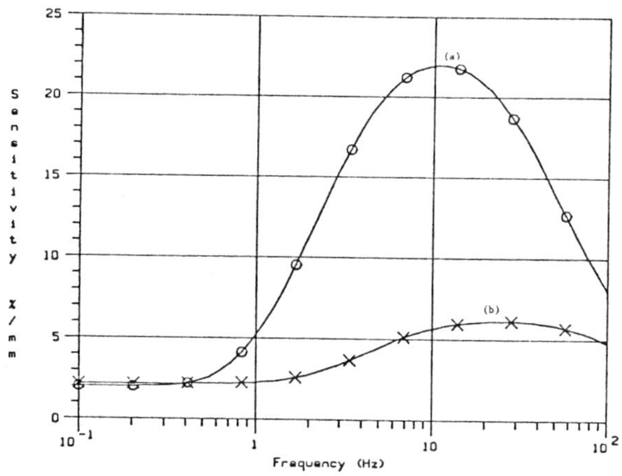

Fig. 7 - Frequency response of SAW (a) and GMAW (b) processes

Although both processes exhibit similar sensitivity for steady-state and low-frequency variations, the low-current-density process shows a much greater sensitivity increase beginning at a significantly lower frequency. The SAW curve begins to break up at 0.5 Hz, and has comparable sensitivity at 1 Hz to the GMA peak at 20 Hz. The 0.5-3 Hz range is an important band of interest for cross-seam oscillation, and the low-current density process exhibits several times the sensitivity at the upper end of this range. Above 20 Hz, further sensitivity improvements in both processes are masked by the power source inductive rolloff. These examples represent two extremes for achieving the desired deposition rate, but illustrate well the effect of current density on signal amplitudes for through-the-arc sensing.

SUMMARY, CONCLUSIONS AND FUTURE RESEARCH

A limited number of SA welds were deposited in 19.1 mm thick plain-carbon steel plate with two different joint designs: (1) single-vee groove weld with 45-degree included angle, and (2) narrow gap, single-vee groove weld with 15-degree included angle. The experimental welding equipment was instrumented to digitally record the following instantaneous welding parameters: (1) voltage, (2) current, (3) wire feed rate, and (4) cross-seam position of the torch.

A controlled series of welds was made representing different spatial relationships between the welding electrode and the weld joint as they affect the dynamic welding arc signals produced. Welds were made with the axis of weld torch travel aligned with the weld joint and other welds skewed to the weld joint. Welds were made to study the effect of inclining the weld joint on CTWD variation.

Preliminary work in dynamic modeling was undertaken, and computer simulation was used to obtain results based on the models developed. The simulation models predicted the experimental phenomena observed.

Albeit limited in scope, the research program reported has led to several conclusions which are listed below. Continuing research efforts are described when appropriate.

1. The current and voltage signal variations that result from the lateral cross-seam oscillation are very robust and quite indicative of weld joint geometry and electrode placement relative to the joint. Research is continuing to extend the data base to a wider range of welding procedures.
2. The dynamics of the SAW process are substantially different than the dynamics of the other consumable electrode processes. As a result, the voltage and current variations as a function of changes in the CTWD are several times greater than with the other processes.
3. The lateral cross-seam oscillation of the electrode has not been observed to have a detrimental effect on the SA weld; instead, it may offer advantages in terms of bead geometry control, sidewall fusion, and perhaps metallurgy as well. Research is continuing to develop models that accurately reflect the presence of lateral oscillation. The experimental welding data base is likewise being expanded.
4. The dynamics of mechanisms which change arc voltage and current in response to changes in CTWD or arc length have been modeled for small signal variations with a relatively simple pair of coupled first-order nonlinear differential equations. Research in modeling and simulation is continuing, with emphasis on expanding the simulation to more accurately model process dynamics and power source regulation dynamics under all conditions of current and electrode extension variations.
5. Digital data acquisition and analysis of the current and voltage signals have proven to serve as an excellent diagnostic tool for better understanding, monitoring, and control of the SAW process, as well as for other arc welding processes. Research is ongoing to conduct statistical analysis of the acquired signal waveforms for the purpose of detecting and correcting the occurrence of process faults, such as might occur due to surface contaminates, moisture in the flux, etc. Models are also being developed to predict the signal variations observed.

REFERENCES

1. Cook, George E., IEEE Transactions on Industrial Electronics, IE-30,252-268 (1983)
2. Cook, George E. and Peter C. Levick, "Developments and Innovations for Improved Welding Production", Welding Institute, Cambridge, England, (1985)
3. Thompson, Jr., Floyd M., Conference on Automation in Welding, Houston, American Society for Metals, (1-3 Nov. 1983)
4. Laing, Brian, R. Heid, and A. Pollack, Welding Journal, 64, 38-42 (1985)
5. Butler, C.A., R.P. Meister, and Milton D. Randall, Welding Journal, 48, 102-108 (1969)
6. Yizhang, Liu, "Through-The-Arc Sensing and Control for Robotic Submerged Arc Welding", M.S. Thesis, Vanderbilt University (1986)
7. Cook, George E. and Mark E. Shepard, "Robotic Welding: International Trends in Manufacturing Technology", IFS Publications Ltd., Bedford, England, (1986)
8. Wu, Gwo-Dong; "The Dynamic Response of Self-Regulation of the Welding Arc"; M.S. Thesis, Ohio State University (1984)
9. Richardson, Richard W., Ohio State University, personal communication
10. Halmoy, E., "Proc. Int'l. Conf on Arc Physics and Weld Pool Behavior", p. 49, The Welding Institute, Cambridge, England (1979)
11. Wilson,J.L., G.E. Claussen, and C.E. Jackson, Welding Journal,35, 1s-8s (1956)
12. Lesnewich, A., Welding Journal, 37 (8), 343s-353s (1958)
13. Amson, J.C., Report A1/47/64, British Welding Research Association, Abingdon Hall, Cambridge (1964)
14. Manz, A.F., WRC Bulletin 223; Welding Research Council, New York, (Jan. 1975)

ACKNOWLEDGEMENTS

This research was supported in part by the Office of Naval Research and the National Science Foundation. Any opinions, findings, and conclusions or recommendations expressed in this paper are those of the authors and do not necessarily reflect the views of either of these agencies.

NUMERICAL SIMULATION OF FEEDBACK CONTROL OF ARC WELD COOLING RATE

Dave F. Farson, Richard W. Richardson, Robert J. Mayhan
The Ohio State University
Columbus, Ohio

ABSTRACT

The development of the coaxial viewing welding torch has made feasible the measurement of weld pool and bead dimensions and geometry in "real time" (i.e. as welding progresses). This capability facilitates the implementation of feedback controls which regulate either weld pool dimensions and shapes or other parameters which can be derived from them. This paper describes a proposed control which regulates weld centerline cooling rates using estimates of cooling rate obtained from weld pool width measurements. The effects which this control has upon the welding thermal cycle are studied using numerical simulation. The simulation predicts that such a control would tend to regulate the centerline cooling rate to nominal values in spite of variations in preheat temperature.

THE AUTOMATION of arc welding processes has become widespread due to the increased application of industrial robotic systems. Experience with these automated systems has indicated that in many cases, some form of feedback control (also generally referred to as adaptive control in the welding literature) is desirable or even necessary for the production of satisfactory weldments. The feedback controls which have been applied to the welding process to date have mostly been of a relatively simple nature (e.g. seam tracking type controls). However, due to advances in sensing technology, the practical implementation of feedback controls which provide a greater degree of regulation of the mechanical properties of the completed weld is now becoming possible.

One sensing method which has the potential to facilitate more complete control over weld mechanical properties utilizes weld pool visual analysis with the co-axial viewing weld torch. Details concerning this sensing concept are available in [1]. In short, the coaxial viewing apparatus utilizes a digital camera situated so as to view the area of welding from directly above, coaxially with the electrode, usually via a fiber optic bundle. Co-axial torches have been developed for use with the gas-tungsten arc (GTA) and the gas-metal arc processes. This sensing approach allows weld pool dimensions (e.g. width and length), joint preparation dimensions and the positional relationship between the pool and joint preparation to be measured using computer vision.

In this paper, a method of using weld pool width measurements to implement a weld cooling rate control is described. The basis of the control is a relationship between weld pool width and cooling rate which is developed from classical heat flow relationships and tested via simulation. A feedback control which utilizes this cooling rate measurement method is then simulated.

ESTIMATION OF CENTERLINE COOLING RATE

In order to implement a feedback control of welding cooling rate, a means of measuring this variable is necessary. In this section, an equation which expresses centerline cooling rate in terms of weld pool width measurements and welding heat input is developed. Such a relationship allows estimates of cooling rate to be calculated using pool width measurements, thus providing the basis for a feedback control. This equation is derived from ideal heat flow relationships.

Prior to the development of a mathematical relationship, a connection between weld pool width and cooling rate was established by a simple experiment. Bead-on-plate welds were produced on 0.023 cm (.09 in.) thick AISI 4130 sheet steel. The thermal conditions were made to differ for each weld by varying preheat and clamping conditions; the heat input was held constant for all welds. In one configuration, the room-temperature specimen was mounted between aluminum bars (see Fig. 1) to provide a large amount of heat sinking.

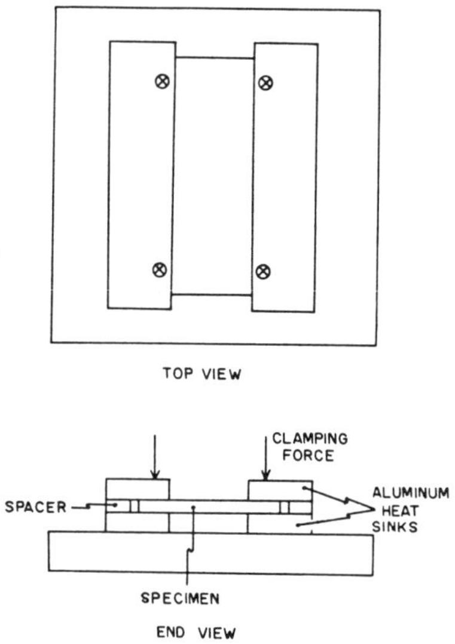

Fig. 1: Weld specimen fixturing

In another variation, the aluminum heat sinks were replaced by an insulating material which allowed relatively little heat transfer. A third sample was mounted in the insulating material and preheated to a temperature of 200 C. Welding conditions for these tests are summarized in Table 1.

Arc current	100 amp.
Arc voltage	10 volts
Travel speed	0.212 cm./sec.(5.0 in./min.)
Shielding gas	32 L/min. Ar (15 cfh)
Backup gas	64 L/min. Ar (30 cfh)
Electrode	0.318 cm.(1/8 in) 2% Th
	120 deg. included angle

Table 1: Welding Conditions

The results shown in Fig. 2 indicate that, roughly speaking, HAZ hardness is inversely correlated to weld bead width for constant welding heat input but varying thermal conditions. Since HAZ properties of the material in question are in the main determined by welding thermal cycles, it follows that thermal cycles are correlated to weld pool width for a given heat input. This relationship will now be expressed in a more concrete, quantitative way.

The following approximate relationships from Wells [2] and Adams [3] respectively, show weld pool width and centerline cooling rate as a function of base metal thermal properties, welding arc energy input, travel speed and preheat temperature. These relationships are obtained from quasi-steady state

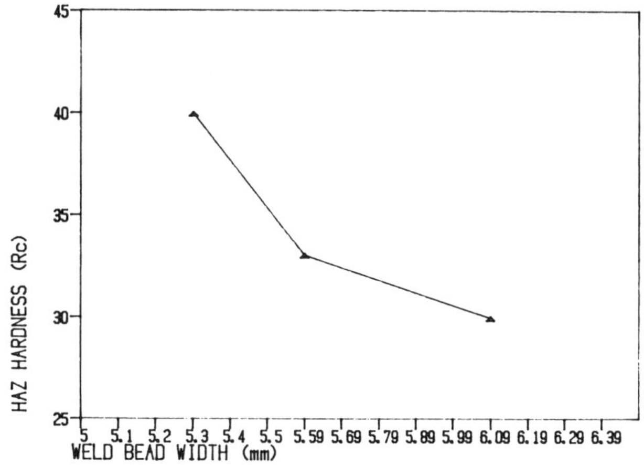

Fig. 2: HAZ Hardness vs. Bead Width

analysis of two-dimensional heat flow from a line source in material with constant thermal properties, thickness and initial temperature.

$$d = \left(\frac{4\alpha}{v}\right)\left(\frac{q}{8Kh(T_m - T_0)} - \frac{1}{5}\right) \qquad (1)$$

$$CR = \left(\frac{2\pi K^2}{\alpha}\right)\left(\frac{vh}{q}\right)^2 (T' - T_0)^3 \qquad (2)$$

where

α = thermal diffusivity
K = thermal conductivity
v = travel speed
q = arc energy input
  = η E I
E = arc voltage
I = arc current
η = arc efficiency
T'= temperature of interest

T_0 = preheat temp.
d = pool width
CR = cooling rate at T'
h = material thickness
T_m = melting temperature

A quasi-steady state relationship which expresses cooling rate as a function of weld pool width can be derived from Eqs. 1 and 2. There are several ways of accomplishing this. In general, Eq. 1 may be rearranged to express a selected parameter (e.g. material thickness, h) in terms of pool width and remaining parameters. This relationship may then be used to eliminate the selected parameter from Eq. 2, yielding an expression for cooling rate in terms of weld pool width and the remaining parameters. For the expression to be a useful means of sensing cooling rate, the remaining parameters must be known with a reasonable degree of accuracy. Hence, in forming the cooling rate equation, one should seek to eliminate a parameter which is considered to be most likely to vary or most difficult to measure.

In this work, the preheat (or interpass) temperature is selected as the parameter to be eliminated in forming the cooling rate relation. Equation 1 may be rearranged to show initial temperature, T , as a function of d,v,q and constants (Eq. 3).

$$T_0 = \frac{-q}{8Kh\left(\frac{vd}{4\alpha}+\frac{1}{5}\right)} + T_m \qquad (3)$$

This value may then be substituted into Eq. 2 to yield cooling rate as a function of d,v,q and constants (Eq. 4).

$$CR = \left(\frac{2\pi K^2}{\alpha}\right)\left(\frac{vh}{q}\right)^2 \left(T' + \frac{q}{8\,Kh\left(\frac{vd}{4\alpha}+\frac{1}{5}\right)} - T_m\right) \qquad (4)$$

The cooling rate as calculated from Eq. 4 will be referred to in the following text as the estimated cooling rate.

Several comments concerning the above method of cooling rate estimation are in order. First, the relationships involved are valid for the quasi-steady state case only. Thus the estimate obtained from them is valid only in the steady state, and in general will have a different dynamic response than the actual cooling rate. The dynamic response of the estimated cooling rate will be the subject of later study.

A limitation of the proposed cooling rate estimation method is that it is based upon relationships which only approximately describe welding heat flow. The equations are based upon the assumption of two-dimensional linear heat flow from a line source in material with constant,known thermal properties and thickness. It was also assumed that no heat is transferred from the plate surfaces to the surrounding environment. These assumptions are not in general satisfied by actual welding heat flow. For example, heat sinking by fixturing and tooling may cause heat flow to be non-two-dimensional. Furthermore, the equations represent approximations even of the ideal case; the exact solutions contain Bessel functions and derivatives of Bessel functions.

In spite of the considerations set forth above, it was found that the proposed method of cooling rate estimation, with a minor modification, performed acceptably in the simulation tests described below. How well such a method is capable of performing in an actual welding situation will be the subject of future investigation.

PROPOSED COOLING RATE CONTROL

A block diagram of a cooling rate control for an automated welding process which uses the preceeding method of cooling rate estimation is presented in Fig. 3. Here, weld pool width is measured by a sensing system, which uses the measurement to obtain an estimate of centerline cooling rate via Eq. 4. The estimated cooling rate is compared to the desired cooling rate and the welding travel speed is adjusted so as to maintain the two equal. The P-I-D control algorithm referred to in the figure is a standard industrial control method which allows the transient and steady state performance of the system to be adjusted. A more sophisticated control algorithm could be used instead to provide more flexibility in the design of the system response.

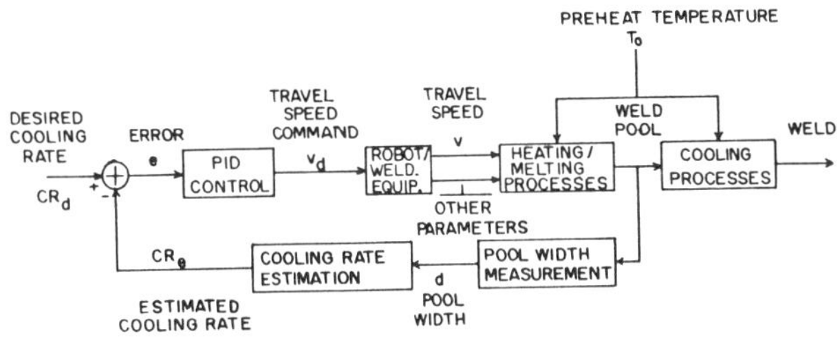

Fig. 3: Control system block diagram

SIMULATION OF WELDING HEAT FLOW

A simulation method similar to that described in [4] was used to test the performance of the proposed feedback cooling rate control algorithm. The simulation employed an integral solution of the linear two-dimensional heat flow problem with gaussian arc heat distribution and infinite boundary conditions (Eq. 5).

$$T(x,y,t)-T_0 = \frac{q}{\pi h\rho c}\int_0^t \frac{\exp\left[-\frac{x-vt^{2}-y^2}{2\sigma^2+4\alpha(t-t')}\right]}{2\sigma^2+4\alpha\, t-t'}\,dt' \qquad (5)$$

where

(x,y) = fixed coordinate frame
t = time
c = volumetric heat capacity
σ = arc heat concentration factor
(w,y) = moving coordinate frame

The integral was approximated using three-point Gaussian quadrature integration [5] over time steps of one second. The temperatures of a rectangular array of points beside the heat source center and a linear array of points behind the source was simulated. The locations of these arrays were fixed with respect to the (w,y) coordinate frame which traveled along with the heat source as shown in Fig. 4.

The weld pool width was determined from the temperatures of the points in the rectangular array as follows: each row of the rectangular array was searched for the location (with respect to the centerline) of the point

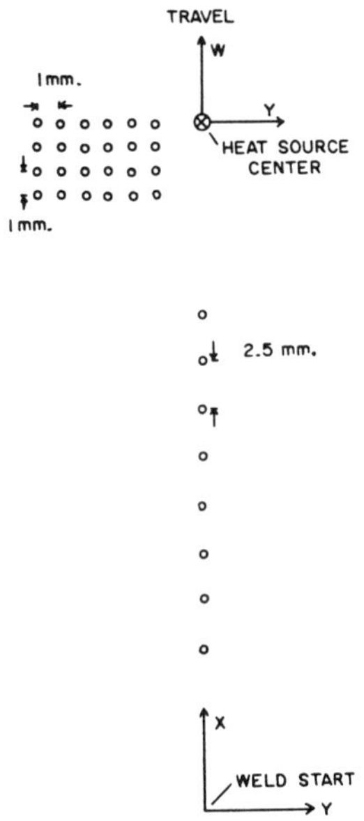

Fig. 4: Locations of points at which temperature was simulated.

having temperature equal to melting temperature. Linear interpolation was used to calculate this distance when it fell between the distances of points in the row. After finding this melting point distance for each row, these values were then compared to determine the maximum, which was taken to be maximum weld pool width.

For comparison with the estimated cooling rate obtained from the weld pool width via Eq. 4, the centerline cooling rate was calculated from the simulation using the relationship:

$$\frac{dT(x,y,t)}{dt} = -\frac{\partial T(w,y,t)}{\partial w} v + \frac{\partial T(w,y,t)}{\partial t} \quad (6)$$

The linear array of points (see Fig. 4) on the centerline behind the heat source center was searched for the point having the temperature of interest (e.g. T' = 500 C), using linear interpolation if necessary. The spatial and time derivatives indicated in the above formula were then approximated numerically at this point. The value of cooling rate obtained from Eq. 6 represents the true simulated cooling rate at T = T', and was used to evaluate the accuracy of the cooling rate estimates obtained from Eq. 4. Hereafter, the cooling rate calculated via Eq. 6 will be referred to as the simulated cooling rate to distinguish it from the estimated cooling rate as calculated from Eq. 4.

The simulation was calibrated using pool width data collected from autogeneous bead on plate welds on .236 cm (.093 in.) thick AISI 1035 and AISI 4130 sheet. The simulation was run and the heat source efficiency and concentration factor were adjusted until the maximum weld pool widths predicted by the model reasonably well agreed with the actual data. An arc efficiency of .58 and gaussian heat concentration factor of .14 cm (.056 in.) were found to produce satisfactory results.

SIMULATION RESULTS

The simulation described in the preceeding section was used to test the accuracy of the cooling rate estimation equation (Eq. 4). The relationship was modified slightly, as will be explained below, to increase its accuracy. Next, the improved relationship was used in the simulation of a cooling rate control, the performance of which was studied.

To investigate the accuracy of the cooling rate estimation relationship (Eq. 4), simulation runs were performed for a variety of preheat temperatures and welding parameter values. The steady state values of estimated cooling rate and simulated cooling rates were compared.

Initial results indicated that the estimated cooling rate was somewhat higher than the simulated cooling rate. This was found to be due to the fact that the estimates of preheat temperature obtained from Eq. 3 were about 40 C lower than the preheat temperature actually input to the simulation. The magnitude of the difference varied somewhat depending upon the welding parameters and actual preheat temperature. To compensate for this difference, the estimated value of initial temperature (as expressed by Eq. 3) was corrected by the addition of a constant value of 40 C before it was substituted into Eq. 2 to obtain the the cooling rate estimation formula. This yielded the corrected cooling rate estimation formula (Eq. 7) which was used in place of Eq. 4 in obtaining the results that follow.

$$CR = \left(\frac{2\pi K^2}{\alpha}\right)\left(\frac{vh}{q}\right)^2\left(T' + \frac{q}{8Kh\left(\frac{vd}{4\alpha} + \frac{1}{5}\right)} + 40 - T_m\right) \quad (7)$$

Comparisons of cooling rates estimated using the corrected formula (Eq. 7) and simulated cooling rates are shown in Fig. 5. The correlation between the two is quite acceptable, the maximum error being about 0.5 C/sec.

Next, the proposed feedback cooling rate control depicted in Fig. 3 was studied in a series of simulation runs. The tests were designed to investigate the effectiveness of a weld cooling rate control implemented using the estimated cooling rate. This was accomplished by varying the preheat temperature input to the simulation, and measuring the extent to which the cooling rate control was able to regulate the value of simulated cooling rate.

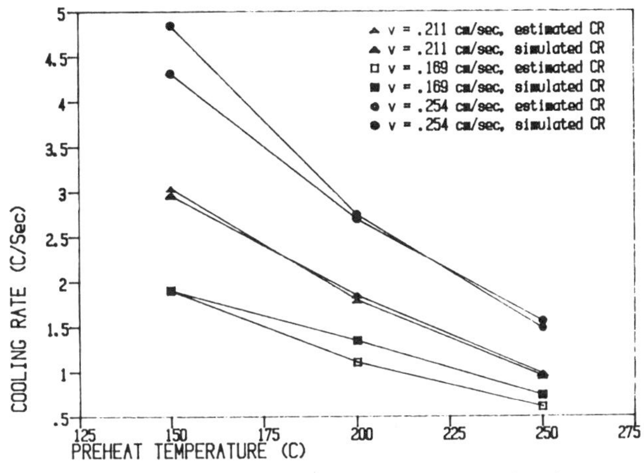

Fig. 5: Estimated and Simulated Preheat vs. Preheat Temperature

Prior to running this series of simulations, a set of nominal welding parameters and a nominal preheat were selected. The nominal welding parameters were 100 amps, 10 volts and heat was 200 C. The first simulation was run without cooling rate control using these nominal conditions and the simulated quasi-steady state cooling rate was found. This value of simulated cooling rate was used as a set point for subsequent control simulation runs. Simulation runs at other preheats (i.e. 150 C, 175 C, 225 C and 250 C) were made with and without cooling rate control. When the control was used, all welding parameters were initially set to nominal and the control subsequently adjusted travel speed so as to maintain the estimated cooling rate equal to the set point cooling rate.

The results of the cooling rate control simulations are shown in Fig. 6. The cooling rate control regulated simulated cooling rate to a nearly constant value, in spite of the preheat variations. The maximum error in cooling rate over all the runs was approximately 0.2 C/sec. which compares favorably to a maximum error of 1.2 C/sec. in the uncontrolled case.

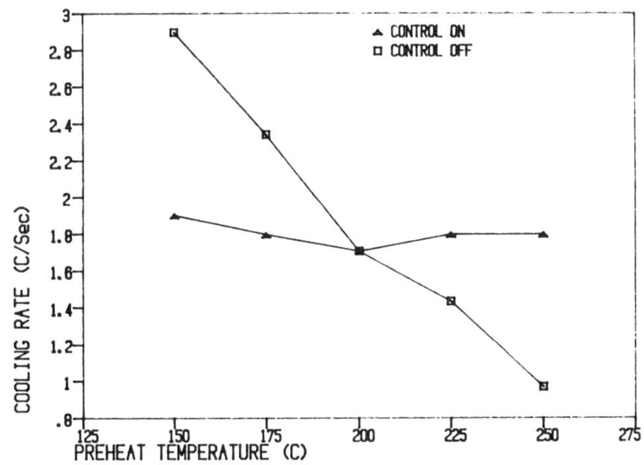

Fig. 6: Simulated Cooling Rate With and Without Cooling Rate Control

The transient response of the feedback controlled system varied somewhat depending upon welding conditions, but was in general quite slow. Settling times on the order of one minute were common. These results could conceivably be improved marginally by refinement of the PID controller gain settings which were used in the simulations. The settings which were used were arrived at based upon qualitative evaluations of control response. Refined controller settings will require modelling of the transient response of the controlled system.

Another possible means of improving the dynamic response is an adaptive control of the system, wherein the controller mechanism analyzes the dynamic response of the system and sets controller parameters for desirable system response. Such a control would be capable of operating over a wide range of welding conditions without modification.

Another improvement in the control which is under investigation is the implementation of a control which simultaneously regulates weld bead width and weld cooling rates. Note that in the tests which were described above, the weld pool width was allowed to fluctuate while the travel speed was adjusted to regulate cooling rate to a desired value. By adding a control loop which simultaneously adjusts weld current, it is conceptually possible to configure a system which allows both cooling rate and pool width to be set to desired values.

SUMMARY

A novel method of implementing feedback control of welding cooling rates was described, and the performance of the control was evaluated using numerical simulation. The unique feature of the proposed control was that the cooling rate measurements needed for the feedback control implementation were obtained from weld pool width measurements. An equation which expressed centerline cooling rate in terms of weld pool width, welding heat input and material thermal constants was obtained from ideal two-dimensional heat flow relationships. The relationship was tested using a welding heat flow simulation and was found to yield cooling rate estimates approximately equal to simulated cooling rates after minor adjustment. A feedback control which adjusted welding travel speed so as to cause the estimated welding cooling rate to equal a desired cooling rate was proposed and simulated. Simulation of the control demonstrated that effective regulation of weld cooling rate was possible, although the transient response of the system was quite slow.

REFERENCES

[1] Richardson, R.W., Gutow, D.A., Anderson, R.A., Farson, D.F., Coaxial Arc Weld Pool Viewing for Process Monitoring and Control. Welding Journal 63(3), 43-50 (1984)

[2] Wells, A.A., Heat Flow in Welding. Welding Journal. 31(5), Res. Supp., 263-s to 267-s (1952)

[3] Adams, C.M. Jr., Cooling Rates and Peak Temperatures in Fusion Welding. Welding Journal. 37(5), Res. Supp., 210-s to 215-s (1958)

[4] Eagar, T.W., Tsai, N.-S., Temperature Fields Produced by Traveling Distributed Heat Sources. Welding Journal. 62(12), Res. Supp., 346-s to 355-s (1983)

[5] Gerald, C.F., "Applied Numerical Analysis", 2nd Ed., p.222, Addison-Wesley Publishing Co., Reading, Mass. (1980)

MULTIVARIABLE CONTROL OF ARC WELDING PROCESSES

Charalabos Doumanidis, Michael Hale, David Hardt
Laboratory for Manufacturing and Productivity
Massachusetts Institute of Technology
Cambridge, Massachusetts

ABSTRACT

A multivariable non-linear model of GMAW is presented that is intended for use in a comprehensive feedback control system. The system includes both the control of the weld pool geometry (width, depth and reinforcement area) and of thermally activated properties (as reflected in the heat affected zone width and the centerline cooling rate). The models that are developed are based on analytical results and on numerical simulation. The basic problem is shown to involve a highly coupled, non-linear dynamic model with four inputs and five outputs. Current work is extending the number of inputs to five and developing both linear and non-linear control schemes based on these models to eventually achieve independent regulation of all output quantities.

INTRODUCTION

The ultimate goal of welding is to join two or more materials with sufficient strength and fracture toughness for the assigned task without excessive distortion of the base materials. These properties are the integration of many individual aspects of the welding process, yet, the task of welding control is to achieve high quality by controlling the melting and resolidification processes involved. With this indirect control authority, it is imperative that a comprehensive control strategy be pursued to prevent a fragmented, redundant, and ultimately low performance solution from resulting. With respect to quantities that can be controlled during the welding processes, the dominant ones are : fusion zone geometry, heat affected zone properties, and gross weldment thermally induced stresses. (Transient imperfections such as porosity or inclusion are of great importance but are assumed at this point to be best addressed by material and weld schedule selection).

This paper concentrates on development of dynamic models and a control strategy for the weld geometry and local material properties for Gas Metal Arc Welding (GMAW). The models demonstrate the truly coupled and non-linear nature of the welding process and are constructed to permit a study of the dynamic interaction of the variables of concern. They also facilitate analysis of the control problem and eventual control system design and experimentation.

CONTROL OBJECTIVES

The goal of most control systems is to achieve stable regulation (i.e. setpoint maintenance) when the process under control changes or is subjected to unpredictable outside influences. This description certainly captures the essence of welding, but a more concise objective can be stated. In this research, we are concentrating on achieving <u>simultaneous</u> regulation of five quantities, three related to fusion zone geometry and two related to heat affected zone properties.

The geometry is characterized by the weld surface width, maximum depth and the area of reinforcement provided by the filler material. These quantities determine the basic mechanical properties of the weld and are key to insuring the basic stress handling capabilities of the joint. The solid material properties are determined (in differing ways depending upon the material in use) by the extent of some critical isotherm (heat affected zone width) and the rate of cooling at some critical temperature. These control objectives constitute a five output control system (see Fig. 1), and since each of these quantities is affected by each of the available controls on a GMAW torch, a fully coupled control strategy must be adopted.

To pursue such a control study, a model of the input-output characteristics of this "block" must be understood, both to identify what inputs to the process (e.g. heat rate, wire feed, etc) will be needed, and to permit a range of controller designs to be explored so as to optimize performance and assess the impact of various feasible feedback measurement schemes.

The remainder of this paper will detail models necessary for the characterization of welding to permit full multivariable control of the process. While ultimately there must be a single, five input, five output model as described above, two separate analyses are presented, one for geometry and one for thermal effect. However, the coupling of the two will be obvious and is reviewed at the end.

FUSION ZONE GEOMETRY CONTROL

Geometry control of the GMA welding process requires control of three output variables, weld bead width, w, bead depth, d, and reinforcement area, A_r. There are three control variables available for regulation of the output variables; open circuit voltage, E_o, wire feedrate, f, and torch speed, v. The dynamic relationship between the control variables and the output variables is expressed by a control model.

The basic system to be modelled is shown in Fig. 2. Of concern to this analysis is the power supply - arc - electrode interaction, the dynamics of the weld melting process, relationships between the weld pool width and depth, and the dynamics of the servos used to control the electrode wire feedrate and torch travel speed.

The first part of the control model that must be established is the dynamic relationship between arc voltage and arc length. This model is modified from a similar development in [1]. The arc dynamics can be established by using conservation of mass on the fill wire. The change in arc length, l, is the difference between the melting rate of the wire and the wire feedrate

$$\frac{dl}{dt} = \frac{dm}{dt} - f$$

where dm/dt is the melting rate and f is the feedrate, a regulated variable. The melting rate is the rate of heat input to the wire divided by the heat required to melt the wire, Q_{rl}. Heat input is given by

$$\frac{dQ_{in}}{dt} = E_1 I + \eta_1 E_2 I$$

where I is the current, E_1 is the voltage drop across the stickout, E_2 represents the voltage drop across the arc, and η_1 is the portion of the heat generated by the arc voltage drop which is transferred to the wire. E_1 can also be written

$$E_1 = R(h-l)I$$

where R is the resistivity of the wire and h is the contact tip to workpiece distance. We need to express the current, I, in terms of open circuit voltage, E_o, which is a control variable. The ideal voltage-current relationship for GMAW power supplies is given by

$$E = E_o - nI$$

where E is the operating voltage and n is the sloping factor. E can be expressed as the sum of voltage drops across the stickout and the arc

$$E = E_1 + E_2$$

or

$$E = R(h-l)I + E_a + E_c + gl$$

where E_a is the voltage drop at the anode, E_c is the voltage drop at the cathode, and g is the slope of the arc voltage across the arc. Solving for I yields

$$I = \frac{E_o - (E_t + gl)}{n + R(h-l)}$$

where

$$E_t = E_a + E_c$$

Substituting into the dynamic arc length equation above yields

$$\frac{dl}{dt} = \frac{R(h-l)I^2 + n_1(E_t + gl)I}{Q_{rl}} - f$$

Notice that although both E_o and f appear as control inputs to the state variable l, the feedrate f is typically provided by a velocity servo, which also has dynamics to be considered. In other words the feedrate cannot be specified and changed instantaneously, and the actual control available to us is f_c, the feedrate command sent to the wirefeeder. The relationship between the command and the actual feedrate can be expressed by a simple first-order equation since the wirefeeder is a velocity servo.

$$\frac{df}{dt} = \frac{-f + f_c}{\tau_f}$$

where τ_f is the time constant which determines how quickly the state f responds to the command f_c.

A model of the weld pool melting dynamics can be developed by applying conservation of energy to the pool. The following model is a modification of GTAW pool dynamics presented in [2]. The pool in this analysis refers to liquified base metal and does not include metal added by the consumable electrode. The change in pool volume, V, is equal to the net energy added to the pool divided by the energy required to melt the base material, Q_{r2}. The net energy added is equal to the difference between the heat input to the pool and the heat conducted away from the pool. Heat input is given by

$$\dot{Q}_{in} = \eta_2 EI$$

where η_2 is an efficiency factor. The heat conducted away from the pool is given by

$$\dot{Q}_{out} = kA_i \frac{d\theta}{dr}$$

where k is the conductivity of the metal, A_i is the area of the solid-liquid interface, and $d\theta/dr$ is a temperature gradient across the interface. The temperature gradient is not the same in all directions and also varies with torch speed. For simplicity we will assume that

$$\frac{d\theta}{dr} = K_v v$$

which states that the average temperature gradient is some linear function of torch speed, v. A dynamic equation can be written for v that is similar to that for f

$$\frac{dv}{dt} = \frac{-v + v_c}{\tau_v}$$

where τ_v is the time constant for the torch velocity servo.

In order to express the pool melting dynamics in terms of the geometry variables of interest, we must assume some shape so that

$$V = F_1(w,d) \quad \text{and} \quad A_i = F_2(w,d)$$

If we assume that the weld pool could be described as half an oblate or prolate spheroid then

$$V = \frac{\pi w^2 d}{6}$$

A_i is simply the surface area of the spheroid. Taking the derivative of V results in

$$\frac{dV}{dt} = \frac{\pi w^2}{6}\frac{dd}{dt} + \frac{\pi w d}{3}\frac{dw}{dt}$$

This is still an indeterminate form, but we can solve the problem by finding some relationship between w and d. Experiments show that width-to-depth ratio can be described as a linear function of arc voltage or arc length (see [3]). If we assume a relationship of the form

$$\frac{w}{d} = 1.0 + \frac{l}{l_m}$$

where l_m is some maximum arc length, then the pool melting dynamics can be written as

$$\frac{dw}{dt} = \frac{2(l+l_m)}{\pi Q_{r2} l_m w^2}\left[\frac{\pi Q_{r2} w^3 l_m}{6(l+l_m)^2}\frac{dl}{dt} + \frac{E_o - (E_t+gl)}{n+R(h-l)} \cdot \eta_2(E_t+gl) - kA_i K_v v\right]$$

So far we have developed a model with four states; l, f, w, v, and three controls; E_o, f_c, and v_c. One output, width, is a state and a second output, depth, is a nonlinear function of two states, w and l. The third output, reinforcement area can be determined from the states by considering a volume balance. The volume of metal added must be equal to the volume of the bead reinforcement

$$fA_e = A_r v$$

where A_e and A_r are the cross-sectional areas of the electrode and the reinforcement respectively. Then we can express reinforcement area as a function of two states

$$A_r = \frac{fA_e}{v}$$

This simple model, then, has four state equations that describe the relationship between the physical welding process including actuator dynamics and the control inputs. There are also three equations which show the relationship between the states and the outputs. With this model we have some indication of how the outputs will respond to changes in the inputs. The model shows very clearly the nonlinear nature of the process and also the strong coupling between the states.

Implementation of feedback control requires real-time measurement of the states or the outputs or both. If the relationship between the states and the outputs is well defined, then measurement of the states is adequate. For the welding process, which is a noisy, nonlinear, poorly-defined process, measurement of the outputs is necessary for good control. This means that, for the model developed above, we must have real-time measurements of the width, depth, and reinforcement area. This requires very complicated sensing techniques and equipment, some of which is not yet available. The sensing problems have a direct effect on the control development. Although there is a good chance of achieving satisfactory control using simple models, there is very little hope of achieving good control of a complicated, nonlinear, multivariable process if the measurements of the desired variables are noisy and inaccurate. In any case, sensors introduce additional dynamics which must be modelled and incorporated into the control analysis.

For required measurements that are not yet available, such as weld pool depth, it is possible to apply techniques of estimation (see for example Somers [3]). Estimation, though, requires more accurate modelling. The tradeoff, then is between modelling and measurement. Those items which cannot be accurately modelled must be accurately measured and vice versa.

We must also be aware of the interaction of the geometry control on other aspects of the weld. For instance very fast, high-quality control of the geometry will require large control power. The large increase in heat input which would be required to increase the width and depth of a weld pool will also increase the heat affected zone. The amount of control which can be achieved over the geometry will certainly be affected by the amount of control which is required for good thermal properties of the weld. In addition, the following section will show that control inputs for thermal and geometry control are not independent.

THERMAL PROPERTY CONTROL

The following analysis focuses on the thermal control problem, i.e. thermal variables will be used to control the material microstructure characteristics of the weld joint. Before selecting the suitable control inputs, the appropriate quantities characterizing the state of the process and determining the important metallurgical features of the final structure must be specified. As covered in detail in [4], these will be the weld nugget cross section area, NS, the heat affected zone width, HZ, and the centerline cooling rate at a critical temperature, CR. Roughly speaking, the weld nugget cross section area, defined by the solidus isotherm T_m, represents a collective measure of the extent of solidification defects in the weld bead, and may also be important in controlling the dilution of the base material with filler material. The heat affected zone width, defined by an envelope isotherm T_H, may indicate the extent of weak areas adjacent to the bead (such as the recrystallization zone), or areas in which some undesirable phase nucleates (such as the sensitization zone in which Cr_4C precipitates), or areas vulnerable to contamination because of increased reactivity with the environment. Finally the centerline cooling rate at the critical temperature T_c gives a measure of the effects of thermal stresses (such as hot cracking or the extent of nucleation of undesirable phases such as martensite). The degree of importance of each of the three features of the temperature fields is different for different types of welding material and is reviewed in [4] for a wide range of materials.

Steady State Modelling

In the following the dependence of these three variables on the process inputs, such as heat imput, travel speed etc, initially at the steady state of welding is examined. The purpose is to identify those thermal quantities that have the greatest effect on the NS, HZ and CR, and thus can serve as inputs for the control problem. Furthermore, it is desirable to select as inputs quantities that have a separate effect on one of the three variables only, leading to independent control of that variable and decoupling the control problem. The steady state relations presented here were selected from a broad survey of the literature on analytical, numerical and empirical modelling of the welding process [4].

An expression for the nugget cross section area may be derived from the classical steady-state conduction model [5,6,7,8]. It can be shown that if the heat input can be modelled by a moving line source (2D case), the cross section of the weld bead will be a rectangle of area:

$$NS_2 = \frac{2\ c}{(2\pi \exp(1))^{1/2}\ \rho C_p} \cdot \frac{Q}{v(T_m - T_o)}$$

where ρ is the density of the material, C_p is the specific heat, T_o the initial preheat temperature, T_m the temperature at the melting point (solidus for alloys), v is the torch travel speed, Q the total heat input supplied by the source, and c is an experimentally determined constant, depending mainly on the efficiency of heat input, n.

The corresponding case with a moving point source (3D case) gives a cross section which is a semicircle and has an area of:

$$NS_3 = \frac{\pi}{2} \left[\frac{2\ \alpha\ c}{v} \left(\frac{Q\ v}{2\alpha k\ (T_m - T_o)} \right)^{1/n_3} \right]^2$$

where k is the thermal conductivity, α is the thermal diffusivity of the material and n_3 is the 3D-conduction exponent = 1.70.

This analysis, which is based on the idealized conditions of the Rosenthal conduction model [6,8] and completely neglects the material transfer in the case of GMAW, may be supplemented by the addition of the reinforcement cross section area:

$$A_r = \frac{fA_e}{v}$$

Note that this implies that A_r is proportional to Q/v since f is proportional to Q.

It is clear from the above relationships that the main dependence of NS is on the heat input Q, the torch velocity v and the preheat temperature T_o. The dependence of NS on Q and v, which is slightly different in the two cases of line and point source, has been extensively investigated experimentally, and various empirically derived relations were proposed. Surprisingly enough, in all these experimental relations [9,10,11,12], the ratio of the exponents of Q and v is 1 over 2/3, and it agrees with the point source model prediction. So it seems reasonable to adopt for the nugget cross sectional area a relationship of the form:

$$NS = C_1 \left(\frac{Q}{v^{2/3}}\right)^{n_a}$$

where the constants C_1 and n_a must be determined experimentally for the particular material and preheat temperature, with the expected value $n_a = 1.25$ [9,11].

The steady state conduction model of welding can also be used to derive expressions for the heat affected zone width, measured from the edge of the weld bead. For the 2D-case of a line source the estimate is:

$$HZ_2 = \frac{2\,\alpha\, c}{v\left(\frac{kt}{Q}\right)^{1/n_2}} \left(\frac{1}{(T_H - T_o)^{1/n_2}} - \frac{1}{(T_m - T_o)^{1/n_2}}\right)$$

where t is the thickness of the plates, c is a constant depending on the efficiency n, n_2 is the 2-D conduction exponent ($n_2 = 0.96$) and T_H is the critical temperature defining the heat affected zone. The relationship for the 3-D case of a point source is:

$$HZ_3 = \frac{2\,\alpha . c}{v\left(\frac{2\alpha k}{Qv}\right)^{1/n_3}} \left(\frac{1}{(T_H - T_o)^{1/n_3}} - \frac{1}{(T_m - T_o)^{1/n_3}}\right)$$

where now n_3 is the 3D-conduction coefficient ($n_3 = 1.70$). As the two constant coefficients n_2 and n_3 imply, the relation of HZ and T_m-T_o is linear on a logarithmic plot for both the 2-D and 3-D cases, a fact that was experimentally verified. [7,8]. It is again reasonable to assume a linear variation of the slope of this line for the intermediate (or 2-1/2-D) case of a real source acting on welded plates of finite thickness [8].

To derive an expression for the centerline cooling rate, CR, the steady-state conduction welding model may again be used, with the condition that $(\partial T/\partial t)_{Tc} = 0$, in which case the cooling rate is:

$$CR = -v(\partial T/\partial x)_{Tc}$$

The line source 2-D model predicts for the cooling rate:

$$CR_2 = -2\pi\, k\rho C_p\, (T_c - T_o)^3 \left(\frac{vt}{Q}\right)^2 \quad \text{at } T_c$$

where T_c is the critical temperature at which the cooling rate is measured. The 3-D-case of the point source gives the relation:

$$CR_3 = -2\pi\, k\, (T_c - T_o)^2\, \frac{v}{Q} \quad \text{at } T_c$$

From the above analysis the nature of the dependence of the variables NS, HZ and CR on the inputs v and Q should be clear. The dependence on the preheat temperature T_o was not emphasized and actually was included in the coefficients c, since a number of problems is associated with the in-process direct control of T_o. An implementation involving a leading preheat torch would provide a nonuniform T_o and introduce considerable transportation lags to the control problem. These difficulties, together with the coupled effect of T_o on all variables NS, HZ, and CR, make T_o inappropriate as a direct input of the control problem. However, since the exponent of the $(T-T_o)$ term in the expressions of the cooling rate is much higher than the respective exponents in the expressions of the heat affected zone width and nugget cross section area, T_o exerts a selective effect on CR which will be exploited by the control scheme of the next section.

The above analysis and discussion makes it clear that the single-torch configuration only provides two distinct inputs (Q_1 and v), to the control problem. A third input must be found in order to ensure steady state controllability. It would also be desirable that the additional input exert a decoupled influence on the three variables, i.e. affect only one of them. With these requirements in mind and since a leading preheat torch, as discussed in the previous paragragh, would not be a wise solution, a trailing secondary torch exploiting the preheating effect of the main torch to define the final CR value may be used. This configuration does not involve transportation lags in the control problem and also satisfies the decoupling requirement if the weld pool and heat affected zone of the secondary torch do not exceed the corresponding quantities of the main torch (Fig. 3).

To see that this can be true, consider the steady-state equations for NS, HZ and CR for the secondary torch, assuming a uniform preheat temperature from the main torch, and the corresponding equations for the main torch, for which there is no preheat. If the values of either NS or HZ are set the same for these two equations, it is concluded that the ratio $(Q/v)/(T-T_o)$ should be the same for the two torches, i.e. a decrease in $T-T_o$ for the secondary torch would mean a proportional increase in v/Q. Since in both the 2D and 3D relations of the CR the exponent of T_c-T_o is higher by one than the exponent of v/Q, this suggests that CR will be less for the secondary torch.

Let us now drop the assumption of uniform preheat from the primary torch and consider the problem of minimization of the cooling rate after the secondary torch, with the requirement that the HZ resulting from the main torch (HZ_1) is not increased by the secondary torch (ie. $HZ_1 = HZ_2$). This also ensures that $NS_2 < NS_1$, as it can be easily shown, and thus the effect of the secondary torch on CR will be completely decoupled.

By using the previously derived equations for the cooling rate, it can be shown (see [4] for details) that with a single trailing torch, the maximum reduction in cooling rate is achieved (for the 3D case) when the secondary torch trails by x_{c1} and the heat input $Q_2 = 0.5\ Q_1$. Under these conditions the CR is reduced by 25%. Similarly, for the 2D case, the trailing distance is $2x_{c1}$, the optimum heat input $Q_2 = 0.42\ Q_1$, and the CR is reduced by 38%.

Although these percentages can be improved if $T_H > T_c$ (which is unlikely for many materials in practice), the effect of a single secondary torch in reducing the CR may not be adequate for control purposes. Although it may not be always realistic in practice, the same optimization problem can be considered for a multiple torch configuation consisting of n distinct torches (including the primary torch), hoping to obtain a further reduction of the cooling rate. In this case it is found that the CR is reduced at most by 50% for the 3D case and in the limit, as the number of torches gets large, the 2D case predicts that the CR can approach 0. Of course, the use of a single distributed secondary heat input is the sensible alternative to the distinct multitorch configuration. Although it was not considered in detail, the constant optimum distance between successive torches in the distinct multitorch approach (x_{ci} in the 3D, $2x_{ci}$ in the 2D) is an indication that the optimum heat distribution will decay slowly from the position of the main torch to a considerable length behind it, i.e. it may not be very practical.

Having always the possibility of a more efficient distributed heat input from a secondary torch in mind as an alternative, at this point the configuration will be limited to one secondary torch only, with a concentrated heat input Q_2. It is assumed that the maximum obtainable reduction of CR by this configuration is adequate for control purposes, and thus the ratio $S = Q_2/Q_1$ may be defined as the third input of the control problem. The steady-state relationships for CR_2 are quite involved, and only their linearized forms are shown here:

2D: $CR_2 = CR_1\ (1 - 0.5\ S)$ near $S = 0$

$CR_2 = CR_1\ (0.91 - 0.33\ S)$ near $S = 0.50$

3D: $CR_2 = CR_1\ (1 - 0.93\ S)$ near $S = 0$

$CR_2 = CR_1\ (0.61 - 0.59\ S)$ near $S = 0.42$

In general, the dependence of CR on the third input, S, may be adequately represented by a relationship of the form:

$$CR_2 = CR_1\ c\ (1 - \beta\ S)$$

where c and β are constants to be determined experimentally.

The effect of S on the variables of the control problem in the steady state has thus been determined. It should be borne in mind that this effect is decoupled, i.e. S affects CR only, as long as S is less than the value at the optimum point (0.50 for the 3D case, 0.42 for the 2D case).

<u>The steady-state model</u>

Now the steady state relationships for the three variables NS, HZ, and CR can be combined to form a complete steady-state model, where the dependence of the variables on the three inputs Q_1, v and S is explicitly expressed:

$$NS = C_1\ \left(\frac{Q_1}{v^{2/3}}\right)^{n_a}$$

$$HZ = C_2\ \left(\frac{Q_1}{v^f}\right)^{1/n_b}\ ;\ f = n_b - \frac{n_b - n_2}{n_3 - n_2}$$

$$CR = C_3\ \left(\frac{v}{Q_1}\right)^{n_g}\ (1 - \beta\ S)$$

This is a steady-state, partially decoupled model of the dependence of NS, HZ and CR on Q_1, v and S. Note that Q_1 is any measure of the heat input provided by the primary source, such as voltage (if the current is kept constant) or current (if the voltage is kept constant) or the product of voltage and current, since the conversion factors are incorporated in the constants c. As it is shown in [4] the constants C_1, C_2 and C_3 and the exponents n_b and n_g can be determined by information available in-process, while the exponent n_a and the coefficient β must be known by previous experiments.

Transient Modelling - Numerical Simulation

In order to model the dynamic dependence of NS, HZ and CR on the selected inputs Q_1, v and S, a solution of the time-dependent temperature field described by the transient conduction model, which is based on the integration of the appropriate Green's function [4,13] turns out to be impractical and limited in scope. Instead, a versatile and general-purpose numerical simulation of the temperature field of the multitorch welding process was developed. The computer program, which integrates the unsteady heat transfer equations in discrete time and space elements by using an explicit backward finite difference formulation is documented in [4]. Its main features are given below:

The geometrical configuration of the simulation is that of butt welding two identical, infinitely long and wide plates with finite thickness (2-1/2-D configuration). This case covers with acceptable accuracy the finite geometries where the width of each plate and the distance from a longitudinal edge at the beginning of welding are greater than ten times the plate thickness [5]. For even more restricted geometries where this condition is not fulfilled, the simulation is still applicable with minor modifications. Another requirement is the symmetry of the heat flow problem about the centerplane, which permits studying the temperature field in one of the plates only. For asymmetric geometry or other heat flow conditions, there is a provision for heat exchange at the interface of the plates, which will permit concurrent simulation of the temperature field in different adjoining plates.

The geometry near the torches is simulated by two rectangular constant-spacing grids moving together with the torches, as in Fig. 4. The coarse grid serves in the study of the temperature field at remote areas (HZ,CR) while the fine grid ensures the required accuracy near the weld pool (NS). The configuration of the process is a multitorch one, with all of the torches aligned on the centerline. The number and distances between the torches are user-defined. The heat input from each torch is spatially distributed according to a Gaussian distribution, which is alterable to any arbitrary form, permitting even, continuous distributions along the centerline. The heat outflow from the welded pieces includes convection and radiation, with different coefficients for the two sides of the pieces.

The properties of the material are temperature-dependent, with the corresponding function defined by the user for each material. These include solidification-fusion latent heat to take into account the thermal effects of this transformation between the solidus and liquidus temperature in the general case of an alloy. The weld pool heat transfer consists of equivalent conduction mechanisms only. The equivalent conduction coefficient used for the weld pool material can be adjusted to make any characteristic feature of the pool (such as NS) match experimentally determined values.

The dynamic thermal model

The numerical simulation of multitorch welding was applied to the investigation of the dynamics of the thermal system. The specific case concerns the welding of two 2mm thick plates of stainless steel 304, without preheat. For the heat affected zone it is assumed that T_H = 870° C. Above this temperature the chromium carbide precipitation proceeds very fast and leads to an excessive amount of sensitization. Finally, for the cooling rate it is specified that T_c = 1224° C. At this range of temperatures the δ-ferrite transforms to austenite, and a low cooling rate is desired to obtain a fully austenitic structure. A set of nominal input values is also specified, which will ensure the desired steady-state values of the variables NS, HZ and CR. These inputs are:

Primary heat input:	Q_1^* = 1500W
Travel speed:	v^* = 5mm^2
Secondary heat input:	Q_2^* = 0

resulting in NS^* = 3.04 mm^2, HZ^* = 0.85 mm, and CR^* = -495° C/s.

The steady state values NS^*, HZ^*, CR^* are obtained by simulating the initial transient response of these variables to the nominal inputs.

The transient responses of these variables to various step inputs were then studied by simulation. Typical step changes in the operating point were ±20% of the operating point or:

Primary heat input:	1800 W and 1200 W
Travel speed:	6mm/s and 4mm/s
Secondary heat input:	S=0.2 and back to S=0.

Typical transient responses are shown in Figure 5. Of note first is the nonlinearity of the system as one compares the step responses to positive and negative steps of the same input, which give obviously different transients. This will become more evident by quantitative comparisons of gains and time constants. However, as a first approach, linearity is assumed in a small neighborhood of the nominal point and the quantitative responses are examined in terms of transfer functions between the variables and the inputs. The range of inputs in which this linearized model is acceptable will have to be determined separately.

The response of the nugget cross-section area NS is generally as expected, i.e. the typical overdamped second order response to a step in either Q_1 or v. The fact that the time constant of the one mode is far smaller than that of the other (i.e. the respective pole is much farther to the left on the s-plane) justifies the simplification of this relationship to a first order dependence, which keeps the dominant mode only. The respective transfer functions may be written as:

$$\frac{NS(s)}{Q_1(s)} = \frac{K_a}{T_a s+1} \; ; \quad \frac{NS(s)}{v(s)} = \frac{K_a'}{T_a' s+1} \; ; \quad \frac{NS(s)}{S(s)} = 0$$

where the K's and T's are gains and time constants respectively.

The response of the heat affected zone width again looks very much like the response of a nonminimum overdamped second order variable to a step in either Q_1 or v. To include the undershooting effect realized soon after the step change, both the positive zero and the second pole are included in the respective transfer functions. The transfer function can be written as:

$$\frac{HZ(s)}{Q_1(s)} = \frac{K_b\,(T_b s+1)}{(T_1 s+1)(T_2 s+1)}$$

$$\frac{HZ(s)}{Q_1(s)} = \frac{K_b'\,(T_b' s+1)}{(T_1' s+1)(T_2' s+1)} \; ; \quad \frac{HZ(s)}{S(s)} = 0$$

where $T_b = (K_1T_2 - K_2T_1)/(K_1 - K_2)$, $K_b = K_1 - K_2$, etc. (see Table 1)

Finally the transient responses of the cooling rate appear to be irregular and rather unrelated between positive and negative steps to the same input. This is explained in terms of transitions of a point on a temperature hill at a constant temperature level T_c and moving on the centerline as the temperature hills go up and down. This topological changes of the temperature hills which are dramatic on the centeline and come as a result of step changes of the inputs cause rather irregular changes of CR or even discontinuous changes in case of transition of the point between two surfaces of different slope. Also, the observed oscillation of the back surface of the weld pool (solidification front) where solidification proceeds with alternating slow and rapid rates, affects the cooling rate on a very fine time scale. However, there is still a dominant time constant, to which the above effects may be superimposed. So a proposed transfer function for CR, similar for all inputs Q_1, v and Q_2 (or S), which considers an oscillating mode and the undershooting effect of a positive zero, as well as the basic exponential mode, is:

$$\frac{CR(s)}{Q_1(s)} = \frac{K_c\,\omega_n^2\,(T_c s+1)}{(T_o s+1)(s^2+2\zeta\omega_n s+1)} \; ;$$

$$\frac{CR(s)}{V(s)} = \frac{K_c'\,\omega_n'^2\,(T_c' s+1)}{(T_o' s+1)(s^2+2\zeta'\omega_n' s+1)} \; ;$$

$$\frac{CR(s)}{S(s)} = \frac{K_c''\,\omega_n''^2\,(T_c'' s+1)}{(T_o'' s+1)(s^2+2\zeta''\omega_n'' s+1)} \, .$$

From the data taken from the simulation of the step responses, the gain and time constants of the transfer functions of NS, the gain and time constants of the transfer functions of HZ and just the gain and the T_k-time constant for the CR-transfer functions have been calculated and are summarized in Table 1.

The linearized response of the thermal variables $\underline{Y}(s)$ to the inputs $\underline{U}(s)$ may be written in the form of a transfer matrix G(s) as follows:

$$\underline{Y}(s) = G(s)\underline{U}(s)$$

$$\text{with } \underline{Y}(s) = \begin{vmatrix} NS(s) \\ HZ(s) \\ CR(s) \end{vmatrix}, \quad \underline{U}(s) = \begin{vmatrix} Q_1(s) \\ V(s) \\ S(s) \end{vmatrix},$$

$$G(s) = \begin{vmatrix} NS(s)/Q_1(s) & NS(s)/V(s) & 0 \\ HZ(s)/Q_1(s) & HZ(s)/V(s) & 0 \\ CR(s)/Q_1(s) & CR(s)/V(s) & CR(s)/S(s) \end{vmatrix}$$

where the elementary transfer functions of G(s) are as above and NS(s), HZ(s) and CR(s) denote deviations from the nominal conditions NS^*, HZ^* and CR^* respectively in all the above relationships.

Finally, at this point it is possible to examine the agreement of the gains caluculated above by the simulation with the steady-state model predictions. This is done by comparing the calculated values of the exponents n_a, n_b and n_g of the steady-state model relations to the respective expected values by the 2-D line source and 3-D point source conduction models.

Nugget cross section area NS:

n_a = 1.061...1.634 (expected: 1.25)

Heat affected zone HZ:

n_b = 0.918...0.949 (expected: ~ 1)

Cooling rate on centerline CR:

n_g = 1.271...1.413 (expected: 1 ~ 2)

This indicates that the (nonlinear) dependence of the gains K on the values of the inputs is in reasonable agreement with the predictions of the steady-state model, the small scatter of values being due to the assumptions of the latter [6] as well as the use of experimental data from different literature sources by the simulation.

CONCLUSIONS

Welding control systems must be considered a comprehensive system regulating all relevant weld properties in real-time. In this paper the process of GMAW is treated as a dynamically coupled, multivariable system in order to achieve optimal welding control system performance. A system description has been forwarded that defines the geometric and thermally activated properties of the weld in a fashion that is amenable to control system analysis. The basis for this analysis, multiple input-multiple output dynamic model, has been derived for both the case of detailed pool geometry and temperature field control. The two models are at this time separate, linked only by the fact that the latter considered the pool area as a single, lumped variable. Both models indicate the inherently non-linear nature of the process and pose particularly challenging control problems.

These models are currently being used to develop specific linear and non-linear control algorithms, the former following the standard Linear Quadratic Gaussian Regulator design for an optimal controller and the latter involving the use of sliding mode control.

However, one problem still must be resolved. While the total number of output variables is five, there exist in the above models only four independent inputs (primary and secondary torch heat input, travel speed, and wire feedrate. Thus additional degrees of freedom must be identified to allow the fully independent multivariable control system to be developed.

ACKNOWLEDGEMENTS

This work was supported by the US Department of Energy, under contract no. DE-FG02-85ER13331.

REFERENCES

1. Lund, Richard Alan, "Multivariable Computer Control of a Gas Metal Arc Welding Process", Ph.D. Thesis, University of Wisconsin-Madison, 1977.

2. Hardt, D.E., Garlow, D.A., and Weinert, J.B., "A Model of Full Penetration Arc-Welding for Control System Design", Trans. of ASME, Journal of Dynamic Systems Measurement, and Control, Vol. 107, March 1985, pp. 40-46.

3. Somers, T. "Control of GTA Weld Penetration Based on Infrared Temperature Measurement", S.M. Thesis, M.I.T., May 1986.

4. Doumanidis, C.C., "Real-Time Control of Thermal Properties in Welding". LMP Research Report, LMP/FAR 86-05.

5. Wells, A.A., "Heat Flow in Welding", Weld Journal, Vol. 31, No. 5, May 1952, pp. 263s-267s.

6. Rosenthal, D., "Mathematical Theory of Heat Distribution During Welding and Cutting", Welding Journal, 20(5), 1941, pp. 220s-234s.

7. Barry, J.M., Paley, Z., and Adams, C.M., "Heat Conduction from Moving Arcs in Welding", Welding Journal, 42, 1962, pp. 97s-104s.

8. Myers, P.S., Uyehara, O.A., and Borman, G.L., "Fundamentals of Heat Flow in Welding", Welding Research Council Bulletin, 123, July 1967, pp. 1-35.

9. Jackson, C.E. and Shrubsall, A.E., "Control of Penetration and Melting Ratio with Welding Technique", Welding Journal, 32, 1953, pp. 172s-178s.

10. Cristensen, N., Davies, V.L. and Gjermundsen, K., "The Distributions of Temperatures in Arc Welding", British Welding Journal, 12, Feb. 1965, pp. 54-75.

11. Bradstreet, B.S., "Effect of Welding Conditions on the Cooling Rate and Hardness in the Heat-Affected Zone", Welding Journal, 48, 1969, pp. 499s-504s.

12. Thorn, K., Feenstra, M., Young, J.C., Lawson, W.H.S., and Kerr, H.W., "The Interaction of Process Variables - Their Influence on Weld Dimensions in GMA Welds on Steel Plates", Metal Construction, 14(3), March 1982, pp. 128-133.

13. Carslaw, H.S., Jaeger, J.C., Conduction of Heat in Solids, 2nd ed., Oxford Press, London, U.K. 1959.

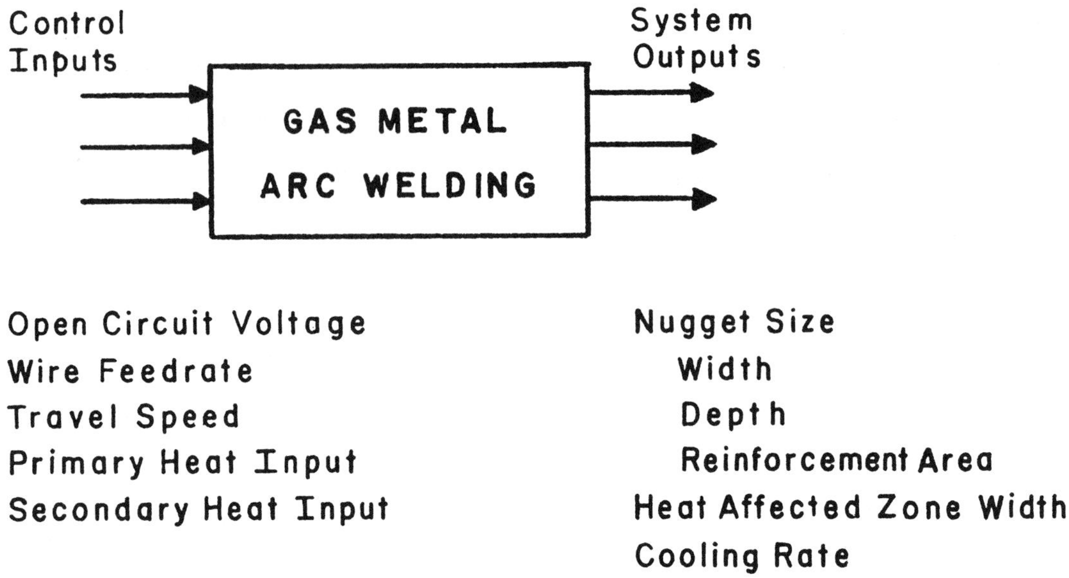

Figure 1. A Block Diagram Model of the GMAW Process

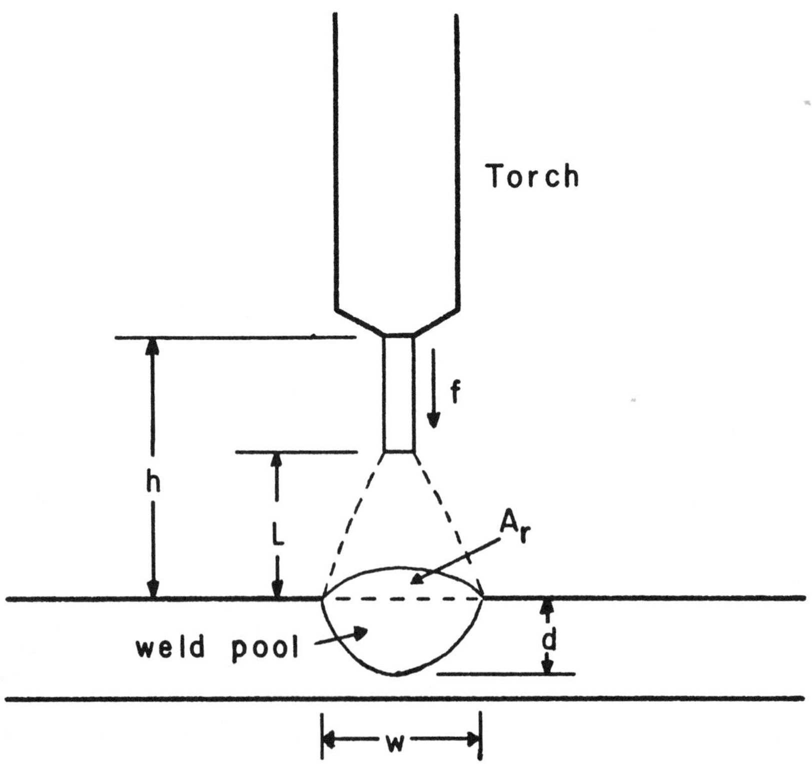

Figure 2. Schematic of GMAW with Definitions for the Geometry Control Model

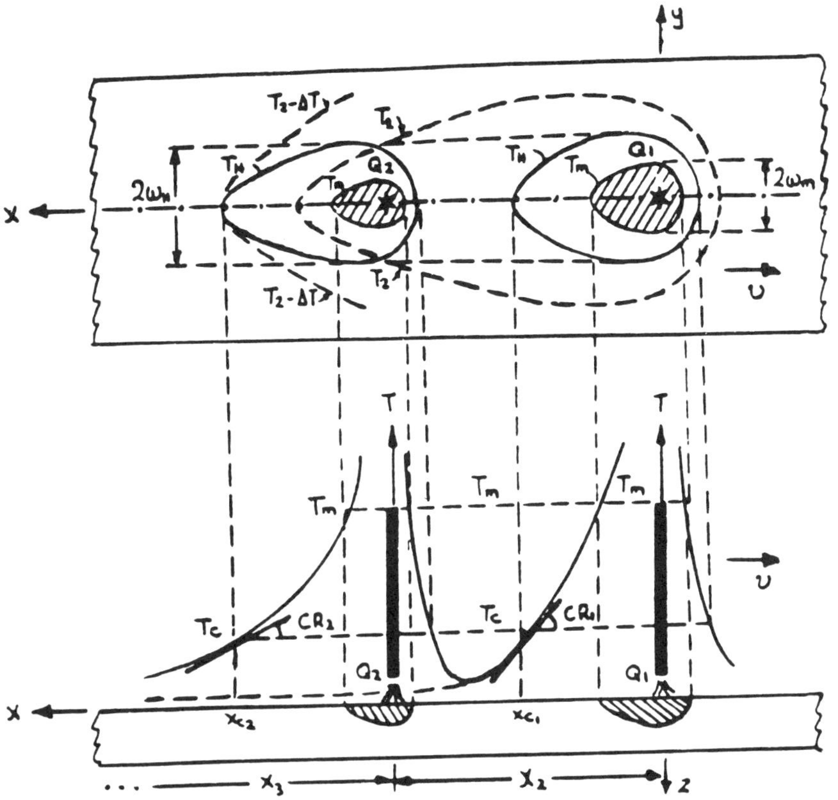

Figure 3. The Double Torch Welding Configuration

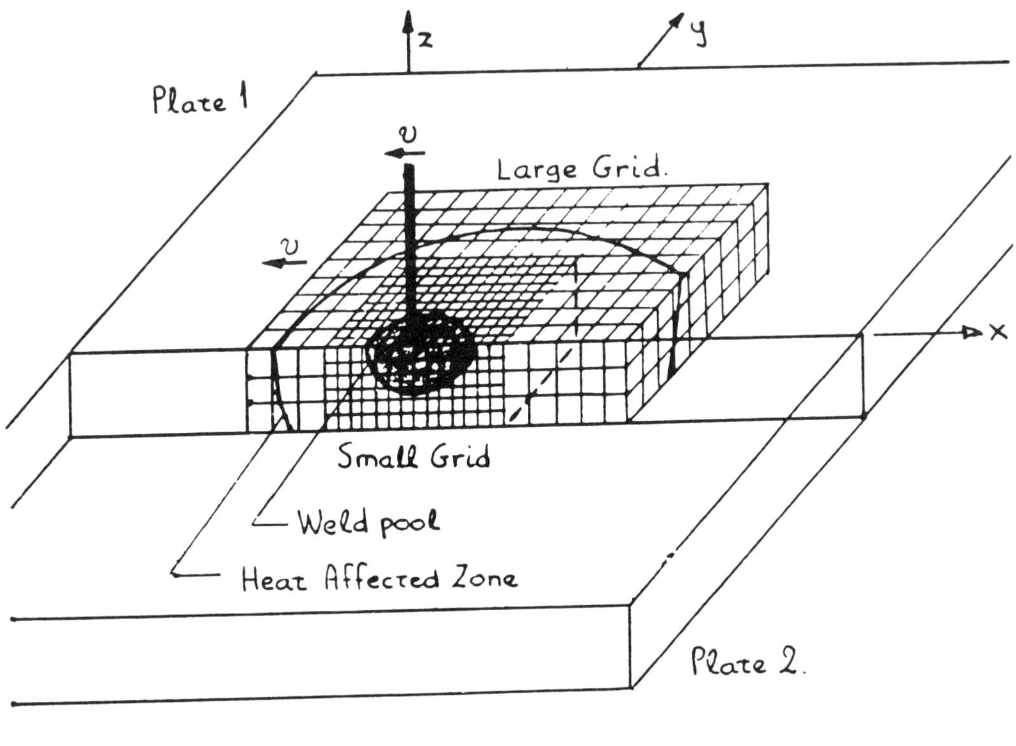

Figure 4. Arrangement of the Numerical Simulation

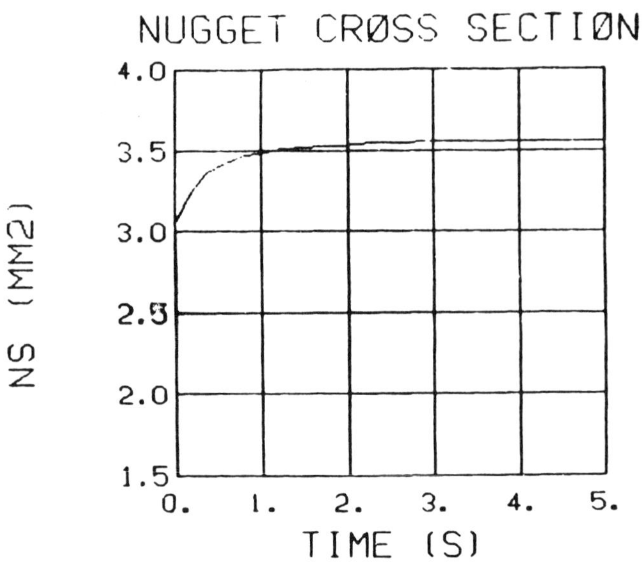

a.

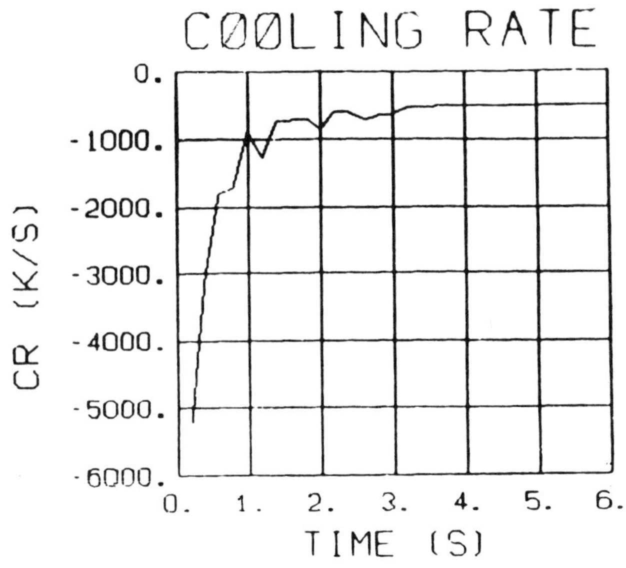

c.

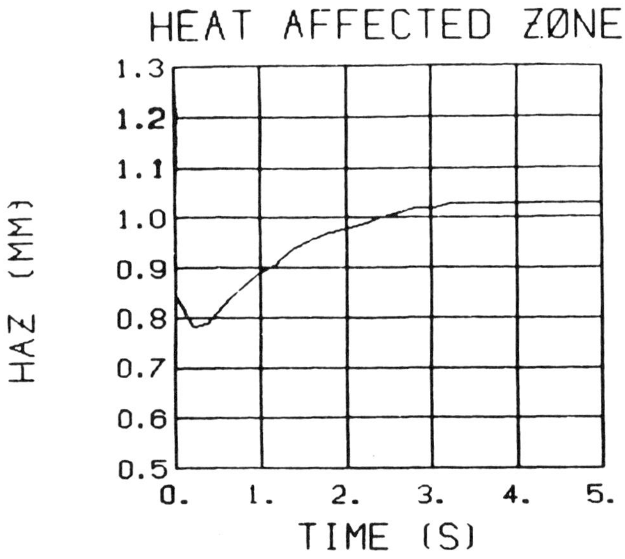

b.

Figure 5. Simulation Results: Controlled Variable Responses to a Step input in Heat Input Q_1.
a) NS b) HZ c) CR

Step Responses		Nugget Section NS		Heat Affected Zone HZ				Cooling Rate CR	
I	Step	K_a	$T_a(s)$	K_1	$T_1(s)$	K_2	$T_2(s)$	K_c	$T_o(s)$
Q	$0.2Q_1^*$	$1.73 \cdot 10^{-3}$ mm²/W	0.67	$1.16 \cdot 10^{-3}$ mm/W	0.91	$0.56 \cdot 10^{-3}$ mm/W	0.42	0.343 K/J	1.09
	$-0.2Q_1^*$	$3.30 \cdot 10^{-3}$ mm²/W	0.44	$1.30 \cdot 10^{-3}$ mm/W	0.74	$0.69 \cdot 10^{-3}$ mm/W	0.38	0.826 K/J	0.80
v	$0.2v^*$	−0.86 mm·s	0.32	−0.40 s	0.48	−0.17 s	0.29	−305 K/mm	0.87
	$-0.2v^*$	−0.52 mm·s	0.75	−0.42 s	1.12	−0.15 s	0.60	−123 K/mm	0.99
S	$0.2Q_1^*$	—	—	—	—	—	—	−285 K/s	1.20
	$-0.2Q_1^*$	—	—	—	—	—	—	−285 K/S	0.37

Table 1. Gains and Time Constants of the Thermal Transfer Functions

GAS METAL ARC PROCESS SENSING AND CONTROL

H.B. Smartt, C.J. Einerson, A.D. Watkins
Idaho National Engineering Laboratory
Idaho Falls, Idaho

R.A. Morris
David Taylor Naval Ship
Research and Development Center
Bethesda, Maryland

ABSTRACT

Welding automation is normally concerned with directly controlling such factors as current, arc voltage, travel speed, and filler wire speed. At least two levels of control beyond present technology can be identified. The first would involve direct control of weld heat input, cooling rate, and bead geometry (bead width, penetration, and reinforcement). The second would involve direct control of weld defects, microstructure, and properties, a level of control which will be achieved in the future.

This work describes the evaluation of a relatively simple steady-state model of the gas metal arc (GMA) welding process which considers wire melting and heat input to the base metal, allowing independent control of the heat and mass input to the base metal. This allows direct, independent control of the weld bead reinforcement area and cooling rate to be achieved.

IN RECENT YEARS there has been a significant growth in the use of mechanized welding systems as cost effective substitutes for shielded metal arc welding. Perhaps the most common process for such substitutions is gas metal arc (GMA) welding and variants, especially for arc welding applications using robots. Yet the GMA process is subject to the same fundamental limitations which virtually all arc welding processes have, and control of the process is limited to those factors which machine builders are accustomed to, such as electrode wire speed, welding speed, current, and voltage. Those factors which the welding engineer would prefer to control, such as reinforcement area, weld heat input, microstructure, and mechanical and physical properties of the weld are not directly controllable except in isolated cases. Yet it has been demonstrated by Lukens and Morris (1) that the mechanical properties of certain alloys are directly related to the welding conditions and that control of such properties by proper selection of welding parameters is feasible.

This work is an effort to independently control two of the above factors, namely the reinforcement area (defined as the transverse cross-sectional area of the material added during welding) and the cooling rate of the weld bead.

BACKGROUND

The GMA welding process employs a consumable wire electrode passing through a copper alloy contact tip. Electrical current, imposed on the wire by a voltage drop between the contact tip and the metal to be welded (base metal), supports an arc between the wire end and the base metal. The electrode is melted by internal resistive power and heat transferred from the arc. Droplets of molten metal are detached and transferred from the wire to the weld pool by the combination of gravitational, Lorentz, surface tension, and plasma forces (2). Heat is transferred to the base metal directly from the arc and also by the molten metal droplets. The electrode wire, molten droplets, weld pool, and solidified weld bead behind the weld pool are protected from oxidation by a shielding gas which may be Ar, CO_2, or mixtures such as Ar with O_2, H_2, He, or CO_2 in various combinations.

An important aspect of GMA welding is that heat input to the base metal and mass addition to the fusion zone are coupled by virtue of the fact that the current is normally controlled by setting the electrode wire speed at the desired value. The current is maintained by the power supply consistent with the contact tip to base metal resistance, which is determined by the wire stickout and arc resistance, and the contact tip to base metal voltage.

GMA welding may be performed with the electrode wire at a positive or a negative potential with respect to the base metal, using either a constant current or a constant voltage power supply, with either pulsed or constant electrode wire speed and/or welding current. However, the most common operating conditions are electrode wire positive, using a constant voltage power supply, with constant wire speed and current. This work is limited to such conditions.

MODEL DESCRIPTION

A model of the heat and mass transfer to the base metal was developed in prior work (3) for use in welding of thick section carbon or alloy steel in the spray metal transfer mode. The model incorporates power supply open circuit voltage and slope; contact tip to work piece distance; as well as the electrode wire type, diameter, and speed. These factors have all been identified as affecting the wire melting process dynamics, and importantly, they may all be measured in real time either prior to or during welding.

The model is a steady-state description of the wire melting rate and the heat input to the base plate. As such it does not contain information regarding the formation of molten droplets on the electrode wire, the detachment of the droplets, or the mode of metal transfer from the electrode wire to the weld pool.

A set of six equations form the basis for the model (3); the terms in the following equations are defined in Table 1.

	Equation	Description
(1)	$H = EI\eta/R$	Heat input to the base metal
(2)	$G = \dfrac{\pi d\ S}{4R}$	Reinforcement area
(3)	$E = E_o + nI$	Power supply characterization
(4)	$IE = IV_w + IV_a$	Power balance between wire and arc
(5)	$IV_w + \eta' IV_a = \dfrac{\pi d \delta S H_m}{4R}$	Power required to melt the wire
(6)	$IV_a = I^2 R_a$	Ohm's Law

These equations may be solved to give (3):

$$S = \frac{E_o \eta\ 4GI + n\eta\ 4GI^2}{\pi\ d^2 H} \tag{7}$$

Table 1 - Definitions of Welding Terms

Term	Definition
H	= heat input per unit length (J/mm)
E	= contact tip to workpiece voltage (V)
I	= current (A)
η	= overall heat transfer efficiency
R	= welding speed (mm/s)
S	= wire feed speed (mm/s)
d	= wire diameter (mm)
E_o	= open circuit voltage (V)
n	= Slope of power supply (V/A)
V_w	= voltage drop across wire length (V)
V_a	= voltage drop across length of welding arc (V)
δ	= average wire density (g/mm^3)
C_p	= specific heat of the welding wire (J/g·K)
T_o	= ambient temperature (°C)
T_m	= melting temperature of wire (°C)
h_f	= heat of fusion of wire (J/g)
η'	= fraction of total power transferred from anode spot to solid wire
H_m	= heat required to melt wire (J/g)
CT	= contact tip to base metal distance (mm)
R_a	= arc resistance
$\dot{Q}$	= power transferred from the anode spot to the solid wire
G	= reinforcement area
$\dot{m}$	= mass flow rate

and

$$R = \frac{S\pi d^2}{4G} \tag{8}$$

where S is the wire speed, R is the welding speed, G is the reinforcement area, and H is the heat input per unit length of weld. The current (I) is given by (3):

$$I = \frac{E_o + \left[E_o^2 + 4\ n-(1-\eta')\ Ra\ S\pi d^2/4\ H_m\right]^{1/2}}{2[(1-\eta')\ Ra - n]} \tag{9}$$

$$\text{where:}\quad H_m = \int_{To}^{Tm} C_p dT + h_f \tag{9'}$$

SENSING REQUIREMENTS

The only operating parameter which is measured by the system during welding is the current. The current is measured by obtaining the voltage drop across a calibrated shunt placed in the secondary current loop. The resulting voltage signal is very noisy, with about a ±1 to 10 A peak-to-peak variation. This variation is too great to allow direct measurement of the voltage to be useful for controlling the process. There are two solutions to this problem. One is to use a hardware filter to remove the high frequency components of the signal. The other is to filter the signal digitally in either the frequency or the time domain. This work employs a digital, linear difference technique in the time domain.

The shunt voltage is measured in sets of 30 values taken at the rate of 3600 per second. The last 20 of these values are processed by a finite impulse response filter (4). The first ten values are discarded to remove any artifacts associated with the relay switching used for multiplexing. At approximately one-half second intervals, the present and the last two output values from the finite impulse response filter are processed using an infinite impulse response filter (4).

SYSTEM PERFORMANCE

Two terms in Eq. (9) require evaluation, η' and R_a. The first of these, η', is the efficiency of heat transfer from the arc to the electrode wire, which for our purposes is defined as:

$$\eta' = \frac{\dot{Q}}{EI} \tag{10}$$

Results from Waszink and Van Den Heuvel (5) give the relationship

$$\dot{Q} = 6I - 0.7 \times 10^6 \dot{m} - 125 \tag{11}$$

for a wire diameter of 1.2 mm, where

$$\dot{m} = \delta S \frac{\pi d^2}{4}$$

The wire diameter used in this work was 0.89 mm, but Eq. (11) is used as the best available approximation. Substituting Eq. (11) into Eq. (10) gives (3):

$$\eta' = \frac{6I - 3415S - 125}{(E_o + nI) \cdot I} \tag{12}$$

A statistical analysis program, SAS (6), was used to determine the arc resistance (R_a) values which give the best agreement between actual and calculated currents; values ranged between 0.10 and 0.15 Ω. In addition, the best fit R_a values were obtained for $\eta' = 0.092$, this fit being of the functional form (3):

$$R_a = C_o + C_1 \cdot CT + C_2 \cdot E_o + C_3 \cdot I + C_4 \cdot CT \cdot E_o$$

where C_o, C_1, C_2, C_3, and C_4 are constants. Using the average η' and the associated R_a values, Eq. (9) was evaluated. The standard deviation of the calculated current was ±7.5 A, for the wire speed it was ±5.1 mm/s. However, certain welds resulted in a negative value for the quadratic term in Eq. (9). Arbitrarily eliminating these terms gave a standard deviation for the current of ±5.8 A, and ±3.9 mm/s for the wire speed.

The initial welding trials indicated that the model performance is not adequate for open loop control of the GMA welding process. Consequently, a feedback control scheme was developed in which the actual welding current is measured and compared to the value predicted in Eq. (9). The difference in these values is used as the basis for correcting the model to give the desired current. The welding speed is simultaneously adjusted to maintain the correct reinforcement area. A block diagram of the control system is given in Figure 1.

Welding trials using the model to control reinforcement area and heat input were made using 0.89 mm (0.035 in.) diameter Type E70S-6 wire, with $Ar\text{-}2O_2$ shielding gas, on 12.7 mm (1/2 in.) thick Type A-36 steel plate in a bead-on-plate configuration. The contact tip to work piece distance was held constant at 15.9 mm (5/8 in.) and the power supply open circuit voltage was 32 V. All welding involved spray transfer of the molten electrode wire material to the weld pool.

A series of seven welds were made to determine the feasibility of controlling the cooling rate of the weld bead using the model. The heat input was varied from 1050 to 1500 J/mm. Three reinforcement areas were used: 30, 35, and 40 mm^2. These parameters and the associated weld bead cooling rates are listed in Table 2.

The accuracy of the system may be seen by examining Figures 2 and 3, where the measured vs desired reinforcement areas and calculated heat inputs are shown for various machine settings. The reinforcement areas are within ±5% of the desired values; the heat inputs are well within ±1% of the desired values. The reinforcement areas were measured from macrophotographs of transverse weld sections using an image digitizing program. The heat inputs were calculated from current, voltage, and travel speed values by the control computer during welding; a 75% heat transfer efficiency was assumed.

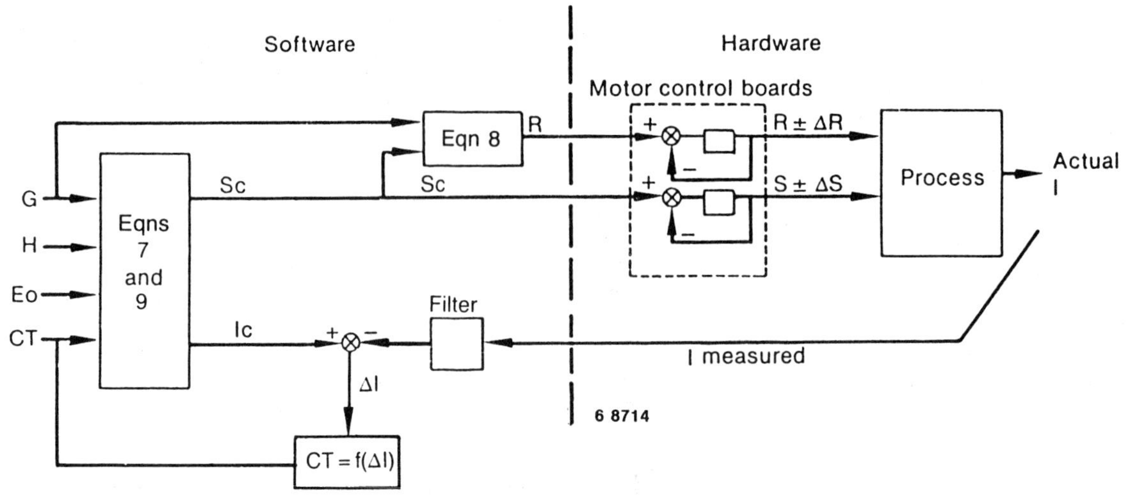

Fig. 1 - Control system block diagram.

Table 2 - Weld Bead Cooling Rate Parameters

Reinforcement (mm^2)	Heat Input (J/mm)	Cooling Rate (°C/s)
30	1050	26.8
30	1100	28.5
35	1200	21.6
35	1300	19.1
40	1300	19.8
40	1400	18.6
40	1500	16.9

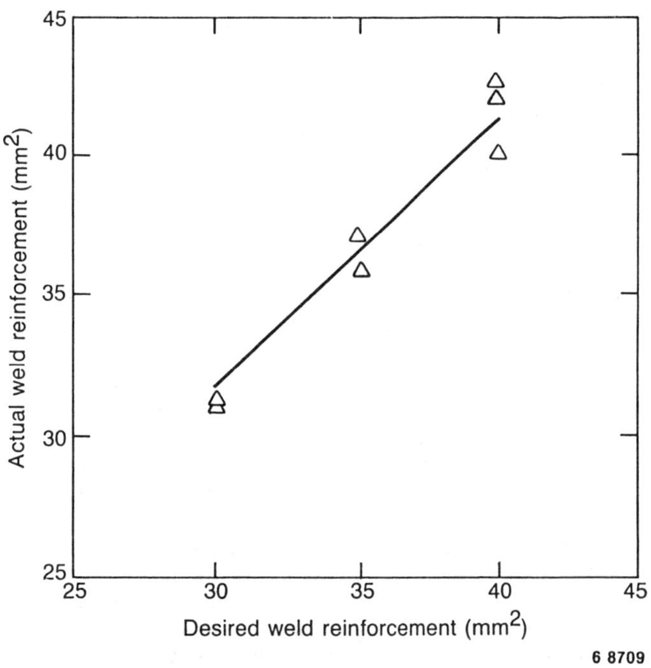

Fig. 2 - Actual reinforcement areas as a function of the desired reinforcement area.

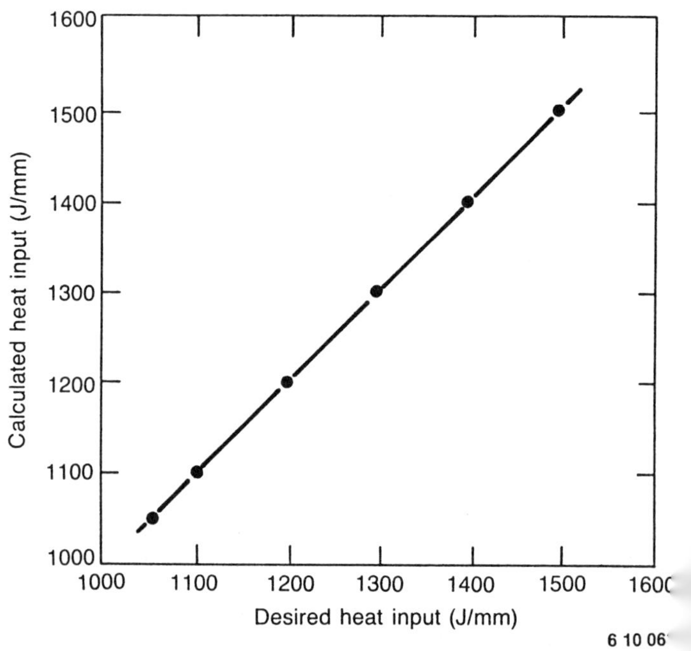

Fig. 3 - Calculated actual weld bead heat input per unit length as a function of the desired heat input.

The weld bead cooling rates were measured using Type K thermocouples plunged into the weld pool at the top center of the weld bead. Cooling rates at 538°C (1000°F) are plotted as a function of the calculated heat input in Figure 4. Although the range of values included in this study is limited, it is clear that the weld bead cooling rate is related to the heat input. Future work on this problem will involve real-time measurement of the weld bead cooling rate with an infrared camera in a manner similar to that of Lukens and Morris (1), and direct closed-loop control of the cooling rate with the error in cooling rate being used to determine the required change in weld bead heat input.

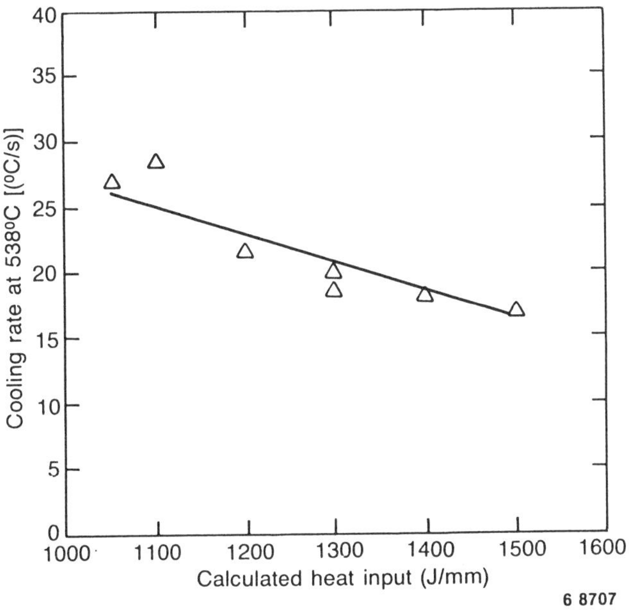

Fig. 4 - Weld bead cooling rate as a function of heat input per unit length.

CONCLUSIONS

This work presents a steady-state model of GMA welding for process control. The following conclusions have been made:

1. A steady-state model of GMA wire melting and base metal heat input has been developed which is suitable for real-time control of welding in the spray metal transfer mode.

2. The model is able to independently control the mass and heat inputs to the base metal.

3. The weld bead cooling rate may be controlled by adjusting the heat input independently of the reinforcement area.

ACKNOWLEDGMENTS

This work is supported by the U. S. Department of Energy, Office of Energy Research, Office of Basic Energy Sciences under DOE Contract No. DE-AC07-76ID01570. Appreciation is expressed to U. S. Wallace for technical assistance.

REFERENCES

1. Lukens W. E. and R. A. Morris, "Infrared Temperature Sensing of Cooling Rates for Arc Welding Control," Welding Journal, Vol. 61, No. 1, (1982), pp. 27-33.

2. C. M. Perrott, "The Role of Electrode Vaporization in GMA Welding," The Australian Welding Institute Conference on "Welding Technology 82," Hobart, Tasmania, October 1982.

3. Smartt H. B., C. J. Einerson, and A. D. Watkins, "Modelling and Control of Gas Metal Arc Welding," in preparation. To be submitted to the Welding Journal.

4. K. G. Beauchamp and C. K. Yuen, "Digital Methods for Signal Analysis," GEORGE ALLEN 7 UNWIN, London, 1979.

5. Waszink J. H. and G. J. P. M. Van Den Heuvel, "Heat Generation and Heat Flow in the Filler Metal in GMA Welding," Welding Journal, Vol. 61 No. 8, (1982), pp. 269-s to 282-s.

6. SAS Institute, Inc., "SAS User's Guide": Statistics, 1982 edition. Cary, N.C.: SAS Institute, Inc., 1982.

DIGITAL SIGNAL PROCESSING AS A DIAGNOSTIC TOOL FOR GAS TUNGSTEN ARC WELDING

Carl D. Sorensen, Thomas W. Eagar
Materials Processing Center
Massachusetts Institute of Technology
Cambridge, Massachusetts 02139

ABSTRACT

One of the major difficulties involved with closed-loop control of welding processes is a lack of a suitable sensors of the welding process. A proposed welding sensor is the welding arc itself. The sensory information is obtained through digital signal processing of the arc voltage and current. Experiments performed to test the suitability of digital signal processing as an arc monitor indicated that the technique had promise as a laboratory sensor, but that it had neither the sensitivity nor the selectivity to act as a production monitor of the welding process.

IN RECENT YEARS there has been a great deal of interest in automating the arc welding process as a means of improving productivity, lowering costs, and improving reliability of welding processes[1]. One of the major difficulties with automation of arc welding is the lack of a suitable weld quality sensor. Typical automatic welding systems may be capable of controlling torch position, travel speed, arc voltage, and/or arc current[2]; however, these variables are generally controlled only according to some predetermined optimum schedule, rather than according to the actual weld condition. In order to use a feedback control system with arc welding, it is necessary to develop better weld quality sensors. If such sensors were available, the welding process could be monitored; and the torch position, travel speed, arc voltage, or arc current could be modified to improve the weld quality.

Several investigators have examined possible weld quality monitors[3-9, e.g.]. Most of these techniques are not yet useful for production systems, in many cases because the sensing systems add bulk to the welding torch or are adversely affected by the intense heat and light of the welding arc. To avoid these problems it is desirable to use the welding arc itself as the weld quality sensor.

The arc is a dynamic electrical element, and thus has electrical properties that vary with frequency. One method of examining the electrical properties of the arc is digital signal processing. This paper is a brief description of an analysis made at M.I.T. of the use of digital signal processing of the arc voltage and current as a diagnotic tool for GTA welding. It describes the analysis techniques and equipment used to measure the electrical properties of the arc. Experiments to test the ability of this analysis technique to measure voltage oscillations are discussed, correlation of the experimental results and theoretical predictions is examined, and the limitations of this technique as a practical weld quality monitor are discussed.

ANALYSIS TECHNIQUE

In order to study the electrical characteristics of the arc/puddle system, an impedance based measurement technique is used. The arc is treated as an electrical "black box" with the weld current as an input and the weld voltage as an output. The electrical characteristics of the arc were measured by means of the frequency response function[10], which is obtained by Fast Fourier Transform (FFT) analysis of the input (current) and output (voltage) of the system. The frequency response function is comprised of the gain, phase, and coherence functions, which measure the amplification, phase lag, and noise as a function of frequency. Of particular interest is the behavior of the system frequency response function near a resonant frequency, where the gain will increase sharply and the phase will shift by 180 degrees. In addition, the coherence is expected to rise, since the magnitude of the output signal will increase due to the increased gain. In practice, the peak in the gain function is easiest to detect and will therefore be used most often to indicate a resonant frequency in the welding arc such as that of the puddle oscillations.

EXPERIMENTAL EQUIPMENT

The equipment used to perform this study consists of a gas-tungsten arc (GTA) welding apparatus combined with a computer-driven measurement system, shown schematically in Figure 1. Welding power is supplied by a 24V, 600 A-h battery. The current is controlled by a solid state linear D.C. current regulator designed by Alexander Kusko, Inc.[11]. This regulator is capable of providing 600 A RMS or 1000 A peak current regulated to within 1% of full scale at frequencies of up to 25 kHz. The desired current waveform is selected by means of an external reference supplied by a Wavetek model 166 signal generator.

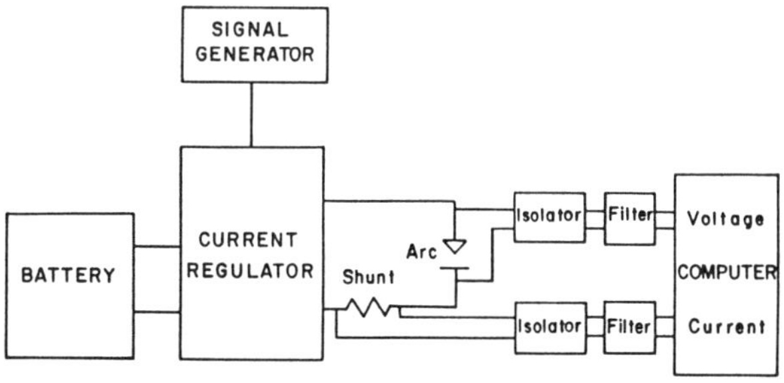

Figure 1. Schematic Diagram of Experimental Equipment

The arc voltage and current are measured with a computer based data collection system. Current is measured by means of a shunt placed in the lead connecting the work to the current regulator. The voltage across the shunt, which is proportional to the weld current, and the arc voltage are fed to a compensated optical isolator to protect the computer. Each of these isolated signals is passed through a Frequency Devices model 901 variable frequency low-pass filter, which serves to prevent aliasing effects due to high frequency components in the signal. The output signals from the filters are connected to a Digital Equipment Corporation MINC/DECLAB-23 computer equipped with ADAC model 1601 timer, 1023 A/D and 1620 DMA modules. This system is capable of sampling both voltage and current waveforms at user-selected frequencies of up to 45 kHz each. These signals are then analyzed off-line to determine the frequency response of the arc-puddle system.

EXPERIMENTAL PROCEDURE AND RESULTS

Three different experiments were performed to evaluate the effectiveness of this new method for sensing the welding process. The first tested the sensitivity of the analytic technique to arc length when measuring weld pool oscillations, the second compared the oscillation frequencies of stationary weld pools with those predicted by a model, and the third investigated the ability of this analysis technique to dectect changes in the welding process.

SENSITIVITY ANALYSIS - As part of the sensitivity analysis, the minimum detectable amplitude of weld pool oscillations was estimated and the input current waveform was optimized. These experiments are discussed in reference 12. An important part of the sensitivity analysis tested the effect of arc length on the ability to measure oscillations in stationary weld pools. As previous experiments with control of puddle size through examining puddle dynamics had been successful only for very short arcs, a major concern with this analysis technique was whether it would work at moderate or large arc lengths. A test was performed to measure the effect of arc length on the sensitivity of the measurement technique. The experiments were performed on a 9 mm (3/8") thick plate of A36 steel. The welding current was a 100 A background with a 300 A pulse of 5 ms duration occurring every 145 ms. A 2.4 mm (3/32") 2% thoriated tungsten electrode with a 60° tip angle was used. Helium was used as a shielding gas. Various arc voltages ranging from 9 to 15 volts were used, corresponding to arc lengths of up to 13 mm (1/2").

Results - Figures 2 and 3 show the system gain function for welds made at 9V and 15V, respectively. The peak in each graph is at the natural frequency of the puddle. The natural frequency is different in each case because the puddle size is different. Figure 4 shows the observed correlation of puddle frequency with weld pool diameter.

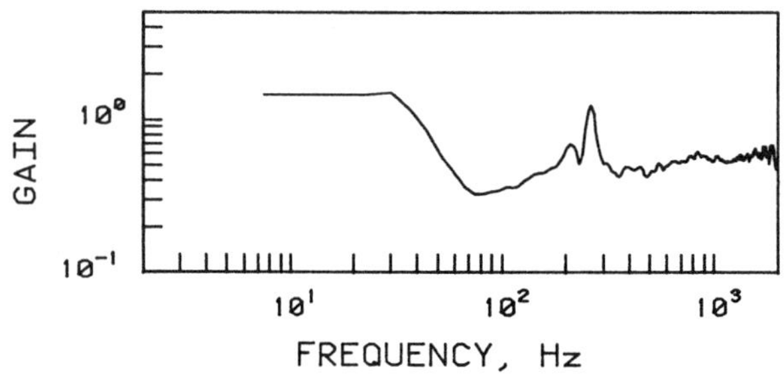

Figure 2. Gain function for 9.5 V arc. Note resonance peak at 280 Hz.

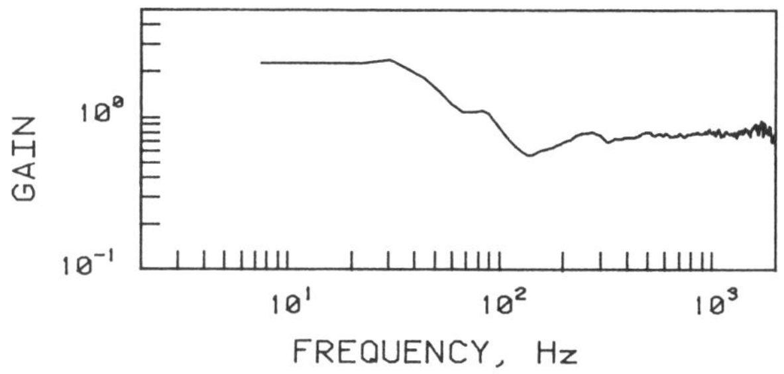

Figure 3. Gain function for 15 V arc. Barely visible resonance peak is at 90 Hz.

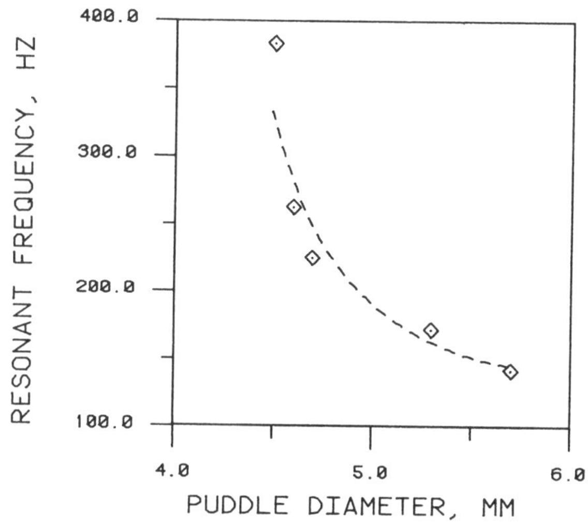

Figure 4. Variation in resonant frequency as a function of weld pool diameter for stationary welds in steel.

OSCILLATION FREQUENCIES OF STATIONARY WELD POOLS - Two different models of a partially penetrated weld pool were developed, a lumped parameter model and a distributed parameter model. These models are explained in more detail in reference 13. In order to verify these models of weld pool oscillation, a series of stationary gas-tungsten arc (GTA) welds were made on steel and copper. These welds were made at varying currents and voltages in order to achieve a variety of weld pool sizes and shapes in each material. As each weld was made, the arc voltage and current were recorded and analyzed as described above. The system transfer functions were evaluated in order to reveal the resonant frequency of the weld pool.

In order to obtain a predicted natural frequency, each weld was sectioned, polished, and etched and the weld width and depth were measured. The surface tension of liquid steel was assumed to be 1800 dyne/cm and that of copper 1300 dyne/cm[14]. The densities of steel and copper were assumed to be 7.0 g/cm^3 and 7.8 g/cm^3 respectively[15]. These data were then substituted into the weld pool models to give predicted frequencies for the lumped and distributed parameter models.

Results - Curves showing the behavior of the weld pool models as the weld pool width changes are plotted in Figure 5. The measured frequencies of each of the welds were plotted against the predicted frequencies for each of these welds in order to determine the agreement between theory and experiment. Figure 6 compares the observed frequencies with the predicted frequencies for steel and copper using both the lumped and the distributed parameter models.

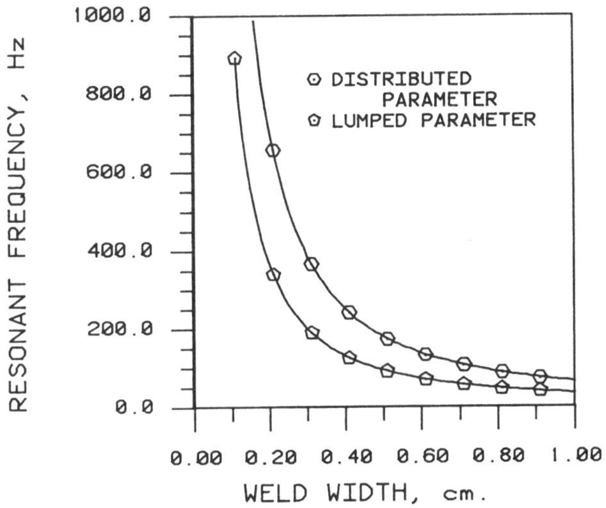

Figure 5. Predicted resonant frequency versus weld pool width at constant width/depth ratio.

DETECTION OF WELD DEFECTS - A suitable weld quality sensor must reveal certain information about the welding process. In order to be used in a closed-loop control system, it must be able to provide signals which indicate the presence of defects in the welding process. Further, such a sensor should be able to differentiate between different types of defects so that the proper corrective action can be taken. These experiments evaluated the ability of signal processing to detect and differentiate various types of welding discontinuities. The weld discontinuities that were investigated were chosen to be representative of actual welding defects.

In these experiments, welds were made using the equipment and procedure described above. However, in each of these welds, a defect was made to occur. The defects or discontinuities that were studied were traveling weld size, humped bead formation, steel to aluminum base metal transition, and partial to full penetration weld transition. Upon completion of these welds, the system frequency response functions were examined to look for signatures of the various defects.

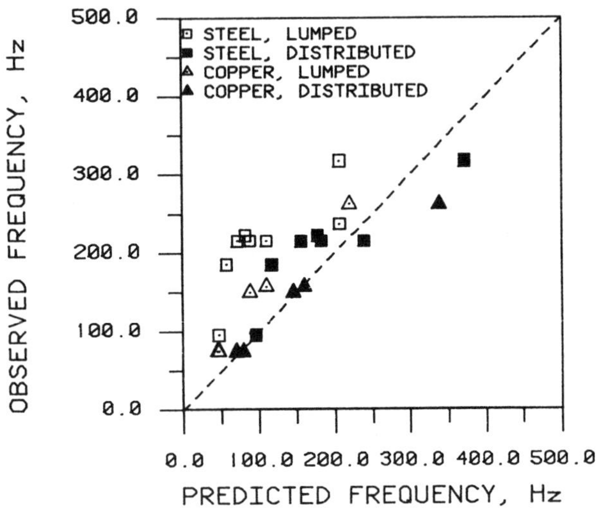

Figure 6. Measured frequency for stationary welds on steel and copper compared to frequency predicted by each of the models.

Results - While the results for these experiments are too numerous to be included in detail here, they can be found in reference 13. However, one of the most important results is included. A comparison was made between observed and predicted weld pool frequency for traveling welds in steel and titanium. The predicted frequency of the steel welds is compared to the observed frequency in Figure 7 for both the lumped parameter and the distributed parameter model, while Figures 8 presents the same results for the titanium welds. Note that the general trend of increasing measured frequency with increasing predicted frequency is followed, but the scatter in the data prevents any prediction of the frequency of a particular weld pool.

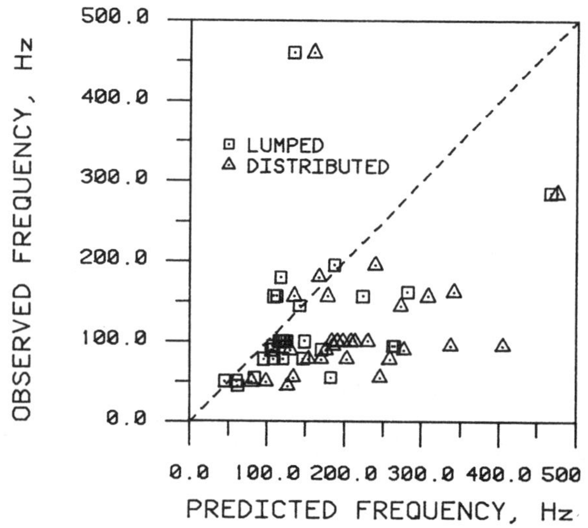

Figure 7. Predicted frequency compared with measured frequency for traveling welds in steel.

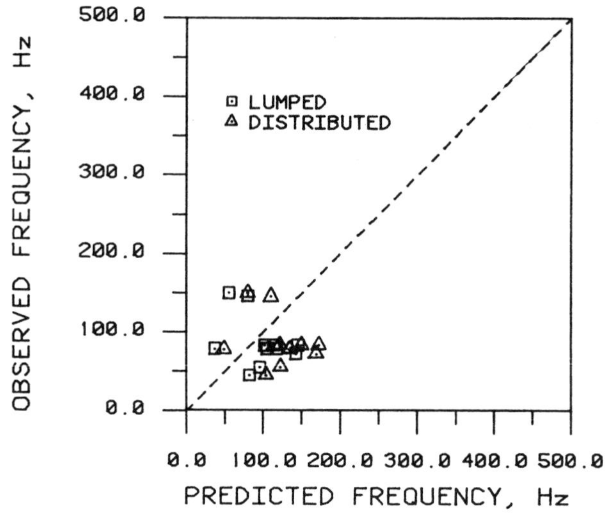

Figure 8. Predicted frequency compared with measured frequency for traveling welds in titanium.

DISCUSSION

SENSITIVITY ANALYSIS - It was anticipated that as arc length increased, the relative amplitude of voltage oscillations would decrease due to the increase in average voltage. This could increase the difficulty of detecting the puddle oscillations. However, experiments performed showed that puddle oscillations could be detected up to the limits of arc stability. Thus, the technique described is effective for measuring the frequency of puddle oscillations of stationary welds at typical arc lengths with helium shielding gas.

As expected, the puddle frequency changed with weld pool size. As the pool size increased, the frequency decreased, which is consistent with the findings of Renwick and Richardson[10]. A detailed analysis of the dependence of natural frequency on weld pool size and shape is found in reference 13.

OSCILLATION FREQUENCIES OF STATIONARY WELD POOLS - As can be seen in Figure 6, good agreement between the predicted frequency and observed frequency was found for both models and both materials. The data for copper show less scatter than the data for steel. The values predicted by the distributed parameter model show better agreement than the values predicted by the lumped parameter model.

The scatter in the experimental data is due in part to the uncertainty in identifying the resonant frequency exactly. In many cases the resonance peak is more than five samples wide, which corresponds to a frequency uncertainty of more than 35 Hz. Thus, there will be a certain amount of scatter simply due to the frequency measurement. In addition, there is some uncertainty in measuring the physical size of the weld pool. This is due to the necessity of sectioning the stationary weld exactly on the diameter in order to measure the correct width and depth. Thus, it is possible that the physical measurements of the welds are smaller than the actual size of the welds, thus leading to a predicted frequency which is too high. Nevertheless, the trend is established.

As can be seen from these figures, the distributed parameter model accurately predicts the weld pool frequency for weld pools of different geometries, densities, and surface tensions. This provides strong evidence for the correctness of this model.

These models predict that there will be significant problems in using signal analysis as a weld size monitor for welds of typical size. As can be seen in Figure 5, there is a very small change in weld pool frequency for welds varying in width from 0.5 to 1 cm. Therefore, in order to use weld pool frequency as an indicator of weld pool size, it is necessary to measure the weld pool frequency very precisely. As described above, it is difficult, if not impossible, to locate the weld pool resonsant frequency this precisely using FFT techniques. The technique, therefore, is not likely to yield satisfactory results in real welds.

DETECTION OF WELD DEFECTS - The correlation between predicted and observed weld pool frequency is not very high, as is shown in Figures 7 and 8. The scatter in the experimental data is due in large part to the difficulty of measuring the puddle frequency in traveling welds. This is due to the measurement technique which is used to detect puddle frequencies.

The measurement technique used to detect puddle frequencies is optimized for use with underdamped systems. The longer the weld pool oscillations continue, the easier they are to measure. The oscillations will continue only if the system is underdamped. In a traveling weld, new material is continually being added to the system through melting, and old material is leaving the system through solidification. The material which is added to the system has no momentum at the time it melts; therefore it must be accelerated by the rest of the weld pool in order to oscillate. This has a damping effect on the oscillations, as the energy of the liquid metal is dissipated in accelerating this new material. Since the oscillations are damped out more rapidly, the resonant frequency is much more difficult to detect. This, in turn, leads to the scatter in Figures 7 and 8. Because of this problem, it is unlikely that signal analysis can serve as an adequate monitor of weld pool size in traveling welds.

In regard to the other discontinuities studied, changes in the frequency response function could generally be observed with some difficulty for each of these defects. However, there was no means of distinguishing between the various defects, and hence, no effective method of controlling the welding process to correct a particular defect. This fact, together with the difficulty of measuring puddle oscillations in traveling welds, indicates that this analysis technique will not serve as an effective weld process monitor. Other monitors must be found which show more selectivity between the various defects if adequate feedback control is to be instituted on the arc welding process.

CONCLUSION

The ability to use frequency spectral analysis as a means of detecting puddle oscillations in the GTA welding process has been demonstrated for stationary welds under carefully controlled conditions. The sensitivity of this technique to variations in arc length has been studied. For stationary welds on steel plate, with helium as a shielding gas, puddle oscillations were measured at arc lengths up to 13 mm. In addition, the change in oscillation frequency with change in puddle size was observed.

A model was developed to predict the resonant frequencies of the weld pool given the size of the weld pool. Experiments performed on steel and copper baseplates with stationary welds verified the ability of the model to predict weld pool frequencies. The model predicts that controlling weld pool size using digital signal analysis will be extremely difficult for typical weld sizes.

This analysis technique works much less well with traveling welds, due to the higher damping of the traveling weld pool. Although many changes in the weld cause changes in the frequency response of the arc-puddle system, the changes are not unique, and hence the defects cannot be distinguished from one another.

Difficulties still remain with measuring puddle oscillations with argon shielding gas. In addition, the input waveform affects calculations of the system transfer function due to the sequential nature of voltage and current sampling. Future work with simultaneous rather than sequential sampling may eliminate this problem, and lead to improved ability to sense puddle oscillations.

Digital signal processing has shown some promise as a laboratory diagnostic tool for gas tungsten arc welding. However, due to the careful setup required to obtain adequate results and the poor performance when applied to real welding problems even in the controlled laboratory environment, it seems unlikely that digital signal processing can serve as a practical sensor for GTA welding.

Acknowledgements

The authors wish to thank the Office of Naval Research for their support of this project under contract N00014-C-80-0384.

REFERENCES

1. Cary, H.B., "Modern Welding Technology", pp. 18,19, Prentice-Hall, Englewood Cliffs, N.J. (1979)
2. Boughton, P., G. Rider, C.J. Smith, "Advances in Welding Processes" ed. J.C. Needham, 203-209, The Welding Institute, Cambridge, England (1978)
3. Chin, B.A., N.H. Madsen, and J.S. Goodling, Welding Journal 62 (9): 227-s to 234-s (1984)
4. Richardson, R.W., D.A. Gutow, and S.H. Rao, "Measurement and Control for Batch Manufacturing" ed. D.E. Hardt, pp. 65-75. ASME, New York (1982)
5. Zacksenhouse, M., and D.E. Hardt, D.E., Journal of Dynamic Systems, Measurement, and Control 105, 179-184 (1983)
6. Arata, Y., K. Inoue, M. Futamata, and T. Oh, Transactions of JWRI 9, 25-30 (1980)
7. JuiHai, Z., W. Quilong, C. Hongjian, and Y. Shuyan, Proc. International Conference on Quality and Reliability in Welding, D-33-1 to D-33-6, Hangzhou, China (1984)
8. Zacksenhouse, M., S.M. thesis., Massachusetts Institute of Technology, Cambridge, Massachusetts (1983)
9. Renwick, R.J., and R.W. Richardson, Welding Journal 62, 29-s to 35-s (1983)
10. Bendat, J.S., and A.G. Piersol, "Engineering Applications of Correlation and Spectral Analysis", Wiley, New York (1980)
11. Galler, D., A. Kusko, and J. Converti, "A Fast Response Transistor Current Regulator for Welding Research, Paper presented at IEEE-IAS Annual Meeting, Philadelphia, PA. (1981)
12. Sorensen, C.D., and T.W. Eagar, Submitted to Welding Journal (1986)
13. Sorensen, C.D., Ph.D. Thesis, Massachusetts Institute of Technology, Cambridge, Massachusetts (1985)
14. Weast, R.C. ed., "CRC Handbook of Chemistry and Physics", F-25 to F-28, CRC Press, Boca Raton, Florida (1980)
15. Ibid. B-223 to B-225.

PROPERTIES OF WELDMENTS

KEYNOTE ADDRESS

STRUCTURE AND PROPERTIES OF WELDMENTS

C.D. Lundin
University of Tennessee
Knoxville, Tennessee

Abstract

This keynote lecture describes some of the difficulties in assessing the properties of weldments on a local scale. The discussion is intended to set the stage for the presentations by the others in this "Properties of Weldments" session of the conference. Several examples of HAZ behavior are discussed and attention is called to some newer techniques for assessing weld properties.

THE FINAL CAPABILITY FOR UTILIZATION of a weldment depends on the level of the properties developed in each of the zones throughout the weld. While the weld region is often simplisticlly divided into the fusion zone and heat affected zone, property variations often occur on a scale coincident with the microstructural variation resulting from the process used to accomplish the weld. In most fusion welds the process thermal distributions result in steep temperature gradients, elastic and plastic strains in the material and rapid heating and cooling rates. The nonequilibrium nature of the microstructural changes attendant upon the thermal/strain excursions create a complexity of structure not often encountered in other metallurgical processing. Thus fusion welds, by their very nature, are complex microstructural continuums which present unique problems in the characterization of local or fine scale properties.

The objective of this introductory talk on structure and properties of welds is to discuss some of the attributes of the characterization of the weld structure, to highlight certain of the problems in determining fine scale property variations and to present some of the solutions for gaining insight into weld properties. In the subsequent discussions the HAZ will be treated almost exclusively but it should be noted that the weld fusion zone also requires careful study and that weld metal HAZ's are created in multipass welds--these aspects should not be dismissed as unimportant. Some recent steam piping failures in the US have also given impetus to the concern for fine scale property definition as crack propagation along the weld "fusion line" is evident in at least one of these incidences.

Probably the single greatest contribution to the technology of weld property determinations on a fine scale was the development of the time-temperature simulator permitting controlled heating and cooling of specimens according to actual weld thermal cycles--GLEEBLE (1). The Gleeble device permits the expansion of a single microstructural zone into a volume of sufficient size to permit mechanical, physical and corrosion testing. Thus, by varying the thermal history point-by-point through the HAZ, a profile of properties can be obtained. While this technique has helped shed considerable light on the HAZ response of a material to welding and aided in the development of alloys with suitable HAZ properties, it has not been without criticism. The principle criticism has been the lack of a strain history to supplement the thermal history during the duplication of the microstructure. This problem was addressed in some earlier models of the device by providing for simultaneous temperature/strain control and the latest model Gleeble is replete with a unique strain capability. (It should be noted at this juncture that the strain history of the various regions in the weld HAZ is not known with any precision and thus incorporation into the simulation cycle is an approximation in any event). An additional criticism involves the testing of unique zones not constrained or surrounded in the same way as the particular zone situated in the weldment. This lack of surrounding constraint has left doubts, on occasion, as to the applicability of the results to actual weldments. Never-the-less the properties measured by the simulation technique have enabled great strides in the technological understanding of the composite weld structures.

HEAT AFFECTED ZONE SENSITIZATION

In recent years a cracking problem has come to light in Boiling Water Reactor piping systems....HAZ intergranular stress corrosion cracking (IGSCC). In the non-transformable austenitic type 304 materials utilized, the effect of weld HAZ thermal cycling results in only minimal metallurgical changes; grain growth and carbide precipitation. Unfortunately the precipitation of the chromium carbides during the thermal cycle (sensitization), when combined with the weld residual stresses and an oxygenated water environment gave rise to the cracking. As part of a study sponsored by the Welding Research Council and the Tennessee Valley Authority, The University of Tennessee under took to develop a simple HAZ profiling test for HAZ sensitization and to compare the results from such a test with Gleeble simulation and measurements of HAZ sensitization in actual welds (2).

The autogeneous bead-on-edge weld test sample, shown in Figure 1 was developed to permit profiling of the HAZ because it gave rise to a HAZ which was straight and easily sectioned. Further, the thermal distribution could be readily measured and related to actual weld thermal cycles. When the HAZ of this sample is evaluated with both ASTM A262 Practice A and the General Electric EPR test (3) a realistic HAZ sensitization pattern results as shown by the EPR number distribution as a function of the distance from the weld fusion zone in Figure 2. This distribution matches that obtained from field welds on the same heat of material.

When Gleeble simulation was attempted to define the maximum degree of HAZ sensitization, it was found that the EPR number was less than that determined from actual welds. When a strain sequence was incorporated into the thermal simulation a better correlation resulted. Concurrent, with the above work, the researchers at Battelle Northwest Laboratories (4) were measuring the actual weld thermal cycle in concert with a strain history at the sample point. These data were determined on a pass by pass basis and made available for Gleeble simulation on the Model 1500 which permits compressive/tensile strain duplication together with the thermal history. Figure 3 shows the data typical of a single weld pass for the combined thermal and strain determinations. When these combined thermal/strain histories were used in the HAZ smulation test a very close match was obtained between the insitu EPR tests and the Gleeble test. It is to be recognized that the data obtained at Battelle Northwest represents the first concerted attempt to obtain a simultaneous thermal/mechanical history of a point in the weld HAZ. The obtaining of this type of information for many weld situations is necessary if the modeling efforts now underway at various laboratories is to be at all successful.

HAZ TRANSFORMATIONS

Diagrams describing the transformational characteristics of a steel under conditions of continuous cooling are considered essential in predicting the microstructure and subsequently the properties of the steel after a thermal treatment. Unfortunately Continuous Cooling Transformation (CCT) diagrams are not readily available for the heating, cooling and peak temperature conditions pertaining to welding of steels (only recently have they become generally available for heat treating conditions). Thus, the starting point for microstructural interpretation is missing. To this end I believe it is imperative that suitable CCT diagrams be developed for steels designed for welding and that a concerted effort by steel producers be instituted to develop these diagrams.

Take for example the two steels used almost exclusively in the containment vessels of nuclear reactors; SA 533 Grade B & SA 508 Class 2. Little work was accomplished to document CCT behavior of the base metal and even less effort was expended on the HAZ structure and on understanding the transformations occurring under fusion welding conditions. Figure 4 presents a CCT diagram derived by both electrical resistivity and dilatometric methods for SA 508 C1.2 over a wide range of cooling conditions. Under the condtions pertaining to welding the HAZ CCT transformation diagrams were developed for the grain coarsened region (adjacent to the fusion line) and the grain refined region (outer reaches of the HAZ) as shown in Figures 5 and 6 respectively. The diagrams are superimposed on the same scale in Figure 7. It is to be noted that the bainite nose shifts for the different austenitizing conditions experienced in the grain coarsened region as opposed to the grain refined region. The weld cooling conditions can be described on the diagrams by the cooling curves and the as-welded hardnesses can be estimated from the appropriate cooling conditions. Thus if diagrams of this type are available for a steel a great deal of insight can be had before the first arc is struck.

Another useful form of the CCT diagram is shown in Figure 8, for A 517 G, high strength quenched and tempered steel. This diagram is shown with the 800°C to 500°C cooling time displayed along the abscissa and permits easier interpolation of cooling time and energy input. When this diagram is compared with other steels within the same specification a welding response can be easily ascertained.

HEAT AFFECTED ZONE TOUGHNESS

By far the majority of the effort expended to measure HAZ properties has been in the area of impact toughness. In the late 1950's and 60's it was common to utilize Charpy samples extracted from actual welds, etch the sample to define the weld fusion line and place the notch in the coarse grained region. The coarse grained region was relatively wide in the C, Mn, Si steels utilized at the time and thus the evaluation of the zone of minimum toughness could be clearly defined. However, in recent years, steel making practice has improved and with the development of microalloyed steels and steels with Ti, N, B, and O as elements added to inhibit grain growth, the occurrence of wide grain coarsened regions in steel welds is not necessarily a forgone conclusion. In fact, in many of the newer grain growth controlled steels a single grain width coarsened region is not uncommon. In addition, the utility of high energy density processes such as Electron Beam and Laser together with Narrow Gap arc welding techniques further limits the total HAZ width and the width of the grain corasened region. Therefore notch placement is difficult. Narrow HAZ regions also compound the difficulty of confining the fracture path to the region of interest as fracture initiates from the notch. In such cases, softer surrounding material may form "plastic hinges" and cause the crack to wander from the zone of interest. While this may appear to be a desirable situation, in that the actual weld contains these same regions, it must be remembered that a Charpy sample is a small sample without restraint where as a large weldment has constraints that may keep a crack propagating in a specific zone. Thus it may often be difficult to perform viable tests with samples extracted from actual welds as well as with simulation samples.

The Gleeble simulation technique which became viable in the early 1950's enabled the documentation of the toughness of the various weld HAZ regions (5). The technique was used extensively to aid in the design of steels for optimum toughness and for the evaluation of the effect of welding conditions on HAZ toughness. Figure 9 represents the toughness of the grain coarsened region for three lots of A 517 quench and tempered steel at 80 KJ/in. energy input, 2" plate at 400°F preheat. It is clear by inspection of this figure that one lot stands above the others in regard to HAZ toughness. This steel was a low S (0.003%) clean product with the lowest carbon content of the group (0.11%). Even with the low carbon content the steel made the strength levels for the grade. Of equivalent interest is the effect of weld energy input on the HAZ toughness which can be readily defined by the Gleeble simulation method. The same three lots of A 517 are shown in Figure 10 with the toughness at 32°F shown as a function of weld energy input. The effect of high energy input is evident on the high toughness lot A and on Lot B. With information of this type, restriction can be placed on welding such that a heat input limitaton could help preserve HAZ toughness.

In the traditional steels the intercritical region was often another potential area of low toughness. However, this is not always the case and some of the current HSLA steels will exhibit excellent toughness when cycled in the intercritical region, also A 516, A 588 and A 36 often fall in this category with a microstructure of martensite and ferrite in the intercritical region yielding good toughness. Some of the microalloyed steels, especially those containing Nb and V, show different responses to multiple weld thermal cycles characteristic of multipass welds. Figure 11, from the work of Eftekhar (6,7), on a low C-Mn-Mo-Nb steel shows that the simulated toughness for the grain refined region (1600°F) and below the critical region (1150°F) lie at lower transition temperatures than the base plate, whereas, the grain coarsened region (2400°F) and the intercritical region (1350°F) have poorer toughness than the base plate. The multiple cycle points reveal a very interesting facet of behavior, in that no improvement of the toughness is evident when a coarse grained sample is subsequently cycled to simulate the grain refined region (it was noted that the grain size was refined). Thus, there are apparently additional metallurgical changes involving the minor alloying elements which control the toughness even though the grain size has been reduced.

Toughening of HAZ regions by multiple cycles has been noted in testing of Drop Weight samples for base metal characterization (8). In Drop Weight testing a crack starter weld bead is employed to produce a running crack which the base metal must arrest at the test temperature to exhibit a no-break condition and thus to yeild a low Nil-Ductility Transition Temperature. However, on some materials and with some crack-starter weld bead techniques a HAZ with a band of material exhibiting improved toughness results. This toughened zone tends to inhibit crack propagation and gives rise to the determination of a Nil-Ductility Transition temperature which is artifically low (8). The fracture face of a Drop Weight sample which shows such a toughened zone is presented in Figure 12. The HAZ zones labeled 1,2,3 are the crack-starter bead, diluted zone, and grain coarsened region respectively. Zone 4, the grain refined region is dark because it contains microvoid coalescence fracture features and is of a fibrous texture. Zone 5 is the base metal. The occurrrence of the toughened zone tends to restrict or slow crack propagation and thus absorb energy in the test. This yields, as indicated above, nonconservative NDTT temperatures.

The ability of the microfractographic features to reveal the apparent toughness of materials has long been a criterion for transition temperatures based on fracture appearance. Several investigators (7,9,10) have employed fracture tests of welds in which the fracture is made to traverse through the weld metal and HAZ and the fracture appearance is evaluated as a function of location in the HAZ (e.g. a "cross weld Charpy test" or Dynamic Tear test with a weld in the fracture path). This evaluation avoids the restriction of having to confine the crack propagation from the notch in one unique microstructural region and aids in the evaluation of narrow coarse grained regions of the HAZ or the HAZ's of high energy density welds as discussed above.

MICROHARDNESS EVALUATION OF THE HAZ

Microhardness measurements have long been a traditional way to evaluate weld HAZ's and gain insight into the strengths of the various regions of the HAZ. Microhardness is one of the few techniques that enables a very fine scale (easily confined to a particular zone regardless of zone size) characterization of the ultimate tensile strength across the HAZ in a "continuous" manner. An illustrative hardness traverse is shown in Figure 13 for an N-TUFF 33 HSLA steel with grain refining additions. The hardness generally decreases through the HAZ with a definite decrease in hardness in the coarse grained region of the HAZ and in the region which experienced temperatures just above the A_{c1} temperature at the outer boundary of the HAZ. The coarse grained HAZ softening in the low carbon HSLA steels is not uncommon and in fact was a characteristic of the A 517 (lot A) shown with the good HAZ toughness in Figure 9 and 10.

Numerous instances of HAZ softening have been revealed in different steels and the concern over the occurrence of the softened regions is related to transverse weld tensile and creep strength. In testing of transverse samples containing a weld fuson zone, HAZ and base metal the softened zone revealed by hardness testing does not always manifest itself. However, there is still cause for concern in full scale weldments where the constraint of the surrounding weld regions may exacerbate the effects of the zone, especialy in the creep regime.

Softening in the outer reaches of the HAZ is common in the Cr-Mo steels used for fossil energy and petrochemical applications and has generally been related to "overtempering" of the region just beow the lower critical temperature, A_{c1}. This loss in hardness in a 3Cr, 1 1/2Mo, 1/2 Ni steel is illustrated in Figure 14 for a mildly tempered base material. The hardness dip is quite evident. Although the majority of investigators have related the hardness dip to the overtempering effect (carbide coarsening) another possible explanation lies in the possibility that the softened region lies "just above" the A_{c1} rather than "just below" the A_{c1} and evidence for this has been gained by weld HAZ hardness characterizations and Gleeble studies isolating the peak temperature for maximum softening (11,12).

The mechanism for softening above the A_1 temperature involves the formation of high carbon austenite and ferrite in the intercritical HAZ region. At temperatures above the A_1, austenite will begin to form at grain boundaries and subgrain boundaries while carbides, primarily concentrated in the same locations, will undergo dissolution. The higher solubility of the alloying elements, particularly carbon, leads to rapid partitioning of alloying elements which stabilize the austenite.

The equilibrium solubility of carbon in the first austenite to form just above the A_1 is approximately 0.8% and therefore the driving force for partitioning of carbon is high. Further, carbon can easily diffuse over the inter-carbide distances during the relatively short times permitted by the welding thermal cycle. With carbon and the austenitizing elements concentrated in the low temperature austenite ($A_1 + \Delta T$); the carbon concentration is lower in the ferrite. Upon cooling of the microstructural consitutents so formed, the austenite transforms to high carbon martensite and the ferrite remains unchanged. The possibility of retained austenite, a soft phase, also exists due to the high carbon content of the initial austenite. Upon PWHT the low carbon ferrite would not experience a change but the high carbon martensite would temper rapidly. Thus a decrease in hardness would be anticipated due to the presence of the low carbon ferrite and softer temper products derived from the high carbon martensite as opposed to the constituents present in a normalized and tempered base metal structure.

At higher temperatures in the intercritical region of the HAZ the carbon content of the austenite would be less and the amount of free ferrite would decrease. Further, just above the A_3, the austenite will be of the same carbon content as the base material and no ferrite will be present. After cooling the tempering of the transformed structures during PWHT would be less dramatic than for the high carbon transformed austenitic regions just above the A_1 and a harder region would exist after PWHT. Thus, a softened zone would result at the outer boundaries of the HAZ (but on the weld side of the HAZ having experienced temperatures "just above" the A_1 during welding). Gleeble testing of Cr-Mo samples tends to show this effect and permits an evaluation of the temperature regime which results in maximum softening (11, 12).

A softened zone has occasionally been revealed adjacent to the fusion line in the Cr-Mo materials. This softened zone is a reflection of carbon migration during heat treatment (annealing or normalizing) or PWHT. The driving force for the carbon migration is primarily a difference in Cr content of the base material versus the weld metal. The carbon migration is toward the higher Cr material where the carbon activity is the lowest (other alloying elements may also affect carbon activity but Cr is the most commonly encountered).

Within the SA specifications for 2 1/4 Cr-1Mo weld and base materials a Cr difference of 0.60% is possible. Such a case was studied in The University of Tennessee Welding Laboratories and is illustrated by the photomicrographs at the fusion line in Figures 15a and b. The Cr content is 0.60% higher in the weld metal than in the base metal and the weldment was annealed and PWHT before examination.

Figure 15a, a transverse section at 50X adjacent to the fusion line (center of micrograph) reveals a pearlite and carbide free zone along the fusion line on the base metal side of the joint. A carbon enriched zone is noted on the weld metal side of the interface (dark pearlitic regions). The carbon denuded zone is more clearly seen in Figure 15b which is an angle slice through the same HAZ/Fusion zone. Angle sectioning enables a "magnification" of the HAZ/Fusion line region and permits more discrete determination of hardness and composition. Naturally, this zone is equivalent to the carbon denuded zone so clearly evident in austenitic stainless steel to ferritic steel transition joints and the same mechanism is applicable.

If the Cr content different is intensified by purposely choosing different Cr content base metals and filler metals the effect can be enhanced even without annealing. The carbon diffusion reaction goes forward equally well just below the lower critical temperature, A_1, as would be the case for tempering, PWHT or service exposure. The HAZ micrograph in Figure 16 represents a 9Cr-1Mo weld deposit on 2 1/4Cr-1Mo base metal. The 9Cr material is at the top in Figure 16 with the fusion line orientated horizontally (scalloped) at the top of the micrograph. A broad 0.030" wide ferrite band exists on the base metal side of the interface after a 1350°F - 2 hours PWHT. A feature that will not be erased by subsequent service at temperatures below the A_1.

The implication of this phenomenon is that a lower hardness, lower strength zone has been created which can be expected to affect transverse weldment properties especially in the creep range. The low hardness region is sandwiched between harder zones and can become particularly troublesome if fracture is forced, by the stress state, to be confined to this low strength region. This is a potential condition certainly worthy of note when selecting filler metals to join the Cr containing steels.

The foregoing discussions of several HAZ phenomena and methods of evaluation of the structure and properties of weldments was constituted to illustrate the complexity of the task of defining the structure/property relationships in weldments and not to specifically provide new data on weldment characterization.

Session K, "Properties of Weldments," at this 1986 International Conference on Trends in Welding Research in Gatlinburg, Tennessee contains papers which discuss investigations of HAZ properties and structures. One particularly noteworthy contribution is the Micro Shear Test by Professor Lutz Dorn of the Technical University of Berlin. The Micro Shear Test appears to offer a new and novel way to determine HAZ properties (strength and ductility) on a fine scale. The test consists of shearing increments (0.5mm spacing) from a small cylindrical specimen (1.5mm diameter) extracted transverse to and containing the fusion zone/HAZ region of a weldment. The instrumented shearing operation permits the quantitative measurement of properties. This methodology is certainly unique and is sure to find utility in the evaluation of HAZ structures on a fine scale.

Another technique, not reported on at this conference, but sure to be employed in future weld zone evaluations is a computerized ultra-fine scale hardness test utilizing an indentation size (depth) in the nanometer range (13). Information on hardness and strength can be derived from the measurements. One of the attributes of the technique is the programed control of the location, hardness indentation parameters and the acquisition of data in a retrievable and subsequently readily analyzable fashion.

The strain history data described above in the HAZ sensitization illustration (4) is to be considered a "first" because it provides simultaneous thermo-mechanical history of various points in the HAZ. More work in this area of weld data acquisition is definitely necessary if the modeling research on welds, currently in vogue, is ever to reach full fruition.

The current research in weld property and structure determination is considered to be on a high plane and conferences of this sort are to be commended as they bring the new work to the forefront quickly where it can be utilized by other researchers in the field.

References

1. "Development of Specimen Simulating Weld Heat-Affected Zones," Nippes, E., Savage, W., Welding Journal, 28, November 1949.

2. "Austenitic Stainless Steel Weld HAZ Sensitization: A method of Evaluation," Lundin, C., Lee, C., Stansbury, E., and Menon, R., Proceedings of JDC University Research Symposium 1985, International Welding Congress, ASM October 1985.
3. "Detection of Sensitization in Stainless Steel Using Electrochemical Techniques," Clark, W., Romero, V., Danko, J., GEAP 21382 (NUREG-025-1) August 1976 or STP 659, ASTM, 1978.
4. "Evaluation of Welded & Repair-Welded Stainless Steel for LWR Service," Atteridge, D., Bruemmer, S., Charlot, L., and Page, R., NUREG/-/cr 3613, PNL 4941 Vol. 3., No. 1.
5. "Impact Performance of Synthetically Reproduced Heat-Affected Zone Microstructures in T-1 Steel," Nippes, E., Sibley, C., Welding Journal 35, October 1956.
6. "The HAZ Impact Toughness of a Mn-Mo-Nb Linepipe Steel," Eftekhar, R., Lundin, C., Welding Journal, April 1977.
7. "An Electron Fractography (SEM) Investigation of the Heat-Affected Zone in the Newly Devloped Climax Low-Carbon Mn-Mo-Nb Steel," Eftekhar, R., Masters Thesis, University of Tennessee, August 1974.
8. "The Influence of the Crack-Starter Bead Technique on the Nil-Ductility Transition Temperature of ASTM A 572 Grade 55 Hot Rolled Plate," Lundin, C., Merrick, E., Kruse, B., To Be Published as ASTM STP-Conference November 1984.
9. "A Fundamental Difficulty in Charpy V-Notch Testing Narrow Zones in Welds," Goldak, J., Nguyen, D., Welding Journal April 1977.
10. "Evaluation of Heat-Affected Zone Toughness by the Cross Weld Charpy Test," Boyd, J., Godden, M., Proceedings of a Symposium-What Does The Charpy Test Really Tell Us? Denver, February 1978, ASM 1978.
11. "Transformation, Metallurgical Response and Behavior of the Weld Fusion Zone and HAZ in Cr-Mo Steel for Fossil Energy Applications," Lundin, C., Henning, J., and Richey, M., Quarterly Progress reports for the periods ending September 30, 1983, December 31, 1983.
12. "Transformation, Metallurgical Response and Behavior of the Weld Fusion Zone and Heat Affected Zone in Cr-Mo Steels for Fossil Energy Application," Lundin, C., Richey, M., and Henning, J., Final Technical Report for period April 1983 to December 1984, US DOE Contract No. De-AC05-84OR21400 April 1986.
13. "Hardness Measurement at Penetration Depth as Small as 20nm," Pethica, J., Hutchings, R., and Oliver, W., Philosophical Magazine A, vol. 48, No. 4, 1983.

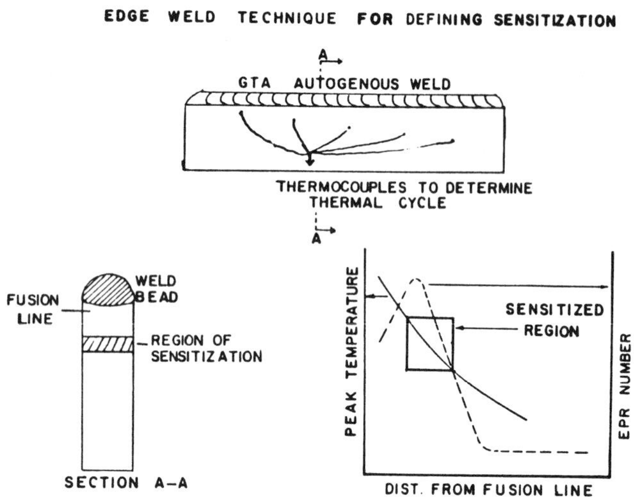

Figure 1. Edge weld sample for HAZ sensitization.

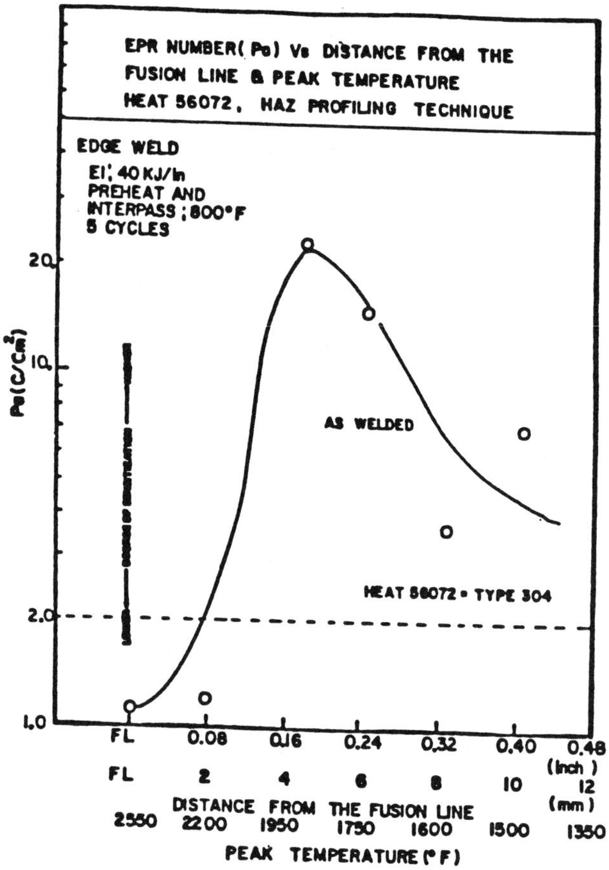

Figure 2. HAZ sensitization of 304 stainless steel.

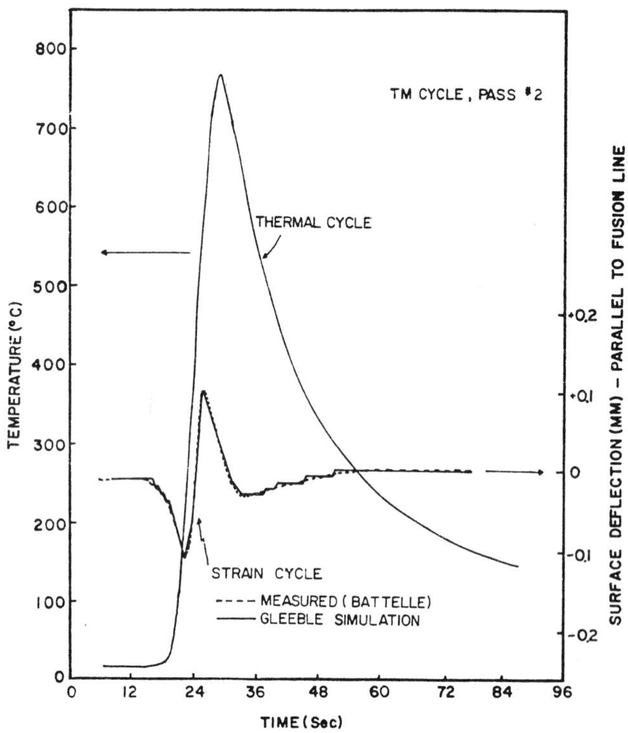

Figure 3. Combined thermo-mechanical history of a point in the HAZ in 304 stainless steel.

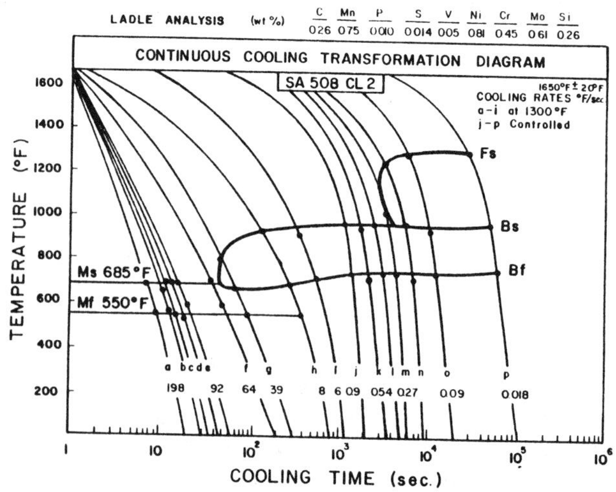

Figure 4. CCT diagram for SA 508 Cl.2 steel.

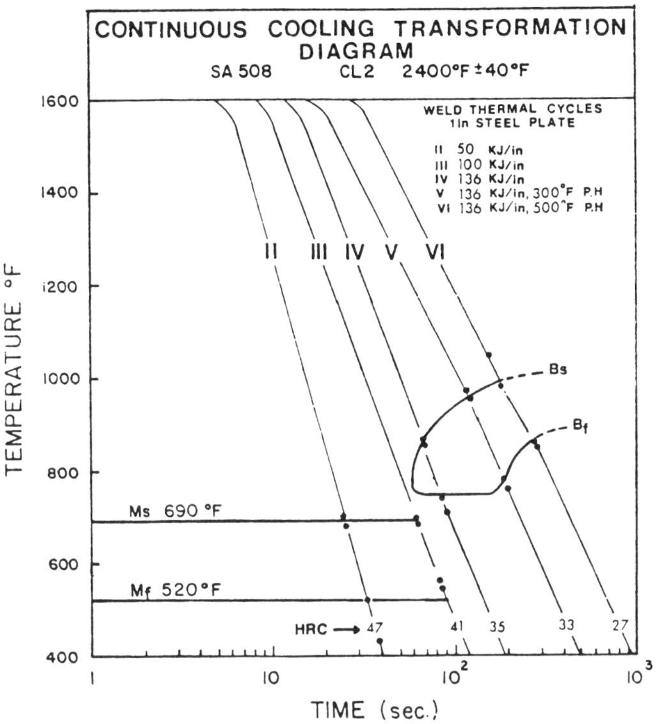

Figure 5. CCT diagram for the grain coarsened region of the weld HAZ of SA 508 Cl.2 steel.

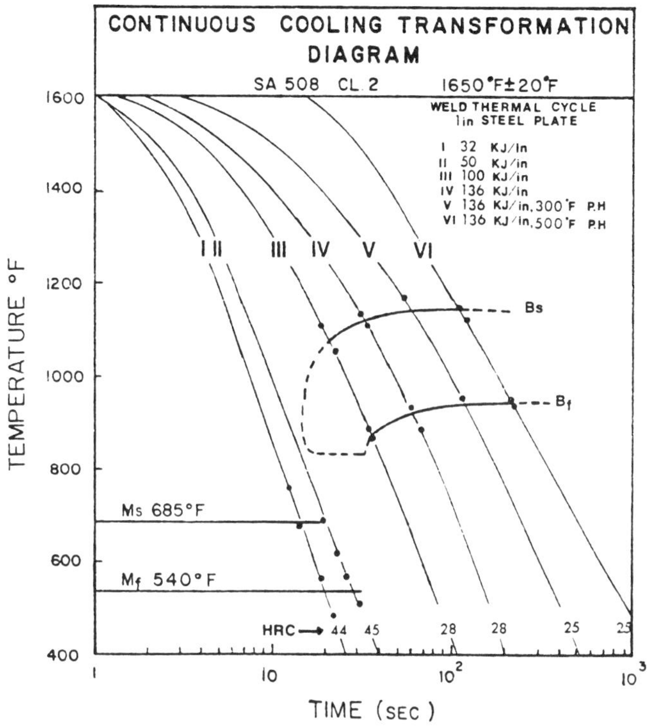

Figure 6. CCT diagram for the grain refined region of the weld HAZ of SA 508 Cl.2 steel.

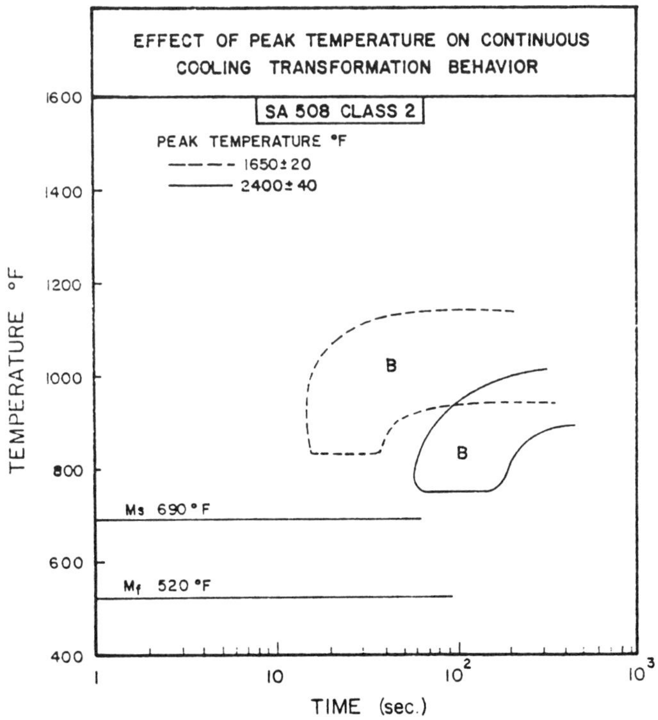

Figure 7. Comparison of the CCT diagrams for the coarse grained and fine grained regions of the HAZ of SA 508 Cl.2 steel.

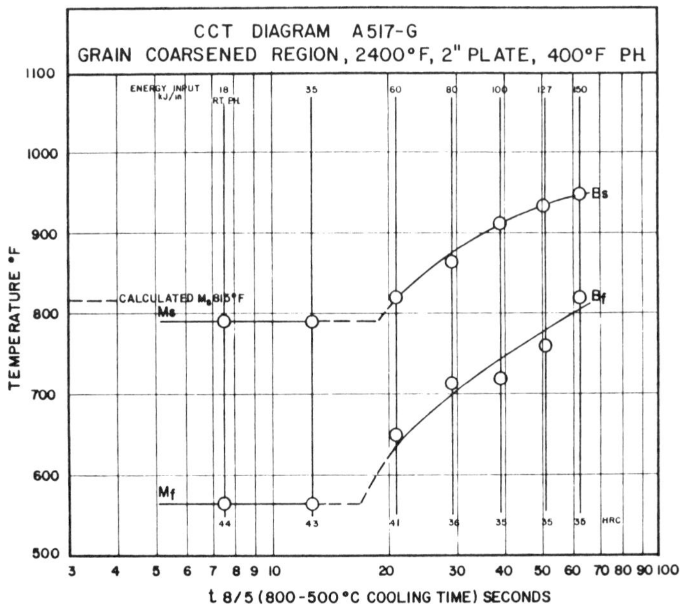

Figure 8. CCT diagram for the grain coarsened region of the HAZ of A 517-G steel.

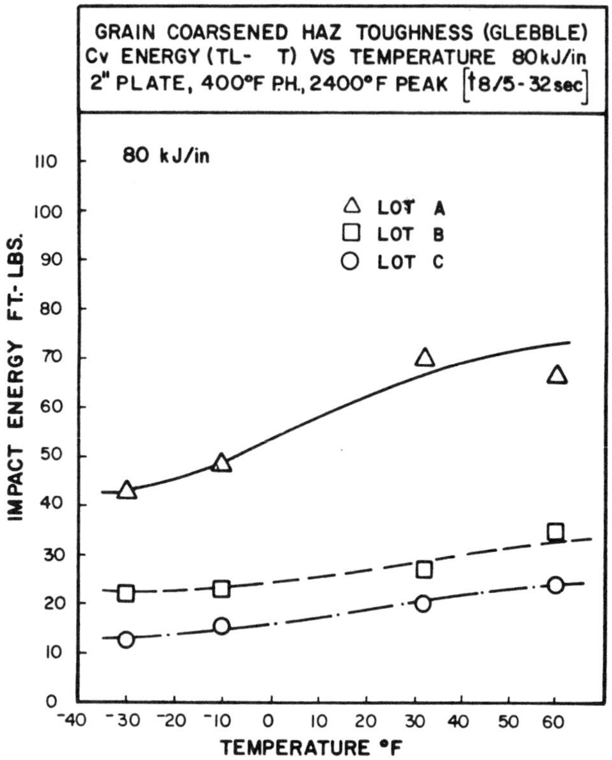

Figure 9. HAZ toughness of the coarse grained region of A 517 steel.

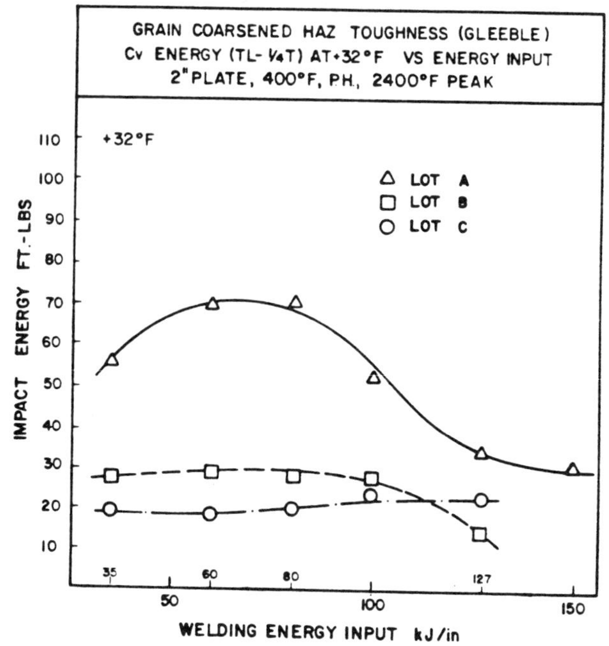

Figure 10. HAZ toughness of the coarse grained region of A 517 steel as a function of weld energy input.

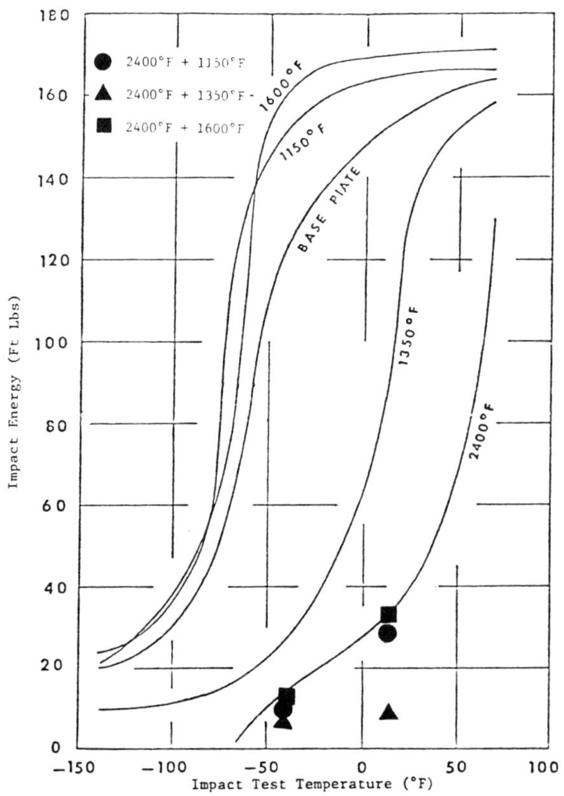

Figure 11. Simulated HAZ toughness of a low C, Mn-Mo-Nb steel.

Figure 12. Fracture face of Drop weight sample tested below NDTT.

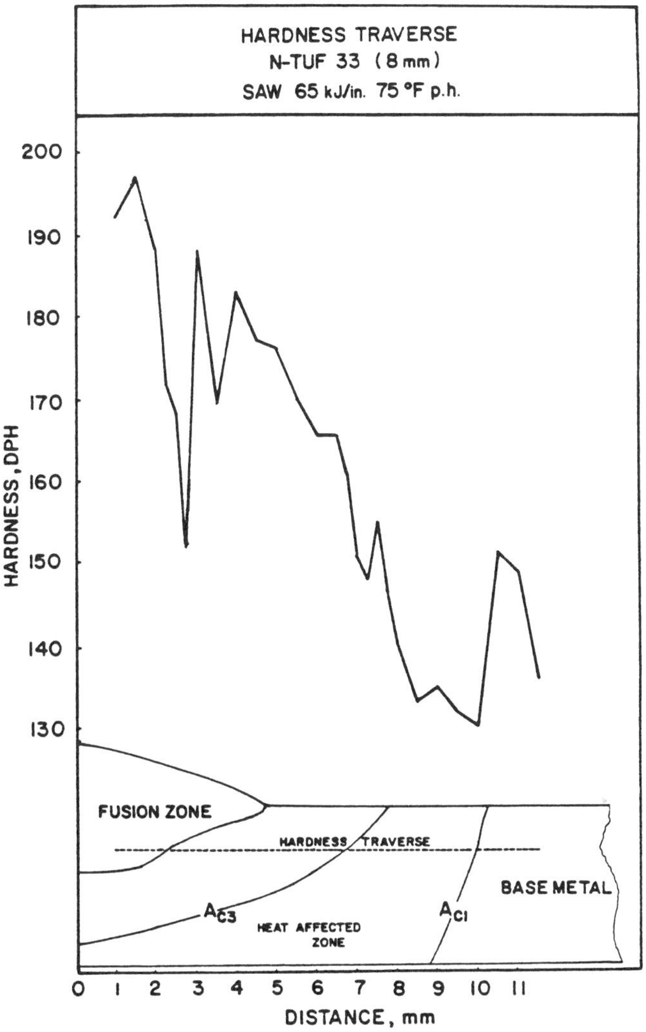

Figure 13. Typical hardness traverse across the weld HAZ in N-Tuff 33 steel.

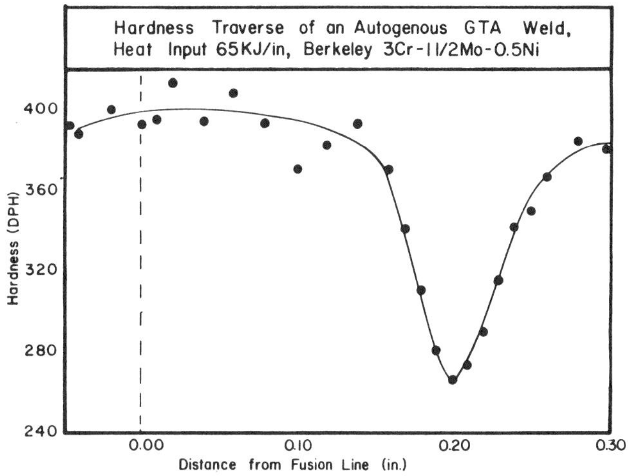

Figure 14. Illustration of the occurrence of a soft zone in 3Cr-1 1/2Mo pressure vessel steel.

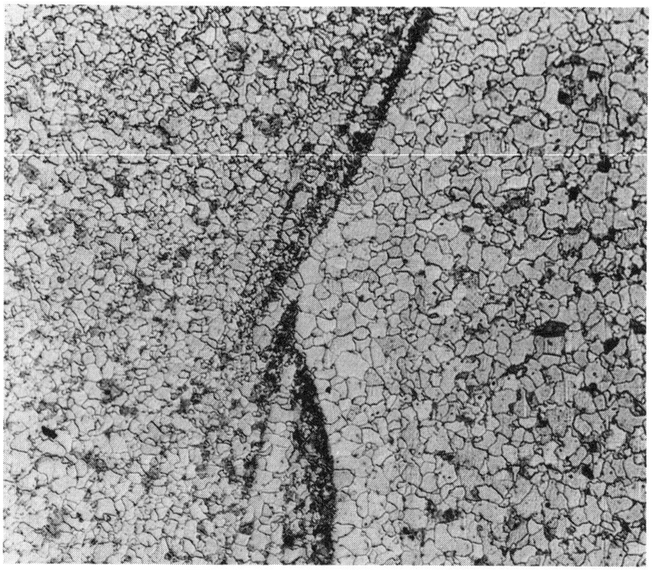

a.

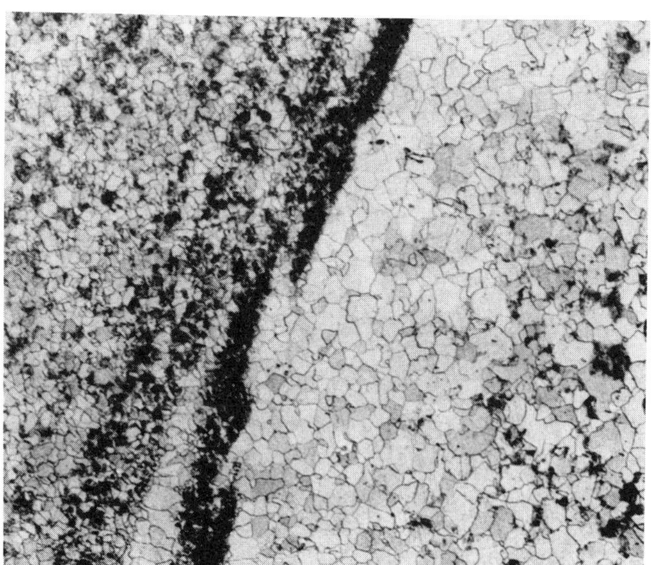

b.

Figure 15. Carbon migration in the HAZ of 2 1/4Cr-1Mo steel (annealed and tempered after welding).
a. Transverse section, 50X
b. Angle section, 50X

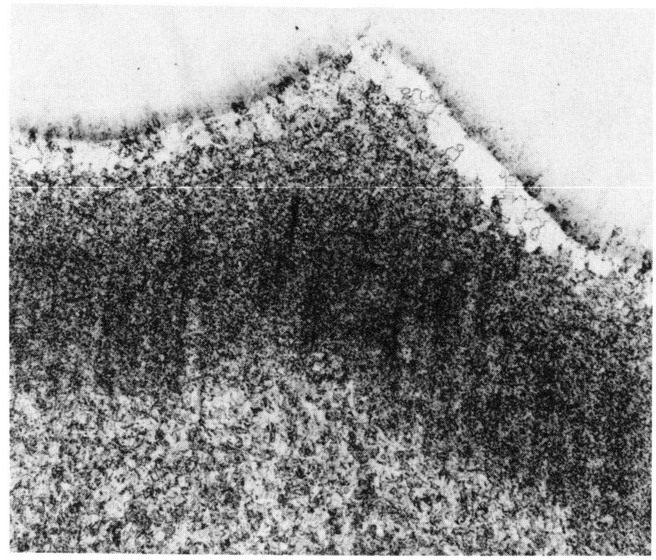

Figure 16. Soft zone at fusion line of a 9Cr to 2 1/4Cr weld. 50X.

DETERMINATION OF WELD STRUCTURE PROPERTIES BY MEANS OF THE MICRO SHEAR TEST

Lutz Dorn
Technical University
Berlin (West), Federal Republic of Germany

Abstract

This paper describes a test method, to examine the different structure properties of welds. The test combines comprehensive information about strength and ductility with simple performance.

THE STRUCTURE OF WELDS consists of several zones of different strength and ductility. To measure the pattern of properties, testing methods are necessary, which combine the following features:

1. Possibility of testing narrow zones
2. Determination of strength and ductility
3. Reliability of test results
4. Simple test equipment and easy test performance

The micro test /1/ seems to be especially promising to fulfil these requirements. The aim of our experiments was to study different welding problems and to prove the qualification of the micro shear test for examination of weld structure properties.

PRINCIPLE OF MICRO SHEAR TEST

The test specimens are 30...50 mm long and have a diameter of 1,5 mm. Instead of cylindrical test specimens, rectangular ones can also be used. One end of the specimen is fixed in a clamp, and a shear tool cuts through the outstanding part. Therewith the force and displacement of the tool are measured so that a force-displacement-diagram can be established (Fig. 1). From the maximum shear force F_m, the original cross-section of the sample S_o and the cross-section in the moment of fracture S_F the following characteristic values can be determined:

Maximum shear strength

$$\tau_m = \frac{F_m}{S_o} \quad (N/mm^2) \qquad (1)$$

Portion of sheared area

$$\alpha = (1 - \frac{S_F}{S_o})\ 100 \ (\%) \qquad (2).$$

Micro shear tests using different steels show that the scattering of the values is low.

On the premises of careful machining, the finishing process by turning or grinding has no influence on the test results.

Each specimen can be submitted to several tests. Only if the distance between the locations of the tests is smaller than 0,5 mm, an influence on the test results can be observed.

A comparison of tensile tests and micro shear tests using the same material showed linear relations between tensile and shear strength

$$R_m = 2,04 \cdot \tau_m - 228 \ (N/mm^2) \quad (3)$$

and also between reduction in area and the portion of sheared area

$$Z = 1,50 \cdot \alpha - 1,45 \quad (\%) \qquad (4).$$

Table 1: Chemical composition of steel 23 CrNiMo 747

% C	% Si	% Mn	% P	% S	% Cr	% Mo	% Ni
0,22	0,24	0,64	0,015	0,015	1,87	1,05	0,61

INFLUENCE OF DIFFERENT JOINT-GEOMETRY AND DIFFERENT WELDERS ON WELD STRUCTURE /2/

The plates of a low alloyed Cr-Ni-Mo-steel 23 CrNiMo 747 (table 1) formed two different joints with three members (Fig. 2). After preheating to 250°C two different welders performed the weld using TIG-method for the root run and basic-coated rods for the filler and cover runs, followed by a post heat treatment (2 hours at 580°C).

Specimens for micro shear tests were taken out of the upperside and underside of the seam. The modification of strength and ductility shows the influence of different joint geometries and different welders in a distinct manner.

This study leads to the conclusion that the micro shear test is suitable to detect also slight differences in the properties of weld structure. Furthermore, due to the joint geometry, other methods for testing ductility (notch impact test, bend test) can hardly be used.

INFLUENCE OF SPECIFIC HEAT INPUT AND POST HEAT TREATMENT CONDITIONS ON WELD STRUCTURE /3/

The plates 1000 x 400 x 40 mm, consisting of a high strength steel W St E 460 (table 2), were arranged corresponding to Fig. 3.

For submerged arc welding a Ø 4 mm filler wire (S 2 Ni 2,5) and a low hydrogen flux were used. The temperature of preheating was 150°C, the interpass temperature 180...200°C. The welds were performed with three different specific heat inputs of 20, 40 and 55 kJ/cm.

Micro shear specimens were taken out of the welds, transverse to the seams. Decreasing heat input increases shear strength in the HAZ and lowers ductility. The molten weld metal shows only a slight change in strength, but ductility is remarkably reduced using 55 kJ/cm heat input. This fact would not be found with the hardness test only.

CONCLUSIONS

This report presents a new method of finding the properties of weld structures: the micro shear test. Compared with conventionals methods the following advantages can be achieved:

a) The area to be tested can be predetermined with high accuracy, so that the properties of each zone of interest can be determined. Contrary to this, the rupture in tensile test occurs always in the weakest region, so that the properties of other zones cannot be determined.

b) Only one cylindrical specimen is necessary to determine the properties of all weld zones transverse to the seam. Therefore, costs can be reduced as compared with conventional methods which require a different specimen for each zone to be tested.

c) Tests can be made within tiny regions and in small distances from each other. For that reason, also very narrow zones with unfavourable properties can be detected.

d) While the information of the hardness test is restricted on strength and that of the notch impact test on toughness, the micro shear test indicates both strength and ductility.

e) According to the correlation between the micro shear test and the tensile test, the shear strength and portion of the sheared area can be converted into tensile strength and reduction in area with satisfactory accuracy. Therefore, the results of the micro

Table 2: Content of steel W StE 460 and filler wire S 2 Ni 2,5

	% C	% Mn	% Si	% Ni	% P	% S	% V
W StE 460	0,155	1,63	0,35	0,35	0,017	0,004	0,17
S 2 Ni 2,5	0,08.. 0,14	0,8.. 1,2	0,05.. 0,12	2,5			

shear test can serve as a basis for designing purposes.

f) The low scattering of test results enables the micro shear test to detect very small differences in weld structure properties, whereas conventional methods often require numerous samples to be tested.

g) Due to small diameter of the specimen thin sheets can also be tested whilst other methods, for instance notch impact, require higher thicknesses.

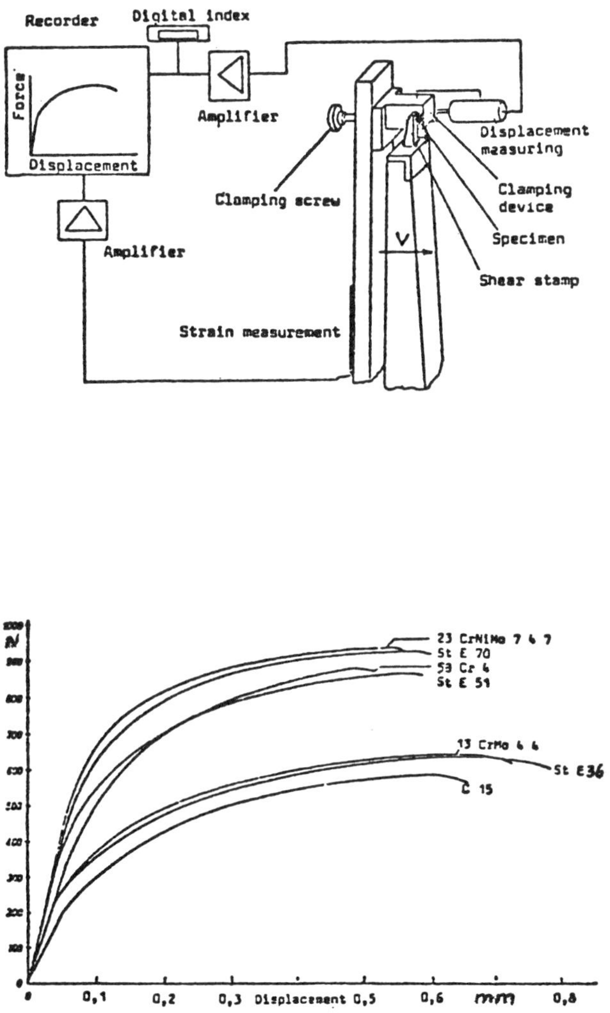

Fig. 1: Principle force-displacement-diagram of the micro shear test

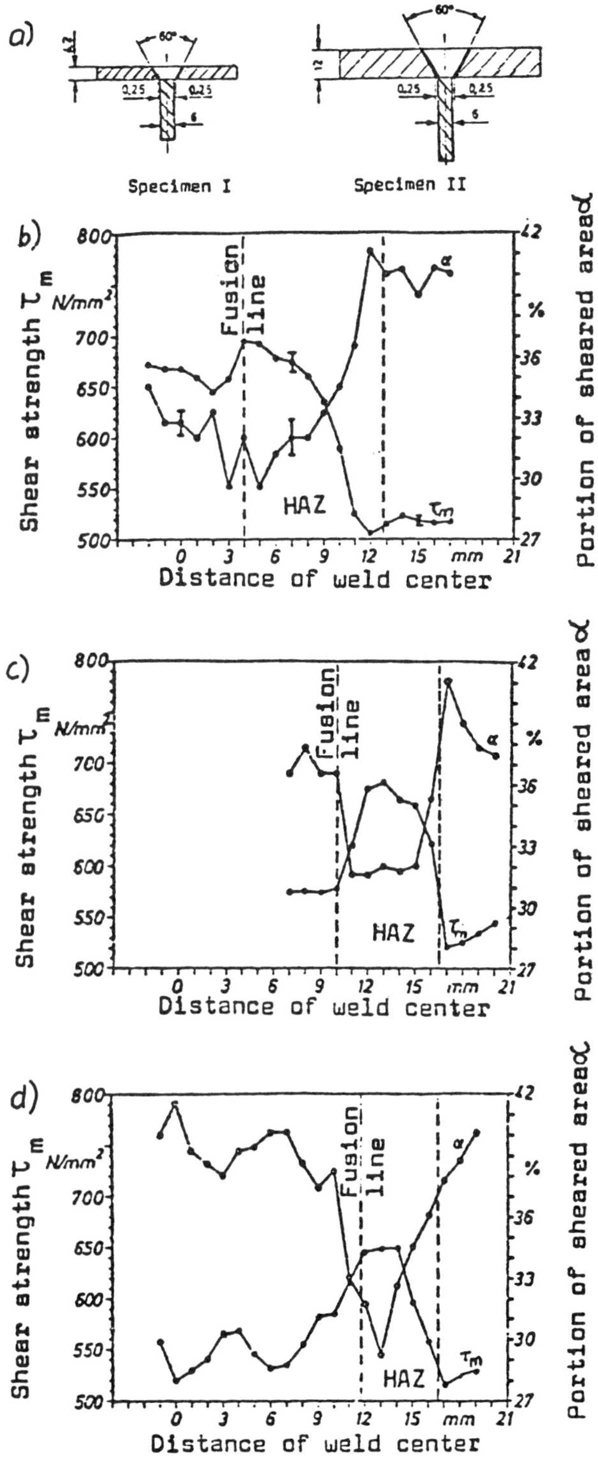

Fig. 2: SMAW-Welding of steel 23 CrNiMo 747
a) Micro shear test - Principle of weldment
b) Micro shear test - Specimen I , welder A
c) Micro shear test - Specimen II , welder A
d) Micro shear test - Specimen II , welder B

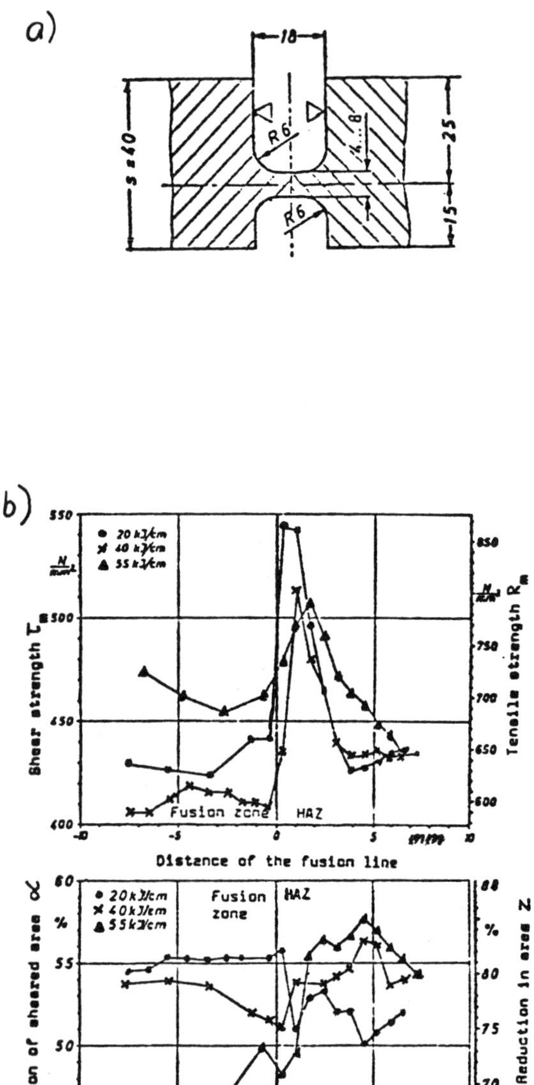

Fig. 3: Submerged arc welding
high strength steel W StE 460
a) Edge preparation
b) Influence of specific heat input according to micro shear tests

REFERENCES

/1/ Ermittlung der Festigkeits- und Verformungseigenschaften kleiner Gefügebereiche mittels Mikroscherprüfung
EX: Dorn, L. and G. Niebuhr, Härterei-Technische Mitteilungen 34, No. 5, P. 227-232 (1979)

/2/ Der Einfluß von Schweißbedingungen auf die Gefügeeigenschaften von 23 CrNiMo 7 4 7; Untersuchungen durch Mikroscher- und Mikrozugprüfung
EX: Dorn, L and G. Niebuhr, Verbindungstechnik - Thermisches Fügen (Ingenieur-Digest-Publishers), P. 7-11 (1979/80)

/3/ Untersuchungen zum Einfluß unterschiedlicher Energiezufuhr und Wärmenachbehandlung auf die Schweißgefügeeigenschaften von WSt E 47 mittels Mikroscherprüfung
EX: Dorn, L. and G. Niebuhr, Schweißen und Schneiden 32, No. 6, P. 216-220 (1980)

MECHANICAL PROPERTIES OF LASER WELDED ASTM A710

E. A. Metzbower
Naval Research Laboratory
Washington, DC 20375-5000

ABSTRACT

Steel plates from two heats of ASTM A710, Grade A, Class 3, were laser welded using a 15kW, continuous wave, CO_2 laser. The heats differed in the amounts of sulfur and oxygen. Acceptable laser welds were made from both heats of steel. The welds and base plate material from the heat of steel with the lower sulfur and oxygen content had CVN read-ings of 60 Joules or greater than material from the heat with higher levels of oxygen and sulfur.

THE LASER BEAM WELDABILITY of several high strength, low alloy (HSLA) steels has been the object of a recent study. One of the high strength, low alloy steels studied was the alloy A710, Grade A, Class 3. Three types of welds were produced; autogenous, Inconel 600 alloyed, and pure nickel alloyed. The alloyed welds were made in an attempt to improve the toughness of the weld material. Comparisons were made between welds from different heats of steel to determine if variations in chemistries could affect the properties of the laser welds. The largest chemical variations were in the sulfur and oxygen content of the steel.

PROCEDURE

All welds were made in 12.7 mm thick plate which was 152 mm x 305 mm in size. Welding parameters ranged from 11 kW to 13 kW at speeds of 11 mm/sec to 14 mm/sec. Welds were produced in a single pass resulting in very narrow welds. The alloyed weldments were produced by the insertion of a shim between the plates. The shim materials were Inconel 600 and pure nickel. The thickness of the shims was 0.13 mm.

RESULTS

Table 1 indicates the sulfur, oxygen, and nitrogen levels of the A710 material as reported by the manufacturers. The results indicate that there were major differences in the amounts of sulfur and oxygen between the two different heats of steels. The largest differences were in the amounts of sulfur where the value was 8 ppm in one heat and 60 ppm in in the other. Oxygen values varied less with values being 21 ppm and 44 ppm, whereas the nitrogen values were 80 ppm and 114 ppm.

Table 1 - Sulfur, Oxygen, and Nitrogen Levels

	A710 Heat Compositions, wt%		
Material	S	O	N
Hi S/O	0.006	-	-
Retest	60 ppm	44 ppm	114 ppm
Weld	60 ppm	98 ppm	127 ppm
Lo S/O	0.004	-	-
Retest	8 ppm	21 ppm	80 ppm
Weld	8 ppm	101 ppm	99 ppm

Fusion zone material was also analyzed for sulfur, oxygen, and nitrogen. Because there were no filler metals involved in the laser welds, the amounts of sulfur did not change. However, because the weld pool is in a molten state and exposed to atmosphere the gas content in the welds did increase. For oxygen the increase was to approximately 100 ppm. For nitrogen the final amounts in the welds were 99 ppm and 127 ppm.

In the A710, Grade A, Class 3, the base plate microstructure, which was identified based on standards produced by The Welding Institute (1,2), was a fine grain polygonal ferrite with some small carbide aggregate regions located along the grain boundaries. The HAZ was a mixture of ferrite with second phase (FS), acicular ferrite (AF) and intergranular polygonal ferrite (PF). The microstructure of the fusion zone of the autogenous weld was a combination of the same microstructures (FS, AF, PF) (Figure 1). In the Inconel 600 and pure nickel insert welds the microstructure was an extremely large grain martensitic structure. As expected there were no major differences in microstructure between plates from the two heats of steel which differed in chemistries.

The microhardness profiles indicate the variations in the microhardness measurements of the different heats of A710 and the influence that the alloy additions of nickel had on the hardness. In both heats of A710 the base plate hardness was approximately 200 Vickers Hardness Number (VHN) with the hardness increasing through the HAZ to the fusion zone. The fusion zone hardness readings varied with the autogenous welds increasing in hardness to a 300 VHN in the fusion zone. With the Inconel and nickel additions the fusion zone hardness increased to approximately 350 VHN.

The weld specimens were machined to AWS specifications with the tip of the V-notch centered in the weld and aligned along the welding direction. Base plate properties were taken in the T-L (rolling) and L-T (transverse) directions.

Figures 2, 3, 4, and 5 contain the toughness results of A710 laser weldments from two different heats of steel. Figure 2 indicates the base plate CVN values of the low sulfur/oxygen steel is approximately 60 J higher in both directions than the high sulfur/oxygen steel. For both heats of steel there is no ductile-brittle transition above -51°C (-60°F) and there is approximately 55 J difference between the L-T and T-L directions. The autogenous welds of the two different heats also showed the superior toughness of the low sulfur/oxygen alloy (Figure 3). The results for the Inconel and nickel alloyed welds are shown in Figures 4 and 5.

Many of these specimens did not break but stopped the Charpy hammer and registered a value of 353 J which is the maximum energy produced by the device used. In these specimens the cracks did not propagate through the weld but split and ran out into the base plate material. Similar results were reported by Moon and Metzbower [3] in their work with Inconel alloyed HY-80 welds.

DISCUSSION

Based on the mechanical tests alone it appears that the A710 material can be successfully laser welded. The microhardness test indicated higher hardness values in the fusion zone which signifies that the fusion zone has a higher strength than the base plate.

The addition of nickel to the A710 is to prevent the copper from causing hot-shortness in the steel [4] and to lower the CVN transition temperature [4,5]. The lowering of the transition temperature is the reason that additions of nickel were made to the laser welds. The exact method by which the nickel accomplishes this is unclear but it has been suggested [5] that it is related to the fact that nickel additions increase cross slip of screw dislocations or by spreading of shear misfit of <100> edge dislocations and thereby making them less effective as crack nuclei.

In the literature there are many references given to the effects of sulfur and oxygen on the toughness of HSLA steels [5-10]. The influence of sulfur and oxygen is that their presence in the form of inclusions act as nucleation sites for the formation of acicular ferrite. In many of these papers it is also stated that acicular ferrite is the preferred microstructure for high CVN results [9,10]. Others have determined that lower sulfur levels increase the toughness especially when the inclusions are shape controlled [6-8].

From the CVN results of the low sulfur/oxygen A710 it is apparent that either an optimum level of sulfur and oxygen has been reached or that the act of laser welding has in some way altered the recommended inclusion levels and still produced a microstructure that has a high impact shelf energy. A possible explanation for this higher toughness value in the laser welding material could be related to the microstructure which is predominantly parallel laths of ferrite with aligned second phase. This structure is randomly oriented throughout the fusion zone and appears to be an interlocking structure similar to normal acicular ferrite.

The nickel [5] reduces the transition temperature in the CVN curves and does so for both chemistries of A710. The most effective addition apparently was in the form of pure nickel but increases in the CVN values with the Inconel 600 inserts were also noted. Microprobe work indicated that the level of nickel in the fusion zone increased from approximately 1% to 6%-8% total weight of the weld material depending on the size of the melt pool, the exact thickness of the insert, and the type of insert used.

CONCLUSIONS

To date the following conclusions can be made concerning the laser welding of high strength low alloy steels: The alloy A710, Grade A, Class 3 can be laser welded with mechanical properties that are comparable to conventionally welded A710. The addition of nickel to the laser weld of A710 increased the toughness of the material. Base plate and welds made from heats containing extremely low concentrations of sulfur and oxygen were found to have better toughness properties than plates and weldments produced from heats having higher amounts.

REFERENCES

1. R.E. Dolby, Metals Tech. 10, 349 (1983).
2. IIW Document No. IXJ-102-85.
3. D. W. Moon and E. A. Metzbower, Second Inter. Conf. on Application of Lasers in Mat. Processing, Jan. 1983, "The Effects of Inconel 600 on the Toughness of HY Steel Laser Welds," ASM, Metals Park, OH (1983).
4. R. E. Kelley, Master of Science Thesis, University of Texas at El Paso, "Optimization of Mechanical Properties of "In-787" Nickel-Copper-Columbian Age Hardenable Alloy" (1974).
5. W. C. Leslie, "The Physical Metallurgy of Steel," Hemisphere Publishing, Washington, 122-124, 204-209, 300-303 (1981).
6. R. Blondeau, L. Cadiou, J. Maniere and G. M. Pressouyre, "HSLA Steel with High Weldability and Improved Characteristics Obtained Through Vacuum Heated Ladle Refining Technique," Int. Conf. on Tech. and Appl. of HSLA Steels, ASM, Metals Park, OH (1983).
7. K. Norring, H. Harvig and B. Lindwall, "Improvements in Thick-Plate Steel for Heavy Structures," Proc. of the Int. Symp. on HSLA Steels, Microalloying '75, ASM, Metals Park, OH, 684 (1975).
8. A. D. Wilson, "Characterizing Inclusion Shape Control in Low Sulfur C-Mn-Cb Steels," Int. Conf. of Tech. and Appl. of HSLA Steels, ASM, Metals Park, OH, (1983).
9. R. A. Ricks, P. R. Howell and G. S. Barritte, J. of Materials Science 17, (1982).
10. P. L. Harrison and R. A. Farrar, J. of Materials Science 16, 2218 (1981).

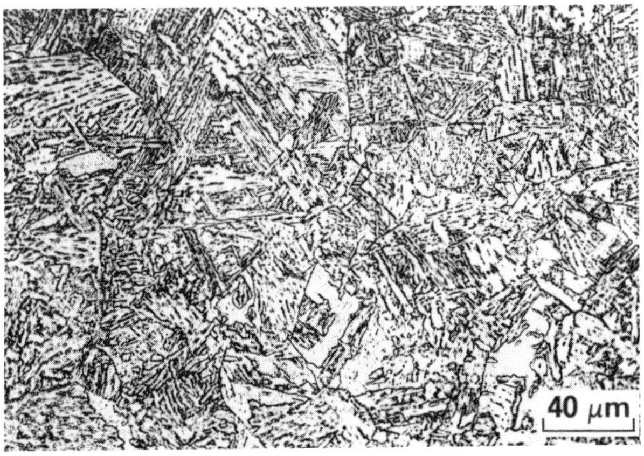

Fig. 1 - Micrograph of fusion zone material from HI S/0 A710 autogenous weld. Present are ferrite with aligned second phase, acicular ferrite, and intergranular polygonal ferrite. (Nital etch).

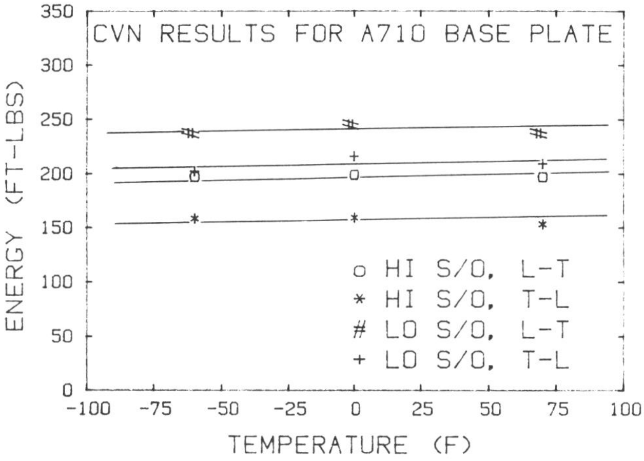

Fig. 2 - Temperature versus Charpy V-notch toughness results for the base plate of both HI S/0 and LO S/0 A710 steel plates.

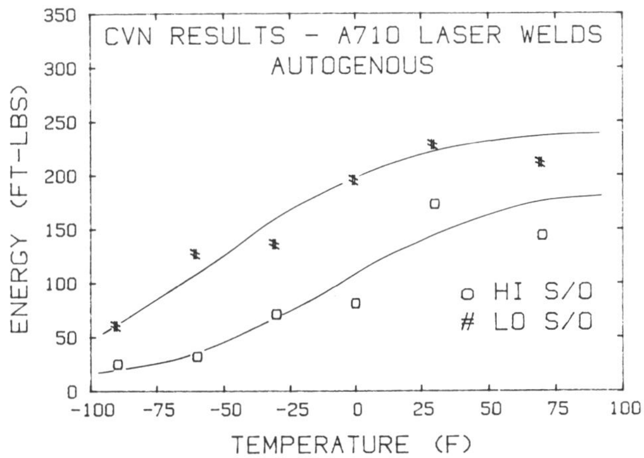

Fig. 3 - Temperature versus Charpy V-notch toughness results for the autogenous welds from HI S/0 and LO S/0 steel plates.

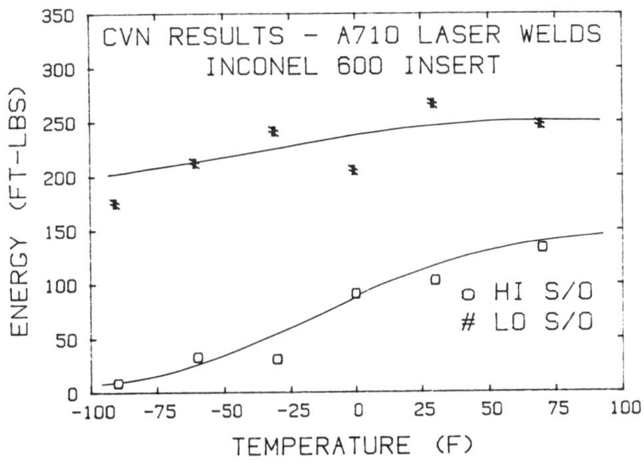

Fig. 4 - Temperature versus Charpy V-notch toughness results for the weldments of HI S/0 and LO S/0 A710 with Inconel 600 inserts.

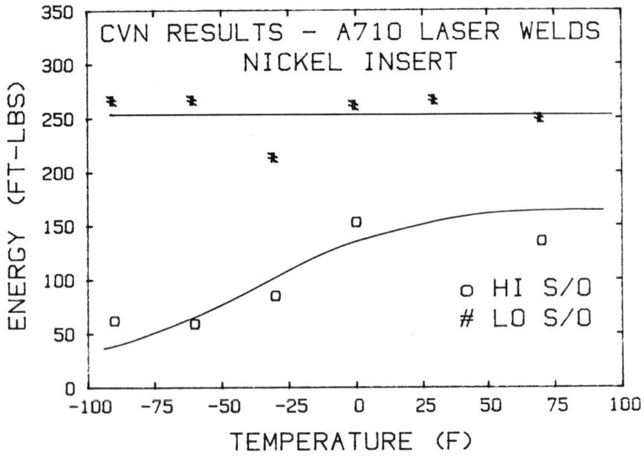

Fig. 5 - Temperature versus Charpy V-notch toughness results for the weldments of HI S/O and LO S/O A710 with pure nickel inserts.

MICROCRACK FORMATION AND FATIGUE INITIATION IN THE HEAT AFFECTED ZONE OF NICKEL ALLOY WELDS

R.G. Thompson
University of Alabama at Birmingham
Birmingham, Alabama

M.L. Callaghan
McDonnell Douglas
St. Louis, Missouri

A.C. Nunes
NASA, Marshall Space Flight Center
Huntsville, Alabama

ABSTRACT

Stress rupture tests have been used in our lab to characterize the fracture process and fracture surface of wrought Inconel 718. This characterization was carried out from temperatures well below the solidus up to the solidus temperature. This allowed us to characterize intergranular fracture at the transition from solid state intergranular cracking to liquid phase intergranular cracking. Surface features were observed in the scanning electron microscope (SEM) which allowed the fracture surface to be identified as either solid state separation or liquid state separation.

The fracture characterizations developed in the stress rupture tests were applied to microfissuring developed in alloy 718 in the varestraint test. Specimens were subjected to a varestraint test cycle to introduce intergranular heat affected zone (HAZ) cracks. These cracks were then mechanically opened for SEM examination. Such examination allowed them to be correlated with fracture surfaces from the stress rupture tests. This evaluation showed the the HAZ cracks were characteristic of a liquid phase cracking mechanism.

The potential for a HAZ crack to initiate a fatigue crack was evaluated by precracking a specimen in the varestraint machine and then running a fatigue test. A strain of 1% was used in the varestraint test to minimize the HAZ precracking. These samples were then tested in tension-tension fatigue to minimize surface damage during the test. These tests were carried to failure and the fracture surfaces examined in an SEM. It was found that an intergranular fatigue crack initiated from the HAZ liquid-phase crack and intergranular fatigue striations were clearly visible.

These tests have shown the potential for an intergranular hot tear, produced in the presence of a liquid phase, to initiate an intergranular fatique crack. It has also provided fracture surface characterization of the material near the solidus temperature where hot cracks of different origins tend to develop.

MANY QUESTIONS ARISE concerning the causes of intergranular hot tears during welding and their relationship to fatigue crack initiation. The questions concerning hot tears is whether they initiate due to intergranular liquation or solid state creep mechanisms. The questions regarding fatigue initiation are usually related to the potential of the microcrack to initiate a larger fatigue crack.

The formation of intergranular hot cracks near or above the solidus of a metal can result from mechanisms such as creep, stress rupture, or grain boundary liquation. During the welding process, this type of crack can form in the heat affected zone (HAZ) adjacent to the molten weld pool. A severe crack which produces a significant stress intensity according to linear elastic fracture mechanics (LEFM) through an equation of the form

$$K_t = Y \sigma_d \sqrt{\pi a} \qquad \text{Eq. (1)}$$

(K_t = crack tip stress intensity; σ_d = design stress, a = 1/2 the crack length, Y = geometrical factor) will have obvious consequences and must be repaired. The same is true for cracks which produce pressure or gas leaks.

When the crack is very small, about equal to the grain size, it may not be of immediate importance. Such hot cracks in the HAZ are often termed microfissures. The concern with microfissures is their ability or inability to develop into larger fatigue or creep-fatigue cracks. In the case of fatigue, the microfissure might be expected to affect the fatigue threshold stress (ΔK_{th}) since

$$\Delta K_{th} = Y \Delta \sigma \sqrt{\pi a} \qquad \text{Eq. (2)}$$

(ΔK_{th} = the minimum level of alternating crack tip stress intensity required to initiate fatigue crack growth; $\Delta \sigma$ is the alternating stress associated with ΔK.)

The size of the microfissures will have an obvious effect on K_t through the crack length term, a. It is less obvious how the microfissures will affect terms such as ΔK_{th} as a result of the fatigue crack initiation path and the sharpness of the microfissure crack tip. One previous study (1) on fatigue crack growth from microfissures found that:

- microfissures act as fatigue initiation sites
- the fatigue crack initiates transgranularly from the intergranular microfissure
- microfissures reduced the fatigue endurance limit and the fatigue life

It could be concluded from this study that the primary effect of microfissures on fatigue initiation is related to their effect as a stress concentrator associated with their crack length (a). This conclusion is reached because the fatigue crack initiated and propagated transgranularly just as it would if microfissures were not present.

The present study examines fatigue initiation from microfissures in wrought and cast alloy 718 while the previous work examined forged alloy 718. The present work also examines the mechanism of hot crack formation in alloy 718 and this is related back to the fatigue initiation process.

EXPERIMENTAL PROCEDURES AND RESULTS

HOT CRACK FORMATION - The mechanism of hot crack formation in alloy 718 was studied using hot tensile tests. Material for the hot tensile tests was Haynes alloy 718, heat HT29C4-EK, solution annealed at 1750F with an initial grain size of ASTM# 7. The 24" x 1" x 0.25" tensile specimens were vacuum annealed at 2000F for 90 minutes. The resultant material had an ASTM# 4 grain size.

The temperature profile of the jacket furnace which surrounded the specimen showed a three inch long hot spot from which data could be taken. Temperature was monitored by a Pt-Pt10Rh thermocouple mechanically held against the specimen. Specimens were brought to test temperature at approximately 200F/minute and soaked at the test

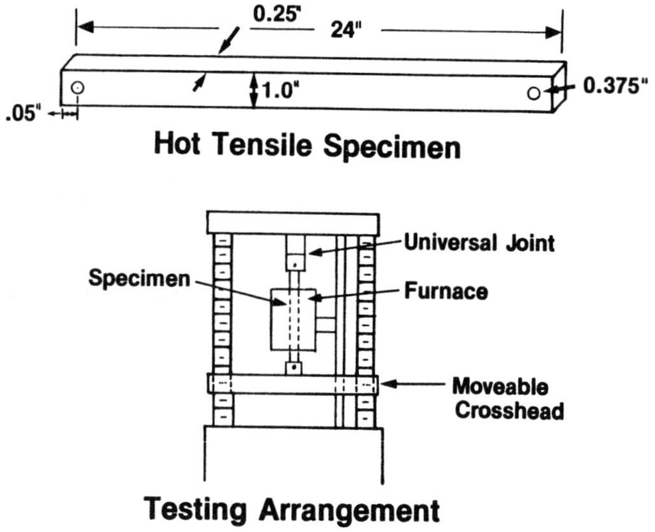

Fig. 1 - Hot tensile specimen and testing arrangement used in the hot tensile tests.

temperature for five minutes before mechanical testing. A cross head displacement rate of 1 inch/minute was used in all tests (Figure 1).

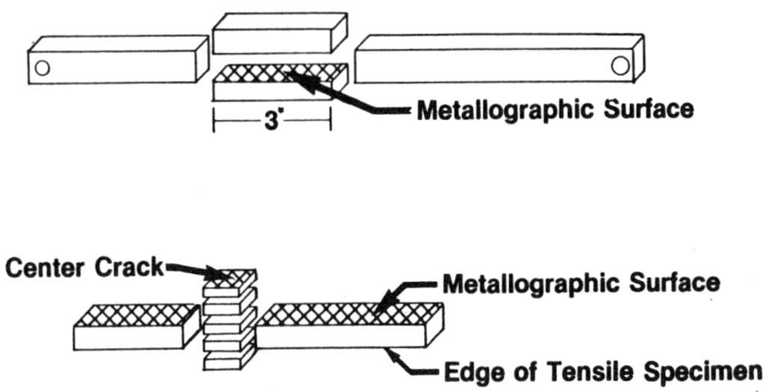

Fig. 2 - Schematic of the sectioning method used to evaluate cracking in the hot tensile test.

The percent of plastic strain was measured by comparing the length of gage marks along the specimen surface before and after the test. Gage marks were approximately 0.14 inches in length and were measured to $\pm$ 0.00025 inches. The permanent plastic strain was thus measured to $\pm$ 0.3%. The constant temperature portion of the bar was sectioned as in Figure 2 for metallographic inspection. Metallographic surfaces were ground, polished and lightly etched in oxalic acid to reveal cracks. A typical micrograph of a specimen with multiple cracks is shown in Figure 3. The cracks in each sample were counted and this data is tabulated in Table 1.

Table 1 - Hot Tensile Test Data

Specimen	Test Temperature	Maximum Strain (% εp)	Number of Cracks Large	Medium	Small
20	2100	5 %	0	1	3
22	2100	1.4%	0	0	0
13	2150	26 %		Fracture	
18	2150	4.5%	1	3	4
19	2150	1.9%	0	0	3
14	2175	11 %		Fracture	
16	2175	5.6%	6	6	11
17	2175	4 %	3	1	4
21	2175	1.7%	0	0	4
4	2200	0 %		Fracture	
9	2200	3 %		Fracture	
15	2200	0.7%		Fracture	
10	2300	0 %		Fracture	
2	2350	4.5%		Fracture	
11	2350	3 %		Fracture	

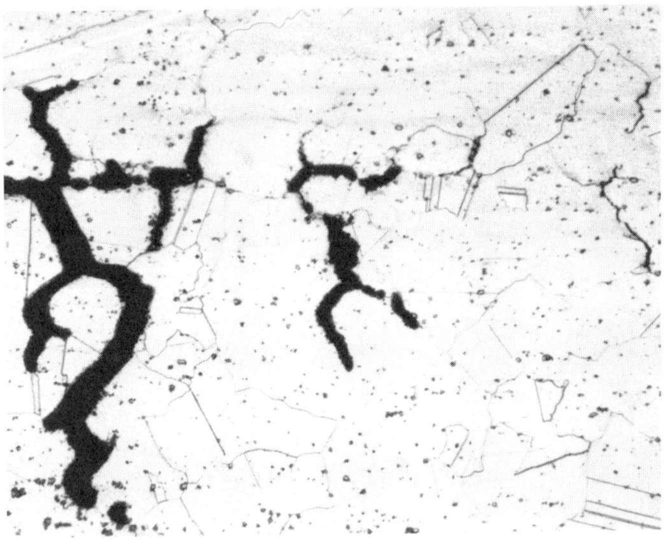

Fig. 3 -Photomicrograph of a section from the hot tensile test showing typical range of intergranular cracks in the multiple-crack temperature range. 200X.

FATIGUE TESTS - The specimen preparation for fatigue testing involved preparing a sample with microfissures in the spot varestraint machine (Tigamajig). The parameters of the spot varestraint test are given in Table 2. Following the spot varestraint test, the weld pool was drilled out of the specimen as illustrated in Figure 4. This produced a fatigue sample that can be evaluated as an edge crack emanating from a hole. The samples were tested in tension-tension fatigue using the parameters given in Table 3.

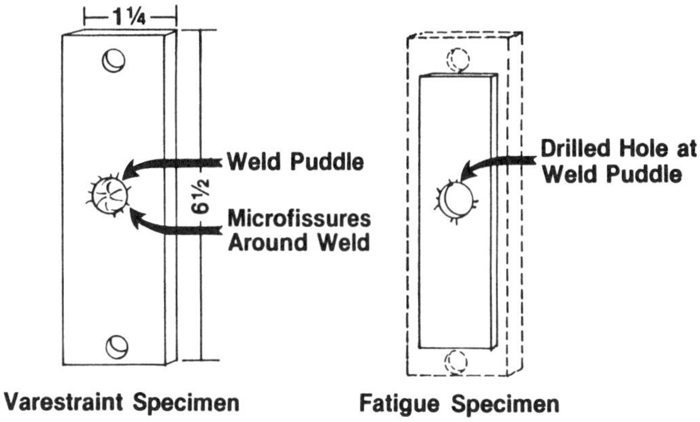

Fig. 4 - Schematic of the fatigue specimen and its relationship to the varestraint specimen from which it is taken.

The fracture surfaces were examined in an SEM following fatigue failure. Figures 5 and 6 show the fatigue initiation from a microfissure in a wrought alloy 718 and Figure 7 in a cast alloy 718.

Fig. 5 -SEM micrograph of wrought alloy following fatigue fracture. The weld zone has been left in place to show its relationship to the HAZ microfissure. 20X.

Table 2 - Varestraint Test Parameters

Welding Current, A	65
Welding Voltage, V	12.5
Arc on Time, sec	10
Argon Shielding, Cfh	30
Electrode to work distance(a)mm(in.)	6.2 (.245)

Table 3 - Fatigue Test Parameters

Cast Alloy Frequency 40 cycles/sec.	Nominal Stress max. (PSI)	Nominal Stress min. (PSI)	Cycles (x 10^6)
	10,000	1,000	0.3.5
	16,000	1,600	3.5-5.5
	22,000	2,200	5.5-7.5
	28,000	2,800	7.5-Failure

DISCUSSION

The hot tensile tests provided valuable insight into the fracture surface characteristics that were found when a sample failed with a solid state boundary as compared to grain boundaries in the liquid state. One question which arose early in the study was, with what confidence can an intergranular fracture be classified as solid state or liquid state failure? Following a correlation of hot tensile fracture data with fracture surface characteristics, it was believed that the change from solid state to liquid state intergranular fracture could be identified with confidence.

The hot tensile fracture data of Table 1 is presented in Figure 8 as fracture ductility versus temperature data. Figure 8 shows the transition from a condition where multiple intergranular cracks form and grow (Figure 3) to a condition where a single intergranular crack forms and propagates to failure. Figures 9a and 9b show the intergranular fracture surface associated with each condition. Figure 9a shows that slip bands and surface tearing, characteristic of solid-to-solid bonding across the grain boundary interface, are present on the intergranular fracture surface in the multiple crack growth temperature

Fig. 6 - SEM micrographs of wrought alloy showing transition from liquid phase microfissure to intergranular fatigue initiation to transgranular propagation.

Fig. 6a - Intergranular microfissure and fatigue initiation zone. 100X.

Fig. 6b - Intergranular microfissure zone next to fatigue initiation. Note fracture surface is typical of liquid phase hot crack. 1000X.

Fig. 6c - Intergranular fatigue initiation zone. Note grain boundary edge in upper third of figure. 2000X.

range. Figure 9b shows a very smooth, glassy intergranular surface which was characteristic of the transition to single crack growth. These features were interpreted in the following way:

- While solid-to-solid bonding existed across the grain boundary, intergranular cracks are easily initiated near the solidus temperature. The ease of this crack initiation increases with increasing temperature, however, the propagation of these cracks is somewhat inhibited by plastic blunting of the crack tips. This leads to multiple crack initiation and a fracture surface characterized by surface tearing and slip bands.
- When intergranular liquid wets the grain boundary, a transition occurs which allows the first intergranular crack that initiates to run to complete fracture of the test specimen. This implies that all of the grain boundaries are immediately wet by liquid at the transition temperature and the propagating crack is not interrupted by solid-to-solid bonding. This liquid wetting leads to the glassy smooth appearance of the intergranular surface in the single crack growth temperature range.

These observations of hot tensile fracture in alloy 718 have implications to the analysis of intergranular hot cracks in the HAZ of alloy 718 welds. Consider the situation shown in Figure 10 for the distribution of intergranular liquid as a function of temperature. The hot tensile data suggests that the temperature range between no wetting and complete wetting is no more than a few degrees. Given the steep thermal gradients typically used to weld superalloys, the dimension in the HAZ between the condition of complete wetting and no wetting would be very small for the wrought alloy tested. A reasonable assumption would be that liquid intergranular wetting stops at one grain boundary and solid-to-solid bonding begins at the next boundary.

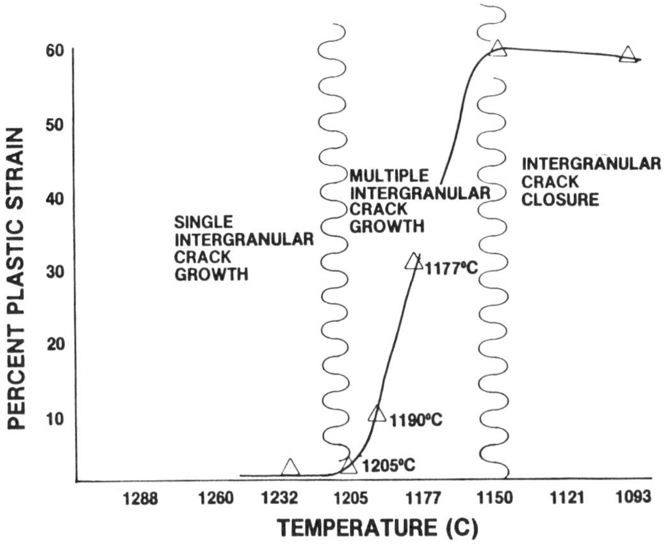

Fig. 8 - Hot tensile data of fractured specimens showing plastic strain at failure as a function of temperature. The division between multiple crack growth and single crack growth is associated with initial liquation on the grain boundaries.

Fig. 7a - SEM fracture surface of cast alloy near fatigue initiation area. Note change in orientation of fatigue striations on different grain surfaces.

Fig. 7b - SEM fracture surface from cast alloy at a great distance from initiation zone, i.e., during intergranular crack propagation. Note dendrite arms and large precipitates. 1000X.

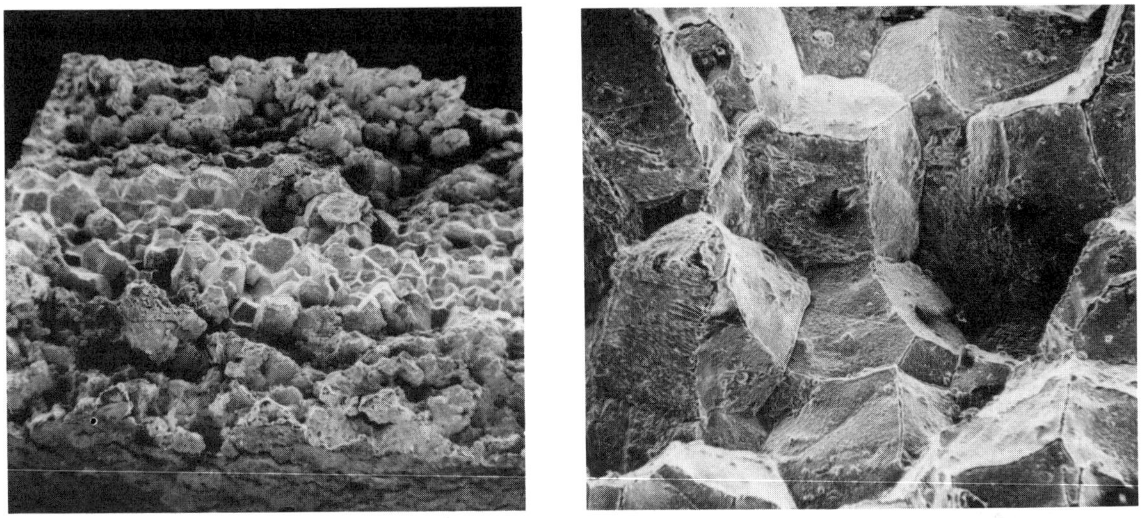

Fig. 9a - Typical SEM fracture surfaces from the multiple crack temperature range of the hot tensile test. Note deformation markings on grain boundary surfaces. 10X and 150X.

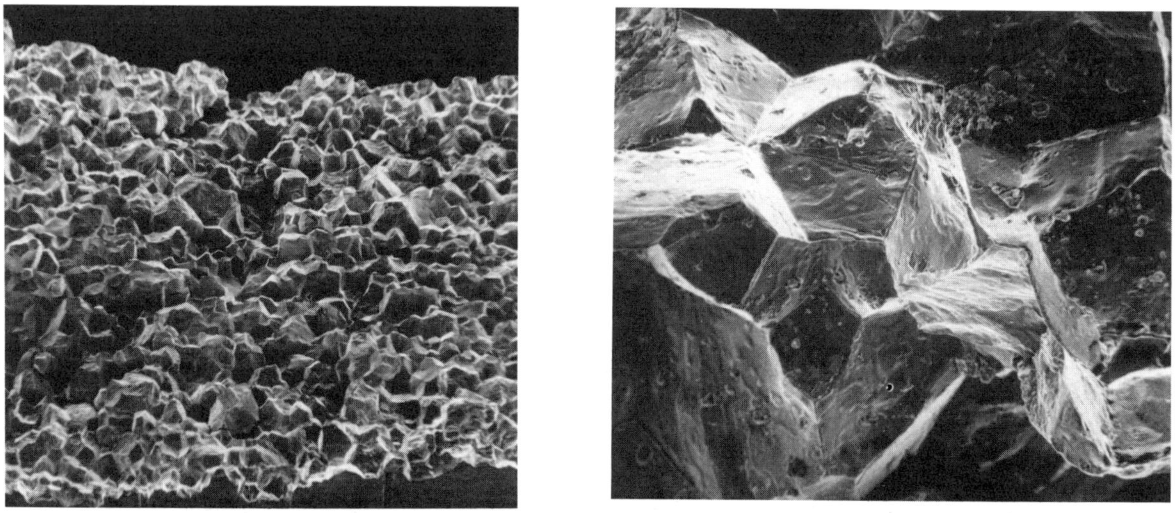

Fig. 9b - Typical SEM fracture surfaces from the single crack growth temperature range. Note smooth grain boundary surfaces. 10X and 150X.

A hot crack, initiated in the thin liquid film of a HAZ grain boundary, would run immediately to the grain containing the interface between the liquid grain boundary film and the solid-to-solid bonding. As will be shown later, the hot crack was found to stop at the solid bonding in wrought 718. This then limits the maximum microfissure length to the length of liquated grain boundary in the HAZ which is established by grain boundary chemistry and weld thermal gradient.

FATIGUE PROPAGATION FROM HAZ HOT CRACKS - Figures 6a, 6b, and 6c show that the fatigue initiation process in the wrought alloy of this study was intergranular in nature. However, the intergranular fatigue initiation zone was only a few microns long at which point the crack moved to a transgranular propagation mode. This is contrary to a previous study on forged alloy 718 in which room temperature fatigue initiation from a HAZ microfissure was transgranular. This is probably indicative of the reluctance of wrought or forged alloys to propagate an intergranular fracture at low temperature.

It also suggests that the fatigue initiation path in wrought alloy is sensitive to test conditions and/or material properties such as grain size. However, no special significance was associated with the differing observations in these two studies.

The results from the fatigue test of the cast alloy were very different from those of the wrought alloy. The cast alloy initiated an intergranular fatigue crack (Figure 7A) and propagated the crack intergranularly (Figure 7B) until failure. While the wrought alloy was very resistant to intergranular crack propagation, the cast alloy showed no resistance to intergranular microfissures and fatigue propagation.

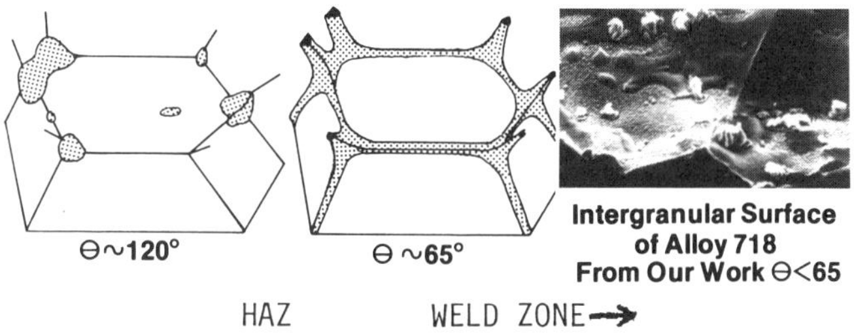

Fig. 10 - Schematic of microfissure propagating in HAZ until it stops at point where liquid no longer wets the grain boundary.

MICROFISSURES AND FATIGUE - The following discussion is based on interpretations of chapters 12 and 13 from Hertzberg's (2), "Deformation and Fracture Mechanics of Engineering Materials".

FATIGUE GROWTH OF PHYSICALLY SHORT CRACKS (MICROFISSURES) - It has been found (3) that very small cracks can grow at ΔK levels well below the ΔK_{th} predicted from LEFM concepts. A HAZ microfissure, being approximately equal to the material's grain size, would certainly qualify as a very small crack. From a design standpoint, one problem with a microfissure is the lack of data on the behavior of small cracks in fatigue. Another problem is how to define a small crack. Most data (4) points to the definition of a small crack in terms of some appropriate microstructural dimension such as grain size. This type of definition would qualify a microfissure as a very small crack relative to larger cracks which are used in LEFM testing which is used to determine da/dN versus ΔK curves.

When very small cracks exist in a fatigue situation, the fatigue behavior may be altered as depicted in Figure 11. As the crack gets large, it behaves just as predicted from LEFM. However, in the early stages of growth the threshold, or fatigue limit, is reduced relative to what would be found in a standard LEFM fatigue test. Thus, a specimen containing microfissures would be expected to behave differently from a standard LEFM fatigue test.

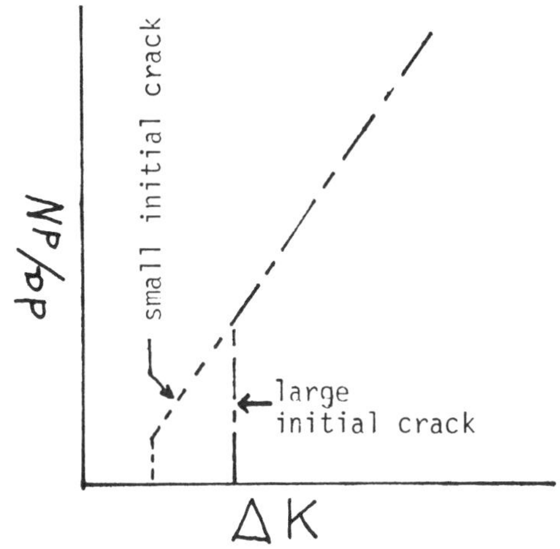

Fig. 11 - Schematic interpretation of how microfissures could alter fatigue threshold.

FATIGUE INITIATION AND PROPAGATION PATH - The observations of intergranular fatigue initiation (wrought and cast 718) and intergranular propagation (cast) would be expected to produce fatigue behavior different from transgranular fatigue. Many observations (5) have been made on the effect of microstructure on fatigue behavior. Microstructure affects both the fatigue threshold and the fatigue propagation rate. It has also been shown (5) that fatigue crack propagation rate was dependent on the microstructural path that the crack follows. The intergranular path seen in the cast alloy (Figures 7A & 7B) was obviously more favorable than the transgranular path under the conditions of this test. This could lead to a reduced threshold stress and higher crack propagation rate for the intergranular path as compared to the transgranular path in the cast alloy. This is depicted schematically in Figure 12.

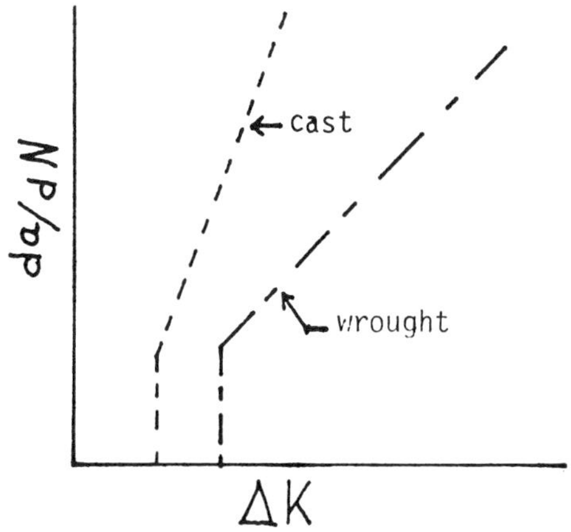

Fig. 12 - Schematic interpretation of how easy intergranular initiation and propagation in cast alloy could alter fatigue properties compared to wrought alloy.

SUMMATION

COMBINED EFFECTS OF MICROFISSURES ON FATIGUE - The following summation is based on very little experimental data and is largely an interpretation of how HAZ microfissures might affect fatigue behavior based on other studies.

The length of the longest HAZ microfissure will be equal to the distance over which liquid wets the grain boundaries at the time the microfissure is formed. These cracks will be small and intergranular.

Since the intergranular microfissures initiated intergranular fatigue cracks, it is expected that the fatigue initiation time is completely eliminated. This would be a significant change from smooth fatigue specimens and possibly from microfissures that initiate transgranular fatigue cracks.

The fatigue strength ($\Delta\sigma_{fat}$) and threshold stress intensity range (ΔK_{th}) have both been shown to be influenced by very small cracks such as microfissures. They have also been shown to be influenced by the crack path related to microstructure. The conditions that exist with microfissures and intergranular fatigue crack initiation are consistant with an unexpectedly low fatigue threshold stress.

Transgranular fatigue crack propagation as seen in wrought alloy 718 would be expected to be typical of fatigue propagation reported elsewhere. However, the intergranular fatigue crack propagation found in the cast alloy could be expected to show an increased propagation rate over transgranular propagation.

The above interpretations, taken together, are shown schematically in Figure 13.

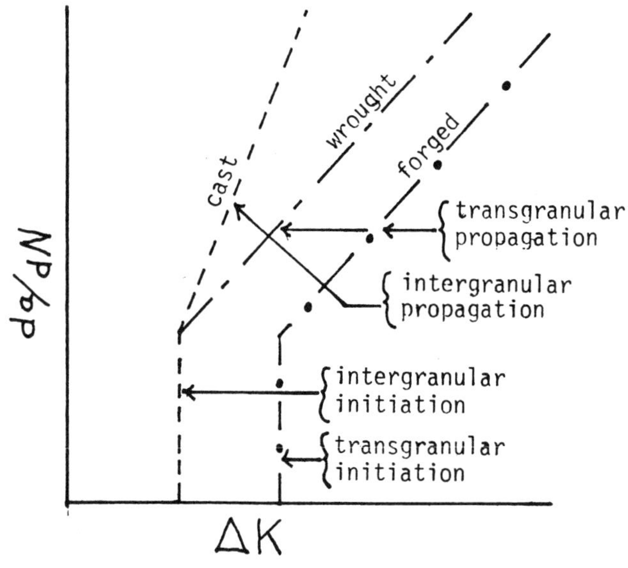

Fig. 13 - Schematic summary interpretation of fatigue differences between cast (intergranular initiation and propagation), wrought (intergranular initiation but transgranular propagation), and forged (transgranular initiation and propagation) conditions in alloy 718.

BIBLIOGRAPHY

1. "High Cycle Fatigue Testing of Inconel 718 EB Weldments with Microfissures", Lassila, D. H., Rockwell International, MPR81-0401,SSME81-0996.

2. R. W. Hertzberg, "Deformation and Fracture Mechanics of Engineering Materials", Second Edition, John Wiley & Sons, 1983 (ISBN 0-471-08609-6).

3. ibid, pg. 548, Section 13.4.2.

4. ibid, pg. 557, Section 13.4.3.

5. ibid, pg. 581, Section 13.7.

ACKNOWLEDGEMENTS

The authors would like to thank NASA, Marshall Space Flight Center; the National Science Foundation; and General Electric (Gas Turbine Division) for their support of various aspects of this research. The authors would also like to thank Mr. Philip Wolf and Mr. Edwin Bradley for their assistance in the study. Finally, the authors express their appreciation to Mrs. Pat Hamaker for her work in preparing this manuscript.

MICROSTRUCTURE AND MECHANICAL PROPERTIES OF SELF-SHIELDED FLUX-CORED ARC WELD METAL

H.D. Mejías,
División Soldadura,
INTEMA-CONICET Mar del Plata;
(7600) Mar del Plata, Argentina

T.E. Pérez
División Soldadura, Comisión Nacional de Energí
Atómica (CNEA); Avda. del Libertador 8250
(1429) Buenos Aires, Argentina

L. De Vedia
Escuela de Soldadura, CONARCO
Calle 18, Nº4113
(1672) Buenos Aires, Argentina

ABSTRACT

In the last years self - shielded arc process has been increasingly used as a site welding process. This paper describes a characterization of mechanical and microstructural properties of self - shielded AWS E61T8-K6, E71T-8, E71T8-K6 and E71T8-Nil type electrodes. They were selected because they are widely used in structural welding and they have high toughness. All weld metal specimens were prepared with each electrode. To study the influence of the heat input welds were made at three heat input levels. Welds were examined in respect of tensile strength Charpy V impact strenght, hardness and microstructure. Crack opening displacement (COD) test was also used to evaluate fracture mechanical properties. The four electrodes presented good toughness; in each case the highest values correspond to the lowest heat input. The E71T8-K6 electrode has the better combination of tensile strenght and fracture toughness and the lowest sensitivity to heat input. The microstructures were fundamentaly primary ferrite and ferrite with aligned second phase. The presence of acicular ferrite as the majority microphase does not seem to be the condition to get good toughness as it is usual in manual metal arc or submerged arc welds. The mechanism suitable to this case is not clear.

IN THE LAST YEARS A SIGNIFICANT INCREASE in the use of welding processes which employ self shielded flux cored arc welding electrodes as consumables has been noticed (1). In these arc processes the protective atmosphere is provided by the flux contained in the wire, thus avoiding the costs and problems involved in the use and transport of the shielding gases commonly employed in semiautomatic welding. This makes self shielded electrodes adequate for their use in "in situ" fabrication where the environmental conditions affect the performance of other processes. Since the alloying elements may be directly added to the flux, it is possible to obtain in an economic way a wide variety of compositions and electrode types, employing the same low carbon sheath.

In the case of self shielded flux cored arc welding electrodes the filler components must insure the production of a weld free of porosity, and with suitable mechanical properties. The alloying elements must be those necessary to provide a desirable microstructure and, at the same time, warrant an adequate protection against the atmospheric gases. The deoxidizers most commonly used are: aluminium, titanium, silicon, magnesium and calcium. Aluminium, titanium and silicon are partially incorporated to the weld metal affecting its toughness. Calcium and magnesium have a low boiling point which makes them less effective in the arc (2).

The dissolved nitrogen is the main cause of porosity, so that strong nitride formers must be added: aluminium is the most frequently used, since titanium tends to reduce the toughness (3). Aluminium has also some deleterious effects, and for this reason its addition should be controlled.

The effect of aluminium on the mechanical properties can be attributed to the following mechanisms:

a. The aluminium affects the decomposition of the austenite, promoting the formation of ferrite with aligned second phase structure. It acts on the microstructure in two ways:

- Its presence in solid solution modifies the transformation curves
- Al has influence on the type, shape, size and distribution of non metallic inclusions. The role of the inclusions in the control of the austenitic grain size (4), as well as their effectiveness as nucleation sites of acicular ferrite (5, 6, 7) has been studied by several authors.

b. Aluminium stabilizes the ferrite phase: in the binary Al-Fe diagram, 1.2% Al is enough to fully suppress the austenite field. In a steel, its effect will depend on the concentration of other alloying elements, such as carbon, manganese and nickel. The presence of a high content of aluminium can lead to a solidification in the δ-phase, without subsequent solid state transformations. These zones will present a final coarse grained microstructure since refinement due to successive solid state transformations is not possible.

c. It eliminates the interstitial nitrogen, improving the impact properties, an excess of aluminium can lead to hardening by solid solution; about a certain level (aprox. >1%)(8), this effect can suppress the advantage achieved by nitrogen elimination.

Slag and gas formers like fluorides, metallic lithium and CO_3Ca are also used to avoid the nitrogen incorporation of the weld pool.

This paper describes a characterization of mechanical and microstructural properties of self shielded AWS E61T8-K6, E71T-8, E71T8-K6 and E71T8-Ni1 type electrodes (9, 10). These electrodes were selected taking into account their application in the structural welding and their high toughness.

EXPERIMENTAL PROCEDURE

In the present work four commercial self shielded flux cored arc welding electrodes conforming to AWS E61T8-K6, E71T-8, E71T8-K6 and E71t8-Ni1 type were used.

All weld metal specimens were prepared for each electrode according to AWS A5.20-79 and A5.29-80 specifications (9, 10). Welding parameters are indicated in Table 1. Voltage and current settings were those recommended by the manufacturer. Angular movement of the base plates was restrained by means of clamps bolted to a rigid welding jig specially built for that purpose. The joint was completed with two passes per layer using a maximum interpass temperature of 150°C. All weld metal tensile and impact specimens were machined and a full Charpy V energy curve between -80°C and 20°C was obtained for each electrode.

Table 1

Pass	Heat Input (HI)(KJ/mm)			
	E61T8-K6	E71T8	E71T8-K6	E71T8-Ni 1
1	1.8	1.6	1.5	1.5
2	2.8	2.2	2.2	2.1
3	2.1	2.3	2.3	2.0
4	2.9	2.2	3.3	2.3
5	2.2	2.7	2.9	2.2
6	3.3	2.7	4.0	2.8
7	2.7	2.7	2.6	2.7
8	3.3	2.2	-	2.8

Wire E61T8-K6 V = 20.5 volt I = 235 Amp
E71T8 V = 20 volt I = 220 Amp
E71T8-K6 V = 20 volt I = 230 Amp
E71T8-Ni 1 V = 20 volt I = 230 Amp

To study the influence of the heat input on the fracture toughness production type welds were used. Plates 19 mm thick were bevelled to form a 70 deg. single V. For each electrode three samples with low, medium and high heat input were produced in the flat position. In all cases a backing pass was carried out and then it was back gouging. The samples with low heat input were made with stringer beads. To obtain the higher heat input a weaving technique was used in order to minimize the welding speed. The medium heat input was obtained using a limited weaving. Interpass temperature was similar to that employed in the all weld metal samples. The welding parameters used are shown in Table II.

Table II

Wire	Pass	HI	Wire	Pass	HI
E61T8-K6 (hHI)	1	2.26	E71T8 (hHI)	1	2.02
	2	3.18		2	2.59
	3	3.07		3	2.65
	4	-		4	3.01
(mHI)	1	2.26	(mHI)	1	1.49
	2	2.86		2	2.01
	3	2.8		3	2.19
	4	2.7		4	1.38
	5	-		5	1.76
(lHI)	1	1.2	(lHI)	1	1.14
	2	2.25		2	1.86
	3	1.16		3	1.01
	4	2.01		4	1.29
	5	1.61		5	1.5
	6	2.07		6	1.79
	7	-		7	1.41
E71T8-K6 (hHI)	1	2.2	E71T8-Ni1 (hHI)	1	2.76
	2	3.21		2	3.06
	3	3.78		3	3.93
(mHI)	1	2.4	(mHI)	1	2.23
	2	2.5		2	2.6
	3	2.44		3	2.36
	4	2.87		4	2.78
(lHI)	1	1.69	(lHI)	1	1.2
	2	1.96		2	1.84
	3	1.28		3	1.28
	4	1.49		4	1.41
	5	1.33		5	1.29
	6	1.52		6	1.38

Wire: E61T8-K6 V = 19.5 volt I = 220 Amp
E71T8 V = 20 volt I = 230 Amp
E71T8-K6 V = 22 volt I = 250 Amp
E71T8-Ni1 V = 20 volt I = 230 Amp
HI = heat input (KJ/mm) hHI = high heat input
mHI = medium heat input lHI = low heat input

From each weld, subsurface Charpy V standard specimens and three full thickness COD specimens were machined. The transition curve was obtained for each experimental condition. COD tests were performed at -10°C, using a three point bend type specimen according to the BS 5762-79 (11). A notch in the through thickness direction, with a fatigue precrack and an initial crack length specimen width ratio (a/w) equal to 0.5 was used.

The diamond pyramid hardness was measured at a load of 5 Kg on the top bead of polished transverse section of the welds. Transverse sections of the weldments were conventionally prepared for metallographic examination. Macrostructure was revealed using 10% Nital. The percentage of heat affected weld metal zone was determined using line intercept technique. The microstructure of the top bead was quantified according to (12, 13). These analysis were performed in all weld metal specimens and also in production type welds.

EDAX system was used for inclusion qualitative analysis in all weld metal specimens. The samples were previously deep etched in a mixture of bromine and methanol. The size distribution and area percentage of inclusions were also determined.

RESULTS

The chemical analysis of all weld metal deposits are given in Table III.

Table III

Element	E61T8-K6	E71T8	E71T8-K6	E71T8-Nil
P	0.023	0.007	0.005	0.006
S	0.007	0.006	0.005	0.006
Sn	0.007	<0.01	<0.01	<0.01
C	0.08	0.08	0.08	0.09
As	<0.01	<0.01	<0.01	<0.01
W	<0.01	<0.01	<0.01	<0.01
Si	0.06	0.22	0.04	0.31
Cu	0.02	0.01	0.02	0.02
Co	<0.01	<0.01	<0.01	<0.01
Ni	0.57	0.02	0.73	0.86
Cr	0.02	0.03	0.02	0.03
Mo	0.02	0.02	0.02	0.04
Mn	0.80	1.48	0.87	1.14
V	<0.01	<0.01	<0.01	<0.01
Nb	<0.01	<0.01	<0.01	<0.01
Ti	<0.01	<0.01	0.02	<0.01
Al	0.91	1.06	0.66	1.00
O_2	0.0094	0.0065	0.0136	0.0113
N_2	0.0353	0.0333	0.0306	0.0288

The main components of columnar region for the different all weld metals are: primary ferrite (PF)(intergranular poligonal, PF(I), and grain boundary ferrite, PF(G)), ferrite with second phase (FS) and acicular ferrite (AF), Fig. 1. The percentage of different components for all weld metals and samples with different heat inputs are summarized in Table IV. The percentages of weld metal heat affected zone (R) are also shown.

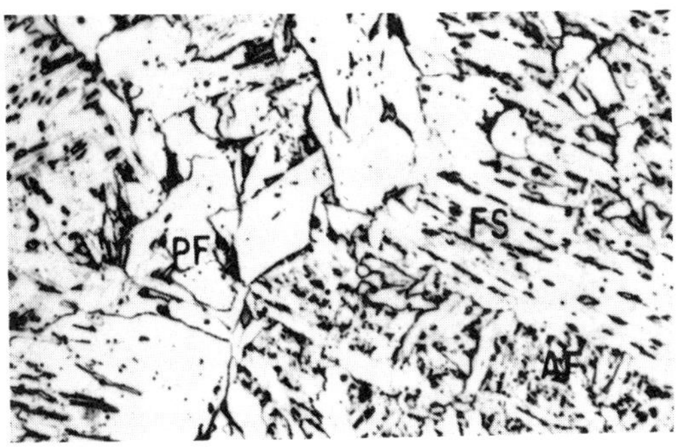

Fig. 1 Typical components of columnar regions.

Table IV

Wire	Weld. proc.	Microstructures PF(I)	PF(G)	AF	FS	%R	DPHN (5Kg)
A	awm	69	2.5	15.5	13	30	167
	hHI	78.7	2.6	12.7	6	30	185
	mHI	67	2	15.5	15.5	32	196
	lHi	67	5	12.5	15.5	27	188
B	awm	49	4	28	19	38	196
	hHI	73.5	2	15	9.5	30	169
	mHI	49	17	15	19	35	191
	lHI	47	12.5	28	12.5	47	207
C	awm	43	9	27	21	53	200
	hHI	77.4	1.2	15	6.4	25	162
	mHI	78.8	0.4	16.4	4.4	36	168
	lHI	46	10.5	31	12.5	32	192
D	awm	54	9	6	31	32	214
	hHI	61	3	21	15	25	201
	mHI	81	5.3	5.7	8	29	198
	lHI	51	5	30	14	34	207

awm = all weld metal
%R = percentage of weld metal heat affected zone
A = E61T8-K6 B = E71T8
C = E71T8-K6 D = E71T8-Ni 1

In every case the structures are fundamentally composed by gross primary ferrite and ferrite with aligned second phase. The percentages of acicular ferrite are low and tend to increase as the heat input decreases.

The majority of the inclusions were principally composed by Al or Al-Mg. The mean diameter and the percentage of inclusions area are presented in Table V.

Table V

Sample	E61T8-K6	E71T8	E71T8-K6	E71T8-Nil
% Area	0.5	0.83	0.22	0.6
D (μ)	0.9	1.1	0.6	0.7

The all weld metal tensile tests results are presented in Table VI. The electrode E61T8 showed the largest reduction in area but a lower yield strength and ultimate tensile strength. The highest strength values corresponded to the E71T8 electrode.

Table VI

Electrode	YS N/mm^2	UTS N/mm^2	RA %	Elong. %
E61T8-K6	367	475	70	30
E71T8	466	505	38	43
E71T8-K6	423	493	58	33
E71T8-Nil	458	544	66	21

Figure 2 shows the all weld metal Charpy V transiton curves obtained for the different electrodes. It is seen that E61T8-K6 and E71T8-K6 electrones showed the highest levels of absorbed energy. The results of hardness tests are summarized in Table IV.

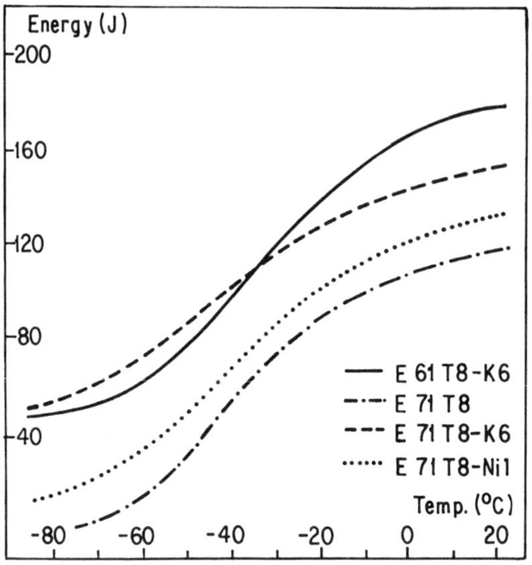

Fig. 2 All weld metal Charpy V transition curves.

The Charpy V notch impact curves obtained for different heat inputs are shown in figure 3.

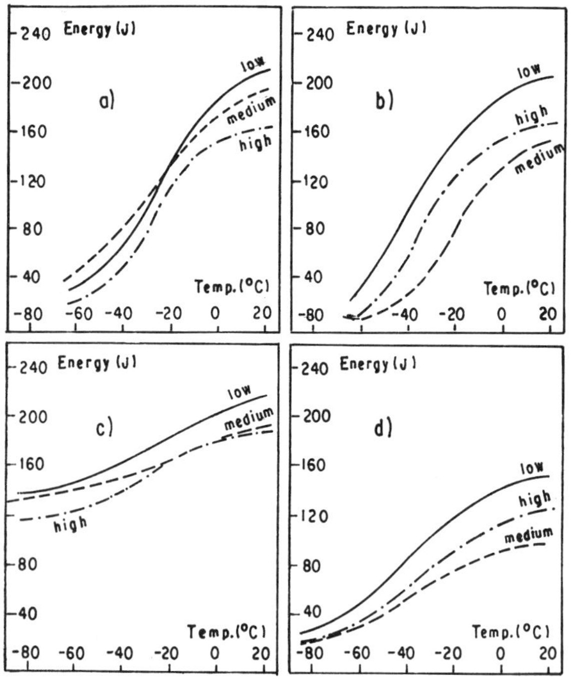

Fig. 3 Charpy V transitions curves for different heat inputs.
a) E61T8-K6 b) E71T8
c) E71T8-K6 d) E71T8-Nil

The transition temperatures corresponding to absorbed energy levels of 50 J and 100 J are indicated in Table VII.

COD test results are given in Table VIII. In most cases critical COD values are referred to the first attainment of a maximum load plateau (δm); only eight values over thirty three correspond to fracture after a certain amount of stable tearing (δu).

Table VII

Electrode	Absorbed energy	Transition temperature (°C)			
		awm	1HI	mHI	hHI
E61T8-K6	50	-75	-46	-54.2	-40
	100	-39	-30.1	-34	-27.3
E71T8	50	-41.3	-54	-29	-42.3
	100	-10.7	-39	-15.3	-29
E71T8-K6	50	-83.7	<-80	<-80	<-80
	100	-41.3	<-80	<-80	<-80
E71T8-Nil	50	-48.7	-58	-41.7	-46.3
	100	-20.7	-31.7	+15.7	-14

Table VIII

Electrode	Welding procedure	Average COD
E61T8-K6	high HI	1.064
	medium HI	0.864
	low HI	0.94
E71T8	high HI	1.157
	medium HI	0.638
	low HI	1.388
E71T8-K6	high HI	1.48
	medium HI	1.459
	low HI	1.453
E71T8-Nil	high HI	1.12
	medium HI	1.286
	low HI	1.04

DISCUSSION

In general, the use of a low heat input procedure seemed to favour the achievement of low transition temperatures and high absorbed energies in the upper shelf.

The E71T8-K6 electrode presented the best combination of tensile strength and fracture toughness. This electrode showed the lowest transition temperatures in the all weld metal specimens and in the production type joints. It is also worth noting that the E71T8-K6 electrode exhibited the lowest sensitivity to heat input. Average COD values were all of the m type and virtually unaffected by changes in the heat input, in addition transition temperatures were mantained at a very low value in all cases for this electrode.

The E71T8-Nil electrode also presented very low transition temperatures at low heat input coupled with good average COD values at -10°C. The COD values for this electrode were with one exception, of the m type and the heat input did not show a significant influence. For these reasons this electrode represents a suitable combination of tensile and fracture properties, particularly if a low heat input procedure can be employed.

Weld deposits produced with the E61T8-K6 electrode, although exhibiting better upper shelf toughness at low heat input than the E71T8-Nil type, presented somewhat higher transition temperatures than this electrode. At the same time, average COD values and tensile strengths were lower for the E61T8-K6 than for the E71T8-Nil electrode.

The E71T8 electrode exhibited very low transition temperatures and high COD values with low heat input. Average COD values were markedly reduced in the case of medium heat input, that resulted in a change of the fracture mode from the δm to δu type. Besides, these results are in agreement with the transition temperature values that for this electrode were maximum at intermediate heat input. All the studied electrodes presented gross structures composed principally by primary ferrite and ferrite with aligned second phase. The latest constituent is promoted by Al when it is present in amounts near 1%. It is worth noting that in spite of this structure the samples exhibed very good impact properties. As another authors pointed out (14), it is seemed that the presence of acicular ferrite as the majority microphase is not the condition to get good toughness as it is usual in manual metal arc (MMA) or submerged arc welding (SAW). The relationship between structure and impact properties that have been well stablished for these processes (15,16) is not clear for self - shielded arc process deposits. The low free nitrogen content that would be expected because of the high Al tenor is benefit but it is unlike that this alone would account for the very good impact properties (14). The different effects of Al will be studied to explain the relation structure - impact properties in this case.

CONCLUSIONS

1. All the wire analysed showed a high toughness in COD tests as in Charpy. The best results were obtained with the lowest heat input.

2. The E71T8-K6 electrode presented the highest values of toughness in both tests. Its behaviour is virtually unaffected by the heat input.

3. As the heat input decreases the acicular ferrite content and the hardness tend to increase.

4. The microstructures were fundamentally primary ferrite and ferrite with aligned second phase. The presence of high percentages of acicular ferrite does not seem to be the condition to get good toughness as it is usual in MMA or SAW. The relation structure - toughness is not clear.

REFERENCES

1. Otegui, J.L. and L.A. De Vedia, Boletín Técnico CONARCO 81, 2-14/27-29 (1985).
2. Dorling, D.V., PhD Thesis Cranfield Institute of Technology (1980).
3. Kotecki, D.J. and A. Moll, Welding Journal 49, 1575-1655 (1970).
4. Cochrane, R., Welding in the World 21, 16-24 (1983)
5. Abson, D. R. Dolby and P. Hart, Int. Conf. on Trends in Steel and Consumables for Welding, London, 75-101 (1978).
6. Ricks, R., P. Howell and G. Banitte, J. Mater. Sci. 17, 732-740 (1982).
7. Cochrane, R. and P. Kirkwood, Int. Conf. on Trends in Steel and Consumables for Welding, London, 103-121 (1978).
8. Irvine, K.J., F.B. Pickering and Gladman, T., IISI 205, part 2, 161-168 (1967).
9. AWS A5.20-79, "Specification for Carbon Steel Electrodes for Flux Cored Arc Welding" American Welding Society (1979).
10. AWS A5.29-80, "Specification for Low Alloy Steel Electrodes for Flux Cored Arc Welding" American Welding Society (1980)
11. BS 5762, "Methods for Crack Opening Displacement (COD) Testing", British Standards Institution (1979).
12. Pérez T. and M. Solari, Proc. V Congreso Argentino de Soldadura (1984).
13. Doc. Int. Welding Institute, IX J 102-85.
14. Haeck R., B. Phelps and A.A. Smith, Welding and Metal Fabrication 48, 441-449 (1980).
15. Evans, G.M., Welding J. 59, 675-755 (1980).
16. Solari, M., T.E. Pérez and J.P. Esperón, Deutscher Verl. Für Schweisstechnik, 27-32 (1982).

DISSIMILAR WELD FAILURE ANALYSIS AND DEVELOPMENT

R.H. Ryder, D.I. Roberts
GA Technologies Incorporated
San Diego, California

R. Viswanathan
Electric Power Research Institute
Palo Alto, California

M. Prager
Metal Properties Council Incorporated
New York, New York

ABSTRACT

Failure of ferritic-austinetic dissimilar welds (DMWs) is a major cause of forced outages in fossil boilers. A program was undertaken by EPRI whose objectives were to develop (1) a clearer understanding of the underlying causes of dissimilar weld failures, (2) a methodology for estimating the residual life of service welds, (3) a critical discriminatory test to predict relative performance of welds, and (4) improved dissimilar welds. The program is now complete and the objectives achieved.

Results of the program show that the durability of dissimilar welds is a function of the filler metal employed and the condition to which the weld is subject in service in terms of time, temperature, cycles and system loads. Accelerated testing methods were developed which involved applying bending and cyclic thermal loads to internally pressurized full-size tubular specimens. Tests had failure times of about 400 hours for stainless filler DMWs and 1500 for nickel-base filler DMWs - about the same ratio observed in service involving significant thermal cycling. A life prediction methodology applicable to stainless filler DMWs was developed. This method (Prediction of Damage in Service - PODIS) uses service conditions as inputs to predict remaining life. Guidance is offered on the potential importance of weld filler selection, weld geometry, heat treatment, etc., in relation to expected service conditions (e.g., severe cyclic vs. normal cyclic) and on how to locate DMWs to optimize service performance. Finally new filler metals were developed which appear to offer the potential for substantial improvements in DMW performance.

DISSIMILAR METAL WELDS (DMWs) are widely used to join ferritic steel to austenitic stainless steel tubing and piping in many high temperature applications in energy conversion systems. The most widespread application is in the superheaters and reheaters of fossil fired electric power generation boilers. The materials joined, in this application, are usually T11, T22 or T9 to 304H, 316H, 321H or 347H. The welds may be made either by induction pressure welding the two materials directly together or by shielded metal arc welding using austenitic stainless steel filler metals or nickel base filler metals. Tube sizes typically lie in the range of 1 in. to 3 in. o.d. Welds of this type have been in use for a great many years. During the late 1950s failures were encountered with a number of welds made using austenitic stainless steel filler. Typically, failure occurred by the development of low ductility cracking in the low alloy steel very close to the weld fusion line. These experiences prompted research into finding better methods of joining the dissimilar materials which led to the use of nickel base filler metals.

The use of nickel base fillers to make DMWs effected a considerable life improvement. However, by the mid-1970s failures were also beginning to occur at a significant rate in DMWs made with nickel base filler. These failures were generally macroscopically similar in appearance to the failures that occurred in DMWs made with austenitic stainless steel filler metals. The forced outage costs resulting from these increased failure rates caused increased utility attention, world wide. In response to this growing concern, a program (RP1874) aimed at identifying problem causes and finding long term solutions was developed. The program, which is funded by EPRI and managed by The Metal Properties Council, was initiated in September of 1980 and is now essentially complete. The intent of this paper is to summarize and review the results of the program.

RESEARCH APPROACH. The objectives of RP1874 were:

1. To obtain a clear understanding of the underlying causes of dissimilar weld failures.

2. To develop a methodology for estimating the residual life of service welds.
3. To identify operating guidelines that would provide maximum weld durability.
4. To devise a critical discriminatory test to predict the relative performance of welds.
5. Based on the discriminatory test, to develop and evaluate optimized weld designs and filler metals.

The program was divided into several tasks which included the following activities: examination of failed and unfailed DMWs from boiler service ranging from 50,000 to 200,000 hours; simulation of the metallurgical structures that develop in DMWs after long time boiler service; analysis of the operating conditions (temperature, stress, cycling, etc.) that contribute to DMW failure; development of accelerated discriminatory testing methods; development of methods for evaluation of remaining life of DMWs in boiler service, and evaluation of the DMW life benefits to be obtained from the use of different weld geometries and improved filler metals. Detailed results of project RP1874 have been provided elsewhere. This paper will, therefore, summarize only the major results and observations.

FAILURE MECHANISMS. Based on the analysis of numerous failed and unfailed DMWs from a total of 25 boilers (owned and operated by 11 utilities), coupled with the results of accelerated discriminatory tests and a thorough review of other programs world wide, it is concluded that while DMW failures are macroscopically similar, appearing as shown in Fig. 1, there are in fact, three distinct failure modes.

1. Failure occurring by the formation and propagation of cracking along prior austenite grain boundaries in the low alloy steel about one to two grains away from the weld fusion line. This failure mode is commonly observed in DMWs with stainless steel filler and is occasionally observed in nickel filler DMWs.
2. Failure occurring by the formation and propagation of cracking along a planar array of globular carbides that develops, in service, along the weld fusion line. This failure mode is commonly observed in DMWs made with nickel base filler.
3. Failure occurring due to the propagation, through the tube wall, of the oxide notch formed on the outside of the tube at the weld/low-alloy steel junction. Oxide notches are commonly observed in all DMWs after service, but in most cases the notches do not propagate. However, failure by this mechanism can occur in both stainless and nickel filler DMWs independently of damage modes 1 and 2.

The terminal failure mechanisms generally involve the formation of creep cavities adjacent to carbides present at prior austenite grain boundaries and/or the fusion line. Accordingly, there is a strong correlation between carbide morphology and cracking. This is particularly notable in the case of interfacial cracking.

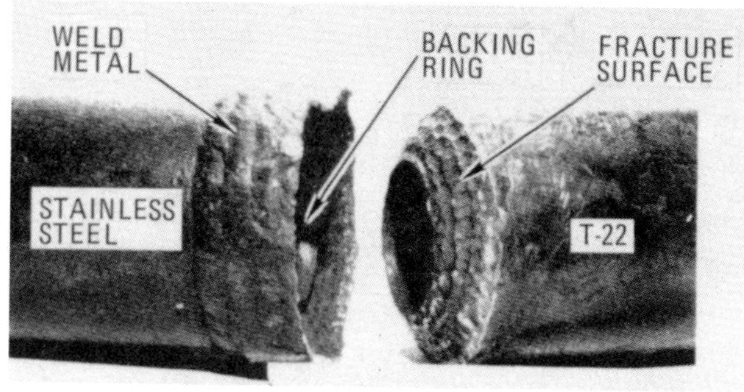

Fig. 1 - Dissimilar weld failure

Two carbide morphologies are commonly observed at DMW interfaces: the first, designated type 1, comprises a planar array of globular carbides, which appear to provide a preferred failure path. The other carbide morphology, designated type 2, consists of a diffuse array of generally smaller carbides. Type 1 carbide is prevalent in nickel filler DMWs after service. In stainless steel filler welds type 2 carbide is the predominant morphology. Details of these metallographic features may be found in Ref. 1.

During RP1874 considerable attention was devoted to determining the kinetics of type 1 carbide formation since CEGB studies have shown (Ref. 1) that the evolution of this carbide at the weld interface has a major influence on the rupture strength and load carrying capability of DMWs made with nickel base filler. Based on those studies and confirmed by experimental activities of RP1874, a mathematical expression was determined to characterize carbide evolution. The details are provided in Ref. 1.

Carbon migration occurs, in service, from the low alloy steel to the weld metal due to the differences in carbon activity of the metals joined. However, it is normally not sufficient to cause a coarse grained, ferritic region to be formed in the low alloy steel. All the evidence observed in this study indicates that such gross decarburization of the low alloy steel region is not a factor in DMW failure.

With respect to the influence of welding processes and materials on performance it is observed that weld defects are not normally significant contributors to DMW failures occurring after the first five years of service. It is evident from both accelerated testing and service observation that DMWs made with nickel base filler give significantly different performance than those of DMWs made with austenitic stainless steel filler. Under testing or service conditions where interfacial carbides in nickel base welds do not grow to a very significant size, welds made with nickel base filler will last approximately five times as long as welds made with austenitic stainless steel filler. However, once the type 1 carbide at the interface between the nickel base filler and the low alloy steel has developed to significant size, the performance differences between the two types of welds are significantly reduced.

LIFE PREDICTION TECHNIQUE AND DAMAGE ASSESSMENT. Based on the examination of the service conditions and damage levels in a large number of welds from many operating boilers it became clear that the performance of DMWs in service is a function of the operating loads and temperatures, the extent of thermal and load cycling, and the metallurgical characteristics of the DMW.

The loadings to which a DMW is subjected can be divided into two categories. In the first, as a result of the expansion differences between the stainless steel tube, the weld filler, and the ferritic steel tube, local stresses are induced at, and near, the DMW as temperature is changed. Damage may develop in elevated temperature service as a result of these stresses, particularly when temperature is cycled as a result of plant startup and shutdown. Such damage is identified as "self stress" damage and is strongly affected by temperature and the type of filler metal used in the DMW. Self stress damage with nickel base filler is less than that in DMWs with stainless steel filler since the expansion coefficient of the nickel base filler is closer to that of the low alloy steel and the stresses are, therefore, lower in the vicinity of the less creep resistant low alloy steel. Since these stresses are strain controlled, they will relax with time in a manner dictated by both the creep properties of the materials and the local stress multiaxiality. However, they may regenerate as a result of temperature cycling.

The second category loads arise because the DMW is part of a large structural system. Some of these loads are applied continuously during service. Dead weight and steam pressure loads lie in this category. Such loads are identified as primary system loads. Other loads to which the DMW is subjected result from constrained thermal expansion of the piping system. These loads are again generally strain controlled and relax at elevated temperatures, but may be regenerated by temperature cycling. Loads of this kind are designated secondary system loads. The stresses induced by the temperature differences across the wall of the tube are also included in this category.

The quantitative technique developed for predicting the level of damage which may be expected in DMWs subjected to a particular set of service conditions is called Prediction of Damage in Service (PODIS) (Ref. 2). The formulation of PODIS is designed to capture the essential characteristics of the previously described damage processes in dissimilar welds. In the method it is assumed that a theoretical estimation of damage may be obtained by the linear addition of three components derived from the loading history. The components are:

D_I - intrinsic (or self) damage caused when temperature changes are applied to the dissimilar weld in the presence of the normal pressure induced hoop stress but in the absence of any other external load.

D_P - damage caused by the primary (load controlled) components of the "system" loads on the weld. "System" loads in this context are such items as dead weight and axial pressure loads.

DS - damage caused by the secondary (strain controlled) components of the "system" loads on the weld. Loads in this category are those due to restrained thermal expansion of the tube assembly and those induced by the tube wall temperature differences. (Note: since the latter is rarely observed to be a significant factor in DMW damage development it is not explicitly included as a parameter in the current model.

The total damage D_{TOT} is then given by the summation of the total values of all three damage components. The value of D_{TOT} is a direct prediction of the amount of metallographically detectable interfacial damage to be expected. Damage at failure is, by definition, equal to unity. The detailed equations describing each of these damage components are fully described Ref. 2. It is sufficient, for the purpose of the present paper, to note that the equation forms and parameters were derived empirically, using data from service and from tests simulating service conditions.

In the case of dissimilar welds made with stainless steel filler, the D_I term is an important parameter due to the significant difference in thermal expansion coefficient between the weld filler metal and the adjacent low alloy steel. For the prediction of behavior of DMWs made with nickel base filler, the D_I term is considered not significant and neglected since, in general, the thermal expansion characteristics of the nickel base filler closely approximate those of the adjacent low alloy steel. However, for the nickel base filler prediction formulation, it is necessary to incorporate an appropriate algorithim to account for the development of type 1 interfacial carbides during service, and the associated degradation in load carrying capability of the weld. The last feature introduces a complication in that the damage done to a weld by any given loading cycle will depend upon when in life that event occurred. As a result, the final damage computation is path dependent, i.e., it depends on the sequence of loading and temperature. A closed form solution is not, therefore, possible and each case must be evaluated by numerical integration.

An important capability of any residual life estimation technique is that it should be able to forecase behavior under loading conditions which are not necessarily the same as those experienced by the weld in the past. That is, it should be capable of accounting for such differences as modification to assemblies, differences in operating temperatures and changes in cycling behavior. The technique developed under RP1874 recognizes parameters such as temperatures, number of cycles, system loads, self loads and metallurgical evolution, which may change as a result of assembly modification and/or differing operating conditions. The general method of remaining life assessment developed for dissimilar welds consists of the following steps:

1. Preliminary review and analysis.
2. Inspection.
3. Sampling.
4. Metallurgical examination and damage assessment.
5. Load history analysis and damage estimation.
6. Damage correlation.
7. Future operating analysis.
8. Residual life estimation.

The amount of sampling and metallurgical examination required generally depends on the level of knowledge of and confidence derived in the determination of temperature distributions and operating history. Details of the steps listed above are given in Ref. 2.

OPTIMIZED WELD GEOMETRY/FILLER METAL. One of the difficulties encountered in earlier research programs on dissimilar metal weld behavior was that of successfully reproducing, in the laboratory, fracture modes typically observed in service. This was particularly true for DMWs made with nickel base filler. An effective test developed in RP1874 program involved the testing of full size tubular DMWs while subjected to internal pressure, axial or bending loads, and cyclic temperature conditions. The tests, which have been described in detail elsewhere (Ref. 1) are typically run in the temperature range 1050° to 1150°F and involve varying degrees of creep rupture, creep fatigue and thermal cycling damage. This test technique has been used to evaluate a variety of commercially available weld filler materials and to assess the effect of weld geometry on performance.

The results of this testing have shown that, as previously noted, as-welded test samples of dissimilar welds made with commercially available nickel base fillers offer a life improvement over DMWs made with stainless steel filler of about four to five times. However, when the nickel base filler welds are aged under conditions that produce large interfacial type 1 carbides, their subsequent life is essentially equivalent to the life of stainless steel filler DMWs. This strongly indicates, as noted before, that, the improved performance of nickel filler welds is in part due to the relatively slow kinetics of the development of the interfacial carbides.

The discriminatory test procedure has also been used to examine the effects of weld geometry. The driving force behind this evaluation was the observation, from the field, that damage in dissimilar welds was far more intense adjacent to weld interfaces lying in a plane normal to the tube axis and damage was often very minor on interfaces lying closer to parallel to the tube axis. Experimental evaluation of the effects of such geometries, confirmed the benefits of a low interface angle. Specifically, in the case of welds made with stainless steel filler, an approximately 50% to 60% increase in life is obtained through using a weld in which the interface, on the ferritic side, makes an angle of 60 deg to a normal to the tube axis (Fig. 2). In the case of DMWs made with nickel base filler the most marked improvement in life was obtained in welds featuring a wide cap (Fig. 3). Welds of this configuration extended the life of nickel filler DMWs by about 50%.

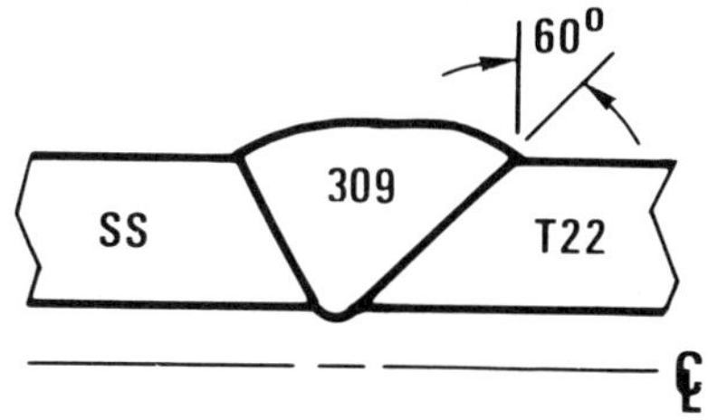

Fig. 2 - Optimized weld geometry for stainless steel filler material

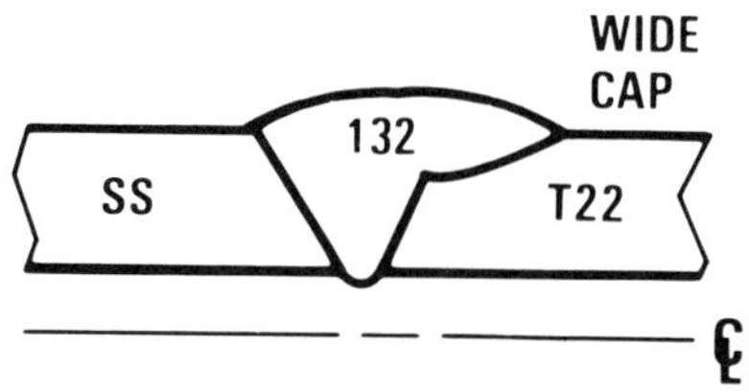

Fig. 3 - Optimized weld geometry for nickel base filler material

IMPROVED FILLER METAL DEVELOPMENT. Notwithstanding the greatly improved ability to predict weld behavior and the ability to enhance BMW behavior through geometry improvements, there still remain a strong interest, on the part of the utilities, in the development of a filler metal that would provide maximum possible life for dissimilar metal welds of this kind. It was viewed as highly desirable that such a filler material be appropriate for use by the shielded metal arc process and that it not require the use of unusual or impractical pre- and post-weld heat treatments.

In response to this concern RP1874 initiated a program for filler metal development in 1983. Based on the program knowledge at that time it was concluded that the desirable attributes of an improved filler metal would be: (a) a thermal expansion coefficient close to that of the low alloy steel, (b) resistance to the formation of type 1 carbides at the weld interface, (c) creep rates intermediate between the low alloy steel and the stainless steel (this allows favorable redistribution of thermally induced stresses), (d) rupture strength better than the low alloy steel, (e) characterisics that discourage carbon transport, (f) adequate corrosion resistance, (g) adequate thermal stability, (h) adequate operability and weldability.

Based upon these criteria a computer aided evaluation of possible alloy formulations was undertaken. From this evaluation a series of eight experimental alloy compositions were defined. Of these, four were nickel base, four were iron base. The alloys were produced by Telledyne McKay for evaluation in the program. In addition, an extended evaluation was made of a wide variety of other commercially available potentially effective filler materials. Those evaluations are listed in Ref. 1. In all, a total of eight experimental and 12 commercial nickel and iron alloys were evaluated.

The evaluation process was such as to allow early screening of the materials down to a few most promising alloys. This first screening process involved assessment of thermal expansion, the popensity for carbon transport, the growth of interfacial carbides, and preliminary evaluations of operability (in the case of the experimental alloys). At the end of this first screening phase the alloy list was reduced to six. The next steps in the evaluation process included more extensive aging testing to determine both interfacial evolution and thermal stability characteristics of the weld deposit creep and rupture testing, and a thorough exploration of the weldability and operability characteristics - particularly of the experimental alloys. This part of the program revealed that some potentially interesting alloys showed excessive thermal instability. A few showed the development of undesirable type 1 interfacial carbides and others showed a marked tendency for creep rupture failure to occur at the interface. Some alloy compositions were concluded to be less desirable as a result of the need for pre- and post-weld heat treatment practices which would be difficult to apply in boiler manufacture - and particularly in boilers requiring repair or replacement.

As a result of all these tests, two experimental filler metal compositions emerged as having the most desirable characteristics. These compositions are shown in Table 1. The weld deposits showed very satisfactory thermal expansion characteristics, a minimal tendency to form type 1 interfacial carbides, excellent thermal stability and good cross weld rupture behavior. At the present time these weld filler metals are being used to develop specimens for accelerated discriminatory testing. Samples of these electrodes are also under evaluation by utilities, on a trail basis, to assess performance under boiler operating conditions.

Table 1 - Desirable Filler Metal Compositions

Element	HFS 5 Lot 6569 HT P1950	HFS 6 Lot 6568 HT P1950
C	0.024	0.033
Mn	1.15	1.26
Si	0.55	0.54
P	0.005	0.005
S	0.003	0.003
Cr	4.10	8.26
Ni	43.69	43.36
Mo	1.81	1.87
Ti	0.05	0.03
Al	0.01	0.01
V	NA	NA
Co	NA	NA
Fe	BAL	BAL

CONCLUSIONS. The RP1874 program has resulted in greatly improved understanding of the mechanisms of DMW failure in elevated temperature service and of the interplay between the metallurgical and mechanical factors that contribute to DMW performance. Several different failure modes have been identified and it is evident that their occurrence is dictated by microstructural features, loadings, and weld geometry. Failures are generally creep damage dominated and, since the creep voiding that precedes fracture is influenced by carbide morphology, microstructure has been found to play a critical role.

It is evident that the performance of DMWs is strongly influenced by the system loadings, temperatures, and temperature cycles imposed on the DMW in boiler service. This knowledge has been used to develop a technique for estimating remaining life of boiler tube dissimilar welds, which takes into account the key boiler operational factors contributing to damage. The technique (which uses the damage computational method designated PODIS) can be used to provide quantitative answers to questions associated with units containing both stainless steel and nickel base filler dissimilar welds. It is simple to apply but requires data to be collected and analyzed which is specific to the unit being evaluated.

Use of the technique enables determination of whether the dissimilar weld metal problem in a particular unit is, or is likely to be, widespread or confined to a specific region. Since the damage estimation method relates to loading conditions at the DMW, it is possible to identify the probable causes of weld distress. Action can, therefore, be taken accordingly and quantitative evaluation can be made of the potential benefits of proposed repairs or modifications. Thus, use of the technique can improve the accuracy and cost effectiveness of run and repair decisions.

The program has also defined and quantified various approaches that can be used to optimize weld filler materials and weld geometry. In particular, it has been shown that certain weld geometries offer substantial life improvements using existing practice filler metals. In addition, experimental filler materials have been developed which appears, at this time, to offer the potential for very substantial improvements in dissimilar weld performance. Final evaluation of this material is currently ongoing.

ACKNOWLEDGMENTS. This work was supported under Electric Power Research Institute contract RP1874-1 and managed by Metal Properties Council.

REFERENCES.

1. Dissimilar-Weld Failure Analysis and Development Program. Volumes 1-4, EPRI Report CS-4252.
2. Roberts, D. I., R. H. Ryder, and R. Viswanathan, Performance of Dissimilar Welds in Service, Journal of Pressure Vessel Technology, Vol 107, p. 247, August 1985.

THE WELDABILITY, MICROSTRUCTURE AND PROPERTIES OF PRECIPITATION STRENGTHENED MARTENSITIC STAINLESS STEELS

Wendy R. Cieslak
Sandia National Laboratories
Albuquerque, New Mexico

John A. Brooks
Sandia National Laboratories
Livermore, California

Warren M. Garrison, Jr.
Carnegie-Mellon University
Pittsburgh, Pennsylvania

ABSTRACT

We have investigated the influence of welding on the mechanical properties and corrosion resistance of two precipitation strengthened martensitic stainless steels: Custom 450 and PH 13-8 Mo. The two alloys exhibited very good weld cracking resistance, although the formation of a low melting NbC eutectic constituent did cause some solidification cracking in Custom 450. The effects of aging temperature on tensile and impact properties were similar in the base metal and Gas-Tungsten-Arc (GTA) weldment for both alloys. The similarity of weld and base metal properties is attributed to the uniform precipitate distribution throughout the weld, as shown by Transmission Electron Microscopy (TEM) and electron microprobe analyses. The susceptibility to pitting corrosion was not impaired by aging or by welding, with the possible exception of a narrow band in the heat-affected zone (HAZ), and both materials were more resistant than 304 stainless steel. The narrow band is identified by electrochemical reactivation testing as a region of increased matrix activity likely caused by localized Cr and/or Mo depletion.

THE PRECIPITATION STRENGTHENED martensitic stainless steels with yield strengths three to four times those of the austenitic stainless steels are excellent candidates for applications requiring high strength, good weldability and corrosion resistance. However, while the variation in properties associated with the fusion zone (FZ) and HAZ microstructures is well established for the austenitics, few data are available for the martensitic stainless steels. We have compared the tensile properties, impact toughness, aging response, microstructures and corrosion resistance of martensitic stainless steel weldments to those of the base metal.

The study has concentrated on two alloys: Custom 450 and PH 13-8 Mo. Based on the alloy content of Custom 450, the strengthening precipitate is probably Cu (1), although the existence of Laves phase in this alloy has also been shown to contribute to strengthening (2). The PH 13-8 Mo is strengthened by the precipitation of the intermetallic NiAl (3). The weldability, as determined by Varestraint testing (4,5), and the localized corrosion resistance, as determined by anodic polarization, have been judged in comparison to the properties of 304 and 304L stainless steels.

EXPERIMENTAL

Compositions for the Custom 450 (Carpenter Fibralloy 450) and PH 13-8 Mo (Armco) are listed in Table 1, along with compositions of two heats of 304L and one heat of 304 stainless steel used as comparison alloys. Before testing or aging the Custom 450 was austenitized at 927°C and the PH 13-8 Mo at 1038°C for one hour and oil-quenched. Both alloys were aged between 300° and 600°C in salt pots for 1 to 3.3 hours and then water-quenched.

The solidification cracking susceptibility was evaluated in the unaged condition using sub-size Varestraint testing (4,5). Welding of Varestraint, mechanical, and corrosion testing samples was performed according to Table 2. Mechanical test specimens were cut from welded plate oriented with the welding direction transverse to the length of the specimens and the weldment centered in the specimen. The notch of the Charpy bars was placed at the root of the weld. Tensile bars (5 mm diameter and 20 mm long) contained a composite of FZ (~50%), HAZ, and base material. The tensile specimens deformed uniformly when tested, thus no one zone dominated the tensile response.

Welded material was sectioned to allow corrosion testing of solely FZ metal, or base metal, or HAZ bordered by small amounts of weld and base metals. Critical pitting poten-

Table 1
Alloy Compositions (wt.%)

Alloy	C	Cr	Ni	Cu	Al	Mo	Si	Mn	Nb+Ta	S	P
PH 13-8 Mo	0.03	12.63	7.9	-	1.08	1.73	0.04	0.05	-	0.005	0.005
Custom 450	0.03	14.75	6.44	1.5	-	0.76	0.4	0.33	0.75	0.005	0.008
304L-Heat 1	0.017	18.6	10.7	-	-	0.21	0.54	1.9	-	0.001	0.024
304L-Heat 2	0.020	19.1	10.9	-	-	0.19	0.59	1.7	-	0.008	0.020
304	0.060	18.7	8.9	-	-	0.12	0.33	1.4	-	0.016	0.033

Table 2
Nominal Weld Parameters*

Samples	Thickness	Amps	Volts	Travel Speed	Wire Feed**	No. Passes
Tensile	0.95 cm	200	10	0.25 cm/sec	~ 0.85 cm/sec	14-16
Charpy	1.27 cm	200	10	0.25	~ 0.85 cm/sec	20-24
Corrosion	1.11 cm	180	10	0.25	-	1
Varestraint	1.11 cm	180	~12	0.25	-	1

* Argon shield gas
** 0.11 cm diameter

tials were determined by potentiodynamic polarization at 0.5mV/sec in 1.0M NaCl + 0.5M Na_2SO_4 solution at 30°C. Potentials were measured and are reported vs. the Saturated Calomel Electrode (SCE). Sensitization, which is due to localized Cr-depletion and is commonly manifested as intergranular corrosion susceptibility, was studied using the double-loop electrochemical potentiokinetic reactivation (DL-EPR) procedure developed by Akashi et al (6) and characterized by Majidi and Streicher (7) for use on austenitic stainless steels. The test electrolyte, 0.4M H_2SO_4 + 0.01M KSCN, was mixed immediately prior to use, and the experiments were conducted in N_2-purged solution at 30 C at a polarization rate of 6V/h.

RESULTS AND DISCUSSION

MECHANICAL PROPERTIES - Figure 1a shows that the tensile and impact properties as a function of aging temperature of the Custom 450 weldments are almost identical to those of the base material. Mechanical properties data are reported as averages of 2-4 tests. Aging at 450°C produces the peak strength of ≈175 ksi and overaging occurs above ≈500°C. Accordingly, the impact energy exhibits a minimum for the 450°C aging treatment and recovers in the overaged condition.

The average peak strength of ≈195 ksi for the PH 13-8 Mo alloy, along with the minimum in impact energy, occurs for aging at 475°-525°C, Figure 1b. Again, the weldment exhibits tensile and impact properties and aging response that differ little from the base material properties.

In both alloy systems, the as-welded condition exhibits properties very close to those of the austenitized-and-quenched condition. These results imply that little

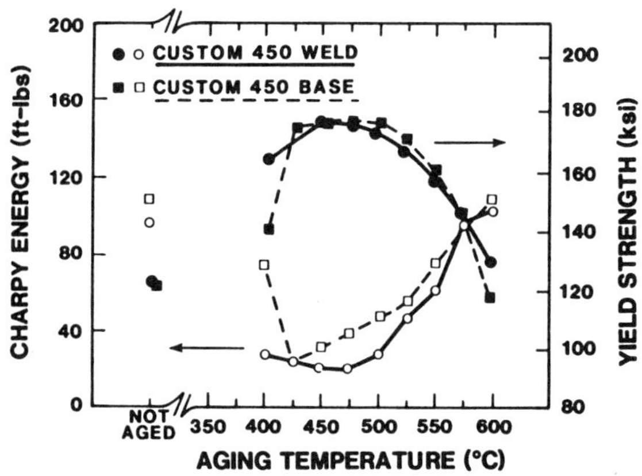

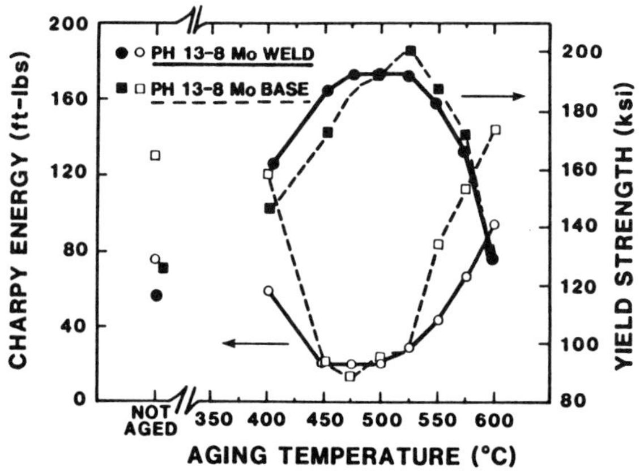

Figure 1. Impact energies and yield strengths as a function of aging temperature for base metals and weldments. Austenitized-and-quenched conditions and as-welded conditions are plotted as Not Aged.
a. Custom 450
b. PH 13-8 Mo

aging occurs during multiple pass welding, consistent with the observed insensitivity of tensile properties to the rate of quenching from the austenitizing temperature (8). The similarity of weld and base metal properties is attributed to the uniform precipitate distribution throughout the weld, as observed in TEM and electron microprobe analyses. Optimization of strength and impact properties for both alloys, whether welded or not, is achieved in the slightly overaged condition that maintains the desired strength for the application, yet allows some recovery of toughness from the minimum associated with the maximum strength.

SOLIDIFICATION CRACKING SUSCEPTIBILITY - Varestraint data are reported as total crack length vs. augmented strain. The experimental results are interpreted by comparison with data for 304L stainless steel. Welds in 304L Heat 1 (Table 1) solidify as primary austenite and contain little or no ferrite, while welds in Heat 2 solidify as primary ferrite and contain ≈4% retained ferrite (F.N. = 4.5.). The Varestraint data in Figure 2 show the excellent hot-cracking resistance of Heat 2, as is typical of 304L that solidifies as primary ferrite. Heat 1, however, typifies the lower resistance known to be associated with 304L that solidifies as primary austenite (9-13). The solidification cracking resistance of the alloys under investigation are also plotted in Figure 2, showing that both are substantially more resistant than the 304L Heat 1. Furthermore, the PH 13-8 Mo exhibits hot-cracking resistance comparable to the 304L Heat 2.

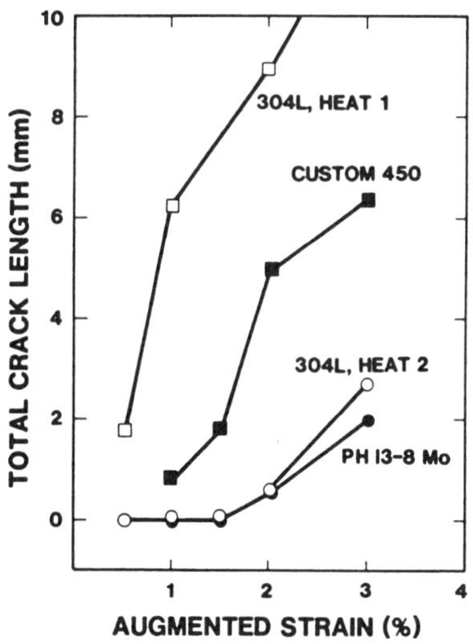

Figure 2. Varestraint test results

Solidification cracking of the Custom 450, Figure 3, is clearly associated with a second phase in the grain boundary ahead of the crack tip. This phase, analyzed at points B and C in Figure 3, is greatly enriched in Nb. The microprobe analysis at these points totals 90-95%, mainly because it cannot account quantitatively for carbon. Analytic Electron Microscopy analyses of particles extracted from the grain boundary regions show that the phase consists almost totally of Nb and lighter elements. The diffraction patterns index as Nb(C,N). Therefore, the low solidification temperature of the NbC/austenite-matrix eutectic constituent is responsible for hot-cracking in the Custom 450. This mechanism has been demonstrated in several austentitic stainless steels (14-16), and in dissimilar metal welds of the martensitic precipitation-hardenable systems HP 9-40-20 to 15-5 PH (17). Nonetheless, the Custom 450 has fairly good hot-cracking resistance, possibly related to relatively low Nb and C concentrations, little wetting of the grain boundaries by the eutectic, or a narrow solidification temperature range. Details of this mechanism are being further studied.

The solidification cracking in PH 13-8 Mo at 3% augmented strain in the Varestraint test does not appear to be associated with any secondary phase of solidification resulting

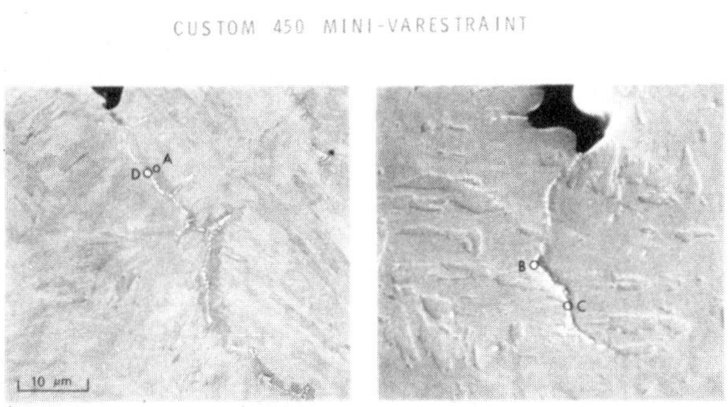

Composition (wt%)

Position	Cr	Fe	Ni	Cu	Si	Nb	Mo	Mn	P	Total
Matrix	14.9	75.7	6.4	1.5	.37	.55	.84	.45	.02	
A	14.5	70.1	9.6	2.5	.44	1.33	.94	.53	.06	
B	11.7	51.3	6.9	1.7	.38	21.7	.90	.43	.04	(95.4+C)
C	8.1	26.8	4.0	1.0	.30	46.0	1.0	.27	.11	(89.8+C)

Figure 3. SEM micrographs and electron microprobe analyses of Custom 450 Varestraint tested at 3% strain.

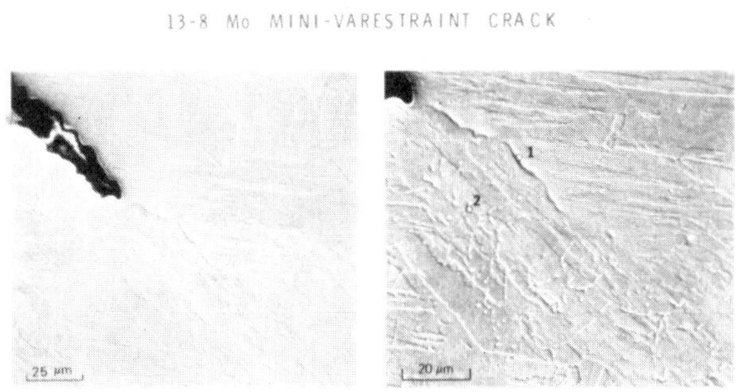

Microprobe Analysis
Composition (wt%)

Area	Cr	Ni	Mo	Mn	Si	P	S	Fe
Base	12.8	8.1	2.3	0.1	0.00	.00	0.00	Bal.
1	15.4	11.6	4.5	0.1	-	(.03)	-	
2	15.5	4.6	3.9	0.1	-	-	-	

Figure 4. SEM micrographs and electron microprobe analyses of PH 13-8 Mo Varestraint tested at 3% strain.

from the segregation of alloying elements or impurities. Microprobe analysis of the region of the grain boundary near the crack tip, point 1 in Figure 4, reveals some increases in Cr, Ni, Mo, and P, but no distinct second phase. The microsegregation in this region likely suppressed the melting temperature of final solidification, thus allowing hot-cracking. Also, it has been shown by Brooks and Baskes (this volume) that the absence of a second phase at room temperature does not necessarily mean that one was not present during the final stages of solidification. However, it could not have been present in great quantities. Cracking is apparently not related to the small amount of lacy-appearing second phase, point 2 in Figure 4. This phase, enriched in Cr and Mo and depleted in Ni, may be delta-ferrite retained from solidification. Therefore, the lack of significant amounts of low-melting second phase promotes the excellent hot-cracking resistance of the PH 13-8 Mo. This study has not addressed heat-to-heat variations in cracking susceptibility, however, such variations are more likely to occur in the Custom 450 than in the PH 13-8 Mo due to the clear evidence of a second phase associated with cracking in the Custom 450 alloy.

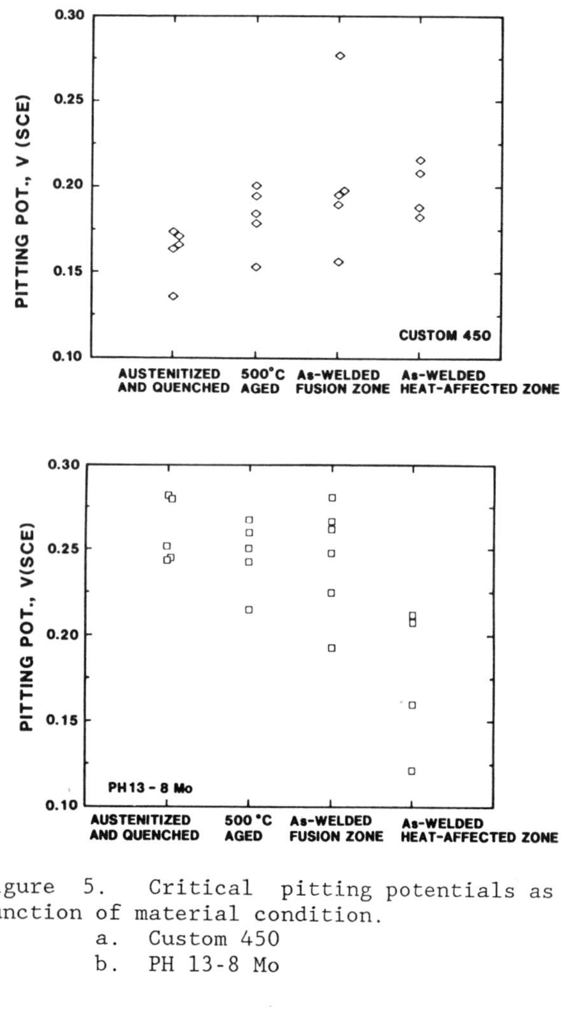

Figure 5. Critical pitting potentials as a function of material condition.
a. Custom 450
b. PH 13-8 Mo

CORROSION RESISTANCE - For both alloys, the 500°C-aged condition and the FZ of the weld exhibit essentially the same resistance (within 2x the standard deviation) to pit initiation as the austenitized-and-quenched condition, Figure 5. The HAZ behaves differently and will be discussed later. The PH 13-8 Mo is the more resistant material, as evidenced by an average pit initiation potential of 0.25V (SCE) vs. 0.18V (SCE) for the Custom 450, with a pooled standard deviation of 0.02V. Microprobe analysis, reported elsewhere (18), reveals that the Custom 450 base metal suffers pitting at sites containing Cu, Nb, C, and trace S, while pitting in the PH 13-8 Mo does not appear to be associated with inclusions. The lack of inclusions, such as sulfides, that initiate pitting and the high Mo content (19) promote the superior resistance of the PH 13-8 Mo to pitting. Both alloys are more resistant than the typical 304 stainless steel, which has a critical pitting potential in this solution of 0.12V(SCE) with a standard deviation of 0.01V.

The lack of an effect of aging treatment on pitting resistance implies that the finely dispersed strengthening precipitates do not influence pitting resistance. This behavior has also been noted for 17-4PH stainless steel (20). The equivalence of fusion zone and base metal pitting resistances implies that the microstructure created by welding does not present any corrosion-susceptible interfaces that had not existed previously, unlike welded austenitic stainless steels (21,22). It has also been shown that regions of major alloying element segregation (particularly Cr and Mo depletion) created by welding reduce the pitting resistance of some austenitic stainless steels (23,24), but such regions are not formed in welds of high-purity ferritic stainless steels (25). Based on pitting resistance, the weld FZ's of the Custom 450 and the PH 13-8 Mo must also be free of substantial Cr or Mo depleted regions. This is not necessarily the case for the HAZ, which, at least for the PH 13-8 Mo, exhibits slightly reduced pit initiation resistance, Figure 5b. The pits in these samples are aligned along the low-temperature HAZ, Figure 6. For Custom 450, this phenomenon has been postulated as caused by Cr-depletion of the matrix as a result of Cr-carbide precipitation (26). However, the phenomenon was not investigated further.

Critical pitting potential is not very sensitive to mild degrees of sensitization. The DL-EPR test has been developed as a sensitive method of quantifying mild degrees of sensitization with very good reproducibility (7). The test sequence, Figure 7, is to anodically polarize the sample through the active dissolution potential range until passivity (sharp reduction in current) occurs. The scan is reversed, and the stable passive film resists active dissolution while Cr-depleted regions depassivate. Therefore, the ratio of the peak current on the reverse (reactivation) scan to the peak current on the forward scan is a measure of sensitization.

For austenitic stainless steels, it has been shown that samples entirely free of sensitization exhibit ratios of $\sim 10^{-3}$ or less (7). A ratio of $>5 \times 10^{-2}$ generally corresponds to a ditch (sensitized) structure in ASTM A262-A.

By using the same criterion for the alloys in the present study, we can show, Figure 8, that the Custom 450 does not appear to undergo sensitization, while the PH 13-8 Mo undergoes sensitization during the 500°C aging and in the HAZ. The numbers for the HAZ do not reflect the maximum sensitization present in this zone because the attack is not uniform. Only a band removed from the FZ (the same band which suffered preferential pitting corrosion) is more severely etched following DL-EPR testing or following standard chemical etching. The etching does not follow prior austenite grain boundaries. This band, therefore, is probably Cr-depleted as a result of general Cr-carbide precipitation. The general rather than intergranular precipitation of Cr-carbides, with the probability of associated matrix Cr-depletion, has been reported following tempering at 566°C of Type 410 martensitic stainless steel (27) and, as noted above, following welding of Custom 450 (26).

Although the HAZ in Custom 450 does not exhibit a significantly reduced critical pitting potential or increased reactivation current in the DL-EPR test, the occurrence of pitting preferentially in the low-temperature HAZ indicates a change in corrosion susceptibility. For the PH 13-8 Mo, the DL-EPR results clearly show that both 500°C aging and welding cause some depletion of those elements - primarily Cr and possibly Mo - which are responsible for the alloy's corrosion resistance. Because the bulk alloy Cr-content is so close to the 11-12% necessary to promote passivity in most aqueous environments, the PH 13-8 Mo is particularly susceptible to destabilization of passivity with very little alloy depletion. As has been shown for Custom 450 (26), heat-treatment after welding, either to diffuse Cr into the depleted zones or to solutionize the carbides, should relieve this problem.

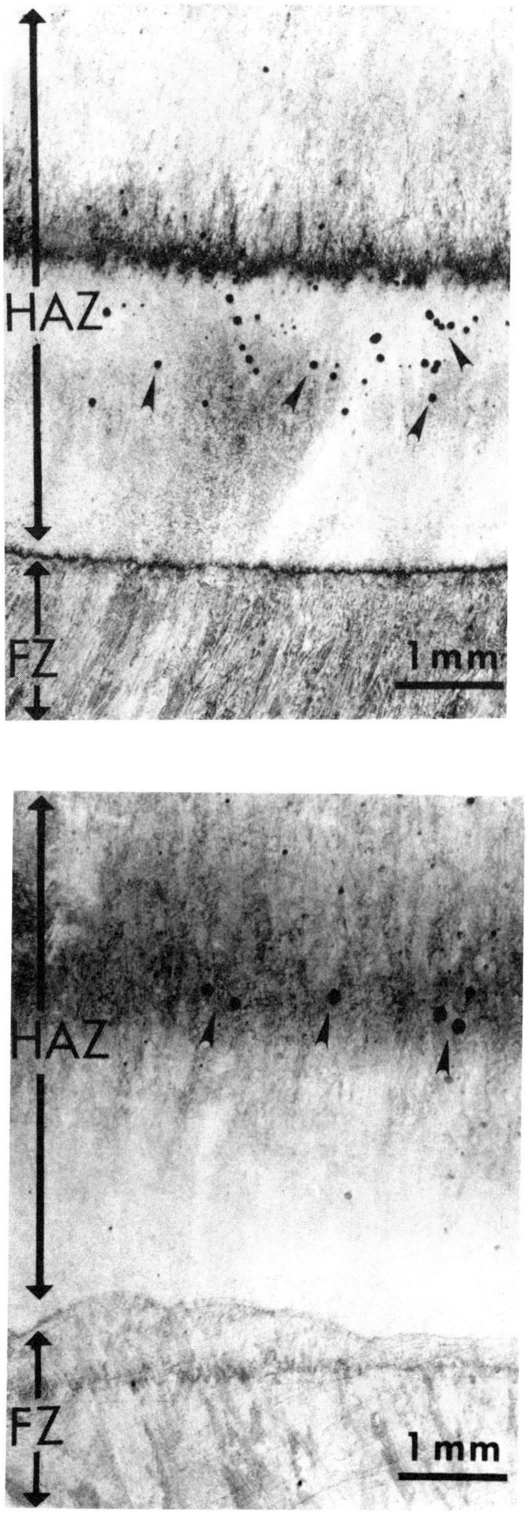

Figure 6. Preferential pitting (arrows) in low-temperature heat-affected zone. Etched with Marble's reagent after pitting corrosion test.

a. Custom 450
b. PH 13-8 Mo

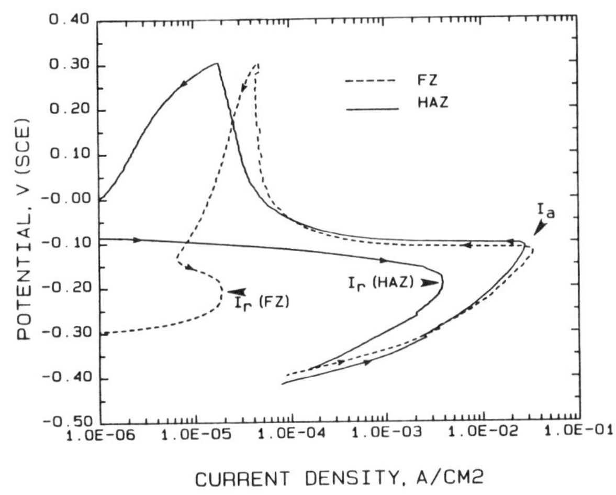

Figure 7. Typical DL-EPR scans performed on PH 13-8 Mo.

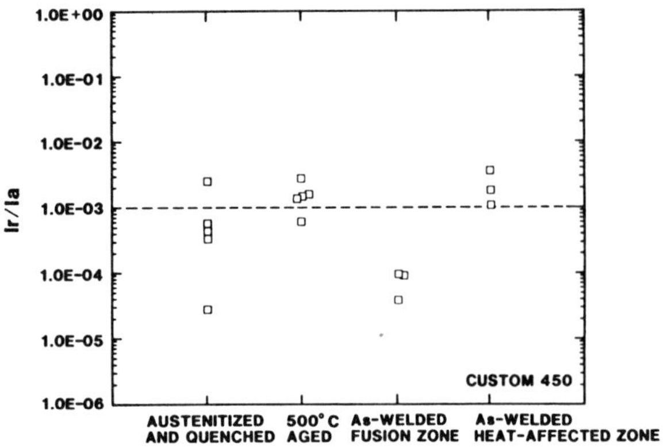

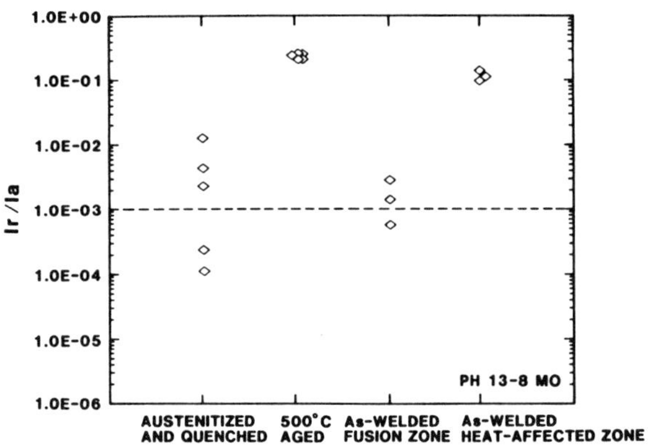

Figure 8. Reactivation ratio (I_r /I_a) as a function of material condition.
a. Custom 450
b. PH 13-8 Mo

CONCLUSIONS

The influence of welding on the mechanical properties and corrosion resistance of two precipitation strengthened martensitic stainless steels has been characterized as follows:

1. Uniform precipitate distribution throughout the weld enables base metal and welded material to exhibit close to identical tensile and impact properties as a function of aging temperature.

2. The lack of a low melting temperature second phase in the PH 13-8 Mo weld promotes excellent resistance to solidification cracking. The existence of a small volume fraction of a NbC-eutectic constituent in the Custom 450 weld metal increases the hot-cracking susceptibility of the alloy relative to a primary delta-ferrite solidifying heat of 304L stainless steel. However, the Custom 450 is more resistant to hot-cracking than a primary austenite solidifying heat of 304L stainless steel.

3. Because the weld fusion zone microstructure is primarily single-phase and free of Cr and Mo segregation, this zone is as resistant to localized corrosion as the base metal. However, a narrow band in the heat-affected zone is slightly more susceptible, probably because of sensitization associated with sub-microscopic Cr-carbides.

ACKNOWLEDGEMENTS

F. S. Bovard, S. Malene, V. V. Pestanas, and D. G. Ruddle provided laboratory assistance. Electron microprobe analysis was performed by C. W. Karfs. This work was performed at Sandia National Laboratories, supported by the U.S. Department of Energy under contract number DE-AC04-76DP00789.

REFERENCES

1. Pickering, F. B., "Iron and Steel Inst. Publ. 114," London, p. 131, (1963).
2. Bride, J. E., Jr. and Maniar, G. N., "Metallography as a Quality Control Tool," Plenum Publishing Corp. (1980) 279.
3. Seetharaman, V., Sundararaman, M., and Krishnan, R., Materials Science and Engineering, 47, 1 (1981)
4. Savage, W. F. and Lundin, C. D., Welding J., 44, 435s (1965).
5. Savage, W. F. and Lundin, C. D., Welding J., 45, 497s (1966)
6. Akashi, M.,Kawamoto, T, and Umemura, F., et.al., Corrosion Engineering (Boshoku Gijutsu), 29, (4), 163 (1980).
7. Majidi, A. P., and Streicher, M. A., Corrosion, 40, 584, (1984)
8. Garrison, W. M., "Heat Treatment Response of Several Martensitic Precipitation Strengthened Stainless Steels," Sandia National Laboratories official report, to be published.
9. Masumato, I., Tamaki, K., Kutsuna, M., Trans. JWS, 41, 1306 (1972).
10. Arata, Y., Matsuda, F., and Katayama, F., Trans. JWRl, 6, 105 (1977).
11. Kujanpaa, B., Suutala, N., Takalo, T., Moisio, T., Welding Res. Inter., 9, 55 (1979).
12. Cieslak, M. J., Ritter, A. M., and Savage, W. F., Welding J., 61, 1s (1982).
13. Brooks, J. A., Thompson, A. W., and Williams, J. C., Welding J., 63, 71s (1984).
14. Ogawa, T. and Tsunetomi, E., Welding J., 61, 82s (1982).
15. Ritter, A. M. and Savage, W. F., Met. Trans. A, 17A, 66 (1986).
16. Jolley, G. and Geraghty, J. E., "Solidification and Casting of Metals," p. 411, The Metals Society, London (1979).
17. Cieslak, M. J., Hills, C. R., and Headley, T. J., to be published in "Microbeam Analysis - 1986."
18. Cieslak, W. R., Semarge, R.E, and Bovard, F. S., to be published in "Microbeam Analysis - 1986."
19. Kolotyrkin, Ja. M., Corrosion, 19, 261t (1963).

20. Syrett, B. C., Viswanathan, R., Wing, S. S., and Wittig, J. E., Corrosion, 38, 273 (1982).
21. Manning, P. E., Duquette, D. J., and Savage, W. F., Weld. J., 59, 260s (1980).
22. Cieslak, W. R. Duquette, D. J., and Savage, W. F., "Trends in Welding Research in the U.S.," p. 361, American Society for Metals (1982).
23. Garner, A., Corrosion, 35, 108 (1979).
24. Garner, A., Weld. J., 62, 27 (1983).
25. Cieslak, W. R., and Duquette, D. J., "Environmental Degradation of Materials in Nuclear Power Systems - Water Reactors," p. 438, National Association of Corrosion Engineers (1984).
26. Henthorne, M., DeBold, T. A., and Yinger, R. J., "Custom 450-A New High Strength Stainless Steel," Corrosion/72, paper #53, NACE, Houston (1972).
27. Lula, R. A., and Davis, J. A. "Intergranular Corrosion of Stainless Alloys, p.233, ASTM STP 656, R. F. Steigerwald, Ed., ASTM, Philadelphia, Pennsylvania, (1978).

PROGRESS IN COMPUTING RESIDUAL STRESS AND STRAIN IN WELDS

J. Goldak, A. Oddy, M. McDill
Carleton University
Ottawa, Canada

A. Chakravarti
AMCA International
Ottawa, Canada

M. Bibby, R. House
Carleton University
Ottawa, Canada

ABSTRACT

Three dimensional transient stress, strain, displacement, and temperature fields around a tack weld in a butt joint are computed with nonlinear transient thermo–elasto–plastic finite element analysis (FEA). The AISI 1020 carbon steel plate is 1.27 cm thick. The arc travels 2.3 cm at 0.508 cm/s in 29 time steps with a power of 1.7 kW. Both starting and stopping transients are analyzed. The assumption of plane strain, which many analysts have assumed, is shown to be invalid.

THE GOAL of computational weld mechanics is to compute the stress, strain, displacement, temperature, microstructure, thermal and mechanical properties in the welding process. Thus the computer should provide a mirror or virtual image of the real welding process that would enable the welding engineer to predict and optimize the welding procedure before an arc is ever struck. The detailed three dimensional (3D) graphic images of the temperature, displacement, strain, and stress fields presented in this paper illustrate progress towards this goal.

Recognizing that the residual strain and stress in a weld depend on the history, this paper follows the only known approach that could yield accurate results; that is to compute the transient strain and stress over the entire history of the weld. Since the space available was not adequate to present both a literature review and the results of this new research, this paper will focus on the results of recent analyses. A few of the more important references are listed to provide an entry into the literature [1] [2] [3] [4].

THEORY

In arc welding, the heating due to the plastic deformation is small compared with the heating due to the arc. Therefore the effect of plastic work on the temperature field can be neglected. This assumption enables the temperature field to be computed before the stress field. The temperature field is computed from the heat equation which is simply a statement of the conservation of energy in the form of a differential equation. In this paper, FEA approximates the exact temperature field in space with a piecewise trilinear polynomial approximation in such a way that energy is conserved in an average (L_2) sense. The elements must be chosen so that this approximation of the geometry, initial conditions, material properties and boundary conditions is adequate. The resulting set of ordinary differential equations is integrated in time with a two point Crank–Nicolson scheme. The time steps must be small enough to accurately integrate the load and temperature. The question of accuracy will be discussed in detail later.

After computing the temperature field, a 3D thermo–elasto–plastic analysis of the weld is performed. The exact displacement field is approximated by piecewise trilinear polynomials (elements). In simple terms, this assumes that matter remains continuous and mass is conserved; i.e, the compatibility equations. Hooke's law is assumed to be true; i.e., $\sigma = E\epsilon$. The conservation of momentum or equilibrium equation is imposed in an average sense.

Elasto–plasticity assumes that the von Mises yield criterion and the flow rule relating the plastic strain rate to the stress rate are obeyed. Isotropic hardening assumes that the von Mises yield function changes size, but does not translate in stress space. Kinematic hardening assumes that the von Mises yield criteria translates but does not change size in stress space. Isotropic–kinematic hardening combines both effects.

The change in the temperature field adds a thermal expansion, $\alpha\Delta T$, to the strain field. The theory, which is given in detail in [1] and [5] closely follows the FEA theory in Bathe's textbook [6]. The most significant differences are the addition of rate–dependent elasticity terms [5] and the 3D mesh grading elements [7].

A number of effects are ignored. Convection and radiation, which were considered in previous FEA analyses, have not yet been implemented for the 3D grading elements. The latent heats of transformation are added to the specific heat. Creep effects are neglected.

DATA

GEOMETRY – One symmmetric half of a $10.16 \times 10.16 \times 1.27$ cm ($4 \times 4 \times 0.5$ in.) bar was analyzed (Fig. 1).

MATERIAL – AISI 1020 carbon steel. Temperature–dependent thermal conductivity and specific heat values were taken from [9]. The temperature dependent Young's modulus, Poisson's ratio, yield strength and hardening modulus were taken from [5].

THE INITIAL temperature was 20 C. Initial strain and stress were assumed to be zero.

BOUNDARY CONDITIONS – In the thermal analysis, the power input was 1700 Watts. The welding speed was 0.508 cm/s (12 ipm). The double ellipsoidal heat source used is described in detail in [1], [8]. All semi-axes were set to 4.77 mm. In this case, the ellipsoid was actually a hemisphere. The arc was moved in 29 time steps ranging from .037 s to .3125 s. The arc position at various times is shown in Fig. 1.

In the thermo–elasto–plastic analysis, the displacements normal to the plane of symmetry were prescribed to zero. The displacements normal to the bottom surface (x–y plane) were set to zero to simulate a plate that could not lift but could expand freely in the x–y plane. The thermo–elasto–plastic analysis used the same time steps as the thermal anlysis.

METHOD

The first step is to create a FEA mesh with as few elements as possible to minimize computing costs and still achieve useful accuracy. In the authors' experience, the size of the smallest elements are controlled by the need for roughly 2 to 10 elements spanning the heat source to approximate both the heat source and the temperature field adequately. Far from the heat source, the temperature field varies linearly and elements should be as large as possible while still approximating the geometry and material properties. The 3D grading elements shown in Fig. 2, which were developed by McDill et al. [7] to meet this requirement, are remarkably effective for computational weld mechanics.

Patel [5] used different meshes and time steps for the thermal and stress analyses. In this paper the same mesh and time steps are used for the thermal analysis and the thermo–elasto–plastic analysis. With the 3D grading elements, the thermal analysis is roughly twenty times cheaper than the stress analysis. Thus by designing the thermal analysis to meet the needs of the stress analysis, the need for interpolating or projecting temperatures from the thermal analysis to the stress analysis was eliminated.

There is one point in the thermal–stress analysis that deserves attention. Within a 4 noded quadrilateral, the 2D equivalent of the 8 noded bricks used in this paper, the polynomial describing the temperature field and thus the thermal strain is $T(x,y) = a_0 + a_1x + a_2y + a_3xy$. The polynomials describing the displacement field are $u(x,y) = b_0+b_1x+b_2y+b_3xy$ and $v(x,y) = c_0+c_1x+c_2y+c_3xy$. The polynomials describing the strain field are $\epsilon_{xx} = b_1 + b_3y$, $\epsilon_{yy} = c_1+c_3x$ and $\epsilon_{xy} = b_3x+c_3y$. The fact that the strain does not have a xy term corresponding to the xy term in the temperature field can cause difficulties. To avoid this problem, the thermal strain field has been integrated with one point integration. The polynomial describing the thermal strain becomes $T(x,y) = a_0$. The strain has been integrated with a $2 \times 2 \times 2$ rule in the 3D thermo-elasto-plastic stress analyses.

In the thermal analyses, it is necessary to move the heat source from node to node if the power input to the FEA is to remain constant.

RESULTS

The temperature fields at various times are plotted in Fig. 1. The corresponding displacement fields are plotted in Fig. 2. The σ_{xx} and σ_{yy} stresses are plotted in Fig. 3 and 5. The σ_{xx} stress on a cross–section through the arc center is plotted in Fig. 4 and 6. To save space, the shear stresses are not plotted. They are an order of magnitude lower.

DISCUSSION

This analysis is a first attempt, an exploratory run, to identify difficulties and refine strategies for 3D FEA analysis of welds. To minimize cpu time, the heat source was broadened to 4.77 mm. The weld length was limited to 2.5 cm. The convergence in the energy norm was set to 0.05 when 0.01 would have been preferred. All of these steps sacrifice accuracy to reduce cpu costs.

In spite of these approximations, the temperatures are believed to be more accurate than 2D analyses previously preformed. There is one test that can and should be applied to results of a thermal analysis. The maximum principle states that the maximum and minimum temperatures must be found on the boundary of a region provided no sinks or sources exist. Therefore, the minimum temperature in this problem should not be less than 20 C. The minimum temperature is actually 18 C. This should not be interpreted to mean that the error in the temperature is 2 C but that the error is at least 2 C. When meshes are too coarse, this error can be much larger.

The thermal strain due to temperatures above 600 C was set to zero. Patel [5] used a cutoff temperature of 1100 C. Without a cutoff temperature an elasto–plastic analysis must fail at some point as the yield strength approaches zero. The effect of increasing the cutoff temperature from 600 to 1100 C remains to be investigated. Ultimately, it will be necessary to invoke a constitutive model that can analyze behavior near the melting point.

The deformation shown in Fig. 2 is clearly 3D. If the deformation were plane strain the x–displacement would be zero. In fact the x–displacement is comparable to the y and z displacements. Clearly the assumption of plane strain, which has often been invoked in the past, is not valid.

In the near future, the computed stresses will be compared to experimental data to evaluate the accuracy. However, Chakravarti et al. [10] computed residual stresses with Patel's [5] methodology and measured residual stresses with a blind hole drilling technique. The results shown in Fig. 7 give some confidence in the current numerical method.

Some FEA theory is needed to fully understand the stress plots in Fig. 3 to 6. Continuum mechanics dictates that the stresses in the exact solution are smooth except possibly at internal boundaries such as those between austenite and ferrite. However, the stresses computed by FEA are not smooth. In general, the stresses are discontinuous across element boundaries. Most FEA programs smooth these stresses by projecting element stresses to nodes and then averaging contributions from the elements that share the node. Usually this smoothing has no rigorous mathematical basis. It is simply a filter to make the stresses easer to visualize. Smoothing algorithms are being developed for the 3D graded brick elements.

In choosing time steps, it was found that convergence was rapid if the maximum temperature change at any node was less than 100 C in any increment. Convergence was slow and could be divergent if the maximum temperature change at any node exceeded 200 C. In particular, small time steps should be used when starting and stopping the arc. This was particularly important when temperatures broke through 600 C and the yield strength dropped sharply. In effect, this highly stressed region suddenly lost its strength. On turning off the arc, the temperature falls rapidly because of the steep gradients. Turning the arc on and off instantaneously is a square wave that requires short time steps for accurate integration.

The welding community has little or no experience in interpreting or applying the stress field data shown in Fig. 3 and 5. The stresses are an internal response to the thermal expansion, with the no displacement in y on the top face and no displacement in z on the back face. The other four faces are entirely free. It may be useful to imagine the bar is entirely free and then force it back into position. How and if stress fields can be applied to understand welding phenomena such as defect formation and fracture remains to be explored.

COMPUTING COST — The computing cost, C, with most current computers (von Neumann) and FEA programs (Gaussian elimination) varies as $C = ANF^2$, where A is a constant for a particular computer and program, N is the number of degrees of freedom (dof) and F the front width or bandwith in the problem. In simple terms, N is the number of equations to be solved and F is the number of terms in each equation. Thermal analyses have 1 dof per node. Stress analyses have three dof per node. Therefore a stress analyses is expected to be 27 times as expensive as a thermal analysis for the same mesh and the same number of increments and iterations. In this case the ratio is actually 20 for the same number of increments but the number of iterations differed. This argument provides a rough guideline that should suffice to show that the thermo–elasto–plastic analysis is roughly 25 times as expensive as the thermal analysis.

One may ask if it is feasible to analyze a complete pass in a girth weld all around the circumference in a 60 cm (24 in.) diameter pipe with the fine mesh shown in Fig. 1. The answer is no; not with current software and hardware. The mesh would have roughly 72 times as many dofs and the same frontwidth. Thus each increment would be 72 times as expensive. For the arc to travel the 190 cm (75 in.) circumference would require 72 times as many increments. Thus the total cost would be 5000 times greater or 33 years of cpu time on a 1 mips, 0.1 mflops computer. If a Cray 2 is 800 times faster and the code were optimized to gain a further factor of ten, the Cray 2 would still require 36 hrs. of cpu time. A 36 pass weld would require 1300 hours of Cray 2 cpu time. In the next decade, the falling price and increasing performance of computers will help but cannot be expected to solve this problem.

Although it is not feasible to analyze a complete weld with current technology, the analysis of short sections of multipass welds could be done. It would be far more realistic than 2D models.

Clearly, the greatest challenge of computational weld mechanics is to develop algorithms that reduce this cost by at least three orders of magnitude. The authors' believe this to be entirely feasible. The analysis presented here contains most of the starting and stopping transients. In between there is a slow change in the quasi–steady state.

SUMMARY

1. The feasability of 3D transient thermo–elasto–plastic analysis of short welds has been demonstrated. The computing costs for long welds are not practical.

2. It is shown that the assumption of plane strain is invalid.

3. The thermo–elasto–plastic analysis requires approximately 25 times as much computer time as the thermal analysis using the same mesh and time steps.

ACKNOWLEDGEMENT

The financial support of NSERC A2992 (J. Goldak), NSERC Strategic Grant G1529 (J. Goldak), NSERC A4601 (M. Bibby) and AMCA International are gratefully acknowledged. Post–graduate scholarships from NSERC (A. Oddy) and Ontario Graduate Scholarship (M. McDill) are acknowledged. Our thanks to Professor N. Holtz for his assistance with computer methods. D. Downey implemented TAURUS, the post–processor, on our Apollo network. Special thanks to John Hallquist, Lawrence Livermore National Laboratory, for providing the post–processor, TAURUS, used to plot results.

REFERENCES

[1] J. A. Goldak, B. Patel, M.J. Bibby and J.E. Moore, *Computational Weld Mechanics*, Invited Opening Paper for AGARD Workshop — Structures and Materials 61st Panel Meeting, Oberammergau, Germany, September 8-13, 1985.

[2] L. Lindgren, *Deformation and Stresses in Butt–Welding of Plates*, Ph.D Thesis, Lulea University of Technology, Sweden, 1985.

[3] Y. Ueda and H. Murakawa, *Applications of Computer and Numerical Analysis Techniques in Welding Research*, Transactions of JWRI, 1984, vol. 13, No. 2, p. 337-346.

[4] J.H. Argyris, J. Szimmat and K.J. Willam, *Computational Aspects of Welding Stress Analysis*, Computer Methods in Applied Mechanics and Engineering, 1982, vol. 33, p. 635–666.

[5] B. Patel, *Thermo-Elasto-Plastic Finite Element Analysis of Deformations and Residual Stresses due to Welds*, Ph.D Thesis, Carleton University, Ottawa, Canada, Jan., 1985.

[6] K.J. Bathe, *Finite Element Procedures in Engineering Analysis*, Prentice–Hall, 1982.

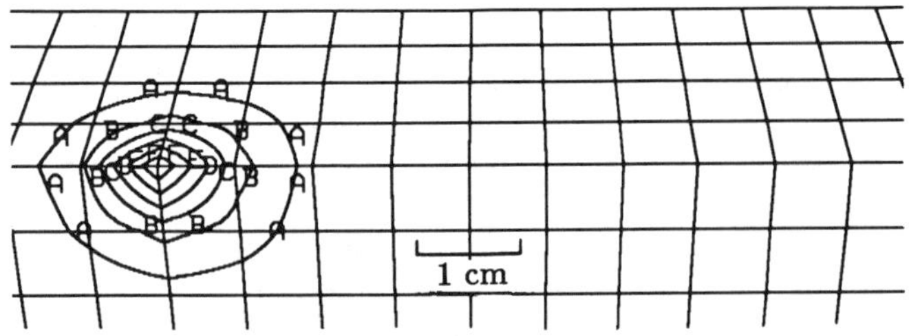

a) Temperature field at 0.52 s. Contour $A = 200$ C, contour $B = 400$ C, etc.

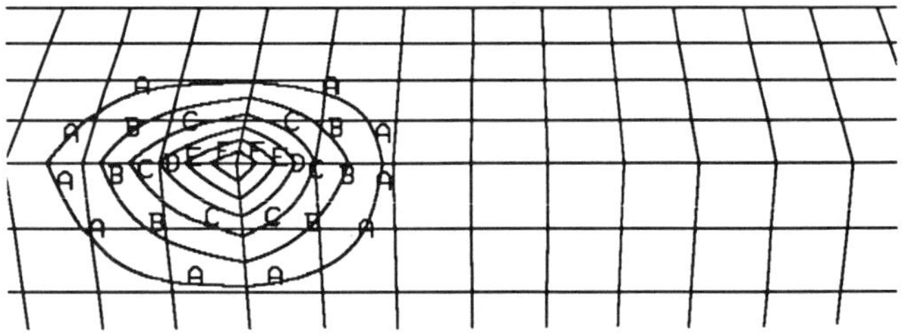

b) Temperature field at 1.04 s. Contour $A = 200$ C, contour $B = 400$ C, etc.

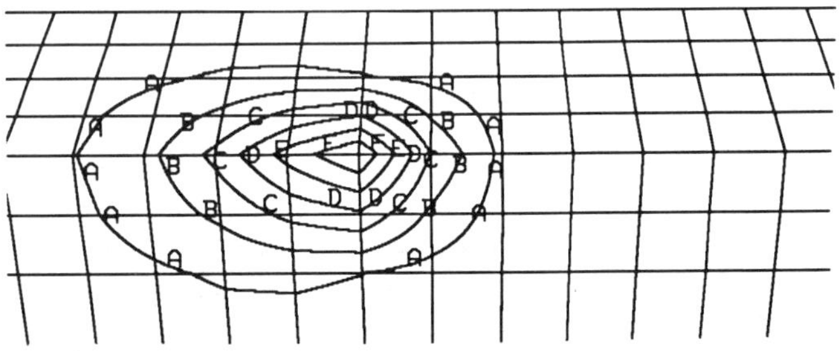

c) Temperature field at 2.20 s. Contour $A = 200$ C, contour $B = 400$ C, etc.

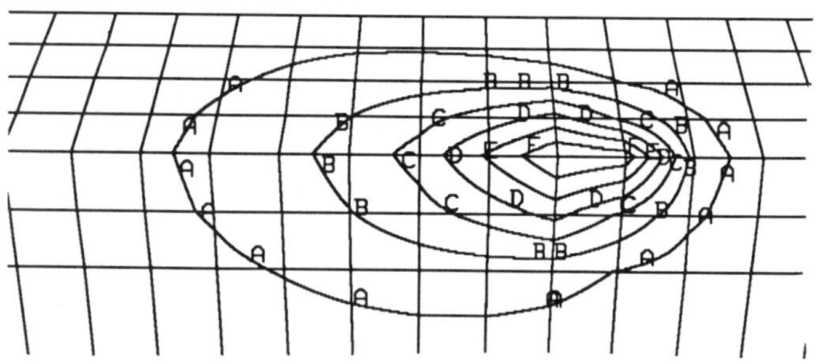

d) Temperature field at 4.40 s. Contour $A = 200$ C, contour $B = 400$ C, etc.

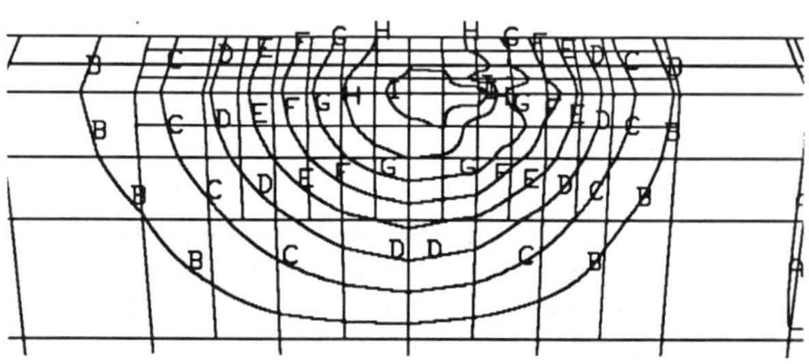

e) Temperature field at 7.75 s. Contour $A = 20$ C, contour $B = 40$ C, etc.

Fig. 1. The temperature fields at various instants of time.

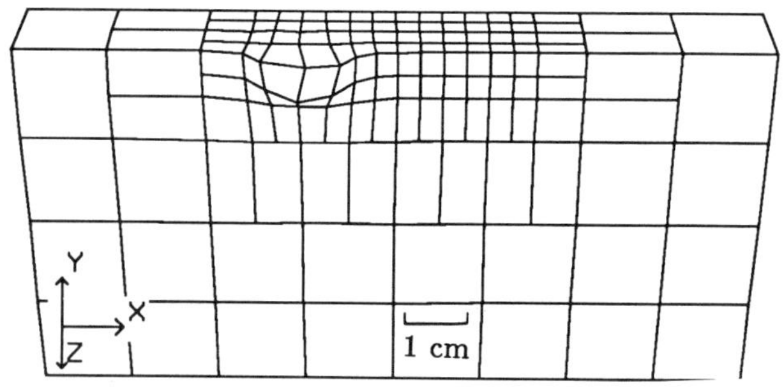

a) Displacement field at 0.52 s.

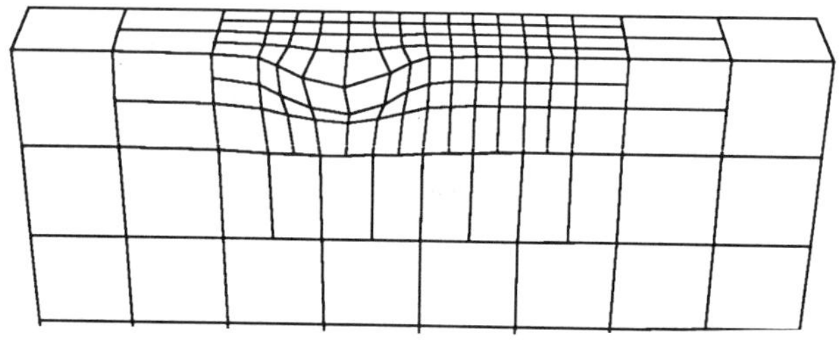

b) Displacement field at 1.04 s.

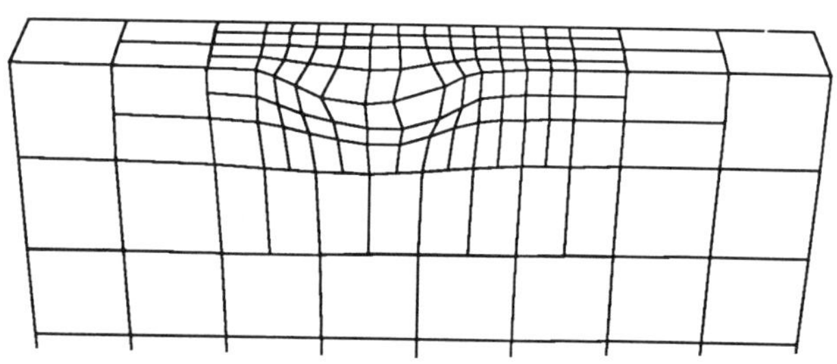

c) Displacement field at 2.20 s.

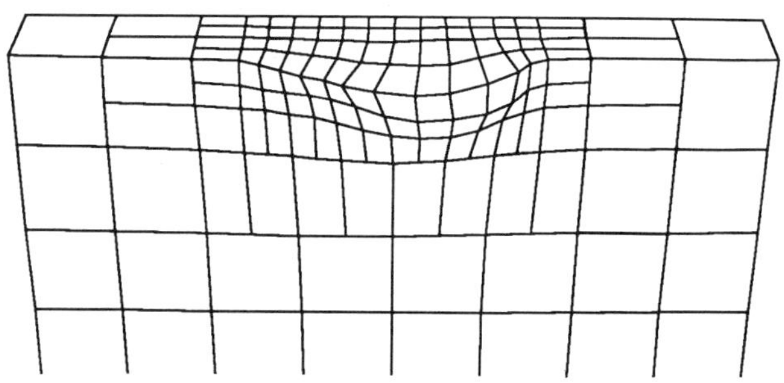

d) Displacement field at 4.40 s.

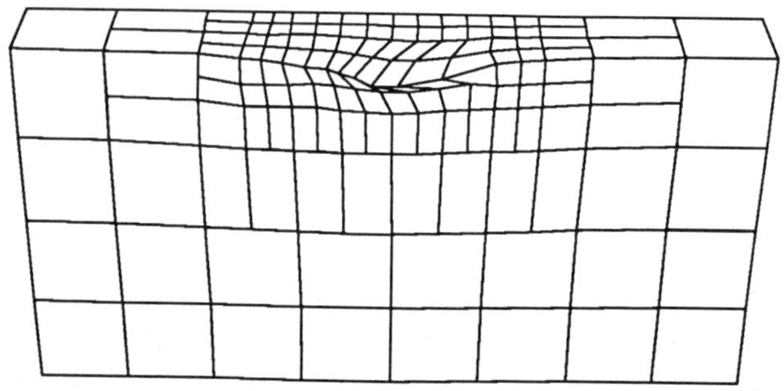

e) Displacement field at 7.75 s.

Fig. 2. The displacements, magnified 100X, at several instants of time.

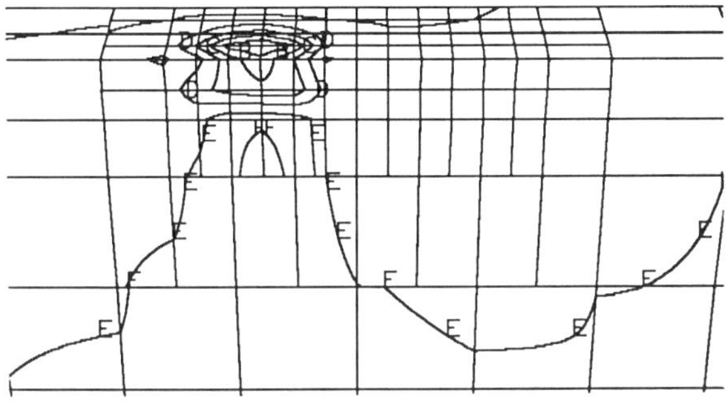

a) σ_{xx}.

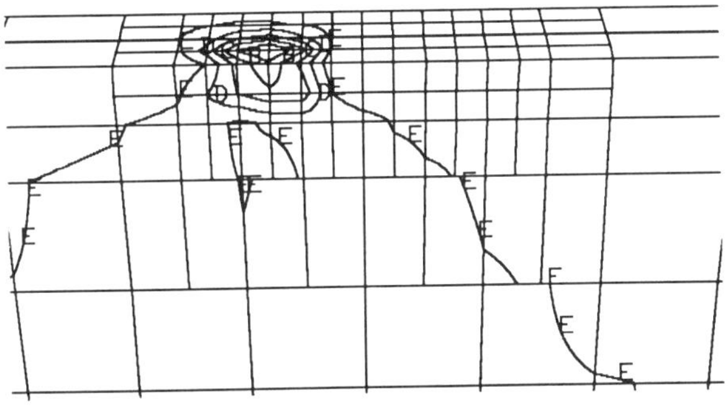

b) σ_{yy}.

Fig. 3. The σ_{xx} and σ_{yy} stress fields at 1.04 s. Contour $A = -800$ MPa, Contour $B = -600$ MPa, etc.

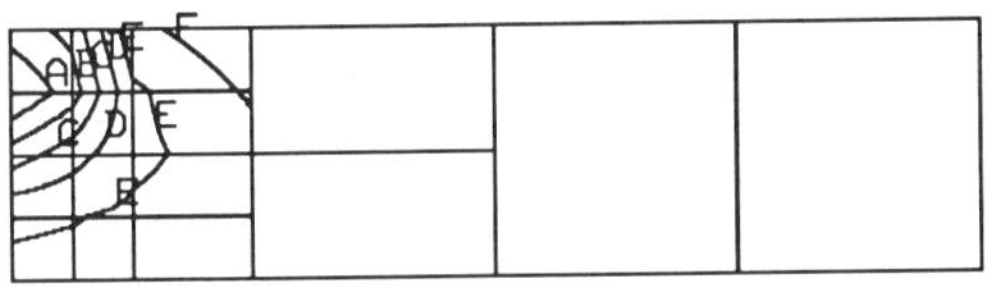

Fig. 4. The σ_{xx} stress field at 1.04 s on a cross-section through the arc. Contour $A = -800$ MPa, Contour $B = -600$ MPa, etc.

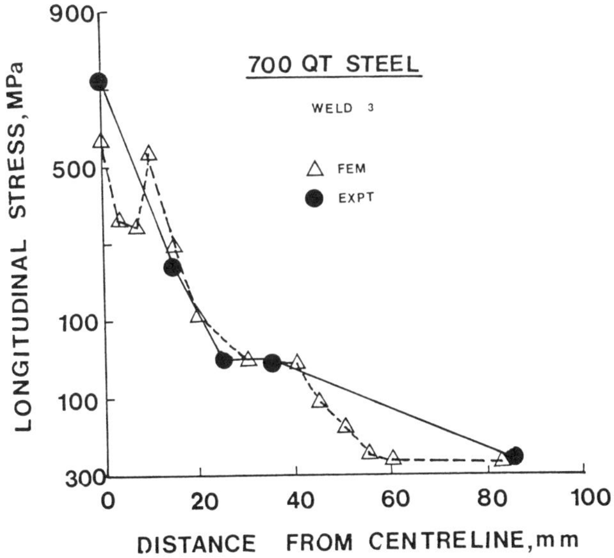

Fig. 7. Distribution of longitudinal residual stress in a bead-on-plate SAW weld in an 8 mm thick quenched and tempered low alloy (800 MPa) steel [10].

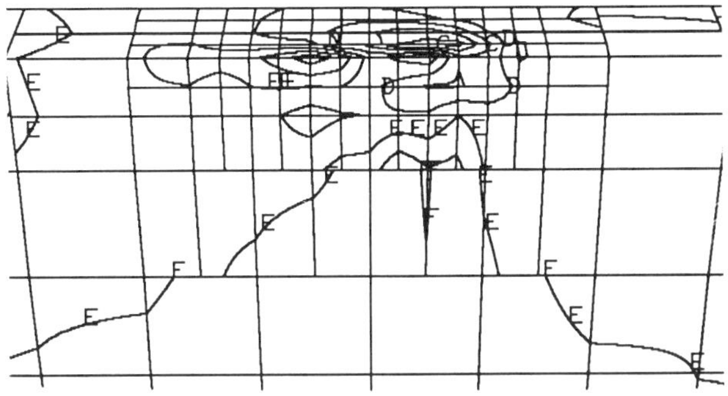

a) σ_{xx}

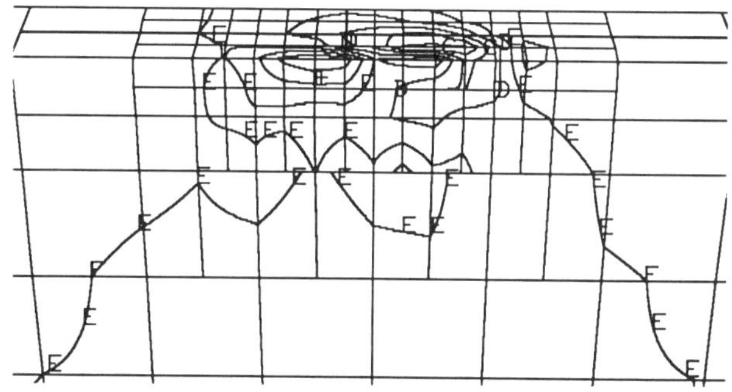

b) σ_{yy}

Fig. 5. The σ_{xx} and σ_{yy} stress fields at 4.40 s. Contour $A = -800$ MPa, Contour $B = -600$ MPa, etc.

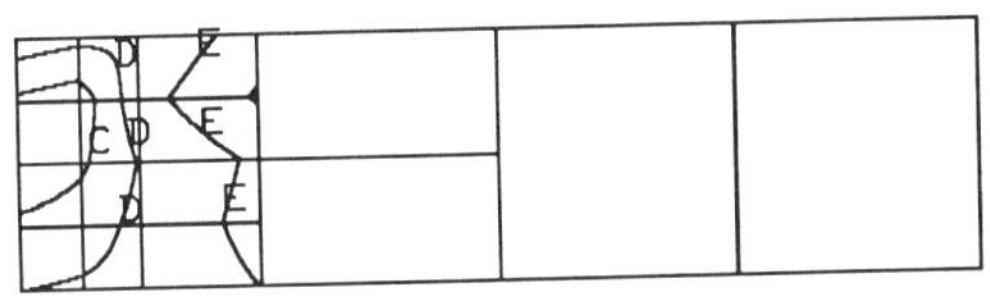

Fig. 6. The σ_{xx} stress field at 4.40 s on a cross-section through the arc. Contour $A = -800$ MPa, Contour $B = -600$ MPa, etc.

[7] M. McDill, J. A. Goldak, A. Oddy, M. J. Bibby, *Isoparametric Quadrilaterals and Hexahedrons for Mesh-Grading Algorithms*, Communications in Applied Numerical Methods.

[8] J.A. Goldak, A. P. Chakravarti and M.J. Bibby, *A New Finite Element Model for Welding Heat Sources*, Trans. AIME., 1984, vol. 15B, p. 299–305.

[9] The British Iron and Steel Research Association, *Physical Constants of Some Commercial Steels at Elevated Temperatures*, London, Butterworths Scientific Publications, 1953.

[10] A.P. Chakravarti, L.M. Malik, and J.A. Goldak, *Prediction of Distortion and Residual Stress in Panel Welds*, Ed. J. Too, Symposium on Computer Modelling of Fabrication Processes and Constitutive Behavior of Metals, Ottawa, Canada, May 15–16, 1986.

NON-CONTACT MEASUREMENT OF OUT-OF-PLANE DISTORTION AS A MEANS FOR CONTROLLING DISTORTION IN WELDED STRUCTURES

Koichi Masubuchi, Andrew DeBiccari, Walter J.C. Cook
Massachusetts Institute of Technology
Cambridge, Massachusetts 02139

ABSTRACT

This paper discusses a novel and simple method of non-contact measurement of out-of-plane distortions of welded structures. This method utilizes an optical method referred to as laser interferometry, where two wide beams of phase-locked monochromatic light are directed onto the specimen surface. When a minute deviation from a perfect plane exists on the surface, the pattern of interference fringes is distorted. The shape and amount of distortion is determined by analyzing the fringe pattern. After a short introduction, this paper presents (a) description of the laser interferometer developed at M.I.T., (b) some basics of analyzing out-of-plane distortion, and (c) results obtained on various types of specimens. The laser interferometer has been successfully used for measuring distortions of various types of specimens including: (1) distortions of fillet welds, (2) angular distortions of butt welds, (3) bending of plates by line heating with an oxyacetylene torch of a high-power laser beam, (4) distortion of a rail segment due to repair welding, and (5) distortions of pipes due to girth welding.

SINCE JUNE 1983, a three-year program entitled "Improvement of Fusion Welding Through Modeling, Measurement, and Real-Time Control" has been performed at the Laboratory for Manufacturing and Productivity of M.I.T. for the Department of Energy. As a part of this program, the research team under Professor K. Masubuchi has studied means for controlling residual stresses and distortions in weldments. A key activity in this study has been to develop an experimental technique for measuring out-of-plane distortions. To be useful for this purpose, the technique should satisfy the following requirements:

(1) The technique requires minimum preparation and instrumentation of a specimen being studied
(2) If possible, it is not necessary to touch the specimen during measurement
(3) Measurement can be made, if possible, in real time during welding
(4) The technique should be applicable to various types of weldments including large structures
(5) The technique should be relatively inexpensive and easy to operate.

After evaluating several possible alternatives, the research team has selected the laser interferometry as a method to be studied further, since it satisfies all of the above-mentioned requirements. Since there was no adequate equipment on the market, the team has designed and constructed equipment that has been successfully used for various experiments. This paper presents:

(a) Description of the laser interferometer developed at M.I.T.
(b) Methods for analyzing fringe patterns
(c) Experiences with the interferometer.

DESCRIPTION OF THE LASER INTERFEROMETER DEVELOPED AT M.I.T.

Further details of the laser interferometer developed at M.I.T. may be found in References [1] and [2]. Figure 1 schematically shows its optical system. A low power (10 milliwatt) helium neon laser is used as a source of monochromatic light having a wavelength of 6328 Å. A laser source was chosen for reasons of wavelength purity and long maintenance-free operation. Within the laser interferometer, the laser beam is expanded, columnated, and split into two phase-locked beams by the action of two concave mirrors. This produces a beam overlap causing a system of parallel interference fringes composed of alternating bright and dark lines to appear on a diffuse surface placed anywhere in the overlap zone.

FRINGE SPACING EQUATION - The spacing of interference fringe planes in the overlapped zone, d, is given by (see Figure 1):

$$d = \frac{\lambda}{\theta} \qquad (1)$$

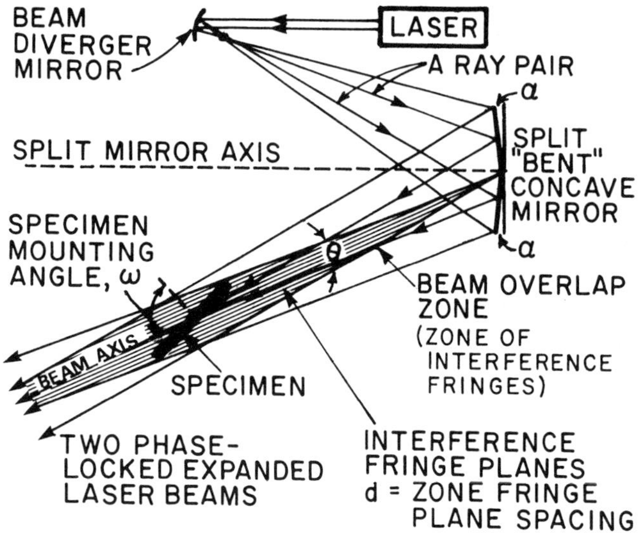

Fig. 1 - Laser interferometer optics [1]

where,

d = bright-to-bright, or dark-to-dark fringe plane spacing

λ = laser beam wavelength (6328 Å for the helium neon gas laser)

θ = 4α

α = tilt angle of each split mirror half.

Then, the spacing of interference fringes on the specimen, s, is given by:

$$s = \frac{d}{\cos \omega} \tag{2}$$

where,

ω = specimen mounting angle.

Normally, an ω approaching 90 degrees is used so that the value of s is in the 1 to 2 mm (0.04 to 0.08 in.) range. Figure 2 is a photograph of a typical fringe pattern on a flat surface. Fringes are planer and equally spaced.

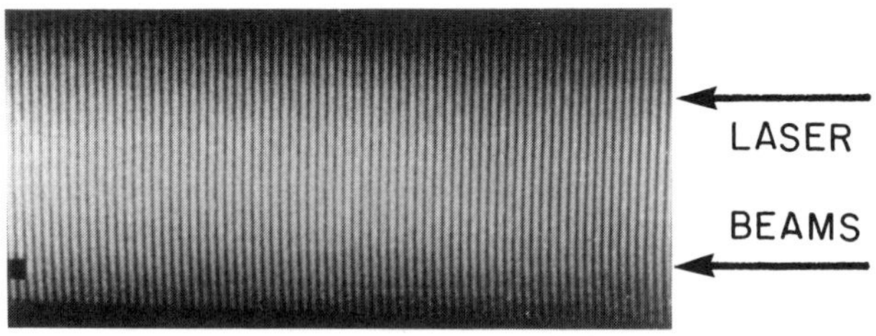

Fig. 2 - Typical fringe pattern on a flat surface

FRINGE PATTERNS PRODUCED BY OUT-OF-PLANE DISTORTIONS

A brief discussion is presented here on the characteristics of fringe patterns caused by the following three types of distortions:

(1) Contour deformation
(2) Lateral bending distortion
(3) Longitudinal bending distortion.

These types of distortions are commonly observed in actual applications.

CONTOUR DEFORMATION AND LATERAL BENDING DISTORTION - Figure 3 shows a typical fringe pattern observed on the back surface of the bottom plate of a fillet weld having both contour deformation and lateral bending distortion. Note that the lower part of the bottom plate is held in the vertical position by the specimen holder. Then the upper part of the bottom plate is bent into the page. The effect of a ridge along the plate center line is to produce fringes on the ridge which bow to the right (toward the laser). Similarly, a depression would cause fringes which bow to the left. The lateral bend angle and the height of the centerline deformation anywhere on the specimen may be easily determined by use of the following two equations:

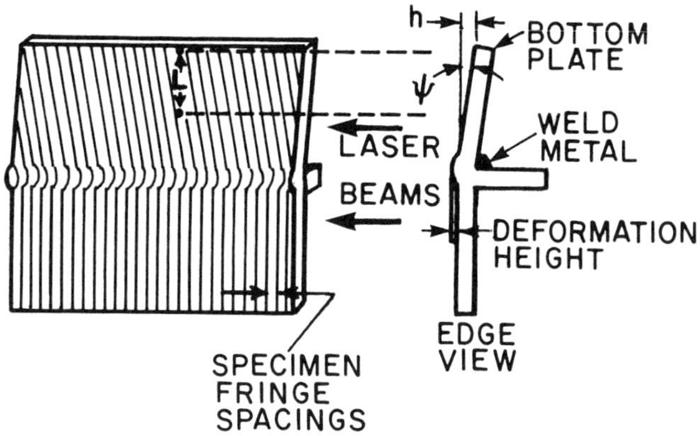

Fig. 3 - Fringe pattern observed on the back surface of a fillet weld having both contour deformation and lateral bending distortion

Deformation Height - Deformation height, H, may be determined as follows:

$$H = d \times N \tag{3}$$

where,

d = zone fringe-plane spacing (approximately 0.07 mm)

N = number of fringes shifted

A deformation height as small as 0.01 mm may be measured.

Lateral Bending Distortion - Bend angle, ψ, may be determined as follows:

$$\psi = \tan^{-1} \frac{h}{L} \tag{4}$$

where,

h = d (Note that ω must be greater than 80 degrees for this simplification to be reasonably accurate)

L = vertical spacing between two adjacent fringes on the specimen surface, as shown in Figure 3

A bend angle as small as 0.1° may be measured.

LONGITUDINAL BENDING DISTORTION - Longitudinal bending of a plate is shown in Figure 4. The upper figure shows the fringe pattern on the plate, while the lower figure shows the top edge view of the bent plate. The spacing between two adjacent fringes is measured for each of the two points on the specimen between which the angle is to be determined. For each fringe spacing, the effective specimen mounting angle

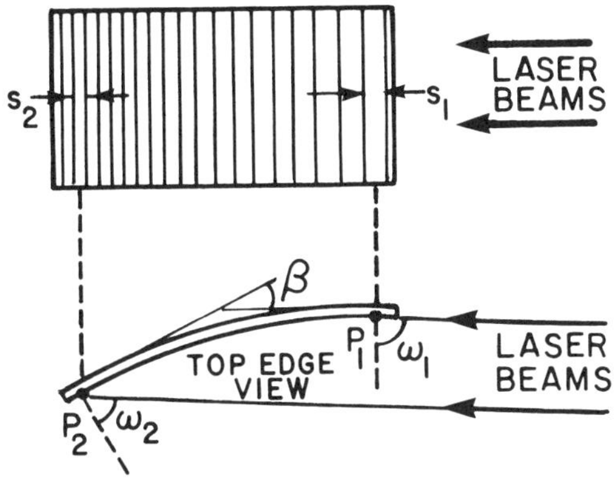

Fig. 4 - Fringe pattern on a plate having longitudinal bending distortion

(angle between the specimen normal at that point and the laser beam axis) is then computed. The longitudinal bend angle, β, between the two points P_1 and P_2, is the difference between the effective mounting angles at these two points, as follows:

$$\beta = \omega_1 - \omega_2 \qquad (5)$$

where,

ω_1 = mounting angle at $P_1 = \cos^{-1}\frac{d}{s_1}$

s_1 = specimen fringe spacing at P_1

ω_2 = mounting angle at $P_2 = \cos^{-1}\frac{d}{s_2}$

s_2 = specimen fringe spacing at P_2

Although Figure 4 shows the fringe pattern on a plate having gradual longitudinal distortion, the basic principle presented here can be used for analyzing a fringe pattern obtained on a specimen having sharply changing longitudinal distortion.

EXPERIENCES WITH THE LASER INTERFEROMETER

The laser interferometer developed at M.I.T. has been successfully used for measuring distortions of various types of specimens including:

(1) Distortions of fillet welds
(2) Angular distortions of butt welds
(3) Bending of plates by line heating with an oxyacetylene flame or a high-power laser beam
(4) Distortion of a rail segment due to repair welding
(5) Distortions of pipes due to girth welding.

Results obtained in cases (1) through (4) have been reported in Reference [2]; therefore, their discussions are brief.

DISTORTIONS OF FILLET WELDS - Figure 3 shows a typical fringe pattern observed on the back of a fillet weld. Two characteristic distortions include (a) lateral bending distortions or angular distortion and (b) contour deformation in regions below the weld metal.

ANGULAR DISTORTIONS OF BUTT WELDS - It has been found that the laser interferometer is capable of determining angular distortions of butt welds.

BENDING OF PLATES BY LINE HEATING - An oxyacetylene flame is widely used for forming plates in various shapes and also straightening distorted structural members. For example, Japanese shipyards frequently use the line heating technique for bending steel plates[3,4]. However, mechanisms of plate bending by line heating are complex, and long-term experience is needed for a worker to master the art of flame bending. M.I.T. researchers have made a movie which shows the transient movement of a plate during heating by an oxyacetylene flame. They have also made video tapes that show real-time distortions of steel plates while they were heated by a high-power laser beam[5].

RAIL TRACK WELD - M.I.T. researchers also used the interferometer to determine the amount of a minute angular distortion of a rail segment caused by a simulated repair welding performed at the head. By use of the interferometer, it was possible to determine the amount of angular distortion as small as 0.2° that could not be detected by a dial gage.

DISTORTIONS OF GIRTH-WELDED PIPES - Efforts have been made to determine distortions of inner and outer surfaces of pipes during girth welding. Details of experimental conditions and results may be found in DeBiccari's thesis [6]. Presented below is a brief discussion of distortion of a girth-welded pipe measured on the outer surface.

Experimentation - Two sections of low-carbon steel pipes, each 230 mm (9") long were welded together. The inner diameter was 305 mm (12"), and the wall thickness was 8 mm (0.313"). The weld was made in two passes using the gas tungsten arc process. The welding conditions were:

Pass #1: Arc voltage - 13.0 Volts
Welding current - 205 Amperes
Travel speed - 0.146 cm/sec (3.46 in/min

Pass #2: Arc voltage - 13.0 Volts
Welding current - 185 Amperes
Travel speed - 0.104 cm/sec (2.46 in/min)

Both passes used a shielding gas of argon at a flow rate of 19 litre/min (40 cubic feet/hour) and a filler wire of 0.9 mm (0.035") in diameter fed in at a rate of 2.59 cm/sec (61.2 in/min).

After the weld was allowed to cool, the radial deflection was measured mechanically by use of a dial gage. After the mechanical measurements were taken, the weld pipe was placed in front of the interferometer and the phase-locked laser beam was projected onto the same area where the mechanical measurements were taken, as shown in Figure 5. The resulting fringe pattern was then photographed.

Results and Discussion - Figure 6 is a photograph of the fringe pattern of the pipe before welding. Fringes are equally spaced with an average spacing of 3.676 mm. Since the measured surface is not flat but curved in a cylindrical form, fringe patterns are parabolic even without distortion.

Figure 7 shows the fringe pattern of the

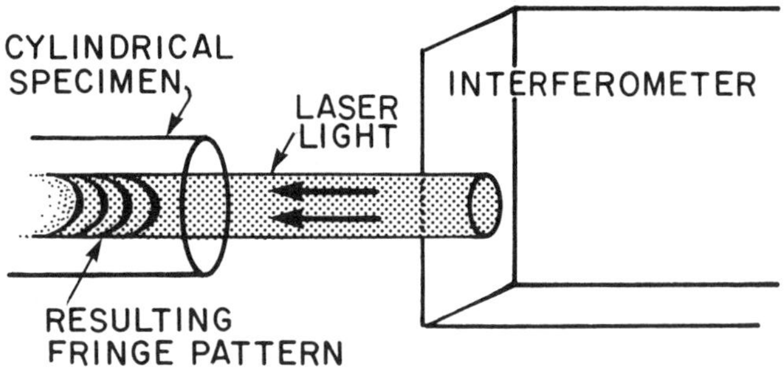

Fig. 5 - Set-up for measuring distortion of a pipe

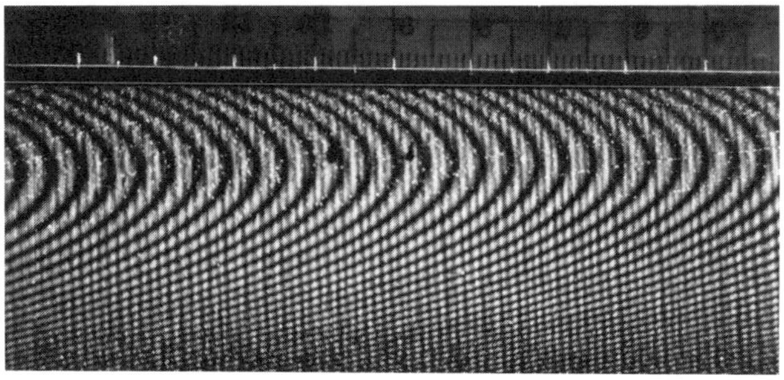

Fig. 6 - Fringe pattern on an unwelded pipe

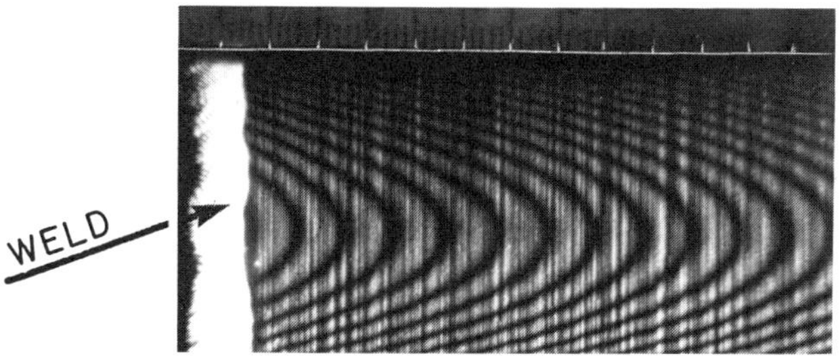

Fig. 7 - Fringe pattern on a girth welded pipe

pipe after welding. As a result of radial distortions that have occurred in regions near the weld fringe, spacings are no longer equal. The distribution of radial distortion can be determined by proper uses of Equation (3). The fringe spacing at a certain location (the distance from the weld) gives the incremental increase (or decrease) in radial contraction at that location. By knowing changes of fringe spacings at various locations, it is possible to determine the amounts of radial distortions as a function of the distance from the girth weld.

Figure 8 shows distributions of radial distortion determined by the interferometer and a dial gage. The results show that regions near the weld have shrunk inward. The results obtained by the two measurement techniques are in a very good agreement.

CONCLUSIONS

It has been found that the laser interferometer can be successfully used to measure out-of-plane distortions produced by welding and heating of various types of specimens. The equipment

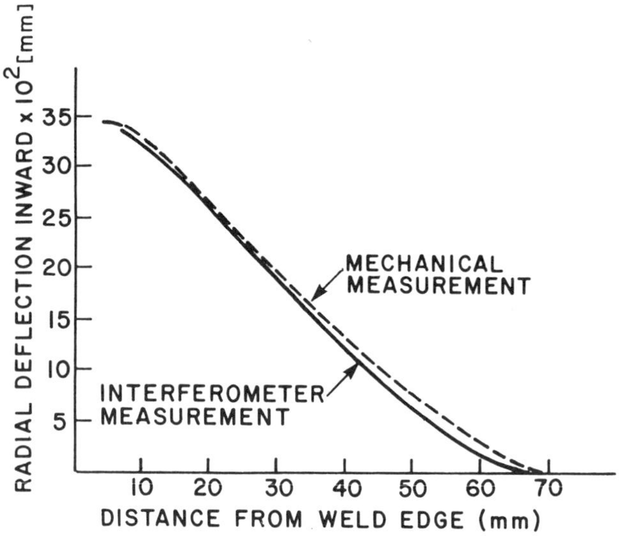

Fig. 8 - Distributions of radial distortion determined by the interferometer and a dial gage of a girth-welded pipe

developed at M.I.T. is rather inexpensive and easy to use. We believe that the laser interferometer can be used not only in laboratories as a research tool but also in fabrication shops for production control and training.

ACKNOWLEDGMENT

The authors are grateful to the DOE project monitor, Dr. Oscar P. Manley of the Division of Engineering and Geoscience, Office of Basic Energy Sciences, for his kind understanding and support of their research efforts. The authors are also grateful to M.I.T. staff members for providing various types of assistance. Special thanks are given to Mr. Anthony J. Zona for preparing welding specimens and Ms. Muriel Bernier for typing and technical illustrations.

REFERENCES

1. Cook, W. J. C., "Laser Interferometer for Weld Distortion Measurement", a report prepared under Subject 13.713 Special Problems in Ocean Engineering, M.I.T., December (1982)
2. Masubuchi, K., and Cook, W. J. C., Proc. Spring Meeting/STAR Symposium of the Soc. of Naval Arch. and Marine Eng., 225-243, May (1983)
3. Hashimoto, T., and Fujishiro, Y., Journal of the Soc. of Naval Arch. of Japan, 104, 201-225, (1959)
4. "Line Heating", The National Shipbuilding Prog., U.S. Dept. of Transp., MARAD in cooperation with Todd Pacific Shipyards Corp., (1982)
5. Research on Laser Heating, research contract for Todd Pacific Shipyards Corp. under Purchase Order PL-67103
6. DeBiccari, A., "Control of Distortion and Residual Stresses in Girth Welded Pipes", Ph.D. Dept. of Ocean Eng., M.I.T., (1986)

USE OF MOIRE INTERFEROMETRY ON WELDS

Susan A. Chavez, Vance A. Deason, Jonathan S. Epstein
EG&G Idaho
Idaho Falls, Idaho

ABSTRACT

Moire interferometry is used to obtain full-field strain data across welds in stainless steel plates with ferralium or nickel filler. Stress-strain curves from strain gage and moire data are compared to verify the moire data. Continuous strain profiles across a section of the weld with no apparent surface defects are obtained using moire data. A defect in the weld is identified by visual examination of the moire fringe pattern and the strain profile across this defect is quantified.

MOIRE INTERFEROMETRY is an optical technique used to measure full-field surface displace ment with resolution <1 μm. It is a valuable tool for detection and analysis of regions where strain gradients are very high or irregular. Such regions include the heat affected zone around a weld, which may experience a high microstructural gradient, and the region surrounding a surface defect.

This paper briefly introduces moire interferometry and its use in determining full-field strain data across welds. Moire results are verified with strain gage data by using both techniques to obtain stress-strain curves in the fillers and base. The full-field data are used to: (1) obtain strain profiles across the base, heat-affected, and fusion zone interfaces, (2) detect surface defects, and (3) obtain strain profiles across a defect.

BACKGROUND

The moire effect can be created by mechanical or optical superposition. Mechanical moire is a very simple phenomena which almost everyone has seen in two parallel snow fences or when two layers of window screen are placed in contact. It occurs when two similar but not quite identical arrays of equally spaced lines are superposed, or when one superposed array is deformed relative to the other. This relative deformation creates a series of fringes, called moire fringes, which is the moire pattern. So, if one repetitious pattern is deformed while the other remains unchanged, the moire pattern records the full-field effect of the deformation. Displacements are calculated from a simple relationship involving the number of fringes and the grating frequency [1].

The moire interferometry effect is created by diffraction and interference of light. However, many prefer to think of optical moire interferometry in terms of mechanical moire, because optically produced fringes have the same interpretation as mechanical moire fringes [2]. A brief review of moire interferometry follows, for a more complete treatment see Reference 3.

When a beam of light is cast upon a reflective diffraction grating (Fig. 1), the incident beam is diffracted back in a multiplicity of lesser intensity beams emerging at angles defined by Eq. (1).

$$\sin\beta_m = m\lambda F + \sin\phi \qquad (1)$$

Where β is the diffraction angle; m, diffraction order; λ, wavelength; F, grating frequency; and ϕ, the angle of incidence.

In the sign convention here, angles that are counterclockwise with respect to the z axis are positive for light propagating in the +z direction and negative for light propagating in the -z direction. For simplicity only the first order (m = +1 and m = -1) diffraction beams will be considered. They have the strongest intensity and are the ones primarily used for data collection.

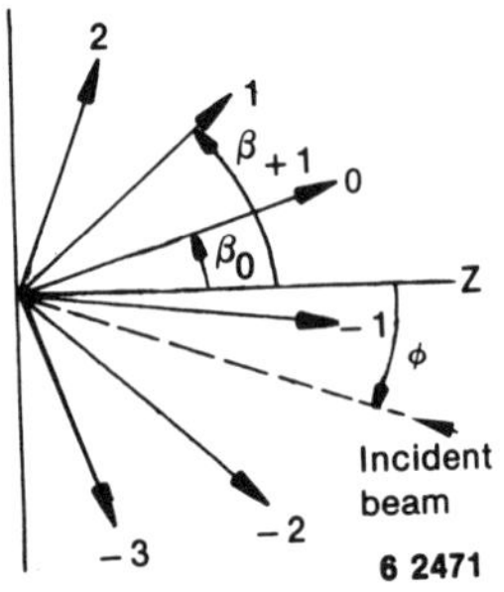

Fig. 1 - Light beam diffracting off reflective phase grating.

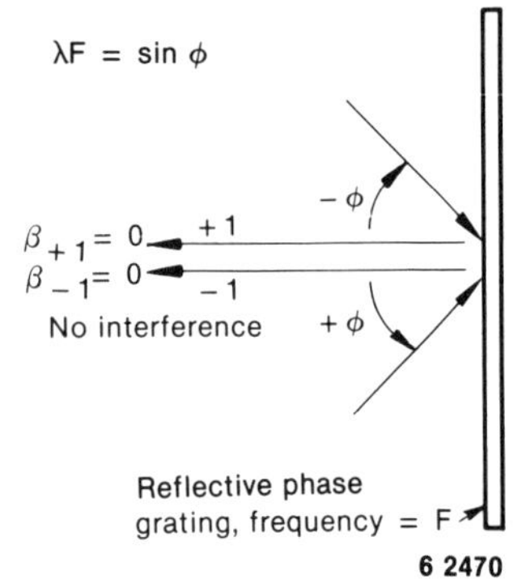

Fig. 2 - Initial undeformed specimen. Incident beams set at specific angles ±ϕ to create no interference of first order diffraction beams.

One can solve for the incident angles (+/-ϕ) which would cause the +1 diffraction order of an incident beam and the -1 diffraction order of another (symmetric) incident beam to emerge perpendicular to the grating (parallel to each other, Fig. 2). This is generally the unloaded, null state. No interference is taking place. If the grating frequency (F) is changed by loading and deforming the specimen, the first order beams will no longer emerge parallel to each other (Fig. 3). Interference of the first order beams is taking place.

Light amplitude can be represented as a sinusoidal wave. When two beams of light are superposed, interference results. Constructive interference occurs when maximum amplitudes of one beam intersect maximum amplitudes of the other. Destructive interference occurs when maximum amplitudes of one beam intersect minimum amplitudes of the other, cancelling the two. As shown in Fig. 4, the interfering beams of light create fringes of constructive and destructive interference. Each fringe represents a contour line of equal in-plane axial displacement. Displacements are calculated from essentially the same equation used in mechanical moire [2,3]:

$$u = n/(2F) \tag{2}$$

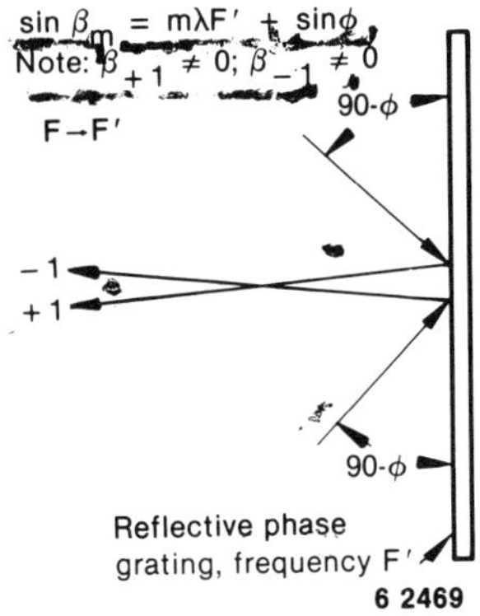

Fig. 3 - Specimen deforms. First order diffraction beams interfere, creating fringe pattern.

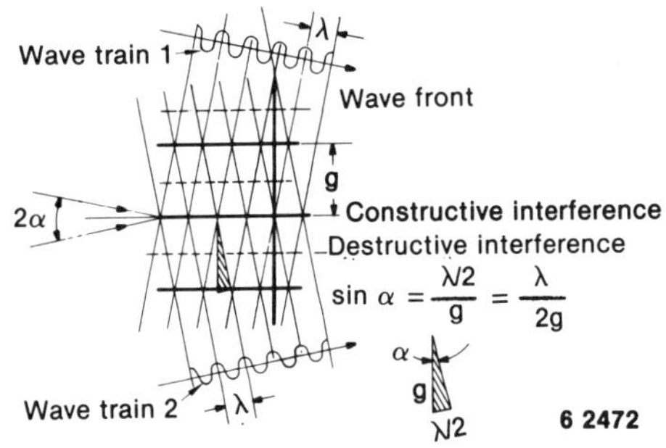

Fig. 4 - Interference of two light wave trains.

Where u is displacement; n, fringe number (counted from arbitrary reference point); and F, original grating frequency.

It is important to note that fringes are lines of constant displacement. They record localized increases (or decreases) in displacement, as well as discontinuities, in the same way a contour map shows changes in elevation.

Hence the two interfering first order beams shown in Fig. 3 create a fringe pattern which can be recorded on a photographic plate and interpreted to full-field displacement data. As the grating frequency changes with the deforming specimen, so does the moire fringe pattern.

Although moire interferometry fringes are created solely by the interference of the first order diffraction beams, many prefer to think of moire interferometry in terms of mechanical moire: superposition of the deforming specimen grating and a constant "virtual grating" created in space by the interference of the two incident beams [4].

Strain (e) can be calculated by measuring the displacement at two points (u_1,u_2) and the distance between the two points ($\Delta\ell$).

$$\epsilon = \Delta u/\Delta\ell \tag{3}$$

From Eq. (2) and (3) the strain between two adjacent fringes is

$$e = 1/(2F\Delta\ell) \quad (4)$$

For a given original grating frequency, then, strain is inversely proportional to the distance between fringes. (Broad fringes indicate low strain).

EXPERIMENTAL

Two Type 304 stainless steel plates were butt welded together with a nickel filler using the gas tungsten arc welding (GTAW) process. Similarly, two other Type 304 plates were welded using a ferralium filler. Two tension specimens were cut from each welded plate with the weld transverse to the tensile axis per AWS QW-462, except that a longer gage length than specified was used. One specimen of each material was mounted with five strain gages (1 in the filler, 1 in each HAZ and 1 on each side of the filler in the base material, see Fig. 5). The other two specimens were covered with a reflective phase gratings (600 l/mm for ferralium and 1200 l/mm for nickel filler) (Fig. 5). Localized strain data from corresponding locations on the specimens were compared.

Gratings were produced by exposing a photographic plate to a "virtual grating" formed in space by the interference of two beams of light (Fig. 4). The plate was developed, bleached, and dried causing localized shrinkage of the constructive interference zones. The resulting regular corrugated surface was dipped in wetting solution before coating with a thin film of aluminum. (The wetting solution aids separation when transferring the aluminum mold to the specimen).

To transfer the aluminum phase grating to the specimen, the specimen was cleaned and

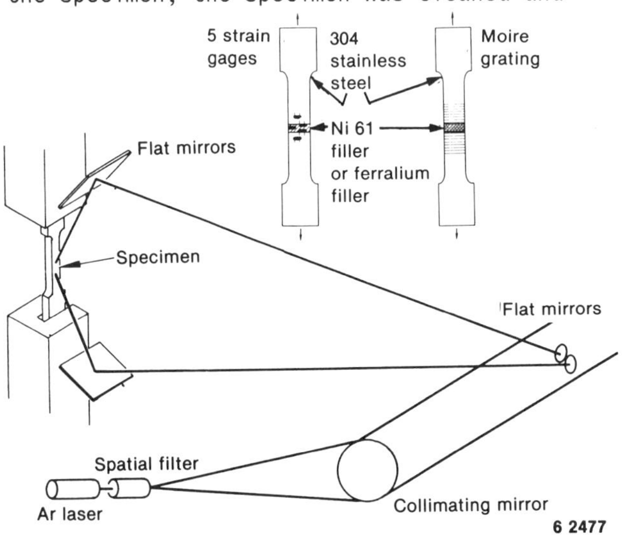

Fig.5 - Schematic moire test configuration.

coated with a suitable epoxy. The aluminum mold was pressed against it until the epoxy polymerized. Extreme care was taken not to introduce contaminants or bubbles into the epoxy layer. The aluminum mold adhered to the epoxy and the specimen, releasing at the wetting solution interface. A complete description of this procedure can be found in Reference 5.

Loads were monitored as specimens were pulled to failure on a 196 N (20 kip) screw-driven test machine. The loading range was "static", 7.5 to 11.0 MPa/min. (1.1 - 1.6 ksi/min.). Strain gage data for all five strain gages were taken simultaneously on a strip chart recorder.

The optical arrangement of the moire test is shown schematically in Fig. 5. Note that the two incoming beams must form a plane parallel to the tensile axis and the grating lines must be perpendicular to the tensile axis. Optical parameters were: grating frequencies, 600 l/mm ferralium specimen, 1200 l/mm nickel specimen; orientation, grating lines perpendicular to tensile axis; laser, argon; wavelength, 514 x 10^9 m; optics, $\lambda/10$ throughout. Photographs of the fringe patterns were taken at known specific loads.

RESULTS AND DISCUSSION

Stress-strain curves from moire and strain gage data are compared in Fig. 6. to verify the moire results. The stress-strain curves of the base material correlate extremely well. Elastic moduli are within 5% of each other and stresses at the 1% offsets are within 10% of each other. Data from filler materials are not as consistent, possibly due to inhomogeneity in the as-laid material. In the ferralium filler data, the strain gage glue appears to be slipping at stresses above 250 MPa.

Approximate elastic moduli for 304 SS, nickel filler, and ferralium filler taken from both methods are 180, 190, and 215 MPa, respectively. Stresses at the 1% strain offsets are 303, 245, and >350 MPa, respectively. This indicates that the nickel filler yields before 304 SS, which yields before the ferralium filler. The differences in yield stresses are reconfirmed by moire strain profiles of the welds which will be discussed.

Moire interferometry is not best suited to finding stress-strain response at one point, but rather, to evaluating full-field displacements with very high sensitivity. Fig. 7 shows a progression of moire patterns taken while testing a weld with nickel filler. The higher density of fringes (higher strain) in the nickel filler at 245 MPa could have been predicted by the stress-strain curves. However, the growing discontinuity between filler and base, probably due to lack of

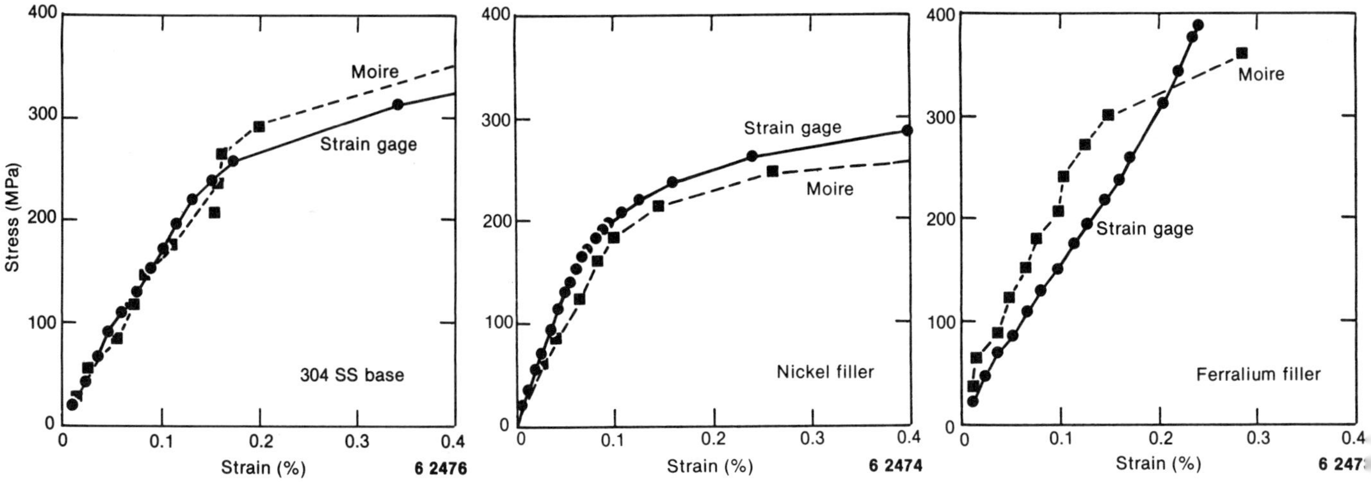

Fig. 6 - Stress-strain curves of 304 SS base, nickel filler and ferralium filler.

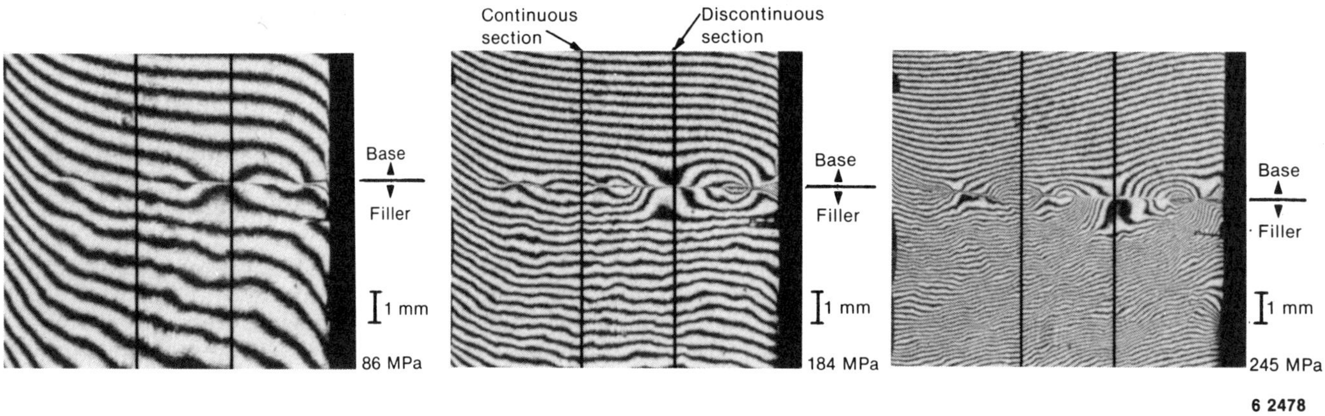

Fig. 7 - Moire fringe patterns of nickel filler/304 SS base interface.

fusion, is detected and quantified only with a full-field technique.

Strain profiles of three weld traverses obtained by moire interferometry are plotted at increasing stresses for ferralium filler weld, nickel filler weld with no discontinuity, and nickel filler weld with discontinuity in Figs. 8, 9, and 10.

The ferralium filler weld showed no surface cracks or defects. Data were taken only to the weld centerline, so symmetry is used to complete the graph for ease of comparison with Figs. 9 and 10. The ferralium has a slightly higher elastic modulus than the 304 SS, resulting in slightly higher strains in the base for stresses below the 304 SS yield stress. At 236 MPa, the 304 SS base begins to yield and at 354 MPa a steep strain gradient between the two materials is formed.

Fig. 9 shows the strain profiles of a section which was continuous across the nickel filler/304 SS base interface (i.e. no surface cracks or defects). At 214 MPa and below, the strain profiles of the filler and the base materials are constant, indicating similar elastic moduli. At 245 MPa and above, the nickel filler yields plastically, forming a very high strain gradient between the nickel filler and 304 SS base. This was observed to be true from the stress strain curves. Because the filler yields before the base, this profile is inverse to the ferralium profile where the base yields before the filler.

Strain profiles of the nickel filler weld across a section with a surface defect are plotted in Fig. 10. At high stresses, the defect exhibits an extremely steep strain gradient from the strain relieved region (adjacent to the defect) to the plastically yielding filler material. A surface crack or hole is indicated by the discontinuity of moire fringes. Also, large displacements and low strains are calculated from the moire fringes around the region.

SUMMARY

Moire interferometry was applied to nickel filler and ferralium filler welds to obtain full-field strain data. Strain profiles from continuous traverses across the welds are consistent with stress-strain data from the

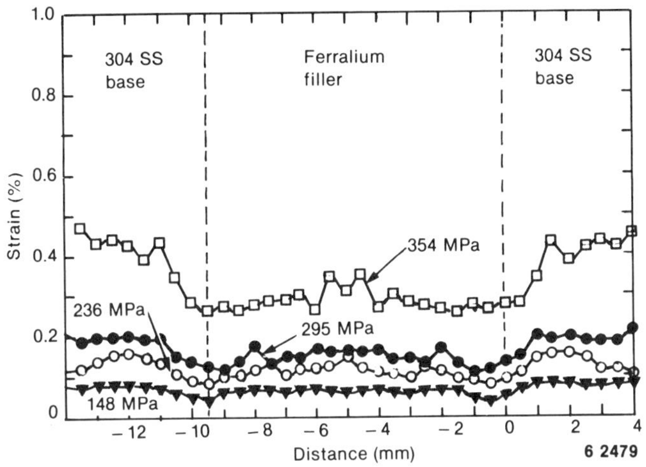

Fig. 8 - Moire strain profiles of ferralium filler, 304 SS base weld.

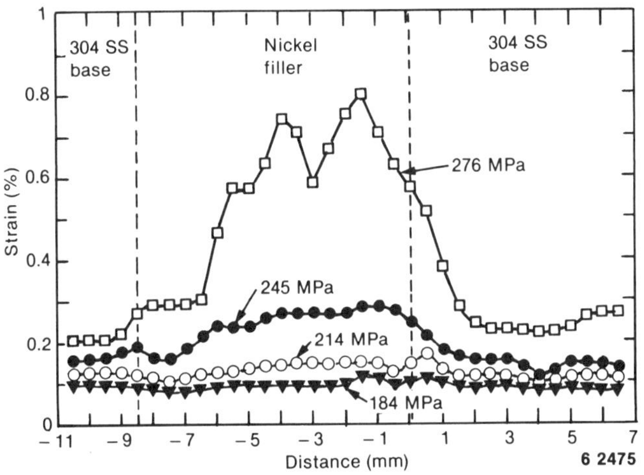

Fig. 9 - Moire strain profiles of nickel filler, 304 SS base weld through continuous section.

304 SS base and filler materials. A surface defect is identified and strain relief around the defect is quantified.

Regions with very high or irregular strain gradients can be analyzed using moire interferometry. High frequency gratings and higher magnification, high resolution optics would be necessary for a more detailed investigation of any particular region. Future work will include such an analysis in the heat-affected zone.

ACKNOWLEDGMENT

The authors gratefully acknowledge Dr. Herschel Smartt for helpful discussions and Terry Pierce for technical assistance. This work was supported by the U. S. Department of Energy Office of Energy Research, Office of Basic Energy Sciences under DOE Contract No. DE-AC07-76ID01570.

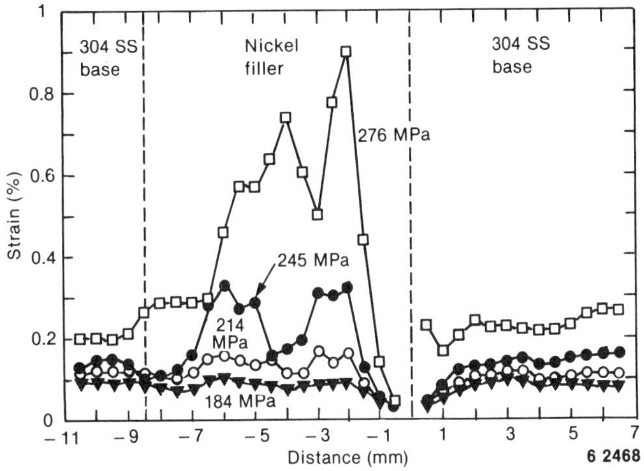

Fig. 10 - Moire strain profiles of nickel filler, 304 SS base weld through defect.

REFERENCES

1. J. W. Dally and W. F. Riley, "Experimental Stress Analysis," New York, McGraw Hill, (1978), p. 384.

2. A. Livnat and D. Post, "The Governing Equations For Moire Interferometry and Their Identity to Equations of Geometrical Moire," Experimental Mechanics, 25, 2, Dec. (1985), pp. 360-366.

3. D. Post, "Moire Interferometry," Ch. 7, SESA Handbook on Experimental Stress Analysis, A. Kobayashi ed., (1984).

4. A. McDonach, et al., "Improved Moire Interferometry and Applications in Fracture Mechanics, Residual Stress and Damaged Composites," Experimental Techniques, (1983), pp. 20-24.

5. M. L. Basehore and D. Post, "High Frequency, High Reflectance Transferrable Moire Gratings," Proceedings of the 1982 Joint Conference on Experimental Mechanics, SESA and JSME, Oahu-Maui, Hawaii, (1982), pp. 156-158.

A METHODOLOGY FOR ANALYZING WELDING DISTORTION OF TUBULAR FRAMES

Chien-ann A. Hou
Department of Mechanical Engineering
Howard University
Washington, D.C.

Chon L. Tsai
Department of Welding Engineering
The Ohio State University
Columbus, Ohio

INTRODUCTION

ONE OF THE MOST troublesome problems to the welding industry is that of welding distortion. The non-uniform heating and cooling cycles, which occur in the weld and the adjacent base metal cause the development of complex strains during welding. The stress resulting from these strains produce internal forces that cause permanent bending and rotation; these are called distortion.

Tubular frames have been used for housing mechanical and electrical devices in today's industry. For some applications, dimensional fitness of the assembled frames is critical. For example, the premachines holes in the tubular members need to be aligned within specified tolerance in the frames for electronic instruments. Therefore, assembly procedures for minimum distortion need to be developed.

Distortion corrections after welding completes have been the common practice to control the dimensional fitness of the finished frames. This postwelding correction procedure is usually troublesome and difficult to assess. In order to avoid these problems, a methodology for analyzing welding distortion of tubular frame needs to be developed. The optimum welding and jigging conditions could be determined from computer modeling and analysis at design stage before manufacturing.

The distortion model used a one-dimensional thermal elastroplastic analysis to get a close-form solution for the thermally induced stresses and strains fields in the joint areas [1]. In order to study the constraint effects from jigging and other frame members on frame distortion caused by welding of a joint, a finite element beam program was linked to the model and the principle of superposition was applied.

METHODOLOGY

THERMAL MODEL - The model of heat flow through a thin wall square tube for a tee joint was developed, as shown in Figure 1. The weld was short and welding speed was fast enough that the heat could be assumed an instantaneous deposition along the joint. Therefore, the temperature distribution would be symmetric about Y = 0. Also the heat loss from the structural surfaces was neglected, assuming that the surface boundaries were insulated. It was then possible to cut the tube along its middle plane of the bottom surface and expanded it into a plate. The thermal problem was therefore reduced from a tubular model to a plate model. In order to avoid mathematical complexity in solving a nonlinear equation, the thermal properties of the tube material were assumed constant.

The solution of the thermal model was obtained by integrating the solution for the instantaneous point source with regard to appropriate space variables and time [2]. The Green's function assumed constant thermal properties, adiabatic boundary conditions and infinite plate.

In the thermal model, the tube was treated as an infinite plate with a weld bead, which had a length equivalent to the tube width, along the joint. The heat distribution was assumed a line source in y-direction over the welding time [3]. Thus the solution became

$$T(x,y,z,t) - T_0 = \int_{-1/2}^{1/2} \int_{0}^{t_s} \frac{\dot{Q}k}{4\lambda[\pi k(t-t')]^{1.5}} \exp\left\{\frac{-R}{4\sqrt{kt-t'}}\right\} dt'\, dy'$$

$$= \frac{\eta_a \dot{Q}}{2\pi\lambda} \int_{-1/2}^{1/2} \frac{1}{R} \left[\mathrm{erf}\left(\frac{R}{2\sqrt{k(t-t'')}}\right) - \mathrm{erf}\left(\frac{R}{2\sqrt{kt}}\right)\right] dy' \quad (1)$$

where $t'' = \begin{cases} t_s & t \geq t_s \\ t & t < t_s \end{cases}$

$$R^2 = x^2 + (y-y')^2 + z^2$$

and T_0: room (reference) temperature

l : weld length

t_s: welding time

λ : thermal conductivity

k : thermal diffusivity

$\dot{Q}$: heat input per unit length of weld

η_a: modification factor

erf : error function

x,y,z: Cartesian coordinates

t: time

DISTORTION MODEL - The distortion model assumed that the distortion was caused by two mechanisms [1]. The first mechanism was the nonlinear distribution of the inelastic strains, which existed due to a nonuniform temperature distribution in the tube. The second mechanism was the shrinkage in the weld nugget area due to the contraction of weld metal during the cooling cycle.

The distortion model divided the flange into many sections as shown in Figure 2. Each section had its own transient temperature field at any given time. During welding, the lcoal temperature of the tube raised up nonuniformly. The resulting nonuniform thermal strains were forced to be compatible to each other by the internal rigidity of the tube section, and therefore caused thermal stresses and deformation. Yielding occurred in the area with high noncompatible strains. In the high temperature areas, yielding usually existed due to drastically reduced elastic modulus upon cooling, the accumulated inelastic strains caused the final distortion and residual stresses in the tube. In the model only the stress component normal to the weld was considered. A narrow band of the ring element was cut from the base member, and the plane of the cross-section of the base member was assumed to remain a plane during the thermal loading process (simple beam theory assumption). Also the thermal loading was assumed monotonic during each time step. The stress-strain relationship can then be written as [4]

$$\begin{aligned} \varepsilon_x &= \frac{S}{H} + \tau + \varepsilon_p \\ \varepsilon &= \varepsilon_x - \tau \\ \varepsilon_p &= \Sigma\Delta\,\varepsilon_p + \Delta\varepsilon_p \end{aligned} \tag{2}$$

where

ε_x : dimensionless total strain

τ : dimensionless thermal strain

ε : dimensionless engr. strain

ε_p : dimensionless total plastic strain

$\Delta\varepsilon_p$: dimensionless plastic strain increment

$\Sigma\Delta\varepsilon_p$: dimensionless sum of previous plastic strain increment

S : dimensionless normalized stress in X direction

H : dimensionless elastic modulus

In each element, the stress component S is a function of the dimensionless surface coordinate η, measured from the midpoint of the top surface to the midpoint of the bottom surface. By assuming zero other stress components, the compatibiliity equation yields a linear total strain distribution across the section.

$$\varepsilon_x = m+2(m-n)[U(\eta-\tfrac{3}{4})(\eta-\tfrac{3}{4})-U(\eta-\tfrac{1}{4})(\eta-\tfrac{1}{4})] \tag{3}$$

where m and n are constants and $U(\eta)$ is the step function. The reaction forces equilibrium equations must be in a condition

$$\int_0^1 S\,d\eta = P^* \qquad \int_0^1 S\,g(\eta)\,d\eta = M^* \tag{4}$$

where P* and M* are the reaction forces and moments in the constrainted joints. The weld metal shrinkage forces can also be represented by P* and M*. For a 1" x 1" square tube function g (η) is defined as

$$g(\eta) = \begin{cases} 1 & 0 \le \eta < 0.25 \\ 2-4\eta & 0.25 \le \eta < 0.75 \\ -1 & 0.75 \le \eta \le 1 \end{cases} \tag{5}$$

The dimensionless total strain can be written as

$$\begin{aligned} \varepsilon &= -\tau + D_1(\tau)+D_2(\Sigma\Delta\varepsilon_p) + D_3(\Delta\varepsilon_p)+D_4(P^*)+D_5(M^*) \\ D_1(\tau) &= B_1 \int_0^1 \tau H\,d\eta - B_2 \int_0^1 \tau H\,g(\eta)\,d\eta \\ D_2(\Sigma\Delta\varepsilon_p) &= B_1 \int_0^1 \Sigma\Delta\varepsilon_p H d\eta - B_2 \int_0^1 \Sigma\Delta\varepsilon_p H\,g(\eta)d\eta \\ D_3(\Delta\varepsilon_p) &= B_1 \int_0^1 \Delta\varepsilon_p H\,d\eta - B_2\int_0^1 \Delta\varepsilon_p H\,g(\eta)d\eta \\ D_4(P^*) &= B_1 P^* \\ D_5(M^*) &= B_2 M^* \end{aligned} \tag{6}$$

where B_1 and B_2 are constants and their magnitudes depend on the integrations of the elastic modulus with respect to the η.

The effects of shrinkage of weld metal was modeled by springs. This is shown in Figure 3. For a weld leg size a, the hypotenuse is $\sqrt{2}$ * a, and the metal contraction Δ from the melting temperature Tm to current temperature T is

$$\Delta = \alpha(T_m-T)\sqrt{2}\,a \tag{7}$$

where α is the thermal expansion coefficient of weld metal.

The shrinkage produces a force Fs directed at an angle of 45 degrees as shown in Figure 3. The magnitude of the force Fs is

$$F_s = K\Delta \tag{8}$$

where K is the temperature dependent spring constant of the weld metal. If we consider the force component in the Z direction, the force will be

$$F = \frac{\alpha E(T_m-T)a l}{\sqrt{2}} \tag{9}$$

This force can be treated as an external force and included in the equilibrium equations.

The distortion model also considers the temperature dependent mechanical properties such as elastic modulus and yielding stress. Figure 4 shows the bilinear temperature dependent mechanical properties for mild steel used in the current study.

MODEL FOR THE FRAME DISTORTION - In order to study the welding distortion of tubular frame structure with jigging constaints, a finite element beam program was used to link the jigging restraints and the frame rigidity to the individual joint distortion analysis. The finite element program contains beam elements, which is an uniaxial element with tension, compression, and bending capabilities. Every element has six degrees of freedom at each node: translations and rotations in the x, y, and z directions.

The frame distortion analysis consists of three parts; the thermal analysis, the distortion analysis and the linkage of the frame tubes and jigs. It was assumed that temperature distribution can be determined independently of the deformation of the structure. Therefore, decoupling of the thermal analysis from the distortion analysis was made in the temperature calculations.

Welding distortion modeling of frame structures can be divided into five programs, with each of the programs performing a specific function. A "preprocessor" needs to be developed to set up the initial conditions, such as the global and local coordinates, the joint geometry and fit-up, the jigging and fixturing conditions, the welding conditions and sequence and the assembly procedures. A numbering system is needed to define the initial conditions and to arrange these variables for subsequent calculations.

The "thermal program" calculates the temperature history of the frame elements caused by sequential welding at various joint locations. The complete thermal history of the frame is saved on magnetic tape and can be retrieved later for distortion calculations.

The "distortion program" calculates the incremental thermal stresses and reaction strains in every joint, assuming that all joints are free of restraining from jigging or frame rigidity.

The "finite element structure program" enforces the compatibility conditions at the fixturing points and applies the frame rigidity to the free joints. The calculated strains (i.e. assuming all elastic) are superimposed to the thermal induced strains. The added strains indicate the curvature of the cross sections, unless yielding occurs. The transient displacements of the joint members can be determined by accumulating the curvatures. For those cross sections exhibiting yielding, the redistribution of stress needs to be considered. The new curvature of these sections is determined by equilibrating the force and moment, and using the temperature dependent stress-strain relationships. For each time increment, a procedure is needed to iterate the calculations until the displacements at the fixturing points become compatible. The calculations move on to the next time increment. The relaxation process due to the removal of the fixtures can be modeled, using the finite element program, by applying the negative restraint forces at the fixturing points. The final distorted shape of the frame can be determined.

To present the calculated results a "postprocessor" needs to be developed. The transient frame distortion can be viewed at any given instant and the final shape of the distorted frame can be obtained at the end of the calculation process.

In this study, a two-dimensional frame structure was investigated to verify the general concept. The frame distortion calculations started when the welding arc initiated and ended when the finished frame cooled to a temperature low enough to cause insignificant thermal strains. The final distorted shape of the finished frame was predicted.

VERIFICATION

EXPERIMENT - An experiment was conducted to investigate the welding distortion of a tubular frame structure to verify the model [3]. The gas metal arc welding (GMAW) process was used to make welds using the Hobart Motoman welding robot. The structure was chosen as a two dimensional square frame with two extra members fixed in the frame (Figure 5). The frame was composited of 1" x 1" mild steel square tubing with a wall thickness of 1/8 inch. Total twelve constraints was applied (two on each member) in order to reduce the welding distortion. The frame had eight joints and sixteen welds. The voltage and current used on each weld were 19.5 V and 100 A, respectively. The welding speed was 15.7 in/min. The distortion was measured by the differences of the relative distance between the premachined holes in the frame member before and after welding. Each member had two holes drilled before assembly. The sequence chosen was according to the restrictions of robot's moveability, accessibility and its speed. The welding sequence and other conditions are listed in Table 1.

DISCUSSION - Comparison of the computed frame distortion shape and the experimental measurement is presented in Figure 6. The magnitude of the frame deformation was enlarged as indicated by the scale in the figure. Reasonable agreement between the prediction and the measurement is observed.

There are several possible sources for errors in the analytical model. Constant thermal properties were used in the thermal analysis. This would cause some errors in the temperature calculations and lead to errors in predicting a distortion. Another source of error is from the computation. In the computer program, the convergence criterion for current plastic strain increment is one percent of the accumulated plastic strain. In addition, in the calculation of thermal strains, numerical integration using Gaussian quadrature method may cause some error.

CONCLUSIONS AND RECOMMENDATIONS

An analytical model for predicting welding distortion of tubular frames is presented in this paper. This model was developed for predicting distortion of three-dimensional frames composited of square tubings. The model used several approximations, but still contained the significant physical characteristics of the frame structure with regard to welding distortion. The model represented a thermal analysis with constant thermal properties, but an elastic-plastic, temperature dependent constitutive behavior. The model also considered the effect of the jigging constraints, the frame rigidity and the welding sequence. Comparison between the computed frame distortion and the laboratory measurement on a two-dimensional frame showed reasonable agreement.

While the ability of the analytical model to predict frame distortion is verified, the limited data base from the present study needs to be expanded before any extrapolation to other tubing sizes and welding conditions can be made. However, the methodology appears to have the potential to be an effective tool for obtaining a better understanding of the mechanism of welding distortion of any frames. The model can be used to study the effects of various parameters including welding conditions, jigging conditions and welding sequences for developing a new and efficient welding procedure to reduce distortion in the design stage before manufacturing.

REFERENCES

1. Hou, C.A., "Modeling and Control of Welding Distortion in Tubular Frame Structures," Ph.D. Dissertation, Ohio State University, Jan. 1986.
2. Carslaw, H.S. and Jaeger, J.C., "Conduction of Heat in Solids", 2nd Edition, Oxford Press.
3. Tsai, C.L., Project Report on "Optimization of Robot arc Welding Procedures for Minimization and Control of Distortion in Tubular Frame Structures", Ohio State University, Jan. 1986.
4. Mendlson, A., "Plasticity: Theory and Application", R.E. Kreger Publishing Company, Malabur, Fl., 1983.

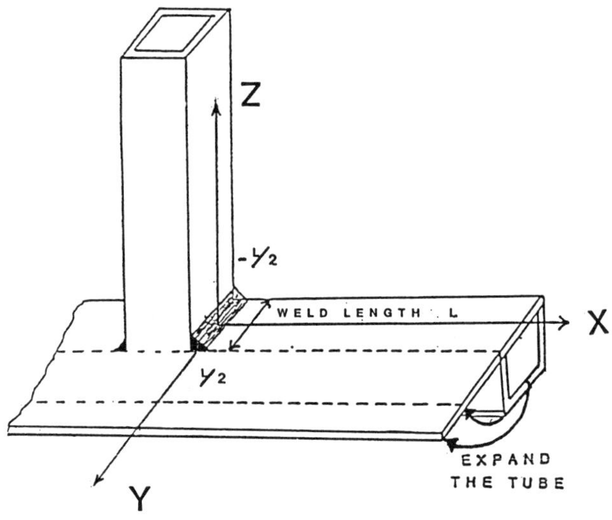

Fig. 1 - Welding thermal model of a tubular joint

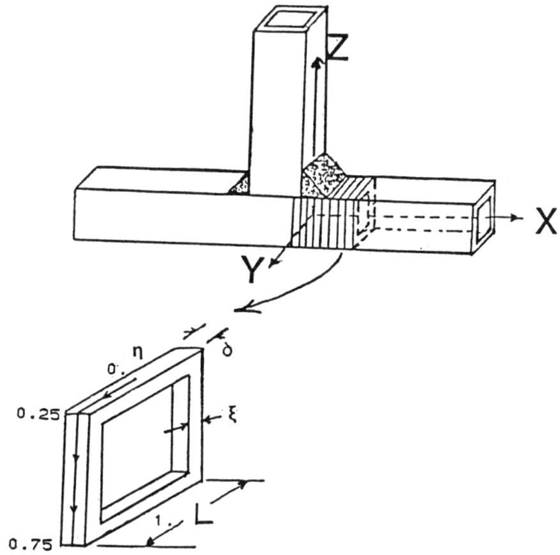

Fig. 2 - The model for welding inelastic strain analysis

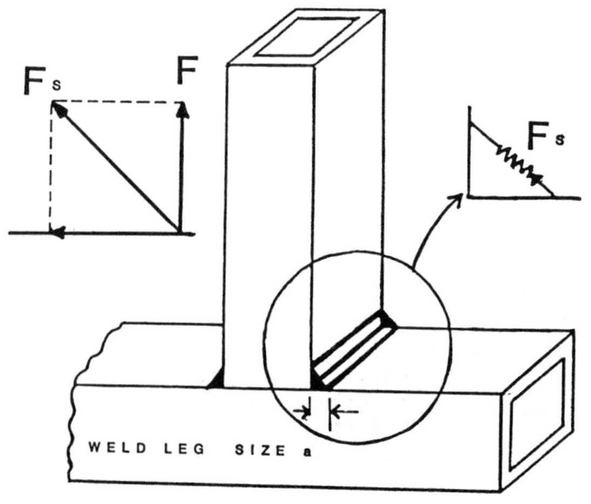

Fig. 3 - The model for weld metal shrinkage in a tee joint

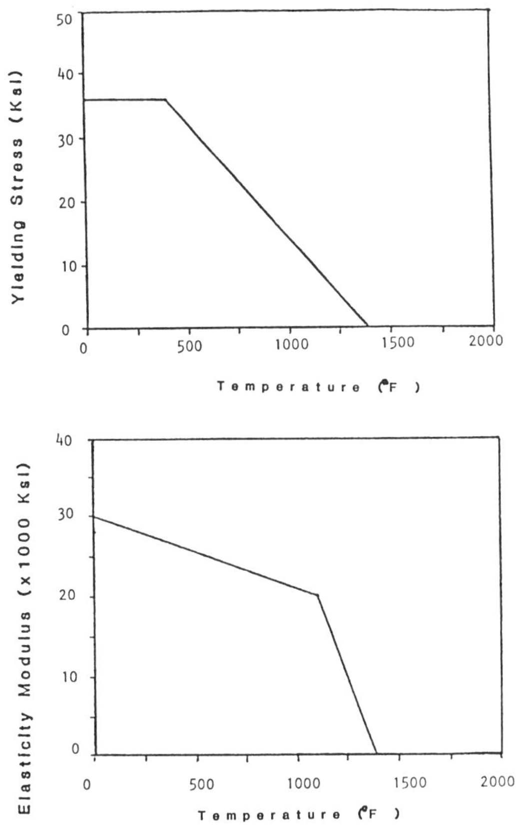

Fig. 4 - Mechanical properties of mild steel

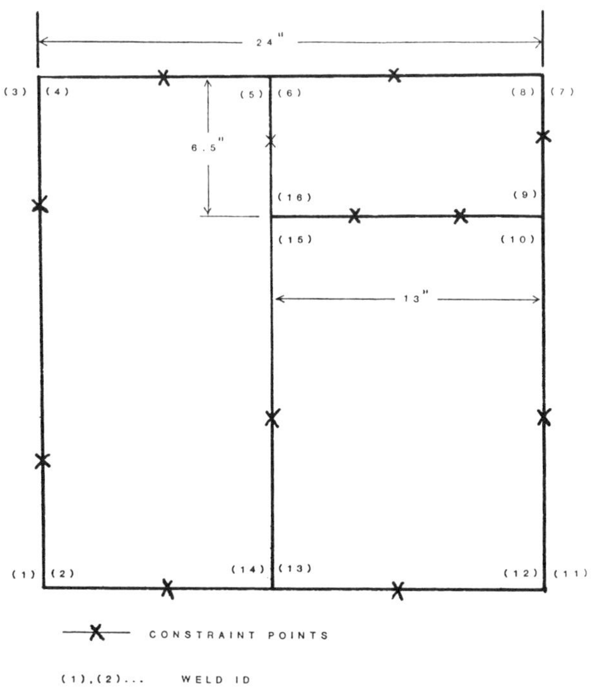

Fig. 5 - Configuration of a 2-D square frame

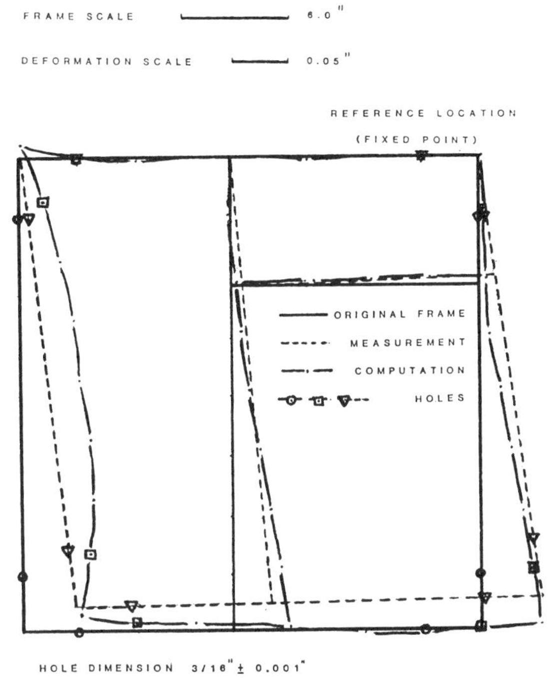

Fig. 6 - Comparison with measurement and computation results of welding distortion of a square tubular frame

Table 1 - Welding Sequence of a Square Tubular Frame

Welding sequence	Weld ID number	Weld starting time	Welding time	Travel time
1	3	0.0	4.0	31.0
2	7	35.0	4.0	31.0
3	11	70.0	4.0	31.0
4	1	105.0	4.0	16.0
5	4	125.0	4.0	36.0
6	8	165.0	4.0	22.0
7	12	191.0	4.0	22.0
8	2	217.0	4.0	29.0
9	5	250.0	4.0	21.0
10	9	275.0	4.0	33.0
11	13	312.0	4.0	33.0
12	6	349.0	4.0	33.0
13	10	386.0	4.0	22.0
14	14	412.0	4.0	35.0
15	16	451.0	4.0	33.0
16	15	488.0	4.0	0.0

(Time is in seconds)

ALTERING RESIDUAL STRESSES IN WELDED AUSTENITIC STAINLESS STEEL PIPING

Joseph C. Danko
American Welding Institute
Louisville, Tennessee

ABSTRACT

Welding austenitic stainless steel pipes by conventional processes results in tensile residual stresses on the inside surface and partially through-wall of the weldment. These residual stresses, operating in the weld sensitized heat-affected zone, have been found to be a major contributor to the intergranular stress corrosion cracking of piping in boiling water reactors. Processes have been developed to alter the stresses by either reducing the tensile residual stresses or imparting compressive stresses in the pipe weldment. With this alteration of the weld residual stresses, the cracking problem is mitigated or eliminated. These processes, heat sink welding, last pass heat sink welding, induction heating stress improvement, and weld overlay, are reviewed.

THE PURPOSE OF THIS PAPER is to review the processes for mitigating or eliminating the intergranular stress corrosion cracking problem in austenitic stainless steel piping in boiling water reactors by altering the weld residual stresses.

During the past decade, the incidents of intergranular stress corrosion cracking in welded austenitic stainless steel piping in boiling water reactors has increased significantly.[1] The majority of the domestic boiling water reactors with conventional AISI types 304 and 316 stainless steel have suffered from this problem. While this is not considered an issue of safety, it has seriously affected plant reliability, availability, economics, and radiation exposure of skilled personnel involved in the inspection and repair operations.

A model that describes the phenomenon of intergranular stress corrosion cracking has been developed[2] and requires the coexistence of three factors: (1) weld sensitization, (2) tensile residual stresses, and (3) the boiling water reactor coolant. With this model, a number of engineering solutions[3] for the problem have been identified. Those based upon processes that alter the weld residual stresses in the pipes are reviewed below.

ALTERING WELD RESIDUAL STRESSES

Processes for altering the weld residual stresses in austenitic stainless steel piping include heat sink welding, last pass heat sink welding, induction heating stress improvement, and weld overlay. To demonstrate how these processes alter the as-welded residual stresses, a brief description of the as-welded stresses is provided.

AS-WELDED RESIDUAL STRESSES[4] - The measurement of weld residual stresses and their role in the intergranular stress corrosion cracking was presented by the author during the first International Conference on "International Trends in Welding Research."[5] Peak axial residual stresses on the internal surface were found to be the highest for 100 mm (4-inch) pipe, intermediate for the 250 mm (10-inch), and the lowest for 660 mm (26 inch) diameter pipe (Figure 1). Through-wall residual stress measurements showed axial tensile stresses from the inner surface to approximately 50 to 75 percent of the wall thickness for the 10 cm pipe, whereas the stresses were strongly compressive at a depth of approximately 25 percent of the wall thickness for the 660 mm pipe (Figure 2). Because of these weld residual stress variations, the large diameter pipes (660 mm) are less susceptible to stress corrosion cracking. Measurements of weld residual stresses of pipes removed from service after 10 years of operation were compared to those of other pipes of the same diameter, and there were no significant differences. This indicates a permanence of weld residual stresses.

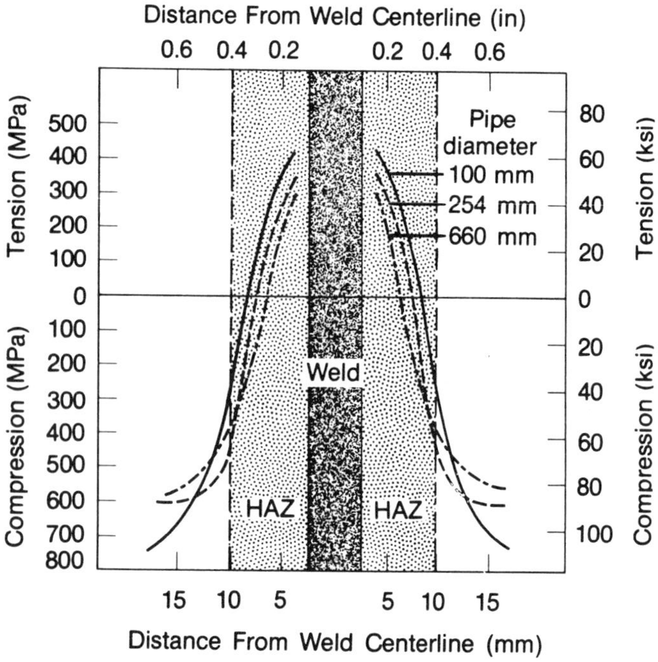

Fig. 1 - Peak Axial Weld Residual Stresses Measured On The Inside Surfaces Of 304 SS Pipes

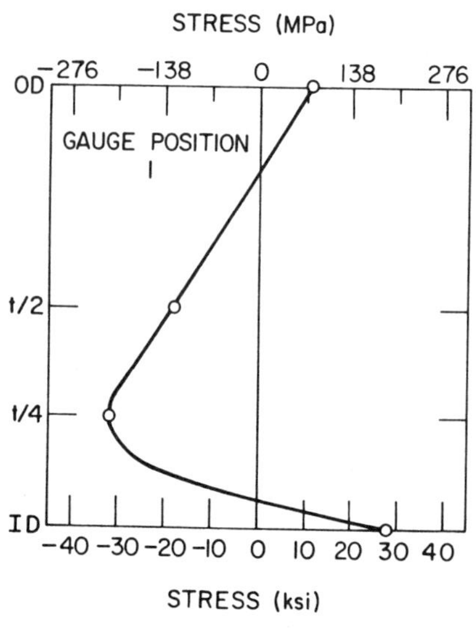

Fig 2 - Through-Wall Distribution of Weld Residual Stresses in 660 mm (26 in) 304 SS Pipe

HEAT SINK WELDING - This process involves the welding of the root pass and the first several layers to provide structural support of the pipe weld.[6] This operation usually is performed by manual gas tungsten arc welding (GTAW). The subsequent welds are currently deposited with machine welding while flowing or sprayed water is applied on the inside of the pipe. With this process, compressive residual stresses result on the pipe inside surface and partially through-wall. The peak axial residual stresses of an as-welded and heat sink welded pipe of 250 mm (10 inch) diameter 304 stainless steel is shown in Figure 3.[7] These results illustrate the alteration of the weld residual stresses produced by heat sink welding. The through-wall residual stresses are also altered with the magnitude and depth of compressive stresses increasing relative to the as-welded condition. Heat sink welding also tends to reduce the level of sensitization,[8] a contributor to the intergranular stress corrosion cracking. However, the sensitization level is not below the critical level to avoid stress corrosion cracking.

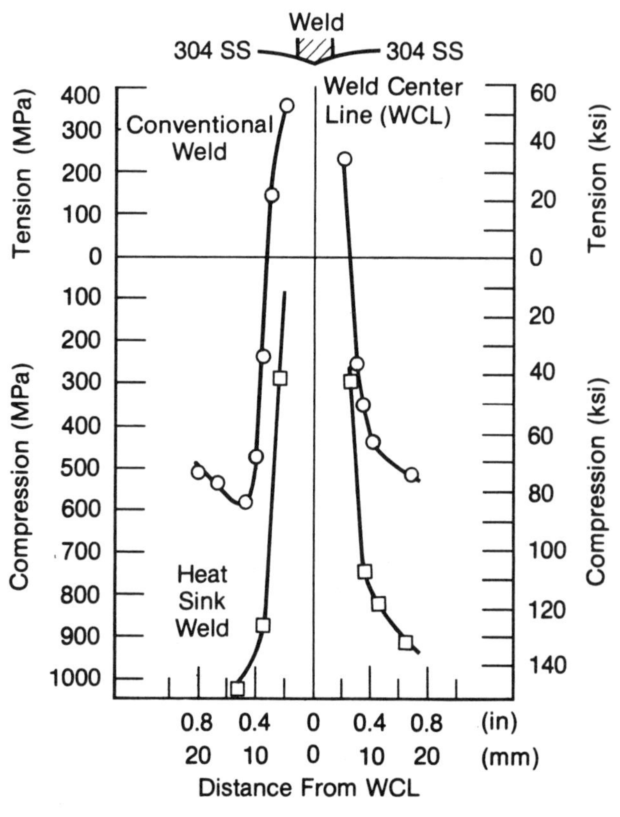

Fig. 3 - Peak Axial Residual Stresses In 250 mm (10 in) 304 SS Pipe Measured On The Inside Surface

LAST PASS HEAT SINK WELDING - This process[9] is a slight modification of heat sink welding. The difference is the introduction of water cooling when the last pass or crown weld deposit is applied. A high heat input is used to establish a proper temperature gradient across the wall thickness of the pipe. A plot of weld residual stresses for a 610 mm (24 inch) diameter 304 stainless steel pipe processed by last-pass heat sink welding is illustrated in Figure 4. These results are very similar to those obtained for heat sink welding. It is important to note that last pass heat sink welding does not reduce the level of weld sensitization.

INDUCTION HEATING STRESS IMPROVEMENT - This process, conceived, developed, and first implemented in Japan,[10,11] involves induction heating the outside surface of the pipe weldment to approximately 500 to 550°C (930 to 1020°F), while maintaining the inside surface temperature at 100 C (212 F) with flowing water. This temperature gradient across the wall thickness results in a small plastic strain in the stainless steel and the original weld tensile residual stresses are altered and compressive residual stresses are produced. Typical compressive residual stresses imparted by induction heating stress improvement for a 406 mm (16 inch) diameter 304 stainless steel pipe weldment are illustrated in Figure 5.[12] An analysis of the relaxation of the residual stresses was performed and the results showed some relaxation with prolonged time, but the compressive residual stresses remained.

WELD OVERLAY - Welded pipes of types 304 and 316 stainless steel with intergranular stress corrosion cracks have been repaired with the weld overlay process.[13] Weld overlays are designed to provide structural reinforcement thicknesses and also produce favorable residual stress distribution. However, weld overlay will not result in compressive residual stresses in all cases. Recent experimental data has shown axial tensile residual stresses in a 716 mm (28 inch) diameter 304 stainless steel pipe after eight layers of weld deposit 22 mm (0.875 inch).[14] In a 406 mm (16 inch) pipe, localized tensile residual stresses were found following weld overlay; compressive stresses were present elsewhere in the root area. This information indicates that weld overlay will not result in compressive residual stresses in all cases, particularly in large diameter thickwalled pipes.

A 308L filler is used in the weld overlay. To achieve a minimum ferrite level of 8FN in the as-deposited material, a starting FN of 11-17 is selected. Low heat input is used on the initial deposit to minimize dilution and higher heat inputs are used for subsequent layers. The reason for a minimum delta ferrite content of 8 FN is that at these delta ferrite levels, intergranular stress corrosion cracking in the pipe will be arrested by the weld overlay. In the application of weld overlay, remote automatic GTAW is used and in most cases reactor coolant remains in the pipe. Under these conditions, a heat sink welding operation is performed.

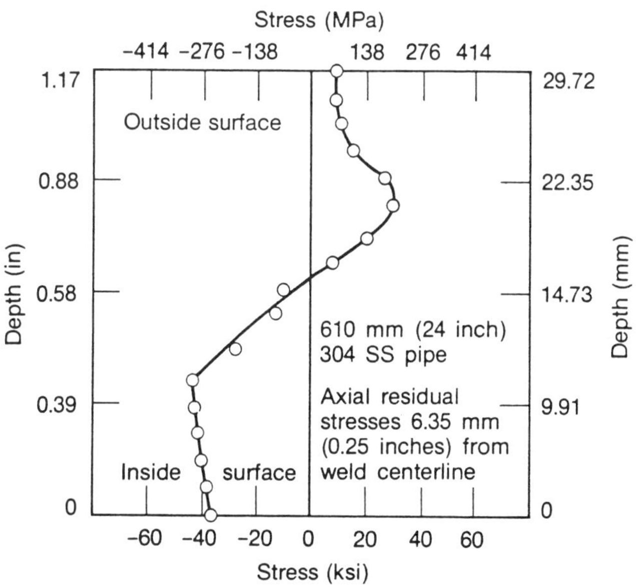

Fig. 4 - Through-wall Weld Residual Stresses In 610 mm (24 in) 304 SS Pipe Processed By Last Pass Heat Sink Welding

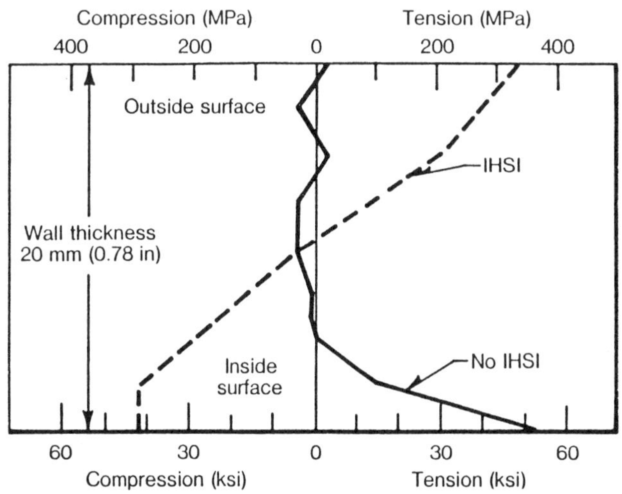

Fig. 5 - Residual Stresses in 406 mm (16 in) 304 SS Pipe With and Without Induction Heating Stress Improvement

Weld overlays also result in shrinkage and distortion that must be considered in the piping system stresses. Measurements have shown diametral and axial shrinkage in the pipe wall after the first several layers of overlay. These generate compressive residual stresses near the inner wall. With additional weld passes, further diametral shrinkage results along with local bending that tends to reverse the initially axial compressive stresses while the circumferential stresses are retained. By increasing the axial length of the overlays, local bending stresses are reduced and the residual stress distribution is improved. However, the longer axial length may produce greater axial shrinkage and increase the pipe system stresses.

Stress analysis of pipe weld overlays have demonstrated good correlation with experimental results. Weld overlays were applied on 305 mm (12 inch) diameter 316 stainless steel pipes with a full structural and mini-overlay. Analysis predicted favorable residual stress patterns. Stress measurements did confirm the residual stresses and in fact these were equal or better than the predictions. This stress analysis is used in the design of the overlay and also applies to the stresses resulting on the rest of the piping system.

Weld overlay repairs have saved the utility industry millions of dollars since the repairs are performed rapidly with reactor coolant in the lines and pipe removal is not required. This is normally accomplished during the scheduled maintenance outages. Typical applications of the weld overlay are shown schematically in Figure 6.

IMPLEMENTATION

The various processes that alter the residual stresses in welded 304 and 316 stainless steel pipes have been implemented in operating and new boiling water reactors. The implementation of these processes is presented in Table 1.

TABLE 1

Implementation

Process	Operating Plants	New Plants
Heat Sink Welding	*	
Induction Heating Stress Improvement	*	*
Weld Overlay	*	

Induction heating stress improvement has found wide application in operating plants, new plant construction, and in some operating plants that have replaced their piping with nuclear grade 316 stainless steel.

Weld overlay has become a popular method of repairing pipe welds in operating boiling water reactors because of the low cost and minimum outage time required for the application. This process is not currently considered an alternative to replacement of piping.

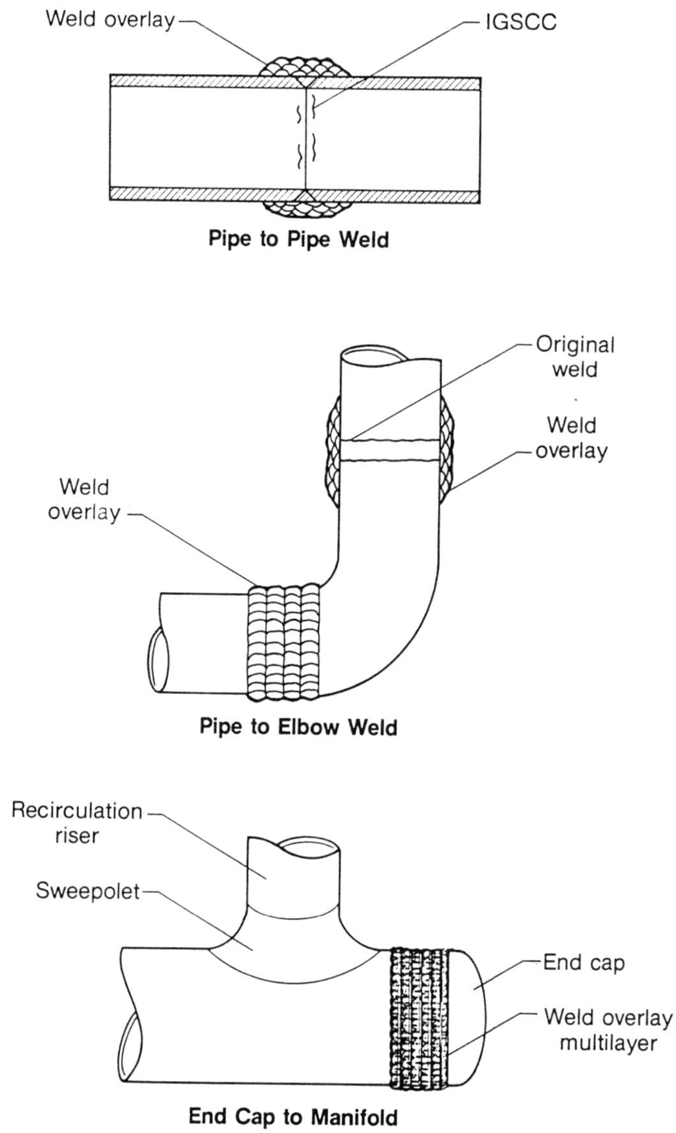

Fig. 6 - Schematic of Typical Weld Overlay Applications

SUMMARY

Incidents of intergranular stress corrosion cracking in welded pipes of AISI types 304 and 316 stainless steel in boiling water reactors have adversely affected plant reliability and economics. The mechanism of cracking is well understood, and one of the major contributors to the problem is the tensile welding residual stresses. A number of processes for altering the weld residual stresses have been developed. These include heat sink welding, last pass heat sink welding, induction heating stress improvement, and weld overlay.

After application of any of these processes, the tensile residual stresses on the inside surface as well as near surface through-wall residual stresses of the as-welded pipe are significantly reduced or become compressive. These alterations will influence the initiation and propagation of intergranular stress corrosion cracking. Cracking will not initiate on the inside surface of the pipe in the presence of compressive residual stresses. For pipe welds with cracking initiated, the cracks may be completely arrested or crack growth retarded, depending upon the depth of the crack and the magnitude of the compressive residual stresses.

Heat sink welding, induction heating stress improvement, and weld overlay have been implemented in operating plants. Induction heating stress improvement has also been used in new plant construction and replacement of piping in operating plants. Use of weld overlay has been implemented in many operating plants and has resulted in significant economic benefits to the utility industry.

ACKNOWLEDGEMENT

The results of the work reported in this manuscript were generated under the Electric Power Research Institute Boiling Water Reactor Owners Group Research Program on Pipe Cracking.

REFERENCES

1. Danko, J. C., Material Performance, Volume 24, pp. 14-7 (1985)
2. Hanneman, R. E., EPRI-WS-79-174, 1 , Paper 3 (1980)
3. Hughes, N., EPRI-WS-79-174, 1 , Paper 4 (1980)
4. Glannuzzi, A. J., EPRI RP 449-2 (1979)
5. Danko, J. C., Conference 820544, ORNL, pp. 379-394 (1982)
6. Hughes, N., EPRI PPT 109-1 (1981)
7. Crenko, R. M., EPRI WS-79-174, 2 , Paper 21 (1980)
8. Danko, J. C., and Smith, R. E., EPRI NP-1222, 2 , 1979
9. Horn, R. M., Herrera, M. L., and Offer, H. P., EPRI NP-3684-SR, 2 , Paper 4 (1984)
10. Shimizu, T., Enomoto, K, Arak, S., and Ikegam, T., EPRI-WS-174, 1 , Paper 10 (1980)
11. Tanaka, S., Umemoto, T., EPRI-WS-174, Paper 13 (1980)
12. Bertossa, D. C., EPRI-WS-174, 1 , Paper 16 (1980)
13. Pitcairn, D. R., EPRI NP-3684-SR, 3 (1984)
14. Gilman, J. D., AWS Conference Proceedings, November, pp, 69-86 (1985)
15. Edwards, J. A., Godby, R. K., AWS Conference Proceedings, pp. 87-114 (1985)

WELD JOINT PROPERTIES AT HYPERBARIC MIG/MAG WELDING

Lutz Müller, Gregor Terlinde, Karl-Heinz Schwalbe
GKSS Research Center
D-2054 Geesthacht, Federal Republic of Germany

ABSTRACT

Hyperbaric unmanned MIG/MAG welding tests at pressures up to 30 bar (300m water depth) were performed. The study focused on the effects of pressure, shielding gas, electrode and welding parameters on the weld metal chemistry, cooling time, microstructure, strength and especially toughness. In order to achieve good toughness at high pressure the conventional welding conditions have to be changed. These changes include less active shielding gases, modified electrodes and welding with negative polarity at high pressure.

INTRODUCTION - The rapid growth of the offshore oil and gas production has led to an increasing demand for underwater welding both for installation and repair of offshore structures like platforms and pipelines. As a result, several welding techniques have been developed which include wet welding, dry or hyperbaric welding and 1 bar-welding; they are described in numerous studies [1-6]. Compared to welding at atmospheric pressure there are two additional factors that have to be considered for underwater welding, namely the increasing pressure at greater water depth (1 bar/10m) and - especially in the case of wet welding - the water. A variety of technological problems related to diving, equipment, handling etc. has to be solved. In addition, poor weld joint properties can result, since the increased pressure as well as the water can strongly influence the welding arc (geometry, stability, temperature, drop transfer) [7-14] and this can result in changes in the metallurgical reactions and the corresponding weld metal chemistry [7, 14-16]. Furthermore, the increased gas density and the water can lead to pronounced effects on the temperature-time cycles, and high hydrogen contents can be observed in the weld joints especially for wet welding, since because of the water dissoziation more hydrogen is available and also soluble, since the pressure is increased. This can lead to severe hydrogen embrittlement (cold cracking), and therefore, code quality welding so far is only carried out by dry or hyperbaric welding.

But also for dry hyperbaric welding problems remain because of the changes in the metallurgical reactions and cooling times, since both are known to influence the microstructures and properties of the weld joints. Since the selection of electrodes, welding parameters and shielding gases is optimized for atmospheric pressure with regard to microstructure and properties, one has to expect inferior properties at higher pressures when using the conventional welding conditions.

Therefore, as a part of the GKSS marine technology program a study was initiated which investigates the effects of pressure, electrodes, shielding gases and welding parameters on weld metal chemistry, cooling times, microstructures, strength and toughness. It is directed towards an improved understanding of the variables controlling the fracture of underwater weld joints, and based on a microstructure control it should lead to improved properties at high pressures.

Hyperbaric, mechanized MIG/MAG welding tests were performed at pressures between 1 bar and 30 bar, (≙ 300m water depth). The base materials were a pipeline steel StE 445.7 TM (≙ X65TM) and a structural steel StE 36 (≙ 50D), both 15mm in thickness. Two commercial flux-cored wires served as electrodes, one being a lower strength C-Mn-wire, the other a slightly higher strength C-Mn-1%Ni wire. In a later stage also a modified version of the C-Mn-1%Ni wire was tested. Various (Ar + O_2)- and (He + CO_2)-shielding gases were used with O_2-contents from 0.1% to 20% and CO_2-contents from 2% to 10%. Since it can become necessary to change the welding parameters with increasing pressure (voltage especially), also voltage, current and polarity were varied.

Preliminary tests had shown that no pronounced effects of pressure on the cooling times were observed which, could affect also the heat affected zone. Therefore, the study has mainly been concerned with the weld metal properties.

EXPERIMENTAL METHODS

The unmanned welding experiments were carried out in a $10m^3$ pressure chamber using automatic MIG-welding equipment. A power source with flat characteristics (I≦600 A, U≦65 V) was employed. Argon was used as pressurizing gas. The shielding gas flow was adjusted to a roughly constant mass flow-rate for all pressures rather than to a constant volume flow-rate.

Butt welds were produced on plates 400m long and 320mm wide with a 60° V-groove and backing strips. The heat input was kept constant at all pressures in most cases (∿13-16kJ/cm).

The butt welds were tested in the as-welded condition. The investigations concentrated mainly on the weld metal and included X-ray examination, chemical analysis, hardness tests, light microscopy and, for some conditions, electron microscopy (SEM and TEM). The microscopy included a quantitative microstructure characterization for selected conditions. In addition, tensile tests with cylindrical specimens (5mm diameter x 25mm gage length) were carried out at room temperature at a strain rate of $8 \cdot 10^{-4} s^{-1}$. Charpy impact tests were performed with V-notch specimens. The specimens were taken from the middle of the plate with the notch perpendicular to the plate surface. For fracture mechanics testing, preliminary tests showed that for our combination of yield stress, specimen thickness (10mm) and toughness, an elasto-plastic analysis was appropriate. The CTOD-values were determined as a function of stable crack growth, with the crack growth measured by the DC potential drop method. The tests were carried out at various temperatures on CT-specimens with a width of 50mm. The details of the test method are described elsewhere [17].

For this study the critical CTOD-values were determined according to British Standard BS 5762, since these values are preferred for offshore applications.

RESULTS AND DISCUSSION

HEAT AFFECTED ZONE - The heat input was kept constant as far as possible. With increasing pressure, however, the voltage had to be increased to maintain a stable arc and a sufficient arc length (28 V at 1 bar, 36 V at 30 bar). This increase in heat input was probably at least partially compensated by the pressure effect leading to slightly faster cooling at high pressure values. Thermocouple measurements at least do not show a significant variation of $t_{8/5}$ at different pressures. Therefore pronounced changes in the heat affected zone microstructures and toughness were not expected. The hardness measurements taken as an indication show only a slight increase of HV10 from 240 at 1 bar to 264 at 30 bar for the pipeline steel and no pressure effect for the structural steel [17]. In addition, Charpy tests with the notch positioned in the heat affected zone gave HAZ impact values lying in a scatterband for all pressures. The level of values was between that of the weld metal and that of the base metal. Therefore, the main effect to be studied was the change in the weld metal chemistry, the corresponding effects on microstructure and toughness and the question how optimum microstructures can be established at high pressures.

WELD METAL - Three series of tests were performed to study individual effects. In the first series conventional welding conditions for atmospheric pressure were taken (commercial wire, Ar + 5% O_2) except for a voltage adjustment at higher pressures. The second series investigated the effect of different shielding gases and an electrode modification on the weld metal chemistry, microstructure and toughness. In the third series the effects of welding parameters were studied. Each series describes the general trends and particular conclusions about a toughness improvement.

The results of the first test series have already been partly presented elsewhere [17] , they will only be summarized as a basis for the subsequent series. When increasing the pressure the weld metal toughness was reduced, while the tensile properties (R_{eL}, R_m, RA) were not significantly affected. The charpy-result and the CTOD-results are illustrated in figures 1 and 2 as a function of temperature for various pressures. At higher pressures the charpy impact toughness is lowered both in the upper shelf and in the ductile/brittle transition, the CTOD-values only show a significant shift in the transition regime. The values of the transition temperatures themselves, however, are lower for the CTOD-values. A detailed analysis of the weld metal chemistry, the microstructure and fracture process revealed that the toughness reduction was mainly due to an increase in the weld metal oxygen content (∿370 ppm at 1 bar, ∿ 750 ppm at 30 bar). The oxygen mainly forms non-metallic inclusions, which can have a pronounced effect on the microstructure development [18-20]. In summary, it was shown, that with increasing weld metal oxygen content the inclusion density increases leading to a smaller prior austenite grain size, a decrease in the acicular ferrite and an increase in grain boundary ferrite and side plates. Cleavage fracture initiates at the inclusions and preferentially propagates through the large grain boundary ferrite so that increased amounts of inclusions and grain boundary ferrite favor the cleavage process resulting in the observed toughness reduction in the ductil/brittle transition. The reduced toughness in the upper shelf region where a dimple type fracture at inclusions occurs can be explained by the higher amounts of inclusions leading to a higher void density.

As a conclusion from this first test series it seemed useful to vary the activity of the shielding gas because it is known to influence the weld metal oxygen content. For a toughness improvement less active gases would be appropriate, but in order to get a more complete picture of the combined effects of pressure and shielding gas a wide selection of gases including also more active gases was taken. From the weld metal chemistry results, two observations are described in figure 3 and 4, since they have a particular effect on the toughness behavior. They apply to the weld metals produced with the C-Mn-wire using different (Ar + O_2) shielding gases. Most of the other elements like C, Ni, N, S, P, V, Nb did not vary significantly, only Al showed a slight increase with pressure. The Ti content increased, decreased or didn't change at all at higher pressures depending on the shielding gas activity. Higher Ti-contents were observed for less active shielding gases. Further studies are underway to clarify possible effects on the microstructure and toughness, since they can be quite sensitive to Ti variations [24]. In figure 3 the weld metal oxygen content is plotted versus the partial pressure of the oxygen in the shielding gas. This partial pressure combines the effect of the oxygen content in the shielding gas and the effect of the absolute pressure, since it is the product of these values. For shielding gases with oxygen contents between 0.1% and 5% a linear log/log relationship has been found between weld metal oxygen content O % and the oxygen partial pressure pO_2 in the shielding gas, it can be described by $[O][wt\ \%] = 0.067 \cdot pO_2^{0.15}$ [bar]. For more active gases separate relationships are found showing higher oxygen levels and different slopes. Figure 4 shows the corresponding Mn- and Si-contents as a function of pressure. With increasing activity of the shielding gas lower Mn- und Si-contents are observed, which is a well known burn-off effect. The extent of burn-off, however, is strongly depending on the pressure. While for shielding gases with up to 5% O_2 no or only a small pressure effect is observed, the more active gases lead to a considerably increased Mn- and Si-reduction at higher pressures. It is interesting to note that this pressure dependent burn off starts at shielding gas activities where the weld metal oxygen contents deviate from the partial-pressure relationships (figure 3). One of the general explanations of the oxygen-effects (figure 3) and the corresponding effects on Mn and Si (figure 4) are certainly the increased solubility of oxygen in the droplet and weld pool at higher pressures, but for a full understanding one will have to take into account possible changes in the arc temperature, the metal transfer, droplet size etc. . Unfortunately only little is known about such correlations. With regard to the oxygen Coe and Moreton [16] have derived a partial pressure relationship $[O][\%] = 0.0726\ pO_2^{0.5}$ based on theoretical considerations. Our results show lower values for the exponent (0.15 e.g.), and the experimental results of the above study fit a relationship with an exponent of 0.15 better than the proposed correlation with 0.5. Generally more detailed investigations including high speed cinematografie appear necessary.

With regard to the selection of shielding gases for hyperbaric welding two points can be made: First, less active gases should be used for welding at high pressures in order to reduce the weld metal oxygen content, and from figure 3 direct information can be gained which oxygen content can be achieved with a certain shielding gas. In order to keep the welding process stable, however, a minimum oxygen content of 0.3% O_2 seems to be necessary based on our tests. The second point applies to the Mn--content. The commercial electrodes are designed for more active gases, giving a certain Mn- and Si-burn off. When using very lean shielding gases, less burn off is observed (figure 4) and the Mn-content can be higher than desired. It is known that for C-Mn-wires Mn-contents above 1.4% and for C-Mn-1%Ni wires Mn-contents above 1.2% lead to a reduced toughness either by increases in yield stress and/or by increased amounts of embrittling martensite/austenite phase [23, 25, 26]. On the other hand, Mn-values below the above mentioned contents also can lead to reductions in toughness - though for different reasons - [23-25] so that generally a control of this element is necessary for toughness optimization. Therefore, based on the results in figures 3 and 4 two steps were taken to improve the toughness at higher pressures: First, less active shielding gases were used (Ar + 0.3% O_2, He + 5% CO_2), and modified Mn-reduced wires were chosen which were fabricated in cooperation with an electrode manufacturer. Figure 5 illustrates the effect of a less active shielding gas on the charpy impact toughness for the C-Mn-1%Ni-wire. The toughness is only improved at lower temperatures, while above around 5°C a lower toughness is observed for the lean shielding gas compared to the more active gas (Ar + 5% O_2). A microstructure analysis revealed less inclusions and less grain boundary ferrite and side plates so that a toughness improvement had to be expected. But on the other hand, the yield stress had markedly increased (630MPa compared to 527MPa), which generally lowers the impact energy. The reasons for the yield stress increase is the higher Mn-content leading to higher amounts of acicular ferrite which has a high intrinsic yield stress because of its fine grain size. In addition, the area fraction of grain refined zones between the passes which are less strong than the as welded areas had decreased for less active gases. In order to reduce the yield stress a Mn-reduced C-Mn-1%Ni-wire was tested. The charpy impact values of the commercial and modified wires are shown in figure 6 as a function of temperature. For the major part of the curve the Mn-reduction from 1.6% to 1.22% has led to a significant increase in the impact toughness, below 40J the values are comparable.

As expected the weld metal yield stress was reduced by the Mn-reduction, the value of 506MPa for the low Mn-content is, however, still sufficient for the pipeline steel. For the C-Mn--wire similar, but less pronounced trends were found.

In a third test series the welding parameters were varied. From earlier studies it is known that at high pressures the welding parameters can affect the metal transfer and the metallurgical reactions in a different way compared to atmospheric pressure [8-12]. Studying such effects is not only worthwhile for an improved understanding of the phenomena, but also for information about allowed tolerances in the welding parameters with regard to microstructure and toughness. Figures 7 to 9 illustrate some of the effects of the welding parameters on weld metal chemistry. Only those results are presented which are thought to have major effect on the microstructures and properties. In figure 7 the effect of voltage and polarity on the weld metal oxygen content is described (C-Mn-wire, Ar + 5% O_2). When welding with straight polarity (+), the oxygen content is not affected by the voltage neither at 1 bar nor at 30 bar. For welding with negative polarity two interesting observations can be made: First, welding with negative polarity results in lower oxygen contents at higher pressures, and second, an increase in voltage further lowers the oxygen content. This trend is found also for other shielding gases (Ar + 0.3% O_2 e.g.), and it follows that welding with a lean shielding gas and negative polarity can result in oxygen contents quite comparable to those at 1 bar. The corresponding results for the weld metal Mn-content are illustrated in figure 8. At (+)-polarity the Mn-content is continously reduced from $\sim$1.5% at 22 V to $\sim$ 1% at 50 V, a particular effect of pressure is not visible. Welding with (-)-polarity shows hardly any Mn--variation. Figure 9 presents the effect of current on the weld metal Mn-content. While for straight polarity (+) no significant Mn-variation is found, a pronounced reduction of the Mn-content from $\sim$1.5% at 300A to $\sim$ 1.0% at 180 A, takes place at 30 bar when reducing the current. An explanation for the effects of voltage, current and polarity can not be given based on the available evidence, unfortunately, more detailed analysis of the metal transfer in the welding would be necessary. Some conclusions, however, can already be drawn for controlling microstructure and toughness at hyperbaric welding. It is apparently attractive to take negative polarity, since it lowers the weld metal oxygen content. On the other hand, in this case the Mn-content depends on the welding current, and in practice it is quite possible that the whole current range studied here is used with the lower values for manuel welding, and the upper values for mechanized welding, for example. It is known from our own results and from other work [23-25] that Mn-values lower than 1.4% for the C-Mn-wire and 1.2% for the C-Mn-1%Ni-wire led to toughness reductions. Therefore it could become necessary to further modify the electrodes depending on the welding parameters. Based on the effects of the welding parameters the last step of the third welding series was to find out if by using negative polarity the toughness can be improved due to a reduction in weld metal oxygen content. The toughness results are presented in the figures 10-12. Figure 10 exhibits the charpy impact values for the C-Mn-wire as a function of temperature. The change from (+)-to(-)-polarity leads to a toughness improvement especially in the ductile/brittle region, as expected. Compared to the toughness curve for 1 bar as a reference (Ar + 5%) (dashed line), the combination of a less active shielding gas and negative polarity leeds to a toughness at 30 bar, which is almost as good. The corresponding CTOD-results are illustrated in figure 11, and again the change in polarity and the use of an O_2-reduced shielding gas has resulted in CTOD values, which especially in the ductile/brittle region are as good as those for 1 bar. Only the upper shelf values are slightly lower at high pressure, the reasons are not clear so far. The charpy impact results for the C-Mn-1%Ni-wire are shown in figure 12. In this case He + 5% CO_2 was taken as shielding gas, since this type of gas is often used for hyperbaric welding. Also for this wire and shielding gas a marked increase in toughness by changing the polarity from (+) to (-) can be observed. The toughness of the 1 bar weld metal is not quite reached, this is probably due to the fact that the weld metal oxygen content at 30 bar is still higher than that at 1 bar. Further improvement should be possible by using (He + CO_2) mixtures with less CO_2(e.g. 2% CO_2).

From a practical viewpoint the combination of less active shielding gas, wire modification and the change of the polarity has resulted in very attractive weld metal toughness values at high pressures. According to British Standard BS 4515 (1984) for example, the minimum required toughness at the design temperature of -10°C is 47 J. In this study much higher values of about 125 J were achieved at this temperature, and the temperature at which the weld metal still had an impact energy of 47 J, was at about -60°C, so that a high margin of safety even for welding at high pressure is guaranteed.

SUMMARY

Hyperbaric, mechanized MIG/MAG welding tests have been performed at pressures up to 30 bar (≙ 300m water depth). Both a pipeline steel (StE 445.7 TM ≙ X65 TM) and a structural steel (StE 36 ≙ 50 D) were welded with flux-cored wires of the C-Mn- and the C-Mn-1%Ni type. Besides the pressure the shielding gases ((Ar + O_2)- and (He + CO_2)- mixtures) as well as the welding parameters were varied. The study was mainly concerned with the weld metal, it focused on the influence of the above parameters on weld metal chemistry, microstructure, strength and toughness including fracture mechanics tests. The results can be summarized as follows.

(1) WELD METAL CHEMISTRY - a) With increasing pressure and increasing O_2 or CO_2-content in the shielding gas, the weld metal oxygen content increases. Systematic correlations between the partial pressure of the oxygen in the shielding gas and the weld metal oxygen content were observed. The Mn- and Si-contents decrease with increasing activity of the shielding gas. At low activitis no pressure effect was found, at higher activities (>5% O_2) the pressure increasingly reduces the weld metal Mn- and Si-content.

b) The influence of the welding parameters on the weld metal chemistry is complex. At high pressure for example the polarity has a pronounced effect on the weld metal oxygen content giving lower oxygen values for negative polarity. In addition for negative polarity the Mn-content is depending on the current while for positive polarity no current effect was found.

(2) STRENGTH AND TOUGHNESS - Hyperbaric welding under conventional atmospheric conditions results in a toughness reduction of the weld metal. Based on the results of the weld metal chemistry and a detailed analysis of the microstructures and fracture processes several steps were taken to reduce the loss in toughness.

a) In order to reduce the undesired, high weld metal oxygen contents at high pressures less active shielding gases should be taken.

b) In case that by this measure the Mn-burn-off is too low, modified, Mn-reduced wires are suggested.

c) At high pressures welding with negative polarity appears attractive since it further reduces the weld metal oxygen contents and gives a stable welding process with little spattering. When welding with negative polarity, attention has to be paid to the effect of current on the Mn-content, since it might necessitate further wire modification.

d) A combination of these steps in several welding series has resulted in improved charpy impact and CTOD-values at 30 bar which are almost as good as for 1 bar and which far exceed to values required by offshore standards.

ACKNOWLEDGEMENTS

We would like to thank W. Puneßen, P. Loebel, A. Dobernowski and A. Schikorr for their help in the welding experiments and V. Schmitz for her support in the metallographic work.

REFERENCES

(1) Underwater Welding: Proc. Int. Conference, Trondheim, Norway, 27-28 June, 1983, Pergamon Press, 1983.

(2) C.L. Tsai, K. Masubuchi; WRC (Welding Research Council) Bulletin 224, 1977.

(3) R. Schaefer, H.G. Schafstall; Schweißen und Schneiden 33 (1981), No.3, p. 1.

(4) H.G. Schafstall, R. Schaefer; Schweißen und Schneiden 31 (1979), No. 9, p. 374.

(5) A.V. Gaudiano, D. Groves; Welding Inst. Seminar "Underwater Welding for Offshore Installations", 9-10 March 1976, Welding Inst. Report 1977, p. 49.

(6) H.D. Knagenhjelm in "Underwater Technology", Proc. Int. Conference, Bergen, Norway, April 14-16, 1980, p. 351.

(7) G. Valland, H. Heiro; Det Norske Veritas Report, 1978.

(8) A. Matsunawa, K. Nishiguchi: in "Arc Physics and Weld Pool Behaviour" Proc. Int. Conf. in London, 8-10 May, 1979, p. 123.

(9) C.J. Allum: in "Underwater Welding", Proc. Int. Conf., Trondheim, Norway, 27-28 June, 1983, Pergamon Press, 1983, p. 171.

(10) D. Lythall, D. Gibson ; Proc. 8th Offshore Technology Conference, Houston 1976, Vol.3, p. 283.

(11) M. Perlman, A.W. Pense, R.D. Stout: Welding J. 49 (6) (1969), p. 231/s.

(12) G.R. Salter: Welding J., 44(3), 1965, p. 107/s.

(13) E.C. Burril, M.L. Levin; Proc. Intern. Conference on Gas Dischanges, IEE, 1970, p.595.

(14) H. Hoffmeister, G. Huismann, H.G. Schafstall: in Proc. Int. Conf. "Underwater Welding", 27-28 June, 1983, Trondheim, Norway, p. 239.

(15) N. Christensen: in Proc. Int. Con."Underwater Welding", 27-28 June, 1983, Trondheim, Norway, p. 71.

(16) F.R. Coe, J. Moreton: in Proc. Int. Conf. "Underwater Welding", 27-28 J, 1983 Trondheim, Norway, p. 293.

(17) G. Terlinde, L. Müller, K.-H. Schwalbe, P.A. Beaven: in Proc. 15th Offshore Technology Conference, Houston, Texas, May 2-5, 1983, Vol. 3, p. 153.

(18) B. Ahlbohm: IIW Doc. IX-1322-84.

(19) M. Ferrante, R.A. Farrar: J. of Materials Science 17 (1982), p. 3293.

(20) R.A. Ricks, P.R. Howell, G.S. Barritte: J. of Materials Science, 17 (1982), p. 732.

(21) W.E.W. Ruckdeschel: IIW Doc. XII-B-174-74.

(22) Y. Ito, M. Nakamishi, Y. Komizo: Met. Constr., Sept. 1982, p. 472.

(23) G. Terlinde, L. Müller, P.A.Beaven, K.-H. Schwalbe: in Proc. of 5th European Conference on Fracture (ECF 5), Sept. 17-21, 1984, Lisbon, Portugal, p. 257.

(24) N. Mori et. al.: J. Jap. Weld. Soc. 50(1981), p. 174.

(25) G.M. Evans: Welding J., Res. Supplement, 59(3) 1980, p. 675.

(26) D. Taylor, J.G. Garland, D. Yates: in Proc. 15th Offshore Technology Conf. Houston, Texas, May 2-5, 1983, Vol. 3, p. 161.

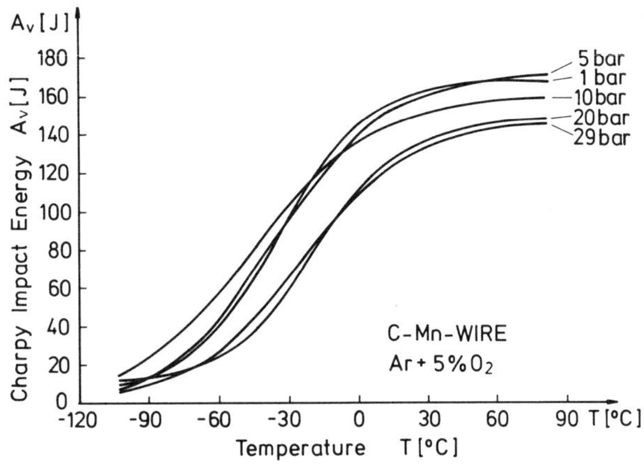

Fig. 1 - Charpy impact energy as a function of temperature for different pressures

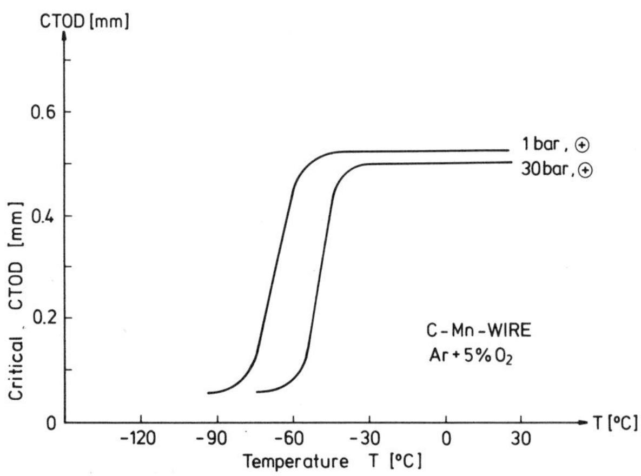

Fig. 2 - Critical CTOD-values as a function of temperature for 1 bar and 30 bar

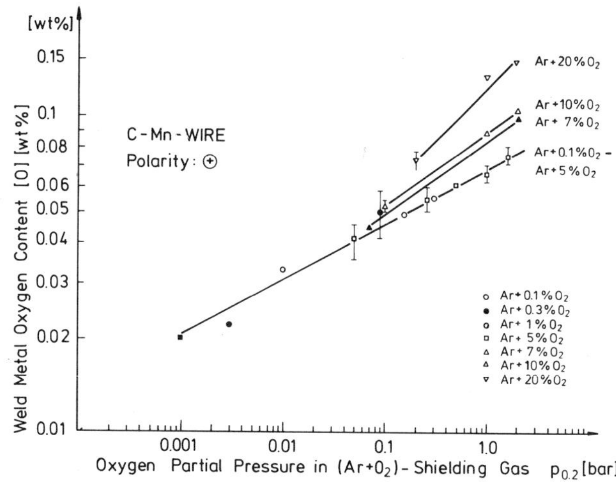

Fig. 3 - Weld metal oxygen content as a function of oxygen partial pressure in the (Ar - O_2)-shielding gas.

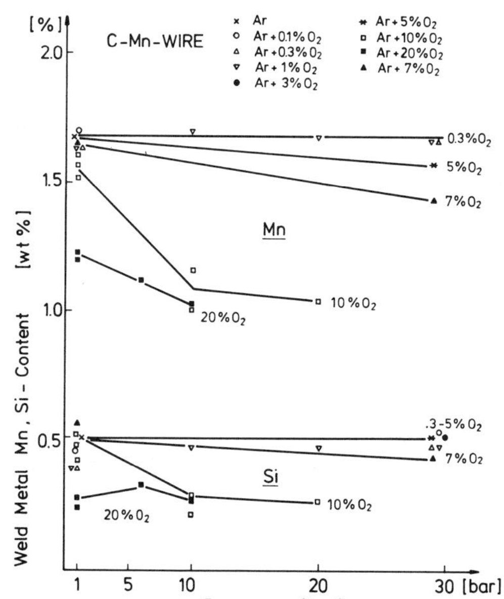

Fig. 4 - Weld metal Mn- and Si-content as a function of pressure for different shielding gases

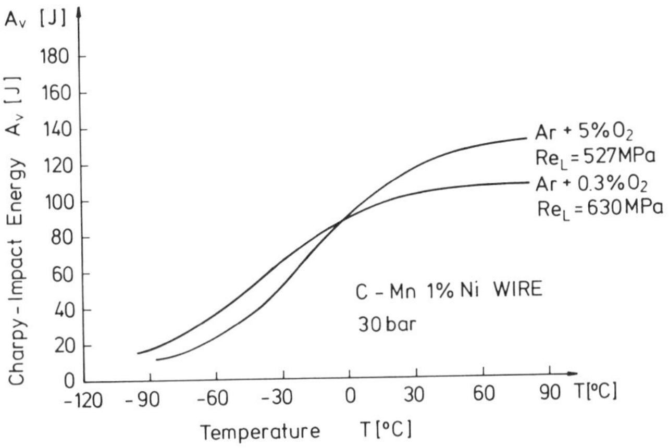

Fig. 5 - Charpy impact energy as a function of temperature for two different shielding gases at a pressure of 30 bar

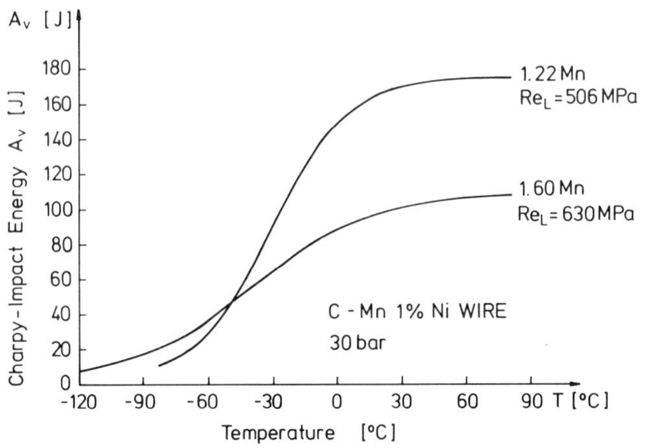

Fig. 6 - Charpy impact energy as a function of temperature for a conventional and a Mn-reduced C-Mn-1%Ni-wire, 30 bar pressure

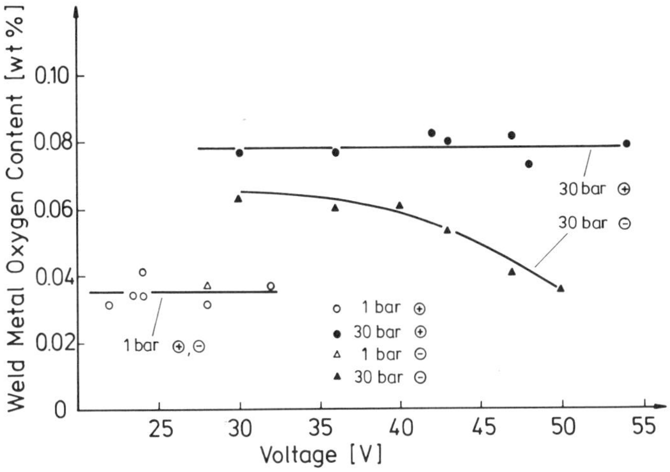

Fig. 7 - Weld metal oxygen content as a function of voltage and polarity at 1 bar and 30 bar

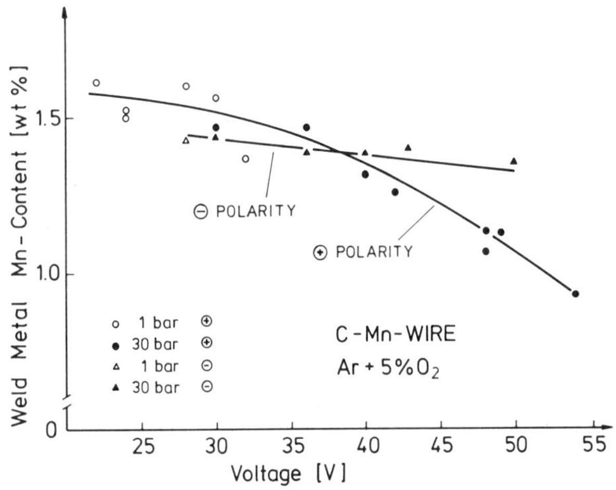

Fig. 8 - Weld metal Mn-content as a function of voltage and polarity at 1 bar and 30 bar

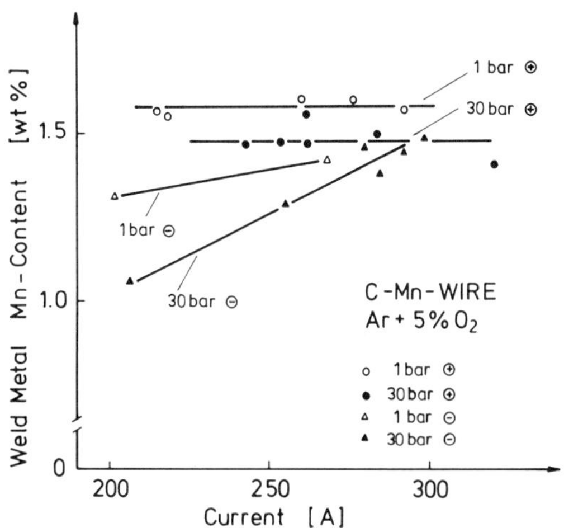

Fig. 9 - Weld metal Mn-content as a function of current and polarity at 1 bar and 30 bar

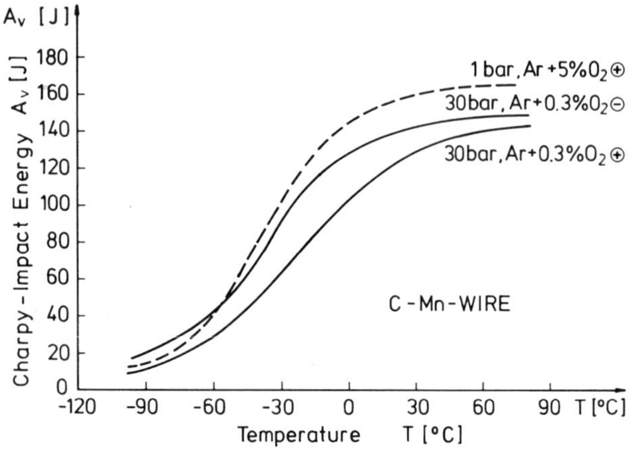

Fig. 10 - Combined effect of less active shielding gas and (-)-polarity on the charpy impact energy, C-Mn-wire

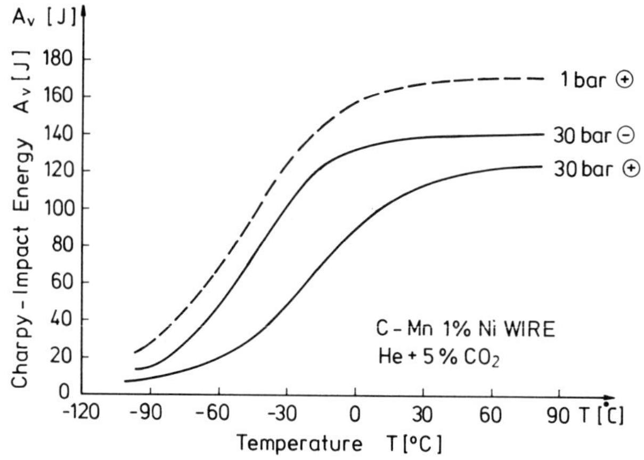

Fig. 12 - Effect of polarity on the charpy impact energy for the modified C-Mn-1%Ni wire welded with He - 5% CO_2

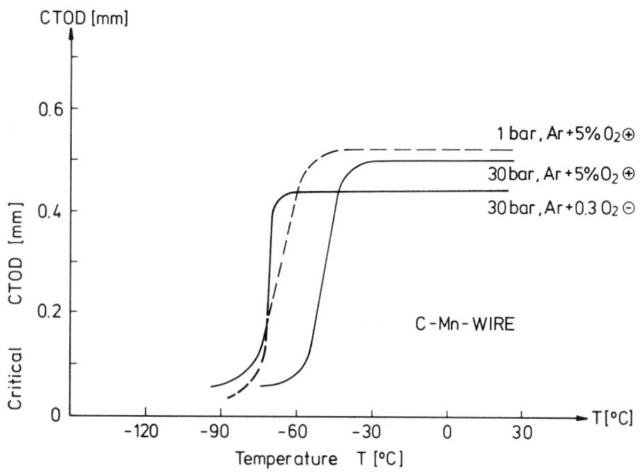

Fig. 11 - Combined effect of less active shielding gas and (-)-polarity on the critical CTOD-values, C-Mn-wire

POST WELD HEAT TREATMENT CRACKING IN A 710 HSLA PRECIPITATION STRENGTHENED STEEL

C.D. Lundin, R. Menon
University of Tennessee
Knoxville, Tennessee

Z. Chen
Dalian Railway Institute
China

Abstract

The evaluation of the post weld heat treatment (PWHT) cracking sensitivity of a heat of ASTM A 710 precipitation hardening HSLA steel is described. The tests utilized were Gleeble simulation, Y-Groove restraint and a C-ring test. The C-ring test has been adapted by the Welding Research Group at The University of Tennessee from one that is conventionally used to evaluate stress corrosion cracking susceptibility and is described in ASTM Standard G-38. The results of the tests indicate that the grain coarsened heat affected zone (HAZ) of the A 710 steel is susceptible to PWHT cracking at post weld heat temperatures of 1050°F, 1150°F and 1250°F. This susceptibility, evident in the Gleeble simulation tests, was reproduced in the Y-groove and C-ring tests where an actual HAZ is produced in the sample. The susceptibility to PWHT cracking is probably related to the intragranular precipitation of Cu-rich precipitates leading to grain matrix strengthening and concomitant strain accumulation in the prior austenite grain boundaries during PWHT.

POST WELD HEAT TREATMENT (PWHT) cracking is defined[1] as cracking that occurs in the heat affected zone (HAZ) during the exposure of welded assemblies to post weld heat treatment or high temperature service. This type of cracking has been found to be predominantly located in the coarse grained HAZ and is intergranular in nature. The occurrence of PWHT cracking has usually been attributed to the precipitation of fine carbides within the grains and the formation of a precipitate denuded zone along the grain boundary. Creep strains developed during PWHT or high temperature exposure are forced into the weakened grain boundary region and this coupled with grain boundary sliding, produces intergranular cracks. The occurrence of PWHT cracking has usually been related to the presence of certain alloying elements in the steel. Some of the elements contributing to PWHT have been reported[2] as chromium, copper, sulfur, phosphorus, arsenic, antimony and tin. Considerable work has been conducted at various research laboratories in evaluating the PWHT cracking susceptibilities of various grades of low alloy steels. However, the behavior of precipitation hardened low alloy steels such as A 710 is relatively unknown.

Thus, the objectives of this research program were to evaluate the PWHT cracking susceptibility of HSLA steel A 710. A 710 is a Ni-Cu-Cr-Mo-Cb bearing steel which derives its strength from precipitation hardening. Copper is the primary precipitation strengthening element. The A 710 steel is now being looked at with renewed interest for possible use as a high strength structural material. The low carbon content of the steel (0.07% max) leads to good weldability, toughness and formability. Mechanical properties can be varied using relatively simple heat treatments to cover a wide range of applications.

The chemical composition and mechanical properties of the A 710 material are given in Table 1. The properties were attained through precipitation hardening which involved austenitization at 1670°F for 1-1/4 hours, water quenching and aging at 1190°F for 3/4 hour. The base material microstructure is shown in Figure 1. The yield and tensile strengths of the steel are characteristic of a Grade A Class 3 steel with impact toughness (140ft-lbs at -120°F) far exceeding those mandated in supplementary requirement S1 (50ft-lbs at -80°F). As far as weldability characteristics are concerned, the steel would be expected to be immune from hydrogen-assisted cracking (HAC) in the HAZ due to its low carbon content. The precipitation hardening nature of the steel, however, would make it a likely candidate for PWHT cracking and it is to this end that this research work has been directed. If is to be noted here that although the various manufacturers of A 710 steel recommend no PWHT, the welding of thick sections (>2") may require some kind of post weld stress relief. Further, in-service high temperature exposure can also lead to a situation similar to PWHT.

EXPERIMENTAL PROCEDURES

Three unique test methods were utilized to evaluate the PWHT cracking susceptibility of the A 710 steel. The Gleeble simulation test was utilized to make an initial assessment of the cracking susceptibility and the Y-groove and C-ring tests were used to verify the results of the Gleeble simulation tests.

GLEEBLE SIMULATION - The Gleeble was used to simulate conditions leading to PWHT cracking in 1/4" diameter round samples of A710. The thermal cycle utilized simulated a heat input of 50 KJ/in in 1" steel plate with a room temperature preheat. The peak temperature was varied to simulate the grain coarsened HAZ (2400°F), the grain refined HAZ (1750°F) and the partially refined HAZ (1450°F). In addition, samples were also thermal cycled to simulate multiple thermal cycling (2400°F + 1750°F). Subsequent to the thermal cycle, the sample was reheated to the PWHT temperature (three PWHT temperatures were utilized: 1050, 1150 and 1250°F) and a tensile stress was simultaneously applied to simulate residual stresses. A schematic representation of the thermo-mechanical cycle is shown in Figure 2.

The stress levels empoyed were selected to cause sample rupture in 60 minutes. If no rupture occurred in 60 minutes, the test was stopped and a higher stress level selected.

The susceptibility to PWHT cracking is determined by the reduction in area (ductility) and the appearance of the fracture surface. To facilitate evaluation in the SEM, tests were also run in an inert gas chamber to prevent surface oxidation.

C-RING TEST - The C-ring test, normally employed to evaluate stress corrosion cracking, has been adapted at The University of Tennessee to study PWHT cracking[3]. This test was a suitable small scale test specimen for verification of the results obtained from the Gleeble testing program. ASTM basic standard G-38 was followed for design of the C-ring specimens and for determination of the deflection necessary to achieve the initial stress state.

To incorporate a weld in the C-ring specimens, a 1 inch by 1 inch square bar is machined with a flat bottomed groove parallel to the rolling direction at the plate surface. Groove geometry can be modified in order to accommodate different diameter shielded metal arc welding electrodes. A single pass shielded metal arc weld deposit of matching strength was used to fill the groove. The welded bars were machined into cylinders and surface ground to a fine, 32 RMS, finish. The cylinders have a nominal 1 inch outside diameter and 0.125 inch wall thickness. The rings are etched with 2% Nital in order to reveal the HAZ and fusion zone. A Charpy V type notch, 0.035 inches deep is broached along the fusion line and a hole drilled through the ring 90 degrees from the axis of the notch. The back of the ring, away from the notch, is removed in order to allow compression of the ring by tightening of a bolt placed through the drilled holes. The bolts were machined from the A 710 steel. Figure 3 schematically shows the successive steps in production of the C-ring specimen from the welded bar to the finished C-ring. Tightening of the bolt through the C-ring results in a constant strain specimen with tensile stresses produced on the exterior of the ring. The amount of deflection of the outside diameter (OD) necessary to produce the desired stress is determined using the equation given in ASTM standard G-38. This equation is modified for the presence of the notch by considering the outer fiber of the ring to be at the root of the notch rather than the outside surface. This results in the OD of the ring and thickness of the ring (t) being reduced by the depth of the notch. The deflection thus calculated is subtracted from the "true" OD to obtain the OD, through to the bolt location, to which the ring has to be deformed to obtain the nominal stress level desired at the root of the notch.

The C-ring specimens were stressed to yield (89 ksi) and PWHT at 1150°F. Periodic examination of the notch root was carried out to determine the onset of cracking. Sections encompassing cracked regions were mounted for metallographic examination. Also, the crack surface was examined optically and in the SEM often fracturing the samples in liquid nitrogen. To facilitate evaluation in the SEM, C-ring samples were post weld heat treated in evacuated quartz tubes to prevent oxidation of the crack surfaces.

Y-GROOVE TEST - The Y-groove test, developed in Japan, has been used extensively to evaluate material susceptibility to both HAC and PWHT cracking. The test simulates the first pass in a large weldment and provides a high restraint at the root of a weld. Ito and Nakanishi[4] have used it extensively to establish parameters to evaluate PWHT cracking susceptibility. The Y-groove specimen is shown in Figure 4. The restraint welds are first deposited using matching strength filler. The test weld is then deposited. The test assembly is held at room temperature for 24 hours and then the test weld is examined for cracking using liquid penetrant testing. After establishing that no HAC has occurred, the test assembly is PWHT at 1150°F for 2 hours. The test weld is cross sectioned into five pieces which are mounted for metallographic examination. Due to the large amount of material required to run this test and the extensive labor involved, only limited testing was performed using the Y-groove test.

RESULTS AND DISCUSSION

GLEEBLE SIMULATION TESTS - The results of the Gleeble tests utilizing various peak and PWHT temperatures are summarized in Table 2. The results indicate that with a peak temperature of 2400°F (grain coarsened HAZ), the reduction in area (RA) is zero at all PWHT temperatures utilized. The fracture mode observed was also entirely intergranular. This portends PWHT cracking problems in the grain coarsened HAZ. When the peak temperature was decreased to 1750°F (grain refined HAZ), the RA increased to about 60% and the fracture mode changed to a ductile dimple mode. A similar result was obtained when the peak temperature employed was 1450°F, to simulate the partially refined HAZ. When a multiple cycle was utilized (peak temperature 2400°F + 1750°F), the RA averaged 21% indicating a decreased susceptibility when compared to the case of a single cycle with a peak temperature of 2400°F. The tests therefore reveal that the grain coarsened HAZ is the region of maximum SRC susceptibility. As a base line test, the as-received material was heated directly to the PWHT temperature of 1150°F. The average RA of 60% indicates that, in the quenched and precipitation hardened condition, the base material is not susceptible to PWHT cracking.

Extensive optical (OLM) and scanning electron microscopy (SEM) was conducted to evaluate the nature and mode of rupture in the Gleeble simulation samples. Figures 5,6 and 7 show fractographs of Gleeble specimens that were subjected to a grain coarsening thermal cycle and PWHT at temperatures of 1050, 1150 and 1250°F respectively. At all three PWHT temperatures, the specimens exhibited zero reduction in area as indicated in Table 2. The fracture features are essentially similar with rupture occurring almost entirely in an intergranular mode. However, the micro-features of the fracture change as the PWHT temperature is raised from 1050°F to 1250°F. At 1050°F the grain faces are "smooth" indicating no ductility in the fracture process (Figure 5b). At 1250°F, the grain faces show evidence of extensive precipitation probably indicating an advanced stage of precipitation hardening (Figure 7b). It is clear that although the age hardening process is furthered during the PWHT, it does not appear to have any decipherable effect on the rupture ductility. A typical longitudinal section of a zero ductility Gleeble sample is shown in Figure 8. The occurrence of PWHT cracking along prior austenite grain boundaries is clearly evident from the location of the secondary cracks. With the peak temperatures lowered to simulate the grain refined HAZ (1750°F) and the partially refined HAZ (1450°F), a dramatic improvement is observed in the rupture ductility accompanied with a ductile dimple fracture morphology (Figure 9).

The results of the Gleeble simulation tests as well as the fractography clearly indicate that the coarse grained HAZ in the heat of A 710 tested is likely to be susceptible to PWHT cracking. To verify these results, the C-ring and Y-groove tests were undertaken so that an actual HAZ could be evaluated.

C-RING TESTS - The C-ring test was conducted as described earlier. The onset of cracking was detected by observing the notch root at periodic intervals during the PWHT. A typical cross section of a sample PWHT at 1150°F is shown in Figure 10. PWHT cracking has occurred in an intergranular manner in the coarse grained HAZ along prior austenite grain boundaries. The cracked surface in a C-ring sample is shown in Figure 11. The morphology of the crack shown in Figure 10 is very similar to the fracture morphology of the Gleeble specimen shown in Figure 6b. Clearly, the results of the C-ring test justify concerns regarding PWHT cracking predicted by the Gleeble simulation tests. The C-ring test accurately simulates real-life situations: it uses an actual weld and HAZ as opposed to the Gleeble test which simulates a HAZ. In the C-ring test, the stress decreases with PWHT which is the situation in a real weld whereas the Gleeble test utilizes a constant load. The notch in the C-ring sample adequately represents stress concentration effects (toe of weld, internal discontinuities), whereas the Gleeble test does not simulate stress concentration effects. Both the Gleeble and C-ring tests are easily evaluated and should therefore be used in a complimentary manner.

Y-GROOVE TESTS - The Y-groove test was conducted to produce PWHT cracking in a highly restrained weldment. Typical cracking in a specimen PWHT at 1150°F is shown in Figure 12. The cracks follow an intergranular path and are confined to the coarse grained HAZ. The crack surface is shown in Figure 13. The cracking mode is intergranular, however, the crack surface is obscured due to extensive oxidation. This oxidation occurred during PWHT. Thus, although the Y-groove test reproduces the susceptibility to PWHT cracking, little information is gained about cracking morphology since the crack surface is oxidized during PWHT. As contrasted to the Gleeble simulation or C-ring tests, the test assembly is too large to be easily accomodated into a vacuum or inert gas chamber. Furthermore, the test is material and labor intensive. Gleeble simulation and the C-ring tests can quite adequately evaluate the susceptibility of a material to PWHT cracking with more facility and economy than the Y-groove method.

The results from all three test methods indicate that the coarse grained HAZ of A 710 Grade A, Class 3 is susceptible to PWHT cracking. The location of the susceptibility, in the coarse grained HAZ, can essentially be ascribed to the following: At the high temperatures (≃ 2400°F) experienced in the coarse grained HAZ, carbides and Cu-rich precipitates go into solution in the austenite. The dissolution is accompanied by austenite grain growth at high temperature. During cooling, the alloying elements remain in solid solution and bainite is formed. During PWHT, intragranular precipitation of Cu-rich precipitates causes grain matrix strengthening in the bainite. Creep strains developed during PWHT concentrate primarily in the prior austenite grain boundaries which are weaker because of more rapid precipitate growth. Cracking occurs when the total grain boundary strain exceeds the grain boundary ductility. The mechanism described above explains the absence of PWHT cracking in the grain refined and partially refined HAZ's in that temperatures are not high enough to take all alloy carbides and precipitates into solution reducing the hardening effect and the grain boundary area is significantly larger and will accomodate more creep strains.

It is anticipated that future work in the program will study the precipitate composition and morphology in greater detail.

CONCLUSIONS

i) A 710 Grade A Class 3 precipitation hardenable HSLA steel containing copper as the principal strengthing element is susceptible to PWHT cracking in the coarse grained HAZ

ii) The susceptibility to PWHT cracking was evident in the Gleeble simulation tests and verified using C-ring and Y-groove tests

iii) The C-ring test appears to be a simple and viable test to evaluate PWHT cracking susceptibility. The test enables the testing of an actual HAZ which can be examined using simple metallographic techniques. The ability to position a notch at any location in the HAZ is an added advantage

iv) It is proposed that the Gleeble simulation and C-ring tests be used in a complimentary manner to evaluate PWHT cracking susceptibility and the Y-groove test be utilized only when a scaled up restraint test is warranted.

ACKNOWLEDGEMENT

The authors wish to thank the Welding Research Council University Research Committee for providing support for conduct of this research program.

References

1. W. Debray, "Reheat Cracking in Low Alloy Steels," Metal Construction, February 1976.
2. J. M. Brear and B. L. King, "Stress Relief Embrittlement in Mn-Mo-Ni Pressure Vessel Steels (ASME SA533/SA508)." Grain Boundaries, presented at the Institution of Metallurgists Conference, (1976).
3. C. D. Lundin and J. A. Henning, "Transformation, Metallurgical Response and Behavior of the Weld Fusion Zone and HAZ in Cr-Mo Steels for Fossil Energy Applications," Fossil Energy Material Program, Quarterly Progress Report for period ending September 1985, ORNL/FMP-85/4.
4. Y. Ito and M. Nakanishi, "Study of Stress Relief Cracking in Welded Low Alloy Steels," Sumitomo Search, No. 7, 1972.

Table 1

Composition and Mechanical Properties of A 710 Grade A Class 3

Chemical Composition (wt %)

C	Mn	P	S	Si	Cu
0.05	0.53	0.008	0.003	0.24	1.13
Ni	Cr	Mo	Al	N	Cb
0.98	0.72	0.21	0.027	0.008	0.034

Mechanical Properties

Yield Strength (0.2 % offset) = 89.3 ksi
Tensile Strength = 98.2 ksi
Elongation (in 8") = 21.5 %

Average Charpy V-notch toughness (ft-lbs)

At $0^{o}F$	185
At $-120^{o}F$	144

Table 2

Gleeble Simulation Test Results

Peak Temperature (^{o}F)	PWHT Temperature (^{o}F)	Average Reduction in Area (%)
2400	1050	0
2400	1150	0
2400	1250	0
1750	1150	60
1450	1150	63
2400 + 1750	1150	21
--	1150	55

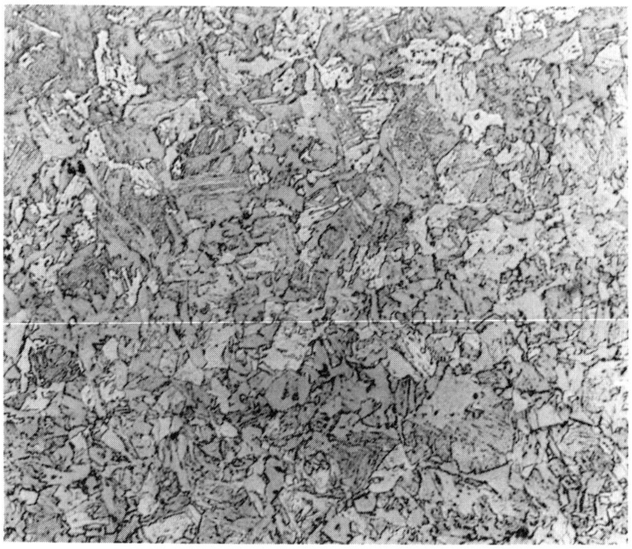

Figure 1. Microstructure of precipitation hardened A 710 Grade A Class 3 HSLA steel. OLM, 500X.

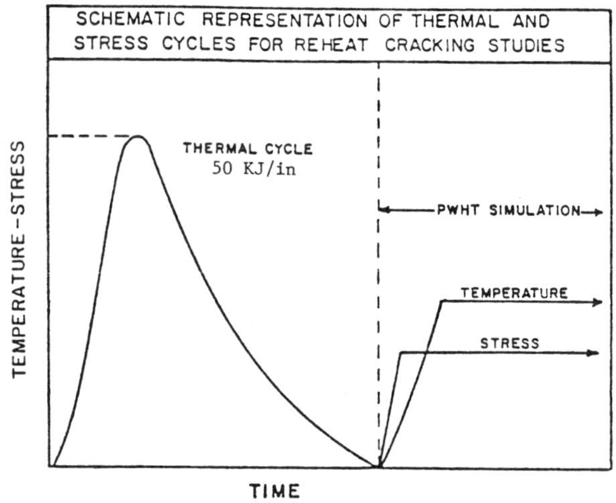

Figure 2. Schematic representation of the thermal treatment-stress cycle employed in Gleeble simulation.

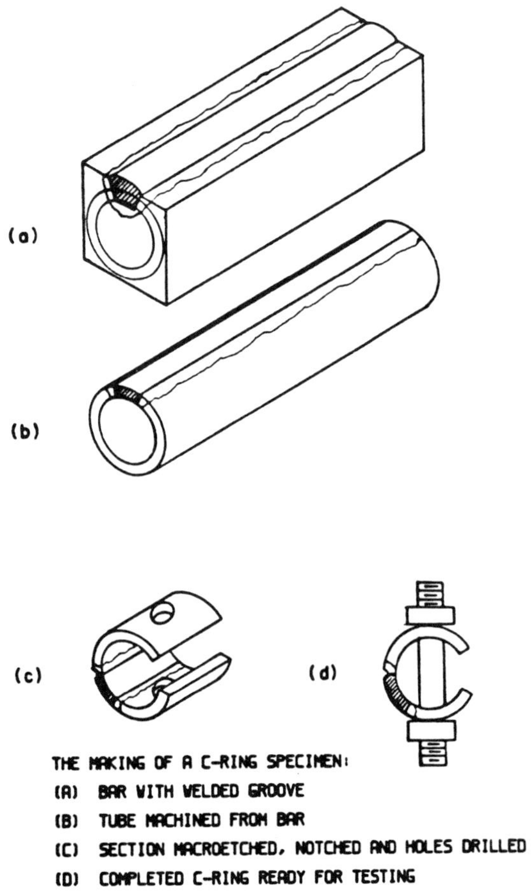

Figure 3. Various stages of C-ring sample preparation.

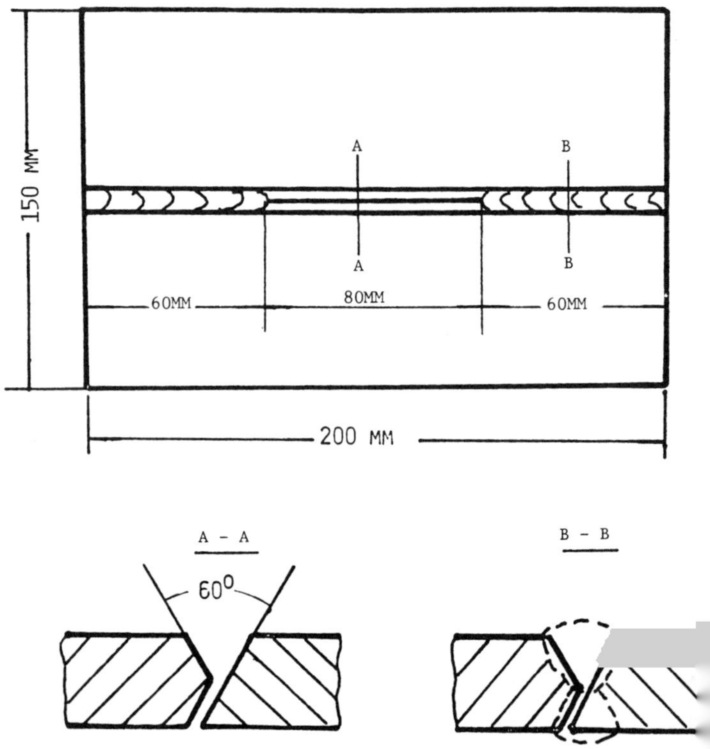

Figure 4. Y-groove test specimen.

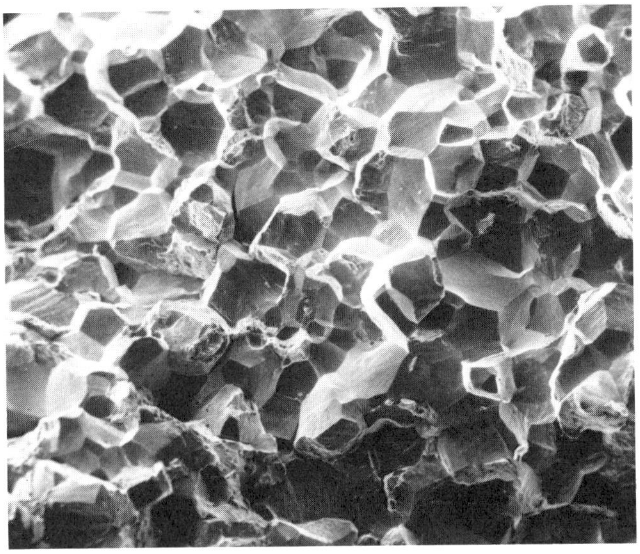

(a)

(b)

Figure 5. SEM fractographs of Gleeble simulation sample, peak 2400°F, PWHT 1050°F. a) 50X, b) 500X.

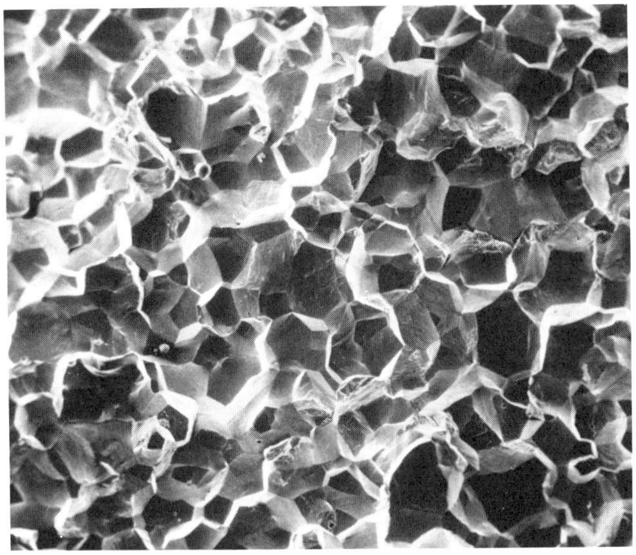

(a)

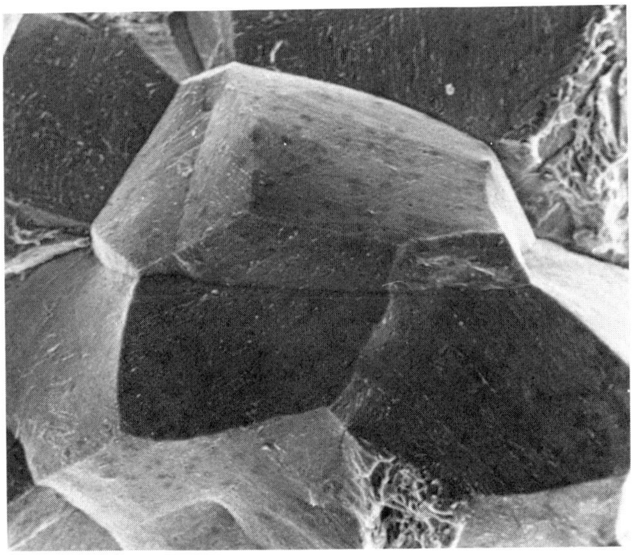

(b)

Figure 6. SEM fractographs of Gleeble simulation sample, peak 2400°F, PWHT 1150°F. a) 50X, b) 500X.

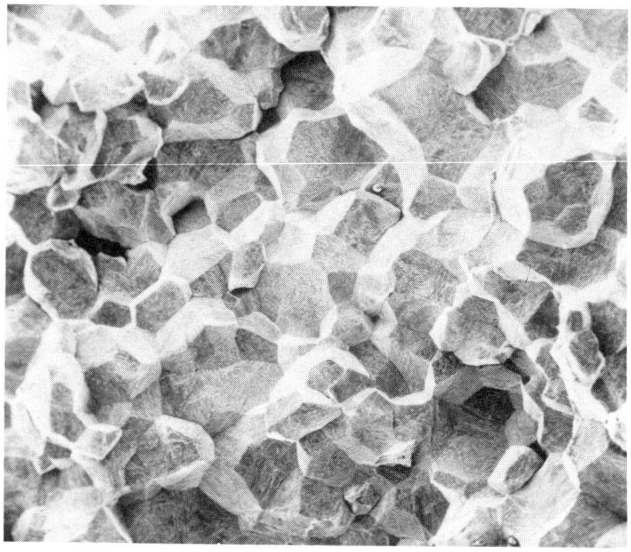

(a)

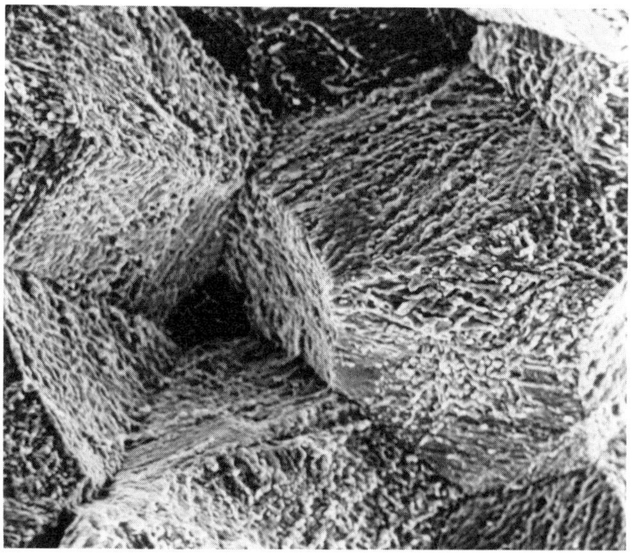

(b)

Figure 7. SEM fractographs of Gleeble simulation sample, peak 2400°F, PWHT 1250°F. a) 50X, b) 500X.

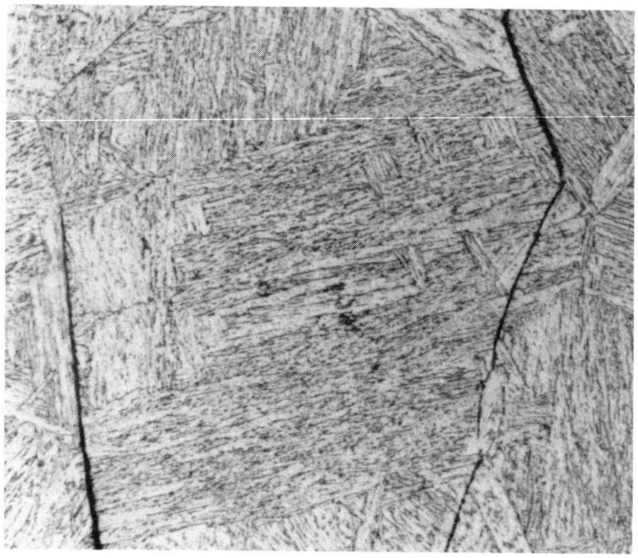

Figure 8. Longitudinal section of Gleeble simulation sample, peak 2400°F, PWHT 1050°F, OLM, 500X.

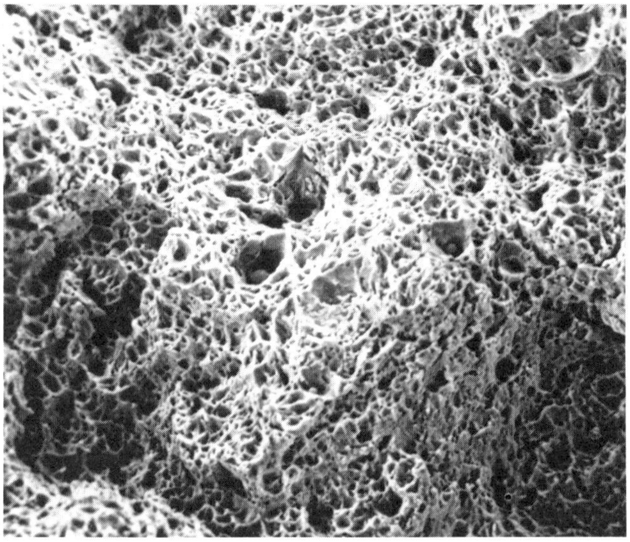

(a)

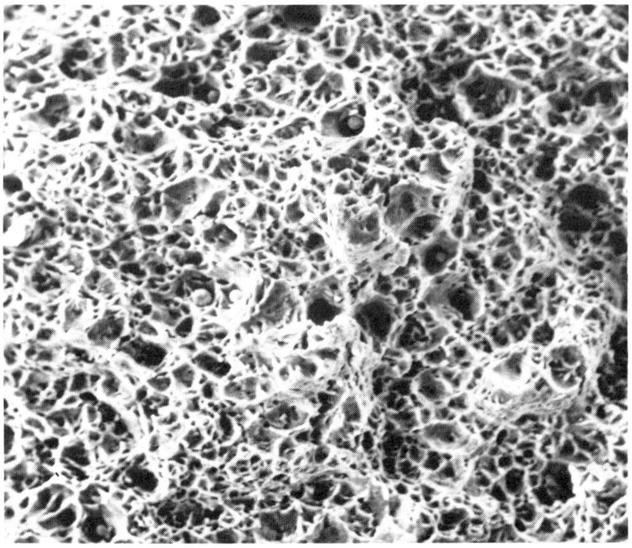

(b)

Figure 9. SEM fractographs of Gleeble simulation samples.
a) peak 1750°F, PWHT 1150°F, 500X
b) peak 1450°F, PWHT 1150°F, 500X

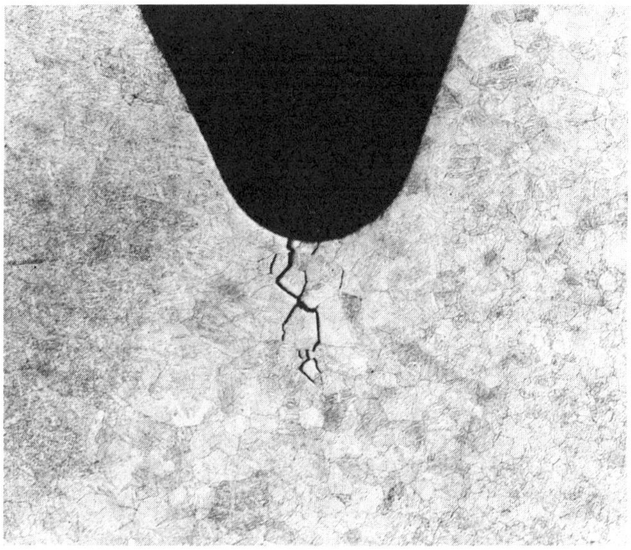

(a)

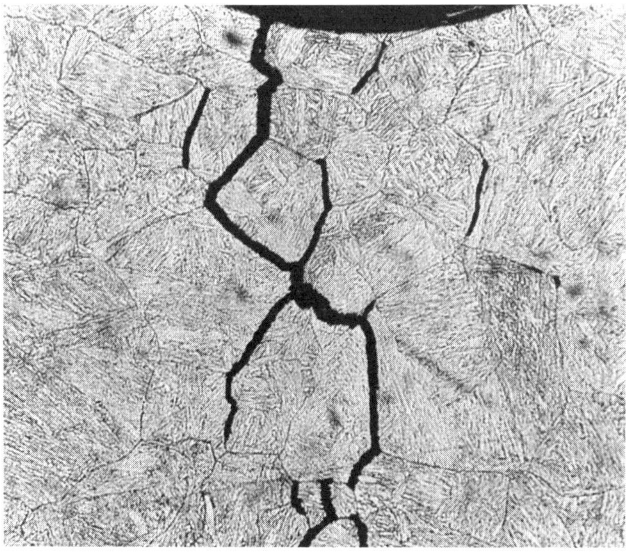

(b)

Figure 10. PWHT cracks in C-ring specimen. PWHT 1150°F-2hrs. OLM,
a) 50X, b) 200X.

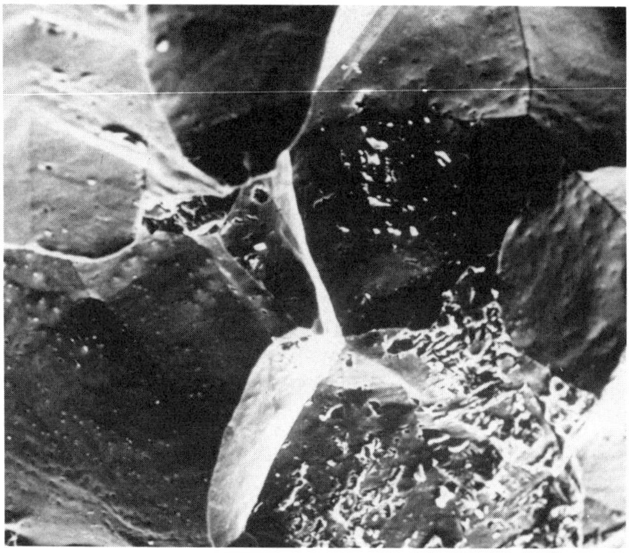

Figure 11. SEM fractograph of crack surface in C-ring specimen. PWHT 1150°F 2 hrs, 500X.

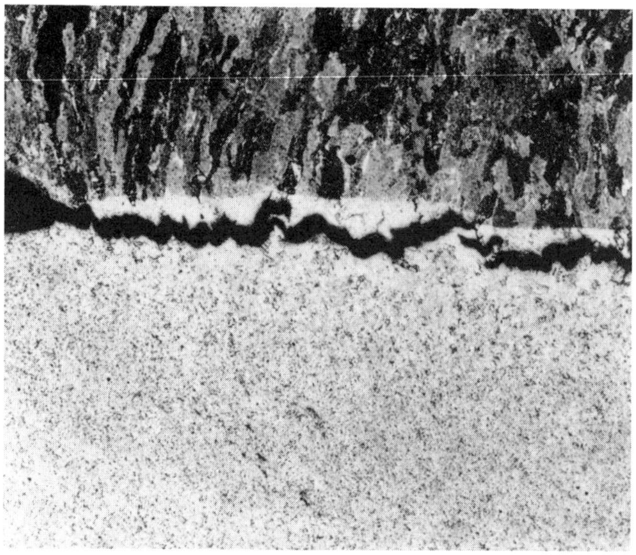

(a)

(b)

Figure 12. PWHT cracking in Y-groove specimen. PWHT 1150°F-2hrs, OLM, a) 22.5X, b) 200X.

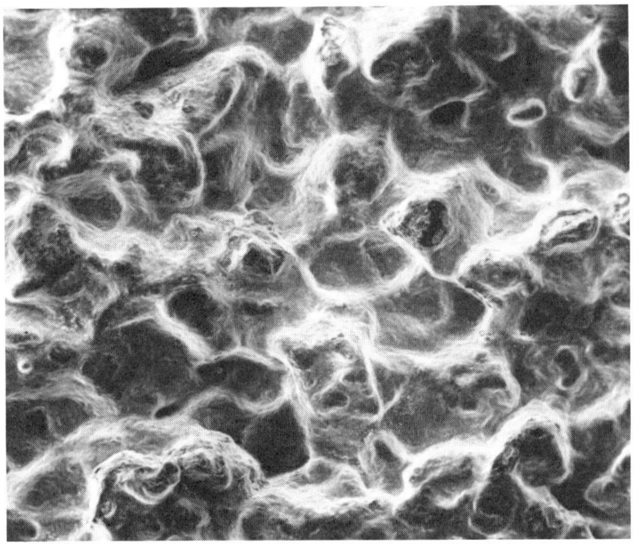

(a)

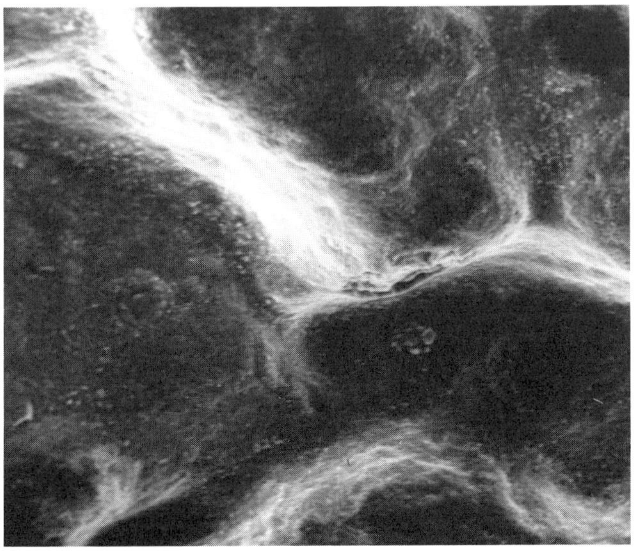

(b)

Figure 13. SEM fractographs of crack surface in Y-groove specimen. PWHT 1150°F-2hrs. a) 100X, b) 500X.

THE INFLUENCE OF MOLYBDENUM ON THE LOCALIZED CORROSION RESISTANCE OF AUSTENITIC STAINLESS STEEL WELD METAL

Silvia N. Simison, Susana R. de Sánchez
Intema
Universidad Nacional de Mar del Plata
Juan B. Justo 4302-7600 Mar del Plata
Pcia. de Buenos Aires-Argentina

Luis A. de Vedia
Conarco, Alambres y Soldaduras S.A.
Calle 18 Nº4113 - 1672 Villa Lynch.
Provincia de Buenos Aires, Argentina

INTRODUCTION

Austenitic stainless steels (ASS) suffer from pitting (PC) and crevice corrosion(CC) in chloride media. When any stainless steel is used in an agressive environment, the corrosion of weldments becomes an important consideration(1).

In order to reduce the risks of localized corrosion high levels of Cr and Mo are necessary. In recent years, much attention has been given to PC and CC of weldments in high alloy steels, and cases have been noted where the corrosion resistance of the weld metal has been lower than that of the wrought base metal.

ASS welds have a lower PC resistance in chloride environments when no filler metal is used or when the filler metal is close in composition to the base metal. Lower pitting resistance of welds is associated with microsegregation of Mo and Cr in the dendritic structure(2-4). However, CC resistance of welds is about equal to that of base materials with nearly the same overall composition(5).

In order to assess the susceptibility to localized attack, immersion tests and electrochemical polarization methods are used.

The breakdown or critical pitting potential methods are used to rate alloys with regard to their resistance to pit initiation in halide media. Although there have been some criticisms of the aplicatibility of those methods, Wilde and Williams (6) have demonstrated both their long and short term aplicability to naturally occurring pit initiation on stainless alloys in chemical and marine environments.

Although the critical pitting potential is qualitatively related to the resistance of a material to passivity breakdown by pit initiation, it is of questionable value in predicting the corrosion performance of a structure containing crevices (7). The propagation of PC and CC responds to the same mechanism. However, initiation mechanisms are different in both processes. CC can be considered as a special case of pitting with a ready - made initiation site (e.g. a geometrical discontinuity).

Since the presence of crevice is very difficult to avoid in actual weldments, CC is the most important form of attack in marine environment.

The resistance to CC depends upon a number of interrelated metallurgical, geometrical an environmental factors (8). This interrelationship makes mandatory to develop a testing technique that allows to change in a controlled and reproducible manner the parameters involved(pH concentration of O_2, Cl^- and $S^=$, surface condition, crevice dimentions, metallurgical structure,etc) creating a crevice similar to those formed in actual service.

The object of the present work is:

- To analyze the influence of Mo content on the susceptibility to localized corrosion of ASS welds.
- To develop suitable techniques for the assessment of localized corrosion susceptibility.

EXPERIMENTAL PROCEDURE

ELECTROCHEMICAL POLARIZATION METHODS - All weld metal samples were prepared on AISI 304 base plate using the joint design shown in Fig.1.

Experimental electrodes of a near 308 base analysis with varying Mo additions to the coating

*Intema - Universidad Nacional de Mar del Plata Juan B. Justo 4302-7600 Mar del Plata, Pcia.de Buenos Aires-Argentina.
**Conarco, Alambres y Soldaduras S.A. Calle 18 N°4113 - 1672 Villa Lynch. Provincia de Buenos Aires - Argentina.

were used. The coating was of a DC-lime type -- (AWS-15 coating).

The welding parameters employed and the chemical composition of the weld deposit are shown in Tables I and II respectively. No postweld -- heat treatments were used.

Cilindrical all weld metal samples of approximately 100 mm^2 cross section were extracted and used to prepare the working electrodes (WE) for the polarization electrochemical studies (Fig.2). The surface of the specimens were finished by -- polishing with alumina powder.

In order to reproduce actual service conditions tests were made in aereated artificial sea water and pH value adjusted to 8. The chemical composition of the artificial sea water was that recommended by Navy Department Specifications -- 44T27B, July, 1940, USA. The design of the cell employed is shown in Fig.2. The removable lens, of 300 mm curvature radius, creates a crevice -- ranging from 0 to 47 μm.

Potentiostatic anodic polarization curves and current vs. time curves were first obtained in the crevice free specimens (without using the -- lens). The potentiostatic curves were used to determine the attack potential (AP), that is the less noble potential for which the current increases steadily.

In order to determine the pitting initiation sites the samples were maintained at the AP during 3 minutes.

Cyclic potentiodynamic anodic polarization -- curves where then made on the specimens with --- crevice at 1 $mVsec^{-1}$ using the lens, begining at -200 mV(SCE). The reversal of the potential ---- sweep was made when the current density reaches 0,40 μAmm^{-2} and the potential of this point was considered as the AP.

Metalographic observations with the light microscope were made in all cases in order to determine the type and nature of the attack.

IMMERSION CREVICE CORROSION TESTS IN $FECL_3$. All weld metal pads as shown in Fig.3 were made with each electrode using the same welding parameters as before (Table I). Samples of dimensions 20x20x5 mm were extracted as indicated in Fig. 3 and abraded with silicon carbide paper to a 120 grit finish, washed in acetone, rinsed with etilic alcohol and air dried. The weight of the samples was determined to within a 10^{-4} mN accuracy.

The method consist of holding two TFE-fluorcarbon blocks, one on either side of the specimen, using two rubber bands at right angles to one another; this gives four rubber band to specimen crevices and two TFE-fluorcarbon to specimen crevices. A 10% $FeCl_3$ solution according to ASTM G48 was used.

Two different types of test were made:

Weight loss - The assemblies were immersed in individual containers and maintained in a thermostatic bath at 10 C during 48 hours. The assemblies were then removed from the bath, cleaned, dried and weighted.

Critical crevice temperature - This is a "go-no go" test in which the assemblies were exposed in the solution begining at 0 C with 2.5 C stepwise increases in temperature every 24 h. After each isothermal exposure period, the assemblies were disassembled and examined for CC. Corroded coupons were discarded and remaining coupons --- were polished again and reassembled for exposure at a 2.5 C higher temperature.

RESULTS

Fig.4 shows the AP corresponding to the crevice free specimens. Typical photomicrographs are shown in Fig. 5 and 6 where two different kinds of attack (pitting and cellular) can be observed.

Fig. 7 corresponds to a photomicrograph showing that PC begings at the δ - γ interphase and progresses into the γ phase.

The AP in the presence of crevice as determined by the cyclic anodic tests are given in Table III.

Fig. 8 shows some typical cyclic polarization curves for different samples.

The weight loss and the critical crevice temperature measured in the immersion tests are indicated in Table IV.

DISCUSSION

The results obtained with the crevice free -- specimens show an increasing resistance to localized attack with increasing Mo contents. For Mo level over about 4% a change from pitting to cellular attack was observed.

The results obtained from the specimens with crevice in artificial sea water indicate that -- two parameters are important to assess the susceptibility to CC. Both are linked with the quality of protective film and they are: The AP, that is related to the breakdown of passivity, and -- the amplitude of the hysteresis loop that is an indication of the repassivation ability. According to the rupture potential, only two widely -- different behaviours corresponding to samples 1-2 and 3-4-5-6 respectively could be observed. However Fig.8 shows that samples with similar rupture potential exhibited very different repassivation abilities, thus allowing a more accuraty selection of the optimum Mo level from the point of view of corrosion resistance.

The effect of Mo on corrosion resistance is also shown by the results obtained from the immersion tests. However, no significant differences could be observed in high Mo levels. On -- the other hand, immersion in $FeCl_3$ does not reproduce actual service conditions. The cell employed in this work allowed the use of sea water under selected conditions, i.e. conditions closer to those in service.

CONCLUSIONS

- Increasing Mo contents raises the corrosion resistance of ASS welds in sea water.

- A marked increase in corrosion resistance is obtained when the Mo level is over about 2% wt.

- Crevice containing samples having the same rupture potential exhibited different repassivation abilities which stresses the importance of cyclic polarization studies.

- A cell was designed and built that allowed the obtention of reproducible results under controlled crevice conditions (pH, O_2, $S^=$, Cl^- concentration, geometry, etc).

- In the absence of crevice a change from pitting to a less localized cellular attack was observed.

ACKNOWLEDGMENTS

The authors would like to acknowledge: CONICET and Conarco, Alambres y Soldaduras, S.A. to have sponsored this work within a research program on welding consumables development; Dr.D.Schiffrin for his help in designing the crevice cell and the assistance of F.Gómez in preparing the samples.

REFERENCES

1) A.P.Bond, H.J.Dundas, S.Ekerot and M. Semchyshen, Stainless Steel 77, N°18.

2) A.Garner, Corrosion/78, Paper N°20, 1978.

3) A.Garner, Pulp and Paper Research Institute of Canada, private comunication, 10/81.

4) A.Garner, Corrosion, 35(3), 1979.

5) N.Suutala and M.Kurkela, Stainless Steels '84, Göteborg, Sweden, 3-5 Sept. 1984.

6) B.E.Wilde and E.Williams, Electrochimica Acta, 1971, Vol.16, pp.1971 to 1985.

7) B.E.Wilde. Corrosion, 28(8), 1972.

8) J.W.Oldfield, Paper presented at the International Colloquium on Stainless Steels, held at Saint-Etienne, France, 7-9 May, 1980.

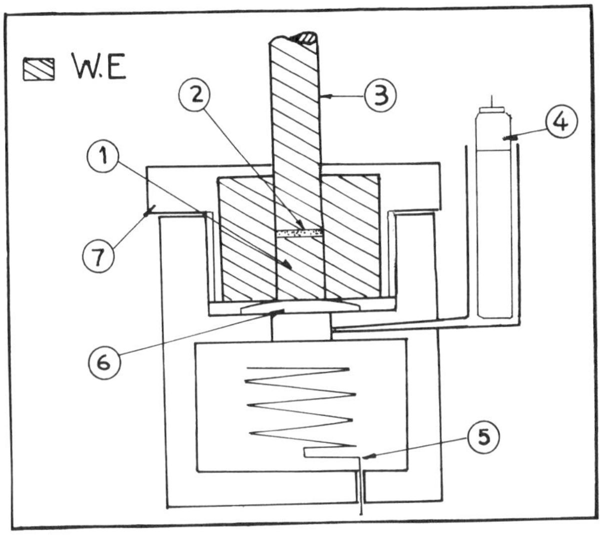

Fig.2: Schematic diagram of the test cell. 1) Test material. 2) Silver-Epoxy resin. 3) Bronze rod. 4) Reference electrode (SCE). 5) Platinum counter electrode. 6) Removable lens. 7) Working electrode holder.

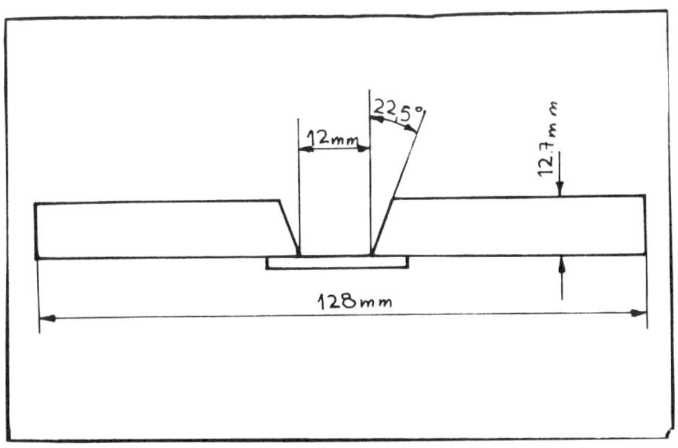

Fig.1: Schematic diagram of the welding joint.

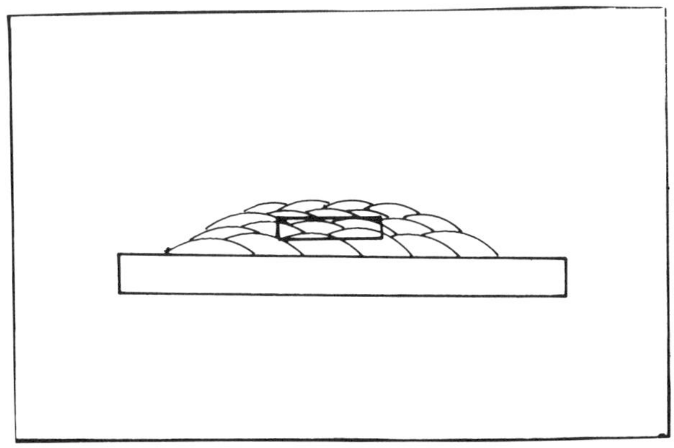

Fig.3: Schematic diagram of the samples extraction zone.

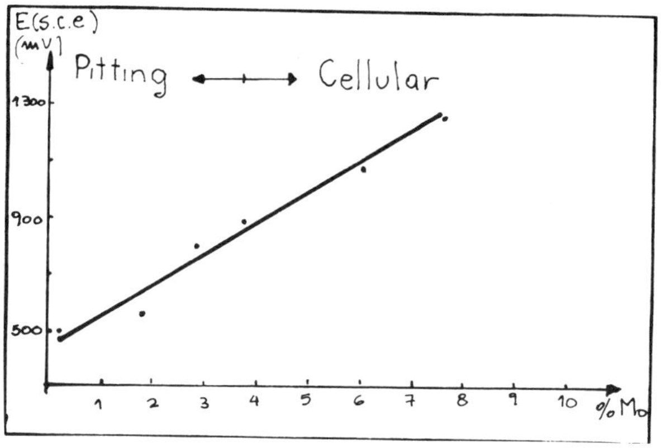

Fig. 4: Attack potential Vs Mo content
(crevice free specimens)

Fig. 5: Pitting attack (material 1)

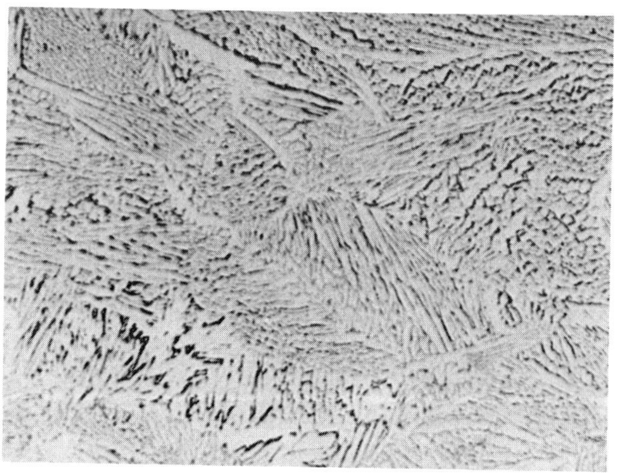

Fig. 6: Cellular attack (material 6)

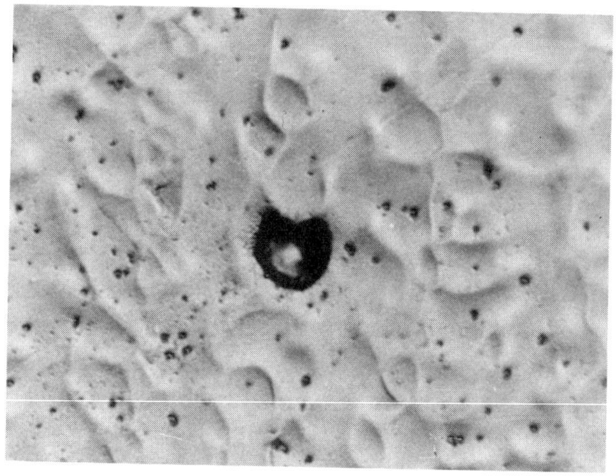

Fig.7: Pitting initiation site (material 2)

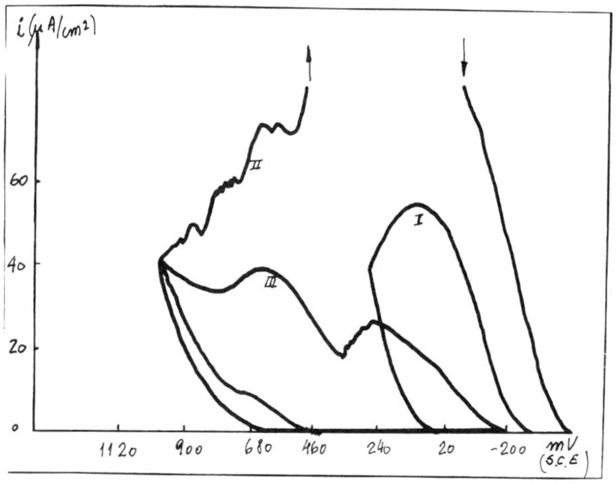

Fig.8: Typical cyclic polarization curves.
I) Material 1.
II) Material 4.
III) Material 6.

Table I : Welding para meters

Electrode diameter [mm]	Weld runs	Current [A]	Voltage [V]	Heat Input $[KJmm^{-1}]$	Travel Speed $[mm\ sec^{-1}]$	Interpass Temperature [°C]
4	9	145	25	1,5	2,4	100

Table II : Chemical composition of the weld metal.

Material	C	Mn	Si	Cr	Ni	Mo	N	O
1	0.05	1.67	0.52	19.6	11.7	0.18	0.048	0.066
2	0.05	1.78	0.59	19.6	11.7	1.88	0.047	0.064
3	0.05	1.66	0.59	19.6	11.5	2.61	0.043	0.060
4	0.055	1.59	0.39	19.5	11.5	3.70	0.046	0.062
5	0.055	1.65	0.39	19.5	11.6	5.50	0.044	0.070
6	0.06	1.54	0.41	19.2	11.3	7.40	0.047	0.064

Table III : Crevice attack potential values [mV]

%Mo	0.18 (1)	1.88 (2)	2.61 (3)	3.7 (4)	5.5 (5)	7.4 (6)
Crevice AP [SCE]	+250	+450	+850	+950	+870	+950

Table IV : Results of the $FeCl_3$ tests.

%Mo	0.18 (1)	1.88 (2)	2.61 (3)	3.7 (4)	5.5 (5)	7.4 (6)
weight loss (% wt)	0.805	0.360	0.230	0.200	0.310	0.165
Critical Crevice temperature °C	0	2.5	7.5	10	12.5	12.5

COPPER CONTAMINATION CRACKING IN THIN STAINLESS STEEL SHEET

R.K. Holbert, Jr., A.G. Dobbins, R.K. Bennett, Jr.
Martin Marietta Energy Systems
Oak Ridge, Tennessee

ABSTRACT

One of the recurring problems experienced in the gas tungsten arc (GTA) welding of thin austenitic stainless steel sheet is copper-contamination cracking. The copper originates from the tooling used in fixturing the sheet parts. This investigation not only has positively identified copper contamination as the cause of the cracking problem, but has derived two solutions to minimize the problem. First, lowering the heat input will reduce the thermal stresses experienced during cooling and minimize the cracking tendency. A robotic welding system will help to ensure that welding technique with the reduced heat input is more consistent from weld to weld. Second, eliminating the copper tooling will eliminate the source of the copper contamination.

THE WELDING OF THIN STAINLESS STEEL SHEET AT THE OAK RIDGE Y-12 PLANT is usually performed using the gas tungsten arc (GTA) welding process. The welding is done by a team of two people, i.e. the sheet metal worker (the person that does all of the weld setup work) and the weld operator (welder). The sheet used is generally Type 304L stainless steel in thicknesses ranging from 0.25 mm (0.010 in.) to 0.38 mm (0.015 in.).

The standard weld joints on the thin sheet are the butt, lap, and edge.[1] The weld process on all of these weld joints is semiautomatic. Another weld joint used frequently on the thin sheet is the tube/eyelet/sheet weld by which a tube can be joined to a sheet part through an eyelet or grommet. This weld is totally manual.

The requirements for the parts indicate that the welds are to be leak tight and full penetration with no surface defects. The inspection methods are a leak check for through crack or defects and a dye penetrant test for surface defects.

One of the problems that can cause the parts to fail the weld requirements is weld cracking. The most frequent types of weld cracking are surface cracking and heat-affected-zone (HAZ) cracking. The surface cracks are shallow defects which can be removed by light sanding, and therefore do not present a major problem. The HAZ cracking is typically deeper and is indeed a major problem.

The HAZ cracking is generally a liquation crack caused by environmental contamination. The most common contaminant is copper originating from the copper tooling used to hold the part. With its high thermal conductivity, copper is not only used to hold the part but also to chill it. Unfortunately, the copper dust that can be deposited on the weld area from worn or chipped fixtures can reach its melting temperature of 1083°C (1982°F) in the HAZ. Liquid copper has a tendency to penetrate the stainless steel grain boundaries which can result in copper-contamination cracking.[2 3 4 5 6]

EXPERIMENTAL EVALUATION

The HAZ cracking problems are not typical for the edge and lap weld joints. The weld fixturing on these welds do not constrain the welds very tightly; therefore, the cooling stresses are not excessively high. The copper clamps confining the butt joint hold the weld area very rigidly, resulting in occasional HAZ cracking (Fig. 1).

Most of the HAZ cracking problems occur in the tube/eyelet/sheet welds because this type of weld experiences more cooling stresses than the other types of sheet welds. The typical diameter of this two-pass weld is less than 8 mm (0.3 in) (Fig. 2). The weld area is surrounded by copper tooling. Also, a small copper bolt, called a "button", is used to hold the eyelet in place during welding. The first step in the tube/eyelet/sheet weld is to center the hole in the sheet (where the tube will be located over the hole in the bottom copper fixture). Then, the top copper fixture is placed on top of the weld area. The eyelet is slipped into the hole in the sheet part and held in place with the copper button. The eyelet is joined to the sheet part and the button is removed. Finally, the tube is placed in the eyelet, held in place by the sheet metal worker, and welded to the eyelet.

HAZ cracking in the tube/eyelet/sheet weld has been a problem for years, but it has caused only an occasional rejection of the welds. Cleanliness of the work environment helps minimize the problem. However, the problem becomes very serious whenever there are deviations from the precise sequence of the approved setup procedure. During a period when there were many welds rejected for HAZ cracking, several of the welds were evaluated. For example, scanning electron microscopy (SEM) of a HAZ crack in Fig. 3 and Fig. 4 resulted in the copper map in Fig. 5 that presents evidence of a copper particle at the start of a crack. On a second tube/eyelet/sheet weld, the weld crack is highly visible under a copper particle in Fig. 6 and its respective copper map in Fig. 7. Fig. 8 shows SEM of a HAZ crack in a third tube/eyelet/sheet weld that has copper on the crack face and in the area around the crack (shown in the copper map in Fig. 9). Furthermore, an electron microprobe of the metallographic cross section of HAZ cracking in a fourth tube/eyelet/sheet weld also presents evidence of copper along the crack face (Fig. 10).

This SEM and electron microprobe data provided a good indication that copper can cause HAZ cracking, but a controlled experiment was necessary for a conclusive evaluation. Therefore, two teams of welding personnel were each given ten tube/eyelet/sheet samples to weld. The sheet in half of the samples were purposely contaminated with copper particles and the welds were done in a random order. The completed welds were subjected to a dye penetrant test to determine acceptance of the weld, with the criterion for acceptance being the absence of surface defects. The first weld performed by the first team was killed because of a problem fixturing the weld. Of the remaining nine welds, all of the copper-contaminated samples were rejected because of HAZ cracking, whereas the non-contaminated samples were acceptable. The welding current used on all of the first team's samples was 21 A. The first sample of the second team was a copper contaminated sample welded at 20 A. The welder then changed the current to 17 A for the remaining tests, which resulted in a lower heat input into the weld. The first weld and another copper contaminated sample were rejected, the remaining welds were acceptable.

RESULTS AND DISCUSSION

The results of the designed experiment and the discovery of copper in and around the HAZ cracking has identified the defects as copper-contamination cracking. To minimize the problem, a reduction in the heat input of the weld can be used to decrease the cooling stresses and the width of the HAZ. However, a very skillful welder is needed to maintain the lower heat input on the manual tube/eyelet/sheet application weld after weld. Thus, an automated weld system is needed for a more consistent weld.

At the Oak Ridge Y-12 Plant, a welding robot has been introduced into the production welding of thin sheet. Presently, only the procedures for a butt weld have been established; however, the tube/eyelet/sheet weld parameter are being developed. Experience on the butt joint indicated that a low heat input can be maintained in the tube/eyelet/sheet weld more uniformly than with an above-average weld operator. A two-pass weld sequence on this tube/eyelet/sheet weld would not be efficient if the robotic system is used, especially if the sheet metal worker has to hold the tube in place while it is being welded. Therefore, to accommodate a change to a one-pass sequence, the end of the tube to be welded was flared so it would not pass through the eyelet. The other end of the tube was gripped with a spring clamp to apply an even tension on the tube. The tube then held the eyelet in place during welding to replace the copper button. This step allowed the eyelet/sheet portion of the weld to be completed, followed by a spiral inward to do the tube/eyelet portion without extinguishing the arc.

The automation of this weld will allow the weld operator to use a low heat input for a uniform weld. Unfortunately, this change will still only minimize, not eliminate the problem of copper-contamination cracking.

Elimination of the copper tooling is essential for this goal.

The production engineering staff tried substituting a silver-copper alloy for the copper. The first attempt was to butter a BAg-8 (AWS A5.8) brazing alloy onto the surfaces of the copper tooling that would be in contact with the sheet parts. This silver-copper surface was machined to the correct dimensions. Later, the discovery was made that there was sufficient chill from the buttered surface to allow mild steel, rather than copper to be used as the base material. The silver-copper alloy not only provided a good chill for minimizing part distortion, but also was not expected to penetrate the stainless steel grain boundaries.[7]

The new tooling eliminated copper-contamination cracking in production for a period of one year, with only one exception. The production staff had trouble with the dimensions on several tubes. To solve that problem, the staff members started using the copper buttons again with the old technique of a two-pass weld sequence. This change led to the rejection of several parts because of copper-contamination cracking. When the new one-pass technique was reinstated, the problems with copper-contamination cracking were eliminated.

CONCLUSIONS

The following conclusions resulted from this investigation:

1) Copper particles originating from copper weld fixtures can cause HAZ cracking in the welds on thin stainless steel sheet.
2) Welding with the minimum heat input necessary for full-penetration welds will decrease the tendency for copper-contamination cracking.
3) A robotic welding system for gas tungsten arc welding will increase the uniformity of the welds performed at the minimum heat input.
4) Copper-contamination cracking can be eliminated by eliminating the sources of copper. In the tube/eyelet/sheet weld application, a silver-copper alloy was successfully substituted for the copper tooling.

ACKNOWLEDGMENTS

The authors wish to thank Jim Somers and Bert Tyree for their tooling ideas. Also, the authors are grateful for the analytical contributions from Larry Walker and Bill Bolinger. Finally, the authors wish to express their gratitude to the production personnel in the Can Shop of the Oak Ridge Y-12 Plant for their support and efforts in this investigation and to Polly Morris for her aid in preparing the manuscript.

REFERENCES

1. D. G. Scott, "Joint Design for Welding Thin Materials", Paper presented at the SME Welding and Brazing Workshop, Chicago, Ill., October 28-30, 1975.
2. W. F. Savage, E. F. Nippes, and R. P. Stanton, Welding Journal, 57 (1), pp. 9s-16s (1978).
3. W. F. Savage, E. F. Nippes, and M. C. Mushala, Welding Journal, 57 (5), pp. 145s-152s (1978).
4. W. F. Savage, E. F. Nippes, and M. C. Mushala, Welding Journal, 57 (8), pp. 237s-244s (1978).
5. E. F. Nippes, Physical Metallurgy of Metal Joining, Proceedings of AIME symposium, pp. 53-73 (1980).
6. E. F. Nippes and D. J. Ball, Final INCRA Research Report on Project No. 287B, International Copper Research Association, June 1980.
7. W. Rostoker, J. M. McCaughey, and H. Markus, Embrittlement by Liquid Metals, Reinhold Publishing, New York (1960).

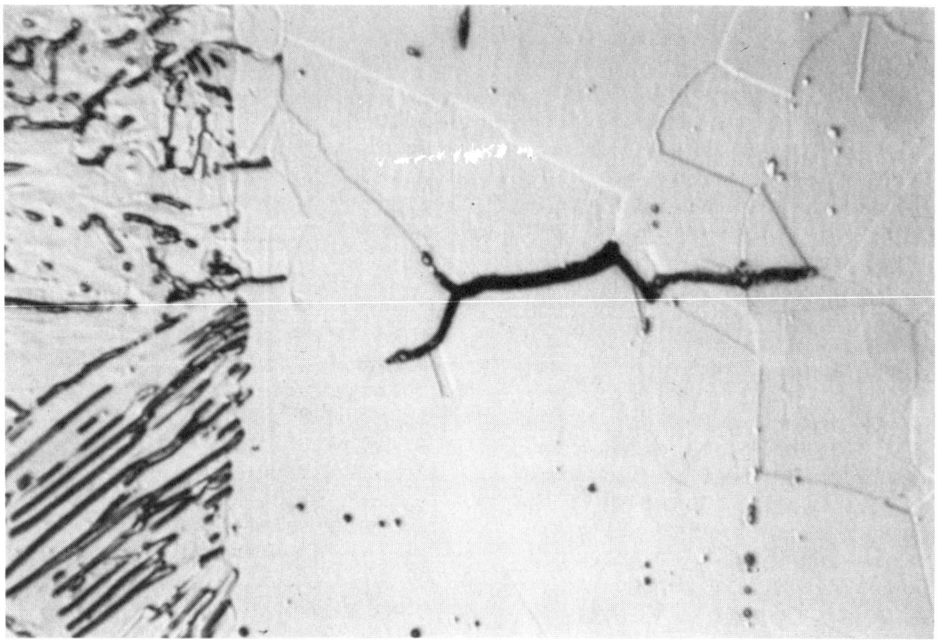

Figure 1. HAZ cracking in a butt weld.

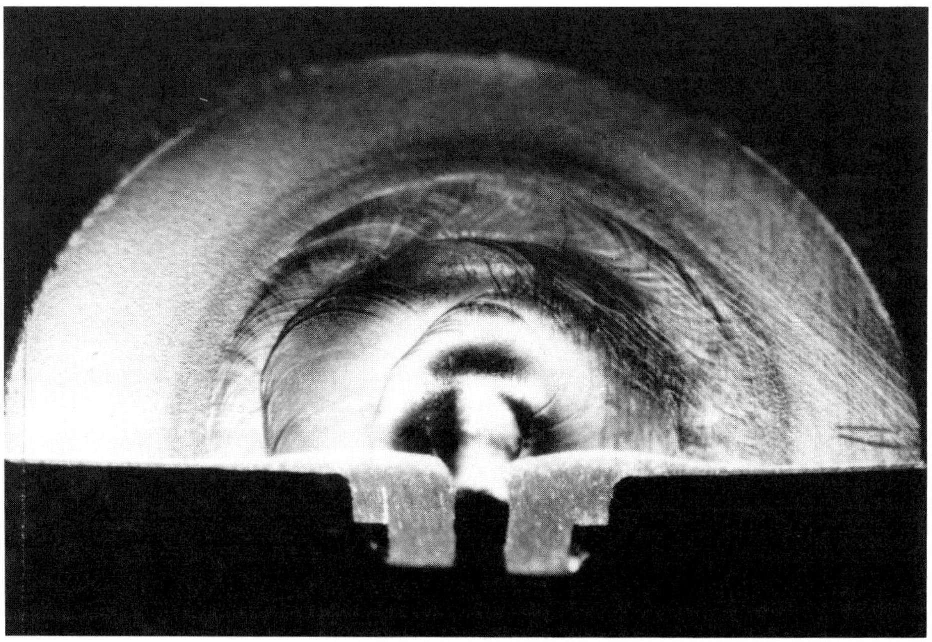

Figure 2. Typical tube/eyelet/sheetweld.

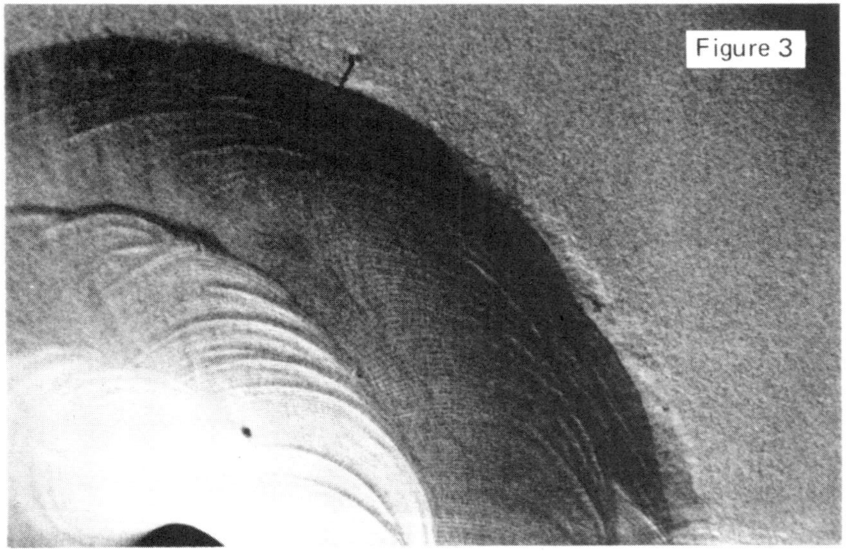

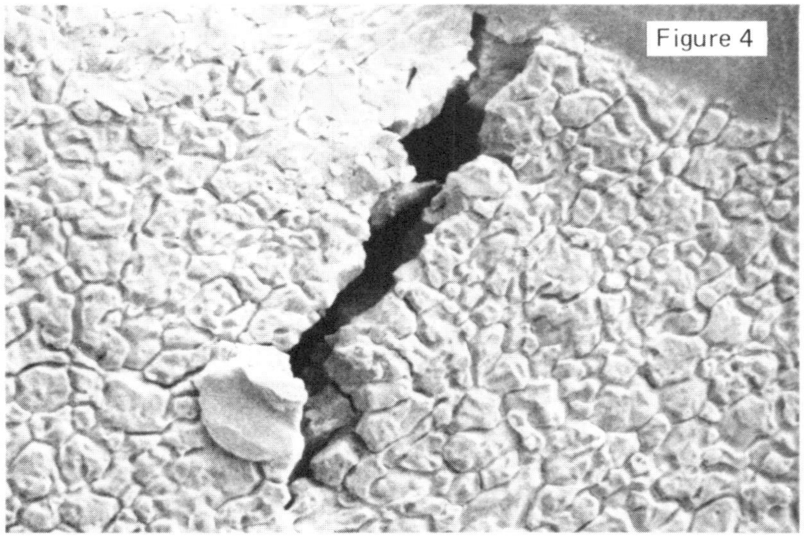

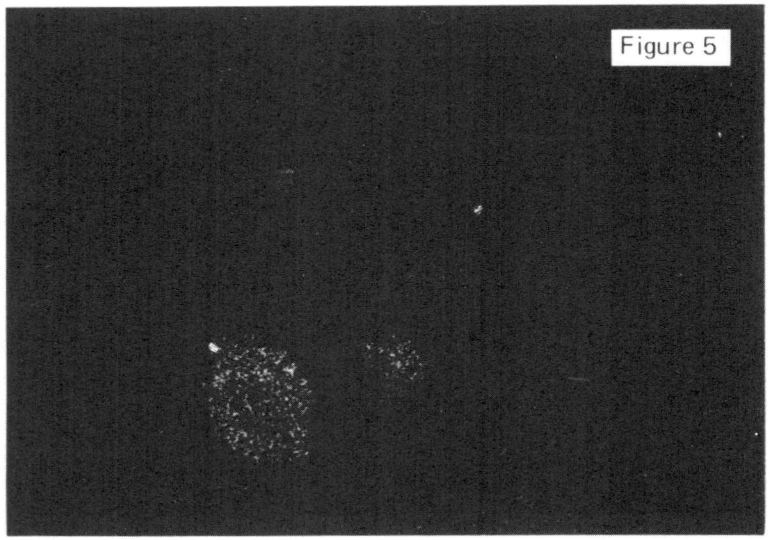

Figure 3. SEM of crack in first tube/eyelet/sheet weld.

Figure 4. Higher magnification of HAZ crack in Figure 3.

Figure 5. SEM copper map of Figure 4.

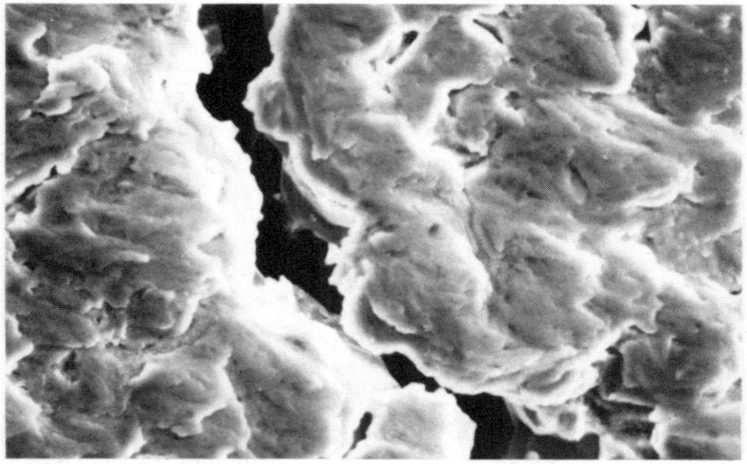

Figure 6. SEM of copper particle and associated crack on second tube/eyelet/sheet weld.

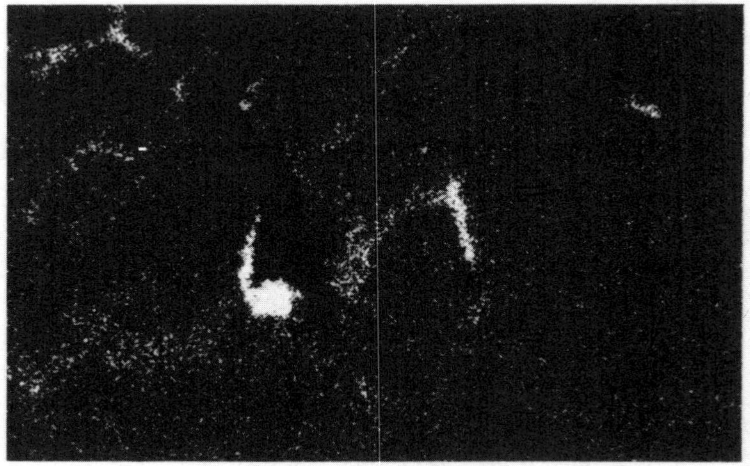

Figure 7. SEM copper map of Figure 6.

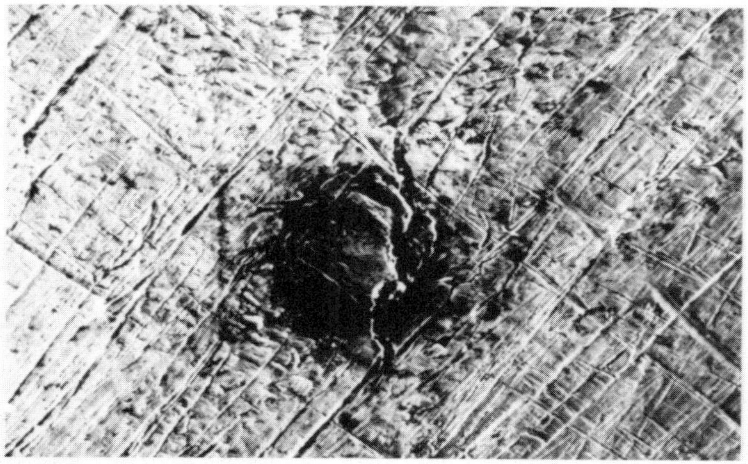

Figure 8. SEM of HAZ crack in third tube/eyelet/sheet weld.

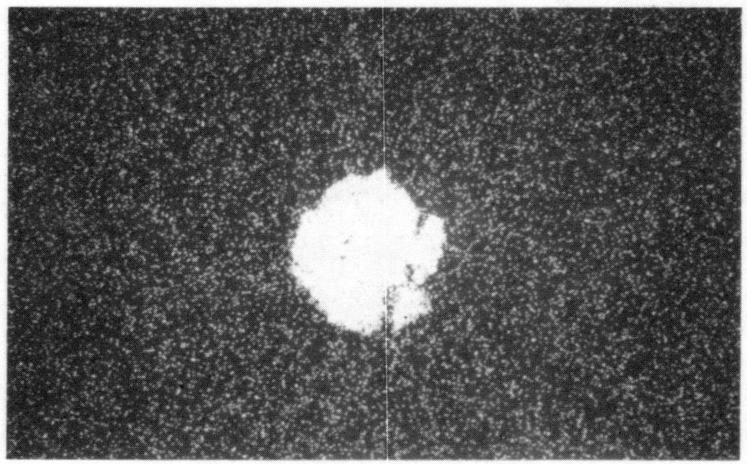

Figure 9. SEM copper map of Figure 8.

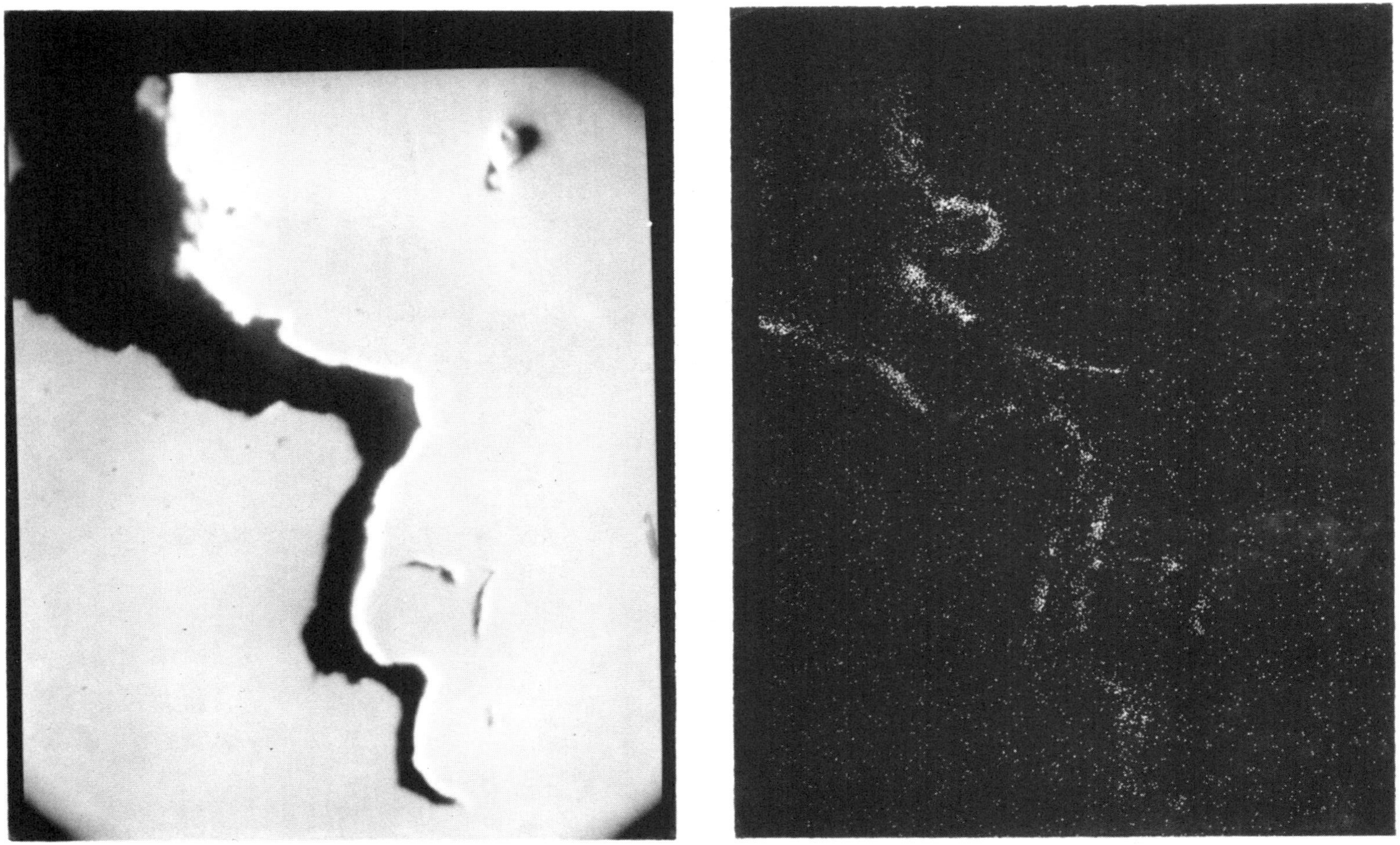

Figure 10. Electron microprobe analysis of metallographic cross section of fourth tube/eyelet/sheet weld.

HYDROGEN-ASSISTED CRACKING SENSITIVITY OF Cr-Mo STEELS DURING WELDING

C.D. Lundin, R. Menon, J.A. Henning
University of Tennessee
Knoxville, Tennessee

M.W. Richey
Martin Marietta Energy Systems
Oak Ridge, Tennessee

Abstract

Hydrogen-assisted cracking (HAC) sensitivity in the weld fusion zone and the heat affected zone (HAZ) of standard and modified Cr-Mo steels with chromium contents ranging from 2 1/4 to 12% has been investigated. An augmented strain test (based on the work of Savage et al. [5,6,7]) called the UT Modified Hydrogen Sensitivity Test (UT-Mod.HST), was utilized to evaluate the potential for HAC. The results of the tests indicate that the HAC susceptibility is primarily related to the composition of the steels. Higher carbon and silicon contents lead to increased susceptibility. Susceptibility is also increased with increasing potential for martensite formation and lower austenite to martensite transformation temperatures. The modifications of the 2 1/4, 3 and 9Cr steels with the additions of Ti, V and B did not significantly affect the HAC response. The addition of nickel to the 3Cr-1 1/2Mo steels appears to have a beneficial effect as far as the potential for HAC is concerned.

HYDROGEN-ASSISTED CRACKING (HAC) is a well known and well documented problem in the welding of ferritic steels. This form of cracking is refered to by various terms; cold cracking, delayed cracking and underbead cracking to state but a few. HAC is a manifestation of hydrogen embrittlement and requires essentially three factors[1] to be present: a sufficient amount of hydrogen, a susceptible microstructure and a sufficient tensile stress level. Hydrogen in welding can result[1] from improperly baked or stored electrodes, moisture in the flux or the presence of moisture and other impurities on the components to be welded. The microstructure is perhaps the most influential variable on the susceptibility to HAC. Increased propensities to HAC results[2] with a microstructural change from softer ferrite + carbides to harder bainitic and martensitic constituents. The microstructure produced depends on the composition and the cooling rate in the weld which in turn depends upon the energy input and the preheat temperature. The grain coarsened HAZ is generally the region with the greatest susceptibility to HAC since it experiences the fastest cooling rate and also possesses a high hardenability. The microstructure in the HAZ depends on the composition of the steel and in Cr-Mo steels is generally:

1. Bainitic and/or martensitic (0.5 to 5% Cr)
2. Martensitic-predominantly martensitic with some delta-ferrite present (5 to 12% Cr)
3. Duplex-martensitic-ferritic (9 to 12% Cr, 2% Mo)

2 1/4Cr-1Mo and 3Cr-1Mo will commonly have bainitic structures in the HAZ whereas 9Cr-1Mo and 12Cr-2Mo steels will exhibit martensitic microstructures. Delta ferrite can be present[3] with the martensite in the higher chromium steels and the potential for delta ferrite formation can be estimated from the chromium equivalent (Cr. E)[4],

$$Cr.E = Cr + 6Si + 4Mo + 1.5W + 11V + 5Nb + 8Ti + 12Al - 40C - 30N - 4Ni - 2Mn - Cu$$

When the Cr.E. is less than ten, delta-ferrite is usually absent. If the Cr.E. ranges from ten to twelve, some delta-ferrite may be present. For Cr.E greater than twelve, delta-ferrite may be present in significant amounts. The influence of delta-ferrite on the HAC susceptibility is not clear. Some effects are hypothesized in the analysis of the results of this investigation. For HAC to occur,the third factor necessary is the presence of a tensile stress. In welded fabrication, joint restraint as well as externally applied stress can contribute to the tensile stress required to cause HAC. Thicker sections will possess greater restraint and therefore are at more risk to undergo HAC.

There is considerable information available on the HAC behavior of plain carbon and low alloy steels. However, the HAC sensitivity during welding of the Cr-Mo steels has not been extensively researched. This paper presents the results of HAC evaluations of a wide range of Cr-Mo steels (2 1/4, 3, 9 and 12Cr) in the standard and modified grades.

MATERIALS AND EXPERIMENTAL PROCEDURES

MATERIALS - The steels tested in this research program encompass a wide range of Cr-Mo steels ranging from 2 1/4Cr to 12Cr ferritic steels. The 2 1/4Cr-1Mo steels tested included three heats of the standard grade (0.10-0.12C) and one heat modified with the addition of Ti-V-B. The alloy additions are made to improve the creep strength and the hardenability. The 3Cr-1 to 1 1/2Mo steels tested include one unmodified grade of 3Cr-1 1/2 Mo (0.08C), one heat of 3Cr-1 Mo (0.14C) modified with Ti-V-B and one heat of 3Cr-1 1/2Mo (0.16C) modified with the addition of 0.5% Ni. The 9Cr-1Mo steels tested comprise two standard heats containing 0.06 and 0.11%C and seven modified heats ranging in carbon composition from 0.08 to 0.11%C. In the 9Cr-2Mo grades, one unmodified heat (0.06%C) and one V-Nb added heat (0.09%C) were tested. Three lots of 12Cr-2Mo steel containing 1,3 and 6% Ni were also evaluated for their susceptibility to HAC. The compositions of all steels together with the calculated chromium equivalents are given in Table 1.

EXPERIMENTAL PROCEDURE - All weldability tests that are employed to evaluate the HAC susceptibility of steels essentially differ in the nature by which they introduce the three important factors responsible for HAC i.e. microstructure, hydrogen content and stress. The microstructure can be controlled with control on the heat input and the hydrogen can be controlled through control on the electrode/flux/shielding gas. The factor that is difficult to control is the stress. In self-restraint tests which are commonly used to evaluate HAC susceptibility, it is difficult to quantify the stress. Self restraint tests also involve extensive sample preparation and analysis. Bearing all these factors in mind, an augmented strain test was used to evaluate the HAC susceptibilities of the Cr-Mo steels. This test allows independent control over strain at constant microstructure and hydrogen levels. The UT-Modified Hydrogen Sensitivity Test (UT-Mod. HST) is described below. This test is a variation of the RPI augmented strain test [5,6,7] and includes the following modifications:

1. Reduced specimen size.
2. Autogenous GTA weld.
3. Hydrogen introduced directly into weld pool.
4. Materials evaluated on a pass-fail basis.
5. Effect of preheat, hold time and straining temperature determined.
6. Acoustic emission determination of crack initiation.
7. No cutting or polishing steps required for crack observation.

The specimen shown in Figure 1 is used in the UT-Mod HST. The specimen is held in a copper fixture which is preheated to the desired temperature. An autogenous gas-tungsten arc (GTA) weld is made along the length of the sample employing the parameters shown in Table 2. Hydrogen is added directly to the shielding gas, saturating the fusion zone. After welding, the sample is allowed to cool to room temperature and is strained in the fixture shown in Figure 2. The nominal augmented strain on the surface is given by the approximation:

$$\varepsilon \simeq t/2R$$

where ε = nominal augmented strain, t = specimen thickness (0.125in (3mm)) and R = bending radius. Die blocks are available to produce strain levels of 1%, 2% and 4%. Susceptibility to cracking is determined by observing if cracks form on the specimen face strained in tension. Cracking is evaluated as a function of preheat, strain level and straining temperature. Two straining temperatures were employed in this investigation (32°F (0°C) and 75°F (24°C)). The 32°F (0°C) straining temperature is considered to be below the ductile to brittle transformation temperature (DBTT) of the HAZ in most Cr-Mo materials, and provides the most conservative assessment of HAC.

After straining, the specimen and fixture are placed in a padded fixture and an acoustic emission (AE) sensor is attached to the die block. The output of the AE instrument is recorded on a strip chart recorder. The specimen is monitored for 24 hours. The time for crack initiation is taken as the time at which sustained AE bursts above the background level are recorded. If the specimen does not crack in 24 hours the test is terminated. For any strain, the preheat level at which at least two tests remain crack free is designated as the preheat required to prevent cracking. After testing, specimens are prepared for metallographic examination.

This testing method is intended to provide a comparative ranking of the HAC sensitivity of Cr-Mo materials based on augmented strain, hydrogen content, microstructure, preheat and welding conditions. Although the augmented strain levels employed in the sensitivity test have no direct correlation with actual measurable restraint encountered during welding, the 4% strain level is analogous to the restraint encountered when welding thick multipass joints, while the lower strain levels (2% and 1%) are analogous to restraint produced during the welding of thinner sections.

The preheat temperatures determined are based on relatively high levels of diffusible hydrogen (17-18 ml per 100gms of fused metal) and restraint, and as a result the preheats determined are conservative in nature and generally higher than commonly employed in welded fabrication.

RESULTS AND DISCUSSION

The preheat temperature was varied from 32°F (0°C) to 750°F (400°C). The augmented strain levels were 1,2 and 4%. The straining was conducted with the samples either at 32°F (0°C) and 75°F (24°C).

9Cr-1Mo AND 9Cr-2Mo MATERIALS - The results for the 9Cr-1Mo and the 9Cr-2Mo materials are shown in Figure 3 as the preheat required to prevent HAC vs. the nominal augmented strain. The augmented strain was applied with the samples at 32°F (0°C) to provide a conservative estimate of the HAC susceptibility. The data indicate that the susceptibility to HAC, as indicated by the increasing level of preheat temperatures required to prevent HAC, increases as the carbon content increases. As the level of augmented strain is increased, the required preheat temperature is increased. The standard 9Cr-1Mo materials with 0.06 and 0.11%C form the lower and upper bounds within which the HAC susceptibilites of the other materials lie. Among the 9Cr-1Mo materials modified with the addition of V and Nb, there appears to be an increased HAC susceptibility for the heats with higher silicon contents. Ranking of the standard and modified 9Cr-1Mo steels can be conducted at the augmented strain level of 1%. The standard 9Cr-1Mo-0.11C is the material with the greatest susceptibility requiring a preheat of 750°F (400°C) to prevent cracking. The standard 9Cr-1Mo-0.06C heat shows the lowest susceptibility with no HAC occurring at a preheat of 75°F (24°C). The susceptibilites of the modified 9Cr-1Mo materials lie in between with the higher silicon heats exhibiting a greater susceptibility. The HAC behavior of the 9Cr-2Mo-0.06C and the 9Cr-2Mo-V-Nb fall essentially within the band, however, the 9Cr-2Mo-V-Nb heat required preheats in excess of 750°F (400°C) to prevent HAC at an augmented strain of 4%. All materials exhibited no cracking when tested at 4% strain, a room temperature preheat and with no hydrogen in the shielding gas. This result confirms that the cracking is hydrogen assisted.

Optical microscopy was conducted to determine the relationship between the microstructure and the susceptibility to HAC. Typical longitudinal and transverse sections of the standard 9Cr-1Mo-0.11C material tested using a preheat of 75°F (24°C) and an augmented strain of 4% are shown in Figure 4. The longitudinal section shows cracks in both the fusion zone and the HAZ. The microstructure is essentially martensitic with some delta ferrite. The cracks appear to run normal to the direction of the maximum tensile stress on the surface of the specimen. Longitudinal and transverse sections of 9Cr-1Mo-V-Nb (Heat 394) and 9Cr-2Mo-0.06C tested under similar conditions as the standard 9Cr-1Mo are shown in Figures 5 and 6 respectively. The greater amount of delta ferrite in the fusion zone of the 9Cr-2Mo-0.06C as compared to the 9Cr-1Mo steels is predicted from its higher Cr.E (Table 1). The influence of delta ferrite on HAC susceptibility is not clear. On the one hand, the formation of delta ferrite can partition the carbon to the austenite resulting in a martensite of higher hardness leading to an increased susceptibility to HAC. On the other hand, delta ferrite may aid in blunting crack advancement as shown in Figure 7. The 9Cr-2Mo-V-Nb material has the highest potential for delta ferrite formation (Cr.E=18.53). Its response to HAC at the 1% and 2% strain levels is approximately equivalent to that of the 9Cr-2Mo-0.06C material. At the 4% strain level, the 9Cr-2Mo-V-Nb material exhibits a greater susceptibility than the 9Cr-2Mo-0.06C material. This is probably related to the difference in potential for ferrite formation and the carbon content.

2 1/4Cr-1Mo AND 3Cr-1 1/2Mo MATERIALS - Results for the 2 1/4Cr-1Mo and the 3Cr-1 to 1 1/2Mo materials are summarized in Figure 8. For comparison, the upper and lower limit of preheat temperature required to prevent HAC determined for the 9Cr-1Mo and the 9Cr-2Mo materials have also been plotted. The augmented strain was applied at 32°F (0°C) to give a conservative assessment of the HAC susceptibility. The results show that there was little change in the preheat temperature required to prevent HAC between heats with increasing augmented strain. On a relative basis the unmodified 3Cr-1 1/2Mo-0.08C steel showed the greatest susceptibility. The modified and the unmodified heats of 2 1/4Cr-1Mo and 3Cr-1 1/2Mo behave within the limits of HAC susceptibility of the standard 9Cr-1Mo materials. The standard 2 1/4Cr-1Mo steels required preheats in the range of 390-570°F (200-300°C) to avoid HAC at the 4% level of augmented strain. The 2 1/4Cr-1Mo heat modified with Ti-V-B showed a lower susceptibility requiring a preheat of only 212°F (100°C) to remain free from HAC. Among the modified 3Cr-1 1/2Mo materials, the nickel modified heat showed the lowest susceptibility with a preheat 390°F (200°C) needed to prevent HAC. Figure 9 shows the test results as bands representing HAC behavior. Comparing the behavior with that of the 9Cr-1Mo materials, the HAC susceptibility of the low carbon-low silicon 9Cr materials is equivalent to that of the 2 1/4 and 3Cr materials at 4% augmented strain. The high carbon-high silicon heats of the modified 9Cr-1Mo appear to be more susceptible to HAC than the 2 1/4Cr and 3Cr materials. This is probably due to the fact that the martensitic structures in the low carbon 9Cr-1Mo steels are less susceptible to HAC than mixed bainitic-martensitic structures in the 2 1/4Cr and 3Cr steels. The situation is reversed with higher carbon martensite resulting in the HAZ of the higher carbon 9Cr-1Mo steels.

Typical HAC in the 3Cr-1 1/2Mo-0.08C material is shown in Figure 10. In general, the cracks run normal to the direction of maximum tensile stress. The cracks are transgranular in nature and extend from the fusion zone to the HAZ. Some transgranular cracks which were delayed in nature were observed oriented parallel to the welding direction. Such cracks may have initiated at a hot crack and then propagated along the centerline in the presence of hydrogen. A typical example is shown in Figure 11.

12Cr-2Mo-Ni MATERIALS - All three 12Cr-2Mo-Ni steels (1,3 and 6% Ni) exhibited cracking when welded using a preheat of 750°F (400°C) and strained at 75°F (24°C) to 1% i.e. cracking occurs even at a relatively high preheat and low strain. Figure 12a shows cracking in the fusion zone and HAZ of a longitudinal section of a sample of 12Cr-2Mo-3Ni steel. Figure 12b shows a transverse section. The relatively high HAC susceptibility of the 12Cr-2Mo-Ni steels is probably related to the potential for martensite formation and on-cooling transformation behavior. The M_s and M_f of these steels are, on the average, 300-350F° (160-190C°) lower than those for the 9Cr and 2 1/4Cr steels. The lower transformation temperatures provide little opportunity for autotempering and self diffusion of hydrogen, thus increasing the cracking susceptibility.

CONCLUSIONS

Based on the results of the UT-Mod.HST, the following conclusions can be arrived at:

1. The HAC susceptibility in the weld fusion zone and the HAZ of standard and modified 9Cr-1Mo steels and unmodified 9Cr-2Mo steel appears to

be largely governed by the carbon content, increasing carbon content leading to increased susceptibility. Increasing levels of silicon in the modified 9Cr-1Mo steels also appear to increase susceptibility to HAC although the reasons for this are not clearly apparent at this time.

2. The 2 1/4Cr-1Mo and 3Cr 1 to 1 1/2Mo standard and modified materials have HAC susceptibilities within those bounded by the susceptibilities of the low and high carbon 9Cr materials. Low carbon 9Cr materials behave similar to the 2 1/4Cr and 3Cr materials.
3. The 12Cr-2Mo materials have a greater susceptibility to HAC than the 9Cr and the 2 1/4Cr and 3Cr materials. This is probably related to the lower austenite to martensite transformation temperatures of the 12Cr-2Mo materials.
4. The role of delta ferrite in affecting the HAC susceptibility could not be clearly defined. The formation of delta ferrite could partition the carbon to the austenite thus leading to an increased susceptibility. However, delta ferrite can also blunt crack propagation. Both these effects apparently act synergistically.

ACKNOWLEDGEMENT

This work was supported by the Department of Energy, Advanced Research and Development Fossil Energy Materials Program, through the Oak Ridge National Laboratory (ORNL), under Subcontract 7685X77 with The University of Tennessee, Knoxville, under Martin Marietta Energy Systems Contract DE-AC05-84OR21400 with the U.S. Department of Energy.

REFERENCES

1. F. R. Coe, "The Avoidance of Hydrogen Cracking in Welding," The Welding Institute (1973).
2. G. E. Linnert, "Welding Metallurgy," Vol. 2,
American Welding Society (1967).
3. F. B. Pickering and A. D. Vassilious, "Effect of Austenitizing Temperature on Constitution, Transformation and Tempering of 9Cr-1Mo Steels," Metals Technology, 7(10), 409-413 (1980).
4. G. C. Bodine, B. Chakravarti, C. M. Ownens, B. W. Roberts, D. M. Vandergriff and C. T. Ward, "A Program for the Development of Advanced Ferritic Alloys for LMFBR Structural Applications: Topical Report," 1977, ORNL/Sub-4291/1 (1977).
5. W. F. Savage, E. F. Nippes and E. S. Szekeres, "Hydrogen Induced Cracking in a Low Alloy Steel," Welding J., 57(9), 276s-283s (1976).
6. W. F. Savage, E. F. Nippes and H. Homma, "Hydrogen Induced Cracking in HY-80 Steel Weldments," Welding J., 57(11), 376s-386s (1976).
7. W. F. Savage, E. F. Nippes and E. S. Husa, "Hydrogen Assisted Cracking in HY-130 Weldments," Welding J., 63(8), 233s-242s (1982).

Table 1
Compositions of Steels Tested (wt%)

Designation	Heat No	C	Mn	P	S	Si	Ni	Cr	Mo	V	Nb	Ti	B	Cr.E
$2\frac{1}{4}$Cr-1Mo-0.10C	BP3	0.10	0.44	0.006	0.026	0.19	0.16	2.18	1.02	0.01	<0.01	<0.01	<0.001	1.99
$2\frac{1}{4}$Cr-1Mo-0.12C	646363D	0.12	0.53	0.006	0.003	0.10	0.17	2.36	0.98	0.01	<0.01	<0.01	<0.001	0.45
$2\frac{1}{4}$Cr-1Mo-0.12C	3443	0.12	0.41	0.006	0.003	0.22	0.02	2.27	1.05	0.01	<0.01	<0.01	<0.001	2.09
$2\frac{1}{4}$Cr-1Mo-V-Ti-B	-	0.12	0.48	0.007	0.006	0.06	0.15	2.31	0.98	0.26	-	0.033	0.0022	3.35
3Cr-$1\frac{1}{2}$Mo-0.08C	P2904	0.08	0.86	0.006	0.009	0.16	0.21	3.02	1.43	0.12	<0.01	<0.01	<0.001	5.26
3Cr-1Mo-V-Ti-B	-	0.14	0.46	0.007	0.008	0.06	0.20	3.11	0.97	0.28	<0.01	0.02	0.003	3.27
3Cr-$1\frac{1}{2}$Mo-$\frac{1}{2}$Ni	-	0.16	0.52	0.010	0.003	0.28	0.51	3.05	1.48	0.007	<0.01	<0.01	<0.001	1.25
9Cr-1Mo-0.06C	K48603F	0.06	0.46	0.018	0.005	0.52	0.20	9.28	0.98	0.01	<0.01	<0.01	0.001	11.90
9Cr-1Mo-0.11C	15965-2C-5	0.11	0.59	0.017	0.008	0.63	0.07	9.38	0.99	0.05	0.01	<0.01	0.001	11.29
9Cr-1Mo-V-Nb	148	0.09	0.47	0.021	0.006	0.28	0.16	9.24	0.96	0.21	0.05	<0.01	0.001	11.12
9Cr-1Mo-V-Nb	176	0.08	0.37	0.010	0.003	0.11	0.09	8.61	0.89	0.21	0.07	<0.01	0.001	9.59
9Cr-1Mo-V-Nb	383	0.08	0.46	0.010	0.004	0.41	0.09	8.46	1.02	0.20	0.07	<0.01	0.001	11.52
9Cr-1Mo-V-Nb	394	0.08	0.46	0.010	0.003	0.40	0.09	8.57	1.02	0.20	0.07	<0.01	0.001	11.61
9Cr-1Mo-V-Nb	A-231001	0.10	0.41	0.004	0.004	0.41	0.01	8.36	0.98	0.22	0.07	<0.01	<0.001	11.70
9Cr-1Mo-V-Nb	82631	0.10	0.41	0.010	0.007	0.42	0.09	8.76	1.00	0.31	0.09	<0.01	<0.001	13.42
9Cr-1Mo-V-Nb	OYYC-9082	0.10	0.43	0.014	0.006	0.40	0.10	8.50	0.98	0.20	0.07	<0.01	<0.001	10.96
9Cr-2Mo-0.06C	592	0.06	0.66	0.015	0.004	0.25	0.06	8.60	1.93	0.03	<0.01	<0.01	<0.001	14.16
9Cr-2Mo-V-Nb	HZ9650/1	0.09	0.93	0.020	0.002	0.42	0.16	9.43	1.95	0.29	0.37	-	-	18.53
12Cr-2Mo-1Ni	2321	0.07	0.82	0.011	0.009	0.31	1.03	11.86	2.04	0.20	<0.01	<0.01	<0.001	15.52
12Cr-2Mo-3Ni	2807	0.06	2.27	0.011	0.014	0.30	3.03	11.91	2.08	0.22	0.10	<0.01	<0.001	5.89
12Cr-2Mo-6Ni	2773	0.07	0.74	0.012	0.014	0.29	5.92	11.96	2.09	0.21	0.09	<0.01	<0.001	-11.50

Table 2. UT-Modified Hydrogen Sensitivity Test Welding Parameters

Variable	Setting
Current	80–100 amps
Voltage	10 volts
Travel Speed	5ipm (12.7 cmpm)
Shielding Gas	5% H_2 - 95% Ar
Energy Input	10.8 KJ/in (0.43 KJ/mm)

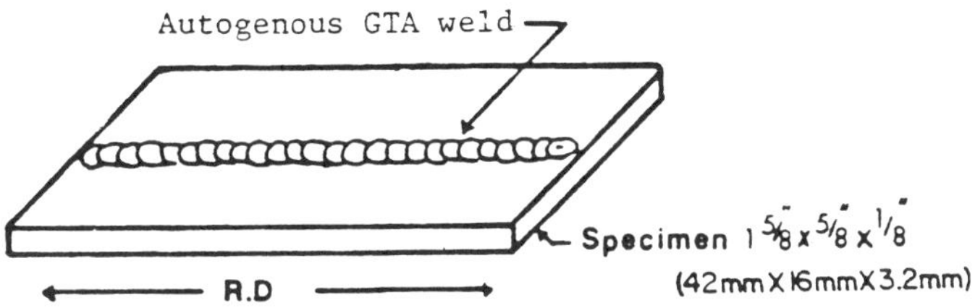

Figure 1. UT-Mod.HST specimen

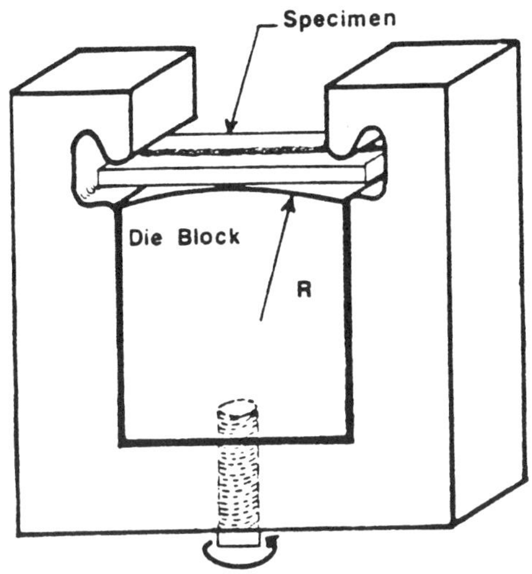

Figure 2. UT-Mod.HST straining fixture

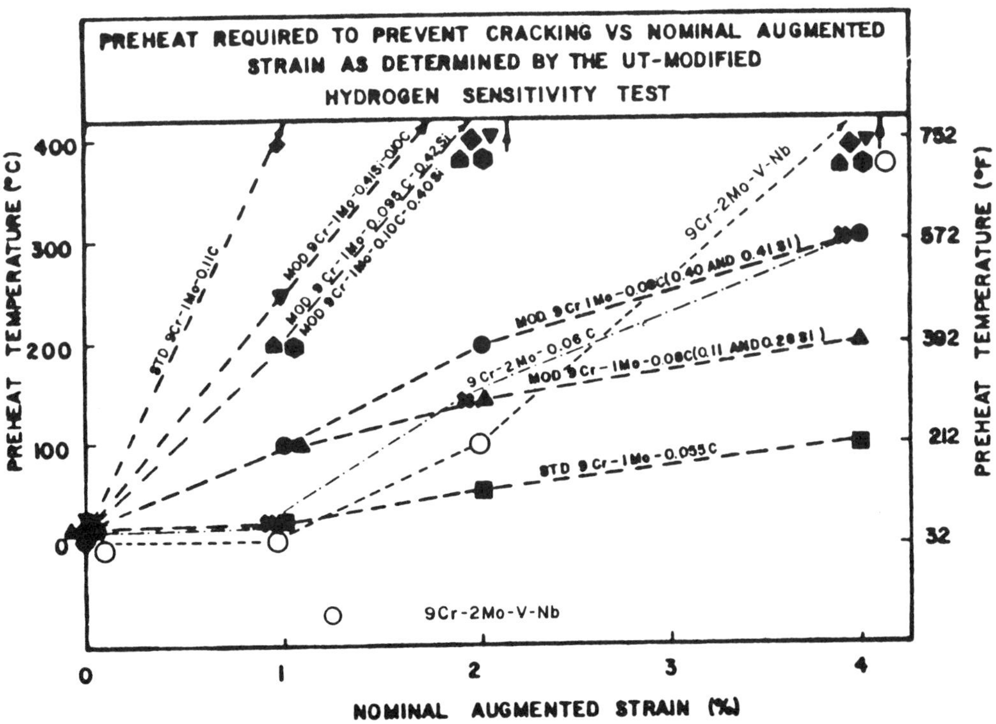

Figure 3. UT-Mod.HST results for 9Cr-1Mo and 9Cr-2Mo steels

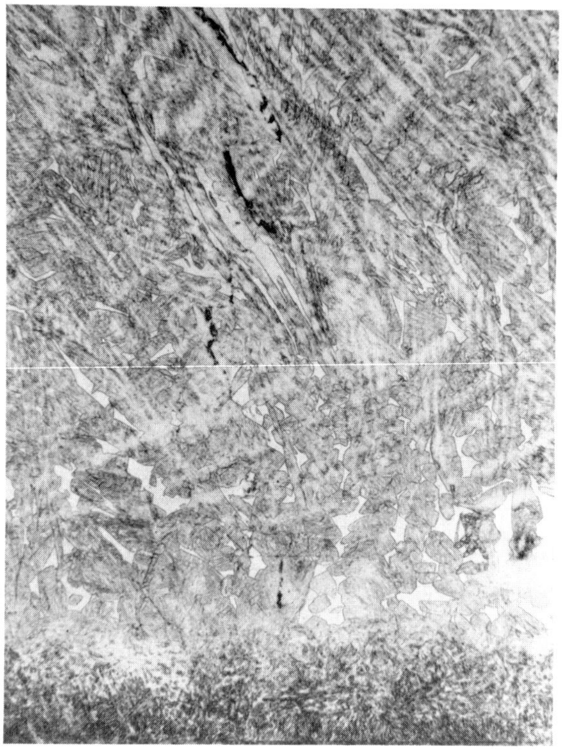

(a)

(b)

Figure 4. Longitudinal (a) and transverse (b) sections of 9Cr-1Mo-0.11C UT-Mod. HST specimen, 75°F (24°C) preheat, 4% strain. OLM, 100X. Modified HCl-Picral

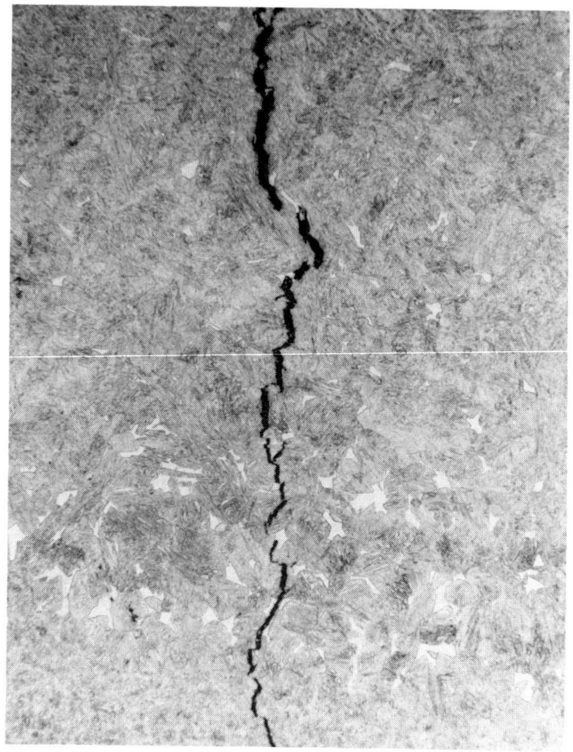

(a)

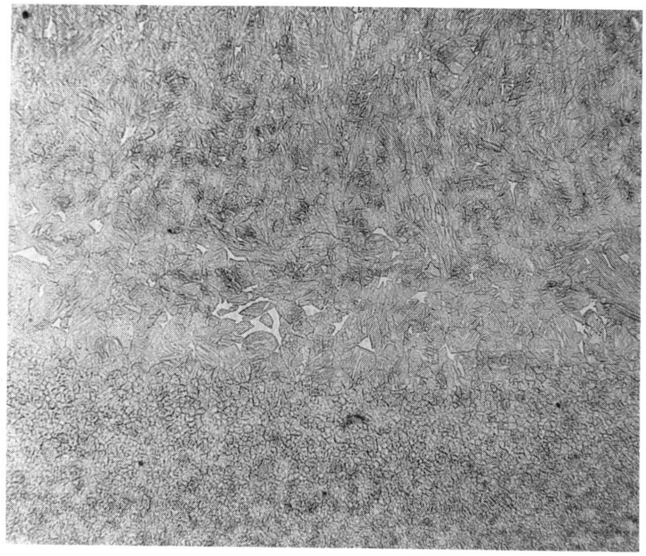

(b)

Figure 5. Longitudinal (a) and transverse (b) sections of 9Cr-1Mo-V-Nb (394) UT-Mod.HST specimen, 75°F (24°C) preheat, 4% strain. OLM, 100X. Modified HCl-Picral.

(a)

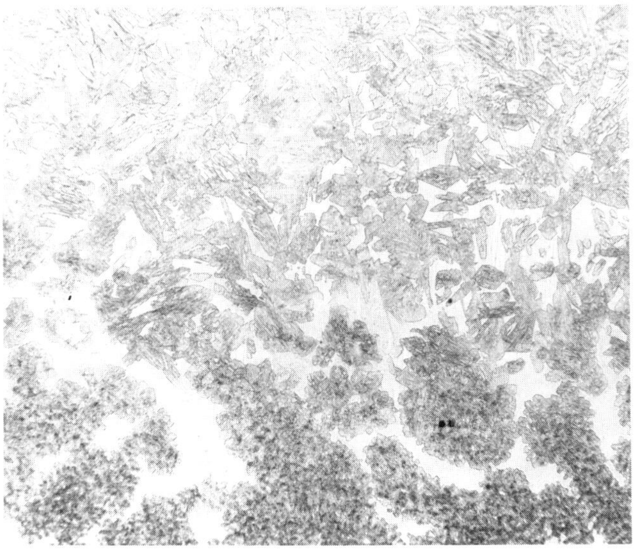

(b)

Figure 6. Longitudinal (a) and transverse (b) sections of 9Cr-2Mo-0.06C UT-Mod.HST specimen, 75°F (24°C) preheat, 2% strain. OLM, 100X. Modified HCl-Picral.

Figure 7. Fusion zone crack in UT-Mod.HST sample of 9Cr-2Mo-0.06C, 75°F (24°C) preheat, 4% strain, OLM, 900X. Modified Nital.

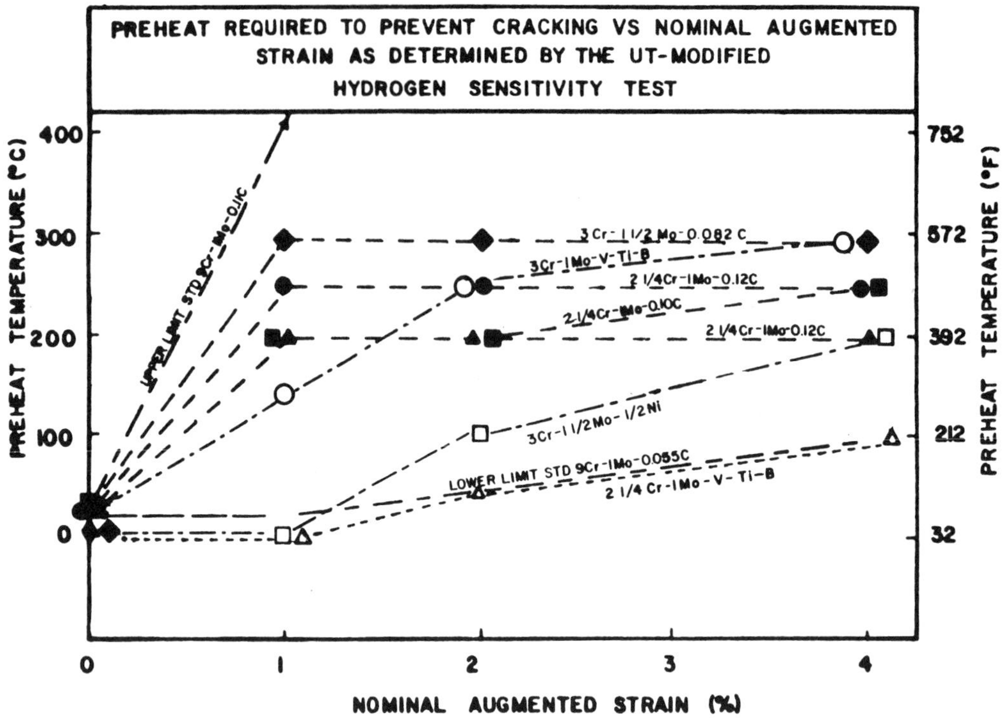

Figure 8. UT-Mod.HST results for 2-1/4Cr-1Mo and 3Cr steels

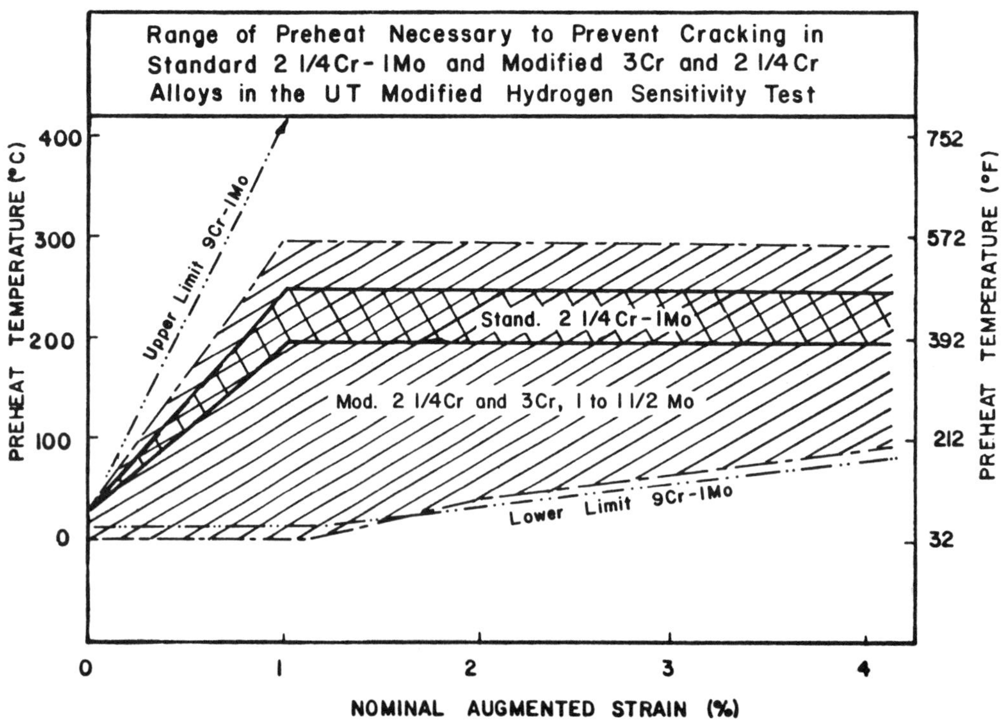

Figure 9. Range of preheats required to prevent HAC in 2-1/4Cr and 3Cr steels determined in the UT-Mod.HST

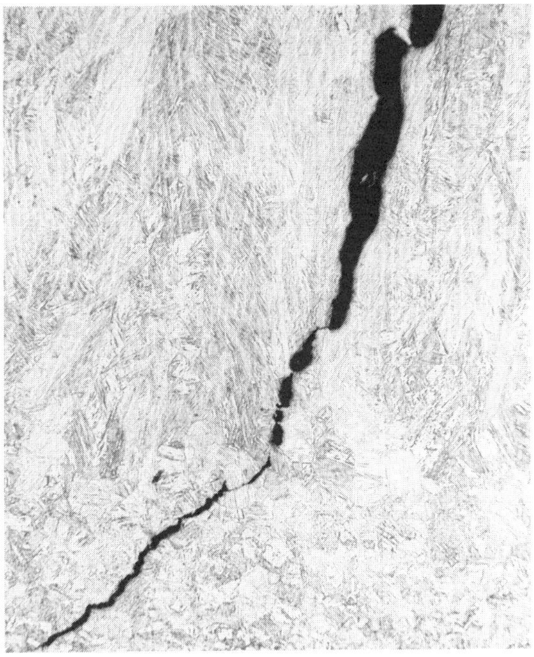

Figure 10. Longitudinal section of 3Cr-1 1/2Mo-0.08C UT-Mod.HST specimen, 390°F (200°C) preheat, 4% strain. OLM, 100X. Picral-Nital.

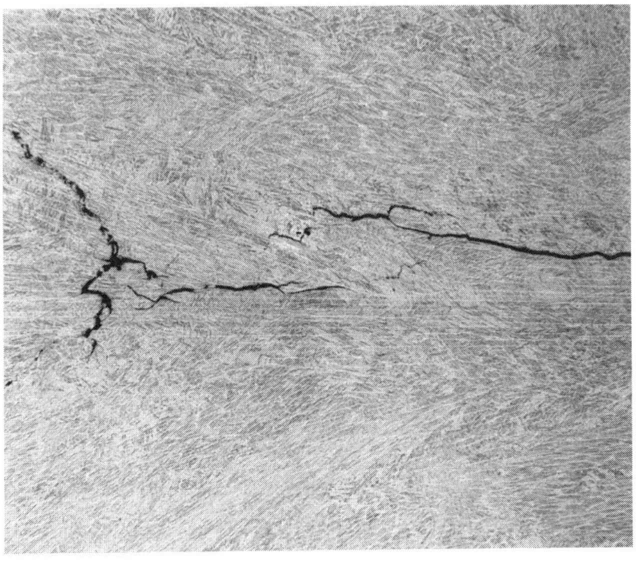

Figure 11. Longitudinal section of 2 1/4Cr-1Mo-0.12C (646363D) UT-Mod.HST specimen, 300°F (150°C) preheat, 4% strain. OLM, 100X. Picral-Nital.

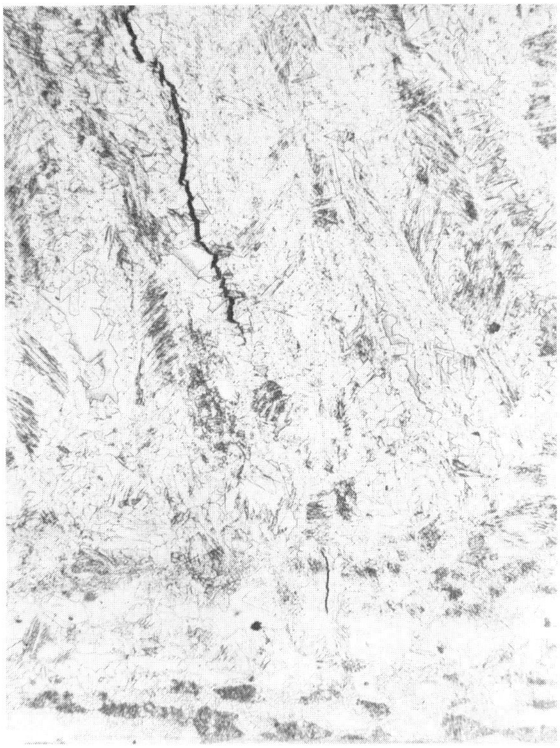

(a)

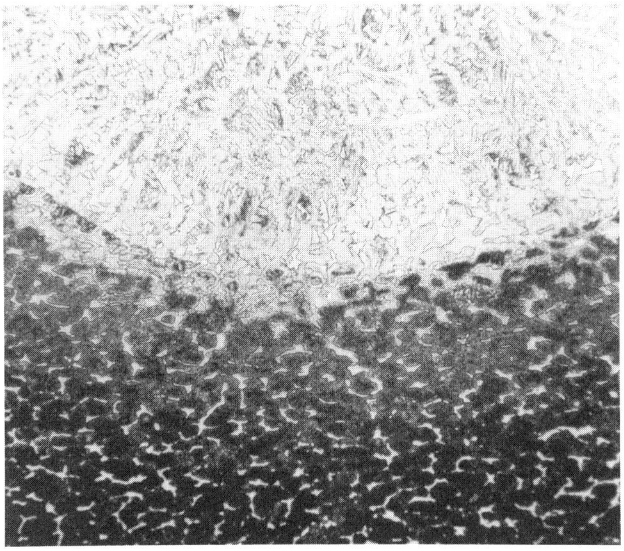

(b)

Figure 12. Longitudinal (a) and transverse (b) sections of 12Cr-2Mo-3Ni UT-Mod.HST specimen, 750°F (400°C) preheat, 1% strain. OLM, 100X. Modified HCl-Picral.

CREEP EMBRITTLEMENT OF HEAT AFFECTED ZONE AND ITS EVALUATION

Moriyasu Nagae, Nakatsugu Abe
Technical Research Center
Nippon Kokan K.K.
Kawasaki-ku, Kawasaki 210, Japan

CRACKS are sometimes detected in 1/2Mo and/or 1-1/4Cr-1/2Mo steel chemical reactor vessels after a long duration service. Such cracks are generally observed to occur in the weld-HAZs, especially in the coarse grain region adjacent to the fusion line and are characterized by a typical intergranular fracture mode.

When conducting post weld heat treatment (PWHT) for those steel weldment, a similar intergranular fracture can occur in the coarse grain region and propagate along the fusion line. Such a fracture has been defined as reheat crack.

The detected cracks mentioned above , however, are somewhat different from reheat crack, because they are detected in the weldments where no cracks had been observed by the nondestructive inspection after PWHT.

It has been reported that 1/2Mo and 1-1/4Cr -1/2Mo steels tend to cause a certain embrittlement in the creep rupture test (referred to creep embrittlement) at higher temperatures. Such a trend is significant in the creep rupture test conducted on the notched simulated weld-HAZ specimens [1]. This investigation was carried out in order to clarify whether or not there is a close relationship between the cracks in service and the embrittlement in the creep rupture test.

Firstly, the cracks recently detected in the reactor vessel for oil refinery equipment was examined with the optical and scanning electron microscope (SEM).

Secondly, the creep rupture characteristics of the same steel was carefully evaluated conducting the creep rupture test on the smooth and notched simulated weld-HAZ specimens. The creep rupture stress obtained was compared with stress actually applied to the reactor vessel in terms of the duration and the running temperature. In addition, the possibility of simulating the fracture mode was assured using slow extension rate technique (SERT) test. Finally, the improvement of creep rupture characteristics of 1-1/4Cr-1/2Mo steel weld-HAZs to avoid cracking in service in the actual chemical reactor vessel were discussed on the basis of the test result for various kinds of steels including laboratory scale melted steels.

MORPHOLOGY OF CRACKS

Fig. 1 shows a schematic configuration of the reactor vessel for oil refinery. Cracks were detected in both the flange and the shell plates as shown in Fig. 1. The number of cracks, however, is predominant in the flange of the forged material.

Table 1 shows the chemical composition of forged flange and shell plate. The contents of impurities such as P, S, As, Sn and Sb are higher in the flange than in the shell plate. The crack modes were classified into longitudinal, transverse and honeycomb-like cracks (see Fig. 2). Fig. 3 shows an example of honeycomb-like cracks detected in the flange. It was observed that those cracks spread widely in the coarse grain region adjacent to the fusion line. The scanning electron microscopy indicated that those cracks were typical

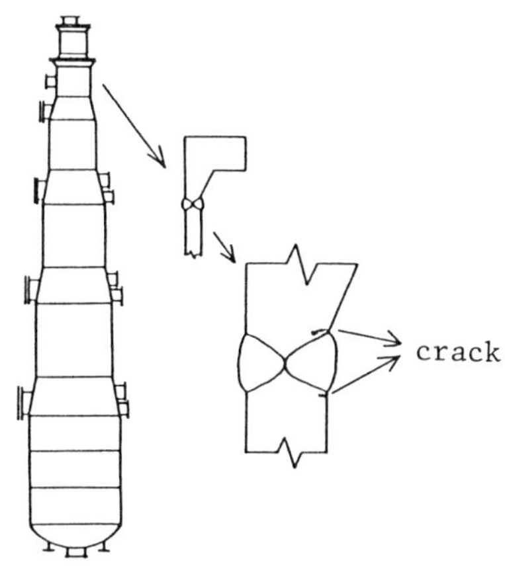

Fig. 1 Schematic configuration of reactor vessel

Table 1 Chemical composition of steels used (wt.%)

Steel	C	Si	Mn	P	S	Cu	Ni	Cr	Mo	As	Sn	Sb
1	0.13	0.74	0.65	0.016	0.032	0.20	0.06	1.04	0.49	0.011	0.043	0.002
2	0.14	0.69	0.47	0.008	0.007	0.03	0.04	1.40	0.58	0.009	0.002	0.001

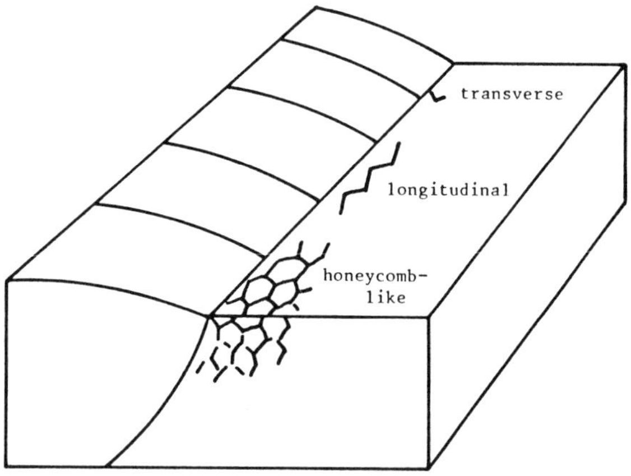

Fig. 2 Schematic appearance of cracks

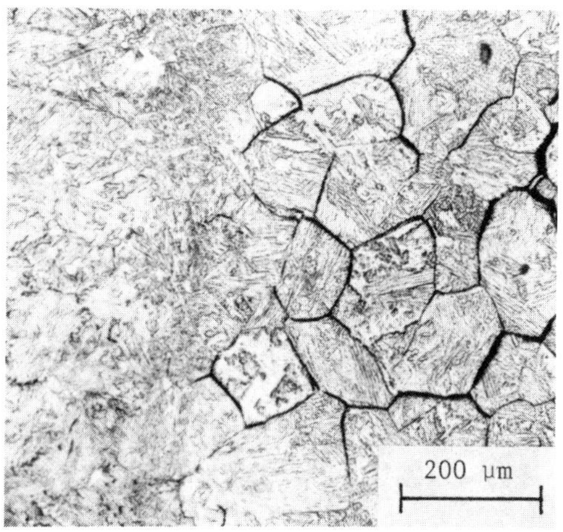

Fig. 3 Cracks detected in the reactor vessel

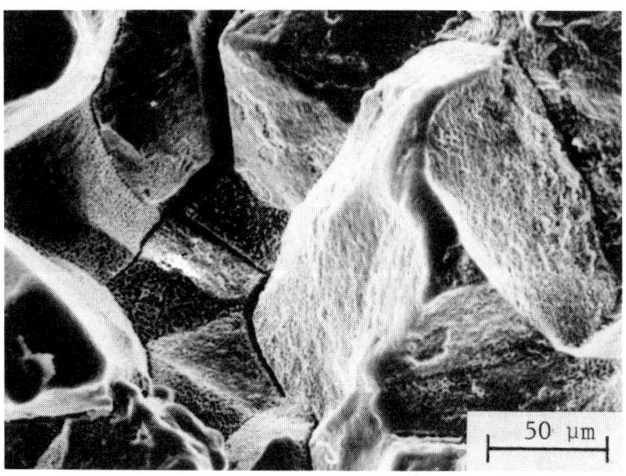

Fig. 4 Scanning electron micrograph of fracture surface from HAZ crack

intergranular ones as shown in Fig. 4. Fine dimples supposed to be creep voids were also observed on the intergranular surface. According to the actual records as to the running conditions of this reactor vessel, those cracks were detected after 58,500 hour duration and the average operation temperature was 504°C. In addition, the stress applied to the cracked flange was estimated to be 75.5 MPa by the numerical analysis. Impurities such as Sn and S are apt to segregate at the grain boundaries, thereby, resulting in the intergranular fracture. Fig. 5 shows the result of Auger electron spectroscopy (AES) analysis conducted on the fracture surfaces. AES analysis was carried out on the intergranular fracture surfaces after a deoxidization treatment. For comparison, AES analysis was also performed on the cleavage fracture surface obtained by splitting the specimen in the liquid nitrogen. AES analysis clearly revealed the segregation of Sn and As at the grain boundaries.

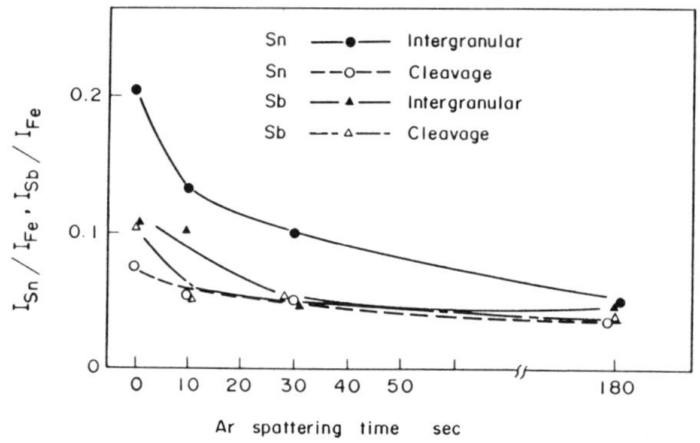

Fig. 5 AES analysis on intergranular and cleavage fracture surface

CREEP-RUPTURE CHARACTERISTICS

The legal inspection had been regularly conducted every one year (8,750 hours) on the reactor vessel in which the cracks were detected. It was confirmed that the cracks were discovered the portion of vessel where any crack had never occurred until last inspection. It is reasonably considered that the cracks are not associated with PWHT, but attributed to creep embrittlement occurred at the high temperature and for the long duration in service. In this study, the creep rupture test was carried out for the specimens sampled from the forged flange and the shell plate of the vessel. Then, the creep rupture stress obtained was compared with

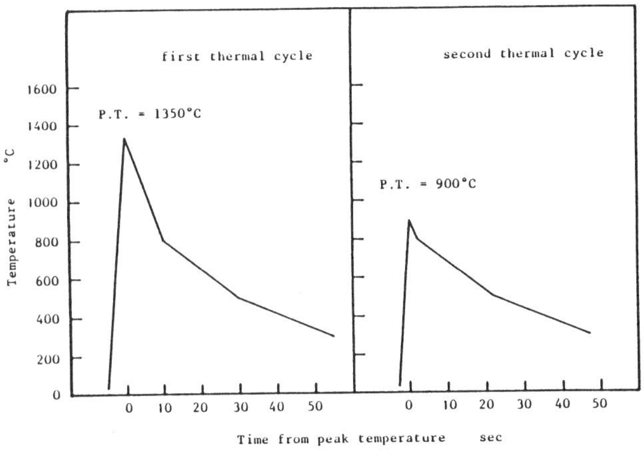

Fig. 6 Thermal cycles for HAZ simulation and refining heat treatment

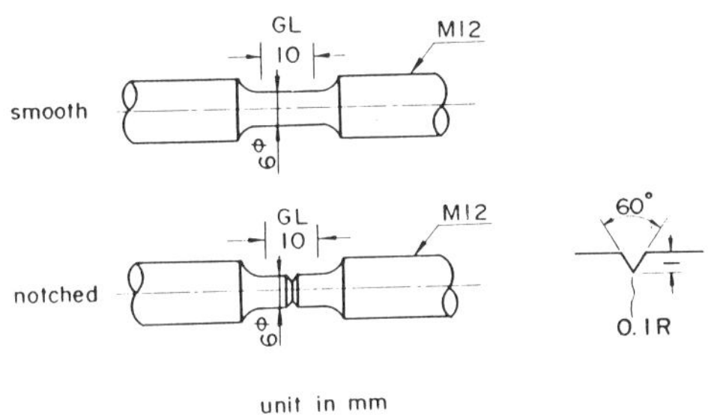

Fig. 7 Test specimens

the stress actually applied to the vessel as a function of running condition.

The first thermal cycle as shown in Fig. 6 was given to the specimen (12 x 12 x 60 mm) in order to eliminate thoroughly the change in material properties during service and simulate the microstructure relative to the coarse grain region of weldment as well. The second thermal cycle as shown in Fig. 6 was provided to half specimens to refine the microstructure. Heat treatment which is equivalent to PWHT was conducted on the rest of half specimens according to the following condition : 650°C x 4 hour (furnace cooling).

An actual welded structure generally involves various factors which bring about a stress concentration such as discontinuity, reinforcement and undercut. The creep rupture test, therefore, was performed for the notched specimens (Fig. 7) of which the stress concentration factor was 5. For comparison, the creep rupture test was also carried out on the smooth specimens.

Figs. 8 to 10 show the creep rupture test results at 500 and 550°C. These figures indicate the creep rupture stress as a function of P (Larson-Miller parameter) which is expressed by the testing condition, namely the holding temperature and time as follows.

$$P = T\,(20 + \log t)$$

where t is the time in hours spent at the testing temperature, T (K).

The data of Fig. 8 show that the creep rupture stress for the notched specimens tends to decrease abruptly as P value becomes large and the tendency is more remarkable than that of the smooth specimens. From this fact, it becomes clear that the forged flange with high impurities used is a material which readily gives rise to creep embrittlement.

As far as the reactor vessel is concerned, P value comes to 19.24 according to the running condition (504°C x 58,500 hours). Therefore, the creep rupture stress for the P value of 19.24 is 66.8 MPa which is relatively in good agreement with the stress 75.5 MPa applied to the flange of vessel. The fact suggests that the detected crack was resulted from creep embrittlement for a long duration in service.

In the case of shell plate material as shown in Fig. 9, a similar tendency associated with the creep embrittlement was obtained at larger P values. However, the notched rupture stress obtained here is higher than that of the flange (Fig. 8) because of low contents of impurities, thereby resulting in small amount of cracks in the shell plate.

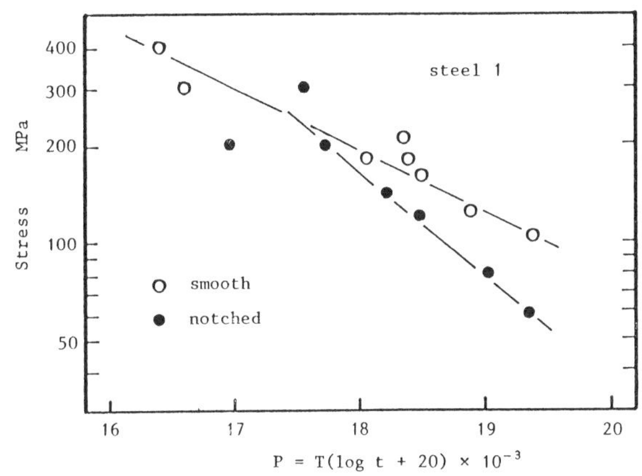

Fig. 8 Creep rupture property of simulated HAZ after PWHT (Steel 1)

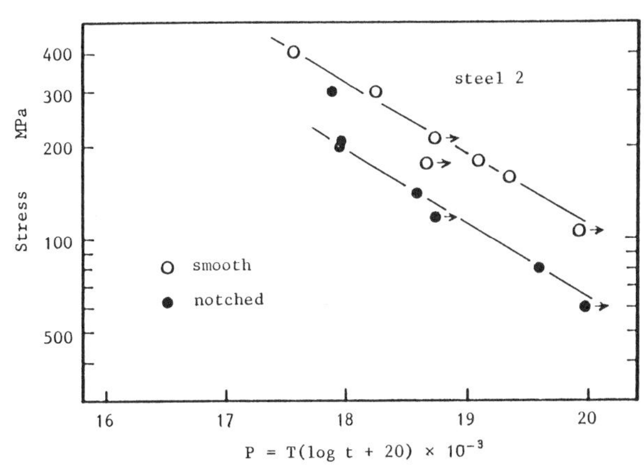

Fig. 9 Creep rupture property of simulated HAZ after PWHT (Steel 2)

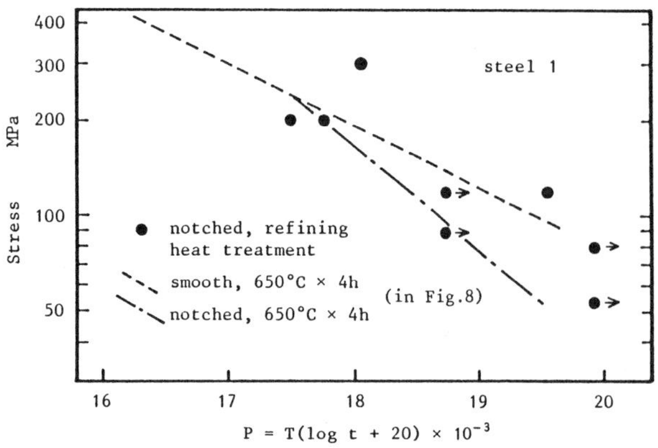

Fig. 10 Creep rupture property of simulated HAZ after refining heat treatment (Steel 1)

Fig. 10 shows the creep rupture test result for the notched simulated weld-HAZ specimens after the refining heat treatment as the 2nd cycle in contrast to after the heat treatment equivalent to PWHT (Fig. 8). As is evident from the figure, microstructure refinement brings about a remarkable improvement of the rupture strength without creep embrittlement in comparison to the simulated PWHT specimens. It should be also suggested that the effectiveness of microstructure refinement has a good relation to the fact that the cracks detected in the reactor vessel occurred only in the coarse grain HAZ.

SIMULATION OF INTERGRANULAR FRACTURE BY TENSION TEST OF SLOW EXTENSION RATE TECHNIQUE

The cracks detected in the actual reactor vessel exhibits a typical intergranular fracture mode as indicated in Fig. 3. It also became clear that the cracks occurred in service could be associated with creep embrittlement (Fig. 8). However, it is not always possible to distinguish the fracture mode of creep ruptured specimens because of the oxidation of fracture surface.

It is considered that the intergranular fracture (Fig. 3) resulted from the embrittlement in terms of both strength and ductility of the grain boundary. It is also considered that the creep embrittlement shown in Figs. 8 and 9 came from the strain concentration to the grain boundaries of lower strength and inferior ductility for a long duration. It seems, therefore, to be possible to simulate such an intergranular fracture when the slow extension rate technique is used in tension test instead of the creep rupture test, because slower stressing brings about the strain concentration to the weak grain boundary. Such a trend is expected to be intensified when the associated grain boundaries are deteriorated by the segregation of impurities.

Then, SERT test was conducted on the simulated weld-HAZ specimens sampled from the forged flange and the shell plate at 0.0005

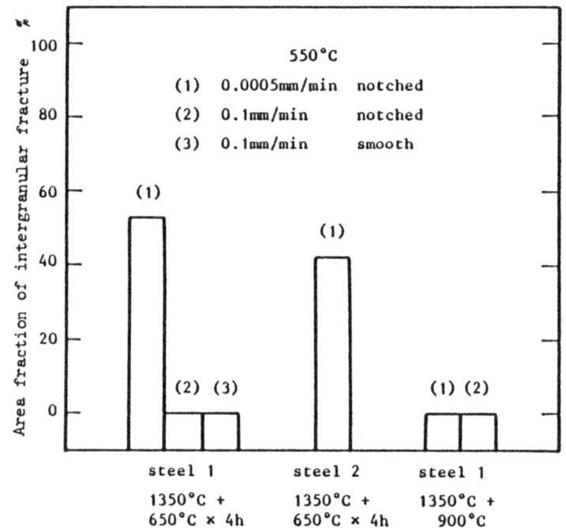

Fig. 11 Effect of thermal cycle and extension rate on fracture mode

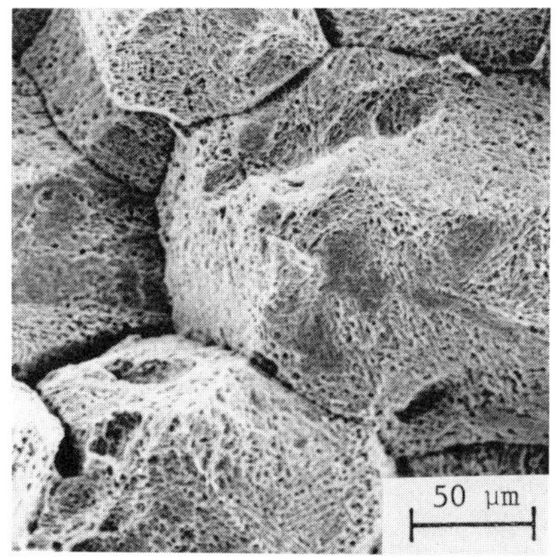

Fig. 12 Scanning electron micrograph of fracture surface obtained SERT test (Steel 1, 1350°C + 650°C x 4hr)

mm/min extension rate. For comparison, the conventional tension test was also performed at 0.1 mm/min extension rate using the Instron testing machine. The configuration of specimen employed was the same as that of the creep rupture test as shown in Fig. 7.

Fig. 11 indicates the SERT test results in terms of the area fraction of intergranular mode. As can be seen from the figure, the simulated Weld-HAZ specimens subjected to the thermal cycle equivalent to coarse grain region exhibit a remarkable intergranular fracture with more than 40 percent. In contrast, the simulated weld-HAZ specimens after refining heat treatment as the 2nd cycle show no longer the same fracture mode. Fig. 12 shows examples of fracture appearance. The data have a good correspondency to the creep rupture test results as shown in Figs. 8 to 10. Namely, the sample which has a tendency of notched embrittlement in the creep rupture test exhibits intergranular fracture to a large extent in SERT test as well. On the contrary, the sample not occurring the creep embrittlement shows no intergranular fracture. This fact could guarantee the possibility that the cracks characterized by intergranular fracture mode in the actual reactor vessel took place due to creep embrittlement occurred in service.

Table 2 Chemical composition of steels used (wt.%)

Steel	C	Si	Mn	P	S	Ni	Cr	Mo	Nb	B	As	Sn	Sb
3	0.14	0.71	0.59	0.004	0.0016	-	1.41	0.56	-	-	0.002	0.001	0.001
4	0.07	0.72	0.59	0.004	0.0019	0.29	1.43	0.58	-	0.0006	0.002	0.001	0.001
5	0.06	0.73	0.59	0.004	0.0019	0.30	1.48	0.57	0.033	0.0005	0.002	0.001	0.001

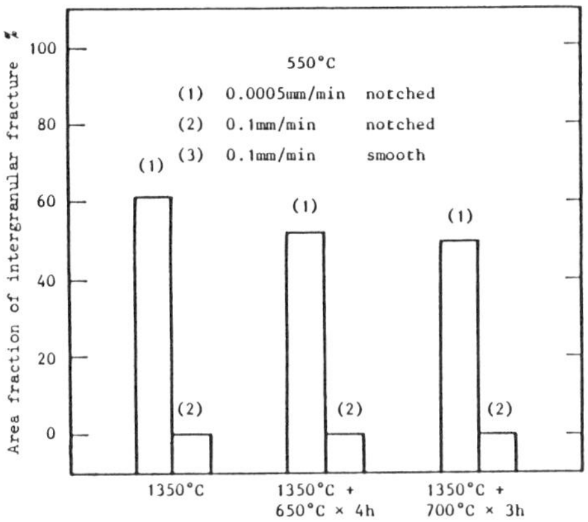

Fig. 13 Effect of PWHT condition and extension rate on fracture mode

As far as test result obtained by the conventional tension test is concerned, no intergranular fractures were observed for all of specimens. Therefore, it is possible to predict whether or not notched embrittlement occurs by the creep rupture test.

Fig. 13 indicates the effect of heat treatment equivalent to PWHT on a tendency of intergranular fracture in SERT test. Although the intergranular fractures were observed for all of specimens tested at slower extension rate, no intergranular fractures occurred at higher extension rate. As a result, it is impossible to reduce creep embrittlement for the material with high impurities.

IMPROVEMENT IN CREEP-RUPTURE PROPERTY

In order to improve the creep rupture property from a metallurgical standpoint, the creep rupture test was conducted on the simulated HAZ of the laboratory melted steels (Table 2) which were subjected to heat treatment equivalent to the PWHT. Fig. 14 shows test result for the individual steels as compared to steels 1 and 2. In this figure, the creep rupture property is represented by the rupture time when the applied stress is 137 MPa and the dotted lines indicate the range of test result obtained on a standard base metal [2]. It was demonstrated that the decrease in both impurities and carbon content is effective to increase the creep rupture strength for the notched specimens.

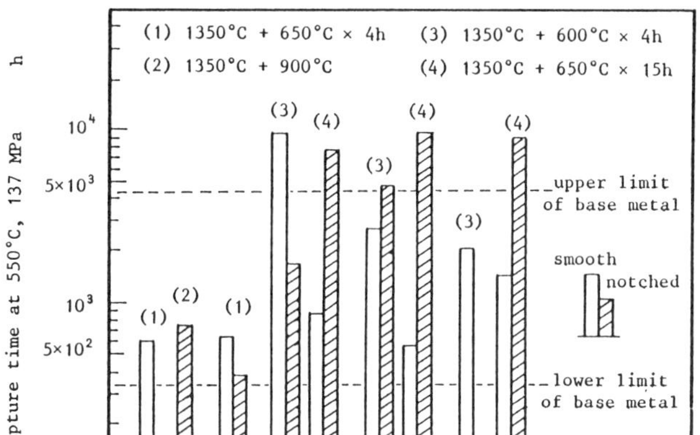

Fig. 14 Effect of chemical compositions on creep rupture property

CONCLUDING REMARKS

In this investigation, the cause of the cracks in service after a long duration recently detected in the reactor vessel for oil refinery was carefully examined by conducting both the creep rupture test and SERT test using the simulated weld-HAZ specimens sampled from the same materials. The cracks were detected in the coarse grain region adjacent to the fusion line of both the forged flange and the shell plate welded joints and had a typical intergranular mode. The creep rupture test conducted on the simulated weld-HAZ specimens clearly revealed that creep embrittlement occurred in both the forged flange and the shell plate. Such a trend is significant in the forged flange material with high impurities. In SERT test performed the same samples mentioned above, the intergranular fracture was successfully simulated especially in the forged flange material. The results of creep rupture test and SERT test indicated a good correspondency each other. As a result, it is reasonably considered that the cracks detected in the reactor vessel resulted from creep embrittlement in service after a long duration. Recently, it has been reported that such cracks were also detected in the materials for piping, tubing and steam turbine casing in addition to chemical reactor vessels. As is evident from the results of this investigation, there remains a possibility of improvement in steel materials by alloying elements modification. It needs much future work to avoid the cracks in service.

REFERENCE

[1] Takamatsu , T., Otoguro, Y., Shiotsuka, K and Hashimoto, K.; Journal of The Iron and Steel Institute of Japan, 65,(1979),851-860

[2] Ishiguro, T., Murakami, Y., Tsukeda, T. and Ohnishi, K.; Journal of The Iron and Steel Institute of Japan, 70,(1984),1421-1428

RECENT DEVELOPMENT IN HIGH TOUGHNESS HEAT-TREATED LARGE DIAMETER PIPES

Shigeru Ohkita, Hiroyuki Homma, Motonori Tamura, Hiroshi Naganuma
Nippon Steel Corporation
Sagamihara, Kanagawa
229 Japan

ABSTRACT

New welding material has been developed for high quality bent pipe, TLP tether and other fittings. And a series of research works, consisting of metallurgical aspect of weld metal toughness after quenched and tempered(QT), development of a new type of submerged arc welding consumable and assessment of QT condition have been carried out. It was found oxygen content should be 200 ppm or under to get good toughness of QT welds. It was also found high toughness of low oxygen welds is attributed to the microstructure. And based on this informations, low oxygen, low nitrogen and low hydrogen flux was developed.

HIGH-STRENGTH STEEL PIPES are used for the tether of tension leg platforms (TLP) and bend pipelines; typical levels of yield strength are more than 80 ksi and 52-70 ksi, respectively. One of the various processes by which they are made comprises the steps of forming, welding, and heat treating. It is necessary to develop such weld metal and base metal that will have satisfactory tensile strength and toughness after the heat treatment has been applied. However, quenched and tempered (QT) weld metals did not show sufficient toughness when conventional welds were applied. This seems to be caused by the high oxygen content of the weld metals compared with the low oxygen content of steel plates whose properties were improved by the heat treatment. In this paper, the metallurgical factors controlling the mechanical properties of QT weld metal are studied. And the development of fluxes of submerged-arc welding (SAW) are described.

EXPERIMENTAL PROCEDURES

The properties of steel plates after heat treatment are known to be controlled by the chemical composition and heat treatment conditions. In this experiment, the influences of oxygen, nitrogen, Al, Ti content, carbon equivalent, and heat treatment conditions on weld metal toughness were investigated. Also, the development of fluxes was simultaneously done by considering the content of gas elements, such as O, N, and H, and welding workability.

The weld metals obtained by SAW using test fluxes were heat treated, and their properties and microstructures were examined. To the fluxes, different quantities of calcium fluoride to control oxygen content and of calcium carbonate to lower hydrogen content were added. Welding wires were the type containing Ni, Cr, and Mo. Deposits were made by using the three-electrodes SAW process with heat input of 53 kJ/cm and using the two electrodes SAW with 45 kJ/cm for 52-70 ksi and 80 ksi steel, respectively.

Table 1 and 2 shows typical chemical compositions of the steel plates used and the weld metals obtained. The heat treatment of weldments consisted of austenitizing at 850-1000°C for 30 minutes and then, after water or oil quenching, tempering at 550-650°C for 30-60 minutes. A simulated heat treatment using induction heating equipment was also conducted to determine the austenitizing behavior.

Table 1 Chemical composition of steel plates.

Steels	element (wt %)										
	C	Si	Mn	P	S	Ni	Cr	Mo	Al	B	N
Base Steels	0.12	0.28	0.97	0.006	0.002	1.41	0.46	0.36	0.07	0.0012	0.0036
	0.10	0.26	0.53	0.011	0.001	2.54	0.40	0.36	0.07	0.0008	—

Table 2 Example of chemical composition of weld metals.

	element (wt %)											
	C	Si	Mn	P	S	Ni	Cr	Mo	Al	B	N	O
weld metals	0.097	0.16	1.30	0.006	0.003	1.01	0.44	0.35	0.010	0.0003	0.0061	0.0339
	0.078	0.13	1.33	0.006	0.003	2.47	0.33	0.30	0.004	0.0004	0.0069	0.0237
	0.090	0.14	2.10	0.007	0.003	0.33	0.11	0.08	0.003	0.0004	0.0058	0.0295
	0.092	0.07	1.61	0.006	0.003	0.38	0.12	0.43	0.005	0.0003	0.0047	0.0324
	0.098	0.13	1.77	0.006	0.003	0.51	0.18	0.12	0.015	0.0007	0.0057	0.0122
	0.099	0.14	1.80	0.008	0.002	0.46	0.16	0.12	0.012	0.0006	0.0058	0.0118
	0.101	0.20	0.70	0.006	0.002	2.55	0.49	0.35	0.025	0.0005	0.0042	0.0163

TEST RESULT

EFFECT OF ALLOYING ELEMENT AND IMPURITIE'

Figure 1 shows the effects of oxygen content on the toughness of 80 ksi QT weld metal. It is clear from this figure that the lower the oxygen content, the higher the toughness. This tendency is the same for the 52-70 ksi QT welds[1]. The microstructure of the metal containing less oxygen (120 ppm) consists mainly of tempered martensite; however, the microstructure of the weld with higher oxygen content consists of coarse-grained bainite in 80 ksi welds. This

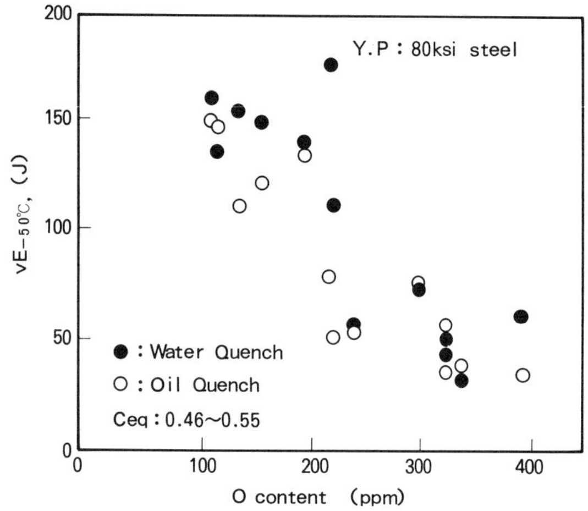

Fig. 1 Effect of O content on toughness.

is in contrast with the fine grained ferrite which shows high toughness in 52-70 ksi QT welds[1].

These differences in microstructure and toughness seem to be due to the behavior of Al in the weld metal and the amount of oxide inclusions, as will be shown later.

Nitrogen is generally regarded as an impurity that deteriorates the toughness of weld. As shown in Fig. 2, with the increase of N, toughness decreases both in 80 ksi and 60 ksi QT welds. However, high toughness can be obtained even if N content is as high as approximately 60 ppm in 60 ksi welds when Al content is high enough. This seems to be related to the behavior of Al and the microstructure.

Al is effective to reduce the vTrs of 60 ksi QT welds, but no remarkable effect can be observed in the vTrs of 80 ksi welds, as shown in Fig. 3. At low strength welds, the main components of microstructure are fine-grained ferrite, perlite, and fine-upper bainite in high Al welds. However, the sizes of each component increase in low Al welds because the austenitic grain size before austenite-to-ferrite transformation seems to increase with the decrease in Al content. On the other hand, the microstructures were almost the same in high strength of 80 ksi welds when the content of Al varied from 0.02 to 0.05%.

Figure 4 shows the effects of Ti content on toughness. It shows small amount of Ti do not affect toughness. But, at 0.023%, Ti in weld deteriorates toughness. As shown in Fig. 5, many fine precipitates can be observed in weld at 0.023% Ti.

Nickel is added to welds to increase the hardenability of steel without reducing its toughness. As shown in Fig. 6. vTrs decrease with increases in Ni content. The microstructure becomes tempered martensite at higher Ni

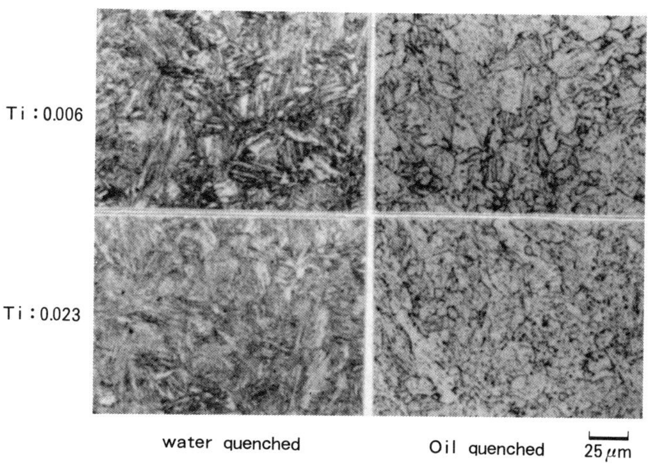

Fig. 5 Effect of Ti content on microstructure of QT welds.

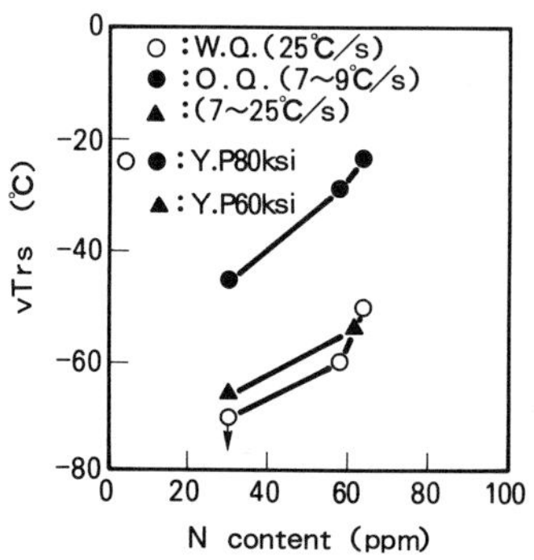

Fig. 2 Effect of N content on toughness.

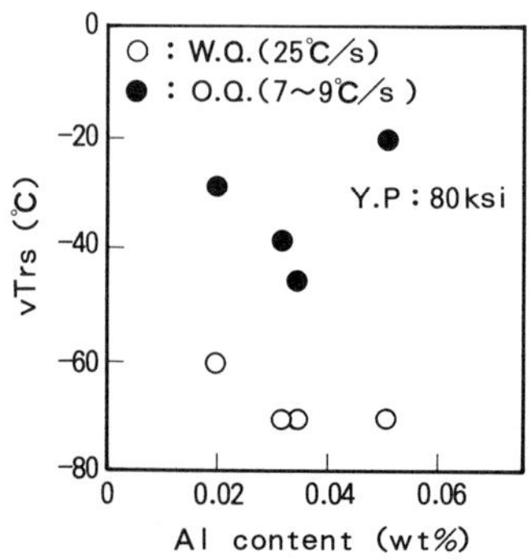

Fig. 3 Effect of Al content on toughness.

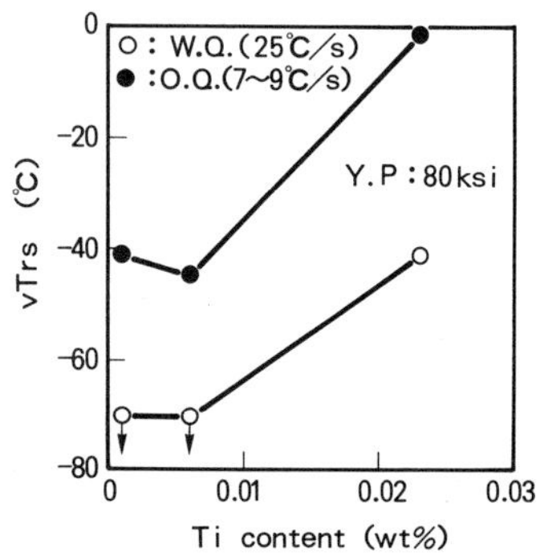

Fig. 4 Effect of Ti content on toughness.

contents, and when tensile strength is high enough. However, excess additions of Ni reduce the shelf energy of welds.

Figure 7 shows the effects of carbon equivalent (CEQ) on toughness of QT welds which are made by using low oxygen type flux and various types of wires. There are two optimum ranges of CEQ which show excellent notch toughness. Namely, excellent vTrs can be obtained at lower CEQ ranges of less than 0.38 and from 0.45 to 0.57 in the case of water quenching. These results can be explained by its microstructures as shown in Fig. 8. The main conponents of microstructure are ferrite and tempered martensite at the CEQ ranges of up to 0.38 0.45-0.57, respectively. However, it becomes upper bainite at the CEQ range of 0.38-0.45.

Alloying design should be changed according to the aimed tensile value.

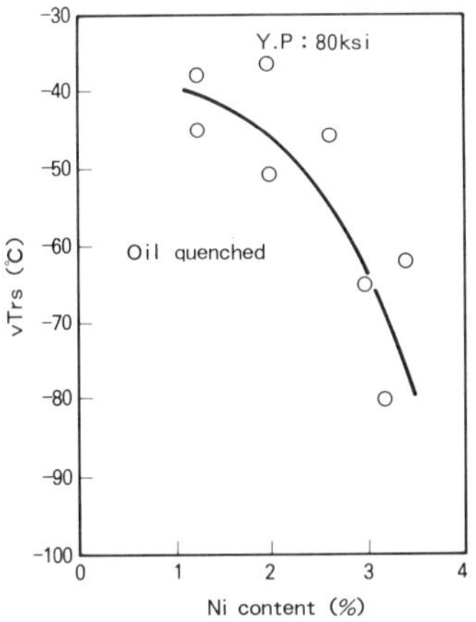

Fig. 6 Effect of Ni content on toughness.

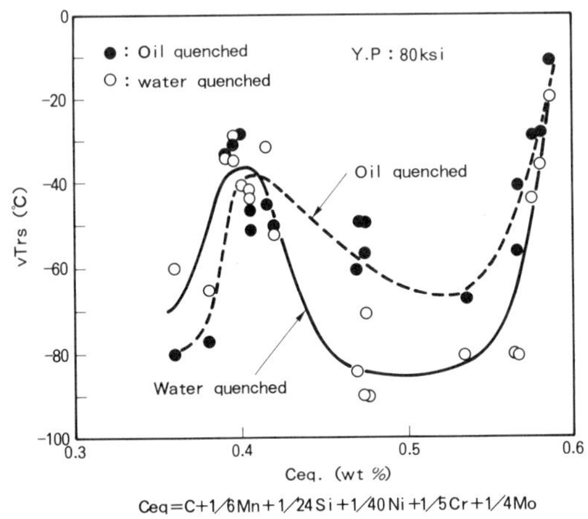

Fig. 7 Effect of carbon equivalent on toughness of QT welds.

CONDITION OF HEAT TREATMENT

Effects of austenitizing temperature, time, quenching rate, and tempering temperature on weld metal toughness and tensile properties have been tested.

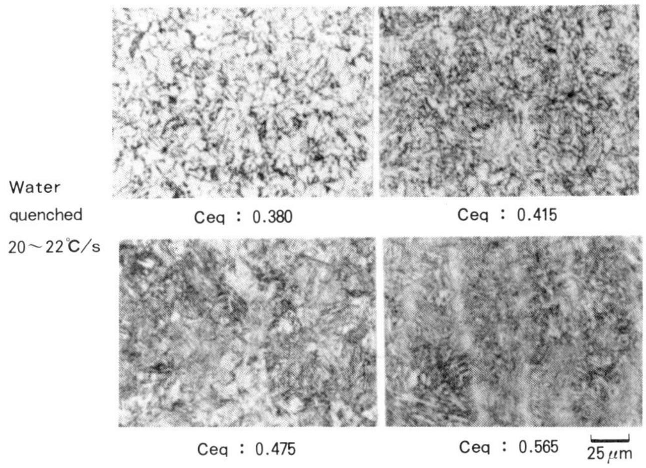

Fig. 8 Effect of carbon equivalent on toughness of QT weld (80ksi).

Austenitizing temperature is very important for toughness because the size of ferrite and martensite are decided by the prior austenite grain size which is controlled by temperature. If the austenitizing temperature is too high, the austenite grain size becomes coarse. If the temperature is too low, perfect austenitization does not occur to leave some as-welded microstructure. For high strength weld, the allowable ranges of austenitizing temperature are wider compared to low strength weld, as shown in Fig. 9. The high strength weld can be austenitized even at low temperatures. And the microstructure austenitized at high temperature easily becomes martensite. Compared with the low CEQ welds need high temperature for perfect austenitization and the microstructure becomes upper bainite at high austenitizing temperatures.

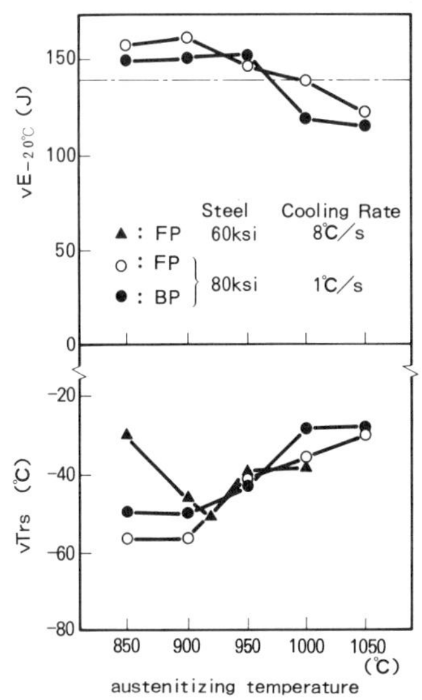

Fig. 9 Effect of austenitizing temperature on toughness of QT welds.

Figure 10 show the effects of tempering temperature on toughness and TS. The toughness does not change remarkably; however, tensile strength decreases with increases of tempering temperature. It should also be noted that as welded weld and base metal show almost the same tensile strength, and the TS of quenched welds are drastically reduced when compared with base metals. This seems to be caused by the difference in oxygen content which is generaly higher in weld metal than in base metal.

The cooling rate is also an important factor to determine the toughness because of its microstructure. Figure 7 also shows the effects of the cooling rate. Oil quenched and water quenched welds have the same tendency in toughness as a function of CEQ. But water quenched welds show high toughness at the CEQ range of 0.45-0.55 as compared to oil quenched welds at CEQ up to 0.38. This is because the microstructures tend to become fine-grained ferrite quenched at low speed of low CEQ (<0.38) weld, and tempered martensite quenched at high speed of high CEQ (0.45-0.55) welds.

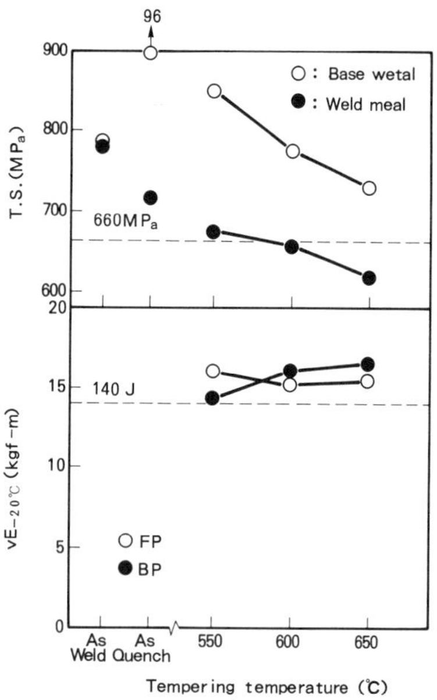

Fig. 10 Effect of tempering condition on toughness and tensile strength of quenched welds.

DEVELOPMENT OF SAW FLUX FOR QT WELDS

Generally, gas composition in SA welds were determined by fluxes[2)3)]. The results described in the previous section clearly indicate oxygen content should be limited less than 200 ppm in all CEQ ranges. Also, it was found the yielding of nitrogen from air to weld should be limited less than 10 ppm/1 pass if 2 or 3 pass welding is to be carried out. And the susceptibility to form such weld defects as cold cracking, pittings, blow wholes, and under cut over a wide range of welding conditions should be taken into account to optimize SAW flux chemistry.

Generally, oxygen content in SAW can be reduced by an increase in the basicity (B) of flux as shown in Fig. 11. Considering weld metal toughness, the basicity should be kept higher than 2.0. However, the absorption of nitrogen and hydrogen[4)] is increased with an increase in basicity. Therefore, calcium fluoride, calcium carbonate, and potassium oxide are used in flux to reduce hydrogen and nitrogen and to improve workability. Consequently, high efficiency, low oxygen, low nitrogen, and low hydrogen SAW flux has been successfully developed as shown in Fig. 12.

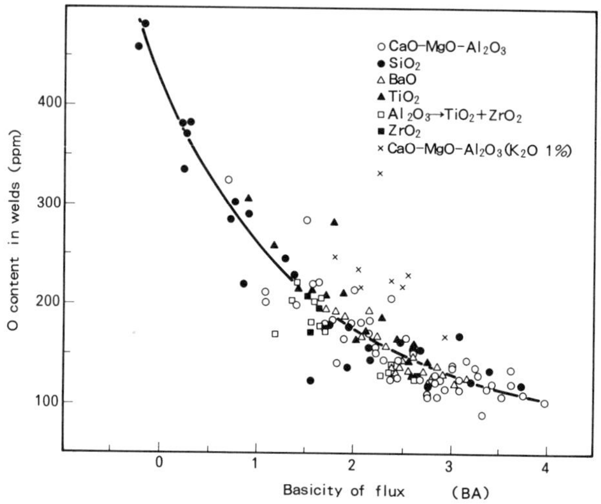

Fig. 11 Effect of flux basicity on oxygen content of weld metals.

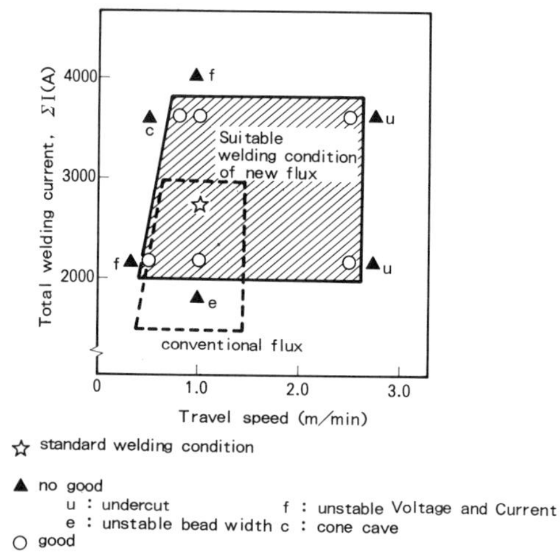

Fig. 12 Suitable welding condition range of welding current and travel speed.

DISCUSSION

EFFECT OF OXYGEN

The higher the oxygen content the higher the toughness of QT weld; and tensile strength of QT weld is lower than that of base metal. Generally, it is well known that ferrite is heterogeneously nucleated at austenite grain boundary. Table 3 shows the results of hardness tests which compared QT base metal and QT weld metals of 80 ksi steel. The hardness of base metal is higher than weld metals. This means oxygen decreases the hardenability of steels and, as described at Fig. 1, the microstructure of high CEQ weld is changed from martensite to upper bainite. This seems to show inclusion acts as a nucleus of ferrite.

Table 3 Effect of oxygen content on hardness (Hv 10kgf).

Sterel	CEQ (Ilw)	As Welded	As Quenched	Q&T.	Oxygen Content (ppm)
Base Metal	0.43	246	285	244	26
Fusion Line	—	253	244	223	—
Weld Metal	0.44	244	217	200	172

cooling Rate : 1°C/s
Austenitizing Temperature : 900°C
Tempering Temperature : 600°C

Besides the above effect, the higher content of oxygen itself seems to deteriorate the QT weld metal toughness both in high and low CEQ welds. The oxygen in weld forms oxide inclusion and the average size and number increase with the oxygen content. Therefore, the interface between oxide inclusion and matrix is increased with the increase of oxygen content. Namely, the interface between oxide inclusion and matrix should be easily separated when stress is engaged at the interface. This means oxide inclusions deteriorate the toughness of QT welds.

EFFECT OF ALUMINUM

The toughness of weld metals can be improved by making the main component of microstructure fine grained ferrite and tempered martensite. Fine grained ferrite is formed when oxygen content is decreased partly because of an increase in the yield of aluminum in weld metals. Figure 13 shows the microstructures obtained by quenching the specimens that were austenitized at various temperatures. In specimens of low oxygen content, fine grained ferrite remained at 950°C. By contrast, grain coarsening occurred around 950°C in high oxygen specimens containing less sol. Al. This seems to be caused by the precipitation of aluminum nitride. The equilibrium solubility product of aluminum and nitrogen can be expressed by the empirical equation[5].

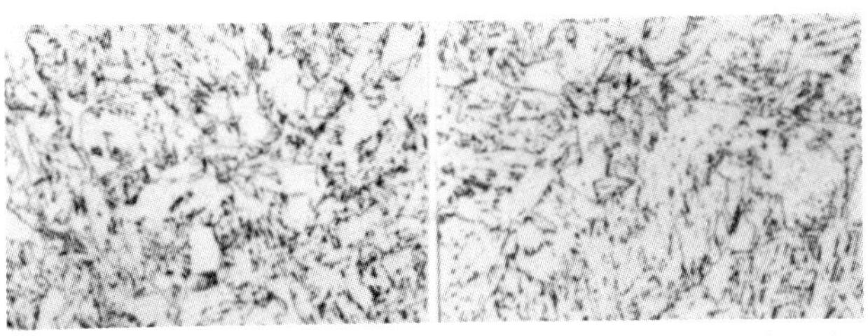

Fig. 13 Effect of austenitizing temperature on microstructure of QT weld (1°C/s.).

From the equation, aluminum nitride is found to be fused completely at a temperature of approximately 970°C when aluminum and nitrogen contents stand at 0.02 and 0.007%, respectively. Fig. 9 shows the toughness of weld austenitized at temperatures higher than 950°C. The test results agreed well with the calculated data.

In high CEQ welds, the microstructure was to be martensite, however, not affected by the change of Al from 0.02 to 0.05%. Usually, to improve the hardenability of steel, boron is added to the steel. Soluble B segregates at prior austenite grain boundary and lowers the austenite to ferrite transformation temperature. The base metal used in this study also contained boron and Al, and it yielded to weld metal. The reason why the microstructure was not changed by the Al content is attributed to the high Al content and the coexistance of B in welds.

Although not covered in this paper, the effects of austenitizing and tempering time and working should be studied.

Austenitizing is mostly accomplished by high-frequency induction heating followed by immediate cooling. Austenitizing time is therefore very short and the material austenitized by this method has higher toughness than that heated in a batch-type furnace as was used in this study. Tempering temperature usually does not have such a great influence on toughness as it has on tensile strength. The effect of working on the austenitized material was not given particular attention; but no remarkable effect can be observed when the pipe was made.

CONCLUSIONS

Considering the factors controlling the properties and weldabilities of QT welds, much important information for development of welds has been obtained. The main results can be summarized as follows:

(1) Using steels of 60 ksi and 80 ksi class, high toughness can be obtained with heat treated weld metals through the control of the microstructure to fine grained ferrite and tempered martensite.

(2) The oxygen content of weld must be lower than 200 ppm to keep the aimed toughness of $vE_{-46°C}>50$ ft-lbs and $vE_{-20°C}>14$ kgf-m at 60 ksi and 80 ksi steel weld, respectively.

(3) While the oxygen content of the weld metal is governed by the basicity of flux, the action of carbon dioxide, potassium oxide, and calcium fluoride should be taken into consideration for the reduction of nitrogen and hydrogen contents.

REFERENCES

1) H. Naganuma, H. Homma, S. Okita, M. Tamura and A. Yamaguchi: Trans. ISIJ of Japan (1986) (to be published).
2) K. Mori: Tetsu to Hagane. 46 (1960) 466 (in Japanese).
3) K. Mori: J. Japan Inst. Metals 24 (1961) 383 (in Japanese).
4) J M Uys, T.B. Kingi: Trans. Met. Soc. AIME 227 (1963) 492.
5) T. Gladman and F.B. Pickering: JISI, (June) 1967, 653.

THE EFFECT OF LOW LEVEL CONCENTRATIONS OF SOME ELEMENTS ON THE TOUGHNESS OF SUBMERGED-ARC WELDED CMn STEEL WELDS

L. Devillers, D. Kaplan, C. Donadille, P.V. Riboud
Institut de Recherches de la
Siderurgie Francaise
France

INTRODUCTION

Weld metal is usually considered as a preferential site for the initiation and propagation of brittle fracture in high strength low alloy (HSLA) steel structures[1]. Weld metal toughness requirements are especially hard to meet in the high heat input single-pass processes such as submerged-arc (SA) welding. This process is characterized by a high dilution of the parent metal (60 to 70%) and low cooling rates which have a tendency to promote low toughness microstructures[2]. Nevertheless it appears that satisfactory toughness values can be obtained through a proper selection of the plate, wire, and flux compositions[3]. The weld metal resistance to brittle fracture is very dependent upon microstructure[2,4] ; it is improved by a large percentage of fine interlocking ferrite grains usually referred to as 'acicular ferrite'[5,6]. Acicular ferrite results from nucleation at numerous sites within the prior austenite grains[5,7,8], and the weld metal microstructure depends upon a balance between grain boundary nucleation and intragranular nucleation of ferrite. These two mechanisms can be affected by low level concentrations of some elements (Al, Si, Ti, S, B, C, Nb, N) to be found either in nonmetallic inclusions or within the steel matrix[8,9]. The effects of these elements on the toughness of the as-welded weld metal are presented and discussed here.

EXPERIMENTAL METHOD

The results have been obtained with 20 mm thick HSLA plates, one molybdenum-bearing wire, and various fluxes. The heat input was 38 kJ/cm (two electrodes in tandem).

Various elements at low concentrations were intentionally brought into the weld pool either through parent plate dilution or through the flux.

Charpy impact tests and J_{Ic} measurements were carried out at different temperatures to assess the fracture toughness of the weld metal from the brittle (cleavage) up to the fully ductile regime.

Thermal treatments were carried out on a 'Gleeble' RPI machine to investigate the behaviour of weld metal at high temperatures and during the austenite to ferrite transformation.

The weld metal microstructure and inclusions were examined with the help of TEM and X-ray microanalysis.

THE EFFECTS OF SOME ELEMENTS ON WELD METAL TOUGHNESS

Oxygen and aluminium

According to the concentration of less stable oxides in the flux such as MnO and SiO_2,[10,11] the amount of oxygen in weld metal varies from 20 up to 100×10^{-3} W %, Fig. 1. This oxygen combines with various deoxidants, most of which are brought into the weld pool through parent plate dilution. The oxide inclusions thus formed have an effect on the resistance to both ductile and brittle fracture.

Results published elsewhere[9] show that the oxide inclusions reduce both the Charpy upper shelf energy and the tearing modulus in the crack propagation regime.

Figure 2 shows that the resistance to brittle fracture does not increase continuously when decreasing the oxygen content : it reaches a maximum at a critical oxygen level. It appears that the weld metal transition temperatures can be much lower than those of heat-affected zones (HAZs) submitted to the same thermal cycles and having similar compositions but for their oxygen content ($\leqq 5 \times 10^{-3}$ W %). The results presented in Fig. 2 were obtained with one plate, one wire, and various fluxes. The welding of other HSLA plates yielded similar results but different critical oxygen levels[3].

Indeed the critical oxygen level, associated to the lowest transition temperatures, appears to depend upon the amount of deoxidants brought into the weld pool through parent plate dilution.

Figure 3 shows that the critical oxygen level in the weld metal increases with the parent plate aluminium content. According to these results, this aluminium content has to be taken into account when selecting the flux. K_{Jc} fracture toughness tests conducted at different temperatures on fatigue precracked specimens confirm the influence of the oxygen content as observed with Charpy specimens[8].

Boron and titanium

Further improvements of the weld metal resistance to brittle fracture can be obtained through boron additions to weld metal[3,7,12]. Small amounts of boron oxide in the flux provide a significant boron concentration in weld metal (up to 10×10^{-3} W %). For boron to be effective it has to be protected from nitrogen[3,7] by some titanium (around 20×10^{-3} W %) brought into the weld pool either from the flux or from the wire.

Figure 4 shows the effect of boron on Charpy transition temperatures. It appears that the transition temperatures can be reduced by 50 deg.C when the proper amount of boron is added to the weld metal. Increasing the weld metal oxygen increases the required level of boron as more boron reacts with oxygen.

Apart from attending to boron transfer, it appears that minute amounts of titanium ($\geqq 2 \times 10^{-3}$ W %) are required to obtain low transition temperatures in weld metal. This was demonstrated with laboratory-prepared fluxes[9,13] free of titanium or with low level titanium oxide additions (3 W %). Around the critical amount of oxygen, a 40 deg.C increase in transition temperatures can be observed when bringing the titanium in weld metal to below $2x10^{-3}$ W %. Other results are presented in Fig. 5 (multipass welding[15]).

Carbon, nitrogen, niobium, sulphur

The weld metal carbon concentration is mainly governed by the parent plate carbon. The effect of carbon on the weld metal resistance to brittle fracture was observed by welding laboratory-prepared plates with different carbon contents. Other parameters being kept constant, the lowest transition temperatures were obtained around $100x10^{-3}$ W % carbon[9].

Increasing the nitrogen level has either no effect or a detrimental effect on toughness according to the weld metal microstructure (Fig. 6).

A comprehensive state-of-the art review has been written on the effect of niobium in weld metal[14]. Our results show that low levels of niobium have a rather beneficial effect on the resistance to brittle fracture (up to $20x10^{-3}$ W % Nb)[15].

Increasing the sulphur level in the weld metal (from 2 to $7x10^{-3}$ W %) appears to the rather detrimental when the copper level is low ($15x\ 10^{-3}$ W %) and has no or little effect when the copper level is higher ($50x10^{-3}$ W %)[15].

DISCUSSION

Weld metal oxygen combines with various deoxidants brought into the weld pool through the parent plate dilution to form oxide inclusions. The mechanisms of ductile crack extension, as described in the literature[16], account for the reduction of weld metal upper shelf toughness as the amount of oxide inclusions increases.

Weld metal resistance to brittle fracture is closely related to its microstructure. Around the critical oxygen concentration ($45x10^{-3}$ W % for these experimental conditions) we observe a high percentage of acicular ferrite. Lower or higher oxygen levels yield coarser microstructures. This effect can be ascribed to a balance between grain boundary nucleation of ferrite which promotes proeutectoid ferrite veins or lathlike ferrite grains, and intragranular nucleation of ferrite which yields fine acicular ferrite grains.

Grain boundary nucleation of ferrite

The amount of grain boundary nucleated ferrite can be decreased either by reducing the nucleation rate at austenite grain boundaries or by increasing the austenite grain size. Boron is known to decrease the nucleation rate of ferrite at austenite grain boundaries[17]. This accounts for the effect of boron as described above.

Figure 7 shows the austenite grain size (15sec at 1400°C) of Gleeble-treated specimens for different oxygen levels. Below $50x10^{-3}$ W % oxygen the austenite grain size is of the order of 100 to 150 µm. Above $50x10^{-3}$ W % oxygen, it is of the order of 50 µm. Therefore large amounts of oxide inclusions may impede the austenite grain growth and promote large amounts of coarse grain boundary nucleated ferrite[18].

Intragranular nucleation of ferrite

Weld metal samples were submitted to heat treatments up to 1400°C to simulate welding. The controlled cooling was interrupted by quenching to room temperature from 575°C. This procedure makes it possible to observe the nucleation sites for ferrite at different oxygen levels. Below the critical oxygen level nucleation occurs mainly at austenite grain boundaries. Around the critical oxygen level nucleation occurs both at austenite grain boundaries and at numerous sites within the austenite grains (intragranular nucleation). Figure 8 shows that these nucleation sites are oxide inclusions. This enhanced nucleation results in fine interlocking ferrite grains impeding each other's growth. At much higher oxygen levels this effect is overcome by an increase of grain boundary nucleation (reduction of the austenite grain size). It appears therefore that intragranular nucleation occurs around particular oxide inclusions resulting from well-defined deoxidation conditions. TEM thin foils examination[9], shows that acicular ferrite is composed of elongated ferrite grains interspersed with carbide-ferrite aggregates or in some cases retained austenite. These aggregates would account to brittle fracture when carbon content increases, and some carbon may have a beneficial effect on the ferrite grain size through a carbon partitioning reaction.

To predict the amount and nature of oxide inclusions we have resorted to a thermodynamics model supported by direct TEM and X-ray microanalysis. We present here some results corresponding to the fracture toughness measurements in Fig. 2. The thermodynamics model describes the weld metal behaviour as it cools down to 1500°C. It takes into account data of the Fe-Al-Si-Mn-O phase diagram, assuming that equilibrium is reached at successive decreasing temperatures during cooling[3]. Table 1 indicates the results obtained in particular conditions for different oxygen levels. Aluminium oxide inclusions should form up to $40x10^{-3}$ W % oxygen, aluminium-manganese silicates above. This change in inclusion nature occurs around the critical oxygen level, Fig. 2. The nature of weld metal oxide inclusions was investigated directly by TEM or carbon replica with X-ray microanalysis of extracted inclusions. The inclusion sizes range from 0.1 to 3 µm in all samples, with no great difference from one sample to the other[19]. Some examples are shown in Fig. 9. At lower oxygen levels most inclusions are pure aluminium oxides ; they are sharp-cornered since they occur as solid precipitates within the liquid metal. About $40x10^{-3}$ W % oxygen, all inclusions are aluminium-manganese silicates ; they are globular since they occur as liquid precipitates within the liquid metal. A lower aluminium level in the base plate yields a lower critical oxygen level i.e. all weld metal inclusions are aluminium-manganese silicates down to a lower oxygen level.

From the results above it appears that intragranular nucleation occurs at aluminium-manganese silicates and not around alumina inclusions. This accounts for the oxygen and aluminium effects described earlier. Furthermore, the efficiency of aluminium-manganese silicates appear to depend on trace amounts of titanium as quoted above. Some oxide inclusions are covered by a thin outer coating of copper sulphide (Fig. 10) which seems to have been rejected from the inclusion core during its solidification. This outer coating does not seem to reduce the ability of the inclusion to act as a nucleation site.

Various mechanisms have been proposed to account for ferrite nucleation at oxide inclusions :

1 - Composition variations in the steel matrix around inclusions. This could not be substantiated by experimental evidence[18],

2 - Small lattice disregistry between ferrite and some oxides (TiO)[7], enhancing any heterogeneous nucleation effect.

We have proposed an other mechanism[9] :

The matrix straining around oxide inclusions during cooling owing to their much lower thermal expansion coefficients. The strained austenite accelerates its transformation into ferrite[20]. Figure 11 shows the thermal expansion coefficients of some inclusions[21]. It appears that some aluminium-manganese silicates display the lowest coefficient and are therefore the most likely to create significant strains during cooling. This interpretation may account for the beneficial effect of a low level concentration of niobium as it is likely to prevent recrystallization of the strained austenite[22].

Oxides usually display a "transition temperature" below which they undergo a drastic change in physical properties[23] (reduction of the thermal expansion coefficient). This temperature is of the order of 2/3 of the melting temperature (°K) i.e. between 600 and 700°C for aluminium-manganese silicates. To act as nucleation sites, oxide inclusions must have a "transition temperature" above the AR_3 temperature of the steel matrix. This could explain the beneficial effect of lowering the AR_3 temperature (Mn, Mo, ...). Furthermore minute amounts of titanium might increase the "transition temperature" of silicates and bring it above the AR_3 temperature. Indeed titanium is a well known catalyst for the crystallization of glasses.

CONCLUSIONS

Weld metal resistance to brittle fracture can be increased by favouring microstructures with large amounts of acicular ferrite (intragranular nucleation) and reduced amounts of large ferrite veins or lathlike ferrite grains (intergranular nucleation).

Acicular ferrite is made of fine interlocking ferrite grains interspersed with ferrite-carbide aggregates. These grains are nucleated at particular aluminium-manganese silicates which contain small amounts of titanium. Aluminium oxide inclusions do not favour intragranular nucleation. Intergranular nucleation can be restrained by avoiding too high oxide inclusion contents which would reduce the austenite grain size. The intergranular nucleation rate can be reduced by boron additions. The effect of oxide inclusions on intragranular nucleation may be ascribed to localised straining of the steel matrix around the inclusions during cooling.

These mechanisms account for the effects of low level concentrations of various elements such as aluminium, oxygen, silicon, boron, titanium, carbon, niobium and sulphur. They could be taken into consideration, along with the effects of higher level concentration elements (Mn, Mo), when parent plates and welding products are to be selected.

ACKNOWLEDGEMENTS

Sincere thanks are here expressed to Dr. B. THOMAS, Mr C. ROUX and Mr D. BOULEAU.

This work was part of an ECSC project.

RERERENCES

1. MASUBUCHI, K. 'Analysis of welded structures'. Pergamon Press, Vol. 33, 1980.
2. BERNARD, G. and FAURE, F. ECSC Final Report, project 6710-93/3/304, EUR 5866.
3. DEVILLERS, L. and RIBOUD, P.V. IIW Doc. IX-1218-81.
4. BONNET, C. Soudage et Techniques Connexes, July-August 1980.
5. ABSON, D.J. and DOLBY, R.E. Welding Institute Conference 'Trends in Steels and Consummables for Welding', London, November 1978.
6. COCHRANE, R. and KIRKWOOD, P. Welding Institute Conference 'Trends in Steels and Consummables for Welding', London, November 1978.
7. HOMMA, H. et al. IIW Doc. IX-1196-81.
8. DEVILLERS, L. et al. Procs 4th ECF Conference, Leoben, Austria, 1982.
9. DEVILLERS, L. et al. Procs WI Conference, London, November 1983.
10. RICHARDSON, F.D. and JEFFES, T.H.E. JISI, 160 261, 1948.
11. CHAI, C.S. and EAGAR, T.W. Welding J., July 1982.
12. HOMMA, H. et al. IIW Doc. IX-1229-82.
13. BONNET, C. and CHARPENTIER, J.P. PROCS WI Conference, London, November 1983.
14. DOLBY, R.E. IIW Doc. IX-1175-80.
15. DEVILLERS, L. Final Report ECSC project, 7210.KA.309.
16. ROESCH, L. et al. Mémoires Scientifiques de la Revue de Métallurgie, November 1966.
17. MAITREPIERRE, Ph. et al. TMS-AIME Meeting, Chicago, October 1973. (IRSID - external report 186).
18. BARRITTE, G.S. et al. Metals Society, book 227, 1981.
19. FERRANTE, M. and FARRAR, R.A. J. of Materials Science, 17-1982.
20. DESALOS, Y. et al. INSTN Conference, Saclay, France, June 1981 (IRSID - external report 863).
21. BROOSBANK, D. and ANDREWS, K. 'Production and Application of clean steels' - Balatonfüred, Hungary, publication of the Iron and Steel Institute, 1970.
22. LE BON, A. and ROSSARD, C. ECSC Final Report, project 6210-82/3/301, EUR 6150.
23. TURKDOGAN, E.T. Metals Society, London, Book 298, 1983.

Table 1 - Prediction of the nature and amounts of oxide inclusions as a function of oxygen content (base plate $Al=53x10^{-3}W$ %, $Si=250x10^{-3}W$ %, Mn=1,4 W %).

Total oxygen content (10^{-3} W %)	Dissolved oxygen at equilibrium at 1500°C (10^{-3} W %)	Dissolved aluminium at equilibrium at 1500°C (10^{-3} W %)	Inclusion types at equilibrium at 1500°C	Amount of inclusions (10^{-3} W %)
10	0.18	26	Aluminium oxide	21
20	0.26	14.8	Aluminium oxide	42
30	0.60	0.40	Aluminium oxide	62
40	2.7	0.41	Aluminium oxide + compound oxides of Al,Si,Mn	78
50	2.9	0.37	Aluminium oxide + compound oxides of Al,Si,Mn	96
60	4.2	0.17	Compounds oxides of Al,Si,Mn	113
70	4.2	0.16	Compounds oxides of Al,Si,Mn	132

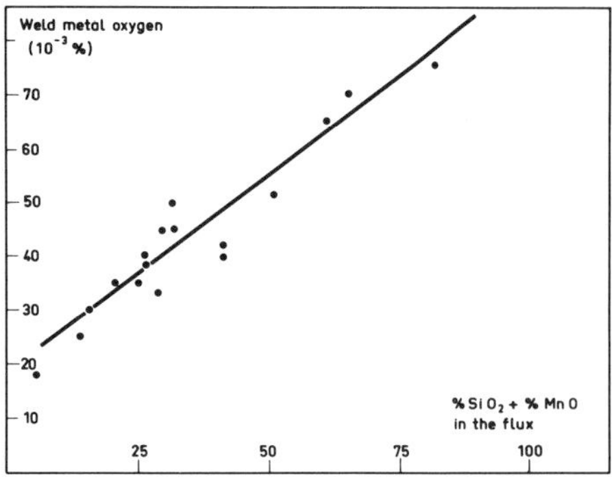

Figure 1 - Variation of weld metal oxygen with SiO_2% + MnO% in the flux.

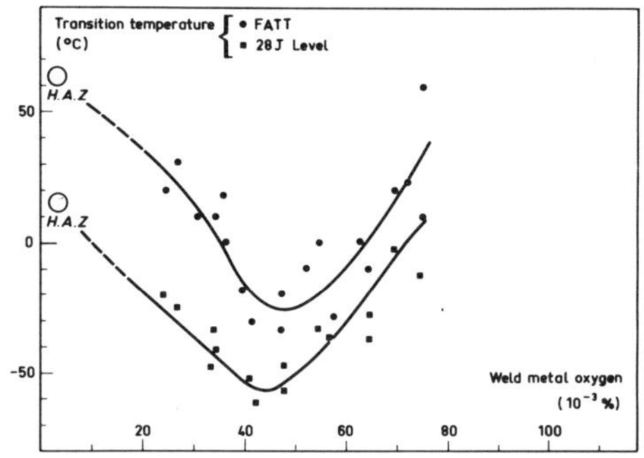

Figure 2 - Charpy V transition temperatures v. oxygen content (base plate Al content 53×10^{-3} W %) - . FATT, 28 J level.

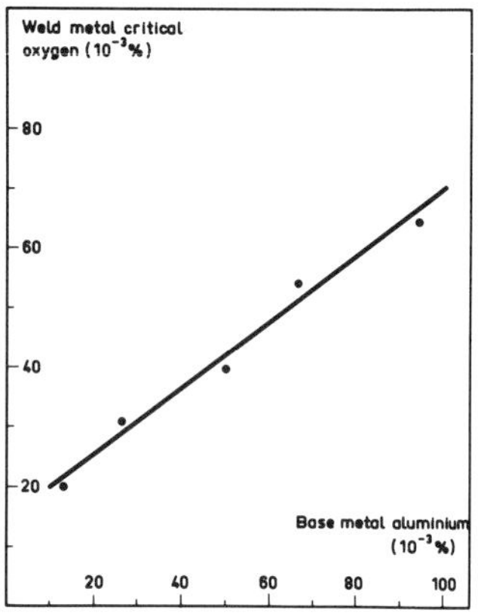

Figure 3 - Weld metal critical oxygen content versus base metal Al content.

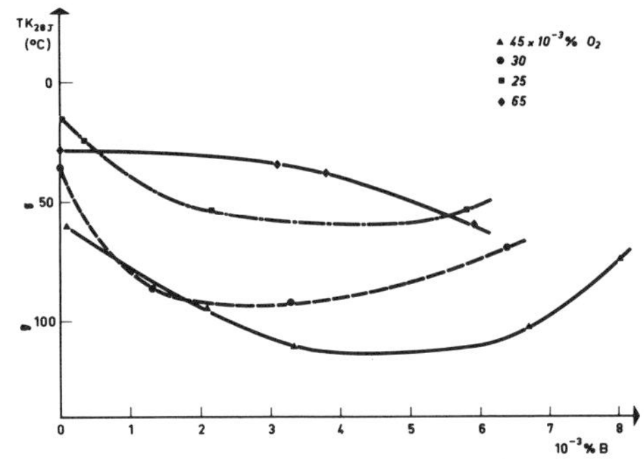

Figure 4 - Charpy V 28J transition temperature v. boron content for different oxygen levels (Ti bearing weld metals).

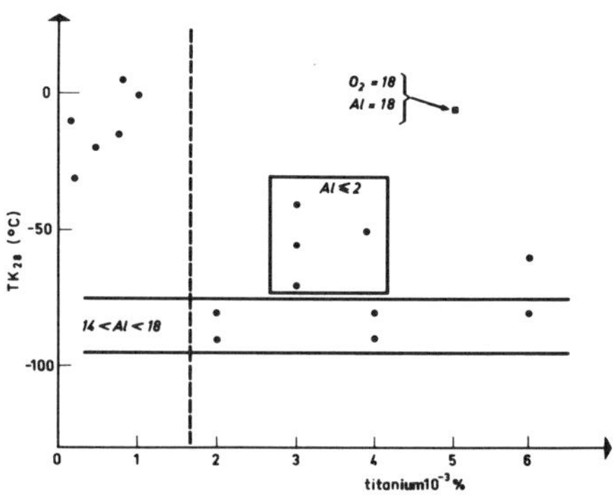

Figure 5 - Charpy V 28J transition temperature v. titanium and aluminium content (10^{-3} W %).

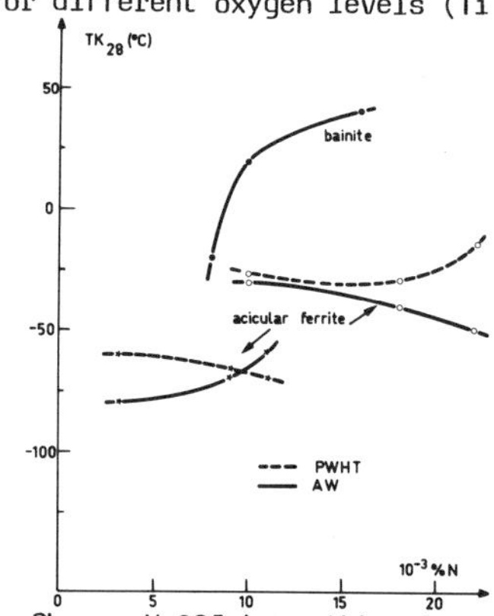

Figure 6 - Charpy V 28J transition temperature as a function of nitrogen content.

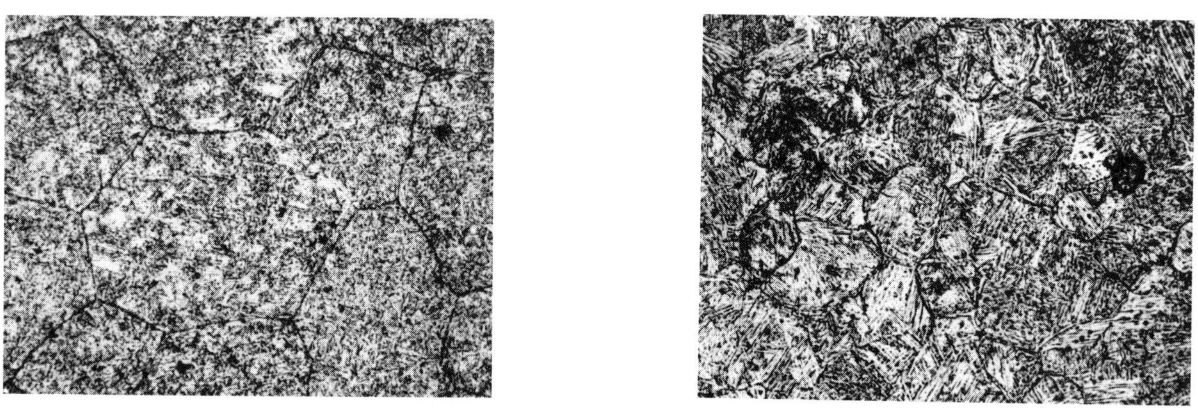

Figure 7 - Weld metal austenite grain size at 1400°C for two different oxygen levels 27×10^{-3} W % (left) and 67×10^{-3} W % (right) - 200 X.

Figure 8 - Nucleation of ferrite at oxide inclusions TEM X 4400

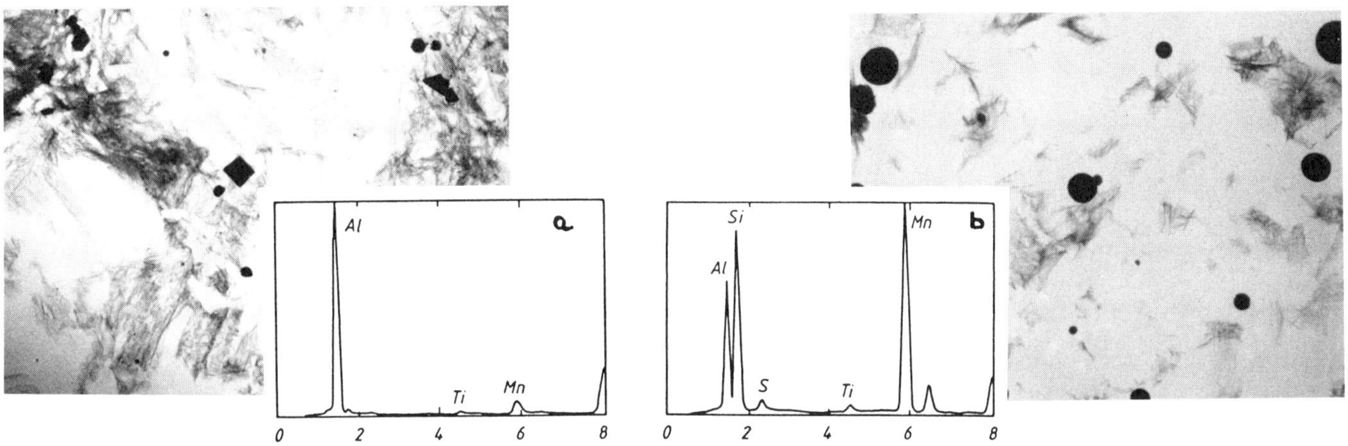

Figure 9 - TEM and microanalysis of extracted inclusions for two oxygen levels. (a) 27×10^{-3} W % O_2 - (b) 68×10^{-3} W % O_2.

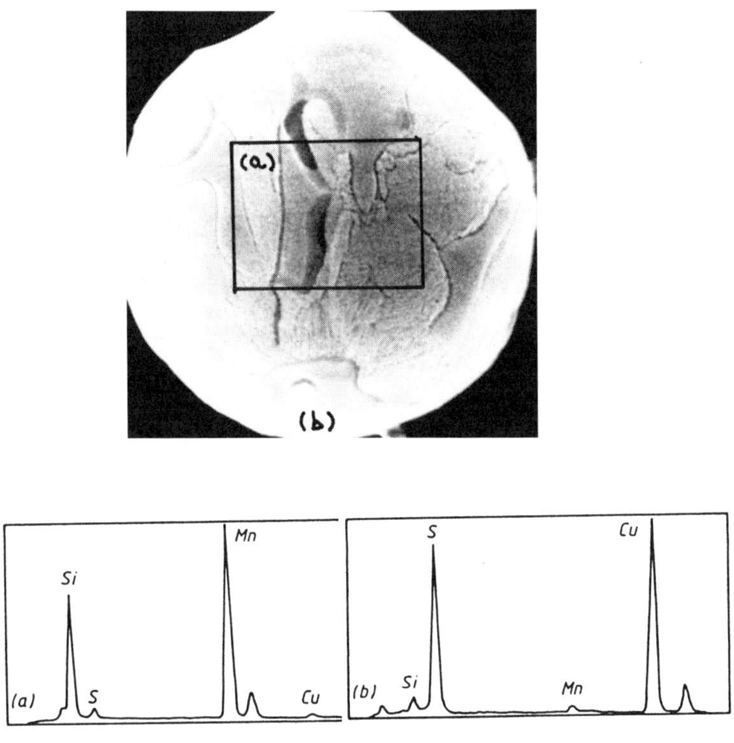

Figure 10 - Aluminium manganese silicate covered by copper sulphide layer X 50000.

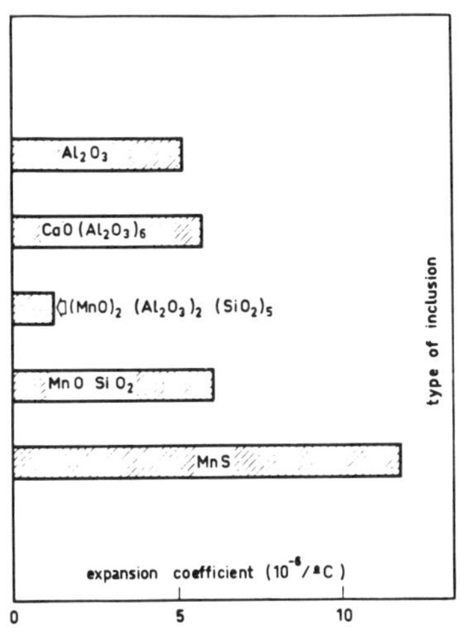

Figure 11 - Thermal expansion coefficients of some inclusions (after Ref. 21) - Austenite coefficient = 23×10^{-6}/°C.

FACTORS AFFECTING WELD METAL MICROSTRUCTURES AND PROPERTIES OF ELECTROSLAG WELDMENTS

H.S. Ann, D.W. Yu, J.H. Devletian, W.E. Wood
Oregon Graduate Center
19600 N.W. Von Neumann Drive
Beaverton, Oregon 97006-1999

ABSTRACT

Welding variables, such as joint gap, flux composition, welding parameters, and alloy content were found to have significant effects on the macro/microstructure and mechanical properties of consumable guide electroslag welds deposited on 2" (50 mm) thick A36 and A588 steel. From these studies, a new classification of characteristic grain structures for electroslag welds has been developed. This study demonstrated the influence of these welding variables on the solidification mechanics, solid state phase transformations and toughness of as-deposited weld metal.

CONSUMABLE GUIDE ELECTROSLAG WELDING (ESW) has been proven to offer the highest deposition rate and cost effectiveness of any welding technique for joining thick section steel. However, ESW requires a large heat input, typically of 100 kJ/mm (for a 50 mm thickness plate), which results in a high base metal dilution and a coarse weld grain structure. Multizoned weld structural features ranging from a coarse to fine columnar grains and variable grain orientations lead to nonuniform weld properties. These structural variations combined with centerline segregation may also result in hot cracking with sufficient restraint. In addition, the typically coarse weld metal microstructure inhibits reliable nondestructive examination of these weldments.

The high oxygen potentials associated with ESW made with the commercial manganese silicate flux lead to many inclusions, which help to nucleate ferrite early in the austenite-ferrite transformation resulting in poor toughness. Acicular ferrite was reported to be the best structure for highest toughness because of its interlocking nature with randomly oriented short ferrite plates [1]. It was demonstrated that an adequate number and appropriate dispersion of weld metal oxide inclusions (around 0.3 μm dia.) were an essential requirement for the nucleation of acicular ferrite [1] and that only an intermediate oxygen content (200-300 ppm) produced not only a high volume fraction of acicular ferrite, but also excellent weld metal toughness [2]. Other significant efforts have been made to better understand the metallurgical factors that control toughness of electroslag welds [3-6].

Furthermore, the electrical resistivity of the molten slag used in ESW is a primary factor in determining the heat generation efficiency. Although wide variations in the pattern of heat generation may be accomplished by changes in voltage, current, and slag volume, the operating limits are set by the slag resistivity [7].

The main objective of this research was to study the effects of process variables, flux chemistry, and minor alloying additions (Ni and Mo) on the solidification structure, macro/microstructure and toughness of consumable guide electroslag welds deposited in A36 and A588 plates. These process parameters include the joint gap, current, voltage, welding speed, and guide tube/plate designs.

EXPERIMENTAL PROCEDURE

MATERIALS - Both ASTM A36 and A588 structural steel plates in 50 mm (2") thick sections were welded by ESW using 2.4 mm (0.093") dia. solid and tubular filler metals. The 12.5 mm (1/2") dia. standard consumable guide tube and the rectangular shaped modified consumable guide plates shown in Fig. 1 were applied. The basicity indices of the commercial and experimental fluxes were calculated by the equation given by Tuliani, et al. [8] except for the omission of CaF_2. The compositions of base metal and consumables are given in Table 1.

WELDING - A commercial constant potential DC welding power supply was used in the reverse polarity mode. Base materials were joined vertically with gaps of 19 mm (3/4"), 25 mm (1") and 32 mm (1-1/4"). Slag depth was maintained constant at about 38 mm (1-1/2") by a continuous flux addition during the welding.

METALLOGRAPHY - Macrostructures of each electroslag weld were prepared in longitudinal and transverse sections after etching with 10% nital solution. The microstructure of each weldment was extracted from longitudinal section specimens. Solidification structures were revealed by a reagent with the following composition: 5 g $CuCl_2$, 5 g $MgCl_2$, 10 ml HCl, 400 ml H_2COOH, and 100 ml H_2O. The solid state transformation products were obtained by etching in a 2% nital solution.

CVN IMPACT TOUGHNESS - Standard CVN impact specimens having the V-notch perpendicular to the welding direction were tested at -18°C (0°F) in accordance with ASTM E23 and AWS D1.1 Structural Welding Code. In addition, impact tests on selected welds were conducted over a temperature range of -73°C to +65°C (-100°F to +150°F) to determine the temperature transition characteristics.

RESULTS AND DISCUSSION

FACTORS AFFECTING THE STRUCTURE OF WELD METAL - Factors affecting weld metal structures included: voltage, current, gap size, welding speed, flux variables (oxygen potential, basicity, and conductivity), and minor alloying additions (up to 2% Ni and 0.4% Mo in the filler metal). A new classification system representing five different types of grain structures were produced in A36 and A588 electroslag weldments as a result of varying process factors. Detailed parametric studies have been presented by the authors' previous work [3,5,6,9].

CLASSIFICATION OF WELD STRUCTURES - As a result of these parametric studies summarized in Table 2, the following characteristic weld metal metallurgical structures have been determined and are shown in Fig. 2: Type I weld structure consisted of only fine grained thin columnar crystals throughout the weld. Type II welds contained an outer zone of coarse columnar grains and an interior zone of thin columnar grains. Type III welds consisted of only coarse columnar grains throughout the weld. Type IV welds possessed not only the coarse and thin columnar grains (as in Type II), but also a zone of equiaxed grains in the weld center. And Type V welds contained an interior zone of equiaxed grains and a zone of thin columnar grains in the weld edges.

Type I. This weld structure was obtained by applying high oxygen potential fluxes resulting in weld metal oxygen concentrations exceeding about 600 ppm as illustrated in Fig. 3. For example, the increased oxygen potential made with a 40 wt.% SiO_2 (manganese silicate) flux produced abundant weld metal inclusions which tended to inhibit austenite grain growth. The results of this work are in full agreement with those of Matsuda, et al. [10] and Fleck, et al. [11] in that oxide segregation at dendritic boundaries hinders the migration (or grain growth) of austenite grain boundaries. Thus, thin columnar austenite grains dominated the entire weld. These oxides, in turn, nucleated copious amounts of blocky, Widmanstatten and grain boundary ferrite as shown in Fig. 2.

Type II. This weld structure was generated under normal welding conditions (including commercial welding practice) using a mild steel electrode, neutral flux and either standard or narrow gaps (Table 2). The sizes of the coarse and fine columnar grain zones varied with changes in welding parameters. For example, for low current, low welding speed and high voltage settings, the form factor exceeded 2.0 and the weld metal macrostructure consisted of a small outer zone of coarse columnar grains and a large interior zone of thin elongated columnar grains. With increasing welding current and travel speed, the form factor decreased and the proportion of coarse columnar grains increased at the expense of the fine columnar grains near the weld center. For high current and welding speeds, the lowest form factor for Type II (about 0.9) was accompanied with centerline hot cracking. The weld metal macrostructure consisted of a large outer zone of coarse columnar grains and small interior zone of thin columnar grains.

The solidification of Type II welds reported by the authors in a previous study [9] was characterized by cellular dendrites dominating near the weld edges and columnar dendrites near the weld center (Fig. 4). The cellular dendritic structure was associated with coarse columnar austenite grains which transformed to large amounts of acicular ferrite. The columnar dendritic morphology prevented lateral grain growth of austenite resulting in the thin columnar structure (Fig. 2) which transformed predominantly into blocky and grain boundary ferrite. Increasing welding speed increased the size of the cellular dendritic zone which resulted in a high acicular ferrite content. Similarly, the amount and fineness of acicular ferrite increased with increasing welding speed. In addition, pearlitic structures along with the proeutectoid ferrite were observed in these steel weldments.

Type III. This fully coarse grained weld structure was produced by ESW with low oxygen potential and high basicity fluxes containing CaF_2-CaO-Al_2O_3. The weld metal oxygen and sulfur content were typically only 80 ppm and 0.012 wt.%, respectively (a standard weld using a neutral flux would contain O_2:250 ppm and S:0.025 wt.%). Improved welding efficiency was also observed because of the increased conductivity of the basic slag (compared to commercial PF-201 flux) as shown in Fig. 5. All of these effects provided larger dendrite arm spacing due to the improved ohmic slag heating and development of coarse columnar grains. The low oxygen and impurity levels in the interdendritic regions were insufficient to retard the gain boundary migration of austenite. Thus, virtually unrestricted lateral austenite growth during weld cooling produced a fully coarse grain structure, as shown in Figs. 2 and 3. The reduced oxygen content decreased the amount of acicular ferrite nucleation while increasing blocky and grain boundary ferrite.

Type IV. This weld was developed using a mild steel electrode, neutral flux and the highest values of current and welding speeds to produce an extremely deep weld pool with a form factor of < 0.9. This solidification structure was characterized by a small zone of equiaxed dendrites at the weld center. The transition from columnar to equiaxed dendrites was believed to be associated with a fast growth rate, high amount of solute segregation and low thermal gradient at the weld center. The high degree of constitutional supercooling provided the driving force for dendrite arm remelting for nucleation of equiaxed dendrites at the weld center. Subsequent solid state transformations produced equiaxed grains containing large amounts of acicular ferrite and proeutectoid ferrite at prior austenite grain boundaries (Fig. 2). The equiaxed grain structure of the weld was so resistant to hot cracking that small radial type hot cracks (usually less than 6.4 mm (1/4") in length) developed intergranularly in the thin columnar grain zone.

Type V. When the low alloyed filler metal (AX90 and TW8544) containing Ni and Mo was added to the medium and high oxygen welds, equiaxed dendrites resulted without the aid of rapid welding speeds. This was due to the extensive degree of constitutional supercooling provided merely by additional alloying. It is believed that the fragile tips of growing dendrites are broken away from the main dendrites and carried into the supercooled region to provide nuclei to form equiaxed dendrites. Unlike Type IV, the equiaxed dendritic structure in Type V welds can occur at much lower growth rates and high temperature gradients. The solid state microstructure also showed equiaxed grains at the weld center with a high volume fraction of acicular ferrite (Fig. 2).

The role of the Ni-Mo alloy addition to the weld metal appears to promote the following: (a) equiaxed dendritic and equiaxed grain structures around the weld center, (b) grain boundary and blocky ferrite reductions, (c) acicular ferrite increases up to about 90%, and (d) elimination of the coarse columnar austenite grain structure near the fusion line. The Ni-Mo alloying additions tended to reduce the nucleation rate of high temperature transformation products (blocky, grain boundary and Widmanstatten ferrite). The retardation of austenite grain growth due to solute interdendritic segregation made the development of thin columnar grain structures across much of the weld possible with a high volume fraction of acicular ferrite.

The proportions of the coarse columnar, thin columnar and equiaxed zones varied with alloy content and oxygen concentration in the weld (Fig. 6). Generally, as these variables increased, the size of the thin columnar zones increased, and with further increase of alloy content and oxygen concentration in the weld, the size of the equiaxed zone increased at the expense of the thin columnar zone.

Increasing oxygen content promoted the nucleation of the higher temperature ferrite transformation products. The effect of oxygen concentration on the ferrite content in electroslag welds is shown in Fig. 7. Welds made with CaF_2-rich basic fluxes showed the lowest weld metal oxygen contents. With increasing silica content in the flux, the weld metal oxygen and silicon contents were found to increase. The best microstructures for maximum acicular ferrite and good toughness were obtained with fluxes having a basicity index of about 1.2, which corresponds to around 180 ppm of oxygen and 50% acicular ferrite at weld center of A36 welds using mild steel electrodes. Up to 90% acicular ferrite and the best toughness values obtained in this study resulted with neutral flux and minor additions of Ni and Mo (AX90 or TW8544) to the weld.

MECHANICAL PROPERTIES - The CVN impact energies at -18°C (0°F) for each structure type are compared in Fig. 8. Type I structures exhibited low impact energy due to the large amount of oxygen inclusions. A588 welds with the Type II developed slightly lower toughness than similar A36 welds, despite the high amount of acicular ferrite. In both A36 and A588 welds, the minor additions of Ni-Mo (AX-90 or TW8544 filler) substantially increased fracture toughness and lowered the ductile/brittle transition temperature of the weld metal as shown in Figs. 8-10. The Ni-Mo alloying additions provided the further benefits of substantially increased resistance to hot cracking because of the development of equiaxed grains in the weld center zone. An important achievement in Types I, III and V structures was the uniformity of CVN impact toughness and grain structure across the entire weld.

SLAG CONDUCTIVITY - The electrical conductivity of the slag was calculated using an empirical formula [12]. Fig. 5 illustrates the marked influence of the slag conductivity on the ohmic heat produced during ESW and also clearly demonstrates the optimum conductivity of slag for the highest welding efficiency which is around 3 $\Omega^{-1}cm^{-1}$ at 1700°C. Excessive slag conductivity produced a small exposed arc between the guide tube and surface of molten slag, while highly resistive slags tended to develop a submerged arc between the electrode and weld pool. In either case, arcing wastes energy and reduces efficiency. Thus, there was an optimum slag conductivity that produced the most efficient ohmic heating of the slag as shown in Fig. 5.

CONCLUSIONS

From this comprehensive study of ESW A36 and A588 steels, the following can be concluded:

(1) A new classification of macrostructures for electroslag welds has been developed.

Five characteristic types of weld metal grain structures were produced as a function of welding parameters, alloy content, oxygen potential and basicity of the slag.

(2) Thin columnar structures were a result of high weld metal oxygen concentrations.

(3) Fully coarse columnar structures occurred with highly basic, low oxygen potential slags.

(4) Equiaxed grains at the weld center were promoted by Ni-Mo addition, high oxygen content, and/or fast welding speeds.

(5) Minor additions of Ni and Mo substantially increased the volume fraction of acicular ferrite and increased the weld metal's resistance to hot cracking. The presence of about 90% acicular ferrite not only increased fracture toughness generally, but also lowered the ductile/brittle transition temperature of the weld metal.

(6) The best toughness and the highest amount of acicular ferrite occurred in welds containing a minor 2% Ni and 0.4% Mo addition to the filler metal (which was the highest alloying level used in this investigation.

ACKNOWLEDGMENT

The authors acknowledge the research support of the DoT/Federal Highway Administration, Contract DTFH61-83-R-00062. The authors also thank Mr. Robert Turpin and Mr. David Roberts for the helpful work in welding processes.

REFERENCES

1. Dolby, R. E., Metals Tech., 10, 349, (Sept. 1983).
2. Cochrane, R. C., et al., Intl. Conf. on Trends in Steels and Consumables for Welding, The Welding Institute, London, England, 103 (1978).
3. Ann, H. S., et al., "Narrow Gap, High Deposition Rate Electroslag Weld", presented at 66th Annual AWS Convention, Las Vegas, NV, Paper 27-D (1985).
4. Venkataraman, S., et al., Proc. of Conf. on Trends in Welding Research in the U.S., ASM, 381 (Nov. 1981).
5. Yu, D. W., et al., "Influence of Low Alloy Filler Metal on Solidification and Subsequent Microstructures", presented at Int'l Conference on Trends in Welding Research, Gatlinburg, TN, (1986).
6. Ann, H. S., et al., "Grain Refinement and Toughness in Electroslag Weldments", presented at 67th Annual AWS Convention, Atlanta, GA, Paper 23-D (1986).
7. Mitchell, A., et al., Met. Trans., 2, 3361 (Dec. 1971).
8. Tuliani, S. S., et al., Welding and Metal Fab., 327 (Aug. 1969).
9. Yu, D. W., et al., "Solidification Study of Narrow Gap ESW", published in Welding Research: The State of the Art, E. F. Nippes and D. J. Ball (eds.), ASM, Metals Park, OH, 21 (1986).
10. Matsuda, F., et al., Trans. of JWRI, 14 (2), 135 (1985).
11. Fleck, N. A., et al., Welding Journal, Welding Research Supplement, 1135 (May 1986).
12. Hara, et al., Trans. of Iron and Steel Inst. J., 23, 1053 (1983).

Table 1 - Chemical Compositions of Base Metal and Consumables

Wt%, BAL: Fe

MATERIAL	C	Mn	Si	Cr	Ni	Mo	Cu	S	P	V	Al
BASE METAL											
A588	0.18	1.20	0.37	0.56	0.16		0.32	0.02	0.01	0.08	0.03
A36	0.17	0.92	0.20	0.02	0.10		0.20	0.02	0.01	0.06	
ELECTRODE											
LINDE WS	0.09	0.50	0.30	0.55	0.50		0.30	0.03	0.02		
HOBART 25P	0.11	1.12	0.50					0.02	0.02		
PS 588	0.07	1.10	0.50	0.50	0.80		0.45				
AX 90	0.08	1.27	0.38	0.06	1.90	0.43		0.007	0.005	0.02	0.01
TW 8544*	0.03	1.20	0.40		2.04	0.40		0.01	0.01	0.04	0.02
GUIDE PLATE											
1010	0.08 /0.13	0.3 /0.6						<0.05	<0.04		

FLUX	HOBART PF201 100% CaF2 MIXTURES of CaF2-Al2O3-CaO

* STOODY TUBULAR FILLER METAL

Table 2 - Significant Variables Promoting Each Type of Characteristic Structure in A36 and A588 Electroslag Welds

	VOLTS (V)	AMP (A)	GAP (mm)	WELDING SPEED (mm/s)	FLUX OXYGEN (ppm)	FLUX BASICITY INDEX	REMARK
TYPE I	35 - 45	600	25 - 32	>0.28	HIGH >600	ACIDIC <0.7	HIGH O
TYPE II	35 - 45	500 - 1100	19 - 32	0.25 - 0.75	250	NEUTRAL 0.7	COMMERCIAL PRACTICE
TYPE III	35 - 45	600	19 - 32	0.25 - 0.30	LOW 80	BASIC 3.6	LOW O LOW IMPURITY
TYPE IV	35 - 40	>1100	19	>0.75	250	NEUTRAL 0.7	HIGH SPEED
TYPE V	35 - 45	600 - 1100	19 - 32	>0.28	<600	NEUTRAL >0.7	Ni - Mo STEEL FILLER

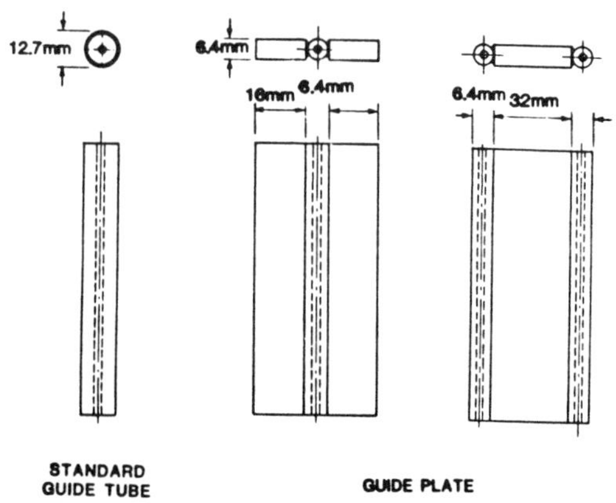

Fig. 1. Guide Plate and Tube Designs for Consumable Guide ESW.

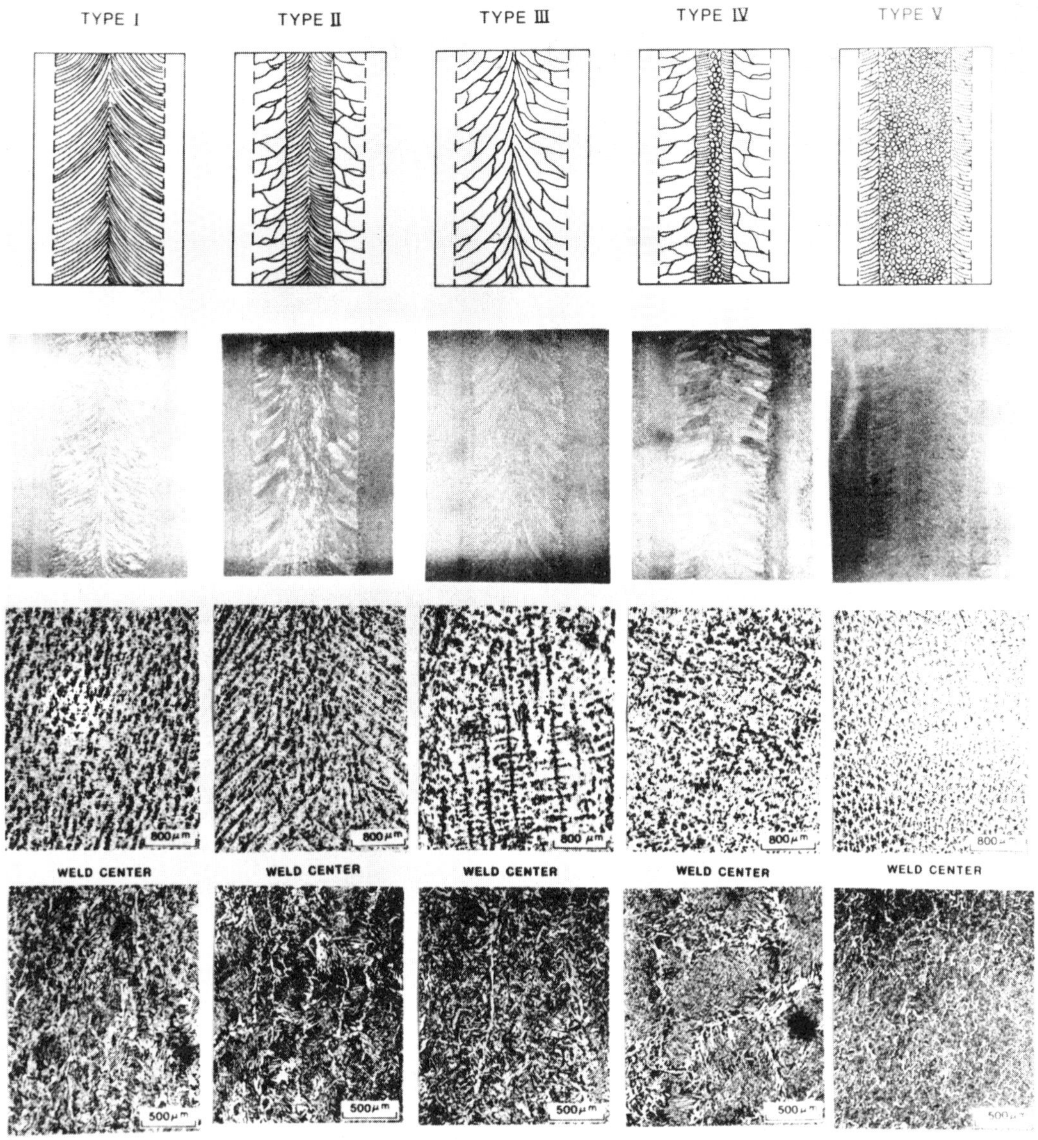

Fig. 2. Classification of Weld Metal Structures Deposited by ESW:
(a) Schematic Representation of Macrostructure,
(b) Macrostructure,
(c) Dendritic Structure, and
(d) Solid State Microstructure at Weld Center.

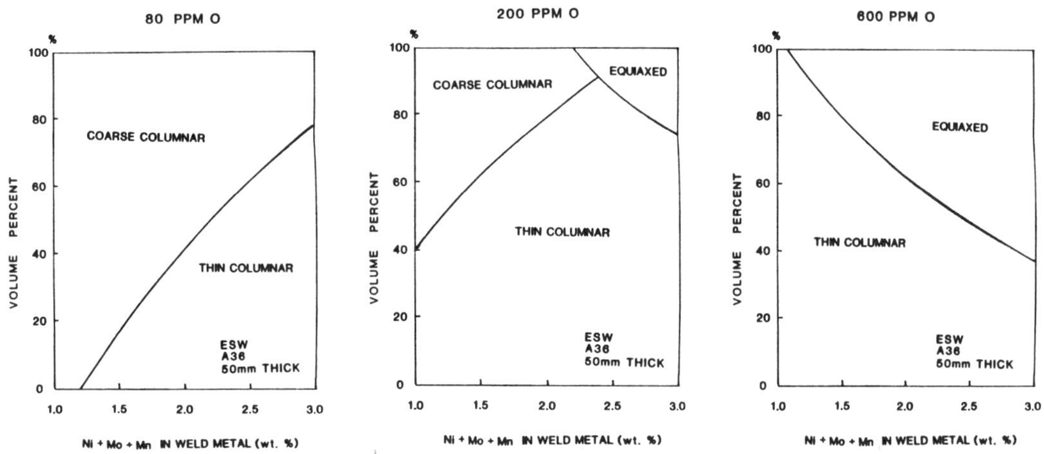

Fig. 6. Effect of Alloy and Oxygen Contents on the Volume Percent of Coarse Columnar, Thin Columnar and Equiaxed Zones in Weld Metal.

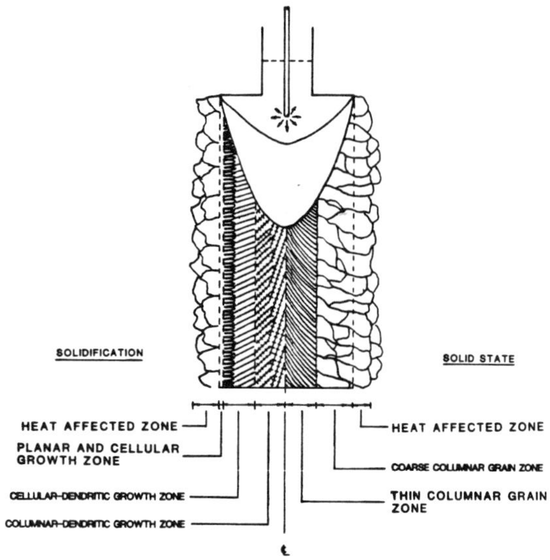

Fig. 4. Relationship Between Weld Metal Solidification Structure and Subsequent Solid State Structure.

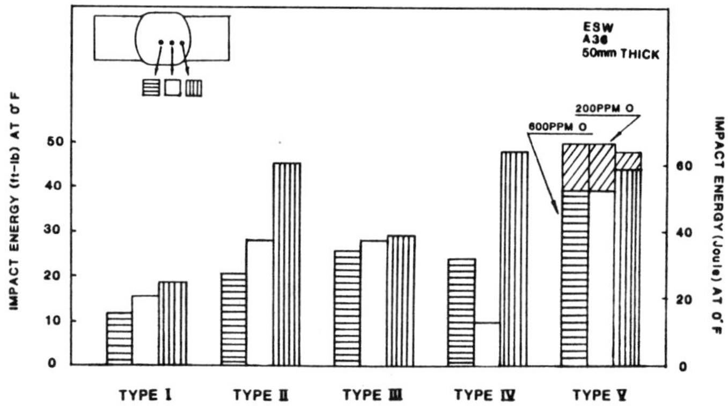

Fig. 8. CVN Toughness of Each Type of Weld Structures.

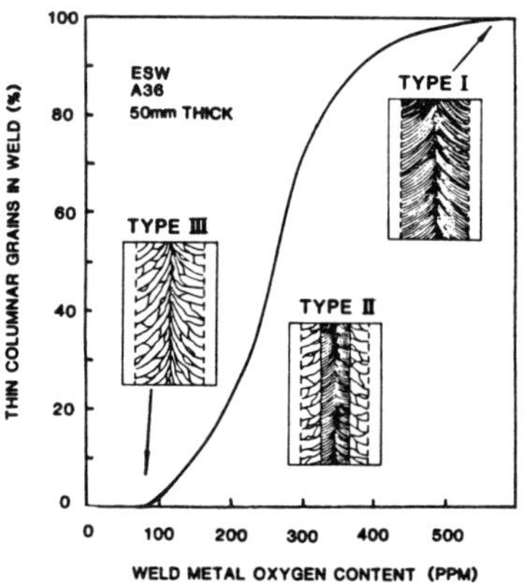

Fig. 3. Macrostructure as Affected by Weld Metal Oxygen Content.

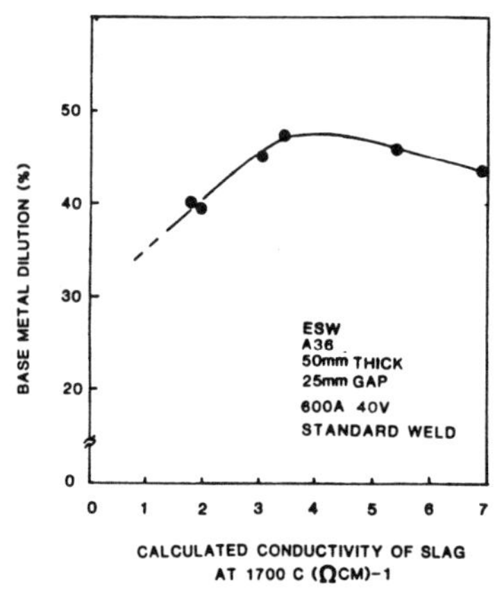

Fig. 5. Relationship Between Slag Conductivity and Base Metal Dilution.

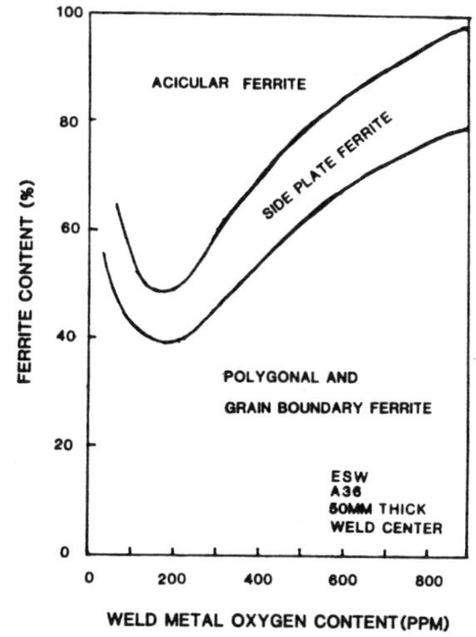

Fig. 7. Ferrite Content at Weld Center as a Function of Weld Metal Oxygen Content.

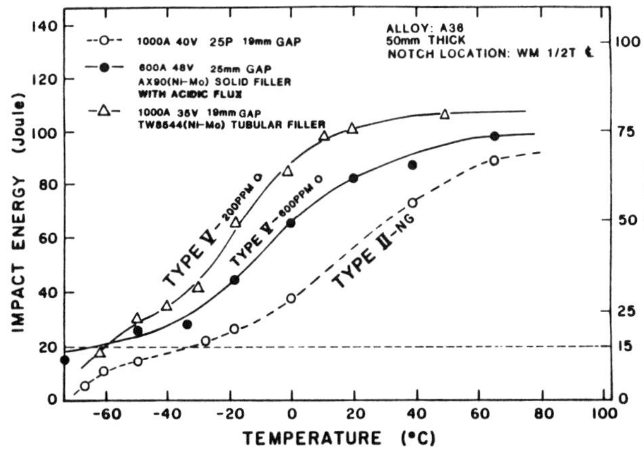

Fig. 9. CVN Toughness Transition-Temperature Curves in A36 Electroslag Welds.

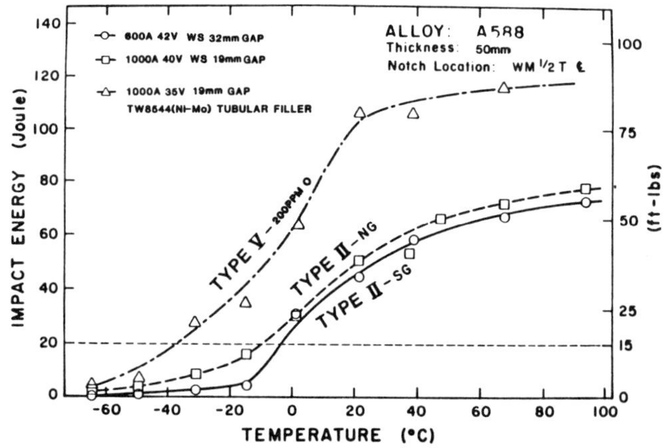

Fig. 10. CVN Toughness Transition-Temperature Curves in A588 Electroslag Welds.

WELDING AND WELDABILITY

INVESTIGATION OF ELEMENTAL EFFECTS ON THE WELDABILITY OF CAST NICKEL-BASED SUPERALLOYS

T. Kelly
General Electric Company
Evandale, Ohio

INVESTIGATION OF ELEMENTAL EFFECTS ON THE WELDABILITY OF CAST NICKEL-BASED SUPERALLOYS

ABSTRACT

Nickel-based superalloys are being investment cast in greatly increased amounts for todays gas turbine engines and will see even greater use in the future. Not only are these alloys being used in increasing amounts, but the thermal and mechanical property requirements are also being increased. Therefore, problems such as weld micro-fissuring in the heat affected zone are becoming a concern in the repair and fabrication of investment cast superalloy parts. Since almost all larger investment castings require some weld repair and many are welded during the fabrication of gas turbine engines, it is important that the cause of HAZ micro-fissuring be identified so that micro-fissures can be eliminated or minimized.

The microstructure of an investment cast superalloy is highly segregated, with major alloying elements frequently partitioning into low melting phases within a NiCr matrix. Due to the possible effect of this partitioning on the weldability of investment cast superalloys, a matrix of six alloying elements was constructed around a generic alloy 718 chemistry. A Plackett-Burmann 12 run design was used to analyze the effect of Fe, Co, Ta, Si, Mn and B on the weldability of cast superalloys. All heats were vacuum induction melted, cast into 3" x 2" x 3/4" slabs and heat treated prior to gas tungsten arc and electron beam welding. The independent variables used for statistical evaluation were the number of cracks per inch of weld, total crack length, and average crack length. During this analysis, B at the 100 ppm level was the single most detrimental element on the weldability of cast superalloys.

INTRODUCTION

The superalloy 718 invented by H.L. Eiselstein was the answer to many of the wrought superalloy welding problems of the 1960's. It is hardened by the sluggish reaction of γ'', which prevented the strain age cracking problem due to γ' which was employed by all of the high temperature superalloys of that time. However, 718 was not designed to be used in the as cast condition. Initially, when 718 was used for structural castings it was only used in noncritical parts. Due to its early success it is now being used in more demanding aerospace applications. In these more demanding applications, problems such as microfissuring in the Heat Affected Zone (HAZ), which could previously be ignored, became potentially serious. Therefore, a program to better understand the origin of these microfissures was undertaken at General Electric.

Cast alloy 718 is a composite structure of γ based dendrites and a mixture of laves, carbide and delta in the interdendritic spaces.(1) Without some means of hot working, little can really be done to homogenize such a structure. Therefore, though these phases may be put into solution, the elemental constituents remain segregated and ready to precipitate the original phases present in the as cast structure. Pease's(2) work, primarily with wrought superalloys, produced a table of generalized elemental effects on weldability. His work indicates that Fe, Cb, Co, and Mn have either no effect or a beneficial effect on welding. While Si is considered variable, it is generally controllable by addition of Cb to Ni-Cr alloys. A ratio of 4 or 5 to 1 is sufficient to prevent hot cracking. Boron is seen as harmful down to 300 ppm in nickel base alloys with variable effects below that. Much of this information is still accurate, though it was generated before alloy 718 was invented.

After the invention of alloy 718, Owczarski(3-5) and Shira(6) performed much of the quality work on weldability of this system. Owczarski found that constitutional liquation of metal carbides (NbC) contributed to HAZ cracking. Additionally, he found a synergistic effect between the presence of carbon and boron in some superalloys, but did not acknowledge the effect of boron in alloy 718. Shira found a concentration of Cb and Ti in the HAZ microfissures and also concluded that Mn additions reduced the HAZ cracking. Shira also pointed out that minor alloying elements were difficult to analyze with existing equipment; therefore, C, B, and S in grain boundaries would not be detectable.

Later work by Thompson(7-8) and Baeslack (9-10) continued to identify constitutional liquation of metal carbides, tramp element segregation, and the presence of low melting laves phase as causes of HAZ microfissuring. Thompson, working with wrought 718 and using Gleeble specimens, identified metal carbide constitutional liquation as the cause of microfissures. Later he was able to demonstrate that tramp element segregation, namely sulfur, to grain boundaries was responsible for some microfissuring. Additionally, he has shown that the segregation of tramp elements can be controlled by proper heat treatment. Baeslack, working with cast 718, has found microfissures associated with both constitutional liquation of metal carbides and the liquation of low melting laves phase.

Cremisio(11) discusses the various effects of heat treatment, segregation, and filler metal quality on the weldability of 718 castings. He points out that the proper homogenization heat treatment prior to Hot Isostatic Pressing (HIPing) can reduce the segregation in a casting and thus reduce HAZ microfissuring due to microsegregation. However, he is addressing fabrication type welds after relatively high temperature HIPing cycles, as high as 1200°C (2200°F). In these studies, he is citing the liquation of low melting phases (probably laves) and the attendant resolidification as boride eutectics. No specific cause of microfissuring in repair welds of castings is addressed.

After reviewing the history of 718 welding metallurgy research, it is somewhat unclear as to the specific cause of HAZ microfissuring. It appears that metal carbides (NbC), laves phase $(Fe, Ni, Cr)_2$ (Nb, Ti) and the presence of borides can all result in reduced weldability. Though all of these mechanisms may be acting in concert, it would be useful to know which is the controlling mechanism. Therefore, this study looks at the effects of alloying elements separated from their phase morphology. In order to accomplish this, the elements associated with the presence and suppression of the phases cited were examined in a statistical matrix. Included in the statistical matrix were Fe, Co, Ta, Si, B and Mn. Once the statistical results were analyzed, the individual phases were viewed using a hot stage scanning electron microscope to verify the findings.

PROCEDURE

The experiment was designed to measure the first order effects of six of the elements permitted in alloy 718. To do this a Plackett-Burman 12 run statistical design matrix was selected. A range for each independent variable was selected, and by superimposing the compositional variables on the matrix, 12 alloy chemistries were developed as shown in Figure 1.

CHEMICAL RANGES USED TO EVALUATE THE WELDABILITY

NUMBER GTA WELD RESULTS OF CRACKS	Co	Fe	Ta	Si	B	Mn
13	13.1	18.5	0.05	0.29	0.0048	0.21
14	12.9	0.05	2.99	0.28	0.0091	0.004
3	0.05	18.3	2.93	0.28	0.0005	0.004
2	12.9	18.4	3,03	0.04	0.0005	0.004
1	13.1	18.4	0.06	0.03	0.0005	0.2
25	13	0.05	0.03	0.05	0.006	0.004
0	0.05	0.05	0.07	0.27	0.0005	0.2
28	0.05	0.05	2.98	0.04	0.0127	0.2
43	0.05	18.2	0.2	0.29	0.0075	0.004
0	13	0.05	3.08	0.28	0.0005	0.2
36	0.05	18.1	3.11	0.05	0.0093	0.21
0	0.05	0,05	0.02	0.04	0.0005	0.004

REMAINING ELEMENTAL ANALYSIS 19%Cr, 5%Nb, 3%Mo, 1%Ti, .5%Al AND 0.04%C

Figure 1: Matrix of six alloy variables used to determine effect on weldability of 718.

Each heat was vacuum induction melted using virgin material for alloy make-up. The heats were then poured in vacuum into graphite crucibles (3" X 2" X 3/4") and allowed to cool in vacuum. After the ingots were removed from the mold, the ingot surfaces were machined to remove any mold reaction product prior to heat treatment. Since two ingots were produced from each heat, one was heat treated for repair welding while the second was heat treated for fabrication welding. The Gas Tungsten Arc (GTA) repair welding was performed after a 1180°C (2150°F)/4 hour thermal treatment of the casting while the Electron Beam (EB) fabrication welds were accomplished after 1180°C (2150°F) HIP and 1090°C (2000°F) solution treatments.

Bead on plate welds were made on surface ground surfaces using both GTA and EB welding processes. Initial evaluation of the weldments was accomplished using 20X binoculars. After the initial evaluation, laydown metallographic specimens were prepared to better evaluate the grain sizes involved in the study and to obtain a better evaluation of the number of HAZ cracks. The number of cracks and the total crack length per each inch of weld were used as the quantitative data for evaluation in the statistical matrix.

RESULTS

Statistical Analysis

Only the GTA welds had visually detectable surface cracks in the HAZ. Boron was the only element to produce a significant response using the visually detected cracks in the statistical analysis. This is shown graphically in Figure 2 where B appears at the 99.5% confidence level when the results of visually detectable crack count are plotted.

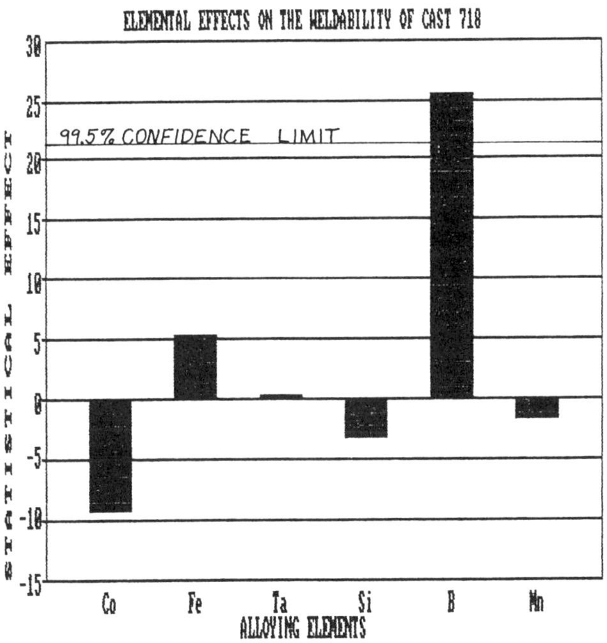

Figure 2: Results of statistical analysis of GTA weld cracking, showing B with 99.5% confidence of being the cause of microfissuring.

Since microfissures in alloy 718 are generally not a surface connected phenomena, all welds were inspected after polishing into the surface of the weld. After metallographic examination, both EB and GTA welds were found to have microfissures. Using total crack count per inch of weld in the statistical analysis, boron again was the only element to appear as significant in this experiment. Even though the maximum boron level was set at only 100 ppm, it appears certain that boron is the major cause of HAZ microfissuring in cast 718.

Using total crack length as a dependent variable in the calculation resulted in boron again being identified as the most harmful alloying element. However, both iron and silicon become noticeable as to their effect on HAZ microfissuring. Both of these elements are known to be associated with laves phase formation and laves phase has been reported to be responsible for HAZ microfissuring. It is interesting that these elements only appear detrimental when absolute crack length is analyzed.

Matrix Chemical Analysis

Figure 1 contains the actual chemical analysis matched to the number of cracks per inch of weld. Note that with zero boron many of the weldments had zero or less than five microfissures, while with as low as 50 ppm B the crack count goes up to greater than ten. It appears that the microfissuring varies almost directly to the boron content.

Additionally, it appears that some synergism exists between B and Fe. In this regard it is unfortunate that the B varied as it did. Since it appears at an unintentional lower level several times when Fe is at a high level, the effect of the iron may have been masked.

HOT SEM OBSERVATIONS

A hot stage Scanning Electron Microscope was used to view samples of the matrix alloys. Specimens with and without boron were viewed at 10,000X while rapidly heating through an approximate HAZ thermal cycle. Specimen temperature was only approximate, but at a temperature in excess of 1090°C (2000°F) a grain boundary reaction was observed in the boron-containing alloys. The non-boron-containing alloys failed to exhibit the same reaction. Additionally, known laves phase pockets were observed melting on grain boundaries, but the molten laves phase only wets the grain boundaries of boron-containing heats. Non-boron-containing heats experienced no grain boundary wetting by molten laves phase.

DISCUSSION

The results of this work point to boron as the most damaging alloying element for weldability of all the alloying elements used in cast superalloys. Data presented here shows that boron in quantities as low as 50 ppm in versions of cast 718 has caused an increased incidence of microcracking in both high voltage EB and GTA welding.

These results tend to rule out constitutional liquation of metal carbides as a cause of microcracks in cast 718. Since carbon was an obvious omission from the experimental matrix it is fair to say it wasn't analyzed; therefore, no conclusion on this theory should be made. However, the NbC phase has a melting point of approximately 1980°C (3600°F). Even though the constitutional liquation theory predicts a series of lower melting eutectics of C-Nb-Ni, it does require quite a significant amount of diffusion in a relatively short time to achieve this in the HAZ. Figure 3 is a photomicrograph of a carbide surrounded by molten laves phase and delta pockets with no appreciable sign of liquation around the carbide. Laves and delta phases in this alloy liquate at 1160°C (2125°F) and 1180°C (2150°F) respectively.

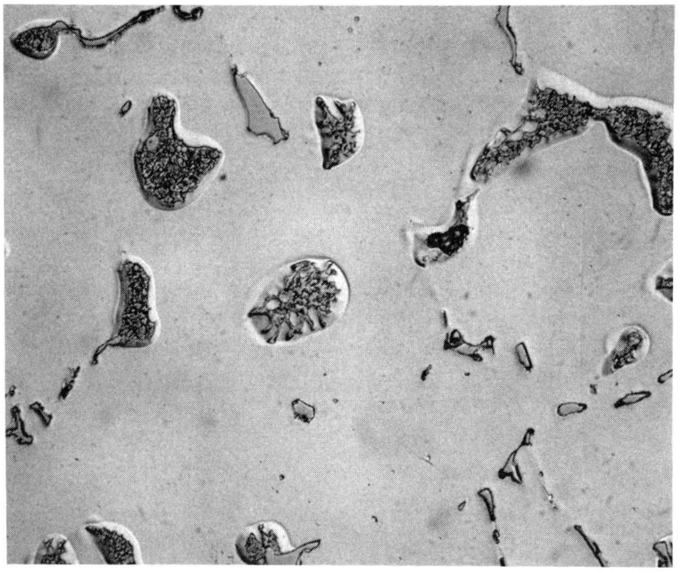

500X

Figure 3: Microstructure of the HAZ of a cast 718 weldment near the fusion zone, weld above top of the field, note the carbides shown no melting while laves pockets were already completely liquated.

It seems that an argument is being made for laves phase liquation as a prime cause of microcracking, but actually it is just the pecking order that is important. In this cast alloy there is always an abundance of laves, delta and carbide phases, as opposed to wrought 718 which is usually completely free of laves phase and, if properly heat treated, free of delta phase. Therefore, in wrought alloys constitutional liquation of carbides may be important but in cast alloys it is unlikely to be a major problem. Thus laves is the lowest melting phase present in 718, but it is not the root cause of microfissuring.

It is easy to understand why the low melting laves phase has received so much attention, but, as mentioned in the hot stage work, laves phase will not wet closed grain boundaries. The only time laves phase was observed to wet grain boundaries was in boron-containing heats. Therefore, though Fe and Si are known laves phase promoters they do not cause microcracking by themselves. It seems that the presence of boron in cast superalloys is the foundation from which all other microcracking mechanisms take root.

Because of the gross segregation found in cast 718, it is unlikely that even greatly reduced amounts of boron will completely eliminate microcracking in the HAZ. Only the complete removal of boron can be expected to have appreciable impact on the microcracking tendencies of cast superalloys. But this comes at the price of high temperature notch ductility; therefore, unless a replacement for boron in found it is unlikely that microcracking can be avoided in the HAZ of cast superalloys.

CONCLUSIONS

1. Boron is the single most detrimental element on the weldability of cast superalloys.

2. Iron and silicon have a strong effect on the size of microcracks in boron-containing cast superalloys.

REFERENCES

1. Carlson, R.G., Radavich, J.F., Microstructural Characterization of Cast 718, TM #85-512, 1985, General Electric, Evendale, Ohio.

2. Pease, G.R., The Practical Welding Metallurgy of Nickel and High-Nickel Alloys, Welding Journal, July 1957, p 330-S-334-S.

3. Owczarski, W.A., Duvall, D.S. and Sullivan, C.P., A Model for Heat-Affected Cracking in Nickel-Base Superalloys Welding Journal, April 1966, p 145S-155S.

4. Owczarski, W.A., Duvall, D.S., Further Heat Affected-Zone Studies in Heat Resistant Nickel Alloys, Welding Journal, Sept. 1967, p 423S-32S.

5. Owczarski, W.A., Some Minor Element Effects on Weldability of Heat Resistant Nickel-Base Superalloys, Proceedings of the Symposium - Effects of Minor Elements on the Weldability of High-Nickel Alloys, Welding Research Council, 1967, p 6-24.

6. Shira, C.S., Morrison, T.J. and Weisenberg, L.A., The Influence of Minor Elements on Alloy 718 Weld Microfissuring,

Proceedings of the Symposium Effects of Minor Elements on the Weldability of High-Nickel Alloys, Welding Research Council, 1967, p 47-68.

7. Thompson, R.G., and Genculu, Microstructural Evolution in the HAZ of Inconel 718 and Correlation with the Hot Ductility Test, Welding Journal, Dec. 1983, p 337S-345S

8. Thompson, R.G., Relationship Between Microstructure and Microfissuring In Alloy 718, To be published in the Welding Journal 1986.

9. Baeslack, W.A. III, Tsay, Y.L., An Investigation of Heat Affected Zone Cracking In Alloy 718, October 4, 1984, Ohio State University.

10. Baeslack, W.A. III, Nelson, D.E. Morphology of Weld Heat Affected Zone Liquation in Cast Alloy 718, To be published in Metallography.

11. Cremisio, R.S., Simon, W., Impact of Metallurgical Cleanliness on Repair Welding of Superalloy Castings, Presentation AIME, Toronto, 1985.

WELDABILTIY OF AN Ni_3Al Alloy

M.L. Santella, S.A. David, J.A. Horton
Oak Ridge National Laboratory
Metals and Ceramics Division
P.O. Box X
Oak Ridge, Tennessee 37831

ABSTRACT

Since joining by conventional welding processes is an important means of fabricating structural components, weldability has become a key issue in the development of these new Ni_3Al alloys. The results of an initial evaluation of the weldability of Ni_3Al containing 0.1 at. % B and 0.5 at. % Hf will be reported. Plates were prepared by conventional methods and used to make full penetration electron beam and gas tungsten arc welds. Initial results indicate that hafnium improves the weldability of Ni_3Al alloys although they are still susceptible to cracking. Examination of microstructures indicated that a distinct microsegregation pattern developed in the welds and affected the ordering behavior of fusion zones. Room temperature tensile testing suggested that welds can have strength and ductility values comparable to base materials, and that postweld heat treatment can improve tensile properties.

THERE HAS BEEN INTEREST IN THE DEVELOPMENT OF Ni_3Al for engineering applications since the discovery that its yield strength increases with temperature up to about 600°C.[1] Besides unusual strength behavior, good creep, oxidation, and corrosion resistance have also been identified with this intermetallic compound.[2] Typically, Ni_3Al crystals exhibit very good room temperature ductility because of their face-centered cubic crystal structure.[2] However, other studies[3,4] of mechanical behavior established that polycrystalline Ni_3Al generally failed by intergranular fracture with little or no ductility. Current opinion is that grain boundaries in polycrystalline Ni_3Al are intrinsically weak and thus are responsible for its low ductility and brittle behavior.[5,6] In the past, this characteristic of Ni_3Al has been the major impediment to its development as a structural alloy.

More recent studies[5,7] have found that adding a few hundred parts per million of boron to polycrystalline Ni_3Al can dramatically improve its ductility by segregating to grain boundaries and

ductility by segregating to grain boundaries and improving their cohesion. This discovery has stimulated a vigorous program of Ni_3Al alloy development[2,5,8] because of the apparent advantages of this material over conventional alloys, especially for high temperature applications including steam and gas turbine components, Stirling and adiabatic diesel engine components, and a variety of structural components in energy conversion systems. Some of the issues being addressed by this alloy development program are hot workability, mechanical properties, oxidation and corrosion behavior, and weldability. Our work has been aimed at assessing the weldability of Ni_3Al alloys, and this paper describes the welding behavior of an Ni_3Al alloy considered to have some commercial value. We will show that the weldability of Ni_3Al can be affected by composition, that defect-free welds can be made in Ni_3Al by electron beam (EB) and gas tungsten arc (GTA) welding, and that these welds exhibit reasonably good tensile properties at room temperature.

MATERIALS AND EXPERIMENTAL DETAILS

The alloy used for this study is known as IC-50, and its nominal composition is given in Table 1. Three different forms were used.

Table 1 - Alloy Composition

Alloy	Ni	Al	Hf	B
IC-50	balance	24.0	0.5	0.1 at. %
	balance	12.5	1.7	0.02 wt %

One form was recystallized sheet either 0.7-mm or 2.0-mm thick, produced from an arc melted ingot by repeated cold rolling with intermediate annealing at 1050°C. The second was a bar produced by hot extrusion of a canned powder. The finished diameter of the bar was 38-mm, and 5-mm-thick welding coupons were cut longitudinally from this bar. The last form used was a 12.5-mm-thick plate casting. In each form the base material was single phase with a fully ordered crystal structure.

Most of the welds used for this study were autogenous, i.e., made without a filler metal. However, some GTA welds were made on the 2-mm sheet using a filler metal made by shearing strips from the rolled sheet. The size of this filler was approximately 2-mm square. All GTA welds were made with argon shielding and standard welding practices. The EB welds were made in a Leybold-Heraeus 15 kW machine.

Specimens for optical metallography were etched by swabbing with glyceregia to show general microstructure or were etched electrolytically with an oxalic acid solution to show solidification structures. Specimens for transmission electron microscopy (TEM) were polished using a perchloric acid solution.

Postweld heat treatments were done in a vacuum muffle furnace. Tensile tests were conducted at room temperature on an Instron testing machine with a crosshead speed of 42 μm/s.

RESULTS

Initially, the welding behavior of IC-50 was assessed by making a series of full penetration EB welds on 0.7-mm sheet. The welds were made at speeds of 2.1, 4.2, 10.6, 21.2, and 42.5 mm/s. Examination of the as-welded surfaces at 100× showed the welds to be free of cracks except for a few crater cracks at the termination point of the high speed welds. This suggested that IC-50 had reasonably good weldability. Figure 1 is taken from the work of David and co-workers[9] and shows that crack-free autogenous welds were never obtained in Ni_3Al + B alloys. Comparison with our data indicates that adding hafnium to the Ni_3Al + B alloys improves their welding behavior.

Following this initial screening, the weldability of IC-50 was studied in more detail. Full penetration EB welds were made in the consolidated powder product and a cross sectional view of the microstructure of one of them is shown in Fig. 2. The powder product base material had a very small grain size (ASTM 9-10), and there was no visible heat-affected zone (HAZ). Some evidence of solidification substructure was found in the fusion zone, but there was no indication of second phase formation. This weld and others similarly made were free of defects except for some isolated porosity.

Several GTA butt welds were made in the 2-mm sheet with the IC-50 filler metal described previously. These welds were found to be defect-free by visual and radiographic examination. The optical microstructure of the GTA weldments was indistinguishable from that of the autogenous EB welds in both the HAZ and the fusion zone. A fusion zone region, etched to reveal solidification structure, is shown in Fig. 3. The dendrite core regions were found to etch quite differently

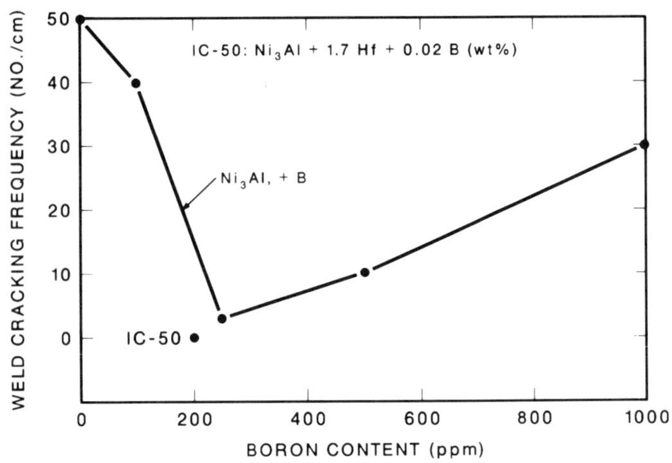

Fig. 1 Cracking frequency plotted against boron concentration for Ni_3Al + B alloy welds. Behavior of IC-50 is also indicated.

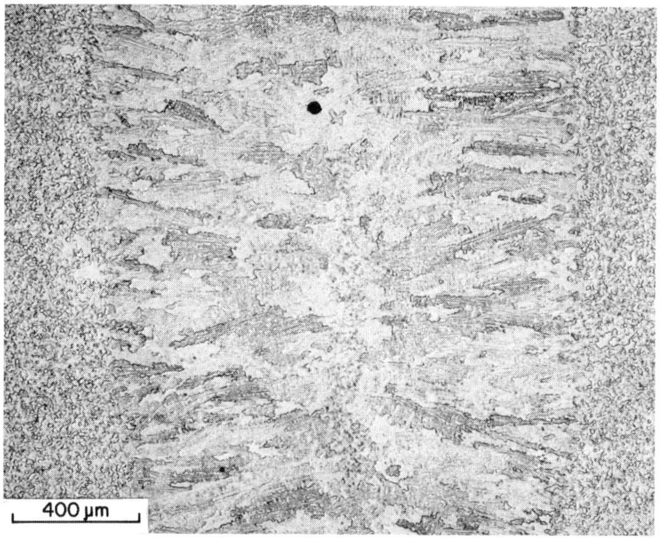

Fig. 2 Microstructure of electron beam weld in hot extruded powder coupon of IC-50.

from the interdendritic areas. Standardless electron microprobe analysis indicated that while the aluminum concentration was fairly constant throughout the fusion zone, hafnium was highly segregated to the interdendritic regions, reaching concentrations up to 4 wt %. Dendrite cores were found to be enriched with nickel.

Specimens of the GTA and EB welds were also examined in the TEM and a fusion zone microstructure is shown in Fig. 4. The dendritic nature of the microstructure is clearly evident. Diffraction analysis of the fusion zone microstructure indicated that both the dark dendrite core regions (A) and the light interdendritic regions (B) are probably completely ordered and of the same crystallographic orientation. The difference in their appearance is due to a variation in ordered domain size in each region. In the dendrite core regions the domain size is on the order of 0.1 to 0.2 μm while the interdendritic areas are essentially free of domain boundaries.

The microstructure of an autogenous GTA weld made on the thin plate cast material is shown in Fig. 5. In this case a large intergranular crack is seen extending from the fusion zone into the HAZ. The cracking frequency was high on welds of this type and the cracks were generally similar in nature to the one shown in Fig. 5. The thin plate cast material had an equiaxed grain structure with a grain size of ASTM 8-9. Although the cast material cracked upon welding, it could still easily be cold reduced without cracking about 40% in thickness at room temperature. This indicates there was enough boron segregation to grain boundaries to make the casting ductile.

The results of room temperature tensile tests of IC-50 weldments are given in Table 2 along with typical tensile properties of IC-50 base material. The orientation of the specimens was such that the weld axis was either perpendicular (transverse) or parallel (longitudinal) to the tensile axis as indicated in Table 2. The EB weld specimen had gage dimensions of 19.0 × 6.4 × 4.4 mm. The transverse GTA weld specimens had gage dimensions of 19.0 × 4.8 × 1.3 mm. The EB weld specimen and the transverse GTA weld specimens all fractured across the fusion zone. The fracture appearance of the EB weld specimen is shown in Fig. 6. The dimpled appearance of the fracture surface indicates that the fusion zone failed by ductile void coalescence. The longitudinal GTA weld specimens had gage dimensions of 12.7 × 3.6 × 0.7 mm. The bead width of these welds was slightly less than the tensile specimen gage width so that the entire weldment was tested. One specimen was given a postweld heat treatment of 1 h at 950°C before testing. All of the longitudinal GTA weld specimens broke in the specimen gage section.

Fig. 3 Solidification structure in fusion zone of gas tungsten arc weld made with addition of filler metal.

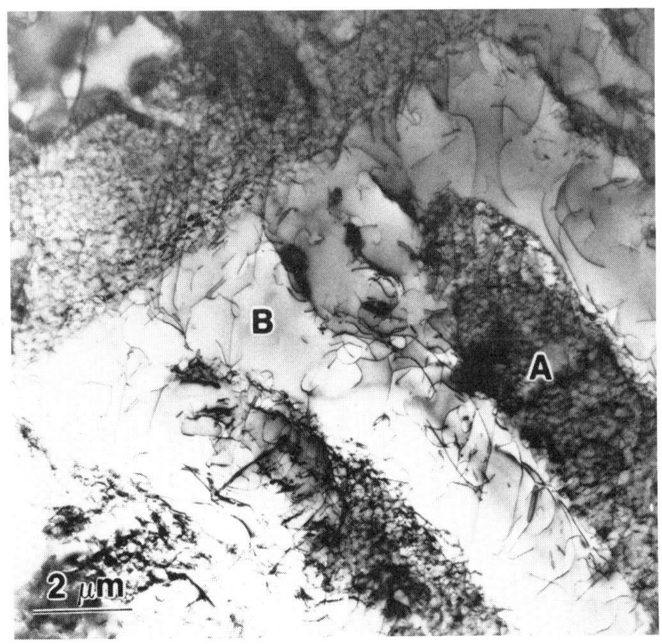

Fig. 4 Transmission electron micrograph in fusion zone of electron beam weld made in hot extruded powder material.

DISCUSSION

Hafnium additions of 1.7 wt % clearly improve the weldability of Ni_3Al + B alloys. Hafnium was added to the base Ni_3Al alloy because

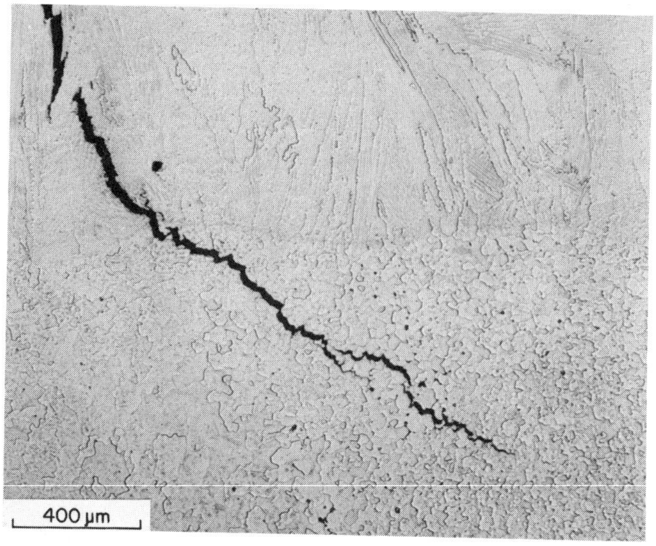

Fig. 5 Microstructure of gas tungsten arc weld showing large intergranular crack in heat affected zone extending into fusion zone.

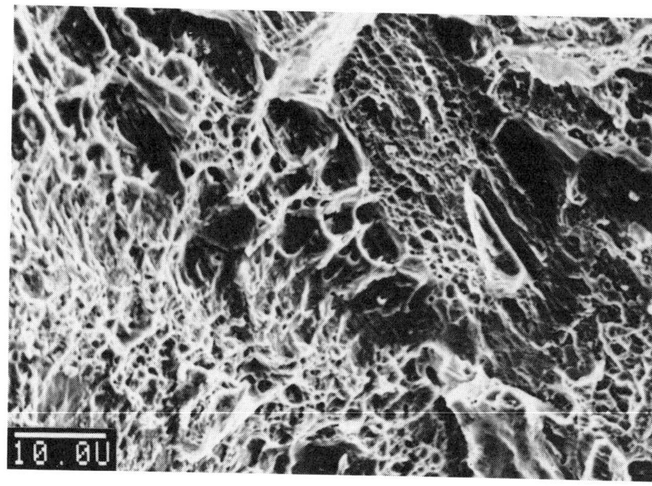

Fig. 6 Scanning electron micrograph of fractured weld showing evidence of ductile void coalescence in fusion zone.

Table 2 - Results of Tensile Testing IC-50 Weldments

Specimen	Condition[a]	Proportional limit (MPa)	Tensile strength (MPa)	Elongation (%)
EB, ⊥ axis	AW	140	1057	24
GTA, ⊥ axis	AW	142	1294	28
GTA, ∥ axis	AW	318	1292	34
GTA, ∥ axis	PWHT	311	1413	43
Base		200–350	1380	40–50

[a]AW = as-welded, PWHT = heat treated after welding.

it improves the high temperature strength by solid solution hardening.[2] Good strength at high temperature also appears important for reducing the tendency for HAZ cracking of Ni_3Al alloys,[10] and apparently the hafnium addition to IC-50 provides enough strengthening to accomplish this.

The variation of ordered domain size in the weld fusion zones can be explained by considering the microsegregation pattern that develops during solidification. Based on the binary Ni-Al phase diagram,[11] we expect the dendrite cores to be enriched with nickel and the interdendritic areas to be enriched with either aluminum, or hafnium, because this element tends to occupy aluminum-type sites in the ordered structure.[12] As discussed by Horton,[13] nickel enriched regions in solidified Ni_3Al alloys will undergo ordering at a lower temperature than regions not so enriched. At low temperatures domain growth is restricted even though it may be rapid at high temperatures. The result of solidification is that domains tend to form at low temperatures and not grow appreciably in nickel enriched regions of Ni_3Al alloys, while domains in aluminum (or hafnium) enriched regions form at relatively high temperatures and may grow to be very large. The microstructure observed in the fusion zone of IC-50 welds agrees with this analysis.

Of the three material forms studied, (1) rolled sheet, (2) hot extruded powder, and (3) thin plate casting, the latter was by far the most susceptible to cracking. In fact, no crack-free welds were made on the casting. Because the grain size of the thin plate casting material was in the same size range as the sheet and hot extruded material (ASTM 7-9) we conclude that grain size was not a major factor in the low crack resistance of the cast material. However, based on our examination of fusion zone microstructure and microprobe results that indicate hafnium enrichment of interdendritic regions, it is reasonable to speculate that grain boundaries in the cast material are enriched with hafnium. The effects of segregation in Ni_3Al alloys on mechanical properties are still being studied, but it is possible that enriching grain boundary regions in Ni_3Al with hafnium could cause their embrittlement through excessive solution hardening. Also, hafnium forms eutectic with nickel[14] so that hafnium enrichment at Ni_3Al alloy grain boundaries may increase their susceptibility to cracking if such phases are present. Data are not presently available to adequately explain the low crack resistance of the cast material.

The tensile properties of the IC-50 weldments compared quite well with the typical properties of recrystallized sheet. The elongation values of all the weld tensile specimens were very good indicating that the weld fusion zones were ductile. This point was confirmed by examination of the fractured specimens in the scanning electron microscope (SEM) as shown in Fig. 6. Also, the ultimate tensile strength of the welds was within a few percent of the base metal properties except for the transverse EB weld. In this case, the lower strength may be characteristic of hot extruded powder product. The yield strength of the transverse weld specimens was significantly lower than base material values of yield strength. The yield strength of Ni_3Al alloys varies inversely with grain size[5] and this may explain the low yield strength of the transverse welds because the grain size was very large in the fusion zones of all of them. Microsegregation in the fusion zones probably influences weld properties also, but more testing will be needed to clearly define the role of these factors on mechanical behavior. The longitudinal welds had yield strength values within the range of the base material properties. However, the fusion zones did not occupy the full cross section of these specimens, making it difficult to form conclusions about these data relative to the base material. In spite of this complication, the longitudinal specimens do indicate that the tensile strength and ductility of IC-50 welds can benefit from postweld heat treatments.

SUMMARY

The welding behavior of the Ni_3Al alloy IC-50 that contains 1.7 Hf + 0.02 B (wt%) has been studied, and the most important findings of this work can be summarized as follows:

1. Hafnium improves the resistance of Ni_3Al + B alloys to cracking compared with alloys without hafnium.
2. Defect-free autogenous welds can be made in IC-50 by EB and GTA welding. Also, defect-free welds can be made by GTA welding with the addition of a filler metal of similar composition to the base material.
3. Microsegregation in weld fusion zones causes dendrite core regions to order differently from interdendritic regions.
4. At room temperature the tensile strength and elongation of IC-50 welds compare favorably with base material properties. Postweld heat treatment of IC-50 welds at 950°C for 1 h improves their tensile strength and ductility.

ACKNOWLEDGMENT

The authors wish to thank R. L. Klueh and R. W. Swindeman for reviewing the manuscript and K. Gardner for typing the manuscript. The work was jointly sponsored by the U.S. Department of Energy AR&TD Fossil Energy Materials Program and Conservation Energy Conversion and Utilization Technologies Program under contract DE-AC05-84OR21400 with Martin Marietta Energy Systems, Inc.

REFERENCES

1. Flinn, P. A., Trans. AIME 218, 145 (1960).
2. Liu, C. T. and J. O. Stiegler, Science, 226, 636 (1984).
3. Westbrook, J. H., Trans. AIME 209, 898 (1957).
4. Grala, E. M., "Mechanical Properties of Intermetallic Compounds," p. 358, ed. J. H. Westbrook, John Wiley and Sons Inc., New York, NY (1960).
5. Liu, C. T., C. L. White, and J. A. Horton, Acta Metall 33, 213 (1985).
6. Ogura, T., S. Hanada, T. Masumoto, and O. Izumi, Met. Trans. A 16A, 441 (1985).
7. Aoki, K. and D. Izumi, Nippon Kinzaku Gakkaishi, 43 1190 (1979).
8. Liu, C. T., W. A. Jemian, H. Inouye, J. V. Cathcart, S. A. David, J. A. Horton, and M. L. Santella, "Initial Development of Nickel and Nickel-Iron Aluminides for Structural Uses," ORNL-6067 (1984).
9. David, S. A., W. A. Jemian, C. T. Liu, and J. A. Horton, Weld. J. 64(1), 22-s–28-s (1985).
10. Santella, M. L. and S. A. David, submitted to Weld. J. for publication.
11. Hansen, M., "Constitution of Binary Alloys," McGraw-Hill Book Co., New York, NY (1958).
12. Ochiai, S., Y. Oya, and T. Suzuki, Acta Metall. 32, 289 (1984).
13. Horton, J. A. and C. T. Liu, Acta Metall. 33, 2191 (1985).
14. Nash, P. and A. Nash, Bull. Alloy Phase Diagrams 4, 250 (1983).

PROGRESS IN WELDING STUDIES FOR CANADIAN NUCLEAR FUEL WASTE DISPOSAL CONTAINERS

P.Y.Y. Maak, M.J. Tinkler
Ontario Hydro Research Division
Toronto, Ontario

Abstract

The progress in the development of closure-welding technology for Canadian nuclear fuel waste disposal containers is described in this paper. Titanium, copper and Inconel are being investigated as candidate materials for fabrication of these containers. Gas-tungsten-arc welding, gas-metal-arc welding, resistance-heated diffusion bonding and electron beam welding have been evaluated as candidate closure welding processes. Characteristic weldment properties, relative merits of welding techniques, suitable weld joint configurations and fit-up tolerances, and welding parameter control ranges have been identified for various container designs. Furthermore, the automation requirements for candidate welding processes have been assessed. Progress in the development of a computer-controlled remote gas-shielded arc welding system is described.

AN IMPORTANT AREA of research in the development of corrosion-resistant disposal containers for the Canadian Nuclear Fuel Waste Management Program concerns the identification of suitable fabrication and closure welding procedures/1/. Since used fuel is highly radioactive, handling, final closure welding, inspection and, if necessary, repair of a loaded and sealed container, must be done remotely. Containers must be capable of withstanding the hydrostatic pressure of a flooded vault (98 MPa at a depth of 1000 m), at temperatures up to 150°C for periods of at least 500 years/1,2/. Several container designs based both on supported-shell and stressed-shell concepts have been developed to meet these requirements/2,3,4,5/. The details of one of the container designs are shown in Figure 1. ASTM grade 2 and grade 12 titanium, pure copper and Inconel 625 are being considered as possible container materials/6/. The progress in the development of welding procedures for these candidate materials is described in this paper/7/.

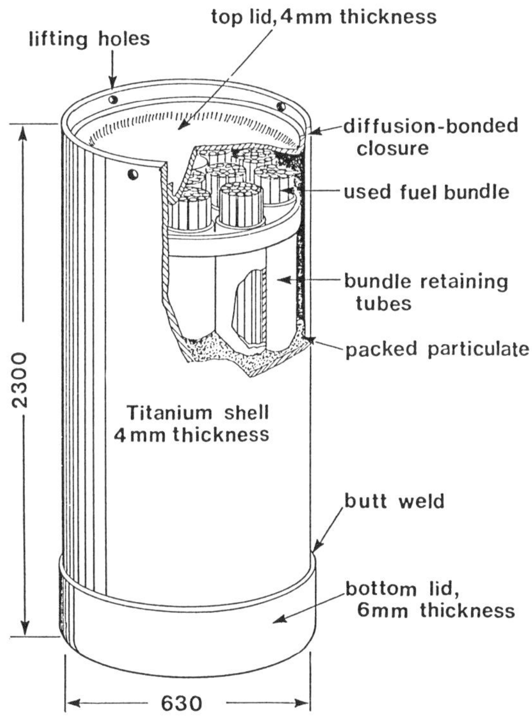

Fig.1 - Design details of a packed-particulate supported-shell fuel isolation container.

WELDING OF TITANIUM

It has been demonstrated that pulsed current, gas-tungsten-arc (GTA) welding, and resistance-heated diffusion bonding are feasible and promising for the final closure of titanium thin-walled supported-shell containers. Demonstration tests were done on titanium pipe and ring mock-ups. Suitable weld joint configuration and fit-up tolerances, allowable ranges of welding parameters, acceptance standards for gas shielding and mechanical properties, and automation needs for remote operation have been identified for these processes/8,9,10,11,12/.

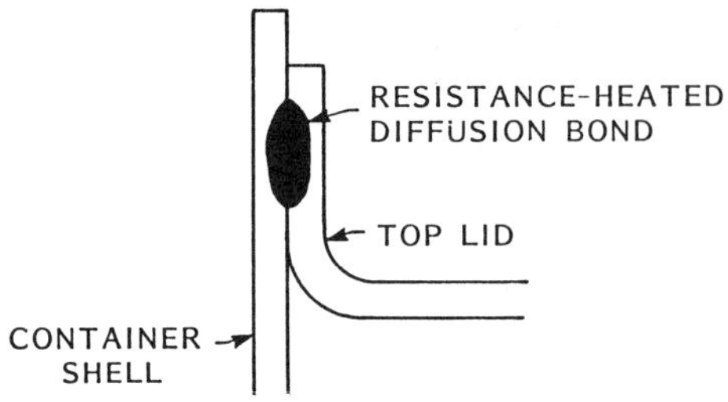

Fig.2 - Recommended weld joint design for the final closure of supported-shell titanium containers by resistance-heated diffusion bonding.

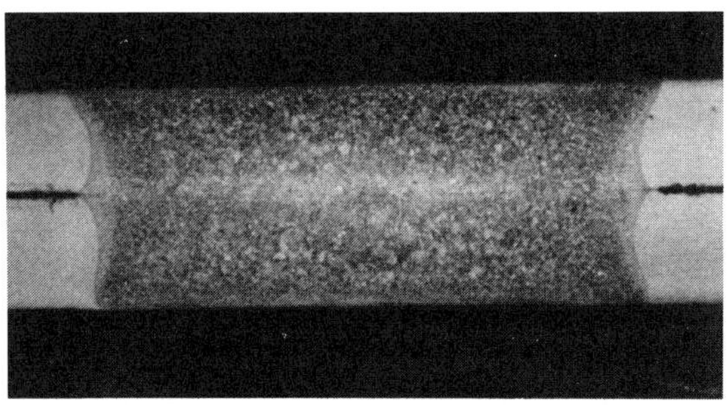

Fig.3 - Cross-section of a typical titanium resistance-heated diffusion bond, mag 5X.

Fig.4 - A typical titanium resistance-heated diffusion bonded ring mock-up.

The recommended joint design for closure by resistance-heated diffusion bonding and a cross-section of a typical titanium resistance-heated diffusion bond are shown in Figures 2 and 3 respectively. A typical titanium resistance-heated diffusion bonded ring mock-up is shown in Figure 4. This solid-state bonding process has been shown to be particularly attractive for remote joining of thin titanium (< 4 mm). The process is simple, rugged and easily controllable. In contrast to fusion welding, no gas shielding or accurate seam tracking is required. However, results to date show that 6.4 mm thick titanium bonded samples display a considerable surface oxidation and deformation. The severity of surface oxidation and deformation can be controlled by adjusting the bonding parameters, but the scope for adjustment is limited by the requirement for high bonding current and relatively long heating time to obtain sound diffusion bonds in titanium of this thickness. Significant improvement in surface quality can likely be achieved by modifying the electrode wheel configuration, and thereby lowering the current density at the electrode wheel/titanium contact surface. The development of an edge bonding technique is in progress to close the crevice present near the top edge of the joint and thus eliminate any risk of crevice corrosion.

In comparison to resistance-heated diffusion bonding, GTA welding appears to more versatile for joining titanium over a wide range of thicknesses. However, the welding parameters must be closely controlled and adjusted for variations in welding conditions, such as weld joint geometry and fit-up, to ensure satisfactory weld quality. The recommended joint design for closure by GTA welding and a cross section of a typical GTA weld are shown in Figures 5 and 6 respectively. Test results to date show that a relatively large root opening (> 2.0 mm) and a small included angle ($\sim 45^{\circ}$) are necessary to ensure complete root fusion and a satisfactory weld bead surface profile.

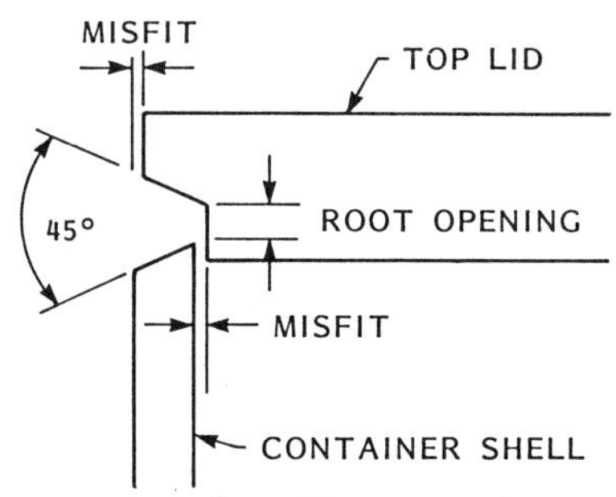

Fig.5 - Recommended weld joint design for the final closure of supported-shell titanium container using the gas-tungsten-arc welding process.

Fig.6 - Cross-section of a typical titanium gas-tungsten-arc weld.

Electron beam (EB) welding appears to be feasible and suitable for joining thick titanium (>12.7 mm thick) but requires high capital equipment cost and close weld joint fit-up tolerances/13/. The cross-section of a titanium EB-welded sample is shown in Figure 7. Among the various welding techniques under investigation, EB welding appears to be least attractive for use on thin-walled containers. However, it is still considered a viable process for the final closure of thick-walled stressed-shell containers.

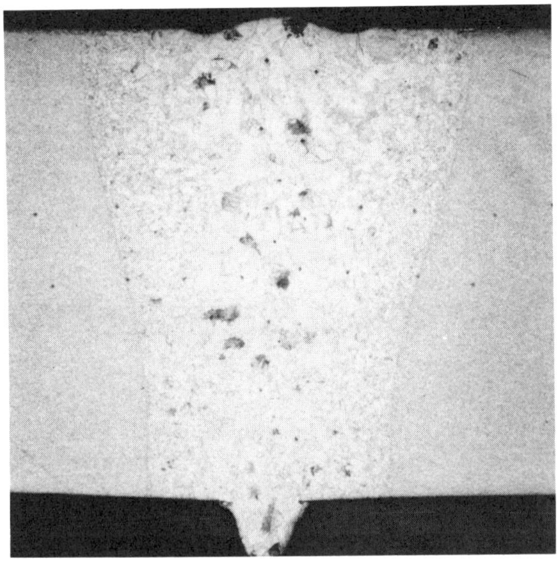

Fig.7 - Cross-section of a typical titanium electron beam welded sample, mag 5X.

Fig.8 - A typical copper electron beam welded pipe mock-up, mag 4.3X.

WELDING OF COPPER

EB welding has been shown to be an ideal process for joining thick copper without preheating/14,15,16,17,18/. EB welding studies have been carried out on copper plates and pipe mock-ups. A typical EB welded copper pipe mock-up is shown in Figure 8. High quality single- pass square butt EB welds with adequate weld penetration (>30 mm) can be readily made in oxygen-free (OF) copper using a wide range of beam power/welding speed combinations. The recommended joint design for closure by EB welding and a cross-section of a typical OF copper EB weld are shown in Figures 9 and 10 respectively. Surface defects such as blow holes, undercut and erratic metal overflow are promoted by high power/high welding speed combinations and appear to be the limiting factor in welding parameter selection for OF copper. It is extremely difficult to control gas porosity in electrolytic tough-pitch (ETP) copper EB welds due to its relatively high and variable level of dissolved oxygen.

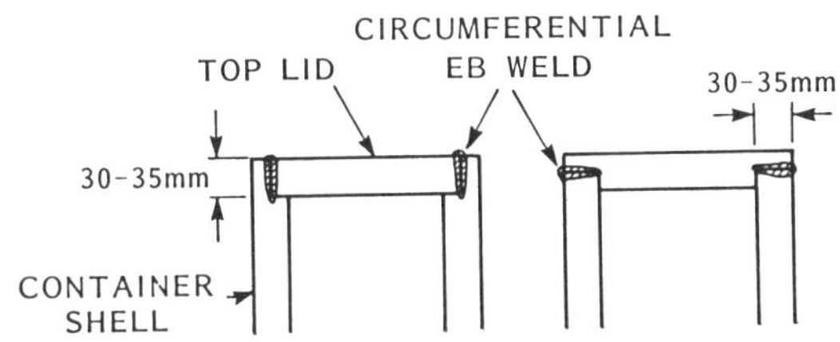

Fig.9 - Recommended weld joint design for the final closure welding of thick-walled supported-shell copper containers using the electron beam welding process.

Fig.10 - Cross-section of a typical oxygen-free copper electron beam welded sample, mag 3.5X.

Sound ETP copper EB welds can only be made using a narrow range of beam power/welding speed combinations. Future EB welding studies will focus on the development of proper weld initiation and termination procedures to ensure sound quality in the overlap area of circumferential welds. Repair welding procedures will also be investigated. Since EB welding is conducted under high vacuum, the possible adverse effects of out-gassing from fuel bundles, will also be examined.

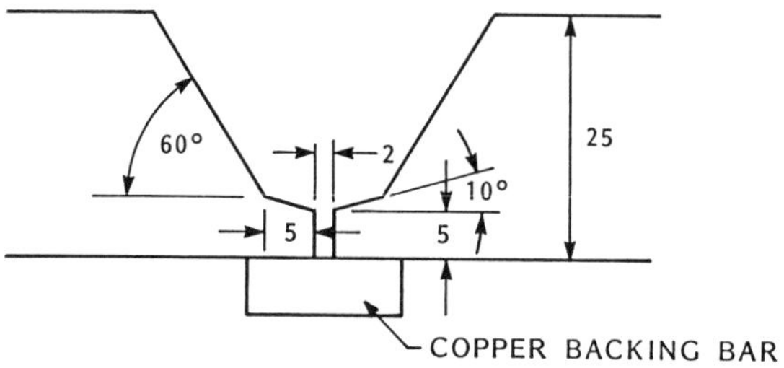

Fig.11 - Recommended weld joint design for the final closure welding of thick-walled supported-shell copper containers using the gas-metal-arc welding process.

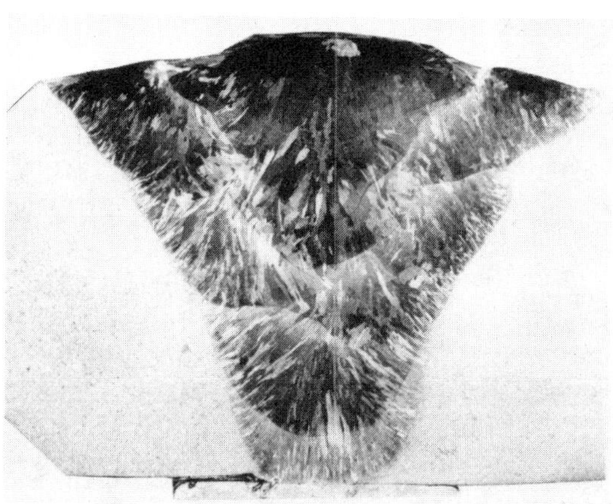

Fig.12 - Cross-section of a typical oxygen-free copper gas-metal-arc welded sample, mag 2X.

Suitable gas-metal-arc (GMA) welding procedures have been developed for the final closure of 25 mm thick copper containers/19,20,21/. Test results to date show that high welding current (>800A) and thus high heat input may increase the hot cracking tendency in copper GMA welds, particularly when high restraint and relatively low preheat (<350°C) are imposed on the weld joint. High current densities can be obtained with the use of small diameter (1.6 and 2.4 mm) electrode wire, which improves the molten metal flow into the groove and wetting to the previous weld passes and the side wall. Sound copper welds can then be made at relatively low current (500 to 600A) and thus at lower heat input. However, suitable selection of preheat in the range of 400° to 600°C is required to prevent formation of lack-of-fusion defects. The recommended weld joint design and a cross-section of a typical copper GMA weld are shown in Figures 11 and 12, respectively.

WELDING OF INCONEL 625

Resistance-heated diffusion bonding has been shown to be unsuitable for joining 6.4 mm thick Inconel 625 plates due to the rapid formation of an oxide film at the bonding interface. EB welding is a feasible process for joining thick plates of Inconel 625 (>12.7 mm)/13/; the cross section of a welded sample is shown in Figure 13. GTA welding procedures for nickel-based alloys are already established and are commonly used in industry, ie there is much more experience of GTA welding Inconel 625 than titanium. Further studies will focus on the optimization of existing GTA welding procedures for Inconel 625.

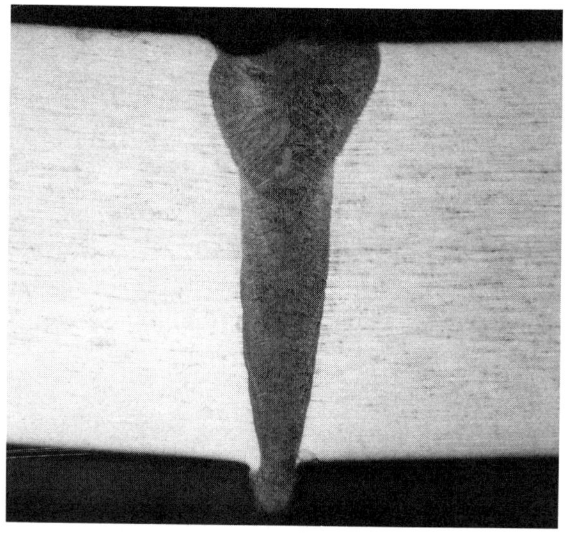

Fig.13 - Cross-section of a typical Inconel 625 electron beam welded sample, mag 5X.

DEVELOPMENT OF REMOTE CLOSURE-WELDING SYSTEMS FOR NUCLEAR FUEL WASTE DISPOSAL CONTAINERS

The automation needs for gas-shielded arc welding, resistance-heated diffusion bonding and EB welding for the final closure of nuclear fuel waste disposal containers have been identified.

In collaboration with Ontario Hydro's automated welding program/22/, the development of a computer-controlled, remote-arc welding capability is in progress, particularly concerning GTA welding of titanium and GMA welding of copper. Areas being investigated include computer-controlled seam and surface tracking, and closed-loop feedback control of welding parameters.

Commercially available resistance welding and EB welding machines have proven very precise and reliable. Various commercial control systems are available for in-process monitoring of welding parameters. Automatic seam tracking, which is available for EB welding, provides the ability to follow the joint accurately and minimizes the occurrences of missed joints, ie the electron beam being misaligned with respect to the centre of the weld joint. Test results to date show that, for resistance-heated diffusion bonding and EB welding, an advanced welding system (ie one with in-process quality monitoring and automatic seam tracking capabilities), combined with complete ultrasonic inspection of finished joints, would ensure the soundness of final closure welds. For these two processes, the development of automatic quality correction systems to adjust bonding and welding parameters according to changes in external conditions does not seem necessary. In future work, these processes will be fine-tuned to determine the welding/bonding parameter control range for in-process quality monitoring. The end results will be used to select the most suitable commercially available welding system.

SUMMARY REMARKS

Feasibility studies of welding/bonding processes have been completed for the fabrication and closure welding of various nuclear fuel waste disposal container designs under consideration in the Canadian Nuclear Fuel Waste Management Program. Resistance-heated diffusion bonding and pulsed-current gas-tungsten-arc welding have been shown to be feasible and promising for the final closure of thin-walled titanium containers. Electron beam welding appears to be an ideal process for the final closure of thick-walled copper containers. Fine-tuning of welding procedures, and the development of repair welding techniques are continuing on small-scale container mockups. The automation needs for various welding/bonding processes under study have also been identified. The development of a remote-controlled arc welding capability is in progress, and includes seam tracking and closed-loop feedback for welding parameter control, particularly for gas-tungsten-arc welding of titanium and gas-metal-arc welding of copper.

REFERENCES

1. Cameron, D.J., 1982, 'Fuel Isolation Research for the Canadian Nuclear Fuel Waste Management Program', Atomic Energy of Canada Ltd Report, AECL-6834.

2. Cameron, D.J., Crosthwaite, J.L. and Nuttall, K., 1983, 'The Development of Durable, Man-Made Containers for Fuel Isolation', Canadian Metallurgical Quarterly, Vol 22, No 1, pp 87-101.

3. Crosthwaite, J.L., Barrie, J.N. and Nuttall, K., 1982, 'Design, Fabrication and Testing of a Stressed-Shell Fuel Isolation Container', Atomic Energy of Canada Ltd Report, AECL-6823.

4. Mikasinovic, M. and Hoy, R., 'Design of Packed Particulate Supported Containment for Irradiated Fuel Immobilization', Proceedings of International Conference on Radioactive Waste Management, Winnipeg 1982, September 12-15.

5. Teper, B., 'Preliminary Evaluation of the Design of a Particulate-Packed Thin-Wall Container for the Disposal of Irradiated Fuel Bundles', Proceedings of International Conference on Radioactive Waste Management, Winnipeg 1982, September 12-15.

6. Nuttall, K. and Urbanic, V.F., 1981, 'An Assessment of Materials for Nuclear Fuel Immobilization Containers', Atomic Energy of Canada Ltd Report, AECL-6440.

7. Maak, P.Y.Y., 1985, 'Progress in Welding Studies for Canadian Nuclear Fuel Waste Disposal Containers', Ontario Hydro Research Division Report 85-98-K.

8. Maak, P.Y.Y., 1982, 'Effect of Air Contamination in the Argon Shielding Gas on the Mechanical Properties of Titanium Gas-Tungsten-Arc Welds', Atomic Energy of Canada Ltd Technical Record, TR-282.

9. Maak, P.Y.Y. and Shui, R., 1984, 'Titanium Resistance/Diffusion Bonding Studies - Phase One', Atomic Energy of Canada Ltd Technical Record, TR-254.

10. Maak, P.Y.Y., 1984, 'Titanium Resistance/Diffusion Bonding Studies - Phase Two', Atomic Energy of Canada Ltd Technical Record, TR-296.

11. Maak, P.Y.Y., 1985, 'Resistance/Diffusion Bonding of Titanium Rings', Ontario Hydro Research Division Report 85-126-K.

12. Cooper, M., Bristol Aerospace Ltd, 1984, 'Diffusion Bonding Development for the Lower Joint of Spent Fuel Disposal Container', Atomic Energy of Canada Ltd Technical Record, in preparation.

13. Maak, P.Y.Y., 1981, 'Bristol Aerospace Ltd - Titanium and Inconel 625 Electron Beam

Welded Coupons', Ontario Hydro Research Division Report, B81-124-K.

14. Maak, P.Y.Y., 1984, 'Electron Beam Welding for Copper Used Fuel Disposal Containers - Phase One', Atomic Energy of Canada Ltd Technical Record, TR-252.

15. Maak, P.Y.Y., 1984, 'Electron Beam Welding of Thick-Walled Copper Containers for Nuclear Fuel Waste Disposal - Phase Two', Atomic Energy of Canada Ltd Technical Record, TR-253.

16. Maak, P.Y.Y., 1985, 'Electron Beam Welding of Thick-Walled Copper Containers for Nuclear Fuel Waste Disposal - Phase Three A', Ontario Hydro Research Division Report 85-100-K.

17. Maak, P.Y.Y.,1985, 'Electron Beam Welding of Thick-Walled Copper Containers for Nuclear Fuel Waste Disposal - Phase Three B', Ontario Hydro Research Division Report 85-125-K.

18. Sanderson, A., Saluka, T.F., Turner, J.L. and Leggatt, R.H., 1983, 'Feasibility Study of Electron Beam Welding of Spent Nuclear Fuel Containers', SKBF Technical KBS report, TR83-25.

19. Ko, K., Bryan, B. and Ditschun, A., 1983, 'Welding of Copper for Radwaste Program', Welding Institute of Canada Report, RC-110 to Atomic Energy of Canada Ltd.

20. Maak, P.Y.Y., 1982, 'Preliminary Study of Gas-Metal-Arc Welding of 25 mm Thick Oxygen-Free Copper', Ontario Hydro Research Division Report, 81-459-K.

21. Maak, P.Y.Y. and Galloway, J.G., 1986, 'Gas Metal Arc Welding Studies for Copper Nuclear Fuel Waste Disposal Containers', Ontario Hydro Research Division Report 86-35-K.

22. Tinkler, M.J. and McNabb, S.C., 1985, 'Development of a Robotic System for the In-Situ Repair of Hydraulic Turbine Runners', Ontario Hydro Research Division Report 85-121-H.

WELDABILITY OF LANTHANIDE-METAL-TREATED 2¼Cr-1Mo STEELS

E.F. Nippes, J.P. Madeka
Rensselaer Polytechnic Institute
Troy, New York 12180-3590

G.A. Ratz
Molycorp, Incorporated
Pittsburgh, Pennsylvania

ABSTRACT

Investigators have shown that the susceptibility of 2¼ Cr - 1 Mo steels to temper embrittlement (TE) is reduced with additions of lanthanide (rare-earth) metals (LnM). This investigation is designed to compare the weldabilities of 2¼ Cr - 1 Mo steels with and without La and LnM additions. Preliminary results of Varestraint tests indicate that LnM additions lower segregation associated hot-cracking. Additions of 0.069 wt % LnM were found to be optimum for the steels used in this investigation.

2¼ Cr - 1 Mo STEELS are widely used in petroleum hydroprocessing pressure vessels. Typical process environments subject these steels to high hydrogen pressures, at temperatures reaching 454°C (850°F). Some reactors have operated for well over a decade, with only periodic shutdowns.[1] However, 2¼ Cr - 1 Mo steels are susceptible to temper embrittlement (TE) and hydrogen-induced cracking (HIC). These phenomena, together with stresses associated with rapid cooling of these reactors to ambient temperatures, may lead to the onset of brittle fracture. Weldments are particularly susceptible to these problems[1] because of the segregation and microstructural changes associated with welding.

Temper embrittlement is manifested by an increase in the ductile-to-brittle transition temperature (DBTT). This occurs if 2¼ Cr - 1 Mo steels are held within, or slowly cooled through, a temperature range of about 371 to 566°C (700 to 1050°F).[1] Although the toughness of these steels is sufficient at operating temperatures, cooling to near-ambient temperatures, may result in a ductile-to-brittle transition. Research into the mechanisms of TE continues, but it is generally accepted that TE is associated with segregation of residuals; mainly P, and to a lesser extent Sb, Sn, and As, (hereafter referred to as 'tramp' elements) to prior austenite grain boundaries. This segregation is enhanced with common alloying additions, namely; Mn, Si, and to a lesser extent Ni, and Cr. Also, the reduction in matrix Mo content due to Mo carbide formation reduces the Mo available to 'getter' P.[2,3].

Previous investigations have shown that LnM additions to 2¼ Cr - 1 Mo steels reduce TE susceptibility as follows:

1.) LnM formed complex, high melting-point, oxi-sulfide compounds with P, and tramp elements.[4-6]
2.) Additions of La alone was the most beneficial.[4-6]
3.) Optimum LnM amounts were well below those dictated from theoretical calculations for a given set of conditions.[4]
4.) Additions of LnM above the optimum amounts lowered CVN upper-shelf energies.[4-6]

Hydrogen from the process environment, and stresses associated with reactor shutdown, may cause HIC. Also, hydrogen and stresses associated with welding may lead to HIC. Investigators have observed a synergistic effect of TE and HIC in 2¼ Cr - 1 Mo steels.[7,8] Additionally, Savage, et al.,[9] found that hydrogen-induced cracks can be nucleated by previously liquated grain boundaries and/or liquated nonmetallic inclusions in HY-80 steel weldments. They determined that small LnM additions to HY-80 steels reduced HIC susceptibility. Therefore, it may be possible to reduce the HIC susceptibility of 2¼ Cr - 1 Mo steels, and their welds, with LnM additions.

The object of this research program is to compare the weldabilities of 2¼ Cr - 1 Mo steels, with and without La and LnM additions. The specific areas of research will include: hot cracking, heat-affected zone (HAZ) toughness, and HIC.

Because 2¼ Cr - 1 Mo weldmetal is most susceptible to TE; the Varestraint[10] test will be used to see if a relationship exists between hot cracking, and TE associated elements such as P, Sb, Sn, and As. Previous investigators[10] have observed a reduction in hot-cracking susceptibility of alloy steels with LnM additions. This is generally attributed to the scavenging of solute by LnM. This reduces the amount of low surface-energy (low melting-point) constituent that segregates or forms at sub-grain interfaces and/or grain boundaries.

This progress report will only be concerned with the areas of hot cracking and HAZ toughness, because of the limited work thus far performed in the area of HIC.

EXPERIMENTAL PROCEDURE

MATERIALS - 13-mm (0.5-in.) thick plates from five 227-kg (500-lb.) experimental heats were investigated. These plates were supplied by the University of Pittsburgh, where a concurrent program is being conducted on base-material properties.[4] Chemical compositions of the five heats are given in Table 1. Ingots were hot-rolled to approximately 178-mm wide by 914-mm long (7-in. x 36-in.) plates, and air cooled. The plates were then austenitized at 954°C (1750°F) for 1 hr., water quenched, tempered 704°C (1300°F) for 1 hr., and air cooled. Next, the plates were divided, and one-half of each plate was subjected to a long isothermal tempering treatment of 520°C (968°F) for 1000 hrs., followed by air cooling. The purpose of this treatment was to induce TE. The average ASTM grain size number for both heat-treatment conditions was approximately 8.5. Approximate mechanical properties for these steels are given in Table 2. Heat C was found to have the optimum Lanthanide metal addition (0.023 w/o La) for the reduction of TE.

HOT CRACKING - The Varestraint Test [10] has been extensively used in determining weld hot-cracking susceptibility of different alloys. Test specimens, 51-mm (2 in.) wide, were cut transverse to the rolling direction of the parent plates. These speicmens were ground, lapped, and finally ground with 600 grit paper. Prior to welding, the specimens were de-magnetized and cleaned with acetone. Gas-tungsten arc (GTA) welding parameters used are given in Table 3. Two augmented strains were chosen, nominally 2 and 4%. Corrected strains, calculated from average material thicknesses, were 1.9 and 3.8%, respectively. A minimum of two specimens per strain were tested. A stereomicroscope with a filar eyepiece was used to measure as-welded hot cracks at 60X.

HAZ TOUGHNESS - A Model 1500 Gleeble thermomechanical simulation device was used to simualte the coarse-grained HAZ in CVN specimens. A thermal cycle for a 20 kJ/mm (50 kJ/in.) weld on 12.7 mm (0.5 in) plate was used. The peak temperature of this thermal cycle was 1300°C (2372°F). A 150°C preheat was incorporated into this thermal cycle. However, no PWHT, typically 690°C (1275°F) for a minimum of 11 hrs, was used. Oversize CVN blanks were cut in the transverse-longitudinal direction. They were thermally cycled in air, and then ground and notched to ASTM specifications. The limitations of this program precluded the development of entire impact-toughness curves. Therefore, impact tests were performed at three significant temperatures: -40°C (-40°F), and 10°C (50°F), test temperatures specified in the ASME boiler code, and 100°C (212°F), the expected upper-shelf temperature. A minimum of three specimens per temperature were tested at -40 and 10°C, and a minimum of two specimens were tested at 100°C.

RESULTS AND DISCUSSION

VARESTRAINT - Varestraint results as measured on as-welded specimens, are shown in Figs. 1 and 2. All hot cracks were located in the weld metal. In general, the propensity for hot cracking, as measured by total crack length (TCL), was lower for heats with LnM additions. Maximum crack lengths were on the order of 400 μm (16 mils), and did not vary significantly from heat to heat. There was an excellent agreement in the trends observd at 1.9 and 3.8% strain, except for TCL in heats A and B. The optimum level of LnM was about 0.069 w/o (Heat D), as cracking susceptibility increased with higher LnM contents. Hot cracks in as-welded specimens of heats A and B were morphologically different to those formed in LnM containing heats. A hot crack from a polished Varestraint specimen is shown in Fig. 3. Energy dispersive spectroscopy (EDS) of the white-etching regions (Fig. 3) revealed the presence of P in higher-than-matrix concentrations.

These results indicate that the Varestraint test may be a means of evaluating the propensity of detrimental P to segregate in weldmetals. The reduction in hot-cracking susceptibility, as measured by TCL, is explained as mentioned in the Introduction. This may also explain the different hot-crack morphologies.

No obvious differences in weld-bead width and geometries were observed. The high currents used (∿320 Amps) may have dominated any surface-tension effects of LnM additions. Therefore, pool geometry differences which may lead to different stress configurations around the weld pool (during Varestraint loading[11]), probably did not exist. An adherent, dark-gray oxide formed on weld beads on heats D and F, but not on weld beads on heats A and B. Heat C weld beads contained an intermediate amount of this oxide. Preliminary auger-electron spectroscopy (AES) was performed on an oxide found on a Heat F weld bead. AES spectra were observed before and after minor surface sputtering, and indicated an absence of LnM. This observation is presently unexplained, however, it may result from an oxide 'doping' phenomenon.

HAZ TOUGHNESS (Preliminary Results) - The as-welded (no PWHT) HAZ produced by synthetic thermal cycles was bainitic and had an average ASTM grain size number of 6 ± 0.5. Preliminary CVN results are shown in Table 4 and Fig. 4. In general, the 100°C impact energy was lower for heats with LnM additions. Heat B had the highest 100°C impact energy, while Heats D and F had the lowest impact energies at this temperature. Heat B had higher impact toughness than Heat A at all temperatures. This probably results from the lower C content (Table 1) and lower resulting hardness of Heat B.(Table 4) The reduction of as-welded (no PWHT) 100°C HAZ toughness is associated with LnM-oxi-sulfide inclusions. Previous investigators have shown this effect in base-metal toughness of similar steels.[4-6] SEM fractographs of CVN specimens from Heats B and F, impact tested at 100°C, are shown in Fig. 5. The increased inclusion contents and dimple sizes, which result from LnM additions are clearly shown.

The as-welded (no PWHT) HAZ impact toughness, and quenched-and-tempered (Q&T) base plate toughness for Heat F, are shown in Fig. 6. A 90°C (194°F) increase in the 54J (40 ft-lb) transition temperature results from thermally cycling the base plate. However, there is no significant difference in impact toughness of the synthetic HAZ produced from Q & T and TE plates. This probably results from the resolutionizing of P and tramps previously located at austenite grain boundaries. This would produce a reversal of TE. A continuing investigation will explore the effects of PWHT, step cooling, and higher HAZ peak temperatures, i.e. 1350°C (2462°F), on HAZ toughness.

CONCLUSIONS

The conclusions of this preliminary investigation are:

1.) Hot-cracking susceptibility of 2¼ Cr - 1 Mo steels, as measured by the Varestraint test, decreased with LnM additions.
2.) Additions of 0.069 wt % LnM were found to be optimum for the reduction of hot cracking in the steels used in this investigation.

ACKNOWLEDGEMENT

This research is funded by Molycorp, Inc., a subsidiary of Union Oil of California. It is being conducted in association with a primary program conducted by C. I. Garcia and G. A. Ratz at the Univeristy of Pittsburgh. The authors also wish to thank J. P. Balaguer for his technical advice and assistance.

REFERENCES

1. Erwin, W. E., and J. G. Kerr, Welding Res. Council Bul. 275 (2), 4 and 32-39. (1982).
2. Jin Yu and C. J. McMahon, Jr., Parts 1 and 2, Met. Trans. A 11 (2), 277-300 (1980).
3. IBID
4. Garcia, C. I., G. A. Ratz, M. G. Burke, and A. J. De Ardo, J. of Metals 37 (9), 22-28 (1985).
5. Knight, R. F., W. R. Tyson, and G. I. Sproule, Metals Tech. 11 (7), 273-279 (1984).
6. Seah, M. P., P. J. Spencer, and E. D. Hondros, Metal Sci., May, 307-314 (1979).
7. Murakami, Y., T. Monura, and J.

Watanabe, ASTM STP 755, 383-417, Edited by G. S. Sangdahl and M. Semchyshen (1982).

8. Takahashi, E., et al., Preprints of Nat. Meeting of Japan Weld. Soc. 26 (1980), as Referenced by: G. R. Prescott, Metals Progress, July 24-30 (1981).

9. Savage, W. F., E. F. Nippes, and H. Homma, Weld. J. (Res. Suppl.) 55 (11), 368s - 376s (1976).

10. Savage, W. F., C. D. Lundin, Weld J. (Res. Suppl.) 44 (10), 433s to 442s (1965).

11. Private Communication; C. D. Lundin, University of Ten. (4/8/86).

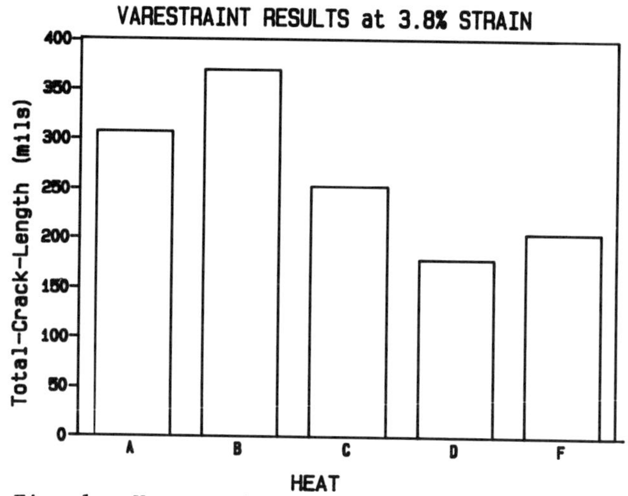

Fig. 1 - Varestraint results at 3.8% augmented strain.

TABLE 1

CHEMICAL COMPOSITIONS OF STEELS USED IN THIS INVESTIGATION (W/o)(1)

HEAT	C	P	Sb	Sn	As	Ce	La	TOTAL(2) LnM
A (base, 0.006P)	0.13	0.006	<0.002	<0.002	<0.002	<0.002	<0.006	<0.008
B (0.021P + Tramps)	0.103	0.021	0.007	0.016	0.024	<0.002	0.009	0.009
C (0.015P + T + 0.023La)	0.14	0.015	0.007	0.011	0.025	<0.002	0.023	0.023
D (0.020P + T + 0.069LnM)	0.13	0.020	0.007	0.010	0.026	0.048	0.021	0.069
F (0.023P + T + 0.085LnM)	0.104	0.023	0.006	0.016	0.025	0.043	0.028	0.085

(1) In addition, all steels contained: S-0.003/0.006, Mn-0.59/0.63, Si-0.14/0.24, Cr-2.21/2.44, Ni-0.11, Mo-0.94/1.03, Al-0.013/0.032, N_2-0.004/0.005

(2) LnM denotes Lanthanide (Rare Earth) Metal (50% Ce, 32% La, 14% Nd, 4% Pr, and balance other Lanthanides) added as michmetal (∿96%LnM), or added as Lanthanide (Rare Earth) Silicide (∿30%LnM).

TABLE 2

APPROXIMATE MECHANICAL PROPERTIES OF STEELS USED IN THIS INVESTIGATION(1)

HEAT TREATMENT	YS, N/mm^2(ksi)	UTS, N/mm^2(ksi)	RA(%)	e(%)	HV
Q&T	600 (87)	758 (109.9)	73.1	21	244
TE	639 (92.6)	742 (107.6)	69.6	23	249

(1) From C. I. Garcia, University of Pittsburgh

TABLE 3

VARESTRAINT GTA WELDING PARAMETERS

Electrode:	3.2mm (1/8 in.) W - 2% Th
Electrode Tip:	Conical, 90° included angle
Electrode Extension:	23.8mm (0.938 in.)
Working Distance:	1.98mm (0.078 in.)
A_r Gas Flow Rate:	16.5 L/min (35 CFH)
Travel Speed:	10.2 cm/min (4 in./min)
Arc Current:	∿320 Amps
Arc Voltage:	∿11.7 Volts
Approx. E.I.(1):	22.1 kJ/mm (56.1 kJ/in)

(1) GTAW efficiency not considered

TABLE 4

AS WELDED (NO PWHT) HAZ CVN DATA

HEAT(1)	Fracture Area %Shear -40°C	-10°C	100°C	Approximate 54J(40 ft-lb) TT (°C)	100°C Impact Energy, J (ft-lbs)	Hardness (HV)
A	13	36	100	-35	128 (94)	372
B	15	46	100	-38	152 (112)	334
C	15	37	100	-28	110 (81)	381
D	12	38	100	-18	94 (69)	372
F	0	30	100	14	88 (65)	336
F(E)	0	27	100	14	96 (71)	353

(1) All specimens from Q&T base steel, except those from heat F(E) which was TE base steel.

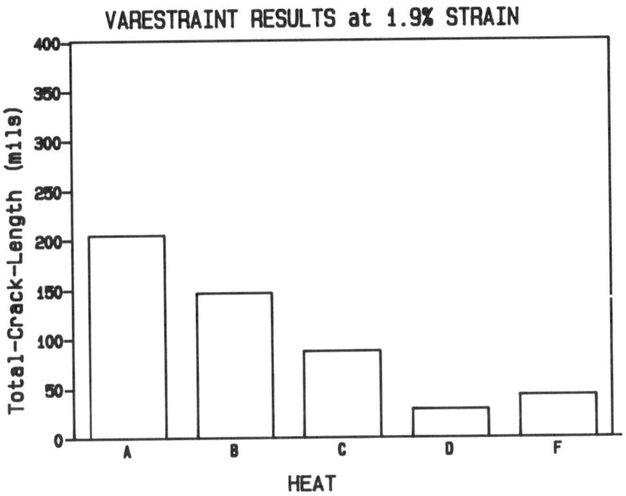

Fig. 2 - Varestraint results at 1.9% augmented strain.

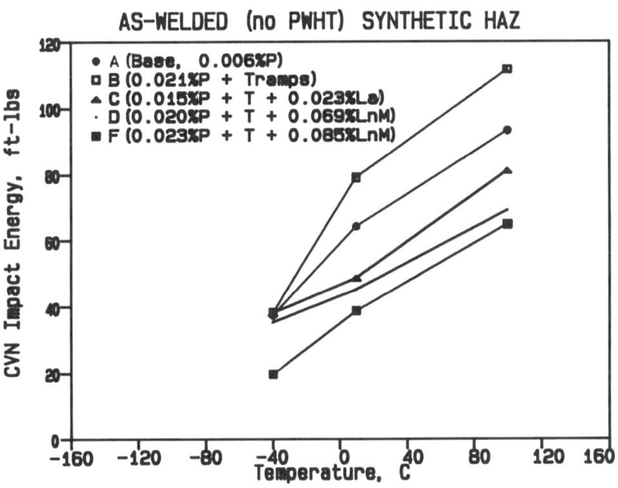

Fig. 4 - CVN results for synthetically produced, as welded, HAZ.

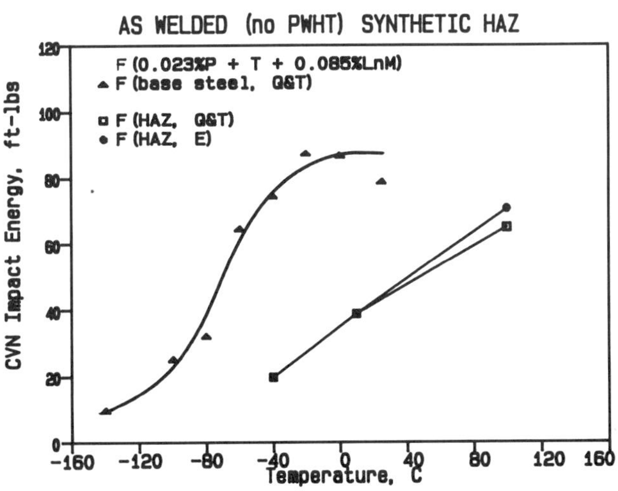

Fig. 6 - Heat F, CVN results for the base steel (provided by Ref.4), and as-welded HAZ. F(HAZ, E) - previously temper-embrittled base steel.

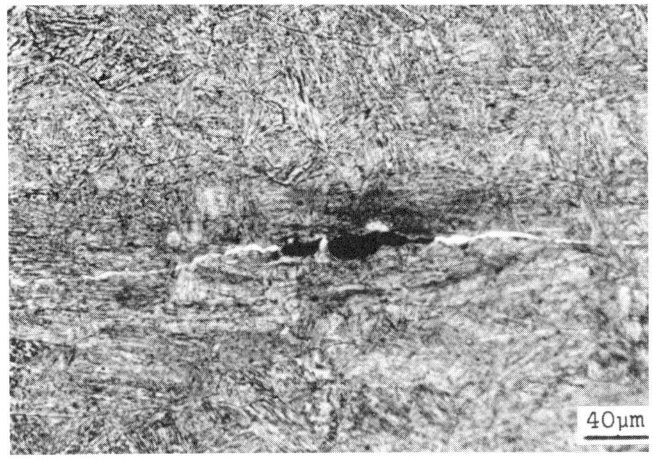

Fig. 3 - Optical micrograph of a polished and etched (2% Nital) hot crack, Heat F.

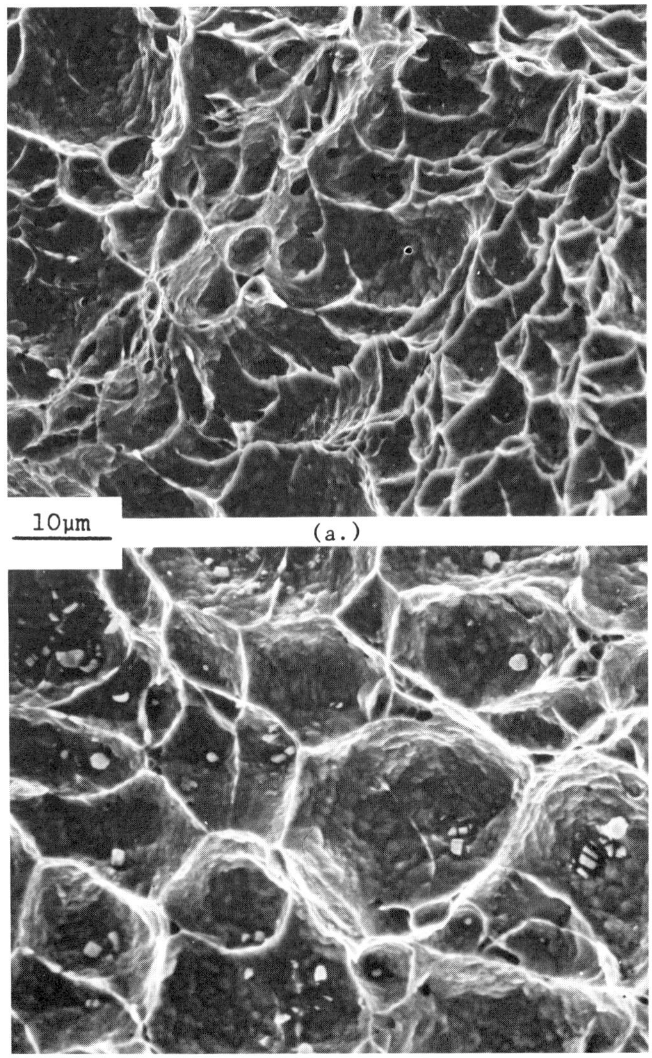

(a.)

(b.)

Fig. 5 - SEM fractographs of as-welded HAZ (no PWHT) CVN specimens, tested at 100 °C . (a.) - Heat B, (b.) - Heat F.

WELD BEAD GEOMETRY PREDICTION FOR PULSE MIG WELDING

R. Senthil Kumar
Hindustan Motors Limited
Calcutta (West Bengal)

R.S. Parmar
Indian Institute of Technology
New Delhi (India)

ABSTRACT

Mathematical models have been developed, for pulse mode on Synergic MIG welding power source, to predict weld bead geometry and shape relationships as affected by peak current and its time period, wire feed rate, welding speed, and the plate thickness. The equations can be utilized for predicting the weld bead geometry or conversely for predicting control parameters for the required bead profile. The relationships can be usefully employed for automated or robotic welding systems.

METAL INERT GAS (MIG) welding introduced to the industry in 1940's[1] has a number of advantages to its credit e.g. high reliability,all position capability, ease of use, low cost,and high productivity [2]. The process can be used with nearly the same ease both for ferrous and non-ferrous materials. However, to get high quality welds it is essential to have complete control over the mode of metal transfer.

Conventionally three modes of metal transfer are recognised viz., short-circuit, globular, and spray [1]. Spray mode is preferred because it can be used for all position welding, gives lower spatter, and keeps the power source evenly loaded. This mode for a given wire diameter however is achieved only if the welding current is above a certain value termed as threshold current [3]. This puts a limitation on the process in that it cannot be used for thin sheets thus necessitating the use of short-circuit mode of metal transfer which can be effectively used for the purpose as well as for the root runs but it is accompanied by high spatter and fluctuating load condition on the power source. Globular transfer which is essentially an inefficient mode of metal transfer is used only for increasing the deposition rate in welding thin sections.

To overcome the above mentioned difficulties of use of spray mode for welding thin sheets and root runs,pulse MIG welding was introduced in 1960's [4]. In this process the background current is kept low and detachment of metal predominantly through pinch effect is obtained by increased current at regular intervals. This makes the process achieve spray mode at a mean current as low as 50A[5] making it a very versatile process for all position welding and a wide range of material thickness even with poor fit-up. The process can be used both for semi-automatic and automatic welding.

With the present trend towards automation [6] and robotic welding it is imperative to use only that process which can cover a wide range of material thickness and be applicable for all position welding with least spatter. Obviously, the Pulse MIG welding with spray mode of metal transfer seems an attractive choice.

However, to get the desired quality welds it is essential to have a complete control over the relevent welding parameters to obtain the required weld bead geometry and shape relationships on which is based the stress carrying capacity of a weldment [7]. For the automatic welding system to use this data it must be available in the form of equations. This paper presents the results of investigations to develop mathematical models for predicting weld bead geometry for pulse mode using one of the latest welding power sources called the Synergic MIG welding system.

SYNERGIC MIG WELDING POWER SOURCE

In conventional pulse MIG welding the droplet transfer frequency is some simple multiple of the mains frequency and the metal transfer is during the pulse period [8,9]. Depending upon the wire feed rate the mean current level is adjusted by setting the background current, the pulse current and their time periods. These settings are required to be accurately controlled to achieve the desired mode of metal transfer. Thus the number of pulses required per unit time can be anything depending upon the droplet size. But because the pulses per unit time are fixed it becomes quite cumbersome to set up an optimal set of conditions to get the desired mode of metal transfer.

In fact for a given material and wire dia. there is a set of unique pulse parameters, called the unit pulse, which are required to detach one droplet per pulse. It is obviously very difficult to set the unit pulse for varying requirements by using the conventional pulse MIG welding power source [10].

To overcome these difficulties a new power source called synergic MIG has been developed which provides what is known as synergic pulsing. This is a transistorized power source in which the pulse current and the background current levels and their time periods can be controlled independently over a frequency range of 15 to 500 Hz making it a very versatile power source to achieve the desired mode of metal transfer [8,10,11]. Block diagram of one such power source, the control circuitory of which is based on Z-80 microprocessor, is shown in Fig. 1.

DESIGN OF EXPERIMENTS

The design of experiments included the following steps.

1. Selection of control variables
2. Selection of limits of variables and the constants of the experiments.
3. Development of design matrix.

These steps are described briefly as follows.

SELECTION OF CONTROL VARIABLES - Welding procedures are quite often established by trial and error within narrow ranges of parameters involved and by varying one parameter at a time.

Superior to trial and error method is the tolerance box technique which can be valuable for selecting process settings but does not provide any data on effefts of interaction between parameters. It is also expensive to carry out [12].

Statistically designed experiments based on 'fractional factorial technique' not only reduce the cost but also gives the required information about the main and the interaction effects on the response factors. It was therefore decided to use this technique to design the experiments [12,13].

In synergic pulse MIG welding process there are many variables affecting the weld geometry. However, it was planned to study the effects of those parameters which affect more definitely the mode of metal transfer and the amount of heat input to the workpiece and thus decide the bead geometry and shape relationships.

Based on their importance the welding variables were put in two groups as follows.

Primay Variables - Pulse current, pulse time, wire feed rate (WFR), welding speed, feed wire diameter, shielding gas composition.

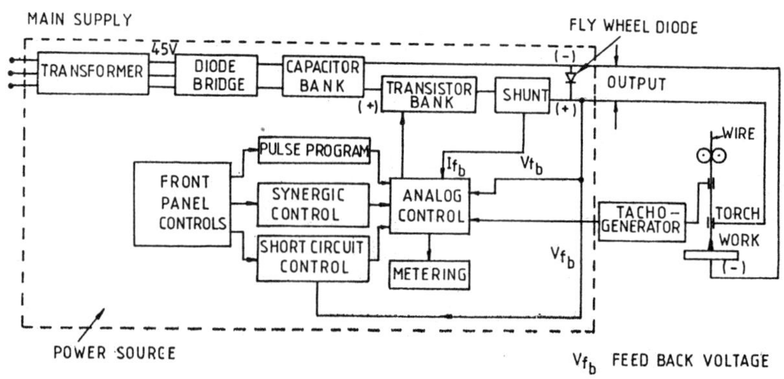

Fig. 1 Synergic power source block diagram

Secondary Variables - Background current level and time period, mean current, volt-ampere ratio, shielding gas flow rate, torch angle, electrode polarity, preheat condition, burn back, rise time, short circuit detect, inductance level, short-circuit current, arc length, and electrode stickout.

Out of these parameters only those primary variables were chosen which had no dependence on each other. Based on this consideration pulse current, wire feed rate, pulse time, welding speed and plate thickness were selected as the control variables.

LIMITS OF VARIABLES AND CONSTANTS OF EXPERIMENTS - The limiting values of the welding parameters and constants of the experiments were selected on the basis of the trial runs - the criterion being an apparently good bead without water cooling the torch. The limits of plate thickness were selected to cover the upper runge of medium thickness plate which are widely used in the industry. Table-1 shows the units, notations and the limits of the control variables.

TABLE 1 - Control Variables And Their Limits

Parameters	Units	Notations	Limits Low (-1)	Limits High (+1)
Pulse current	Amp	X_1	325	400
Wire feed rate	m/min	X_2	2.5	3.0
Pulse time	ms	X_3	3.5	4.5
Welding speed	cm/min	X_4	40	60
Plate Thickness	mm	X_5	10	20

Selected constants of the experiments were as follows.

Feed wire diameter	1.2 mm
Torch-to-work angle	90°
Electrode polarity	+ve
Shielding gas	Argon+5%CO_2
Gas flow rate	30 lit/min
Volt-ampere	9.5 (cc)
Burn back	6.0
Short circuit detect	3.0
Rise time	0.0 ms
Inductance level	0 Amps.
Short circuit current	450 Amps.
Nozzle-to-plate distance	20 mm
Preheat	Nil

DEVELOPMENT OF DESIGN MATRIX - Due to narrow ranges of the variables it was decided to use a two level fractional factorial design. Taking into account the main and second order interactions plus the constants of the equations a two level, 5 variable, half fractional factorial design provided $2^{5-1} = 16$ degrees of freedom which were adequate to find all the coefficients of the mathematical models [14,15].

For selecting the generating relation, a maximum possible resolution and confounding properties were considered.

The generating relation $X_5=X_1X_2X_3X_4$ has a maximum possible resolution of 5. Further, it also confounds the interacting parameters in such a way that main and second order interactions are confounded with cubic interactions.

The coefficients were determined by the following confounded relations [16].

$B_1=b_1+b_{2345}$	$B_{15}=b_{15}+b_{234}$
$B_2=b_2+b_{1345}$	$B_{23}=b_{23}+b_{145}$
$B_3=b_3+b_{1245}$	$B_{24}=b_{24}+b_{135}$
$B_4=b_4+b_{1235}$	$B_{25}=b_{25}+b_{134}$
$B_5=b_5+b_{1234}$	$B_{34}=b_{34}+b_{125}$
$B_{12}=b_{12}+b_{345}$	$B_{35}=b_{35}+b_{124}$
$B_{13}=b_{13}+b_{245}$	$B_{45}=b_{45}+b_{123}$
$B_{14}=b_{14}+b_{235}$	

Satisfying the orthogonality laws of matrix for statistically designed experiments and for a 2^{5-1} condition, the matrix which was evolved for conducting the experiments is given in Table-2.

EXPERIMENTAL PROCEDURE AND RESULTS

Welding was carried out on 220mm x 110 mm low carbon steel 'flats' with the help of a linear manipulator and using bead-on-plate technique. The weld runs, based on the design matrix, were made at random to avoid any systematic error. Selection of random numbers for experimental runs was made through the computer and are indicated in Table-2. The control variables were reset after each experimental run but the constants of the experiments were not disturbed.

Two test pieces were cut from the central zone of each of the 16 welded specimens by using a metallurgical cut-off wheel. For measuring the weld bead dimensions, the test pieces were polished by the standard metallurgical technique and etched using 2 per cent nital.

The polished and etched testpieces were examined with the help of shadowgraph. The weld bead contours, at 10X magnification were traced from the shadograph screen and one set of such bead profiles is shown in Figure 2. Weld

Table 2 - Design Matrix

Run No. Design	Run No. Exptl.	X_1	X_2	X_3	X_4	X_5	Code
1	8	-	-	-	-	+	e
2	2	+	-	-	-	-	a
3	10	-	+	-	-	-	b
4	4	+	+	-	-	+	abe
5	15	-	-	+	-	-	c
6	9	+	-	+	-	+	ace
7	1	-	+	+	-	+	bce
8	13	+	+	+	-	-	abc
9	16	-	-	-	+	-	d
10	5	+	-	-	+	+	ade
11	11	-	+	-	+	+	bde
12	14	+	+	-	+	-	lbd
13	3	-	-	+	+	+	cde
14	12	+	-	+	+	-	acd
15	6	-	+	+	+	-	bcd
16	7	+	+	+	+	+	abcde

penetration (P), width (W), reinforcement height (H) were measured with the help of travelling microscope whereas the reinforcement area and the total bead area were determined with the help of a planimeter. From these observations all the six response factors were determined and are given in Table-3.

DEVELOPMENT OF MATHEMATICAL MODELS

Mathematical models were developed by the method of regression and analysis of variance technique was utilized to determine the regression coefficients. The significance of each coefficient was checked by F-test. Models were brought to the final forms by dropping the insignificant coefficients. All these steps were carried out by developing the required computer software based on the flow chart given in Figure 3. The final models for the six response factors are as follows.

$$P = 0.95+0.066X_2-0.072X_4+0.097X_5-0.094X_{23} -0.069X_{45}$$

$$W = 5.756+0.491X_2-0.566X_4-0.269X_{34} -0.398X_{45}$$

$$H = 1.848+0.039X_2-0.152X_4-0.042X_{15} -0.113X_{45}$$

$$D = 29.63+1.09X_1-2.625X_{25}-1.057X_{34} -1.753X_{45}$$

$$WPSF = 6.228-0.069X_1+0.152X_2-0.044X_3 -0.213X_4-0.479X_5+0.265X_{13}+0.125X_{15} +0.669X_{23}-0.354X_{45}$$

$$WRFF = 3.118+0.196X_2+0.094X_{14}+0.119X_{23} +0.106X_{24}-0.15X_{34}-0.216X_{45}$$

where X_1, X_2...... are the control variables, and X_{12} the quadratic interaction between X_1 and X_2, and so on.

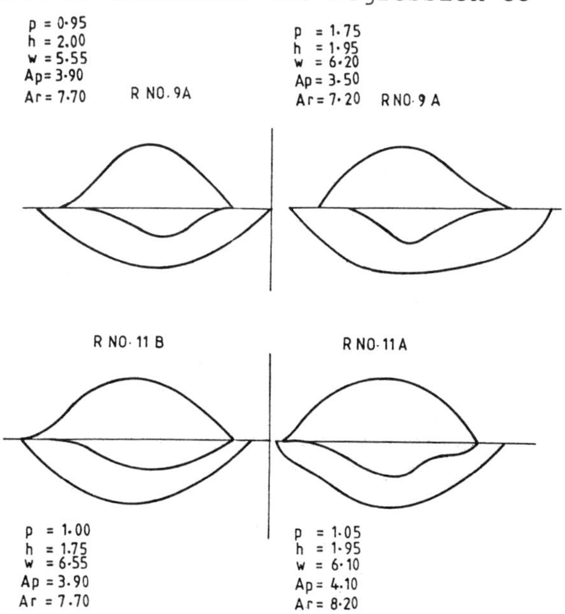

Fig. 2 Weld bead profiles

Table 3 - Calculated Mean Values of Response Factors

Design Run No.	P	W	H	D	WPSF	WRFF
1	1.100	6.225	1.800	28.05	5.659	3.458
2	0.675	5.325	2.050	21.14	7.889	2.598
3	0.900	6.925	2.075	24.84	7.694	3.337
4	1.525	6.125	2.050	36.90	4.016	2.988
5	0.975	6.125	1.875	29.63	6.282	3.267
6	1.050	5.875	1.975	33.76	5.595	2.975
7	1.025	6.775	2.125	29.68	6.610	3.188
8	0.925	7.200	2.050	27.39	7.784	3.512
9	0.750	5.075	1.625	29.66	6.767	3.123
10	0.775	5.200	1.575	26.50	6.710	3.302
11	1.025	6.325	1.850	33.48	6.170	3.419
12	1.125	5.625	1.675	35.60	5.000	3.358
13	1.025	4.300	1.875	27.39	4.195	2.293
14	0.725	4.000	1.700	36.44	5.517	2.353
15	0.750	5.250	1.600	25.59	7.000	3.281
16	0.850	5.750	1.650	23.02	6.765	3.432

All the above mentioned equations are for control variables in coded form which is based on the following relationship [16].

$$X = \frac{X_n - X_{no}}{J}$$

where, X = coded value of the factor
X_n = natural value of the factor

X_{no} = natural value of the factor at the middle level

J = variation interval with respect to middle level.

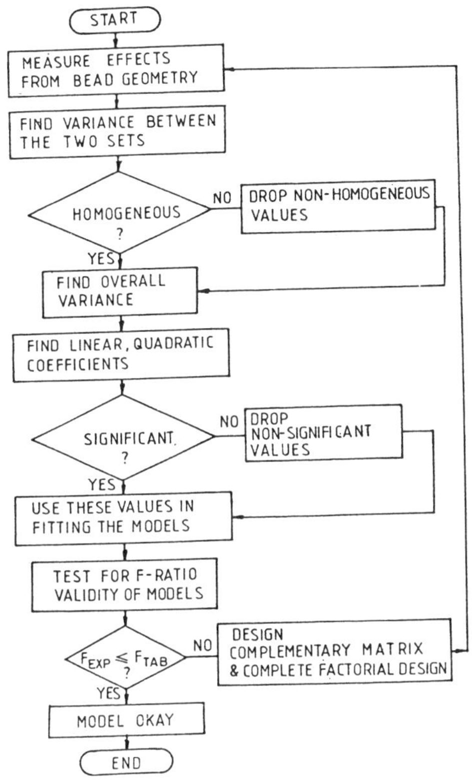

Fig. 3 Modelling Technique

ANALYSIS OF RESULTS

The evolved mathematical models can be utilized to predict the required weld bead geometry dimension or shape relationship for the range of parameters used in the investigations or conversely the required control parameters can be predicted for a given bead dimension. The results can as well be optimized for a specific response factor. Based on the evolved set of equations the response factors were computed and the results plotted. Figure 4 shows one such set of results for all the six response factors as affected by the wire feed rate and figure 5 shows the effect of interaction between wire feed rate and pulse time on penetration.

Figure 4 shows that the penetration, width, reinforcement height and percentage dilution increased with the increase in wire feed rate while WPSF and WRFF showed the opposite trends. These can be considered as fairly well expected results and good representation of the observed data.

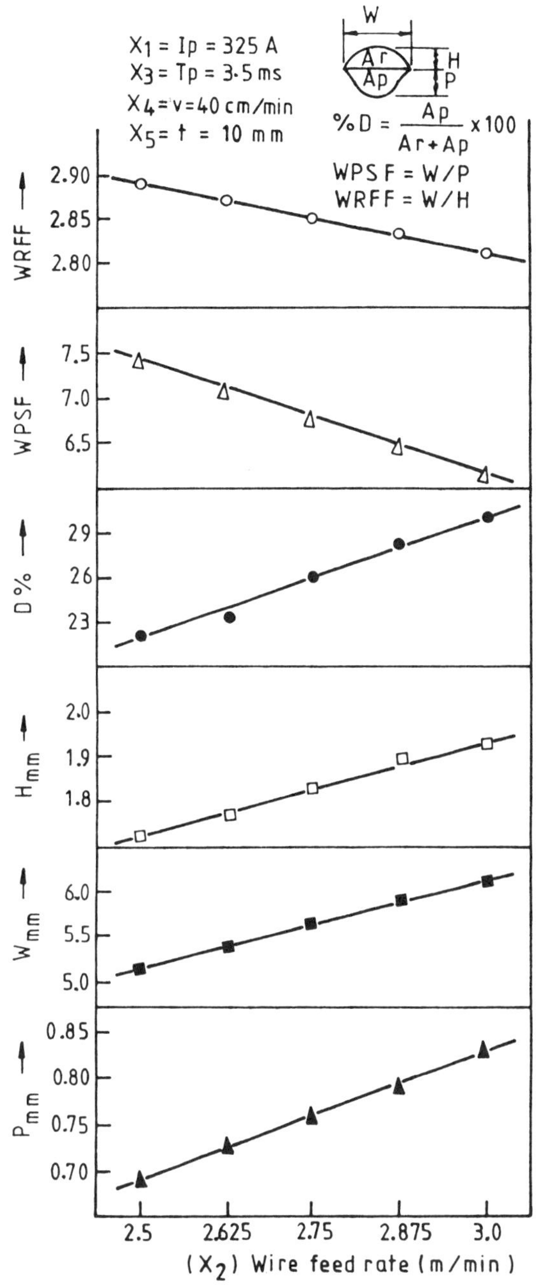

Fig. 4 Wire feed rate (WFR) vs P, W, H, D, WPSF, WRFF

Figure 5 shows the effect of wire feed rate (X_2) and pulse time (X_3) on penetration. It is evident that the penetration increased with the increase in pulse time below the WFR of 2.75m/min but had an opposite effect above that value. This could be explained on the basis of pulse current level and its time period at that level required to detach the molten metal from the tip of the feed wire and the volume of the droplet detached.

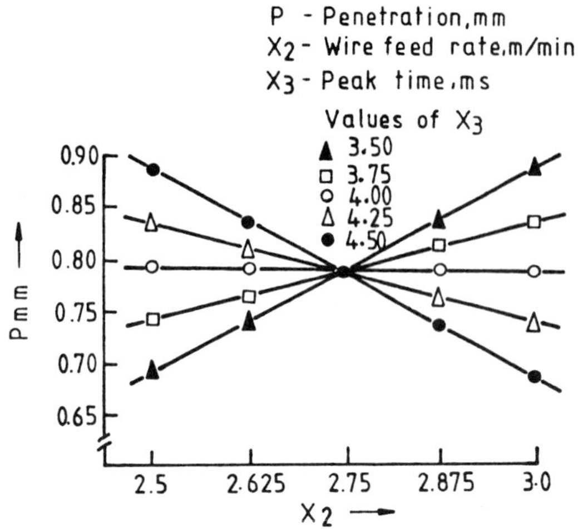

Fig.5 X_2X_3 Intraction vs penetration

CONCLUSIONS

From the development of mathematical models and the analysis of results the following conslusions were arrived at.

1. Fractional factorial technique can be effectively utilized to design the experiments for pulse MIG welding.
2. Mathematical models based on regression and analysis of variance technique can be an effective tool for prediction of weld bead geometry and shape relationships.
3. The models developed can also be utilized for predicting the values of control variables for achieving a desired weld bead profile.
4. These equations can be usefully employed for control of weld bead contours in mechanised or robotics welding systems.
5. The models can provide a vivid picture of interaction between control parameters and the response factors.
6. The low range of wire feed rate used for the experiments provide a useful data which can be utilized for making root runs in butt welds of medium thickness plates.

REFERENCES

1. Allum, C.J., Metal Construction 15, 347-353 (1983).
2. Lyttle, K.A., Welding Journal 62 15-22 (1983).
3. P.T. Houldcroft,"Welding Process Technology", p. 108, Cambridge University Press, Cambridge (1977).
4. Ma, Jilong and R.L. Apps, Welding and Metal Fabrication 51, 119-128 (1983).
5. Lucas, W., Welding Institute Research Bulletin Cambridge (1979).
6. Thorn,K., M. Feenstra, J.C.Young, W.H.S. Lawson and H.W. Kerr, Metal Construction 14, 128-133(1982).
7. Jackson, C.E.,Welding Journal 39, 225S-230S (1960).
8. Lucas, W., Welding Institute Research Bulletin, Cambridge, 304-310 (1977).
9. Amin, M., Metal Construction 15, 272-278 (1983).
10. Amin, M., Ibid., 13, 349-353 (1981).
11. Lucas, W., Ibid., 17,30-34 (1985).
12. Harris, P. and B.L. Smith, Ibid., 15, 661-666 (1983).
13. McGlone, J.C., Ibid., 14, 378-384 (1982).
14. Box,G.E.P., W.G. Hunter and J.S. Hunter,"Statistics for Experimenters", John Wiley and Sons, New York (1978).
15. O.L. Davis,"The Design and Analysis of Industrial Experiments", Longman Group Ltd., New York (1978).
16. Adler, Yu.P., E.V. Markava and Yu. V. Granovsky,"The Design of Experiments to Find Optimal Conditions", MIR Publishers Moscow (1975).

ARC AND METAL TRANSFER CHARACTERISTICS DURING THE GMA WELDING OF RARE EARTH TREATED C-Mn-Nb HSLA STEELS USING SOLID AND FLUX CORED WIRES

Madapura S. P. Murthy, Carl R. Loper, Jr.
Madison, Wisconsin

ABSTRACT

Metal transfer and arc characteristics during the GMA welding of rare earth treated C-Mn-Nb HSLA steels were studied using a high speed photographic technique and utilizing solid wire and flux cored wire containing varying amounts of rare earths. Rare earth content of steels ranged from 0.000 to 0.156%, and that of flux cored wire ranged from 0.000 to 0.061%.

Both short circuiting and globular/spray modes of metal transfer were studied with solid wires. Only the globular/spray mode was used in the flux cored wire study of the rare earth effect.

Significant differences in the arc and metal transfer characteristics were observed in the solid wire welding of the various rare earth content steels. With increasing rare earth content in the steel, the welds became ropey and generally unsatisfactory. With flux cored wires the effects of rare earths in the steel plates were moderated by both the presence of rare earths in the wire and the presence of flux per se.

A VARIETY OF RARE EARTH TREATED STEELS ARE available which offer a valuable combination of mechanical properties and technical advantages [1-3]. The optimum rare earth content of these steels is a function of the composition of the steel and the specific properties desired.

The weldability of rare earth treated steels has been studied by a number of investigators [4-10] and has included a variety of welding processes, i.e., SMAW, GMAW, GTAW and SAW. The advantages of using these steels have been tempered, however, by limitations which arise in welding them, e.g., arc instability, lack of adequate penetration, improper bead shape (ropey welds), undercutting, etc. Accordingly, many industries are concerned about specifying rare earth treated steels even though recent studies have attempted to understand the fusion zone and arc characteristics of a number of rare earth treated steels joined using the short circuiting mode of metal transfer [11,12].

GMA welding has been widely used in pipeline and automobile industry applications attractive to rare earth treated steels. The weld quality to GMAW type processes is due to the combined effect of arc and metal transfer characteristcs; and any adverse effect on the arc and metal transfer behavior is correspondingly reflected in the weld. It is well known that certain minor elements play a major role in the arc behavior and penetration attained [13].

Information relative to the effects of rare earths in this respect is limited. This study was initiated to examine the arc and metal transfer characteristics of rare earth treated steels GMA welded using solid and flux cored wires. High speed photography was utilized to document these studies. The results should assist in understanding the role played by rare earths in steel during welding and in reaching a solution to the problems encountered when these steels are welded.

MATERIALS AND EXPERIMENTAL PROCEDURE

The experiments to be described were conducted on a laboratory scale using hot rolled plates of 0.08C-1.04 Mn-0.08 Nb HSLA steels containing from 0.000 to 0.156% rare earths and E70 grade (AWS-specification), 1.6 mm diameter solid wire and flux cored wire containing from 0.000 to 0.061% rare earths.

Automated GMA bead on plate (BOP) welds were made on surface ground steel plates using a variable inductor, rectifier type power supply. All welds were made in the direct current reverse polarity mode (DCRP). A high speed FASTAX camera was used to observe the arc and metal transfer phenomenon during welding. The observations were recorded on Kodak Ektochrome film at 4000 frames/sec., using three quartz-tungsten lamps for backing light.

RESULTS AND DISCUSSION

The films thus obtained were studied carefully to document information on short circuiting frequency (SCF) or drop frequency, short circuiting time, arc length, maximum diameter of droplets, direction of metal transfer, arc stability and amount of spatter. The quality of welds were judged from the appearance of the weld beads.

E70S-1B SOLID WIRE, SHORT CIRCUITING MODE--The results of welding using the short circuiting mode of metal transfer are summarized in Table 1, and the salient features of the arc and metal transfer behavior are sequentially shown in Figure 1. These data indicate significant differences in SCF, arc length, maximum diameter of droplets and the direction of metal transfer as the rare earth level of the steel is raised to 0.156%. It is interesting to note that even though the welding parameters were pre-set for the short circuiting mode, the presence of 0.156% REM in the plate results in the growth of large globules in a non-axial direction. Subsequently, arcing is established between globules and the puddle. Transfer of these globules occurs when short circuiting is established as the globules bridge. The flaring and wander of the arc due to the presence of rare earths in the plate is evident in Figure 1.

In contrast, a steel without rare earths exhibits excellent short circuiting behavior, and relatively quiescent arcing conditions.

E70S-1B SOLID WIRE, GLOBULAR/SPRAY MODE--Table 2 summarizes the results of welding in the globular/spray mode at high as well as low welding speeds; while Figures 2 and 3 illustrate corresponding selected frames taken from the high speed film documentation. The following observations may be made: (1) the effect of welding speed on metal transfer characteristics is significant in non-rare earth containing steels. Low welding speeds promote greater spray than globular transfer at the same power input; (2) A lower welding speed also promotes a more axial type of transfer and reduces the arc length appreciably; (3) In the case of high rare earth containing steel, welding speed has little influence on the arc and metal transfer behavior; (4) The presence of rare earths in the steel reduces the drop frequency, increases the drop size, promotes a non-axial globular type of transfer, promotes growth of globules away from the arc and introduces heavy arc flaring and wander.

E70T-5 FLUX CORED WIRE, GLOBULAR/SPRAY MODE--Data gathered from welding at high and low speeds is summarized in Table 3. Figures 4 and 5 show corresponding selected photographs representing the arc and metal transfer behavior.

No discernible difference is observed due either to the welding speed or to the presence of rare earths in the steel plate. The globular/spray mode of transfer is observed in all cases, and the metal transfer is dominated by the heavy growth of globules in a non-axial direction, with those globules detaching as a result of short circuiting due to bridging. An interesting feature is that globules always grow out of the arc. These forementioned characteristics result in flaring of the arc, the amount of which appears to be somewhat greater in the case of rare earth treated steels.

E70T-5 FLUX CORED WIRE CONTAINING RARE EARTHS, GLOBULAR/SPRAY MODE--Appreciable improvement in metal transfer is observed with the introduction of 0.061% rare earth metal (REM) into the flux formulation of the flux cored wire. These results are listed in Table 4. The representative frames from the high speed film are shown in Figure 6. These results indicate the following: (1) When the rare earth level of the flux cored wire exceeds that of the plate, free flight transfer dominates and the drop size decreases; (2) Where the rare earth level of the plate is greater, the metal transfer is similar to flux cored wire welding without rare earths in the wire. However, considerable dipping of the electrode into the weld pool is observed; (3) When rare earths are present in the flux cored wire, the growth direction of the globules approaches 90 degrees to the wire axis, unlike that of conventional flux cored wires; (4) The arc length appears to be more stable in welds produced where the rare earth level in the flux cored wire is greater than that in the plate.

SUMMARY AND CONCLUSION

1. Rare earths are shown to affect the arc and metal transfer behavior in both short circuiting and globular/spray modes during solid wire GMA welding.

2. Where flux cored wires are employed, the effects of rare earths in the steel are less prominent.

3. The presence of greater levels of rare earths in the flux formulation appears to counteract the effects of rare earths in the steel being welded.

4. Welding speed has a significant effect in solid wire welds. No appreciable changes are seen with flux cored wires.

5. The use of flux cored wires appears to be

an effective means to reduce the deliterious effects of rare earths in steels during welding.

6. The utilization of rare earths in the welding wire appears to be an effective means to counteract the effect of rare earths in the base plate on the welding arc and resulting metal transfer. The best way to understand the interaction of rare earths is to isolate the effect of the flux by using solid wires/metal cored wires of varying rare earth content. Research is presently underway in this direction.

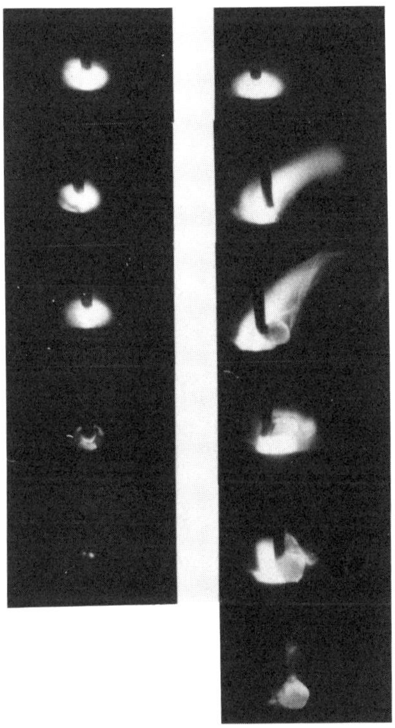

LEFT
No REM in plate

RIGHT
0.156% REM in plate

Figure 1 - Short circuiting mode; I:190A, arc voltage: 20V, S:400 mm/min., shielding: argon + 24.5% CO_2, E70S-1B solid wire of 1.6 mm dia.

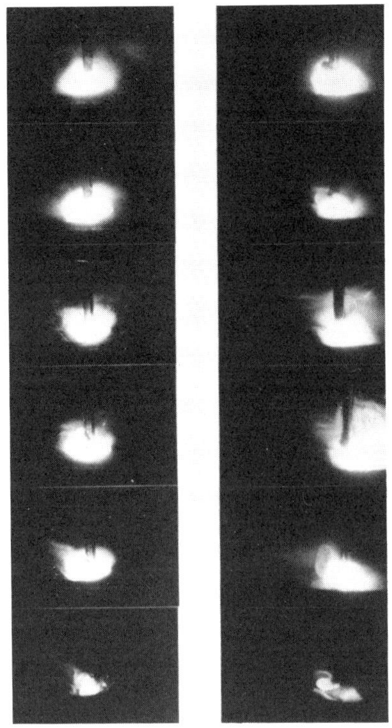

LEFT
No REM in plate

RIGHT
0.156% REM in plate

Figure 2 - Globular/spray mode; I:325A, arc voltage; 26V, S:1000 mm/min., shielding: argon + 24.5% CO_2, E70S-1B solid wire of 1.6 mm dia.

ACKNOWLEDGEMENTS

The assistance of Mr. William Heitmann of Inland Steel Company is greatly appreciated.

REFERENCES

1. Waudby, P. E., Intl. Met. Rev., No. 2, 74-98 (1978).
2. Zavyalov, A. S. and Sandomirskii, M. N., FTD-MT-24-1194-71, Foreign Tech. Div., Wright-Patterson AFB, Ohio, AD-743, 682 (1971).
3. Kepka, M. and Skala, J. Weld. Res. Abd. 19, 3, 26-38 (1973).
4. Banks, E. E. and Gunn, K. W., Intl. Conf. Weld. of HSLA Struc. Steel, Italy,

Table 1 - Short Circuiting Mode of Metal Transfer

Observations	No REM in Plate	0.156% REM in Plate
Short circuiting frequency/sec.	103	6 (Globules)
Short circuiting time. sec.	0.003	0.009
Arc length, mm	2.450	3.400
Max. dia. of drops, mm	2.000	3.2-4.8
Direction of metal transfer	Axial	Non-axial
Arc stability	Smooth and stable	Heavy flaring/ wander
Spatter	Nil	Very small
Quality of weld	Smooth and nice beads	Undercuts and ropey

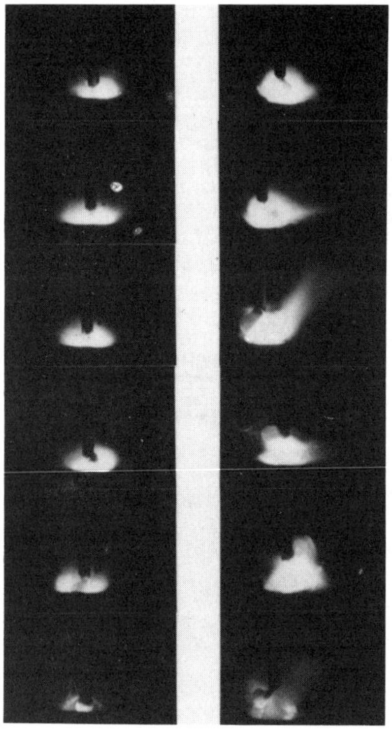

LEFT
No REM in plate

RIGHT
0.156% REM in plate

Figure 3 - Globular/spray mode; I:325A, arc voltage: 26V, S:500 mm/min., shielding: argon + 24.5% CO_2, E70S-1B solid wire of 1.6 mm dia.

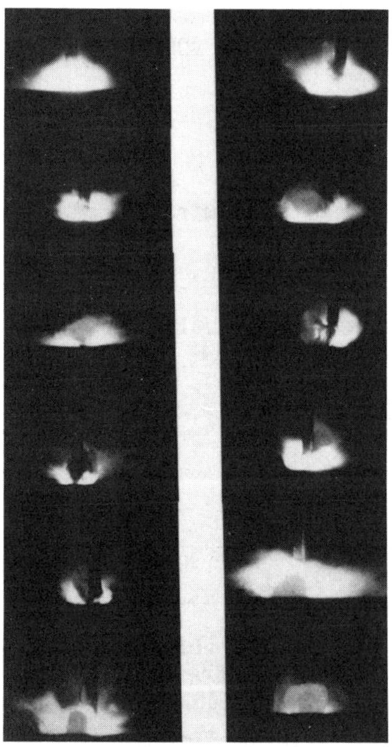

LEFT
No REM in plate

RIGHT
0.156% REM in plate

Figure 4 - Globular/spray mode; I:325A, arc voltage: 26V, S:1000 mm/min., shielding: argon + 24.5% CO_2, E70T-5 flux cored wire of 1.6 mm dia.

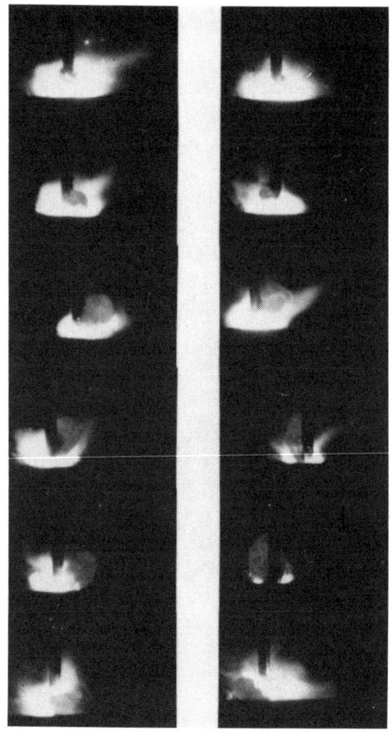

LEFT
No REM in plate

RIGHT
0.156% REM in plate

Figure 5 - Globular/spray mode; I:325A, arc voltage: 26V, S:500 mm/min., shielding: argon + 24.5% CO_2, E70T-5 flux cored wire of 1.6 mm dia.

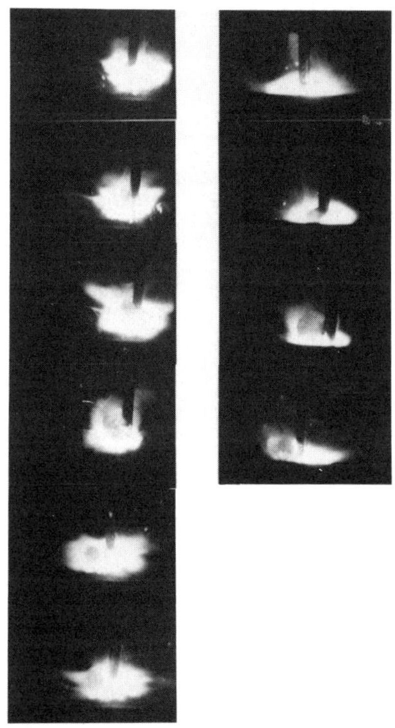

LEFT
No REM in plate

RIGHT
0.156% REM in plate

Figure 6 - Globular/spray mode; I:325A, arc voltage: 26V, S:1000 mm/min., shielding: argon + 24.5% CO_2, E70T-5 flux cored wire containing 0.061% REM of 1.6 mm dia.

Table 2 - Globular/Spray Mode of Metal Transfer With Solid Wire

Observations	High Welding Speed		Low Welding Speed	
	No REM in Plate	0.156% REM	No REM in plate	0.156% REM
SCF/sec.	37	8	64	4
Drop Frequency/sec.	77	21	-	27
Arc length, mm	3.8-4.0	3.0-5.0	1.5-2.0	≥ 4.0
Short circuiting Time, sec.	0.002	0.002	0.002	0.002
Max. Dia. of Droplets, mm	1.6	2.5-4.4	1.6	2.5-4.4
Direction of metal transfer	Axial	Non-axial	Axial	Non-axial
Growth Direction	Into the arc	Out of arc	Into the arc	Out of arc
Arc stability	Stable	Heavy flaring	Stable	Heavy flaring
Spatter	nil	Moderate	nil	Moderate
Quality of weld	Nice beads	Ropey	Smooth beads	Ropey

Table 3 - Globular/Spray Mode of Metal Transfer with Flux Cored Wire

Observations	High Welding Speed		Low Welding Speed	
	No REM in plate	0.156% REM	No REM	0.156% REM
Drop frequency/sec.	13	10	11	9
Arc length, mm	≤4	≤4	≤4	≤4
Max. Dia., of droplets, mm	3.0-4.8	3.0-4.8	3.0-4.8	3.0-4.8
Direction of transfer	Non-axial	Non-axial	Non-axial	Non-axial
Growth of direction	Out of arc	Out of arc	Out of arc	Out of arc
Arc stability	Flared	More flared	More flared	More flared
Spatter	Less	More	Less	More
Quality of weld	NO discernible differences in weld bead shapes			

Table 4 - Globular/Spray Mode with Flux Cored Wire Containing REM

Observation	No REM in Plate	0.156% REM in Plate
Drop frequency/sec.	22	16
Arc length, mm	≤4	≤2
Max. Dia. of Droplets, mm	3.0-4.0	3.0-4.0
Metal transfer	Free flight	Dipping type
Direction of metal transfer	Non-axial	Non-axial
Growth direction	Into arc	Out of arc
Arc stability	Stable	Less stable
Spatter	Less	More
Quality of weld	Uniform Bead Shape	

ASM/AIM, 467-492 (1976).

5. Sasaki, H., Akahide, K. and Tsubi, J. Trans. Japan Weld. Soc. 7, No. 2 (1976).
6. Kazennov, Y. I. and Raviznikov, L. I., Weld. Prod. 25, No. 11, 27-30 (1978).
7. Sorokin, L. I., Weld. Prod. 24, No. 10, 12-15 (1977).
8. Abralov, M. A. and Sergeev, V. G., Weld. Prod. 24, No. 12, 8-10 (1977).
9. Laz'Ko, V. E. and Koval'Chuk, V. G., Weld. Prod. 25, No. 8, 20-22 (1978).
10. Cassel, C. E., Weld. J., 48, 167-170s (1969).
11. Ratz, G. A., Nippes, E. F., Mathew, J. and Baek, W. H., Weld. J. 63, 11, 333s (1984).
12. Li, S. and Patchett, B. M., AGA Contract PR-152-121 (1982).
13. Glickstein, S. S. and Yaniscavich, W., WRC Bull. 226, 1-8 (1976).

WELDMETAL CRACKING

EFFECT OF CENTER LINE SEGREGATION ON THE COLD CRACKING IN CONCAST STEEL WELDMENT

H.J. Kim, M.R. Kim, D.H. Park
Hyundai Heavy Industries Company Ltd.
Ulsan, Korea

ABSTRACT

The dark line revealed at the mid-thickness of the concast steel plates was investigated to clarify its nature and its effect on cold cracking susceptibility. Metallographic observation and chemical analysis demonstrated that the dark line was resulted from the center line segregation associated with the dense distribution of thick pearlite band structure. Because of alloy segregation and thus a localized increase in carbon equivalent along the mid-thickness, the steel plate having a dark line is quite susceptible to cold cracking. Based on the above informations, it can be concluded that the crack occurred at the mid-thickness of the concast steel weldment or plate is colsely related with alloy segregation along the dark line.

I. INTRODUCTION

AFTER etching the steel plate produced by continuous casting, a dark line is often revealed along the mid-thickness. One typical example is represented in the sample plate A shown in Fig.1. Referring to recent research (Ref.1), it was reported that this dark line led to very low resistance to hydrogen induced cracking (HIC). In addition, the dark line has been considered as a quite susceptible region to cold cracking since such problems were often faced during ship-building and construction of off-shore steel structure (Ref.2).

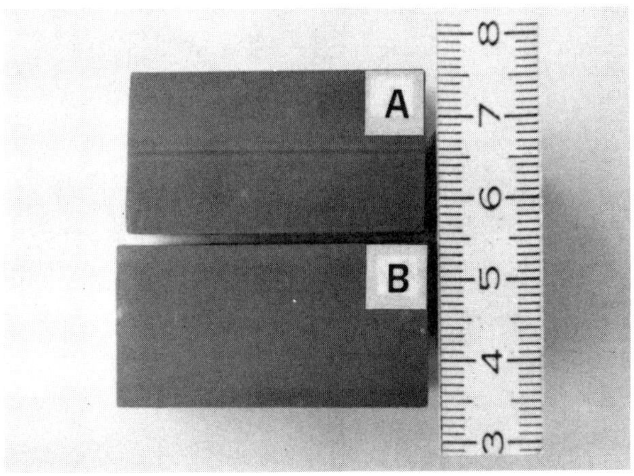

Fig.1 — Macrostructures, of (a). steel A containing the dark line along the mid-thickness and (b). steel B which does not contain the dark line.

An example of such cracks occurred along the dark line is shown in Fig.2.

Therefore, the present research is mainly directed to investigate the nature of the dark line and its effect on cold cracking susceptibility of the steel plate containing a dark line.

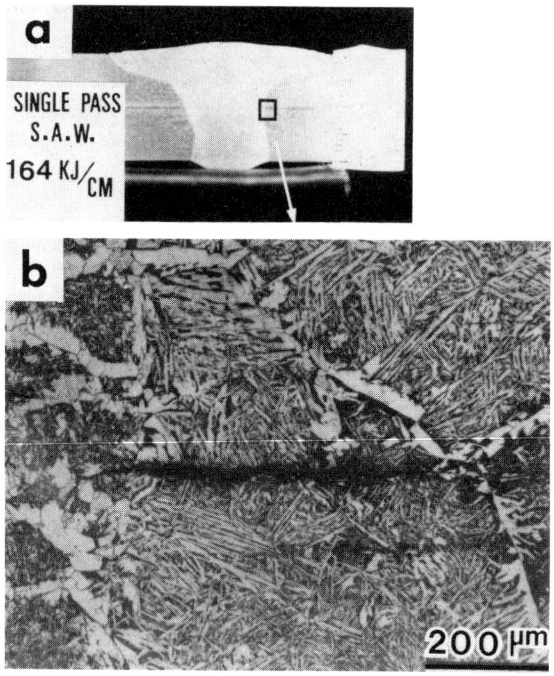

Fig.2 —a). Macro-view of the SAW weldment containing a crack along the dark line at the mid-thickness and b). magnified view of cracked area.

II. EXPERIMENTAL PROCEDURE

MATERIALS — Through-thickness macrostructures of sample plates (ABS AH32 grade) used for this study are shown in Fig.1. In this figure the dark line is clearly observed along the mid-thickness of the steel plate A, whereas is not observed in the plate B. The chemical compositions and the tensile properties of both plates are given in Table 1 and 2. No significant difference between plate A and B was noted except the reduction of area(RA) in thickness (Z-) direction.

Table 1 - Chemical Composition of Steel Plates (wt.%)

Plate	C	Si	Mn	P	S	Al	Fe
A	0.16	0.32	1.41	0.019	0.009	0.029	Bal.
B	0.14	0.35	1.36	0.018	0.008	0.044	Bal.

Table 2 - Mechanical Properties of Steel Plate A and B

Plate	Dir.	YS (Kg/mm^2)	UTS (Kg/mm^2)	Elong. (RA %)	Cv at 0^oC (Kgf.m)
A	L	38	55	25	12.5
	T	•	•	31	0.5
B	L	37	52	25	14.1
	T	•	•	50	1

MECHANICAL TESTING — Charpy V-notch impact tests and Vickers hardness tests were carried out to characterize the dark line. The Charpy impact test specimens were made in longitudinal and through-thickness directions as shown in Fig.3. Tab-pieces were welded on both sides of steel plate for making through-thickness impact specimens. Fig.3 also shows the two different notch locations of through-thickness Charpy specimens, i.e. notched at the center of the plate (1/2t) and at the quarter thickness (1/4t). The impact values obtained with specimens having their notch at 1/4t can be regarded as the bulk toughness while those obtained at 1/2t are the toughness values affected by the dark line.

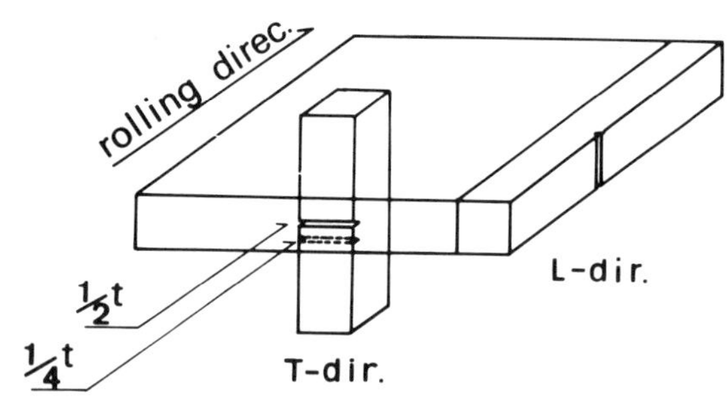

Fig.3 — Locations of V-notch in Charpy impact specimens with respect to plate rolling direction.

WELDING — In order to investigate the effect of dark line on cold cracking susceptibility of steel plate, the plate A containing a dark line was machined into 50° single V-groove and then welded by one side-tandem submerged arc welding (SAW) with a heat input of 164KJ/cm. In addition, bead-on-plate welding with lower heat input was carried out to determine the heat input effect on the maximum hardness in the HAZ.

CHEMICAL AND METALLOGRAPHIC ANALYSIS — In order to measure the variation of alloying contents across the dark line, a full-thickness specimen (20mm) was taken and then ground the surface parallel to the plate surface successively. The chemical analysis was conducted using a X-ray fluorscence spectrometer on the exposed surface after each grinding. By sequential grinding and analysis, full profiles of alloying contents across the dark line could be obtained.

Specimens for optical microcopy were cut from the weldment or the broken Charpy bars to reveal the microstructure below the fracture surface. The samples for microstructural analysis were prepared by standard metallographic procedure.

III. RESULTS AND DISCUSSION

MICROSTRUCTURES — Detailed examination on the microstructure of sample plate A and B showed that the bulk structures are indentical but the structures in the mid-thickness are quite different between two specimens. Fig.4 shows the microstructure taken in the area of mid-thickness of each plate.

The specimen A having a dark line shows a dense distribution of pearlite band structure at the center of the dark line (Fig.4-a), while the specimen B has an uniform distribution (Fig.4-b). A consequence is that the specimen A has a high volume fraction of pearlite at the mid-thickness of steel plate. Therefore, an attempt was made to measure the pearlite volume fraction across the mid-thickness quantitatively using an image analyzer. This result is shown in Fig.5 along with that obtained from sample plate B. As expected, the pearlite formed in the steel plate B distributed uniformly in the form of band structure, i.e. uniform thickness of bands and uniform distance between bands throughout the thickness.

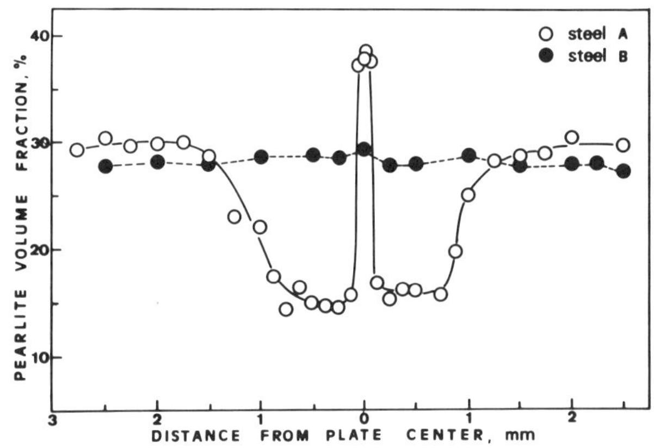

Fig.5 — The variation of pearlite volume fraction of steel plate A and B as a function of distance from plate center.

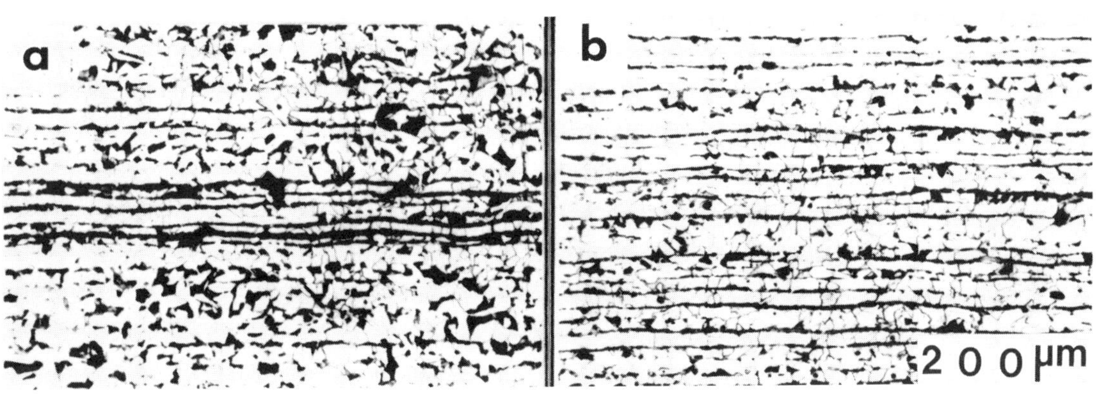

Fig.4 — Microstructures of the mid-thickness region in steel plate A and B

However, the pearlite distribution shown in the plate A exhibits maximum value (38%) at the center of the dark line and minimum values (15%) at the both sides of the peak. The occurrence of minimum content is quite surprising and the difference between the maximum and minimum is quite drastic. Detailed microstructural examination made in the areas having a minimum pearlite content demonstrated that pearlite phase formed in this area did not develop the band structure but distributed finely in the form of isolated islands. As a result of this investigation, it can be concluded that the dark line revealed along the mid-thickness is due to the microstructural inhomogenity which results in dark contrast after etching.

SEGREGATION PROFILES — Chemical analysis on the specimen A was conducted to investigate the segregation profile of specific alloying elements by X-ray fluorscence spectrometer. As shown in Fig.6, the dark line is highly segregated with alloying elements such as Mn, P and Si, and has negative segregation of such elements at the both sides of the plate center. It is quite interesting to note that all the segregation profiles have a same trend to that of pearlite volume fraction. The correlation between the microstructural variation and segregation profile has been reported by H. Tamehiro et.al. (Ref.3).

MECHANICAL PROPERTIES OF BASE METAL — Fig.7 shows the hardness profile across the plate center. Hardness profile of plate A is also consistent to the pearlite and segregation profiles. The hardness profile obtained, therefore, was able to be explained with an associated change in volume fraction of pearlite phase, i.e. linear correlation between pearlite content and hardness could be obtained.

The Charpy impact test results obtained with the plate A are presented in Fig.8. As shown in this figure, the orientation and notch location of specimen have a profound effect on

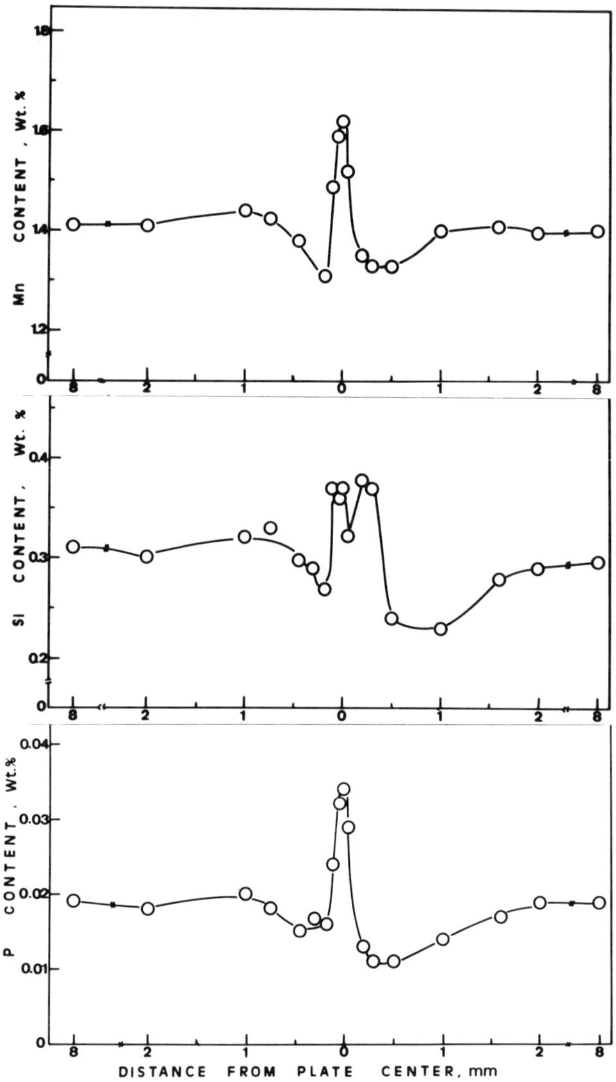

Fig.6 — Segregation profiles of the steel plate A

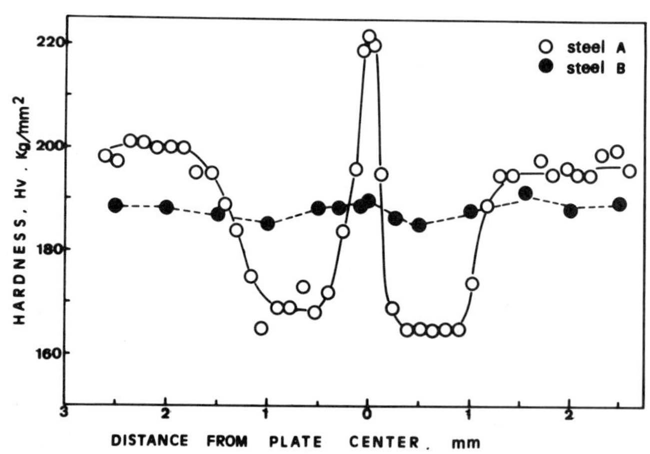

Fig.7 — Hardness variation of steel A and B as a function of distance from plate center.

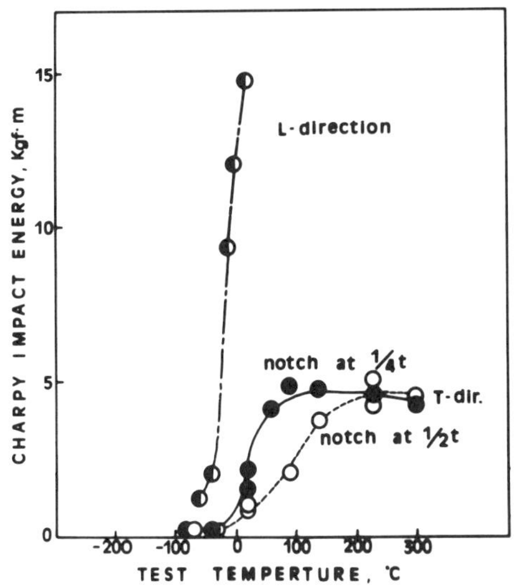

Fig.8 — DBTT-curves of steel A depending on the various notch locations.

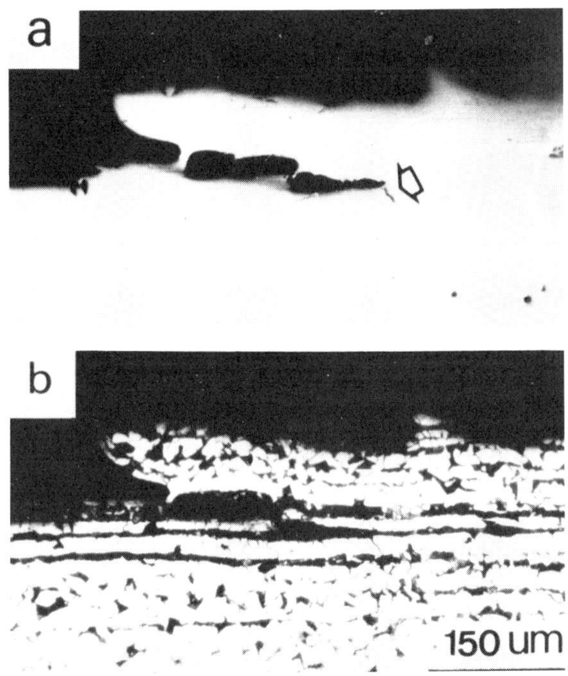

Fig.9 — Cross-section of the broken Charpy specimen notched at the center of the plate. a) was taken before etching, b) was taken after etching.

the resultant impact values and the ductile-brittle transition temperature (DBTT). Especially, the through-thickness impact toughness varies significantly depending on the notch locations, i.e. 1/2t and 1/4t. The DBTT obtained from 1/2t specimen is much higher by about 70°C than that from 1/4t. To explain this difference, the broken Charpy specimens were sliced to investigate the fracture path. Fig.9 shows crack propagation path in the specimen notched at 1/2t and then tested at the room temperature. It is clearly seen in this figure that the crack mainly propagates following the pearlite band structure. Considering the brittleness of pearlite phase and the thick band morphology at the mid-thickness, it is reasonably supposed to result in the poor toughness at 1/2t. It also can cause the deterioration of RA value in Z-direction tensile test as already shown in Table 2.

HARDNESS IN HAZ — Hardness test results in HAZ of SAW weldment were shown in Fig.10 with a schematic illustration of joint geometry. As show in this figure, the hardness value recorded along the dark line (1/2t) is quite higher than that measured along 1/4t. For example, the maximum hardness values at 1/2t and 1/4t are Hv 265 and Hv 230 respectively. It should be noted here that the crack shown

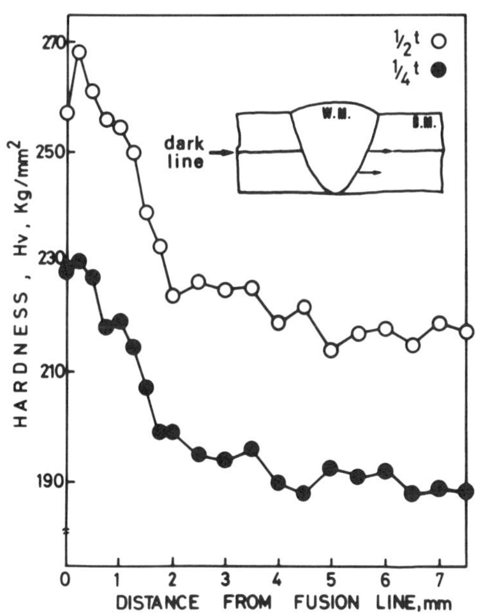

Fig.10 — HAZ hardness profiles along a dark line (1/2t) and a line of 1/4t in a SAW weldment.

in Fig.2 was found in this weldment.

On the other hand, bead-on-plate welding for specimen A was conducted to determine the maximum hardness with a lower heat input at 1/2t and 1/4t respectively. For this purpose, two specimens were prepared. One is for the fusion line of weld bead to be adjacent to the plate center and the other is to the quarter thickness as shown in the schematic drawings inserted in Fig.11. From this figure, it is noted again that hardness value along the dark line is significantly high comparing with that of quarter thickness by about Hv 60.

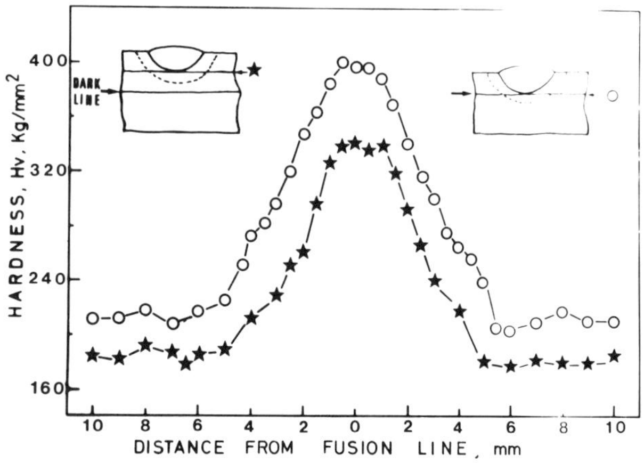

Fig.11 — HAZ hardness profiles along the dark line (1/2t) and a line of 1/4t in the bead-on-plate weldment.

Moreover, the maximum hardness recorded along the dark line is over Hv 400. Correlating the segregation profiles with the HAZ hardness, it can be concluded that the high hardness is attributed to the increased hardenability imposed by alloy segregation along the dark line. It must be borne in mind, therefore, that the maximum hardness checked on the bulk as a requirement criterion for cracking susceptibility has no meaning in the materials having a dark line. Similarily, considering the difference in chemical composition at the bulk and the dark line (Fig.6), the carbon equivalent calculated based on the base metal composition is no longer valid in the dark line area. As is seen from Table 3, the carbon equivalent of the

Table 3 — Chemical Composition of the Bulk Material and the Dark Line Region.

	Bulk	Dark line
Mn	1.41	1.62
Si	0.32	0.37
C	0.16	0.21*
Ceq.	0.41	0.50

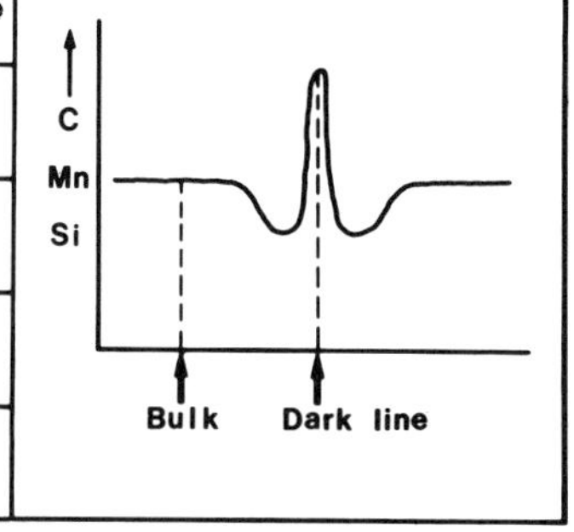

$$C^{*} = \text{Bulk C} \times \frac{\text{Pearlite volume fraction on dark line}}{\text{Pearlite volume fraction of bulk}}$$

$$C_{eq.} = C + \frac{Mn}{6} + \frac{Si}{24} \ (\%)$$

bulk is 0.41 while that of the dark line is 0.50. Therefore, it is also necessary to reconsider the carbon equivalent of the steel plate when it contains a dark line. This is because the carbon equivalent calculated with the chemical composition of the bulk does not consider the localized increase in carbon equivalent at the mid-thickness which in turn results in the increased susceptibility to cold cracking during welding. From the above results and discussions, it is supposed that for the improvement of cold crack resistant it is important to prevent the formation of center line segregation during continuous casting process, that is, it is necessary to reduce the alloy segregation at the mid-thickness of the slab.

IV. CONCLUSION

On the basis of this study, the following conclusions were made.

1) The dark line revealed is due to the microstructural inhomogeonity at the mid-thickness of steel plate.

2) Microstructural inhomogenity occurred is originated from the alloy segregation during continuous casting process.

3) Because of the alloy segregation, the dark line area is quite susceptible to cold cracking because of its higher carbon equivalent than that of base metal.

4) It is necessary to re-evaluate the carbon equivalent and maximum hardness requirement in case of the steel plate containing a dark line, i.e. having a center line segregation.

V. REFERENCES

1. H. Tamehiro, N. Yamada and H. Masuda, Transactions of ISIJ, Vol.25 (1985) 54
2. M. R. Kim, D. H. Park and H. J. Kim, Hyundai Technical Review, Vol.5, No.4 (1985) 22
3. H. Tamehiro, T. Takeda, S. Masuda, K. Yamamoto and N. Okumura, Transactions of ISIJ, Vol.25 (1985) 982

EXAMINATION OF WELD HOT CRACKING IN A TRANSITION JOINT BETWEEN HP 9-4-20 STEEL AND 21-6-9 STAINLESS STEEL

M.W. Richey, M.W. Doughty
Martin Marietta Energy Systems, Inc.
Oak Ridge Y-12 Plant
Oak Ridge, Tennessee 37831

C.L. White
Martin Marietta Energy Systems, Inc
Oak Ridge National Laboratory
Oak Ridge, Tennessee 37831

ABSTRACT

A statistically designed, full-scale, weldment test series has been employed to appraise the possible influence of weld production irregularities on the cracking tendency of a gas-tungsten arc/cold wire feed weld between HP 9-4-20 steel and 21-6-9 stainless steel. Optical light microscopy, scanning electron microscopy, and Auger electron spectroscopy were utilized to characterize the metallurgical nature of weld cracking. The HP 9-4-20 filler material was found to be the most important production variable. Weld cracking was contingent on a particular heat of weld wire. The 21-6-9 material is apparently a prerequisite for the cracking process. Weld wire and weld metal chemical analysis affords no clarification of the cracking problem. Fractographic evidence suggests liquation as the most plausible cause of weld cracking. Enrichment of boron, nitrogen, and sulfur along weld cracks was detected by Auger electron spectroscopy. Data gained from an investigation of Auger free-surface segregation shows that the potential for boron and nitrogen enrichment is greater in the crack-susceptible weld metal. The full implications of the Auger observations with regards to weld cracking are presently unknown.

DURING PRODUCTION AT THE OAK RIDGE Y-12 PLANT, weld cracking has been encountered in a weld between HP 9-4-20 steel* and 21-6-9** stainless steel. The HP 9-4-20 material is a readily weldable, high-strength, high-toughness steel [Aerospace Material Specifications (AMS) 6525], and the 21-6-9 material is a nitrogen-strengthened, austenitic stainless steel (AMS 5656A). The gas-tungsten arc/cold wire feed (GTA/CWF) welding process is employed to produce the HP 9-4-20 and 21-6-9 weld. The weld filler wire material is 0.063-in. dia (1.59mm) HP 9-4-20 (AMS 6468). No serious weld fabrication cracking problems have been reported in the 9Ni-4Co alloy series(1,2,3).

Approximately nine weld passes are needed to complete the weld. The weld fabrication parameters are summarized in Table 1. Because of the self-tempering nature of HP 9-4-20 steel, and the temper heat-treatment thermal cycle associated with multipass welding, no postweld heat treatment is required (3).

Table 1 - Welding Parameters
For The GTA HP 9-4-20/21-6-9 Weld

Voltage	8.5-10.5 volts
Current	250-290 amps
Travel speed	7.625 ipm (19.4 cmpm)
Energy input	16.7-24.0 KJ/in.(6.6-9.5 KJ/cm)
Electrode	0.125 in. (0.32 cm) dia Thoriated Tungsten, 15° included angle
Gas	Argon 40 cf/h (18.9 L/min)

The occurrence of weld cracking in the HP 9-4-20 and 21-6-9 weld during production was indeed unexpected because an earlier series of crack-free laboratory development welds had been made. Between laboratory development welding and production welding, several potentially important procedural modifications which could influence weld cracking occurred:

1) Substitution of a different heat of HP 9-4-20 weld wire;

2) Change in welding deposition sequence from staggered (development) to stacked (production); and

3) Variation in interpass temperature due to the unavoidable constraints encountered in a production environment (e.g., diversity in weld operators, break-time interruptions, inconsistency in brushing sequence, etc.).

*Republic Steel Corporation -
9% nickel, 4% cobalt and 0.2% carbon.
**Armco Steel Corporation -
21% chromium, 6% nickel, 9% manganese, and 0.2%-0.4% nitrogen.

This paper examines the nature of weld cracking in the HP 9-4-20 and 21-6-9 weld and analyzes the influence of weld wire composition, welding sequence, and interpass temperature on the weld-cracking tendency. Optical light microscopy (OLM), scanning electron microscopy (SEM), and Auger electron spectroscopy (AES) were employed to characterize the metallurgical nature of the weld cracks. A statistically designed weld-cracking experiment analyzed the direct effect of production procedures on the weld cracking problem.

EXPERIMENTAL PROCEDURE

METALLOGRAPHIC AND FRACTOGRAPHIC EXAMINATION - Transverse and longitudinal metallographic specimens were extracted from a production weld exhibiting a high degree of weld cracking. Standard metallographic techniques were utilized in the grinding and polishing process. Optical light microscopy was employed to locate an area which promised to yield an AES tensile specimen suitable for Auger fracture analysis.

The longitudinal weld section immediately adjacent to the metallographic specimens yielded an impact specimen which was notched in the root pass, immersed in liquid nitrogen, and fractured. The fracture surface was then examined in the scanning electron microscope.

STATISTICALLY DESIGNED WELD-CRACKING STUDY - Four weldments which were fabricated from full size components were employed in the study of weld cracking. A completely randomized half-replicate 2^3 factorial experimental design was selected for the investigation. The three welding factors under examination were:

1) HP 9-4-20 weld wire (two wires designated Development and Production respectively);

2) Welding deposition sequence (staggered-development and stacked-production); and

3) Interpass temperature (5 minutes and 20 minutes interpass time). Interpass temperature was not measured directly, instead, the interpass time was varied (5 and 20 minutes) to simulate the interpass temperature range which might be encountered during development and production welding. An interpass time of 5 minutes correlates to a "high" interpass temperature, conversely an interpass time of 20 minutes corresponds to a "low" interpass temperature. During the weld-cracking study, the welding variables other than those under examination were controlled closely to the production parameters outlined in Table 1. The point runs for the weld-cracking study are enumerated in Table 2.

Table 2 - Test Runs For Weld-Cracking Study

Run	Weld Wire	Interpass Time (min)	Welding Sequence
1	Development	5	Staggered
2	Development	20	Straight
3	Production	20	Staggered
4	Production	5	Straight

The welding, cutting, and inspection orders were fully randomized. Following welding, four transverse metallographic specimens (spaced at 90° intervals with 0° selected arbitrarily) were produced from each weldment. The specimens were ground, polished, etched, and the linear length of each weld crack was measured and recorded.

AES EXAMINATION OF A PRODUCTION WELD CRACK SURFACE - An AES specimen [~0.08 in. x 0.12 in. x 0.71 in. (~2 mm x 3 mm x 18 mm)] was extracted from the center portion (forth to sixth pass region) of a production weld. This was the area of the weld exhibiting the greatest crack density. The axis of the AES tensile specimen and the welding direction were both located in the same plane. The tensile specimen was introduced into the Auger tensile fracture device, fractured under vacuum, and transferred to the analysis position. A series of Auger elemental maps and Auger spectra were collected from the as-fractured weld crack surface. Composition profiles were obtained by alternate argon ion-beam sputtering and Auger reanalysis.

The sputtering conditions used here (5-keV beam, 12-μ A/cm^2 current density) are estimated to result in an etching rate between 0.5 and 2.0 nm/min. All sputtering was carried out in the normal incidence mode, with the ions impinging upon the fracture surface from the same perspective as the electron beam during AES analysis.

AES FREE-SURFACE SEGREGATION STUDY - Auger analysis of free-surface segregation entailed AES examination of a thin [0.010-0.015 in. (255-380μm)] polished specimen during a thermal-cycle treatment. A surface segregation specimen was sectioned from both development and production welds. Following thinning and polishing to a 0.25-μm finish, the specimen for analysis was mounted on a small specimen heater and introduced into the Auger system. The specimen was then transported under vacuum to a heating stage in the analysis area, argon ion-beam sputter cleaned, and subjected to a step profile thermal cycle. The segregation specimen was initially heated to 800°C. After approximately 160 minutes (total elapsed heating time 184 minutes), the temperature of the segregation specimen was increased to 1000°C. The specimen temperature was maintained at 1000°C for approximately 75 minutes, after which time the power was switched off, and the segregation specimen was allowed to free-cool on the heating stage.

Auger spectra were obtained from the as-received surface, after argon ion-beam sputter cleaning, during the heating cycle, and following the free-cool period. The boron, nitrogen, and sulfur peak height intensity ratios (I and I_{Fe}) were determined for the Auger spectra taken during the heating cycle.

RESULTS AND DISCUSSION

CHARACTERIZATION OF WELD CRACKS - Photomicrographs from transverse and longitudinal weld sections are combined in Fig. 1 to form a spatial three-dimensional view of a typical family of production weld cracks. The weld cracks follow the former austenite grain-boundaries, seemingly ignoring the cellular and cellular-dendritic solidification structure. The weld cracks are generally located between the second and seventh weld passes, and are evenly distributed within each weld pass. The weld cracks are confined to the weld pass in which they lie.

Fig. 1. Three-dimensional representation of production weld cracks.

A series of SEM fractographs from the impact specimen are shown in Fig. 2. The fusion zone exhibits a mixed intergranular and fibrous transgranular fracture morphology (Fig. 2a). Microvoid coalescence is responsible for the fibrous appearance of the transgranular areas (Fig. 2b). The intergranular fracture areas follow the prior austenite grain-boundaries and display two distinct types of topography (Points A and B, Fig. 2c). The microductility visible at point A apparently occurred during the impact fracture. The preexisting weld crack surface is present at point B and is decorated with a network of stepped striations, which are most likely the remains of a dendritic structure. A similar fracture surface has been noted with liquation cracking(4).

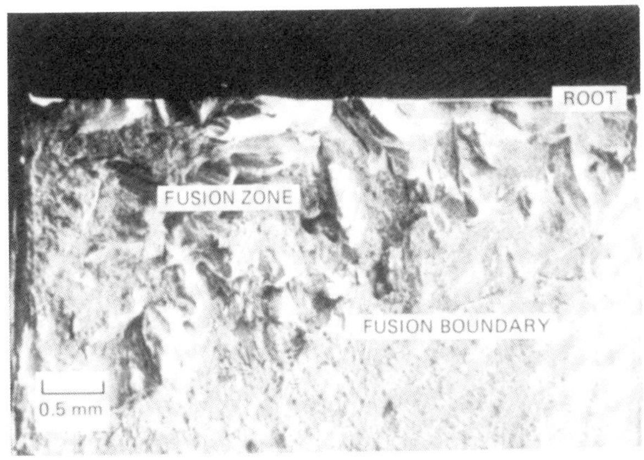

a) 20X

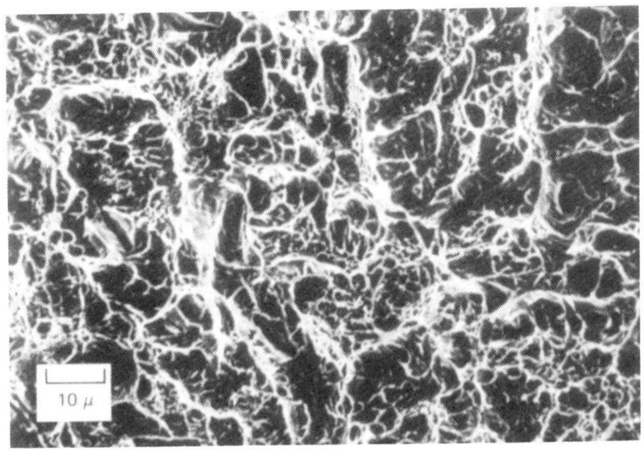

b) 1000X

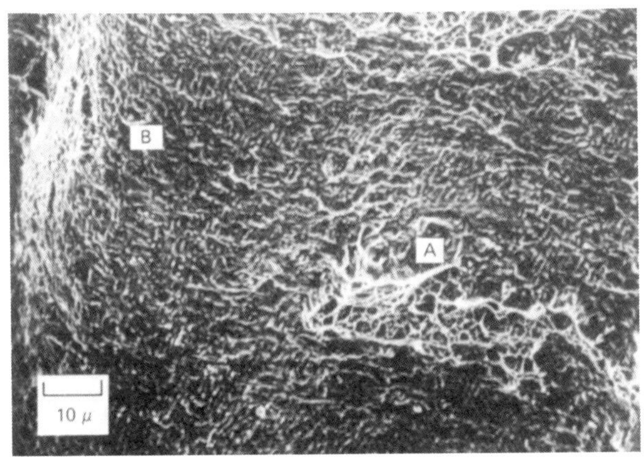

c) 1000X

Fig. 2. Fractographs from the impact specimen.

WELD-CRACKING STUDY - The Fig. 3 histogram summarizes the results of the weld-cracking study. The Production wire welds exhibited substantial cracking while the Development wire welds exhibited virtually no cracks. Clearly, the onset of the production weld cracking problem can be directly linked to the production weld filler wire.

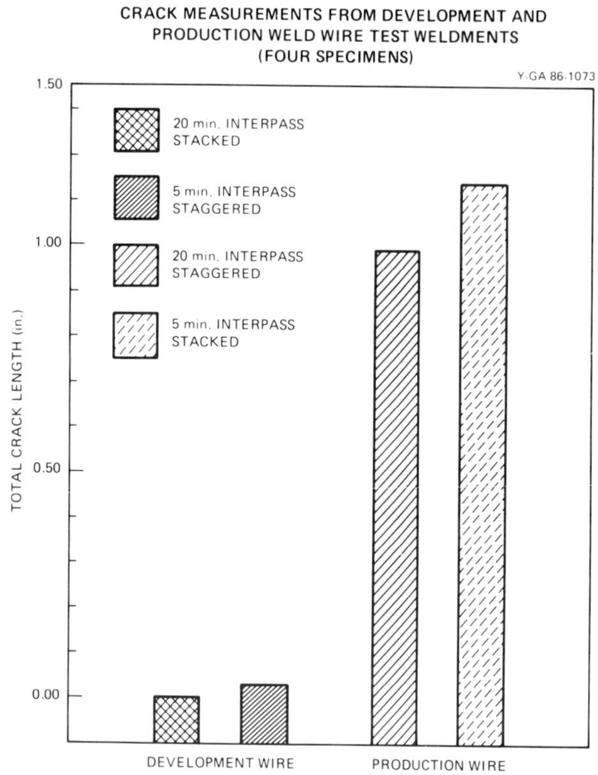

Fig. 3. Results of the weld-cracking study.

Many researchers have established that weld hot cracking is a direct result of certain corruptive impurity elements (viz., phosphorus and sulfur) whose potency is enhanced or diminished by various alloying elements (e.g., carbon, chromium, nickel, manganese, and silicon). Chemical analyses for the Development and Production wires are found in Table 3. The AMS 6468 specification is also included in Table 3.

Table 3 - Chemical Analysis For Development and Production HP 9-4-20 Weld Wire

Element	Development Wire	Production Wire	AMS 6468
Carbon	0.14	0.14	0.14- 0.17
Manganese	0.41	0.48	0.40- 0.55
Silicon	0.25	0.22	0.15- 0.25
Phosphorus	0.004	0.001	--- - 0.008
Sulfur	0.010	0.010	--- - 0.008
Chromium	0.95	1.01	0.90- 1.05
Nickel	10.08	8.98	9.75-10.25
Cobalt	3.81	3.81	3.50- 4.00
Molybdenum	0.46	0.50	0.40- 0.50
Vanadium	0.09	0.08	0.06- 0.10
Copper	0.04	<0.01	--- - 0.10
Boron	0.002	0.001	----------
Aluminum	0.01	0.01	----------
Titanium	<0.01	0.02	----------

With the exception of sulfur and nickel, both HP 9-40-20 weld wire compositions are within the specification limits. Upon comparison, there is no readily discernable difference in the nominal chemical composition of the weld wires that could easily account for their contrasting welding behavior. The vacuum-induction melt and vacuum-arc remelt (VIM and VAR) process utilized in the production of the HP 9-4-20 weld wire resulted in a relatively low level of phosphorus and sulfur. The combined phosphorus plus sulfur level for both wires is well below the minimum level (P + S <0.020) which other investigators have shown necessary to prevent hot cracking in 6% nickel and high-strength steel welds (5,6).

During production, the Production wire had successfully been employed to make sound welds between two HP 9-4-20 steel components that were geometrically similar to the HP 9-4-20 and 21-6-9 weld. Evidently, base metal dilution from the 21-6-9 material contributed to the weld cracking problem. Weld metal chemical analyses from both development and production welds are presented in Table 4. Again chemical analysis offers no indication of a clear composition trend which might be metallurgically associated with the weld cracking problem.

Table 4 - Chemical Analysis of Development and Production Wire Weld Metal

Element	Development Wire	Production Wire
Carbon	0.14	0.14
Manganese	1.28	0.89
Silicon	0.26	0.23
Phosphorus	0.009	0.009
Sulfur	0.002	0.003
Chromium	2.94	1.93
Nickel	9.55	8.95
Cobalt	3.49	3.67
Molybdenum	0.51	0.52
Vanadium	0.07	0.06
Copper	0.07	0.02
Boron	0.001	0.001
Aluminum	0.02	0.03
Titanium	<0.01	0.02

Perhaps weld cracking resulted from a subtle interaction between intrinsic (intentionally added) and extrinsic (residual impurity) elements which enriched the concentration of latent corruptive elements along the weld metal grain-boundaries. The potential for this interaction might not be evident from the bulk chemical analyses. Auger electron spectroscopy was employed to examine production weld metal for evidence of solute enrichment along crack surfaces.

AES INVESTIGATION OF A PRODUCTION WELD CRACK - A SEM fractograph of the production weld AES tensile specimen is shown in Fig. 4. The tensile specimen fracture surface is very similar to the fracture surface that was broken out of the production weld (Fig. 2). Auger elemental maps over approximately the same field of view as Fig. 4 are assembled in Fig. 5. The intergranular

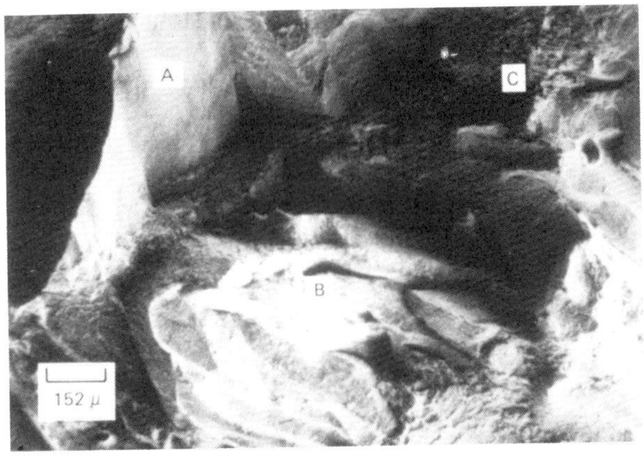

Fig. 4. SEM fractograph of AES Tensile specimen, SEM 66X.

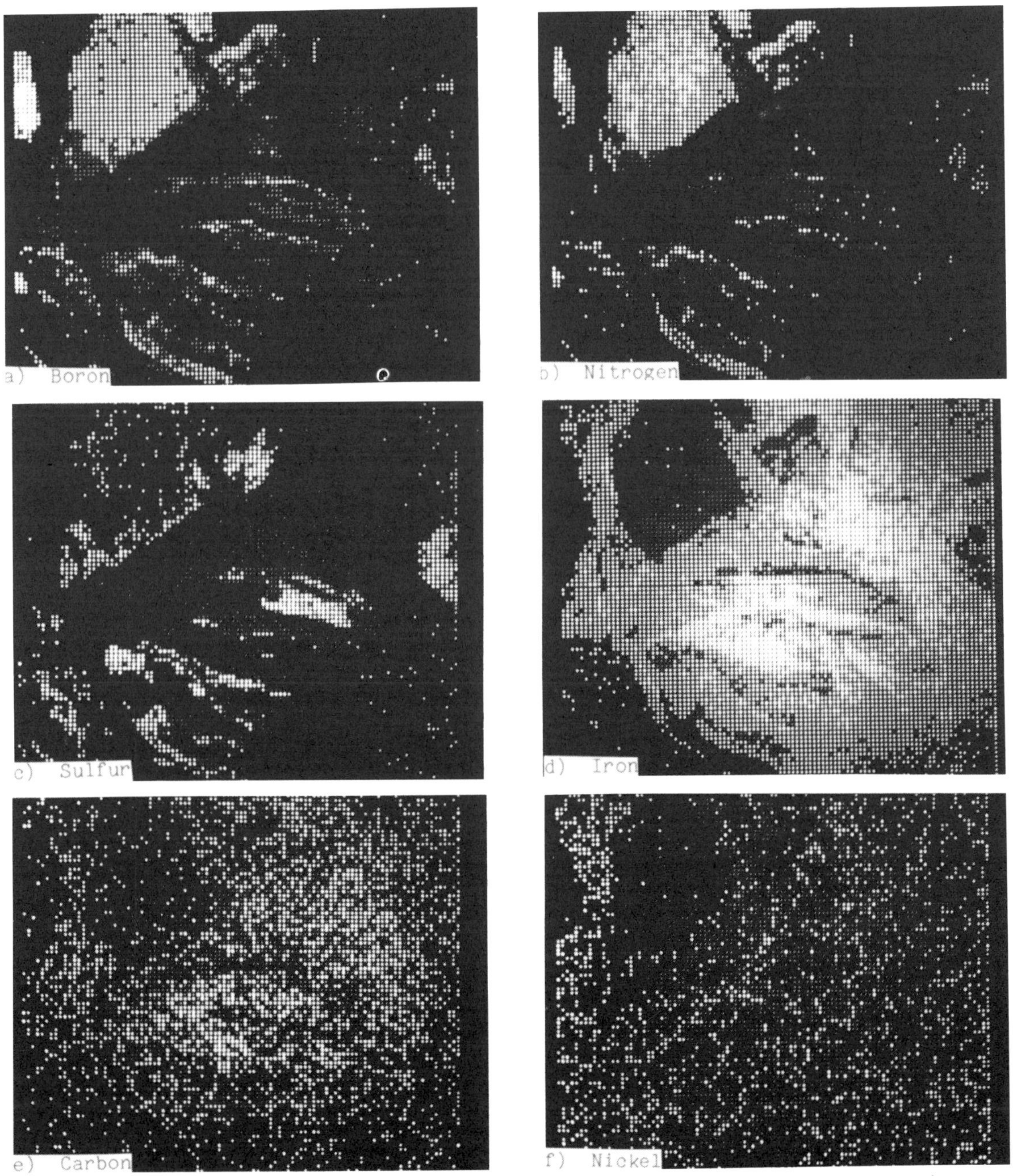

Fig. 5. Auger elemental maps over the fracture surface shown in Fig. 4.

fracture areas are enriched in boron, nitrogen, and lesser amounts of sulfur (Figs. 5a, 5b, and 5c). The grain-boundary surfaces are denuded of iron (Fig. 5d), and no significant variations in nickel or carbon enrichment were observed over the fracture surface (Figs. 5e and 5f).

Auger spectra obtained at various locations on the fracture surface generally confirmed the compositional variations indicated in Fig. 5. Figure 6a shows a high-magnification SEM micrograph near point A of Fig. 4. Figures 6b-6d show Auger spectra from this region in the as-fractured condition (6b), after sputter etching for 2 minutes (6c), and after sputter etching for a total of 7 minutes (6d). Comparison of these spectra clearly indicates extensive enrichment of boron, nitrogen and sulfur on the fracture surface, and that the solute enriched layer is only a few atom layers thick. Kujanpaa et al have found similar evidence of sulfur enrichment along a weld crack surface from an austenitic stainless steel Spot-varestraint specimen (7). A substantially smaller amount of boron, nitrogen, and sulfur enrichment can be found along a ductile dimple grain-boundary area produced during the tensile fracture (Fig. 7). No evidence of enrichment was found along a transgranular fracture

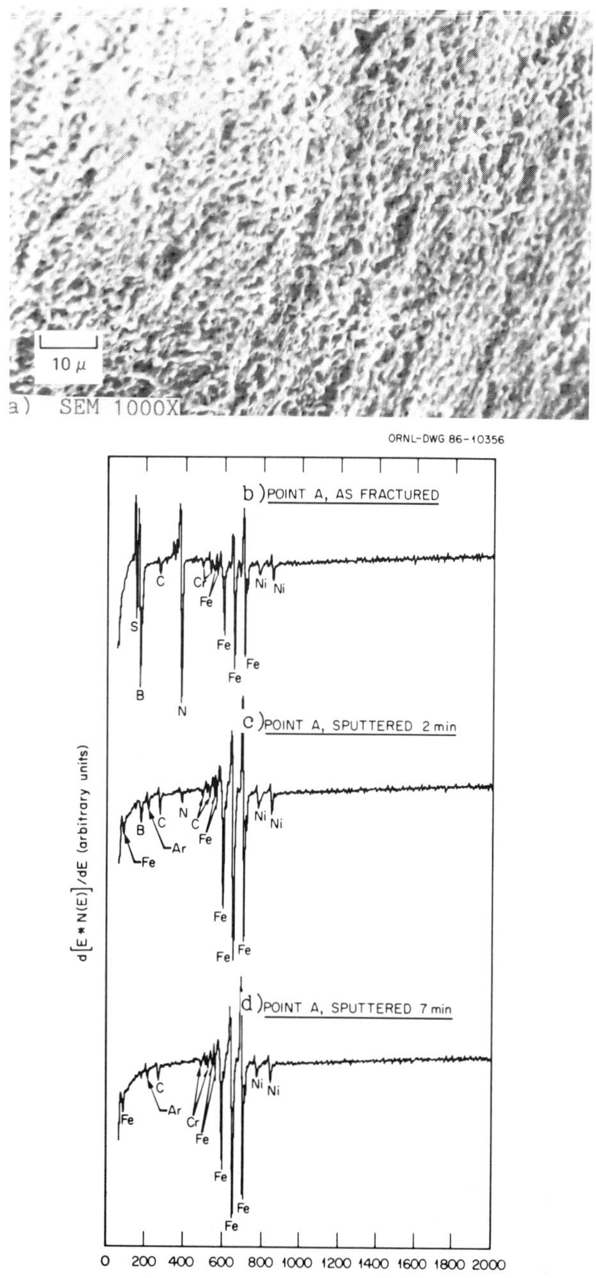

Fig. 6. SEM fractgograph and Auger Spectra from point A, Fig. 4.

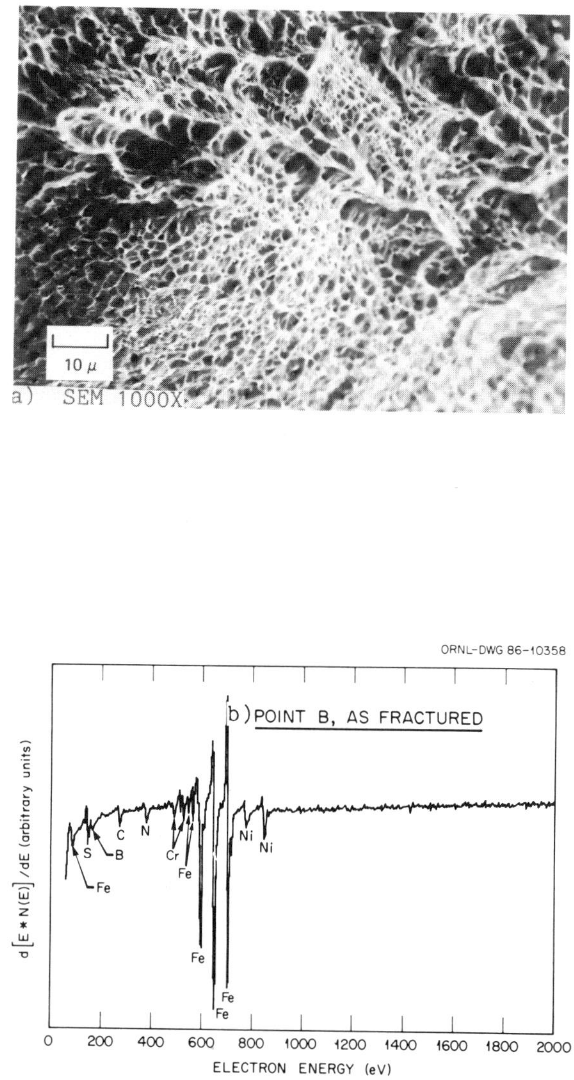

Fig. 7. SEM fractograph and Auger spectrum from point B, Fig. 4.

region (Fig. 8). Evidently, enrichment was limited to the preexisting crack surfaces.

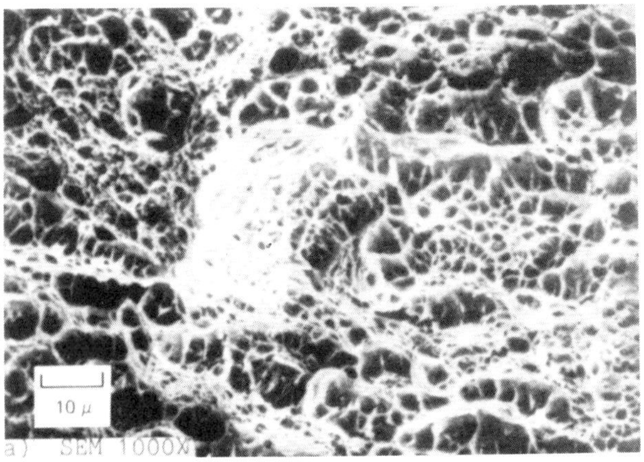

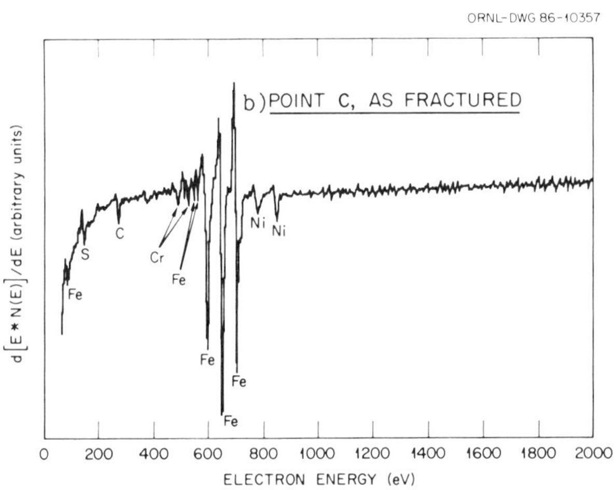

Fig. 8. SEM fractograph and Auger spectrum from point C, Fig. 4.

The next step in the AES investigation was to compare a production weld grain-boundary crack surface with a grain-boundary surface from a crack-free development weld. Providing an appropriate fracture surface for direct Auger comparison proved to be the main obstacle fronting the AES investigation. No expeditious method could be found to promote intergranular fracture in an AES tensile specimen from a sound development weld. The inability to produce a suitable AES fracture surface in a sound material is a problem frequently faced by Auger investigations. Because free-surfaces and grain-boundaries are thermodynamically similar, comparison of free-surface segregation in the development and production welds was carried out to assess the relative tendency for trace element segregation in these two weld metals.

AES FREE-SURFACE ANALYSIS - Boron, nitrogen, and sulfur peak height intensity ratios are plotted versus heating time in Fig. 9. Comparison of results for the development and production weld metals indicates a striking difference in free-surface segregation behavior. The production weld metal exhibited rapid and extensive segregation of boron and nitrogen upon heating to 800°C. A slow increase in sulfur was also observed during that period, with some indication that sulfur was displacing the boron and nitrogen. Upon heating to 1000°C, the sulfur segregated rapidly to the free-surface displacing boron and nitrogen.

In contrast to the production weld metal, the development weld metal exhibited no detectable boron segregation and relatively weak nitrogen segregation at 800°C. The time and temperature dependence of sulfur segregation was qualitatively similar to that observed in the production weld metal. The intensity of sulfur segregation was, if anything, greater in the development weld metal. The free-surface analysis results, taken with the AES weld crack surface analysis, suggest a connection between boron and nitrogen segregation and the tendency of these welds to exhibit cracking.

Boron has been associated with weld heat-affected zone (HAZ) cracking in nickel-base alloys (8) and in ferritic steels (9). No evidence of any previous boron connection was found for weld cracking similar to that discussed previously. The investigation sighted in Reference 9 indicates that nitrogen plays a critical role in the HAZ cracking process in ferritic steels. This observation may well explain why only welds incorporating the 21-6-9 nitrogen-strengthened alloy exhibited cracking.

SUMMARY

The results of the investigation into the HP 9-4-20 and 21-6-9 weld-cracking problem are most intriguing. While the development and production welds are similar in composition, their cracking behavior is a complete contrast. The cracking observed in the production welds cannot be simply explained in terms of nominal chemical composition. The weld cracking process is quite complicated, cracking seems to commence only when the 21-6-9 material is present. Apparently a rather subtle cracking mechanism (possibly liquation) is responsible for weld cracking.

The boron, nitrogen, and sulfur detected on the grain-boundary cracks in the production weld are consistent with a mechanism involving weld metal partitioning and liquation cracking. Enrichment of these elements could also result from segregation subsequent to cracking; however, the results of the surface segregation study strongly implicate boron in the cracking process. The strong correlation between boron and nitrogen segregation observed in both weld-crack and free-surface segregation examinations explain why the 21-6-9 material was required for cracking. The

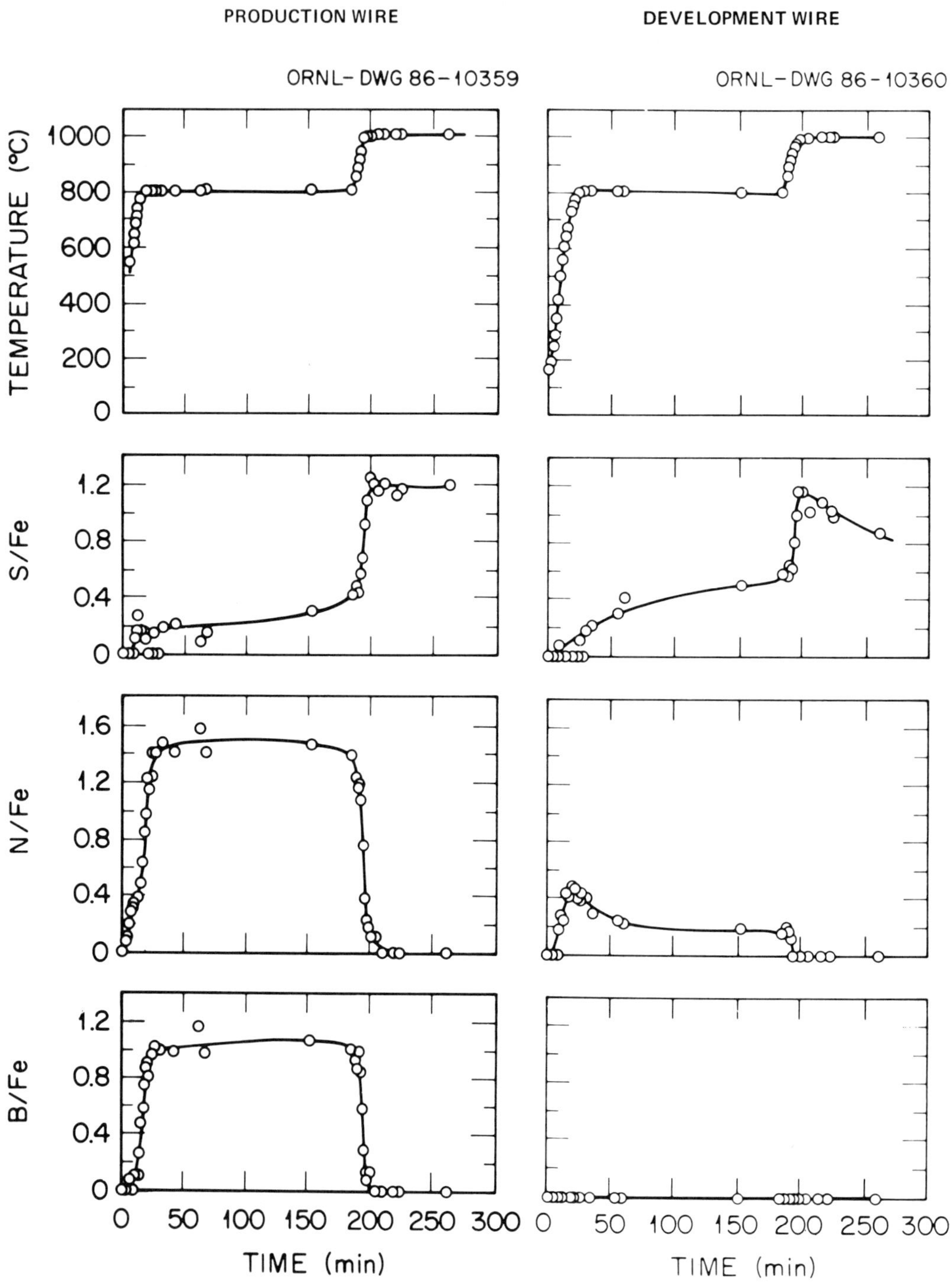

Fig. 9. Plots of the peak height intensity ratios for AES free-surface segregation study.

weld-cracking test results strongly implicate the weld wire as the source of the problem; however, the boron level in the Development wire was reported to be higher (0.002%) than in the Production wire (0.001%). This is most puzzling and indicates that additional bulk chemistry on these materials may be in order. Beyond the compositional factors responsible for this cracking phenomenon, little is known about the finer details of the mechanism. Clearly, additional research, preferably using alloys having controlled boron and nitrogen contents, is in order.

CONCLUSIONS

The following conclusions emerged from this investigation:

1) Weld cracking occurred in the HP 9-4-20 and 21-6-9 weld when the HP 9-4-20 Production wire was substituted for the HP 9-4-20 Development wire;

2) An unknown dilution contribution by the 21-6-9 stainless steel is apparently necessary to initiate weld cracking;

3) The mechanism of weld cracking is rather subtle and is not evident from bulk chemical analysis;

4) An enrichment layer of boron, nitrogen, and sulfur is associated with the weld cracks; and

5) The potential for grain-boundary enrichment is greater in the crack-susceptible production weld metal than the crack-free development weld metal.

REFERENCES

1. Reis, G. D. and S. W. Poole, Weld. J. 45 (10), 465s-476s (1966)
2. Savas, J., Becker, P. C. and S. J. Matas, Weld. J. 48 (11), 479s-492s (1969)
3. Centra D'information Du Cobalt, "Cobalt-Containing High-Strength Steels," Brussels, Belgium (1974)
4. Honeycombe, J. and T. G. Gooch, Met. Constr. Br. Weld. J. 2 (9), 375-380 (1970)
5. Mishler, H. W. Monroe, R. E. and P. J. Rieppel, Weld. J. 40 (1), 1s-7s (1961)
6. Kenyon, N. and A. L. Epstein, Weld. J. 50 (1), 46s-48s (1971)
7. Kujanpaa, V. P., David, S. A., and C. W. White, to be published, Weld. J.
8. Vincent, R., Acta. Metall. 33 (7), 1205-1216 (1985)
9. McMahon, C. J., Z Metall. 75, 496-509 (1984)

ACKNOWLEDGEMENTS

The authors are grateful to C. W. Houck and R. A. Padgett for their assistance in the experimental investigation.

EMBRITTLEMENT PHENOMENA IN THE INTERCRITICAL HAZ OF BORON CONTAINING LOW CARBON STEELS

M. Ramberg, O.M. Akselsen
SINTEF
N-7034 Trondheim-NTH
Norway

Ø. Grong
The Norwegian Institute of Technology
N-7034 Trondheim-NTH
Norway

ABSTRACT

In the present investigation, the formation of the M-A (martensite- austenite) constituent in the intercritical HAZ of two boron containing low carbon steels has been studied by optical, scanning and transmission electron microscopy of thermally cycled CVN specimens. The extent to which boron affects the HAZ properties was found to depend both on the weld thermal programme (T_p, $\Delta t_{6/4}$) and the base metal boron content. At the highest boron level (26 ppm B), no significant improvement of the HAZ toughness could be observed within the specific peak temperature or cooling time ranges examined, reflecting the fact that boron promotes the formation of the M-A constituent during the austenite decomposition. This observation is in sharp contrast to steels of lower boron contents (11ppm B), where an increase in either T_p or $\Delta t_{6/4}$ resulted in a general enhancement of the CVN impact energy due to the development of higher transformation products such as pearlite.

THE MAJOR IMPETUS for developments in low carbon microalloyed steels for offshore applications has been provided by the need for an improved heat affected zone (HAZ) resistance against hydrogen-induced cold cracking, stress corrosion cracking and brittle fracture initiation, without sacrificing base metal strength, ductility or low temperature toughness /1-3/. Despite an increased emphasis on weldability, experience shows that procedure test plates of such steels often fail to meet the stringent HAZ toughness requirements representative of welded offshore structures /4/. In addition to the grain coarsened region adjacent to the fusion boundary, low impact properties are frequently observed in the intercritical HAZ where the peak temperature has been between the Ac_1 and the Ac_3 level /5,6/. Although the mechanisms involved are not yet fully understood, it is reasonable to assume that this type of embrittlement is closely related to the decomposition of pearlite during the initial heating cycle, and subsequent retransformation of the austenite to high carbon (twinned or lath) martensite on cooling.

The present study was undertaken with the objective of investigating the specific effects of boron on the formation of the M-A (martensite-austenite) constituent in the intercritical HAZ of low carbon microalloyed steels, based on the weld simulation technique. In particular, attempts have been made to correlate the resulting HAZ toughness with the microstructure, as determined by light and transmission electron microscopy.

EXPERIMENTAL PROCEDURE

MATERIALS - The chemical composition and mechanical properties of the two boron containing low carbon steels examined are outlined in Tables 1 and 2. It is apparent from the composition data presented in Table 1 that the overall chemistry of the two base plates (in the following designated A and B) is reasonably similar, with the exception of the boron content (11 and 26ppm B, respectively). However, there is a marked difference in the base metal microstructure and inclusion morphology between the two steels. In addition to a significantly coarser ferrite grain structure of steel B as compared to steel A (14.7 μm compared to 6.9 μm), islands of martensite and bainite could also be observed within the pearlite bands of the former steel as illustrated by the optical micrographs in Fig. 1. A closer metallographic examination of this steel further revealed stringers of manganese sulphide, aligned in the base plate rolling direction (indicated by the arrow in Fig. 1B). Both factors will give rise to a low toughness of steel B, particularly in the transverse direction as shown by the CVN data contained in Table 2.

Table 1. Steel composition in weight percent. (Plate thickness: 30mm).

Steel	C	Mn	Si	P	S	Cr	Ni	Cu	Mo	V	Al	Nb	Ti	N	B	O
A	0.11	1.44	0.37	0.009	0.004	0.02	0.15	0.15	0.02	0.04	0.036	0.017	0.004	0.0040	0.0011	0.0024
B	0.13	1.38	0.30	0.013	0.004	0.03	0.18	0.11	0.07	0.04	0.057	0.002	0.003	0.0035	0.0026	0.0025

Table 2. Mechanical properties (transverse specimens).

Steel	Yield strength (MPa)	Tensile strength (MPa)	Elongation (%)	Notch toughness at -40^0C(J)
A	385	514	32	242
B	330	510	28	35

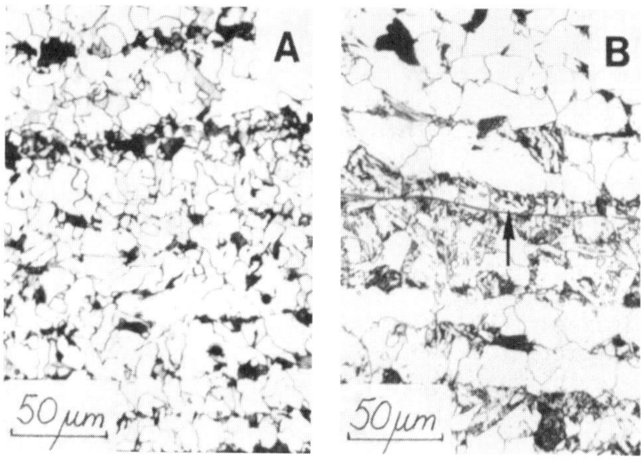

Fig. 1. Base metal microstructures: (a) Steel A, (b) Steel B.

WELD SIMULATION - The intercritical HAZ microstructures were produced by resistance heating of 11 mm x 11 mm x 100 mm steel bars in a computer controlled weld thermal simulator. In these experiments two dissimilar peak temperatures were applied ($T_p \approx 770$ and 840^0C) in combination with four different cooling programmes, corresponding to $\Delta t_{6/4}$ (the cooling time from 600 to 400^0C) equal to 5, 10, 20 and 50 s, respectively for Rosenthal thickplate heat cycles. In addition, one series of experiments was also carried out at varying peak temperatures (at a constant cooling time of $\Delta t_{6/4} \approx 6s$) within the range from 600 to 1350^0C to obtain comparable information on the impact properties of both the subcritical, the grain refined and the grain coarsened regions.

MECHANICAL TESTING - The thermally cycled steel bars were machined down to the appropriate CVN dimensions according to the ASTM specifications /7/, prior to impact testing at -20^0C. All CVN data reported in the present investigation are the average of two parallel tests. Both steels were assessed in the transverse (T-L) direction (i.e., the notch parallel to the plate rolling direction). In addition, steel B was tested in the longitudinal (L-T) direction (i.e., the notch perpendicular to the plate rolling direction) to account for an expected anisotropy in the HAZ mechanical properties caused by the presence of elongated manganese sulphide stringers in this particular base plate.

METALLOGRAPHIC EXAMINATION - The metallographic examination of the dual phase microstructures included both optical and transmission electron microscopy (TEM) as well as microhardness testing ($HV_{0.05}$) for identification and characterization of the M-A constituent. The thin foils for the transmission electron microscope were prepared by mechanical grinding down to a thickness of about 0.1mm, with subsequent thinning in an automatic jet-polishing apparatus (operating at $+20^0C$ and 36V), using a 95% 2-butoxy-ethanol/ 5% perchloric acid solution as electrolyte. In addition, selected broken CVN specimens were examined with a scanning electron microscope (SEM) to reveal the fracture mode.

RESULTS

BASIC CVN DATA - The effect of peak temperature on the CVN energy absorption at a constant cooling time of $\Delta t_{6/4} \approx 6s$ is shown in Fig. 2. From the experimental data presented it is clear that embrittlement in the HAZ of steels A and B mainly is associated with the intercritical region during low heat input weld simulation. Since the observed reduction in toughness at $T_p \approx 770^0C$ is approximately the same in both cases, it means that the pertinent variations in ferrite grain size, inclusion morphology and chemical composition between the two base plates only have marginal effects on the resulting HAZ toughness at high cooling rates.

In Fig. 3, the results obtained for the two weld simulation test series carried out at a constant peak temperature ($T_p \approx 770$ and 840^0C, respectively) and varying cooling programmes have been plotted versus the parameter $\Delta t_{6/4}$. An inspection of the figure reveals significant differences in the HAZ impact properties of steels A and B when the cooling time from 600 to 400^0C is extended beyond about 6 seconds. In the case of steel A (11 ppm B), the HAZ toughness increases steadily with increasing values of $\Delta t_{6/4}$ at both peak temperatures. In contrast, the measured CVN energy of steel B (26 ppm B) is low and virtually independent of both T_p and $\Delta t_{6/4}$ within the ranges examined. It is interesting to note that the data of steel B indicate no discernible difference in the impact properties between the (T-L) and the (L-T) direction under the prevailing circumstances. This result is to be expected if the HAZ toughness is controlled by the dual phase microstructure developed during intercritical weld simulation.

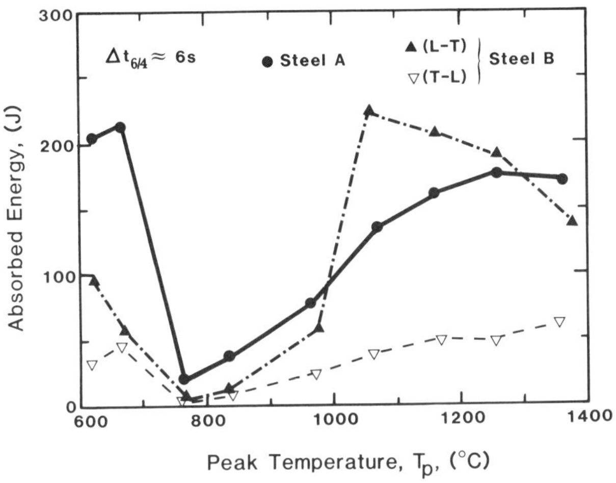

Fig. 2. Effect of peak temperature on the HAZ toughness at -20°C.

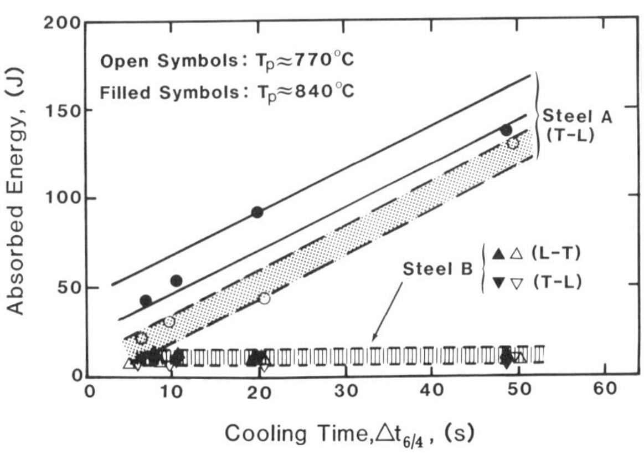

Fig. 3. Effect of weld thermal programme on the HAZ toughness at -20°C.

HARDNESS AND MICROSTRUCTURE - The results from the microhardness measurements are presented graphically in Fig. 4. At high cooling rates (i.e., $\Delta t_{6/4}$ less than about 10s), the microhardness of the M-A constituent formed along the prior base metal pearlite bands is approaching 500 $HV_{0.05}$. A decrease in the cooling rate beyond this threshold results in a dramatic fall in the hardness level. A comparison between the two steels reveals, however, that the pertinent drop in the hardness is less pronounced in the case of steel B, indicating that boron promotes the formation of the M-A con-

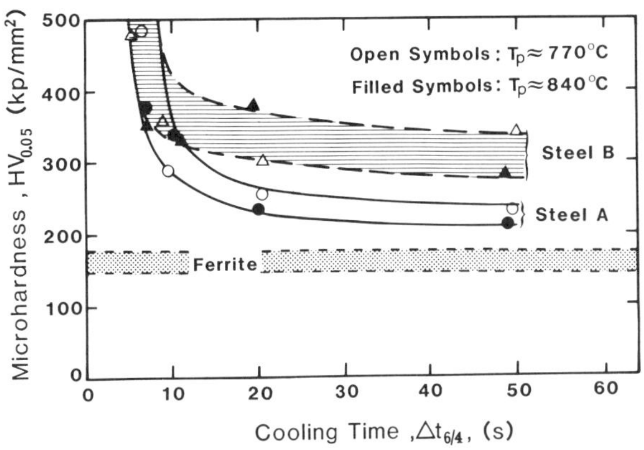

Fig. 4. Effect of weld thermal programme on the microhardness of M-A islands. The corresponding bulk hardness of the ferrite is indicated by the horizontal scatter band in the graph.

stituent during relatively slow cooling. This assumption is further substantiated by the TEM micrographs contained in Figs. 5 and 6. At high cooling rates , the austenite formed within the HAZ of both steels due to dissolution of pearlite is seen to retransform to high carbon (twinned or lath) martensite on cooling, as would be expected from the hardness data presented in Fig. 4. In contrast, during relatively slow cooling, the transformation product is mainly pearlite in the case of steel A (11 ppm B), while lath martensite still can be observed within the HAZ of steel B (26 ppm B) even at $\Delta t_{6/4}$ close to 50 seconds.

DISCUSSION

BORON HARDENABILITY MECHANISMS - The primary effect of boron on the steel hardenability is to raise the energy barrier to ferrite nucleation at the austenite grain boundaries. Although the mechanisms involved are not yet fully understood, free boron is known to segregate extensively to the austenite grain boundaries /8/ where it reduces the grain boundary energy /8-10/. Based on transformation kinetics theory it can be argued that a decrease in the austenite grain boundary energy will suppress the nucleation of ferrite at these sites /11/. Further enhancement of the energy barrier to ferrite nucleation due to an increase in the total strain energy of the embryo is possible if the free grain boundary volume becomes filled with either boron atoms /12/ or borocarbide precipitates, $Fe_{23}(B,C)_6$. In addition, these precipitates have been proposed to inhibit ferrite nucleation at the early stages of the transformation by virtue of their semicoherent nature /13/. The inhibition of ferrite nucleation at the coherent side of the austenite grain boundary has been verified experimentally /14/. On this basis it is not surprising to find that the conditions for martensite formation during intercritical weld si-

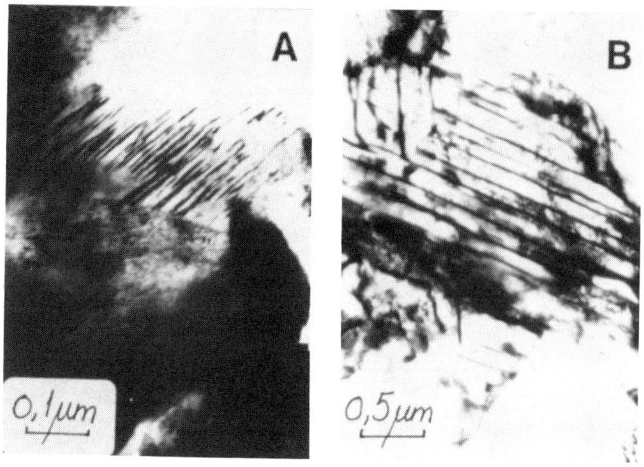

Fig. 5. Steel A. TEM micrographs of different transformation products at observed $T_p \approx 770^0C$; (a) Twinned martensite ($\Delta t_{6/4} \approx 6s$), (b) Pearlite ($\Delta t_{6/4} \approx 50s$).

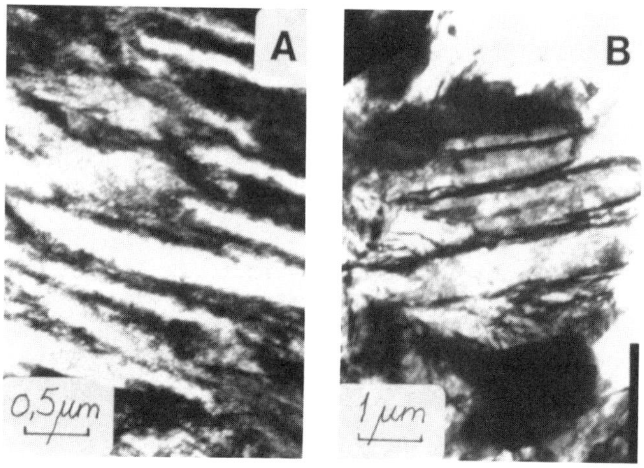

Fig. 6. Steel B. TEM micrographs of different transformation products observed at $T \approx 770^0C$; (a) High carbon lath martensite ($\Delta t_{6/4} \approx 6s$), (b) Low carbon lath martensite ($\Delta t_{6/4} \approx 50s$).

mulation generally are seen to be more favourable in steel B than in steel A. In fact, a theoretical analysis of the extent of boron segregation which occurs to the austenite grain boundaries during various thermal programmes shows that the base metal boron content must be close to (or above) 26ppm B in order to significantly affect the HAZ hardenability under the prevailing circumstances /15/.

KINETICS OF AUSTENITE FORMATION - In general, the formation of austenite during intercritical annealing of low carbon microalloyed steels can be separated into the following three main stages /16/:

(1) Rapid growth of austenite into pearlite until the dissolution process is completed.

(2) Slower growth of austenite into ferrite at a rate which is either controlled by carbon diffusion in the austenite or by diffusion of substitutional elements such as manganese in the ferrite.

(3) Very slow final equilibration of solute concentration gradients within the austenite through diffusion.

At peak temperatures above 770^0C, the rate of pearlite transformation would be expected to be sufficiently high for complete dissolution of the phase during weld simulation /16/. Further growth of the austenite into the ferrite requires, however, that the peak temperature of the applied thermal cycles is raised to about 840^0C in order to reach the kinetic (paraequilibrium) stage where the reaction is controlled by diffusion of carbon in austenite /16, 17/. Under such conditions, the growth rate of the austenite increases by several orders of magnitude compared with that observed at lower peak temperatures, which, in turn, gives rise to a significant increase in the amount of martensite formed within the HAZ of steel B as illustrated by the optical micrographs in Fig. 7. The trend observed in Fig. 7 will probably be reversed in the absence of extensive boron segregations to the grain boundaries (steel A), particularly at lower cooling rates, due to partial retransformation of the austenite to ferrite and pearlite. This may explain why the HAZ toughness of steel A is significantly improved by an increase in the peak temperature from 770 to 840^0C during medium and high heat input weld simulation as previously shown in Fig. 3.

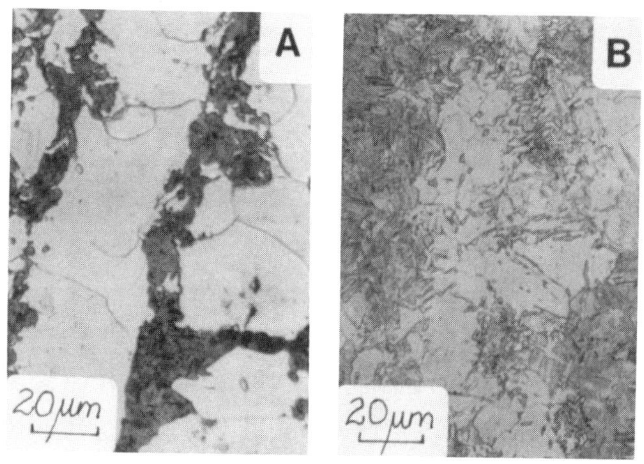

Fig. 7. Steel B. Examples of microstructures developed during intercritical weld thermal simulation ($\Delta t_{6/4} \approx 6s$); (a) $T_p \approx 770^0C$, (b) $T_p \approx 840^0C$. The martensite is indicated by the dark areas in the micrographs.

STRUCTURE - PROPERTY RELATIONS - As already stated, the characteristic low toughness of the intercritical HAZ is closely related to the existence of islands of high carbon (twinned or lath) martensite within the ferrite matrix. From recent literature on dual phase steels, it can be rationalized that the formation of such martensite islands will result in the development of significant strain in the adjacent ferrite due to the high transformation stresses involved /18/. During subsequent plastic deformation of the steel, the lower flow strength of the ferrite compared with that of the martensite gives rise to a further enhancement of the stress concentrations in these areas /18, 19/. At room temperature, the dislocations formed are mobile, which explains the excellent tensile ductility of dual phase steels /20/. However, when mechanical testing is performed at lower temperatures, their mobility becomes strongly reduced, and hence, a brittle fracture can develope /18, 21/. Accordingly, it is likely that embrittlement in the intercritical HAZ of ferrite-pearlite steels arises from an extensive build-up of dislocations at the ferrite/martensite interface, with consequent initiation of a cleavage type of fracture in the ferrite. This conclusion is rather surprising, since it has been generally believed that failure of dual phase microstructures is caused by fracture in the martensite region /18/.

Based on the above discussion, one would expect a relation between the HAZ toughness on the one hand and the microhardness of the M-A constituent on the other. Such a plot is shown in Fig. 8, using data from Figs. 2, 3 and 4. It should be noted that for this correlation, the measured CVN energies have been normalized with respect to the base metal toughness, as observed after a short tempering heat cycle at $T_p \approx 620^0C$ (Fig. 2), to account for the pertinent differences in both the microstructure

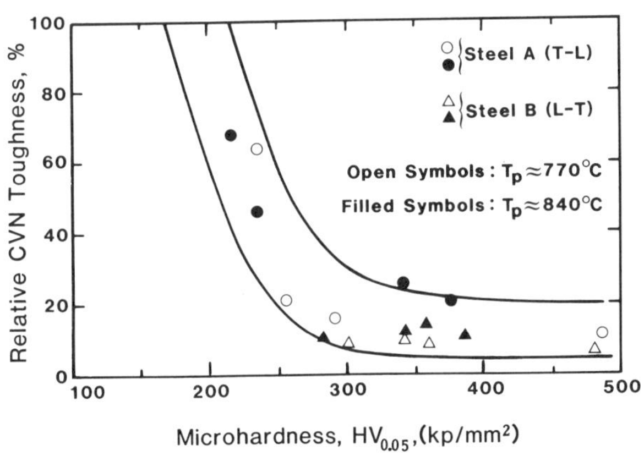

Fig. 8. Correlation between the microhardness of M-A islands and the relative CVN toughness of steels A and B.

(ferrite grain size, martensite/pearlite banding) and the inclusion morphology between steels A and B. From the results presented in Fig. 8 it is apparent that the relative CVN data for both steels fall within the same scatter band. When the microhardness level exceeds about 300 $HV_{0.05}$, the transformation product is predominantly twinned or lath martensite, which gives rise to a quasicleavage fracture mode (Fig. 9) and a significant reduction in the HAZ impact properties. Below this threshold, the HAZ toughness of steel A rapidly approaches that of the base metal due to the formation of higher transformation temperature products such as pearlite.

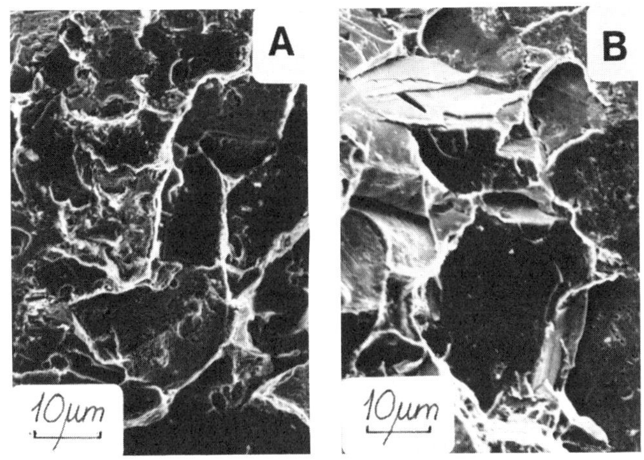

Fig. 9. SEM fractographs of CVN specimens; (a) Steel A - $T_p \approx 770^0C$, $\Delta t_{6/4} \approx 6s$. (b) Steel B-$T_p \approx 770^0C$, $\Delta t_{6/4} \approx 6s$.

CONCLUSIONS

The basic conclusions that can be drawn from this investigation are as follows:

- Embrittlement in the intercritical HAZ of low carbon microalloyed steels is closely related to the decomposition of pearlite during the initial heating cycle, and subsequent retransformation of the austenite to high carbon (twinned or lath) martensite on cooling.

- The detrimental effect of boron on the HAZ toughness observed in the present investigation can mainly be attributed to the fact that boron enhances the formation of the M-A constituent during relatively slow cooling.

- Although the mechanisms involved are not yet fully understood, indications are that the characteristic low toughness of the intercritical HAZ arises from an extensive build-up of dislocations at the ferrite/martensite interface during the γ to α' transformation, with consequent initiation of a cleavage type of fracture in the ferrite.

REFERENCES

1. F.B. Pickering: The Metallurgist and Materials Technologist 5 375-379 (1975).
2. E.E. Fletcher: "High Strength Low Alloy Steels - Status Selection and Physical Metallurgy", Ohio, Battelle Press (1979).
3. H. Suzuki: Welding in the World 20, 121-148 (1982).

4. V.G. Marshall: Met. Constr. 14, 528-532 (1982).
5. C. Thaulow, A.J. Paauw, Å. Gunleiksrud, O.J. Næss: Met. Constr. 17, 94R-99R (1985).
6. Ø. Grong, O.M. Akselsen: "On the HAZ Toughness of Low Carbon Microalloyed Steels for Offshore Applications", Met. Constr., (in press).
7. ASTM Standard: E23-83 (1983)
8. E.P. Hondros, M.P. Seah: "Physical Metallurgy", 3rd ed., p. 855-931, Elsevier Science Publisher B.V., Amsterdam (1983).
9. J.C. Fisher: Trans. AIME 200, 1146-1147 (1954).
10. D.A. Mortimer, M.G. Nicholas: Met. Sci. J. 10, 326-332 (1976).
11. D.A. Porter, K.E. Easterling: "Phase Transformations in Metals and Alloys", 1st Ed., p. 272-273, Van Nostrand Reinhold (1981).
12. C.R. Simcoe, A.R. Elsa, G.K. Manning: J. Metals, 193-200 (1955).
13. R.C. Sharma, G.R. Purdy: Met. Trans. A 4A, 2303-2311 (1973).
14. Ph. Maitrepierre, D. Thivellier, R. Tricot: Met. Trans. 5A, 287-301 (1975).
15. O.M. Akselsen, Ø. Grong, P.E. Kvaale: "A Comparative Study of the HAZ Properties of Boron Containing Low Carbon Steels", Met. Trans. (in press).
16. G.R. Speich, V.A. Demarest, R.L. Miller: Met. Trans. 12A, 1419-1429 (1981).
17. X-L. Cai, A.J. Garratt-Reed, W.S. Owen: Met. Trans. 16A, 543-557 (1985).
18. N.J. Kim, G. Thomas: Met. Trans. 12A, 483-489 (1981).
19. X-L. Cai, J. Feng, W.S. Owen: Met. Trans., 16A, 1405-1415 (1985).
20. C.A.N. Lanzilotto, F.B. Pickering: Met. Sci. 16, 371-382 (1982).
21. J.H. Chen, Y. Kikuta, T. Araki, M. Yoneda, Y. Matsuda: Acta Met. 32, 1779-1788 (1984)

HOT-CRACKING STUDIES OF TURBINE ROTOR Cr-Mo-V STEELS

S.Y. Yoo, A.L. Lu, A.W. Pense
Lehigh University
Bethlehem, Pennsylvania

ABSTRACT

An investigation has been made of four rotor materials, three CrMoV steels and a NiCrMoV steel, both with respect to solidification and HAZ liquation cracking. It involved Varestraint testing using autogenous gas tungsten arc welding and metallographic examinations using scanning and/or transmission electron microscopy. The detrimental effects of S and P on both the solidification and the HAZ liquation cracking have been confirmed. The harmful effect of Ni on solidification cracking has also been confirmed in a modern vacuum poured NiCrMoV steel which contains relatively small amounts of S and P, 0.008 wt% and 0.005 wt%, respectively. On weld metal fracture surfaces, eutectic type sulfides of MnS-CrS rather than those of MnS-FeS have been found in an old CrMoV steel. The precipitation of square or triangular shaped sulfides has been observed on the HAZ hot crack surfaces. A number of precipitates hitherto undetected have been found in the steel: a) iron phosphide of dendritic form on the weld metal fracture surfaces, b) precipitation of eutectic-like vanadium carbides on the HAZ hot crack surfaces, c) iron phosphide on the HAZ hot crack surfaces even at high heat input welding (i.e. slow cooling rate).

LOW ALLOY STEELS containing chromium, molybdenum, vanadium, and/or nickel are used extensively as the rotor material of steam turbines in power generating plants. The rotors can often be damaged during service due to creep, fatigue, or stress corrosion[1]. Until the last few years, weld repair of the damaged rotors has not been usually considered feasible because of their high carbon content, impurities, alloying elements, and the complicated geometry of the rotor area to be repaired. This was especially true of old 1950's air melted rotors.

As welding technology has progressed, however, it is becoming clear that these problems can be overcome. In fact, Brown Boveri Corporation (BBC) has been actively pursuing the turbine rotor weld repair market, and has reported that a typical weld repair can be accomplished in five weeks while rotor replacement would take eight months and would be 6 or 7 times more costly than weld repairs [2].

When considering weld repairs, two types of weldability are of concern: fabrication weldability and service weldability. From the fabrication weldability viewpoint, hot, cold and stress relief cracking must be avoided, along with other more common weld defects. From the service weldability viewpoint, the strength, toughness, and creep resistance of the weld metal must match that of the base material. Although all are worthy of study, the investigation here is directed to the problem of hot cracking in both weld metal and the heat affected zone (HAZ) using the Varestraint test [3] developed at Rensselaer Polytechnic Institute.

Weld metal hot cracking in rotor steels usually occurs in the root pass of a weld repair where dilution of the weld metal from base material is extensive. Considerable research has been undertaken in order to correlate hot cracking in weld metal with chemical compositions or welding parameters [4,5,6]. The effect of phosphorous, however, does not seem to be explained clearly in terms of metallurgical evidence [6,7,8]. Most studies on the HAZ hot cracking have been performed using a thermal cycle simulation test and have showed that liquated boundaries due to in-situ melting of sulfides contributed to the HAZ hot cracking [9,10]. Japanese workers [11,12], however, reported that the segregation of alloying elements such as Ni, Cr, and Mo at the bands associated with rolling caused the HAZ liquation cracking.

In this paper, an emphasis is placed on investigating metallurgical causes of hot cracking in both the weld metal and the HAZ using scanning and/or transmission electron microscopy.

EXPERIMENTAL PROCEDURE

MATERIALS - The materials used in this study were sectioned from steam turbine rotor forgings. The chemical compositions of the rotors are listed in Table 1. Two rotors, the Buck and the Gallatin, were 1950's air-melted 1 Cr - 1.25 Mo - 0.25 V steels, which had been retired after about 20 years service. The former was due to temper embrittlement and the latter was due to the catastrophic failure in 1974 [13]. The other two (labeled 2A and 3A) were modern electric furnace vacuum poured rotor forgings. The 2A is a conventional 1 Cr - 1.25 Mo - 0.25 V steel forging used for the high pressure section and the 3A is a 2½ NiCrMoV steel forging used for the low pressure (LP) section of steam turbines. With these four materials, the differences in base metal chemistry which exist could be expected to produce variations in hot cracking susceptibility for study.

VARESTRAINT TEST - The hot cracking susceptibility of each material was evaluated by the Varestraint test using autogeneous Gas Tungsten Arc Welding (GTAW). The device for Varestraint testing is schematically shown in Figure 1. The rotor was cut through cross section in 50 mm interval and specimens were cut from the resulting 50 mm wide discs. The dimensions of the specimen were 50 x 300x 12.7 mm and the face of 50 x 300 mm was parallel to the longitudinal axis of the rotor (disc). Specimens were ground with Blanchard grinder to the required thickness. Specimens were degreased with acetone before welding. Welding was performed at two different heat inputs, 2.7 kJ/mm and 1.2 kJ/mm, respectively. The welding parameters used are listed in Table 2. Augmented strains were varied from 0.25% to 3.7% by changing the die block radii. The strain is calculated using the expression, $\varepsilon = t/2R$, where t is the specimen thickness and R is the radius of the die block. Since the most clear ranking of hot cracking susceptibility was found at 3.7% augmented strain, most of the testing was performed at this strain level. In order to evaluate the effect of welding speed on hot cracking of weld metal, three different travel speeds, 100 mm/min, 150 mm/min, and 175 mm/min at the same low heat input were used. Three specimens were tested at each condition.

For the 3A forging, a Mn addition was tried using a steel tube of 3.2 mm O.D. packed with pure Mn powder. The tube was laid down in a groove, 3.3 mm wide, 125 mm long, and 2 mm deep, made in the center of the 50 mm x 300 mm face of the Varestraint specimen. The purpose of the Mn addition to the 3A forging was to increase the Mn/S ratio because the low Mn/S ratio of the 3A, 35, would obscure the effect of Ni on solidification cracking. Welding was carried out two times at high heat input: one for premelting Mn powder and one for the Varestraint test.

The length of cracks was measured on the as-welded surface of each specimen under an optical microscope with a calibrated reticle at 80X magnification. The total combined crack length was plotted as a function of augmented strains. All data presented here represent the average of three test values.

METALLOGRAPHY - Microsections where hot cracks formed were removed from the Varestraint specimen for optical and electron microscopy. These were polished and etched with a 2% nital or 2% sodium bisulfite ($NaHSO_3$) in water. The latter reveals the weld metal solidification structure very nicely. Elemental segregation in both weld metal and the HAZ was studied with Jeol Superprobe 733. Qualitative analyses using X-ray area scanning with a Wavelength Dispersive Spectrometer (WDS) were performed. Quantitative x-ray microanalyses included interdendritic areas and the HAZ liquated boundaries.

Cracks were opened by carefully making notches near the tip of the crack and then impact loading at liquid nitrogen temperature. Opened hot crack surfaces were examined in scanning electron microscope (SEM) with an Energy Dispersive Spectrometer (EDS) system to characterize the liquid films or precipitates on the surfaces. Direct carbon extraction replicas were made on both the HAZ hot crack surfaces and the weld metal fracture surfaces (not high temperature crack surfaces) by depositing carbon in a vacuum chamber. The replicas were released by immersion in a 10% nital in methanol and supported on a copper grid. Examination was carried out in a Philips 400 electron microscope with an EDS system. The phases on the HAZ hot crack or weld metal surfaces were identified by EDS, selected area diffraction pattern and convergent beam diffraction pattern (CBDP). By these means, the nature of the impurity segregation associated with the cracking could be determined.

RESULTS AND DISCUSSION

SOLIDIFICATION CRACKING - The results of Varestraint test are summarized in Tables 3, 4, 5, and 8 and Figures 2,3,4,6, and 13. Figure 2

Table 1 - Chemical Analysis of Rotor Steels Studied (wt%)

Elements / Rotors	C	Mn	P	S	Si	Ni	Cr	Mo	Cu	V	Sn	Mn/s
Buck	.31	.72	.028	.036	.29	.14	1.01	1.25	.13	.22	.09	20
Gallatin	.36	.84	.012	.011	.23	.06	1.07	1.34	.11	.26	.03	76
2A	.31	.71	.008	.006	.25	.45	1.12	1.28	.057	.25	.004	118
3A	.27	.28	.005	.008	.05	2.81	1.73	0.37	.067	.08	.004	35

Table 2 - Welding Parameters Used

Parameters	High Heat Input, 2.7 KJ/mm	Low Heat Input, 1.2 KJ/mm
Direct Current, amp	310	170
Voltage, volts	14.5	11.5
Travel Speed, mm/min	100	100
Arc Length, mm	3	2.5
2% Thoria Tungsten Electrode dia., mm	3.2	2.4
Electrode Tip Angle	90°	90°
Argon Gas Flow, cfh	40	40

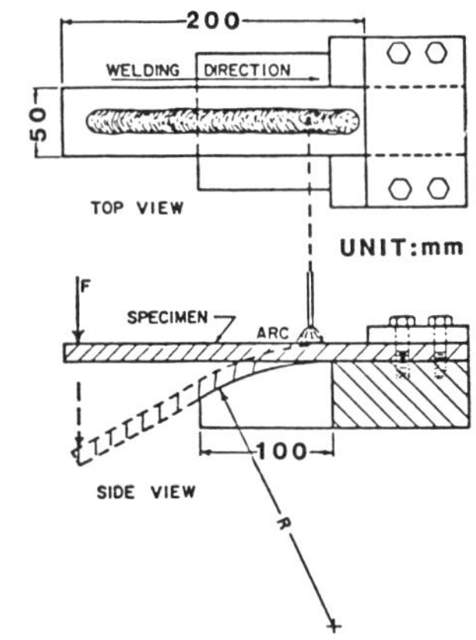

Figure 1. Simplified sketch of the Varestraint testing device.

Table 3 - Weld Metal Total Crack Length, mm of Varestraint Test Results For Four Rotor Materials

Augmented Strain, %	Buck 2.7 KJ/mm	Buck 1.2 KJ/mm	Gallatin 2.7 KJ/mm	Gallatin 1.2 KJ/mm
0.25	0.3	-	-	-
0.33	3.4	0	0	-
0.50	8.6	1.0	1.8	-
1.0	10.4	1.7	2.2	-
2.0	18.9	7.5	-	-
3.7	27.8 ± 0.8	8.6 ± 1.7	12.5 ± 0.6	5.3 ± 0.6

Augmented Strain, %	2A (CrMoV) 2.7 KJ/mm	2A (CrMoV) 1.2 KJ/mm	3A (NiCrMoV) 2.7 KJ/mm	3A (NiCrMoV) 1.2 KJ/mm
0.25	-	-	-	-
0.33	-	-	0.4	-
0.50	-	-	1.3	-
1.0	0	-	5.1	-
2.0	-	-	-	-
3.7	3.7 ± 1.9	2.4 ± 0.8	18.2 ± 2.5	7.7 ± 1.2

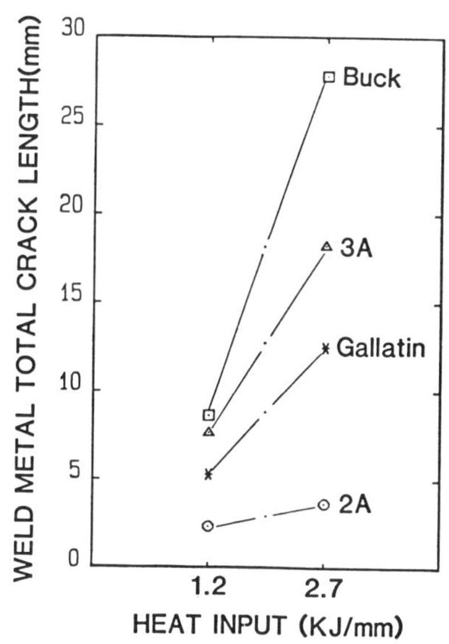

Figure 2. Effect of heat input on Varestraint test results at 3.7% augmented strain.

shows the weld metal total combined crack length determined at 3.7% augmented strain for all materials at both 2.7 kJ/mm and 1.2 kJ/mm heat inputs (cf. Table 3). Of the three CrMoV materials, i.e. the Buck, the Gallatin, and the modern CrMoV, 2A forging, it is clear that the historic rotors are more susceptible to solidification cracking than the modern 2A forging. The cracking as might be expected, followed the sulfur level of the materials, with the Buck rotor having the highest sulfur content, 0.036 wt%, and the greatest hot cracking tendency whereas the modern CrMoV, 2A, forging has the lowest sulfur content, 0.006 wt%, and the least hot cracking susceptibility. There is widespread agreement that sulfur promotes cracking and the effect of sulfur is often expressed by Mn/S ratios [14,15] due to the interaction between them. As Mn/S ratio increases, the solidification cracking susceptibility decreases. Figure 3 shows that the ranking of materials is in good agreement with the Mn/S ratios, i.e. the ratio of 35 for NiCrMoV forging, 3A, is better than the Buck ratio of 20 and worse than both the Gallatin ratio of 76 and the CrMoV forging, 2A, ratio of 118.

The 3A forging which is rated as having the second highest hot cracking susceptibility also contains high Ni content (2.8 wt%). Nickel is known [16,17] to affect hot cracking susceptibility adversely by favoring peritectic reaction, which causes the molten weld metal to start solidifying as more primary austenite. The maximum solubilities of sulfur and phosphorus are 0.15 wt% and 2.55 wt%, respectively in ferrite, which are much larger than those of S and P in austenite, 0.065 wt% and 0.3 wt%, respectively [18]. Therefore, more segregation of S and P occurs as the solidification mode changes from the primary ferrite to the primary austenite. Although the modern CrMoV forging, 2A, has 0.45 wt% Ni, this Ni level does not seem to affect the hot cracking susceptibility at all.

Since the low Mn/S ratio of the 3A forging obscures the effect of Ni on cracking, Varestraint test with the Mn addition was performed. Table 4 summarizes the results. Even though the Mn/S ratio was increased to the value of '180' which is much higher than the 2A Mn/S ratio, 118, the total combined crack length (TCL) of the 3A with Mn added, 14.3 mm, was still much longer than that of the 2A, 3.7 mm. Therefore, the modern forging, 3A, does not necessarily guarantee good hot cracking resistance. When Ni content is high in a steel, the permissible impurity levels should be reconsidered.

Heat Input - As shown in Figures 2 and 4, the high heat input welding produced more cracks than the low heat input welding did. First of all, the width of the weld bead (15.5 mm) at high heat input was much wider than that of the low heat input weld bead (7 mm). This means there is more area for hot cracks to form at high heat input. Secondly, solidification microstructures at low heat input (Fig. 5B & D) are finer than those (Fig. 5A & C) at high heat input. The coarser microstructures induce more segregation at interdendritic areas thus resulting in more cracks. Thirdly, the smaller temperature gradient in the molten weld metal at high heat input increases the cracking temperature range [19] thus resulting in the increased amount of cracking.

Travel Speed - The results of Varestraint test with three different travel speeds, 100, 150, and 175 mm/min are summarized in Table 5 and Figure 6. These show that the cracking susceptibility tends to increase as the travel speed increases. The results may be correlated with the change in both solidification substructure and its size, and the transverse thermal stress at the trailing edge of the weld pool due to different welding speeds as proposed by John et al. [20].

SOLIDIFICATION STRUCTURES

The NiCrMoV Forging, 3A Fig. 7A shows a very pronounced cellular dendritic microstructure revealing a hot crack formed during Varestraint test. The crack follows along a prior austenite grain boundary. The back scattered electron image in Fig. 7B clearly reveals that inclusions (arrowed), mostly sulfides, form only at the prior austenite grain boundaries. This explains the occurrence of the cracking along austenite grain boundaries. This also indicates that segregation of impurities is more dependent on the austenite grain size than the substructure size of weld metal in the NiCrMoV type steel. This point was shared by Thompson [21] who reported that cracking was not related to cell size, but is dependent on solute segregation and structural features of grain boundaries. Elongated and slightly deformed sulfides arrowed in Fig. 7C and a debris of the dendritic shaped sulfide in Fig. 7D were observed on the opened weld metal hot crack surfaces.

In addition to the dendritic shaped sulfides, a network of sulfides and chrysanthemum-like patterns were found (Fig. 8). Closer examination revealed that the pattern consisted of arrays of rod-like inclusions and their morphology was indicative of a eutectic solidification structure. A qualitative analyses of an inclusion within the pattern and the sulfide network (labeled 1 in Fig. 8B) by EDS show the local concentration of sulfur, chromium and manganese (Fig. 8C).

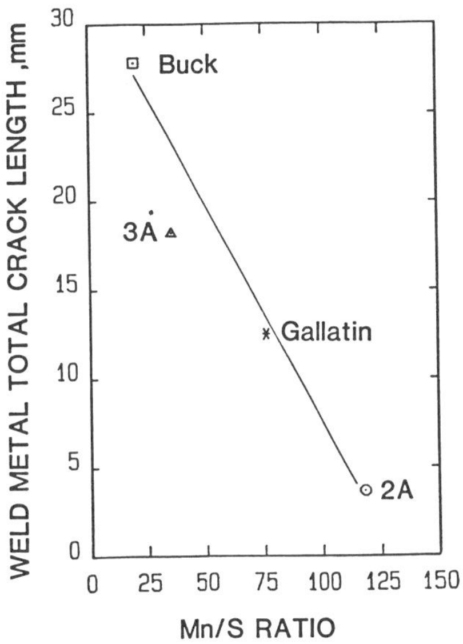

Figure 3. Weld metal total crack length vs. Mn/S ratio for four rotor materials at 3.7% augmented strain and high heat input (2.7KJ/mm).

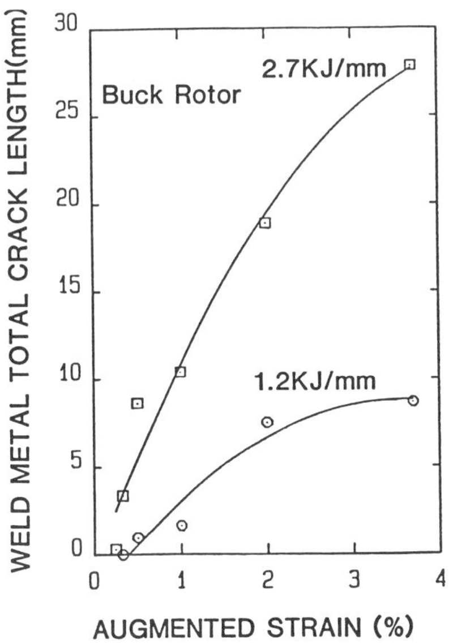

Figure 4. Effect of heat input on the Buck rotor Varestraint test results.

Table 4 - Weld Metal Total Crack Length (mm) of the 3A Forging With Mn Addition, Varestraint Test Results at High Heat Input (2.7 KJ/mm)

Augmented Strain, %	3A Mn/s = 35	3A + Empty Tube Mn/s = 42	3A + Tube With Mn Mn/s = 180
3.7	18.2 ± 2.5	16.6 ± 0.9	14.3 ± 0.8

Table 5 - Weld Metal Total Crack Length (mm) of Four Rotor Materials at Three Different Travel Speeds, Varestraint Test Results at 3.7% Augmented Strain and Low Heat Input (1.2 KJ/mm)

	Buck			Gallatin	
100	150	175	100	150	175
8.6 ± 1.7	12.1 ± 0.9	12.3 ± 0.8	5.3 ± 0.6	5.9 ± 0.6	7.8 ± 1.4
	2A			3A	
100	150	175	100	150	175
2.4 ± 0.8	-	8.7 ± 1.2	7.7 ± 1.2	6.9 ± 0.1	11.1 ± 3.7

(mm/min)

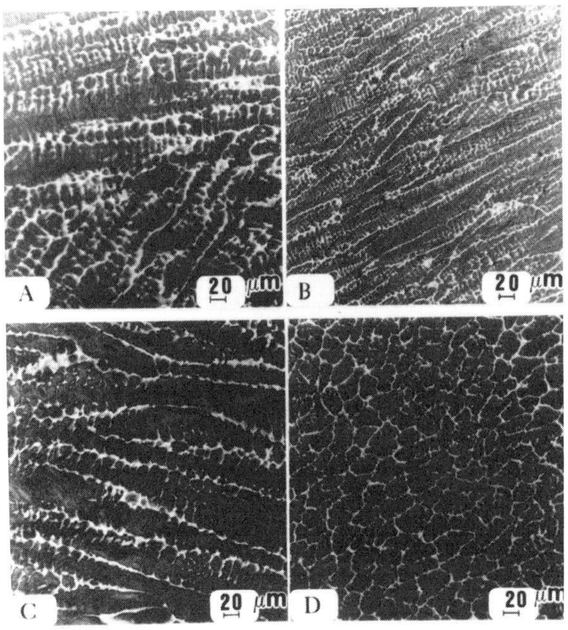

Figure 5. A) & B) Weld metal microstructures of bead top surface at high and low heat input welding respectively. C) & D) Microstructures of weld cross section at high and low heat input welding respectively. The Buck Rotor. 2% Aqueous Sodium Bisulfite Etch.

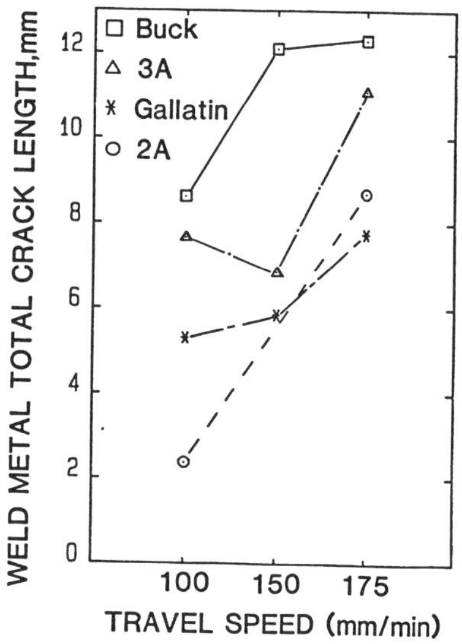

Figure 6. Effect of travel speed on Varestraint test results at 3.7 % augmented strain and low heat input (1.2KJ/mm).

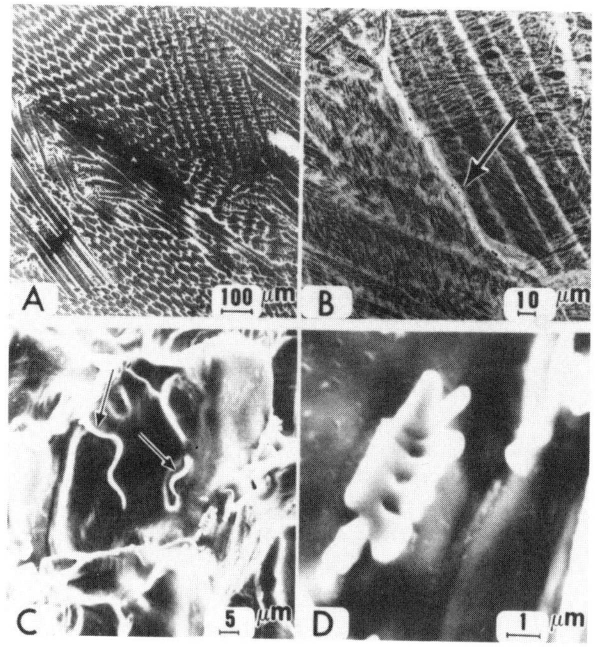

Figure 7. A) Solidification cracking along grain boundaries at 3.7% augmented strain. B) BSE image of weld metal showing sulfides formed only on an austenite boundary. C) & D) Weld fracture surface showing sulfides. The 3A Forging.

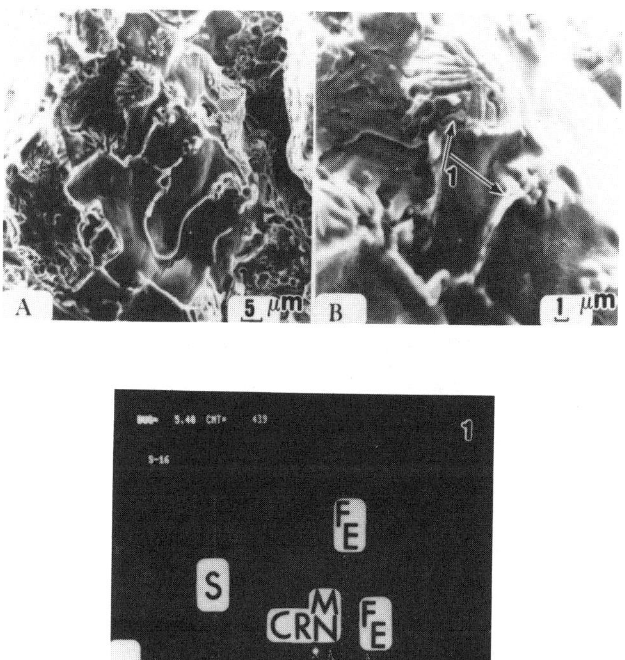

Figure 8. A) Sulfides network and chrysanthemum-like pattern on hot crack surfaces of the high heat input 3A weld metal. B) Magnified view of A. C) EDS spectrum obtained from the area labeled 1 in B.

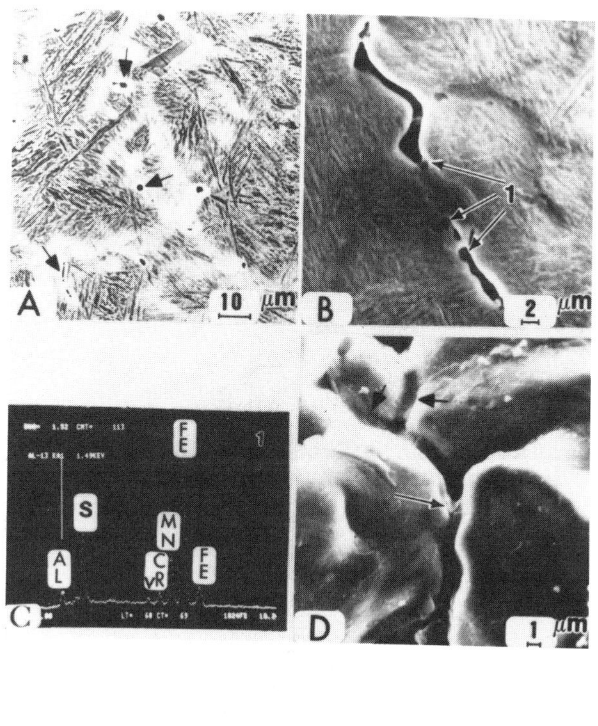

Figure 9. A) BSE image of weld metal showing sulfides at interdendritic areas. B) Hot cracks along the dendritic boundary. C) EDS spectrum from the sulfides labeled '1' in B). D) Dendritic appearance of the fracture surface showing sulfide liquid films at boundaries. The Buck Rotor.

These clearly show metallurgical evidence of solidification cracking caused by low melting point sulfides rich in S, Cr, and Mn formed at the grain boundaries. Carbon extraction replicas on the polished and etched bead top surface and thin foil TEM specimens of the weld metal failed to detect the distribution of sulfides. These inclusions can therefore only be characterized in detail by examination of fracture surfaces in SEM or TEM using carbon extraction replicas.

Buck Rotor - Fig. 9A shows a back scattered electron image of the Buck weld metal where globular or elongated sulfides (arrowed) formed at the interdendritic areas. In Fig. 9B, hot cracks follow along the interdendritic areas and sulfides can be seen at the tip of the cracks (labeled 1). The chemical composition of the sulfides were obtained with EDS in SEM as shown in Fig. 9C. The sulfides have local concentration of S, Mn, Cr, V, and Al. In Fig. 9D typical hot crack surfaces of pronounced dendritic structures are shown, which indicate a weak interdendritic condition at the moment that the augmented strain was introduced. This supports Borland's theory [22] of solidification cracking, which states that low melting point liquid caused by the elemental segregation at the growing tip of the dendrites is not able to accommodate the shrinkage or the applied strain and cracking results.

Fig. 10A and B show interdendritic sulfides extracted from the polished and nital etched bead top surface of the Buck weld metal with their EDS spectra. Surprisingly, the sulfides labeled 1, 2, and 3 do not have iron but are rich in S, Mn, Cr, and V. The precipitates labeled 4 and 5, which formed at the edge of globular sulfide, show the local concentration of Mo, V, Mn, Cr, and Fe. From the chemical composition it could be inferred that these would be carbides. Direct carbon extraction replica was tried on hot crack surfaces but it was not successful in pulling out previous liquid films etc., partly due to the heavy oxidation of crack surfaces.

A Buck weld metal of high heat input welding far away from the strained area was broken open assuming that a crack would propagate along the weakest planes on which sulfides or phosphides were present. Direct carbon extraction replicas on these fresh weld metal fracture surfaces revealed several interesting features. Fig. 11 shows eutectic-like sulfides (A and B) and EDS spectrum (D) of S, Mn, Cr, V, and very little Fe and Mo peaks. A typical CBDP is reproduced in Fig. 11. A number of CBDP's obtained from the eutectic-like sulfides were determined by calculating interplanar spacings and the sulfides were identified as α-MnS; α-MnS posesses a fcc structure and a lattice parameter of 5.224A. It is surprising that the sulfides have almost no iron but high Cr in contrast to very common eutectic type sulfides Fe-Mn-S. Comparing the EDS spectrum of the sulfides in the SEM (Fig. 9C) with the one in the STEM (Fig. 11D), the difference is in the iron peak: the high iron peak in the SEM and almost no iron peak in the STEM. This is due to the large interaction volume of x-ray emission in the case of the bulk materials in the SEM [23]. Therefore, x-rays are produced from the depth of the matrix even though the particles being analyzed are sitting on the surface. More correct chemical composition of the sulfides can be obtained with EDS using extraction replicas in the STEM.

Apart from the eutectic type sulfides a number of hitherto undetected dendritic shaped thin films (Fig. 12A & B) were observed in replicas removed from the weld metal fracture surfaces. EDS analysis showed that they contained Fe, P, Mo, Cr, and V (Fig. 12D). From the analysis of electron diffraction patterns, they were identified as iron phosphide, FeP. Calculated interplanar spacing "d" values corresponded to those given for the compound, FeP (Table 6).

Although eutectic type sulfides (Fig. 11) and dendritic shaped iron phosphides (Fig. 12) were observed on weld metal fracture surfaces, not on the hot crack surfaces, it is clear that these low melting point phases formation could lower the freezing temperature of the liquid thereby causing solidification cracking when the restraint is present. The role of phosphorous in promoting hot cracking was understood as having formed low melting point Fe-Fe_3P eutectic in the Fe-P phase diagram [4,18] by Mishler and coworkers. Wilkinson and coworkers [16] reported that phosphorous would assist segregation of sulfur but there was no supporting evidence for this fact. In the present study, phosphorous was found to form the iron phosphide, FeP.

Quantitative x-ray microanalysis was accomplished on the Buck rotor weld metal to study elemental segregation (Table 7). Elements such as Sn, P, Mo, V, Cr, and Si segregate to the cell boundaries in decreasing order. In the case of Mn and S they interacted with each other to form sulfides. Therefore, most Mn and S were consumed and were not

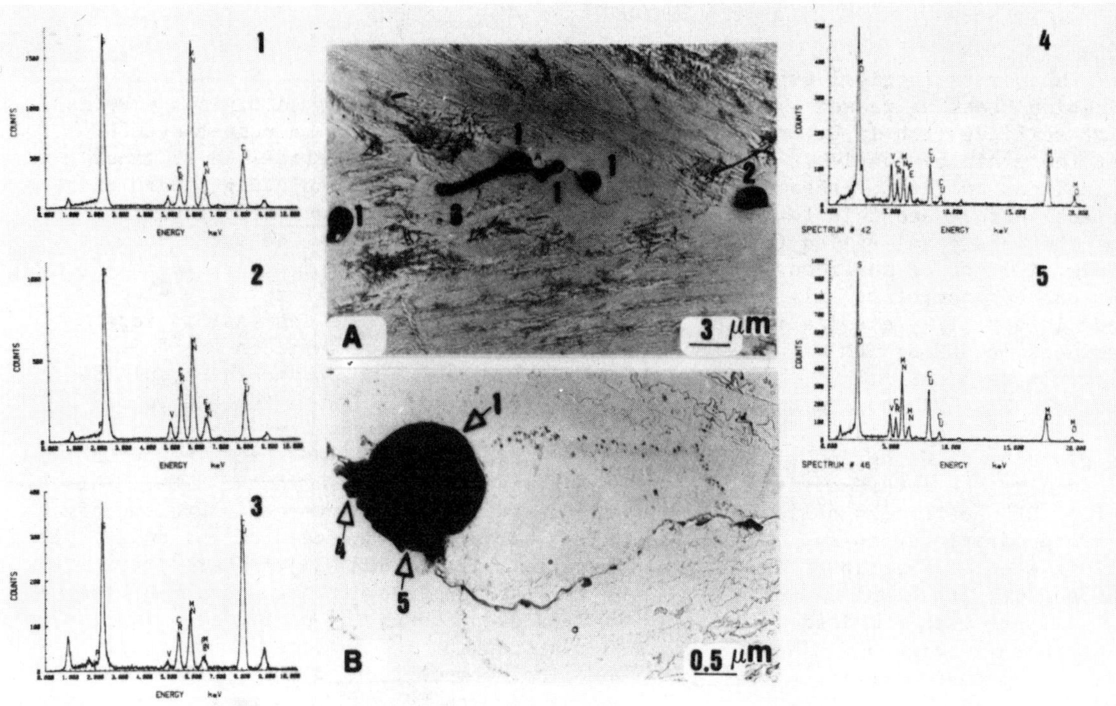

Figure 10. A) and B) Sulfides extracted from the polished and 2% nital etched high heat input Buck weld metal bead top surface. Carbon extraction replica.

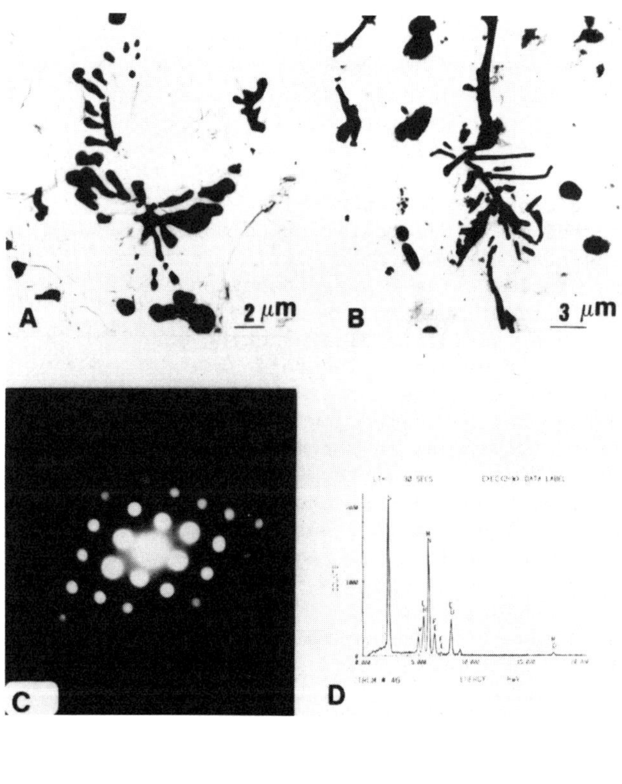

Figure 11. A) and B) Eutectic sulfides extracted from the high heat input Buck weld metal fracture surfaces. C) CBDP obtained from the eutectic sulfides, Beam direction B = [001]. D) EDS spectrum obtained from the eutectic sulfides. Direct carbon extraction replica.

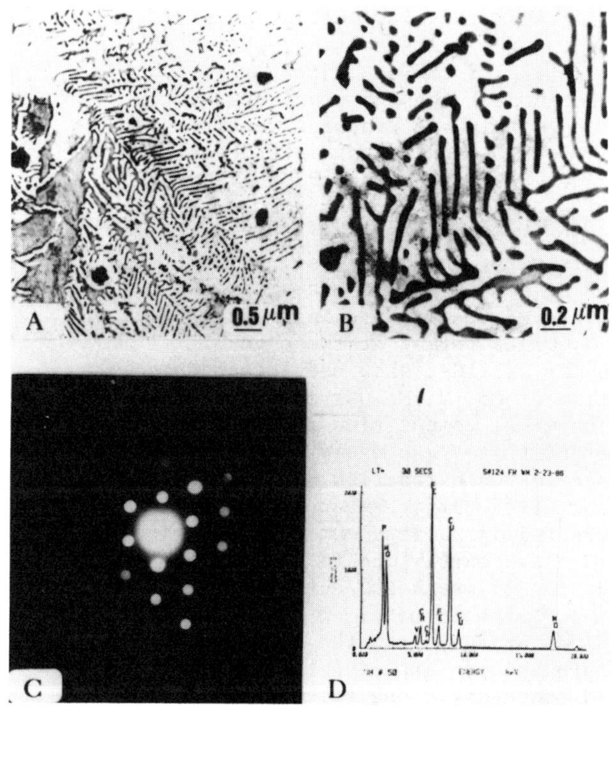

Figure 12. A) and B) Dendritic iron phosphides extracted from the high heat input Buck weld metal fracture surfaces. C) CBDP obtained from the iron phosphides, FeP. D) EDS spectrum obtained from the precipitates in A and B. Direct carbon extraction replica.

Table 6 - Determination of Diffraction Patterns From Precipitates in Photo 10.

FeP d values, A Orthohombic a = 5.782 b = 5.177 c = 3.089	Calculated d values, A Pattern 1	Pattern 2	Pattern 3	Pattern 4	Pattern 5
3.38			3.44	3.55	
3.15	3.11			2.94	
3.00					
2.62			2.59		2.39
2.52		2.50			
2.39				2.26	
2.28					2.22
2.20					
2.15		2.18			
2.11	2.11				
2.04		2.07	2.05		1.88
1.98					
1.88	1.88				
1.84					
1.80					
1.77			1.70		
1.69					
1.64					
1.59	1.58				
1.55				1.51	
1.51					
1.48					1.45
1.44			1.42		
1.42		1.40			
1.40					
1.36	1.36			1.34	
1.33					
1.30					
1.28					
1.27			1.20		
1.25					

Table 7 - Quantitative Electron Probe Microanalysis Results with Jeol 733 (wt%)

Elements	WM Matrix -1	WM Matrix -2	WM Matrix -3	Avg.WM Matrix	Cell Bdy -1	Cell Bdy -2	Cell Bdy -3	Avg. Cell Bdy*
Ni	0.12	0.13	0.12	0.12	0.15	0.15	0.17	0.16
Mo	1.09	1.04	1.01	1.05	2.43	2.63	2.52	2.53
Cr	0.95	1.01	0.95	0.97	1.32	1.25	1.29	1.29
Si	0.54	0.54	0.57	0.55	0.67	0.65	0.82	0.71
V	0.26	0.25	0.25	0.25	0.52	0.53	0.57	0.54
Mn	0.73	0.82	0.75	0.77	0.66	0.54	0.67	0.62
Sn	0.05	0.07	0.06	0.06	0.53	0.68	0.60	0.60
S	0.05	0.04	0.05	0.05	0.06	0.09	0.06	0.07
P	0.02	0.02	0.02	0.02	0.13	0.14	0.08	0.12
Fe	95.49	96.45	95.10	95.68	94.46	92.91	93.0	93.46
TOTAL				99.52				100.1

* Avg. Cell Bdy: Average Cell Boundary

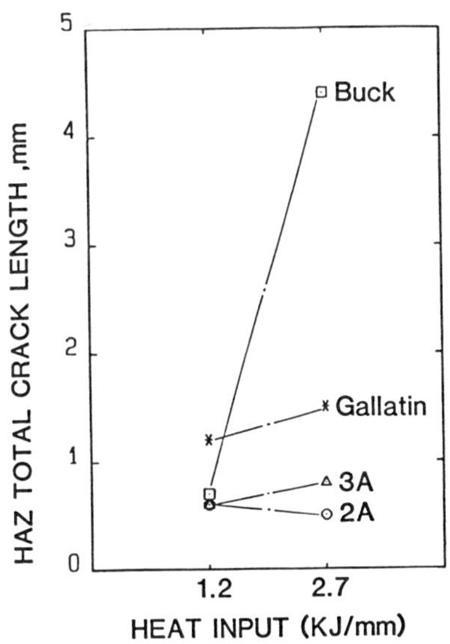

Figure 13. Effect of heat input on Varestraint test results at 3.7% augmented strain.

detected at the interdendritic areas. Ni did not show any segregation. In the authors' opinion, these elemental segregations could lower the melting point of interdendritic liquid or could form a low melting point eutectic compound, thus increasing hot cracking susceptibility. It is not surprising that Mo, Cr, and V are present in iron phosphide, FeP, (Fig. 12D) and Cr and V are present in MnS type sulfides (Fig. 11D) because these are the elements which segregate to the interdendritic areas.

HAZ LIQUATION CRACKING - HAZ cracking generally occurred only in Varestraint specimens subjected to relatively large augmented strains and the threshold strain for the onset of HAZ cracking is always greater than the threshold strain for solidification cracking (Table 8). The HAZ total crack length is plotted as a function of heat inputs at 3.7% augmented strain for four rotor steels (Figure 13). The ranking of the materials at high heat input is the Buck, the Gallatin, the 3A, and the 2A in that order. In contrast with the solidification cracking results, the ranking of the Gallatin and the 3A rotors was reversed in the HAZ liquation cracking. For low heat input welding, the HAZ hot cracking sensitivity of four rotors remained almost the same. The Buck rotor showed 6 times increase in the HAZ total crack length at high heat input than at low heat input welding (Table 8). The other three materials, the Gallatin, the 3A, and the 2A, did not show significant HAZ hot cracking sensitivity to different heat inputs.

Buck Rotor - Fig. 14 shows a HAZ liquated boundary close to the fusion line with a crack on it and x-ray mappings of the boundary for elements such as V, Mo, P, Cr, Mn, S, and Ni. It is found that Mo, V, Cr, and P segregate at the liquated boundary but Ni does not. This is contradictory to the work by Tamura et al. [12] who reported that liquated boundaries are caused mainly by the enrichment of Ni and Cr on the boundaries. They employed x-ray line scanning across the boundary using thermally simulated high strength Ni-Cr-Mo steels. From the results of Mn and S x-ray mappings (Fig. 14) it can be inferred that the MnS type sulfides precipitate at the liquated boundary. Fig. 15 shows that a HAZ crack propagates along the interface between the matrix and the manganese sulfides. At the edges of the crack (arrowed), Mn and S coating has occurred, which indicates in-situ partial melting of the sulfides.

In an effort to determine if the liquated boundary microstructure resulted primarily from a crack back-filling mechanism as originally described by Savage and coworkers [24], a Varestraint test was made at high heat input without bending. Quantitative Microprobe analysis of the liquated boundaries (Table 9) in the unstrained specimen revealed a localized enrichment in Sn, Mo, P, V, Cr, and Si in decreasing order. The degree of segregation in the HAZ liquated boundary is almost the same as interdendritic segregation reported in Table 7. Therefore it is clear that elemental segregations at liquated boundaries can occur in the absense of any backfilling.

Opened HAZ hot crack surfaces were examined in the SEM to characterize precipitates or liquid films formed on them. Dendritic shaped sulfides which contain S, Mn, Fe, Cu, Cr, and V were observed (Fig. 16). Fig. 16C clearly shows the evidence of the origin of the sulfide films coming from in-situ melting of a sulfide inclusion which happens to sit in the HAZ very close to the fusion line. In addition to the sulfide films, surface "ferns" were observed on the HAZ fracture surfaces (Fig. 17 A and B) and their morphology was indicative of an eutectic solidification. They exhibited three interesting features: a) rod-like or globular inclusions in surface fern areas, b) a hole in the center of the fern, c) an inclusion sitting in the center of another fern. The EDS spectrum both from the rod-like inclusions within the ferns (labeled 1B) and the lower part of the inclusion (labeled 1 in 17A) in the center of the ferns shows localized concentration of S, Cu, Cr, and Mn. It is interesting to note that sulfides in the Buck rotor HAZ at high heat input contain copper (Fig. 16D and 17D). The prsence of Cu, Cr, and V in MnS type inclusions would further reduce the melting point of the inclusions thus making the material more susceptible to the HAZ liquation cracking. Another interesting finding was that the EDS spectrum from the upper part of the inclusion (labeled 2 in Fig. 17A) showed the localized concentration of P only. This indicates that an iron phosphide has a lower melting point than that of the sulfides and is the last constituent to solidify during cooling. Additional features of the HAZ crack surfaces were the presence of iron phosphide sheets (arrowed) at grain boundaries (Fig. 18 C and D) and MnS type sulfides (Fig. 18 A & B) with square or triangular morphology.

In Fig. 19A, SEM micrograph shows the HAZ liquation cracks on the polished and sodium bisulfite (2% aqueous solution) etched

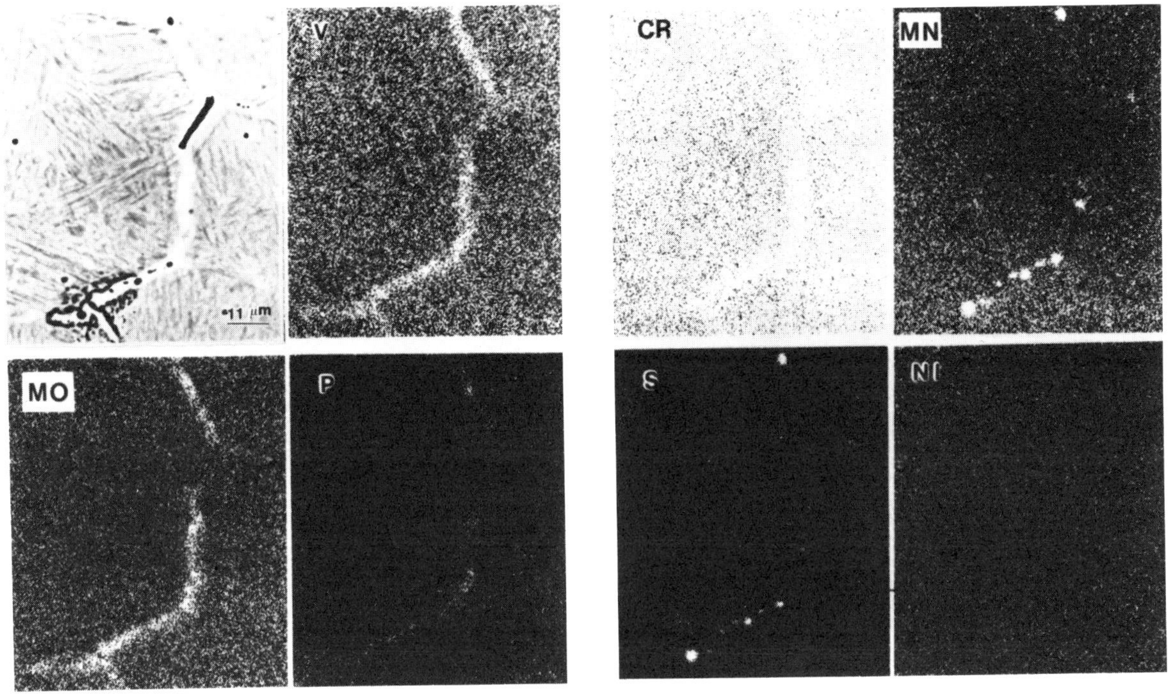

Figure 14. X-ray area scans of the high heat input Buck HAZ liquated boundary.

Table 8 - HAZ total crack length (mm) of Varestrait Test Results at Two Different Heat Inputs

Augmented Strain, %	Buck		Gallatin	
	2.7 kJ/mm	1.2 kJ/mm	2.7 kJ/mm	1.2 kJ/mm
1.0	0	0	0	-
2.0	0.8	0.2	-	-
3.7	4.4 ± 2.2	0.7 ± 0.3	1.5 ± 1.6	1.2 ± 0.6

Augmented Strain, %	2A (CrMoV)		3A (NiCrMoV)	
	2.7 kJ/mm	1.2 kJ/mm	2.7 kJ/mm	1.2 kJ/mm
1.0	0	-	0	-
2.0	-	-	-	-
3.7	0.5 ± 0.4	'0.6 ± 0.2	0.8 ± 0.4	0.6 ± 0.1

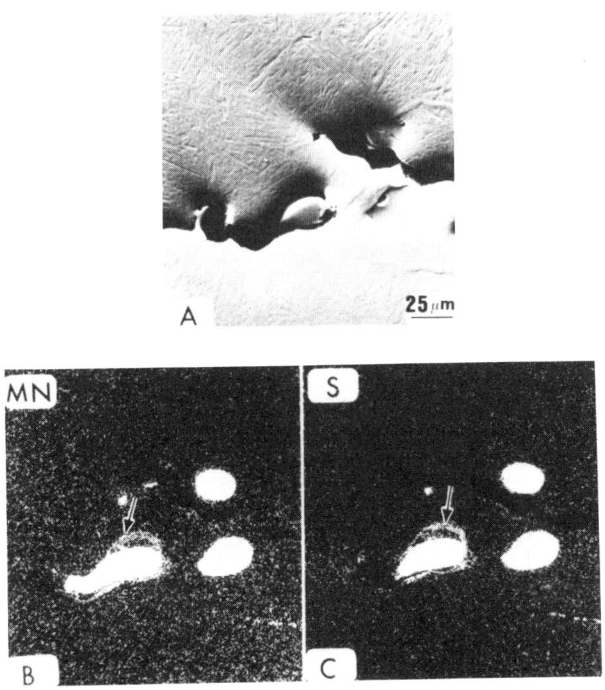

Figure 15. X-ray area scans of the high heat input Buck HAZ hot crack associated with sulfide inclusions. A) HAZ close to the fusion line. B) Mn scan. C) S scan.

Table 9 - Quantitative Electron Probe Microanalysis Results with Jeol 733 (wt%)

Elements	HAZ Matrix -1	HAZ Matrix -2	HAZ Matrix -3	Avg HAZ Matrix	Liq. Bdy -1	Liq. Bdy -2	Liq. Bdy -3	Liq. Bdy -4	Avg. Liq. Bdy*
Ni	0.14	0.14	0.15	0.14	0.15	0.14	0.14	0.14	0.14
Mo	1.21	1.13	1.26	1.20	2.12	2.64	1.91	2.39	2.27
Cr	1.00	1.04	1.08	1.04	1.28	1.35	1.27	1.33	1.31
Si	0.58	0.58	0.60	0.59	0.72	0.71	0.72	0.72	0.72
V	0.28	0.27	0.30	0.28	0.45	0.57	0.43	0.53	0.50
Mn	0.78	0.84	0.92	0.85	1.07	0.44	1.08	1.09	0.92
Sn	0.13	0.12	0.13	0.13	0.33	0.56	0.30	0.48	0.42
S	0.05	0.04	0.05	0.05	0.05	0.06	0.05	0.06	0.06
P	0.04	0.04	0.04	0.04	0.04	0.13	0.07	0.10	0.09
Fe	95.03	94.38	95.19	94.87	92.82	92.85	94.70	93.08	93.36
TOTAL				99.19					99.79

* Avg. Liq. Bdy: Average Liquated Boundary

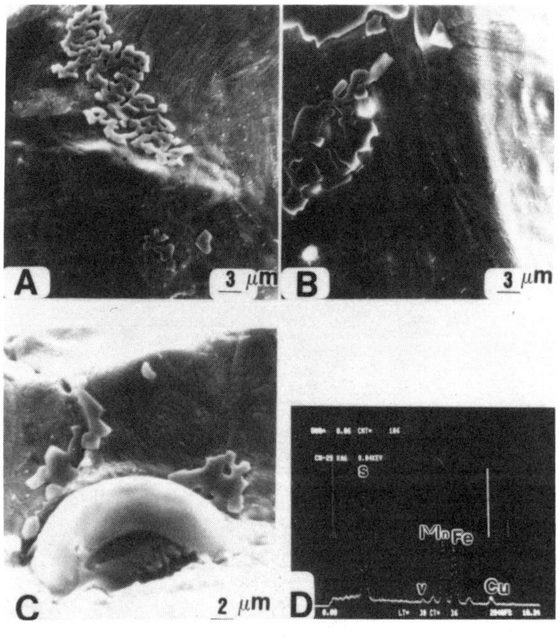

Figure 16. A), B) and C) SEM images of the sulfide films on the HAZ hot crack surfaces of the Buck rotor at high heat input. D) EDS spectrum obtained from the sulfide films in A, B, and C.

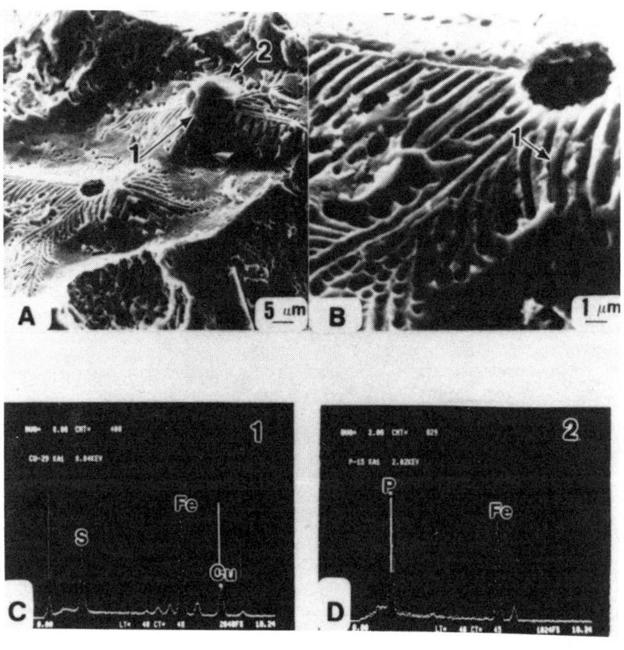

Figure 17. A) and B) SEM images of fern-shape sulfides on the HAZ grain faces of the Buck rotor at high heat input. C) EDS spectrum obtained from the area labeled 1. D) EDS spectrum obtained from the area labeled 2.

Varestraint specimen. A higher magnification image of the area labeled 'a' in Fig. 19A revealed the presence of precipitates on the liquated boundaries as shown in Fig. 19B. The EDS spectrum obtained from the precipitates (Fig. 19C) shows the peaks of Mo, V, Cr, and Fe. Since sulfur K line peaks overlap with Molybdenum L line peaks, it is not easy to distinguish them from each other. One way to solve this problem was to operate SEM at 30 kV and to compress the energy span to 20 kV. Then the very small Mo $K\alpha$ peak appears at 17.40 kV. Also from experience in the STEM, whenever vanadium content was relatively high Mo was always present. From the chemical compositions, the precipitates in Fig. 19B could be carbides. To confirm this point further, direct carbon extraction replicas were made on both the HAZ hot crack surfaces and the HAZ fracture surfaces.

The dendritic shaped precipitates hitherto undetected were observed in HAZ hot crack surfaces (Fig. 20). EDS analysis showed they were enriched in V and Mo. The chemical composition indicates them to be carbides precipitated during welding. Electron diffraction patterns (Fig. 21C and D) obtained from the same type of precipitates were determined as vanadium carbides, VC (Table 10). They seem to be brittle because numerous cracks are present in them (Fig. 21A). Thin sheets of precipitates around a round particle were also observed in the replica (Fig. 23). A qualitative analysis by EDS shows that they contain P, Mo, Fe, V, and Cr. The round particle could be an iron oxide because only an iron peak was observed. The typical electron diffraction patterns obtained from the thin precipitates are reproduced in Fig. 22C and D. They were identified as iron phosphide, FeP, as determined in Table 6. Thin sheets of iron phosphides precipitated in dendritic manner were also observed in the HAZ fracture replicas (Fig. 23A and B). Apart from the thin iron phosphides, rather thick ones were observed (Fig. 23C and D). This particular one has a localized concentration of Sn in addition to the other elements such as P, Mo, Cr, and V. The plates of iron phosphide at the grain boundaries will almost certainly contribute to the brittleness of the material.

Other Rotor Steels - In the case of the Gallatin rotor, liquation of sulfides along the grain boundary and precipitation of square-shape sulfides as shown in Fig. 24A and B, respectively, also occurred on the HAZ hot crack surfaces. A hole (arrowed) in the grain face of Fig. 24A indicates that the liquid film resulted from in-situ melting of inclusions, here MnS type inclusions, in the manner proposed by Pepe and Savage [25] in their constitutional liquation theory.

The 3A forging, NiCrMoV steel, also showed the precipitation of square-shape MnS type sulfides on the HAZ crack surfaces (Fig. 24C) but the degree of precipitation was not as extensive as the Buck rotor.

The metallurgical causes of HAZ liquation cracking can be summarized as follows: a) The infiltration of extensive dendritic MnS type sulfides into grain boundaries due to in-situ melting of inclusions, b) the formation of fern shaped sulfides, c) the precipitation of MnS type cuboidal sulfides, d) the precipitation of thin sheets of iron phosphides, e) the precipitation of vanadium carbide, VC.

The role of sulfur in promoting cracking is evident. The MnS type inclusions partially melt and penetrate into the grain faces due to the capillary action. This results in crack formation and embrittlement under the action of the tensile stresses resulting from the augmented strain and the shrinkage stress of the solidifying weld metal. It is well established that intergranular precipitates make the grain boundaries brittle by raising stress concentrations at the edges of the precipitates and resulting in small crack formation. It can be argued qualitatively that as more intergranular sulfides are precipitated, the total stress required to propagate cracks along grain boundaries would be reduced. This leads to a greater portion of intergranular fracture.

In contrast to the work by Brammar [26] who reported that iron phosphides in burnt steels precipitated only at very high cooling rates, i.e. brine quenching, but redissolved in the matrix at slower cooling rates, thin sheets of iron phosphides on the HAZ fracture surfaces at high heat input were observed in the present study. Although Philips and Jordan [27] did not observe phosphides in the HAZ crack surfaces, they reported that phosphouous is 2.8 times more potent than sulfur in promoting the HAZ cracking in low alloy steels.

The ranking of materials according to HAZ hot cracking susceptibility in Fig. 13 can be explained by their impurity contents. The Buck rotor with the highest S and P, 0.036% and 0.028%, respectively, showed the worst cracking

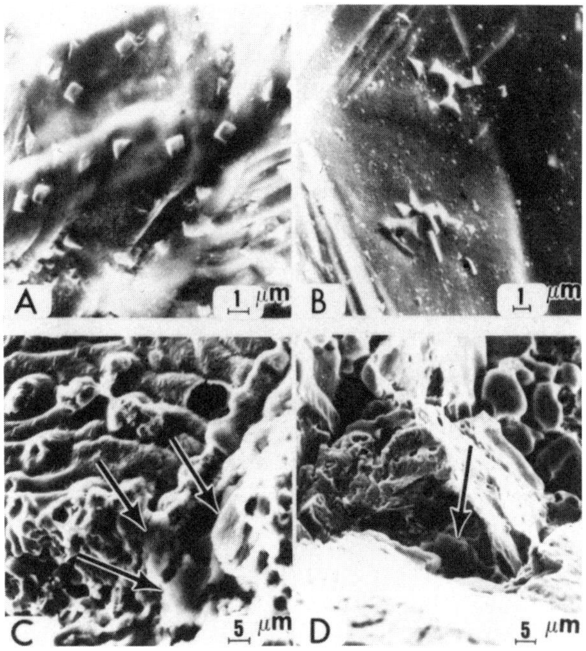

Figure 18. A) & B) Weld HAZ fracture surface showing the sulfides precipitation with square or triangular morphology. C) & D) Weld HAZ fracture surface close to fusion line, showing sheets of iron phosphide at grain boundaries. The Buck Rotor.

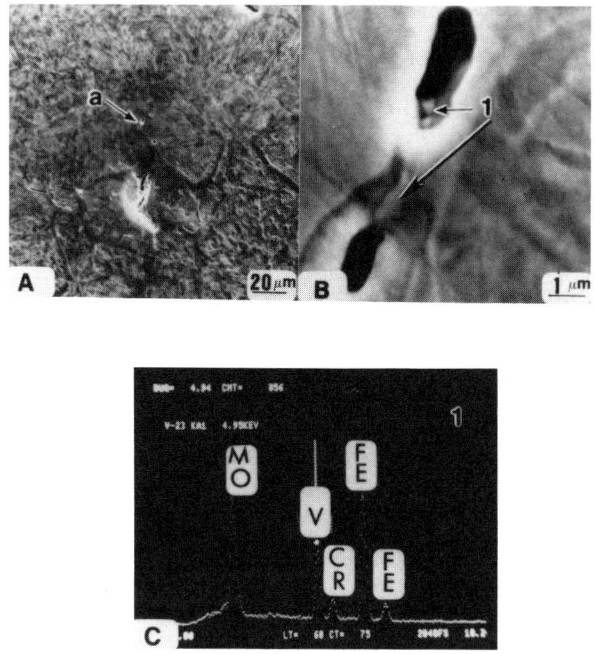

Figure 19. A) SEM images of the fusion line of the Buck rotor at high heat input showing the HAZ hot cracks (labeled a). B) Magnified view of area 'a' in A showing precipitates formed on the liquated boundary. C) EDS spectrum obtained from the precipitate labeled 1 in B.

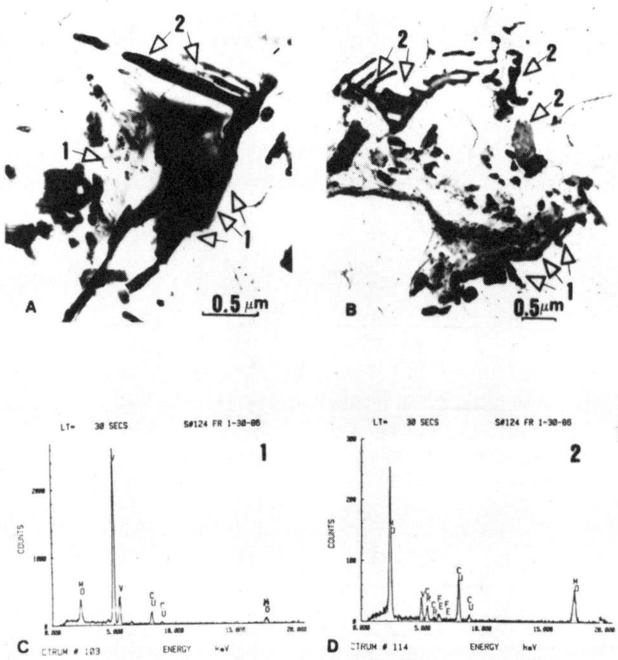

Figure 20. A) and B) TEM micrographs of the HAZ fracture surfaces showing the dendritic precipitates. C) EDS spectrum from the region '1'. D) EDS spectrum from the region '2'. The Buck rotor at high heat input. Direct carbon extraction replica.

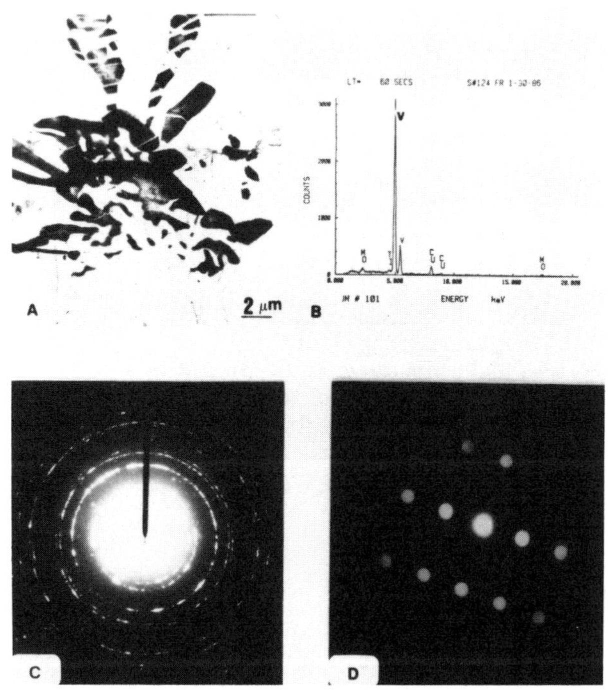

Figure 21. A) TEM image of thin precipitates on the HAZ fracture surfaces of the Buck rotor at high heat input. Direct carbon extraction replica. B) EDS spectrum obtained from the precipitates in A. C) Selected area diffraction pattern (SADP) obtained from the precipitates in A. D) CBDP of the precipitates in A, Beam direction, B = [112].

Table 10 - Electron Diffraction Results from Vanadium Carbides

VC FCC, a = 4.16A		Pattern 1 B = [112]	Pattern 2 B = [011]	Ring Pattern
d(A)	hkl	d	d	d
2.40	111	2.38	2.39	2.44
2.07	200		2.09	2.13
1.47	220	1.48	1.43	1.49
1.25	311	1.26	1.25	1.27
1.20	222			1.22
1.04	400			
0.95	331			
.93	420	0.93		0.94
.85	422			
.80	333			

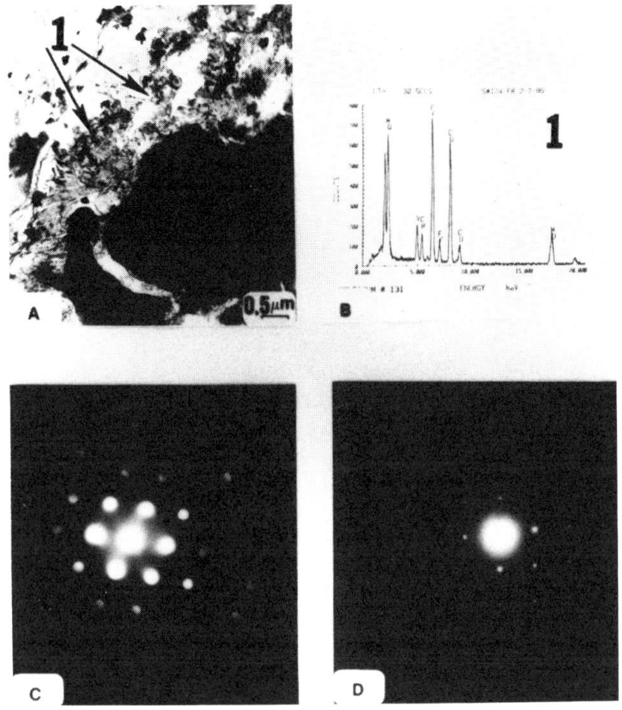

Figure 22. A) TEM image of iron phosphides in thin plate form on the HAZ fracture surfaces of the Buck rotor at high heat input. Direct carbon extraction replica. B) EDS spectrum obtained from the area '1' in A. C) CBDP obtained from the thin precipitates in A. D) µµ diffraction pattern from the thin precipitates in A.

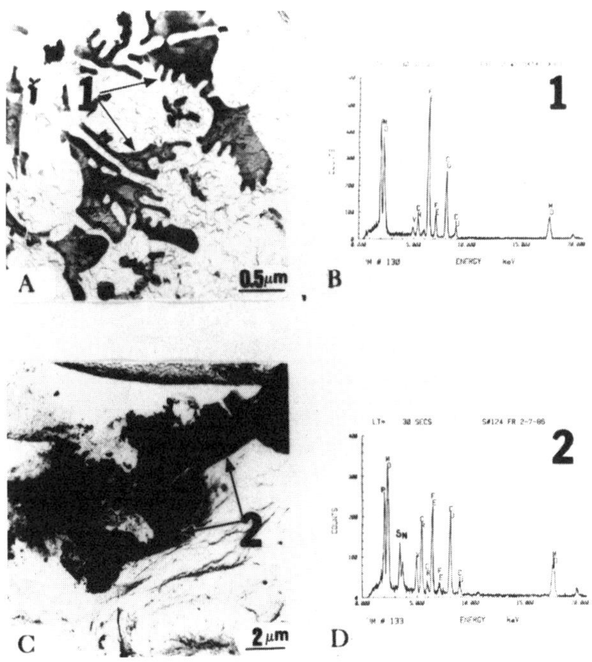

Figure 23. A) and B) TEM image of iron phosphides on the HAZ fracture surfaces and EDS spectrum obtained from them respectively. C) and D) TEM image of complex compounds on the HAZ fracture surfaces and EDS spectrum obtained from them respectively. The Buck rotor at high heat input. Direct carbon extraction replica.

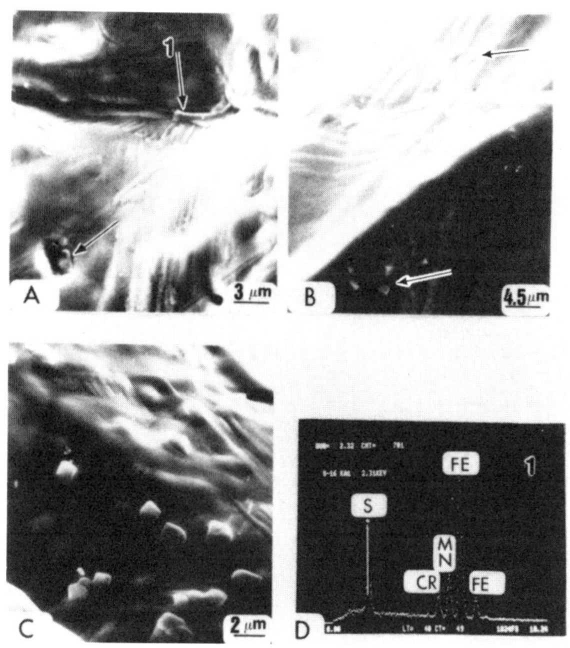

Figure 24. A) and B) SEM micrographs of the HAZ hot crack surfaces of the Gallatin rotor at high heat input showing sulfide films (labeled 1) along the grain boundary and cuboidal sulfides precipitation respectively. C) SEM image of the HAZ hot crack surfaces of the 3A forging at high heat input showing cuboidal sulfides precipitation.
D) EDS spectrum from the sulfides in A, B, and C.

and the Gallatin with S and P, 0.011% and 0.012%, respectively, was the second worst. The 3A and the 2A forgings stand very close together because S and P contents are almost the same. Another finding is that the Mn/S ratio does not seem to affect the HAZ hot cracking because the 3A forging has the Mn/S ratio, 35, which is close to that of the Buck rotor, 20. The results are contrary to those of Meitzner and Stout [28] who reported that the HAZ microcracking increased as the Mn/S ratio decreased. The lack of correlation with the Mn/S ratio has also been reported by Phillips and Jordan [27].

The significant finding in the present work is the role of V in promoting the cracking. The precipitation of eutectic-like vanadium carbides in association with Mo (Fig. 19, 20 and 21) could make grain boundaries brittle; thus producing intergranular fracture. The degree of contribution by VC to the HAZ cracking would be much smaller than the contribution by S and P.

Increase in heat input/unit length did not affect the cracking for the Gallatin, the 2A and the 3A. The Buck rotor crack length at high heat input was 6 times longer than low heat input crack length. Increase of heat input/unit length has been reported as having no effect on cracking by Masubuchi and Martin [29], and Boniszewski and Watkinson [30]. But Meitzner and Stout [28] reported that it caused an increase in cracking. It appears that steels with high impurity levels such as the Buck rotor show an increase in the HAZ hot cracking with increasing heat input/unit length.

CONCLUSIONS

1. The detrimental effect of Ni on solidification cracking has been confirmed in a modern NiCrMoV forging. When Ni content is high, here 2.8 wt%, the permissible impurities level, especially sulfur, should be reconsidered.
2. The harmful effects of sulphur and phosphorous on solidification cracking have been confirmed. In particular, direct evidence has been found that dendritic shaped iron phosphide, FeP, promotes cracking.
3. An increase in heat input/unit length or in travel speed at low heat input causes an increase in solidification cracking susceptibility.
4. The major HAZ hot crack-promoting effects of sulfur and phosphorus have been confirmed in the Buck rotor. Direct evidence of MnS type inclusions being liquated thus spreading along grain boundaries has been observed. Even at high heat input welding (i.e. slow cooling rate) iron phosphides have been observed in the HAZ fracture surfaces.
5. Dendritic vanadium carbide precipitation in the HAZ has contributed to the HAZ hot cracking, but the contribution by VC is smaller than the sulfur and phosphorous effects.
6. No evidence has been found that nickel contributes to the HAZ liquation cracking. In fact, Ni has neither segregated at the interdendritic areas nor to the HAZ liquated boundaries.

REFERENCES

1. Bush, S.H., "Failures in Large Steam Turbine Rotors," Workshop Proceedings: Rotor Forgings for Turbines and Generators, EPRI Report WS-79-235, Sept., 1-1 - 1-27, (1981).
2. Freedlander, G.D., "New Life Available for Old T/G;s and Boilers," Electrical World, No. 5, 87-96, (1983).
3. Savage, W.F. and Lundin, C.D., Weld. J. 44, 433-s-442-s (1965).
4. Mishler, H.W., Monroe, R.E., and Reippel, P.J., Weld. J, 40, 1-s - 7-s, (1961).
5. Morgan-Warren, E.J., and Jordan, M.F., Metals Technology, 1, 271-278, (1974).
6. Morgan-Warren, E.J., and Jordan, M.F., Metals Technology, 3, 29-39, (1976).
7. Huxley, H.V., Weld. and Metal Fabrication, No. 1, 29-32, (1963).
8. Brockhurst, P.J., and Muir, H., Weld. Fabrication and Design, 9, 80-86, (1965).
9. Vinckier, A.G., Weld. J., 50, 19-s - 24-s, (1971).
10. Pilows, V., Kovove Materialy, 21, 60-67, (1983).
11. Watanabe, T. and Okane, I., Trans. Jap. Weld. Soc. 13, 2-12, (1982).
12. Tamura, H., Kato, N. and Watanabe, T., Trans. Jap. Weld. Soc. 5, No. 2, 72-83, (1974).
13. Kramer, L.D., Randolph, D.D., and Weisz, D.A., "Analysis of the Tennessee Valley Authority, Gallatin Unit No. 2 Turbine Rotor Burst," Winter Annual Meeting of ASME, New York, Dec. (1976).
14. Jones, P.W., Brit. Weld. J., 6, 282-290, (1959).
15. Senda, T., Matsuda, F., Takano, G., Watanabe, K., Kobayashi, T. and Matsuzaka, T., Trans. Jap. Weld. Soc. 2, 141-162, (1971).
16. Wilkinson, F.J., Cottrell, C.L.M., and Huxley, H.V., Brit. Weld. J., 5, 557-562, (1958).
17. Kihara, H. and Matsuda, F., Trans. Jap. Weld. Res. Inst., 2, 83-95, (1973).
18. American Society for Metals, Metals Handbook, Vol. 8 (8th Edition), 304-305, (1973).

19. Lundin, L.D., Lingenfelter, A.C., Grotte, G.E., Lessmann, G.G., and Matthews, S.J., Welding Research Council Bulletin 280 Aug. (1982).

20. John, R., and Richards, W.G., Metal Constr. and Brit. Weld. J., 19, 127-134, (1972).

21. Thompson, R.E., U.S. Navy Marine Engineering Lab. R & D Report 13/67, March, (1967).

22. Borland, J.C., Brit. Weld. J., 7, 508-512, 1960.

23. Goldstein, J.I., Newbury, D.E., Echlin, P., Joy, D.C., Fiori, C., Lifshin, E., Scanning Electron Microscopy and X-ray Microanalysis, p. 59, Plenum Press, New York and London, (1981).

24. Savage, W.F., and Dickinson, D.W., Weld. J., 55 (6), 181-s - 188-s, (1972).

25. Pepe, J.J., and Savage, W.F., Weld. J., 46 (9), 411-s - 422-s, (1967).

26. Brammar, I.S., J. of the Iron and Steel Inst., 201, 752-761, (1963).

27. Phillips, R.H., and Jordan, M.F., Metals Technology, 3, 571-579, (1976).

28. Meitzner, C.F. and Stout, R.D., Weld. J., 45, 393-s - 383-s, (1966).

29. Masubuchi, K., and Martin, D.C., Weld. J., 41, 375-s - 383-s,(1962).

30. Boniszewski, T., Watkinson, F., Brit. Weld. J., 11, 610-619, (1964).

EFFECT OF NITROGEN ON THE AUSTENITIC STAINLESS STEEL WELD METAL HAZ CRACKING

Chang-Pin D. Chou
National Chiao Tung University
Taiwan, Republic of China

Peir-Shin Wu
Chun-Shan Institute of Science and Technology
Taiwan, Republic of China

Abstract

Autogenous GTA welds with nitrogen in the shielding gas were made on the austenitic stainless steel plates 304, 316, and 310S. The Fissure Bend Test was utilized to evaluate the hot cracking resistance of austenitic weld metals. Optical metallography, SEM fractography and energy dispersive X-ray analysis were employed to determine the cause and mechanism of hot cracking in the austenitic weld metals. It was found that the addition of nitrogen into the argon shielding gas during welding can reduce the ferrite content in the welds and therefore increase the hot cracking tendency of ferrite-containing austenitic welds. While in the fully austenitic stainless steel weld metal, the addition of nitrogen can reduce the hot cracking propensity. It was also found that multiple thermal cycles enhance the fissuring propensity in austenitic weld metals. Fissuring occurred in the weld metal "Hazard HAZ" region with low ferrite content. Liquid film evidence was seen in the fissure surface which had higher silicon and manganese content than the bulk weld metal.

WITH THE INCREASING USE OF HEAVY SECTION NITROGEN-CONTAINING AUSTENITIC STAINLESS STEEL WELDMENTS, repair after fabrication and repair in service become more common. Thus, the understanding of the effect of nitrogen and the effect of multiple thermal cycles on weld metal cracking becomes more important and need intensive study.

Many austenitic stainless steels develop a duplex, austenite plus delta ferrite, microstructure after welding. The retained delta ferrite form a discontinuous network in a matrix of austenite. Early experience showed that retention of 5-10 vol.% ferrite in the weld bead could be used to control hot cracking of welded austenitic steels (1,2). The amount of delta ferrite retained from high temperature can be predicted from special phase diagrams (3,4) and measured by metallographic and magnetic measurements. While the retained ferrite is associated with reduced hot cracking, this phase can be detrimental to the aqueous corrosion (5) and stress corrosion cracking (6). In high temperature service, ferrite may promote rapid transformation to the embrittling sigma phase (7). Thus, for critical applications, fabricators are frequently required to use fully austenitic stainless steel weld metals.

It is well known that nitrogen dissolves interstitially in austenite and is a strong austenite stablizer. The stablizing effect of nitrogen is 20-30 times that of nickel (4,8). Therefore, nitrogen is a potent element which can be used as substitute for nickel. Nitrogen additions are valuable in increasing the tensile strength at all temperatures (9-11) although this tensile strength increase is sometimes accompanied by reduced ductility and toughness. The weldability studies of nitrogen-strengthened stainless steels become more important and draw more attention (12-14). There are three basic ways of alloying weld metal with nitrogen: by using weldable nitrogen-containing base metal, by incorporating nitrogen in the welding materials (wire, flux, electrode coating), or by introducing nitrogen in the gas shield. The most economical and effective way to obtain a fully austenitic weld is adding nitrogen to the shielding gas during welding fabrication (15,16).

The effect of multiple weld metal HAZ thermal cycles on the hot cracking resistance of austenitic stainless steel weld metal has been extensively investigated using the Fissure Bend Test and the Gleeble Hot Cracking Test at the University of Tennessee (2,15,16). It was found that multiple thermal cycles enhance the fissuring propensity in both the fully austenitic weld and the ferrite-containing weld.

The main object of this study is to investigate the effect of nitrogen content in the shielding gas and the number of thermal cycling

Table 1 - Chemical Composition Analysis of Austenitic Stainless Steel Weld Metal

Weld Metal	wt.%								FN
	C	Cr	Ni	Mo	Mn	Si	P	S	
304	0.045	18.01	8.51	--	1.45	0.561	0.021	0.007	5.8
316	0.085	16.99	9.95	2.22	1.39	0.573	0.026	0.009	4.5
310S	0.074	25.12	18.51	--	1.79	1.020	0.021	0.006	0.0

on the microstructure and fissuring sensitivity of austenitic stainless steel weld metals.

EXPERIMENTAL PROCEDURES

Austenitic stainless steel plates of 304, 316, and 310S with thickness of 6 mm were used in this study. The chemical composition and measured Ferrite Number (FN) are shown in Table 1. FN measurement were made on all weld metals using the Magne Gage.

Autogenous Gas Tungsten Arc (GTA) welds with several number of remelting passes were made on the stainless steel plates. The dimensions of tested plate and remelting sequence are schematically shown in Figure 1. In this manner, five regions were obtained after final remelting pass. Regions A and E are base metal Heat Affected Zone (HAZ). Regions B and D are weld metal. Region C, where most cracks occurred is the weld metal HAZ. The letter N represents the number of remelting pass of weld metal D which produces N thermal cycle excursions in region C. The welding parameters used are; arc voltage - 12 V, welding current - 170 A, welding speed - 150 mm/min, gas flow rate - 35 cfh. The shielding gas used in the first pass (Region B) was argon plus nitrogen ranging from o.5% up to 15%. Pure argon was then used in the subsequent passes (Region D).

The Fissure Bend Test device was utilized to determine the fissuring tendency of the welded stainless steel plates. The bent surface was then liquid penetrant tested and examined under binocular microscope up to 60X. The number of verified fissures were counted and recorded.

Selected samples were cut, mounted, polished, and etched in the solution of 10 ml HNO_3, 20 ml HCL, and 30 ml H_2O. Fractographic examination was performed using a scanning electron microscope. Energy dispersive X-ray analysis was also applied in order to determine elemental distributions in selected regions.

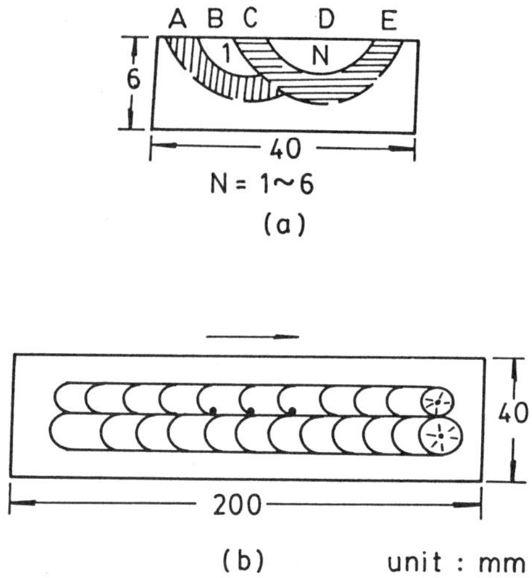

Fig. 1 - Dimensions and Remelting Pass Sequence of Weld Plate. (a) Transverse View, (b) Top View

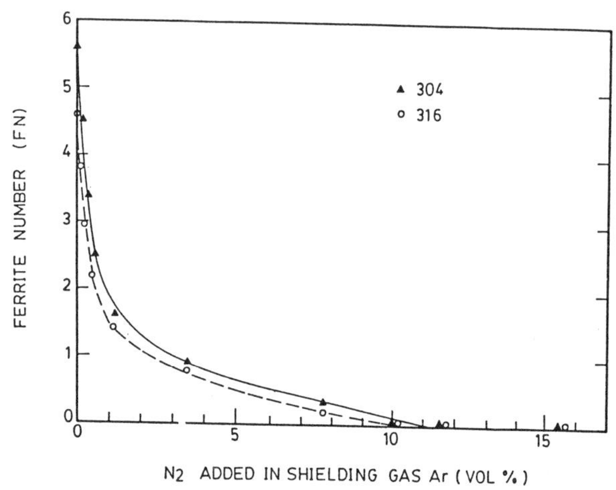

Fig. 2 - Measured FN vs. % N_2 in Shielding Gas

RESULTS AND DISCUSSION

EFFECT OF NITROGEN CONTENT ON THE MICROSTRUCTURE - Figure 2 shows the results of measured FN as a function of volume percentage of nitrogen in the shielding gas for 304, 316, and 310S. A rapid drop in measured FN is observed as the volume of nitrogen in the shielding gas is increased. With 10% nitrogen, the FN is effectively reduce to zero for both 304 and 316 weld metals. The chemical analyses for nitrogen of these welds, presented graphically in Figure 3, shows that weld metal becomes saturated with nitrogen after being exposed to a shielding gas containing approximately 10%. It is also found that the solid solubility of nitrogen decreases with increasing chromium content in the austenitic stainless steel weld metals.

The microstructures of weld metals 304, 316, and 310S are shown in Figure 4. Typical vermicular ferrite morphology is clearly seen in the ferrite-containing welds 304 and 316. The ferrite phase are uniformly distributed in the austenite matrix. In the fully austenitic weld 310S no ferrite is observed and the cellular dendritic solidification mode and segregation pattern are seen. With the increasing of nitrogen in the

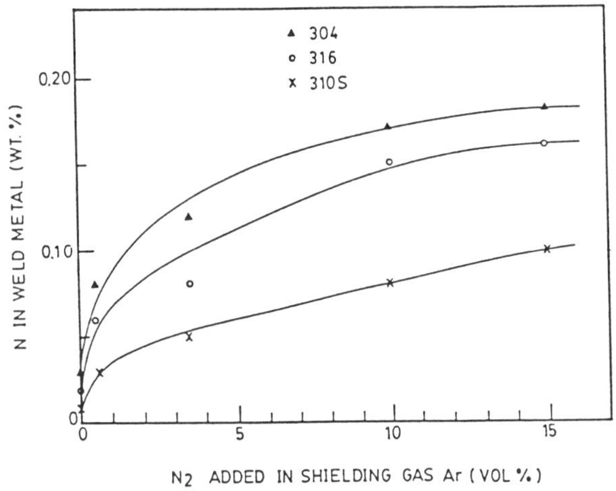

Fig. 3 - Weld Metal Nitrogen Content vs. % N_2 in Shielding Gas

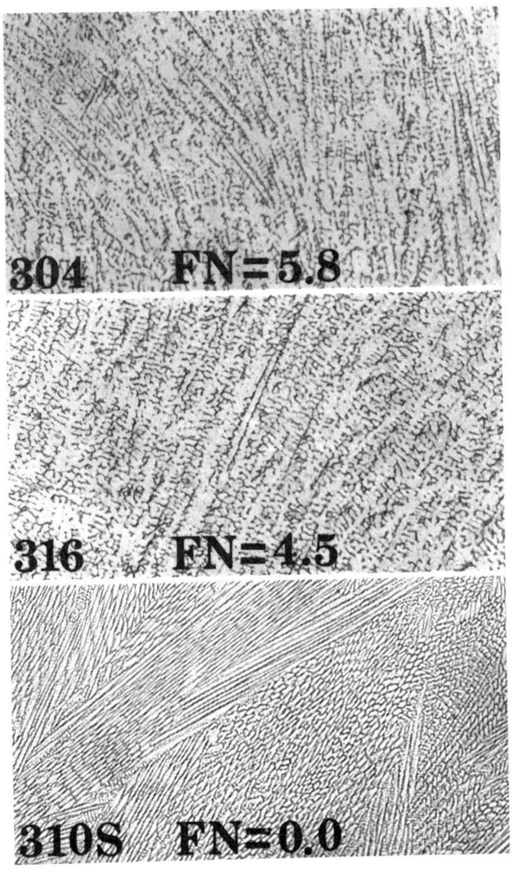

Fig. 4 - Microstructures of Austenitic Stainless Weld Metals

shielding gas, the ferrite number reduced from 5 to near zero. This can been seen in Figure 5, which shows the microstructures of weld metal 316 with different nitrogen content in the weld metal. Therefore, nitrogen is a strong austenite stabilizer. Adding a small amount of nitrogen into the shielding gas is an economical and effective way to reduce the residual ferrite content in the austenitic stainless steel weld metals.

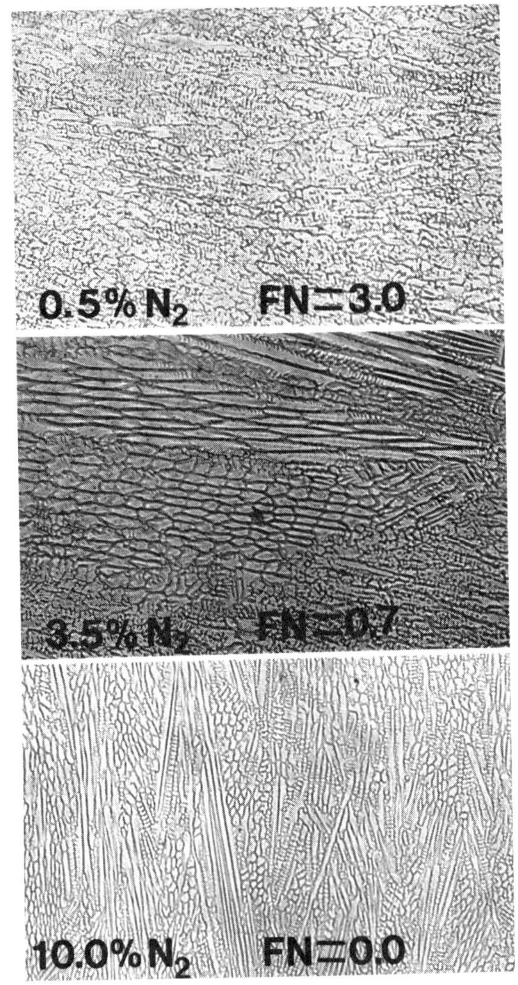

Fig. 5 - Microstructures of Weld Metal 316 with Different Nitrogen Content in the Weld Metal

EFFECT OF NITROGEN CONTENT ON THE HOT CRACKING RESISTANCE - The Fissure Bend Test was utilized to determine the effect of nitrogen and number of thermal cycles on the hot cracking resistance of austenitic stainless steel weld metals. Figure 6 shows the results of data obtained as a function of the number of HAZ exposures and volume percentage of nitrogen in the shielding gas of weld metal 304. The ferrite number was measured in the weld metal HAZ where fissures are located. This low-ferrite region is adjacent to the fusion boundary and is termed "Hazard HAZ" as proposed by Lundin and Chou (17). From these data it is clear that nitrogen has an adverse effect on the hot cracking resistance. It is also found that high nitrogen welds are more sensitive to cracking than the low nitrogen welds as the number of HAZ exposure is increased. This is probably due to the reduced ferrite content in the weld metal HAZ of high nitrogen welds. While the number of HAZ exposure has little effect on the measured FN in the "Hazard HAZ" region, the number of fissuring increased as the number of HAZ exposure increased. This phenomenon indicates other mechanisms are also dominated for the fissuring occurrence during multiple thermal cycling. Similar results are also obtained for the weld metal 316 as shown in Figure 7.

The Fissure Bend Test results of fully austenitic weld metal 310S are shown in Figure 8. It is shown that thermal cycling is detrimental to the weld metal HAZ at all levels of nitrogen content. Nitrogen can effectively reduce the number of fissures in fully austenitic weld 310S. The probable causes of the increase of cracking resistanceby the addition of nitrogen are due to grain refining, prevention of grain growth, and increasing of toughness on the grain boundaries.

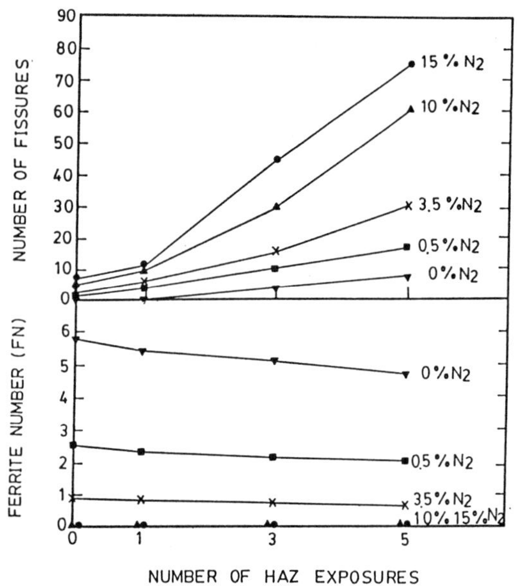

Fig. 6 - Number of Fissures vs. Number of HAZ Exposures and Volume Percentage in Shielding Gas of Weld Metal 304

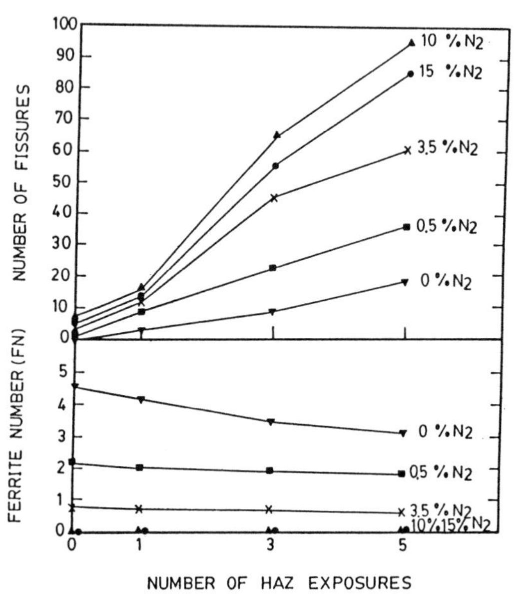

Fig. 7 - Number of Fissures vs. Number of HAZ Exposures and Volume Percentage in Shielding Gas of Weld Metal 316

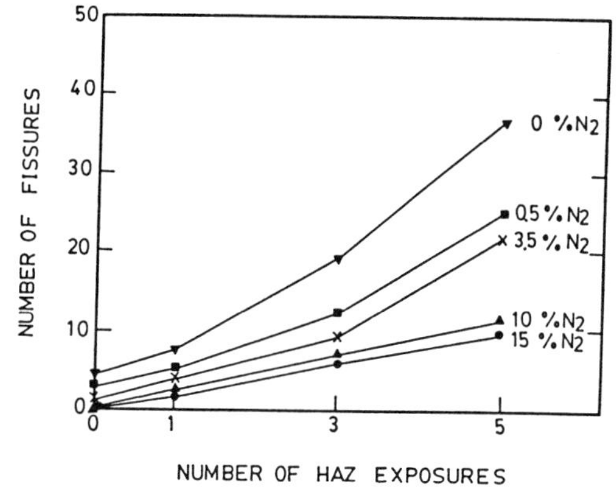

Fig. 8 - Number of Fissures vs. Number of HAZ Exposures and Volume Percentage in Shielding Gas of Weld Metal 310S

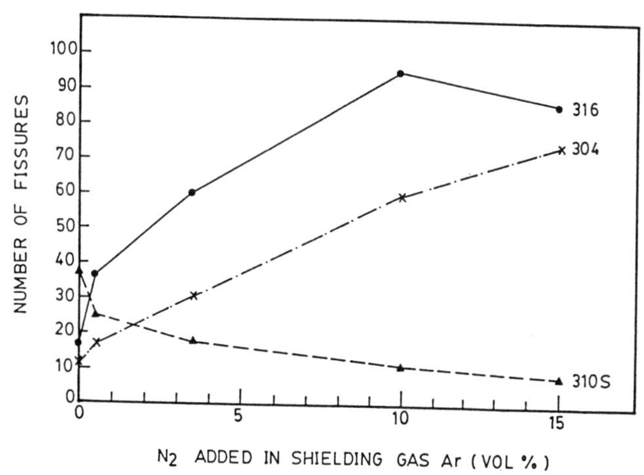

Fig. 9 - Effect of Nitrogen Addition in the Shielding Gas on the Number of Fissures in the Austenitic Stainless Steel Weld Metals

The effect of nitrogen addition in the shielding gas on the number of fissures in the weld metal HAZ is summarized in Figure 9. Nitrogen plays an important role in cracking resistance of fully austenitic welds and ferrite-containing welds. The mechanism of fissuring in nitrogen-strengthened welds is unsolved and need more intensive investigation.

FRACTOGRAPHIC EXAMINATION OF FISSURE SURFACE - The results of the SEM fractography studies of fissures of weld metal HAZ are shown in Figure 10. The weld metal is stainless steel 316 autogenously remelted by GTA process with 10% nitrogen in the shielding gas. Several fissures are observed in the weld metal HAZ which was exposed five thermal cycles by the subsequent remelting passes. It is to be noted that these cracks are in the "Hazard HAZ" region adjacent to the fusion zone of the weld bead producing the HAZ exposure. The high magnification of crack at the bottom of Figure 10(a) is shown in Figure 10(b). Area A, which is away from the fusion line (i.e., reheated to a lower temperature during welding), shows a typical ductile overload region with sheared dimples as shown in

Figure 10(c). Typical liquated appearance is clearly seen in area B. The EDAX results showed that the Si and Mn content is higher in the liquated region than that found in the bulk metal. A partial explanation would revolve around liquation of the segregated grain boundary region produced upon solidification and/or those

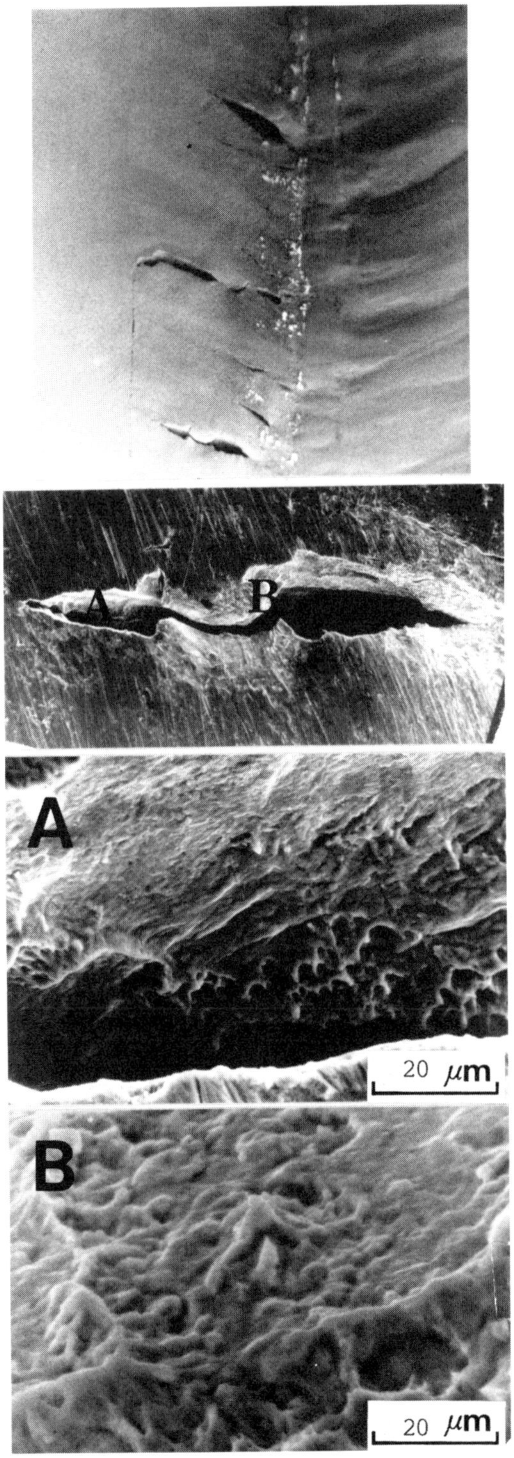

Fig. 10 - (a) Fissures in the Weld Metal HAZ
(b) Fissure Surface under SEM
(c) High Magnification of Area A
(d) High Magnification of Area B

boundaries with enhanced segregation produced during thermal cycling. In addition, the continued HAZ straining by subsequent weld passes no doubt plays an important role.

SUMMARY

Based on the Fissure Bend Test, weld metal HAZ cracking has been investigated on the nitrogen-containing weld metal obtained by adding nitrogen to the shielding gas during autogenous GTA welding fabrication. Fissures observed are in the weld metal "Hazard HAZ" region located near the fusion zone of a subsequent weld bead. Fissuring is enhanced by multiple thermal cycling in the "Hazard HAZ" regardless of ferrite content. In the ferrite-containing austenitic welds, nitrogen addition can increase the hot cracking tendency. In the fully austenitic welds, the addition of nitrogen can reduce the hot cracking tendency. Liquid film evidence was seen in the fissure surface which had higher silicon and manganese content than the bulk metal had.

ACKNOWLEDGMENT

The research described in this paper was sponsored by the National Science Council of Republic of China.

REFERENCES

1. Hull, H. C., Welding J., 46, p399s (1967).
2. Lundin, C. D., Delong, W. T., and Spond, D. F., Welding J., 54, p241s (1975).
3. Schaeffler, A. L., Metal Progress, 56, p680 (1949).
4. Delong, W. T., Ostrom, G. A., and Szumachowski, Welding J., 35, p521s (1956).
5. Gooch, T. G., British Welding J., 15, p345 (1968).
6. Baselack, W. A., Duquette, W. A., and Savage, W. F., Corros-on, 35, p45 (1979).
7. Willingham, D. C. and Gooch, T. G., The Welding Inst. Res. Bulletin, 12, p19 (1971).
8. Hull, F. G., Welding J., 52, p193s (1973).
9. Cullen, T. M. and Davis, M. W., ASTM STP 522, p60 (1973).
10. Janzon, B., JISI, 207, p826 (1971).
11. Blake, P. D., Metal Const., 11, p196 (1979).
12. Espy, R. H., Welding J., 61, p149s (1982).
13. Enjo, T., Kikuchi, Y., Kobayashi, T., and Kuwana, T., Trans. of JWRI, 9(2), p173 (1980).
14. Brooks, J. A., Welding J., 54, p189s (1975).
15. Lundin, C. D., Chou, C.-P. D., and Sullivan, C. J., Welding J., 59, p226s (1980).
16. Lundin, C. D. and Chou, C.-P. D., WRC Bulletin, 289 (1983).
17. Lundin, C. D. and Chou, C.-P. D., Welding J., 64, p113 (1985).

POSTER PRESENTATIONS

WELDABILITY STUDIES OF Fe-Mn-Al STAINLESS STEELS

Chang-Pin D. Chou, Chien Hsun Lee
National Chiao Tung University
Taiwan, Republic of China

Abstract

Based on the composition of 30% Mn, 10% Al, and balanced Fe, Fe-Mn-Al alloys with different carbon content and thus different ferrite content were studied. Tensile test and U-bend test on the autogenous GTA welded specimen were utilized to evaluate the weldability. FN measurement, optical metallography, and microhardness measurement were made to study the microstructural characteristics of the weld metal and the base metal HAZ. It was found that the carbon content has a strong influence on the amount of residual ferrite in the Fe-Mn-Al weld metals studied. The tensile strength are excellent for all welds in both the longitudinal and transverse welding directions. The elongation of butt joint welded specimen is heavily reduced from that of base metal, especially for the fully austenitic weld metal. U-bend test results give a positive indication of good weldability of Fe-Mn-Al alloys.

FE-MN-AL ALLOYS have recently been considered to be a promising candidate for an austenitic stainless steel. Aluminum and manganese replace the expensive and heavy alloying elements in conventional Fe-Cr-Ni system. This new alloy system also yield a long range strategic advantages (1).

Aluminum plays a major role in the oxidation and corrosion resistance which is characteristic of the binary Fe-Al alloy. Manganese is the austenite stabilizer as the combination of Mn-C is believed to have the capability to retain the face-centered-cubic austenitic structures (2). Hence a suitable composition of Fe-Mn-Al-C should give an austenitic alloy with good mechanical properties and oxidation resistance.

The Fe-Mn-Al alloys were first developed in 1930s. Korter and Tonn established a partial phase diagram for this system (3). Dean and Anderson presented a series of complex Fe-Mn-Al alloys (4). Later. Ham and Carins reported a composition of Fe-34.4Mn-10.2Al-0.76C (5). Recently, Inoue et al. presented a series of results concerning rapid quench in the Fe-(Ni-Mn)-Al-C system (6). With the addition of silicon, the marine corrosion resistance and cavitation erosion resistance were increased and can be used for propellers, especially for seagoing vessels (7).

Based on the composition of Fe-10Mn-30Al-1.0C alloy, a well organized research group headed by Wan has conducted a series of systematical research projects at National Tsing Hua University, Taiwan, R.O.C. for the past few years (8-12). Wan et al. found that, combined with significant cost and density advantages, good mechanical properties, excellent corrosion and oxidation resistance, Fe-Mn-Al stainless steels are good candidates for replacing at least part of the conventional Fe-Cr-Ni stainless steels. In order to enhance the applicability, the weldability of Fe-Mn-Al alloys was studied and preliminary results are reported in this paper.

EXPERIMENTAL PROCEDURES

Three kinds of Fe-Mn-Al-C alloys were prepared from pure iron (99%), electrolytic manganese (99.96%), high purity aluminum (99.99%), and high purity carbon with a vacuum induction furnace (L-H IS 8/III) and casted under argon atmosphere. The as-cast alloys were hot forged at 1200°C, homogenized at 1050°C for 12 hours, then cold rolled and annealed at 950°C for one hour in an argon atmosphere furnace, followed by air cooling. The composition of these alloys were analyzed by wet chemical method and listed in Table 1. The measured Ferrite Number (FN) of base metal and weld metal are also listed in Table 1. FN measurements were made using the Magne Gage.

Autogenous GTA welds were made using the welding parameters listed in Table 2. The dimensions and bead configuration of welded plates

Table 1 - Chemical Composition Analysis of Fe-Mn-Al Alloys Studied

Alloy	Mn	Al	C	S	Fe	Ferrite Number(FN)	
						Weld Metal	Base Metal
A	28.44	7.96	0.53	0.006	Bal.	13.0	3.5
B	29.28	8.59	0.81	0.004	Bal.	5.0	0.0
C	29.05	8.03	1.17	0.001	Bal.	0.5	0.0

Table 2 - Welding Parameters Used in Autogenous GTA Welding

Welding Condition	Voltage (Volt)	Current (Ampere)	Travel Speed (mm/min)	Flow Rate (cfh)	Electrode (in.)	Heat Input (KJ/mm)
1	8.5	80	150	45	1/16	0.27
2	9.8	105	150	50	1/16	0.41
3	10.5	140	150	50	1/16	0.59
4	13.8	220	150	50	3/32	1.21
5	14.0	260	150	50	3/32	1.46

are shown in Figure 1. In Figure 1(d), two parallel, overlapped remelting passes were made so that regions of weld metal, weld metal HAZ, and base metal HAZ can be tested and evaluated at the same time.

Standard tensile test and U-bend test were utilized to determine the effect of composition and welding conditions on the weldability of the Fe-Mn-Al alloys. Metallographic samples were prepared and etched with 10% nital. SEM and EPMA were also applied to determine elemental distributions in selected regions.

RESULTS AND DISCUSSION

MICROSTRUCTURAL STUDIES - The microstructures of weld metals A, B, and C are shown in Figure 2. Typical columnar dendritic solidification mode is clearly seen in the fusion zone. The vermicular type ferrite is located at the core of the originally-solidified primary ferrite dendrites. The ferrite phase can be qualitatively identified using concentration profiles of Mn and Al across the ferrite in weld metal A as shown in Figure 3. The effect of heat input on the measured FN is presented in Figure 4. The ferrite is increased as the heat input increases up to 0.6 KJ/mm. Since carbon is a very strong austenite-forming element, the addition of carbon in the Fe-Mn-Al alloys improves the stability of austenite. Increasing the carbon content in the alloys A, B, and C from 0.5%, 0.8%, to 1.2%, the ferrite number decreases from 13, 5, to near zero. After solution heat treatment, only alloy A remains a duplex structure. This can be seen in Figure 5.

TENSILE TEST RESULTS - Tensile tests were conducted for the autogenously welded specimens in both longitudinal and transverse directions. In order to determine the joint strength, the butt joint welded specimens were also tested in the transverse direction. All the welding conditions were tested in triple and the averag values were calculated. The results are presented graphically in Figure 6. The conventional

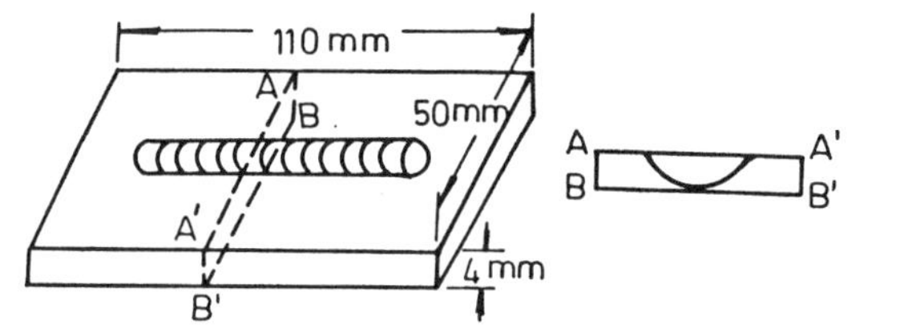

(a) Autogenouly welded, longitudinal sample

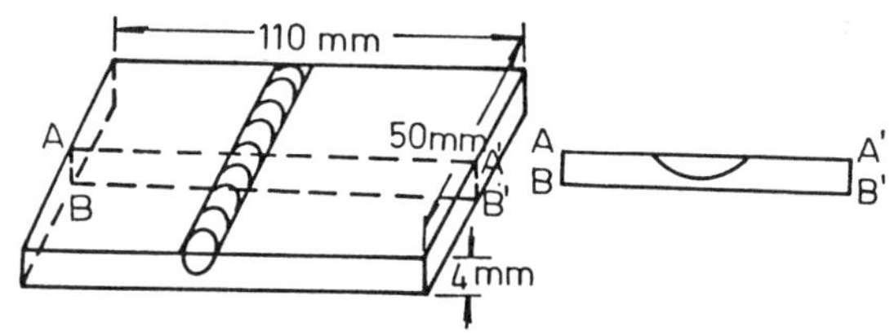

(b) Autogenously welded, transverse sample

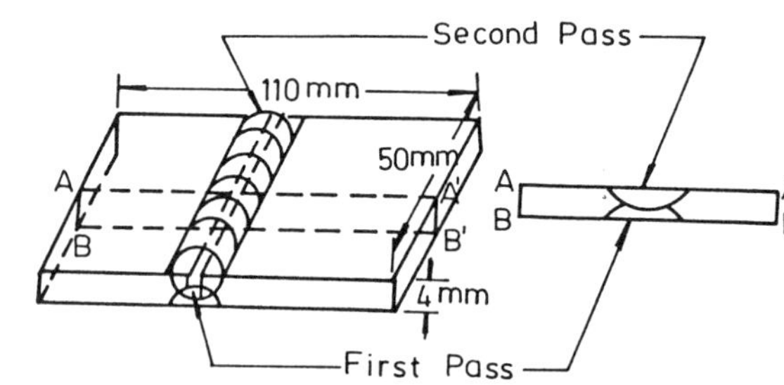

(c) Butt joint welded, transverse sample

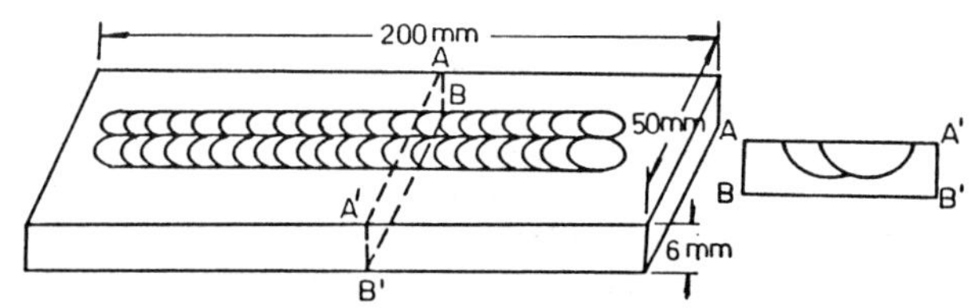

(d) Overlapped remelting, longitudinal sample

Fig. 1 - Dimensions and Bead Configuration of Samples for the Tensile Test and U-Bend Test

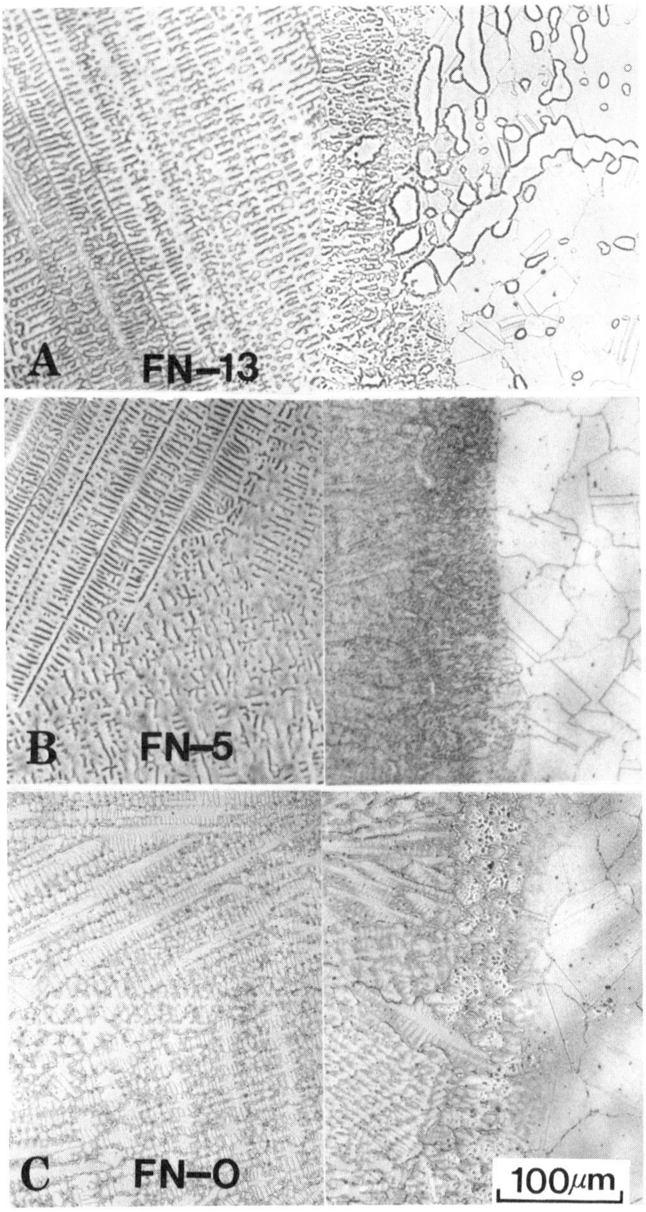

Fig. 2 - Microstructures of Fe-Mn-Al Weld Metals A, B, and C

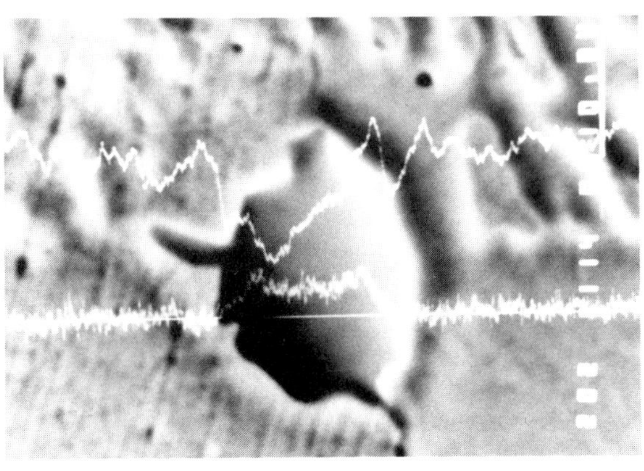

Fig. 3 - Concentration Profile of Manganese and Aluminum Across the Ferrite Phase in Weld Metal A

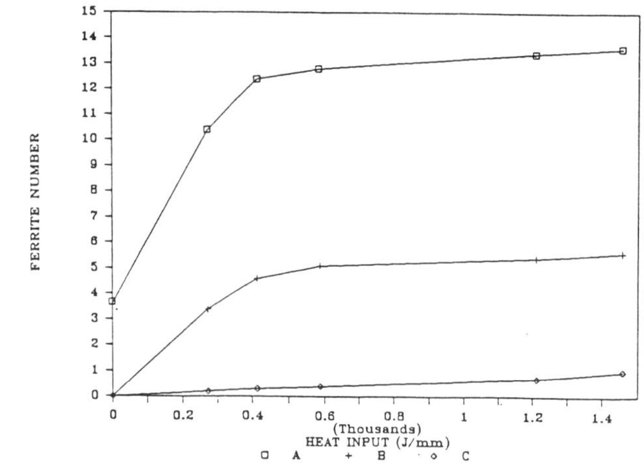

Fig. 4 - Effect of Heat Input on the Measured FN of Fe-Mn-Al Weld Metals A, B, and C

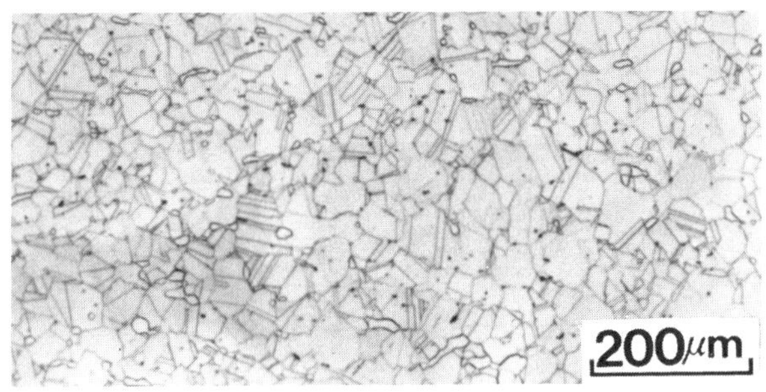

Fig. 5 - Microstructure of Fe-Mn-Al Base Metal A

Fe-Cr-Ni austenitic stainless steels test results are also presented for comparison. All the three Fe-Mn-Al alloys show promising strength and ductility in the autogenously welded conditions. Alloy C is the strongest one. While in the butt joint welded condition, the alloy C shows an inferior result. The elongation is reduced from 54% in the unwelded condition to only 19% in the butt joint welded condition. Nearly all the tensile specimens were fractured at the center line of the weld metal. This is probably due to the enhanced level of solute and impurity segregation at the central part of a weld bead. An SEM fractograph of the fractured surface of weld metal A is shown in Figure 7. Typical ductile fracture surface with uniform dimples is clearly seen. There are appreciable amount of Al-riched particles identified by EPMA in the dimples.

U-BEND TEST RESULTS - A promising results were obtained for the butt joint welded specimens when tested in the transverse direction. No crack was observed for all welding conditions tested. The pictures of butt joint weldment of alloy A are shown in Figure 8. There was a few fissures appeared in the unaffected base metal after U-bend testing for some welding conditions. It was examined and concluded that these fissures are probably due to the slag inclusion cotained in the base metal during the material preparation.

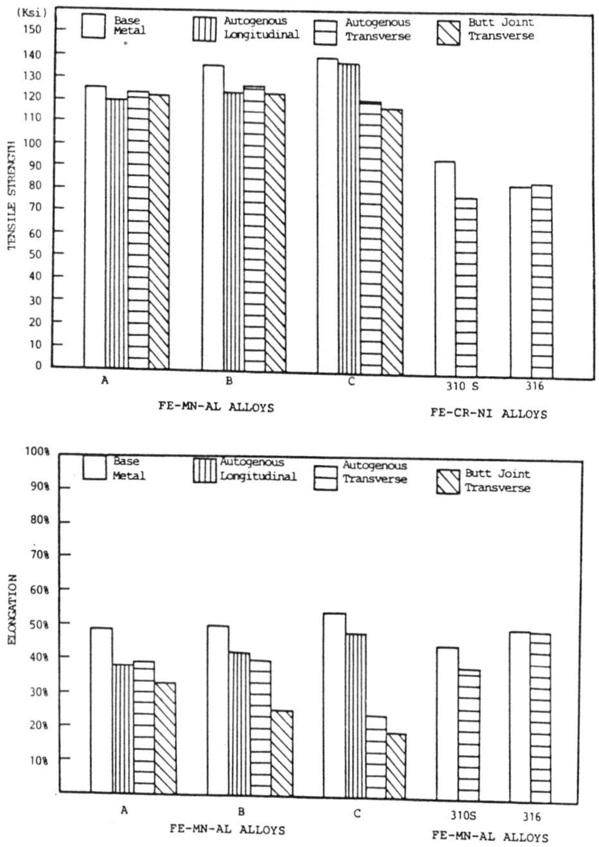

Fig. 6 - Tensile Test Results of Fe-Mn-Al Weld Metals A, B, and C

Fig. 8 - U-Bend Tested Weldment of Alloy A

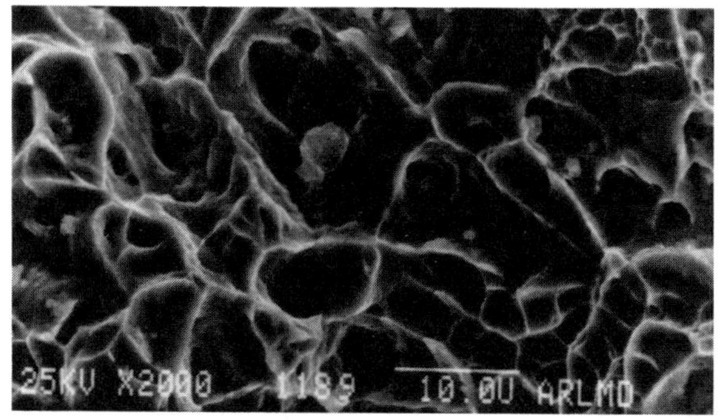

Fig. 7 - SEM Fractograph of Tensile Tested Fractured Surface of Weld Metal A

SUMMARY

The Fe-Mn-Al alloys with basic composition of 30% Mn, 10% Al, balanced Fe were prepared, fabricated, and tested in several directions. The preliminary results of weldabilty testing shows that the Fe-Mn-Al weld metal have very good strength and ductility. The alloying elements such as carbon, silicon, etc. may have strong effect on the weldability of this new alloy system. More intensive investigation need to be done in the future.

ACKNOWLEDGMENT

The research described in this paper was sponsored by the National Science Council of Republic of China under grant NSC 74-0405-E009-002.

REFERENCES

1. Charles, J., Berghezan, A., Lutts, A., and Dancoisne, P. L., Metal Progress, May, p71 (1981)
2. Banerji, S. K., Metal Progress, April, p59 (1978)
3. Korter, W. and T-nn, W., Archiv fus das Eisenhuttenwesen, 7, p365 (1933)
4. Dean, R. S. and Anderson, C. T., U.S. Patent 2,329,186 (1943)
5. Ham, J. L. and Carins, R. E., Product Engineering, 29(52), p50 (1958)
6. Inoue, A., Minemura, T., Kitamura, A., and Masumoyo, T., Met. Trans., 12A, p1245 (1981)
7. Wang, R. and Beck, F. H., Metal Progress, March, p72 (1983)
8. Tarn, C. L., Hu, C. T., and Wan, C. M., Chinese J. of Materials Science, 15A(2), p75 (1983)
9. Liu, T. F., Wan, C. M., and Lee, B. K., Chinese J. of Materials Science, 15A(2), p11 (1983)
10. Duh, J. G., Huarng, S. H., and Wan, C. M., Chinese J. of Materials Science, 16A(1), p14 (1984)
11. Liu, T. F. and Wan, C. M., Scripta Metallurgica, 19, p727 (1985)
12. Sheu, T. S.., Chang, S. C., and Wan, C. M., Chinese J. of Materials Science, 17A(2) p27 (1985)

GAS TUNGSTEN-ARC WELDING REFRACTORY METALS WITH LOCAL SHIELDING

Sherwood Goldstein
Reading Research & Development Laboratory
Cabot Corporation
Cabot Beryllium Products
Reading, Pennsylvania 19603

ABSTRACT

This report describes room-temperature properties from gas tungsten-arc welding using local shielding on 0.15- to 0.030-inch thick strip of columbium, tantalum, Cb-1%Zr, Ta-2.5%W, Ta-10%W and Ta-40%Cb alloys. Welding procedures were developed for full penetration butt welds and the resulting joints were characterized by hardness tests, metallography, and room-temperature tensile and bend tests.

It was determined that these metals fall into three ductility groups after welding with local shielding, as follows:

1. No weld ductility impairment:
   Cb, Cb-1%Zr, Ta-2.5%W

2. Slight ductility impairment:
   Ta - powder grade
   Ta - metallurgical grade
   Ta-40%Cb

3. Substantial ductility impairment:
   Ta-10%W

THE HIGHEST QUALITY gas tungsten-arc welds in tantalum and columbium and their alloys are produced in vacuum-purged chambers backfilled with inert gas to prevent contamination from nitrogen, oxygen and hydrogen. Common practice to further minimize weld contamination is to acid clean joint members before welding by pickling in a solution of nitric, sulfuric and hydrofluoric acids.

In some applications, however, such as in welded tubing or for large chemical processing vessels, it is necessary to weld outside of a chamber. The objective of this work was to evaluate the quality and room temperature properties of welds in several columbium and tantalum alloys made in the open with inert gas trailing shields and with minimum preweld cleaning.

TECHNICAL APPROACH

MATERIAL - Annealed sheet between 0.015-inch and 0.030-inch thickness of unalloyed tantalum (originating from both electron-beam cast ingot and sintered powder), Ta-2.5%W, Ta-10%W, Ta-40%Cb, unalloyed columbium and Cb-1%Zr were evaluated.

PREWELD PREPARATION - Sample pieces were sheared and degreased and either chemically etched or wire-brushed.

WELDING - Test pieces were clamped in a fixture to produce a square butt joint and were welded in the open with secondary shielding from a trailing shield and from a drilled backup bar underneath the joint. Full penetration welds were made without filler material using pulsed current gas tungsten arc welding and mechanically controlled travel speed.

WELD EVALUATION - Weld quality was judged by overall assessment of nondestructive test examination (visual, liquid penetrant, and Xray), room-temperature tensile strength, room-temperature bend ductility, microhardness, and metallographic examination of weld cross section.

RESULTS

WELD QUALITY - Visual, liquid penetrant and X-ray inspection indicated that all welds were crack-free regardless of preweld cleaning (etched or wire-brushed). Welds that had edges etched prior to welding contained no or few spots of porosity compared with only slightly more scattered porosity in the wire-brushed samples. Normally, welding is not recommended for powder-grade tantalum because the weld would be expected to be very porous, but the samples in this investigation contained only scattered porosity.

TENSILE STRENGTH - Tensile strength of the welded material was judged by the acceptance criteria of the ASME Boiler and Pressure Vessel Code.

The tensile test results for the welded samples, in order of decreasing strength, are listed in Table 1. The weld strength in all cases was higher than the minimum tensile strength for the annealed base metal and therefore all samples would pass the tensile test requirement of the ASME Boiler Code.

Also note that with one exception, the welded samples failed either outside the weld and heat-affected zones, or when in the weld zone at a joint efficiency of at least 97%.

METALLOGRAPHY - Full penetration, sound welds (except for isolated spots of porosity) were obtained in all test welds. The microstructure typically showed epitaxial growth from coarsened grains in the heat-affected zone; in some samples, the grain size in the weld extended to the full thickness of the sample. Figure 1 shows the weld, heat-affected zone and base metal structures and corresponding hardness survey in Cb-1%Zr.

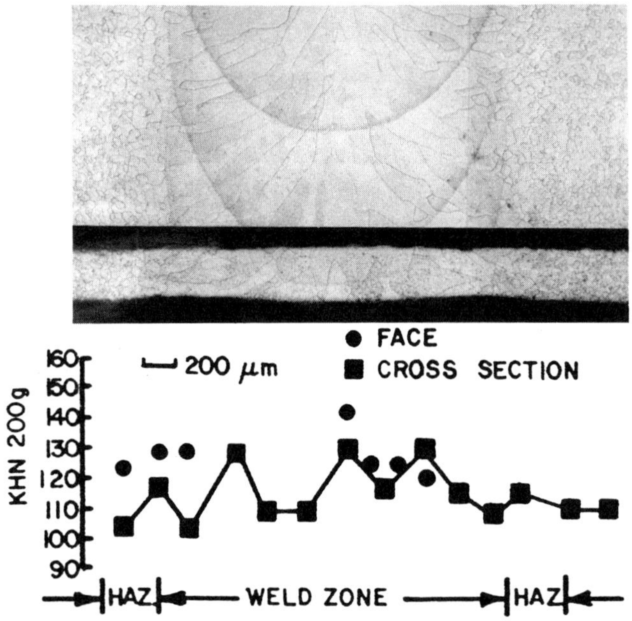

Fig. 1 - Polished and etched top surface and cross-section of weld and corresponding hardness survey across weld in Cb-1%Zr.

Table 1 - Tensile Strength of Gas Tungsten-Arc Welds (Base Metal and As-Welded Transverse Weld Tensile Strength)

	UTS (ksi)	YS (0.2%) (ksi)	Elongation (%)	Fracture	Joint Efficiency % (UTS Weld/UTS BM)
Ta-10%W					
Min. Base Metal	70.0	55.0	20		
Actual Base Metal	108.7	97.7	21		
Weld	105.8	89.4	6	Weld	97
Ta-2.5%W					
Min. Base Metal	40.0	28.0	20		
Actual Base Metal	55.5	38.8	29		
Weld	55.4	39.7	29	BM	
Tantalum Powder-Grade					
Min. Base metal	30.0	20.0	20		
Actual Base Metal	50.6	36.9	35		
Weld	50.1	37.0	26	Weld	99
Ta-40%Cb					
Min. Base Metal	40.0	28.0	25		
Actual Base Metal	44.5	28.5	59		
Weld	43.8	28.3	20	Weld	98
Cb-1%Zr					
Min. Base Metal	35.0	20.0	20		
Actual Base Metal	41.7	25.6	29		
Weld	41.6	25.2	30	BM	
Tantalum Met-Grade					
Min. Base Metal	30.0	20.0	20		
Actual Base Metal	42.7	28.3	44		
Weld	38.2	28.4	17	Weld	89
Columbium					
Min. Base Metal	30.0	20.0	25		
Actual Base Metal	36.5	23.9	32		
Weld	34.1	22.2	27	BM	

MICROHARDNESS SURVEY OF WELD - It was assumed that welding in the open would result in interstitial pickup of oxygen and/or nitrogen which - in some alloys - could be detected by an increase in hardness.

Knoop hardness measurements were taken at mid-thickness of welded cross-sections through base metal, heat-affected zone and weld. Hardness measurements were also taken on the polished weld face and root and in some instances, through the thickness of the weld and heat-affected zone.

For the mid-thickness hardness measurements only, the mean hardness measurements in the weld and base metal and the percent change in hardness between the weld and base metal are listed in Table 2. Interstitial hardening in the weld zone varied from virtually no change to about a 28% increase from base metal hardness; in general, being greater in the higher strength alloys.

BEND TESTS - The weld samples were bent with the bend axis perpendicular to the weld (transverse weld bend) and with the bend axis centered in the weld (longitudinal weld bend); in some cases, the bend axis was also centered in the heat-affected zone (longitudinal weld bend). For each bend orientation, one coupon was run with the weld face in tension, and one coupon was run with the weld root in tension. Bend test results are listed in Table 3.

It was expected that an increase in weld hardness would be associated with a corresponding decrease in weld ductility, but this did not occur. There was no correllation between interstitial hardening in the weld zone and loss of weld ductility. In some instances, no or a small increase in hardness was associated with cracking in the bend sample, and in others, a large increase in hardness produced satisfactory bend test results.

As a point of reference, the ASME Boiler Code requires an 180° bend R/T equal to 2, where R is the bend radius and T is the sample thickness, to demonstrate adequate weld ductility. Columbium and Cb-1%Zr passed all bend tests at an 180° bend R/T = 1.1 and Ta-2.5%W passed all bend tests at an 180° bend R/T = 1.6, indicating these materials retain excellent ductility after welding with local shielding. Note that each material experienced a different degree of interstitial hardening in the weld, from virtually no change, to the maximum detected.

Tantalum metallurgical-grade, tantalum powder-grade and Ta-40%Cb each developed slight heat-affected zone cracks in some bend tests and passed others at an 180° bend R/T = 1.1. An increase in hardness in the vicinity of the cracks indicated that oxygen pickup in the heat-affected zone occurred. These materials can be welded in the open but require better local shielding than the previous group.

Gross weld and heat-affected zone cracks occurred in the transverse weld bend specimens in Ta-10%W, but no interstitial weld hardening was detected. This alloy requires substantially better local shielding than was provided in these tests or it should not be welded outside of an inert gas-atmosphere chamber.

SUMMARY

With the exception of Ta-10%W, these materials can be welded outside of an inert gas chamber using secondary local shielding - similar to techniques used to weld titanium - and retain room temperature strength and ductility in the weld and heat-affected zones comparable to unaffected base metal.

The best weldability was exhibited by columbium, Cb-1%Zr and Ta-2.5%W. Each had the following properties in common: fracture of the tensile coupons occurred outside the weld or heat-affected zones, and bend ductility was good. These materials appear to be highly tolerant to welding in the open with local shielding.

Tantalum metallurgical-grade, tantalum powder-grade, and Ta-40%Cb were less tolerant to welding in the open than the previous group. Welded tensile coupons fractured in the weld, and slight heat-affected zone cracking occurred in bend coupons. These materials appear to be amenable to welding in the open but require better secondary shielding than the previous group.

Ta-10%W exhibited appreciable loss of bend ductility. This alloy is not a good candidate for welding outside of an inert gas atmosphere chamber.

Table 2 - Hardness Survey of Gas Tungsten-Arc Welds in Several Refractory Metals (Knoop Hardness Number - 200g load)

Alloy	Base Metal Mean Value	Weld Mean Value	Percent Change In Hardness
Ta-10%W	278	268	-3.6
Ta Met.-Grade	94	97	3.2
Columbium	81	87	7.4
Cb-1%Zr	107	121	11.6
Ta-40%Cb	108	122	13.0
Ta Powder-Grade	116	144	24.1
Ta-2.5%W	141	181	28.4

ACKNOWLEDGMENT

The author acknowledges the valuable technical assistance of Mr. Joseph Wertz in developing the welding parameters.

Table 3 - Bend Tests in Gas Tungsten-Arc Welds in Several Refractory Metals (Base Metal and Welded 180° Bend Test R/T Values (R = Radius of Plunger, T = Base Metal Thickness)[1]

Alloy Thickness (Inch)	Base Metal	Transverse Weld: Face in Tension	Transverse Weld: Root in Tension	Longitudinal Weld, Bend Axis in Weld: Face in Tension	Longitudinal Weld, Bend Axis in Weld: Root in Tension	Longitudinal Weld, Bend Axis In Heat-Affected Zone: Face in Tension	Longitudinal Weld, Bend Axis In Heat-Affected Zone: Root in Tension
Columbium 0.030	1.1	1.1	1.1	1.1	1.1	1.1	1.1
Ta Met-Grade 0.030	1.1	1.1	slight HAZ cracking at R/T = 1[2]	1.1	1.1	1.1	slight HAZ cracking at R/T = 1[2]
Cb-1%Zr 0.0155	1.0	1.0	1.0	1.0	1.0		
Ta-40%Cb 0.030	1.1	1.1	1.1	1.1	slight HAZ cracking at R/T = 1[2]		
Ta Powder-Grade 0.025	1.1	HAZ cracks at R/T = 2	1.1	HAZ cracks at R/T = 2	1.1[3]		
Ta-2.5%W 0.020	1.6	1.6	1.6	1.6	1.6		
Ta-10%W 0.020	1.6	1.6	gross weld & HAZ cracks at R/T = 2.4	1.6			

(1) Acceptance criteria - no cracks at 30X examination in binocular microscope.

(2) Slight cracks detected with binocular microscope at 30X, but none were detected in cross-section examined at 200X.

(3) No cracks detected with binocular microscope at 30X, but cross-section revealed cracks at inside surface.

STATISTICAL EXPERIMENTATION IN ARC WELDING

Michel Galopin
Ecole de Technologie Supérieure
Montréal, Canada

Elie Boridy
Université du Québec à Montréal
Montréal, Canada

ABSTRACT

A Computer Assisted Experimentation system dedicated to arc welding and based on statistical principles has been designed. The system, composed of hardware and software, assists the experimenter in the definition of the scope of the experiment, the choice of the variables among the welding parameters, the design of the experimental plan, the statistical analysis of the collected data and the tabular or graphic display of results.

THE DETERMINATION of an acceptable arc welding procedure which satisfies various criteria and requirements follows generally a more or less structured approach widely based upon subjective appreciations. This approach, although generally used, has major drawbacks. It is subject to human error; it is not of universal reach; it contributes little to the understanding of the problem and yields a solution which, if acceptable, is not necessarily optimum in terms of cost or quality. A more structured methodology is therefore highly desirable.

ANALYTICAL METHODS

Analytical studies in the welding field progress rapidely with the spreading use of computers [1]. The existing programs deal mostly with heat flow, metallurgical and mechanical behaviour of the weld at high temperatures, solidification process etc. However, these analytical techniques are affected by the lack of knowledge of the basic phenomena and oversimplification is necessary to solve even common problems. The design of a system based upon such analytical methods and which satisfies the industrial needs, constitutes at most a long term objective.

SYSTEMIC METHODS

Systemic methods take into consideration the whole welding process. Computers are used to store all the available empirical information obtained with the joint of various materials. A great deal of work has been done to design algorithms [2] for the determination of the most suitable welding conditions based on existing empirical laws. However, the practical application of such results can be done only in the very restrictive context in which the laws have been established.

Another methodology based upon systemic experimentation has been proposed in reference [3]. According to this method, domains are defined with the desired precision in the space of variables which obey to the requested criteria for a given welded joint. Such a partition is obtained following an inductive and recurrent choice of tests. The latter methodology leads to a complete and explicit solution of the welding problem. It preserves all the compiled information, allows a rigorous determination of the effects on the quality of any modification of the welding parameters and finally, offers a well-informed choice of the welding parameters in terms of the constraints imposed by the production process (tolerances, cost...). Its application requires however a very large number of tests which cannot be justified on an practical basis.

We propose a more efficient approach based on the statistical design and analysis of experiments.

STATISTICAL DESIGN OF WELDING EXPERIMENTS

A trial-and-error approach to the design and execution of experiments cannot only be time consuming and expensive but even self defeating. Whatever the objectives of an experiment are, no analysis will ever overcome the handicap of poorly designed experiments.

Statistical design of experiments has proved useful and efficient in the planning of experiments and their sequential execution. These methods have found widespread uses in disciplines like agricultural science, biology, medecine, engineering and physical sciences. We applied these methods to the selection of an arc welding procedure and more generally to experimentation in welding.

By statistical design, we refer to the planning of the welding experiment so that appropriate data can be collected and analyzed by statistical methods resulting in valid and objective conclusions. The design of the experiment and the statistical analysis of the data are closely related, since the method of analysis depends directly on the design chosen. The three basic principles of experimental design are replication, randomization, and blocking.

We designed a Computer Assisted Experimentation system based on statistical techniques to solve classical problems in the determination of welding procedures. Three common problems are identified.

1- Determination of relations between responses (some measures of the quality of the weld) and variables associated to the base materials, the geometry of the joint, the characteristics of the welding process and the operating mode (Fig. 1.a).

2- Determination of a subset of the variable space for which some specifications are satisfied (Fig. 1.b).

3- Comparative studies to evaluate or classify some factors (Fig. 1.c).

COMPUTER ASSISTED EXPERIMENTATION

The system designed includes software and hardware (custom-built programmable controller) to manage the various stages of experimentation which are illustrated in figure 2.

WELDING PARAMETERS - All the welding parameters related to the experiment have to be defined. These parameters define the joint, the preparation, the welding process and the operating mode. A typical list is proposed in figure 3.

CHOICE OF RESPONSES - Selection of the responses or criteria (dependent variables). The responses to be measured should provide meaningful information about the quality of the weld. These responses could refer to the geometry of the weld, the mechanical properties of the joint, the results of some NDT tests...

CHOICE OF WELDING VARIABLES and levels. The experimenter must select the independent variables or factors to be investigated in the experiment. The factors may be qualitative (gas composition, wire type, flux type, etc.) or quantitative (welding current, welding speed, voltage, etc.). Six variables distributed over the three passes, were chosen for the experiment presented at figure 4.

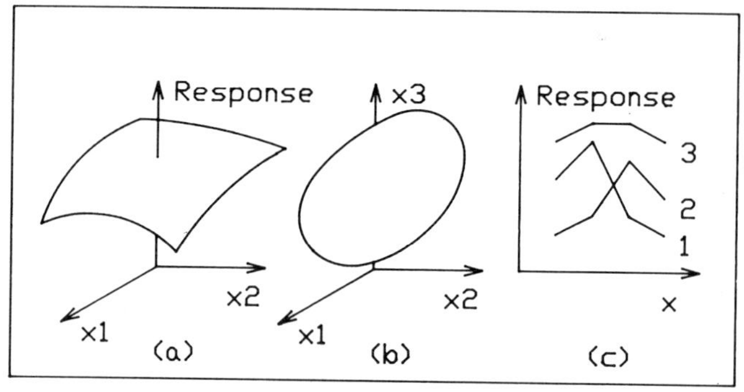

Fig. 1 - Three classes of problems in welding experimentation.

```
EXPERIMENTATION MENU                                                   date:   04-14-1986
EXPno  1  CFGno 01  PASno  0  PLNno  1  VARno     REPno      TSTno 12  BLCno  1
EXPnb  3  SEQno     PASnb  1  PLNnb  1  VARnb  6  REPnb   12 TSTnb  16 BLCnb  1
PARno     Procs GMA PLNty 2N  statusD   LVLnb  2  CBNnb   64 TSTtn  16 OBSnb  1

EXPERIMENTATION MENU

   <1> CHOICE OF VARIABLES                    <S> SAVE
   <2> CHOICE OF EXPERIMENTAL DESIGN          <D> DEBUG
   <3> PERFORMING THE EXPERIMENT
   <4> DATA COLLECTION
   <5> STATISTICAL ANALYSIS
   <6> REGRESSION SURFACES
   <7> SPECIFICATIONS
   <8> JOINT DRAWING
   <9> OTHER DESIGN / EXPERIMENT
   <0> END

Enter choice
< ESC >: Previous entry
```

Fig. 2 - Menu of the experimentation software.

```
CHOICE OF VARIABLES                                                    date:   04-14-1986
EXPno  1  CFGno 01  PASno  1  PLNno  1  VARno     REPno      TSTno 12  BLCno  1
EXPnb  3  SEQno     PASnb  1  PLNnb  1  VARnb  6  REPnb   12 TSTnb  16 BLCnb  1
PARno     Procs GMA PLNty 2N  statusD   LVLnb  2  CBNnb   64 TSTtn  16 OBSnb  1

Parameters
1 JOINT            2 PREPARATION     3 PROCESS         4-5 OPERATING PARAMETERS

JOINT B            WType         2   GAS   A02         Speed      45   Xinit     0
PENET L            PREPA MACH        Flow    15.00     Feed     1300   ISpee   120
MatA  A36          Fixtu 1           Wire  S101                        WX       90
                   GAP        1.00   Size     0.90     uarc     21.0   RX        0
MatB  A36          ANGL1        60   Stick   20.00                     RSpee   120
                   ANGL2         0   Lead       90
tA.mm      9.50    S1         8.00   Work       90
tB.mm      9.50    S2(mm      0.00
Backg N            R(mm)      0.00
                   GAMMA         0
                   Misma      0.00                                     Pre P     5
                   ALPHA         0                                     PostP     5

< ENTER >: Next
<PRINT>: Printout
```

Fig. 3 - A welding procedure: starting point of experimentation.

```
CHOICE OF VARIABLES                                                    date:   04-14-1986
EXPno  1  CFGno 01  PASno  0  PLNno  1  VARno     REPno      TSTno 26  BLCno  1
EXPnb  1  SEQno     PASnb  3  PLNnb  1  VARnb  6  REPnb   12 TSTnb  16 BLCnb  1
PARno     Procs GMA PLNty 2N  statusD   LVLnb  2  CBNnb   64 TSTtn  16 OBSnb  1

Review table
Pass  Variable

   1  GAP
   1  Mismatch
   2  Speed
   2  Feed
   3  Speed
   3  Feed
```

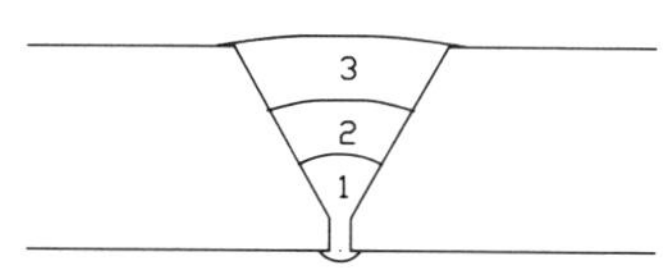


```
< ENTER >: Next
```

Fig. 4 - Choice of variables for a three pass butt weld.

CHOICE OF EXPERIMENTAL DESIGN - In selecting the design, it is important to keep in mind the experimental objectives. We will classify welding experiments in three groups.

Measuring the efects of variables - Factorial designs are extremely useful for this purpose. Screening experiments are usually performed in the early stages of the study to discriminate between the effect of many factors on the response and the 2^k design, with high fractional replication, is particularly useful. A one-quarter fraction of the 2^6 design has been chosen to investigate the effect of 3 preparation-related variables and 3 welding variables on the geometry of the root pass of a butt joint. The experimental plan and the analysis of variance are presented in figure 5. The experimenter may be interested in producing an empirical model that describes how a given set of input variables affect the response over some specified region of interest. The two and three levels factorial designs are appropriate to estimate the regression model. With two levels for each factor, we must assume that the response is approximately linear over the range of the factor levels chosen. Quadratic terms are introduced in the model for the 3^k design. Figure 6 presents a one-third fraction for a 3^4 design of a single pass butt weld.

Specification problems - Similar designs are used to spotlight the region of the variable space in which complex requirements may be met, or to indicate in which direction process changes should be made if the characteristics of the joint stay outside specifications.

Comparative experiments - The evaluation of materials (base metal, preparation, heat treatment...) or consummables (wire, electrode, gas mixture, flux, etc.) consists generally in the comparison of several treatments which are randomly assigned to the experimental material. The analysis of variance is the appropriate procedure for comparative designs.

From here, all the referred results correspond to the design of figure 5.

PERFORMING THE EXPERIMENT - This is the actual execution of the welding tests in accordance to the experimental design and the collection of the response data (micrography, mechanical testing, NDT, etc.). Rigorous control of the operating variables is achieved by the hardware.

STATISTICAL ANALYSIS - Statistical analysis provide guidelines as to the reliability and validity of results. Plots of the effects of different factors on normal probability paper (figure 7) provides a simple way to analyse unreplicated factorial design. Negligible effects are normallly distributed and will tend to fall along a straight line. They can be combined to form an estimate of the experimental error. The resulting analysis of variance is presented in figure 8 for the width of the weld.

CHOICE OF EXPERIMENTAL DESIGN date: 04-14-1986
EXPno 1 CFGno 01 PASno 0 PLNno 1 VARno REPno TSTno BLCno 1
EXPnb 3 SEQno PASnb 1 PLNnb 1 VARnb 6 REPnb 12 TSTnb 16 BLCnb 1
PARno Procs GMA PLNty 2N statusD LVLnb 2 CBNnb 64 TSTtn 126 OBSnb 1

Review table Pass	Variable	/ Random Order of tests 0	1	2	Confounding (a=AA)
1	GAP	0.00	1.00		
1	ANGL1	45	60		I
1	Mismatch	0.00	1.50		ABCE
1	Speed	45	55		ACDF
1	Feed	1200	1400		BDEF
1	uarc	20.0	22.0		
12 15 39 29 19 26 59 53 46 36 1 41 24 64 6 50					I

< ENTER >: Next
<PRINT>: Printout

Fig. 5 - 2^k factorial design with fractional replication.

CHOICE OF EXPERIMENTAL DESIGN date: 04-14-1986
EXPno 1 CFGno 01 PASno 0 PLNno 1 VARno REPno TSTno BLCno 1
EXPnb 1 SEQno PASnb 1 PLNnb 1 VARnb 4 REPnb 12 TSTnb 27 BLCnb 1
PARno Procs GMA PLNty 3N statusD LVLnb 3 CBNnb 81 TSTtn 137 OBSnb 1

Review table Pass	Variable	/ Random Order of tests 0	1	2	Confounding (a=AA)
1	GAP	0.00	0.50	1.00	
1	Mismatch	0.00	0.75	1.50	I
1	Speed	68	73	78	AbCD
1	Feed	2100	2200	2300	aBcd
13 17 46 38 20 30 72 68 56 42 1 43 31 61 35 54					I
25 24 76 50 75 9 80 64 12 60 5					

< ENTER >: Next
<PRINT>: Printout

Fig. 6 - 3^k factorial design with fractional replication.

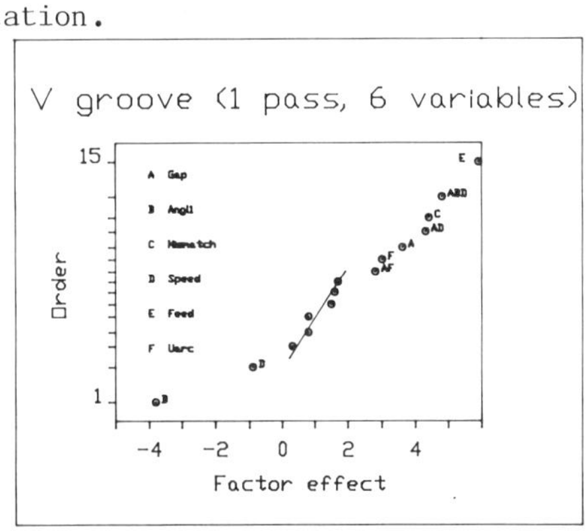

Fig. 7 - Normal plot of effects for the design of figure 5.

STATISTICAL ANALYSIS date: 04-14-1986
EXPno 1 CFGno 01 PASno 0 PLNno 1 VARno REPno TSTno 12 BLCno 1
EXPnb 3 SEQno PASnb 1 PLNnb 1 VARnb 6 REPnb 12 TSTnb 16 BLCnb 1
PARno Procs GMA PLNty 2N statusD LVLnb 2 CBNnb 64 TSTtn 16 OBSnb 1

ANOVA Table Response: Penet

SOURCE OF VARIATION				SS	DF	MS	Fo	alpha
A	BCE	CDF	ABDEF	0.53	1	0.53	8.63	0.03
B	ACE	ABCDF	DEF	0.57	1	0.57	9.36	0.02
AB	CE	BCDF	ADEF	Neglect				
C	ABE	ADF	BCDEF	0.78	1	0.78	12.86	0.01
AC	BE	DF	ABCDEF	Neglect				
BC	AE	ABDF	CDEF	Neglect				
ABC	E	BDF	ACDEF	1.38	1	1.38	22.67	0.00
D	ABCDE	ACF	BEF	0.03	1	0.03	0.53	1.00
Error				0.37	6	0.06		
Total				5.94	15			

Scrolling:
< ENTER >: Next

Fig. 8 - Anova table for the design of figure 5.

DISPLAY OF RESULTS - The experimenter may draw conclusions or inferences about the results. The statistical inferences must be physically interpreted, and the practical significance of these findings evaluated. Recommendations may include a further round of experiments, as experimentation is usually an iterative process. The use of graphical displays is a very effective way to present important results; an example of a response curve for the width of the bead is presented in figure 9. Contour plots of interesting sections of the k-dimensional response surface are also specially useful. In effect, specifications problems require generally the simultaneous consideration of several responses. Contour overlays as illustrated in figure 10 for the previous fractional factorial design at two levels, clearly show how and whether complex requirements may be met. It allows new conditions to be determined if specifications change, and indicates in which direction process changes should be made if the characteristics of the manufactured products stray outside specifications.

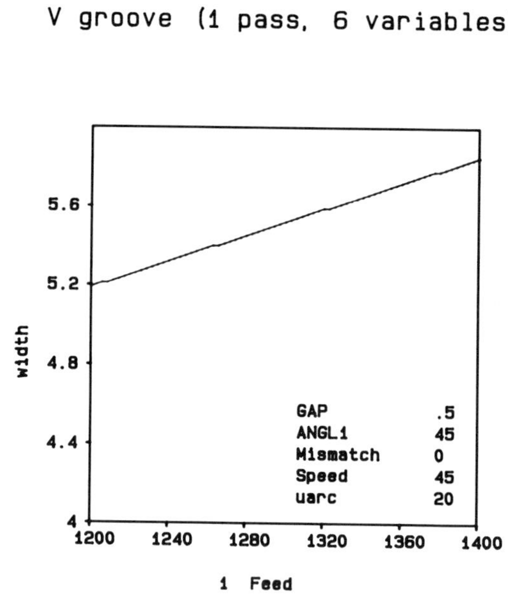

Fig. 9 - A response curve for the width of the bead.

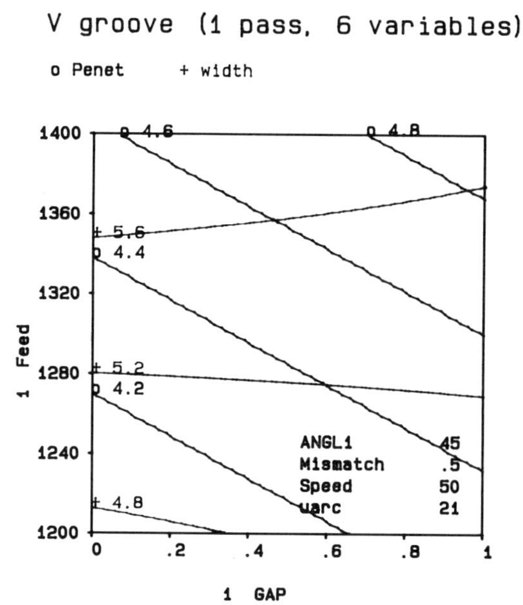

Fig. 10 - Contour overlays of the penetration and the width of the weld.

CONCLUSIONS

The statistical design approach is an effective solution to procedure selection. It provides meaningful data on the particular problem studied without any explicit reference to the fundamental mechanisms underlying the welding process. There are no limitations in terms of the factors and responses investigated and no oversimplification of the problem is necessary.

Our Computer Assisted Experimentation system permits a straightforward implementation of this approach. The next step consists in linking this system to a welding data base manager to ensure the integration of optimized data in production.

REFERENCES

(1) Masubuchi, K., Soudage et Techniques Connexes, 34, 80-94 (1980)

(2) Jones, S.B., J. Doherty and G.R. Salter, Welding Journal, 6, 19 (1977)

(3) Galopin, M. and E. Boridy, Soudage et Techniques Connexes, 37, 403-412 (1983)

(4) Box, G.E.P., W.G. Hunter and J.S. Hunter, "Statistics for experimenters", John Wiley, (1978)

(5) Montgomery, D.C., "Design and analysis of experiments", John Wiley, (1984)

HOT-DUCTILITY BEHAVIOR OF TWO ALUMINUM ALLOYS

J.P. Balaguer, E.F. Nippes
Rensselaer Polytechnic Institute
Troy, New York

D.W. Walsh
California Polytechnic State University
San Luis Obispo, California

ABSTRACT

The weld HAZ hot-cracking susceptiblity of two aluminum alloys, Al-4Mg-2Li and Al-3.5Mg-4Zn, was evaluated with hot-ductility testing. Specimens of each material were subjected to high-speed tensile testing on heating and on cooling. The nil-strength temperatures of the Al-Mg-Li and Al-Mg-Zn alloys were 540°C and 570°C, respectively. The Al-Mg-Li alloy experienced a loss in ductility near 500°C, and was judged to be the more susceptible of the two alloys to HAZ hot cracking. The hot tensile strength of both alloys was independent of cooling rate from the peak temperature, with the Al-Mg-Li alloy somewhat stronger at lower temperatures. The hot ductility of the Al-Mg-Li alloy was sensitive to cooling rate; the ductility was greater for the faster cooling rate at 400°C and below and greater for the slower cooling rate above 400°C.

THERE HAS BEEN CONSIDERABLE interest recently in the area of aluminum-lithium alloys. It has been well documented that additions of lithium to aluminum alloys increase the elastic modulus and strength and decrease the density and ductility. The high strength/weight and stiffness/weight ratios of Al-Li alloys make them excellent materials for aerospace applications. Previous investigations have shown that Al-2Li-Mg alloys exhibit maximum tensile ductility at Mg levels of 4%[1].

There has been little recent work examining the weldability of Al-Li alloys. A study by Cross[2] has shown that the maximum hot-tearing susceptibility, as measured by the Varestraint test, was observed for Al-2.6Li binary alloys. The present investigation will examine the hot ductility of Al-4Mg-2Li, a precipitation-hardened alloy.

Aluminum alloy 7039, an Al-3.5Mg-4Zn precipitation-hardened alloy designed for armor applications, will also be examined. Although little information on this alloy is available in the literature, previous studies at Rensselaer[3] have shown it to be susceptible to hot cracking in Varestraint tests. Hot-ductility testing has been successfully used to evaluate the heat-affected zone (HAZ) hot-cracking susceptibility of many alloys[4,5]. The object of this study will be to compare and contrast the hot-ductility behavior of two precipitation-hardened aluminum alloys in order to evaluate their susceptibility to HAZ hot cracking.

MATERIALS AND PROCEDURE

MATERIALS. The materials used in this investigation were an experimental Al-4Mg-2Li alloy and a commerical Al-Mg-Zn alloy (7039-T61). The chemical compositions of these alloys are given in Table I. The Al-4Mg-2Li alloy, fabricated from pre-alloyed powders, was rolled to 12.5-mm (0.5-in.) thick plate. The Al-Mg-Zn alloy was obtained in the form of 37.5-mm (1.5-in) thick plate. The room-temperature mechanical properties of these materials are listed in Table II.

EXPERIMENTAL PROCEDURE. Hot-ductility testing was performed using a Gleeble Model 1500 thermomechanical simulation device. High-speed elevated-temperature tensile tests were performed on 6.35-mm (0.25-in.)

diameter cylindrical specimens. Temperature control was provided by a two-wire, chromel-alumel, thermocouple percussion welded to the specimen. The nil-strength temperature (NST) of each alloy was determined by placing a stress of 10 MPa (1.5 ksi) on the specimen and heating at 25°C/sec (45°F/sec) until failure occurred. The temperature at which failure occurs is the nil-strength temperature.

Two synthetic thermal cycles were then designed, based on temperature measurements in aluminum plate by Masubuchi[8], to simulate the weld HAZ. Both thermal cycles consisted of heating at 50°C/sec (90°F/sec) to 500°C (932°F). The two thermal cycles differed only in cooling rate; the "fast" cycle had a cooling rate of 66°C/sec (119°F/sec) and the "slow" cycle had a cooling rate of 33°C/sec (59°F/sec).

High-speed tensile tests, with a stroke rate of 5 cm/sec (2 in./sec), were performed on the heating and cooling segments of the synthetic thermal cycle. On-heating tests at 500°C were performed immediately upon reaching peak temperature, while on-cooling tests at 500°C were performed after holding at peak temperature for 1 sec. The resulting force data were gathered using a high-speed data acquisition memory (Datalyzer). Reduction-in-area data were gathered using an optical microscope with a filar eyepiece. The fracture surfaces of several Al-Mg-Zn specimens were elliptical as a result of the anisotropic properties of the plate from which they were machined. The area of these fracture surfaces was estimated using the formula for the area of an ellipse. The accuracy of this technique was confirmed using an image analyzer.

RESULTS

SLOW CYCLE. The nil-strength temperatures of the Al-Mg-Li and Al-Mg-Zn alloys were 540°C (1004°F) and 570°C (1058°F), respectively. The results of hot-ductility tests on the Al-Mg-Zn alloy, for the slow cooling rate, are shown in Figs. 1 and 2. The hot tensile strength of this alloy decreased to a minimum measured value of 41 MPa (6 ksi). As shown in Fig. 1, the hot tensile strengths measured on heating (OH) were slightly greater than those measured at the same temperature on cooling (OC). The maximum difference in measured hot tensile strengths was observed at the lowest common test temperature, 300°C (572°F).

The hot-ductility response of the Al-Mg-Zn alloy is shown in Fig. 2. Similar to the hot tensile strengths, the maximum difference in hot ductility was observed at the lowest common test temperature, 300°C. No true ductility dip was observed up to the peak temperature of 500°C; this alloy had a reduction in area of 90% at this temperature.

The results of hot-ductility tests on the Al-Mg-Li alloy are shown in Figs. 3 and 4. The results shown in Fig. 3 indicate that there is little difference in the on-heating and on-cooling hot tensile strengths of this alloy. The minimum measured hot tensile strength of the Al-Mg-Li alloy, 34 MPa (4.9 ksi), was observed at 500°C.

The hot-ductility response of the Al-Mg-Li alloy is shown in Fig. 4. The maximum hot ductility was observed at 450°C for on-heating and on-cooling tests. Ductilities measured on cooling were approximately 10 to 15% less than those measured on heating. Tests performed at 500°C, however, produced conflicting results. Two tests performed on heating produced ductilities of 14% and 56% and tensile strengths of 33 MPa (4.8 ksi) and 34 MPa (4.9 ksi), while two tests performed on cooling produced ductilities of 7% and 41% and tensile strengths of 36 MPa (5.2 ksi) and 44 MPa (6.4 ksi). (The data points in Fig. 4 represent the average of two tests). Despite the aforementioned scatter, the tensile strength/ductility ratio of this material at 500°C appeared considerably higher than that of the Al-Mg-Zn alloy. This indicates that the Al-Mg-Li alloy would have the greater susceptibility to HAZ hot cracking.

FAST CYCLE. The results of hot-ductility tests, performed on cooling, for the fast cycle are shown in Figs. 5 and 6. Also included in these figures are the previously discussed results from the slow-cycle tests. Figure 5A shows that the hot tensile strength of the Al-Mg-Zn alloy was slightly greater for the slow cycle than for the fast cycle. The Al-Mg-Li alloy results, shown in Fig. 5B, indicated somewhat superior hot tensile strength for the slow cycle at 400°C and below, with the reverse true at temperatures of 450°C and 500°C. The greater slow-cycle tensile strength of both alloys probably results from enhanced precipitation on cooling.

The hot-ductility response of the Al-Mg-Zn alloy appeared to be independent of cooling rate, save that no data were collected at 450°C for the fast cycle (Fig. 6A). The results for the Al-Mg-Li alloy, again, were mixed. The hot ductilities measured for the fast cycle appear superior at 400°C and below, and inferior at 450°C (Fig. 6B). Hot-ductility tests at 500°C gave mixed results similar to those obtained for the slow cycle: ductilities of 67%, 24%, and 10% were measured with corresponding tensile strengths of 52 MPa (7.6 ksi), 37 MPa (5.3 ksi), and 21 MPa (3.1 ksi). As shown in Fig. 6B, the average hot ductility of the Al-Mg-Li alloy at 500°C was significantly less for the fast cycle. The poor ductility recovery of this material probably resulted from a network of oxides at prior-particle boundaries, a factor which plagued previous investigations of this material[6]. Because the tensile strengths were identical for the two cycles, it appears that the HAZ hot-cracking susceptibility is higher for material subjected to the fact cycle.

SUMMARY

The results of hot-ductility testing on two age-hardened aluminum alloys have been used to provide an indication of their relative HAZ hot-cracking susceptibilities. While the Al-Mg-Li alloy had generally higher high-temperature strength, it suffered from ductility loss at approximately 500°C. Both the Al-Mg-Zn and Al-Mg-Li alloys displayed lower ductilities for the slow thermal cycle as a result of enhanced precipitation on cooling. On the basis of hot-ductility test results, it appears that the Al-Mg-Li alloy will be more susceptible to HAZ hot cracking than the commercial Al-Mg-Zn alloy. The apparent increase in hot-cracking susceptibility of the Al-Mg-Li alloy with increased weld HAZ cooling rates will be investigated further in future work.

TABLE I

COMPOSITION OF ALLOYS USED IN THIS INVESTIGATION

Alloy	Composition (w/o)						
	Mg	Zn	Li	Si	Cr	Fe	Al
Al-Mg-Li[6]	4.24	0.004	2.13	0.10	<0.002	0.01	bal.
Al-Mg-Zn	3.44	3.99	--	0.047	0.163	0.115	bal.

TABLE II

MECHANICAL PROPERTIES OF ALLOYS USED IN THIS INVESTIGATION

	UTS MPa (ksi)	YS MPa (ksi)	% Elong. (in 25 mm)
Al-Mg-Li[6]	413 (60)	351 (51)	12
Al-Mg-Zn[7]	414 (60)	345 (50)	14

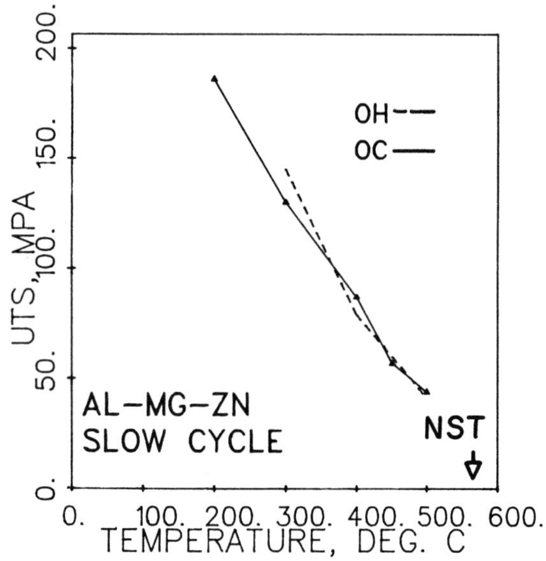

Fig. 1: Hot-tensile strength results for the Al-Mg-Zn alloy.

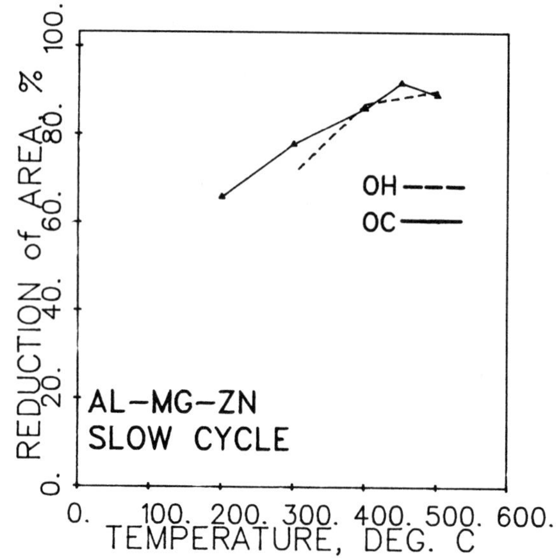

Fig. 2: Hot-ductility results for the Al-Mg-Zn alloy.

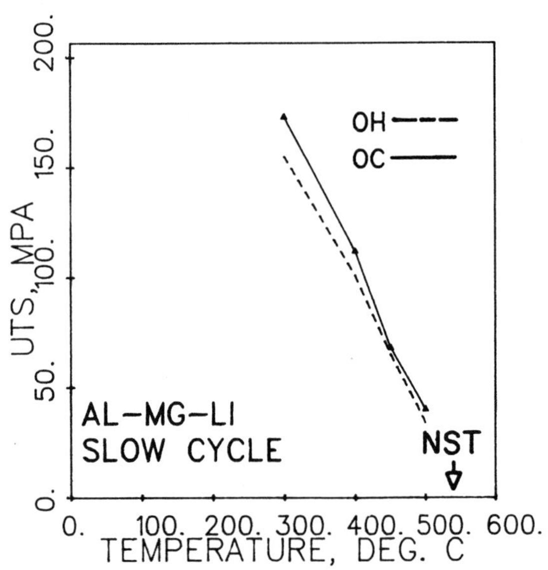

Fig. 3: Hot-tensile strength results for the Al-Mg-Li alloy.

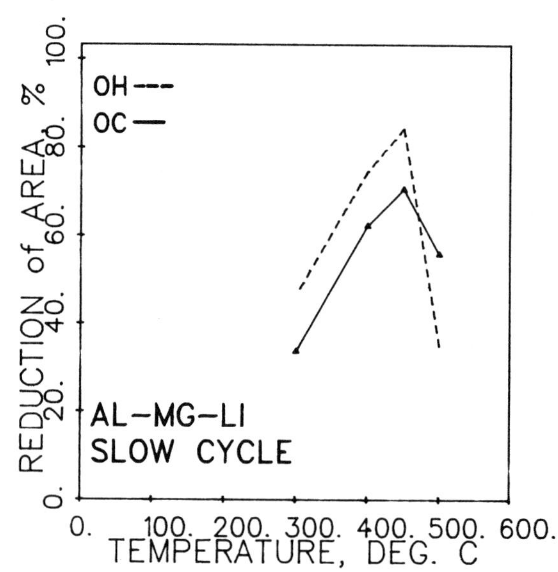

Fig. 4: Hot-ductility results for the Al-Mg-Li alloy.

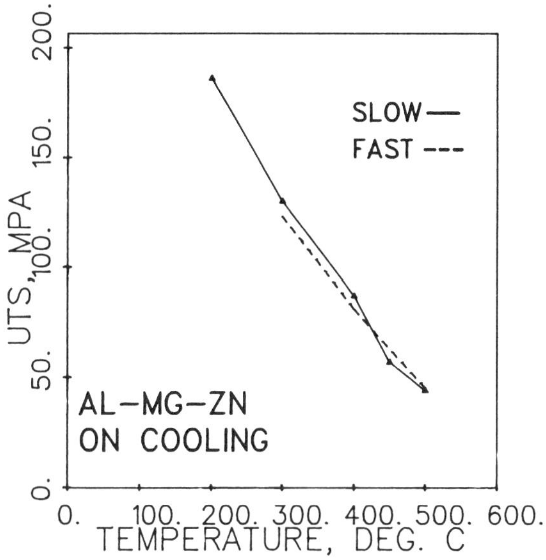

(a) Al-Mg-Zn

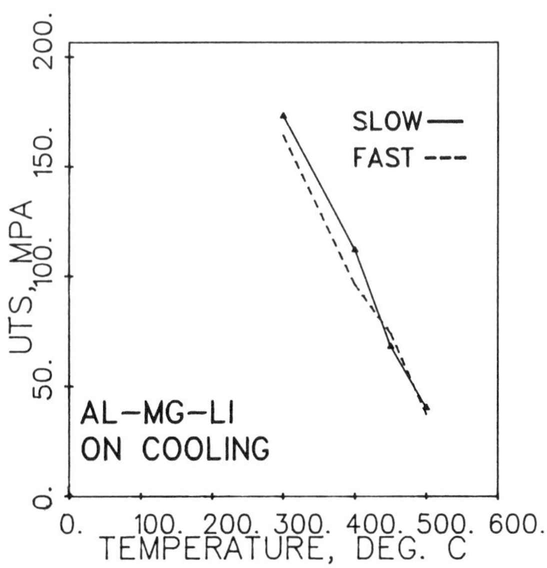

(b) Al-Mg-Li

Fig. 5: On-cooling hot-tensile strength results for the fast and slow cycles.

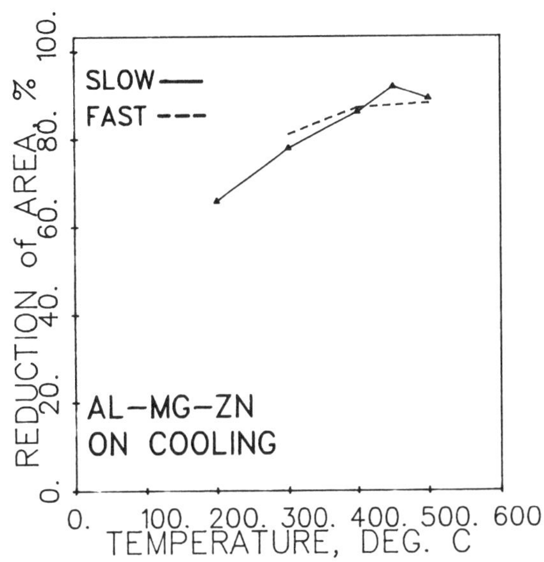

(a) Al-Mg-Zn

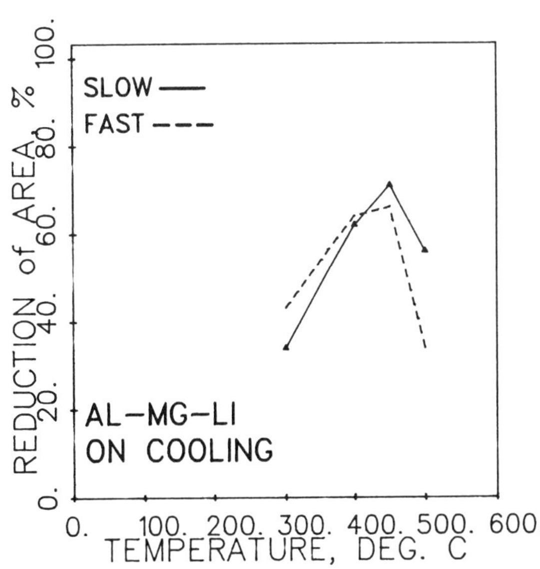

(b) Al-Mg-Li

Fig. 6: On-cooling hot-ductility results for the fast and slow cycles.

ACKNOWLEDGEMENTS

The authors would like to acknowledge the materials supplied for this investigation by FMC Corporation, San Jose, California, and the Naval Air Development Center (NADC), Warminster, Pennsylvania. The authors also wish to thank David Sims and Roger Hatfield for their assistance in the experimental portion of this investigation.

REFERENCES

1. Dinsdale, K., S. J. Harris and B. Noble, "Proceedings of the First International Aluminum-Lithium Conference", TMS-AIME, pp. 101-118 (1981).

2. Cross, C. E., D. L. Olson, G. R. Edwards, and J. F. Capes, "Proceedings of the Second International Aluminum-Lithium Conference", TMS-AIME, pp. 675-682 (1983).

3. Lippold, J. C., "An Investigation of Weld Hot Cracking in 5083-0 Aluminum", M.S. Thesis, Rensselaer Polytechnic Institute (1975).

4. Nippes, E. F., W. F. Savage, H. F. Mason, B. J. Bastian, and R. Curran, Welding Journal (Research Supplement), 34, 183s-196s (1955).

5. Stout, R. D. and W. D. Doty, "Weldability of Steels", Welding Research Council, pp. 312-315 (1978).

6. Ricker, R., "Corrosion Fatigue of an Al-Zn-Mg Alloy and an Al-Mg-Li Alloy", Ph.D. Thesis, Rensselaer Polytechnic Institute, Troy, NY (1983).

7. "Aluminum Construction Manual", The Aluminum Association (1972).

8. Masubuchi, K., "Analysis of Weld Structures", Pergamon Press, Oxford, p. 263 (1980).

RESULTS OF RESEARCH ON WELDING RESEARCH

C.B. Shaw, Jr.
Idaho National Engineering Laboratory
EG&G Idaho Incorporated
P.O. Box 1625 - ILF
Idaho Falls, Idaho 83415

IN 1981 THE INTERNATIONAL Institute of Welding (IIW) authorized establishment of Provisional Commission VII, "Welding Research Strategy and Collaboration," now Study Group RES, with the charter to exchange information among interested organizations on their structure, identify and define welding research problems, assist IIW in opening channels of research cooperation, promote transfer and application of results of welding research, identify needs for improved training of welding personnel at all levels, and assist in the organization of conferences. One of the first actions of the new organization, under its founding chairman Prof.-Dr. Y. Arata, was to draft a questionnaire for worldwide distribution to establish baseline data on the current status and perceived needs of the welding research community. The form was based on one circulated in the United States by Dr. John Gross. The National Councils of IIW were enlisted to aid in seeking out appropriate organizations to be solicited for information. Responses were received from 273 organizations in 17 of the 36 nations represented in IIW. The author has used them to build a data base in Ashton-Tate's dBASE III language, which can be queried by anyone having a personal computer compatible with the IBM PC. While the attempt to computerize the responses to the questionnaire revealed certain deficiencies which are expected to be corrected when the data base is updated, this first survey has been analyzed to yield the findings presented here. Those interested should note that the data base includes not only the names and addresses of the responding organizations, but of specific individuals who can be contacted for further information or to make arrangements for cooperative activity of mutual interest in the field of welding research.

ANALYSIS OF QUESTIONNAIRE RESPONSES

Responses to the IIW Questionnaire came from organizations which identified themselves as universities (23%), research institutes (22%), government laboratories (10%), manufacturers (33%), and other (13%). Most of those in the last category were public utilities. As Table 1 shows, however, categorization of the type of organization does not fully define the nature of its activities.

WELDING RESEARCH RESOURCES AND SUBJECTS - The responses analyzed indicate that at least 9,300 individuals are engaged in welding research, including 3300 professionals, 4100 technicians, almost 800 students, and 1100 support personnel. Their research institutions occupy facilities totalling 4.2 million square feet and have annual operating budgets of $730 million. The latter amounts to $78 thousand per person, $98 thousand per technical person (professionals plus technicians), or $220 thousand per professional.

The percentage of effort devoted to various fields of welding research is shown in Table 2.

NEEDS TO ADVANCE WELDING TECHNOLOGY - The respondents were asked to rate various suggested needs as unimportant (given a score of 0), somewhat important (scored 1), important (2) or very important (3). Existence of a central (national) welding research organization, and national-level coordination of welding research had average scores above 2, the latter by only a narrow margin. The other suggested categories fell slightly below 2; in sequence they were increased government funding, increased funding by industry, better training of personnel, and improved mechanisms for technology transfer. No write-in candidate received a significant score.

A suggested needs rating of 3, very important, was given by 27% of the respondents to the central welding research organization, and by 13% to national coordination. No other suggestion was considered 'very important' by as many as 9%. It may be noted in passing that central welding research organizations were among the most faithful respondents.

ATTITUDES TOWARDS COOPERATIVE RESEARCH - Fully 92% of respondents state that they encourage some form of cooperative welding

Table 1 - Percentage of Various Activities Carried out by Different Types of Organizations.

Organization	Activity: Education	Research	Technical Assistance	Fabrication	Other
University	64	34	1	0	0
Research Institute	6	65	29	1	0
Government laboratory	2	84	3	3	8
Manufacturer	3	20	10	59	8
Other	5	23	20	6	45

Table 2 - Percentage of Effort Devoted to Various Fields.

Field	Percent
Physics of Welding	3
Welding Processes	19
Power Supplies	17
Consumables	9
Cutting Processes	3
Welding Design	4
Stress and Distortion	4
Metallurgy	9
Weldment Properties	9
NDT	3
Fracture and Fatigue	7
Instrumentation/Control	3
Automation	7
Other	5

research. Cooperation with individual research organizations in the same country is practiced by 82%, in sharp contrast to cooperation with foreign organizations, whether affiliated (15%) or not (11%). However, 33% serve on international committees fostering research cooperation, not greatly below the 56% involved with domestic committees.

MECHANISMS FOR SHARING RESEARCH ACCOMPLISHMENTS - Table 3 shows respondents' opinions of various actual or potential mechanisms for sharing the results of welding research on a national or international level. Many of those disapproving Staff Exchange stipulated that considerations of national security made such exchanges impractical for their organizations. Conventions (such as those of the American Welding Society) which include equipment exhibitions were indicated in several comments as drawing disapproval for a less-than-serious atmosphere. Generalized conferences draw significantly less approval, and much less strong approval, than seminars on specialized topics, the latter rating second in endorsement only to the international meetings of IIW. The topic most often suggested for special seminars was robotic welding, which was later the subject of the international conference held at the beginning of the 1985 Annual Assembly of IIW.

TECHNOLOGY TRANSFER METHODS IN ACTUAL USE - Despite the sentiments expressed in favor of wide cooperation in welding research, and in the sharing of research results, the fact remains that 63% of respondents reported that their own research results go into internal reports only, and 23% make their results known only (at least initially) to the financial sponsors of the work. Internal reports are thus a very close second to technical journals (65%) and are tied with conference proceedings. The IIW is the forum for reporting 56% of results, theses account for 27%, and other types of reports total 11%. Clearly, most welding research is reported more than once, but not always shared with outsiders.

An important message contained in the responses studied here is difficult to quantify, with the data base in its current form. Scanning the list of specific journals used for technology transfer shows not only the expected preference for publication in one's own country and language, but in a journal read primarily by those trained in one's own specialty. In view of the broad spectrum of research being covered by welding organizations, as indicated above, this means that a large fraction of welding research is being communicated only to a small fraction of welding researchers. As one respondent commented, 'Welding research may fall through the cracks between more recognized disciplines.'

CONCLUSIONS

Welding research is widely pursued and adequately funded in many quarters of the world. It is highly interdisciplinary in nature, which may account in part for the widespread desire for cooperation among institutions in its pursuit and in the reporting of results. National and international research seminars on specialized topics within the field are perceived as highly valuable. While most organizations report practicing some form of research cooperation, more needs for cooperation are perceived than are being addressed.

ACKNOWLEDGEMENTS

"Work supported by the U.S. Department of Energy, Office of Basic Energy Sciences, under DOE Contract DE-AC07-761D01570."

Table 3 - Percentage Scores of Modes of Exchange

Exchange Mode	Score: No Opinion	Strongly Disapprove	Disapprove	Approve	Strongly Approve
Group Research					
National	25	0	1	44	30
International	48	0	1	33	18
Convention					
National	38	0	19	22	15
International	37	0	18	37	8
Conference					
National	51	1	6	38	5
International	27	1	8	54	10
Seminar					
National	30	0	0	22	48
International	25	0	0	25	49
IIW					
National	62	0	0	32	6
International	29	0	0	48	22

THEORIES OF LOW TEMPERATURE SOLID STATE BONDING

R.A. Nichting, G.R. Edwards, D.L. Olson
Colorado School of Mines
Golden, Colorado 80401

B.B. Rath
Naval Research Laboratory
Washington, D.C. 20375

ABSTRACT

Various theories for low temperature solid state bonding are identified. These theories address both the attainment of ultimate bond strength and the mechanistic models for plastic deformation and enhancement of transport during bonding to achieve as high as possible metallic contact across the bonding interface.

Fabrication of sophisticated modern structures requires extremely low temperature joining processes to avoid undesirable phase transformations and distortion. For example, rapidly solidified materials, refractory and reactive metals, and metal matrix composites must all be joined to make the high precision assemblies currently being designed. In numerous instances, the joining of metals to non-metals must be accomplished. In all these examples, low temperature bonding is desirable, if not essential, for the production of assemblies with the required properties (1-7).

Solid state bonding may be classified into two categories, as shown in Figure 1. These categories are: 1) the bonding of dissimilar metals directly and 2) the bonding of dissimilar metals with the use of a interlayer metal. This interlayer may be used to guarantee a clean mating surface for bonding, and to reduce the effect of surface asperities. Nickel, gold, and silver have been used as interlayer metals (8-14). Silver as an intermediate layer has been used very successfully in achieving high-strength diffusion bonds at very low bonding temperatures (9). Nickel interlayers have been successfully used but require high bonding temperature (7,15). The Group IVA, VA and VIA elements have also been evaluated as 4nterlayer materials (16).

Solid state bonds made using a thin intermediate layer of a second material may perform in a manner mechanically similar to the base metal for uniaxial loading in the elastic range even though the yield strength of the intermediate metal may be much lower than that of the base metal. This behavior results from the triaxial stress state generated in the intermediate layer during loading. The weaker intermediate layer is restrained from contracting in the directions normal to the loading direction by the stronger base metal. The mechanical strength of these solid state joints with a thin intermediate layer of silver are found to go through a maximum with decreasing joint thickness (9). The deformation and fracture characteristics of joints encompassing thin intermediate layers, specifically braze joints, have been treated in detail by several investigators (17-20). The interlayer depositions for diffusion bonding have been applied by electroplating (15,21), vacuum-evaporation, and hollow-cathode sputtering-evaporation techniques (22-24,25).

O'Brien et. al. (9) have characterized the specific influence of bonding parameters on the bonding of silver-coated maraging steel tensile specimens. They found that the silver solid state bond strength can be as high as 690 MPa (100,000 psi). This is an excellent indication of the strength of the silver-silver bond and

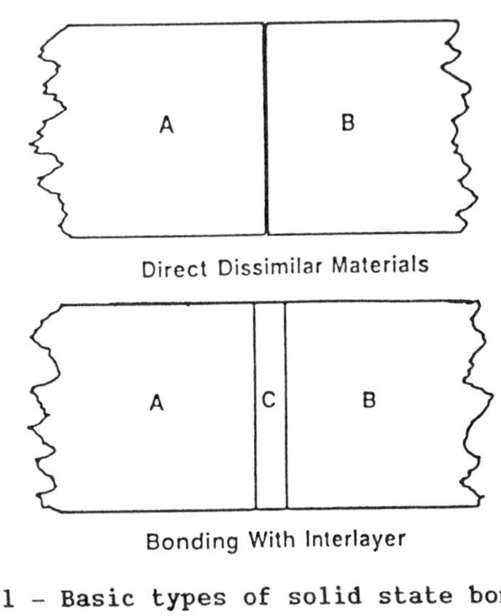

Fig. 1 - Basic types of solid state bonding.

the magnitude of restraint to plastic flow experienced by the thin intermediate silver layer. O'Brien et. al. (9) found that strength decreased drastically as the bonding temperature dropped below 150°C, suggesting that thermally activated processes were required to achieve a viable bond. If the bonding process is thermally activated as the temperature dependence suggests, the kinetics must be extremely fast. A definite threshold in bonding pressure essential to achieve an acceptable bond was also reported. For a bonding temperature of 200°C, the threshold bonding pressure was found to be approximately 35 Mpa (5,000 psi). Increasing the bonding pressure increased the bond strength.

Numerous models have been developed to account for low temperature bonding of materials. These models involve mechanisms which are related to: 1) the attainment of ultimate bond strength, 2) influence of plastic deformation and 3) the enhancement of transport during bonding. The traditional low temperature solid state bonding models have concentrated on the relationship of deformation and surface preparation to bond quality. These models were based on empirical studies which did not consider principles of chemical thermodynamics, plastic deformation, mass transport or solid state physics in any comprehensive way.

This paper introduces a number of fundamentally based concepts which potentially can assist either in increasing the ultimate bond strength or in improving the rate at which the bond is established.

The Ultimate Bond Strength

The ultimate interfacial bond strength between two materials is achieved when they are in complete contact. Conrad and Rice (26) demonstrated that bonding of clean surfaces of silver, aluminum, copper and nickel that the bond strength was almost equal to the compressive load applied. Similar results have been reported by Bay (27) and Upit and Manik (28). These results and the model of Kulagin (29) suggest that the bond strength is directly related to the amount of metal-to-metal contact area. Current bonding technology addresses the principles by which contact is made, but does not address any means by which the ultimate bond strength between two given materials can be predicted. It would be very valuable in making joint designs and material selections to have a means to make a quantitative calculation of ultimate bond strength for various combinations of materials.

Concepts related to the attainment of ultimate bond strength include: 1) enthalpy of mixing, 2) electronic contact potential, 3) chemical bonding, and 4) surface film barriers. Considerations of bulk thermodynamic phase equilibria based on enthalpy of mixing provides a basis for materials and alloying addition selection. Small additions of an element known to have a positive enthalpy of mixing in the solid state with the element to be bonded should have a negative effect on joint integrity, while elements known to have a negative enthalpy of mixing should increase joint integrity. Lison and Stelzer (31) have classified the solid state bonds by the physical and chemical properties of the metals making the combination. They considered the potential tendency for given phase reactions, i.e. eutectic reaction.

The concept of how contact potential should affect joint integrity is illustrated in Figure 2. Any compositional change which elevates the Fermi level on one side of the interface should promote a electron transfer, thereby establishing a coulombic force across the interface, and increasing the ultimate bond strength (32). Chemical bonding, as quantified by interfacial surface energies, can in concept be strengthened by alloying. The surface energy of adhesion, $\Delta\gamma$, is given by

$$\Delta\gamma = \gamma_1 + \gamma_2 - \gamma_{12} \qquad (1)$$

where γ_1 and γ_2 are the surface energies of the two metals and γ_{12} is the energy of the interface between them (33). The "film barrier breaking" model assumes that the initial oxide layer present on metallic surfaces must be displaced and nascent surfaces placed into intimate contact before attainment of a high integrity solid state bond is possible (34).

Miedema, et. al.. (35-38) have developed alloying theories which they have extended to the contact surfaces between metals. These theories are thus directly applicable to the interfacial interactions that occur during solid state bonding. Miedema calculated the ultimate bond strength, γ_{AB} as follows:

$$\gamma_{AB}^{Bond} = -0.85\,(\gamma_A^S + \gamma_B^S) + \gamma_{AB}^{Chem} \qquad (2)$$

$$\text{where } \gamma_{AB}^{Chem} = k\,\frac{\Delta H_{sol}}{V_A^{2/3}} \qquad (3)$$

Here, γ_A^S and γ_B^S are free surface energies of metal A and B respectively, V_A is the atomic volume of metal A, k is a constant and ΔH_{sol} is the change in the enthalpy when mixing A atoms into solvent B.

ΔH_{sol} is given by:

$$\Delta H_{sol} = \frac{2V_A^{2/3}}{(n_{ws}^A)^{-1/3} + (n_{ws}^B)^{-1/3}} \qquad (4)$$

$$= [-Pe\,(\Delta\phi)^2 + Q_o\,(\Delta n_{ws}^{1/3})^2 - R] \qquad (5)$$

where $\Delta\phi$ is directly related to the work function, Δn_{ws} is the difference in electron density at the Wigner-Seitz boundary, V_A is the specific atomic volume, and R, e, Q_o and P are constants. Oberg et. al. (39) utilized these concepts to consider bond strength in explosively bonded couples. They were able to select which metal-metal pairs should explosively bond with high joint integrity. Diebold et. al. (40) have also used the Miedema equation to calculate adhesive strengths for nickel and silver interlayers. Their results are illustrated in Table I and indicate that nickel interlayers have significantly higher adhesive strengths than silver interlayers. Because the Miedema expression uses terms based on solution thermodynamics, interfacial energies and contact potentials to calculate a theoretical bond strength, the model is very useful as a conceptual tool.

Attainment of the Ultimate Bond

Solid state bonding is achieved only when one free metal surface is brought to within the atomic distance of surface atoms for another metal surface. A common major obstacle is an oxide film on the bonding surface. To get the necessary intimate contact between the metal surfaces, this film must be broken. Much of the effort devoted to developing diffusion-bonding processes has focused on procedures to minimize the oxide layer.

Bryant (41) classified the various metal oxide films according to their expected influence on diffusion bonding. The oxides were classified in groups according to solubility for oxygen and the standard free energy of oxide formation, ΔF^o. The metals with high ΔF^o values (-100 kcal/mole) were found also to be metals for which the least difficulty in diffusion bonding was experienced. Metals with ΔF^o values lower than -350 kcal/mole demonstrated significantly less tendency to diffusion bond. Using bend testing to determine bond strength, Bryant (41) found for each of his oxide classes a direct relationship between the minimum bonding pressure and the minimum bonding temperature.

Munir (42) and Jellison (43) considered the effect of the dissolution of the surface oxide film on the solid state bonding of various

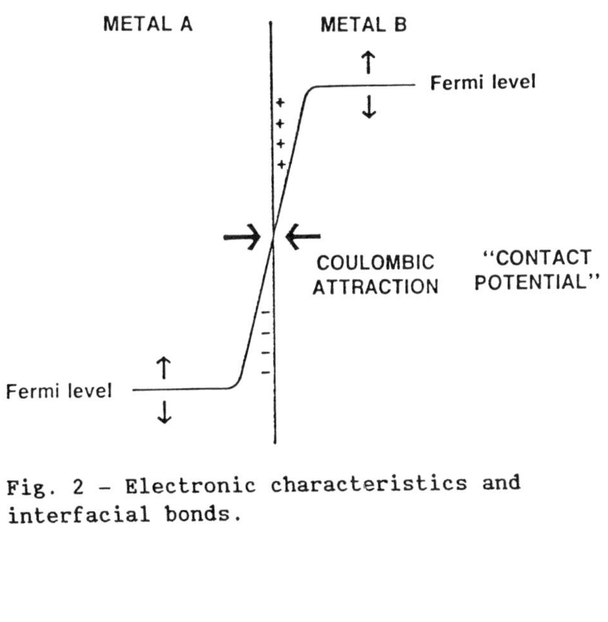

Fig. 2 - Electronic characteristics and interfacial bonds.

metals. Munir (42) introduced an analytical method to calculate the time necessary for dissolution and thus determine the minimum bonding time.

The necessary application of pressure during bonding can contribute in two ways to oxide film breakup. The first is the removal of the oxide by the initial shear displacements as the two microscopically rough surfaces come into contact and move relative to one another. The plastic yielding of surface asperities has been reported (44,45) essential in achieving bond integrity. Holmes (46) has studied diffusion bonds made by applying the load at various angles to the joint surfaces. He found that much stronger bonds were obtained under sliding conditions, especially with small deformations. If the bonding process requires oxide rupture by shear displacements of the surfaces during initial stages, there should be an optimal surface microroughness required to achieve maximum bonding strength. With oxide free gold surfaces, it has been demonstrated that microscopically rough surfaces are essential to produce diffusion bonds (47).

Mohamed and Washburn (47) have discussed a second way in which pressure contributes to the initiation of bonding. This effect involves a complex mechanism of oxide film fracture, caused by aligned dislocation pileups on each side of the bonding interface. They suggest that, with the onset of plastic flow, the dislocations most favorably oriented with respect to the applied

TABLE I - Calculated Adhesion Strengths Using Miedema Equation for Bond Strength (40)

Metal Interfaces	Silver-Metal Bond Strength ($-MJ/m^2$)	Nickel-Metal Bond Strength ($-MJ/m^2$)
Ta	3400	5300
Re	3400	5000
W	2900	4900
Ti	3000	4700
Ru	3100	4600
Pt	3200	4300
Cr	2500	4200
Fe	2500	4200
Ni	2700	----
Co	2700	4200
Si	2200	3700
Y	2500	3600
Mn	2100	3600
Al	2100	3600
Cu	2500	3400
Ge	2000	3200
Au	2400	3100
Zn	1900	3100
Ga	1800	3100
Mg	1900	2800
Sn	1600	2700
Ag	----	2700
Sb	1600	2600
Ca	1900	2600
In	1600	2600

stress start to move on their slip planes. As these dislocations approach the oxide film, they experience a repulsion, reflecting the strain energy of the elastically harder oxide material. The degree to which dislocations will tend to pile up at the oxide-metal interface and produce a stress concentration depends on the relative hardness of the metal and its oxide film. Dislocation pileup has been considered a potential source of crack opening in the oxide. Mohamed and Washburn (47) suggest that the oxide film on the two surfaces should break independently. It can be seen that bonding will not occur unless freshly exposed areas in the open cracks of both oxide films overlap and metal is plastically flowed through these gaps to guarantee intimate contact. If plastic flow controls the amount of bonding area, this model would suggest that metal/metal oxide combinations with large differences in the relative hardness should be more easily bonded (47). It is interesting to note that the metal/metal oxide hardness difference becomes larger with increasing temperature.

The major problem with using nickel as an interlayer is that it takes a high bonding temperature to achieve the necessary mass transport during bonding. Methods to enhance low temperature transport are necessary if nickel interlayers are going to replace silver interlayers for low temperature bonding. Models based on mechanisms which enhance transport kinetics to promote low temperature bonding include: 1) near surface defect structure effects, 2) diffuser enhanced transport, 3) capillarity effects and 4) thermal gradients. Techniques such as ion implantation can be used to produce excessive defects, and to selectively alloy the region adjacent to the bonding surfaces. The defect concentrations in excess of the thermal equilibrium concentrations permit transport processes to occur at very high rates at relatively low temperatures. Diffuser enhanced transport promotes bonding by employing elements on the bonding surface which increase the vacancy concentration at the bonding surface when they diffuse away (48). A similar result can be achieved by establishing a severe thermal gradient across the interface. This could be achieved by bringing two materials, initially at different temperatures instantaneously together.

Considerable deformation occurs at relatively high strain rates at the bonding interface when a diffusion bond is in progress. Such deformation can cause a large, non-equilibrium increase in the concentration of point and line defects. Non-equilibrium defect generation, controlled by the deformation processes and not by the thermal frequency of events, should cause an increase in the mass transport kinetics (49). An increase in mass transport kinetics in the region of the bonding interface in turn should accelerate the process of establishing a strong interfacial bond.

Naimon et. al. (22) have studied the adhesion of silver films on beryllium deposited from a hot hollow cathode source. Naimon et. al. (22) correlated their bond strengths very successfully to the "total" amount of energy used for the ion bombardment cleaning prior to silver deposition. There is an obvious threshold effect, as seen in Figure 3, for the amount of cleaning or surface damage that is necessary to achieve high-strength silver-bonded beryllium joints. These results suggest that there may be tremendeous increase in the mass transport by effective utilization of ion implantation.

The application of a thin layer of mercury on the top of the silver deposition, prior to applying the bonding pressure, will produce reliable and reproducible high-strength joints (50,51). Mercury activation does not appear to increase the maximum strength of the silver solid state bonds, but reduces the scatter in the strength values caused by contamination and roughness of the silver coating. The mercury was found to migrate readily into the silver rather than stay at the bonding surface. The concept of using a diffuser is to take advantage of the vacancy flux created when interdiffusion occurs between two elements of widely different diffusivities. Transport of the diffuser atom away from the interface causes a flux of vacancies toward the interface. These excess vacancies can assist in the completion of the bonding process.

Plastic deformation during bonding can only bring the contact area up to approximately 80 to 90 percent. King and Owczarski (52), Karlinski (53), Munir (42) and Allen and White (54) have been concerned with void formation and shrinkage at the bonded surface. They suggest that solid state bonding is a manifestation of processes very similar to sintering. The applied bonding pressure will push the bonding surfaces together, producing regions of bonding and voids. The initially irregular voids along the bonding interface will be subject to shape changes due to the desire to minimize their surface energy. These capillarity driving forces will promote surface and volume diffusion to produce spherical voids, which, with further bonding time, will shrink and disappear.

King and Owczarski (52) found in their studies of titanium diffusion bonding that removal of the bonding pressure after a so-called "first stage" does not alter the joining process in subsequent stages, although

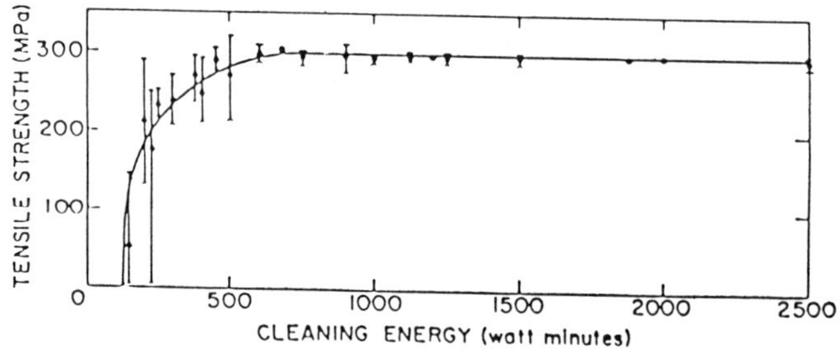

Fig. 3 - Ag/Ag Bond - Argon Ion Bobmardment (22).

premature removal of the bonding pressure before the completion of the first stage is detrimental. This behavior is explained by proposing a model in which voids are formed and eliminated in three separate stages similar to sintering. Each stage is associated with a period of time in which a particular mechanism is controlling. The first stage is defined as the period in which the contact area grows to a large fraction of the joint area by the localized deformation of the contacting surface asperities. A second stage of diffusion bonding is defined as that period when all of the voids in the joint shrink and most are eliminated. It has further been suggested that in this second stage, the interfacial boundary migrates out of the plane of the joint to a lower energy equilibrium configuration. The high-diffusivity paths, such as grain boundaries and free surfaces, may assist in this void elimination. During the third stage the voids are finally filled by volume diffusion.

The influence of surface roughness on bonding strength can also be explained by this void model since it is expected that a rough initial surface produces more and larger voids than a smoother initial surface. Under the same bonding conditions the rough initial surface will require a longer time or higher temperatures for the mass transport process to shrink and eliminate the voids.

Recently, Calderon, Walmsley and Munir (55) evaluated the high temperature mechanical behavior of solid state bonds. They related the long time high temperature properties to rate of intermetallic phase formation at the interfaces of the joint. The intermetallic phases may also form if the bonding times and temperatures are not well controlled.

Summary

General principles of diffusion bonding have been reviewed. These principles reconcile current empirical knowledge with both old and new theoretical concepts, providing a basis for further research. Many materials of high technology require new joining proesses applicable at low temperature. A vigorous research effort to exploit new concepts of low-temperature bonding is therefore imperative.

Acknowledgement

The authors appreciate and acknowledge the research support of the Office of Naval Research through Contract N00014-85-K-0504. The advice and interest of Dr. Bruce MacDonald is also gratefully acknowledged.

References

1. M.J. Albom, solid state bonding, Weld J. 43, 491-504 (1964).
2. P.M. Bartle, Weld J. 54, 799-804 (1975).
3. R.F. Tylecote, The Solid Phase Welding of Metals, pp. 217-218, St. Martin's Press, New York (1968).
4. T.H. Batzer and R.F. Bunshah, J. Vac. Sci. Tech. 4, 19-28 (1967).
5. D.R. Milner and G.W. Rowe, Met. Rev. 7 (28), 433-480 (1962).
6. J.M. Gerken and W.A. Owczarski, A Review of Diffusion Welding, Welding Research Council, New York, Bulletin No. 109 (1965).
7. N.F. Kazakov, "Diffusion Bonding of Materials", Pergamon Press, New York (1985).
8. J.L. Knowles and T.H. Hazlett, Weld. J. 49, 301s-310s (1970).
9. M. O'Brien, C.R. Rice and D.L. Olson, Weld. J. 55, 25-27 (1976).
10. R.S. Rosen, D.R. Walmsley and Z.A. Munir, Weld. J. 65, 83s-92s (1986).
11. E.R. Naimon, J.H. Doyle, C.R. Rice, D. Vigil and D.R. Walmsley, Weld. J. 60, 17-20 (1981).
12. J.A. Hoffman, G.R. Baxter, R.C. Bertossa and B.R. Cottrell, Weld. J. 41, 160s-166s (1962).
13. V.S. Golovchenko and B.A. Gololobov, Welding Prod. 18, 55 (1971).
14. I.M. Barta, Weld. J. 43, 241s-247s (1964).
15. J.W. Dini, Weld. J. 61(1), 33-39, (1982).
16. R. Lison, Z. Werkstofftech. 10(2) 59-67 (1979).
17. J.J. Saxton, A.J. West, and C.R. Burrett, Met. Trans. 2, 999 (1971).
18. W.G. Moffatt and J. Wulff, Weld. J. 42, 115s-125s (1963).
19. N. Bredzs, Weld. J. 33, 545s-563s (1954).
20. B.S. Kasatkin, A.K. Tsaryuk, G.K. Kharchenko, and G.N. Kovab, Weld Prod. 13(7), 4-9 (1966).
21. J.W. Dini, W.K. Kelley, W.C. Cowden and E.M. Lopez, Welding J. 63, 26s-34s (1984).
22. E. Naimon, D. Vigil, J. Villegas, and L. Williams, J. Vac. Sci. Tech. 13, 1131-1134 (1976).
23. D.G. Williams, J. Vac. Sci. Tech. 11, 374-376 (1974).
24. D.T. Larson and H.L. Draper, Thin Solid Films 107, 327-334 (1983).
25. E.R. Naimon, J.H. Doyle, C.R. Rice, D. Vigil and D.R. Walmsley, Weld. J. 60 (11), 17-20 (1981).
26. H. Conrad and L. Rice, Met. Trans. 1, 3019 (1970).
27. N. Bay, Weld. J. 62. 137s-142s (1983).
28. G.P. Upit and J.J. Manik, Wear 13 77-84 (1969).
29. D.P. Kulagin, Avt. Svarka No. 3, pp. 11-15 (1984).
30. E.S. Karakozov et. al., Weld. Prod. 20(2) 83-86 (1973).
31. R. Lison and J.F. Stelzer, Weld. J. 58, 306s-314s (1979).
32. H. Czichos, J. Physics D (Applied Physics) 5(10), 1890-1897, Oct. (1972).
33. M.D. Pashley and J.B. Pethica, J. Vac. Sci. Technol. A3 (3), 757-761 (1985).
34. R.F. Tylecote, D. Howd and J.E. Furmidge, Brit. Welding J. 5 (1), 21-38 (1958).
35. A.R. Miedema, J. Less-Common Met., 46, (1976), 67-83.

36. A.R. Miedema, Z. Metallkde., 69, (1978), 287-292.
37. R. Boom, F.R. deBoer and A.R. Miewdema, J. Less-Common Met., 46, (1976), 271-284.
38. A.R. Miedema and F.J.A. den Broeder, Z. Metallkde., 70, (1979), 14-20.
39. A. Oberg, N. Martensson and J.A. Schweitz, Met Trans. 16A, (1985), 841-852.
40. T. Diebold, G.R. Edwards and D.L. Olson, Met. Trans. 17A, 374 (1986).
41. W.A. Bryant, Weld. J. 55, 433s-435s (1975).
42. Z.A. Munir, Welding Journal 62, 333s-336s (1983).
43. J.L. Jellison, Surface Contamination: gensesis detection and control, Vol. 2, 899-923, ed. K.L. Mittal, New York: Plenum Press (1979).
44. G. Garmong, N.E. Paton and A.S. Argon, Met. Trans 6A, 1269 (1975).
45. C.H. Hamilton, "Titanium Science and Technology", ed. R.I. Jaffee and M.M. Barte. Plenum Press, New York, Vol. 1, 625 (1973).
46. E. Holmes, Br. Weld. J. 6, 29-37 (1959).
47. H.A. Mohamed and J. Washborn, Weld. J. 55, 302s-310s (1975).
48. W. Feduska and W.L. Horigan, Weld. J. 41, 28s-35s (1982).
49. R.W. Balluffi and A.L. Ruoff, J. App. Phys. 34, 1634-1647 and 2862-2872 (1963).
50. D.L. Olson and A.L. Liby, Beryllium Science and Technology, 2, 275-296, eds. D.R. Floyd and J.N. Lowe, Plenum Press, New York, (1979).
51. V.K. Grotzky, C.R. Rice, J.H. Doyle and D.L. Olson, Report RFP-2514, Rockwell International, Rocky Flats Division, Golden, CO (November 19, 1976).
52. W.H. King and W.A. Owczarski, Weld. J. 47, 444s-450s (1968).
53. W. Karlinski, Welding Research International 9 (4), 25-43 (1979).
54. D.J. Allen and A.A.L. White, "Sintering Processes in Diffusion Bonding", Proc. of Inst. of Metallurgist Conf. on "The Joining of Metals: Practice and Performance", Vol. 2 (18), pp. 96-103, (1981), held at Univ. of Warwick Coventry, UK, 10-12 April (1981).
55. P.D. Calderon, D.R. Walmsley and A.Z. Munir, Weld. J. 64, 104s-112s (1985).

DIFFUSION BONDING OF Ti-6Al-4V

Daniel W. Walsh
California Polytechnic State University
San Luis Obispo, California

John P. Balaguer
Rensselaer Polytechnic Institute
Troy, New York

ABSTRACT

The Gleeble, equipped with a high-vacuum chamber, was used to establish suitable parameters for the diffusion bonding of an α/β titanium alloy, Ti-6Al-4V. Bonding experiments were performed at temperatures of 850 C (1562 F), 950 C (1742 F), and 1050 C (1922 F) in order to simulate bonding with low, medium, and high volume fractions of β, respectively. Loads of 68 kg (150 lbs), 100kg (220 lbs), and 227 kg (500 lbs) were used, times were limited to 30 min or less, and a vacuum of 5 x 10^{-5} torr was maintained during bonding. Bonding at 850 C resulted in a low-quality weld interface. Bonding at 1050 C, for short times, resulted in excessive grain growth, large upset, and a slight increase in hardness relative to the base material. Relatively good bonds were formed between specimens welded in the α/β region at a temperature of 950 C for 30 min and a load of 100 kg. Samples bonded at 950 C displayed slighty excessive upset, moderate grain growth, and small changes in hardness relative to the base material.

DIFFUSION BONDING is presently used for the welding of reactive metals, dissimilar metals, and other metals which are not readily joined by fusion welding. This welding process is of particular interest to the aerospace industry, where light, reactive metals must be joined to provide high strength and low added weight. Considerable work has been done on the diffusion bonding of α and α/β titanium alloys (1, 2) and this process is commonly used in the commercial welding of these alloys. Other investigations have studied the diffusion bonding of titanium alloys to dissimilar metals (3). The object of this investigation is to establish suitable specimen geometries and welding parameters for the diffusion bonding of Ti-6Al-4V on the Gleeble. These parameters will then be used to study the diffusion bonding of titanium alloys to dissimilar metals in future investigations.

MATERIALS AND PROCEDURE

The material used in this study was Ti-6Al-4V, an α/β alloy. A pseudo-binary diagram for this alloy (4) is shown in Fig. 1. Cylindrical specimens 9.5 mm (0.375 in.) in diameter and 57 mm (2.25 in.) in length were machined from 13 mm (0.5 in.) thick plate. Diffusion bonding experiments were performed using a Gleeble Model 1500 thermomechanical simulation device equipped with a vacuum chamber. The α transus and β transus temperatures for the alloy used in this investigation were experimentally determined on the Gleeble using dilatometry.

Bonding experiments were performed at temperatures of 850 C (1562 F), 950 C (1742 F), and 1050 C (1922 F) in order to simulate bonding with low, medium, and high volume fractions of β, respectively. All specimens were resistance heated at 48

C/sec to the bonding temperature and maintained at constant temperature during bonding. After bonding was complete, samples were conduction cooled through water-cooled copper blocks. Loads of 68 kg (150 lbs), 100kg (220 lbs), and 227 kg (500 lbs) were used, resulting in compressive stresses of 0.95 kg/mm^2 (1.4 ksi), 1.4 kg/mm^2 (2.0 ksi), and 3.2 kg/mm^2 (4.5 ksi). Bonding times were limited to 30 min or less and a vacuum of 5 x 10^{-5} torr was maintained during bonding. An upset parameter, P_u , was defined to quantify the plastic deformation which occurred upon bonding.

$$P_u = \frac{D_m - D_o}{D_o} \times 100 \qquad \text{(Eqn. 1)}$$

where P_u = Upset Parameter
D_m = Maximum Diameter
D_o = Original Diameter.

RESULTS

The α transus and β transus temperatures of the Ti-6Al-4V used in this investigation were 713 C (1315 F) and 1070 C (1958 F), respectively. The microstructure of the as-received material, shown in Fig. 2, is a fine-grained mixture of α (light) and accicular α formed on cooling from β (dark). The hardness of this alloy was approximately HV 354 (HRC 36) in the as-received condition.

The hardnesses and calculated upset parameters resulting from various bonding conditions are listed in Table I. The microstructure of specimens bonded at 850 C, 100 kg, 30 min is shown in Fig. 3. The resulting microstucture consisted of a small fraction of accicular α (transformed β) in a matrix of prior α . The original interface (bond line) in this sample remained clearly visible, indicating that an incomplete bond was obtained. The microstructure of specimens bonded at 950 C, 100 kg, 30 min is shown in Fig. 4. The bond line in this sample had almost disappeared and the grain size had increased to somewhat greater than that observed in the sample bonded at 850 C. Additionally, the microstructure contained a greater fraction of accicular α (transformed β) as a result of the higher bonding temperature.

The microstructure of specimens bonded at 1050 C, 68 kg, 40 sec is shown in Fig. 5. The bond time for this sample was very short, and the bond line was still visible. Despite the short bond time, however, significant grain growth occurred as a result of the high bonding temperature. The microstructure contained a small fraction of α , as the bonding temperature of 1050 C was still slightly below the β transus, in a matrix of transformed β . The α produced by the on-cooling transformation of β in this sample appeared considerably more accicular than that observed in the sample bonded at 950 C. However, the cooling rate after bonding was too slow to suggest a martensitic transformation. Additionally, there were regions of blocky α , transformed from β , at prior β grain boundaries. As the transformation temperature decreased, kinetic limitations forced a transition in α morphology from a grain boundary allotriomorph to a Widmanstätten structure, resulting in the microstructure shown in Fig. 5. It is probable that, with longer bonding times at this temperature the grain size would be increased and therefore the volume fraction of blocky α would be reduced.

Based on these results, it was determined that 950 C, at which there is a mixture of approximately 50% α and 50% β , was the optimum bonding temperature. All remaining welds were made at this temperature with varying loads and times. Welds made at 950 C, 68 kg, 10 min resulted in upset (P_u = 5.3) similar to that produced in welds made at 950 C, 227 kg, 1 min (P_u = 4.0). The microstructure of the 950 C, 68 kg, 10 min sample, shown in Fig. 6, is a mixture of accicular α and oriented α . The oriented α , not transformed upon heating, is a holdover from the mechanical working of the base material prior to bonding. Similar microstructures were observed in samples bonded at 950 C, 227 kg, 1 min and 950 C, 68 kg, 5 min. However, prior α present in samples bonded at 950 C, 227 kg, 30 min was no longer oriented with the axis of the specimens. Apparently, the limited plastic deformation associated with low loads and/or short times is insufficient to break up the oriented α . Also visible in Fig. 6 is the bond line, indicating the necessity for longer times at these parameters.

Microhardness traverses, for bonding parameters listed in Table I, are shown in Fig. 7. These traverses indicate the variation in hardness as a function of distance from the bond line. With the exception of the specimens bonded at 1050 C, the weld-interface hardness of all samples was somewhat lower than that of the base material. Specimens bonded at 950 C, 1 min, 227 kg and 1050 C, 0.7 min, 68 kg show somewhat elevated hardnesses distant from the weld interface. This probably results from the greater fraction of Widmanstätten α and was not observed in samples bonded for longer times at 950 C.

SUMMARY

The Gleeble, equipped with a high-vacuum chamber, has been successfully used to diffusion bond Ti-6Al-4V. Bonding at 850 C resulted in a low-quality weld interface. Bonding at 1050 C, for short times, resulted in excessive grain growth, large upset, and a slight increase in hardness relative to the base material. Relatively good bonds were formed between specimens welded in the α/β region at a temperature of 950 C for 30 min and a load of 100 kg. Samples bonded at 950 C displayed slighty excessive upset, moderate grain growth, and small changes in hardness relative to the base material. Future investigations will utilize these results to study the effect of bonding parameters on joint efficiency and the diffusion bonding of titanium and dissimilar metals.

ACKNOWLEDGEMENTS

The authors would like to thank Duffers Scientific, Inc. for their assistance in supplying specimens and equipment for this investigation. The authors also appreciate the efforts of J. Murray in the completion of the experimental portion of this program and the constructive comments of Dr. E. F. Nippes.

REFERENCES

1. King, W. H. and W. A. Owczarski, Welding Journal (Research Supplement), 46, 289s-298s (1967).

2. King, W. H. and W. A. Owczarski, Welding Journal (Research Supplement), 47, 444s-450s (1968).

3. Zanner, F. J. and R. W. Fisher, Welding Journal (Research Supplement), 53, 105s-112s (1974).

4. Rausch, J. J., F. A. Crossley, and H. D. Kessler, Journal of Metals, 8, 211-214 (1956).

TABLE I

UPSET PARAMETER AND HARDNESS RESULTS

Weld Parameters			Hardness	P_u
Temp. (C)	Time (min)	Load (kg)	(HV)	
950	1.0	227	360	4.0
950	30.0	227	340	58.3
950	5.0	68	356	4.2
950	10.0	68	344	5.3
1050	0.7	68	352	1.8
Base Metal - As Received			354	---

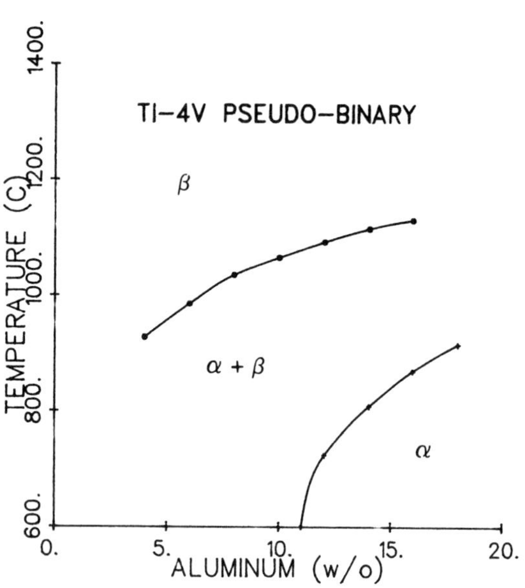

Fig. 1 - Pseudo-binary phase diagram for Ti-4V (after Rausch, 1956).

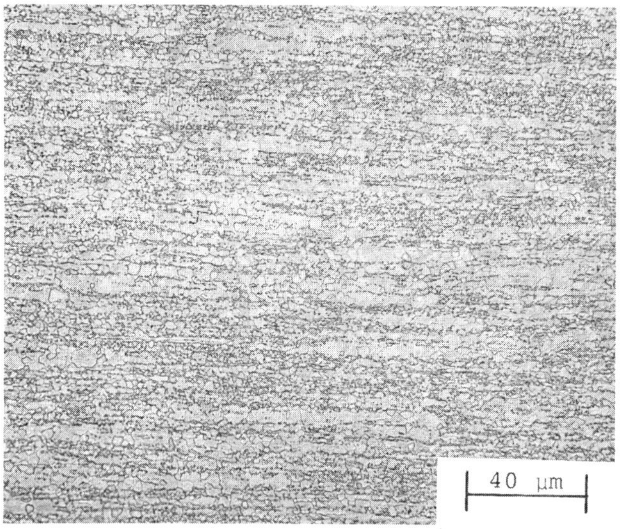

Fig. 2 - Photomicrograph of the as-received material, Ti-6Al-4V, 500X.

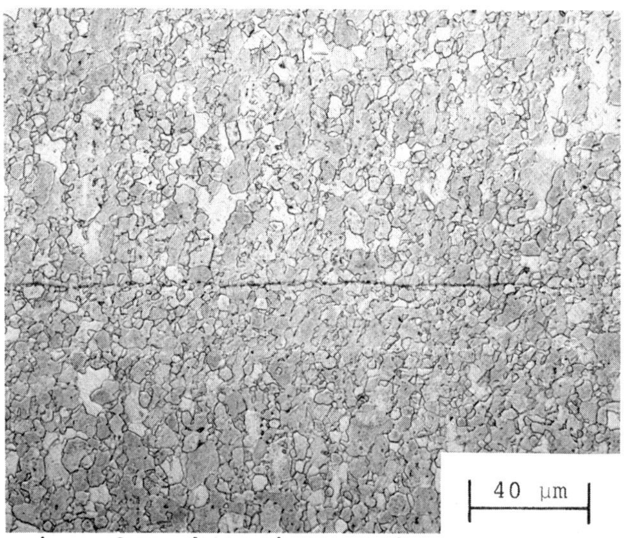

Fig. 3 - Photomicrograph of specimens bonded at 850 C, 100 kg, 30 min, 500X.

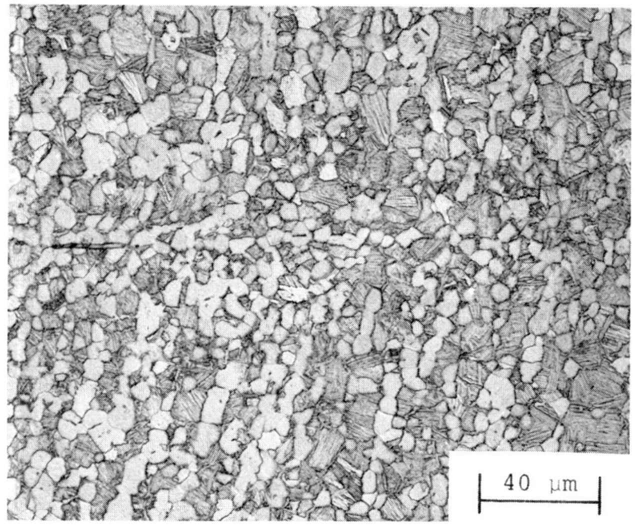

Fig. 4 - Photomicrograph of specimens bonded at 950 C, 100 kg, 30 min, 500X.

Fig. 5 - Photomicrograph of specimens bonded at 1050 C, 68 kg, 40 sec, 100X.

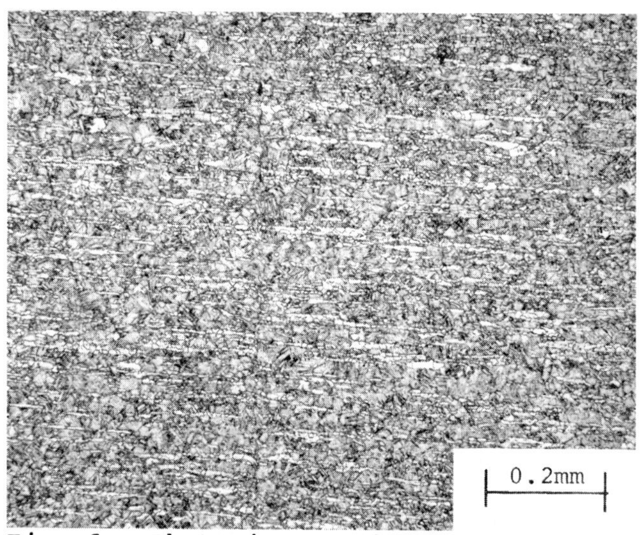

Fig. 6 - Photomicrograph of specimens bonded at 950 C, 58 kg, 10 min, 100X.

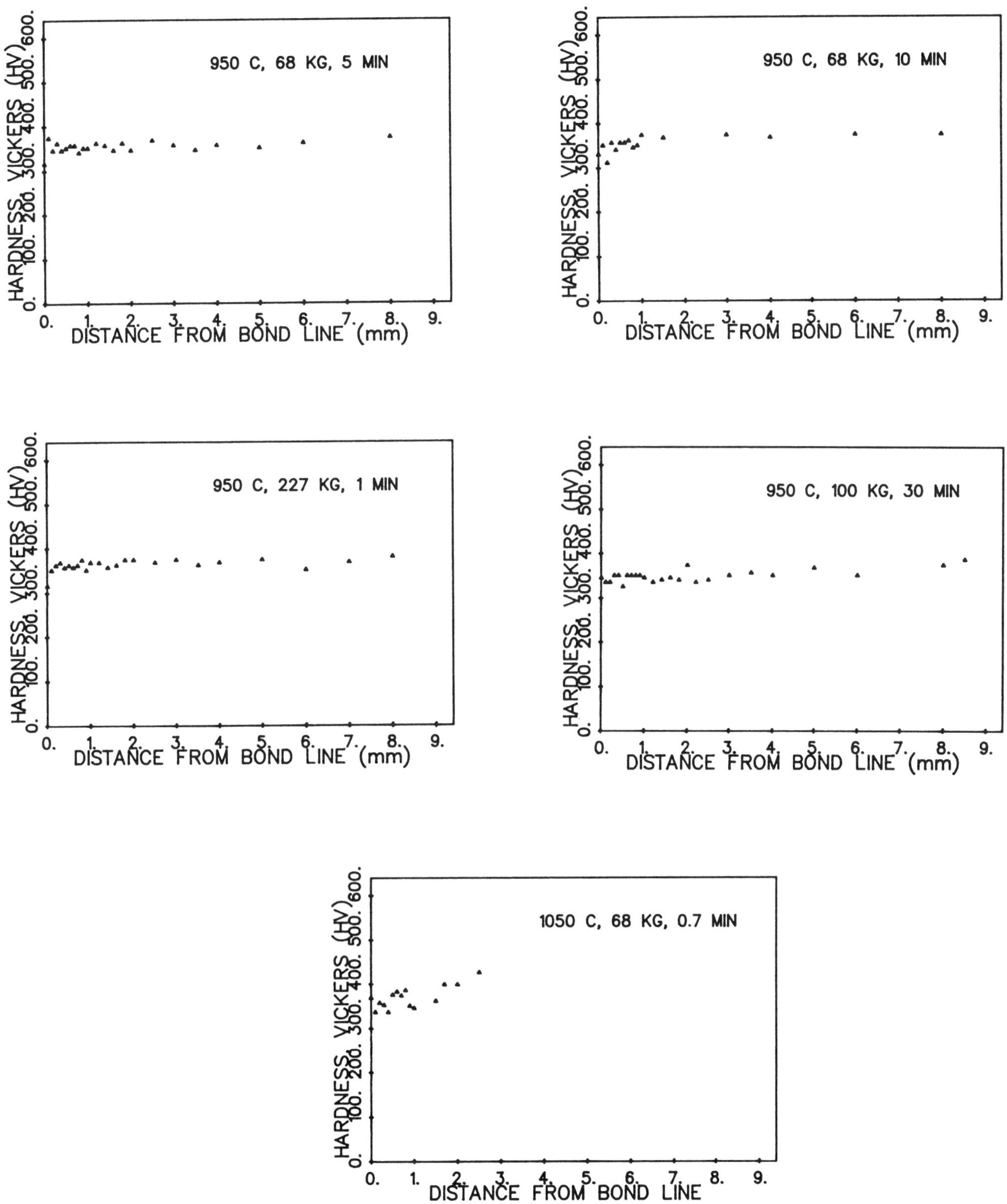

Fig. 7 - Microhardness traverse results for bonds made with the parameters shown in Table I.

SOLID PHASE BONDING OF COPPER-IRON ALLOYS UNDER DEFORMATION

Roman Dubrovsky
New Jersey Institute of Technology
Newark, New Jersey

Abstract

In bimetal or clad technology during joining of disimilar metals designed from copper-iron alloys the main concern is a quality of diffusion bonding.

An experiment was conducted to measure the copper diffusion rate in α and γ irons and the effect of deformation on copper mobility in the area of diffusion zone. The diffusion parameters and their temperature dependence in deformed and non-deformed states were evaluated. The results showed that the diffusion penetration of copper into iron increased sizably with deformation, but the rate of the diffusion increment dropped with an increase in temperature above the transformation line. These results indicate that the probability of diffusion mobility of copper into iron is greater at a temperature range below the critical line when the grain boundary and short circuiting diffusion provide a high diffusion rate.

THE DIFFUSION WELDING PROCESS (dwp) has now become well established in industry for production of dissimilar metal joints in many combinations of materials. This process can take place in solid state due to the relatively low temperature in the interface between joints. One of dwp's major attractions for industry has been its ability to successfully join a wide range of materials in many different combinations using very simple procedures when no additional materials or high operator skills are involved, or when this process can be brought to automation and process monitoring for quality assurance purposes [1,2]. The incentive to manufacture materials with high load-carrying capacity and wear resistance leads one to the creation of new bimetals on a copper-iron base [3]. Because of significant differences in processing techniques and joint materials, however, prediction of processing parameters is not easy, and has been done only on the base of diffusion parameters.

As appeared in Ref. [4], diffusion welding can be performed under limited temperature, pressure and time conditions. In order to bring the asperity tips to deformation, the compressive stresses cannot be higher than the elastic limit of the materials. On the other hand, the effective dwp can be developed only by reduction of diffusion time at reasonable temperatures and pressures. This means that it is necessary to establish diffusion parameters of participants under deformation conditions.

This paper investigates the parameters required for diffusion of copper into iron in non-deformed and deformed states to produce copper-iron alloy bimetals.

EXPERIMENTAL PROCEDURE - The chemical composition of Armco Iron and Copper are shown in Table 1. To cause grain growth, the iron ingots were annealed at 1200ºC temperature during 10-12 hours in neutral atmosphere. Then the specimens were cut to the dimensions: Iron - 12*7*7mm; Copper - 12*7*3mm to prepare a sandwich as shown in Fig. 1. Two parallel flat faces of each iron specimen were carefully ground on waterproof abrasive paper and polished to the nominal surface finishing of 5 μm. The adjacent surface of copper specimens possessed the same surface roughness.

It is to be supposed that the diffusion mobility of copper atoms into the iron matrix can be increased by the deformation in the interface. The diffusion mobility of copper in a Cu-Fe system of materials was studied on an IMASH 5S-65 apparatus, equipped with a specially designed fixture, which could perform non-deformed and deformed (compressive) states of weld interface, Fig. 1.

Table 1 - Chemical Composition

ARMCO-IRON

Elements	C	Mn	Si	Al	S	P	Cr	Ni	Cu
Content in weight %	0.02	0.16	0.35	0.015	0.018	0.024	0.05	0.04	.02

COPPER (FRHC)

Elements	Cu(incl. Ag)	As	Sb	P	Te
Content in weight %	99.90	with unspecified oxygen and other named elements			

All tested specimens were annealed at six different temperatures (700, 800, 900, 1000, 1050°C) during different times at both deformed (compressive stresses 1 kg/mm^2) and non-deformed conditions. After diffusion annealing, all specimens were cut, polished along the direction of diffusion or deformation, and etched. The described method of sectioning was employed to determine the distribution of copper concentration in the bulk of the iron matrix, using optical microscopy, layer-by-layer spectroscopy locally with respect to the depth (ISP-30)[5], and electron probe microanalysis. The layer-by-layer method is an effective means to measure diffusion parameters of the elements since it characterizes the average diffusion permeability of grains over a sparking surface equal to 3-4 mm^2 and allows for the contribution of diffusion occuring along the dislocations as well as along other crystal-lattice defects.

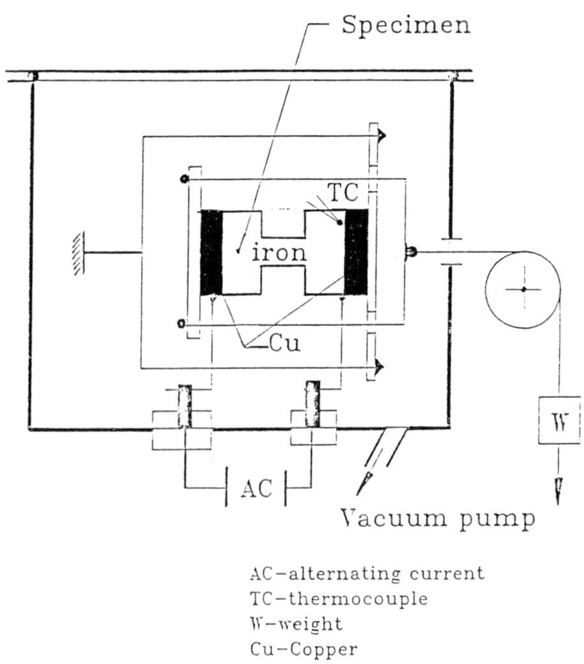

Fig. 1 - Vacuum chamber of IMASH 5S-65 apparatus and scheme of specimen loading and heating

For one dimensional diffusion the solution of Fick's second law for two coupled semi-infinite bodies when the volume diffusion coefficient is constant, or does not depend on concentration, is given in Ref. [6] as:

$$C(x,t)=\text{const.}*\exp(-x^2/4D_V t) \quad (1)$$

where C(x,t) is the concentration of copper at a distance "x" from the original surface; D_V is the volume diffusion coefficient and t is the diffusion time.

Except for the bulk, diffusion runs along grain boundaries and other structural defects. The input of grain boundary diffusion in the diffusion process can be defined using Fisher's approximation [7], when it is possible to find the increase of boundary concentration velocity at any point "x" along the boundary. Then the concentration at any points in the case of layers thickness equal to dx will be

$$C(x,y,t)=C_o*\exp[-x\sqrt{2}/(\pi D_V t)^{\frac{1}{4}} \times (\delta D_B/D_V)^{\frac{1}{2}}]*[1-\text{erf}(y/2\sqrt{D_V t})] \quad (2)$$

where C_o is the concentration at the interface, when x=0; δ is the width of boundary, conditionally is equal to 5.10^{-8} cm; D_B is the boundary diffusion coefficient, and y is the distance from the grain boundary. Layer-by-layer spectroscopy is used to measure the amount of substance in any layer "dx" as a function of its depth "x". From (2), $\ln\bar{C}$ is a linear function of x, and knowing the slope of this straight line allows one to calculate D_B [8] as

$$\delta D_B = m^{-2}*(4D_V/t)^{\frac{1}{2}}*A^2 \quad (3)$$

where m - the slope of straight line $\ln\bar{C}=f(x)$ obtained from the experiment; A-parameter, in the case of Fisher's solution, A=0.751

RESULTS AND DISCUSSION - Penetration of copper into iron strengthens the iron. The microhardness values are the greatest at the interface where the concentration of the copper is the greatest.

Metallographic investigation enabled the depth of the diffusion layers to be evaluated and the special features of the copper penetration into the iron to be established. The diffusion grows with the temperature of anneal; Fig. 2, a,b,c shows that diffusion takes place through the grain volumes, grain boundaries and other structural defects. The

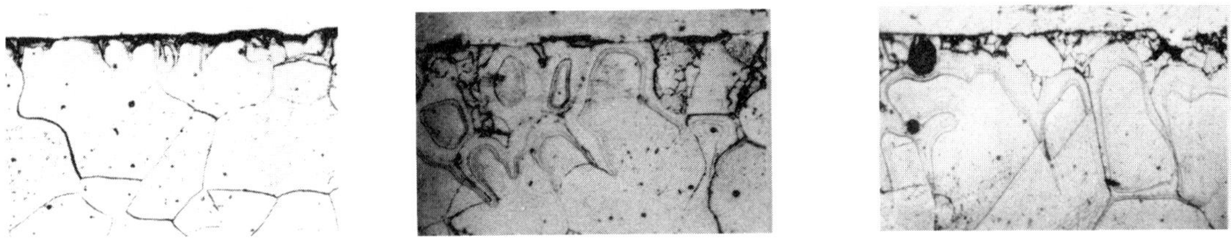

a b c

Fig. 2 - Microstructure of Diffusion Zone

a-Tan=800°C, time=75 hrs., x300;
b-Tan=1000°C, time=30 hrs., x500
c-Tan=1050°C, time=25 hrs., x500.

diffusion zone consists of developed diffusion flux along grain boundaries, where the flux of copper initiates from the grain boundaries into the grain bulk.

Bulk diffusion of copper into Armco-Iron was investigated with X-ray spectroscopy. The surface to be probed was etched, and for the microanalyzer probe, the "corridors" were marked out by a microhardness tester in the volume of the grains at the greatest distance from their boundaries. These "corridors" were parallel to the direction of the copper diffusion from the interface. The results of copper distribution during diffusion into iron in "α" and "γ" states are shown in Fig. 3. The linear relationship observed between logC and X^2 confirms the fact that volume diffusion is being measured.

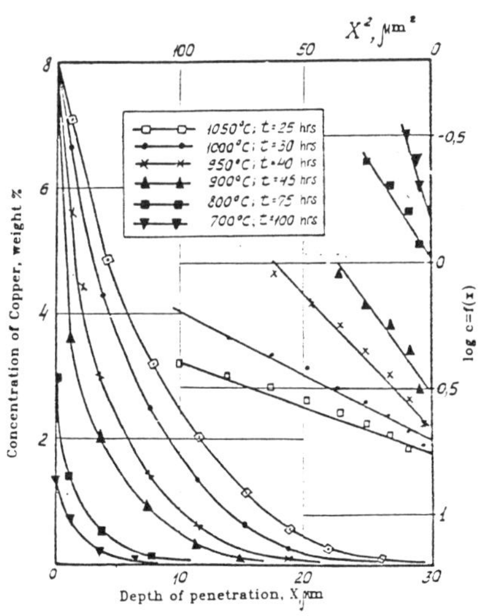

Fig. 3 - Concentration curves of copper distribution in α and γ -irons and relationship between log c=f(x^2).

This procedure was undertaken to eliminate the influence of boundary diffusion. The diffusion coefficient "D_V", the frequency factor "D_{OV}", and the energy of activation "Q_V" were obtained from Arrhenius plots. The temperature dependence of the volume diffusion coefficient is shown in Fig. 4. It is possible to find δD_B from (3) using the slope m of straight line log$\bar{C}$=f(x) obtained from Fig.5. The diffusion parameters are listed in Table 2. The temperature dependence of the boundary diffusion coefficient is shown in Fig. 4.

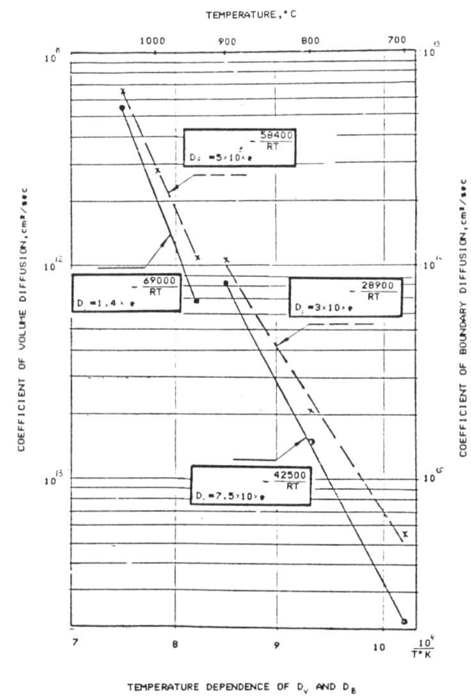

Fig. 4 - Temperature dependence of D_V and D_B.

As can be seen from Fig. 2 a, the boundary diffusion at 800°C plays a minor role, at the same time, in the "γ" state of diffusion anneal, the boundary diffusion is clearly developed (Fig. 2 b,c), and it is possible to see very deep penetration of copper along the grain boundaries, a substantially developed substructure in the diffusion zone,and sizable volume diffusion.

The distribution of effective copper concentration under deformation in "α" and "γ" states obtained by layer-by-layer spectroscopy are shown in Fig. 6 a,b. The small deviation from linearity is observed for the log$\bar{C}$=f(x^2). This may be due to some diffusion along the grain boundaries and short circuiting. The diffusion rate was evaluated by means of the effective diffusion coefficient (Def), frequency factor (Doef), and the activation energy (Qef), and are listed in Table 3. As shown in this table, the contribution of deformation to the total diffusion is sizable (one or two orders of magnitude). The allotropic transformation decreases the diffusion rate during deformation. The influence of the temperature on the effective diffusion coefficient was measured by means of the relative effective diffusive coefficient, given by

$$\Delta ef = \frac{D'ef - Def}{Def}$$

where D'ef and Def-represent effective diffusion coefficients in deformed and non-deformed states correspondingly. In the deformed state, the value of D'ef is between one and two orders of magnitude higher than Def in a non-deformed state. The activation energy, Q'ef, is smaller too, and it shows an easier occurence of diffusion processes in the deformed iron compared with the non-deformed state. The increase in the diffusion rate and the reduction in activation energy are evidently associated primarily with the formation of a developed substructure at the deformed iron surface, consisting of dislocation and low angle boundaries. Comparison of relative effective diffusion coefficients at different temperatures (Table 3) shows that the increment rate of the diffusion coefficient is decreased in the deformed state approximately 50- fold. The main cause of this phenomenon is a reduction of the amount of the available paths for mass transfer because of increased rate of annihilation of dislocation at a higher temperature. The increased value of activation energy in the "γ" state confirms this assumption. The diffusion rate can be 10 times more in the deformed state at the temperature range of allotropic transformation.

Table 2 - Volume and Boundary Diffusion Parameters

Diffusion parameters	α-iron			γ-iron		
Temperature of anneal, °C	700	800	900	950	1000	1050
Time of anneal, hrs	100	75	45	40	30	25
VOLUME DIFFUSION						
$D_V \times 10^{14}$, cm²/sec	2.2	15	35	67	220	550
Q_V, kcal/mole	42.5			69		
D_0, cm²/sec	7.5×10^{-3}			1.4		
BOUNDARY DIFFUSION						
$D_3 \delta \times 10^{6}$, cm /sec	5.51	21.5	100	110	290	600
$D_3 \times 10^{5}$, cm²/sec	1	4	20	22	53	120
Q_3, kcal/mole	28.9			53.4		
D_0, cm²/sec	3×10^{-2}			5×10^{3}		

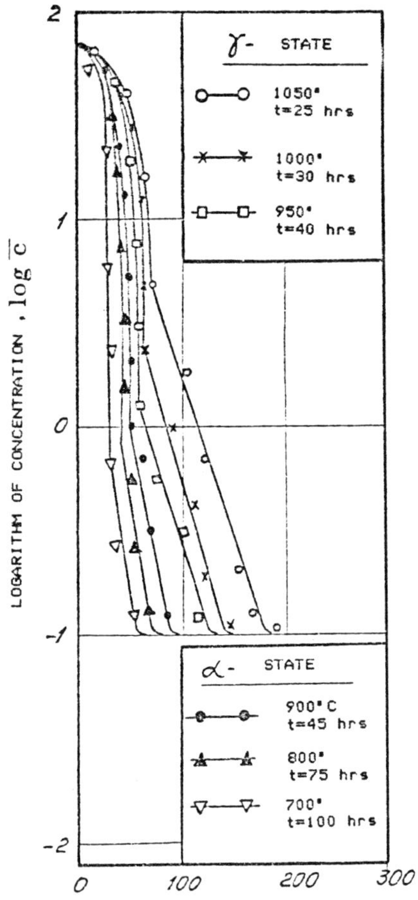

Fig. 5 - Relationship between log $\bar{c}$=f(x) at layer-by-layer analyses of iron-copper system.

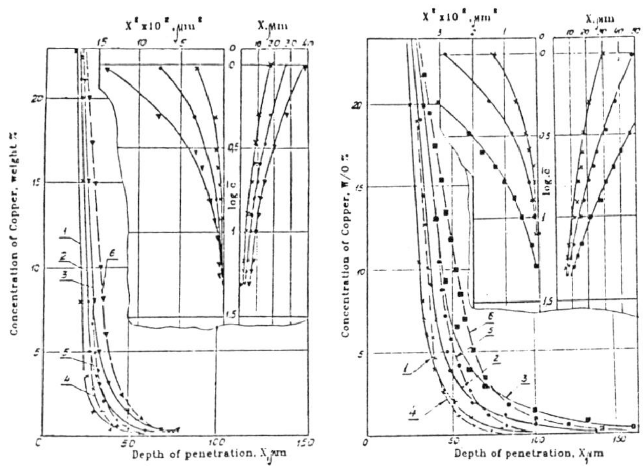

Fig. 6 - Concentration curves of effective copper distribution in deformed and non-deformed states.

a - α - iron state; 1,4-700°C; 2,5-800°C; 3,6-1050°C.

b - γ - iron state; 1,4-950°C; 2,5-1000°C; 3,6-1050°C.

1,2,3 - non-deformed specimens

4,5,6 - deformed specimens

Table 3 - Effective Parameters of Diffusion in Deformed and Non-deformed States

No	Parameters of diffusion	α - Fe			γ - Fe		
1	Temperature of diffusion annealing, °C	700	800	900	950	1000	1050
				Non-deformed State			
2	Time of annealing, hours	100	75	45	40	30	25
3	Def $\frac{cm^2}{sec}$	$2.1x10^{-13}$	$8x10^{-13}$	$1.8x10^{-12}$	$3x10^{-12}$	$5x10^{-12}$	$9.8x10^{-12}$
4	Qef $\frac{Kcal}{mole}$		$2.5x10^4$			$3.9x10^4$	
5	Do,ef $\frac{cm^2}{sec}$		$1.7x10^{-5}$			$2.5x10^{-3}$	
				Deformed State			
6	Time of annealing, hrs	10	10	10	10	10	10
7	D'ef, $\frac{cm^2}{sec}$	$1.6x10^{-11}$	$2x10^{-11}$	$3.2x10^{-11}$	$3.3x10^{-11}$	$4.5x10^{-11}$	$6.3x10^{-11}$
8	Qef, $\frac{Kcal}{mole}$		$0.8x10^4$			$2.1x10^4$	
9	D'o ef, $\frac{cm^2}{sec}$		$6.0x10^{-8}$			$1.8x10^{-5}$	
10	Δef	75	24	17	10	8	5.3

CONCLUSION - By combining the data of diffusion joining of copper with iron in deformed and non-deformed states, it should be assumed that the surface projections are subjected to plastic deformation when two mating parts are brought into contact. Deformation in the area of contact increases the rate of diffusion. The diffusion parameters and their temperature dependence in deformed and non-deformed states were evaluated. The diffusion penetration of copper into iron increases sizably with deformation, but the rate of diffusion increment drops as the temperature increases above the transformation line. The obtained results show that the Δef increases by decreasing the temperature. At 700°C, the D'ef was 100 times higher than the Def; at 1050°C, the D'ef was 7 times higher than the Def.

REFERENCES

1. Baginski, W.A., Study of Dissimilar Metal Joining by Solid-State Welding, NASA Report N66, 327787, Contract NAS 8-20156.
2. Wallach, E.R., Solid-State Joining: Techniques and Applications. The Metallurgist and Materials Technologist, N2, (1984).
3. Kozlov, V., Ph, J. Properties and Structure of Powder Antifriction Material. Poroshkovaja Metallurgia, N6, (1981).
4. Krasulin, Yu.L., Shorshorov M.Kh., J. Fizika i Khimia Obrabotki Materialov, N1, (1967).
5. Krishtal, M.A., Davidov, Yu.I., Korvachev, V.D., J. Zavodskaja Laboratoria, N8, (1964).
6. Shewmon, P.G., Diffusion in Solids. McGraw-Hill Book Co., Inc., New York, (1963).
7. Fisher, I.C., Journal Appl. Phys., 22, N1, (1951).
8. Le Cleire, A.D., Brit. Journ. Appl. Phys., 14, N6, (1963).

AN ANALYTICAL STUDY ON THE DEVELOPMENT OF RESIDUAL STRESS DURING WELDING

S. Suresh, A. Shanmugam
Welding Research Institute
BHEL
Tiruchirapalli, India

ABSTRACT

Through this study the authors have proposed an analytical method to compute longitudinal residual stress distribution in edge welded plates. The method is based on Okerblom's thermomechanics theory modified by incorporating a kinetic analysis of the stress development during welding. The proposed method has also been validated through extensive experimental work conducted on three different steels employing three levels of heat input.

RESIDUAL STRESS is caused by non-elastic strains induced in the regions near the weld, where steep temperature changes occur during welding. Though several attempts have been made to compute & analyse residual stress through various models, a simple and less expensive methodology is yet to be developed.

Through this paper the authors have proposed a simple computation methodology to compute longitudinal residual stress in edge welded plates. The paper also deals with the experimental verification of the methodology and the results obtained thereof.

2. ANALYSIS OF RESIDUAL STRESS DEVELOPED DURING WELDING

The analysis of most of the cases of residual stress development during welding can be separated into two parts. The first part being the calculation of temperature distribution. The second part is the stresses that result out of the temperature distribution. The distinguishing feature of the welding process is the extremely high temperatures involved wherein very little information is available on the structural behaviour of materials at elevated temperatures.

The analysis of the first part is relatively simple as extensive theoretical works have been already done and models are available which may need only little modifications to suit the exact situations that are commonly encountered. For instance the Rosenthal's quasi-stationary model using a line source theory along with the consideration of variation of thermal properties with temperature can be easily applied to many situations of welding.

On the other hand, in the field of stress analysis despite extensive work, very little is directly applicable to welding process. For some of the phenomena that occur during welding, like plasticity, strain hardening, variation of material properties with temperature and other metallurgical changes, only limited data is available. This situation naturally calls for several assumptions to be made during such analysis.

However attempts so far made prove that the validity and exactness of these assumptions can be greatly improved by using computers for analysis. This paper deals with one such approach for analysis of residual stress developed during welding in edge welded plates.

In this work attempt has been made to compute the longitudinal residual stress developed during welding, using a kinetic approach which takes into account cumulative stress build-up with time. It is assumed that the stress developed in the transverse direction is zero and the stress is uniform through the thickness making the analysis one dimensional. However, the thermal analysis is two dimensional. By a simple assumption that plane sections remain plane after deformation, the compatibility conditions also get satisfied. The overall analysis is based on the principle that when free thermal deformations occurring in local areas are restricted by adjacent cooler parts of the welded part, residual deformations, that remain in the welded plate even after cooling and are also of self-equilibrating type. However, the elastic residual deformations and corresponding stresses can be separated once we substract the shape change deformations.

The step by step calculation procedure is as follows:

Step 1:

An initial temperature distribution for a given time is calculated using the appropriate model corresponding to the considered plate thickness. For thicknesses less than 12.5mm a 2D model of heat flow is assumed and temperature at any point is calculated using -

T = To+EXP(-D.V.E1).KO(D.V.R).A/(π.K.G)....(1)

for plate thickness greater than 12.5mm and less than 25mm, while a 2.5D model for heat flow with reflections is assumed given by

T = To+Q/(2.π.K).(Exp(-D.V.E1)
(EXP (-R3)/R3........................(2)

As the variation of the thermal properties with temperature affect the temperature distribution, this phenomenon is also taken into account during calculation using an iteration technique. To compute the heat input Q, an efficiency value of 0.8 is taken for SAW process and 0.6 for SMAW process.

By calculating the temperature over 100 Points across the width using either of the above equations the temperature distribution at any given time is arrived at. A typical temperature distribution for various times can be obtained.

Step 2:

By multiplying the temperature at these points by an appropriate thermal expansion coefficient corresponding to that temperature the free thermal deformation distribution of the fibres contained by these points is calculated.

Step 3:

Some initial shape change deformation is assumed.

Step 4:

Now the elastic deformation at this time is the difference between the free thermal deformation and the shape change deformation limited by the elasto-plastic behaviour of the material with temperature. (Fig.1) The shaded area describes the above statement.

Step 5:

Force and moment equilibrium conditions are now applied to check the validity of the assumption. Iteration process is done until these conditions are met by a relevant shape change deformation. The final values are stored.

Step 6:

Steps 1 and 2 are repeated for the next incremented time.

Step 7:

The transient elastic deformation is then calculated by applying the zonal conditions with these two sets of data (obtained till Step 5 and in Step 6). The zonal conditions get defined by the temperatures attained by the individual fibres. The fibres whose temperatures are greater than 600°C are said to be completely plastic (Zone 1) for which the elastic deformation is zero.

Fibres whose temperatures are less than 600°C but greater than 500°C are said to be in elastoplastic zone for which the elastic deformation is the difference between the previous elastic deformation value less the incremental thermal deformation.

Fibres whose temperatures are less than 500°C are said to be in completely elastic zone for which the elastic deformation is the difference between the previous elastic deformation less the incremental free deformation. While calculating the elastic deformation values as said above, the magnitude is limited by the corresponding yield values at those temperatures in either directions. Typical yield stress variation with temperature for the steels analysed in this report are given in Fig.2.

Step 9:

Equilibrium conditions are once again applied along with iterations to arrive at the right shape change and elastic deformations. The difference between the elastic deformation and its corresponding shape change deformations multiplied by an appropriate elastic modulus gives the transient residual stress for the present time under analysis.

Step 10:

The final residual stress is arrived at by repeating the calculation Steps from 6 to 9 until the time the temperature distribution reaches equilibrium.

As the temperature and deformation calculations are to be done over 100 points repeatedly along with iterations a computer program has been developed for performing these calculations. The program has about 300 statements written in simple BASIC language and can be run on any desk-top computer. The inputs of the program are the welding parameters, plate dimensions, the thermal diffusivity data over 100 discrete temperatures and the yield stress data given over 10 discrete temperatures. The flow chart of the program is given in Fig.3.

3. EXPERIMENTAL PROCEDURE

To validate the calculation methodology experiments were conducted using edge welded plates of dimensions as shown in Fig.4 using MMAW and SAW processes. The welding trials were conducted on a structural steel conforming to IS 226, a high strength weldable high tensile steel HTW Gr.55 conforming to IS 961 and a pressure vessel steel conforming to SA 515 Gr.70, on two thicknesses over three heat input levels. The chemical composition of these steels as analysed and the welding parameters used are given in Table 1 and 2 respectively. During welding the thermal cycles as well as shape change deformations were measured using a fixture and set up as shown in Fig.5. The thermal cycles were recorded using chromel-alumel thermocouples along with appropriate compensating cables using an XY recorder. The shape change deformation with time was measured on two locations using a highly sensitive inductive type LVDT-W5K transducer coupled with a UV-recorder. The plates were also stress relieved prior to welding in a furnace. After welding the residual stress distribution was measured using electrical strain gauges by employing dissection method. The initial and the final strains were recorded using a digital strain indicator with a multipoint scanner in a full bridge mode. These arrangements are shown in Fig.6.

4. RESULTS AND DISCUSSIONS

As the temperature distribution is the base for the calculation method, a good agreement between the measured and computed results is essential. Fig.7 shows a typical comparison of thermal cycles obtained for a case, indicating quite a close agreement between the two. As plotting of temperature distribution is practically not feasible, comparison is done using the thermal cycle plots. However, this does not cause any error since the same model is used for both the calculations. The measured shape change deformation values while compared with computed values show a reasonably good agreement as shown in Fig.8.

The progress of the kinetic step by step calculation method is illustrated in Fig.9 which describes the stress build-up with respect to time. The residual stress values measured for all the three cases along with the computed results shown through Fig.10-12 illustrates the close agreement between the two. The residual stress distribution diagrams indicate that in both the cases, the width over which tensile residual stress occur is larger with higher heat inputs.

5. CONCLUSIONS

An analytical computation methodology has been proposed to calculate the longitudinal residual stress distribution in thin and medium thickness edge welded plates.

Corrobration of experimental and computed results are in very close agreement for both temperature distribution and residual stress on 12.5mm and 20mm plates welded by MMAW and SAW processes which proves the efficacy of the proposed method.

As this is a kinetic approach capable of calculating transient deformations also, this can be used for analysing hot cracking problems along with transformation diagrams. Further, it can also be used for computing residual stress in open corner welded joints.

6. ACKNOWLEDGEMENT

The authors wish to thank the Management of BHEL for having permited to present this paper at this Conference.

7. NOMENCLATURE USED IN THE PAPER

T. = Temperature at any point ^{o}C
To = Initial temperature of the plate ^{o}C
D = Thermal diffusivity Cm^2/Sec.
V = Velocity of the welding source Cm/Sec.
E1 = Moving co-ordinate of time/Sec.
R = SQR $(E1^2 + Y^2)$
K = Thermal conductivity Cal/Cm ^{o}C/Sec.
Ko = Bessel function
g = Thickness of the plate in Cm
R3 = D.Vr3
r3 = SQR $(E1^2+Y^2+(Z+2.n.g.)^2)$
Z = Distance of the point in the thickness direction.
n = integer
Q = V.I.E
E = Efficiency of welding process
I = Welding current
V = Welding voltage

8. MAIN REFERENCES

1. ROSENTHAL,D., W.J.20,P.220S-234S(1941): Journal.
2. MYERS,P.S., UYEHARA,P.A., BORMAN,G.L., Fundamentals of heat flow in welding, WRC Bulletin 123.
3. TALL,L., "The calculation of Residual stress - in perspective", Conference Proceedings on Residual Stress and their effects, Welding Institute Publication 1978.

4. MASUBUCHI,K., Control of shrinkage and distortion in welding, WRC Bulletin 149.
5. RYKALIN,N.N., "Calculation of heat flow in welding", (Translated version) MIR Publishers, Moscow (1955).
6. BOULTON,N.S., LANCE MARTIN, Residual stresses in arc welded plates, Proceedings of Institution of Construction Engineers, 1936.
7. OKERBLOM,N.O., "Calculation of deformations in welded structures", HMSO Publication, London (1958).
8. TALL,L., Welding Journal, 05 P10S-23S (1964).
9. SURESH,S., SHANMUGAM,A., ACHAR,D.R.G., "AN ANALYTICAL STUDY OF RESIDUAL STRESS DEVELOPMENT DURING WELDING", Paper presented at the IIW National Welding Seminar, at Tiruchirapalli, India, 1984.

APPENDIX

TABLE-1

CHEMICAL COMPOSITION OF THE STEELS ANALYSED IN THIS STUDY

Sl. No.	Material Specification.	C	Mn	Si	S	P	Cr	Mo	Yield Stress Kg/Cm^2	Tensile Strength Kg/Cm^2
01.	IS 226 (Parent metal)	0.22	--	--	0.05	0.05	--	--	26	42
02.	SA 515 Gr.70 (Parent metal)	0.27	0.8	0.24	0.016	0.019	0.15	0.12	30	52
03.	HTW Gr.55 (IS 961-62)	0.11	0.66	0.2	0.01	0.011	0.4	0.4	35	53
04.	Automelt Gr.A Wire/4mm dia.	0.08	0.5	0.1	--	--	--	--	--	--

TABLE-2

WELDING PARAMETERS USED IN THE ANALYSIS

Sl. No.	Material	Set No.	Specimen No.	Welding Current in Amp.	Welding Voltage in Volt	Welding Speed Cm/S	Heat Input Kcal/Cm	Remarks
01.	IS 226 (12.5mm thick)	1	A1,A2,A3	180	25	0.25	4.3	MMAW process with 4.0mm dia electrode.
02.	SA 515 Gr.70	2	B1,B2,B3	725	36	0.80	7.8	SAW process with 4.0mm wire.
03.	-do-	3	B4,B5,B6	850	45	1.0	8.6	-do-
04.	-do-	4	B7,B8,B9	750	35	0.6	10.5	-do-
05.	HTW Gr.55/IS 961-62	5	B10,B11,B12	725	36	0.8	7.8	-do-
06.	-do-	6	B13,B14,B15	800	46	1.0	8.8	-do-
07.	-do-	7	B16,B17,B18	750	35	0.6	10.5	-do-

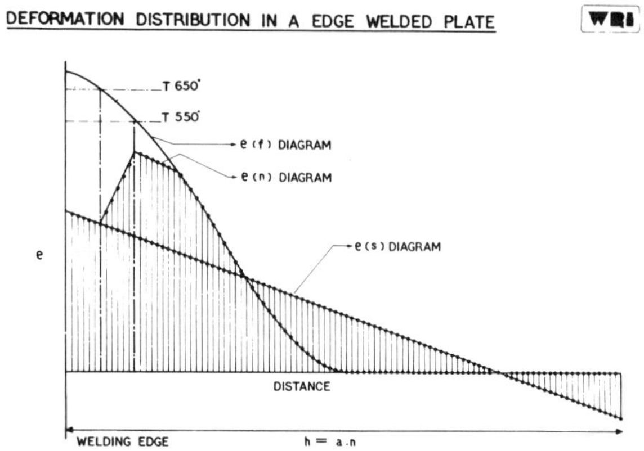

Fig.1 - Initial assumed distribution.

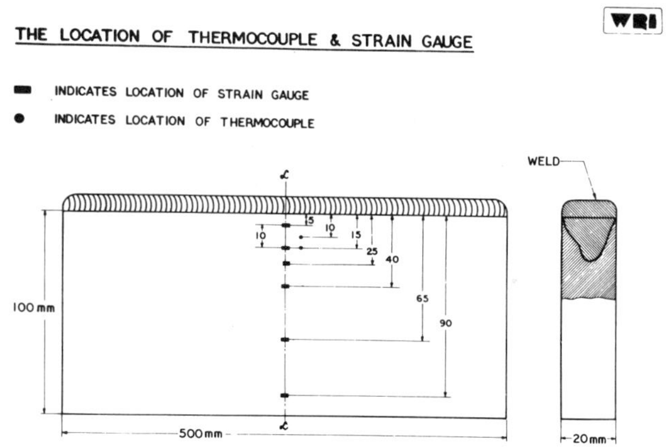

Fig.4 - Plate dimensions used in the study.

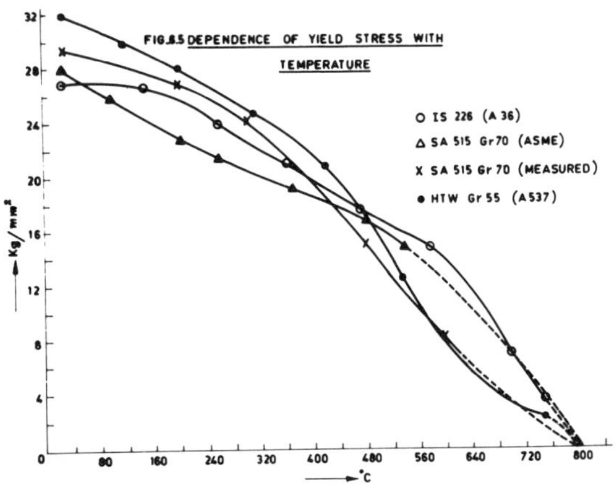

Fig.2 - Variation of yield stress with temperature for the steels analysed.

Fig.5 - Overall view of the measurement arrangement.

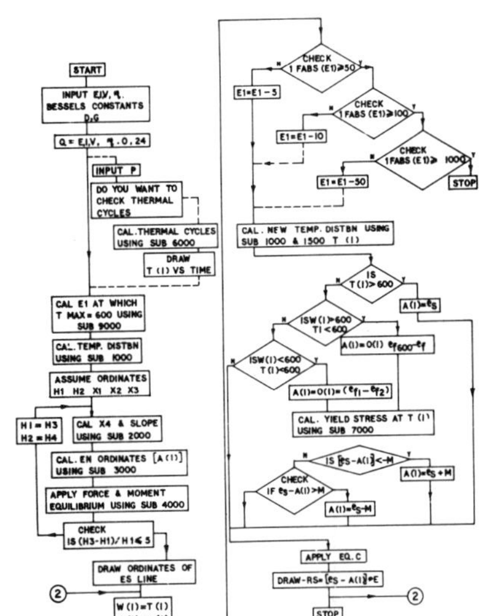

Fig.3 - Flow chart of the program.

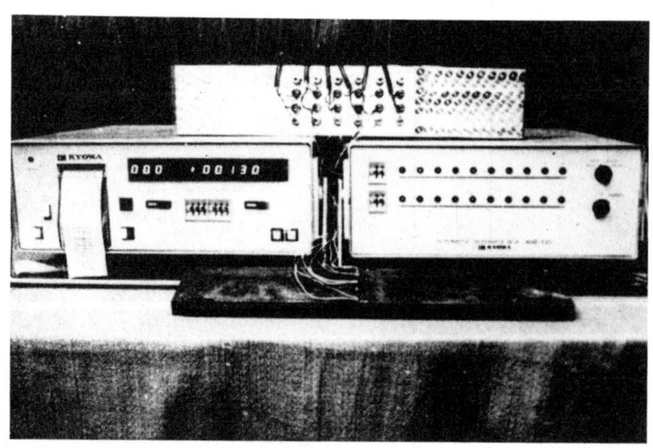

Fig.6 - Residual stress measurement set-up.

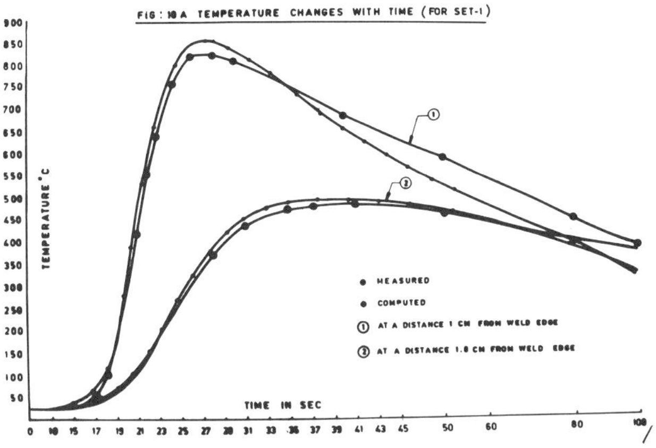

Fig.7 - Typical comparison of thermal cycles.

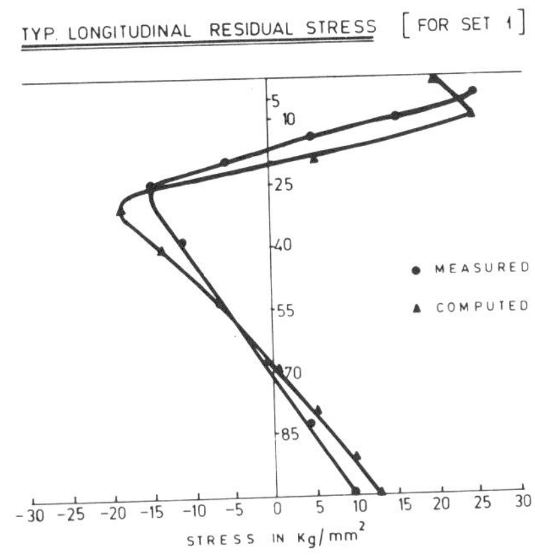

Fig.10 - Comparison of computed residual stress distribution with measured results in thin plate (IS 226)

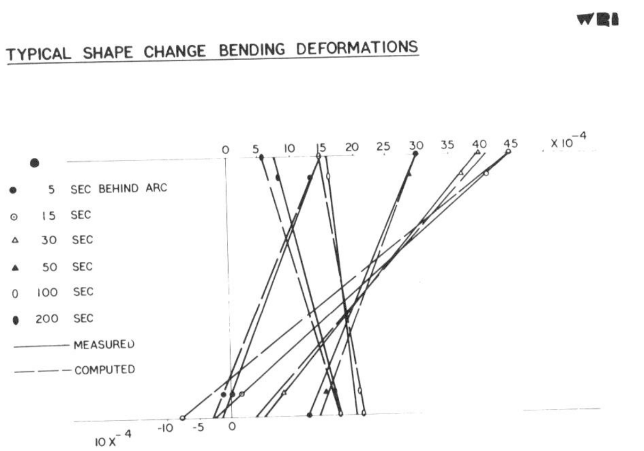

Fig.8 - Typical comparison of shape change deformation.

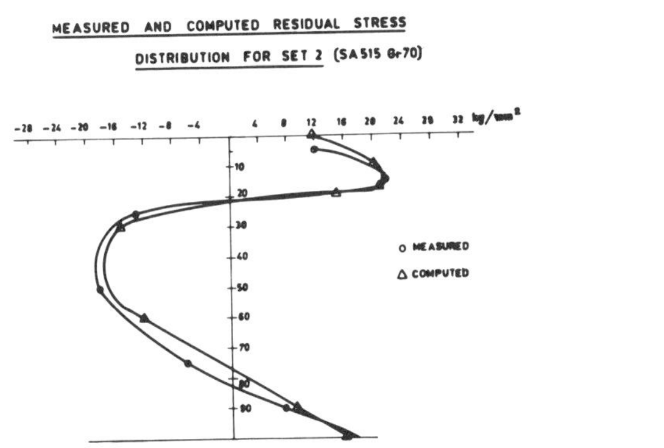

Fig.11 - Typical comparison of computed residual stress distribution with measured results in medium thick plate (SA 515 Gr.70)

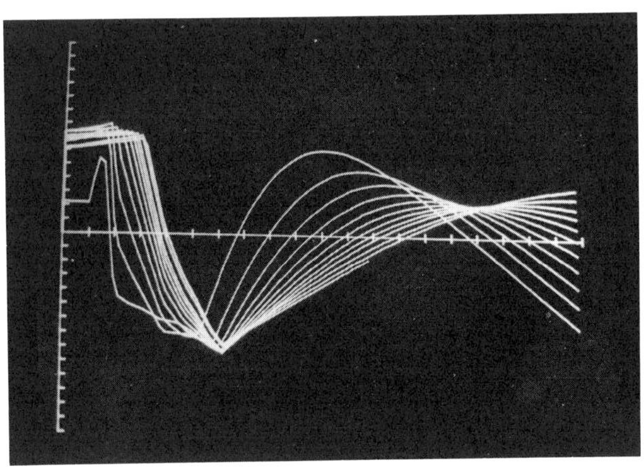

Fig.9 - Typical kinetic stress development during welding.

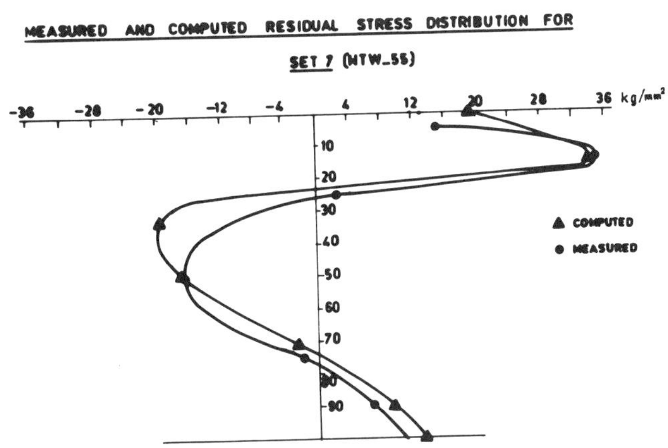

Fig.12 - Typical comparison of computed residual stress distribution with measured results in HTW Gr.55 steel.

INVESTIGATION OF HEAT AFFECTED-ZONE CRACKING OF GMA WELDS OF Al-Mg-Si ALLOYS USING THE VARESTRAINT TEST

M. Katoh
Department of Metallurgical Engineering
Kyushu Institute of Technology
1-1 Sensuicho, Tobatku
Kitakyushu, Japan

H.W.Kerr
Department of Mechanical Engineering
University of Waterloo
Waterloo, Ontario, Canada, N2L 3G1

Abstract

The effect of the filler metals and welding conditions on HAZ cracking was studied for GMA welds using the Varestraint test. Two types of HAZ cracking were observed; longitudinal cracking, parallel to the weld bead, and transverse cracking, normal to the weld bead. In the case of 4043 filler metal only the transverse cracks were observed. In the case of 5356 filler metal, however, both the transverse and the longitudinal cracks were observed. The evidence of liquation was observed on fracture surfaces of both cracks. Both the augmented strain and the heat input must exceed critical values in order to cause transverse cracking. The maximum crack length increased with further increase of these parameters above the critical values. At a given augmented strain and heat input, the maximum transverse crack length was considerably greater in the case of 4043 filler metal than in the case of 5356 filler metal. This phenomenon can be explained using the difference of the solidus temperatures of the weld metals.

Al-Mg-Si ALLOYS are superior to other high strength Al alloys in the weldability and corrosion resistance, but have a problem of high crack sensitivity in the weld heat-affected zone (HAZ) due to liquation cracking. Some reports have been published on liquation cracking of these alloys. Gittos and Scott showed that HAZ cracking occurred parallel to the weld bead (longitudinal crack) in GMA welds using a patch test (Ref.1). They attributed cracking to the base metal solidus being below the weld metal solidus. However, Tsujimoto et al, who performed GMA butt welding with a groove, showed that liquation cracking occurs even when the solidus of the base metal is higher than that of the weld metal (Ref.2).

It is apparent that there is some disagreement in previous papers over the mechanisms of liquation cracking. Part of the difficulty in interpreting these results is that different testing techniques were used. In addition, the comparison of cracking using different filler metals or welding conditions has generally been qualitative.

In this investigation, the effect of the filler metals and welding conditions on HAZ liquation cracking was studied for GMA welds using the Varestraint test.

MATERIAL USED AND EXPERIMENTAL PROCEDURES

The base metal used is a commercial 6061 alloy (thickness 6.3 mm). The Varestraint test was performed to evaluate the sensitivity of the HAZ liquation cracking. GMA welding on plates was performed using welding speeds (v) ranging from 254 to 508 mm/min, welding currents (I) from 110 to 170 A, and arc voltages (E) from 22 to 24 volts. The filler metals used were 4043 and 5356 (diameter 1.6 mm).

The sensitivity of the material to HAZ liquation cracking was mainly evaluated by the maximum crack length in the HAZ. The crack length was observed using a projector (x20) after machining off the reinforcement. The maximum crack length (Lmax) changes when either the width of the specimen or the ram speed of the cylinder for bending the specimen is changed (Ref.3). In this investigation the specimen width was 50 mm and the ram speed was 12.3 mm/s (measured in bending the base metal without welding).

RESULTS AND DISCUSSION

Fig.1 shows a schematic representation of the cracks observed for the two filler metals, 4043 and 5356. In the case of 4043 filler metal only cracks which were normal to the weld bead were observed (transverse cracks). The transverse cracks are HAZ liquation cracks. In the case of 5356 filler metal both the transverse cracks and cracks which were parallel to the bead (longitudinal cracks) were observed. In the case of GTA welds only the transverse cracks were

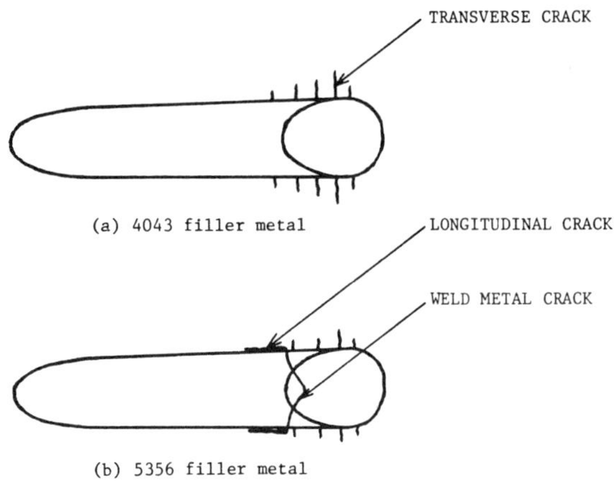

Fig.1 - Schematic representation of the cracks observed

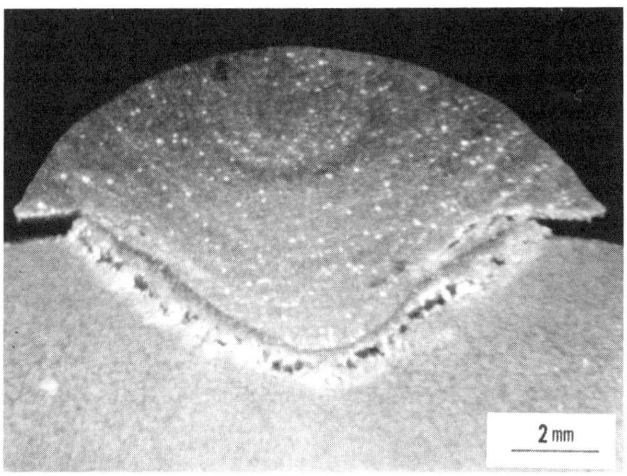

Fig.2 - A macrostructure of the longitudinal crack

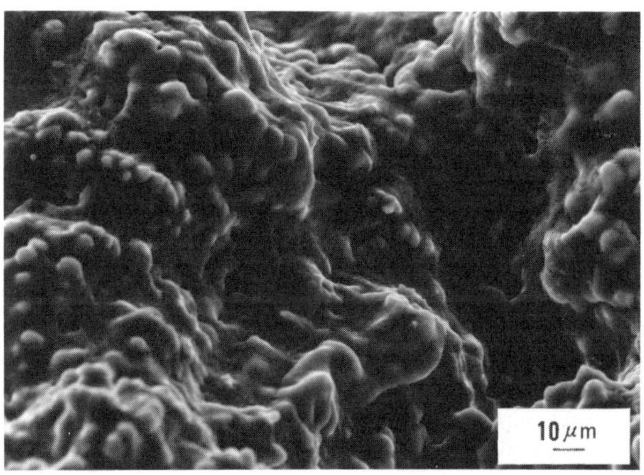

Fig.3 - An example of fractographs of the longtudinal crack

observed.

Fig.2 shows a macrostructure of the longitudinal crack. In the cross section the crack is observed to be parallel to the fusion boundary and is propagating in the partially molten zone. Fig.3 shows an example of fractographs of the longitudinal crack and the evidence of liquation is observed. The crack length increased gradually when the augmented strain increased. At a given augmented strain a critical heat input must be exceeded to cause longitudinal cracking. When the heat input was larger than the critical value the crack length increased linearly (Ref.4).

As was seen in Fig.3 it was clear that the fracture surface was partially melted. The amount of the liquid increased closer to the weld crater. For the longitudinal cracks to be developed, it is necessary that the region near the fusion boundary is melted (probably due to constitutional liquation) and that the weld metal has partly solidified when the augmented strain is applied.

Fig.4 shows the relation between the maximum transverse crack length and the augmented strain for 4043 and 5356 filler metals. No transverse cracks were observed when the augmented strain was less than about 0.5% for both filler metals. Maximum crack length increased linearly as the augmented strain increased above about 0.5%.

Fig.5 shows the relation between the maximum transverse crack length and the heat input. No cracks were observed when the heat input was less than about 4 kJ/cm. When the heat input became larger than 4 kJ/cm the maximum crack length increased linearly with heat input. As is clear from Fig.5, the maximum crack length in 4043 filler metal was generally more than twice that in 5356 filler metal for a given test condition.

It is clear from Figs.4 and 5 that critical values of both the augmented strain and the heat input must be exceeded to cause transverse crack-

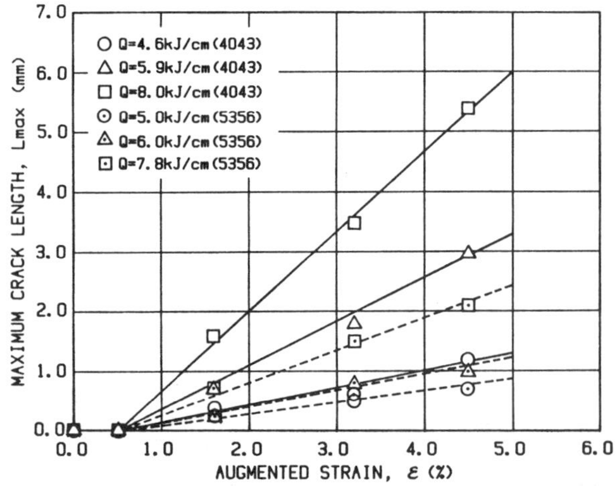

Fig.4 - Relation between the maximum transverse crack length and the augmented strain ing. The filler metal also considerably influences transverse cracking.

It should be noted that the relative cracking tendency is opposite for the two types of cracking with the different filler metals; 4043 filler metal promotes transverse cracking, but prevents longitudinal cracking.

Figs.6 and 7 show examples of fractographs of the transverse cracks for 4043 and 5356 filler metals, respectively. Many globular structures are observed, showing that the region where the transverse cracks were observed was partly melted during cracking.

The most logical explanation for the transverse cracks is the influence of molten metal penetrating from the weld metal along grain boundaries in the partially molten zone. The relative effects of the two filler metals on the maximum crack length can be explained using the difference of the solidus temperatures of the weld metals. That is, the lower solidus temperatures with 4043 filler metal permit liquid to be present to a lower temperature than in the case of 5356 filler metal. However, the effects may also be influenced by fluidity effects (Ref.4).

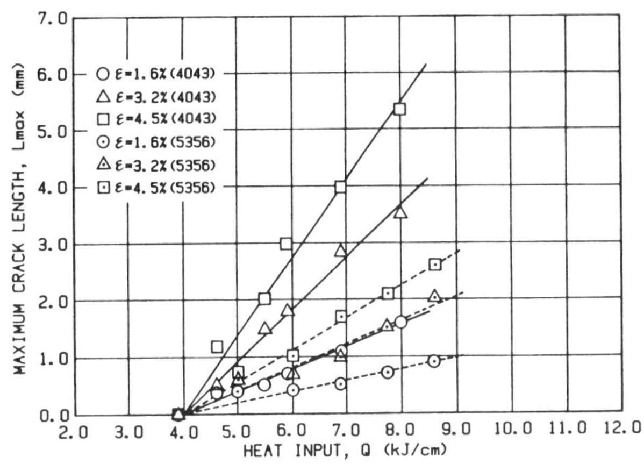

Fig.5 - Relation between the maximum transverse crack length and the heat input

CONCLUSIONS

1. Two types of HAZ liquation cracking were observed; longitudinal cracking, parallel to the weld bead, and transverse cracking, normal to the weld bead.
2. Longitudinal cracking was observed only in the case of 5356 filler metal, but not when 4043 filler metal was used.
3. Transverse cracking was observed for both 4043 and 5356 filler metals, and for GTA welds.
4. Both the augmented strain and the heat input must exceed critical values in order to cause transverse cracking.
5. At a given augmented strain and heat input, the maximum transverse crack length was considerably greater in the case of 4043 filler metal than in the case of 5356 filler metal.

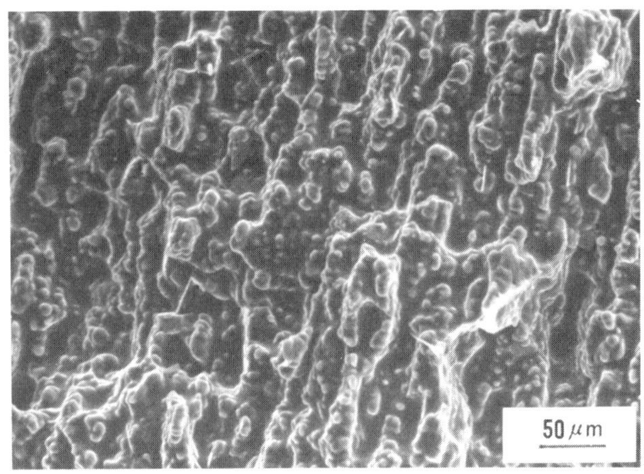

Fig.6 - An example of fractographs of the transverse cracks (4043 filler metal)

ACKNOWLEDGEMENTS

This work was supported by a grant from the Natural Sciences and Engineering Research Council of Canada, and by the University of Waterloo. Alcan graciously supplied the base metals, and also carried out the chemical analyses and differential scanning calorimetry. Messrs. N. Wilhelm, A.Degroot and H.Kamler provided important technical support at the University.

REFERENCES

1. Gittos, N. F. and M. H. Scott, Weld. J. 60, 95s-103s (1981)
2. Tsujimoto, K., A. Sakaguchi, T. Kinoshita, K. Tanaka and S. Sasabe, IIW Doc.IX-1273 1-13 (1983)
3. Katoh, M. and H. W. Kerr, to be published
4. Katoh, M. and H. W. Kerr, to be published

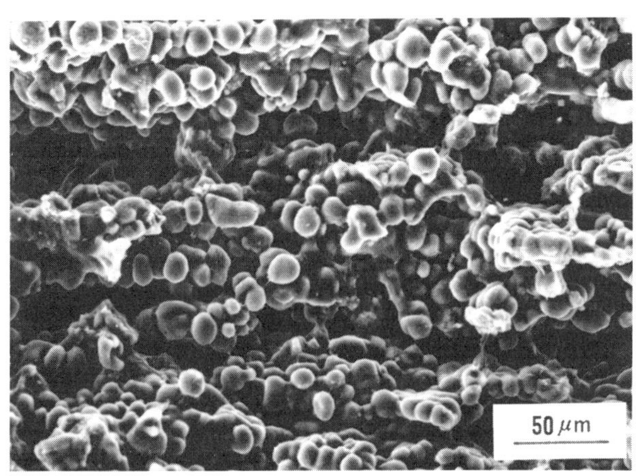

Fig.7 - An example of fractographs of the transverse cracks (5356 filler metal)

ACOUSTIC ARC LENGTH CONTROL

Pellervo Kaskinen, Gregory J. Mueller
Merrick Engineering, Incorporated
Nashville, Tennessee

ACOUSTIC ARC LENGTH CONTROL

BACKGROUND - Arc voltage control is generally used with automatic GTA welding to maintain electrode to work distance while the part deviates from ideal shape or when multipass welding with filler wire addition causes gradual change in the part diameter. Ideally, the electrode to work distance should remain constant with this feedback technique, but in practice some difficulties arise:

Changing Current - Often the welding current must for different reasons be altered during the welding. As the sensing point can not be at the hot tip of the electrode, there is some voltage drop that is dependent on the current. Therefore, the voltage reference cannot remain constant, if the current is changed.

Materials - The arc voltage itself is not a constant even at constant current but depends on the shielding gas composition, electrode and work materials and electrode shape. It also can vary with arc blow (due to turbulent gas flow or magnetic fields).

Negative Resistance - At low currents, below some 30 to 50 A, there is a negative resistance region. This means that the voltage goes up with reduced current. This effect can be quite dramatic, bringing the voltage up to over twenty volts when the current is reduced to below 5 A, as compared to a minimum of about 8 V found at 30 A.

Arc Dynamics - When pulsing the welding current, the arc column keeps changing in cross sectional area. Also the surface area of the electrode actually emitting current increases and decreases. Both of these effects can be thought to affect the cross sectional area of a conductor and thereby its resistance. The resistance is higher with a smaller arc, i.e., with lower current. But the effect is continuously varying with the pulse frequency. The outcome is overshoot in each direction: a high current through a narrow column causes a short high voltage spike at the onset of the high pulse. On the other hand, an undershoot takes place at the beginning of the background current. All this causes difficulties to the arc voltage control system.

Electrode Wear - Loss of Thorium oxide from the surface of the electrode during a continued welding increases the arc voltage. Therefore, the voltage reference must be adjusted every now and then for long welds to keep the gap constant.

CURRENT AVC SYSTEMS - The difficulties have been well known, and generally 0.1 V accuracy is the best specification for commercial arc voltage controllers. Also, if the voltage can be kept at 10 V or higher, there is not too much to worry about. The situation is quite different if very short arcs are necessary, say something that represents 8 V. Then the natural changes in the emittivity and other variations are likely to cause the electrode to hit the puddle sooner or later.

Another problem is the tendency to become unstable with sloped welding current (taper/downslope). This has been a challenge mostly to the weld programmers. Either the voltage control is locked out before the taper, or the reference is increased to compensate for the voltage increase. While this can be done with modern programmers, these methods clearly present possibilities for programming errors.

Instability during pulsing is aggravated by the spikes after every transition. Two basic approaches can be used, filtering, or lock-out, plus delay. Both methods can work reasonably well, until the background current needs to be at the negative resistance region of the arc characteristics. Then the situation tends to get out of hand as far as filtering or time delay is concerned. A separate reference voltage for peak current and for background current is the best concept available for this extreme condition.

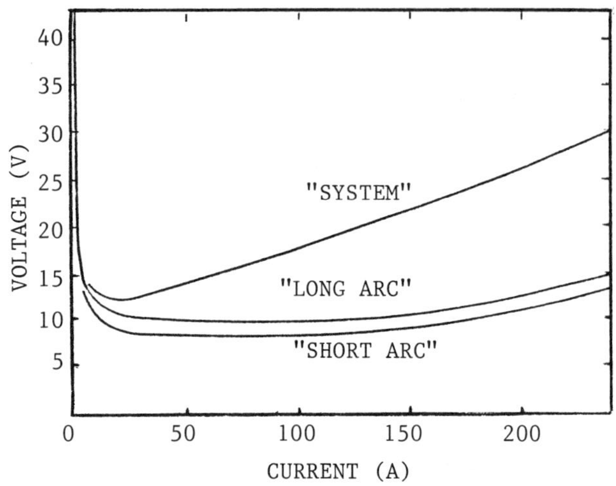

Fig 1 Arc V-A Characteristics

NEW CONCEPT - As was mentioned in the discussion about pulsing effects, the arc expands and contracts with every pulse. Why? Quite obviously, this is due to the instantaneous power in the arc column. An increased heat generation means higher temperature in the column, and according to the laws of thermodynamics, this increases the pressure in the gas, causing expansion against a constant ambient pressure. The larger column both radiates more heat and, with the larger cross sectional area, generates less heat so a temporary balance is achieved. The opposite takes place when the welding current is reduced to the background level. There is cooling and heating, until a new balance between heat generation and dissipation has been achieved in each direction.

Now, this increase and decrease of the plasma volume is audible, if proper frequency of pulsing is used, and the pulse amplitude is sufficient. In fact, there have been attempts to use this phenomenon for a very low mass loudspeaker. They have demonstrated clear and easy trebles, but midrange and bass have not been too convincing. In any case, we have a hint here. The arc sings louder at high frequencies, as well as with higher amplitudes of pulsing. This can be put into a mathematical formula stating that the sound intensity emitted by a pulsed arc is proportional to the rate of change in the instantaneous power input or:

$$S = K \frac{d(V\,I)}{dt}$$

where K is a constant.

In the product term, there may be some relationship between the two terms. A partial derivation yields:

$$S = K \left(I \frac{dV}{dt} + V \frac{dI}{dt} \right)$$

Studies of the normal ranges and variations of the voltage and the current in typical welding suggest that, with constant arc length, the voltage remains practically constant if the current is modulated by a small amount, like a couple of amperes around a reasonably high base level of 20 to 500 A. The term dV/dt approaches zero and the equation for sound pressure becomes

$$S = K\,V \frac{dI}{dt}$$

Therefore, the sound pressure is proportional to dI/dt or rate of current change. On the other hand, the sound pressure is also linearly proportional to the arc voltage! And we should emphasize that here we are talking about the voltage drop in the plasma column, not the anode and cathode drops, and definitely not the drops in the metallic conductors. This has an extremely far reaching consequence: it turns out that this portion of the voltage is also directly proportional to the arc length, rather than having a large offset (some 6 to 8 volts) and superimposed gap-dependent portion.

A PRACTICAL CONTROLLER - A practical arc length controller has been built as a front end attachment to existing Arc Voltage Controllers (AVC's). It modulates the welding current by "stealing" a certain amount of current from the arc through a network that is connected in parallel with the arc. This stealing takes place at a crystal controlled frequency (about 2 kHz presently, but not limited to it). The operation is with 50% duty cycle, and as the peak current shunted is about 4 A, the average reduction of the welding current in the arc is 2 A. This reduction can, of course, be easily compensated in the weld program.

There is an acoustic wave guide from close to the arc to a microphone that is located at a safe distance from the heat and high frequency radiation of the arc. The signal picked up by the microphone is amplified and filtered in a narrow band around the modulation signal. Then it is rectified and sent out to the original AVC unit as a simulated arc voltage. In fact, if the acoustic unit is not powered up, or the voltage in the arc is not within a preset window, the current modulation does not start and, instead, the voltage from the torch (through a HF filter) is sent directly to the AVC.

The absolute position of the sensing tube is not critical, but once put, it should stay in the same position. Any movement of the sensing point causes a change in the attenuation of the sound pressure signal captured. The feedback loop treats this change as any change in the arc length and tries to compensate. We should also note that the operating frequency is high enough to demonstrate direction sensitivity. If the direction of the pick-up tube is changed, there will be a change in the signal level produced. This is particularly obvious from an experiment we ran (unwittingly).

The tube was pointing in the general direction of the arc, but the clamping was rather temporary and the slippery teflon tube moved gradually. We were observing the arc and the wire feeder operation, using handheld, large window welding helmets. We noticed some odd movements of the torch and after several minutes of observations finally correlated the seemingly erratic jumps with our own movements! After that, it was easy to verify the reason: the sound emitted by the arc was bounced off the glass lenses of the helmets and then picked up by the sensor tube. Depending on the position of the helmets, the signals might add to or subtract from the signal arriving directly from the arc to the tube. At this point, the distance of the tube head was some 50 mm (2 inches) from the arc and the direction was more toward one helmet than to the arc. Correcting the aiming and reducing the distance back to the nominal 38 mm (1.5 inches) restored proper operation.

Generally, the sensing tube can be routed along the torch but if there is a bulky clamp, like one holding a wire feeder, there may be difficulties. Then it is advisable to attach the tube to the wire guide instead.

TEST RESULTS - The system is so new (although the basic research was conducted a few years ago and there are patents issued in most principal countries) that actual comparisons to a state of the art conventional AVC unit are not ready for publication. There are some introductory results though.

The simple addition of a unit in front of a Merrick Volt-Trak II produced oscillations. The reason was that the acoustic unit had been optimized for use with another popular AVC head that has a deadband in its response. While this deadband was and is beneficial for a certain kind of circuit (to eliminate hunting around the balance position), the Merrick AVC uses a linear response to achieve the highest possible sensitivity and response. Also, the speed of response in the Merrick unit was higher than the time constant in the acoustic feedback unit could support. These problems were easy to overcome by reducing the sensitivity settings on the Volt-Trak. Later on, the filtering time constant in the Acoustic Arc Length control unit was reduced to one-fifth of the original. The resulting system response is very competitive at any arc length and superior at very short gaps.

The first tests were run on a slow SCR type power supply in which the pulse frequency achievable was limited to 0.1 - 25 pulses per second. Normally, the Volt-Trak AVC is operated in a pulse synchronized mode so that it is locked out during the background current to prevent undue hunting. This lock-out was disconnected and the system was operated under the acoustic feedback. There was practically no hunting at any frequency if the background current was kept above 20 A. Reducing the background current to 9 A produced some hunting. During this time the peak current was run as high as 260 A.

Later on, the system was run with a fast power supply that could pulse beyond the modulation frequency of the Acoustic Arc Length sensing unit. We tried several frequencies and different amplitudes of pulsing, but there was no interference until we got to essentially the same frequency with the measuring modulation. In fact, there was about 1 Hz difference between the two crystal time bases. This could be heard and seen in the head movement. There was a definite 1 Hz movement in the head height due to the interference between the two signals. But, surprisingly, the system did not crash or do anything drastic at all!

With both power supplies the gap control was able to prevent the electrode from diving into the puddle throughout a taper from 50 A to 5 A without any changes to the parameter setting whatsoever. The initial gap corresponded with a measured voltage of about 8 V at the power supply terminals (rather than at the torch). The operation also was steady enough to build some 30 or more layers of wire in a knife edge fashion on a steel cylinder, using 0.9 mm (0.035") stainless steel wire. This kind of buildup is quite a challenge to the precision of the AVC and the wire feed mechanism. Generally, any bumps in the outside diameter magnify from layer to layer, soon making further buildup impossible.

MICROSTRUCTURE AND MICROCHEMISTRY OF TYPE 316L STAINLESS STEEL SUBMERGED ARC WELDS

V. Seetharaman, D.D.N. Verma, I. Gowrisankar
Metallurgy Program
Indira Gandhi Center for Atomic Research
Kalpakkam, 603102 India

D.R.G. Achar
Department of Metallurgical Engineering
Indian Institute of Technology
Madras, 600036 India

ABSTRACT

The influence of heat input used during multipass submerged arc welding of type 316L stainless steel plates on the microstructure and microchemistry of the weld joints has been investigated. Optical and scanning electron microscopy studies show that the spacing between the secondary arms of the ferrite dendrites increases with an increase in heat input. While the ferrite content at the crown of the welds is independent of the heat input, the ferrite content at the root regions increases with heat input. Electron probe microanalysis has clearly established that the extent of solute partitioning between austenite and ferrite increases systematically with the heat input.

SUBMERGED ARC WELDING not only offers high deposition rates, thermal efficiency and weld metal recovery but also becomes economical for welding of thick plates. Despite these advantages, this technique is not used commonly for welding of austenitic stainless steels mainly because of the problems relating to high heat input. These problems include lower delta-ferrite content and the associated risk of hot cracking, loss of alloying elements such as chromium and manganese, increased width of the heat affected zone (HAZ) coupled with excessive grain growth and sensitization. Therefore, a systematic attempt has been made to investigate the effect of heat input on the microstructure and mechanical properties of type 316L weldments made by the SAW technique.

EXPERIMENTAL PROCEDURE

Hot rolled plates of AISI type 316L stainless steel (1000 x 150 x 25 mm), Avesta - 800 flux (basicity = 0.9) and 316ELC filler wire were used for this investigation. After edge preparation corresponding to a double V-groove geometry, the plates were tacked and given a root pass by the TIG process. This was followed by (i) a second pass using the SAW method with a constant heat input of 1.7 kJ/mm for all the plates and (ii) back grinding of the TIG deposit. Subsequently, these pads were welded by the SAW technique to yield four separate weldments corresponding to heat input values of 1.73, 2.58, 4.08 and 5.77 kJ/mm (designated as A, B, C and D). The welding conditions used and the chemical composition of the parent metal, filler wire and the four weld deposits are reported elsewhere [1].

Transverse sections were prepared metallographically and etched electrolytically using an aqueous solution containing 10% $(NH_4)_2S_2O_8$ to reveal the general microstructure of the weld, HAZ and the base metal. Delta-ferrite content was determined using the Magne-gage calibrated according to AWS A4.2-74. Variations in the ferrite content along the thickness of the weld were measured at intervals of 1 mm. Samples used for SEM study for the ferrite morphology were prepared by a special electrochemical technique which involves selective dissolution of austenite [2]. SEM was also used for the determination of the interdendritic spacing between the delta ferrite arms. Electron probe microanalysis was used to determine the compositions of ferrite and austenite phases and to determine the point to point variations of composition across the delta-ferrite dendrites. The intensity ratios obtained were corrected for the atomic number effect, absorption effect and the flourescent effect by standard procedures.

RESULTS

Figure 1 shows the general microstructure of the weld, HAZ and the base metal for plates A and D. The main inferences from the metallographic observations are as follows: (a) the maximum widths of the HAZ were 2.3 and 5 mm for plate A and D respectively, (b) the mean spacing between

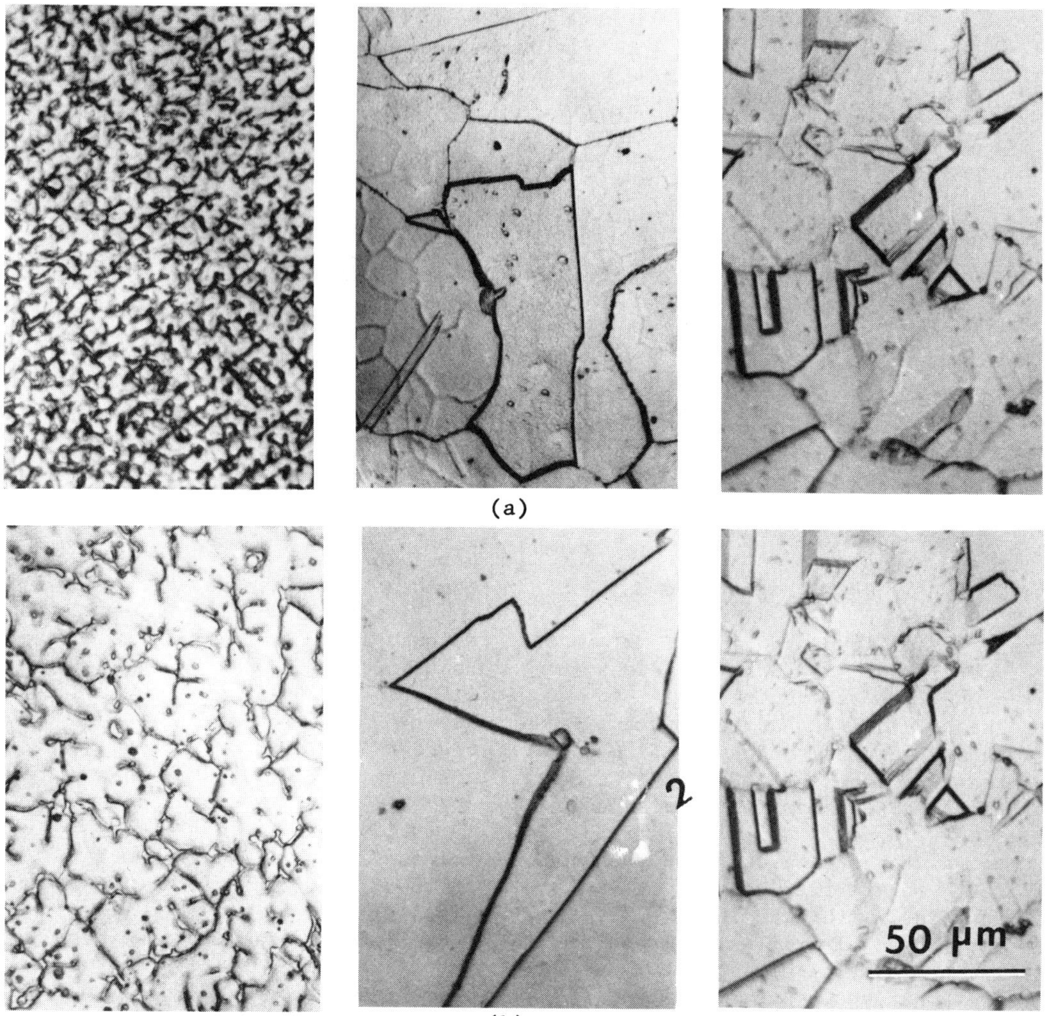

Fig. 1 - Microstructures of the weld, HAZ and base metal: (a) plate A and (b) plate D.

the secondary arms of austenite in the welds A and D were ~ 6 and 10 μm, respectively, (c) the mean values of the austenite grain size in the HAZ are 60 and 150 μm for plates A and D, respectively and (d) the grain size of the base metal was found to be 35 μm.

Figure 2 shows the variation of the ferrite number from the root to the surface of the welds A and D. It is observed that the ferrite number increased substantially as one moves away from the root region towards either surface of the weld for all the heat input values. As the heat input is increased from 1.73 kJ/mm (plate A) to 5.77 kJ/mm (plate D) the ferrite number increased from 5.2 to 6.5 in the root region. However, the ferrite number values near the top of the weldments were found to be nearly independent of the values of the heat input parameter.

Figures 3(a) and (b) illustrate the morphology of the delta-ferrite phase in plate A and D, respectively. It can be seen that the ferrite forms as a continuous network in both these samples, but the mean spacing between the secondary arms of the ferrite dendrites increases from 8 to 12 μm as the heat input is increased from 1.73 to 5.77 kJ/mm. It may be noted that these values determined by SEM are slightly higher than those measured by optical microscopy.

Composition profiles of different elements across the delta-ferrite dendrites were obtained by electron probe microanalysis. Figure 4 shows that the ferrite dendrites are enriched with respect to ferrite stabilizers such as chromium and molybdenum and are depleted with respect to austenite stabilizers such as nickel. This is in agreement with results reported elsewhere [3-7]. Further, the extent of enrichment of Cr and Mo and the depletion of Ni in the delta-ferrite dendrite cores is observed to increase with an increase in the heat input.

DISCUSSION

A decrease in the heat input from 5.77 to 1.73 kJ/mm leads to an increase in the solidification and cooling rates of the weld metal. In addition the weld metal and the heat affected zone experience increased number of thermal

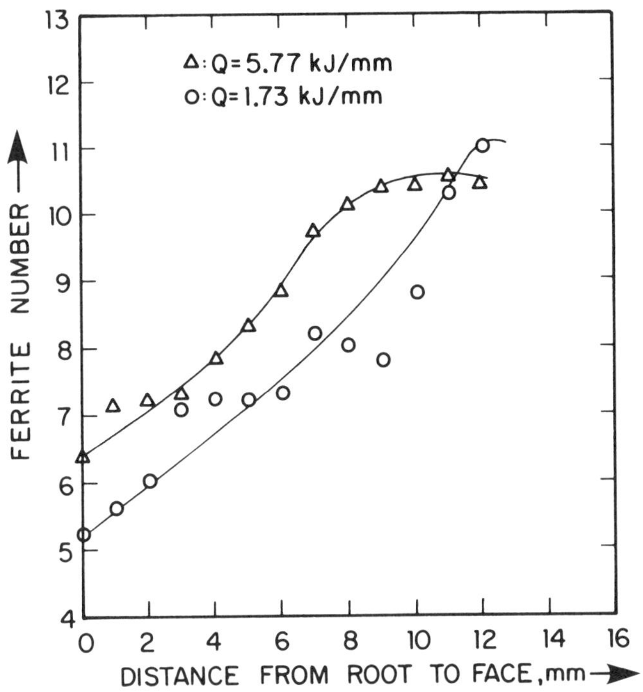

Fig. 2 - Variation of ferrite number from the root to the surface of the welds.

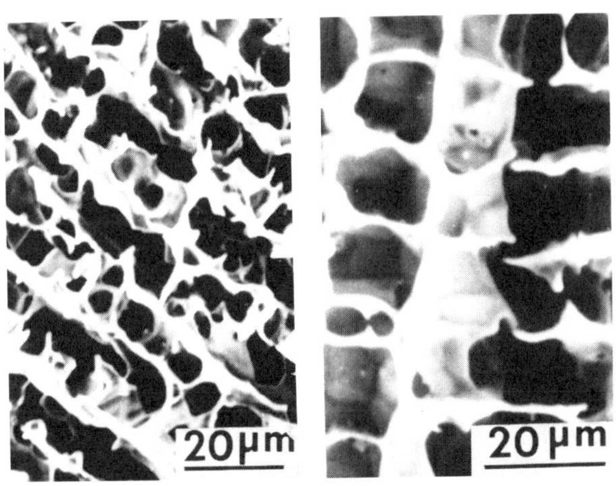

(a) (b)

Fig. 3 - Scanning electron micrographs of the samples which have been selectively etched to remove austenite: (a) plate A and (b) plate D.

cycles resulting from an increase in the number of passes from 5 to 13. It is well established that the spacing between secondary arms of the dendrites decreases with an increase in the solidification rate. Thus the observed increase in the dendrite arm spacing from 8 to 12 μm when the heat input is increased from 1.73 to 5.77 kJ/mm is consistent with the above argument based on the effect of solidification rates.

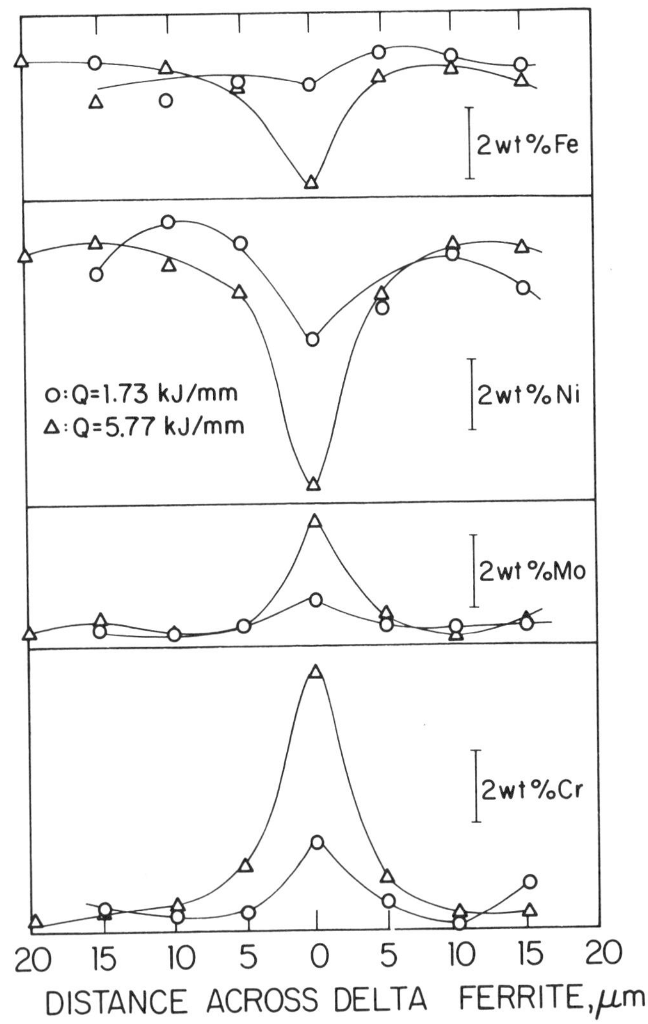

Fig. 4 - Composition profiles of Cr, Ni, Fe and Mo across the delta-ferrite dendrites as determined by electron probe microanalysis.

The values of chromium equivalent and nickel equivalent for this weld metal are 21.91 and 14.37, respectively. On the basis of predictions made by Long and DeLong [8] and by Suutala [9] it can be expected that under submerged arc welding conditions this weld metal will solidify with ferrite as the primary phase and that the net ferrite content at room temperature will be ~ 10 FN. The ferrite content values measured at the surface of the welds and the microstructures observed are in good agreement with these predictions. It is important to note that the differences in the cooling rates per se do not seem to influence the ferrite content at the surface of the weld. This is consistent with the fact that the rate of transformation of ferrite to austenite is very high in the temperature range 1400-1100°C. Massive transformation of ferrite to austenite has been found to be nearly complete in samples quenched from 1400°C at rates approaching 15000°C/s [7].

An important characteristic of multipass welds is the large number of thermal cycles imposed on the initial passes during the deposition of subsequent passes. The number of thermal cycles varies inversely with the heat input used per pass. During these thermal cycles, the residual ferrite can transform to austenite or decompose into a mixture of austenite, $M_{23}C_6$ and intermetallic phases, leading to a net reduction in the ferrite content. Low values of ferrite content observed at the root region and the marked increase in ferrite content from the root to the surface of the submerged arc welds (Fig. 2) clearly confirm the important role played by the thermal cycles. Furthermore, the ferrite content observed in the root region of low heat input welds is significantly lower than that for the high heat input weld. This can be explained on the basis of the increase in the total number of thermal cycles with an increase in the number of passes. Moreover, the morphology of ferrite was found to change from vermicular in the root region to the dispersed morphology at the surface of the welds. This is similar to the results reported by Lai and Haigh [10].

The ferrite formed during welding is enriched in chromium and molybdenum and depleted in nickel [3-7]. The extent of solute partitioning between austenite and ferrite is, however, governed by the rate at which the welds cool in the solid state. The cooling rates in the centerline of these welds at temperatures just below the solidus are estimated as 36 and 120°C/s for heat input values of 5.77 and 1.73 kJ/mm, respectively. At such low cooling rates diffusion in the solid state is very effective in redistributing the solutes between the austenite and the ferrite phases. As cooling rate decreases, the extent of the diffusional redistribution becomes quite significant (Fig. 4). These results are also in agreement with those reported by Vitek and David [11].

The kinetics of grain growth in the heat affected zone is strongly dependent on the heat input used for welding. Using the grain growth diagrams developed by Ashby and Easterling [12], one can predict that the maximum grain size near the fusion line should increase from 80 μm to 150 μm as the heat input is increased from 1.73 to 5.77 kJ/mm. The observed maximum grain sizes of 60 and 150 μm are in reasonable agreement with the above prediction. However, the measured values of the HAZ width were considerably smaller than the predicted values of 8 and 17 mm. Sensitization tests carried out as per ASTM A262 Practice A and E revealed that both the weld and the HAZ were not susceptible to intergranular corrosion. This is due to the fact that low carbon content in the weld metal and in the base metal has precluded the precipitation of $M_{23}C_6$ carbides despite the slow cooling rates involved.

CONCLUSIONS

This investigation on the role of heat input in multipass submerged arc welds of type 316L stainless steel has led to the following conclusions:

1. An increase in the heat input leads to increases in the width of the heat affected zone, maximum grain size near the fusion line and interdendritic spacing in the weld metal.
2. The ferrite content at the surface of the weld is independent of the heat input used whereas the ferrite content at the root region increases with an increase in the heat input.
3. The degree of solute partitioning between austenite and ferrite increases with heat input.

ACKNOWLEDGEMENTS

The authors are grateful to Dr. P. Rodriguez, Head, Metallurgy Programme and Materials Science Laboratory for his continuous encouragement and support for this work. They are also thankful to Dr. T. P. S. Gill and Mr. A. K. Bhaduri for many valuable discussions and assistance in the preparation of the manuscript. One of the authors (V.S.) wishes to acknowledge the support provided by Ames Laboratory, Ames, Iowa, USA, which is operated for the USDOE under contract no. W-7405-Eng-82, during the preparation of this manuscript.

REFERENCES

1. D. D. N. Verma, M.S. Thesis, Indian Institute of Technology, Madras (1983).
2. Gill, T. P. S., R. K. Dayal and J. B. Gnanamoorthy, Weld. Journal 58, 375s-378s (1979).
3. Raghunathan, V. S., V. Seetharaman, S. Venkadesan and P. Rodriguez, Metall. Trans. 10A, 1683-89 (1979).
4. J. C. Lippold and W. F. Savage, Weld. Journal 58, 362s-374s (1979).
5. David, S. A., G. M. Goodwin and D. N. Braski, Weld. Journal 58, 330s-336s (1979).
6. Brooks, J. A., J. C. Williams and A. W. Thompson, Metall. Trans. 14A, 23-31 (1983).
7. Singh, J., G. R. Purdy and G. C. Weatherly, Metall. Trans. 16A, 1363-69 (1985).
8. Long, C. J. and W. T. DeLong, Weld. Journal 52, 281s-297s (1973).
9. Suutala, N., Metall. Trans. 14A, 191-197 (1983).
10. Lai, J. K. and J. R. Haigh, Weld. Journal 58, 1s-6s (1979).
11. Vitek, J. M. and S. A. David, Metall. Trans. 16A, 1521-23 (1985).
12. Ashby, M. F. and K. E. Easterling, Acta Metall. 30, 1969-78 (1982).

A COMPARATIVE EVALUATION OF INCONEL 182 AND 16-8-2 ELECTRODES FOR SHIELDED METAL ARC WELDING OF INCOLOY 800 AND TYPE 304 STAINLESS STEEL

I. Gowrisankar, V. Seetharaman
Metallurgy Program
Indira Gandhi Center for Atomic Research
Kalpakkam 603102, India

ABSTRACT

Transition metal joints between 2 1/4 Cr - 1 Mo steel and Type 304 stainless steel with Incoloy 800 as the transition material are being developed for application in the Prototype Fast Breeder Reactor. As a part of this development program, hot cracking susceptibilities of Inconel 182 (E NiCrFe-3) and 16-8-2 electrodes have been evaluated. Detailed microstructural characterization and evaluation of mechanical properties have been carried out on shielded metal arc welded butt joints between Incoloy 800 and type 304 stainless steel. This paper reports the relative hot cracking tendencies of Inconel 182 and 16-8-2 weld metals and also discusses the microstructural features and mechanical properties of the joints made with these consumables.

DISSIMILAR METAL JOINTS between low alloy ferritic steels and austenite stainless steels are used widely in steam generators of power stations. In fossil fuel fired plants the primary boiler and heat exchanger tubes made of chromium-molybdenum steels, especially 2 1/4 Cr - 1 Mo steel, are welded to austenitic stainless steel tubes used in the high temperature sections such as the superheaters and reheaters. In liquid sodium cooled fast breeder reactors, steam generator tubes made of 2 1/4 Cr - 1 Mo steel are joined to austenitic stainless steel tubes running from the intermediate heat exchanger. These dissimilar metal joints are usually produced by gas tungsten arc welding (GTAW) or induction pressure welding. Initially austenitic stainless steel filler wires were used for the GTAW process; but recently nickel base filler wires have also been employed. The current interest in dissimilar weldments is related to the fact that several failures have occurred in these joints (in fossil fired plants) after 15 to 20 years of operation, well below the design lifetime of the tubings [1]. These failures have invariably occurred in the heat affected zone of the ferritic steel through the propagation of circumferential cracks very close to the fusion boundary. Principal factors contributing to this type of failure can be summarized as follows [1-4]:

(a) Cyclic thermal stresses at the ferritic steel/austenitic weld metal interface due to the differences in the coefficients of thermal expansion between the base metal and the weld metal.

(b) Preferential stress oxidation at this interface and the propagation of oxide notches from the outer surface of the tubes.

(c) Carbon migration from the ferritic steel to the austenitic weld metal leading to accelerated creep in a narrow carbon depleted zone adjacent to the interface.

(d) Microstructural instabilities involving dissolution of $M_{23}C_6$ carbides at regions well away from the fusion line and their reprecipitation along prior austenite grain boundaries very close to the fusion line.

Nickel base weld metals such as Inconel 182/82 not only have coefficients of thermal expansion close to that of the ferritic steel but also reduce the rate of carbon migration from the ferritic steel very effectively. As a result, nickel base welding consumables are considered to be superior to austenitic stainless steel consumables and are now used almost exclusively. Indeed, significant improvements in the service life of the dissimilar metal joints have been achieved by adopting this approach. Nevertheless, premature failures of these joints occur frequently even with the use of nickel base filler metals. Nucleation and propagation of creep-fatigue cracks along a planar array of globular carbides formed on the fusion line during service are considered to be responsible for this type of failure. Thus it is very clear that extensive investigations on the microstructural evolution and the deformation behavior of dissimilar metal joints at elevated temperatures and high stresses will be necessary before optimum performance of these joints can be achieved. Different approaches and specific research efforts required for the successful development of these joints have been outlined by Campbell et al. [5].

DEVELOPMENT OF TRANSITION METAL JOINTS

One of the approaches adopted for circumventing the problems associated with the dissimilar metal welding is the use of trimetallic transition joints [6]. A transition piece is chosen to have a coefficient of thermal expansion (CTE) that is intermediate between those of ferritic steel and austenitic stainless steel. This results in a more gradual change in the CTE and a consequent decrease in the magnitude of cyclic thermal stresses. In addition to the thermal expansion characteristics, the inherent weldability, metallurgical compatibility and stability during long term service must also be considered in the choice of the transition piece and the welding consumables.

Figure 1 illustrates the relationship between the CTE and chemical composition for a range of stainless steels and high nickel alloys

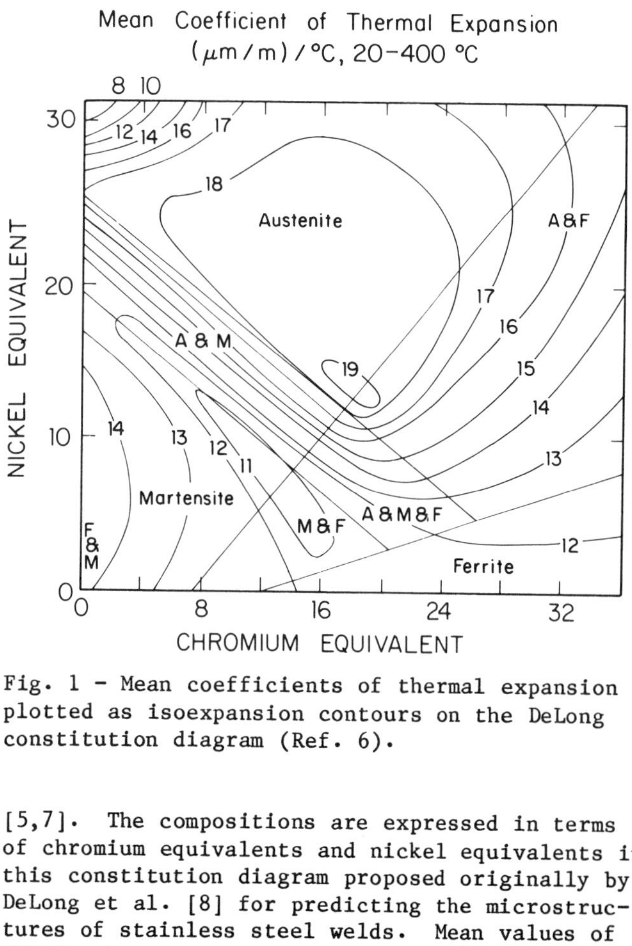

Fig. 1 - Mean coefficients of thermal expansion plotted as isoexpansion contours on the DeLong constitution diagram (Ref. 6).

[5,7]. The compositions are expressed in terms of chromium equivalents and nickel equivalents in this constitution diagram proposed originally by DeLong et al. [8] for predicting the microstructures of stainless steel welds. Mean values of CTE in the temperature range 20-400°C plotted as isoexpansion contours are superposed on this diagram. The region of the highest CTE (19 μm/m/°C) is obtained in the austenite phase field corresponding to conventional austenitic stainless steels and is centered around Ni_{eq} = 14 and Cr_{eq} = 18. The CTE decreases as the composition is changed in any manner with respect to this point. If we increase Cr_{eq} at a constant value of Ni_{eq} = 14, then duplex microstructures containing austenite and ferrite (A&F) are obtained and this leads to significant reduction in the CTE. In contrast, the CTE decreases very gradually with an increase in the nickel equivalent for fully austenitic microstructures. It is interesting to note that nickel rich alloys have approximately the same CTE as ferritic steels.

The mean CTE of 2 1/4 Cr - 1 Mo steel and type 304/316 stainless steel are 14.0 and 18.5 μm/m/°C, respectively. Among the materials which have CTE intermediate to these values, Incoloy 800 which exhibits excellent resistance to oxidation and creep at elevated temperatures is the most attractive choice for the transition piece [6]. Inconel 182 (E NiCrFe-3), which has a CTE of ~ 16 μm/m/°C and is a consumable used commonly for dissimilar metal joints, has been chosen for the joint between Incoloy 800 and 2 1/4 Cr - 1 Mo steel. For welding Incoloy 800 to austenitic stainless steel, austenitic welding electrode containing 16% Cr, 8% Ni and 2% Mo (16-8-2) with a CTE of 17.9 μm/m/°C is the natural choice. However, Inconel 182 electrodes are specified as the alternative welding consumable for this joint mainly because of the large experience base available for the welding of Incoloy 800 with Inconel 82/182 consumables. A typical transition joint configuration adopted for the steam generator circuit of the 500 MWe Prototype Fast Breeder Reactor is shown in Fig. 2. Detailed stress analysis carried out by Dalcher et al. [9]

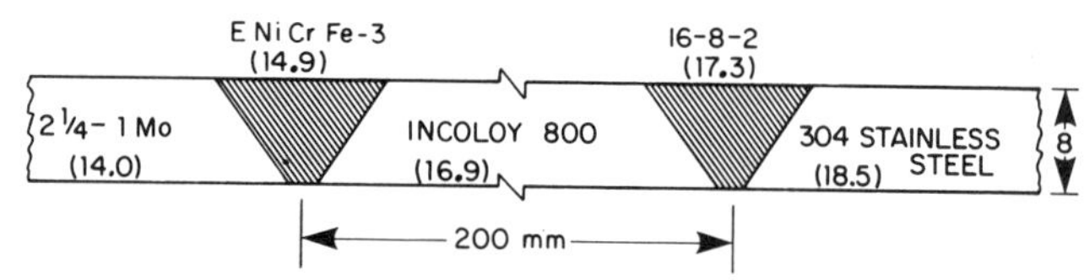

Fig. 2 - Transition metal joint configuration adopted for the prototype fast breeder reactor. The number given below each material denotes the mean thermal expansion coefficient in the temperature range 20-538°C.

has shown that the hoop stress values in the 2 1/4 Cr - 1 Mo steel close to the root region decrease by 37% when Incoloy 800 is used as the transition material. This clearly establishes the improvement in thermal stresses generated in trimetallic transition joints over the (direct) bimetallic joints.

A comprehensive program on the development of transition metal joints to be used in the PFBR has been undertaken at Indira Gandhi Center for Atomic Research, Kalpakkam. This consists of detailed weldability studies, metallurgical characterization and evaluation of both short term and long term mechanical properties of this joint. Since the transition piece made of Incoloy 800 is specified to be 200 mm long, the two welds involved in this transition joint can be considered to be practically independent of each other and can hence be studied separately. This paper deals with a comparative evaluation of the hot cracking susceptibilities of Inconel 182 (E NiCrFe-3) and 16-8-2 electrodes. In addition, it reports preliminary results obtained on the microstructure and mechanical properties of the butt welds between Incoloy 800 and type 304 stainless steel produced by the shielded metal arc welding technique using the two consumables mentioned above.

EXPERIMENTAL PROCEDURE

The hot cracking susceptibility of the joint between Incoloy 800 and type 304 stainless steel was assessed by performing modified Tekkon Y-groove and key-hole tests. These tests were chosen due to their simplicity and reproducibility and also because they approximately simulate conditions existing during the root pass deposition. The Y-groove test specimen consisted of Incoloy 800 and type 304 stainless steel plates joined together to form a Y-groove slot in the center and the weld deposit is made by the GTAW process in the central groove, restrained by anchor welds on both ends. In the key-hole test the center slot extended up to the edge of the plate, the weld restraint increasing as welding proceeded from the edge to the key-hole.

Solution-annealed plates (10 mm thick) of type 304 stainless steel and Incoloy 800 were edge-prepared for butt joints with a singe V-groove geometry and an included angle of 80° and were welded by the shielded metal arc welding (SMAW) process using Inconel 182 or 16-8-2 electrodes. The welds were examined by non-destructive techniques such as X-radiography and liquid penetrant test.

Button head type tensile test specimens were machined from the base materials as well as the welded joints (all weld and transverse orientations). Tensile tests were carried out at 25, 350, 500, and 700°C and at a nominal strain rate of $3.2 \times 10^{-4}\ s^{-1}$. Subsize (55 x 10 x 5 mm) charpy V-notch impact test specimens were prepared from the welded plates in such a way that the notch would propagate entirely through the weld and along the welding direction. Impact tests were carried out at room temperature using an instrumented impact testing machine. Transverse sections of the joints were prepared metallographically and etched electrolytically in 10% ammonium persulfate solution to reveal the microstructural features of the weld and the heat affected zone.

RESULTS AND DISCUSSION

Y-groove specimens welded with Inconel 82 filler wire revealed significant microcracking, especially in the root region, along the fusion boundary with stainless steel, whereas those welded with 16-8-2 filler wire revealed only a few microcracks. This is in agreement with results reported by King et al. [6]. SMAW deposits made on key-hole specimens machined from Incoloy 800 were found to be free from microcracks irrespective of the welding consumables used. It is important to note that no macrocracks were observed on Y-groove or key-hole specimens. This indicates that hot tearing is unlikely to occur under the welding restraints employed in this work. It should also be pointed out that the specimens welded with 16-8-2 filler wire showed better wetting characteristics compared to that welded with Inconel 82 filler wire. This is illustrated in Fig. 3, which shows a concave profile for the 16-8-2 weld metal and a flat profile for the Inconel 82 weld metal. This suggests that welds made with Inconel 82/182 weld metal may contain lack of fusion type defects. Butt welds made with Inconel 182 as well as 16-8-2 electrodes were, in general, free from weld defects. However, Inconel 182 weld metal occasionally exhibited lack of fusion in the root region. This is consistent with the results obtained from the hot cracking tests.

Figure 4 shows the optical microstructures of the weld interior and the fusion line regions of 16-8-2 and Inconel 182 welds. Inconel 182 weld reveals a fully austenitic microstructure

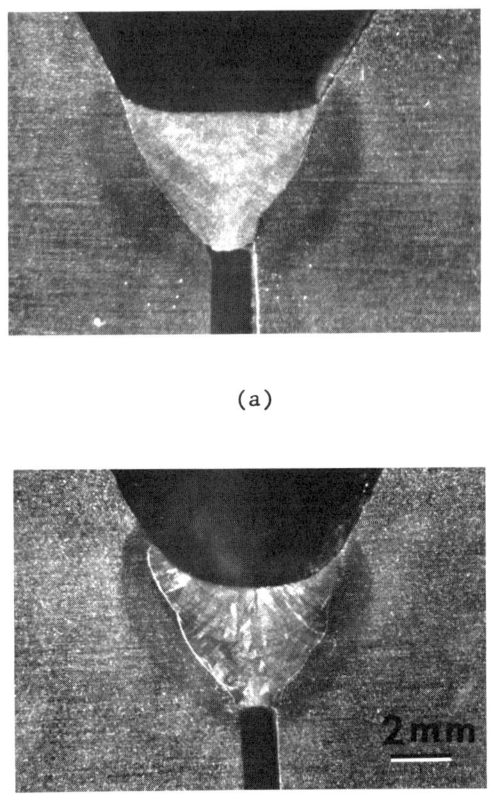

(a)

(b)

Fig. 3 - Macrographs showing the wetting profiles of (a) Inconel 182, and (b) 16-8-2 weld metals.

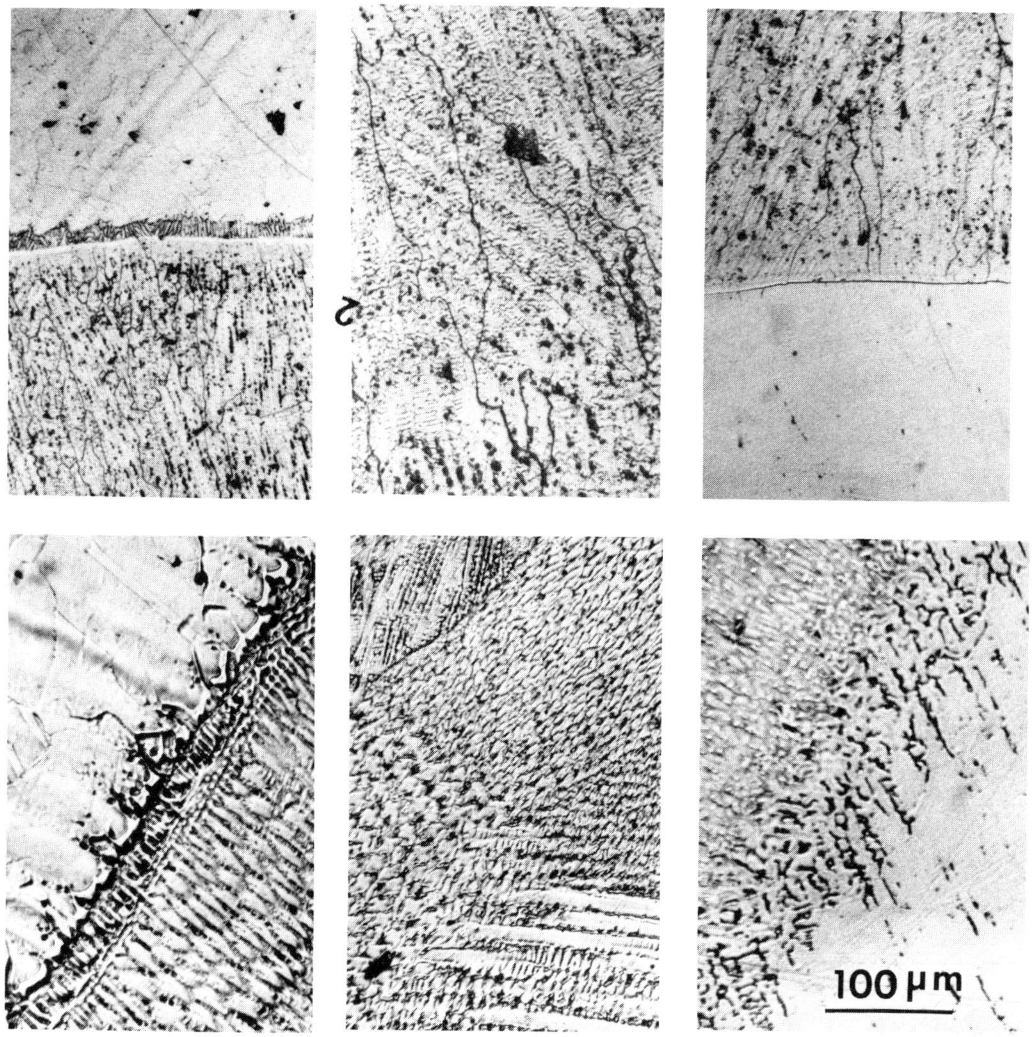

Fig. 4 - Optical microstructures at the centerline and fusion boundaries of (a) Inconel 182, and (b) 16-8-2 weld metal. (Left: Incoloy 800/weld metal; right: weld metal/stainless steel).

with niobium carbide precipitates distributed in the interdendritic regions. In contrast, the 16-8-2 weld reveals a cellular dendritic structure with a small amount of deta-ferrite distributed in the intercellular regions. The ferrite content was found to vary from 1-2%. This can be attributed to the slight increase in nickel content in the weld region close to the Incoloy 800 base metal, presumably due to dilution effects. A narrow zone (~ 50 μm wide) containing solidification microstructure is observed in the Incoloy 800 base metal adjacent to the fusion boundary. This is clearly noticeable at the interface between Incoloy 800 and Inconel 182 weld metal. This type of microstructure results from the localized melting and solidification of Incoloy 800 without any convective mixing with the weld pool. However, no such in situ melted and solidified regions are observed in the stainless steel base metal. This can be rationalized in terms of the differences in the melting points of Incoloy 800 and type 304 stainless steel. The interface between Inconel 182 and stainless steel is very sharp and nearly planar whereas that between 16-8-2 and stainless steel is very ill-defined.

The yield strength of type 304 stainless steel decreased monotonically from 275 MPa to 130 MPa when the test temperature was increased from 25 to 650°C. In contrast, the yield strength of Incoloy 800 decreased steadily with an increase in the test temperature to reach a minimum value of 175 MPa at 500°C and then increased to 200 MPa at 650°C. However, the ultimate tensile strength of both base materials decreased monotonically with an increase in the test temperature.

Figure 5 illustrates the variations in tensile properties of the Inconel 182 and 16-8-2 weld metals as a function of the test temperature. The data for the type 304 stainless steel base material are also included for comparison. The yield and ultimate tensile strength values of the 16-8-2 weld metal at 25°C were significantly higher than those of the Inconel 182 welds.

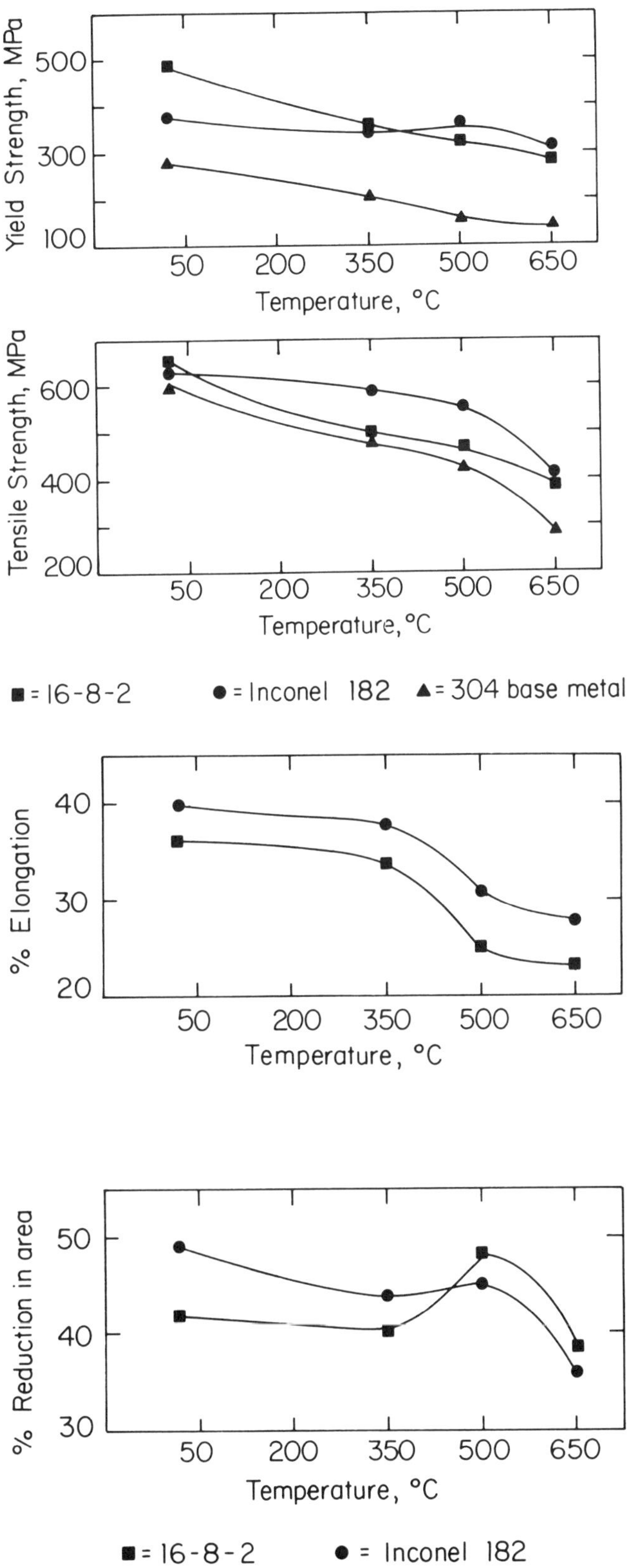

Fig. 5 - Temperature dependence of all weld tensile properties of Inconel 182 and 16-8-2 welds. Data for type 304 stainless steel base metal are also included for comparison.

However, the yield and tensile strength values of 16-8-2 welds exhibited a stronger temperature dependence than those of Inconel 182 welds. The ductility of the 16-8-2 weld metal was found to be marginally less than that of the Inconel 182 weld metal.

The yield and ultimate tensile strength values of the transverse weld specimens decreased monotonically with an increase in the test temperature from 25 to 650°C. None of the transverse specimens failed in the weld metal. This suggests that both Inconel 182 and 16-8-2 weld metals are stronger than the base materials used. While the transverse weld specimens tested at 25°C fractured in the Incoloy 800 side, those tested at 500 and 650°C fractured in the stainless steel side. At 350°C some of the transverse weld specimens fractured in the Incoloy 800 side while the others failed in the stainless steel side. This transition at 350°C of the fracture location in transverse weld specimens, irrespective of the welding consumable used, is consistent with the crossover in tensile strength vs. temperature plots of type 304 stainless steel and Incoloy 800 observed at 350°C.

The impact toughness values obtained from tests conducted at room temperature on subsize charpy V-notch samples are presented in Table 1.

Table 1 - Charpy V-notch Impact Toughness Data

Material	Impact Energy (J)
Incoloy 800	85
Type 304 stainless steel	70
Inconel 182 Weld	65
16-8-2 Weld	60

The corresponding values for Incoloy 800 and type 304 stainless steel base metals are also included in this table for comparison. The Inconel 182 weld metal is marginally tougher than the 16-8-2 weld. It is clear, however, that the impact toughness of both Inconel 182 and 16-8-2 weld metals are lower than those of the base metals.

CONCLUSIONS

Major conclusions from this investigation on the dissimilar metal welding between Incoloy 800 and type 304 stainless steel are:

(1) Inconel 182 weld metal exhibits slightly higher tendency for microfissuring than 16-8-2 weld metal, although the overall hot cracking tendency of both the weld metals is found to be quite low under the weld restraints employed in this work.

(2) Shielded metal arc welding technique can be used to produce radiographic quality welds between Incoloy 800 and type 304 stainless steel, especially with the use of 16-8-2 electrodes.

(3) 16-8-2 welds exhibit higher yield and tensile strength values than Inconel 182 up to about 350°C. At higher temperatures, Inconel 82 welds are stronger than 16-8-2 welds. Both the weld metals are much stronger than both Incoloy 800 and type 304 stainless steel in the temperature range 25-700°C. Room temperature impact toughness values of Inconel 182 welds are marginally higher than those of 16-8-2 welds.

ACKNOWLEDGEMENT

The authors are grateful to Dr. P. Rodriguez, Head, Metallurgy Program, for his constant support and encouragement during the course of this work. The authors have benefited immensely from the discussions they have had with Mr. S. C. Chetal, Reactor Group and would like to thank him sincerely. They also wish to express their appreciation to Messrs. A. K. Bhaduri and S. Venugopal for their kind help and cooperation in the experimental program. One of the authors (V.S.) wishes to acknowledge the support provided by Ames Laboratory, Ames, Iowa USA which is operated for the USDOE under contract no. W-7405-Eng-82, during the preparation of this manuscript.

REFERENCES

1. Roberts, D. I., R. H. Ryder and R. Viswanathan, J. Pressure Vessel Tech. 107, 247-254 (1985).
2. Emerson, R. W., R. W. Jackson and C. A. Dauber, Weld. Journal 41, 385s-393s (1962).
3. Tucker, J. T. and F. Eberle, Weld. Journal 35, 529s-540s (1956).
4. Klueh, R. L. and J. F. King, Weld. Journal 61, 302s-311s (1982).
5. Campbell, G. M., J. W. Elmer, W. S. Gibbs, D. K. Matlock and D. L. Olson, "Trends in Welding Research in the United States, "p. 443-470, S. A. David (Ed.), American Society of Metals, Metals Park, Ohio (1982).
6. King, J. F., M. D. Sullivan and G. M. Slaughter, Weld. Journal 56, 354s-358s (1977).
7. Elmer, J. W., D. L. Olson and D. K. Matlock, Weld. Journal 61, 293s-301s (1982).
8. DeLong, W. T., G. A. Ostrom and E. R. Szumachowski, Weld. Journal 35, 521s-528s (1956).
9. Dalcher, A. W., T. M. Yang and C. L. Chu, J. Eng. Mater. Tech. 99, 65-69 (1977).

PROCESS CONTROL IN GMA WELD OVERLAY CLADDING OF Fe-Cr-Al ALLOYS

Robert H. Frost, Tim R. Vaughan
Department of Metallurgical Engineering
Colorado School of Mines
Golden, Colorado

ABSTRACT

Experimental gas metal arc welds were made with mild steel and iron-chromium-aluminum alloys containing 15 to 17 wt%. chromium and 4 to 6 wt.% aluminum. The objective was to eliminate weld cracking in Fe-Cr-Al weld overlay cladding used for corrosion protection applications. Additional objectives were to investigate the functional relations between welding process variables and to develop process control nomograms.

The results showed the welding potential to be a linear function of the welding current, the current to be a linear function of the electrode velocity, and the slope of the current-electrode velocity plot to decrease with increasing electrical resistivity of the alloy. Nomograms were developed for process control in potential-current coordinates or in voltage tap-electrode velocity coordinates.

Weld cracking was found to be aggravated by the large grain size caused by epitaxial growth and by the segregation of aluminum to the columnar grain boundaries. The use of a high nitrogen cover gas was found to provide aluminum nitride precipitate particles which pin grain boundaries in the heat affected zone to prevent grain growth, and provide heterogeneous nucleation sites for refinement of the solidification structure.

THE USE OF SPECIALIZED ALLOYS for structural, electronic, abrasion resistant, and corrosion resistant applications makes it important to understand and control welding processes parameters. The heat input and deposition rate determine the microstructure and mechanical properties of the weld deposit. They are dependent variables controlled by the welding potential, wire speed, current, and weld travel speed. For a given electrode and weld geometry the independent process variables include: the open circuit potential (V_0), the electrode velocity (W), and the weld travel speed (S). The dependent variables include: the welding potential (V), the welding current (I), the deposition ratio (D), and the heat input (H).

Conventional Gas Metal Arc power supplies operate at constant potential to maintain a constant arc length. The welding current is controlled by the electrode velocity. and it influences a number of process characteristics such as the metal transfer mode, penetration characteristics, wire melting rate, and arc stability, etc.(1).

The objectives of this research were to investigate the influence of GMA welding variables on weld structure and morphology, and to develop a series of nomograms for process control. These nomograms are designed to provide a map of process space which can be used to locate parameters that were to be used in a surface cladding project.

The application of abrasion or corrosion resistant clad layers requires close control of process variables. In the present study the heat input was varied to control cracking in ferritic stainless steel weld deposits. The alloys investigated had compositions in the range of 15-17 percent chromium and 4-6 percent aluminum. They show good resistance to sulfidation and corrosion in coal processing environments; however, their strength is low for structural applications, particularly at elevated temperatures (2,3). The FeCrAl clad layers were deposited on a higher strength 9 percent molybdenum ferritic steel substrate to provide high temperature creep resistance.

Susceptibility of weld metal to solidification cracking appears to be dependent of three main factors: (1) the coarseness of the solidified microstructure, (2) the degree of alloy and tramp element segregation, (3) the geometry of the joint (4,5,6). Epitaxial solidification causes the coarseness of the weld deposit microstructure to be inherited from the grain growth region of the heat affected zone. Thus, higher energy input to the weld causes more grain growth in the heat affected zone, and a coarse microstructure in the solidified weld deposit (7,8).

If the alloy freezes over a large temperature range, liquid may be present along the center line when the remainder of the weld becomes stressed as the weld cools. This leads to hot cracking or hot tearing along the center line of the weld. Liquid films are present in most alloys, but they are not all equally susceptible to hot cracking. The existence of liquid films is a necessary, but not a sufficient condition for hot cracking (4,5).

Ductility and toughness may be increased and cracking minimized by preheating the parent plates before welding. The benefits from preheating are obtained by reducing the thermal gradient and hence the stresses between the weld metal and the base plate, and by reducing the yield strength of the base plate when preheated above 600°F. The reduction of base plate yield strength is beneficial because the maximum stress on the weld bead is decreased to this level (6,9,10).

Refinement of the fusion zone grain structure provides increased resistance to hot cracking. This may be achieved by: mechanical or electrical stirring of the weld pool, by the addition of nucleation agents or inoculants, or by process transients which increase the degree of under cooling and help to nucleate new grains. Grain refiners are added to provide high melting intermetallics or nonmetallics (i.e. oxides, carbides, etc.) which cause the nucleation of new grains. Aluminum, vanadium, titanium, zirconium, and nitrogen act as grain refiners in carbon and low alloy steels. Grain refiners are important to weldability, as decreased columnar grain size improves the ductility and minimizes the tendency for hot-short cracking.

EXPERIMENTAL PROCEDURE

Alloys of compositions: 15%Cr-4%Al, 15%Cr-6%Al, 17%Cr-4%Al, and 17%Cr-6%Al were provided by the Lockheed Corporation in the form of extruded 0.50-in diameter bar as well as 0.30-in and 0.25-thick plate. Welding wire of the four experimental FeCrAl alloy compositions was produced by hot rolling, cold rolling, and swaging to a final wire diameter of 0.0625 inch (1.59 mm).

A gas metal arc welding process was used with a Hobart RC 500 limited droop DC power supply, a 0.0625 inch (1.59 mm) diameter steel or FeCrAl alloy electrode, and a helium or argon cover gas. The welding current and potential were monitored using a Hewlett Packard 7100 BM strip chart recorder. The initial welding experiments were conducted with mild steel in order to conserve the limited supply of experimental material provided, as well as to evaluate the nature and ranges of process variables.

A set of 45 welds was made using 0.0625-in mild steel wire on 0.5-in mild steel plate using an argon cover gas, and another set of 24 welds was made using helium as a cover gas. Emphasis for the FeCrAl alloys was placed on producing crack free welds from the experimental materials provided. A set of 19 FeCrAl welds was made on plate of the same composition as the electrode. Ten of the welds used helium as a cover gas, three used an argon cover gas and preheat temperatures between 304°C and 445°C, and six used a mixed argon-nitrogen cover gas to add nitrogen to the metal for grain refinement.

The metallographic characterization of the welds included: optical metallography, transmission electron microscopy, and scanning electron microscopy. The macrostructures of the carbon steel weldments were examined on the transverse section after grinding through 600 grit paper and etching with 2% Nital. The microstructures of the stainless steel weldments were examined after polishing through 0.05 micron alumina, removing scratches with colloidal silica, and etching with a solution of 15% hydrochloric acid in methanol.

Transmission electron microscopic studies were used to evaluate the nonmetallic particles in the weld metal, particularly in the welds which were produced using nitrogen gas to introduce nitrogen to the weld for grain refinement purposes. Several micrographs and diffraction patterns were obtained from the area around the perforations in order to identify the nonmetallic particles.

The size, shape, and distribution of the particles were investigated by scanning electron microscopy using the same specimens which were prepared for optical microscopy. In addition to information on the nature and distribution of inclusions, line scans were performed to provide information about the degree of segregation, the alloy element segregated, and the composition of the particles. Additional information on the total nitrogen content of the welds was provided using a LECO TC-136 nitrogen/oxygen analyzer.

RESULTS AND DISCUSSION

The objectives of the welding experiments were to investigate the influence of process variables on weld bead cracking, and to develop process control models which can be used to control the size of the weld bead and the heat input per unit length of weld. For a given weld geometry and electrode the independent variables include: the open circuit potential (V_o) which is controlled by the voltage tap; the cross sectional area of the electrode (A_w), the electrode velocity (W); the weld travel speed (S), and the gap between the contact tube and the work. For the present experiments the electrode cross section and the gap between the contact tip and the work were held constant. The dependent variables include the welding potential (V), the current (I), the deposition ratio (D), and the heat input (H). Figure 1 shows a schematic of the GMA welding process.

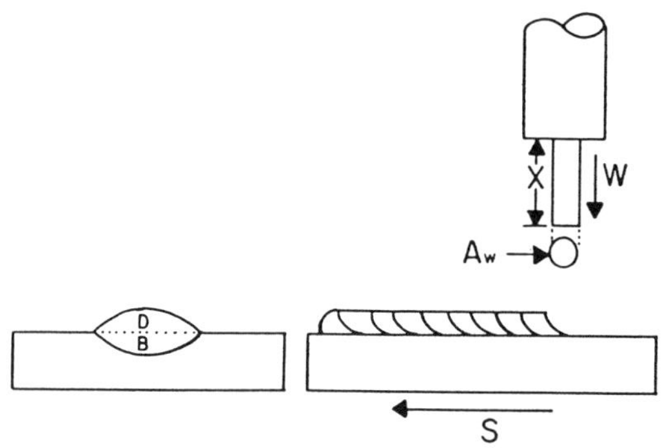

Figure 1 - Schematic of the Gas Metal Arc Welding Process.

The welding potential (V) is a linear function of the welding current controlled by the machine characteristic or droop of the power supply:

$$V = V_o - c(I) \qquad (1)$$

where: V_o = open circuit potential
c = is a machine constant

Figure 2 shows the experimentally determined machine characteristic curves for voltage dial settings from 3 to 6. The data for mild steel

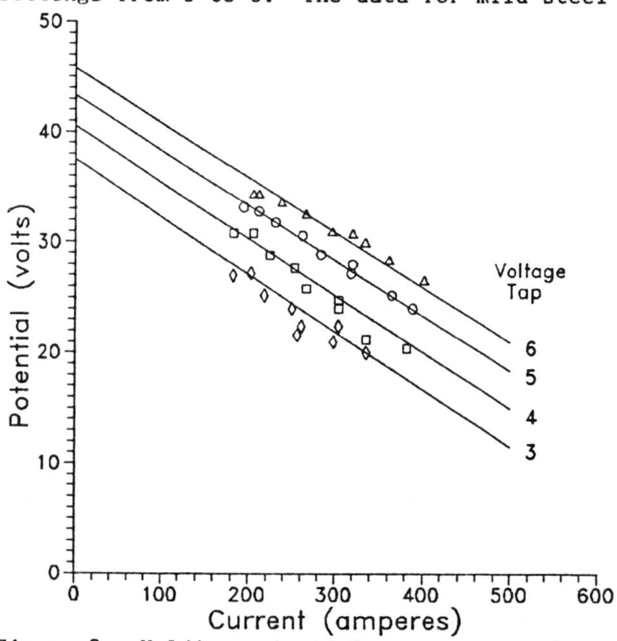

Figure 2 - Welding potential versus current as a function of the voltage tap setting.

welds shows that the linear potential or the intercept at zero current is the open circuit potential (V_o). The open circuit potential for these welds ranges between 37 to 47 volts, and the slope of the machine characteristic represented by the constant (c) in equation 1 was -0.05 volts per ampere.

The welding current (I) is a linear function of the electrode velocity:

$$I = a + b(W) \qquad (2)$$

where: a and b are constants
Figure 3 illustrates the linear dependency of current on electrode velocity for mild steel and FeCrAl alloy electrodes. The difference in the slopes for the two alloys is caused by the difference in electrical resistivity.

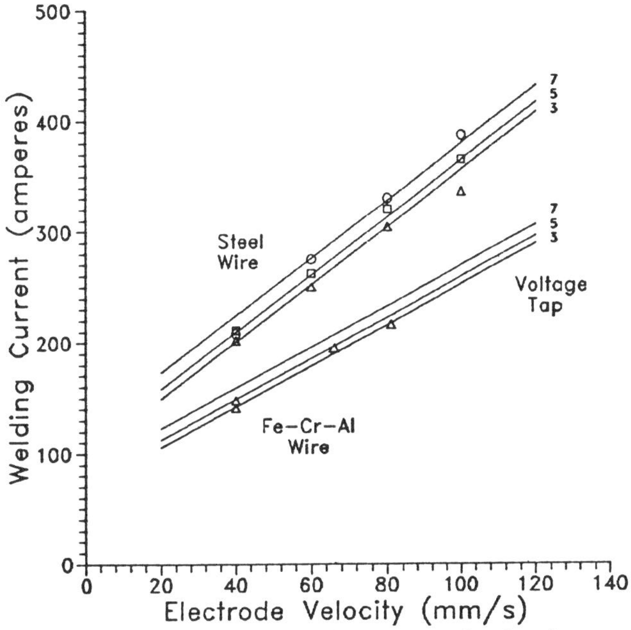

Figure 3 - Welding current versus electrode velocity for mild steel and alloy electrodes as a function of the voltage tap setting.

The deposition ratio (D) is defined as the product of the electrode cross-sectional area (A) and the ratio of the electrode velocity to the weld travel speed. It has units of area (cm^2) and it is proportional to the cross sectional of the deposited metal:

$$D = (W/S)A \qquad (3)$$

The heat input per unit length of weld (H) is the product of the current and potential divided by the weld travel speed:

$$H = VI/S \qquad (4)$$

The heat input controls the superheat in the deposited weld metal, the degree of melting of the base plate and the solidification rate of the weld bead.

This increased resistivity results in a higher electrode velocity for the FeCrAl electrode than for that of the mild steel electrode at constant current. The higher electrode velocity may also be interpreted as an increase in the respective melt rates, due to joule (I^2R) heating of the electrode. This joule heating is taking place in the stickout portion of the electrode, located between the contact tip and the arc, indicated by (x) in Figure 1. The data in Figure 3 shows that the linear current-electrode velocity relation shifts to increasing current with an increase in the voltage tap setting.

The functional relations expressed by equations 1 through 4 can be used to provide an expression for the electrode velocity as a function of the deposition ratio, the heat input, and the constants in equations 1 and 2. Substituting equations 1,2, and 3 into equation 4 the electrode velocity is given by;

$$W^2 + \{(HA/Cb^2D) + Z\}W + Y = 0 \qquad (5)$$

where: $$Z = (2a/b) - (V_0/cb) \qquad (6)$$

$$Y = \{a - (Vo/c)\} + a/b^2 \qquad (7)$$

The electrode velocity determined using equation 5 can be used in equations 1,2, and 3 to find the potential, the welding current, and the weld travel speed required to produce a given deposition ratio and heat input.

Figure 4 shows a welding process control nomogram in potential-current coordinates which is based on the functional relations in equations 1 through 5 and on the experimental data in Figures 2 and 3. The nomogram is for FeCrAl alloy welds with a constant deposition ratio of 0.05 cm. Machine characteristic lines are drawn for voltage tap settings from 2 through 6, and constant heat input contours are drawn for heat input levels from 0.1 to 0.6 MJ/M. An increase in the deposition ratio or size of weld bead would cause the heat input contours to shift to higher currents.

Figure 4 can be used for welding current and welding potential. For operational purposes, however, a control nomogram based on the voltage tap setting (V*) and the wire electrode velocity (W) is more convenient. Figure 5 shows the process control nomogram in voltage tap - electrode velocity coordinates for mild steel welds produced at a constant deposition ratio of 5 mm^2. Constant heat input contours are plotted for heat input levels from .1 to .6 MJ/M. The data points show the locations of welds for which the power was too low, the electrode velocity was too high, and the electrode rammed the parent plate.

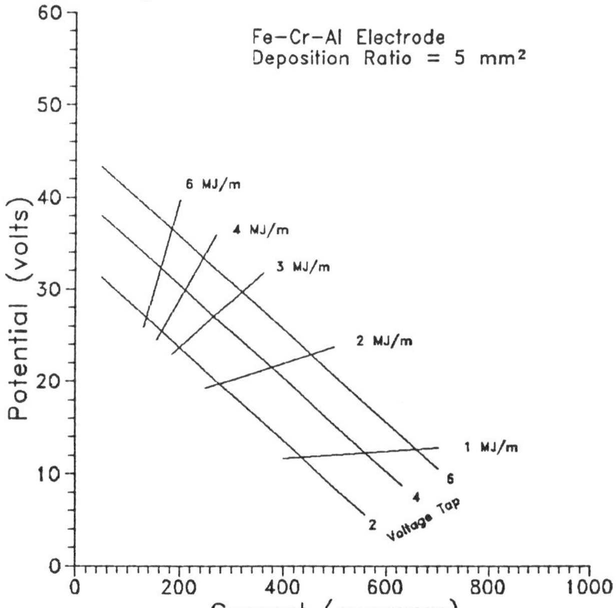

Figure 4 - Welding process control nomogram in potential-current coordinates.

The data in Figure 5 indicates that for steel wire at a deposition ratio of 5 mm^2 a minimum heat input level of about .25 MJ/M is required to produce a successful weld. Other welding problems can also limit the range of process variables for successful welding. At high potentials and low electrode velocities the heat input and parent plate melting become excessive, the arc length may also increase to such a point that the electrode fuses to the copper guide tube. The best welds in Figure 5 were produced at heat inputs of .25 to .4MJ/M and voltage tap settings between 3 and 6.

The process control nomogram for FeCrAl alloy electrodes is shown in Figure 6. This high electrical resistivity of the FeCrAl alloys shifts the constant heat input contours to lower electrode velocities. The area enclosed by the dashed line indicates the maximum possible variation in process variable settings: however, successful welds can not be produced over this entire range. At heat input levels less than .2 MJ/m the welds become intermittent because the electrode rams the base plate. At heat input levels of .4 MJ/m or greater the excessive parent plate melting, under cutting, and weld bead cracking were observed.

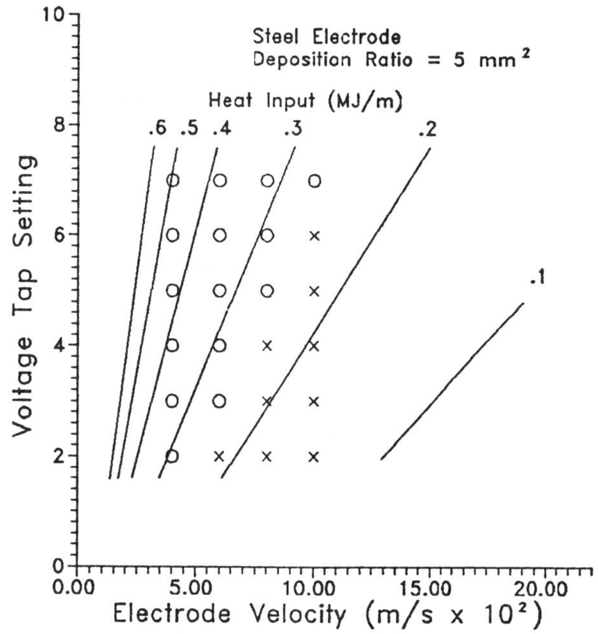

Figure 5 - Acceptable (O) and unacceptable (X) welds as a function of heat input in voltage tap setting - electrode velocity coordinates for mild steel welds.

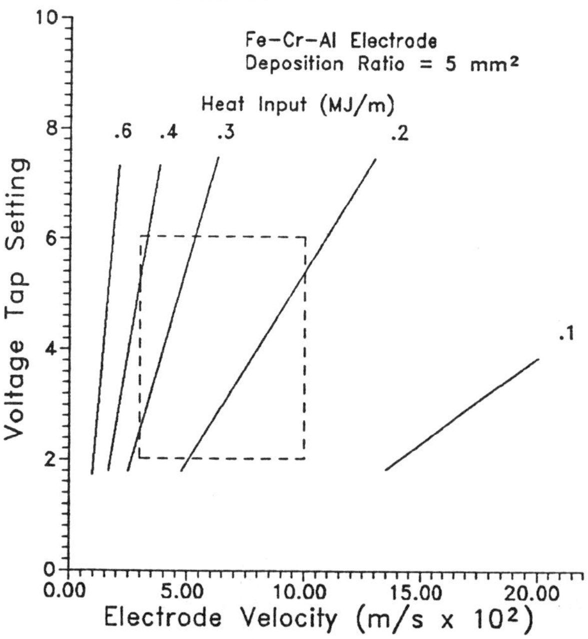

Figure 6 - Process control nomogram in voltage tap - electrode velocity coordinates for Fe-Cr-Al alloy welds.

The experimental FeCrAl alloy welds were produced at heat input levels of .23 and .4 MJ/m. The .2 MJ/m welds were produced at an electrode velocity at about .0657 M/s, and a voltage tap setting of 3 which provided welding potentials of around 26.0 volts. The welds with heat input levels of .3 and .4 MJ/m were produced at electrode velocities of .048 and .030 M/s, and at a voltage tap setting of 5.0 which provided welding potentials from 34 to 36 volts. The initial welds at .4 MJ/m produced using 15%Cr - 4%Al wire showed substantially greater weld cracking at a high heat input.

The degree of cracking is related to the heat input through grain growth and segregation effects. The higher heat input causes more grain growth in the heat affected zone and larger epitaxial columnar dendrites in the fusion zone of the weld. This reduces the surface area of the columnar grain boundaries and increases the segregation.

The scanning electron microscope was used in an attempt to understand the mechanism responsible for cracking in these alloys. An Energy Dispersive Spectral Analysis (E.D.S.) was performed on the cell interior as well as on the cell boundary of the final solidified structure of Weld No. 15-4-4. This analysis revealed a significant amount of aluminum segregation occurred at the cell boundary. The concentration of aluminum in the cell interior was found to be 5.83 wt.% while the concentration on the boundary was significantly higher at 23.98 wt.%. An (E.D.S.) line scan analysis across a typical crack in Weld No. 15-4-4 indicated a high aluminum content at the crack surface similar to that observed at columnar grain boundaries. This segregation resulted in the formation of a low melting constituent at the columnar grain boundaries. The presence of a liquid film leads to hot cracking when thermal stresses develop. The remaining welds were produced at lower heat input levels of .2 to .3 MJ/m to minimize weld cracking.

Grain refiners generally provide high melting point intermetallics or nonmetallics which act as heterogeneous nucleation sites. Weld number 15-4-7 was produced using a 100% nitrogen cover gas in order to control the grain size through aluminum nitride precipitation. The result of nitrogen additions to the weld metal can be shown by a comparison of the fusion zone structures of welds produced with and without nitrogen additions to the cover gas. Figure 7 shows a micrograph of Weld No. 15-4-4 which was produced with a helium cover gas, and Figure 8 shows a micrograph of Weld No. 15-4-7 which was produced under a nitrogen cover gas:

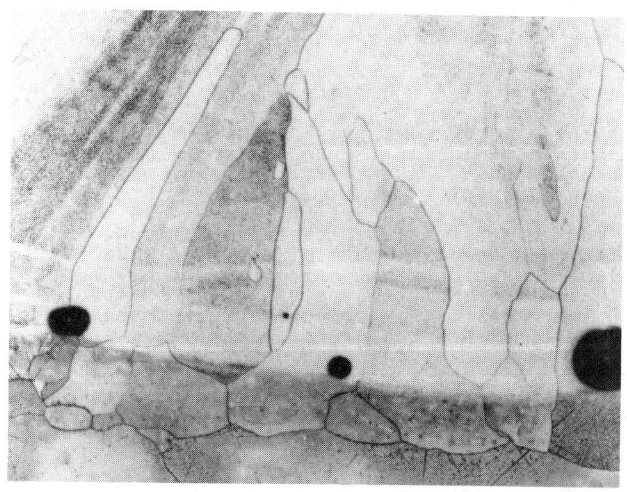

Figure 7 - Epitaxial grain growth and large columnar grains in a Fe-Cr-Al Weld No. 15-4-4 produced using a helium cover gas (45X).

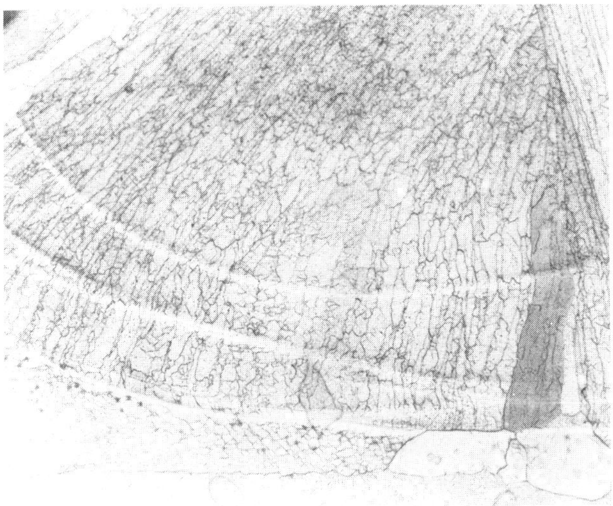

Figure 8 - Grain refinement in Weldment No. 15-4-7 produced with a 100% nitrogen cover glass (45X).

A metallographic examination of the welds produced under helium and nitrogen cover gasses showed that the grain size, as measured by the Mean Linear Intercept (L_3), is refined in both the fusion and heat affected zones of the welds produced using the nitrogen cover gas. Weld 15-4-4, produced under a helium cover gas, showed a mean linear intercept in the fusion zone of $L_3 = 1.266 \times 10^{-2}$ mm. and an L_3 in the heat affected zone of 1.73×10^{-2} mm. Weld 15-4-7, produced under a nitrogen cover gas, showed a fusion zone $L_3 = 3.33 \times 10^{-3}$ mm., and an $L_3 = 2.38 \times 10^{-3}$ mm. for the heat affected zone. These results indicate that the nitrogen diffused into the heat affected zone to produce AlN particles that pinned the grain boundaries and minimized heat affected zone grain growth. The refined solidification structure in the fusion zone is the result of: (1) a refined epitaxial grain size caused by the fine grain size in the heat affected zone, and (2) increased nucleation during solidification caused by the presence of heterogeneous nuclei.

Examination of weld 15-4-7 using scanning and transmission electron microscopy indicated that small particles had precipitated out in the matrix of the weld metal. Energy dispersive analysis (E.D.S.) of the particles showed a high aluminum content of 81.07 wt.% Al. Thin foils examined by transmission microscopy showed the particles to be hexagonal aluminum nitride.

Additional welds were produced using mixtures of argon and nitrogen ranging in composition from 50% Ar - 50% N_2 to 93.75% Ar - 6.25% N_2. Significant refinement of the fusion zone grain structure occurred only in the weld produced using the 50% gas mixture. The refinement in this weld took on the appearance of a cellular type of sub- structure, but the degree of refinement was not as pronounced as that in the weld produced using a 100% N_2 cover gas.

CONCLUSIONS

1. For the gas metal arc welding process the welding potential is a linear function of the welding current, and the welding current is a linear function of the electrode velocity. The slope of the linear current-electrode velocity function decreases with increasing electrical resistivity of the electrode alloy.
2. A set of simple nomograms can be developed for control of the gas metal arc welding process to produce desired deposition ratios and heat input levels. These nomograms can be plotted in potential-current coordinates, or in coordinates based on the independent process variables such as the voltage tap setting and the electrode velocity controls on the power supply.
3. The electrical resistivity of high alloy electrodes for corrosion or abrasion resistant cladding applications can make a substantial difference in the levels of process variables observed during gas metal arc welding.
4. The coarse columnar grain size in the Fe-Cr-Al alloys used in this investigation is closely related to the grain size in the grain growth region of the heat affected zone through the mechanism of epitaxial growth. The grain size in the fusion zone can be significantly reduced by the addition of nitrogen to the cover gas and the formation of aluminum nitride inclusions.

ACKNOWLEDGEMENT

The authors gratefully acknowledge the support of the U.S. Department of Energy and the Lockheed Corporation for this research.

REFERENCES

1. Essers, W.G., and Walter, R., Conference Proceedings on Arc Physics and Weld Pool Behavior, sponsored by The Welding Institute, London, Abington Hall, Abington Cambridge, pp 289-303, 1979.

2. Perkins, R.A., and Nagarajan, V., "Sulfidation Resistant Coatings for Coal Gasification Process Equipment," Lockheed Missiles and Space Co., Palo Alto, CA, 1985.

3. Gordon, B.A., and Nagarajan, V., Oxidation of Metals, Vol. 12, No. 4, pp. 313-322, 1978.

4. Flemings, M.C., "Solidification Processing", McGraw Hill, 1974.

5. Baker, R.G., British Welding Journal, pp. 288-291, June, 1968.

6. Kah, D.H., and Dickenson, D.W., Welding Journal Research Supplement, pp. 135-142-s, 1981.

7. Easterling, K.E., "Introduction to the Physical Metallurgy of Welding", Butterworth, London, 1983.

8. Lancaster, J.F., "Metallurgy of Welding", Third Edition, George Allen & Unwin, London, 1980.

9. Savage, W.F., "Solidification, Segregation and Weld Imperfections," Houdremont Lecture, Welding in the World, Vol. 18, No. 5, 1980.

10. "Fundamentals of Welding", Welding Handbook, Volume 1, Seventh Edition, American Welding Society, 1976.

WELDING EQUIPMENT FOR COMPLEX ARC WELDING PROCESSES

Howard Cary, Steve Barhorst
Hobart Brothers Company
Troy, Ohio

WELDING EQUIPMENT FOR COMPLEX ARC WELDING PROCESSES

By

Steve Barhorst & Howard Cary

HOBART BROTHERS COMPANY

Troy, Ohio

The metalworking industry is currently undergoing a revolution based on advances in automation, communication and improved processing. All of this activity is based on the anticipation for the factory of the future. In this way it is hoped to reduce manufacturing costs by incorporating computer aided design and computer integrated manufacturing at all levels. The welding equipment industry is responding to these requirements by providing three basic things. Improving welding productivity, improving weld quality, and removing the human welder from the immediate area of the welding arc. All Three of these objectives are being met by changing the method of application of welding. This means moving from the manual and semi-automatic method of application to mechanized welding methods; including machine welding, automatic welding and automated welding.

For many years the metalworking industry accepted the fact that there were four ways to arc weld. That is four methods of applying welds which were manual welding, semi-automatic welding, machine welding and automatic welding. These can be considered the man machine relationship in making arc welds.

These are graphically shown along with the activities involved by Figure 1.

We believe that the welding user industry wants three things:

Improved Welding Productivity
Improved Weld Quality
A desire to remove the human welder from the immediate area of the welding arc.

It is possible that all three of these objective can be met by changing the method application of arc welding. This means moving from the manual and semi automatic method of application to mechanized welding methods, including machine welding, automatic welding and automated welding.

For many years industry accepted the fact that there were four ways to weld, that is four methods of applying welds, these were:

Manual welding
Semi Automatic welding
Machine welding
Automatic welding

These are graphically shown along with the activities involved by Figure 1. Automatic welding has most of the advantages mentioned above related to welder/user requirements and overcomes many of the problems associated with the individually controlled manual, semi-automatic, and even machine welding. Automatic welding has two basic types: Standardized equipment for specific types of welds or joints and dedicated equipment for specific welded products. By using automatic welding systems, higher quality welds and improved productivity will result, however, automatic equipment is more expensive and must be kept busy to become economically feasible.

To further improve quality and productivity and remove the welder from the arc area we must consider a new development known as the automated method of applying welding. It is the method of applications that is more advanced than automatic welding and is shown on the right hand column of Figure 1. Automated welding involves complete control of the welding operation, including accommodation for poorly fitted joints and the ability to modify taught travel patterns. The difference between automated welding and automatic welding is the use of programmable controllers with memory plus sensing devices for feed forward and feed back to adaptive controls. These devices along with the proper arc manipulator system and work motion system will produce high quality welds even when the fit up and accuracy of the weldment is poor.

An arc welding robot can be either an automatic or an automated welding system. It is reprogrammable and can weld many different items in the course of a day. A robotic arc welding system is an automatic welding system normally, however, if sensors and an adaptive controller is included it becomes an automated welding system.

Automated welding is defined as welding equipment that performs the entire welding operation without human observation or intervention. Required adjustments are determined by sensing devices which monitor the arc and feed information to the control system. The adaptive controller modifies welding parameters and motion patterns to insure a quality weld.

The control to accommodate changes is shown as closed loop adaptive control. Closed loop means that the value of an element is reported back to the controller so that corrective action can be taken if required. Adaptive control requires sensing devices and computerized controls so that the value of an element can be altered by information furnished by the sensing device. The brain of the human welder provides adaptive control, thus manual and semi-automatic and even machine welding are closed loop systems. Automatic welding does not have this feed back feature and is an open loop system. Automated welding is a closed loop system, but the degree of control depends on the number and type of sensors involved.

The difference between automatic welding and automated welding needs to be thoroughly understood. In automatic welding the apparatus will be programmed to provide the exact taught motion pattern and the weld parameters that are to be used. In many cases the weldment is straight forward and the parts are sufficiently accurate so that changes in the welding conditions or taught motion pattern is not required. Quality welds will result each and every time and the inherent tolerance of the welding process will accommodate minor variations. If the fit up or joint location is beyond minor variations, a defective weld may result.

In automated welding the equipment will provide a quality weld even if there are deviations such as inaccurate piece part preparation or inaccurate location of the joints. This will be done without human operation, supervision, or intervention, it is a closed loop system. There may be however, limits to the amount or degree of inaccuracies that can be accommodated.

There are differences in the degree of automation that are included in the welding equipment. This relates to the number and type of the sensors employed to monitor specific functions. Sensing devices and adaptive controls are expensive and at this time only a few real-time sensors are perfected for shop production use. Automated welding will become more widely used in the future, however, today automatic welding with close control of fit-up will provide high quality welds.

The welding industry is reacting to the needs of higher quality, greater productivity, and removing the welder from the arc area in several different ways. It is obvious that more control must be placed over the welding process. The technology of welding is advancing rapidly and is profiting from the advances in electronics, power solid state device improvements, better control systems and the computer. All of these factors help improve the technology of welding to provide more precise equipment which in turn allows more precise deposition of weld metal which enhances the quality of the weld. The principal advances are being made in the area of the continuous wire systems including gas metal arc (MIG) and flux cored arc welding and the gas tungsten arc (TIG) and plasma arc welding processes. These four welding processes are being refined and equipment upgraded to improve performance primarily in the mechanized welding systems.

The heart of a mechanized welding system designed for high quality welding and high productivity is a welding program. Figure 2 shows a welding program for Gas Tungsten Arc welding. It is a sequence of instructions to be executed by a controller to control the procedure. One of the early uses of a program of this type was the orbital type welding heads for tube-to-tube sheet welding. Mechanization of Gas Tungsten Arc welding used for orbital or rotational type heads for joining tubes to a tube sheet is shown by Figure 3. By observing manual welding it was possible to determine the parameters required to make a successful weld and it was soon learned that some of the parameters needed to be changed during the welding operation. Thus the program shown in Figure No. 2 was developed and shows that the welding current changes based on time. Programs such as this were established for the more simple welds which would be done by merely a stop and start, but would include auxiliary functions such as shielding gas flow, cooling water flow, high frequency current, and motion as required, including filler wire feed. More complex programs would include regulation and change of welding current based on the needs of the particular job.

Welding power sources were then developed that included controllers which could be programmed in accordance to the need of a specific welding application. These machines included solenoid valves for shielding gas and cooling water and also high frequency current superimposed on welding current to aid in arc starting. A machine including these features is shown by Figure 4. The one parameter that is changed continuously is the welding current and it was found that this is needed to control the weld penetration and molten weld pool while rotating around the tubing.

It was found that parts being welded may have a variation in material thickness. The procedure required a change in welding current and the controller would provide this feature. Current reduction is normally required as heat builds up in the weldment. This is obtained by the downslope control which reduces the current as required when joining metals that cannot tolerate too much heat input.

The termination of the weld requires a special attention to avoid the crater. This is done with the program and the controller by continuing the filler metal wire feed after the travel motion is stopped. This is known as the filler metal feed stop delay time. During this period the welding current is gradually decreased to a minimum level at which time the filler wire feed and current are both terminated. Different levels of complexity of controllers are provided in order to provide different welding programs. Different welding controllers/programmers are shown in Figure 5. These give the operation a wide variety of possibilities in establishing welding procedures. They are straight forward and are of the open loop design which means there is no feedback signal to indicated that the selected values were provided.

Programmed, mechanized Gas Tungsten Arc welding equipment provided reproducible high quality welds on joints where parts preparation and joint is uniform. As programmed welding became more popular there was a need to improve its tolerance for misaligned weld joints, for improper weld preparation, for welding thick to thin materials, for welding dissimilar metals, and for other situations where conditions were not continuously uniform. Low frequency pulsing of the welding current added a degree of tolerance to the welding process making these more difficult welds possible with quality deposits.

In pulsed current Gas Tungsten Arc Welding the welding current continuously changes between levels as shown by Figure 6. During the high pulsed time the high current creates deep penetration and increased molten pool size. During the low pulse time, cooling occurs and some solidification takes place so that the molten metal remains under control. The welding machine controller automatically switches from low to high current and holds each value for a specific time. The pulsed current method actually makes the weld seam of overlapping arc spot welds. Torch movement is done during the low pulsed current time.

Pulsed current Gas Tungsten Arc Welding is also used for manual welding. The welder soon learns the rhythm of the pulsing and is able to produce higher quality welds than when using conventional, steady current. There are several other advantages of pulsed current welding. It reduces distortion and warpage when welding on thinner metals. It allows making welds on misaligned joints and allows the welding of light to heavy sections. It helps overcome the variations in heat sink due to variations in joint preparation and tooling conductivity. Out of position welding is also easier with pulsed current welding since the molten pool size is restricted and the effects of gravity thus less severe. Metallurgical improvements of a deposit weld metal may also occur because of pulsed current welding. This relates to weld strength, grain size and reduces cracking tendencies of crack sensitive materials. Combining pulsed current welding with a mechanized programmed controller system improves the quality and reproducability of welds made with mechanized equipment. It allows the welding of thinner materials and also improves quality when fit-up is less than optimum. It provides better control of root penetration and the ability to weld some materials heretofore considered difficult. It eliminates the need to greatly change welding parameters while rotating around the tubing. The power source and controller once properly set up can be operated with minimum attention by the welder. It is truly automatic welding controlled by a start button.

The orbital tube welding heads are successfully employed to product high quality welds on tubing and pipe of carbon steels, low alloy steels, stainless steels, nickel alloys, copper alloys, titanium alloys and other special materials. The successful welding of the aluminum alloys, however, did not become immediately practical.

Generally, welding of aluminum piping systems was done with the gas tungsten arc welding process using AC power. Superimposed high frequency current was used for manual welding of aluminum. However, the superimposed high frequency when used with the orbital welding head created many problems. It tended to cause insulation breakdowns in the miniature drive motors, and caused burn-outs of delicate electronic control circuits.

Aluminum can be welded using direct current with the electrode negative however, the base metal must be well cleaned and free from surface oxide. This oxide coating quickly reforms on the clean aluminum surface so that weld areas must be quickly welded if a quality x-ray clean weld is to be made. It was not, however, an acceptable solution for high volume construction or manufacturing. Cathode etching as a cleaning method has been utilized for TIG welding of aluminum piping. It is done by passing the positive electrode arc over the welding surface. This combined with a new square wave form power supply made orbital welding of aluminum pipe practical.

The square wave waveform eliminates the need for high frequency superimposed current except for very low current applications and for arc starting. The waveform is such that as the voltage passes through zero quickly that arc reignition is instantaneous and rectification does not occur. In addition, the amount of positive and negative half cycles can be varied within limits to aid in welding aluminum as shown by Figure 7. The square wave AC/DC Cyber-Wave machine is shown by Figure 8.

When an extremely high quality deposit is required a different type of power source is used. This is known as a "Variable Polarity" or VP machine and is shown by Figure 9. This machine is actually two machines in one. The amplitude and time of each half cycle can be set. The optimum waveform for high quality aluminum welding is shown by Figure 10. Water clear x-rays of welds in aluminum are obtained when using the variable polarity power source. The VP machine is also used for plasma welding. Keyhole welding of aluminum is possible with the variable polarity power source.

The plasma arc welding process will utilize many of the same programs and controls required for the Gas Tungsten Arc Welding process, Plasma however, requires additional controls since more variables are involved. In addition, with the keyhole plasma technique, special controls are required tc close the keyhole. Plasma provide more energy and a stiffer arc that can be better directed than Gas Tungsten Arc. Keyhole welds in aluminum were made that reduce the overall joint welding time, produces a cleaner deposit and reduces angular distortion than a multipass Gas Tungsten Arc weld. Figure 11 shows a weld made in 13mm thick aluminum in one pass. With Gas Tungsten Arc welding, three passes were required. This procedure has been employed by the National Aeronautical and Space Administration for use on the longitudinal seams of the external fuel tank used for the space shuttle program. Variable polarity plasma aluminum is also finding applications where extremely high quality welds are required.

Conventional welding power sources have served industry well for many years. Automatic applications and specific processes have demanded the development of increasing complex welding power sources and control systems. An analysis of the requirements of various applications shows that they can be grouped as types or families. Each of these families of features can be further subdivided according to the complexities of a given process or application. Increasing complexity required more special engineering and assembly which lead to more expensive equipment. It was decided that a new approach in the design of welding power sources and process controls should be tried. The resultant design is shown in Figure 12. This figure compares a traditional orbital GTAW design to a modular one. The concept is predicated on the idea that the power source, if properly controlled, can be the heart of many different systems. In order to meet the acceptance criteria, a control system should be produced that would provide suitable output current for different types of welding. For instance, voltage or current pulsing would be delivered by the power source under the direction of the controller. Other control circuits can handle wave form control, as in the case of pulsed GMAW. By interconnecting the appropriate controls with the power source, the process requirements can be met.

Welding power sources must accept feedback and control from external sources in order to provide optimum conditions for welding different materials with different process variations. Motion devices control welding parameters in some cases, like GMAW. When using the constant voltage variation, the power source output current is determined by the wire feed speed. Output voltage is held nearly constant by the wire feed speed in the constant current method. Different material types require different parameters; GMAW spray transfer of aluminum requires fewer volts and amperes than steel using the same diameter electrode. In the case of gas tungsten arc welding (GTAW) or plasma arc welding (PAW) using cold wire feed, the motion device (wire feeder) is controlled to deliver the amount of filler wire specified in the welding procedure.

Arc initiation or starting is an important factor especially in automated welding systems because a poor start can cause an unacceptable joint. GTAW has traditionally used a superimposed high frequency current to initiate the arc. This is also used to stabilize the arc during sinusoidal AC welding. However, this superimposed high frequency is potentially damaging to micro-circuits and other components. A relatively new advance in arc starting uses high voltage from a capacitor discharge system.

Shielding gas must be controlled by means of valve timers and flow meters to provide pre and post flow shielding gas. Cooling water is required for certain applications, except for low current work. As welding procedures become more complex, or more complex processes are used, additional controls are required. For example, in plasma arc welding, both shielding and plasma gases must be controlled and there must be a control to switch from the internal arc within the torch to be transferred arc to the work. Pilot arc power supplies are involved and included in the total system.

As mechanized and automatic applications become more popular, the control systems of the total welding system must include motion control devices. These range from a simple linear motion device for use on a side beam carriage or rotary motion required for an orbital type tube head. In other words, one axis of motion. However, other applications require two axes of motion like in an X and Y drive motion system. For many welding applications filler wire is required. A wire feed motor and a motor speed controller is utilized for feeding the cold filler wire. As systems becomes more complex, it is necessary to add oscillation to the arc. This involves the speed of oscillation, the width of oscillation and dwell time at either end of the oscillation stroke. This requires special controls which must be provided in the total control system.

The base for weld control is usually time, however, time may not be sufficiently accurate nor precise in some cases. Motion motors may be required to have positioning feedback to identify exact locations which complicates the system. The automatic arc length control system is a feedback system. It is normally required for gas tungsten arc and plasma welding. It utilizes the arc voltage signal to provide a constant arc length which is required for many automatic applications.

The solution to provide a total welding system is to utilize modular design and construction. In modular design, each function requires a control circuit or panel and a mechanical actuator. They must be matched and employed together. An example would be the cold wire feeder including the motor, gear box, drive rolls and the motor speed control circuit. Modules of this type are now available with standardized mounting hardware so that they can be incorporated in the total welding system as required. The control circuits or panels and built to international standards and mounted in racks as shown by Figure 13.

This concept has advanced to the point where a constant current power source is available as a module in a 150 amp size. These modules are designed in such a manner that they can be paralleled to provide 300 ampre output, or even 450 ampere output. They are compatible from an electrical and mechanical point of view and can be mounted in the international standardized 19-inch rack mount. Figure 14 shows this module and its compact size. The arc starting module can be the newer high voltage type system or the normal high frequency system previously built in the machine, but available as a separate or remote panel.

Cooling water systems and gas shielding systems are required. Two types of modules are available - one, shown in Figure 15 provides both water and gas controls, the other provides gas controls where water cooling is not required. The water circulator has normally been considered as a piece of auxiliary equipment. It is now available packaged for rack mounting. The BTU rating is matched to the welding requirements of the system and must be of sufficient size to accommodate the heat produced.

A basic part of a welding system is a programmable controller. Controllers of different levels of complexity are available. For example, a programmable controller would provide preflow delay and travel delay. It would provide for an initial current which may ramp to a higher current and hold steady for a pre-determined period of time before reducing the current in one or two steps. It would be followed by a postflow delay and a travel delay delay. Programmers allow presetting of specific levels of current and specific ramp schedules for increasing or decreasing current. These have become quite popular for gas tungsten arc and plasma arc welding. Programmers of this type are available as separate modules and are now being incorporated into the total system as a module that can be selected for the system. An example of such a controller is shown by Figure 16. Pulsing characteristics may or may not be incorporated in this delay/current level control systems.

There are three methods of controlling electrode wire feed speed. One is for the CC (constant current) type MIG applications. One is for the CV (constant voltage) application. The third one is for the cold wire feeder used in gas tungsten arc or plasma arc welding. Each type has a different control signal requirement and may have different speed rate capabilities. Each control module is matched to the appropriate electrode wire drive motor system for applying the electrode wire to the arc at the appropriate rate. The speed ratios or gear box ratios is different for different applications and should accommodate different wire sizes. A modular wire drive control panel is shown in Figure 17.

Joint tracking or following control systems are becoming available. In the GMAW process, the welding electrode wire is physically oscillated across the weld joint. The control circuit monitors arc voltage and weld current to provide both vertical and horizontal correction capabilities.

The ultimate in a welding control system is the microprocessor Microprocessor based welding control systems are now available. Figure 18 shows a computerized welding system for repairing the edges on end-mill cutters. Microprocessors are available with different levels of capability depending on the number of axis of motion that must be controlled, the number of input and output circuits that are controlled, etc.

As the price of microprocessors decrease and their capabilities increase, they will be used automated welding stations or cells. Akin to the "handshake" procedures of computers, power supplies and control modules will be integrated. A host central processing unit (C.P.U.) would monitor all facets of motion, weld parameters, part identification and location, gas supply pressures and flow rates, and visual inspection system results. This would require analog to digital conversion of information from sensors and interconnections among the host C.P.U. and slave mini computers responsible for welding parameters, fixture motion and visual inspection systems. Higher production rates, reduction of rejects and lower production costs will result.

Ideally, the whole system would be a modular turn-key operation. If the user required more than one welding process on the same production line, the software would have to recognize each type of slave unit and configure I/O's accordingly; like the handshakes between a personal computer and its data storage units or printers. Self-diagnostics and peripheral tests would be mandatory. Downtime due to system failure would be minimized using the same modular philosophy by replacement of entire control units or printed circuit boards.

With the availability of modules as described above, it is now possible to produce the required welding power source and control the system properly selecting and interconnecting the available modules to provide the features necessary to practice specific welding procedures. Welding equipment manufacturers are developing systems to keep pace with this rapidly changing technology. As the microprocessor continues to evolve, the welding industry will demand more systems with the modular design.

TABLE OF ILLUSTRATIONS

Function \ Method of Application	Manual	Semi Automatic	Mechanized	Automatic	Automated
Start & Maintain The Arc	Manual	Machine	Machine	Machine	Machine
Feed Filler Wire	Manual	Machine	Machine	Machine	Machine
Joint Travel	Manual	Manual	Machine	Machine	Machine
Joint Guidance	Manual	Manual	Manual	Machine	Machine
Overcome Variations	Manual	Manual	Manual	Defect	Machine

FLEXIBILITY ←

PRODUCTIVITY →

Fig. 1 Man Machine Relationship (including automated)

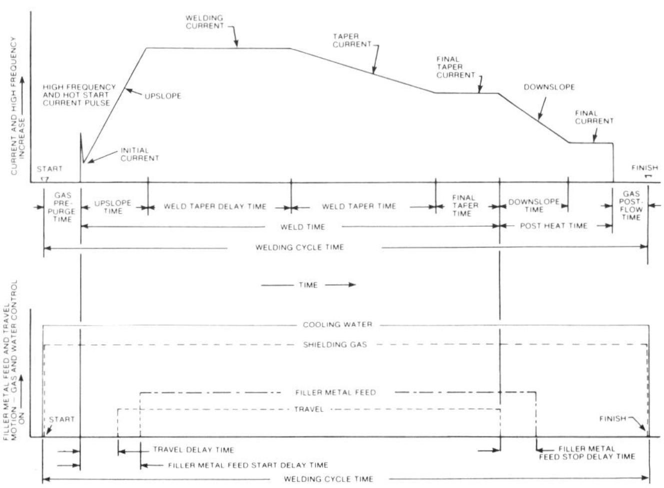

Fig. 2 Welding Program

Fig. 3 Welding Tube-To-Tube Sheets

Fig. 4 Cybertig II Machine

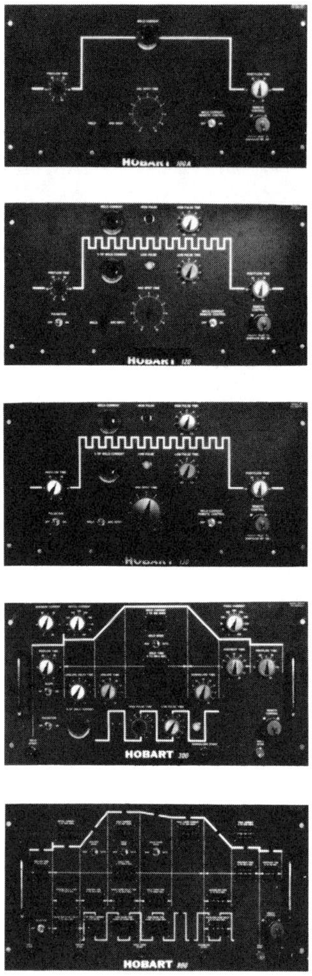

Fig. 5 Different Control Drawings

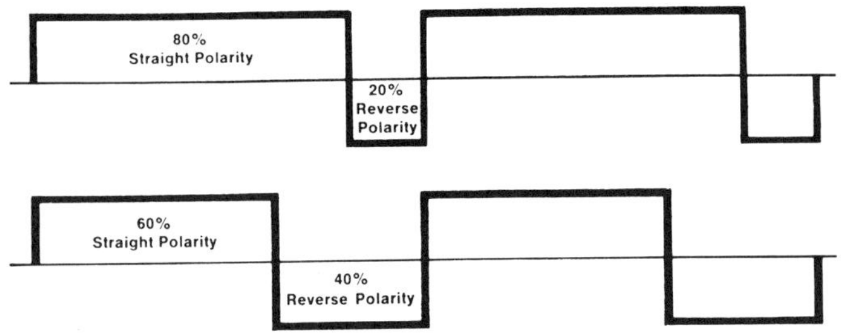

Fig. 7 Square Wave AC Current

Fig. 8 Cyberwave

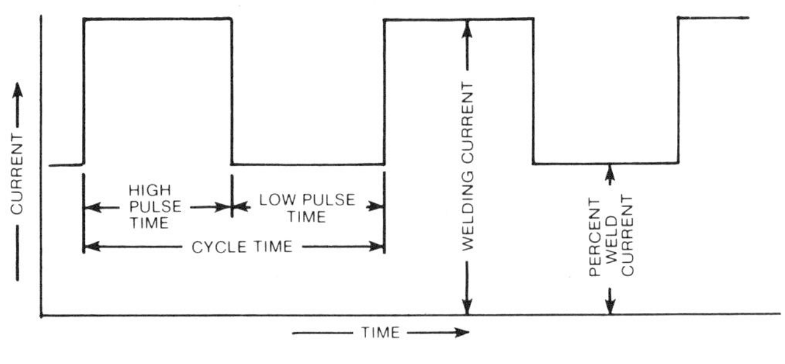

Fig. 6 Pulsed Weld Current

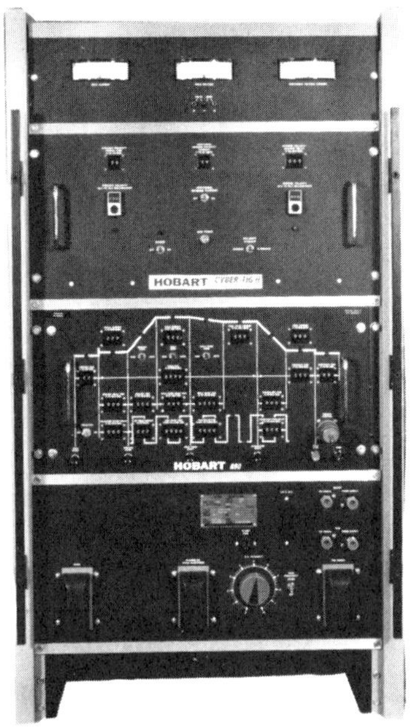

Fig. 9 VP Machine

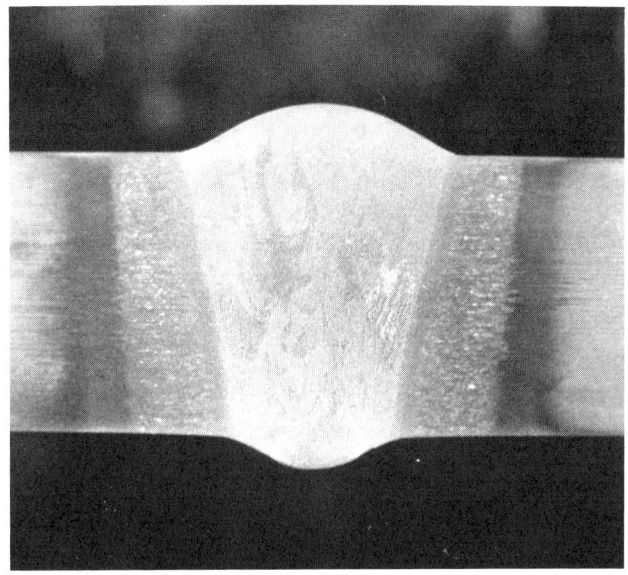

Fig. 11 Cross Section of VP Plasma Weld

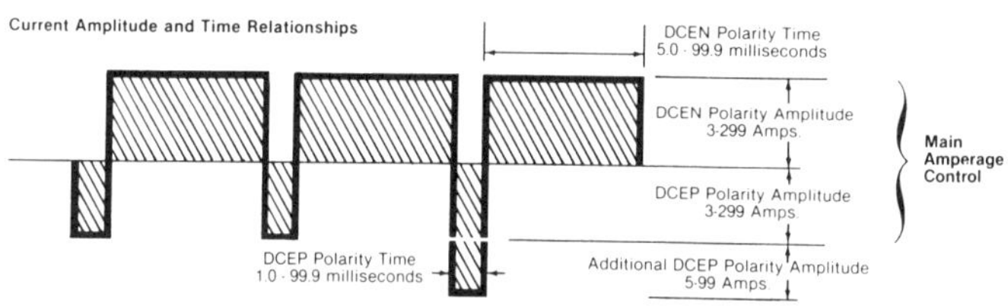

Fig. 10 VP Waveform

Fig. 12 Rack Mounted & Standard Cybertig Machine

Fig. 13 Rack Mounting of Control Circuits

Fig. 15 Gas & Water Cooling Module

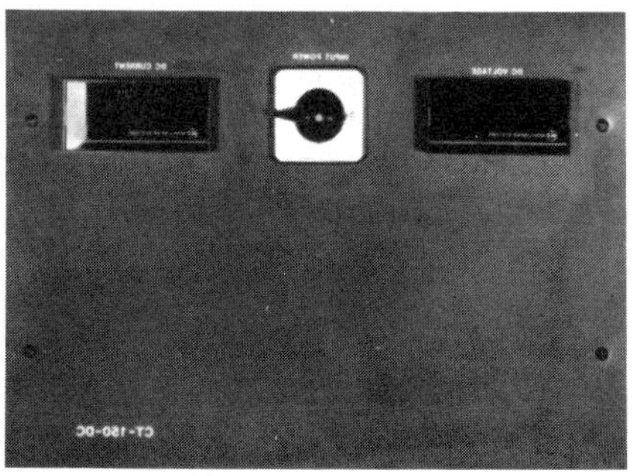

Fig. 14 150CT Power Source

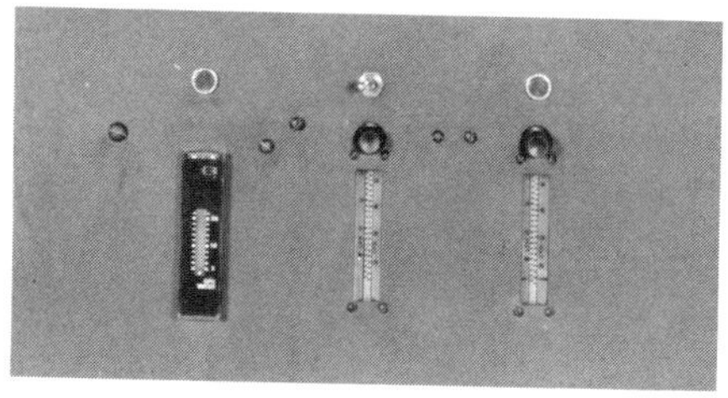

Fig. 16 Rack Mounted Controllers

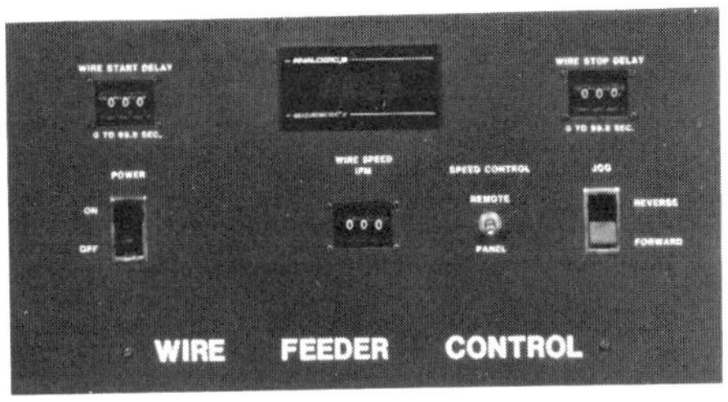

Fig. 17 Modular Wire Drive Control

Fig. 18 Microprocessor Controlled Welding Station

WETTING OF MOLYBDENUM BY COMMERCIAL BRAZING FILLER ALLOYS

M.M. McDonald, D.L. Keller, W.L. Johns
Rockwell International
Golden, Colorado

Abstract

The wetting of molybdenum by several commercial brazing filler alloys has been examined. Wettability tests have been conducted for each candidate filler alloy at several temperatures. A wetting index, which is a function of contact angle and braze alloy contact area, was determined for each filler alloy at each brazing temperature.
The nature and extent of interaction between the brazing alloys and the molybdenum was also investigated. The microstructures of the braze alloys and the base metal were examined with optical metallography, scanning-electron microscopy, and electron microprobe techniques. Results of the present study are compared with results of a previous wettability study on type 304 stainless steel.

AS PART OF A STUDY to determine the best filler alloys to use in joining molybdenum to Monel and to types 304, 316, and 321 stainless steel, the wettability of molybdenum by a number of commercial braze alloys was examined. Considerable information is available on vacuum brazing 300 series stainless steels; information on the brazing of molybdenum appears to be more scarce. The results of a wettability study of many of the same commercial alloys on type 304 stainless steel are discussed herein as a basis of comparison and also as basis for determining compatibility. Compatibility is meant in the sense of the applicability of a given filler alloy for use in joining molybdenum to type 304 stainless steel.

EXPERIMENTAL PROCEDURE

Wettability coupons, 3 mm x 50 mm x 50 mm (0.125 in. x 2 in. x 2 in.), were prepared from commercially pure molybdenum plate stock. A 1.6 mm (0.063 in.) diameter by 12 mm (0.5 in.) deep thermocouple hole was drilled at the edge of each specimen. The specimens were thoroughly degreased in acetone and polished on 400 grit metallographic paper. The samples were then degreased again. Braze filler alloy samples were also thoroughly degreased in acetone. All braze alloys were in wire form of generally 0.50 mm (0.020 in.) or 0.63 mm (0.025 in.) diameter. The wire was formed into coils of varying weights such that the total volume of braze metal used was 7.5 mm^3 (0.0005 $in.^3$). The composition, trade name, liquidus and solidus for each filler alloy are given in Table I.

The molybdenum coupons, with the braze alloys placed on the coupons, were placed in a standard, cold wall vacuum furnace. Chromel-Alumel theromcouples were inserted into each sample. The furnace was evacuated to 5 x 10^{-5} mm of Hg or lower. The samples were heated to the selected brazing temperature and held five minutes. The samples were then static inert gas cooled to 10°C (18°F) below the solidus and then rapid gas cooled to below 100°C (180°F) before removing them from the furnace. After being removed from the furnace, the area of the braze metal on each specimen was measured and the contact angle determined by shadowgraph. Selected specimens were then cross-sectioned for metallographic examination in optical and electron microscopes. Elemental maps of molybdenum and the major alloying elements in the filler alloys were obtained over the cross sections of selected specimens by X-ray microprobe analysis. This type of map allows qualitative analysis of the composition of the phases present.

Table I

Compositions of Candidate Filler Alloys

Alloy No.	Tradename	Composition (wt%)	Liquidus (°C)	Solidus (°C)
1	Palcusil 5	68Ag-27Cu-5Pd	810	807
2	Palcusil 10	58Ag-32Cu-10Pd	852	824
3	Palcusil 15	65Ag-20Cu-15Pd	900	850
4	Palcusil 20	52Ag-28Cu-20Pd	925	875
5	Palcusil 25	54Ag-21Cu-25Pd	950	900
6	Nicusil 3	71.5Ag-28.1Cu-0.75Ni	795	780
7	Nicusil 8	56Ag-42Cu-2Ni	893	771
8	T50	62.5Ag-32.5Cu-5Ni	866	780
9	T51	75Ag-24.5Cu-.5Ni	802	780
10	T52	77Ag-21Cu-2Ni	830	780
11	Cusiltin 5	68Ag-27Cu-5Sn	760	743
12	Cusiltin 10	60Ag-30Cu-10Sn	718	602
13	Braze 580	57Ag-33Cu-7Sn-3Mn	730	605
14	Braze 630	63Ag-28Cu-6Mn-3Ni	850	750
15	Braze 655	65Ag-28Cu-5Mn-2Ni	850	750
16	Nioro	82Au-18Ni	950	950
17	Nicoro 80	81.5Au-16.5Cu-2Ni	925	910
18	Palniro 7	70Au-22Ni-8Pd	1037	1005
19	Gapasil 9	82Ag-9Ga-9Pd	880	845

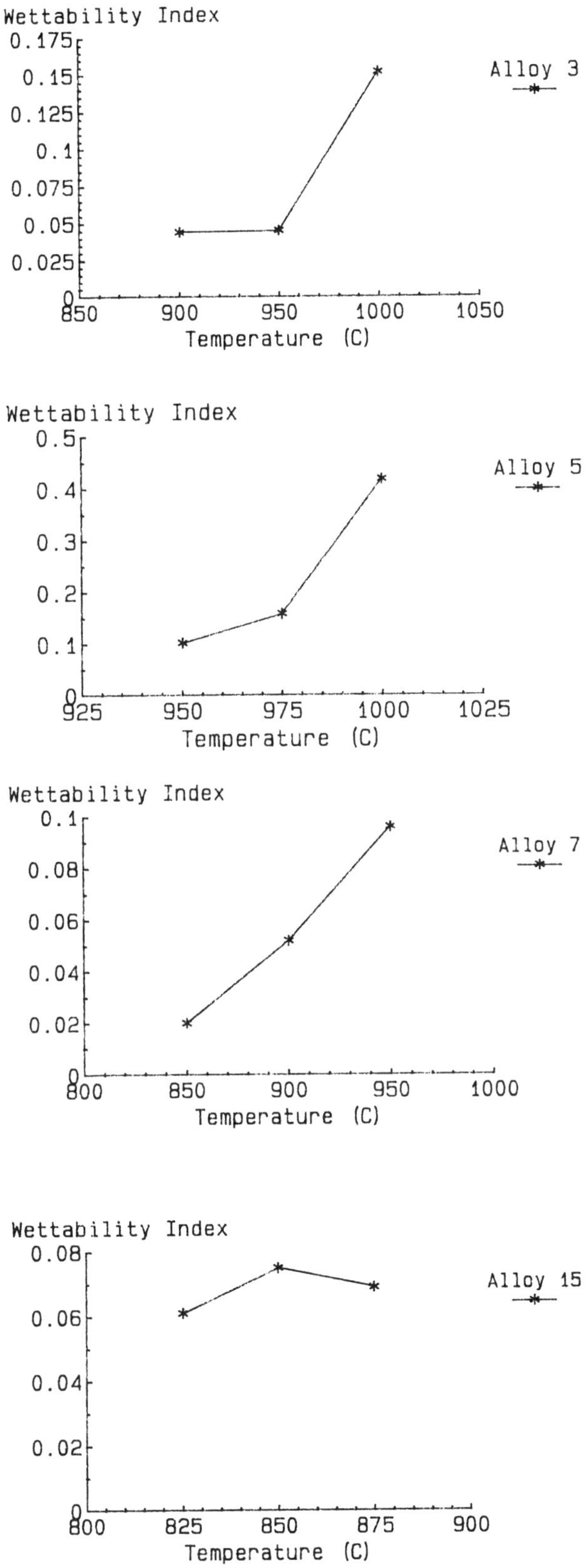

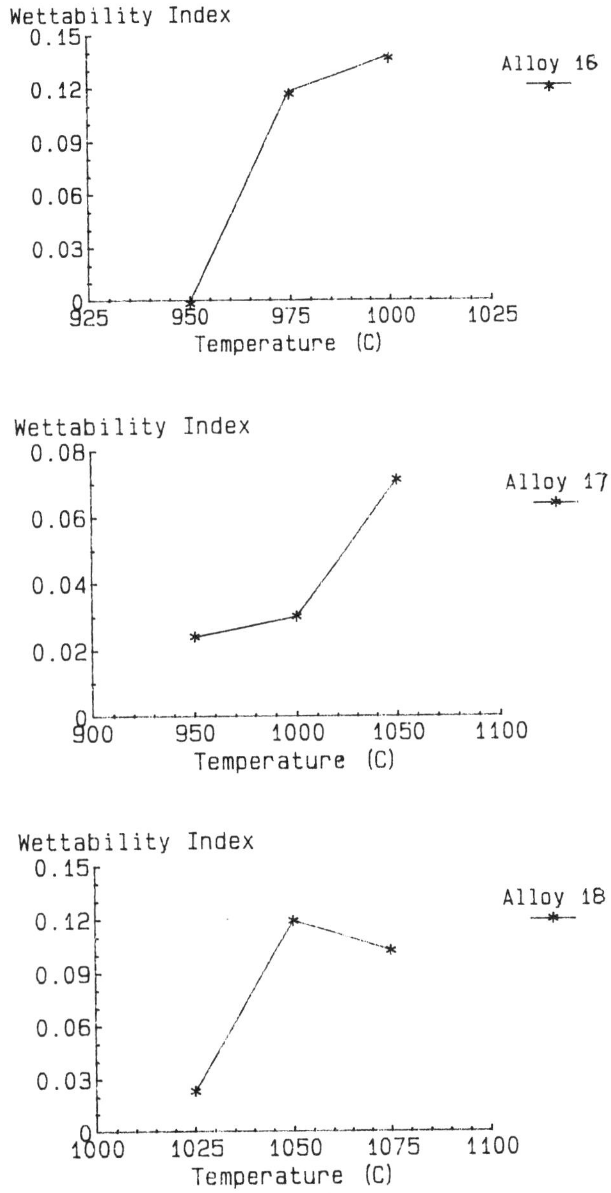

Figure 1. The Wettability Indices as a Function of Temperature for Selected Filler Alloys on Molybdenum.

RESULTS AND DISCUSSION

GENERAL RESULTS - The contact angle and braze alloy flow measurements on each coupon are presented in the form of wetting index. This index, which was developed by Feduska (1), allows condensation of the voluminous raw data into a meaningful parameter. The wetting index (W.I.) equals the area covered by the braze alloy times the cosine of the contact angle. The W.I. for each braze alloy is given in Table II. Note that the W.I. here cannot be compared directly with those of Feduska because of differences in the amount of braze metal used. The W.I. here can, however, be compared with the W.I. determined by Bennett et al. (2) on type 304 stainless steel, as equal volumes of braze alloy were used. The W.I. of this study varied from 0, where no wetting occurred, to a maximum of 0.418. In these tests, W.I. values of 0.05-0.10 are considered indicative of good wettability, and values above 0.10 are considered indicative of excellent wettability.

Nine specimens were chosen for metallographic analysis based on their W.I. and the availability of corresponding data for the braze alloy on type 304 stainless steel. Plots of W.I. versus temperature for the selected alloys are given in Figure 1. (Note that the scales vary from plot to plot.) A summary of qualitative analysis of the phases present in the selected specimens is given in Table III.

Cracks were observed in three of the molybdenum coupons after the braze cycle. The two coupons which exhibited significant cracking were the one which was heated to 1000°C with alloy 16 and the one which was heated to 1050°C with alloy 18. A micrograph of the alloy 16 specimen is shown in Figure 2. Cracking occurred in the basemetal, parallel to the interface, approximately 100 microns (0.004 in.) from the interface. The coupon which was heated to 1000°C with alloy 17 exhibited a small amount of cracking. All of these alloys are gold-based and contained some amount of nickel, and all were heated to temperatures greater than or equal to 1000°C, which is higher than the temperatures achieved with most of the silver-based alloys. No cracks were found in the molybdenum sample heated with alloy 17 after heating to 1050°C. Unlike the two gold-based alloys associated with significant cracking in the molybdenum, no molybdenum-nickel phase was formed at the braze metal-molybdenum interface with alloy 17. This may be related to the low nickel content of alloy 17.

Several possible reasons for the cracking are being investigated. These include embrittlement due to recrystallization or precipitation, or cracking due to thermal constraint. The AWS brazing manual (3) warns of the possibility of embrittling molybdenum when brazing the metal above the recrystallization temperature of 1177°C (2150°F). Impurities can lower the recrystallization temperature

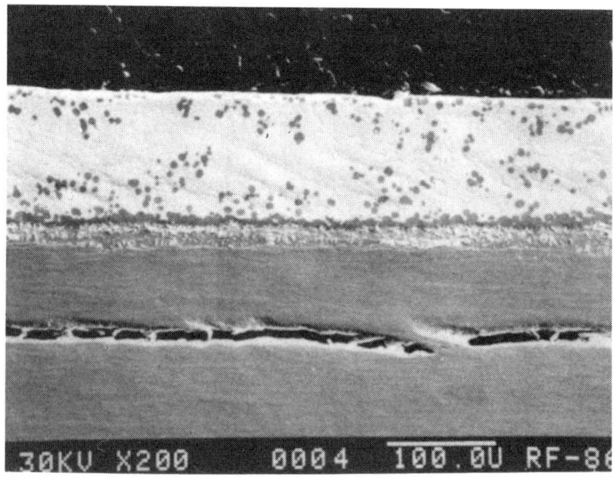

Fig. 2 Alloy 16 1000°C: Gold-Based Filler Alloy Associated with Cracking in the Base Metal.

significantly. The presence of brittle constituents such as oxides, nitrides and carbides are also reported to greatly lower the ductility of molybdenum. A temperature range for carbide and oxide embrittlement of molybdenum has been reported as being between 980-1650°C (1800-3000°F) (4) The present observations are consistent with this observed embrittlement range.

None of the silver-based alloys examined exhibited any evidence of interaction with the base metal, i.e. no grain boundary penetration, erosion or formation of a molybdenum-containing phase in the solidified base metal. The eutectic type structure that was common to the Ag-Cu-Pd based alloys is shown in Figure 3. The constituents present in these alloys were composed of varying amounts of copper, silver and palladium. The silver-based alloys other than the Ag-Cu-Pd based alloys, alloys 7 and 15, contained a silver-rich phase and a silver depleted phase. These two phases can be seen in the micrograph of the alloy 7 specimen heated to 950 C shown in Figure 4.

COMPARISON WITH RESULTS OF 304 STAINLESS STEEL STUDY - The results of the study by Bennett et al. (2) on the wettability of type 304 stainless steel by numerous filler alloys were compared with the results of the present study on the wettability of molybdenum by several of the same filler alloys. The alloy 3 specimen heated to 1000°C and the alloy 5 specimen heated to 1000°C both exhibited excellent wettability on the molybdenum. These same filler alloys heated to the same temperature on type 304 stainless steel also exhibited excellent wettability on the steel. The alloy 15 specimen heated to 850°C and the alloy 16 specimen heated to 1000°C both exhibited good wettability on the molybdenum and the type 304 stainless steel. However, the molybdenum heated with alloy 16 was cracked after the braze cycle as previously discussed.

Table II

Wettability Indices for Commercial Brazing Alloys in Molybdenum

Alloy No.	Wettability Index (Temperature (°C)) 750	800	825	850	875	900	925	950	975	1000	1025	1050	1075
1				.012		.034		.045					
2				.011		.030		.045					
3						.044		.045		.152			
4					.015	.037	.045						
5								.102	.119	.418			
6		0		.024		.053							
7				.020		.052		.096					
8				.023		.052		.069					
9		0		.023		.049							
10				.024		.057		.075					
11				.008	.012	.057							
12			.021	.037	.031								
13	.030	.037		.045									
14		0		.045									
15			.061	.075	.069								
16								0	.118	.138			
17								.024		.030		.071	
18											.023	.119	.102
19						.017		.061		.071			

Table III

Semiquantitative Analysis of Phases Present in Filler Alloys by Microprobe

Alloy No.	Temperature (°C)	Major Phase (wt%)	Second Phase (wt%)
18	1050	70Au-25Ni-5Pd	55Ni-30Au-10Mo-5Pd
17	1050	75Au-20Cu-5Ni	
17	1000	60Au-34Cu-6Ni	
16	1000	85Au-15Ni	75Ni-25Mo
3	1000	75Ag-20Cu-5Pd	60Cu-30Ag-10Pd
5	1000	50Ag-35Cu-15Pd	
7	950	90Ag-10Cu	80Cu-20Ni
3	900	75Ag-15Cu-5Pd	70Cu-15Ag-15Pd
15	850	70Ag-15Cu-10Ni-5Mn	50Cu-25Ni-15Mn

(a) Alloy 5 1000°C

(b) Alloy 3 900°C

Fig. 3 Structure of Ag-Cu-Pd Filler Alloys Containing Two Phases of Varying Amounts of Each of the Elements Present in the Filler Alloys.

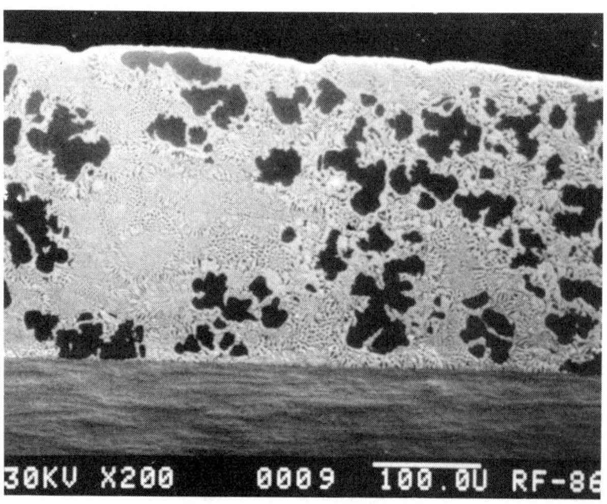

Fig. 4 Nicusil 8 950°C: Structure of Silver-Based Filler Alloy Containing a Silver-Rich Phase and a Silver-Depleted Phase.

CONCLUSIONS

The following conclusions can be drawn from the present study:

(1) Alloy 3, 65Ag-20Cu-15Pd, (1000°C), alloy 5, 54Ag-21Ci-25Pd, (1000°C) and alloy 15, 64Ag-28Cu-5Mn-2Ni, (850°C) are promising candidates for joining molybdenum to type 304 stainless steel.

(2) The silver-based alloys examined did not exhibit any interaction with the base metal (grain boundary penetration, erosion or formation of a molybdenum-containing phase).

(3) The gold-based alloys which exhibited interaction with the base metal in the form of molybdenum-nickel phase at the interface were associated with significant cracking in the base metal. The origin of this cracking is under investigation.

FUTURE WORK

Alloys 3, 5, and 15 will be examined as candidates to join molybdenum and 304 stainless steel. The joints will be characterized metallographically and their strengths determined.

The origin of the cracking in the base metal of the specimens with the gold-based alloys will be investigated. Areas of investigation include the possibility of the embrittlement being due to recrystallization or precipitation or cracking due to thermal constraint.

ACKNOWLEDGMENTS

The authors would like to acknowledge the assistance of W. E. Hofmann for his major contribution to the experimental work, T. O. Wilford for his assistance in determining the wettability index, and D. H. Riefenberg and A. F. Jankowski for their microprobe analysis. The authors would also like to acknowledge the many helpful discussions with C. R. Heiple, T. G. Glenn and C. L. Quimby.

REFERENCES

(1) W. Feduska, "High-Temperature Brazing Alloy-Base Metal Wetting Reactions," Welding Journal, March 1959, p122-s.
(2) W. S. Bennett, R. F. Hillyer, D. L. Keller an D. H. Riefenberg, "Vacuum Brazing Studies on High Manganese Stainless Steels," Welding Journal, November, 1974, p510-s.
(3) Brazing Manual, American Welding Society, 1976, p223.
(4) The Metal Molybdenum, J. J. Harwood, ed., American Society for Metals, 1956, p249

PREDICTION OF HAZ SENSITIZATION DEVELOPMENT IN 300 SERIES STAINLESS STEEL PIPE WELDS

S.M. Bruemmer, D.G. Atteridge
Pacific Northwest Laboratory
Richland, Washington 99352

ABSTRACT

A mathematical model has been developed to predict the development of sensitization in the heat-affected zone (HAZ) of 300 series stainless steel (SS) pipe welds. Stainless steel alloys exhibiting this structure are found to be susceptible to intergranular stress corrosion cracking (IGSCC). Sensitization development prediction in the HAZ is based on the thermomechanical (TM) history induced by the given weld parameters, material chemistry, and the initial state of sensitization and cold work.

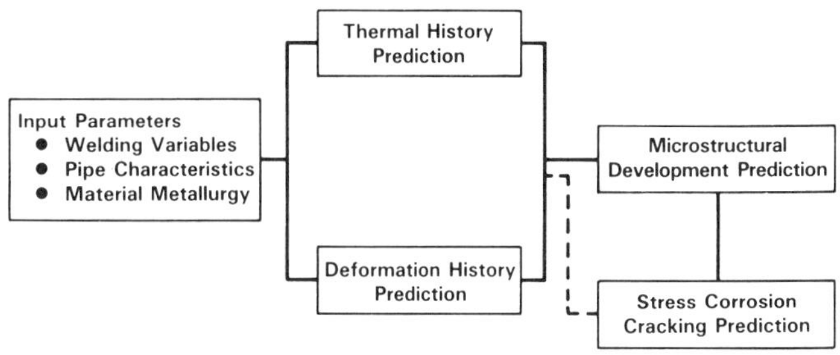

Fig. 1 - Flow diagram illustrating the critical components of HAZ-SCC prediction model

AUSTENITIC STAINLESS STEELS become sensitized and thus susceptible to IGSCC when a chromium-depleted layer forms around grain boundaries due to precipitation of chromium-rich carbides at grain interfaces (1). The composition and extent of growth of this region are controlled by the thermodynamics of carbide formation and the kinetics of chromium diffusion. Stainless steels of moderate carbon content can become sensitized in a relatively short time in a temperature regime where carbide precipitation is thermodynamically stable (<850°C) and chromium diffusion is sufficiently rapid (>500°C), such as in the HAZs of SS pipe welds.

The sensitization prediction work discussed in this paper stems from development efforts towards a model capable of predicting IGSCC susceptibility of as-fabricated weld HAZs in nuclear plant coolant SS piping systems (2,3,4,5). A schematic of the overall program approach is shown in Figure 1.

The prediction of HAZ sensitization level from known thermomechanical treatment is presented here. A model for predicting the change in HAZ sensitization level on a pass-by-pass basis has been developed. These predictions are based on the weld-induced HAZ TM history, as illustrated in Figure 2.

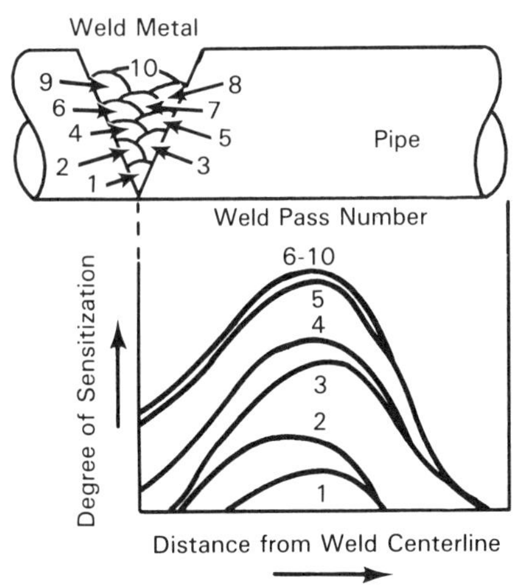

Fig. 2. Schematic of pass-by-pass prediction of HAZ sensitization development

SENSITIZATION PREDICTION MODEL DEVELOPMENT - A model for the prediction of material degree of sensitization (DOS) as a function of bulk composition, initial condition, and TM history is being developed. Basic components of the model include: 1) determination of the equilibrium chromium concentration at the carbide/matrix interface based on the thermodynamics of carbide formation; 2) chromium concentration gradients based on "effective" diffusivities; and 3) an empirical correlation between chromium depletion and DOS as measured by the electrochemical potentiokinetic reactivation (EPR) test. Empirical correlations also present in the model adjust diffusion coefficients and the onset and progression of healing (during extended heat treatment) as a function of bulk composition.

The computation sections of the model include a thermodynamic calculation to define chromium concentrations at the carbide/matrix interface for each temperature and a kinetic calculation to define, through conventional chromium diffusion analysis, the depleted zones as a function of time and temperature. The initially developed model used data for the thermodynamics and diffusivities for austenitic SS appearing in the literature. As the experimental and literature data base has increased, the model has been empirically adjusted.

An important aspect of the model is the prediction of DOS in a form that allows straightforward comparison to experiment; i.e., prediction of expected EPR value. Thus, existing model capabilities and attempts to improve the predictive ability can be evaluated using the EPR technique.

ISOTHERMAL AND CONTINUOUS-COOLING SENSITIZATION PREDICTIONS - The model was originally developed to predict isothermal sensitization development. Continuous-cooling sensitization development is evaluated using small, additive isothermal time steps to approximate the cooling (or heating) curve. Model prediction results for these conditions are presented and compared with experimentally measured results elsewhere (4,5,6).

The isothermal sensitization predictions show excellent agreement with experiment and the continuous-cooling predictions are reasonable. HAZ predictions, however, were found to be consistently low. This is due to the need to consider strain-induced sensitization acceleration taking place in the HAZ. HAZ TM history determination, discussed in a companion paper (7), revealed that straining takes place before and during the time the HAZ is in the sensitization development temperature range. Thus there is a need to quantify the effects of deformation on sensitization.

PREDICTION OF DEFORMATION EFFECTS ON SENSITIZATION - Deformation either before or during heat treatments within the temperature range where sensitization occurs sharply accelerates sensitization development (5,6,8,9). Model predictions at present account for deformation effects by increasing chromium diffusivity with increasing deformation. While this approach does not consider effects of deformation on carbon diffusion, carbide nucleation and precipitate thermodynamics, it does allow a straightforward method of predicting the acceleration in sensitization kinetics observed with deformation. An example of this predictive capability is presented in Figure 3. Chromium diffusivity

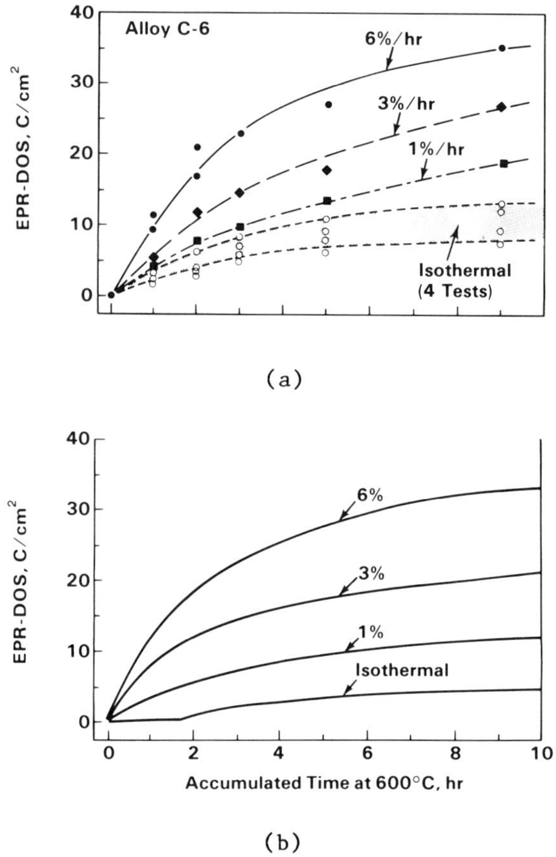

Fig. 3 - Comparison of measured (a) and predicted (b) sensitization development as a function of deformation and annealing time at 600°C

coefficients are increased by a nonlinear function that varies from 1 at 0.5% deformation to approximately 10 at 10% deformation. Deformation effects are predicted to saturate at 10% and to decrease at levels above 20%.

HAZ SENSITIZATION PREDICTIONS - Sensitization development in the HAZ of several 14- and 24-in.-dia (36- and 61-cm-dia) pipe welds has been modeled. Thermally induced model predictions for a 14-in.-dia weld and thermal plus strain predictions for a 24-in.-dia pipe will be discussed.

Thermally induced chromium depletion and EPR development predictions for the HAZ of a 36-cm-dia pipe girth weld are presented in Figures 4 and 5. The experimentally determined thermal history near the fusion line can be seen in Figure 4c. Predictions for development of the chromium depletion zone and corresponding EPR measurements are seen in Figures 4a and b, respectively.

The model predicts that stages of both growth and healing take place during individual weld passes near the fusion line. It also illustrates that the majority of sensitization development for this high heat input weld (nominally 41 kJ/cm) took place during the sixth and

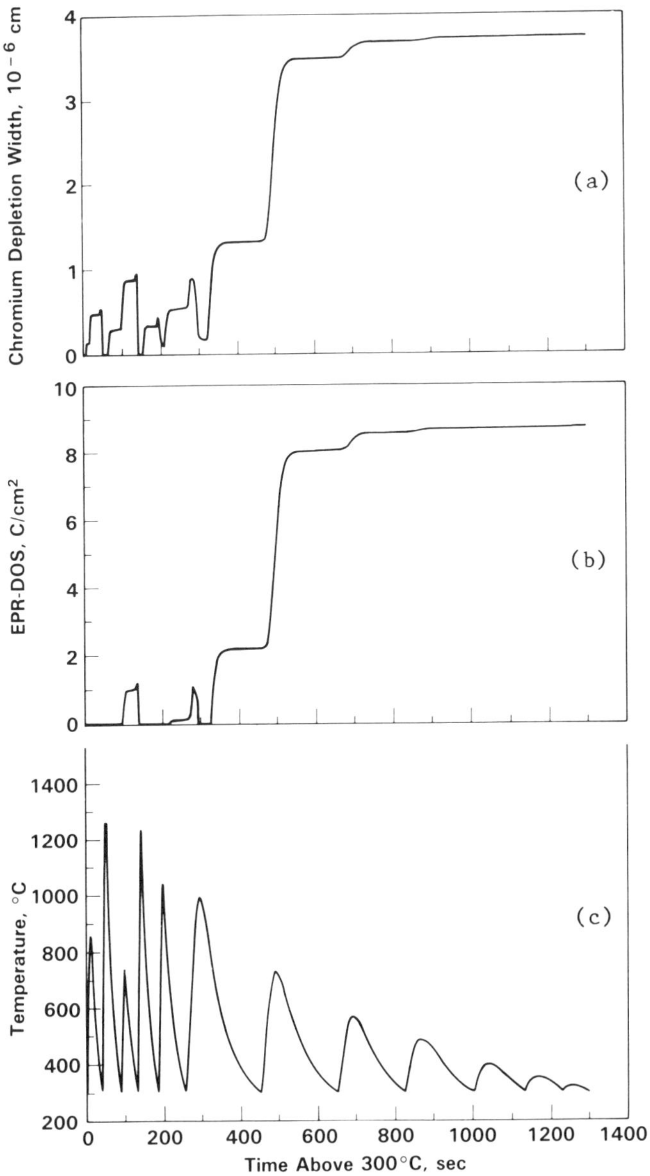

Fig. 4 - Development of the chromium depleted zone 0.2 in. from the weld centerline in a 14-in.-dia schedule 160 SS pipe girth weld

seventh passes. This was where the maximum temperature reached during welding stayed within the upper temperature range of the sensitization development range and heating and cooling rates were reasonably slow.

The thermally induced development of the HAZ as a function of distance from the weld centerline is shown in Figure 5. The EPR values represent sensitization development at the end of each pass. Sensitization remains low for the first several passes, builds rapidly during three to four passes and then remains essentially constant for the rest of the weld. Maximum sensitization takes place 0.5 cm

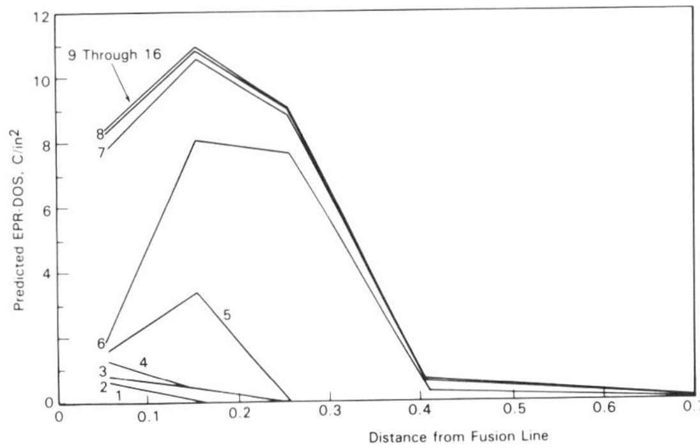

Fig. 5 - Predicted EPR-DOS results as a function of distance from the weld centerline on a pass-by-pass basis for a 14-in.-dia schedule 160 SS girth weld using only thermal data

(0.2 in.) from the fusion line in a HAZ of approximately 1 cm (0.4 in.).

A comparison of measured and predicted DOS will be presented here for a 24-in.-dia TM history weld. The 24-in.-dia weld is unique because it was fully instrumented to map pass-by-pass TM history with DOS measured in the HAZ after each pass by EPR. Thus, an accurate pass-by-pass data set is available for direct comparison of experiment and model prediction.

Thermal history data from HAZ thermocouples were compiled during heating and cooling and used as input for model predictions. A gradual increase in DOS is predicted from passes 1 through 5 as illustrated in Figure 6a. Predictions are only made at thermocouple locations of 0.38, 0.51, 0.64, and 0.77 cm from the weld centerline. Extrapolation of thermal input between these data points was not done, but it appears that maximum DOS occurs between 0.5 and 0.6 cm due to the balance between maximum temperature and cooling rate.

Model predictions were assessed by comparison to the DOS experimentally measured after each pass (Figure 6c). Final measured DOS levels were significantly higher (~2X) than predicted levels based on thermal history alone. Differences might be even greater if the model did not overpredict the first few passes. The overprediction during the early passes reflects the need for a better treatment of precipitate nucleation in the model for continuous-cooling thermal cycles.

The underprediction of DOS results primarily from neglecting the effect of deformation during the thermal cycle. If, for example, empirical correlations based on the isothermal strain effects on sensitization experiments described above are used to accelerate sensitization kinetics, predictions are remarkably close to those measured as shown in Figure 6b.

Good agreement can be seen in the final predicted and measured EPR profiles in the location of maximum DOS and the shape of the DOS

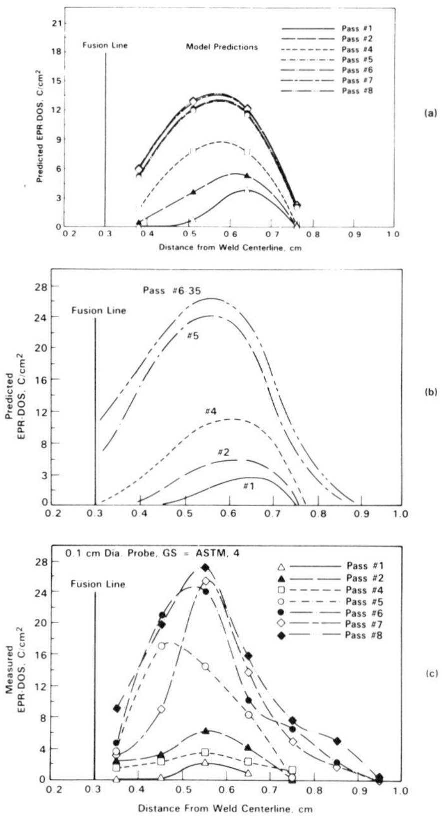

Fig. 6 - Sensitization development in the HAZ of the 24-in.-dia TM history weld: (a) thermal history model predictions per pass, (b) strain-accelerated thermal history model predictions per pass and (c) measured degree of sensitization

versus HAZ location curve. Overall, the correlation between prediction and experiment is promising, considering that the model data base to date is primarily from isothermal sensitization data.

SUMMARY

A model has been developed which enables quantitative predictions of sensitization development in the HAZ of Types 304 and 316 SS pipe welds. The model is based on isothermal sensitization development and has been used to predict isothermal, continuous-cooling, and strain-enhanced sensitization development as well as that found in HAZs. Good agreement between measured and predicted sensitization has been found for a wide range of material compositions and thermomechanical heat treatments.

ACKNOWLEDGMENT

The contributions of L. A. Charlot and A. L. Corrigan and technical assistance of B. W. Arey are gratefully acknowledged. This work was supported by the division of Engineering Technology, Office of Nuclear Regulatory Research, U.S. Nuclear Regulatory Commission.

REFERENCES

1. Bruemmer, S. M. and L. A. Charlot, Scripta Met. 20, 1019-1024 (1986)
2. Atteridge, D. G., S. M. Bruemmer and R. E. Page, NUREG/CR-3613 Vol. 1 (1984)
3. Atteridge, D. G., S. M. Bruemmer and R. E. Page, NUREG/CR-3613 Vol. 2 (1985)
4. Atteridge, D. G., S. M. Bruemmer, L. A. Charlot and R. E. Page, NUREG/CR-3613 Vol. 3, No. 1 (1985)
5. Atteridge, D. G., S. M. Bruemmer, W. E. Anderson, L. A. Charlot, B. W. Arey, B. Norton and R. E. Page, NUREG/CR-3613 Vol. 3, No. 2 (1986)
6. Bruemmer, S. M. and L. A. Charlot, "Measurement and Prediction of Thermomechanical History Effects on Sensitization Development in Austenitic Stainless Steels," No. 242, Corrosion 86, NACE (1986)
7. Atteridge, D. G. and W. E. Anderson, in this proceedings
8. Solomon, H. D. and D. C. Lord, Corrosion J. 36 (1980)
9. Charlot, L. A., B. W. Arey and S. M. Bruemmer, "Strain Effects on Sensitization Development in 304 Stainless Steel as Measured by EPR and STEM/EDS Methods," Proceedings for Symposium on Materials Characterization, Materials Research Society, Pittsburgh, Pennsylvania (1986)

SLAG METAL ELEMENT DISTRIBUTION IN COPPER ALLOY SUBMERGED ARC WELDS

Paul Tews
CRC-Evans Automatic Welding
Houston, Texas

Vjekoslav Pavelic
The University of Wisconsin-Milwaukee
Milwaukee, Wisconsin

Abstract

The law of mass action relates the concentration of oxides in slags to the concentration of elements dissolved in molten metal. This law was applied to three welding slag reactions to determine the law's usefulness in predicting weld metal element concentrations. Submerged arc welds were deposited using two copper alloys; aluminum bronze and copper-1.5% iron. Three fluxes were used; cryolite plus alumina, cryolite plus silica, and cryolite plus manganese monoxide. The slags and welds were chemically analyzed to determine element and oxygen distributions. The law of mass action was found to fit the experimental data for the alumina and silica reactions reasonably well. Apparent equilibrium temperatures derived from the measured values of K ranged from 2100 to 2700 degrees C. These temperatures are higher than the temperatures derived from the values of K measured by previous investigators studying ferrous slag-metal reactions.

PREVIOUS RESEARCHERS have studied the slag-metal element distributions which occur when the submerged arc welding process is used to deposit low carbon steel on low carbon steel base plates.* In this investigation, the distribution of elements between cryolite slags and copper weld metal was studied. Base plate conforming to the specifications of CDA C61300 was selected because of the popularity of this alloy and the need to develop a suitable submerged arc flux for joining and overlaying this alloy. An electrode conforming to the specifications of AWS CuAl-A2 was chosen because this electrode has been used successfully to join CDA C61300 plate using the gas metal-arc welding process. For comparison purposes, welds were made on copper plate using an electrode containing the same amount of iron as in AWS CuAl-A2 but without aluminum. The major constituent of the fluxes used in this investigation was cryolite, because cryolite has been shown to hold promise as a submerged arc flux for aluminum bronze.* The elements selected for study were aluminum, silicon, and manganese, because these three elements play major roles in determining the properties of aluminum bronze.

EQUILIBRIUM EXPRESSIONS

Equilibrium expressions were derived for the three slag-metal reactions by referring to the literature for the Gibbs' free energy of oxide formation and excess Gibbs' free energy of elements dissolving in copper. No reported values were found for the excess Gibbs' free energy of silica and manganese monoxide dissolving in copper. Only one value was found for the excess Gibbs' free energy of alumina dissolving in cryolite.** For this reason, the activity co-efficient for the oxides dissolved in cryolite has been taken as 1 for the present work. Future determinations of the actual activity co-efficients of oxides in cryolite should be used to refine the results obtained here.

Table 1 is a list of the Gibbs' free energy of formation for the three oxides studied. Table 2 is a list of the excess Gibbs' free energy of solution for selected elements dissolved in copper. Table 3 lists the one weight percent activities for selected oxides dissolved in cryolite assuming ideal solution

*Belton, G. R., et. al., The Welding Journal, 42, (7), 289s-297s (1963)
Chai, C. S., et. al., Metallurgical Transactions, 12B, 539-47 (1981)

*Milichenko, S. L., et. al., Automatic Welding, 21, (7), 59-61 (1968)
Gooch, T. G., et. al., W. I. Members Report 120/180

**Dewing, E. W., Canadian Metallurgical Quarterly, 13, (4), 607-18 (1974).

behavior.

Table 1 - Gibbs' Free Energy of Formation

Oxide	ΔG	
Al_2O_3	-378,810 + 67.79 T cal/mole	(1)
SiO_2	-229,700 + 47.7 T cal/mole	(2)
MnO	- 82.400 + 13.64 T cal/mole	(3)

(1) Kubaschewski, Chase, Shpilrain
(2) Kubaschewski, Stull, Hultgran
(3) Hultgran, Turkdogon, Kubaschewski

Table 2 - Excess Gibbs' Free Energy of Solution

Element	wt%	$\Delta H - T\Delta S$	Temp.C	
Al (liq)	1	-8630-13.84T cal/mole	1100	(1)
Al (liq)	8.6	-6174-10.51T cal/mole	1100	(2)
Si (liq)	1	-15000-7.5T cal/mole	1550	(3)
Mn (liq)	1	-1950-8.83T cal/mole	1244	(4)
1/2 O_2 (gas)	1	-20400+4.43T cal/mole	1100	(5)

(1) Hultgren, Sigworth
(2) Hultgren
(3) Chart
(4) Hultgren, Krenzer, Seetharaman
(5) Sigworth

Table 3 - One Weight Percent Activities

Oxide	Activity
Al_2O_3	2.04×10^{-2} cal/mole
SiO_2	3.41×10^{-2} cal/mole
MnO	2.90×10^{-2} cal/mole

Equilibrium constant expressions were derived using the values given in Tables 1 through 3. For aluminum in aluminum bronze:

$$\ln K = 153{,}616/T - 34.11 \quad (1)$$

where:

$$K = \text{wt\% } Al_2O_3 \ / \ (\text{wt\% Al})^2 \ (\text{wt\% O})^3$$

For aluminum in copper + 1.5% iron:

$$\ln K = 151{,}144/T - 37.46 \quad (2)$$

For silicon in both aluminum bronze and copper - 1.5% iron:

$$\ln K = 85{,}498/T - 19.94 \quad (3)$$

where:

$$K = \text{wt\% } SiO_2 \ / \ (\text{wt\% Si}) \ (\text{wt\% O})^2$$

For manganese in both aluminum bronze and copper - 1.5% iron:

$$\ln K = 30{,}218/T - 5.57 \quad (4)$$

where:

$$K = \text{wt\% MnO} \ / \ (\text{wt\% Mn}) \ (\text{wt\% O})$$

PROCEDURE

Twenty-one batches of flux were prepared for this investigation. Seven batches were formulated with alumina in quantities ranging from 0 to 35 weight percent. Seven batches were formulated with silica, and seven batches with manganese-monoxide. The predominant flux ingredient was cryolite, with 2.5 weight percent sodium silicate added for bonding. The powdered ingredients were agglomerated, screened, air dried, oven baked, screened again, and stored in sealed containers until use. The fluxes were preheated to 120 degrees C before welding and maintained at this temperature between passes. Aluminum bronze weld beads were deposited on copper base plate. The welding parameters are listed in Table 4. The beads were wire brushed before depositing successive beads. Fused flux was discarded and unfused flux reclaimed by screening.

Table 4 - Welding Parameters

Amps:	450 to 500
Volts:	32 DCRP
Electrode Dia:	3.2 mm
Electrode Feed:	38 mm/sec
Travel Speed:	4.44 mm/sec
CTWD:	38 mm
Flux Depth:	50 mm

RESULTS AND DISCUSSION

Figure 1 shows a cross-section of a typical aluminum bronze weld bead pyramid. Microscopic examination of the weld metal confirmed that the oxygen was present in the form of finely dispersed oxides, and not in the form of entrapped slag particles. Slag chemistries were determined using x-ray fluorescence spectroscopy. The spectroscopic analysis included oxygen and fluorine. Aluminum bronze weld metal chemistries were determined by emission spectroscopy. Copper-iron weld metal chemistries were determined by wet chemistry methods. Weld metal oxygen contents were determined by vacuum fusion analysis. Figures 2 through 7 are plots of ln K

Fig. 1 - Cross Section of typical weld bead pyramid on Aluminum Bronze Plate.

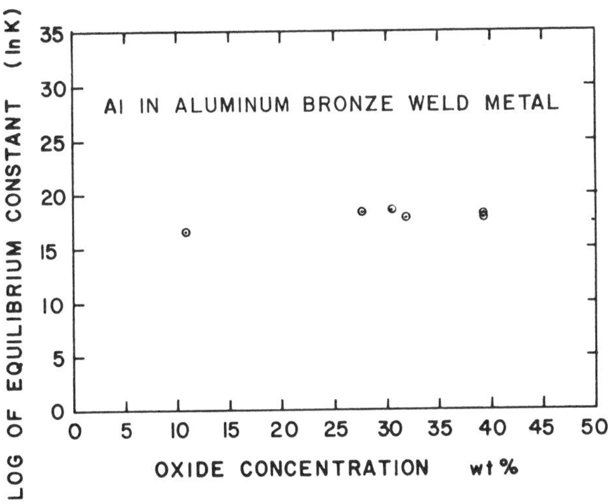

Fig. 2 - ln K for Aluminum in Aluminum Bronze Weld Metal.

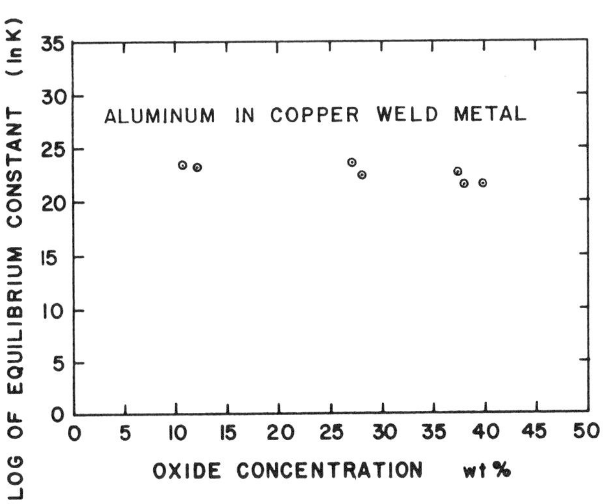

Fig. 3 - ln K for Aluminum in Copper -1.5% Iron Weld Metal.

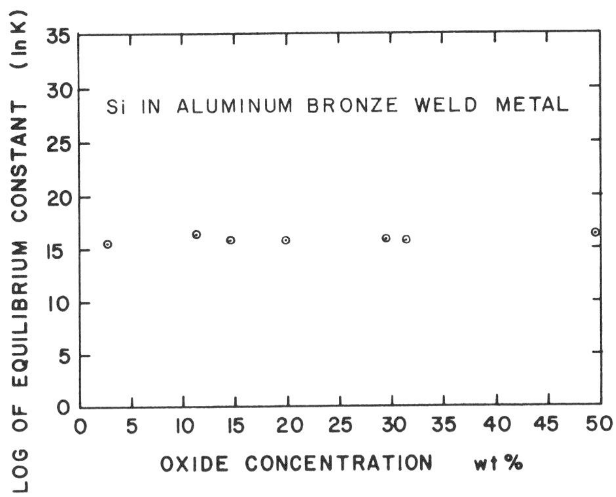

Fig. 4 - ln K for Silicon in Aluminum Bronze Weld Metal.

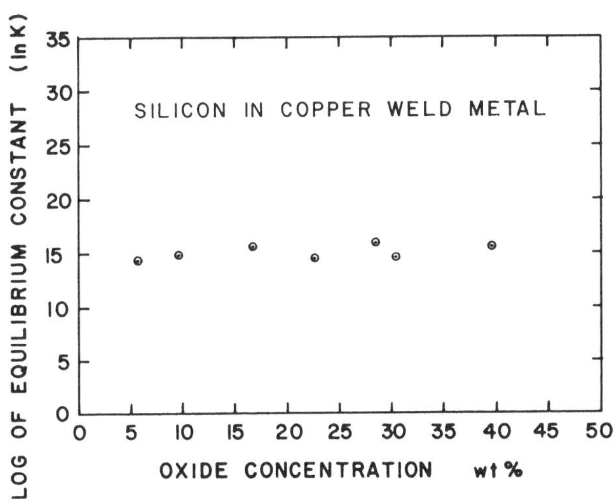

Fig. 5 - ln K for Silicon in Copper -1.5% Iron Weld Metal.

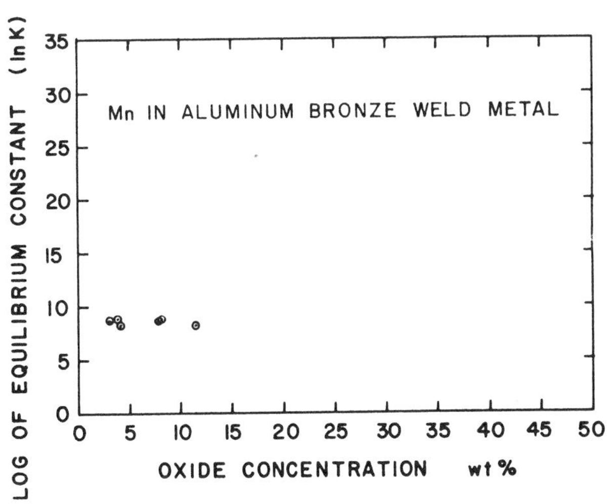

Fig. 6 - ln K for Manganese in Aluminum Bronze Weld Metal.

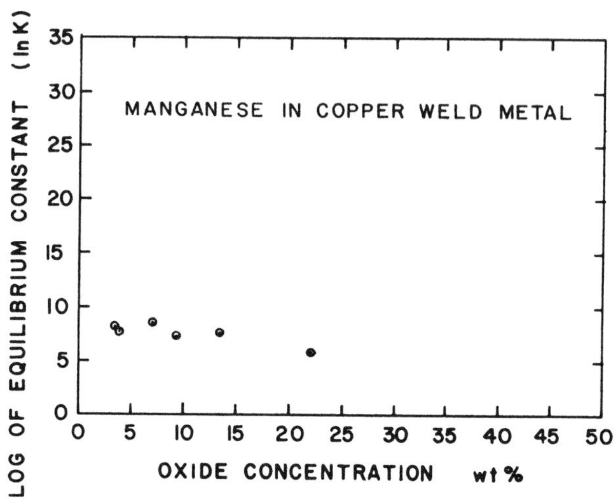

Fig. 7 - ln K for Manganese in Copper -1.5% Iron weld Metal.

for the six combinations of flux and weld metal. The value of ln K is fairly constant for the alumina and silica reactions, but not for the manganese monoxide reaction. When the average values of ln K are substituted into equations 1 through 4, the following apparent equilibrium temperature result:

Aluminum in aluminum bronze: T=2687 deg. C.

Aluminum in copper-iron: T=2236 deg. C.

Silicon in aluminum bronze: T=2118 deg. C.

Silicon in copper-iron: T=2175 deg. C.

CONCLUSIONS

Although the apparent equilibrium temperatures derived from the data gathered in this investigation are higher than the apparent equilibrium temperatures derived by previous researchers investigating slag-metal element distributions for steel, the use of the law of mass action as a mathematical tool for predicting slag-metal element distributions between copper alloys and cryolite slags appears justified for the case of silicon and, to a lesser degree, aluminum.

REFERENCES

Belton, G. R., Moore, T. J. and Tankins, E. S., The Welding Journal, 42, (7), 289s-297s (1963)

Chai, C. S. and Eagar, T. W., Metallurgical Transactions, 12B, 539-47 (1981)

Milichenko, S. L. and Bykovskii, O. G., Automatic Welding, 21, (7), 59-61 (1968)

Gooch, T. G., Davis, M. L. E. and Barlow, I. A., W. I. Members Report 120/180

Dewing, E. W., Canadian Metallurgical Quarterly 13, (4), 607-18 (1974).

Kubaschewski, O. and Alcock, C. B., "Metallurgical Thermochemistry, 5th Ed.", Pergamon Press, Oxford, England (1979)

Chase, M. W., Cornutt, J. L., McDonald, R. A. and Syverud, A. N., Journal of Physical Chemistry Reference Data, 7, 793 (1978)

Shpilrain, E. E., Kagen, D. N. and Barkhatou, L. S., High Temperature and High Pressure, 4 605 (1972)

Stull, D. R. and Prophet, H., "JANAF Thermochemical Tables", NSRDS-NBS 37, U.S. Dept. of Commerce, Washington, DC (1971)

Hultgren, H., Desai, D. D., Hawkins, D. T., Gleiser, M., Kelly, K. K. and Wagman, D. D., Selected Values of the Thermochemical Properties of the Elements, American Society of Metals, Metals Park, Ohio (1973)

Turkdogan, E. T., "Physical Chemistry of High Temperature Technology", p. 15, Academic Press, New York, NY (1980)

Sigworth, G. K. and Elliot, J. F., "The Thermodynamics of Liquid Copper Alloys", Canadian Metallurgical Quarterly, 13 (3) (1974)

Krenzer, R. W. and Pool, M. J., Trans. of The Met. Soc. of AIME, 245 (1), 91-8 (1969)

Seetharaman, S., Abraham, K. P. and Staffanson, L. I., Scandinavian Journal of Metallurgy, 7 176-80 (1978)

Chart, T. G., NPL Report, Chemistry, August 18 (1972)

AUTOMATIC REAL-TIME BEAD HEIGHT CONTROL WITH ARC SENSOR-WELDING SPEED CONTROL IN PULSED GMAW

Hirokazu Nomura, Yuji Sugitani, Naohiro Tamaoki
Tsu Laboratories
Nippon Kokan K.K., Japan

- Abstract -

Arc sensor system based on constant arc length control has been developed. The advantages of this system are not only seam tracking but also optimum weaving width control which gives important information concerning to the groove shape just under the arc.

In the pulsed GMA welding, the variation value of the weaving width with arc sensor control is same as that of groove width. Based on this relation, the welding speed control system has been developed and applied to the development of multi-pass MAG arc welding robot. The robot has been applied to the fabrication of steel box columns.

1. Introduction

The authors already have developed some kinds of seam tracking control process with arc sensor.(1) One of them utilizes the constant arc length control method(AVC or ACC). Advantages of this process are ; The optimum weaving width control can be carried out as well as seam tracking control in variations of the groove width and the groove center position. Moreover, information concerning to variation of the groove area can be obtained from the trace of weaving.(2)

Applying this arc sensor system to the pulsed GMA welding, the adaptive welding speed control method to keep a bead height constant has been developed. The pulsed GMA welding robot for multi-pass welding of butt joint has been developed. The main function of this robot is the adaptive microcomputer control with above mentioned arc senor system.

This paper describes the development and application of the process.

2. Seam tracking control with arc sensor system

2.1 Principle of arc sensor seam tracking control

Principle of automatic seam tracking control process with arc sensor is shown in Fig.1. A turning point of transverse weaving with constant arc length control is decided on the condition that the torch hight (ey) reaches to the constant preset level (eo). Here, the constant arc length control is known as AVC (Automatic Voltage Control) in GTA welding and ACC (Automatic Current Control) in GMA welding, and the preset level (eo) is the distance from the surface of the base plate.

2.2 Waveforms of torch displacement and welding current

Fig.2 shows the displacement of the transverse weaving (X-axis), the torch height (Y-axis) and the waveform of welding current. The torch height (Y-axis) rises influenced by the inclined groove wall. When the torch height reaches to the constant level (eo), the direction of X-axis changes (point A). For a little while after changing the direction of X-axis, Y-axis still rises continuously by the characteristic of ACC motion. Welding current waveform shows the maximum value when Y-axis reaches to eo (point A). Welding current value at the peak point (B) of Y-axis is equal to the reference value (Io=300A) of ACC.

Now, at the central part of transverse weaving, variation of Y-axis is small in comparison with the variation of welding current. This is caused by off-band treatment of ACC motion.

Where, the power source is switching transistor type, the diameter of electrode wire is 1.2mm, the shielding gas is 80%Ar + 20%CO_2, and the average welding current is 300 amperes of 200Hz pulse frequency. The V-shaped groove with wall angle of 25 degrees each side is used.

3. Development of welding speed control process

The basic examinations in the pulsed GMA welding have made it clear that the weaving width with arc sensor control gives an accurate information to detect the root width variation just under the arc.

Accordingly, the authors have the welding speed control process to keep the bead height constant using only the detected value of the weaving width with arc sensor control. Fig.3 shows its principle. In the intial state of the root width Go, the bead height ho and the deposited area Ao, the weaving width with arc sensor control is Wo. Then, in the state where only the root width is changed from Go to G (in this figure, the width is increased), the weaving width with arc sensor control may be changed from Wo to W as shown in the figure. Assuming that the groove angle is constant, the variation of weaving width will become equal to that of the root width.

$$G - Go = W - Wo \quad [1]$$

Therefore, in order to keep the bead height constant, the deposited amount should be increased by ΔA.

$$\Delta A = (G - Go)\, ho = (W - Wo)\, ho \quad [2]$$

So, the optimum welding speed V can be obtained from the following equation.

$$V = \frac{Vf}{Ao + \Delta A} \quad [3]$$

$$V = \frac{Vf}{Vf/Vo + (W - Wo)\, ho} \quad [4]$$

Here, Vf is the wire feeding rate and Vo is the initial welding speed.

It should be noted that the equation [3] and [4] can be used on the condition that the turning level (eo) is set on the basis of the position of base plate surface and, in strictly speaking, the depth of groove is also not changed. The equation [4] can be used to asymmetric groove as a single bevel and to not only 1pass-1layer welding but also to 2pass-1layer welding.

4. Control experiment

Welding speed control has been carried out with continuously varied root width from 6mm to 16mm, where the wire feeding rate (Vf) is 12.5m/min, the desired bead height (ho) is 5mm and the initial welding speed (Vo) is 34 cm/min at 6mm root width.

Fig.4 shows the controlled welding speed and weaving width as against the continuously increased root width. Satisfactory and precisely controlled result is obtained. The weaving width is nearly equal to the root width and the welding speed well corresponds to the calculated value to keep a constant bead height (broken line). The measured bead height at each root width are shown, where the broken line shows the calculated value assuming constant welding speed (Vo). The average bead height is about 6mm. Although it is a little higher than the desired bead height, the bead height is well controlled within 0.8mm.

5. Development of multi-pass arc welding robot

Microcomputer controlled multi-pass GMA welding robot has been developed on the basis of the adaptive arc sensor control system. In this chapter, the outline of the robot mechanism and its control functions will be described.

5.1 Out line of the robot mechanism

The robot mechanism (welding head) is composed of three cartesian axis (X, Y, Z), the torch angle adjusting mechanism (β-axis) and the auxiliary axis for correcting eo level (eo-axis) as shown in Fig.5. X-axis is for seam tracking and weaving, Y-axis for ACC, Z-axis for welding travel and β-axis for adjusting the torch angle in the welding direction. The eo-axis corrects the turning level (eo) in Y-axis by means of detecting the surface of the base plate with the magnetic sensor. All these axes are driven by DC servomotors and use the potentiometers or encorders to detect their positions.

5.2 Control function

Fig.6 shows the block diagram of the control circuit and its configuration. The controller has several 8 bit CPU. The functions of main CPU are the sequence control, memory-play back control and the display on CRT. Sub CPU performs the arc sensor control and the welding speed control. Using CRT and key board, the preset data for the welding conditions control can be set up. The operating box arranged near the welding head is operated by a welder. There are a liquid cristal display, a key board and switches in the front panel of the operating box. A welder operates the key board and switches according to the liquid crystal display by man-machine dialogic style.

In addition, optical fiber cables for transmission of control signals are adopted between the main controller and the operating box. Using the optical fiber cables prevents noise influence and decreases the number of control cables. Pendant for manual operation of all axes in welding head is the wireless remote control type in consideration of handling and preventing disconection.

5.3 Build-up sequence control

The number of passes in one layer and the number of build-up layers for finishing welding are controlled automatically as shown in Fig.7. The number of weaving cycle (n) that the weaving width exceeds the preset critical width (W max) is counted through out the pass. When the ratio of this number and all times of weaving (N) in the pass exceeds the preset ratio (n/N), the next layer will be welded with two passes (2pass-1layer).

The weaving position in 2pass-1layer is controlled as follows. The center of weaving (Wc) in 1pass-1layer welding is memorized at short intervals of the welding length and played back in 2pass-1layer welding. The played back Wc is used as the turning point of weaving at the center side of the groove and the other turning point at the wall side of the groove is controlled by arc sensor.

While, the final layer is decided by comparison between the preset base plate thickness value and the integrated displacement of the torch height from the first layer.

6. Welding experiment

A welding experiment using this robot has been carried out. The groove shape is single bevel of 30mm thickness plate with a varied groove width from 9mm to 15mm and the welding length is 600mm. The critical weaving width (W max) is preset by 15mm and the ratio of n/N is preset by 70%.

On the result of the welding examination it was welded with 1pass-1layer until third layer, with 2pass-1layer from fourth layer and completed by ten passes continually. Fig.8 shows the cross sectional bead shape of three cases of build-up steps at start (G=9.5mm) and end (G=14.5mm) parts of the joint. The bead height at each build-up step is controlled very well by the arc sensor system.

7. Conclusion

The welding speed control method to keep the bead height constant in the pulsed GMA welding has been developed. Microcomputer controlled multi-pass GMA welding robot with adaptive arc sensor control has been developed and applied to the welding for the diaphragms of the box columns. Results obtained in this work can be summarized following .

(1) By using detected weaving width, the following welding speed control equation to keep a bead height constant can be obtained.

$$V = \frac{Vf}{Vf/Vo + (W-Wo)ho}$$

(2) Satisfactory performance of the developed welding control method has been confirmed by experiments carried out for 1pass-1layer and 2pass-1layer welding.

(3) Multi-pass GMA welding robot with adaptive control by arc sensor has been developed and applied to the welding for diaphragms of box columns with satisfactory results.

Reference

1) H. Nomura, Y. Sugitani, K. Kobayashi; "Automatic seam tracking method with an arc as a sensor" (In Japanese) Arc Physics Comm., JWS, 81-453-1981

2) H. Nomura, Y. Sugitani, Y. Suzuki ; "Automatic Real-time Bead Hight Control with Arc Sensor in TIG Welding", IIW Doc. XIIC-108-85

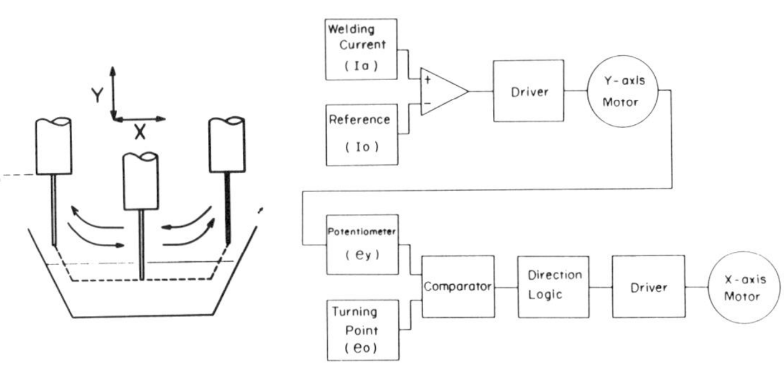

Fig.1 Principle of automatic seam tracking control with arc sensor

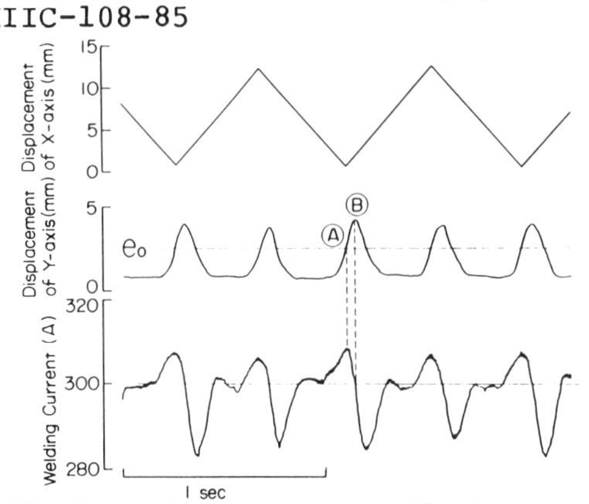

Fig.2 Waveforms of torch displacement and welding current

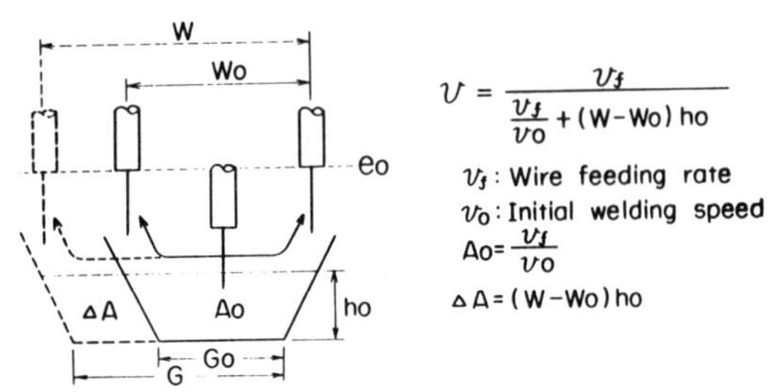

Fig.3 Principle of welding speed control

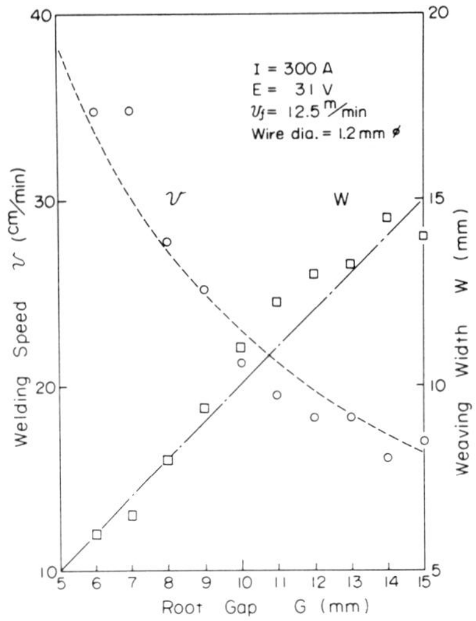

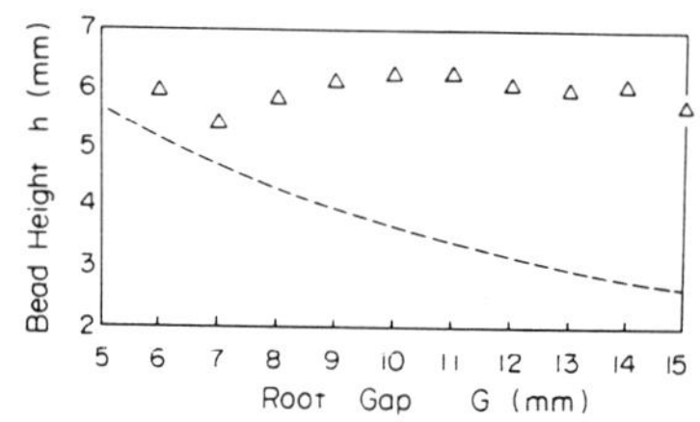

Fig.4 Controlled weaving width and welding speed by arc sensor system in continuously increased root width

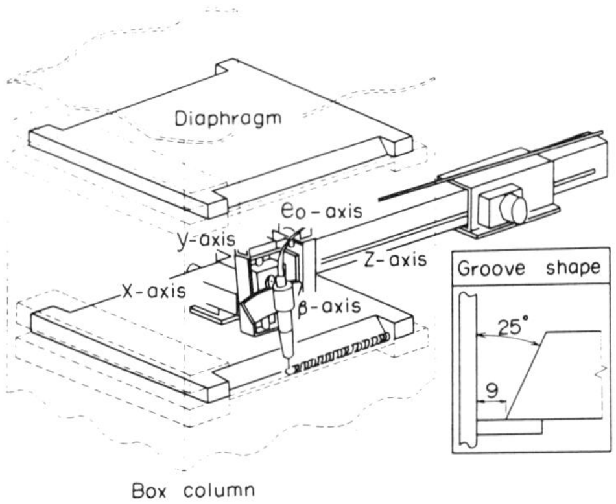

Fig.5 The robot mechanism in application to a steel box column

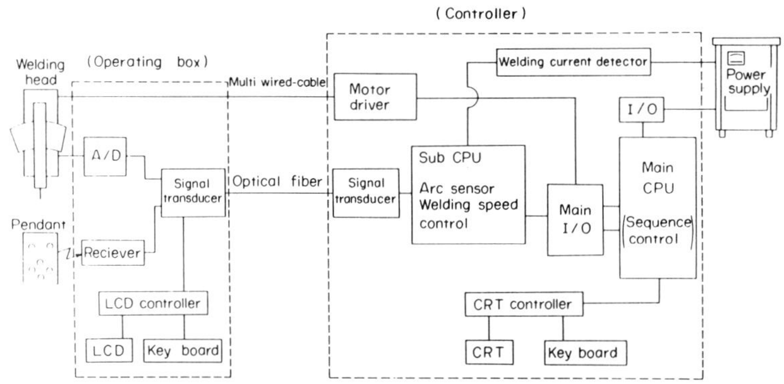

Fig.6 Block diagram of control circuit

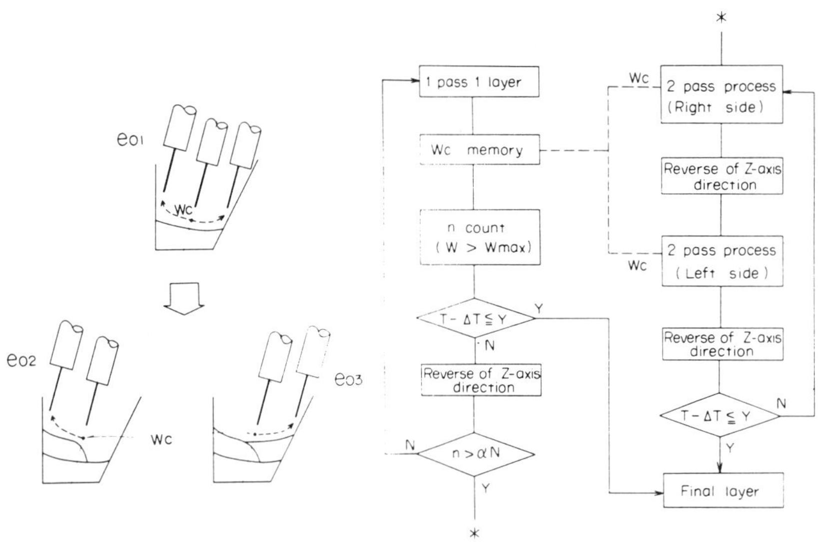

Fig.7 Flow chart of build-up sequence control

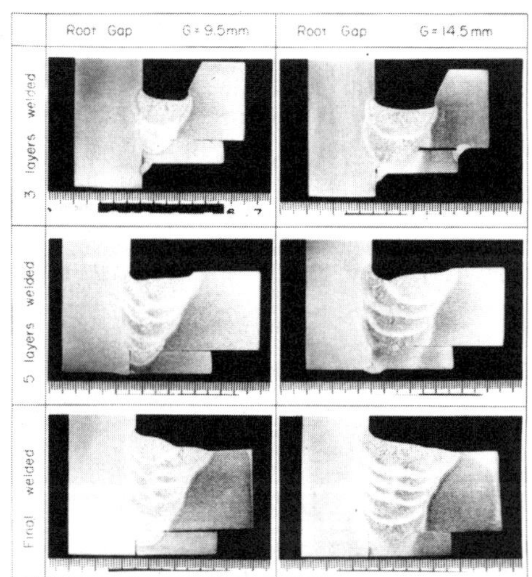

Fig.8 Result of welding experiment multi-pass MAG with micro-computer controlled arc welding robot

THE EFFECT OF MICROSTRUCTURE ON THE DUCTILITY OF 308L WELD METAL

S.J. Goodwin, F.W. Noble
Department of Metallurgy & Materials Science
University of Liverpool
Liverpool, England L693BX

B.L. Eyre
U.K.A.E.A. Risley
U.K

ABSTRACT

A study has been made of the effect of microstructural variations on the tensile ductility of 308L weld metal and 304L plate. The latter material was heat-treated to produce two different δ-ferrite contents while the former was tested 'as-deposited' and after heat-treatments designed either to eliminate the δ-ferrite or to induce its transformation to σ-phase. The results indicate that the presence of δ has relatively small effect on tensile ductility and confirm the detrimental effect of σ formation. In the absence of σ the ductility of the weld metal is controlled by the development of inclusion nucleated voids and transmission electron microscopy (TEM) was used to monitor the nucleation and growth of these voids as a function of tensile strain. The results revealed that void nucleation occurs by decohesion of the inclusion/matrix interface and increased progressively with strain. It was also established that void growth consists largely of elongation in the tensile direction. In the presence of σ the fracture surface was much more brittle in appearance and only isolated regions of ductile dimple fracture were visible. Evidence of extensive σ-phase cracking was found, the cracking occurring through the σ rather than along the γ-σ interface.

VARIATIONS IN MICROSTRUCTURE between 'as-deposited' austenitic weld metals usually involve differences in inclusion content and amount of δ-ferrite present. The δ is metastable and further microstructural variations can result from its decomposition, in subsequent heat treatment, to form carbides and intermetallic phases (σ,χ, Laves and R in 316 (1,2,3); σ in 308 (4,5)), the transformation sequence and products being dependent on composition, thermomechanical history, and, possibly, δ content (4). Each of these microstructural components can affect the mechanical properties of the weld metal though, to date, there has been little (4,6,7,8) systematic work on the magnitude of these effects and the mechanisms underlying them. This paper reports preliminary results of a study of such effects in 308L weld metal and particularly the influence of δ-ferrite, σ-phase and inclusion content on tensile ductility. Data obtained from 304L plate with different amounts of δ are also included for the purpose of comparison.

EXPERIMENTAL PROCEDURE

MATERIALS AND HEAT TREATMENT - The 308L weld metal was deposited using submerged arc welding at 400A and 30V at a speed of 400mm/min. Two different samples were obtained, as a result of using fluxes with different chromium contents, and their compositions are given in Table I. Magnetic measurements showed that W1 contained about 22% δ-ferrite in a more or less continuous network, while W2 contained about 6% in isolated patches. The weld metals were deposited in the form of bead-on-plate welds permitting their properties to be determined without the complications arising from base metal dilution. For the purposes of evaluating the effects of δ on ductility a heat treatment of 3h at 1050°C was employed to eliminate (virtually) the δ-ferrite from W2 (this treatment had little effect on W1). For the purposes of studying the effects of σ-formation, an ageing treatment of 100h at 700°C was applied to W1 resulting in the transformation of about 50% of the δ (as indicated by magnetic measurements) with little or no carbide formation being apparent.

Table I - Composition of the Weld Metals

	Cr	Ni	C	Mn	Mo	Si
W1	22.51	9.21	.027	1.07	.13	.90
W2	19.70	9.79	.031	1.62	.14	.43

Table II - Tensile Properties

Condition	% δ ferrite	0.2% P.S. MN/m^2	U.T.S. MN/m^2	U.T.S. /P.S.	True Strain to Failure	True Uniform Strain	Non-Uniform Strain	Reduction in Area %
304L (1h at 1260°C)	2	257	596	2.32	1.91	0.51	1.40	85
304L (1h at 1375°C)	14	279	618	2.22	1.69	0.45	1.24	81
W1 (as deposited)	22	460	695	1.51	0.73	0.25	0.48	52
W1 (aged 700°C)	11	391	719	1.84	0.15	0.12	0.03	15
W2 (as deposited)	6	365	577	1.58	0.91	0.29	0.62	60
W2 (annealed 1050°C)	1	247	558	2.26	0.74	0.59	0.15	53

The 304L plate material tested in this work was treated to produce either 2% δ (1h at 1260°C) or 14% δ (1h at 1375°C) and was rapidly cooled from the heat treatment temperature.

The materials were tensile tested at room temperature and strain to failure data calculated from the reduction of area at fracture. After testing the specimens were examined both by scanning and transmission electron microscopy (SEM and TEM). In the latter case thin foils were prepared from longitudinal sections cut from deformed specimens and thinned electrolytically in a solution of 10% perchloric acid in methanol at 0°C.

RESULTS

The tensile properties of the 304L plate and the samples of weld metal, for the heat treatments indicated, are shown in Table II. The results for the 304L indicate that its ductility is high and is similar for both δ-ferrite contents. The ductility of all the weld metal samples is much lower than that of the plate but again variations in δ content (in unaged W1 and in W2 before and after annealing) appear to have very little effect on ductility. After ageing W1, however, to produce σ-phase, there is a marked decrease in overall ductility and the non-uniform strain component falls effectively to zero.

Comparison of the inclusion content of weld metal and plate showed that the inclusions were largely spherical in both materials but that the inclusion density was considerably greater in the weld metal. The data obtained from examination of the inclusion population of plate and weld metal (W1) are summarised in Table III.

Table III - The average diameter (d) spacing (λ) and volume fraction (f) of the inclusions

	d (μm)	f	λ (μm)
304L	0.63	3.1×10^{-4}	13
W1	0.70	4.3×10^{-3}	4

The development of voids at inclusions in a sample of W1 weld metal strained to 0.24 is illustrated by the TEM micrograph shown in Fig. 1.

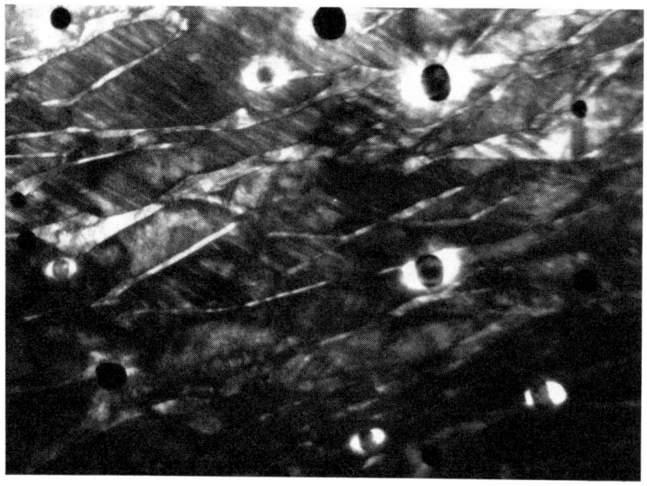

1·5μ

Fig. 1 - Electron micrograph of deformed W1 showing void formation at inclusions ε = 0.24

The voids appeared to have nucleated mainly by end-cap decohesion and foils prepared from samples deformed to lower strains indicated that nucleation commenced at the onset of plastic flow. The progressive increase in void nucleation with deformation for W1 is shown in Fig. 2, in which the percentage of inclusions associated with voids is plotted as a function of strain. The results indicate that nucleation begins to saturate at a strain of about 0.25 when, apparently, only about 50% of the inclusions have nucleated voids. However, this is considered to be a spurious observation and to arise because the population of inclusions observed by TEM includes those inclusions attached to the top and bottom surfaces of the foil from which the surrounding material has been removed during thinning. If due allowance is made for these inclusions the results are consistent with all of the inclusions having nucleated voids at the saturation strain.

The growth of inclusion nucleated voids was such that even at strains close to fracture no lateral growth was observed and the voids developed by longitudinal growth parallel to the tensile axis to form ellipsoids, the minor axes of which coincided with the diameters of the inclusions at which they had formed. Analysis of the void sizes associated with specific, narrow inclusion size intervals showed a range of aspect ratios to be associated with each interval, indicative of a spectrum of nucleation strains and providing no evidence of inclusion size being a significant factor in determining ease of void nucleation at a specific inclusion.

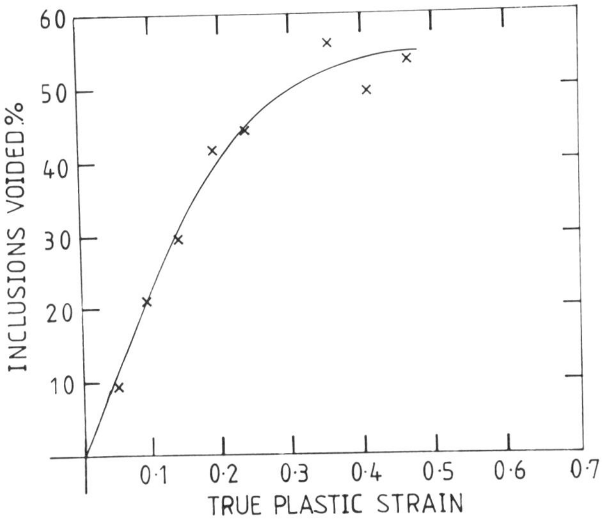

Fig. 2 - Variation in percentage of inclusions associated with voids as a function of strain

Figs 3 and 4 are fractographs taken from tensile specimens of W1 in the unaged and aged condition respectively. Fig. 3 is typical of a ductile-dimple failure resulting from the growth and coalescence of inclusion nucleated voids and the dimple spacing is consistent with the inclusion spacing. Fig. 4 exhibits the brittle facets which characterised the fracture surface of aged W1 and which indicate that much of the fracture has involved cleavage of the σ-phase.

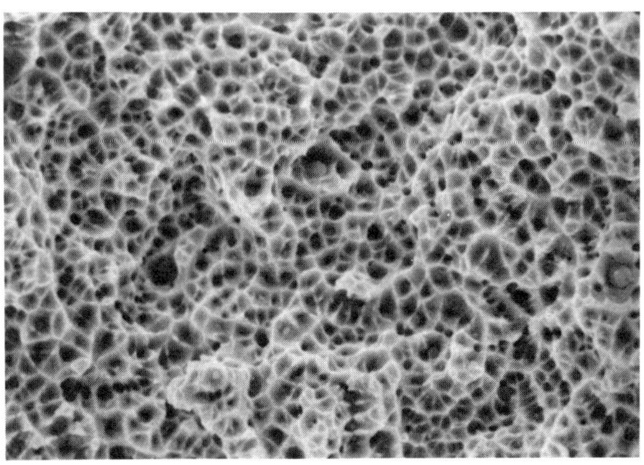

Fig. 3 - Fractograph from tensile specimen of 'as-deposited' W1(x850)

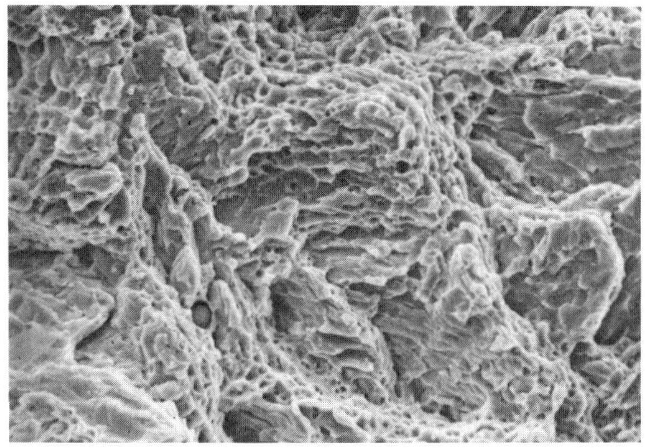

Fig. 4 - Fractograph from a tensile specimen of aged W1(x425)

Carefully prepared metallographic sections of tensile specimens of aged W1 showed that σ-phase cracking was readily observable, and was extensive, especially in ligaments of σ oriented in the direction of the tensile axis. Occasional cracks were also found in large masses of σ lying at right angles to the tensile axis, Fig.5, these cracks being up to 10 μm in length in some cases.

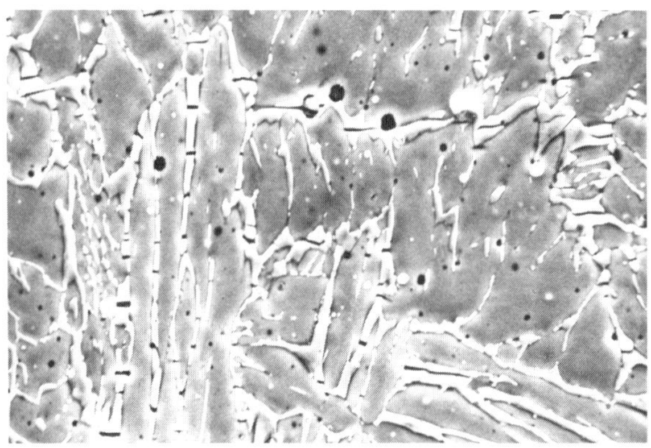

Fig. 5 - Longitudinal metallographic section through a tensile specimen of aged W1 pulled to fracture showing cracking in the σ-phase(x3400)

DISCUSSION

The Effect of δ-Ferrite - The results from previous studies of the influence of δ on the mechanical properties of stainless steels are inconclusive. For example, in studies of 308 weld metal Edmonds et al (4) found that the presence of δ-ferrite (up to 15%) reduced the impact toughness of the material while Hawthorne and Menke (9) concluded that δ had no significant effect on the notch toughness. The latter conclusion was supported by the results of Roberti and Firrao (10) who studied wrought material (18Cr 5Ni 2½Mo) containing much larger amounts of δ (54-79%) and found its fracture toughness to be little affected by the δ content.

While tensile properties are not strictly comparable with impact and toughness data it is reasonable to assume the different properties to be inter-related. In the present work the data for the 304L plate provides unambiguous evidence of the negligible effect of δ on the ductility of this material and the similarity in strain to failure between W1 and W2 provides further confirmation of this - though the two weld metals may have slightly different inclusion contents.

In the case of the results obtained from W2 before and after the 1050°C treatment it is of interest to note that the U.T.S./P.S. ratio and the (uniform) strain at the U.T.S. both increase significantly after the treatment and approach the values for the (annealed) plate material. This suggests that the weld metal is in a work hardened state after deposition (as observed by King et al (11)) and the annealing treatment not only eliminates the δ-ferrite but also relieves the internal stress in the material. The overall strain to failure, however, is only slightly affected because although the uniform component of strain doubles after the anneal there is a sharp drop in the non-uniform component which more than compensates for this. Although these effects complicate the interpretation of the results in terms of the influence of the δ on ductility, the marginal decrease in failure strain resulting from the anneal is difficult to interpret in any other way than that the effect of the δ is negligible.

Inclusion Nucleated Failure - It is widely recognised (12-14) that the growth and coalescence of inclusion nucleated voids can be a major factor in limiting the ductility of metals and, on this basis, the different ductilities of weld metal and plate can be attributed to the higher inclusion content of the weld metal. It is of interest to note, however, that while all the inclusions nucleate voids before fracture the growth of the voids is almost entirely longitudinal suggesting that in plain tensile specimens at least, the accumulation of deformation induced damage of this type is dominated by longitudinal strain rather than stress state. This being so, the attainment of a sufficient void volume fraction to produce, flow localisation, coalescence, and final fracture will be largely determined by the accumulation of sufficient total strain irrespective of whether this arises from a large uniform and small non-uniform component or vice-versa. This is consistent with the comparatively small change in the total strain to failure in W2 before and after the 1050°C heat treatment and the similar value of failure strain for as deposited W1, the total ductility in each case being determined solely by the inclusion content.

σ-Phase Cracking - Clearly σ-phase formation in aged W1 results in fracture that is essentially brittle in character, as evidenced by the fractography, the marked reduction in ductility and particularly the virtual absence of any non-uniform strain. Nevertheless, there is significant plastic strain prior to the U.T.S. resulting in deformation induced damage in the form of a progressively increasing amount of cracking in the σ prior to final failure. In contrast to previous work in which cracking was reported to occur along the σ- γ interface (4), most of the cracking observed here occurs within the σ itself (Fig. 5) and the final stages of fracture must involve a significant component of ductile failure of the intervening austenite to allow interlinkage of the cracks to occur. The very low failure strain associated with aged W1 (∿0.15) is such that the general development of inclusion nucleated voids prior to fracture is small, but the concentration of stress and strain in the austenite adjacent to severely cracked regions of σ must eventually reach sufficient levels to induce its failure by the growth and coalescence of inclusion nucleated voids. In this respect, even though the cracking of the σ plays a critical role in the damage accumulation which leads to fracture, the inclusion content of the material still plays some part in determining the overall strain at which failure occurs.

REFERENCES

(1) Farrar, R.A. and R.G. Thomas, J. Mat. Science 18, 3461-74 (1983)

(2) Lai, J.K. and J.R. Haigh, Welding J. Res. Suppl. 58, 1s-10s (1979)

(3) J.M. Leitnaker, ibid. 61, 9s-12s (1982)

(4) Edmonds, D.P., Vandergiff,D.M. and R.J. Gray, 'Properties of Steel Weldments for Elevated Temperature Applications', p47-61, San Francisco (1978)

(5) F.R. Beckitt, J.I.S.I. 207, 632-38 (1963)

(6) Farrar, R.A. and R.G. Thomas, J. Mat. Science 20, 2828-38 (1985)

(7) Thomas, R.G. and S.R. Keown, 'Mechanical Behaviour and Nuclear Applications of Stainless Steels at Elevated Temperature', p30-38, The Metals Society, London (1981)

(8) Horton, C.A.P., Marshall, P. and R.G. Thomas, ibid, p66-72 (1981)

(9) Hawthorne, J.R. and B.H. Menke, 'Structural Materials for Service at Elevated Temperatures in Nuclear Power Generation', p351-64, Houston, Texas (1975)

(10) Roberti,R. and D. Firrao, 'Advances in Fracture Research', Sixth International Conference on Fracture, p1337-44, New Delhi, India (1984)

(11) King, R.T., Steigler, J.D. and G.M. Goodwin, Welding J. Res. Suppl. 53, 307s-13s (1974)

(12) K.E. Puttick, Phil Mag. 4, 964-69 (1959)

(13) Goods, S.H. and L.M. Brown, Acta Met. 27, 1-15 (1979)

(14) Edelson, B.T. and W.M. Baldwin, Trans. Am. Soc. Met., 55, 230-250 (1962)

(15) Wegrzyn, J. and A. Klimpel, Welding J. Res. Suppl.60, 146s-54s (1981)

WELDABILITY OF RAPIDLY SOLIDIFIED TYPE 304 STAINLESS STEEL

D.E. Clark, C.J. Einerson, R.B. Loop
EG&G Idaho, Incorporated
Idaho Falls, Idaho

RAPID SOLIDIFICATION PROCESSING (RSP) of metals is generally recognized to have four beneficial characteristics as compared with ingot metallurgy processing:[1] (1) the reduction or elimination of solidification microsegregation; (2) the production of fine scale microstructures; (3) the extension of solid solubility of alloying elements; and (4) the retention of metastable phases at ambient temperatures. These characteristics offer potential advantages in new alloy development, and in particular in reducing the dependence of these new alloys on strategic and critical materials such as chromium.

Rapid solidification occurs when a material loses heat to its surroundings during solidification at cooling rates greater than a technologically defined limit of 10^3 K/s. In order to extract heat sufficiently quickly, at least one dimension of the melt must be small, a few tens of micrometers or less. Thus the RSP product has at least one dimension of this size, and has the general form of a ribbon, whisker, or powder.

In order to be useful as an engineering material, the raw product must be consolidated. For some applications, near net shape fabrication is logical; for others, the secondary fabrication of bulk shapes is more sensible. In either case, the thermal cycles associated with consolidation processing must not harm the desirable RSP characteristics. Depending on the material and which of these characteristics are to be preserved, consolidation thermomechanical cycles can range from the very fast, low bulk temperature rise of dynamic compaction (as by explosives), through conventional hot working, such as hot extrusion, to the relatively slow, high-temperature hot isostatic pressing.

* The work described in this paper was supported by the Interior Department's Bureau of Mines under Contract No. J0134035.

Similar criteria apply to joining as a secondary fabrication process: its associated thermomechanical cycles must not exceed the previous thermal history of the material so as not to destroy the desirable RSP properties for which it was chosen in the first place.

SCOPE OF THIS WORK

The present investigation involves experimental and modeling work in welding processes considered potentially suitable for the joining of RSP materials.

The processes investigated experimentally in some detail include laser, capacitor discharge, inertia, and gas tungsten arc welding. Although the latter is clearly not a rapid solidification process, it was used as a well characterized standard for comparing the welding response of rapidly solidified materials with both conventional and other RSP materials.

Modeling efforts were considered important to this study for correlation with the original RSP parameters. Such a correlation is required for determining the suitability of joining techniques. Many joining processes, and in particular those that may be most suitable for RSP joining, take place so quickly or in such small volumes that their parameters are, like those of RSP production techniques themselves. very difficult to measure.

The material chosen for this work is Type 304 stainless steel. It is technologically important, commonly welded, a large user of chromium, and, in ingot metallurgy form, well studied by physical and welding metallurgists.

EXPERIMENTAL AND MODELING PROCEDURE

MATERIALS - The chemical compositions of the RSP materials used in this work are given in Table 1. The centrifugally atomized (CA) material was produced by Pratt & Whitney Government Products Division. In this process, a molten stream hits the center of a shallow,

Table 1 - Chemical Compositions of Type 304 RSP Materials, wt%

Material	C	Mn	Si	P	S	Cr	Ni	Mo	Cu	He at. ppm
CA	0.052	0.80	0.65	0.023	0.002	18.36	9.13	0.60	0.48	7.1
VGA	0.015	0.03	0.04	0.005	0.005	18.97	9.88	0.015	0.005	0.002

rapidly spinning cup, and is atomized at its edge; subsequent rapid radiative and convective cooling occurs in the circulating helium atmosphere. These particles have a very clean surface, with few "satellites" (clinging smaller particles), an average diameter of approximately 40 μm, and a dendritic solidification structure with a secondary dendrite arm spacing corresponding to a cooling rate of approximately 10^4 K/s. The vacuum gas atomized (VGA) powder, from Homogenous Metals, is atomized by releasing a gas-supersaturated melt into a vacuum. These particles have a heavier surface oxide, more satellites, an average diameter of approximately 80 μm, and a cellular solidification structure whose cell spacings suggest cooling rates during solidification of the order of 10^3 K/s.

The consolidation methods used to produce bulk shapes from these powders include hot extrusion at 900°C, hot isostatic pressing at 900°C and 207 MPa for 2.5 h, and dynamic consolidation by an explosively generated shock wave. The baseline RSP 304 material in this study is the centrifugally atomized powder consolidated by hot extrusion.

WELDING PROCEDURES AND MODELING TECHNIQUES - For each process examined, experimental procedures, welding parameters, and the modeling used are described below.

Autogenous Gas Tungsten Arc Welds - Some travelling welds were made, but for material conservation and modeling precision, most of the work was done with spot welds. The parameters were: material thickness, 3 mm; arc voltage, 9-10 V; shielding gas, argon; current, 50 to 80 A; weld times, 3, 6, and 9 s. The gas content of porosity found in these welds was investigated by breaking notched specimens in a scanning Auger microscope with attached residual gas analyzer. The modeling uses CONRAD,[2] a 2D finite element code which assumes a Gaussian heat source, models the transient response from arc ignition through resolidification and cooling, and includes heat of fusion and the temperature dependence of thermal properties, but does not include the effects of fluid flow.

Capacitor Discharge Welds - This process involves the initiation of an arc between rapidly approaching workpieces, superficial melting, and extinguishing of the arc and expulsion of molten material in a short time, approximately 1 ms. Electrical conditions were approximately 90 V at 80,000 μF. Drop heights ranged from 50-70 mm, drop weights from 6-13 kg, and ignition tip length from 0.6-1.0 mm. The model used for this process is SINGLE[3], a 1D version of the same finite element code used to characterize GTA spot welds.

Laser Welds - Autogenous laser welds were made at 400 W average power with a Nd-YAG pulsed laser under argon shielding gas. Travel speeds from 2 to 42 mm/s (5 to 100 in./min.) were used to produce welds of various penetrations and depth-to-width ratios. No attempt was made to model these particular laser welds; their microstructures were examined and compared with those produced by other fusion welding processes.

Inertia Welds - Inertia welds were produced in 22 mm (0.875 in.) diameter material using a 5.09 lb-ft^2 flywheel at 4200 rpm and 1138 psi upsetting force, and in 13 mm (0.500 in.) diameter material using a 0.61 lb-ft^2 flywheel at 5300 rpm and 700 psi upsetting force.

RESULTS AND DISCUSSION

GTA WELDS - The most obvious feature of the GTA welds in the CA material, no matter what the consolidation technique, is the large amount of porosity in the fusion zone. Such porosity persists in material preweld heat treated at up to 1200°C, and was not found in the VGA or conventional 304 materials. The gas content of these pores was determined from welds fractured through the porosity to be helium. Analysis of the bulk material yielded a He content of 7.1 atomic ppm, which corresponded well with the amount of porosity observed. The source of the helium is almost certainly the helium atmosphere in which the original RSP product was solidified, though the precise mechanisms are uncertain.

The most notable HAZ feature in the CA/hot extruded material was the complete lack of grain growth (from an as-extruded grain size of approximately 8 μm) under the welding thermal cycle, see Figure 1. The sluggish grain growth response of this material has been confirmed in furnace annealing treatments, and is still under investigation. The grain size is much smaller than the original powder particle size, and residual surface oxides that might pin the grain boundaries are not in evidence in the extruded product. The helium itself may play a role, either by the direct mechanical action of

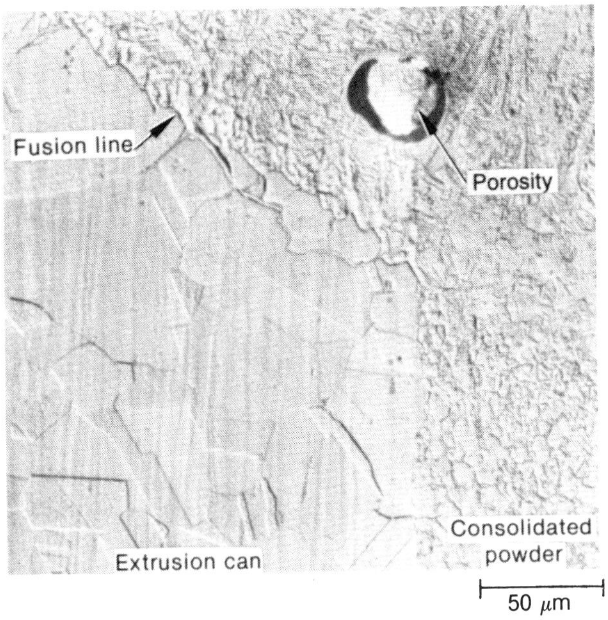

Fig. 1 - The fusion zone and near HAZ of a GTA weld overlapping the consolidated powder material and the conventional ingot metallurgy 304 extrusion can, showing much reduced grain growth in the powder material under identical thermal cycles.

He bubbles, the tying up of vacancies, or the stabilization of another phase.

The predictions of the model and secondary dendrite arm spacing measurements in the fusion zone of the GTA spot welds agreed in indicating cooling rates of the order of 10^3 K/s.

CAPACITOR DISCHARGE WELDS - Porosity in the material produced by centrifugal atomization, again assumed to be due to residual helium from the powder processing, was evident in a number of CD welds.

As can be seen in Figure 2, the CD weld is quite thin (40-100 μm), or roughly the same dimensions as both the original RSP product and the products of splat quenching techniques known to produce high cooling rates and rapid solidification.[4,5] In the case of CD welding, however, since growth is epitaxial, there is no interface, and cooling is ideal rather than Newtonian. Both the finite element model developed in this work and an extrapolation of observed primary dendrite arm spacings[6] (solidification mode is cellular) suggest cooling rates of 10^5 to 10^6 K/s, considerably in excess of those in the original powder products. Although these materials have Cr_{eq}/Ni_{eq} ratios that produce primary ferrite solidification in GTA welds, no ferrite was observed metallographically, and the presumption is that solidification occurred as primary austenite due to the high cooling rates.[5] The model also calculates solidification rates and temperature gradients in the liquid ahead of the solidification front, allowing welding and rapid solidification methods to be mapped onto a plot of G_L (temperature gradient in the liquid at the liquid/solid interface, K/mm) vs. R (solidification rate, mm/s), as shown in Figure 3.

LASER WELDS - The deep penetration laser welds showed a variation in cell spacing from bottom to top, indicating that the cooling rates are different in these regions, as suggested by other work.[5] The shallow laser welds made at higher travel speeds have approximately the same thickness as the CD welds and the original forms of the RSP materials, although primary dendrite arm spacings (again solidification is cellular) indicate somewhat slower cooling rates. Porosity, again assumed to be due to residual helium, was obvious in the laser welds, causing considerable surface eruption in the deeper penetration welds.

INERTIA WELDS - This welding process produced sound welds in RSP and conventional materials. The thermal cycles associated with inertia welding last for a second or more near the melting point and, though associated with severe mechanical work, are thus comparable in

Fig. 2 - Capacitor discharge weld in centrifugally atomized HIP-consolidated Type 304 RSP material. The weld is very thin, of approximately the same dimensions as the original RSP powders.

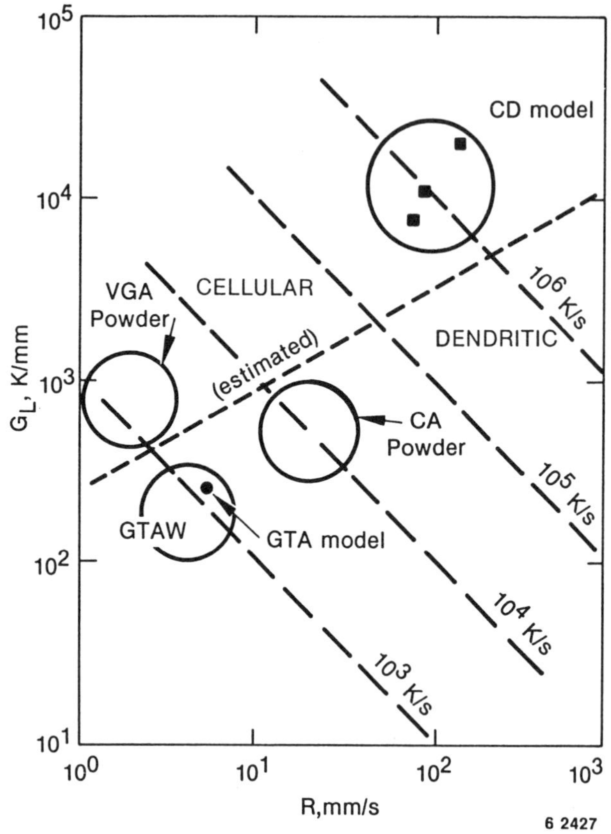

Fig. 3 - Schematic representation of solidification parameters as calculated by thermal models in this work. The cellular/dendritic transition is estimated based on the observed solidification structures. The areas representing VGA and CA powders can be placed on the plot based on known solidification structures and cooling rates estimated from dendrite arm or cell spacings, and the resulting G_L and R estimated.

scale and duration to HAZ thermal cycles in the GTA process. As a solid state process, inertia welding avoids doubt about resolidification rates. The gross helium porosity problem in fusion welds in materials produced by present centrifugal atomization methods is not present, but the possibility of very fine helium porosity in the solid state at higher temperatures is still under investigation.

SUMMARY AND CONCLUSIONS

The criteria for joining rapidly solidified alloys without losing the desirable characteristics imparted by RSP include (1) thermal cycles in the HAZ which do not exceed those experienced in the preceding rapid solidification and consolidation, and (2) the resolidification of any fusion zone in the RSP regime. Capacitor discharge and laser welding involve rapid thermal cycles approaching those of RSP techniques. Inertia welding subjects the material to longer but still relatively short thermal cycles, and avoids melting. Gas tungsten arc welding, while ultimately unsuitable for conserving RSP properties in a weldment, reveals a porosity problem in centrifugally atomized material due to helium entrapped during rapid solidification processing; the helium may also be responsible for the sluggish grain growth kinetics in this material.

REFERENCES

1. N. J. Grant, Rapid Solidification of Metallic Particulates, J. Metals, January 1983, 20-27.

2. H. B. Kraus, "COUPLE/MOD6 General Purpose Two-dimensional Finite Element/Finite Difference Heat Conduction Code," RE-A-84-002, January 1984.

3. H. B. Kraus, "SINGLE/MOD2 - A One-dimensional Finite Element Steady State and Transient Heat Conduction Code with Contiguum Element Capabilities for Use on a CDC Cyber 176/Cyber 173 Computer," 1983.

4. D. M. Kroeger, W. A. Coghlan, D. S. Easton, C. C. Koch, J. O. Scarbrough, A Study of Cooling Rates During Metallic Glass Formation in a Hammer and Anvil Apparatus, J. Appl. Phys., 53:3 1445-1453, 1982.

5. J. M. Vitek, A. Dasgupta, and S. A. David, Microstructural Modification of Austenitic Stainless Steels by Rapid Solidification, Met. Trans. 14A:1833-1841, 1982.

6. T. Okamoto, K. Kishitake, K. Murakami, Solidification Structure and Segregation in Iron-Chromium-Nickel Alloys, Trans. Iron and Steel Institute of Japan, 21:641-648, 1981.

EVALUATION OF WELDABILITY IN 60 Kg/m HEAVY RAILWAY STEEL USING GLEEBLE 1500 SIMULATION TECHNIQUE

Zu Ze Xu
Central Iron and Steel Research Institute
13 Tai Ping Zhuang
Beijing, China

ABSTRACT

Various aspects of weldability of heavy rail steel have been examined by simulation using the Gleeble 1500 technique. Both thermit and flash welding are found to produce acceptable HAZ properties from the viewpoints of strength, fatigue, hardness and toughness.

INTRODUCTION

Transportation in China depends largely on rail traffic and is expected to develop greatly in the near future. The increasing demand has required production of rails suitable for heavy duty services. Since 1976, 60 Kg/m heavy railway steel has been produced at two steel companies in Baotou and Dukou cities to meet this demand. The modernisation of the rail network in this regard has necessitated that welding be used to join rails rather than the mechanical joining which had mainly been used previously. It was therefore essential to examine the weldability of the rail steels to judge their suitability for different types of welding and their mechanical properties in the welded condition (1,2,3).

Welding processes which are relevant in this case are thermit welding and flash welding by electrical resistance heating. It was considered of interest to examine different carbon contents within the normally accepted range (0.67%-0.80%) as well as the influence of phosphorus as an impurity element and niobium as a possible alloying element.

The Gleeble 1500 welding simulator has been applied in this work to assess the suitability of the steels for certain aspects of welding such as hot ductility. The Gleeble has also been used to simulate heat affected zone (HAZ) structures for different conditions of welding. These samples were subsequently evaluated using a series of testing methods for hardness, toughness, tensile and fatigue strength, and related to the continuous cooling transformation (CCT) behaviour.

The experimental programme was as shown in Figure 1. All combinations of conditions were examined for the two commercial steels. A more limited range of testing was carried out on the laboratory steels. Table I shows the chemical analyses of the investigated materials, and Figure 2 shows the experimental lay-out in schematic form.

TABLE I CHEMICAL ANALYSES OF THE INVESTIGATED STEELS

Steel code	%C	%Mn	%Si	%S	%P	%Nb
A	0.67	0.90	0.21	0.01	0.021	-
B	0.77	0.86	0.23	0.012	0.019	-
C	0.70	0.90	0.18	0.017	0.022	0.02
D	0.70	0.92	0.20	0.017	0.020	0.05
E	0.73	0.92	0.20	0.018	0.020	0.08
F	0.71	0.98	0.22	0.017	0.020	0.10
G	0.68	0.96	0.21	0.017	0.007	-
H	0.69	0.96	0.21	0.017	0.012	-
I	0.68	0.95	0.21	0.015	0.033	-
J	0.61	0.89	0.18	0.040	0.054	-

Steels A and B, commercially manufactured by the Baotou Iron and Steel Co.
Steels C-J, laboratory vacuum-induction melts

RESULTS

CONTINUOUS COOLING TRANSFORMATION - the situation for steels of this type is shown in Fig.3. Provided that the cooling rate is less than 5 deg.C/sec no martensite is formed. For higher cooling rates bands of martensite start to appear associated with segregation of the alloy elements. For cooling rates of more than 0.5

degC/sec the amount of ferrite is reduced to below 5%. The conditions typical of flash and thermit welding are, therefore, acceptable from the point of view of correct microstructure development.

HARDNESS TESTS - on simulated welds, Fig.4, show that the values lie between 30 and 40 Rockwell C. The commercial base steels have about 28 RC hardness while the laboratory steels are significantly softer. Weld hardness is increased by increasing carbon content and cooling rate e.g. in the flash weld cycle. However, the hardness is almost independent of position in the rail or content of phosphorus and niobum.

TOUGHNESS - is somewhat reduced in the high temperature simulated HAZ structure but only minor differences exist between the two welding processes. Greatly increased toughness is found in parts of the microstructure which have just exceeded the AC_3 temperature. This appears to be associated with refinement of the pearlite structure.

FATIGUE TESTS - show that in general the HAZ structure has equal or better properties than the base material (Fig.6). Some deterioration is observed with thermit welding of the low carbon variant which may be associated with an increased ferrite content.

HOT DUCTILITY - Fig.7 show that hot ductility is good between 700°C and 1200°C. There is a small decrease around 600°C while nil ductility and nil strength occur at 1300°C. Problems are not encountered with bot ductility in practice.

CONCLUSIONS

The base composition of the steels now being produced in China for 60 Kg/m heavy rail is suitable for both flash welding and thermit welding. With HAZ cooling rates between 0.2 °C/sec and 2°C/sec it is possible to avoid martensite formation and maintain the content of proeutectoid ferrite below 5%. Hardness distributions are quite uniform and can be controlled within the range 30-40 Rockwell C.

There is only minimal deterioration in tensile strength and ductility as a result of the thermal cycles from welding. Toughness is slightly decreased after exposure to high temperature (∿1300°C) but is greatly improved in those regions which reached temperatures of ∿760°C. The laboratory steels showed good toughness in all cases but with a reduced hardness. Fatigue strength is not greatly affected by the welding simulations that have been examined.

Good hot ductility is observed between 700°C and 1100°C. Nil ductility and nil strength temperatures occur close to 1300°C.

The Gleeble 1500 dynamic thermo-mechanical technique has proved to be of great value in the evaluation of weldability of heavy railway steel carried out in this project.

ACKNOWLEDGEMENTS

The present project was performed by the group for studing weldability of the P14-60 kg/m railsteel and suported by the National Science and Technology Commitee, and the Ministry of Metallurgy Industry in China. The actual welding thermal cycles are provided by the Institute for Railway Research in Beijing. Dr Bevis Hutchinson and Mr Alf Sandberg of the Swedish Institute for Metals Research kindly helped with preparation of the manuscript. The author is indebted to members of the above organizations for valuable support and help.

REFEREBCES

1. Zu ze, Xu, Internal report of Central Iron and steel Research Institute in Beijing, China." On the weldability and test methods of heavy railsteel" (1985, April).

2. Fletcher, F.B, Proceedings of the Heavy Railway Conference, Perth, WesAAustralia. "Development of high strength chromium-molybdenum Rail steel with improved weldability" (1977, Sept).

3. Teruaki, Transation ISIJ, 21, 260-269. (1981).

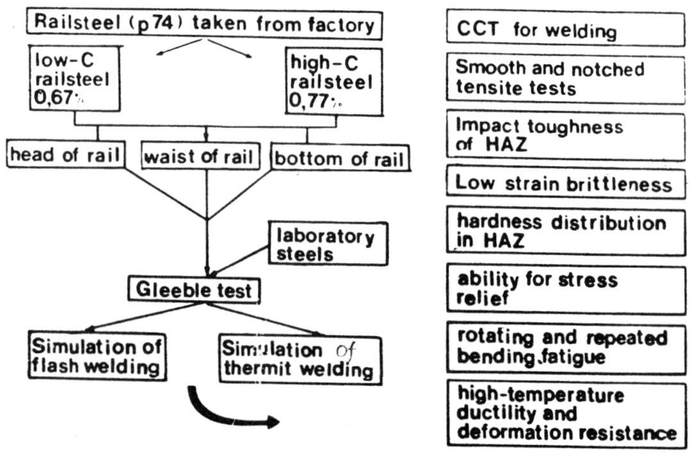

Fig 1

Fig 2

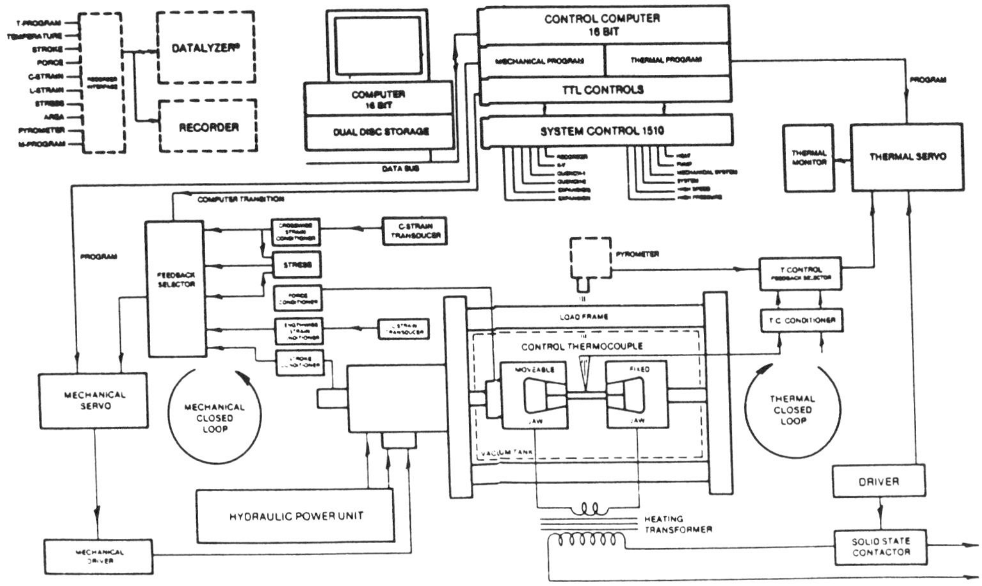

Fig 3

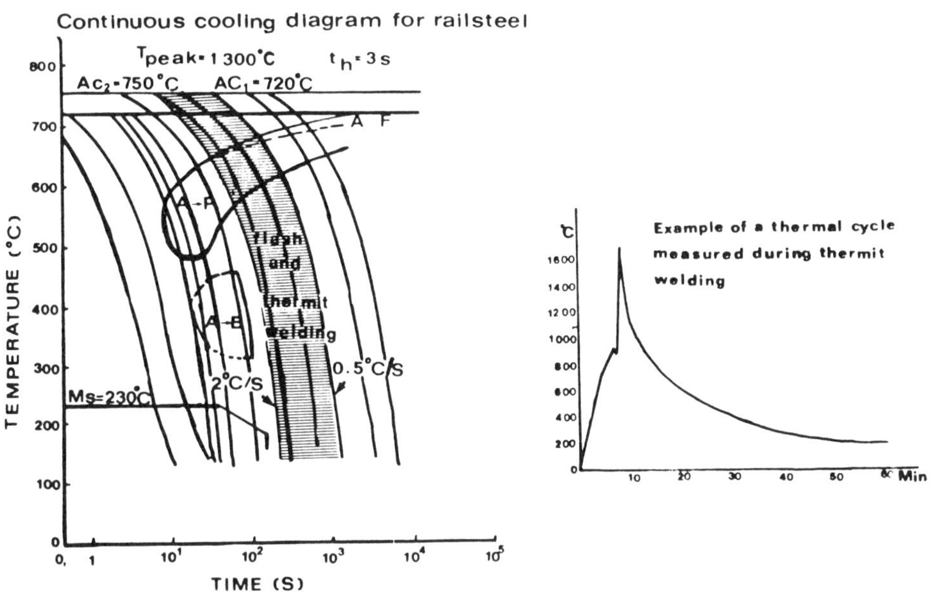

Fig 4

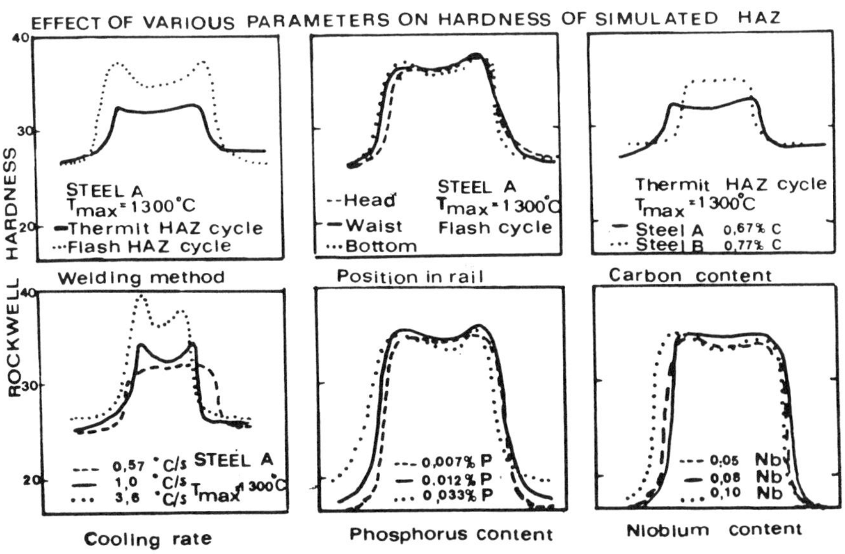

Effect of various HAZ conditions on impact toughness

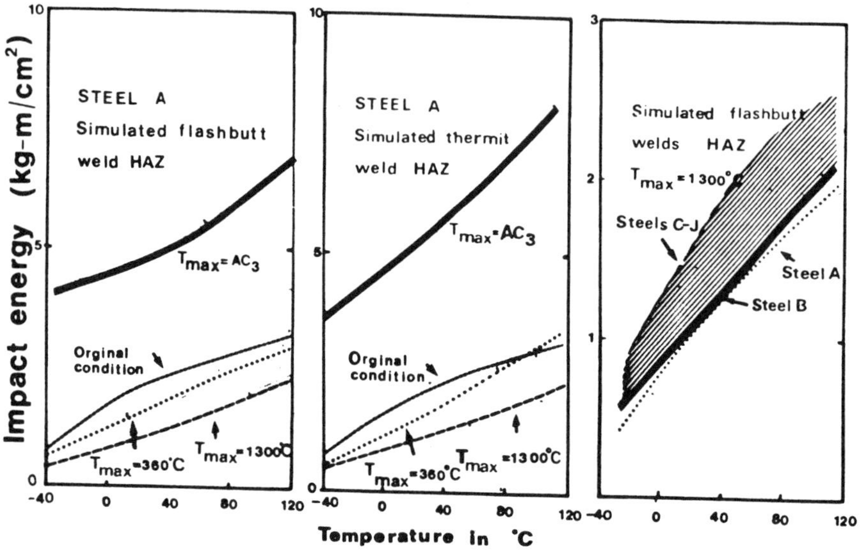

Fig 5

EFFECT OF WELD SIMULATION ON FATIGUE STRENGTH IN ROTATING BENDING

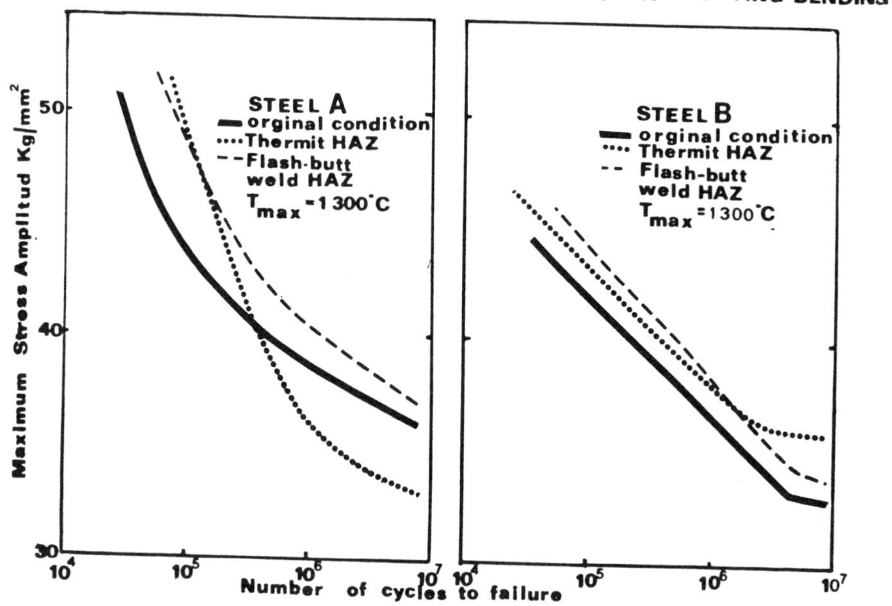

Fig 6

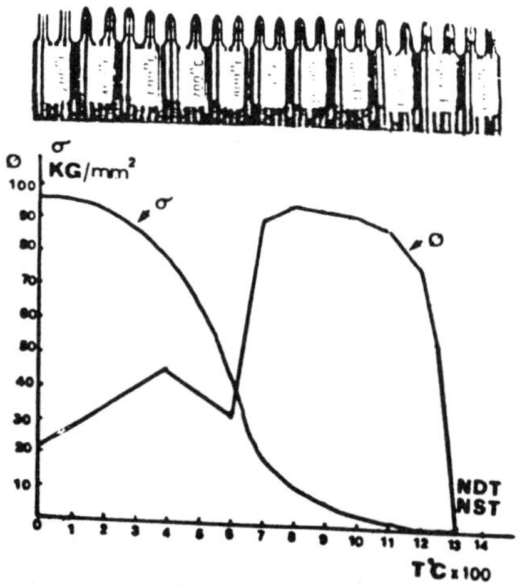

hot-ductility and strength of P17-60KG/M rail steel

Fig 7

HAZ THERMOMECHANICAL HISTORY FOR A LARGE-DIAMETER STAINLESS STEEL PIPE WELD

D.G. Atteridge, W.E. Anderson
Pacific Northwest Laboratory
Richland, Washington 99352

ABSTRACT

Thermomechanical (TM) history during welding was determined for a 24-inch-dia (61-cm-dia) Type 304 stainless steel (SS) schedule 80 pipe. It was measured on the inside counterbore surface as a function of distance from the weld centerline (WCL). Both thermal and strain measurements were obtained. The girth weld was made using a cold-wire-feed, mechanized gas-tungsten-arc pipe welding head. Weld parameters as well as inside counterbore surface temperatures and deflection measurements were recorded during each pass. The weld was completed in 35 passes, with the first half inch of weld metal into a 37-1/2-degree bevel and the second half inch into a 10-degree bevel. The TM history measurements, taken about the WCL, included weld shrinkage measurements parallel to the pipe axis, counterbore surface temperature measurements, counterbore surface deflections parallel and perpendicular to the weld fusion line, and radial deflections of the counterbore surface.

AUSTENITIC STAINLESS STEEL COMPONENTS of commercial boiling water and pressurized water reactors have experienced stress-corrosion cracking (SCC) in the heat-affected zone (HAZ) of SS welds in service. Both the plastic strain and the thermal histories produced during welding contribute to SCC susceptibility of the HAZ, as discussed in a companion paper (1). Development of a method of predicting SS HAZ resistance to SCC has been under way for several years (2,3,4,5). Quantifying weld TM history is part of this larger program oriented toward predicting SCC susceptibility of austenitic SS pipe welds based on pipe chemistry and fabrication parameters.

EXPERIMENTAL

Thermomechanical history measurements on a 24-in.-dia SS pipe girth weld demonstrated that a weld HAZ is subjected to a complicated strain history along with, and induced by, heating and cooling cycles. The TM history was monitored on a pass-by-pass basis while HAZ sensitization development was monitored between passes (5,6,7). The pipe was welded in the 2G position, pipe axis vertical, using mechanized gas tungsten arc welding (GTAW). The TM history data were collected by a computer-based system, and the various TM history sensor signals were electrically and thermally isolated before they entered the analog-to-digital converter in order to keep currents/voltages associated with welding away from the computer (4,5).

Temperatures and strains were recorded as a function of arc-on-time and distance from the WCL. Counterbore surface strains were measured parallel and perpendicular to the weld. Radial surface deflections were measured perpendicular to the counterbore surface. Weld-induced changes in length of a section of the pipe, centered on the WCL, were also monitored (5).

Sensors consisted of chromel-alumel thermocouples and modified crack-opening-displacement clip gages that were used to measure the four different types of pipe strains or deflections. All temperatures and inside pipe surface deflection measurements were confined to the counterbore surface-area region for the first 20 passes. Outside surface thermocouples were then added and monitored for the rest of the weld.

The weld was completed in 35 passes. Weld pass sequence of all 35 passes is shown in Figure 1, along with the actual weld fill placement, which is contrasted with the superimposed initial weld preparation geometry. This difference between actual and initial weld groove geometry provides some indication of how weld joint geometry changes during welding due to weld metal shrinkage.

Contraction during welding, as measured by the internal extensometer, began as soon as welding initiated and before any measured temperature change was recorded at the extensometer location. Most of the contraction takes place

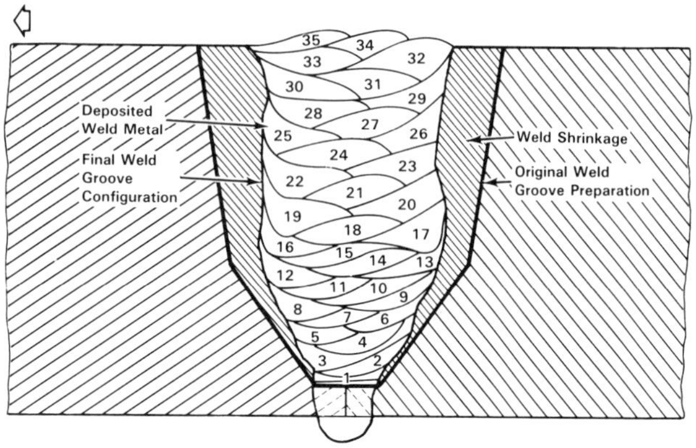

Fig. 1 - Schematic of the original and final weld groove geometry and the sequence of welding

once the arc-induced temperature spike passes the gage. The permanent post-weld contraction measured by an internal extensometer on a pass-by-pass basis is presented in Figure 2. Per-pass contraction decreased as welding continued while the overall pipe shrinkage reached close to 0.4 in. (1 cm) at weld completion.

Change in heat distribution characteristics with increasing pass number is presented in Figure 3, which shows plots of maximum temperature versus distance from the WCL for five different passes where the weld pool was approximately centered over the WCL. These maximum surface temperatures drop off relatively quickly during the early passes but stay relatively constant during later passes. The profiles tend to be flatter and broader with increasing pass number.

Figure 4 illustrates the correspondence between electrode position and counterbore temperatures. Electrode positions and weld groove geometries prior to welding are shown above the

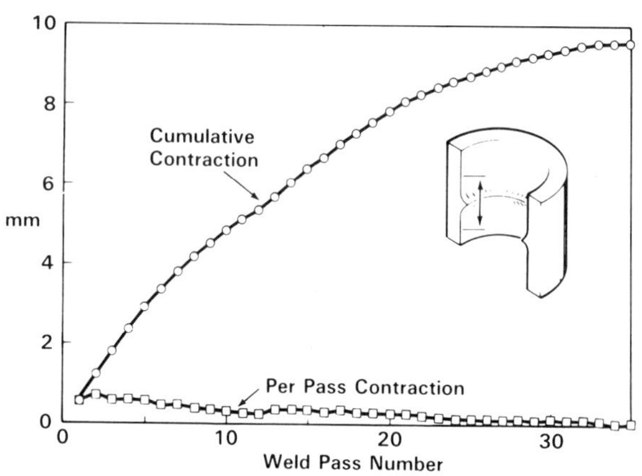

Fig. 2 - Post-pass internally measured axial pipe contraction during welding of a 24-in.-dia Type 304 SS pipe

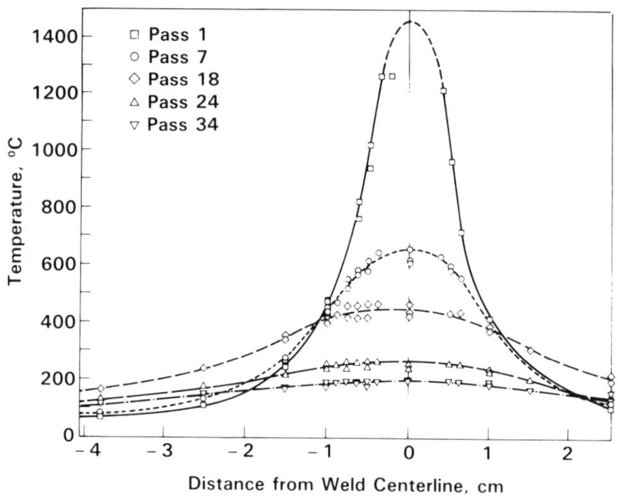

Fig. 3 - Change in maximum inside counterbore surface temperature profile as a function of weld pass number for a 24-in.-dia SS weldment

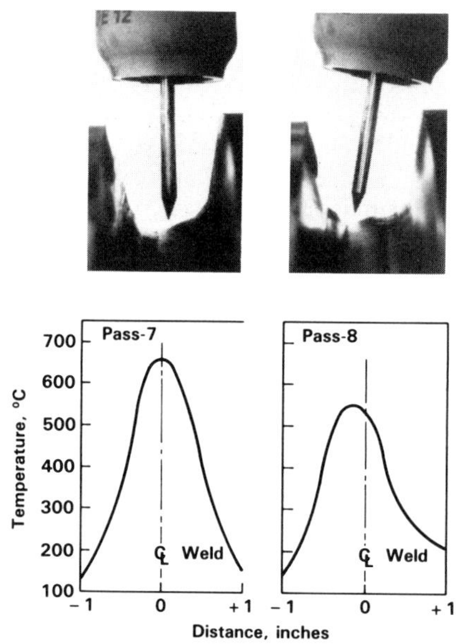

Fig. 4 - Weld-induced groove geometry changes and the effect of electrode placement on maximum temperature profiles

maximum-temperature profiles for passes 7 and 8. Pass 7 was essentially centered in the weld groove, while pass 8 was skewed to the side. Both temperature distributions exhibited the characteristics of their respective electrode position. Passes 7 and 8 were both welded at the same heat input (31 kJ/in. or 12.2 kJ/cm), and yet pass 8 exhibited a peak maximum temperature lower than pass 7 by 110°C. This is attributed to a greater heat sink capacity when the molten pool is directly in contact with the side wall.

An overall picture of the thermal characteristics of the weld aspects can be seen in a plot of the maximum inner-surface temperatures and the actual heat input per pass in Figure 5. Because the temperatures result not only from weld settings but also from the particular bead position, these too are presented as a function of weld bead centroid radius. It is evident that the distance of the weld bead from the inside counterbore surface is a dominant factor, with heat input being a secondary consideration.

Straining of the counterbore surface during welding does not take place until the arc is close to the strain sensor position. A typical surface strain and temperature profile is shown in Figure 3 for a clip gage located 0.40 in. (1 cm) from the original WCL. The surface material experienced compression before a surface temperature change was measured and went into tension once the temperature began to increase. The material then went back into compression as the arc traveled on. A permanent tension offset appeared after pass completion. The magnitude of the strain change, as well as the maximum temperature, changed on a pass-by-pass basis in relationship to the position of the welding electrode and the distance of the molten pool from the counterbore surface.

The motions depicted in Figure 6 can be evaluated in terms of unit strain, Figure 7. These strains can be viewed as consisting of two major aspects: 1) permanent plastic strain, which is the difference between the pre-weld strain and the cooled, post-weld strain; and 2) total "cyclic" strain, which is the sum of tension and compression strains taking place between the initial and final states. Total cyclic strain includes thermal expansion

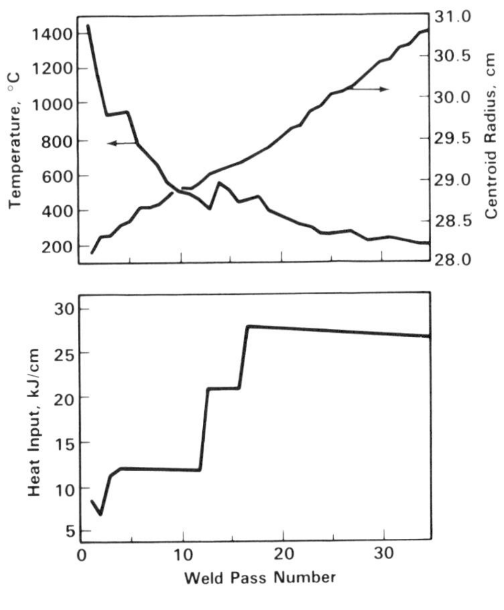

Fig. 5 - Relationship between maximum inner-surface temperatures, actual heat input and weld bead position

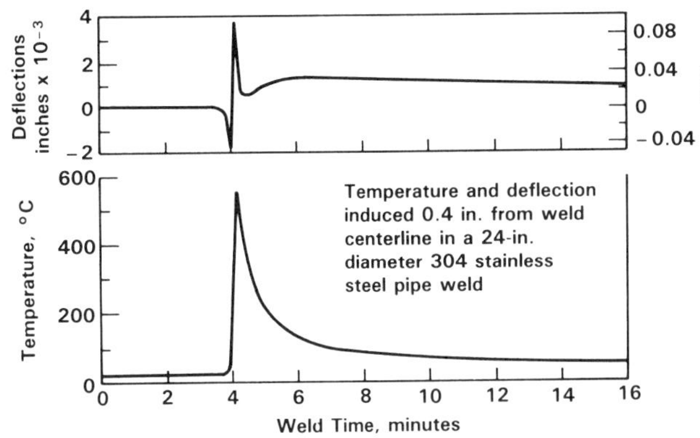

Fig. 6 - Temperature and surface deflection parallel to the weld centerline induced on the counterbore surface 0.40 in. from the weld centerline in pass 3 of the 24-in.-dia pipe weld

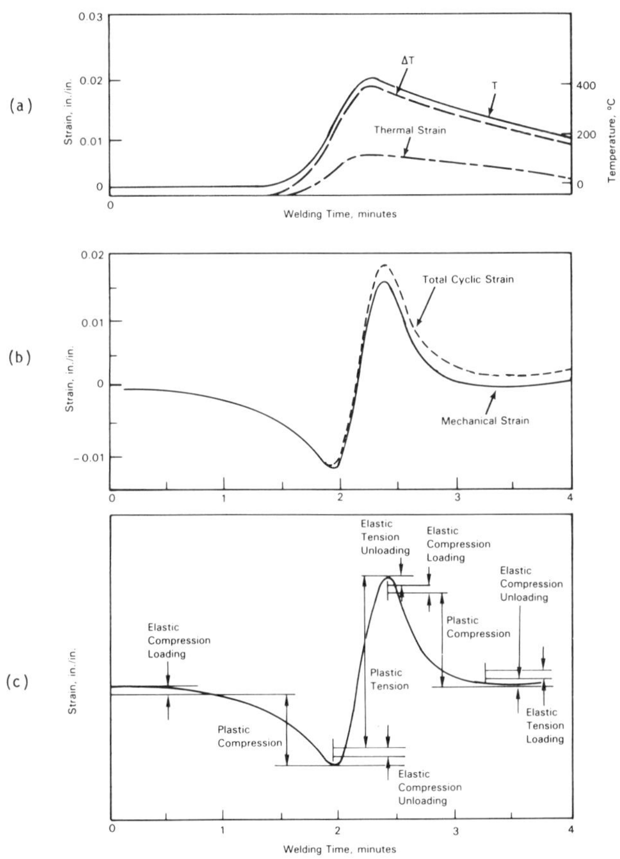

Fig. 7 - Illustration of the separation of strains measured parallel to the WCL into (a) measured temperature cycle and thermal strain component, (b) measured total strain and derived mechanical strain, and (c) separation of the mechanical component into compression and tension elastic and plastic regions - data for pass 18 at the WCL strains, plus elastic strains as well as compression and tension plastic strains.

Thermal strain in the surface "layer" represents metal expansion due to local transient heating; in the tests it caused gage spreading and thus increased apparent tension displacements or decreased apparent compression displacements. Corrections for thermal strains should therefore enlarge indicated compression values and diminish indicated tension values, as shown in Figure 7a and 7b, where a "mechanical strain" (measured minus thermal) has been graphically computed and displayed. Thermal strain was estimated by calculating the expected unrestrained expansion of a unit length of pipe metal. Maximum thermal strain corrections are approximately 0.007 cm/cm (i.e., less than 1%) for a temperature excursion (ΔT) of 400°C.

Mechanical strain includes elastic loading and unloading segments as suggested in Figure 7c; the remainder represents cyclic plastic flow or "warm/cold working" and is numerically approximated as the absolute sum of the two compression squeezings and the one tension stretch indicated in the figure. Elastic loadings are about 0.001 each, so about 0.005 (1/2% strain) should be subtracted from the summed mechanical strain excursions to produce an estimate of cyclic plasticity. For pass 18, on the WCL, Figures 7a and 7b, this results in about 3-1/2% plastic strain.

Counterbore surface deflections, recorded when the weld pool moved past strain measuring devices, imply the existance of a thermally induced deformation bump, nominally traveling under and along with the weld pool. The bump is an indication of cyclic plastic work taking place in the HAZ during each weld pass; it is pictured schematically in Figure 8.

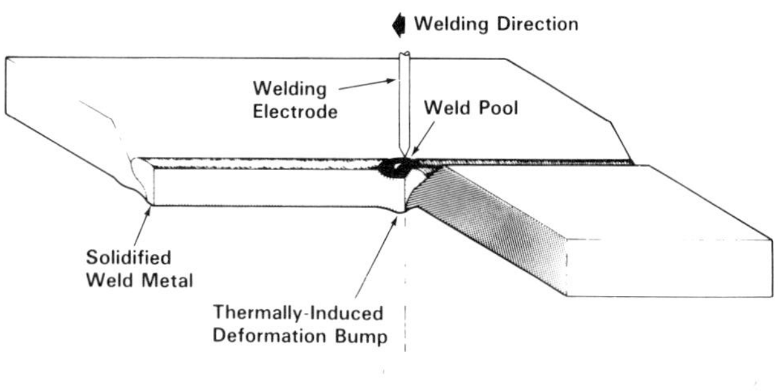

Fig. 8 - Schematic of thermally induced deformation bump

SUMMARY

The HAZ TM history during a multipass SS pipe weld has been quantitatively measured. Results indicate that complex strain and thermal cycles take place in the HAZ. Temperatures reached on the inside counterbore are primarily controlled by the distance of the weld pool from the counterbore surface. Asymmetry of the weld pool in relation to the WCL is mirrored in the spacial placement of the thermal distribution on the counterbore surface.

Counterbore strain history is controlled by expansion of the heated metal around the weld pool as well as shrinkage of the metal in the weld pool. The HAZ is subjected to both tension and compression strain components during each weld pass. These strain cycles induce work hardening in the HAZ and result in elevation of HAZ flow stress above that of the base metal.

ACKNOWLEDGMENT

The contribution of S. M. Bruemmer and R. E. Page and the technical assistance of B. Norton is gratefully acknowledged. This work was supported by the Division of Engineering Technology, Office of Nuclear Research, U.S. Nuclear Regulatory Commission under Contract DE-AC06-76RLO 1830 with Pacific Northwest Laboratory, which is operated for the U.S. Department of Energy by Battelle Memorial Institute.

REFERENCES

1. Bruemmer, S. M. and D. G. Atteridge, in this proceedings

2. Atteridge, D. G., S. M. Bruemmer and R. E. Page, NUREG/CR-3613 Vol. 1 (1984)

3. Atteridge, D. G., S. M. Bruemmer and R. E. Page, NUREG/CR-3613 Vol. 2 (1985)

4. Atteridge, D. G., S. M. Bruemmer, L. A. Charlot and R. E. Page, NUREG/CR-3613 Vol. 3, No. 1 (1985)

5. Atteridge, D. G., S. M. Bruemmer, W. E. Anderson, L. A. Charlot, B. W. Arey, B. Norton and R. E. Page, NUREG/CR-3613 Vol. 3, No. 2 (1986)

6. Bruemmer, S. M., R. E. Page and D. G. Atteridge, Nuc. Eng. and Design 89, 319-331 (1985)

7. Bruemmer, S. M. and L. A. Charlot, "Measurement and Prediction of Thermomechanical History Effects on Sensitization Development in Austenitic Stainless Steels," No. 242, Corrosion 86, NACE (1986)

A COMPARISON BETWEEN ADHESIVE-BONDING AND WELDBONDING OF HSLA STEEL

G. Moniatis, L. Dorn
Technische Universitat
Berlin, West Germany
Fugetechnick / Schweisstechnik

A. ElSawy
Faculty of Engineering
Cairo University
Giza, Egypt

ABSTRACT

The main objectives of this paper is to compare the tensile-shear load to fracture of both adhesive-bonded and weldbonded St E 340 HSLA steel lap joints, with different surface preparations, subjected to verious environmental degredations. Because no standard test for lap joints impact shear test, a special fixture has been designed to be attached to the pendulum of the impact testing machine. Results showed that the joint strength is affected by surface preparation, and degredation tests duration. The strength of the weldbonded joints is superior than the adhesive-bonded joints in withstanding environmental degredation and impact load, especially the degreased and ground joints.

THE REDUCTION OF AUTO-BODY weight dictated the use of HSLA steels to substitute the conventional low-carbon steel because of their higher strength to weight ratio. Because spot welding is the major joining process used in automobile industry, HSLA steels are required to be spot welded. The two major problems facing the weldability of such steels are :1) the restricted range of acceptable welding parameters and 2) the lack of joints ductility under impact loads (1).

Many research works have been made to solve the first problem (2-11). While the second problem, which has its practical significance in car accidents, has not yet been convincingly solved. The alternatives of spot welded joints are either to use adhesive-bonded or weldbonded joints which showed superiority under tensile and fatigue loading in different applications (15). Therefore, this work is addressed to study the reliability of the adhesive-bonded and weldbonded joints under different environmental degredation conditions and impact loading.

EXPERIMENTAL PROCEDURES

The material used in this work is St E 340 HSLA steel (0.06% C, 0.27% Mn, 0.012% P, 0.015% S, 0.02% Si, 0.027% Al, 0.01% W, 0.04% Ni, 0.03% Nb, 0.01% Mo, 0.06% Cr, 0.03% Cu, 0.003% As, 0.001% Sn, 0.005% Sb, and 0.002% Pb), and its mechanical properties are: 365 N/mm^2 yield strength, 452 N/mm^2 ultimate tensile strength, 27% percent elongation, and 70 BHN. The dimensions of the two sheets of the joints are 100x40x1.6 mm each with 12 mm overlap. To date, no single surface preparation has been found to give both optimum welding and adhesive bonding conditions (15). Therefore, prior joining, arbitrary three kind of surface preparations were used: 1) as received, 2) degreasing only with acetone, and 3) degreased and ground with 100 grid sand paper. A one component epoxy adhesive (Terokal 3919 S made by Teroson GmbH) was used because of its easeness of application and its acceptability by car industry. This kind of adhesive cures at 180 deg. C for 70 minutes. The gap thickness used in adhesive bonded joints ranging between 0.2 - 0.3 mm to get the maximum shear strength recommended by the manufacturer of 20 N/mm^2.

In weldbonding, it is not possible to achieve a large gap thickness because that the adhesive will squeeze out under the welding pressure as well as during the solidification and cooling of the spot weld nugget. The spot welding parameters used to get a nugget diameter of 5 mm, according to the following relation (1-10):

$$D = 5\sqrt{S} \quad \text{mm}$$

where: S is the sheet thickness in mm, are: 20 cycles squeeze time, 10 cycles weld time, 15 cycles holding time, 3 KN electrode force, and 7.7 KA welding current. The Cu-Zr-Cr electrodes were used as recommended by DIN 44759 for low alloy steels. For the easeness of squeezing away the adhesive from the nugget region, a hemispherical electrode shape of 750 mm diameter were used. The presence of adhesive between the faying surfaces causes an insulation effect. This was overcomed by spot welding immediately after the application of the adhesive, while its still soft and with low viscosity.

Prior tensile-shear tests, specimens were subjected to the following accelerated environmental degredation tests:

1) heating up to 20, 80, 115 and 150° C prior tensile-shear tests, 2) storage in 0.2% liquid soap diluted with distilled water at 40° C, (as suggested by industry), 3) salted water spray test (DIN 50021), and climate test(at 20° C and 65% relative humidity).

At present, there is no standardized method for spot welds impact testing (3, 12-14). Figure 1 shows the specially designed fixture to be connected to the pendulum of the impact testing machine to measure the joints impact shear strength.

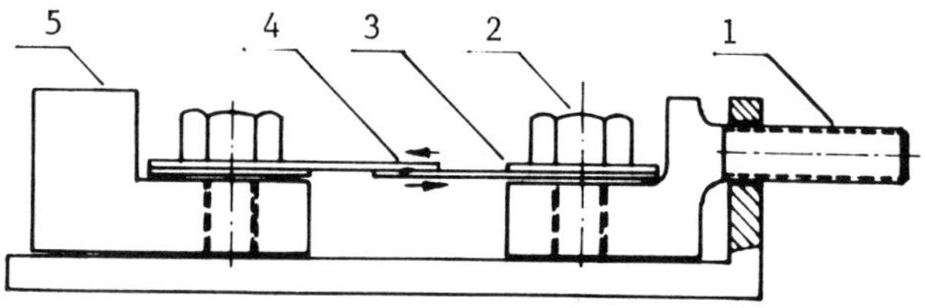

Fig. 1 - Lap joint impact testing attachement (1. end fixed to the pendulum, 2. lap joint fixing bolts, 3. hold pads or washers, 4. lap spot welded joint, 5. striking end)

Because the hammer additional weight, caused by the connection of the fixture, the following corrections have been made to keep the initial velocity v_o = 5.5 m/s according to DIN 53448.

$$V_o = \sqrt{2.g.L_{red}.(1-\cos\alpha)}$$

$$L_{red} = (0.25\ g/\pi^2)(t.n)^2/n$$

where L_{red} is the pendulum length (1.513 m)
g is the gravitational acceleration
t is the time
n is the frequency
α is the angle which makes V_o=5.5 m/s

It was found that $\alpha = 90.5^o$, for the pendulum and fixture weight of 85.503 Kg .

RESULTS & DISCUSSION

Figure 2 shows the load-deformation diagram of a weldbonded joint. The general trend shows that after a charp increase of load, a sudden drop happens succeded with another increase of the load to another peak with more deformation followed by a drop of load with deformation till fracture. This behavior suggesting that at the bignning, the whole force is carried out with both the adhesive and welded spot. Then because that the adhesive used is brittle, it fails suddenly causing the drop of the load. Then, the whole load is carried out by the welded spot only till fracture. The type of failure in the weldbonded joints depends on the combination of base metal yield strength, joint design, shear strength of the weld metal, as well as the number of spots and the adhesive material properties. During the test, the joint interface rotates in a way to be more normal to the tensile direction. For spot welded joints, at that condition, the weld metal experiences shear loading (4). As the rotation proceeds, the stress normal to the weld

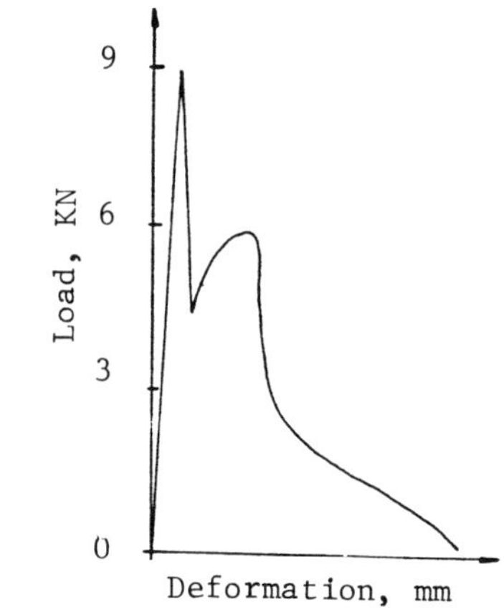

Fig. 2 - Load-Deformation Diagram of a St E 340 Weldbonded Joint

interface increases very rapidely, but the shear stress increases at a decreasing rate. If the weld metal is strong enough to resist the shear stress, the weld-metal interface rotates sufficiently to cause pullout failure, away from the lapped area, if not, interface failure takes place by shearing. With material having a higher yield strength or greater thickness, as well as for weldbonded or adhesive bonded joints,,the rotation is smaller due to the higher rigidity of the joint at the same load. So that, the shear stress increases much faster than for a lower rigidity joints. Therefore, unless the shear st strength of the material withstanding to the shear force is high enough, interface failure tends to occur as shown in Fig. 3 (4). Consequently, the determination of the shear strength of the joint interface materials withstanding to the shear stress became a necessity, in case if it is decided to use weldbonding joints to reduce the car-body weight.

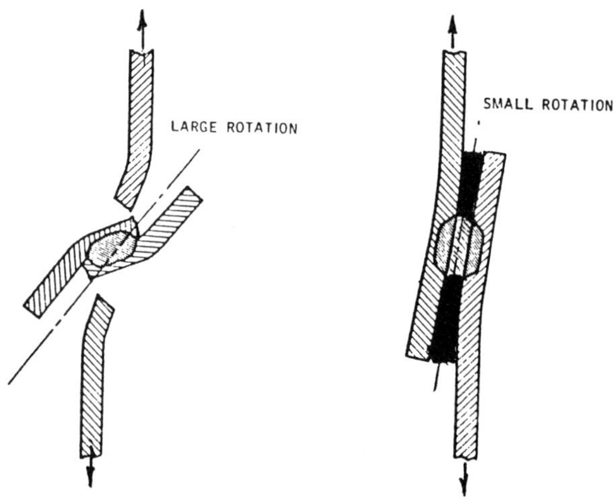

Pull-out Failure

Interface Failure

Fig. 3 - Schematic Diagram of Failure Types in Tensile-Shear Test (4)

Moreover, it was found that the surface preparation has a considerable effect on both adhesive-bonded and weldbonded joints. The strength of the joints made with specimens after degreasing and ground and tested under different environmental conditions, are higher and more consistent (less scatter) than the ones joined in the as received or degreased conditions (see Fig. 4) The reason that the tensile-shear strength of the specimens joined in the as received condition is less consistent is that the surface contaminants cause the random nonuniform distribution of heat between the faying surfaces leading to nonconsistent welding and bonding areas.

The hot tensile-shear test results showed that the testing temperature affects the strength of adhesive-bonding joints more than the weldbonded joints (Fig. 4). Generally speaking, the testing temperature has the major effect on strength for both adhesive-bonded and weldbonded joints. As previously mentioned, the surface preparation has a considerable effect also, in adhesive-bonded joints, while its effect is less noticeable in weldbonded joints. This is because that the adhesive strength decreases with the increase of temperature.

By water storage, the as received and degreased joints are the ones how showed an increase of strength up to 240 h and then decreased again. The degreased and ground weldbonded joints showed superiority among the other tried joints, while with the same surface preparation, the adhesive-bonded showed remarkable inferiority. The reduction of strength along with longer storage duration can be attributed to the degredation of adhesive with time.

Salt spray test showed different trends along with the different surface preparations, except the degreased and ground weldbonded joints who showed superiority and consistency (the reduction of strength is 2.6 KN in 1000 h). The as received adhesive-bonded joints strength have after a duration of 240 h a great reduction, after which the strength increased again. Also, the increase of strength is also remarkable with the degreased and ground adhesive-bonded joints (about 2 KN in 1000 h). The loss of strength is reasonable in that test with time due to the loss of adhesive with the longer salt spray.

Under the normal climate test, the tried joints showed a sudden reduction of strength up to 240 h, then levels down up to 1000 h. The greatest reduction is remarkable for the as received adhesive-bonded joints up to 480 h, followed by an increase of strength up to 1000 h. The higher strength of the weld-bonded joints is mostly based on the spot weld strength.

Impact test preliminary results for spot welded, adhesive-bonded, and weldbonded joints using the designed fixture is shown in Table 1. The weldbonded joints showed higher impact strength over the spot welded and adhesive-bonded joints. The degreased joints also gave the higher strength results among the tried surface preparations.

Table 1 - Preliminary Impact Test Results

Type of Joint	Surface Preparation	Impact Energy (Nm)
adhesive-bonded	as received	12.80
	degreased	13.25
	degreased & ground	9.36
spot welded	as received	31.44
weld-bonded	as received	33.64
	degreased	36.28
	degreased & ground	34.08

CONCLUSION

As a result of this study it may be concluded that the weldbonding is a good method for joining of HSLA steel sheets. It advantage in comparison to spot welding and adhesive-bonding at room temperature is: 1) the higher tensile-shear strength, 2) the resistance against impact loads, and 3) the tightness of the connections. In comparison to adhesive bonding, weldbonding is more reliable under environmental degredation conditions mainly if the joints are degreased and ground prior joining.

REFERENCES

1. Kaiser, J.G., Dunn, G.J. and T.W. Eager, Weld. J., Res. Suppl., 61,167-s to 174-s (1982)
2. Kimchi, M., Weld. J., Res. Suppl., 63, 58-s to 63-s (1984)
3. Sawhill, J.M. Jr., and J.C. Baker, Weld. J., Res. Suppl., 59,19-s to 30-s (1980)
4. Sawhill,J.M. Jr., Watanabe, H., and J.W. Mitchell, Weld. J., Res. Suppl.,56, 217-s to 224-s (1977)
5. Pollard, B., Weld. J., Res. Suppl., 53, 343-s to 350-s (1974)
6. Cappelli, P.G., procedings of the Welding of HSLA (microalloyed) Structural Steels, Rome, Italy, Nov. (1976)
7. Savage, W.F., Nippes, E.F., and F.A. Wassell, Weld. J., Res. Suppl., 56, 365-s to 370-s (1977)
8. Savage, W.F., Nippes, E.F., and F.A. Wassell, Weld. J., Res. Suppl., 57, 43-s to 50-s (1978)
9. Wu, K.C., Weld. J., Res. Suppl., 54, 436-s to 446-s (1975)
10. Williams, N.T., and T.B. Jones, Met. Const., October, 541-546 (1979)
11. Connell, L.D., Met. Const., Jan., 30-32 (1977)
12. Rivett, R.M., Doc. No. III-WG7-24-82, Weld. Inst., May (1983).
13. Dinsdale, W.O. and K.I. Johnson, Doc. No. P/66/74, Weld. Inst., (1974).

14. Scheetz, H.A., Koh, S.L., and J.N. Brennan, J. Appl. Polymer Sc., IV(12), 296-313 (1960)

15. M.M. Schwartz, "WeldBonding", Document No. IV-145-74, Rohr Industries, Chula Vista, CA, Jan. (1974).

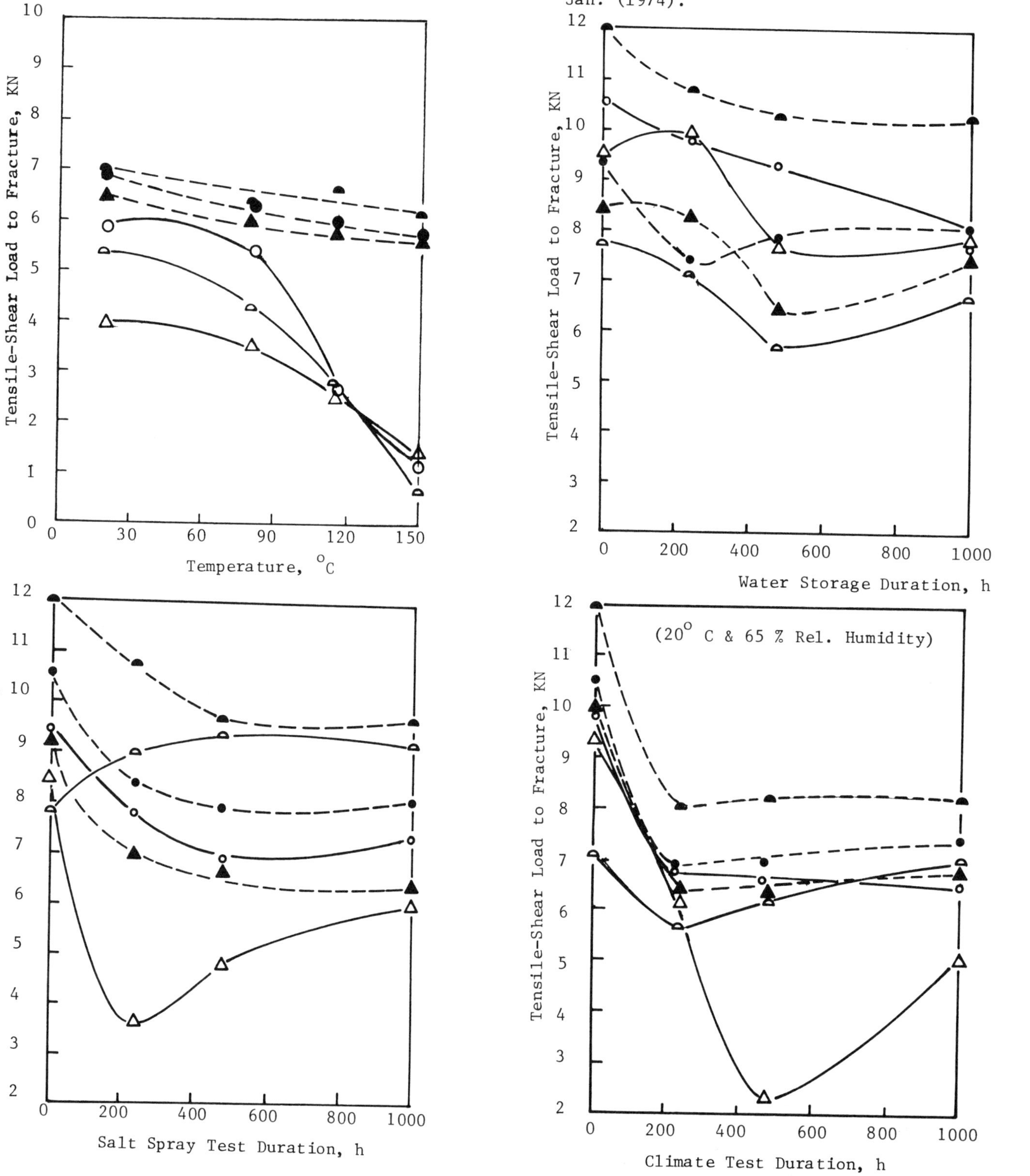

Fig. 4 - Effect of Environmental Degredation and Surface Preparation on the Tensile-Shear Load

As received △ —— adhesive-bonded, ▲ – – – weldbonded

Degreased ○ —— adhesive-bonded, ● – – – weldbonded

Degreased/ground ◓ —— adhesive-bonded, ◒ – – – weldbonded

EFFECT OF STRUCTURES FORMED IN HAZ ON BEHAVIOR OF HYDROGEN DIFFUSION AND PROPERTIES FOR THE ULTRA HIGH STRENGTH STEEL 30CrMnSiNi2A

Chen Zhongxiao, Zhang Zuyiao, Yang Dexin, Liu Gichen
Dalian, People's Republic of China

Abstract

This paper deals with the rules of hydrogen absorption for 3 types of structure formed in HAZ, diffusion coefficient and solubility of hydrogen on overheated region at HAZ, changes in the behavior of hydrogen diffusion after subzero treatment, and the sensitivity to hydrogen brittlement, by investigating the influence of the simulative structure formed in HAZ on the behavior and characteristic of hydrogen diffusion by using the method of transmission electron microscope, cathodic electrolysis and the definite load delayed fracture test, for 30CrMnSi2A ultra-high strength steels. It has been denoted that the structure improvement in overheated regions is the determining condition for the distinct improvement of properties of the post-treated HAZ. The retained austenite which is distributed over between the martensite lath packets in a film-like form plays an important role in raising the property of HAZ.

GREAT WIDESPREAD ATTENTION has been paid to the problem of hydrogen-induced crack in weld HAZ. A large number of tests and studies show that the microstructure constituent of HAZ, the hydrogen content and its distribution, especially the behavior of diffusible hydrogen in HAZ, are very important factors influential in making the hydrogen-induced crack in HAZ (1-3). Thorough studies have been made of the effect of the microstructure upon the hydrogen-induced crack in HAZ, but there are still many problems to be studied further, such as the problem of the effect of the different microstructure constituent formed in HAZ on the action as well as on its mechanism for preventing hydrogen-induced crack, and the problem of how to adjust the welding condition reasonably in order to obtain an expected microstructure in HAZ, etc. All of them and many others are worth our while when medium carbon low alloy ultra high strength steel is welded.

According to the change made in HAZ after immediatly post-heating 30CrMnSiNi2A steel in welding, and from the point of view of toughening of the structure formed in over-heated region, this paper has observed the problems on the following:

1) the effect of microstructure constituent upon the sensitivity of hydrogen-induced crack in HAZ.
2) the effect of the presence of retained austenite upon the behavior of diffusible hydrogen and role for improvement of the properties in HAZ.
3) all-round appreciation the significance of microstructure improving for toughness.

MATERIAL AND EXPERIMENT PROCEDURE

MATERIAL - The material tested is round bar of 46mm diameter, having the following chemical composition: 0.3% C, 1.2% Mn, 1.10% Si, 1.09% Cr, 1.56% Ni, 0.019% P, 0.01% S. After quenching and tempering, the mechanical properties become as follows: tensile strength 164 Kgf/mm^2, yield point 136 Kgf/mm^2, elongation 12.5%, reduction of area 52%, impact value 12.5 Kgf/m.

EXPERIMENTAL PROCEDURE - The test specimens of 11 x 11 x 100 mm are to go through three kinds of thermal cycles, by using CKP-1 type thermal simulative numerical control machine in order to obtain the simulated structure similar to that of actual weld HAZ's in following welding condition: 1) neither pre-heating nor post-heating, 2) pre-heating at 250°C and post-heating at 200°C for one hour, 3) pre-heating at 250°C and post-heating at 260°C for one hour.

The square board test specimens of 11 x 11 mm are cut off from the homoeneous heated zone of the thermal simulated test specimen subject to the above thermal cycles and then penetrated by hydrogen with the method of cathodic electrolysis after the surface been treated strictly. The test specimens has penetrated with hydrogen are to be cleaned and dried in 20 seconds, and then are measured for hydrogen by using the chromategraph apparatus. Microstructure constituent and its morphology are observed by the TEM and Electron Diffraction Pattern. Retained austenite content is measured by x-ray diffraction.

The thermal simulated test specimens are to be machined to 10 x 10 x 100 mm, a V-type groove is made in the middle of homogeneous heated zone and after that the three-point bent definite load delayed fracture test is conducted after penetrating hydrogen separately by cathodic electrolysis in homogeneous heated zone. The fracture toughness K_{1c} in the case of no hydrogen and the $_HK_{1c}$ in the case of 0.25 ml/100 g diffusible hydrogen content are measured by using the specimens of all kinds of thermal simulated structure.

EXPERIMENTAL RESULTS

MICROSTRUCTURE CONSISTUENT AND ITS MORPHOLOGY OF THE THREE KINDS OF OVER-HEATED REGION - The characteristics of microstructure and the amount of retained austenite in three kinds of over-heated region are listed in Table 1 and shown in Photos 1,2,3.

Table 1 Characteristics of Coarse-Grained Zone

WELDING CONDITION	HAZ STRUCTURE	RETAINED AUSTENITE (%)
NO Preheating NO Postheating	Coarse Twinned Martensite + Minor Amount of Retained Austenite	1
Preheat 250°C Postheat 200°C	Coarse Twinned Martensite + Higher Amount of Retained Austenite	5
Preheat 250°C Postheat 260°C	No Twinned Martensite + Highest amount of Retained Austenite distributed in the Martensite Laths.	11

RULES OF HYDROGEN ABSORPTION FOR THREE KINDS OF STRUCTURE FORMED IN OVER-HEATED REGION - The Fig. below shows the relationship between the penetrated time for hydrogen and the quantity of hydrogen absorbed in the three kinds of structure formed in HAZ.

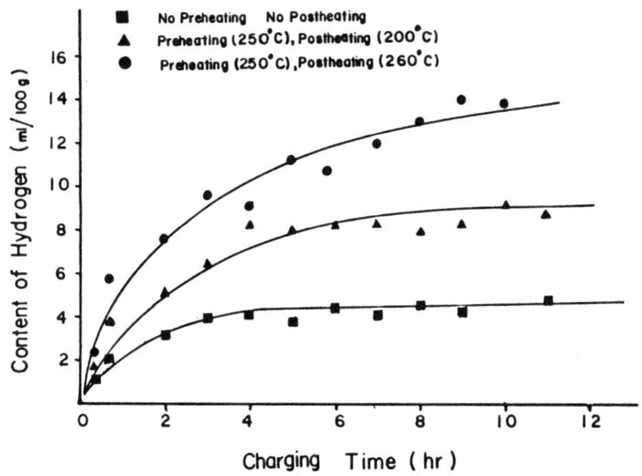

Fig. - Relationship between the penetrated time for hydrogen and quantity of absorbed hydrogen in the three kinds of structure formed in HAZ.

The rules of hydrogen absorption of the three kinds of structure formed in HAZ are all in complete agreement with the equation

$$Y = aX^b$$

where Y - Quantity of Absorbed Hydrogen, X- Penetrated Time for Hydrogen, a,b - Constant.

The rules of hydrogen absorption of the three kinds of structure formed in HAZ, the diffusion coefficient and the solubility of hydrogen in them are listed in Table 2.

Table 2 Relationship Between Welding Conditions and Diffusion Coefficient and Charge Saturation Level

WELDING CONDITON	DIFFUSION COEFFICIENT (CM^2/S)	CHARGE SATURATION LEVEL (ml / 100g)
NO Preheating NO Postheating	6.96×10^{-7}	10
Preheat 250°C Postheat 200°C	2.87×10^{-7}	23
Preheat 250°C Postheat 260°C	1.62×10^{-7}	36

THE CHANGES OF DIFFUSION COEFFICIENT AND SATURATION SOLUBILITY OF HYDROGEN IN STRUCTURE FORMED IN HAZ AFTER SUBZERO TREATMENT - After subzero treatment only the amount of retained austenite in the structure is varied, i.e., the amount decreases from 10.77% to 7.86%. The diffusion coefficient of hydrogen in this structure is $2.2 \times 10^{-7} cm^2/s$, increased 40% as compared with $1.62 \times 10^{-7} cm^2/s$ before subzero treatment. The saturation solubility of hydrogen is about 30ml/100g, reduced in comparison with 36ml/100g before subzero treatment, but greater than 23ml/100g after postheating at 200°C.

Photo. 1- Twin martensite (neither preheating nor postheating) x 59000

a) Twin martensite (x 63000)

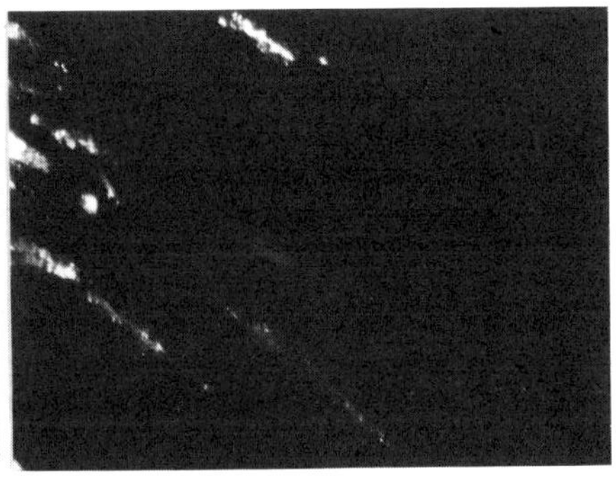

b) Retained austenite (darck field x 13500)

Photo. 2- Preheating at 250 deg.C and Postheating at 200 deg.C

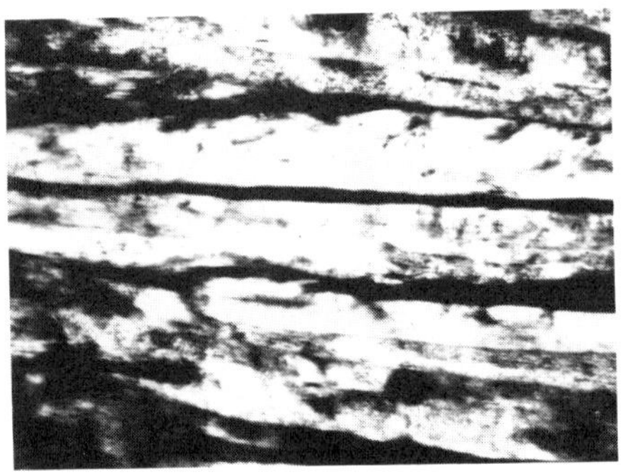

a) Lath martensite (x 21000)

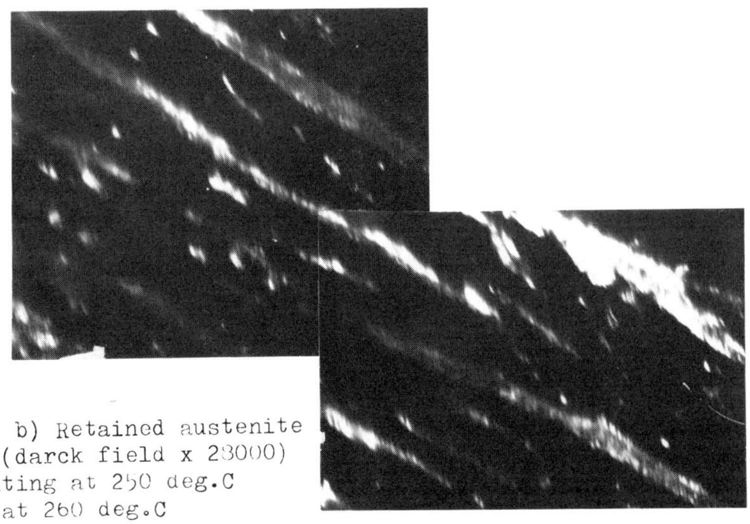

b) Retained austenite (darck field x 28000)

Photo. 3- Preheating at 250 deg.C and Postheating at 260 deg.C

THE EFFECT OF STRUCTURE FORMED IN HAZ ON PROPERTIES

The Results of Three Point Bent Definite Load Delayed Fracture Test - In Table 3 are listed both the delayed fracture critical stress (σcr) and the relevant sensitivity index of hydrogen brittleness of the three kinds of structure formed in over-heated region in the case of different initial contents of hydrogen.

Table 3 Results of Three-Point-Bending Test

Hydrogen (ml/100g) / Properties / Welding Condition	3		2		1		0.25		0
	Rupture Strength (Kg/mm²)	Reduction Ratio (%)	Rupture Strength (Kg/mm²)	Reduction Ratio (%)	Rupture Strength (Kg/mm²)	Reduction Ratio (%)	Rupture Strength (Kg/mm²)	Reduction Ratio (%)	Rupture Strength (Kg/mm²)
NO Preheating NO Postheating	12	96	16	95	34	89	40	87	304
Preheat 250°C Post heat 200°C	20	94	44	88	76	60	126	60	317
Preheat 250°C Postheat 260°C	72	79	88	74	146	57	304	12	344
Base Metal							308	10	300

The Determined Results of Fracture Toughness - The plane strain fracture toughness K_{1c} and K^H_{1c} of the three kinds of structure formed in over-heated region are determined in the case of no containing hydrogen or the content of hydrogen being 0.25ml/100 g. The results are listed in Table 4.

Table 4 Relationship between Welding Condition and Toughness Reduction Ratio

WELDING CONDITION	FRACTURE TOUGHNESS (K_gf/mm)		TOUGHNESS REDUCTION RATIO (%)
	WITHOUT (H_2)	WITH (H_2)	
NO Preheating NO Postheating	242	214	11
Preheat 250°C Postheat 200°C	253	223	8
Preheat 250°C Postheat 260°C	326	319	2

DISCUSSION

THE EFFECT OF POST-HEATING ON THE STRUCTURE CONSTITUENT FORMED IN OVER-HEATED REGION - It has been indicated through the test results that proper post-heating immediately after weld can improve markedly the structure constituent at 260°C and post-heating immediately after weld at 260°C for an hour, the structure of over-heated region has been improved greatly.

First, the morphology of martensite, i.e. plate-type martensite disappear, while the lath-type martensite is in the majority. The distribution of carbon atoms in over-cold austenite tends to be even so that the difference of temperature (M_s) in very zone become less, hence the temperature M_s are relatively elevated. The document (4) points out that the two critical shear-transformation stress which are needed to cause sliding and twinning in over-cold austenite between M_s and M_f is not large for medium-carbon low alloy steel. When transformation temperature of over-cold austenite is higher, and the critical shear-transformation stress required to cause sliding is lower than that of the twinning. Therefore, the elevation of the temperature M_s of over-cold austenite is useful to itself in transformation into lath martensite in cooling course.

The content changes and distribution state of retained austenite in structure post-heating at 260°C are the other remarkable characteristics. The post-heating at 260°C elevates the stability of over-cold austenite and makes hot stabilizing of the austenite because of the diffusion and segregation of C, N atoms of isothermal temperature (5).

Because of the increase of its content, the distribution state of the reained austenite is improved greatly, which distributed continuously along the boundaries of lath martensite and over between the martensite lath packets in film-like form. This is one of the useful distribution states (6,7,8).

It is clear that the changes of structure constituent as well as morphology will directly affect the behavior of hydrogen diffusion in structure and the hydrogen brittleness sensitivity of over-heated region.

THE EFFECT OF MICROSTRUCTURE ON THE BEHAVIOR OF HYDROGEN DIFFUSION - It can be seen from the above mentioned results that the microstructure, especially the presence of retained austenite in film-like form affect the behavior of hydrogen diffusion markedly.

The diffusion coefficient of hydrogen is the biggest ($D = 6.96 \times 10^{-7} cm^2/s$) in the structure formed in over-heated regions with neither preheating nor postheating. But it becomes obviously less in the structure with preheating at 250°C and postheating at 260°C for one hour, $D = 1.62 \times 10^{-7} cm^2$, being the smallest one among three kinds of structure over-heated region.

After postheating at 260°C, the retained austenite in structure increases evidently and it distributes continuously over between the martensite laths and the martensite lath packets in a film-like form. The diffusion coefficient of hydrogen in austenite is very small, i.e., five numeral grades is lower than that of martensite, but the solubility of hydrogen is very remarkable.(1) Therefore the retained austenite in strucutre is, in fact, the resistant layers for diffusible hydrogen and it effieicntly blocks the diffusion of hydrogen. This causes the diffusion coefficient of hydrogen in structure to reduce evidently. When subzero treatment has been taken at -196°C, 27% retained austenite in structure is transformed, and the spread of hydrogen diffusion is elevated to $2.27 \times 10^{-7} cm^2/s$, i.e., 40% higher than before. This shows further that the retained austenite distribution in film-like form can block the diffusion off efficiently. The varied rule of the solubility of hydrogen is just contrary to the diffusion coefficient. The solubility of hydrogen in strucutre postheating at 260°C is the biggest, because the place for absorpting hydrogen in structure increases evidently, while the diffusible hydrogen is more scattered and it is difficult for the segregation of hydrogen to emerge.

On the basis of modern theory about delayed crack (9), the action of the structure to the behavior of hydrogen diffusion must affect the sensitivity of hydrogen-induced crack.

THE ACTION OF STRUCTURE CHANGES IN INCREASE OF ABILITY TO RESIST CRACKING - The critical fracture stress appears a sudden change here under level of hydrogen content. Critical fracture stress of the structure postheating at 260°C is all high, while the sensitivity index of hydrogen crack is very low.

The critical fracture stress of the structure is only 12 Kgf/mm^2 with the hydrogen content being 3 ml/100g. It only increased to 40 Kgf/mm^2, when the hydrogen content decreased to approach the level of hydrogen (0.25 ml/100g) for postheating at 260°C, with neither preheating nor postheating, while critical fracture stress of structure postheating at 260°C is improved clearly at the same hydrogen content, the critical fracture stress is 304 Kgf/mm^2, being 7.6 times than that of no preheating and no postheating. As a result, it can be proved that the improvement of the structure constituent is the basic reason, for which the properties of the structure postheating at 260°C are improved obviously.

The document (10) indicates that the hydrogen-induced crack can not be caused until the hydrogen content in stress focusing field attains to a critical density.

It can be seen that the diffusion rate of hydrogen in structure and toughness of structure are the key influential factors for cracking.

Hence, both the improvement of the toughening after structure changes, and the behavior of hydrogen diffusion, are the main reason to explain the increase of the ability to resisting crack of structure postheating at 260C.

Preheating at 250°C and postheating at 260°C for an hour can, on one hand, improve ductility greatly and make even the degree of deformation of structure when being forced, and make difficult the appearance of partial zone where strain is much heavy; on the other hand, make the diffusible hydrogen to separate evenly in the retained austenite and over between the martensite lath packets as well as a result of much positions able to absorb hydrogen. It is more important that the diffusion rate of hydrogen is very slow because of the block of film-like form of retained austenite and influence of other structure constituent. Therefore the hydrogen is difficult to diffuse and focus towards the partial area where the stress is centrilized, and thus in the case of containing certain amount of hydrogen, the possibility to cause crack reduced, and the critical fracture stress elevated and the properties of resistance to cracking increased.

In order to prove further the effects of retained austenite on the properties of HAZ, the paper below has been studied thoroughly the properties of structure with the amount of retained austenite changed. The results are listed in Table 5.

Table 5 Effect of Retained Austenite on Properties

Treatment Condition	DELAYED FRACTURE (Kg/mm2)			FRACTURE TOUGHNESS (K_gf/mm)		
	WITHOUT [H]	WITH [H]	REDUCTION RATIO (%)	WITHOUT [H]	WITH [H]	REDUCTION RATIO (%)
No Treatment	344	304	12	326	319	2
-196°C 24 hrs.	345	226	35	313	288	8

[H] = 0.25 ml/100g

The results show that under the circumstances of no changing other structure the retained austenite in structure reduces 27%, the K_{1c} 4.1% when the hydrogen content is 0.25 ml/100g, critical fracture stress reduces 26%, $K^H{}_{1c}$ 9.7%. This strongly proves that the retained austenite in film-like form has important effects on the toughening and the ability to resist hydrogen-induced crack.

It can also be seen from above mentioned results that the reduced degree of property caused by the change of retained austenite with diffusible hydrogen is not bigger than that contains no hydrogen. This shows the existence of retained austenite in film-like form has more special significance to elevate the ability of structure to resist the hydrogen-induced crack because of the action of the retained austenite to block the hydrogen diffusion and solid dissolution. It ought to be emphasized that the retained austenite is quite stble above -45°C.

MAIN CONCLUSION

For 30CrMnSiN2A the ultra-high strength steel after preheating at 250°C and postheating at 260°C during welding, the over-heated region can offer a structure, major lath martensite and minor lower bainite with 10.77% retained austenite, which is distributed over between the martensite lath and its packets in a film-like form. The diffusion coefficient of hydrogen in this structure is low, while solubility of hydrogen is high, it is because of the action of the retained austenite in blocking the hydrogen diffusion and improving solid solubility. Thus, its σcr, K_{1c}, K^H_{1c} are all elevated. Precisely because of this reason, especially due to the action of the retained austenite in blocking the hydrogen diffusion, lowering the diffusion rate, increasing the places for hydrogen absorption, hence the hydrogen there is difficult to diffuse towards a partial area. So that the toughening for the over-heated region is elevated obviously.

REFERENCES

1. "Weld Crack of Steels," Japan.
2. "Welding theory and Welding Technology," China.
3. Cold Crack in the Welding of Steels, Weld Res. Abroad, 1972.
4. "Martensite Transformation and Martensite," S.P.C. China, 1981.
5. "Theory of Metal Treatment," MEPC, China 1981.
6. Chemicophysics Inspection (physical volume) No. 1, 1981.
7. E. R. Parker, Met. Trans., Vol 8A (1977).
8. Iron and Steel, Vol. 62 (1976).
9. Atlas for Metallographic Analysis on Welding Crack. H.W.I. China (1981).
10. Foundation of Metal Wedlability. MEPC, China (1982).